Modern Atomic and Nuclear Physics

Revised Edition

Modern Atomic and Nuclear Physics

Revised Edition

FUJIA YANG
Fudan University, China & Nottingham University, UK

JOSEPH H HAMILTON
Vanderbilt University, USA

World Scientific

NEW JERSEY • LONDON • SINGAPORE • BEIJING • SHANGHAI • HONG KONG • TAIPEI • CHENNAI

Published by

World Scientific Publishing Co. Pte. Ltd.
5 Toh Tuck Link, Singapore 596224
USA office: 27 Warren Street, Suite 401-402, Hackensack, NJ 07601
UK office: 57 Shelton Street, Covent Garden, London WC2H 9HE

British Library Cataloguing-in-Publication Data
A catalogue record for this book is available from the British Library.

The first edition was published in 1996 by McGraw-Hill, New York.

MODERN ATOMIC AND NUCLEAR PHYSICS
Revised Edition

ISBN-13 978-981-283-678-6
ISBN-10 981-283-678-0
ISBN-13 978-981-283-679-3 (pbk)
ISBN-10 981-283-679-9 (pbk)

Printed in Singapore

ABOUT THE AUTHORS

Fujia Yang has taught atomic and nuclear physics for over thirty years at Fudan University, Shanghai, China. He was a postdoctoral fellow at the Niels Bohr Institute, Copenhagen, Denmark. He also was a visiting professor at the Niels Bohr Institute, Copenhagen, Denmark, the State University of New York at Stony Brook, NY, and Rutgers University, and an adjunct professor at Vanderbilt University.

Professor Yang is the author of *Atomic Physics*, a book for undergraduate and beginning graduate students (over 20,000 copies sold in China), and of *Ion Beam Analysis*, a book for beginning graduate students. He has authored more than forty research papers in atomic physics, nuclear physics, and ion-beam analysis in leading journals worldwide. His current research is in atomic physics, PIXE, and ion-beam analysis. He was the associate editor of *Nuclear Instruments and Methods A*, *Hyperfine Interactions*, and *Chinese Physics Letters*.

Professor Yang was awarded a prize as the author of one of the three outstanding physics texts published in China in 1987 for his book *Atomic Physics* and a Distinguished Teacher Award in Shanghai in 1987.

He has served as President of Fudan University; Professor and Director of the Institute of Modern Physics, Fudan University; Director of the Institute of Nuclear Research, Chinese Academy of Science; and is a member of the Chinese Academy of Science. Since 2001 he has served as Chancellor of the University of Nottingham in England.

Joseph H. Hamilton has spent over forty years teaching atomic and nuclear physics at Vanderbilt University. He has had guest appointments at the University of Uppsala, Sweden; the Institute for Nuclear Research, Amsterdam, Netherlands; the University of Frankfurt, Germany; and the L. Pasteur University, Strasbourg, France. He is an adjunct professor at Tsinghua University, Beijing, and Honorary Professor at Fudan University.

Professor Hamilton has been a coeditor and/or coauthor of 13 international conference proceedings and reference works and over 950 research publications in nuclear physics and a coauthor and author of one book and twenty-one publications for nonscientists. He has received the following awards: NSF Postdoctoral Fellowship; Jesse Beams Award for Outstanding Research (Southeastern Section of the American Physical Society); Alexander von Humboldt Prize, Federal Republic of Germany; Earl Sutherland Award for Outstanding Research (Vanderbilt University); George Pegram Award for Outstanding Teaching (Southeastern Section of the Americal Physical Society); Guy and Rebecca Forman Award for Outstanding Teaching

(Vanderbilt University); 1991 Professor of the Year in the State of Tennessee by CASE (Council for Advance and Support of Education); Thomas Jefferson Award for Outstanding Service (Vanderbilt University); Award for Development of International Cooperation, American Association for the Advancement of Science; Ilkovic Gold Medal for Career Achievement in Science, Slovak Academy of Science; International Scientific and Technological Cooperation Award of the People's Republic of China; G.N. Flerov Prize for Major Achievement in Nuclear Physics, Joint Institute for Nuclear Physics, Russia; Fellow of the American Physical Society and of the American Association for the Advancement of Science.

Professor Hamilton is the Landon C. Garland Distinguished Professor of Physics at Vanderbilt University and Director of the Joint Institute for Heavy Ion Research, Oak Ridge, Tennessee.

CONTENTS

PREFACE

Our book aims to give students a broad perspective of current understandings of the basic structures of matter as scientists have probed ever deeper levels from atoms, to the nucleus, on to leptons, quarks, and gluons, along with the necessary introductory quantum mechanics. We have interwoven the material with its historical development and challenging future directions. Beyond a broad understanding of atomic, nuclear, and particle physics, we want students early in their studies to appreciate the uncertain path of success and failure, opportunities seized and opportunities missed and the roles of chance and intuition by scientists in the unfolding human drama of scientific discovery. By working through some of the issues and struggles that occurred in the development of modern physics, students better understand the material as well as learn how science works. Several significant historical developments are presented that are not found in any textbooks.

We also believe that it is important for students as early as possible to begin to think about some of the current intellectual challenges of these fields and to glimpse the ever-increasing power of new techniques and facilities to probe these and future challenges. Thus, examples of very recent developments and future plans are described to excite students by allowing them to see how the techniques and ideas of atomic, nuclear, and particle physics have been used and are being used to attack important problems in other basic and applied areas of physics, chemistry, and biology on to major societal problems in medicine, energy resources, new tailor-made materials and environmental pollution, and in areas of wide cultural and historical interest such as dating the Shroud of Turin, the levels of civilization revealed by the compositions of ancient artifacts, and the cause of the extinction of the dinosaurs.

A questioning spirit is at the heart of scientific discovery. Many discoveries began by someone asking "Why is that so?" or "How could we do that differently?" Thus, throughout the book students are encouraged to reflect on problems and to ask questions. One of the important traits of great scientists is the ability to identify, out of the many open questions and problems, the significant ones whose answers would truly make a difference.

The text is an outgrowth of our many years of teaching courses in atomic physics, in nuclear and particle physics and in modern physics, our research in these fields, and our strong interest in the history of physics and the philosophy of science. The book can be used in these different courses by sophomore students to beginning graduate students for a one-semester or a full-year course by emphasizing different chapters and different material within a chapter. Only an introductory calculus-based physics course is a necessary prerequisite. Students in biology, chemistry, engineering, medicine, and other fields in addition to physics will find the material very helpful in their work. We hope students will come to share the challenges and excitement we experience in seeking to understand the basic structures of matter at the atomic, nuclear, and particle levels.

One of the authors (F. Yang) thanks Professor Aage Bohr for his invitation to work in Copenhagen on several occasions and his cultivation in the author of a deep appreciation for the "Copenhagen Spirit" at the Niels Bohr Institute. It was a rare privilege to trace much of the history of the development of quantum mechanics and atomic and nuclear physics in the Niels Bohr Archives there.

Because of the simplicity in stating many equations in electricity and magnetism, we have more often used the Gaussian (CGS) system rather than the SI (MKS) units. The conversion of factors between the two systems of electrical units are given in an appendix. Again, for ease in solving problems, energies are given in electron-volts up to GeV (10^9 eV). Several constants which are composed of different fundamental constants are given to facilitate the working of various practical problems. Other appendices are given to provide more detailed deriva-tions, examples of the applications of the principles to different areas of research, and useful tables.

We thank J. Y. Tang, Ge Ge, F. Q. Lu, J. Y. Li, M. S. Ting, X. M. Li, Y. N. Sun and S. Zhu for their kind help in translating part of the book from Chinese to English. We would also like to express our thanks for the many useful comments and suggestions provided by following reviewers of the first edition: Ellen A. Berning, Loyola University, M. Broyles, University of Texas, Robert Luke, Boise State University, and David Ward, Union University. We also want to thank Professors F. T. Avignone, D. J. Ernst, C. K. Gelbke, F. Q. Lu, W. E. Johns, C. F. Maguire, R. J. Scherrer, A. V. Ramayya, Dr. J. P. VanDevender and Y. S. Wang for their help regarding new material for Chapters 10, 11, 12, 14 and Appendix III. Ms. C. Soren, Dr. J. K. Hwang and S. H. Liu are thanked for their extensive help with new figures, tables and typing for the revised edition.

Fujia Yang
Joseph H. Hamilton

INTRODUCTION

Physics is the discipline of science that concerns itself with the study of the most general laws of matter and energy and the fundamental structure of matter. The branch of atomic physics aims primarily at the study of that structure at what is called the *atomic scale*—a term for sizes between those of molecules and the nuclei of atoms. Figure I.1 gives us a perspective on the dimensions of our universe from clusters of galaxies to the nucleus of the atom. While the concept of atoms has a history of more than 2000 years, atomic physics as we know it today is a discipline that began to form in the present century and parallels the development of modern physics.

As is well known, the word "atom" is derived from the ancient Greek with its original meaning being "the indivisible." In the fourth century B.C., the ancient physicist Democritus advanced this concept and thought of atoms as the smallest unit of matter. In the same period, however, some scholars such as Aristotle, Anaxagoras, and others denied such an atomic view of matter and held that matter was continuous and so was divisible indefinitely. This point of view dominated through the Middle Ages. But, after the sixteenth century, the atomic picture of matter was revived with the new emphasis on experimentation and the development of new experimental techniques. Famous scholars such as Galileo Galilei, René Descartes, Robert Boyle, and Isaac Newton supported such a point of view. In the early 1800s John Dalton gave several rules that formed the basis of an atomic theory in which atoms were the smallest units of matter. Dalton's atomic model included the following basic postulates: (1) All matter is composed of small particles, which are called atoms. (2) The atoms of each element are all alike in weight and in all other properties. (3) The atoms of different elements are of different weights, and their other properties also differ. (4) Atoms are indestructible and can neither be created, destroyed, nor divided. They preserve their identity during chemical reactions and only undergo rearrangement. (5) When two or more elements combine to form a compound, their atoms combine to form identical groups of atoms, which are called molecules.

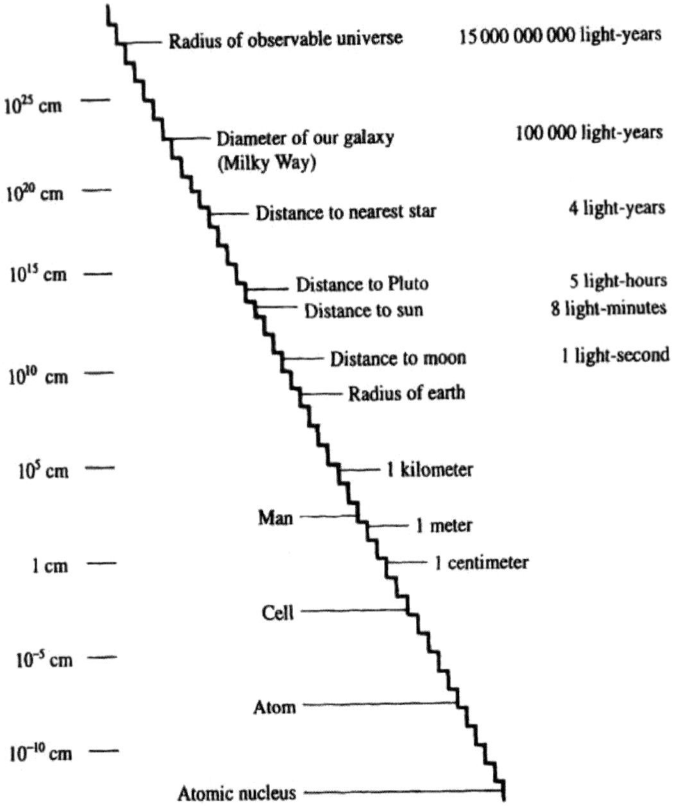

FIGURE I.1
Dimensions of objects in our universe.

It is not as generally well known in the West that the concept of atom goes back equally far in Eastern thought. In China, the two viewpoints mentioned above may be traced back at least to the Period of the Warring States (476–221 B.C.). The most famous school that denied the infinite divisibility of matter was founded by Mo-Di. In the book called *Mo-Jing*, there is the following sentence: "Point: The orderless and foremost of the body." Basically, the so-called "point" is something that is indivisible ("orderless") and primary ("foremost") and from which bodies are composed. At the same time Huishi said: "A thing is so small that it has no interior, and this is called the smallest unity." Here the "smallest unity" is without an inside, so it can not be divided further. It is the most elementary particle. In that period, a book *Zhong-Yung* (The Intermediate) of the Confucian School pointed out more definitely: "As to the small, no body can break it in the world." A later authority on Confucian doctrine, Zhu-Xi of the Sung Dynasty (960–1279), explained: "No one can break it means interiorless. That is to say, suppose there is a smallest thing that one can break into two parts, then there may be some room for something else between those two parts.

Conversely, if it is interiorless, it can never be broken again." When the scholar Yan-Fu of the Qing Dynasty (1616–1911) translated Mill's *Logical System* into Chinese, he introduced the word "atom" into China for the first time. He translated "atom" as "Mo-Po" (the indivisible) and translated "atom theory" as "Mo-Po particle theory".

However, another school of the Warring States period maintained the infinite divisibility of matter as represented by Gung-Sun-Lung, who wrote, "With a rod one foot long, you can never exhaust it by cutting off one half of it everyday." In many ways the speculation of Gung-Sun-Lung two thousand years ago has been pushed continuously to smaller and smaller scales by modern science.

Toward the end of the last century, it began to be realized that atoms are only one of the levels in the structure of the matter. Many important discoveries led to our understanding of atoms as important units of matter and of smaller units and scales including the atomic nucleus, the nuclear constituents, the proton and neutron, and then their substructures of quarks and gluons.

1806	J. Proust of France discovered that molecules of chemical compounds obey the law of definite proportion.
1807	J. Dalton of England discovered the law of multiple proportions and suggested an atomic theory.
1808	J. Gay-Lussac of France discovered the law that in the chemical combination of gases, their volumes are in simple ratios to each other; hence he concluded that, in a gaseous state, the weight of an element in a given volume would be proportional to its atomic weight.
1811	A. Avogadro of Italy made his hypothesis: Under the same temperature and pressure, there are the same number of molecules in the same volumes of gases.
1826	R. Brown of England observed that some corpuscles suspended in liquid were in random fluctuating motion, the so-called Brownian motion.
1833	M. Faraday of England formulated the law of electrolysis, and reduced the chemical affinity to electric forces.
1869	Independently, D. Mendeleev of Russia and L. Meyer of Germany suggested a periodic law of the elements.
1895	W. Roentgen of Germany discovered x-rays.
1896, 1897	H. Becquerel followed by M. and P. Curie of France discovered radioactivity, and J. Thompson of England discovered the electron. These two discoveries gave the first evidence that atoms were not indestructible and indivisible as Dalton had proposed.
1900	M. Planck of Germany introduced the idea of the quantum which, with the previous three great discoveries lifted the curtain of modern physics.
1905	A. Einstein of Germany explained Brownian motion as the result of the bombardment of the suspended particles by the random

motion of the atoms and molecules which make up the liquid, as well as giving us the special theory of relativity and a quantum explanation of the photoelectric effect.

1908 J. Perrin of France carried out careful measurements to confirm Einstein's theory of Brownian motion and so firmly established the existence of atoms.

1911 E. Rutherford of England interpreted the work of his co-workers, Geiger and Marsden, to be evidence that atoms had tiny cores which contained essentially all the mass and all the positive charge of the atom and that the core was almost 100 000 times smaller than the atom itself.

1913 N. Bohr of Denmark proposed the first quantum theory of atomic structure to open the door to new chapters in atomic physics.

1919 E. Rutherford performed the first artificial transmutation of elements by nuclear reactions.

1923 A. H. Compton of the United States completed the particle (photon) nature of light with the discovery of the Compton effect.

1924 L. de Broglie of France proposed that particles of matter have waves associated with them, to complete the duality of particles and waves.

1925 E. Schrödinger of Austria and W. Heisenberg, Germany, developed the new quantum mechanics, and W. Pauli of Switzerland proposed the exclusion principle to give a basis for the Periodic Table of the Elements.

1928 P. Dirac of England predicted the existence of a new companion particle to the electron, the positron, based on his new relativistic quantum theory of the electron.

1930 W. Pauli suggested the existence of another new particle beyond the electron and proton, the neutrino.

1932 Two new particles were discovered, the positron by C. D. Anderson of the United States and the neutron by J. Chadwick of England.

1934 I. and F. Joliot-Curie of France discovered new radioactivities which are not present in nature, and E. Fermi of Italy proposed his theory of beta decay in which the electron and neutrino are created by the weak nuclear force.

1935 H. Yukawa of Japan proposed still another new particle, the meson, to explain the force that holds protons and neutrons together.

1936 The μ lepton was discovered by H. Neddermeyer, C. D. Anderson and co-workers of the United States, but it was not until 1947 that C. Powell and co-workers of England discovered the meson which has properties similar to those proposed by Yukawa.

1939 O. Hahn and F. Strassmann of Germany discovered the fission of uranium induced by neutron absorption.

1940	G. Seaborg, E. McMillan and co-workers in the United States discovered the first new elements beyond uranium, then the heaviest known element and the heaviest to be found in nature.
1949, 1953	The first detailed predictive models of the structure of nuclei were proposed, the shell model by M. G. Mayer of the United States and J. H. D. Jensen and co-workers in Germany, and the collective model by A. Bohr of Denmark and B. R. Mottelson of the United States.
1955	O. Chamberlain, E. Segre, and co-workers in the United States discovered the antiproton. This fueled efforts to build larger and larger particle accelerators and opened the gates to the discovery of multitudes of new particles.
1964	M. Gell-Mann and G. Zweig in the United States proposed a new substructure of several quarks held together by gluons to explain the multitude of heavy particles being observed, including the proton and neutron.
1974	B. Richter and co-workers and S. Ting and co-workers in the United States separately discovered the J/Ψ particle to establish the fourth quark, called the charmed quark. Now, to explain the known heavy particles, one needs six quarks and six antiquarks, each with three colors or types—36 quarks held together by 8 colored gluons in the current Standard Model!
1983	C. Rubbia, an Italian physicist at Harvard-CERN, and S. van der Meer, a Dutch physicist and co-workers at CERN, discovered the massive $W^{\pm}$, Z^0 particles that mediate the weak nuclear force.

More precise experimental data from atoms led to the creation of the new quantum mechanics. Our understanding of the structure of atoms has continued to advance significantly in each decade right on through the present as new and more powerful experimental techniques have been developed and new, improved quantum mechanical theories have come forth. A renaissance is occurring in atomic physics today through the use of high-powered lasers and nuclear accelerators, including very energetic heavy-ion accelerators capable of accelerating atoms up through uranium to relativistic velocities. These are yielding exciting new insights into the structure of atoms. For example, with lasers atoms can be studied in previously inaccessible exotic conditions where the electrons are in orbits with radii up to 1000 times larger than those in normal atoms. Now scientists are very much interested in higher-energy beams of uranium up to several GeV (10^9 electron-volts) and higher-powered lasers to study in detail atoms up to uranium with only one or two electrons surrounding the nucleus. There previously unseen higher-order processes can occur to test and expand our theoretical insights and to probe possible previously unobserved fundamental properties of nature.

Nuclear physics concerns itself with the structure and behavior of the next layer of matter, which forms the tiny inner core of the atom. The atomic nucleus is about 100 000 times smaller than the atom but contains all the positive charge and essentially all the mass of the atom. One of the long-sought goals of nuclear physics is to understand the strong nuclear force which holds protons and neutrons together in the nucleus. Much progress has been made in understanding this force through the study of the radioactive decays of nuclei and through the interactions of nuclei with probes ranging from photons, electrons, and mesons, through light ions (p, n and α) and more recently heavy ions all the way up to uranium at energies from ultracold neutrons ($\ll 1$ eV) to 2 TeV (10^{12} eV) carbon-12 projectiles. In the absence of being able to express the strong nuclear interaction in closed form, models of the nucleus have been developed. The tiny nucleus of the atom can exhibit many different shapes from spherical to various deformed shapes such as prolate and oblate spheroids and triaxial, and now two or more of these shapes are seen in the same nucleus for different energy levels—nuclear shape coexistence. Nuclear matter is being probed in exotic regions of high temperature, high angular momentum, and high density—hotter than the sun and at densities approaching those of neutron stars.

The structure of matter has been taken to an even deeper layer as we probe in high-energy particle physics the substructure of the neutron and the proton. Neutrons and protons are themselves composite particles composed of three types of quarks held together by gluons. The quark structure of matter is now described by the "standard" model of quantum chromodynamics, QCD, in which the quarks have the additional property of flavor or color in addition to mass and charge. These latter two fields of nuclear and particle physics will be sketched in broad outlines to complete our picture of the basic structure of matter.

Chapter 1 presents the basic principles of the special theory of relativity which are necessary for understanding atomic, nuclear, and particle physics.

In Chapter 2, the first models of atoms are introduced. Then it will be shown that these models and classical theories could not account for even the experimental data at the turn of the century. Chapter 3 deals with Niels Bohr's seminal work that originated from his endeavors to avoid the classical difficulties—the introduction of quantum behavior into the atomic region for the first time. He postulated the concept of stationary quantum states to explain the experimental data. To this point only the gross structure of the atom, where the electrons and the nucleus are point charges, was considered. In Chapter 4, starting from experimental facts, the necessity of the concept of "electron spin" is introduced. That concept was essential in understanding the finer structure of atoms and the spectra of light they emitted. The word "spin" was not unfamiliar in classical physics, but this concept represents a totally new form of motion in the microworld that has no classical counterpart at all. At the end of this chapter, we show how our understanding of the hydrogen atom has been deepened step by step. In Chapter 5, we generalize the properties of one-electron systems like hydrogen to many-electron ones and explain the periodic properties of the elements from our modern understanding of atomic structure. Here, one of the important concepts is the exclusion principle introduced by Pauli. In Chapter 6, the production and properties of x-rays that were discovered in 1895 are considered. Various experimental facts are presented to show that x-rays display both wave aspects and corpuscular aspects. The contents of the first chapters will describe various difficulties that classical physics met in the atomic realm. Chapters 7 and 8 complete the necessity of the creation of quantum mechanics. There we try to disclose the essence of quantum mechanics from physical concepts but leave many details to a course on quantum mechanics.

Our understanding of the other constituent of the atom, the atomic nucleus, is introduced in Chapter 9. The basic features of the nucleus are presented there. Radioactivity is presented in Chapter 10. The nuclear force and nuclear models are discussed in Chapter 11. In Chapter 12, reactions between nuclei are considered along with their role in nuclear energy production. The hyperfine interactions between the atomic electrons and the nucleus that arise from the nuclear magnetic and quadrupole moments and electron motion and their fields are treated in Chapter 13. This field has developed into an interdisciplinary science between atomic and nuclear physics with many broad applications. The further substructure of the neutrons and protons in the nucleus and our present understanding of the basic constituents of matter at the next lowest level, as probed in the field of high energy physics, are described in Chapter 14.

Finally, some concrete examples of the practical applications of atomic and nuclear physics are given in Appendix I. These include the techniques of ion-beam analysis, emphasizing the applications of Rutherford scattering (Chapter 2), x-rays (Chapter 6) and nuclear reaction methods (Chapter 12) in material analysis.

Today studies of the structure of the atom, the structure of the nucleus, and the substructure of the nucleon and basic constituents of matter are exciting and expanding frontiers in science. The text seeks to interweave some of the new directions and research opportunities with the basic insights needed to go farther in these fields. Major, new facilities recently completed and under development including, for example, more powerful lasers, giant superconducting accelerators, large detector arrays, and more advanced super computers will continue to open up new research and new insights into our basic understanding of the workings of matter at each of these three levels. As well, they will make possible new opportunities in many other areas of basic and applied research along with new solutions to important societal problems.

In the early decades of the twentieth century when many of the basic discoveries which transformed our understandings of the laws of physics were taking place, international conferences would include the leading physicists in all areas of physics, as illustrated in Fig. I.2. Most of the founders of twentieth-century physics are seen in Fig. I.2, and their monumental works are the subject of much of this book. Today the number of physicists has grown tremendously and, unfortunately, their work has become so specialized that separate conferences are held to consider each of the subspecialties of physics. However, there is held in Lindau, Germany, a regular series of more popular conferences which bring together Nobel Laureates in all areas of physics. Reminiscent of the earlier conferences, as see in Fig. I.2, numbers of the new leaders along with some of the early ones are seen in Fig. I.3 at the 1983 Lindau Conference.

FIGURE I.2
Participants in the 1927 Solvay Conference. From left to right, bottom row: E. Langemeir, M. Planck, M. Curie, H. A. Lorentz, A. Einstein, P. Langevin, Ch. E. Guye, C. T. R. Wilson, O. W. Richardson. Second Row: P. Debye, M. Knudsen, W. L. Bragg, H. A. Kramers, P. A. M. Dirac, A. H. Compton, L. V. de Broglie, M. Born, N. Bohr. Back Row: A. Piccard, E. Henriot, P. Ehrenfest, Ed. Herzen, Th. DeDonder, E. Schrödinger, E. Verschaffelt, W. Pauli, W. Heisenberg, R. H. Fowler, L. Brillouin.

FIGURE I.3
Participants in the 1983 Lindau Conference from left to right: (1) T. C. Koopmanns, (2) Graf L. Bernadotte, (5) F. Bloch, (6) R. Hofstadter, (7) A. Butenandt, (8) I. Giaever, (10) J. Meade, (11) P. Dirac, (12) L. Esaki, (13) A. Penzias, (14) V. Fitch, (15) R. Wilson, (16) P. Samuelson, (17) K. Siegbahn, (19) A. Schawlow, (20) E. Wigner, (21) Walter Brattain, (22) N. Bloembergen, (24) R. Mössbauer, (25) S. Ting, (26) E. O. Fischer, (27) F. von Hayek, (28) J. R. Schrieffer, (29) W. Lamb, (30) H. Alven, (34) J. Schwinger, (35) A. Kastler

CHAPTER 1

THEORY OF RELATIVITY

If I have seen far, it is because I stood on the shoulders of giants.

Isaac Newton

1.1 SPECIAL RELATIVITY

The special theory of relativity was proposed by Albert Einstein in 1905 in one of his three classic papers that year. The other two were on the theory of Brownian motion (the precise verification of which helped win Perrin the Nobel Prize) and the theory of the photoelectric effect for which Einstein won the Nobel Prize. The theory is called "special" because it applies to normal, flat pseudo-Euclidean or Minkowski geometry. General relativity involves curved space. When it was introduced, the special theory of relativity was hailed by many but not accepted by others. Now that the records of the Swedish Academy of Science on the awarding of Nobel Prizes have been opened for those of more than 50 years ago, we know that Einstein was nominated eight times for the Nobel Prize in Physics for his work on the special theory of relativity but was blocked by a member of the physics selection committee who did not believe there was any validity to this theory. This is only one of many examples where even great scientists do not accept, or overlook, important new discoveries which are outside the accepted laws and theories. Today the special theory of relativity is one of the great milestones of the twentieth century. As another interesting aside, after the Russian revolution, Lenin and his followers barred Einstein's theories from being taught for many years in Russia. This was because he felt that Einstein's theory, in which matter and energy could be interchanged, did away with the importance of the material world and so was opposed to Dialectical Materialism.

Before the special theory of relativity, our laws of motion were based on the works of Newton and Galileo. Motion occurred in absolute space and time according to the laws of motion of Newton. All motion could be judged relative to some absolute frame of reference.

Now look at the basic principles as they were seen in 1900. The principle of inertia stated that an object at rest tends to remain at rest unless acted on by an outside force. But this is not true in an accelerating frame of reference. For example, consider a coin sitting on a moving book. It is at rest in the reference frame of the book. But if the book is accelerated by stopping it, the coin continues moving. The same occurs in a rotating (accelerating) frame of reference. Newton's laws hold in nonaccelerating frames ($v = 0$ or constant). So inertial frames of reference are all frames which move with constant (including zero) velocity.

To go from one inertial frame of reference to another requires a Galilean transformation in which velocities are added in the simplest (commonsense) way. Consider a train moving with constant velocity $\boldsymbol{u}$. A person on the train throws a ball with velocity $\mathbf{v}'$ with respect to the train. A person on the ground sees the object with a velocity

$$\mathbf{v} = \mathbf{v}' + \boldsymbol{u} \tag{1.1}$$

Let $\mathbf{v}'$ and $\boldsymbol{u}$ be in the x direction. Then choosing coordinate systems that coincide at $t = 0$, that is, the origins of the coordinate systems are at $x_0 = x_0' = 0$, when $t = 0$ (see Fig. 1.1),

$$\mathbf{v} = \mathbf{v}' + \boldsymbol{u} \tag{1.1}$$

becomes

$$\begin{aligned} \mathrm{v}_x &= \mathrm{v}_{x'} + u_x \\ \mathrm{v}_y &= u_y = 0 \\ \mathrm{v}_z &= u_z = 0 \end{aligned} \tag{1.2}$$

where

$$x = x' + ut, \qquad y = y', \qquad z = z' \tag{1.3}$$

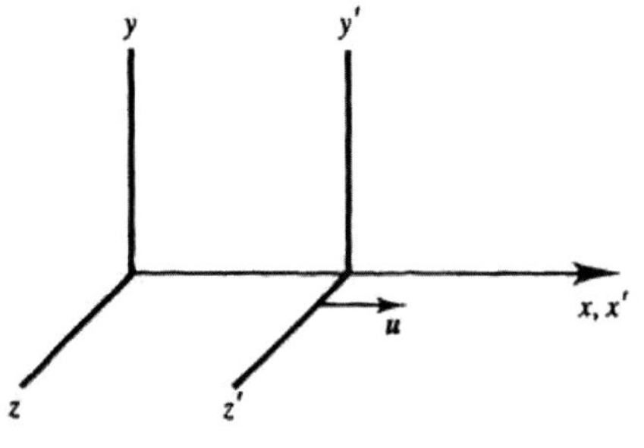

FIGURE 1.1
The x', y', z' reference frame moves with velocity u in the x direction relative to frame x, y, z.

From these transformation equations, we see that the acceleration $d\mathbf{v}/dt$ and thus the forces on the object are the same in both systems:

$$\frac{d\mathrm{v}_x}{dt} = \frac{d\mathrm{v}_{x'}}{dt} + \frac{du_x}{dt} \tag{1.4}$$

But u_x is constant, so

$$a_x = a_{x'} \tag{1.5}$$

$$\boldsymbol{F} = m\boldsymbol{a}_x = m\boldsymbol{a}_{x'} \tag{1.6}$$

Also, both observers will agree on conservation of energy and other fundamental laws. We have had practical, commonsense experience with these laws. For example, running with or against the wind, we easily sense the difference in our motion relative to the ground. World records (v) are not allowed in track events if the wind speed (u) exceeds 2.0 miles/hour. If you fly to Europe, the time of flight is shortened or lengthened depending on whether there is a tail or a head wind blowing to add to or subtract from the plane's velocity, $\mathrm{v} = \mathrm{v}' \pm u$ as shown in Fig. 1.2, and similarly if you are swimming or going by boat upstream or downstream in a river. But there is a subtle problem with this analysis.

As Lord Kelvin noted in 1900, the Michelson and Morley experiment provided evidence that Eq. (1.1) was incorrect! Maxwell's work explained light as an electromagnetic (EM) wave of transverse oscillating electric and magnetic fields. In classical physics all waves need a medium for propagation. So what is the medium on which EM waves propagate? We do not experience any effects of such a medium. So an ether—an invisible, massless medium which permeates all space—was proposed. Its only function was to allow the propagation of EM waves. This was again simple common sense. After all, water waves need water and sound waves need a gas, or liquid, or solid for transmission. However, since the ether permeates all space, it can serve as an absolute coordinate system by which to judge absolute motion. For example, measuring the motion of the earth through the ether would give its absolute motion.

Michelson and Morley set out to search for the ether. They set up an experiment shown in Fig. 1.3. The light beam is split by a half-silvered mirror into two orthogonal horizontal beams. When they recombine, if there is a phase difference, an interference pattern should be observed. The phase difference can arise from an actual difference in path length. In addition, the velocity of the earth through the ether should lead to a different path length for light going in the direction of the earth's motion relative to light going perpendicular to that

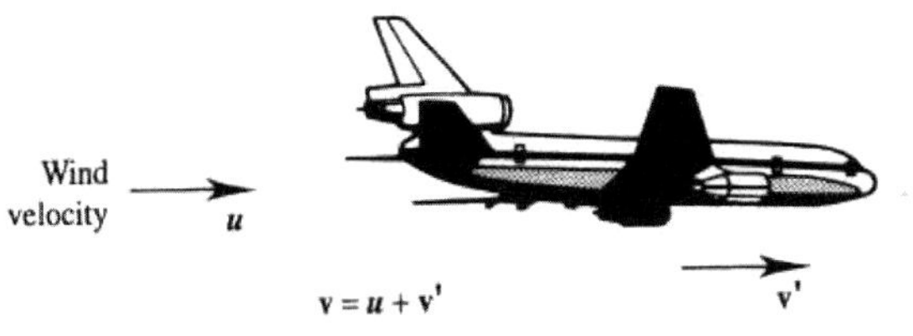

FIGURE 1.2
An airplane is traveling with velocity $\mathbf{v}'$ with respect to still air and experiences a wind blowing with velocity $\boldsymbol{u}$ in the direction of its motion relative to the earth, so its net velocity relative to the earth is $\mathbf{v} = \mathbf{v}' + \boldsymbol{u}$.

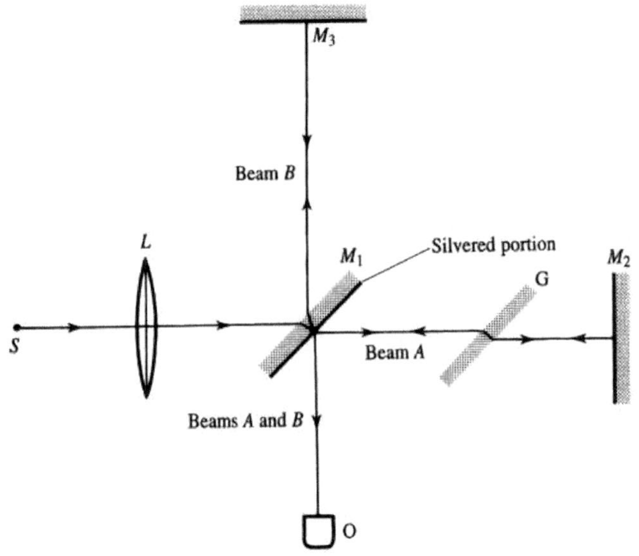

FIGURE 1.3
A schematic diagram of the apparatus used by Michelson and Morley. The apparatus was mounted on a stone slab that was floating in a pool of mercury to reduce vibration. Light from a source, S, was collimated by the lens, L, and split into two coherent beams by the half-silvered mirror, M_1. Beams A and B were reflected back from mirrors M_2, and M_3 to M_1 and then superimposed to produce an interference pattern seen by an observer at O. Actually, multiple mirrors were used on paths A and B to lengthen the paths for increased sensitivity. A glass plate, G, was inserted to insure that beams A and B had the same optical paths.

motion. It is like swimming upstream and then downstream as compared with swimming across stream and then back. So the two light rays when recombined should exhibit an interference pattern. However, it would be impossible to observe the shift in path length directly. If the system is rotated by 90°, then one should see a phase shift in the fringes, since the two beams would have exchanged places as to which is going "across the stream." Michelson and Morley carried out this experiment and, to their surprise, they saw *no* shift in the interference pattern. They thought that perhaps the orbital motion of the earth was canceling the earth's motion through the ether. Good scientists must be sensitive to considering the possible distortions of their results. They repeated the measurement six months later when the earth's direction of motion in its orbit would have reversed direction. They still found no shift, as reported in 1887. This result deeply disturbed scientists for nearly twenty years. The special theory of relativity offered an explanation. However, many scientists still did not accept Einstein's explanation until many years later. In Figs. 1.4(a) and 1.4(b) Michelson and Einstein are seen enjoying their favorite sports.

About five years after Michelson and Morley's results, Lorentz (more than 10 years before Einstein) worked out the coordinate transformation equations

FIGURE 1.4 (a)
Michelson indulging in a game of billiards. (Reprinted with permission of Macmillan Publishing Co. from *The Master of Light* by D. M. Livingston).

necessary to account for the Michelson–Morley result. In 1900 Larmor wrote the Lorentz transformation equations in the following form for the transformation from a frame x', y', z' moving with velocity u in the x direction in reference frame x, y, z (Fig. 1.1):

$$x = (x' + ut')\gamma, \qquad y = y', \qquad z = z'$$
$$t = \left(t' + \frac{ux'}{c^2}\right)\gamma \tag{1.7}$$

$$x' = (x - ut)\gamma, \qquad y' = y, \qquad z' = z$$
$$t' = \left(t - \frac{ux}{c^2}\right)\gamma \tag{1.8}$$

where $\gamma = (1 - \beta^2)^{-\frac{1}{2}}$, $\beta = u/c$, and c is the speed of light.

Einstein was able to derive these equations from two simple postulates, the postulates of relativity:

1. The speed of light in vacuum is the same in all inertial reference frames.
2. The laws of physics are the same in all inertial reference frames.

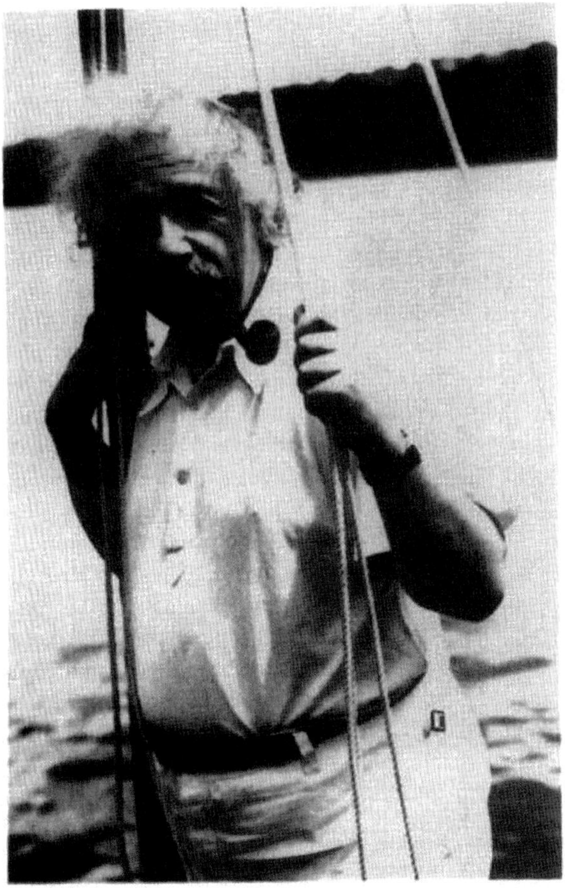

FIGURE 1.4 (b)
Einstein sailing off Long Island, about 1935 (AP/Wide World Photos).

A flash of light at $t = 0$ is seen by an observer in the x frame to expand in a spherical shell and satisfies

$$x^2 + y^2 + z^2 = c^2t^2 \tag{1.8a}$$

The same flash is seen by an observer in the x' frame with

$$x'^2 + y'^2 + z'^2 = c^2t'^2 \tag{1.8b}$$

For the x' frame moving as in Fig. 1.1 with respect to the x frame, we want the relation between (1.8a) and (1.8b). A Galilean transformation would give

$$(x' + ut')^2 + y'^2 + z'^2 = c^2t'^2 \tag{1.8c}$$

We can let

$$y' = y \tag{1.8d}$$

$$z' = z \tag{1.8e}$$

but trouble arises for the x and t components. We write a general linear transformation for these coordinates and then require (1.8a) and (1.8b) to be satisfied:

$$x' \equiv ax + bt \tag{1.8f}$$

$$t' \equiv dx + et \tag{1.8g}$$

and as $u \to 0$ we must have $a, e \to 1$ and $b, d \to 0$. The origin of the x' frame moves at u in the x frame, thus for $x' = 0$ (1.8f) implies $b = -ua$, since we must have $x = ut$. Now substituting (1.8f) and (1.8g) into (1.8b), which we equate with (1.8a), gives algebraic relations that we solve to find

$$a = e = \frac{1}{\sqrt{1 - \dfrac{u^2}{c^2}}}$$

and

$$d = \frac{-u}{c^2}\, a$$

Combining all these results gives the Lorentz transformations (1.7) or (1.8) as promised.

Note that one of the important consequences of these equations is that space and time are mixed together. From these equations, one can derive the equation for length contraction of an object in a moving frame of reference. For example, see Fig. 1.5. As seen from Eq. (1.7) and Fig. 1.5, only the length in the direction of the relative motion contracts. The other two lengths remain the same. An observer on a street corner measures the length of a building along the street as $L_B = x_2 - x_1$. An observer in a car moving at a high speed measures the length of the building to be $L'_B = x'_2 - x'_1$. To obtain this length, the positions x'_2 and x'_1 must be measured at the same time, $t'_1 = t'_2$. By using Eq. (1.7),

$$L_B = x_2 - x_1 = \gamma(x'_2 + ut'_2) - \gamma(x'_1 + ut'_1) \tag{1.9}$$

but $t'_1 = t'_2$, so

$$L_B = \gamma(x'_2 - x'_1) = \gamma L'_B \tag{1.10}$$

or

$$L'_B = \frac{1}{\gamma} L_B = \sqrt{1 - \frac{u^2}{c^2}}\, L_B \tag{1.11}$$

This is the Lorentz–FitzGerald space contraction equation. Remember in using this equation that L' is the length of the object seen to be moving by an observer as measured by this observer.

The length of the building is measured to be shorter to the observer in the car than to the observer on the street. Remember that if the person on the street measures the length of the car, the same equation now applies with the roles reversed. So the observer on the street measures the length of the car to be L'_C, which is shorter than the length L_C measured by the driver of the car. To the

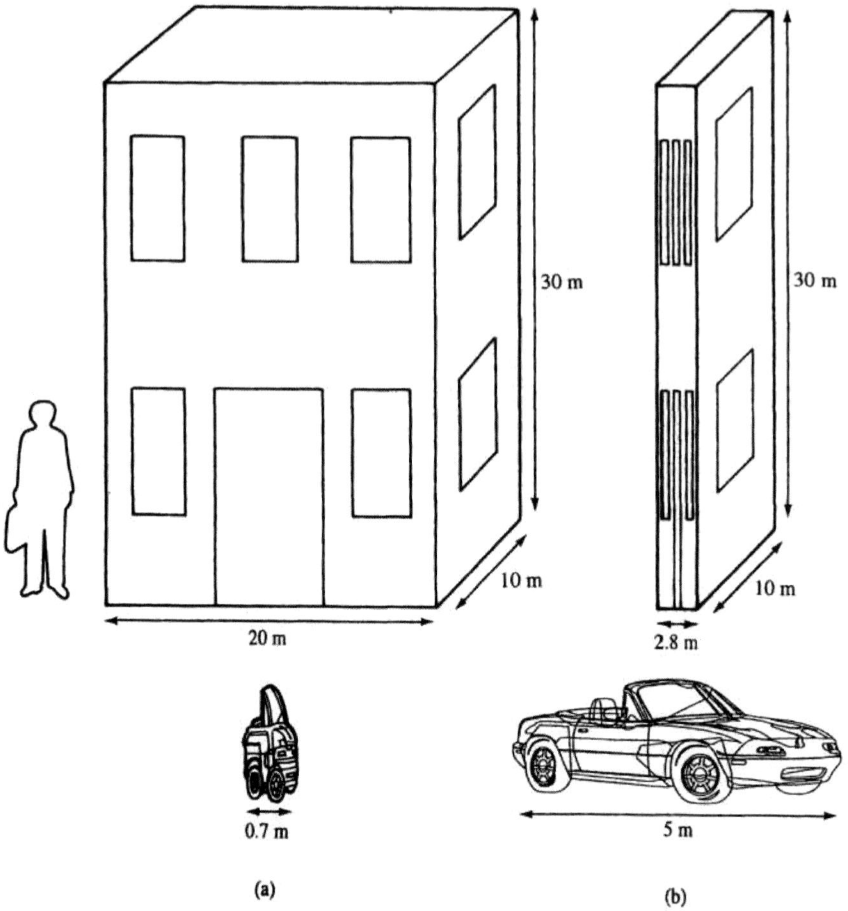

FIGURE 1.5
An illustration of a car passing the front of a building at velocity $u = 0.99c$, (a) as seen by an observer on the corner and (b) as seen by the driver.

person in the car the buildings contract in their direction of motion but the height and depth of the building are unchanged, while to the person on the street the length of the car contracts but its height and width are unchanged. The contraction occurs only in the direction of motion. If the length of the front of the building along the street, as measured by the person on the street, is 20 meters, then the person in a car moving down the street in front of the building at $u = 0.99c$ measures it to be

$$L' = 20\sqrt{1 - (0.99)^2} = 20 \times 0.14 = 2.8 \text{ meters}$$

The driver sees the length of the front of the building to be shorter than the length of the car, which the driver measures to be 5 meters. The driver and the person on the corner both measure the same height for the building and the same depth for the building. The person on the street corner measures the length of the car to be $L' = 5 \times 0.14 = 0.7$ meters, which is thirty times shorter than what the street observer measures for the length of the building. Again, the two both measure the same height for the car of 1.6 meters.

In the same way we can obtain the so-called dilation equation. For a clock fixed (at rest) in the x frame with time Δt between clock ticks,

$$\Delta t = t_2 - t_1$$

then

$$\Delta t' = \gamma\left(t_2 - t_1 + \frac{u}{c^2}[x_2 - x_1]\right) \tag{1.12}$$

but $x_2 - x_1 = 0$ since the clock is at rest in the x frame. Thus,

$$\Delta t' = \gamma(t_2 - t_1) = \gamma\,\Delta t \tag{1.13}$$

From this equation time is seen to run slow in the moving frame to an observer in the stationary frame. For a car moving down the street at $0.99c$ ($\beta^2 = 0.98$ and $1/\gamma = 0.14$), when the car clock showed 60 seconds to have passed to the driver, the driver would see only $60 \times 0.14 = 8.4$ seconds to have passed on the building clock. It would take an additional 368 seconds for a total of 428 seconds on the clock in the car for the driver to observe the building clock's second hand to go around once (60 s):

$$\Delta t'_{\text{car}} = \frac{60\text{ s}}{0.14} = 428\text{ s}$$

Likewise, a person on the street would say that it took 428 seconds on the building clock for the seconds hand on the car clock to go around once.

The driver sees the 5 m long car pass one end of the building in a time

$$\Delta t = \frac{5\text{ m}}{0.99c} \approx 1.7 \times 10^{-8}\text{ s}$$

A person on the street, who sees the car length as 0.7 m, says to the driver, "You're crazy. It took only

$$\frac{5\text{ m}/\gamma}{0.99c} = \frac{0.7\text{ m(car)}}{0.99c} = 2.4 \times 10^{-9}\text{ s}$$

for your car to pass the end of the building." On the other hand, if you asked the driver how long it would take for the front end of the car to go from one edge of the building to the other (a length of 2.8 m as seen by the driver), the driver would say, "According to the car clock

$$(\Delta t_{\text{Driver}}) = \frac{2.8\text{ m}}{0.99c} = 0.94 \times 10^{-8}\text{ s}"$$

while the person on the street corner would say, "According to the street clock

$$(\Delta t_{\text{street observer}}) = \frac{20\text{ m}}{0.99c} = 6.7 \times 10^{-8}\text{ s}"$$

Note that asking how long it takes for the building to pass one end of the car is not the same as asking the question embodied in Eq. (1.13), which asks "when 0.94×10^{-8} s has passed on the car clock as seen by the driver, how much time has passed on the building clock as seen by the driver?" When 0.94×10^{-8} s have passed on the car clock, the driver sees the wall clock as having advanced only $0.14 \times 0.94 \times 10^{-8}\text{ s} = 0.13 \times 10^{-8}$ s according to Eq. (1.13). Time dilates (runs slower) in the moving system compared to a clock in the observer's frame. Always make sure you set up the equation so this occurs. Each observer sees the other's clock run slow, so events seen in the moving frame are like a picture run in slow motion.

Time dilation applies to the biological clocks that determine our age. The fact that each observer sees the others' clock run slow gives rise to the famous twins paradox. In a space expedition one twin, Jan, remains on earth and the other twin, Nell, travels in a spaceship at $0.5c$ to a distant star which is 10 light years away (assuming, for simplicity, that the galaxy is not moving relative to the earth). Jan will calculate that Nell's clock runs slow compared to her own and expect Nell to be younger when she returns, while Nell will calculate that Jan's clock runs slow compared to hers and expects Jan to be younger. The paradox is that only one twin can be younger! Which one is really younger and why?

The laws of special relativity apply only to frames of reference which are moving relative to one another at constant velocity—inertial frames of reference. Suppose Nell is accelerated to $0.5c$ rapidly and then travels at constant velocity relative to Jan to the star. Already Nell has gone from the earth's frame to another frame. Upon arrival, Nell must undergo an acceleration from a frame moving away from earth at $0.5c$ to another frame of reference moving toward earth at $0.5c$. Thus, Nell goes from one inertial frame to another to return. These changes of inertial frames are what lead to Nell being the younger twin upon arrival, as both Jan and Nell agree. The space travelers who go out and return will always be the younger.

An elegant demonstration of this was made by comparing two extremely precise synchronized atomic clocks; one was left in the United States and one was put on an airplane which circled the earth. The airplane test is more complex because of the rotation of the earth and predictions of general relativity about changes that the clock experiences from the change in gravitational field strength with altitude. Nevertheless, taking these effects into consideration, the clock that flew off, circling the globe to return to the same place, ran slower as predicted by special relativity, within the experimental errors, providing an elegant confirmation of the theory, as described by J. C. Hafele and R. E. E. Keating [*Science* **177** (1972) 166].

If the above equations were known before 1905, what was Einstein's contribution in developing the special theory of relativity? The above results were developed as an explanation of the Michelson–Morley experiment. Einstein set forth basic principles from which these equations follow as a natural consequence, as does the result of the Michelson–Morley experiment. In addition, these principles, as summarized in the two postulates of the special theory of relativity, give rise to other new and unexpected phenomena.

While the Michelson–Morley experiment suggests that the speed of light is the same upstream, downstream, and across stream, Einstein's postulates were an inspired, complete break with classical physics. Postulate 1 simply sets forth the constancy of the speed of light in vacuum as a fundamental principle of nature. As an example of the consequence of this postulate, if a spaceship leaves earth with a speed $c/4$ away from earth and sends out a light signal in its direction of motion with velocity c relative to the spaceship, an observer on earth will observe a speed c for the speed of light—not $c + c/4$ as one would expect from ordinary experience in a Galilean transformation. In relativity, velocities no longer add linearly as in a Galilean transformation. Remember that the velocity of light in materials is less than its velocity in vacuum (their relative velocities are expressed in the index of refraction). It is possible for a particle to move faster in a material than light moves in that material. The Cerenkov effect involves visible light radiation being emitted by electrons moving in a liquid or other material at a velocity that exceeds the velocity of light in that material. The emitted light comes out in a shock front very similar to the sound wave shock front (the "sonic boom") generated when an airplane travels faster than the speed of sound in air.

One can derive new velocity relations from the Lorentz transformation equations for two reference frames moving with velocity u relative to each other (Fig. 1.1), where the velocity is u in the x direction. We find (as shown below)

$$\begin{aligned} \mathrm{v}'_x &= \frac{\mathrm{v}_x - u}{1 - \mathrm{v}_x u/c^2} \\ \mathrm{v}'_y &= \frac{\mathrm{v}_y\sqrt{1 - u^2/c^2}}{1 - \mathrm{v}_x u/c^2} \\ \mathrm{v}'_z &= \frac{\mathrm{v}_z\sqrt{1 - u^2/c^2}}{1 - \mathrm{v}_x u/c^2} \end{aligned} \tag{1.14}$$

Note that now v'_y and v'_z both depend on v_x, too. We can derive these equations by noting, for example,

$$\mathrm{v}'_x = \frac{dx'}{dt'} = \frac{dx'}{dt}\frac{dt}{dt'}$$

From Eq. (1.7),

$$\frac{d}{dt}x' = \frac{d}{dt}(x - ut)\gamma = (\mathrm{v}_x - u)\gamma$$

and for Eq. (1.6),

$$\frac{dt'}{dt} = \frac{d}{dt}\left(t - \frac{ux}{c^2}\right)\gamma = \left(1 - \frac{u\mathrm{v}_x}{c^2}\right)\gamma$$

So,

$$\mathrm{v}'_x = \frac{(\mathrm{v}_x - u)\gamma}{\left(1 - \frac{u\mathrm{v}_x}{c^2}\right)\gamma} = \frac{\mathrm{v}_x - u}{1 - \frac{u\mathrm{v}_x}{c^2}}$$

In one dimension, suppose we have two objects moving with velocities v_{x_1} and v_{x_2}. Their relative velocity, v', as measured by an observer in x', which is moving with velocity $u = \mathrm{v}_{x_2}$ relative to the x frame is

$$\mathrm{v}' = \frac{\mathrm{v}_{x_1} - \mathrm{v}_{x_2}}{1 - \frac{\mathrm{v}_{x_1}\mathrm{v}_{x_2}}{c^2}} \tag{1.15}$$

So if two objects move toward each other with $\mathrm{v}_{x_1} = c$ and $\mathrm{v}_{x_2} = -c$, then their velocity of approach as seen by either object is

$$\mathrm{v}' = \frac{c - (-c)}{1 - c(-c)/c^2} = \frac{2c}{2} = c$$

not $2c$. The same would hold if they were moving away from each other with each traveling at a velocity $= c$.

Suppose a spaceship moving at velocity $u = c/2$ in the x direction emits a light ray with velocity $\mathrm{v}'_x = c$ as seen by an observer on the spaceship (see Fig. 1.6). What will an observer on the ground see for the light velocity v_x?

$$\mathrm{v}'_x = c = \frac{\mathrm{v}_x - u}{1 - \mathrm{v}_x u/c^2} = \frac{\mathrm{v}_x - \frac{1}{2}c}{1 - \mathrm{v}_x \frac{1}{2}\frac{c}{c^2}}$$

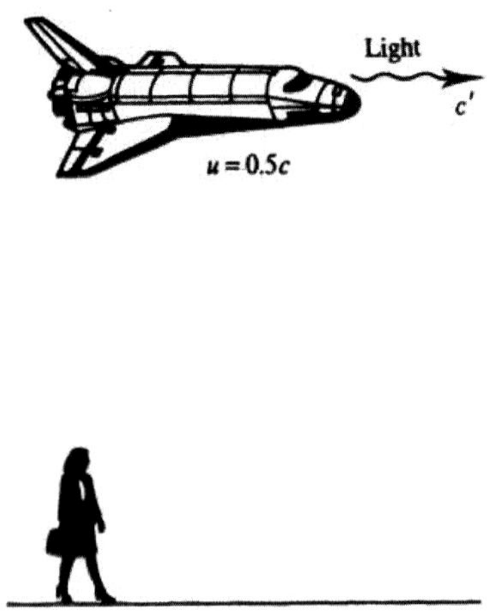

FIGURE 1.6
An observer on the ground sees a spaceship traveling at $c/2$ in the x direction emit a light signal with $c' = c$ in the x direction as seen by an observer on the spaceship. In the ground observer's rest frame, what does the ground observer measure for the light signal's velocity in the x direction?

Multiply each side by the denominator on the right (using c for the left-hand side); then

$$c - \tfrac{1}{2}\mathrm{v}_x = \mathrm{v}_x - \tfrac{1}{2}c$$
$$\tfrac{3}{2}c = \tfrac{3}{2}\mathrm{v}_x$$
$$\mathrm{v}_x = c$$

So the observer on the ground likewise measures a velocity c.

This new velocity addition equation has been well verified experimentally. One of the most accurate tests was that of Alvager et al. [*Phys. Lett.* **12** (1964) 260] who measured, in the laboratory, the velocity of a gamma ray emitted in the decay of a π^0 meson in the direction of motion of the π^0, which had $\mathrm{v} = 0.999\,75c$. They expressed their measured results for the velocity of light in the laboratory, c', in terms of the equation $c' = c + k\mathrm{v}$, where v is the velocity of the moving source (the π^0) and k is a constant. They measured $c' = 2.9977 \pm 0.004 \times 10^8$ m/s to yield $k = (-3 \pm 13) \times 10^{-5}$ confirming that c is the maximum velocity to within a few parts in 100 000.

Postulate 2 of the special theory is called the principle of relativity. It is a complete break with classical physics. Postulate 2 says there is no absolute reference frame by which we can judge motion or measure velocities. We can only measure the motion of one inertial system relative to another—not their absolute motion. Whether there is or is not some grand absolute reference frame, we can never know according to postulate 2 because we can never do an experiment which will demonstrate its existence. So for us it does not exist in our experience. Thus, the Newtonian concepts of absolute space and time no longer exist and space and time are mixed up for different observers.

We have already seen that some known consequences of relativity, namely the Lorentz transformation equations, the Lorentz–FitzGerald contraction $L' = L/\gamma$, and time dilation $\Delta t' = \gamma\,\Delta t$ flow naturally from Einstein's two postulates.

At the time of Einstein's paper, the Michelson–Morley experiment could have been the experimental impetus for his inspirational, imaginative work. (It is still controversial today whether Einstein was actually familiar with the work of Michelson and Morley.) As one began to explore the properties of the newly discovered (1897) particle, the electron (and much later other similarly small particles), the special theory of relativity was found to provide an explanation of their motions and velocities as their velocities approached the velocity of light. In more recent times a nice example of the confirmation of the length contraction and time dilation equations is provided by the decay of μ mesons. Muons are unstable particles which can be created in a collision which involves a high-energy particle. In the rest frame of the muon it decays into an electron and neutrino with a lifetime τ of about 2×10^{-6} s; this is what is measured in the laboratory when muons decay at rest. Muons are produced in our upper atmosphere at a height of about 100 km by the collision of high-energy cosmic rays with atoms. In our rest frame, in a time of 2×10^{-6} s, a muon moving at the velocity of light, 3×10^8 m/s, would travel only 600 m. So, how does the muon reach the surface of the earth 100 km from the point of its creation? In their rest frame muons do live for only 2×10^{-6} s; however, when viewed from our frame of reference, the muon's clock runs much slower. To travel 100 000 m at 3×10^8 m/s takes about 3.3×10^{-4} s. What velocity must the muon have for its time dilation to yield $\Delta t' \simeq 3.3 \times 10^{-4}$ s, for the muon to decay as measured by an observer on Earth, while the muon clock shows only 2×10^{-6} s elapsed?

$$\Delta t' = \frac{\Delta t}{\sqrt{1 - u^2/c^2}}$$

$$3.3 \times 10^{-4}\ \text{s} = \frac{2 \times 10^{-6}\ \text{s}}{\sqrt{1 - u^2/c^2}}$$

Solving for u yields $u = 0.999\,982c$. A muon created at the top of our atmosphere and traveling at this velocity would then reach the Earth's surface about the time it decays. We could have looked at the question from the perspective of the muon by asking what the length of the atmosphere is to the muon? The muon at this high velocity sees the height of the atmosphere contracted to

$$l' = l(1 - u^2/c^2)^{1/2} = (100\ \text{km})(1 - 0.999\,982^2)^{1/2} = 100\ \text{km} \times 1/165 = 606\ \text{m}$$

which is essentially the distance it can travel in its lifetime of 2×10^{-6} s at $u = 0.999\,982c$.

When the first mile-long linear accelerator for electrons was built at SLAC at Stanford University, there was worry about what the intensity of the electrons would be at the end of this long path. Typically, new accelerators do not give the calculated intensity performance during initial operation. Surprisingly, they found the intensity to be much greater than they had expected. How was this possible? It was suggested that it was a relativity effect. The electrons did not see a 1.6 km-long tube; at their injection velocity into the accelerator they saw a tube only 1 to 2 cm long, and in this short distance did not have a chance to disperse out of the beam.

The next consequence of the transformation equations involves the simultaneity of two events at different locations. The time difference between two events is not the same in different inertial systems which move relative to each other. Time differences in the two frames are given by

$$t_2' - t_1' = \left[(t_2 - t_1) - \frac{u}{c^2}(x_2 - x_1)\right]\gamma \tag{1.16}$$

Here time and space are mixed together. The time $t_2' - t_1'$ depends on the spatial positions of x_2 and x_1. Thus, two events which occur simultaneously at times $t_2 = t_1$ in the stationary frame occur at different times in the moving frame, $t_2' \neq t_1'$ unless $x_2 = x_1$. Note that the order of two events in which t_1 occurs before t_2 can never be changed so that t_1' occurs after t_2' (i.e. $t_1' > t_2'$) as long as

$$t_2 - t_1 > \frac{(x_2 - x_1)u}{c^2}$$

This says that as long as the time interval $t_2 - t_1$ is longer than the time it would take to send a signal at the velocity of light c between x_2 and x_1, then one cannot see a reversal in the order of events. So one can never change the order of two causal events; for example, if an event at x_1,t_1 causes an event at x_2,t_2 then there is no reference frame where one sees the event at x_1',t_1' occur after x_2',t_2'. This is because for the event at x_1,t_1 to cause the event x_2,t_2, it must be able to send a signal to event x_2,t_2, and a signal cannot travel faster than the velocity of light. For example, if the events are separated by $t_2 - t_1 = 5\,\text{s}$ and it only takes 3 s to send a signal at a velocity $= c$ from event x_1 to x_2, then event x_1 can in principle cause x_2. In this case the simultaneity equation says that there is no reference frame where event 1 occurs after event 2 to destroy cause and effect. The separation between events which can be connected by a light signal is called timelike event, and the time order of these events cannot be reversed. The separation between events which cannot be connected by a light signal is called spacelike, and the order of two such events can be reversed in different reference frames.

Consider the car–building problem discussed earlier, where the car had $u = 0.99c$. On the corner there is a street sign 0.7 m wide to the person standing on the street. Since the street observer measures the moving car to be 0.7 m long also, then he would have seen the front end of the car line up with one end of the sign at the same time as the back end of the car lined up with the other end of the sign. However, the driver would see the 0.7 m sign as

$$\begin{aligned} L' &= 0.7 \times 0.14\ \text{m} \\ &= 0.098 \approx 0.1\ \text{m} \end{aligned}$$

To the driver the time between when the nose of the 5 m-long car first reached the far end of the sign until the tail of the car reached the first edge of the sign would be

$$\Delta t = \frac{5\,\text{m} - 0.1\,\text{m}}{0.99c} = 4.95\ \text{m}/c = 1.65 \times 10^{-8}\ \text{s}$$

So the driver would not see the times that the ends of the car reached the two ends of the sign as being simultaneous as would the person on the street. In terms of Eq. (1.16) with $t_2 = t_1$ and $x_1 - x_2 = 0.7\,\text{m}$,

$$\begin{aligned} t_2' - t_1' &= 0 + \frac{\left(\frac{u}{c^2} \times 0.7\,\text{m}\right)}{\sqrt{1 - u^2/c^2}} \\ &= \frac{0.7\,\text{m} \times 0.99/c}{0.14} = 4.95\,\text{m}/c = 1.65 \times 10^{-8}\ \text{s} \end{aligned}$$

The Lorentz transformation equations also lead to a new relativistic equation for the Doppler shift in the frequency ν and wavelength λ of a wave when there is relative motion between an observer and a source. The relativistic Doppler shift equations are obtained using the fact that the phase of a light wave, which is related to the number of maxima and minima passing a point in space in a time t, is invariant under Lorentz transformations. So for a traveling wave

$$e^{(\mathbf{k}\cdot\mathbf{x}-\omega t)} = e^{i(\mathbf{k}'\cdot\mathbf{x}'-\omega' t)}$$

or

$$\mathbf{k}'\cdot\mathbf{x}' - \omega' t = \mathbf{k}\cdot\mathbf{x} - \omega t$$

For the x' system moving at $\mathbf{u} = u\hat{x}$ and light moving in the $\hat{x}$ direction, we have (it is possible to derive more general formulas when these vectors are not in the same direction)

$$k'_y = k_y = 0, \qquad k'_z = k_z = 0$$

and

$$k'_x = \gamma\left(k_x - \beta\frac{\omega}{c}\right), \qquad \omega' = \gamma(\omega - uk_x)$$

But for light in vacuum $k_x = |\mathbf{k}| = \omega/c$ and $k'_x = |\mathbf{k}'| = \omega/c$,

$$\omega' = \gamma(\omega - \beta\omega) = \omega\gamma(1-\beta)$$

or

$$\omega' = \omega\sqrt{\frac{1-\beta}{1+\beta}}$$

The angular frequency can be written in terms of the frequency,

$$\nu' = \nu\frac{\sqrt{1 \pm u/c}}{\sqrt{1 \mp u/c}} \tag{1.17}$$

or inverted to give the transformation for the wavelength,

$$\lambda' = \lambda\frac{\sqrt{1 \mp u/c}}{\sqrt{1 \pm u/c}} \tag{1.18}$$

where the upper signs under the square roots (+ in the numerator and − in the denominator in Eq. (1.17)) are used when the observer and source approach each other with relative velocity u. These equations must be used when calculating the relative motions between ourselves and distant stars and galaxies. For example, suppose a distant galaxy is moving away from us. You find that the 434 nm blue line from hydrogen is shifted to 1892 nm in the infrared region. Then the galaxy must be moving away from us with velocity u to produce such a shift, so

$$\lambda' = \lambda\sqrt{\frac{1+u/c}{1-u/c}} = 1892\,\text{nm} = 434\,\text{nm}\sqrt{\frac{1+u/c}{1-u/c}}$$

Rearranging,

$$\frac{1+u/c}{1-u/c} = \left(\frac{1892}{434}\right)^2 = (4.36)^2 = 19.0$$

$$\left(1+\frac{u}{c}\right) = 19\left(1-\frac{u}{c}\right)$$

$$20\frac{u}{c} = 18$$

$$u = 0.9c$$

The galaxy is moving away from us at 90 percent of the velocity of light. This is an example of the famous red shift of light emitted by all distant stars and galaxies.

Next, let us look at unexpected consequences of Einstein's postulates. We have already seen the consequence that velocities add differently. Now in relativity

$$\boldsymbol{F} \neq m\boldsymbol{a}$$

If $\boldsymbol{F}$ were equal to $m\boldsymbol{a}$, then by applying $\boldsymbol{F}$ long enough the velocity of an object would continue to increase until its velocity exceeded the velocity of light in vacuum. This is not possible according to Einstein's postulates, so something is wrong. Einstein solved the problem by noting that

$$\boldsymbol{F} = \frac{d}{dt}(\boldsymbol{p}) \tag{1.19}$$

But now one has a new definition of momentum,

$$\boldsymbol{p} = \frac{m_0\mathbf{v}}{\sqrt{1-\mathrm{v}^2/c^2}} = \frac{m_0\mathbf{v}}{\sqrt{1-\beta^2}} = m_0\mathbf{v}\gamma \tag{1.20}$$

where m_0 is the mass measured when the object is at rest (i.e., its rest mass). This equation is well verified in modern physics. Evidence for this phenomenon had been observed already in the 1890s, when the charge to mass ratio, e/m, of cathode rays was observed to depend on the accelerating voltage across the cathode ray tube (and thus on electron velocity). However, these data were not understood because Thompson had not even proposed the electron yet. Historically, in most beginning physics texts this equation leads to the association of γ with mass, so that the mass of an object is said to increase with velocity and approaches infinity as its velocity approaches c. This idea was introduced earlier by Lorentz in 1899:

$$m = \frac{m_0}{\sqrt{1-\mathrm{v}^2/c^2}} \tag{1.21}$$

Here m is the mass of an object when its velocity is $\mathbf{v}$. Since, in this equation, the mass approaches infinity as its velocity approaches c, no finite force can accelerate an object to $\mathrm{v} = c$. In more advanced texts and some recent beginning texts, mass is treated as an invariant, a constant which does not change, and γ is associated with velocity to convert $\mathbf{v}$ into part of the relativistic velocity four-vector. In this case there is no need to write m_0 since the mass of an object at rest is the only mass that appears. This approach is particularly useful in using velocity four-vectors to treat special relativity. There is an ongoing debate as to which approach to use in introducing relativity. See, for examples, C. G. Adler [*Am. J. Phys.* **55** (1987) 739], L. B. Okun [*Physics Today* (1989) p. 31] and R. Baierlein [*The Physics Teacher* (1991), p. 170] on why not to introduce relativistic mass, and T. R. Sandin [*Am. J. Phys.* **59** (1991) 1032] on why to use relativistic mass. The two approaches do not differ in their physics, but each can lead to some misconceptions or potential difficulties in understanding. Einstein wrote in his later years that he felt it was best to use only the rest mass and not introduce Eq. (1.21). Part of the confusion is in our association of mass with inertia, the resistance offered when a force is used to accelerate an object. The other confusion relates to the equation $E = mc^2$ discussed next.

Equation (1.20) leads to a new relativistic energy equation. In either approach one finds that

$$\text{kinetic energy} \neq \frac{1}{2}\frac{m_0 \mathrm{v}^2}{\sqrt{1-\beta^2}}$$

In classical physics, the kinetic energy is defined in terms of the work done by an external force to give an object a certain speed starting from rest. So, for an initial kinetic energy of zero, the final kinetic energy of an object is

$$K = \int \boldsymbol{F} \cdot d\boldsymbol{x} \tag{1.22}$$

The relativistic force can be written as $\boldsymbol{F} = d\boldsymbol{p}/dt$ So,

$$K = \int \frac{d\boldsymbol{p}}{dt} \cdot d\boldsymbol{x} = \int \mathbf{v} \cdot d\boldsymbol{p}$$

Using integration by parts, one can show the relativistic kinetic energy is

$$K = \int \mathrm{v}\, dp = \int d(p\mathrm{v}) - \int_0^{\mathrm{v}} p\, d\mathrm{v} = \gamma m_0 c^2 - m_0 c^2 = mc^2 - m_0 c^2 \tag{1.23}$$

The total relativistic energy is the kinetic energy plus the rest energy,

$$E = m_0 c^2 + K = \gamma m_0 c^2 = mc^2 \tag{1.24}$$

W. Bertozzi [*Am. J. Phys.* **32** (1964) 551] has published a nice experimental verification of this equation. As taken from that paper, Fig. 1.7 compares the predictions of the relativistic equation for $(\mathrm{v}/c)^2$ with that of classical physics where $K = m_0 \mathrm{v}^2/2$.

Here again if one uses m_0 as an invariant and does not introduce Eq. (1.21), the total energy is written as $\gamma m_0 c^2$ not mc^2. Written as $E = mc^2$, this is the best-known equation of Einstein's by the general public, where the total energy of an object is measured in terms of its mass. (This was the equation that upset Lenin.) However, this is not what Einstein originally wrote. He wrote $E_0 = mc^2$ where E_0

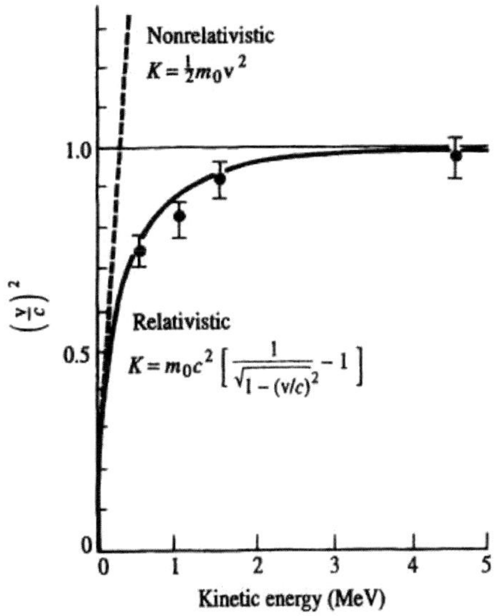

FIGURE 1.7
Experimental verification of the relativistic kinetic energy equation as reported by W. Bertozzi, *Am. J. Phys.* **32** (1964) 551.

is the rest energy of a stationary particle and $\Delta E = \Delta mc^2$ where ΔE is the energy gained or lost in some form when there was a change in mass of Δm. In 1900 Poincaré wrote $E = mc^2$ to indicate the mass of a light pulse with energy, E. However, we now know the photon is massless. Moreover, the gravitational mass of an object is not E/c^2. Nevertheless, $E = mc^2$ has become a part of our culture, but unfortunately its meaning is misunderstood by the public in general. Mass and energy are not equivalent as generally interpreted. In the equation $E = mc^2$, m is inertia (a measure of the property that resists changes in the motion or velocity of an object) or inertial mass not matter. The equation does not mean mass (inertia), resistance to changes in motion, and energy (ability to do work) are equivalent. It means E and m are proportional to one another and the proportionality constant is c^2, as stressed by Baierlein. What $\Delta E = \Delta mc^2$ tells us is that as the energy of a system changes, its inertia changes. It is not a conversion of energy into mass (or matter), e.g. there are not more protons or electrons or other particles when the energy increases. One can convert matter (protons and electrons) into electromagnetic radiation and vice versa, but that is a separate feature of nature.

Is it really true that one can exchange or convert mass to energy or vice versa? In 1905 no one had reported the conversion of matter into energy or energy into matter. Can the energy in light be transformed into particles and vice versa? Yes, since 1932 with the discovery of the positron, the energy in light has been observed under the right circumstances to be converted into a pair of particles, a particle and its antiparticle. Likewise, a particle and its antiparticle can combine and their rest masses and kinetic energies (if not at rest) will be converted into light energy with no mass. We receive sunlight from the Sun because nuclear processes convert some of the rest mass of the Sun into light energy. The Sun is a nuclear fusion reactor. The Sun converts into energy 4×10^{12} grams per second out of its present mass of about 2×10^{33} g. We will also see how exchange of mass and energy play other absolutely essential roles in our existence, for example in combining protons and neutrons to form nuclei, and the binding of electrons to nuclei to form atoms and to form molecules. For example, the binding energy of an electron in hydrogen is 13.6 eV (electron volts). So an electron gives up 13.6 eV of its rest mass when it is bound in hydrogen and this energy must be supplied to it to pull the electron out of the hydrogen atom and set it free (to be free, it must have its full rest mass). The equation $\Delta E = \Delta mc^2$ or $E = mc^2$ can be used to calculate the energy released when mass is converted to energy or vice versa, and the energy required to produce particles with rest mass. Since we will not be using four-vectors in this text, we will use the simple expression involving relativistic mass which is a function of velocity (Eq. 1.21) and use $E = mc^2$ to calculate conversion of mass into energy and vice versa.

The final very useful equation for dynamics is

$$E^2 = p^2c^2 + m_0^2c^4 \tag{1.25}$$

Note that this equation is independent of whether you use Eq. (1.21) to define a relativistic mass or just speak of a rest mass. This has the form of a Pythagorean theorem where pc and m_0c^2 are the sides of a right-angle triangle and E is the hypotenuse. Note that if a particle has no rest mass then

$$E = pc \tag{1.26}$$

The photon, and perhaps the neutrino, are particles with no rest mass, so this simplified equation holds for them. This equation is also the extreme relativistic limit in which the kinetic energy is so great that $pc \gg m_0c^2$ and the m_0c^2 can be neglected. As a rule of thumb, relativistic effects clearly set in for a particle as the relativistic kinetic energy approaches the particle's rest energy. The rest energy of an electron is 511 keV, and for a proton or neutron about 940 MeV.

Let us look at some examples of Eqs. (1.21)–(1.26).

(1) Find the total energy, kinetic energy, and momentum of an electron, whose rest energy $m_0c^2 = 511$ keV, when moving at a velocity of $0.6c$.

$$E = mc^2 = \frac{m_0c^2}{\sqrt{1 - v^2/c^2}} = \frac{511\,\text{keV}}{\sqrt{1-(0.6)^2}} = \frac{511}{0.8} = 639\,\text{keV}$$

$$K = E - m_0c^2 = 639\,\text{keV} - 511\,\text{keV} = 128\,\text{keV}$$

$$E^2 = p^2c^2 + m_0^2c^4$$

$$pc = \sqrt{E^2 - m_0^2c^4} = \sqrt{(639)^2 - (511)^2} = 383\,\text{keV}$$

$$p = 383\,\text{keV}/c$$

In relativistic problems it is common to give momentum in units of keV/c, MeV/c or GeV/c because such units make the computations easier. One could convert to kg · m/s by converting keV to joules and dividing by c. To calculate p, one could also use the equation

$$p = \frac{m_0 v}{\sqrt{1 - v^2/c^2}} = \frac{1}{c}\frac{m_0 c^2 v/c}{\sqrt{1 - v^2/c^2}} = \frac{511 \times 0.6}{0.8}\frac{\text{keV}}{c}$$
$$= 383\,\text{keV}/c$$

(2) Next, what is the total energy, momentum, and velocity of a neutron with 10 GeV of kinetic energy. The rest energy of the neutron is 0.94 GeV. Since the rest energy is small compared to the kinetic energy, we may approximate it as 0.9 GeV for this example.

$$E = K + m_0c^2 = 10\,\text{GeV} + 0.9\,\text{GeV} = 10.9\,\text{GeV}$$
$$pc = (E^2 - m_0^2c^4)^{1/2} = [(10.9)^2 - (0.9)^2]^{1/2}\,\text{GeV} \simeq 10.9\,\text{GeV}$$

Neglecting the rest energy changes the term under the square root by less than 1 part in 120. From Eq. (1.21), squaring and rearranging, we find

$$\frac{v}{c} = \sqrt{1 - \left(\frac{m_0}{m}\right)^2} = \sqrt{1 - \left(\frac{0.94}{10.9}\right)^2} = \sqrt{1 - 0.0074} = 0.996$$

In the above relativistic equations the following concepts of classical physics still hold in any inertial frame:

1. Newton's second law, $\boldsymbol{F} = \frac{d\boldsymbol{p}}{dt} = \frac{d(m\boldsymbol{v})}{dt}$
2. Conservation of energy
3. Conservation of linear momentum
4. Conservation of angular momentum

Finally, the above relativistic equations must all reduce to the classical physics equations, such as kinetic energy $= mv^2/2$, as the velocity approaches to zero. Thus,

$$K = mc^2 - m_0c^2 = \frac{m_0c^2}{\sqrt{1 - v^2/c^2}} - m_0c^2$$

As $v \to 0$ we expand (by Taylor series) the square root and keep terms of order v^2 to find

$$K \to m_0c^2\left(1 + \frac{1}{2}\frac{v^2}{c^2}\right) - m_0c^2 = \frac{1}{2}m_0v^2$$

The special theory of relativity must also be incorporated into the behavior of electric and magnetic fields when the source of these fields is moving at speeds approaching the velocity of light, and into the quantum behavior of matter. As a brief mention of the former, fast-moving charged particles lose energy by their electric fields interacting with the charged electrons and nuclei in matter, a

Coulomb interaction. The electric field around a stationary charged particle is spherically symmetric and falls in intensity with the square of the distance from the particle. The stopping power of matter decreases with increasing energy of the incident charged particles. However, as the velocity of light is approached, the electric field contracts (less intensity with distance) in the direction of motion and extends out further (greater intensity farther out) in the directions perpendicular to the direction of motion. This outward extension enables the charged particles to interact over a larger distance or cross-sectional area in the matter, and the stopping power (energy loss) shows a relativistic rise with increasing energy.

In summary, Einstein's two postulates of the special theory of relativity (a) the speed of light in vacuum has the same value in all inertial systems, and (b) the laws of physics are identical in all inertial systems (the postulate of relativity of motion) provided the underlying physical principles to explain the Michelson and Morley results. From these postulates flow all the equations to describe the motions of systems moving at velocities approaching the velocity of light.

1.2 GENERAL RELATIVITY

From the results of special relativity, it is clear that Newton's theory of gravity is incomplete. In addition, the special theory of relativity cannot treat phenomena in different noninertial frames (frames that are accelerating with respect to each other). Einstein's theory of general relativity, which is one of the triumphs of twentieth-century physics, treats accelerating noninertial frames and generalizes Newton's theory of gravitation. In pursuing to its logical conclusion the train of thought that led to special relativity, Einstein was led further and further into a realm where there was as yet little or no experimental data. However, once general relativity was completed, it made predictions for results of many new experiments. Some of these predictions, like the bending of light rays passing near the Sun, and the precession of the planetary orbits (the one effect of general relativity known before the theory was completed) were borne out by the 1920s, but other predictions, like the existence of gravitational waves, are still being tested. Although gravitational waves are as yet undiscovered, there is great hope that within a decade they will be seen by a new generation of experiments.

What is general relativity? First, consider an observer standing near the center of a rotating disk (or the surface of a rotating Earth). An object thrown toward the outer edge of the disk will follow a curved path in the rotating frame of reference—not a straight path. As another example, a Foucault pendulum on the surface of the Earth changes its plane of rotation in the absence of any external force. Newton's laws of motion do not hold in accelerating noninertial frames of reference. One can restore Newton's laws by assuming the existence of inertial forces in addition to other forces present given by

$$F = \text{inertial mass} \times \text{acceleration}$$

The Coriolis and centrifugal forces are inertial forces in a rotating frame of reference.

Einstein was led to the *principle of equivalence* by noting that since the period of a pendulum is independent of the mass of the pendulum, its inertial and gravitational masses must be equal. So, locally there is no way to distinguish between gravitational and inertial forces. The principle of equivalence states that

> the effect of a homogeneous gravitational field is equivalent to that of a uniform accelerated reference frame

where the acceleration is in the opposite direction to that of the gravitational field. Inside a closed box it is impossible to tell the difference in the acceleration produced in one direction by a uniform gravitational field and uniform acceleration in the opposite direction. The principle applies only to local events where the gravitational field is homogeneous and not in general (for example, large scales).

A straightforward prediction of the principle of equivalence is the change in frequency of light when there is a difference in gravitational potential between the source and detector which is equivalent to an acceleration that gives a relative velocity between the source and detector and thus a Doppler shift. However, the effect is very small for light (the relative shift is less than 10^{-12}) if measured on Earth and is complex if measured from a distant star. As we shall see in Appendix 7C, the Mössbauer effect permitted the gravitational red shift to be measured in the laboratory.

Next, consider the problems special relativity causes Newton's theory of gravitation. For instance, we can calculate the escape velocity for a particle from the surface of a star of mass M and radius R:

$$V_{\text{esc}} = \sqrt{\frac{2GM}{R}}$$

where G is Newton's universal gravitational constant. But special relativity says that no particle can travel faster than the speed of light c, hence we must have

$$V_{\text{esc}} < c$$

or

$$\frac{2GM}{Rc^2} < 1$$

For a star of fixed radius, we would conclude from special relativity that nothing, including light, can escape from a star that has a mass greater than $Rc^2/2G$. Similarly, the critical radius from which light cannot escape is

$$R_s = \frac{2GM}{c^2}$$

where R_s is called the Schwarzschild radius. The existence of black holes, objects from which not even light can escape, was a later prediction of general relativity. R_s is the critical radius for a mass, M, to be a black hole. Our Sun would have to have a radius of about 3 km (rather than its approximate 700 000 km) to be a black hole. Outside R_s a black hole has the same gravitational effect as any other mass.

All this is definitely not a part of Newton's theory of gravity, since that theory has no limiting velocity (however, the velocity of light was known by the seventeenth century, and with this Laplace in 1798 realized that light could not escape from a sufficiently heavy star of a given radius).

Another example where special relativity runs amok is that of a spherically symmetric rotating mass M of radius R. If a stationary observer (with respect to the center of mass of M) sees an object of length L_0 on the equator when M is not rotating, then special relativity tells us that when M rotates, L_0 is Lorentz contracted to $L = L_0/\gamma$. In fact, the equator itself appears smaller, and therefore the circumference C (at the equator) is less than $2\pi R$ for a rotating sphere. In this case what Einstein argued was that a large mass like a star distorts the geometry of space; that is, the space we live in is not just a simple flat Minkowski space, but is curved space where the curvature is caused by the mass (and energy) in the space. The classic example of a system with similar properties is a stretched rubber sheet with a marble or ball bearing sitting on it. The rubber sheet is distorted by the marble, and the indentation it makes can be thought of as a potential well. Furthermore, if two marbles are placed on the sheet, they will attract as each influences the other through its potential. So what general relativity says is that

$$\text{geometric curvature} = \text{mass-energy}$$

Technically, this is written as

$$G_{\mu\nu} = \kappa T_{\mu\nu}$$

and is called Einstein's equation, where κ is 8π times Newton's constant. Here $T_{\mu\nu}$ is called the stress energy tensor. For the simple case of a noninteracting gas of density ρ and pressure p,

$$T_{\mu\nu} = \begin{pmatrix} p & 0 & 0 & 0 \\ 0 & p & 0 & 0 \\ 0 & 0 & p & 0 \\ 0 & 0 & 0 & \rho \end{pmatrix}$$

$G_{\mu\nu}$ is called the Einstein tensor, which is built from the metric tensor and the curvature tensor. The metric tensor is used to measure distances in a space; for instance, the infinitesimal distance ds from x^μ to $x^\mu + dx^\mu$ is given (see Appendix 1A) by

$$ds = \sqrt{g_{\mu\nu}\, dx^\mu\, dx^\nu}$$

For a spherically symmetric geometry, for example the geometry of a non-rotating star, the metric tensor can be shown from Einstein's equations to be

$$ds^2 = \left(1 - \frac{2GM}{rc^2}\right)^{-1} dr^2 + r^2(d\theta^2 + \sin^2\theta\, d\phi^2) - \left(1 - \frac{2GM}{rc^2}\right) dt^2$$

Note the factors of $1 - (2GM/rc^2)$. These factors vanish when the escape velocity $v_{esc} = c$. This allows Einstein's theory to describe the case of a star so dense that light does not escape; in other words, as noted, general relativity predicts the existence of black holes. There are now strong astrophysical candidates for black holes. The stellar system Hercules X1 was the first nearby object to have the right properties—it is a strong emitter of x-rays related to the release of energy by matter infalling into the black hole. Even more dramatic, black holes may have been discovered in the centers of galaxies including our own Milky Way. These black holes seem to have enormous masses on the order of one million solar masses or greater.

Another place where general relativity is indispensable is in the description of big bang cosmology. The uniform geometry of our universe can be described with the metric

$$ds^2 = R^2(t)[dr^2 + r^2\, d\theta^2 + r^2 \sin^2 \theta\, d\phi^2] - dt^2$$

where $R(t)$ is called the scale factor of the universe. An analysis of Einstein's equations in this case clearly shows the existence of an expanding universe solution. One finds that today the universe is expanding as if it was filled with non-interacting dust, and the present scale size of the universe is a function of time:

$$R(t) \sim t^{2/3}$$

In the very early universe, when the temperature was still very high, the universe expanded at a different rate, since the equation of state that enters Einstein's equation is that of a relativistic gas (the same as the equation of state for black body radiation). The present expansion is taking place in such a way that distant objects, for instance galaxies or quasars, are receding from us, and the more distant the object, the faster it recedes. Remarkably, this is what was discovered by Edwin Hubble in his landmark observations at the Wilson Observatory in California in the 1920s. Hubble observed a linear relationship between galaxies' brightness and their red shifts, again indicating that distant objects recede faster and faster at larger and larger distances.

Because of recent technological advances in lasers, interferometers, space platforms, and the like, general relativity has become a very exciting area of research experimentally, with new projects underway and more being planned. To support these new efforts, new and more accurate theoretical analysis has also been undertaken; in particular the theoretical description of gravitational waves from tightly orbiting black holes and neutron stars is being studied intensely with the hope of having predictions for the spectrum of these waves available before they are discovered experimentally.

APPENDIX 1A FOUR-VECTOR NOTATION

The usual geometric definition of a measure of infinitesimal distance in a three-dimensional flat space is

$$ds^2 = d\boldsymbol{x} \cdot d\boldsymbol{x} \equiv \sum_{ij} g_{ij}\, dx^i\, dx^j$$

where $i, j = 1, 2, 3$ and

$$g_{ij} = \begin{pmatrix} 1 & 0 & 0 \\ 0 & 1 & 0 \\ 0 & 0 & 1 \end{pmatrix}$$

is called the metric tensor. All the eigenvalues of g_{ij} are positive in this case, and the space is called Euclidian.

Let us for the moment consider a light signal moving at a constant velocity $\boldsymbol{v} = c$, then its position is given by

$$\boldsymbol{x}_f = \boldsymbol{x}_i + (t_f - t_i)\mathbf{v}$$

or

$$\Delta x = \mathbf{v}\, \Delta t$$

where $\Delta \boldsymbol{x} = \boldsymbol{x}_f - \boldsymbol{x}_i$ and $\Delta t = t_f - t_i$. Squaring the last equation (i.e., dotting it with itself) gives

$$\Delta \boldsymbol{x} \cdot \Delta \boldsymbol{x} = \mathbf{v} \cdot \mathbf{v}\, \Delta t^2 = c^2 t^2$$

The infinitesimal version is

$$d\boldsymbol{x} \cdot d\boldsymbol{x} - c^2\, dt^2 = 0$$

or

$$ds^2 \equiv \sum_{\mu,\nu} g_{\mu,\nu}\, dx^\mu\, dx^\nu = 0$$

where μ and ν run from 1 to 4 and

$$dx^\mu = (dx_1, dx_2, dx_3, c\, dt)$$

and

$$g_{\mu\nu} \equiv \begin{pmatrix} 1 & 0 & 0 & 0 \\ 0 & 1 & 0 & 0 \\ 0 & 0 & 1 & 0 \\ 0 & 0 & 0 & -1 \end{pmatrix}$$

is the metric tensor, which we are describing in what is known as four-vector notation. This flat space with one negative eigenvalue is called *Minkowski space.*

It is standard practice to suppress the summation symbol $\sum_{\mu,\nu}$ in relativity theory, where repeated indices are always summed if not stated otherwise; this is the Einstein summation convention, which is very convenient in practice, and only a few exceptions ever need to be considered.

PROBLEMS

1.1. Calculate the velocity of an electron with kinetic energy of 100 keV, 1 MeV and 10 MeV.

1.2. Calculate the relativistic masses and momenta of the electrons in the three cases in Problem 1.1.

1.3. Calculate the velocity of a proton with kinetic energy of 100 keV, 1 MeV, 10 MeV, and 1 GeV.

1.4. Calculate the momentum, kinetic energy and total energy of a proton ($m_0c^2 = 938$ MeV) moving at $v = 0.90c$.

1.5. Calculate the velocities at which the electron and proton masses are 2 percent greater than their rest energies.

1.6. A space shuttle moving in a vertical direction away from an observer on earth at a velocity of $0.2c$ fires a bullet with a velocity of $0.9c$ relative to the shuttle and in the same direction. What is the velocity of the bullet as seen by an observer on the ground?

1.7. Two spaceships approach each other, one traveling at $0.3c$ and the other at $0.6c$. What is their relative velocity of approach as seen by an observer on either ship?

1.8. For a particle with rest mass m moving with velocity v, show that $v/c = pc/E$ where p is the momentum and E the total energy, which equals γmc^2.

1.9. Event 1 occurs at $x_1 = 10$ m at time $t_1 = 0$. Event 2 occurs at $x_2 = 600\,000\,010$ m at time $t_2 = 1.8$ s. Is there any reference frame where these two events can be reversed so that event 2 occurs before event 1? Prove your answer.

1.10. Calculate the velocity of a distant galaxy when a blue 434 nm line is observed at 1950 nm. Where does this line occur in the electromagnetic spectrum?

1.11. A space shuttle that is 20 m long in its rest frame is passing a space docking station which is 110 m long in the docking station rest frame. If the space shuttle is moving at $0.6c$ relative to the station, what will be the length of the docking station to the person on the shuttle? How long will it take the shuttle from the time the front tip of the shuttle reaches one end of the station until the front tip is seen by the person on the shuttle to reach the other end of the station?

1.12. What would the velocity of the shuttle in Problem 1.11 have to be for the person in the shuttle to see the front tip of the shuttle coincide with one end of the station when the other end of the shuttle coincides with the other end of the station?

1.13. An object from outer space moves past the Earth at $0.8c$. You measure the length of the object as 3.3 m in the Earth's frame. In the object's rest frame, what is its length?

1.14. You are traveling past a station in space at $0.3c$. The clock in your shuttle says it took 3 s for you to pass the station. In your shuttle what time has passed for this event to occur according to what you see on the clock on the wall of the station?

1.15. A 30-year-old female astronaut makes a trip from the Earth to a star that is 6 light years away at an average velocity of $0.9c$. Neglecting the time it takes to turn the spaceship around for the return, how old is the woman when she returns to Earth?

How old is her boss, who was 40 when she left? Explain which one aged more than the other and why it was that person who aged more.

1.16. A particle has a total relativistic energy of 6 MeV and a momentum of 5 MeV/c. What is the rest mass of the particle?

1.17. Suppose you see that light from a particular type of atom in a distant star that should be at 950 nm actually occurs at 525 nm. What is the velocity of the star relative to you, including the direction of the velocity?

1.18. A K^0 particle with rest mass 497.7 MeV decays at rest into a π^+ and π^- each with rest mass of 139.6 MeV. Find the kinetic energies and momenta of the π^+ and π^-.

1.19. Suppose the Concorde flies from Paris to New York, a flying distance of about 3800 miles at a supersonic velocity of 1200 miles/hour. Neglecting the time and distance to reach that velocity, what is the distance as measured by an observer on the plane? What is the time to cover the 3800 miles as measured by a ground observer for a clock on the ground? What is the time as measured by the observer on the ground for the clock on the Concorde?

CHAPTER 2

THE CONFIGURATION OF THE ATOM: RUTHERFORD'S MODEL

> For example, if you're doing an experiment, you should report everything that you think might make it invalid—not only what you think is right about it . . . . If you make a theory, for example, and advertise it, or put it out, then you must also put down all the facts that disagree with it, as well as those that agree with it.
>
> Richard Feynman[1]

2.1 THE BACKGROUND

2.1.1 The Discovery of the Electron

In 1833, Michael Faraday put forward his laws of electrolysis, from which it was deduced that every mole of univalent ions of any element always contains the same amount of unbalanced electric charge. This quantity, the Faraday constant F (for its value see Appendix III at the end of this book), was first determined by

[1] From *Surely You're Joking, Mr. Feynman!*, Bantam Books, New York (1986), p. 311; used with permission of W. W. Norton & Co., Inc., New York.

Faraday from experiments. Remembering that A. Avogadro in 1811 proposed that there is a constant number, N_A, of atoms or molecules per mole of a substance, one can infer from F that there exists a smallest unit of electric charge. A mole is defined such that there are $N_A = 6.023 \times 10^{23}$ (Avogadro's number) atoms or molecules in one mole of a substance. Jumping ahead, the mole is also defined as a gram-atomic or gram-molecular weight of a substance; for example, 4 grams of the isotope ^{4}He is a gram-atomic weight of helium or 28 grams of N_2 is a gram-molecular weight of N_2 where, as we shall see, atomic and molecular weights are related to the number of protons and neutrons in the atom or molecule. Thus, an electric charge consists of some elementary unit or "atom" of electricity ($e = F/N_A$).

However, such a deduction, or inference, was only made as late as 1874 by G. J. Stoney. He pointed out that the charge of an ion was an integral multiple of an elementary charge and he calculated the approximate value of that elementary charge using Avogadro's number. Stoney suggested in 1881 the name "electron" for the smallest unit of electric charge.

Nevertheless, it was J. J. Thomson[2] who established experimentally in 1897 the actual existence of a new particle, named the electron, which carries the basic unit of negative charge. Figure 2.1(a) is a photograph of Thomson and his discharge tube, and Fig. 2.1(b) is the tube shown schematically as taken from his original work. Cathode rays are emitted by the cathode C, accelerated toward the positive potential on A, focused into a beam with velocity v on passing through slits A and B, then passed between two metallic parallel plates D and E, and finally strike the screen, which has a scale on it on the right. An electric field may be applied between D and E, and a magnetic field perpendicular to the electric field can be created by coils located outside the tube. When an electric field E is applied, the point at which the rays strike the screen will be deflected from point P_1 to point P_2. This deflection shows that the cathode rays are electrically (and in fact, negatively) charged. Let e be the charge on each particle. Now, let us further apply a suitable magnetic field H perpendicular to the plane of the figure, so that the mark will return to point P_1. Then we know that the Lorentz force (Hev) and the electric force (eE) are equal in magnitude and opposite in direction. Hence we may deduce the velocity of the rays as $v = E/H$.

When the electric field is removed, the path of the beam will become a circular arc, since the magnetic field is perpendicular to the velocity. Let the radius of the arc be r. The centrifugal force on the particles will be mv^2/r, with m denoting the mass of the particles. This force must be balanced by the Lorentz force Hev. From this equality, the ratio of e/m may be deduced, since the velocity v has been obtained from E/H.

Before J. J. Thomson's work, the history of the study of cathode rays had already lasted several decades. So why had such a simple experiment as Thomson's been postponed to the last years of the nineteenth century? The crucial

---

[2] J. J. Thomson, *Phil. Mag.* **44** (1897) 293.

(a)

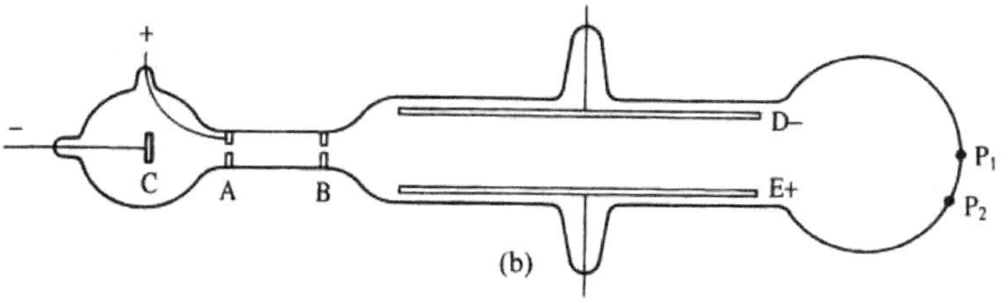

(b)

FIGURE 2.1
(a) J. J. Thomson and his discharge tube. (b) Thomson's tube shown schematically.

point was the attainment of a high vacuum. (The reader can estimate how high the vacuum must be to ensure the success of Thomson's experiment.) When Thomson began his work, vacuums that could be obtained were not very high. So he observed no deflection at all. At that same time, the noted physicist H. R. Hertz, the discoverer of electromagnetic waves, had done experiments similar to Thomson's. He also observed no deflection and made the incorrect conclusion that cathode rays were not charged.

Actually, Thomson may be rightly called "the first man to open the gate leading to elementary particle physics." The reason is not so much that he determined the value of e/m, but rather that he was courageous enough to break with the traditional concepts and to propose the actual existence of a new very low-mass particle, the electron, for the first time.

In fact, A. Schuster (professor of physics at Manchester University in England, a position later occupied by Ernest Rutherford) had already in 1890 studied the deflection of cathode rays in a hydrogen discharge tube. He calculated that the charge/mass ratio of the cathode ray particles was over a thousand times larger than that of hydrogen ions, but he did not dare believe his own result. It was absurd to him to conclude that "the mass of a cathode ray particle is less than a thousandth of the mass of a hydrogen atom." Were atoms not indestructible? How could one break off such a small piece? So he incorrectly assumed that a cathode ray particle weighed about the same as an atom but that its charge was much larger than that of a hydrogen ion.

In 1897, W. Kaufmann also did similar experiments. He determined the e/m ratio much more accurately than had Thomson the same year. Kaufmann's value differed from the modern value by only about 1 percent. Moreover, he observed a velocity dependence of the e/m value. This fact would be explained by Einstein's theory of relativity in 1905; that is, the mass of a body increases with an increase in its velocity as discussed in Chapter 1. But Kaufmann, too, did not dare to publish his results until 1901, since he also did not believe that cathode rays actually consisted of such tiny particles.

Unfortunately these scientists were so committed to the then held world view that they could not see new insights when these new insights were right in front of their faces. As we shall see, even several Nobel Laureates observed but overlooked discoveries that led to Nobel Prizes for others because their results did not fit the then current scientific views. Someone like J. J. Thomson was needed to break with the then universally accepted scientific view put forth by J. Dalton that atoms were the smallest indivisible units of matter and to draw the correct conclusion that there exist particles much smaller than atoms.[3]

2.1.2 The Charge and Mass of the Electron

Less than two years after his determination of e/m_e, Thomson separately determined the charge and the mass of the electron. He noticed that electric charges behaved as nuclei for condensation in a saturated vapor. From measurements of the number and total charge of the droplets, the average electronic charge was calculated. He obtained a value of 3×10^{-10} esu.

As is well known, the first accurate determination of the electronic charge was done in 1910 by R. A. Millikan, when he did his famous "oil-drop experiment." After repeated work over a number of years, Millikan obtained a value of 4.78×10^{-10} esu (1.59×10^{-19} C). This value was taken as the most reliable one until 1929, when it was found that there was an error of about 1 percent. This arose from an error in the determination of the viscosity of air. The modern value of this quantity is

$$e = 1.602176487(40) \times 10^{-19}\ \mathrm{C}$$

[3] For the life and work of J. J. Thomson, see E. Segre, *From X-Rays to Quarks*, Freeman, San Francisco (1980); R. P. Crease and C. C. Mann, *The Second Creation*, Interlink Publishing Group Inc. (1997).

with an accuracy of three parts per ten million. The number 49 in parentheses indicates the error in the last two figures, which means (33 ± 49).

Above all, Millikan found that electric charge is quantized; that is, any charge can only be an integral multiple of e, and e is the smallest charge that any object can have. In the next chapter, we will discuss in more detail the concept of quantization. This concept occurs not only in quantum physics but also in classical physics, but it plays no dominant role in the latter. Why is electric charge quantized and why that particular value? These are really very deep questions that are still quite open today. With the probing of the structure of matter into deeper levels, the "oil-drop experiment" has been repeated to search for fractional charges.[4]

From the experimental values of e/m and e, the electronic mass may be deduced to be

$$m_e = 9.10938215(45) \times 10^{-31}\,\mathrm{kg}$$

Moreover, from Faraday's laws of electrolysis, we may find the quantity of electricity needed to dissociate one mole of hydrogen (i.e., the Faraday constant). Then we may calculate the charge-mass ratio e/m_p of the H ion (which was named the "proton" by Rutherford in 1914). With e/m_e and e/m_p, we may derive the ratio between the mass of the proton and the mass of the electron:

$$\frac{m_p}{m_e} = 1836.15267247(80)$$

This ratio is one of the two most important dimensionless constants in atomic physics (the other one being the fine structure constant α, see Chapter 3). It is this constant that determines the most fundamental characteristics of atomic physics, and indeed of our universe. Were this ratio of the order of magnitude of 1 (or, if e were a factor of 10 greater), our physical world would have an utterly different appearance. Why this ratio has the value it has instead of another one cannot be answered at present. This constant still cannot be derived yet from primary (a priori) physical principles.

From the values of m_p/m_e and m_e, we can obtain

$$m_p = 1.672621637(83) \times 10^{-27}\,\mathrm{kg}$$

The mass of the lightest atom is nearly the same as this value, while the masses of the heaviest atoms are some 250 times larger. It is obviously inconvenient to measure the masses of atoms in kilograms, more absurd than measuring the weight of a human body in tons. In the world of atoms, we use an "atomic mass unit" u. It is agreed internationally to take the mass of the isotope of carbon with six protons and six neutrons, ^{12}C, as exactly 12 u. From this we may deduce

$$m_p = 1.00727646677(10)\,\mathrm{u}$$

In a rough approximation, this may be taken as one atomic mass unit (1 u).

[4] See, for example, G. LaRue, et al., *Phys. Rev. Lett.* **46** (1981) 967.

From the mass–energy relation (see Chapter 1) $E = mc^2$ given in Einstein's special theory of relativity, we may obtain

$$m_e = 0.510998910(13)\,\text{MeV}/c^2$$

$$m_p = 938.272013(23)\,\text{MeV}/c^2$$

$$1\,\text{u} = 931.494028(23)\,\text{MeV}$$

This is the usual method of expressing masses in energy units in microphysics, where c is the velocity of light; sometimes even c^2 is omitted. Here MeV stands for million electron-volts. One electron-volt is the energy acquired by a particle with a unit charge e when accelerated by an electric field across a voltage difference of one volt. The relation between this and the common unit of energy, the joule, is

$$1\,\text{eV} = 1.602176487(40) \times 10^{-19}\,\text{C} \times 1\,\text{V}$$

$$= 1.602176487(40) \times 10^{-19}\,\text{J}$$

As mentioned above, Kaufmann established experimentally that the ratio e/m decreased as the velocity of electrons increased (published in 1901). This is another example of how many of the consequences, including Lorentz space contraction and FitzGerald time dilation, of Einstein's special theory of relativity were already known before Einstein gave these phenomena a physical basis. So, the change in mass with velocity was understood in 1905 using the relativistic mass formula

$$m = \frac{m_0}{\sqrt{1 - \frac{v^2}{c^2}}}$$

where m_0 is a particle's mass when it is not moving (its rest mass), m is its mass when its velocity is v, and c is the velocity of light in vacuum, as presented in Chapter 1.

While the mass of an electron increases with increasing velocity, just as the masses of other bodies do, the most precise experiments still show that the electric charge of an electron never changes with its velocity.

2.1.3 Avogadro's Constant

Avogadro's constant gives the number of atoms in one mole of atoms or the number of molecules in one mole of molecules. We shall show by the following examples that this constant is the physical quantity that is an important link between the macroscopic and the microscopic world.

By the definition of Avogadro's constant, one mole of ^{12}C atoms, that is 12 g of ^{12}C, contains N_A atoms of ^{12}C. Then, the mass of every ^{12}C atom, when measured in grams, must be $(12/N_A)$ g. Now, this mass is defined to be 12 u, so the relation between u and grams will be $12\,\text{u} = 12/N_A$ g, that is,

$$1\,\text{u} = \frac{1}{N_A}\,\text{g} \qquad \text{or} \qquad 1\,\text{g} = N_A\,\text{u}$$

It may be derived from the value of N_A that

$$1\,\text{u} = 1.660538782(83) \times 10^{-24}\,\text{g}$$

Since the gram (g) is a macroscopic unit and the atomic mass unit (u) is a microscopic one, the role of N_A is seen to be a bridge between the macro and the micro. Note that u is smaller than the free proton or neutron mass because each gives up mass to be bound in ^{12}C, so $m(^{12}\text{C}) < 6m_p + 6m_n$.

Note that many fundamental constants are not mutually independent; the relation between u/g and N_A is an example. However, they may be independently measured by experiments. Hence, to obtain a consistent set of constants from experimental data, it is necessary to treat the experimental values and related formulas with the least-mean-square method to obtain an optimal set of values. As a result of incessant improvements in experimental measurements, there was a new international "least-square adjustment" of constants and a new set of self-consistent values given by the Committee on Data for Science & Technology of the International Council of Scientific Unions (CODATA) in 2006 to replace the previous set of 1986. All the constants used in this book are from the 2006 set (see Appendix III). Measurements published later than 2006 will cause these constants to be adjusted in order to be self-consistent.

Next, note that the relation between the Faraday constant F and electronic charge e is

$$F = eN_A$$

Here F is a macroscopic and e is a microscopic quantity, and the bridging of e and F is also through N_A. Moreover, we have

$$R = kN_A$$

where R is the universal gas constant, a macroscopic quantity, while k is the Boltzmann constant, a microscopic quantity, and these two quantities are also connected through N_A.

Whenever we do any experiment to study some quantity in the microscopic world, we will always deal consciously or unconsciously with Avogadro's constant, which acts as a bridge to the macroscopic world where our experiments must be performed. When we derive a microscopic quantity from the measurements of some macroscopic quantities, we need a bridge to go through, and the constant N_A may serve as such a bridge.

2.1.4 The Size of Atoms

What is the physical size of an atom? To answer this, let us try a simple estimation. For any kind of atom ^{A}X, there are N_A atoms of X in A grams of those atoms. Let the mass density of this substance be ρ(g/cm^3); the total volume of A

TABLE 2.1

Element	Mass number A	Mass density (g/cm^3)	Atomic radius r (Å)
Li	7	0.7	1.6
Al	27	2.7	1.6
S	32	2.07	1.8
Cu	63	8.9	1.4
Pb	207	11.34	1.9

grams of X atoms will be A/ρ. Suppose the volume occupied by a single atom is $(4/3)\pi r^3$ (r being the "radius" of the atom), then we have

$$\frac{4}{3}\pi r^3 N_A = \frac{A}{\rho}$$

(Here, once again we see the bridge role of N_A.) From this we get a formula for the atomic radius:

$$r = \sqrt[3]{3A/4\pi\rho N_A}$$

As an example, for ^{63}Cu, $r = (3 \times 63/4\pi \times 8.9 \times 6.02 \times 10^{23})^{1/3} = 1.41 \times 10^{-8}$ cm. Using this formula, we may calculate the radii of various atoms; Table 2.1 gives some examples. Here the angstrom (1 Å $= 10^{-10}$ m) is a unit of length often used in the atomic region. It is named after the Swedish scientist A. J. Ångström.

We see from Table 2.1 that the radii of many different atoms are very nearly the same. Indeed, the radii of most atoms are the same within a factor of 2 or 3. This is a fact that classical physics is unable to explain. We will come back to this point a few chapters later.

So far we have learned that in atoms there are electrons, the mass of which is only a small fraction of that of an atom, and that the electrons are negatively charged while the atoms are electrically neutral. So there must be some positively charged parts that are the site of the larger part of the atomic mass. Inside an atom, i.e. in a region of angstrom dimensions, what will be the distribution and motion of the positive parts and the negative electrons?

2.2 THE EMERGENCE OF THE RUTHERFORD MODEL

After Thomson's discovery of the electron, there appeared a number of different ideas about how the positive and negative charges are distributed in atoms.[5] An important one among the early models was suggested by Thomson himself in

[5] E. N. da C. Andrade, *Scientific American* **195** (Nov. 1956) 53.

1898, and further improved in 1903 and 1907. Thomson considered that the positive charge in an atom was uniformly distributed in a sphere of the same size as the atom, and the electrons were embedded in that sphere. Sometimes Thomson's model is called the "watermelon model" or the "plum-pudding model," where the electrons are the seeds or the plums in the positive melon or pudding. For more about this model, see J. J. Thomson[6] and F. L. Friedman and L. Satori.[7] To explain the periodic table of elements, Thomson further assumed that the electrons were distributed in rings: the first (innermost) ring could contain 5 electrons, the second ring could contain 10 electrons and if there are 70 electrons to be distributed, one would need 6 concentric rings in all. Thomson's model did have some success in explaining the periodic properties of the elements. Moreover, although it was disproved by subsequent experiments, it really contained some very significant and revolutionary new ideas such as that of concentric rings and that of the "limited number of electrons in each ring" (the first suggestion of quantization of electron energies in atoms and the Pauli exclusion principle).

In 1903, P. Lenard discovered in an experiment on the absorption of cathode rays by matter that "the atoms are rather dilute," the first evidence that atoms are mostly empty space. Based on this experiment, the Japanese physicist Hantaro Nagaoko proposed his "Saturn-ring model" of atoms in 1904. He imagined that the positive charge in an atom was concentrated in its center, and that the electrons were orbiting about the center in a ring like the rings of Saturn, but he did not go any further. Then in 1909 came the sensational discovery of Rutherford's student and associate H. Geiger and E. Marsden concerning the backward scattering of α particles when bombarding atoms by α particles. Rutherford discovered in 1899 the general characteristics of α and β rays emitted in radioactive decay. He proved in 1906 that α particles were helium atoms with two positive elementary charges (see Chapter 10). Geiger and Marsden found that α particles striking a gold foil had a probability of about one in eight thousand of bouncing backward at angles greater than 90° to the forward direction of the α beam. Rutherford was amazed by that result. He said, "It was as incredible as if you fired a 15-inch shell into a tissue-paper and it came back and hit you." This surprise is because in Thomson's model, as we shall see, the probability that an atom could exert a force sufficiently great to make the α scatter into an angle greater than 90° is 1 part in 10^{3500}! However, Rutherford would not overlook the experimental facts and, after some rigorous theoretical reasoning, he proposed his "nuclear model of the atom" in 1911.[8] Since Rutherford's model contained some factors that cannot be understood from the classical point of view (let the reader ponder on these, which will be discussed in detail later), his paper was not accepted favorably by authoritative scholars. Even some famous books on atomic

---

[6] J. J. Thomson, *The Corpuscular Theory of Matter*, Constable & Co., London (1970).

[7] F. L. Friedman and L. Satori, *The Classical Atom*, Addison-Wesley, Reading, MA (1965).

[8] E. Rutherford, *Phil. Mag.* **21** (1911) 669.

structure published in 1913 and 1915 do not contain a single word about Rutherford's work.

The main difference between Thomson's model and Rutherford's model is that the former takes the positive charge as uniformly distributed over the whole atomic volume, while the latter takes it as concentrated in a very small region, the dimension of which is 10 000 to 100 000 times smaller than that of the atomic radius. From the experimental point of view, how do you decide between the two atomic models?

Suppose we have two electric charged spheres of the same size, mass and charge, but that the charge density of one sphere is uniform everywhere while the charge of the other is concentrated in its center. Imagine what will happen if we bombard these two spheres with the same charged particle. From Gauss' law in electricity, it is easy to understand that, at a great distance from the spheres of radius R ($r \gg R$, see Fig. 2.2), the charges on each may be taken as concentrated at the center. But when the particle penetrates into the spheres ($r < R$), the situations will become quite different: inside the uniformly charged sphere, the Coulomb force on the particle is proportional to r and so goes toward zero as the particle comes nearer and nearer to the center. Thus, it is comparatively easy for the particle to go through that sphere. On the other hand, in an atomic sphere with its positive charge concentrated in the center, the Coulomb force on the incoming particle is still inversely proportional to r^2, just as before the particle enters the atomic sphere. So the force will become larger and larger as the particle comes nearer to the center of the atom. Thus the force can become sufficiently large for the particle to be stopped and thrown out in the backward direction.

All of these are qualitative and intuitive considerations. Let us now consider Thomson's model somewhat quantitatively. We see from Fig. 2.2 that, in

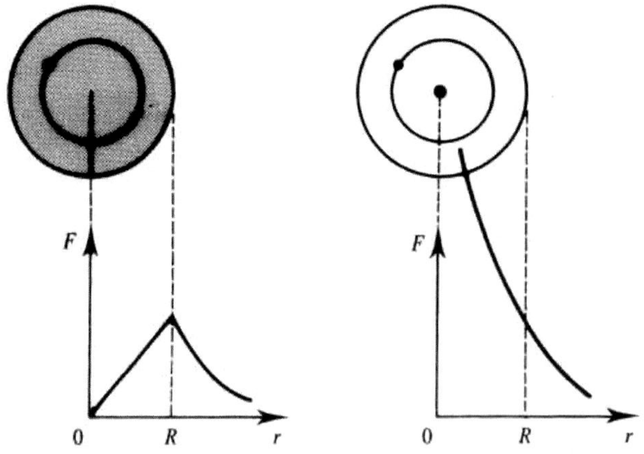

FIGURE 2.2
Two different distributions of electric charge (on the left a uniform charge of radius R and on the right a tiny core of charge) will apply different forces on an incident charged particle. (The vertical scales are different for these two pictures. If they were the same, how would you make the drawings?)

Thomson's model, the maximum force occurs at $r = R$ (R being the radius of the atom), where the force of the positive charge Ze of the atom on the α-particle ($2e$) is

$$F = \frac{(2e)(Ze)}{R^2} \tag{2.1}$$

In the SI (MKS) system, e^2 is replaced by $e^2/4\pi\varepsilon_0$ with no other changes.

To estimate the change produced in the momentum of an α particle by the scattering, it is only necessary to multiply the force by the time interval ($\sim 2R/\mathrm{v}$) during which the particle is in the vicinity of the atom, so we have

$$\begin{aligned} \frac{\Delta p}{p} = \frac{F\,2R/\mathrm{v}}{m_\alpha \mathrm{v}} = \frac{2Ze^2/R}{\frac{1}{2}m_\alpha \mathrm{v}^2} &\simeq \frac{2Z \times 1.44\,\mathrm{fm} \cdot \mathrm{MeV}/1\,\text{Å}}{E_\alpha(\mathrm{MeV})} \\ &\simeq 3 \times 10^{-5}\frac{Z}{E_\alpha} \quad \text{radian} \end{aligned} \tag{2.2}$$

Here we have used a very useful expression for e^2:

$$e^2 = 1.44\,\mathrm{fm} \cdot \mathrm{MeV} \tag{2.3}$$

where fm stands for femtometer, where $1\,\mathrm{fm} = 10^{-15}\,\mathrm{m} = 10^{-5}\,\text{Å}$. In Eq. (2.2), we have taken $R \simeq 1\,\text{Å}$ and E_α is the kinetic energy of the α particle in MeV and for small θ, $\Delta p/p \approx \theta$, see Fig. 2.3.

Formula (2.2) gives the greatest deflection of incident α particles produced by a positive charge in Thomson's model. Even in head-on collisions, the action of the electrons is negligibly small, since the electronic mass is about 8000 times smaller than the mass of an α particle, so

$$\frac{\Delta p}{p} \simeq \frac{m_e}{m_\alpha} \sim \frac{1}{8000} \sim 10^{-4} \tag{2.4}$$

From looking at Eqs. (2.2) and (2.4), a conservative estimation of the deflection angle is

$$\theta < 10^{-4}\frac{Z}{E_\alpha} \quad \text{radian} \tag{2.5}$$

For the scattering of 5 MeV α particles by a gold ($Z - 79$) foil, the greatest deflection angle in any collision will be smaller than 10^{-3} radian. To produce a deflection of 1°, we need many collisions. Since the deflections in different collisions are random in direction, it is very improbable that a large-angle deflection

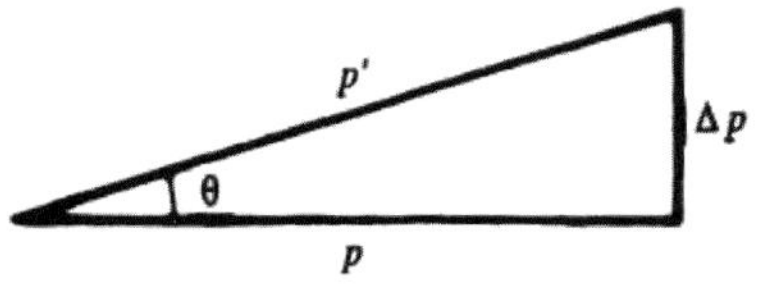

FIGURE 2.3
Momentum change produced by scattering.

will be produced through accumulation. It may be estimated that the probability of producing a 90° deflection is roughly 10^{-3500}, while the experimental value obtained by Geiger and Marsden is 1/8000. The question of how to resolve this problem led to the Rutherford model discussed in the following section.

2.3 RUTHERFORD SCATTERING FORMULA

2.3.1 The Derivation of the Coulomb Scattering Formula

Figure 2.4 illustrates the scattering of particles of energy E and charge Z_1e by a target nucleus of charge Z_2e. First, let us prove a very important formula in physics, the Coulomb scattering formula

$$b = \frac{a}{2}\cot\frac{\theta}{2} \tag{2.6}$$

where the quantity

$$a \equiv \frac{Z_1Z_2e^2}{E} \tag{2.7}$$

is called the Coulomb scattering factor, and b is the impact parameter. As seen in Fig. 2.4, b is defined as the shortest perpendicular distance between the incoming path of the Z_1 and a parallel line through the center of charge Z_2 when there is no interaction between the incident particle and the scatterer at rest, and θ is the scattering angle for impact parameter b. A particle incident within an area πb^2 centered on the nucleus will be scattered into an angle $\geq \theta$. When $\theta = 90°$, the Coulomb scattering factor a is two times the impact parameter b.

Before deriving the Coulomb scattering formula (2.6), we make some assumptions about the scattering process: (1) There are only single scatterings; (2) there is only a Coulomb force interaction; (3) the action of the extranuclear electrons may be negected; and (4) the target nucleus is at rest. After the deriva-

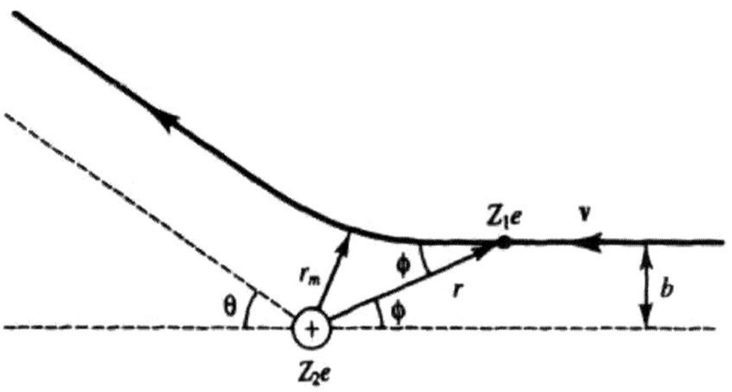

FIGURE 2.4
The Coulomb scattering of charged particles.

tion of the formula, we shall investigate which assumptions may be valid and which ones may not be valid.

The starting point of the derivation of (2.6) will be:[9]

$$\boldsymbol{F} = m\boldsymbol{a}$$

Substituting the actual form of the force, we get

$$\frac{Z_1 Z_2 e^2}{r^2}\boldsymbol{r}_0 = m\frac{d\mathbf{v}}{dt} \tag{2.8}$$

where $\boldsymbol{r}_0$ is the unit vector along $\boldsymbol{r}$. Since the Coulomb force is a central force (see Appendix 2A), and for central forces there exists the conservation of angular momentum, i.e.

$$mr^2\frac{d\phi}{dt} = L\,(\text{constant})$$

we may eliminate the time factor dt from (2.8). We rewrite (2.8) as

$$\frac{Z_1 Z_2 e^2}{r^2}\boldsymbol{r}_0 = m\frac{d\mathbf{v}}{d\phi}\frac{d\phi}{dt}$$

or

$$d\mathbf{v} = \frac{Z_1 Z_2 e^2}{mr^2\dfrac{d\phi}{dt}}\,d\phi\boldsymbol{r}_0 = \frac{Z_1 Z_2 e^2}{L}\,d\phi\boldsymbol{r}_0$$

Integrating,

$$\int d\mathbf{v} = \frac{Z_1 Z_2 e^2}{L}\int \boldsymbol{r}_0\,d\phi \tag{2.9}$$

The integration of the left-hand side is very simple,

$$\int d\mathbf{v} = \mathbf{v}_f - \mathbf{v}_i = |\mathbf{v}_f - \mathbf{v}_i|\boldsymbol{u} \tag{2.10}$$

where $\boldsymbol{u}$ is the unit vector along the direction of $\mathbf{v}_i - \mathbf{v}_f$ (Fig. 2.5), and $\mathbf{v}_i$ and $\mathbf{v}_f$ denote, respectively, the velocities of the incident particle ($Z_1 e$) before and after the collision when the particle is very far from the target. By the conservation of energy,

$$E = \frac{1}{2}m|\mathbf{v}_i|^2 = \frac{1}{2}m|\mathbf{v}_f|^2$$

That is to say, $\mathbf{v}_i$ and $\mathbf{v}_f$ must have the same magnitude (denoted by v), but their directions are different. Hence, as may be seen from Fig. 2.5(a), the vector $(\mathbf{v}_i - \mathbf{v}_f)$ must have a magnitude $|\mathbf{v}_i - \mathbf{v}_f| = 2\text{v}\sin(\theta/2)$, and it makes an angle $\theta/2$ with the y axis.

---

[9] See J. C. Willmott, *Atomic Physics*, Wiley, New York (1975), p. 36.

The right-hand side of (2.9) is the integral of a unit vector. Since $\boldsymbol{r}_0$ is a variable unit vector, it must be expressed in terms of fixed unit vectors $\boldsymbol{i}$ and $\boldsymbol{j}$ before integration. Then we have

$$\begin{aligned}\int \boldsymbol{r}_o \, d\phi &= \int_0^{\pi-\theta} (\boldsymbol{i} \cos \phi + \boldsymbol{j} \sin \phi)\, d\phi \\ &= 2 \cos \frac{\theta}{2} \left(\boldsymbol{i} \sin \frac{\theta}{2} + \boldsymbol{j} \cos \frac{\theta}{2} \right)\end{aligned} \tag{2.11}$$

As can be seen from Fig. 2.5(b), the unit vector $\boldsymbol{u}$ along $(\mathbf{v}_i - \mathbf{v}_f)$ has an x-component $\boldsymbol{i}\,\sin(\theta/2)$ and a y-component $\boldsymbol{j}\,\cos(\theta/2)$. This means that the quantities in parentheses on the right-hand side of (2.11) are just the unit vector $\boldsymbol{u}$, in agreement with (2.10). This is necessarily so, since (2.9) is a vector equation, and the directions of the two sides must be the same.

Substituting the above results into (2.9), we finally obtain

$$\mathrm{v} \sin \frac{\theta}{2} = \frac{Z_1 Z_2 e^2}{L} \cos \frac{\theta}{2} = \frac{Z_1 Z_2 e^2}{mvb} \cos \frac{\theta}{2} \tag{2.12}$$

Because $m\mathrm{v}^2 = 2E$, we find again for (2.6) and (2.7)

$$b = \frac{a}{2} \cot \frac{\theta}{2} \tag{2.6}$$

$$a \equiv \frac{Z_1 Z_2 e^2}{E} \tag{2.7}$$

and this is the Coulomb scattering formula.

What is the impact parameter when 6.0 MeV α particles are scattered into 60° by lead atoms?

$$\begin{aligned}b = \frac{Z_1 Z_2 e^2}{2E} \cot \frac{\theta}{2} &= \frac{2 \times 82 \times 1.44\,\text{fm} \cdot \text{MeV}}{2 \times 6\,\text{MeV}} \cot 30^\circ \\ &= 19.6\,\text{fm} \times 0.577 = 11.3\,\text{fm}\end{aligned}$$

Now let us return to the four assumptions made before our derivation. First of all, let us consider the fourth one—the target being at rest. In experiments, this is in general impossible, for the target will recoil when interacting with the incident particle. Thus we must deal with the general process of a two-body interaction, in which we must conceptually modify some parameters (2.6) to be sure that this equation will be still valid. Let us regard θ as the scattering angle in the center-of-mass system, that is, a coordinate system with its origin fixed on the center of mass (COM) of the two bodies in question, where V_{M} is the velocity of the COM, and regard E as the energy in that system:

$$E = \frac{1}{2} m V_m^2 + \frac{1}{2} M V_m^2 = \frac{P_m^2}{2m} + \frac{P_M^2}{2M}$$

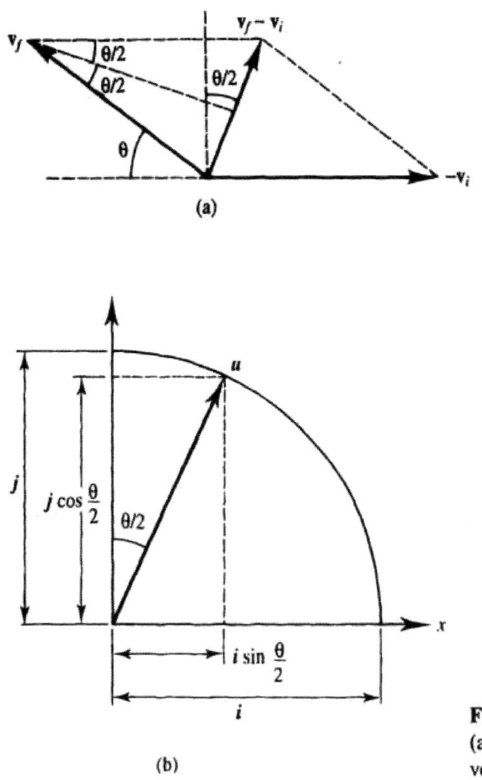

FIGURE 2.5
(a) The vector diagram of the initial and final velocities. (b) The projections of a unit vector.

Since $P_m = P_M$ in COM,

$$E = \frac{P_M^2}{2}\left(\frac{1}{m} + \frac{1}{M}\right)$$

and conservation of momentum in COM gives

$$m(\mathrm{v} - V_M) = MV_M$$

so

$$V_M = \frac{m\mathrm{v}}{M+m} \qquad \text{and} \qquad P_M = MV_M = \frac{mM}{M+m}\mathrm{v}$$

Substituting P_M^2 into E above, one finds

$$E = \tfrac{1}{2}\mu \mathrm{v}^2$$

where the reduced mass is

$$\mu = \frac{mM}{m+M}$$

with m and M denoting the masses of the incident particle and the target, respectively. After such a modification, Eq. (2.6) will hold generally. Nevertheless, when $m \ll M$, such a modification may be neglected.

Example. ^{214}Po (before it was identified as a new element this isotope was called RaC) emits α particles with an energy of 7.68 MeV. When these are scattered by a gold foil, a relation between b and θ follows from (2.6). The results are given when the motion of the target is neglected. Taking this motion into consideration, the reader may calculate the error involved. When calculating with (2.6), we recommend the following method: Use MeV as the unit of energy E of the α particles and $e^2 = 1.44\,\text{fm}\cdot\text{MeV}$; then the value of b in fm (1 fm $= 10^{-15}$ m) will follow immediately. For example, take $b = 100$ fm, then in the above case,

$$100\,\text{fm} = \frac{Z_1 Z_2}{2E} e^2 \cot\frac{\theta}{2} = \frac{2\times 79}{2\times 7.68\,\text{MeV}} \times 1.44\,\text{fm}\cdot\text{MeV}\cot\frac{\theta}{2}$$

$$\cot\frac{\theta}{2} = \frac{768}{79\times 1.44} = 6.75$$

$$\frac{\theta}{2} = 8.45$$

$$\theta = 16.9$$

Impact parameter, b (fm)	Angle of scattering, θ (deg)
10	112
100	16.9
1000	1.7

Consequently we see that, to give a large angle θ, the scattering must occur in a very small region, radius less than or approximately equal to 10 fm. This is why large-angle scattering cannot occur in Thomson's model where the positive charge is spread out over the whole atom while it can occur in Rutherford's model, as seen in Fig. 2.6.

It must be noted that, theoretically, formula (2.6) is very important, but it cannot be used in experiments, because the impact parameter b still cannot be measured experimentally. In order to be able to compare with experiment, we must go one step further.

2.3.2 The Derivation of Rutherford's Formula

We can see from Eq. (2.6) that there is a correspondence between θ and b. When b is large, θ is small, and vice versa. For a given b, there corresponds a definite θ.

Those α particles with impact parameters lying between b and $b + db$ will leave with scattering angle lying between θ and $\theta - d\theta$, as shown in Fig. 2.7(a).

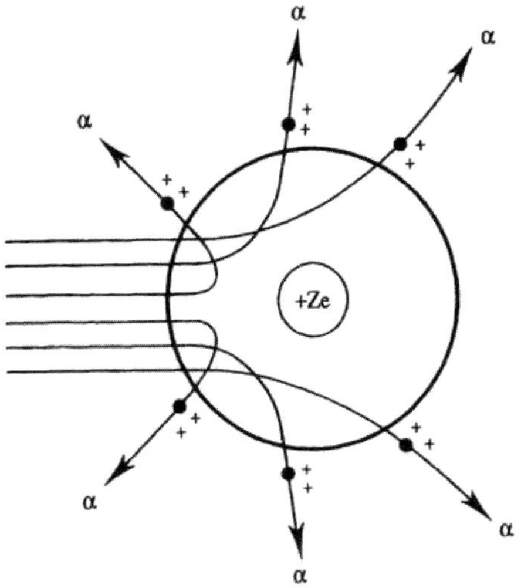

FIGURE 2.6
Rutherford's atomic model: a planetary model.

Thus, we see that α particles flying through the area of a circular ring of inner and outer radii b and $b + db$ will be scattered into a hollow cone with vertex angle lying between θ and $\theta - d\theta$. Now we ask: What is the probability of the particles hitting the above-mentioned ring? Let A be the area of a foil and t be its thickness (where the foil is so thin that to the incident α particles the nuclei in the foil do not screen each other). You should estimate the order of magnitude of the foil thickness to ensure such a condition. The area of the ring is $2\pi b|db|$, so that the probability for α particles to hit this ring is

$$\begin{aligned}\frac{2\pi b \cdot |db|}{A} &= \frac{2\pi}{A}\left(\frac{a}{2}\cot\frac{\theta}{2}\right) \cdot \left| -\frac{a}{2}\csc^2\frac{\theta}{2} \cdot \frac{1}{2} d\theta \right| \\ &= \frac{a^2 2\pi \sin\theta \, d\theta}{16A \sin^4\dfrac{\theta}{2}}\end{aligned} \tag{2.13}$$

From Fig. 2.7(c), it may be seen that the relation between the solid angle of the hollow cone indicated by the shaded area and the angle $d\theta$ is

$$d\Omega = \frac{2\pi r \sin\theta \cdot r\, d\theta}{r^2} = 2\pi \sin\theta \, d\theta$$

Substituting into (2.13), we find

$$\frac{2\pi b|db|}{A} = \frac{a^2 \, d\Omega}{16A \sin^4\dfrac{\theta}{2}} \tag{2.14}$$

As shown in Fig. 2.7(b), there are many such rings in a foil. One such ring corresponds to each atomic nucleus. If the number of nuclei per unit volume is n, the number of nuclei in the foil volume At will be nAt, and there are also this

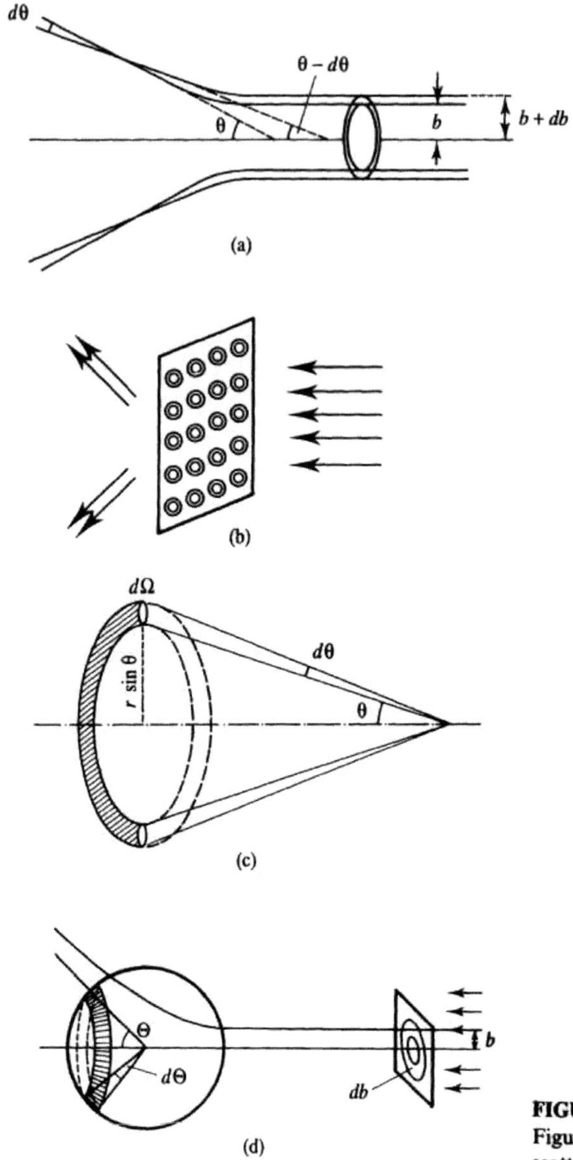

FIGURE 2.7
Figures used in the calculation of the scattering probability.

many "rings." All the α particles hitting such rings will have an angle of scattering θ. So the probability for one α particle hitting the foil and scattering into a region $\theta \to \theta - d\theta$ (that is, scattered along a direction lying in $d\Omega$) must be

$$dp(\theta) = \frac{a^2\, d\Omega}{16A \sin^4 \frac{\theta}{2}}\, nAt$$

where n is the concentration of nuclei in the foil (and assuming that the nuclei in the foil do not screen each other, so that every nucleus is equally effective). Now suppose there are N particles hitting the foil. Then the number of scattered particles measured in the direction of $d\Omega$ must be

$$dN' = N \frac{a^2\, d\Omega}{16A \sin^4 \frac{\theta}{2}}\, nAt = ntN \left(\frac{Z_1 Z_2 e^2}{4E}\right)^2 \frac{d\Omega}{\sin^4 \frac{\theta}{2}} \tag{2.15}$$

We define the differential cross-section as

$$\sigma_c(\theta) \equiv \frac{d\sigma(\theta)}{d\Omega} \equiv \frac{dN'}{Nnt\, d\Omega}$$

It denotes the number of scattered particles per number of atoms per unit area of the target per incident particle per solid angle. Then (2.15) may be written as

$$\sigma_c(\theta) = \left(\frac{Z_1 Z_2 e^2}{4E}\right)^2 \frac{1}{\sin^4 \frac{\theta}{2}} = \frac{a^2}{16} \times \frac{1}{\sin^4 \frac{\theta}{2}} \tag{2.16}$$

This is the famous Rutherford formula. Here $\sigma_c(\theta)$ has the dimension of area. Its meaning is that, when particles are scattered between θ and $\theta - d\theta$, $\sigma_c(\theta)$ represents the effective scattering cross-section of every atom into the solid-angle element $d\Omega$.

Notice that, in the above derivation, the atomic nuclei are assumed to be at rest. If we discard this assumption, formula (2.16) still holds, but it does so only in the center-of-mass coordinate system. Theoretically, this formula is very important, but it is necessary to transform it into the laboratory coordinate system in practical applications (see Appendix 2C).

Let us look at some examples of the use of these equations. First, how many of the N incident α particles of 6 MeV are scattered into an angle equal to or greater than 60° from an aluminum ($Z = 13$) foil of thickness $t = 10^{-3}$ cm and n atoms/cm^3? (Recall that $nt = \rho t \times$ Avogadro's number$/A$). The number scattered into an angle greater than 60° is the sum of all events with impact parameters for $\theta = 60°$ to 180°, which correspond to $b \le b(\theta = 60°)$, e.g. from $b = 0$ to $b(60°)$, which means the number of events in the area $\pi b(60°)^2$, so

$$\Delta N = Nnt\pi b^2$$

$$\Delta N = N \times \frac{\rho t}{A} N_A \pi \times \left(\frac{a}{2} \cot \frac{\theta}{2}\right)^2$$

Thus, the number scattered per number of incident particles is

$$\frac{\Delta N}{N} = \frac{2.7\,\mathrm{g}}{27\,\mathrm{cm}^3} \times 10^{-3}\,\mathrm{cm} \times 6.02 \times 10^{23}\,\mathrm{g}^{-1} \times \pi$$
$$\times \left(\frac{2 \times 13}{2 \times 6\,\mathrm{MeV}} \times 1.44\,\mathrm{MeV} \times 10^{-13}\,\mathrm{cm} \times \cot 30\right)^2 = 55 \times 10^{-6}$$

Equation (2.16) gives us two ways to calculate the differential cross-section at an angle θ. We may know the Z of the projectile and scattering foil and the energy of the particle; for example, for 6 MeV α particles on aluminum at an angle of 30°,

$$\sigma_c(30^\circ) = \left(\frac{2 \times 13 \times 1.44\,\mathrm{MeV} \cdot \mathrm{fm}}{4 \times 6.0\,\mathrm{MeV}}\right)^2 \frac{1}{\sin^4 15^\circ}$$
$$= (1.56\,\mathrm{fm})^2 \frac{1}{(0.259)^4} = 542\,\mathrm{fm}^2 = 542 \times 10^{-26}\,\mathrm{cm}^2 = 5.42\,\mathrm{barns}$$

Alternatively, if we measure $\Delta N/N =$ for $\theta \geq 60^\circ$ but not the energy as above, then we can turn the equation around to solve for a to substitute into Eq. (2.16).

2.4 THE EXPERIMENTAL VERIFICATION OF THE RUTHERFORD FORMULA

2.4.1 Geiger–Marsden Experiment

Rutherford's formula is based on his nuclear model of the atom in which the positively charged parts of the atom are concentrated at the center of the atom in a very small volume (the nucleus) which contains more than 99.9 percent the mass of the whole atom, and on the assumption that the α particles are moving outside the nucleus, subject only to the Coulomb force of the total positive charge of the atom. If the above conditions describe the atom, the experimental results should agree with formula (2.16). From (2.15), we can see four relations: (1) For the same α particle source and the same scatterer, dN' is inversely proportional to $\sin^4(\theta/2)$, i.e., $dN' \sin^4(\theta/2) =$ const. (2) For the same α particle source and the same material of the scatterer, and under the same scattering angle, dN' is proportional to the thickness t of the scatterer. (3) For the same scatterer and under the same scattering angle, dN' is inversely proportional to E_α, i.e., $dN' E_\alpha =$ const. (4) For the same α particle source and with the same scattering angle and the same Nt value, dN' is proportional to Z. In 1913, H. Geiger and E. Marsden verified these four conclusions drawn from (2.16) through their experiments. In 1920, J. Chadwick improved the apparatus and, using Rutherford's formula, directly determined the charge number Z of atoms in his experiments for the first time. It was proven that this charge number Z equals the atomic number of the element in question. This conclusion agrees with other information about atomic structure from other experiments and offered additional strong support for Rutherford's formula. How accurately does Rutherford's formula

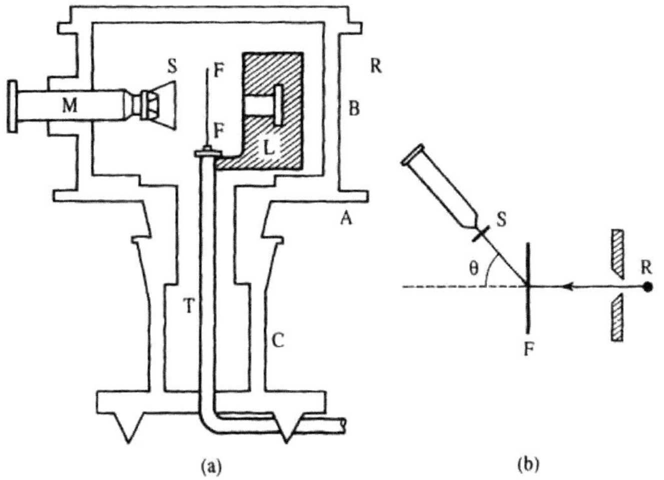

FIGURE 2.8
Apparatus used in the experimental verification of Rutherford's scattering formula: (a) side view; (b) top view. R, particle source; F, scattering foil; S, fluorescent screen. The round metallic box B is fixed on the graduated disk A. A and B may rotate in the smooth tube C. R and F are mounted on the tube T, which is independent of the box. The whole box is evacuated through the tube T. The number of flashes on the screen S is counted by observing with the microscope M.

agree with experiments? This question is still being investigated today. The reason is that, beginning in 1967, the Rutherford formula has become more and more useful in the analysis of materials (see Appendix I). Thus, scientists are concerned more and more with its accuracy. Here, only the cases of small θ and $\theta = 180°$ are considered. For the general case, the article of J. R. MacDonald[10] is easy to read.

For the apparatus and its description, see Fig. 2.8. It should be noted that many formulas that hold in classical physics fail in quantum physics, but Rutherford's formula is one of a few that are derived according to classical physics which retain their original forms in quantum physics.

2.4.2 Rutherford Formula for Small Angles

It follows from Eq. (2.15) that, when θ is very small, the number of particles dN' going into a solid angle $d\Omega$ may become larger than the number N of incident particles; and at an extremely small θ, dN' will even tend to infinity. This is evidently absurd.

A small angle corresponds to a large impact parameter. In this case, the assumption that the action of extranuclear electrons can be neglected will no

[10] J. R. MacDonald, et al., *J. Appl. Phys.* **54** (1983) 1800.

longer hold under the usual experimental conditions. When b increases to the size of an atom, Coulomb scattering will no longer take place at all, because now the atom appears to be electrically neutral. It is, therefore, no longer legitimate to neglect the screening effects of the extranuclear electrons in the Rutherford formula.

However, if we consider a thought experiment in which all the atomic electrons around the target nuclei are stripped away, how should we express the number of α particles at small θ? At what small θ will Rutherford's formula no long hold? We leave this to the reader to answer.[10]

2.4.3 The Estimation of the Nuclear Size

When deriving the Rutherford formula, we took the atomic nucleus as a point and we considered the Coulomb force only. But, in fact, every nucleus has a certain size (see Chapter 9). So when an incident particle comes sufficiently near to a nucleus, the force acting on the particle ceases to be purely Coulomb. There will then arise discrepancies between the Rutherford formula and experimental results.

How near can the incident particle approach to the atomic nucleus? We now find a smallest distance r_m at which the Rutherford formula remains valid, so that the linear dimension of the atomic nucleus must be smaller than r_m. This r_m can at least be used as an upper bound of that linear dimension.

How does one find such an r_m? You might think that since we have the Coulomb scattering formula (2.6), we can put $\theta \to 180^\circ$ and find the smallest value of b and use it as r_m. But this would be a misunderstanding of the definition of b. In fact, b is the smallest linear distance between the incident particle and the scatterer when there is no interaction between them. But what we are considering is the smallest distance at which two interacting particles can approach to each other (see Fig. 2.4). The r_m and the impact parameter b represent two different concepts.

Let the velocity of the α particle be v when the particle is very far from the nucleus, so that its kinetic energy is $\frac{1}{2}m\mathrm{v}^2$. As the α particle experiences the Coulomb force from the nucleus, $Z_1Z_2e^2/r^2$, its kinetic energy will decrease to $\frac{1}{2}m\mathrm{v}'^2$, and the conservation of energy requires that

$$\frac{1}{2}m\mathrm{v}^2 = \frac{1}{2}m\mathrm{v}'^2 + \frac{Z_1Z_2e^2}{r} \tag{2.17}$$

Again, since the α particle is moving in a central field of force, we know that the angular momentum is also conserved, i.e.

$$|\boldsymbol{r} \times m\mathbf{v}| = mr^2\frac{d\phi}{dt} = L(\text{constant}) = m\mathrm{v}b \tag{2.18}$$

Equations (2.17) and (2.18) are the general laws of the motion. Now we take the special character of r_m into consideration. We have the smallest value of r when $r = r_m$. Hence, at this distance the radial velocity will become zero and there remains only the tangential velocity—this is a characteristic of the so-called perihelion. Then we have

$$mvb = mv_m r_m \tag{2.19}$$

At r_m, Eq. (2.17) becomes

$$E = \frac{1}{2}mv_m^2 + \frac{Z_1 Z_2 e^2}{r_m}$$

Substituting (2.18) into (2.19) and solving for v_m to go into this, we get

$$E = \frac{L^2}{2mr_m^2} + \frac{Z_1 Z_2 e^2}{r_m} \tag{2.20}$$

The first term on the right-hand side is the rotational energy of the particle $Z_1 e$ (the α particle), and the second term is the potential energy at the perihelion. From classical mechanics the trajectory of the α particle subject to a repulsive Coulomb force is

$$\frac{1}{r} = \frac{1}{b}\sin\phi + \frac{a}{2b^2}(\cos\phi - 1) \tag{2.21a}$$

(see derivations in Appendix 2A). When $r = r_m$, ϕ is symmetric about r_m and equals $(\pi - \theta)/2$. Substitution of b and ϕ into (2.21a) leads to

$$r_m = \frac{a}{2}\left(1 + \csc\frac{\theta}{2}\right) \tag{2.21b}$$

If the incident particles are negatively charged, we should replace 1 by -1 in (2.21b). Notice that, Eq. (2.6) applies to the cases of both two-body repulsion and two-body attraction.

It follows from (2.21b) that r_m reaches its minimum $r_m = a$ when $\theta = 180°$. This is the smallest distance that two bodies in a field of a repulsive force can approach to each other through a head-on collision. This statement is another expression of the physical meaning of the quantity a in (2.7).

Hence we reach the conclusion that the upper bound of the nuclear size of the scattering material is a, provided the Rutherford formula holds when $\theta = 180°$. Note that if we use incident particles of higher energy, we get a smaller a, and the estimation of the nuclear size will be nearer to reality, provided the formula still holds at $\theta = 180°$.

We also may obtain the above conclusion very simply through physical considerations. Consider a mountain climber who has used up all his energy and changed it into potential energy. Then his position is the nearest point to the summit that he can reach. By analogy, suppose a positive particle $Z_1 e$ moves toward an atomic nucleus with an energy $(\frac{1}{2})mv^2$. When all its energy is changed into potential energy, i.e., when

$$E = \frac{1}{2}mv^2 = \frac{Z_1 Z_2 e^2}{r_m}$$

it will reach its nearest point to the nucleus, since it has no energy to go further. Hence the corresponding smallest distance is

$$r_m = \frac{Z_1 Z_2 e^2}{E} \equiv a \tag{2.22}$$

However, our derivation of the same conclusion in (2.21) has the advantage that (2.21) is valid under arbitrary θ, so the perihelion formula (2.21) is very useful.

As shown by experiment, when the α particles from $^{210}_{84}\mathrm{Po}$ (with energy of 5.3 MeV) are scattered by $_{29}\mathrm{Cu}$ at $\theta = 180°$, the Rutherford formula turns out to be still correct. From Eq. (2.22), $a = 2 \times 29 \times 1.44\,\mathrm{fm \cdot MeV}/5.3\,\mathrm{MeV} = 15.8\,\mathrm{fm}$. Hence the nuclear radius of copper atoms must be smaller than 15.8 fm.

2.4.4 Rutherford Formula at $\theta = 180°$

In some cases, the experimentally determined cross-sections for Rutherford scattering are much larger than the values determined by the Rutherford formula.[11] However, such deviations occur only around 180° and the range of the angle is not larger than 1° (when $\theta = 179°$, the agreement between experiment and theory is very good). Such deviations have been considered elsewhere.[12] But before going to the references, consider the following: What is the difficulty in measuring the cross-section for Rutherford scattering at a scattering angle of exactly 180° and how do you overcome this difficulty? To gain a deeper knowledge of Rutherford scattering, first see the MacDonald paper cited in footnote 10. Then you may consider the experimental problems of 180° scattering and obtain instructive insights from the paper of Oen.[13]

2.5 SUMMARY OF THE SIGNIFICANCES AND DIFFICULTIES OF A NUCLEAR MODEL

2.5.1 The Significances

1. The most important significance is the advance of the "nuclear structure" of atoms, that is, the concept of the concentration of essentially all the mass and all the positive charge of an atom in a nucleus with very small volume and very high density. This concept divides an atom into two parts, the intranuclear and the extranuclear part (in our everyday life, we deal with only the latter).

2. Not only did Rutherford scattering play a very important part in atomic and nuclear physics, but the method of investigating the structure of matter by means

---

[11] A. N. Mantri, *Am. J. Phys.* **45** (1977) 1122.

[12] T. E. Jackman, et al., *Nucl. Instrum. & Meth.* **191** (1981) 527.

[13] O. S. Oen, *Nucl. Instrum. & Meth.* **194** (1982) 87.

of the scattering of particles continues to be important in the present. Whenever we observe in scattering experiments the characteristics of Rutherford scattering (the so-called "shades of Rutherford"), we may deduce that there is a point-like substructure in the object of our investigation.

3. Rutherford scattering provides a means for material analysis. In 1967, the United States sent a spacecraft to the Moon. On that craft was an α source, which was used to analyze the elements present in the surface of moon by means of Rutherford scattering. The results were sent back to Earth. These results were found to be in agreement with the results of later analysis of samples brought back from the Moon in 1969. Since then, Rutherford scattering has been used more and more in laboratories and has become a powerful means for the analysis of materials (see Appendix I). "Rutherford spectroscopes" constructed according to such principles have become commercially available in the market.

2.5.2 The Difficulties

Once the tiny nucleus (radius $\sim 10^{-13}$ cm) was proposed, the question immediately arose as to what the electrons were doing so that atoms had radii of $\sim 10^{-8}$ cm. The electrons could not be standing still or they would be drawn into the nucleus by the attractive Coulomb force. The simplest solution was to consider atoms to be like our solar system, with the electrons like the planets and the nucleus the sun: then both systems are governed by a $1/r^2$ force law which can be balanced by rotational motion, and 99.9% of the total mass of each system is concentrated in its center (the atomic nucleus or the Sun). However, the force acting in the solar system is the force of gravity while that acting in an atom is the Coulomb force. This difference immediately brings up the first of the three fatal difficulties of a classical solar model for the atom.

1. A solar model for atoms is not stable. It is well known that any charged particle in accelerated motion will radiate away energy in the form of electromagnetic waves. The circular motion of an electron around the nucleus is an accelerated one. Since the electron is charged, it should continuously emit electromagnetic waves and lose its energy according to classical theory. Thus, the radius of its orbit would become smaller and smaller, that is, the electron would move toward the nucleus along a spiral, and in a very short time (of the order of 10^{-9} s) the electron would fall into the nucleus. There the positive and negative charges would neutralize each other, and the atom would have collapsed to form a system the size of the nucleus. In the real world, however, no one has ever seen such a collapse. On the contrary, atoms in their normal (ground) state have never been observed to change their size or nature (with the exception of the few radioactive atoms found in nature). Gold atoms in rings made thousands of years ago are still gold today. So atoms are quite stable—a fact the classical planetary model could not explain.

2. The model cannot explain why atoms are identical. We know from the laws of classical mechanics that the solar system of today is determined by its initial conditions when it began to be formed. Different initial conditions cannot lead to exactly the same results. Since the changes and evolutions in the universe are astonishingly unpredictable, it is absolutely unthinkable that there should be two identical solar systems in the universe. But it is a different story for atoms—iron from America, and even from the Moon, is identical with iron in China in its atomic structure. Why atoms of each element are identical cannot be explained by the planetary model.

3. The model cannot explain the regeneration of atoms. In the solar system, once a planet is impacted by a comet, the state of the planet will be perpetually disturbed and the original state can never be regained—this is well-known common sense. But what is the situation in atoms? Suppose an atom interacts with an incoming particle. After the interaction, as soon as the incoming particle has gone away, that atom will return to its original state, just as if nothing had happened at all. This regeneration ability again cannot be explained by Rutherford's model.

2.5.3 Quantifying a Difficulty with Rutherford's Model

For the sake of simplicity, assume that an electron is revolving in a circular orbit around a nucleus (Ze). The Coulomb force on the electron provides the centripetal acceleration needed to keep a moving electron in a circular orbit about the nucleus, so

$$\frac{m_e \mathrm{v}^2}{R} = \frac{Ze^2}{R^2} \tag{2.23}$$

Introducing the angular momentum,

$$L = m_e \mathrm{v} R \tag{2.24}$$

the velocity of the electron can be written as

$$\mathrm{v} = \frac{Ze^2}{L} \tag{2.25}$$

and its acceleration

$$a = \frac{\mathrm{v}^2}{R} = \left(\frac{Ze^2}{L}\right)^2 \left(\frac{m_e Ze^2}{L^2}\right) = m_e \frac{(Ze^2)^3}{L^4} \tag{2.26}$$

According to classical electrodynamics, the energy radiated per unit time is

$$P = \frac{2}{3}\frac{e^2}{c^3} a^2 = \frac{2}{3}\frac{e^2}{c^3} m_e^2 \frac{(Ze^2)^6}{L^8} \tag{2.27}$$

Let τ be the necessary time interval for the electron to radiate out all its kinetic energy, then we have

$$P\tau = \frac{1}{2} m_e v^2 \tag{2.28}$$

Substituting (2.25) and (2.27) into (2.28), and rearranging, we have

$$\tau = \frac{3}{4} \frac{1}{Z^4} \frac{L}{m_e c^2} \left(\frac{Lc}{e^2}\right)^5 \tag{2.29}$$

Using

$$L^2 = Ze^2 m_e R \tag{2.30}$$

Eq. (2.29) may be expressed as

$$\tau = \frac{3}{4} \frac{R^3}{Zcr_e^2} \tag{2.31}$$

where

$$r_e = \frac{e^2}{m_e c^2} = 2.818\,\text{fm} \tag{2.32}$$

is called the classical radius of the electron.

Taking $R = 1\,\text{Å} = 10^5\,\text{fm}$ and $Z = 1$, and using Eq. (2.31) to estimate τ, we find that the time necessary for the electron to fall spirally into the nucleus is

$$\begin{aligned} \tau &= 3(10^5\,\text{fm})^3/4 \times 3 \times 10^8\,\text{m} \cdot \text{s}^{-1} \times 10^{15}\,\text{fm} \cdot \text{m}^{-1}(2.8\,\text{fm})^2 \\ &\simeq 3.2 \times 10^{-10}\,\text{s} \end{aligned} \tag{2.33}$$

This is a critical drawback of the planetary model. Electrons do not fall into nuclei.

APPENDIX 2A
CENTRAL FORCES

In theoretical mechanics, for a body moving in a field or force (or for two bodies interacting with each other, although this problem can always be reduced to a single-body problem), we have for a conservative field

$$\oint \boldsymbol{F} \cdot d\boldsymbol{r} = 0 \qquad \text{and} \qquad \boldsymbol{F} = -\frac{\partial U}{\partial \boldsymbol{r}} = -\left[\frac{\partial U}{\partial x}\hat{\boldsymbol{i}} + \frac{\partial U}{\partial y}\hat{\boldsymbol{j}} + \frac{\partial U}{\partial z}\hat{\boldsymbol{k}}\right]$$

We know that a conservative force is not necessarily a central force, but a central force is necessarily conservative. A central force is a function of the radius vector, i.e., $\boldsymbol{F} = F(r)\hat{\boldsymbol{r}}$, where $\hat{\boldsymbol{r}}$ is a unit vector in the $\boldsymbol{r}$ direction, and the potential function corresponding to it will be a function of the radius r only, $U = U(r)$.

When a particle moves in a central field of force, its trajectories may be divided into two categories, the finite ones and the infinite ones. Among the finite ones, there are the closed and the open ones:

$$\text{Trajectories of a particle in a central field of force} \begin{cases} \text{finite} \begin{cases} \text{closed} \\ \text{open} \end{cases} \\ \text{infinite} \end{cases}$$

It can be shown rigorously that only in two central fields, i.e., $F \sim 1/r^2$ and $F \sim r$, with potential functions $U \sim 1/r$ and $U \sim r^2$, respectively, can the trajectories of a particle be closed. Examples of the former are gravity and the Coulomb force, and an example of the latter is the simple-harmonic (i.e., elastic or pseudo-elastic) force. With the exception of these, all the trajectories in other fields can never be closed. Let us now show the possible trajectories of a particle in the case of a $1/r^2$ force as follows:

Force ($F \sim 1/r^2$)	Energy	Type of trajectory
(1) Gravity attraction and Coulomb attraction	>0	Infinite hyperbola
	$=0$	Infinite parabola
	<0	Finite ellipse
(2) Coulomb repulsion	>0	Infinite hyperbola
	$=0$	Infinite parabola

As an example, consider the trajectory of the α particle in Rutherford scattering. From Newton's law,

$$\boldsymbol{F} = m\boldsymbol{a} = \frac{Z_1 Z_2 e^2}{r^2} = m\left[\frac{d^2 r}{dt^2} - r\left(\frac{d\phi}{dt}\right)^2\right] \tag{2A.1}$$

Remember that the angular momentum is

$$L = mr^2 \frac{d\phi}{dt} = mvb = \text{constant} \tag{2A.2}$$

Let $r = 1/w$ and transform dr/dt into

$$\frac{dr}{dw}\frac{dw}{d\phi}\frac{d\phi}{dt}$$

Then substituting for $dr/dw = -1/w^2$ and Lw^2/m for $d\phi/dt$ from Eq. (2A.2) leads to

$$\frac{dr}{dt} = \frac{dr}{dw}\frac{dw}{d\phi}\frac{d\phi}{dt} = -\frac{1}{w^2}\frac{dw}{d\phi}\frac{Lw^2}{m} = -\frac{L\,dw}{m\,d\phi} \tag{2A.3}$$

Then write

$$\frac{d^2r}{dt^2} = \frac{d}{dt}\frac{dr}{dt}$$

in terms of $d\phi$ and substitute Eq. (2A.2) for $d\phi/dt$ and Eq. (2A.3) for dr/dt to give

$$\frac{d^2r}{dt^2} = \frac{d}{d\phi}\frac{d\phi}{dt}\left(\frac{dr}{dt}\right) = -\frac{L^2w^2}{m^2}\frac{d^2w}{d^2\phi} \tag{2A.4}$$

Now putting (2A.2) and (2A.4) into (2A.1) yields

$$m\left[-\frac{L^2w^2}{m^2}\frac{d^2w}{d\phi^2} - \frac{1}{w}\left(\frac{Lw^2}{m}\right)^2\right] = Z_1Z_2e^2w^2 \tag{2A.5a}$$

or

$$\frac{d^2w}{d\phi^2} + w = -\frac{Z_1Z_2e^2m}{L^2} = -\frac{Z_1Z_2e^2m}{m^2\mathrm{v}^2b^2} = -\frac{a}{2b^2} \tag{2A.5b}$$

where

$$a = \frac{Z_1Z_2e^2}{\frac{1}{2}m\mathrm{v}^2}$$

The general $a/2b^2$ solution of this second-order differential equation is

$$w = A\cos\phi + B\sin\phi - a/2b^2 \tag{2A.6}$$

where A and B are arbitrary constants. Substituting in the initial conditions $\phi \to 0$ as $r \to \infty$ and $dr/dt \Rightarrow -\mathrm{v}$ as $r \to \infty$, we find

$$w = \frac{1}{r} = 0 = A\cos 0 + B\sin 0 - \frac{a}{2b^2}$$

so $A = a/2b^2$.

Then using the condition $dr/dt = -\mathrm{v}$ as $r \to \infty$ and Eq. (2A.3) and differentiating Eq. (2A.6), we find

$$\frac{dr}{dt} = -\frac{L}{m}\frac{dw}{d\phi} = -\mathrm{v} = -\frac{L}{m}(-A\sin 0 + B\cos 0)$$

from which we find

$$B = \frac{mv}{L} = \frac{1}{b}$$

Thus,

$$w = \frac{1}{r} = \frac{1}{b}\sin\phi + \frac{a}{2b^2}(\cos\phi - 1) \tag{2A.7}$$

APPENDIX 2B
ELECTRIC UNITS

The interaction force between two point charges separated by a distance r as shown is

$$Z_1e \cdot \text{-------}_r\text{-------} \cdot Z_2e \qquad F = k\frac{Z_1Z_2e^2}{r^2}r_o$$

For different systems of units, the coefficient of proportionality k is different. If the Gaussian system (cgs) is used, the unit of charge is esu, the unit of distance (between charges) is the centimeter, the unit of mass is the gram, and the unit of force is the dyne. Then the expression for the force will be

$$F = \frac{Z_1Z_2e^2}{r^2}$$

where the coefficient k is

$$k = 1\,\frac{\text{dyne}\cdot\text{cm}^2}{\text{esu}^2}$$

If the SI system (MKS) is used, the unit of charge is the coulomb, the unit of distance is the meter, the unit of mass is the kilogram, and the unit of force is the newton. Then the coefficient in the force equation is

$$k = \frac{1}{4\pi\varepsilon_0} = 8.98755 \times 10^9\,\frac{\text{N}\cdot\text{m}^2}{\text{C}^2}$$

In order to go from the Gaussian system to the SI system, it is necessary only to multiply e^2 in a formula with $1/4\pi\varepsilon_0$.

APPENDIX 2C THE DERIVATION OF RUTHERFORD'S FORMULA IN THE LABORATORY SYSTEM

In the center-of-mass system, Rutherford's formula is

$$\sigma_c(\theta_c) = \left(\frac{Z_1 Z_2 e^2}{4E_c}\right)^2 \csc^4 \frac{\theta_c}{2} \tag{2C.1}$$

where the unit of σ_c is cm^2/(solid angle degree), e is in esu, and E_c is in ergs.

By the invariant relation (the index L indicates "laboratory"),

$$\sigma_c(\theta_c) \sin\theta_c \, d\theta_c = \sigma_L(\theta_L) \sin\theta_L \, d\theta_L$$

which can be transformed into

$$\sigma_L(\theta_L) = \sigma_c(\theta_c) \frac{\sin\theta_c}{\sin\theta_L} \frac{d\theta_c}{d\theta_L} \tag{2C.2}$$

Since

$$\sin(\theta_c - \theta_L) = \gamma \sin\theta_L \tag{2C.3}$$

where γ is the ratio of the masses of the incident particle and target nucleus ($\gamma = m_1/m_2$), we can easily obtain

$$\frac{d\theta_c}{d\theta_L} = 1 + \gamma \frac{\cos\theta_L}{\cos\Delta} = \frac{\sin\theta_c}{\sin\theta_L} \cdot \frac{1}{\cos\Delta}$$

where

$$\Delta \equiv \theta_c - \theta_L$$

Thus Eq. (2C.2) becomes

$$\sigma_L(\theta_L) = \sigma_c(\theta_c) \left(\frac{\sin\theta_c}{\sin\theta_L}\right)^2 \frac{1}{\cos\Delta} \tag{2C.4}$$

The energies in the center-of-mass system and the laboratory system are related by

$$E_c = \frac{1}{1+\gamma} E_L \tag{2C.5}$$

Substituting (2C.1) and (2C.5) into (2C.4), we obtain

$$\sigma_L(\theta_L) = \left(\frac{Z_1 Z_2 e^2}{2E_L}(1+\gamma) \frac{\sin\theta_c}{2\sin\theta_L} \frac{1}{\sin^2 \frac{\theta_c}{2}}\right)^2 \frac{1}{\cos\Delta} \tag{2C.6}$$

Using

$$1 + \gamma = 1 + \frac{\sin\Delta}{\sin\theta_L}$$

we obtain

$$\frac{\sin\theta_c}{2\sin^2\frac{\theta_L}{2}} = \frac{1}{\tan\frac{\theta_c}{2}} = \left(\frac{\sin\theta_L + \sin\Delta}{\cos\theta_L + \cos\Delta}\right)^{-1}$$

or

$$\sigma_L(\theta_L) = \left(\frac{Z_1 Z_2 e^2}{2E_L}\right)^2 \frac{(\cos\theta_L + \cos\Delta)^2}{\sin^4\theta_L \cos\Delta}$$

or

$$\sigma_L(\theta_L) = \left(\frac{Z_1 Z_2 e^2}{2E_L}\right)^2 \csc^4\theta_L \frac{\left[\cos\theta_L + \sqrt{1-\gamma^2\sin^2\theta_L}\right]^2}{\sqrt{1-\gamma^2\sin^2\theta_L}} \tag{2C.7}$$

After rearranging terms and using $e^2 = 1.44\times10^{-13}$ MeV, this becomes

$$\sigma_L(\theta_L) = 5.2\left(\frac{Z_1 Z_2}{E_L\sin^2\theta_L}\right)^2 \frac{\left[\cos\theta_L + \sqrt{1-\left(\frac{m_1}{m_2}\sin\theta_L\right)^2}\right]^2}{\sqrt{1-\left(\frac{m_1}{m_2}\sin\theta_L\right)^2}} \tag{2C.8}$$

where θ_L is the scattering angle in the laboratory system in degrees, σ_L is the cross-section in millibarns (10^{-27} cm^2), m_1 and Z_1 are the mass and charge number of the incident particle, m_2 and Z_2 are those of the target nucleus, and E_L is the energy of the incident particle in MeV.

PROBLEMS

2.1. A nonrelativistic α particle with velocity v collides with a free electron at rest. Show that the largest angle of deflection of the α particle is approximately 10^{-4} rad.

2.2. (*a*) When α particles with kinetic energy of 5.00 MeV are scattered at 90° by gold nuclei, what is the impact parameter?

(*b*) If the thickness of a gold foil is 1.0 μm, in what percentage of cases will the incident α particles be scattered at angles larger than 90° (this is called back-scattering).

2.3. When there occurs a head-on collision between a 4.5 MeV α particle and a gold nucleus, what is the distance of closest approach? What will it be if we replace the gold nucleus by a lithium nucleus?

2.4. (*a*) Suppose the radius of a gold nucleus is 7.0 fm; what is the necessary energy for an incident proton just to reach the surface of the nucleus? Generally the proton size is neglected. What is the difference if the proton is taken to have a radius of 1 fm?

(*b*) Suppose we replace the gold nucleus by an aluminum nucleus the radius of which is assumed to be 4.0 fm. What is the necessary energy of the incident proton in order that the proton can just reach the surface of the aluminum nucleus in a head-on collision?

2.5. A narrow beam of protons with a kinetic energy of 1.0 MeV impinge perpendicularly on a gold foil of mass-thickness 1.5 mg/cm^2. A counter counts the protons scattered at an angle of 60°. The window of the counter has an area of 1.5 cm^2, and the distance from the scattering region in the foil is 10 cm. The window faces and is at right angles to the protons that fall on it. What is the ratio between the numbers of protons that enter the window and that impinge on the foil?

2.6. A beam of α particles impinges perpendicularly on a metal foil. After scattering, what is the ratio between the number of the scattered particles for which the angles of scattering are larger than 60° and 90°, respectively?

2.7. A narrow and homogeneous α particle beam impinges perpendicularly on a tantalum foil of thickness 2.0 mg/cm^2. The ratio between the numbers of scattered and incident particles that are scattered at an angle $\theta > 20°$ is 4.0×10^{-3}. Calculate the differential cross-section of a tantalum nucleus corresponding to the scattering angle $\theta = 60°$.

2.8 (*a*) Incident particles of mass m_1 are elastically scattered by a target nucleus at rest. The mass of the target is m_2 (where $m_2 \leq m_1$). Show that the largest possible scattering angle of the incident particle in the laboratory system $\theta_{L,\max}$ is determined by $\sin\theta_{L,\max} = m_2/m_1$.

(*b*) If an α particle is scattered by a deuterium nucleus which is originally at rest, what is the maximum angle of scattering in the laboratory system?

2.9. A narrow beam of protons with a kinetic energy of 1.0 MeV impinges perpendicularly upon a gold foil of mass thickness 1.5 mg/cm^2. Suppose the foil contains 30% silver. Calculate the relative number of particles that have a scattering angle larger than 30°.

2.10. An accelerator-produced proton beam with an energy of 1.2 MeV and a beam-current of 5.0 μA, impinges perpendicularly upon a gold foil. Find the number of protons that are scattered in 5 min by the foil into the following intervals of angles:

(*a*) 59–61°

(*b*) $\theta > \theta_0 = 60°$

(*c*) $\theta < \theta_0 = 10°$

2.11. Assume Thomson's model is right: that the positive charge in an atom is uniformly distributed over the entire atom. If an α particle has an energy of 5.0 MeV and the radius of a gold atom is 1 Å, show that the largest angle of deflection of the α particle scattered from a gold atom would be about 10^{-4} rad (electrons are neglected).

2.12. We have two identical gold foils and scatter α particles off of each separately. Of course, the results (scattering probability) are the same (call the result A). When the two foils are put together, the scattering probability is measured to be $2.1A$ (with the same scattering angles). If an initial α particle energy of 5.0 MeV is used, calculate the α particle energy loss after passing through the first foil.

2.13. If α particles with kinetic energy up to 7.7 MeV energy are scattered by a gold foil and the Rutherford scattering formula is still correct, estimate the size of the gold nucleus.

2.14. When a proton with energy of 2.0 MeV approaches a gold nucleus, how close can it come in a head-on collision before it is stopped?

2.15. A narrow proton beam of 2.0 MeV impinges perpendicularly on a gold foil of mass-thickness 1.5 mg/cm^2. The beam current is 10^{-9} amp. A counter with a window of 4.0 mm diameter is put 10 cm away from the foil. The window faces the gold foil and is at 160° with respect to the incoming proton beam. How many proton counts would be accepted by this detector within 10 min?

2.16. When the α scattering angle is very small (less than 15°), the experimental data are far from the theoretical values. What is the reason?

2.17. In the Rutherford scattering experiment, if the mass of an incoming particle in the beam is exactly the same as that of the target particle, show that the scattering angle in the laboratory system is one-half of the respective angle in the center of mass system.

2.18. When an α particle is scattered by a helium nucleus at rest, what is the largest scattering angle in the laboratory system?

2.19. The thickness of a thin gold layer on a silicon backing can be measured by Rutherford backscattering (RBS). A 2.0 MeV proton beam from an accelerator impinges perpendicularly on a thin gold layer and a detector with 1.0 mm diameter is placed 5.0 cm away from the sample. The scattered protons at 160° are recorded. When the incoming protons have been accumulated up to 100 μC, the scattered proton count is $N' = 12500$. What is the thickness of this gold layer?

2.20. When protons are scattered by a copper ($Z = 29$) foil, the scattered protons are recorded at 90°. For lower-energy protons, the measured scattered proton number is in agreement with the Rutherford formula, but when the proton energy is larger than 10 MeV, deviations occur. What is the radius of the copper nucleus? (The recoil of the copper nucleus may be neglected.)

2.21. A proton with kinetic energy of 25 keV is scattered by a helium nucleus at rest. The helium recoils in the direction of 60° with respect to the incoming proton. What is the impact distance of the incoming proton?

2.22. Consider what some of the consequences would be if the mass of the electron were more nearly equal to the mass of the proton. What would this do to electron binding energies in atoms and molecules? How would the thermal energy radiation from the Sun influence processes on the Earth like photosynthesis, for example?

CHAPTER 3

QUANTUM STATES OF ATOMS: THE BOHR MODEL

What is a model? A model is like an Austrian timetable. Since Austrian trains are always late, a Prussian visitor asks an Austrian conductor why they bother to print timetables. The conductor replies "If we did not, how would we know how late the trains are?"

Victor F. Weisskopf[1]

3.1 BACKGROUND

In 1900, Max Planck put forward his famous quantum of light hypothesis. But at that time few paid any attention to his article, let alone understood it. Even Planck himself did not like his "quantum" and thought of it only as a mathematical trick. He and many others wanted to bring the quantum hypothesis into the classical fold. However, in 1905, Albert Einstein took the revolutionary concept seriously. He used the concept of the quantum of light to explain the photoelectric effect in the same year in which he put forward the special theory of relativity. Likewise, Einstein's quantum paper was not taken seriously either by leading scientists. Planck said, when introducing Einstein as a new member of the

[1] From H. Frauenfelder and E. M. Henley, *Subatomic Physics*, Prentice-Hall, Englewood Cliffs, NJ (1974), p. 351.

German Academy of Science in 1909: He has done so much good work, we can forgive him for "dabbling in the idea of the quantum." Even in 1913, the four most famous German physicists (including Planck) called Einstein's light-quantum a "losing direction" in a letter.[2] But Niels Bohr, a Danish physicist, at the age of only 28, creatively applied the concept of the quantum to the Rutherford model of the atom, which was in serious difficulty at that time, and explained the 30-year-old puzzle of why atoms have only certain spectral lines. In this chapter, we will introduce Planck's quantum hypothesis, Einstein's concept of the light quantum, and some experimental facts related to the spectra of atoms. Ironically, years later Einstein, who won the Nobel Prize for his quantum theory of the photoelectric effect, came to be the chief opponent of quantum mechanics as a fundamental theory.

3.1.1 Evidence of the Quantum Hypothesis I: Blackbody Radiation

What is blackbody radiation? Sometimes we call someone "black hearted" (for example, Shylock, the loanshark in Shakespeare's comedy "The Merchant of Venice"). By that we mean that the person greedily grabs everything and gives nothing back. Similarly, if a body absorbs all light incident upon it without reflecting any light back we call it an "absolute blackbody" or simply "blackbody." Of course, a blackbody can radiate energy but this is a different process. In fact, there is no "absolute blackbody," but some bodies can be treated as a "blackbody" approximately. As shown in Fig. 3.1, once a beam of light enters into a cavity through a narrow hole, it is very hard for it to be reflected out of the cavity through the hole again. The hole in the cavity can be treated as a blackbody.

As we all know, all bodies emit heat radiation and, just like visible light radiation, heat radiation is an electromagnetic wave that spans a certain

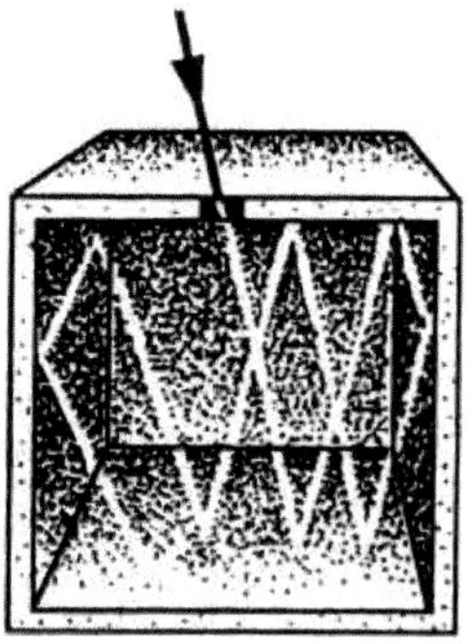

FIGURE 3.1
Simulation of blackbody radiation.

[2] Max Jammer, *The Conceptual Development of Quantum Mechanics*, McGraw-Hill, New York (1966), p. 44.

frequency range. Whether steel-making is good or bad depends on the temperature of the furnace and the temperature is indicated by the color of the flame. Thus we must know the intensity distribution of the heat radiation in a furnace, $u(\lambda)$ (the intensities of each wavelength radiation, that is, the intensities of the different colors), and how it varies with the temperature in order to select the proper time for making steel. Similarly, astronomers determine the surface temperatures of stars by measuring the intensity distributions of the radiations from these stars. The needs of metallurgy and astronomy have promoted much research on heat radiation.

In 1859, G. R. Kirchhoff proved that when a blackbody is in equilibrium with heat radiation, the shape and the location of the curve of u_ν as a function of frequency ν (u_ν is the density of radiation energy, e.g., energy per unit volume) are related only to the absolute temperature T of the blackbody and have nothing to do with the shape and the composition of the cavity. Therefore, using a blackbody, we can investigate the law of heat radiation itself universally while leaving aside the specific characteristics of matter.

Then, in 1879, J. Stefan found that the total energy (total radiant intensity) over all frequencies (or all wavelengths) emitted per unit time per unit area from a blackbody at a temperature T, called the radiancy, is given by $R_T = \int R_T(\nu)\, d\nu = \sigma T^4$. Here $\sigma = 5.67 \times 10^{-8}\, \mathrm{W/m^2 \cdot K^4}$ is called the Stefan–Boltzmann constant.

In 1893, W. Wien discovered the displacement law of blackbody radiation: The frequency ν_m corresponding to the maximum radiation energy density is proportional to the absolute temperature T of a blackbody (see Fig. 3.2).

$$T = \frac{0.2898}{c}\,\nu_m$$

For example, at 1000 K, ν_m is given by

$$1000\,\mathrm{K} = \frac{0.2898\,\mathrm{cm \cdot K}\,\nu}{3 \times 10^{10}\,\mathrm{cm/s}}$$

$$\nu_m = 1.035 \times 10^{14}\,\mathrm{Hz}$$

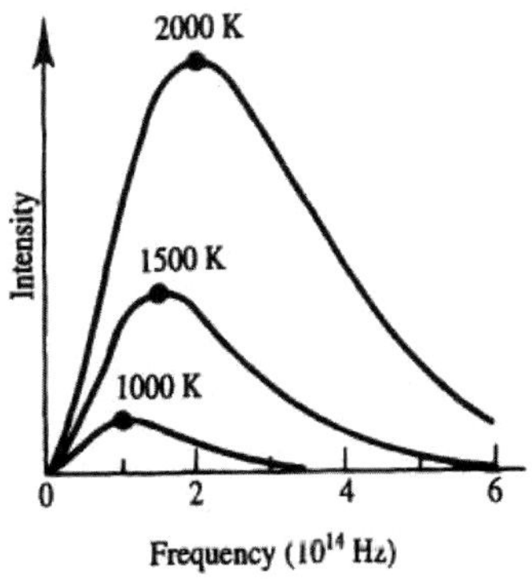

FIGURE 3.2
Illustration of Wien's displacement law of blackbody radiation. A dot indicates ν_m for each T. Note how ν_m increases linearly with temperature.

One can also write that the product of wavelength λ_m at the maximum and T remains a constant

$$\lambda_m T = 0.2898\,\text{cm}\cdot\text{K}$$

Wien derived the following empirical expression for the radiation energy density u_ν in a frequency interval $d\nu$ between ν and $\nu + d\nu$:

$$u_\nu\, d\nu = C_1 \nu^3 \exp(-C_2\nu/T)\, d\nu \tag{3.1}$$

where C_1 and C_2 are empirical parameters and T is the temperature at equilibrium. Except for an obvious deviation at low frequency (long wavelength), the agreement between experiment and this formula is very good.

In 1899, J. W. S. Rayleigh and J. H. Jeans, using classical electrodynamics and statistical mechanics, derived the formula

$$u_\nu\, d\nu = \frac{8\pi}{c^3} kT\nu^2\, d\nu \tag{3.2}$$

where c is the velocity of light and k is the Boltzmann constant. This formula is in good agreement with experiments at low frequency (long wavelength). However, as the frequency increases, the agreement vanishes since Eq. (3.2) predicts an impossible infinite energy density when $\nu \to \infty$. This is the famous "ultraviolet catastrophe," as shown in Fig. 3.3. Thus Wien's empirical formula fit the short wavelength end, and the Rayleigh–Jeans formula fit the long wavelength end of the energy distribution.

In April 1900, Sir William Thomson, Lord Kelvin, said in an article summarizing physics in the preceding few hundred years: "In the scientific mansion which has been basically completed, it seems that the only thing left for younger physicists is to repair it. But in the distant place of the clear sky of physics, there still exist two ominous black clouds." The two black clouds represented the two phenomena in physics which could not be explained at that time. One was related to blackbody radiation, and the other to the Michelson–Morley experiment. These two black clouds were soon to bring about a profound revolution in

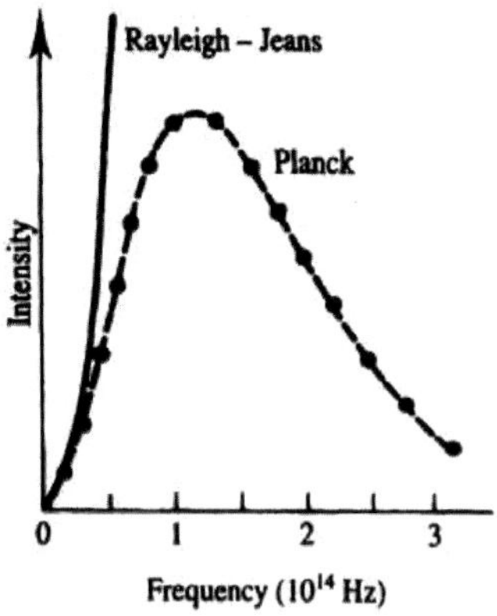

FIGURE 3.3
Comparison of the Rayleigh–Jeans formula and the Planck formula with experiment (points).

physics. One led to the birth of quantum mechanics and the other led to the theory of relativity.

On October 19, 1900, Planck proposed a formula for the energy distribution in blackbody radiation at the German Physical Society:

$$u(\nu) = \frac{8\pi h\nu^3}{c^3} \frac{1}{e^{h\nu/kT} - 1} \tag{3.3}$$

Planck obtained this formula by an inspired guess in order to fit experimental data at all wavelengths[3] but was guided in his guess by the forms of the Wien and Rayleigh–Jeans formulas. On the day the formula was proposed, H. Rubens checked it against the most accurate experimental data then measured by O. Lummer and E. Pringsheim and found the agreement to be extremely good. Rubens told Planck the good news the next day. This agreement made Planck determined to "find out a theoretical explanation at any expense." After two months' hard work, Planck proposed on December 14, at the German Physical Society, that the energy exchange of electromagnetic radiation can only be in the form of quanta, i.e., $E = nh\nu$, $n = 1, 2, 3, \ldots$ where h is called Planck's constant. The value of h was given by Planck[4] as

$$h = 6.55 \times 10^{-34}\,\mathrm{J \cdot S}$$

which is only 1 percent lower than the presently accepted value. Planck also derived the Boltzmann constant:

$$k = 1.346 \times 10^{-23}\,\mathrm{J/K}$$

which is about 2.5 percent lower than the present value. From these values, it was possible to calculate Avogadro's constant N_A and the charge of the electron e quite accurately. It was 20 years later before N_A and e were measured separately by experiments to better accuracy.

In explaining the meaning of the quanta, Planck said initially that electromagnetic radiation both in its emission and absorption could only be in quanta (certain discrete, fixed units of energy $h\nu$), but then he showed that the formula only required emission in quanta while absorption could be continuous as in classical physics. Because this concept deviated seriously from classical physics, Planck regretted very much proposing this "quantum hypothesis." In the following ten years and more, he left no means untried to bring it into classical physics. For example, he proposed the quantization was "false quantization,"—it was just like butter, people could only buy a whole piece from shops, but could cut it bit by bit at home. Only after all classical explanations had failed was the real and profound meaning of the quantum hypothesis realized.[5]

[3] See Jin-Yan Zen, *Quantum Mechanics*, Scientific Publishing Co., Beijing (1981), p. 1; W. Greiner, *Quantum Mechanics*, Springer-Verlag, Berlin (1989) for detailed discussions on blackbody radiation.

[4] M. Planck, *Ann. der Physik* **4** (1901) 553.

[5] M. J. Klein, *Physics Today* **19** (1966) 23.

Because of the great difference between Planck's quantum hypothesis and the concepts of classical physics, essentially no one paid any attention to it during the next five years after Planck's formula was put forward formally. The first serious application of this revolutionary new idea was in 1905 when Einstein put forward a light-quantum hypothesis to explain successfully the photoelectric effect by using $E = h\nu$.[6]

3.1.2 Evidence of the Quantum Hypothesis II: Photoelectric Effect

DISCOVERY OF THE PHOTOELECTRIC EFFECT. In 1887, H. R. Hertz first observed electromagnetic waves in a discharge experiment and showed that the velocity of electromagnetic waves was equal to the velocity of light. Thus, Hertz verified Maxwell's predictions of the existence of electromagnetic waves and that their velocity equaled the velocity of light. Hertz noted in his experiment that it was easy for a discharge to take place when ultraviolet light was incident on the cathode. This was the first sign of the photoelectric effect. The next year, W. Hallwachs made the further observation in connection with this phenomenon that a clean and isolated zinc plate acquired positive charge when illuminated by ultraviolet light while a negatively charged plate lost its negative charge. In 1900, P. Lenard proved experimentally that metals would emit electrons when illuminated by ultraviolet light. After two years, he further concluded that the experimental data on the photoelectric effect could not be explained by a wave theory. In 1905, Einstein put forward a light-quantum hypothesis to explain the photoelectric effect.

In 1916, Millikan measured the relation between the light frequency and the energy of the emitted electrons carefully. He verified Einstein's quantum formula for the photoelectric effect and determined the Planck constant accurately.

EXPERIMENTAL DATA ON THE PHOTOELECTRIC EFFECT. An experimental setup for the study of the photoelectric effect is shown in Fig. 3.4. When monochromatic light with frequency greater than some critical frequency ν_0 is incident on a metal surface which is the positive electrode, electrons are emitted from the surface. A negative voltage V is added to another electrode to stop the electrons (stopping potential). So V is a direct measurement of the electron energy. If we assume that the maximum kinetic energy of the electrons emitted from the positive electrode is $\frac{1}{2}mv_m^2$, no electron will reach the cathode under the condition that $eV = eV_0 = \frac{1}{2}mv_m^2$. Then the photoelectric current i will be zero.

At first sight one might think that the photoelectric effect is very easy to understand. However, careful consideration of the experimental data reveals that there is something deep behind the data. The experimental results indicate:

---

[6] A. Einstein, *Ann. der Physik* **17** (1905) 132; A. Arons and M. Peppard, *Am. J. Phys.* **33** (1965) 367.

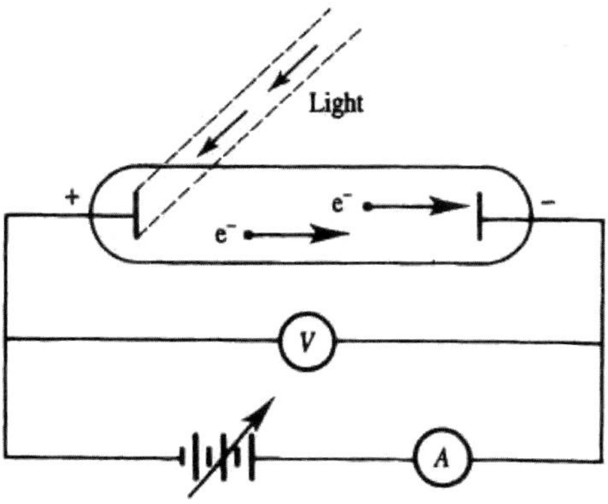

FIGURE 3.4
An experimental arrangement for observing the photoelectric current.

1. For a fixed intensity I and frequency ν of incident light, the relation between the photoelectric current i and time t after the light is incident on the metal surface is as shown in Fig. 3.5(a). The current begins essentially instantaneously (<1 ns). More importantly, the time t is independent of I; that is, the current begins at the same time for a very weak intensity light as for a very strong one.
2. For a fixed stopping potential V and frequency of light ν, the photoelectric current i is proportional to the intensity of the light I; that is, the number of electrons per second emitted from the surface is proportional to the intensity of the light, as shown in Fig. 3.5(b).
3. For a fixed intensity I and frequency of light ν, the photoelectric current i decreases as the stopping potential V increases, as shown in Fig. 3.5(c). When $V = V_0$, $i = 0$, because even electrons with the maximum energy will be stopped (when V is low, only those electrons with low energy will be stopped).
4. For a given surface, the stopping potential V_0 depends only on the frequency of light ν, and has nothing to do with the intensity of the light I, or the photoelectric current i.
5. Finally, a photoelectric current flows only when the frequency of the incident light exceeds a certain threshold frequency for the metal being illuminated. Each metal, such as cesium, potassium or copper, has its own fixed threshold frequency ν_0. No matter how large the intensity of the light, there will be no photoelectric current i if $\nu < \nu_0$, as shown in Fig. 3.5(d). For most metals, ν_0 is in the ultraviolet range. Because a typical value of V_0 is a few volts, the kinetic energy of the photoelectrons emitted is of the order of a few electron-volts. It is clear from Fig. 3.5(d) that for $\nu \leq \nu_0$, $V_0 = 0$, because no electrons are emitted even when no stopping potential is added. This threshold frequency for the emission of photoelectrons is the most crucial datum that cannot be explained in classical physics.

CLASSICAL EXPLANATIONS OF THE PHOTOELECTRIC EFFECT. Classical physics understood light as a type of wave. Electrons across the whole surface

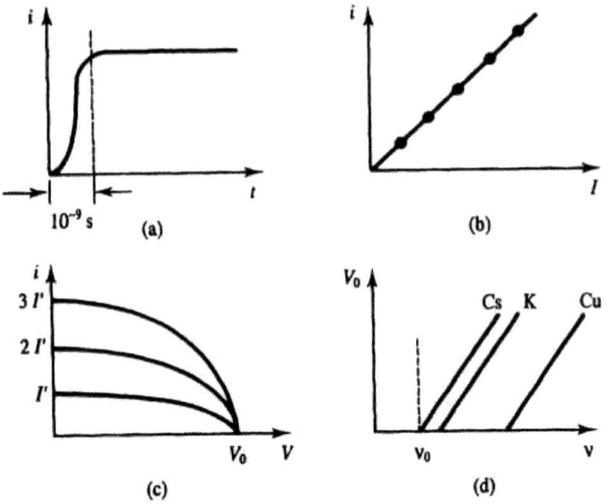

FIGURE 3.5
(a) For a fixed intensity I and frequency ν of light, the relation between photoelectric current i and the time of creating the photoelectric current i. (b) For a fixed stopping potential V and frequency ν of light, the relation between the photoelectric current i and the intensity of light I. (c) For a fixed intensity I and frequency ν of light, the relation between the photoelectric current i and the stopping potential V. (d) For a specific metal surface, the relation between the stopping potential V and frequency of light ν (slope = h, Planck's constant).

of a metal should gain energy continuously when illuminated by a light wave. When the energy gained continuously by an electron reached a certain value, the electron would be ejected from the atom and the metal surface. But how much time does it take for an electron to obtain sufficient energy from light. Classically, this time should depend on the intensity of the light: the lower the intensity the longer the time before emission. Experiments tell us that the photoelectric current can be detected virtually instantaneously even when light with an intensity of $1\,\mu W/m^2$ is incident on the surface of sodium. This is analogous to the situation where a light source with an intensity of 500 W illuminates a plate of sodium 6300 m away from it. Even then photoelectrons are emitted essentially instantaneously. It is easy to estimate that there are about 10^{19} Na atoms per m^2 per one atomic layer and 10^{20} Na atoms per ten atomic layers. (From $\frac{4}{3}\pi r^3 N_A = A/\rho = 23/0.97$, one can estimate that the radius of a sodium atom is $r = 2$ Å. The volume of $1\,m^2$ and one atomic layer is $2 \times 2\,Å \times 100^2 = 4 \times 10^{-4}\,cm^3$, in which there are $4 \times 10^{-4}/\frac{4}{3}\pi r^3 = 1.2 \times 10^{19}$ sodium atoms per m^2 per atomic layer.) If one assumes that the energy of the incident light wave is absorbed uniformly by the atoms in these ten layers, each atom would obtain $10^{-26}\,W = 10\,J/s = 10^{-7}\,eV/s$. This indicates that on a sodium plate of $1\,m^2$ area, the energy received by each atom per second is about $0.1\,\mu eV$. An electron needs about 1 eV of energy to escape from the surface of sodium. Even if only one electron in each atom receives energy, it will take $10^7\,s \simeq 1/3$ year (1 year $= 3.15 \times 10^7$ s) for this electron to obtain an energy of

1 eV to escape. This delay is in serious contradiction with the experimental facts.

In addition, according to classical theory, it is not the frequency of light but the intensity of the light that determines the energy of the electrons, just as the intensity (as measured by the height) of a water wave determines its effect on a beach or pier. But the experimental fact is that the energies of the electrons emitted from a metal illuminated with a weak blue light are much larger than with an intense red light (when ν of both blue and red light exceed ν_0 of the metal). Such a relation between the electron energy and light frequency is impossible to explain by classical physics. Finally, in the classical absorption of light by a surface there should be no threshold frequency for the emission of electrons. Red light with the same energy intensity as blue light should be equally as effective in releasing electrons. But, experimentally, if ν (red) $< \nu_0$ no electrons are emitted even for a very intense red light.

QUANTUM EXPLANATIONS FOR THE PHOTOELECTRIC EFFECT. In 1905, Einstein applied Planck's quantum hypothesis to the photoelectric effect. Planck assumed that the energies of the oscillators in matter that emitted the radiation were quantized. So light was emitted from a source discontinuously in quanta or packets of energy, $h\nu$, but light propagated in the form of a wave and could be absorbed continuously as a wave. But Einstein proposed that light was absorbed by matter in quanta like light is a particle. Such a particle is now called a light quantum or photon. Einstein explained the photoelectric effect successfully using the light-quantum hypothesis.

Einstein proposed that when light is incident on a metal surface, the light energy is not absorbed continuously as it would be for a wave but discontinuously all at once as a photon of energy $h\nu$ is absorbed by an electron. A part of the energy the electron absorbs is used to overcome the binding energy of the electron to the metal, and the remaining part goes into the kinetic energy of the electron after escaping from the surface. This energy relation can be expressed as

$$\tfrac{1}{2}mv_m^2 = h\nu - \phi \tag{3.4}$$

Here the energy of the photon $h\nu$ minus the binding energy (work function) ϕ of the electron in the metal equals the maximum kinetic energy of the electron. As shown in Fig. 3.6, which is Fig. 3.5(d) amplified, when $h\nu < \phi$, no electron can escape from the metal surface because its energy is less than its binding energy. Thus no photoelectron appears. The frequency of light determines the energy of the photons as well as the energy of the photoelectrons. The greater the intensity of the light the more photons there are and the more photoelectrons there are. But the possibility of creating photoelectrons depends on the frequency of the incident light. One can ask what happens to an electron which absorbs light with ν just below ν_0: Why could it not absorb a second photon and the sum of the energies of the two allow it to escape? This does not occur normally because the extra kinetic energy of the electron after absorbing the first photon $h\nu$ is given very quickly to

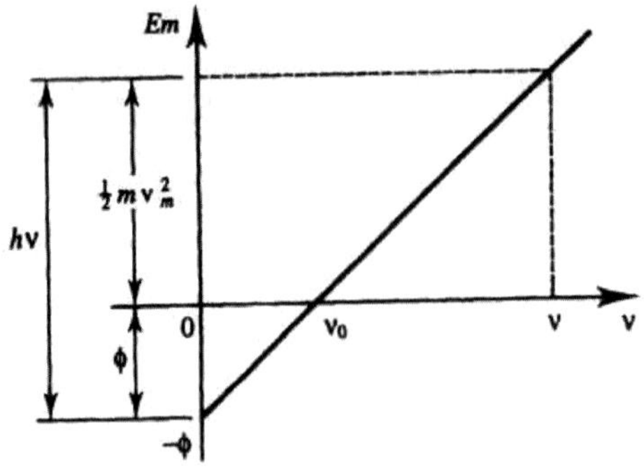

FIGURE 3.6
Einstein's explanation for the photoelectric effect, $E_m = \frac{1}{2}mv_m^2 = h\nu - \phi$ for a given ν, where ν_0 is the threshold frequency and m is the electron rest mass.

other electrons and atoms by collisions to raise the temperature of the metal long before a second photon can be absorbed. In summary, all the data on the photoelectric effect that cannot be understood by classical physics are explained by the absorption of light in quanta $h\nu$.

As an example, suppose a substance had a work function of 1.5 eV. What would be the maximum kinetic energy of the photoelectrons if blue light of 4000 Å were allowed to shine on the substance?

$$\frac{1}{2}mv_m^2 = h\nu - \phi = \frac{hc}{\lambda} - \phi = \frac{4.1 \times 10^{-15}\,\text{eV}\cdot\text{s} \times 3 \times 10^8\,\text{m/s}}{4000 \times 10^{-10}\,\text{m}} - 1.5\,\text{eV}$$
$$\approx 3.1\,\text{eV} - 1.5\,\text{eV} = 1.6\,\text{eV}$$

The straight line in Fig. 3.6 can be obtained from experiments. From its slope, Planck's constant, h, which is a universal constant for all metals, can be measured directly. In 1916, by carefully carrying out measurements as shown in Fig. 3.6, Millikan measured Planck's constant and obtained a value very close to the present value. Professors Einstein and Bohr, the first two scientists to exploit the idea of the quantum of light, are seen in Fig. 3.7.

3.1.3 Spectrum of Light

A spectrum of light shows its intensity distribution as a function of frequency. Measuring this distribution is one of the important ways of investigating the structure of atoms. Newton said as early as 1704 that the only way for us to understand the structure inside matter is to investigate its light spectrum.

A spectrum is measured by a spectrometer. There are many kinds of spectrometers, but their basic principles are essentially the same. They consist of three parts: a light source, a spectroscope (prism or grating) to separate the light into different frequencies, and a device to record the intensities of the different frequencies into which the light has been divided. Figure 3.8 illustrates the principles of a prism spectrometer. When a line-slit is used to define the light source, after separation each frequency appears as a line image of the slit.

FIGURE 3.7
Albert Einstein and Niels Bohr taking a walk.

Different light sources have different spectra. If a hydrogen lamp is used as a light source, the light will be characteristic of hydrogen. The spectrum detected by a spectrograph will be the spectrum of hydrogen, as shown in Fig. 3.9.

By 1885, fourteen spectral lines (14 frequencies) from hydrogen had been observed by using spectrographs. That year, J. J. Balmer put forward an empirical formula that gave the observed wavelengths or wavenumbers (reciprocals of wavelengths) of the hydrogen spectral lines which are in the visible light range:

$$\lambda = \frac{Bn^2}{n^2 - 4} \qquad \tilde{\nu} \equiv \frac{1}{\lambda} = \frac{4}{B}\left(\frac{1}{2^2} - \frac{1}{n^2}\right) \tag{3.5}$$

where $n = 3, 4, 5, 6, \ldots$ and $B = 3645.6$ Å is an empirical constant. The values of the wavelengths calculated by this formula are in very good agreement with the experimental values within the experimental errors. This formula was later called the Balmer formula. The group of hydrogen spectral lines represented by it, all in the visible light range, is called the Balmer series.

J. R. Rydberg (1889) suggested the following universal equation to represent many series:

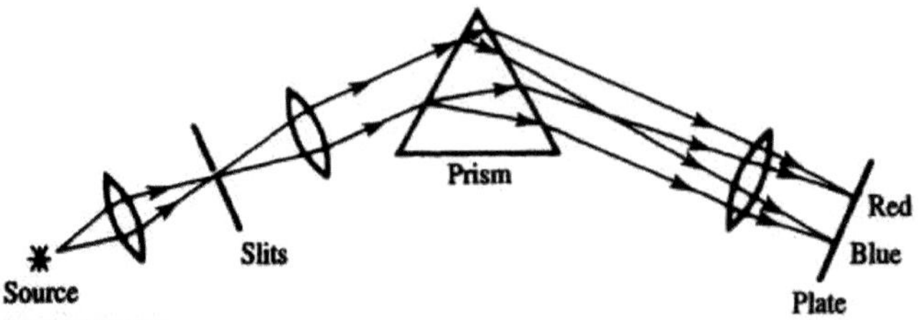

FIGURE 3.8
Schematic diagram of a prism spectrograph.

$$\tilde{\nu} \equiv \frac{1}{\lambda} = R_H \left[\frac{1}{n^2} - \frac{1}{n'^2} \right] = T(n) - T(n') \tag{3.6}$$

where $T(n)$ and $T(n')$ are called term values.

The Balmer formula for hydrogen is a special case of Rydberg's equation. Here, $R_H = 4/B$ is an empirical parameter called Rydberg's constant and $n = 1, 2, 3, \ldots$ where for each n there exists $n' = n+1, n+2, \ldots,$ to give a spectrum series. For example, the $n = 1, n' = 2, 3, 4, \ldots,$ spectrum series for hydrogen is in the ultraviolet range. The first member of this hydrogen series was discovered by T. Lyman in 1906, but was not attributed to hydrogen. This series was predicted by Bohr in his quantum theory of hydrogen and the observation of this series ending in $n = 1$, called the Lyman series, was a great triumph of Bohr's quantum theory of atoms.

The $n = 2, n' = 3, 4, 5, 6, \ldots,$ series for hydrogen is in the range of visible light and is the Balmer series (1885). Among them is the famous red H_α line ($n' = 3, \lambda = 6563$ Å) which was determined first by A. J. Ångström in 1853. The common unit of wavelength the angstrom (Å) was named after him. The year 1853 may be regarded as the beginning of scientific spectroscopy. He, along with G. R. Kirchhoff and R. W. Bunsen, laid the foundations of spectral analysis.

The $n = 3, n' = 4, 5, 6, \ldots,$ series is in the infrared range. It was discovered by F. Paschen in 1908 and is called the Paschen series.

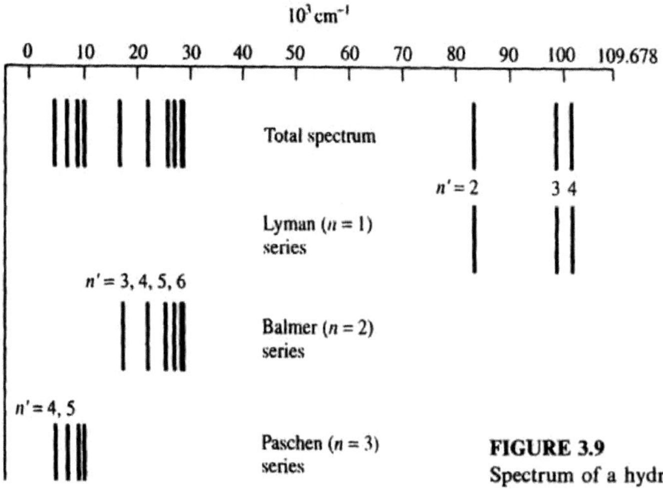

FIGURE 3.9
Spectrum of a hydrogen atom.

The $n = 4, n' = 5, 6, 7, \ldots,$ series is in the infrared range as well. The first three members were discovered by F. Brackett in 1922, so it is called the Brackett series.

The $n = 5, n' = 6, 7, 8, \ldots,$ series is farther in the infrared range. It was discovered by H. A. Pfund in 1924 and is called the Pfund series.

Additional members of the hydrogen spectral series with $n = 4, n' > 7$; $n = 5, n' > 7$; and $n = 6, n' = 7$ were discovered later by C. S. Humphreys.

As an example of Eq. (3.6), calculate the wavelength of the second line in the Paschen series. Here $n = 3$ and $n' = 5$.

$$\frac{1}{\lambda} = \frac{4}{B}\left(\frac{1}{n^2} - \frac{1}{n'^2}\right) = \frac{4}{3645.6\,\text{Å}}\left(\frac{1}{3^2} - \frac{1}{5^2}\right) = \frac{4}{3645.6\,\text{Å}} \times \frac{25-9}{225} = \frac{4 \times 0.071\,111}{3645.6\,\text{Å}}$$

$$\lambda = 12\,816.7\,\text{Å}$$

It is evident from Eq. (3.6) that any spectral line of hydrogen can be expressed by the difference between two spectral terms. The spectrum of hydrogen is the synthesis of the differences between spectral terms. The many spectral lines which are so complicated outwardly can actually be expressed simply by Eq. (3.6). This was a remarkable achievement. However, it was a puzzle for nearly 30 years why the agreement between the Rydberg formula and experimental data was so good. This puzzle was solved when Bohr introduced Planck's quantum hypothesis into the Rutherford model and began to unravel the inner workings of the atom. Professors Bohr and Planck are seen in Fig. 3.10.

3.2 THE BOHR MODEL

Bohr apparently did not know of the existence of the Rydberg equation before February, 1913. Later someone asked Bohr: "How did you not know the Balmer and Rydberg formula?" Bohr replied: "At that time most physicists thought that the spectra of atoms were too complicated and they would never be part of basic physics." In February when he heard of this empirical expression for the hydrogen spectral lines from his friend, he obtained "the last and decisive piece of the jig-saw puzzle" of his theory.[7] In March of that year, Bohr put forward his theory of the hydrogen atom and published three successive monumental works in July, September, and November.[8] Bohr finished his hydrogen theory in three separate stages.

3.2.1 Classical Orbits with Stationary State Conditions

Bohr suggested that an electron in the hydrogen atom was moving in a circular orbit around the nucleus (classical orbit) like planets around the sun, but then laid

[7] L. Rosenfeld and E. Rudinger, *Niels Bohr*, North-Holland, Amsterdam (1968), p. 51.

[8] Niels Bohr, *Phil. Mag.* **26** (1913), 1; **26** (1913) 476; **26** (1913) 857.

FIGURE 3.10
Niels Bohr and Max Planck holding discussions.

down the following absolute (rigid) rules: (a) The electron could only be in certain discrete orbits and it could only rotate in these orbits around the nucleus; (b) in addition, the acceleration associated with this rotation would not produce electromagnetic radiation. These are Bohr's stationary state conditions. It should be pointed out that "an absolute (rigid) rule" is often necessary when a new theory is put forward. Einstein laid down rigid rules when he established the theory of relativity. To determine if a rigid rule is tenable, the level of agreement between the conclusions produced by the rigid rule and the experimental results must be considered.

As shown in Fig. 3.11, an electron with mass m is rotating in a circular orbit of radius r around the proton. According to classical dynamics, the centripetal force on the electron is

$$F = m\frac{v^2}{r}$$

which can only be supplied by the Coulomb attraction between the electron and the proton, namely,

$$F_c = \frac{e^2}{r^2} = \frac{mv^2}{r}$$

From this we can get the energy expression for electrons rotating in circular orbits

$$E = T + V = \frac{1}{2} m\text{v}^2 - \frac{e^2}{r} = \frac{1}{2}\frac{e^2}{r} - \frac{e^2}{r}$$

or

$$E = -\frac{1}{2}\frac{e^2}{r} \tag{3.7}$$

The frequency for electrons rotating in circular orbits is

$$f = \frac{\text{v}}{2\pi r} = \frac{1}{2\pi r}\sqrt{\frac{e^2}{mr}} = \frac{e}{2\pi}\sqrt{\frac{1}{mr^3}} \tag{3.8}$$

3.2.2 Frequency Condition

According to Bohr's hypothesis, an electron would not emit electromagnetic radiation when it was rotating in a stable orbit, so it would not lose its energy and fall into the nucleus. But when would electromagnetic waves be emitted by an atom? Bohr postulated that when an electron changed from one orbit to another, an electromagnetic wave of energy $h\nu$ would be emitted (or absorbed). Its energy would be equal to the difference in energy between the two orbits. In symbols, the energy emitted when an electron jumps from a state $E_{n'}$ to a lower energy state E_n is

$$h\nu = E_{n'} - E_n \tag{3.9}$$

This is the frequency condition or radiation condition Bohr postulated. Here Bohr introduced Planck's constant into the atomic field.

Only shortly before publishing his theory did Bohr notice that, comparing Eq. (3.9) with Eq. (3.6), one immediately finds that

$$E_n = -\frac{Rhc}{n^2} \tag{3.10}$$

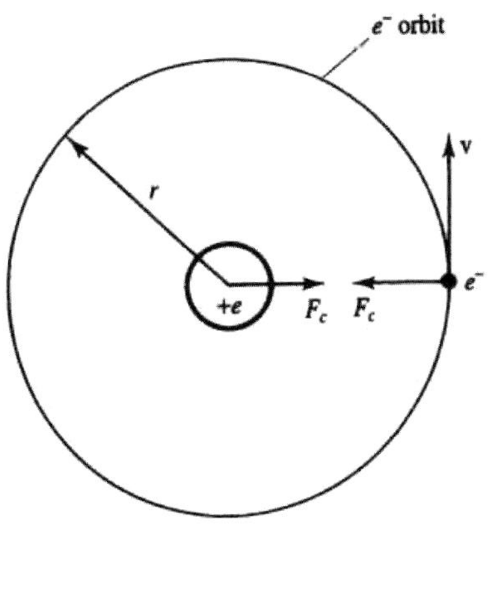

FIGURE 3.11
Classical orbit of an electron in a hydrogen atom.

Once described in this form, the Rydberg formula can be explained as follows: It represents the energy released when an electron jumps from one stationary state n' (with energy $E_{n'}$) to another stationary state n (with energy E_n). The corresponding wavelength is λ and frequency is ν.

From Eqs. (3.10) and (3.7), we find

$$r_n = \frac{e^2}{2Rhc} n^2 \tag{3.11}$$

This gives the allowed orbital radii of electrons in hydrogen atoms in stationary states n, where n can only be positive integers so the orbits are discrete. But these results for r_n were not so remarkable, because r_n could not be determined from experiment to compare with Eq. 3.11. R was still an empirical constant.

3.2.3 Quantization of Angular Momentum

Some authors regard the quantization of angular momentum as Bohr's third postulate. In fact, this idea was derived from what is called the *correspondence principle*. Phenomena in the atomic field and phenomena in the macroscopic domain may follow their respective own rules, but when the rules in the microscopic field are extended to the classical field, the numerical results obtained from such rules should agree with the results from the classical rules. This is main content of the correspondence principle.

Let us first rewrite formula (3.6) as

$$\nu = \tilde{\nu} c = Rc \frac{n'^2 - n^2}{n'^2 n^2} = Rc \frac{(n' + n)(n' - n)}{n'^2 n^2} \tag{3.12}$$

When n is very large, the frequency of the transition between two adjacent states n and $n'(n' - n = 1)$ is

$$\nu \simeq Rc \frac{2n}{n^4} = \frac{2Rc}{n^3} \tag{3.13}$$

According to the correspondence principle, it must be in accordance with the classical relation (3.8), for frequency,

$$\frac{2Rc}{n^3} = \frac{e}{2\pi} \sqrt{\frac{1}{mr^3}} \tag{3.14}$$

From which we obtain

$$r = \sqrt[3]{\frac{e^2}{16\pi^2 R^2 c^2 m}} n^2 = \frac{e^2}{2Rhc} n^2 \tag{3.15}$$

which must accord with Eq. (3.11) as repeated on the right. From this we obtain an expression for the Rydberg constant:

$$R = \frac{2\pi^2 e^4 m}{ch^3} \tag{3.16}$$

Now the Rydberg constant is no longer an empirical constant. It is composed of several basic constants (e, m, h, c) and can be calculated accurately.

Substituting Eq. (3.16) into Eq. (3.11), we obtain

$$r_n = \frac{\hbar^2}{me^2} n^2 \qquad \text{where } \hbar \equiv \frac{h}{2\pi} \tag{3.17}$$

for the radii of the orbits of the electrons.

Substituting Eq. (3.16) into Eq. (3.10), we get the energy expression of the electrons in this system:

$$E_n = -\frac{me^4}{2\hbar^2 n^2} \tag{3.18}$$

Now, in classical theory, the angular momentum, L, of the electrons should be

$$L = mvr = m\sqrt{\frac{e^2}{mr}}r = \sqrt{me^2 r} \tag{3.19}$$

Bohr saw that substituting Eq. (3.17) into (3.19), yields

$$L = n\hbar, \qquad n = 1, 2, 3, \ldots \tag{3.20}$$

which is the quantization condition of angular momentum.

It should be pointed out the formulas (3.16) to (3.20) are all derived under the condition of very large n values. However, we assume that all of these formulas are correct even when n takes a small value. This is another key point of the corresponding principle. Whether the assumption is right or not must be determined by experiments.

The idea of the quantization of angular momentum had been proposed a year earlier by P. Nicholson and also by P. Ehrenfest in 1913. Having seen that a consequence of his quantized energy is that angular momentum is quantized, Bohr tried to use this principle to explain the structure of heavier elements, but with little success. Nevertheless, the idea of the quantization of angular momentum was to prove basic in explaining the periodic table and many other phenomena. It is ironic that the formula (3.20) is wrong, as shown later in the Schrödinger formalism of quantum mechanics. The n quantum number in the energy and radius formula is not associated with angular momentum in the later theory, but angular momentum quantization is still important (see Chapter 7).

So far, we have introduced the basic model of the hydrogen atom put forward by Bohr in 1913. There are three key steps in the Bohr theory: (a) the stationary state condition (concept of quantum state); (b) the frequency condition for radiation (quantum transition); and (c) the correspondence principle. These assume that the conservation of energy is still valid when going down to the scale of the atom. Is the Bohr theory correct? This depends on comparisons between the results obtained from theory and experiment. Before making comparisons, we introduce some calculational methods first.

3.2.4 Numerical Calculational Methods

We now have expressions for the Rydberg constant, and the energy and the radius of an electron in the hydrogen atom, see equations (3.16) to (3.18). It is obvious that, in order to make numerical calculations, one need only substitute all the basic constants $(m, e, h, ..)$ into these formulas. But such a process is not only tedious but also lacking in physical meaning. So let us introduce simple numerical calculational methods.

Introducing the following composite constants (their meanings will be clear later) with accuracy sufficient for normal calculations,

$$\begin{aligned} hc &= 1240\,\text{fm}\cdot\text{MeV} = 12.4\,\text{Å}\cdot\text{keV} \\ \hbar c &= 197\,\text{fm}\cdot\text{MeV} = 1970\,\text{Å}\cdot\text{eV} \\ e^2 &= 1.44\,\text{fm}\cdot\text{MeV} = 14.4\,\text{Å}\cdot\text{eV} \\ mc^2 &= 0.511\,\text{MeV} = 511\,\text{keV} \end{aligned} \tag{3.21}$$

we can easily calculate the radius of the first Bohr orbit of hydrogen (the value of r_n when $n = 1$). From Eq. (3.17), we have

$$\begin{aligned} r_1 \equiv a_1 &= \frac{\hbar^2}{me^2} = \frac{(\hbar c)^2}{mc^2 e^2} \\ &= \frac{(1970)^2(\text{Å}\cdot\text{eV})^2}{0.511\times 10^6\,\text{eV}\times 14.4\,\text{Å}\cdot\text{eV}} \simeq \frac{3.9\times 10^6\,\text{Å}}{7.3\times 10^6} \simeq 0.529\,\text{Å} \end{aligned} \tag{3.22}$$

Note that putting in and canceling units is always helpful in obtaining correct answers. So we can write:

$$\begin{aligned} r_n = a_1 n^2 &= 0.529 n^2\,\text{Å}\,\text{(for hydrogen atoms)} = a_1 n^2 \\ r_n &= a_1 n^2/Z\,\text{(for hydrogenlike atoms)} \end{aligned} \tag{3.23}$$

Now calculate the energies of the electrons in hydrogen. First, rewrite equation (3.22) as

$$E_n = -\frac{me^4}{2\hbar^2 n^2} = -\frac{mc^2}{2}\left(\frac{e^2}{\hbar c}\right)^2\frac{1}{n^2} \tag{3.24}$$

The constant

$$\frac{e^2}{\hbar c} \equiv \alpha \simeq \frac{1}{137} \tag{3.25}$$

is called the fine structure constant for reasons to be made clear later. It is obvious from Eq. (3.24) that α is a dimensionless constant. It and m/m_p are the two most important constants in atomic physics. They are dimensionless constants and cannot be derived from first principles up to now.

After introducing α, Eq. (3.24) becomes

$$\begin{aligned} E_n &= -\frac{1}{2}m(\alpha c)^2\,\frac{1}{n^2} \text{ for the hydrogen atom} \\ &= -\frac{1}{2}\,m(\alpha c)^2\,\frac{Z^2}{n^2} \text{ for any one-electron, hydrogenlike atom} \end{aligned} \tag{3.26}$$

For $n = 1$,

$$\begin{aligned} E_1 &= -\frac{1}{2}\,m(\alpha c)^2 = -\frac{1}{2}\,mc^2\alpha^2 \\ &= -\frac{1}{2}\,(0.511\times 10^6\,\text{eV})\times\left(\frac{1}{137}\right)^2 \simeq -13.6\,\text{eV} \end{aligned} \tag{3.27}$$

$$E_n = -13.6\,\text{eV}\,\frac{Z^2}{n^2} \text{ for hydrogenlike atoms} \tag{3.27a}$$

The energy for an electron in the ground state ($n = 1$) of doubly ionized lithium ($Z = 3$) is

$$E_1(\text{Li}^{2+}) = \frac{-13.6\,\text{eV}\times 9}{1} = -122.4\,\text{eV}$$

Equation (3.27) gives the binding energy of the ground state of an electron in hydrogen. If we define that the energy of the ground state of hydrogen is zero, then

$$E_\infty = \frac{1}{2}\,m(\alpha c)^2 = 13.6\,\text{eV} \tag{3.28}$$

is the energy needed to move an electron in the ground state of hydrogen to infinity. This energy is the ionization energy of hydrogen. Using Eq. (3.27a), we can obtain the ionization energy for an electron in the ground state for any hydrogenlike atom. The ionization energy for an electron in the nth orbit of hydrogen is $(13.6/n^2)\,\text{eV}$. For example, for the third orbit of hydrogen, $n = 3$, it is $13.6/9\,\text{eV} = 1.51\,\text{eV}$.

Thus, we have two important physical parameters characterizing atoms. One is a dimension, Bohr's first radius, and the other is an energy, the ground state energy or the ionization energy of hydrogen. From Eq. (3.27) we also see that

$$\alpha c = \text{v}_1 \tag{3.29}$$

which is defined as the electron velocity in the first Bohr orbit. We obtain the same result from the characteristics of circular orbital motion where

$$\frac{m\text{v}_1^2}{r_1} = \frac{e^2}{r_1^2} \qquad (n = 1) \tag{3.30}$$

From this we find

$$\mathrm{v}_1 = \sqrt{\frac{e^2}{mr_1}} = \frac{e^2}{\hbar} = \frac{e^2 c}{\hbar c} = \alpha c$$

in accordance with Eq. (3.29). Thus we know that the velocity of electrons moving in the first orbit in the hydrogen atom is 1/137 of the velocity of light. At this low velocity, relativity is not needed generally.

Note that from Eq. (3.26)

$$\mathrm{v}_n = \frac{\alpha c}{n} \quad \text{(for H)} = \frac{\alpha c Z}{n} \quad \text{(for hydrogenlike)} \tag{3.31}$$

In addition, we can rewrite the expression for the Rydberg constant in Eq. (3.16) as

$$R = \frac{1}{2} m(\alpha c)^2 \frac{1}{hc} = \frac{E_\infty}{hc} \tag{3.32}$$

Thus we see that the Rydberg constant is proportional to the ionization energy of hydrogen, where E_∞, the energy, and R are connected by hc. From $h\nu = E$, the wavenumber, the reciprocal of the wavelength, is

$$\tilde{\nu} \equiv \frac{1}{\lambda} = \frac{E}{hc} \tag{3.33}$$

Thus the wavelength of a transition of energy E is

$$\lambda = \frac{hc}{E} = \frac{12.4}{E} \text{ Å} \cdot \text{keV} \tag{3.34}$$

in which the energy E must be in keV. From Eq. (3.32) or Eq. (3.33), we see the meaning of the composite constant hc (or $\hbar c$): it is the bridge connecting the two ways of characterizing a system, its energy, or its length. The dimension of hc (or $\hbar c$) is the product of energy and length, which are the most important physical parameters in any system. The product of these two parameters being constant means that smaller lengths connect with higher energies. The e^2 is connected by $\hbar c$ to yield the dimensionless fine structure constant.

3.3 EXPERIMENTAL EVIDENCE I: SPECTRA

3.3.1 Spectrum of Hydrogen

Using Eq. (3.16) or Eq. (3.32) and Appendix III, we obtain

$$R = 109\,737.315\,\text{cm}^{-1} \tag{3.35}$$

which is in reasonable agreement with the experimental value

$$R_H = 109\,677.58\,\text{cm}^{-1} \tag{3.36}$$

Thus for the first time the Rydberg constant was obtained theoretically. But there was still a problem. The difference between the calculated and the experimental result exceeds 5×10^{-4}, but the experimental accuracy of spectroscopy at the time this was first done had reached 1×10^{-4}. This question was raised by A. Fowler and answered by Bohr in 1914. In the original theory the nucleus of hydrogen was taken to be stationary. However, since the mass of the hydrogen nucleus (proton) is not infinite, the nucleus cannot be stationary when an electron is rotating around it. Two-body motion would take place as shown in Fig. 3.12. Thus the reduced mass μ should be substituted for the mass of the electron in the former expression for the energy. The Rydberg constant corresponding to the nucleus with mass A should be written as

$$R_A = \frac{2\pi^2 e^4}{ch^3}\mu = \frac{2\pi^2 e^4}{ch^3} m \frac{1}{1+\frac{m}{M}} = R\frac{1}{1+\frac{m}{M}} \tag{3.37}$$

which is reduced to $R_\infty = R$ when the mass M is infinite.

This shows that the Rydberg constant given by Eq. (3.35) is in fact R_∞, which corresponds to a nucleus with infinite mass. Then R_A with a finite mass A should be calculated by Eq. (3.37). Some specific examples are listed in Table 3.1.

It is clear from Eq. (3.37) that so long as the constant R_A corresponding to an individual atom is determined accurately, R_∞ can be derived from it. It was found that R_∞ derived from experiment was identical with the theoretical value, Eq. (3.35). Thus the Bohr theory gave Rydberg's empirical formula a firm physical basis.

In the Rydberg formula, the wavenumber is given by

$$\tilde{\nu} = \frac{1}{\lambda} = \frac{\nu}{c} = R\left(\frac{1}{n^2} - \frac{1}{n'^2}\right) \tag{3.38}$$

where now n' and n represent the quantum numbers of the allowed states occupied by an electron before and after the transition, respectively. Diagrams of the possible orbits r_n expressed by Eq. (3.17) and the possible energies E_n expressed by Eq. (3.18) are shown in Fig. 3.13 and Fig. 3.14.

Figure 3.14 is the energy level diagram of a hydrogen atom drawn to scale. Each horizontal line in the diagram represents the binding energy of the electron in each n shell and the distance between the lines is the energy difference which an electron emits when it jumps to a lower level. Care must be taken when using the energy level diagram: the larger the energy gap, the shorter the wavelength.

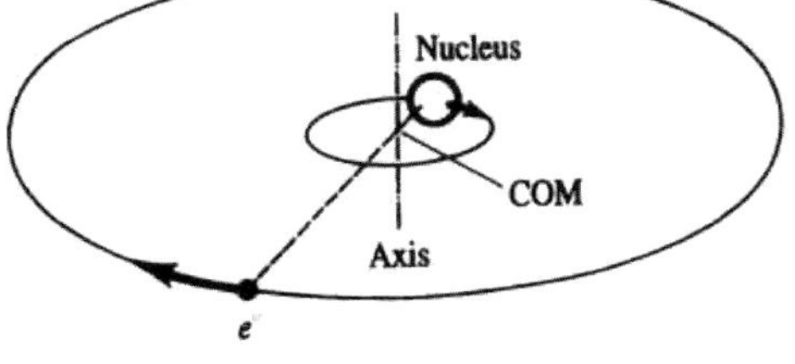

FIGURE 3.12
Rotations of the electron and the nucleus in a hydrogen atom.

TABLE 3.1
Rydberg constant R_A (cm^{-1})

	$R_\infty = 109\,737.315\,685\,27(73)$ cm^{-1}		
$^1_1$H	109 677.58	$^4_2$He$^+$	109 722.27
$^2_1$D	109 707.42	$^7_3$Li^{2+}	109 728.80
$^3_1$T	109 717.35	$^9_4$Be^{3+}	109 730.70

Note: the number at the upper left of the element symbol represents the atomic mass number A and that on the lower left the atomic number Z.

Energies can be added and subtracted from each other directly, but this is not true for wavelengths. For example, when an electron jumps from the $n = 2$ energy level to the $n = 1$ energy level, the energy radiated by the electron is

$$h\nu = E_2 - E_1 = \left(-\frac{hcR}{4}\right) - (-hcR) = \frac{3}{4}\,hcR$$

But the wavelength of this electromagnetic wave is not the difference between the wavelengths calculated from E_2 and E_1, respectively. Another example, the energy difference between $n = 3$ and $n = 2$ is equal to the energy interval of $n = 3 \to n = 1$ minus the interval of $n = 2 \to n = 1$; but the transition wavelength from $n = 3$ to $n = 2$ (H_α line in Balmer series) is not equal to the difference between the transition wavelength of $n = 3 \to n = 1$ (the second line in the Lyman series) and the transition wavelength of $n = 2 \to n = 1$ (the first line in the Lyman series).

Also remember that if one wants to use a particle like an electron or a proton to excite an electron in a hydrogen (or hydrogenlike) atom from the $n = 1$ to $n = 2$ orbit, the threshold (minimum) kinetic energy required for the projectile is

$$E_K = \left(1 + \frac{m_p}{m_H}\right)(E_2 - E_1)$$

where m_p is the mass of the projectile and m_H the mass of hydrogen (or hydrogen-like) atom to account for their relative motion. Note that for an electron neglect of the reduced mass is only a very small correction to its kinetic energy.

Thus the Bohr model explained the spectrum of hydrogen successfully and thereby unraveled "the puzzle of the Balmer formula" which had existed for nearly 30 years. This was a great success for the Bohr theory.

3.3.2 Spectra of Hydrogenlike Ions

Hydrogenlike ions[9] are ions which have only one electron outside the nucleus, which has a positive charge with $Z > 1$. Different values of Z represent different hydrogenlike systems. For example:

[9] H. F. Beyer and V. P. Shevelko, *Introduction to the Physics of Highly Charged Ions*, Taylor & Francis (2002).

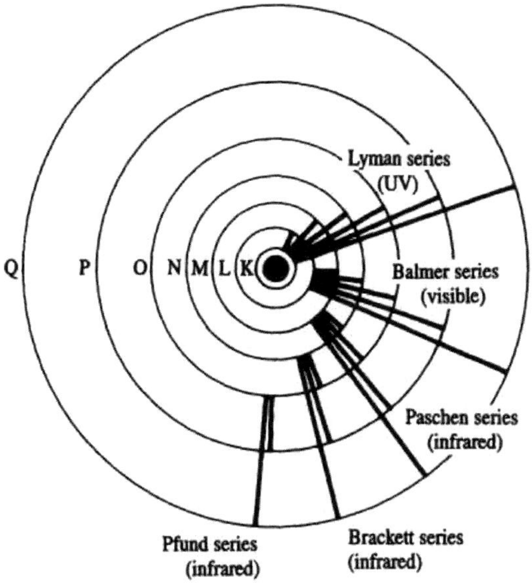

FIGURE 3.13
Orbits and spectral lines of a hydrogen atom.

			Labeled as
$Z = 1$	H	Neutral	HI
2	He^+	with one unit of positive charge	HeII
3	Li^{2+}	with two units of positive charge	LiIII
4	Be^{3+}	with three units of positive charge	BeIV

Hydrogenlike ions of higher Z, such as O^{7+}, Cl^{16+}, Ar^{17+} up to U^{91+} can now be produced by using accelerator techniques.

The extension of the Bohr theory to the spectra of hydrogenlike ions is very simple. Whenever there is an e^2 in the original formula, multiply by Z. For example, the wavenumber of the spectra of hydrogenlike ions is as follows: Since R_A includes e^4, R_A must be changed to $R_A Z^2$:

$$\left(\frac{1}{\lambda}\right)_A = R_A\left(\frac{1}{n^2} - \frac{1}{n'^2}\right)Z^2 = R\left\{\frac{1}{\left(\frac{n}{Z}\right)^2} - \frac{1}{\left(\frac{n'}{Z}\right)^2}\right\} \tag{3.39}$$

From this equation, we see that although n, n', and Z are integers, the ratios of n/Z and n'/Z are not necessarily integers. This is exactly the difference between the spectra of hydrogenlike ions and the spectrum of hydrogen.

For He^+ ($Z = 2$) as an example, Eq. (3.39) becomes

$$\left(\frac{1}{\lambda}\right)_{He^+} = R_{He}\left[\frac{1}{(n/2)^2} - \frac{1}{(n'/2)^2}\right]$$

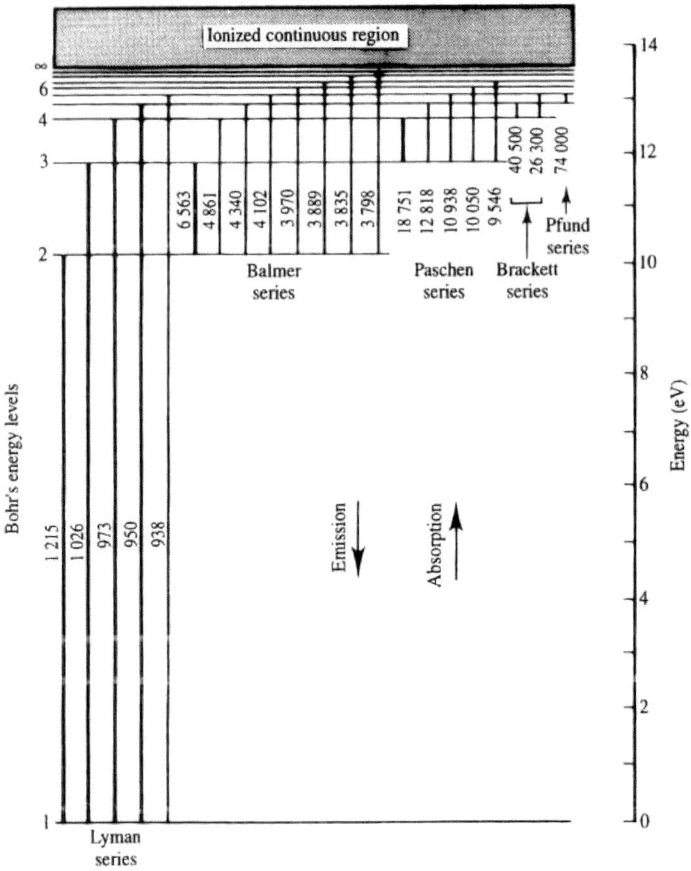

FIGURE 3.14
Energy level diagram of a hydrogen atom. The wavelengths of the transitions shown are given in Å.

For $n = 4$, and $n' = 5, 6, 7, \ldots$, we obtain

$$\left(\frac{1}{\lambda}\right)_{He^+} = R_{He}\left[\frac{1}{2^2} - \frac{1}{n_1^2}\right], \text{ where } n_1 = 2.5, 3, 3.5, \ldots$$

However, the Balmer formula for hydrogen is

$$\left(\frac{1}{\lambda}\right)_H = R_H\left[\frac{1}{2^2} - \frac{1}{n'^2}\right], n' = 3, 4, 5, \ldots$$

There are two major differences between these two formulas: One is that the number of the spectral lines of He is more than that of hydrogen. The spectral lines corresponding to $n = 2.5, 3.5, 4.5, \ldots$ do not exist in the spectrum of hydrogen. The other is the difference between R_{He} and R_H. Thus, even the spectral lines with $n_1(\text{He}) = \text{n}'(\text{H})$ are not exactly at the same place in the spectrum.

In Fig. 3.15, the group of relatively taller spectral lines represents the Balmer series of hydrogen and the shorter ones represents the Pickering series of He^+. The Pickering series was found by the astronomer E. C. Pickering in 1897 when he observed the optical spectrum of the star ζ in the constellation Puppis. From the figure we can see that the spectral lines of the Balmer series nearly coincide with those of every other line of the Pickering series. But there are also some spectral lines between the two neighboring lines of the Balmer series. These correspond to the lines with n_1 not being an integer. For those lines which nearly coincide with each other, the small differences between their wavelengths are related to the differences of R.

At first, some thought that the Pickering series was the spectral lines of hydrogen and that hydrogen on the earth was different from that on other stars. Think what that would do to science. But starting from his theory, Bohr pointed out that the Pickering series was not emitted by hydrogen but belonged to He^+. E. J. Evans upon hearing Bohr's idea, went to the laboratory and studied the spectrum of He^+ carefully. His results verified that Bohr's conclusion was absolutely correct. The successful extension of Bohr's theory to the spectra of hydrogenlike ions further convinced people of the reliability of his theory. When given the news, Einstein also was completely convinced and praised Bohr's theory as a "great discovery."

3.3.3 The Existence of Deuterium

In 1932, H. C. Urey discovered in his experiments that there was another spectral line (6561.00 Å) by the side of the H_α line of hydrogen (6562.79 Å). The difference between them was 1.79 Å. He postulated that this spectral line belonged to an isotope of hydrogen, deuterium, and further that $m_H/m_D = 1/2$. He calculated the different Rydberg constants R_H and R_D, respectively, using Eq. (3.37), and then the corresponding wavelengths. The results showed that the agreement between the theoretical and the experimental values was very good to confirm the existence of deuterium (D, heavy hydrogen).

3.3.4 Note I: Nonquantized Orbits

We have introduced quantized states and discrete line spectra of light from atoms. It is clear from Eqs. (3.17) and (3.18) that the radii of the orbits are proportional

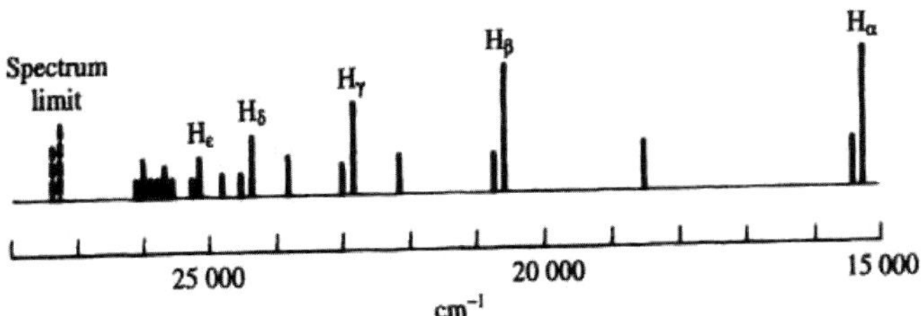

FIGURE 3.15
Comparison between the Pickering series and Balmer series.

to n^2 and the absolute values of the energies E vary inversely as n^2. The larger the n, the larger the radius, while the energy tends to zero; that is, the quantized energy is negative and the maximum quantized energy is zero. There follows the question: Are there any states with positive energy? Such states have been found in experiment. Beyond the series limit of the Balmer series, there is a continuous band produced by atoms with positive energy.

When an electron is far from the nucleus, it has positive kinetic energy and its potential energy is nearly zero, so its total energy is equal to its kinetic energy: $\frac{1}{2}mv_0^2$. When the electron approaches the nucleus, its orbit is a hyperbola, that is, the orbit is infinite, as shown in Fig. 3.16 (see Appendix 2A). At any point of the orbit, the energy is equal to the value when the electron is far from the nucleus, which is positive, and is expressed as

$$E = \frac{1}{2}mv^2 + \left(-\frac{Ze^2}{r}\right) = \frac{1}{2}mv_0^2$$

This energy of the electron in this nonclosed orbit is not quantized and can be any positive value. If an electron jumps from this nonquantized orbit to a quantized orbit, a photon will be emitted by the atom and the energy of the photon is

$$\begin{aligned} h\nu = E - E_n &= \left(\frac{1}{2}mv^2 - \frac{Ze^2}{r}\right) + \frac{me^4Z^2}{2n^2\hbar^2} \\ &= \frac{1}{2}mv_0^2 + \frac{hcR}{n^2}Z^2 \end{aligned}$$

for a bare nucleus of charge Z.

The first term on the right side of the equation can be any positive value starting with zero and the second term corresponds to the energy of the series limit of a certain spectrum series. Thus, the frequency of light emitted varies continuously and its value increases starting with the series limit, extending in the direction of short wavelength.

3.3.5 Note II: Rydberg Atoms

A Rydberg atom[10] is a highly excited atom in which one of its electrons has been excited to a high quantum state (n is very large). So far, hydrogen with $n = 105$ has been produced in the laboratory, and in large atoms states with $n = 630$ have been detected by the observations of radioastronomy.

The history of investigating atoms in highly excited states covers nearly one hundred years. In 1885, after Balmer had put forward the Balmer formula of hydrogen, someone observed the spectral lines of $n = 31$ by way of interplanetary observation and established the connection between a Rydberg atom and astronomy. In 1906, the Rydberg state of sodium of $n = 51$ was observed.

Because of the inherent limit of traditional spectroscopes, the course of investigating Rydberg atoms was always very slow. Such investigations have

[10] Thomas F. Gallagher, *Rydberg Atoms*, Cambridge Univ. Press (2005).

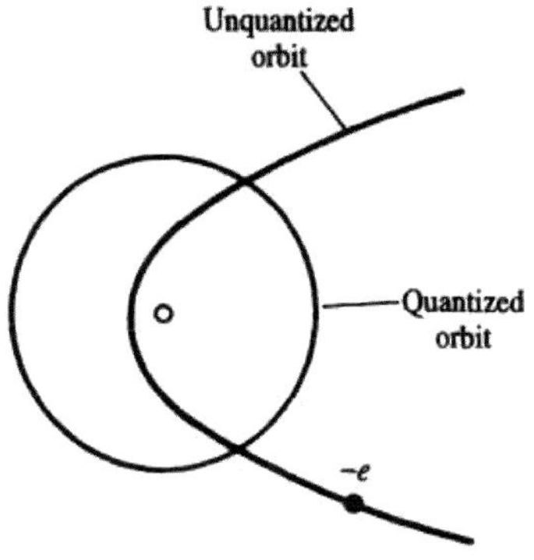

FIGURE 3.16
Nonquantized orbits of hydrogen.

flourished, however, since laser techniques were introduced into this spectroscopy. But why are people interested in Rydberg atoms?

In a Rydberg atom, only one outer electron is in a highly excited state which is far from the closed core (atomic nucleus and other electrons). The electrostatic Coulomb action on it by the closed core is like that caused by a point charge ($+e$). Thus any atom, when it is excited to become a Rydberg atom in a highly excited state, can be treated as a hydrogenlike atom. Regarding the atom as being composed of one outer electron and a closed core, the many-body question can be reduced to the one-electron question and can be treated by the methods of quantum mechanics for one-electron atoms.

Rydberg atoms have many peculiar qualities. According to Bohr's correspondence principle, the motion of the electron in an atom will be close to the condition of classical physics when it is excited to a state of very large n. Its orbital radius is proportional to n^2 (Eq. 3.17); For example, the radius is about 480 Å when $n = 30$, and 3.3 μm when $n = 250$, which is close to the size of bacteria. It is really an exceptionally large atom. We have pointed out in Chapter 2 that the masses of atoms existing in nature may differ more than 250 times while there is hardly any difference between their radii in normal cases (see Table 5.5). But here the Rydberg atom is 100 000 times as big as an atom in its ground state.

When an excited quantum state is reasonably low in energy (n' is a number only a few integers larger than n of the ground state), it is easy for the electron to return to the ground state, so the mean lifetime of an excited state is generally about 10^{-8} s. But when an electron is in a state where n is very large, the radiation lifetime is approximately proportional to $n^{4.5}$. It is very common that the lifetime is as long as 10^{-3} s or even 1 s, if the atom is not impacted by other atoms.

The binding energy of the outer electron of a Rydberg atom varies approximately inversely as n^2. Thus, the energy interval of the two neighboring bound states of a Rydberg atom varies approximately inversely as n^3 (see Eq. 3.13), so it decreases rapidly with increasing n. For example, $\Delta E_n = 10^{-3}$ eV when $n = 30$; $\Delta E_n = 6.3 \times 10^{-7}$ eV when $n = 350$. Such a small energy interval is very difficult to measure, so that high-resolution spectrometric techniques must be used. This also opens up some new phenomena. For example, usually it was thought that the influence of blackbody radiation at room temperature on an atom could be neglected completely, because the frequency of blackbody radiation is inversely proportional to the temperature. At room temperature (300 K), its frequency is much lower than the common radiation frequency of atoms. But for Rydberg atomic states, ΔE_n is low enough to match the frequency spectrum of blackbody radiation at room temperature and therefore the influences of blackbody radiation at room temperature on the lifetimes of highly excited atoms have been observed.[11]

The Coulomb attraction inside the atom is relatively strong in general in the ground states of atoms and the influences of applied electric fields and magnetic fields on atoms are rather small. But in Rydberg atomic states, when the electron is in a highly excited state far away from the atomic center, the Coulomb attraction by the central part of the atom is relatively weak. Then an applied electric or magnetic field may affect it easily and so bring about some interesting phenomena.

3.4 EXPERIMENTAL EVIDENCE II: FRANCK–HERTZ EXPERIMENT

3.4.1 Basic Idea

The Bohr theory was quickly verified by investigations of optical spectra. But any important physical law must be tested by at least two kinds of independent experimental methods. J. Franck and G. Hertz used a different method which was independent of the study of optical spectra to test Bohr's theory soon after its publication.

The main points of the Bohr theory are that there exist stationary quantum states in atoms and that electromagnetic waves will be absorbed or emitted when electrons jump between two quantum states. Experiments involving spectra reflect the discrete characteristics of the emission or absorption of electromagnetic waves to prove the existence of quantum states. But the Franck–Hertz experiment used an electron beam to excite atoms. If electrons in atoms can be only in certain discrete energy states (quantum states), then the experiment should certainly show that only electrons with certain energy can have a large probability of exciting atoms.

Why use electrons as the means to excite atoms? To answer this question, we must review two laws in theoretical mechanics.

A. In the case of an elastic collision, the kinetic energy of one particle can be transferred totally to another particle only when the masses of these two particles are precisely equal. For an elastic collision, consider a head-on collision of par-

---

[11] E. J. Beiting et al., *J. Chem. Phys.* **70** (1979) 3551; T. F. Gallagher and W. E. Cooke, *Phys. Rev. Lett.* **42** (1979) 835.

ticles of mass m_1 with velocity v_{1i} on an object with mass m_2 and velocity v_{2i} approaching each other. Momentum and energy conservation give

$$m_1 v_{1i} + m_2 v_{2i} = m_1 v_{1f} + m_2 v_{2f}$$

$$\tfrac{1}{2} m_1 v_{1i}^2 + \tfrac{1}{2} m_2 v_{2i}^2 = \tfrac{1}{2} m_1 v_{1f}^2 + \tfrac{1}{2} m_2 v_{2f}^2$$

Rewriting the energy equation:

$$m_1 (v_{1i}^2 - v_{1f}^2) = m_2 (v_{2f}^2 - v_{2i}^2)$$

$$m_1 (v_{1i} - v_{1f})(v_{1i} + v_{1f}) = m_2 (v_{2f} - v_{2i})(v_{2f} + v_{2i})$$

Now substitute into this the momentum conservation equation written as

$$m_1 (v_{1i} - v_{1f}) = m_2 (v_{2f} - v_{2i})$$

one finds

$$v_{1i} + v_{1f} = v_{2f} + v_{2i}$$

Substituting this into the momentum equation for v_{2f} gives

$$m_1 v_{1i} + m_2 v_{2i} = m_1 v_{1f} + m_2 [v_{1i} + v_{1f} - v_{2i}]$$

Solving this yields

$$v_{1f} = \left(\frac{m_1 - m_2}{m_1 + m_2}\right) v_{1i} + \left(\frac{2m_2}{m_1 + m_2}\right) v_{2i}$$

and

$$v_{2f} = \left(\frac{2m_1}{m_1 + m_2}\right) v_{1i} + \left(\frac{m_2 - m_1}{m_1 + m_2}\right) v_{2i}$$

So, in an elastic collision, if $m_1 = m_2$ the objects exchange velocities and kinetic energies. For a very heavy particle, m_1, colliding with a very light particle, m_2, at rest, $v_{1f} \simeq v_{1i}$ and $v_{2f} \simeq 2v_{1i}$. An example of this is an α particle colliding with an electron in an atom. The α particle loses only a small amount of energy and continues on in the same direction.

B. In the case of an inelastic collision where momentum is conserved but kinetic energy is not, only when the masses of the two particles are very far apart can the kinetic energy of the light particle be changed totally into internal energy of the heavy one.

In fact, these two kinds of collision processes can be expressed by one equation:

$$\tfrac{1}{2} m v^2 + \tfrac{1}{2} M V^2 = \tfrac{1}{2} m v'^2 + \tfrac{1}{2} M V'^2 + \Delta E$$

in which ΔE is the internal energy and the primes indicate the velocities after the collision. For $\Delta E = 0$, the collision is elastic; for $\Delta E \neq 0$, the collision is inelastic and $\Delta E > 0$ and $\Delta E < 0$ are called the first kind and the second kind of inelastic collisions, respectively.

When an electron collides with an atom, it collides either with its nucleus or with the electrons outside the nucleus. When an electron collides with a nucleus, either energy is transferred to the nucleus as a whole or its internal energy levels are excited (nuclear internal energy levels also are quantized, see Chapter 11).

Since the kinetic energies obtainable for electrons before the 1930s were much lower than needed to excite the nucleus, from 100 keV to several MeV for nuclear excitation, no energy could be transferred from the electron to the internal excitation of the nucleus. Moreover, since the electron mass and nuclear masses differ greatly, no energy is transferred to the nucleus as a whole (law A), so the only energy transfer is to the orbital electrons. Since their masses are equal, the collision of an electron with an electron is elastic by law A. However, since the electrons are bound to the atom, one can treat the atom as a whole. Then according to law B, the electron can change its total kinetic energy into internal energy of the atom to excite an electron to a higher energy state. Thus it is very effective to use electrons as a means to excite electron states in atoms.

3.4.2 Franck–Hertz Experiment

In 1914, the year after Bohr published his theory, J. Franck and G. Hertz did an experiment in which atoms were bombarded with electrons to prove that the internal energy of electrons in atoms is really quantized. A schematic diagram of the Franck–Hertz experiment is shown in Fig. 3.17.

The glass container is filled with the gas to be studied. Electrons are emitted from the thermal cathode K and are attracted toward the positively charged grid G. But between G and the receiver R, a negative voltage of 0.5 V is added. Electrons pass through the KG space and then enter the GR space. If the electrons have sufficiently large energy, they can overcome the negative voltage and reach the receiver R and so become an electric current passing through the ammeter, A. If electrons collide with atoms in the region between K and G, and give some of their energy to the atoms, the residual energy of the electrons may be so small that after passing through the grid the electrons cannot overcome the negative voltage to reach R. If many electrons undergo such collisions, the reading of the ammeter will decrease noticeably.

The initial gas investigated by Franck and Hertz was mercury vapor. In the experiment, they increased the voltage between K and G gradually and observed

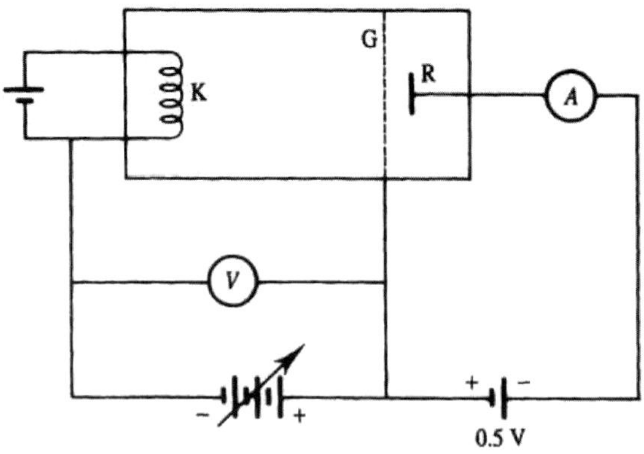

FIGURE 3.17
Schematic diagram of the Franck–Hertz experiment.

the electric current of the ammeter. Figure 3.18 shows the curve of the current to receiver R varying with the voltage across KG: Initially, the current on R increases as the voltage between K and G increases. After increasing gradually from zero, there is a sudden drop in the current and then it increases again. A series of peaks and valleys appear in the current. The interval between two peaks (or two valleys) is nearly same: about 4.9 volts. That is, the sudden drops in the current take place when the voltage between K and G is an integer multiple of 4.9 V. Why does this phenomenon take place? We will answer this in the following.

The experimental phenomena mentioned show clearly that the electrons in the mercury atoms do not accept just any external energy. Electrons in mercury atoms absorb external energy only when it is 4.9 eV, that is, there exists an excited quantum state with an energy of 4.9 eV for electrons in a mercury atom. When the voltage between K and G is lower than 4.9 eV, the energy an electron obtains after being accelerated through KG will be lower than 4.9 eV. If an electron with $E < 4.9\,\text{eV}$ collides with a Hg atom, according to the classical point of view expressed in law **B** it should transfer almost all its energy to the Hg atom since classically the internal energy states of Hg are continuous. Any amount of energy can be transferred classically from an electron to internal energy of a heavy Hg atom. But now the experiment shows that this electron still has enough energy to pass through the grid and overcome the negative voltage to reach electrode R, thereby contributing to the current. So, when its energy is less than 4.9 eV, this electron does not transfer any energy to a Hg atom. For electron energies lower than 4.9 eV, as the energy of the electrons increases, it is easier for the electrons to reach electrode R and the electric current increases. When the voltage between K and G reaches 4.9 V, if the electrons collide with Hg atoms near the electrode G, then it is possible for the electrons to transfer all the energy they have obtained to electrons in the Hg atoms and cause the electrons in the Hg atoms to go into an excited state of 4.9 eV. After losing that energy the electrons cannot reach the electrode R, and the current will decrease sharply. When the voltage between K and G is slightly over 4.9 V, the electrons do not give up all their energy to the Hg atoms as predicted by classical law; they only hand over 4.9 eV. Thus the electrons keep the part of their energy above 4.9 eV. This energy is sufficient for them to overcome the negative voltage to reach electrode R so that the current increases again. When the voltage between K and G is twice 4.9 V, the electrons which have lost an energy of 4.9 eV in one collision in the region between K and G may acquire an energy of 4.9 eV again in this region and incur a second inelastic collision with another atom. Again the electrons exhaust their energy and so the current decreases sharply again. The same principle applies when the voltage between K and G is three times as large as 4.9 V. Then the electrons may use up all their energy through three collisions. Thus, data as in Fig. 3.18 verified the existence of quantum states of electrons in atoms clearly.

The experimental setup Franck and Hertz used in 1914 (Fig. 3.17) had a shortcoming: it was difficult for the energy of the electrons to exceed 4.9 eV because once they were accelerated to reach 4.9 eV, they would collide with Hg atoms and lose their energy. Thus they could not excite electrons in Hg atoms to

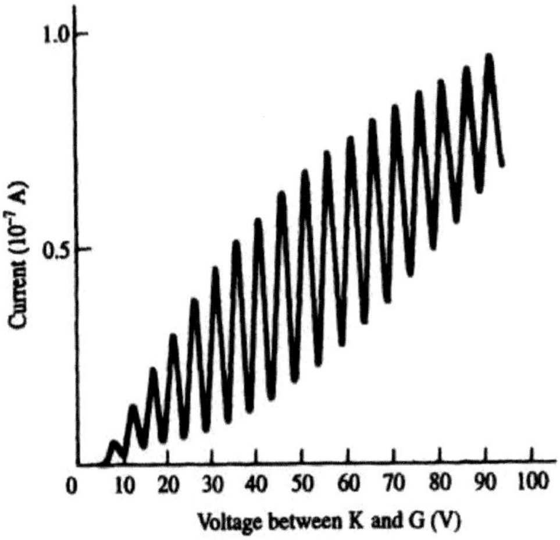

FIGURE 3.18
Measurement of the first excitation potential of a mercury atom. (Provided by the Modern Physics Laboratory of Fudan University, F-H tube produced by Fudan University.) The first peak is not observed at 4.9 eV in the figure for experimental reasons, but the difference in energy between each peak is 4.9 eV as expected. The fact the peaks do not appear exactly at $4.9n$ volts where $n = 1, 2, \ldots$ is related to an experimental artifact which shifts all peaks the same amount. Neighboring peaks are separated by 4.9 V.

higher energy states, so that only one quantum state of 4.9 eV in a Hg atom could be verified.

3.4.3 Improved Franck–Hertz Experiment

In 1920, Franck improved the original experimental setup, as shown in Fig. 3.19. Compared with the former one (Fig. 3.17), three improvements were made: (1) An electrode plate was added in front of the original cathode K to heat the cathode indirectly in order to make the electrons be emitted evenly. The energies of the electrons could then be measured more accurately. (2) A grid G_1 was added near the cathode K and the pressure of the gas in the tube was reduced so as to let the interval between K and G_1 be shorter than the average mean-free path of the electrons in the Hg vapor. To do this required establishing an acceleration range without collision in which the electrons were accelerated but did not collide with Hg atoms. (3) The electrode G_1 and the electrode G_2 close to the electrode R were at the same electric potential; this established an isopotential range as a collision range (G_1–G_2) in which the electrons only collided with Hg atoms and were not accelerated. Thus the acceleration and collision took place separately in two different ranges, to avoid the shortcomings of the former equipment. Then the electrons could obtain higher energies in acceleration. The experimental results did show a series of quantum states existing in the Hg atom, as shown in Fig. 3.20(a). Readers should draw a curve similar to Fig. 3.18 of the current on the

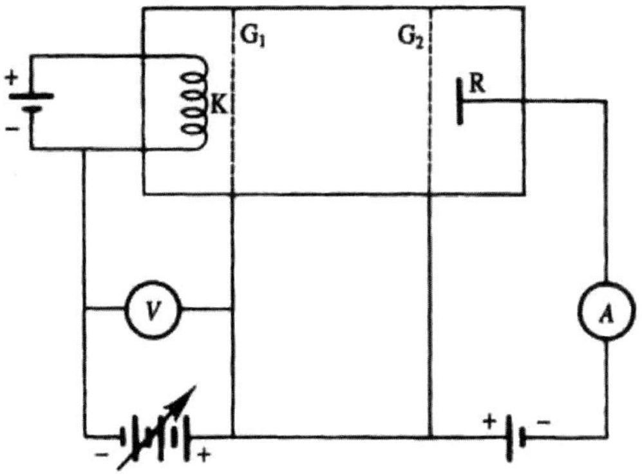

FIGURE 3.19
Schematic diagram of a modified Franck–Hertz experiment.

electrode R as a function of the voltage between K and G_1 and make some illustrations.

In 1924, Hertz did experiments again using the improved equipment described above. He observed the emission spectrum of Hg atoms carefully while measuring the energy levels of Hg atoms. When the voltage between K and G_1 reaches $\gtrsim 8$ V, six lower excited states of those shown in Fig. 3.20(a) can be produced. It is clear from the figure that six transitions may take place, but two of them (1850 Å and 2537 Å) are in the ultraviolet range. Thus four spectral lines will be observed by a visible-light spectrometer and the experimental results show that, as seen in the left part of Fig. 3.20(b). When the voltage between K and G_1 reaches ≥ 9.9 V, thirteen lines can be observed, as shown in the right part of Fig. 3.20(b). The numbers by the side of the spectral lines in the figure represent the wavelengths, expressed in Å. The numerical values in parentheses represent the voltages between K and G_1 at which the corresponding spectral lines are observed.

These results clearly indicate that when an atom is excited to different states, the energy absorbed is not continuous; that is, the internal energy of an atomic system is quantized. Thus, the Franck–Hertz experiments confirm the existence of quantum energy states in atoms.

3.4.4 Concluding Remarks

There are three different types of very important and very famous experiments in the history of the development of atomic physics and of quantum mechanics. The first includes experiments which confirmed the quantum of light, in particular experiments that involve blackbody radiation, the photoelectric effect, and the Compton effect (the latter is discussed in Chapter 6). The second includes the experiments that confirmed the existence of quantum states in atoms, such as the discrete spectrum of light from atoms and the Franck–Hertz experiments. The third type includes experiments that confirmed the wave character of matter, as discussed in Chapter 7.

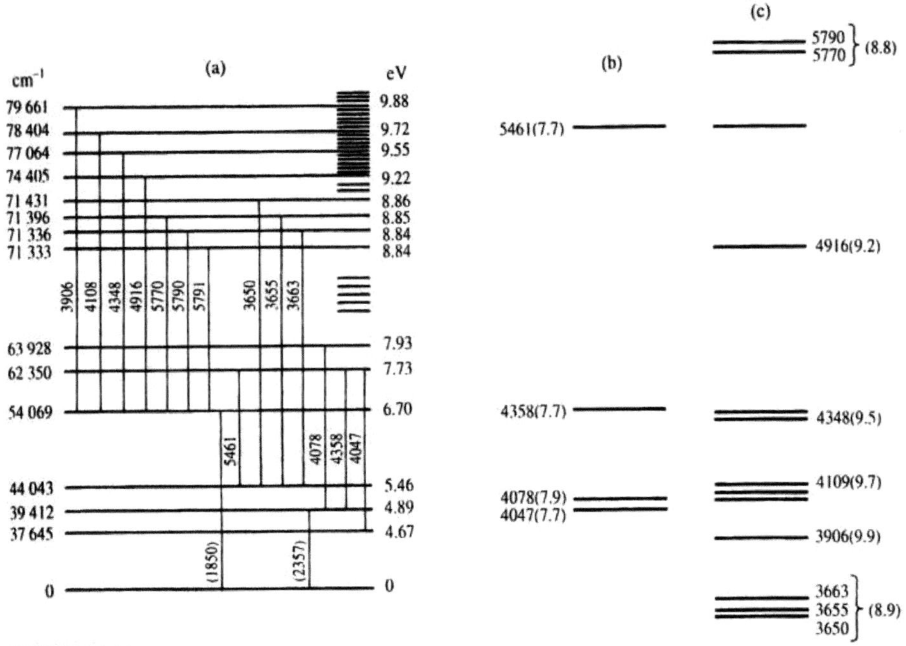

FIGURE 3.20
(a) Energy level diagram of a mercury atom. The levels are identified by energy in eV on the right and $1/\lambda = \tilde{\nu}$ on the left and the transitions by wavelengths in Å. (b) Visible emission spectrum of Hg for voltages $\gtrsim 8$ V but less than 8.8 V in a Franck–Hertz experiment. Transitions are labeled with wavelengths in Å. (c) Visible emission spectrum of Hg for voltages > 10 V in a Franck–Hertz experiment. Transitions are labeled with wavelengths in Å.

The Franck–Hertz experiment occupies an important place in atomic physics. The method of the experiment was independent of the methods of investigating optical spectra. It confirmed the existence of quantum states in atomic systems from a different point of view and achieved controlled excitation of atoms.

In experiments on optical spectra, the energies of the quantum states of atoms are expressed by a wavenumber, $\tilde{\nu} = 1/\lambda$, and in the Franck–Hertz experiment energies are expressed in electron-volts. The observed quantities in the two experiments λ and E are connected by $hc = 12.4\,\text{Å} \cdot \text{keV}$, that is

$$\lambda = \frac{hc}{E} = \frac{12.4\,\text{Å} \cdot \text{keV}}{E(\text{keV})}$$

3.5 EXTENSION: BOHR–SOMMERFELD MODEL

3.5.1 Extension of Quantization

Bohr obtained in his theory of the hydrogen atom that only those orbital motions of electrons which satisfied Eq. (3.20), the condition of quantization of angular momentum,

$$L = n\frac{h}{2\pi}, \qquad n = 1, 2, 3, \ldots$$

may actually exist. Rewriting this equation as

$$2\pi L = nh, \qquad n = 1, 2, 3, \ldots \tag{3.40}$$

shows that the orbits Bohr considered were circular orbits. Here r is constant and the only variable is ϕ. Since integrating an angle variable over one cycle gives 2π, Eq. (3.40) can be written $\oint L\,d\phi = nh$ where L is the one-dimensional momentum. This is the condition of quantization of angular momentum in the case of one dimension. What will happen if we extend it to the case of many dimensions? Here we should recall the Lagrangian equation in theoretical mechanics:

$$\frac{d}{dt}\left(\frac{\partial \mathcal{L}}{\partial \dot{q}_a}\right) - \left(\frac{\partial \mathcal{L}}{\partial q_a}\right) = 0, \qquad a = 1, 2, \ldots$$

Here, $\mathcal{L}$ is the Lagrangian representing the difference between the kinetic energy and potential energy of a mechanical system, that is, $\mathcal{L} = T - V$; $\partial \mathcal{L}/\partial \dot{q}_a$ is the generalized momentum; and q_a is the generalized coordinate.

In the case of two dimensions, there are two generalized momenta, $\partial \mathcal{L}/\partial \dot{\phi} = mr^2\dot{\phi} = L$ and $\partial \mathcal{L}/\partial \dot{r} = m\dot{r} = p_r$, and two generalized coordinates, ϕ and r. Stated simply, by analogy to the one-dimensional case, in two dimensions each dimension has a quantization condition. So there are two quantization conditions corresponding to the two generalized momenta, namely,

$$\oint L\,d\phi = n_\phi h \qquad \text{and} \qquad \oint p_r\,dr = n_r h \tag{3.41}$$

Extending the quantization condition further to many dimensions, we obtain the general expression of the quantization conditions:

$$\oint p_q\,dq = n_q h, \qquad n_q = 1, 2, 3, \ldots \tag{3.42}$$

in which dq is the shift of angle or the shift of position, and p_q is the momentum corresponding to q, that is, angular momentum or linear momentum. The symbol $\oint$ means that the integration is taken over a whole period of motion. It is obvious that this expression includes Bohr's quantization condition of orbital angular momentum of circular motion.

3.5.2 Elliptical Orbits

In 1916, soon after the publication of Bohr's theory, A. J. W. Sommerfeld put forward the theory of elliptical orbits. Two things were done by Sommerfeld in his theory. One was to extend Bohr's circular orbits to elliptical orbits, and the other was to introduce relativistic corrections.

Sommerfeld thought that the electrons performed elliptical motion in a plane around the atomic nucleus. This is two-dimensional motion. The coordinates ϕ and r in plane polar coordinates can be used to describe positions in elliptical motion. The generalized momenta corresponding to the two coordinates are angular momentum L and momentum p:

$$L = mr^2\dot{\phi} \qquad \text{and} \qquad p = m\dot{r}$$

in which $\dot{\phi}$ is the angular velocity of the electron, $r\dot{\phi}$ is the component of the velocity perpendicular to r, and $\dot{r}$ is the component of the velocity in the direction of r. The energy of this system can be expressed as

$$E = \frac{1}{2}mv^2 - \frac{Ze^2}{r} = \frac{1}{2}m(\dot{r}^2 + r^2\dot{\phi}^2) - \frac{Ze^2}{r}$$

It is known from Eq. (3.41) that a system undergoing two-dimensional motion is determined by the two quantization conditions

$$\oint L\, d\phi = n_\phi h \qquad \oint p\, dr = n_r h \tag{3.41}$$

where n_ϕ and n_r are integers and are called the angular quantum number and the radial quantum number, respectively. Then $n = n_\phi + n_r$ is called the principal quantum number.

Starting from the equations mentioned above, and performing mathematical manipulations, Sommerfeld obtained the expressions for the semimajor axis a and semiminor axis b of the elliptical orbits and the energy in his theory:

$$\begin{aligned} a &= \frac{n^2\hbar^2}{\mu Ze^2} = n^2\frac{a_1}{Z} \\ b &= a\frac{n_\phi}{n} = nn_\phi\frac{a_1}{Z} \\ E &= -\frac{\mu Z^2e^4}{2n^2\hbar^2} \end{aligned} \tag{3.43}$$

These results were obtained by Sommerfeld after he extended the circular orbits to elliptical orbits. It is easy to see by comparing them with the case of Bohr's circular motion that the energy expression is unchanged, while the radius of the orbits is divided into a semimajor axis and semiminor axis and the two quantum numbers n and n_ϕ appear. The physical meanings of these results are that the semimajor axis of the orbit and the energy of the system are determined only by the principal quantum number n, and are independent of n_ϕ; that the angular momentum depends on the angular quantum number n_ϕ; and that the shape of the orbits depends on the ratio $b/a = n_\phi/n$ (see Fig. 3.21). This means that for the same n but different n_ϕ, different systems corresponding to different orbits will have the same energy. Such a system is said to be degenerate. For example, for a certain value of n, because n_ϕ may have the values of $1, 2, 3, \ldots n$, there will be n possible orbits. Thus there will be n states, but all these n states will have the same energy—the system is n-fold degenerate. Figure 3.22 shows the cases of energy degeneracy of the hydrogen atom.

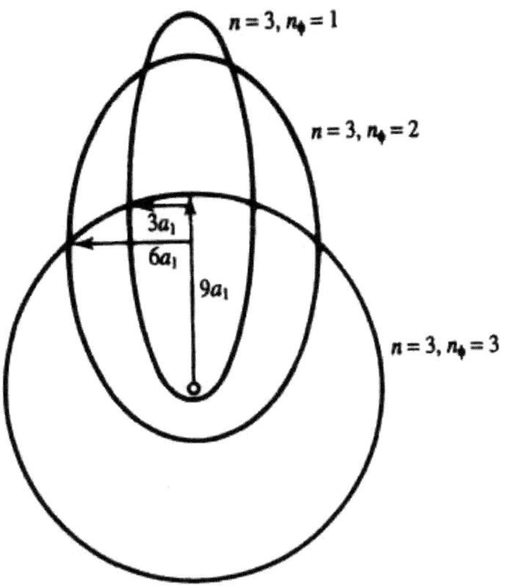

FIGURE 3.21
Orbits corresponding to the same n but different n_ϕ.

The conclusions presented above are derived simply using mathematical means. Next we will verify from an experimental point of view that the concept of elliptical orbits really has some factual basis.

3.5.3 Relativistic Corrections

In the theory of relativity, the mass of a moving body is no longer a constant but is related to its velocity by

$$m = \frac{m_0}{\sqrt{1 - \beta^2}}, \qquad \beta \equiv \frac{\text{v}}{c}$$

in which m_0 is the static mass of the body ($\text{v} = 0$), v is its velocity and c is the velocity of light in vacuum. Here $m \cong m_0$ when $\text{v} \ll c$. When v is close to c, however, m will be much larger than m_0.

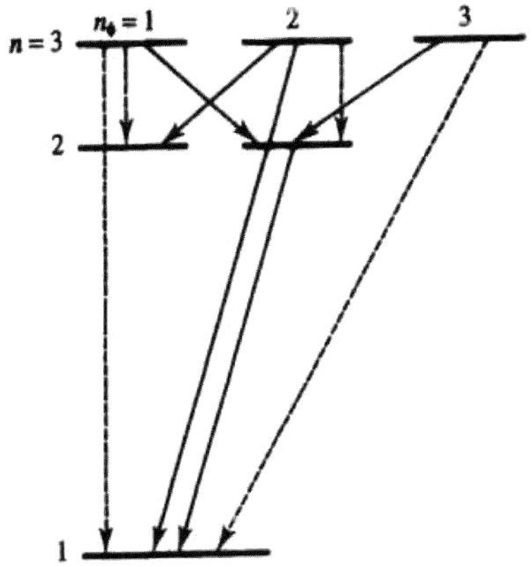

FIGURE 3.22
Degeneracy of the energy levels of hydrogen.

The expression for the kinetic energy of a moving body given by relativity is

$$T = (m - m_0)c^2 = m_0c^2\left(\frac{1}{\sqrt{1-\beta^2}} - 1\right) \tag{3.44}$$

which is different from the classical formula. When $v \ll c$, that is, when β is very small, one can perform a series expansion for the first term on the right-hand side of Eq. (3.44) and neglect the infinitesimally small higher terms. Then the equation can be rewritten as

$$T = m_0c^2\left(1 + \frac{1}{2}\beta^2 - 1\right) = \frac{1}{2}m_0v^2 \tag{3.45}$$

This is the classical expression for kinetic energy.

So, when relativity is applicable, the mass can be expressed as $m = m_0/\sqrt{1-\beta^2}$ and the kinetic energy can be expressed as $T = (m - m_0)c^2$. In relativity, the key constant is the velocity of light c. Next, we will use relativity to modify the circular orbits and elliptical orbits, respectively.

The modification of the circular orbits. Let us discuss this from the energy point of view. Energy can be expressed as the sum of kinetic energy and potential energy

$$E = T - \frac{Ze^2}{r} \tag{3.46}$$

First, let us look at the second term. Because

$$r_n = \frac{\hbar^2}{me^2}\frac{n^2}{Z}$$

holds for general cases, the second term can be expressed as

$$\frac{Ze^2}{r} = \frac{Z^2e^2me^2}{\hbar^2n^2} = \frac{Z^2}{n^2}\frac{e^4}{\hbar^2c^2}mc^2 = \frac{Z^2}{n^2}\alpha^2 mc^2 \tag{3.47}$$

Using Eq. (3.26) and considering that the atomic nucleus has charge Ze, we obtain

$$E = -\frac{1}{2}m\left(\alpha c\frac{Z}{n}\right)^2 \tag{3.48}$$

Remember that we know that $v = \alpha cZ/n$, that is, $\beta \equiv v/c = \alpha Z/n$. Thus after considering relativistic effects and substituting for T and Ze/r, Eq. (3.46) is expressed as

$$\begin{aligned} E &= (m - m_0)c^2 - mc^2\left(\frac{Z\alpha}{n}\right)^2 = -m_0c^2 + mc^2\left[1 - \left(\frac{Z\alpha}{n}\right)^2\right] \\ &= -m_0c^2 + \frac{m_0c^2}{\sqrt{1-\beta^2}}(1-\beta^2) \end{aligned} \tag{3.49}$$

or

$$E = m_0c^2(\sqrt{1-\beta^2} - 1)$$

This is the energy expression. Performing a series expansion for the square root term and neglecting the terms of order higher than β^4, we have

$$\begin{aligned} E &= m_0c^2\left(1 - \tfrac{1}{2}\beta^2 + \frac{\tfrac{1}{2}(\tfrac{1}{2}-1)}{2!}\beta^4 - 1\right) \\ &= -m_0c^2\left(\tfrac{1}{2}\beta^2 + \tfrac{1}{8}\beta^4\right) \end{aligned} \tag{3.50}$$

or

$$E = -\frac{m_0c^2}{2}\left(\frac{Z\alpha}{n}\right)^2\left[1 + \frac{1}{4}\left(\frac{Z\alpha}{n}\right)^2\right] \tag{3.51}$$

In the process of the simple derivation presented above, one sees how the relativistic effects are included into the formula. Clearly, the first term in Eq. (3.51) is the one the Bohr theory had given and the second term is the additional modification term that includes the relativistic effects. Although the simple derivation holds only for circular orbits, the main effects caused by relativistic corrections are included (compare it with Eq. (3.52) next).

It is worth pointing out that when expanding the above equation, we assumed that $\beta \ll 1$ or at least $\beta < 1$. So, what will happen if $\beta = Z\alpha/n \geq 1$, that is, if $n = 1$ and $Z \geq 137$? Since $\beta = \mathrm{v}/c > 1$ is not allowed in relativity, there is a breakdown and the energy dives to negative infinity. Such problems do not occur at present, because in fact the heaviest known nuclide produced in the laboratory is only $Z = 111$ and because the theoretical modifications which take into consideration the finite size of the nucleus indicate that the case of $\beta > 1$ will not occur for $Z \leq 172$. However, heavy-ion accelerators, by bombarding uranium on uranium, may produce nuclides up to $Z = 184$ for times that are long compared to the time for an $n = 1$ electron to orbit the combined nuclei. At the present time, studies of such reactions are a very active research field for probing the quantum electrodynamics of very strong fields and the behavior of electrons under extreme conditions.

The modification of the elliptical orbits. After extending Bohr's circular orbits to elliptical orbits, Sommerfeld considered the relativistic effects in the elliptical orbits again. After complicated mathematical calculations, the energy of the system was found to be expressed as

$$\begin{aligned} E &= -\frac{mZ^2e^4}{2\hbar^2n^2}\left[1 + \left(\frac{Z\alpha}{n}\right)^2\left(\frac{n}{n_\phi} - \frac{3}{4}\right)\right] \\ &= -\frac{1}{2}m(c\alpha)^2\frac{Z^2}{n^2}\left[1 + \left(\frac{Z\alpha}{n}\right)^2\left(\frac{n}{n_\phi} - \frac{3}{4}\right)\right] \end{aligned} \tag{3.52}$$

From (3.52) we get the expression for the spectral term $T(n, n_\phi)$,

$$T(n, n_\phi) \equiv -\frac{E}{hc} = \frac{RZ^2}{n^2} + \frac{RZ^4\alpha^2}{n^4}\left(\frac{n}{n_\phi} - \frac{3}{4}\right) \tag{3.53}$$

in which the first term is the result of the Bohr theory and the second term is the modification term that includes relativistic effects. It is obvious that for the same n but different n_ϕ, the values of the second term are different. Thus, for those orbital motions with the same n but different n_ϕ, the energy is different. Now the original degeneracy of the energy levels with same n and different n_ϕ is eliminated after considering the relativistic corrections. But the value represented by the second term is much smaller than that of the first term and the split yields only a slight difference in energy for the levels.

3.5.4 Comparison with Experiments

SPECTRUM OF HYDROGEN. Sommerfeld then attempted to explain the fine structure of the spectrum of hydrogen observed in experiments. For example, as early as 1896, Michelson and Morley found that the H_α line of hydrogen was a doublet, $0.36\,\text{cm}^{-1}$ apart. Later, three close neighboring lines were observed at high resolution. The introduction of elliptical orbits did not resolve the problem of the degeneracy of the energy levels (see Fig. 3.22). One sees this by comparing the two energy level diagrams of hydrogen corresponding to circular orbits from Eq. (3.24) (corrected for reduced mass) and elliptic orbits from Eq. (3.43), respectively. However, on introducing the relativistic corrections, the energies split. For example, because of the difference for n_ϕ, the energy level of $n = 3$ is separated into three levels, while the energy level for $n = 2$ is separated into two. Using Eq. (3.53), we obtain the energy differences from the splitting and so obtain the transition wavelengths between the three energy levels corresponding to $n = 3$ and the two energy levels corresponding to $n = 2$ (pay attention to the selection rules, a transition must obey $\Delta n_\phi = \pm 1$). Thus three H_α lines are obtained that correspond well with the experimental results. But in Chapter 4 we will find that this "complete correspondence" was a coincidence. Sommerfeld's theory is not complete.

In addition, the theory of Sommerfeld indicated that when an electron was moving in an elliptical orbit, its velocity varied continuously. It moved fast when it was close to the nucleus and moved slowly when it was far from the nucleus, so as to keep the angular momentum constant. Thus the mass of the electron varied continuously. As a result the orbit of the electron was not a closed one. It looked as though the elliptical orbit had a continuous motion, as shown in Fig. 3.23. For those orbits with the same n but different n_ϕ, the variation of velocity and the variation of mass were different, so the energy differed slightly.

THE SPECTRA OF THE ALKALI ATOMS. The spectra of hydrogenlike ions were well explained by the Bohr theory. Here we discuss the spectra of the alkali atoms, which have one electron outside a closed electron shell. Closed electron shells will be discussed fully in Chapter 5. For now, we only state that each electron orbit or

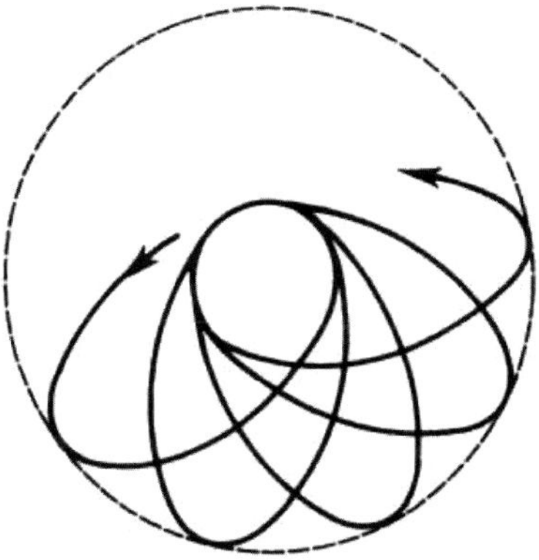

FIGURE 3.23
Precessional motion of the orbit of an electron.

shell can contain only a specific number of electrons. When a shell (orbit) has the maximum number of electrons it is allowed to have, it forms a closed shell (or inert core) that does not interact to first order with electrons in the next orbit. Both atoms with only one electron and atoms with one electron outside a closed shell are similar to hydrogen. Each has its own strong points as to which is more like hydrogen. From the point of view of having only one electron outside the nucleus, obviously the hydrogenlike ions are better than the alkali atoms. But considering the effect of the closed-shell core on the outermost electron, the alkali atoms are superior to the hydrogenlike ions, as the net charge of the alkali atoms seen by the outermost electron is 1. However, as we will see in a full quantum treatment, the outermost electron actually spends various small percentages of its time inside the closed electron shell core so the net charge it feels is somewhat greater than 1. We will shortly see the effect of this perturbation. So after discussing the spectra and structure of the hydrogenlike ions, it is a natural development to extend the established theories next to the more complicated alkali atoms, so as to investigate their structure.

When we discussed the spectrum of hydrogen, we knew that the characteristics of the Balmer series established that the formula (3.6) has two terms. The first term is a fixed term which determines the limit of the series and depends on the final state, and the second term is the variable term which is determined by the initial state. The spectra of the alkali atoms are similar to the Balmer series, that is, with the variation of wavelength from long to short, the intensity of light varies from strong to weak and the intervals between the spectral lines vary from small to large. Figure 3.24 gives the spectral line series of lithium. From such data (coming directly from experiments), we can obtain the energy level diagram of the electron. Figure 3.25 is the electron energy level diagram of lithium. In order to remember the energy level diagrams of alkali atoms easily, we summarize the main characteristics of the diagrams in the following four points.

1. Four sets of spectral lines—the initial position of each set is different so there are four sets of changeable terms.
2. Three terminals—three fixed terms.

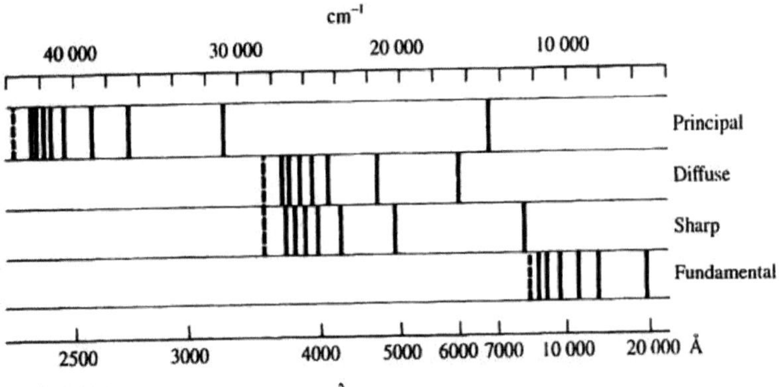

FIGURE 3.24
Spectral line series of lithium. The principal series, the sharp series, the diffuse series, the fundamental series in terms of wavelength (Å) and wavenumber (cm^{-1}).

3. Two quantum numbers—n and n_ϕ. We know that $n_\phi = 1, 2, 3, \ldots n$. First substitute k for n_ϕ, that is, $k = 1, 2, \ldots n$, which is only rewriting the symbol. But then make the significant substitution of l for k, where $l = 0, 1, 2, \ldots n-1$. Pay attention to the fact that originally k starts from 1, but here l starts from 0. The change of the initial l from 1 to 0 comes out of quantum mechanics and cannot be understood by classical physics. We will discuss this change later. Here l is the quantum number of the angular momentum.
4. One rule—the dashed lines shown in Fig. 3.22 represent transitions which are not observed in experiments, because of the selection rule $\Delta l = l_i - l_f = \pm 1$ for the transitions between two atomic energy levels. We can give the following explanation of this selection rule: The difference between the l values of two states is the difference between their angular momenta. Because the intrinsic angular momentum of a photon is 1, emitting a photon for a transition between the two states requires that the difference between the angular momenta must be at least 1. (More than one unit of angular momentum change is possible but this involves a much higher-order effect with the photon having to carry off orbital angular momentum in addition to its intrinsic angular momentum of $1\hbar$, so such transitions are not normally observed.) With this selection rule, we can draw the transitions in the energy level diagram. In Fig. 3.25, the energy levels are classified according to the value of l; the energy levels with the same l are drawn on the same line.

   Because the four series in Fig. 3.25 (ordered by decreasing energy and increasing wavelength) start from $l = 1, 0, 2, 3$, respectively, and the first letters of the English names (given next) of the four series as follows are P, S, D, F, respectively, $l = 0, 1, 2, 3$ are expressed by S, P, D. F. The values of l larger than 3 are listed alphabetically (G, H, I, . . . represent $l = 4, 5, 6, \ldots$ respectively). The n value is given in front of S, P, ... In the figure, the transitions from the various nP states to the 2S state, $n\text{P} \rightarrow 2\text{S}$, are called the

Principal series. The transitions of nS → 2P are called the Sharp series (they are also called the Secondary subordinate series). The transitions of nD → 2P are called the Diffuse series (also called the Primary subordinate series). The transitions of nF → 3D are called the Fundamental series (or Bergmann series).

It should be pointed out that the spectral line series of lithium shown in Fig. 3.24 are the emission spectral lines, namely, the spectrum emitted directly by excited lithium (the means of excitation may be high temperature or collision). If we let light with a continuous distribution of wavelengths pass through lithium gas and observe the absorption spectrum of lithium, we will find that the spectrum

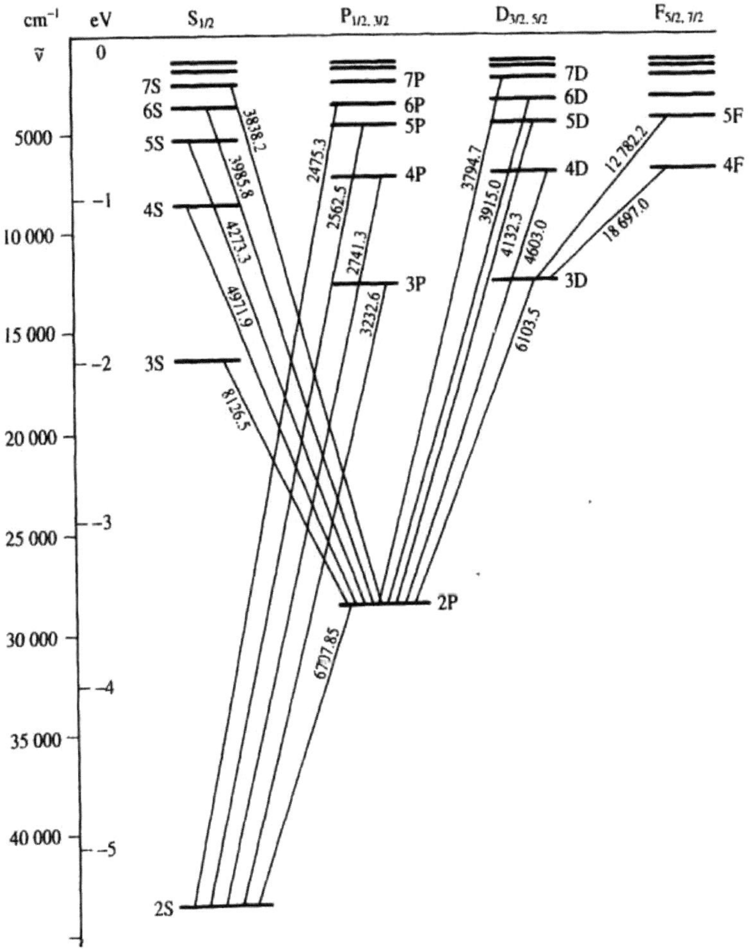

FIGURE 3.25
Energy level diagram of lithium. Transitions are labeled with wavelength in Å.

now has dark lines or dark bands caused by the absorption of light with certain wavelengths by the lithium. In this case, we can only observe the spectrum corresponding to the principal series. This is true for the absorption spectra of all alkali atoms, because only the principal series connects with the ground state of the alkali atoms and the atoms used to create the absorption spectrum are in the ground state.

In Fig. 3.26 the energy level diagrams of hydrogen, lithium, and sodium are compared. It is clear from the figure that these three cases are very similar. For example, for $n = 3$, each has three discrete energy levels: 3S, 3P, and 3D. However, for hydrogen, the splitting of the three energy levels is so small that it cannot be seen in the figure, but for lithium the splitting is reasonably obvious and for sodium a large splitting exists. For the four $n = 6$ energy levels which are shown, all the splittings decrease with sodium still having the largest split. These splittings are sufficiently small that they are readily observable only for the energy levels with small l (6S, 6P).

Although Sommerfeld explained the splitting of the energy levels of hydrogen and hydrogenlike ions reasonably successfully using a theory based on elliptical orbits and relativistic corrections, the splitting of the energy levels of the alkali metals greatly exceed the theoretical predictions given by Sommerfeld.

How do we explain the splitting of the energy levels of the alkali atoms correctly? Here we discuss it only qualitatively. Compared with hydrogenlike ions, the closed core of the alkali atoms is not closed. When the valence electron is far from the closed core (n is large) and the orbit is quite circular (l is large), as shown in Fig. 3.27(a), no major problem occurs. But when the valence electron is close to the closed core (n is small) and the orbit is very elliptical (l is small), two cases which do not exist in hydrogen will appear: One is the penetration of the orbit through the closed core, as shown in Fig. 3.27(b), and the other is the polarization effect of the closed core (Fig. 3.28). Both these phenomena will affect the energy of the electrons. We will discuss them separately.

First consider the penetration of the orbit. It is clear from Fig. 3.25 that the S energy level of sodium is much lower than that of hydrogen. The S corresponds to an orbit with very large eccentricity and very small l. Those orbits which are close to the closed core will penetrate the core and affect the energy of the atoms. In the new quantum mechanics (see Chapters 7 and 8) one no longer speaks of regular orbits of any shape, but the electrons have smeared out, overlapping probability distributions for their radial behavior. Nevertheless the concept of which n, l orbits can be considered as penetrating orbits is still useful. Such "penetration" decreases as l and n increase.

Next look at the polarization of the closed core. The original structure of the closed core is spherical. Inside the core, the nucleus has Ze positive charges and the $(Z-1)$ electrons have $(Z-1)e$ negative charges. When the valence electron is moving near the closed core, the electric field of the valence electron produces a slight relative displacement between the nucleus with positive charge and the electrons with negative charge in the closed core, as shown in Fig. 3.28 by the solid circle. Thus the center of the negative charges is no longer at the center of the

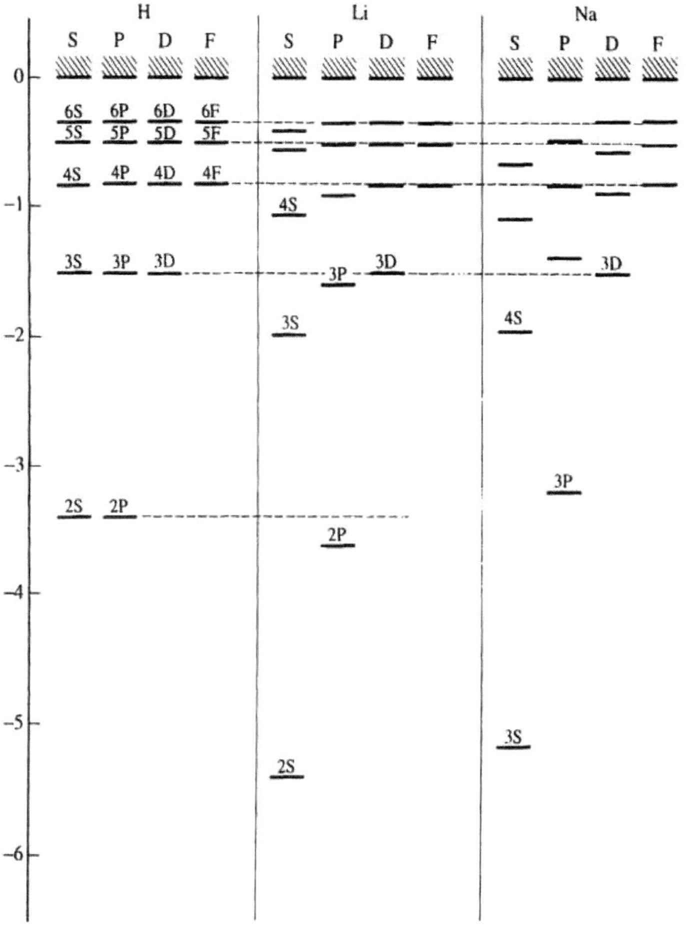

FIGURE 3.26
Comparison of the energy levels of hydrogen, lithium, and sodium.

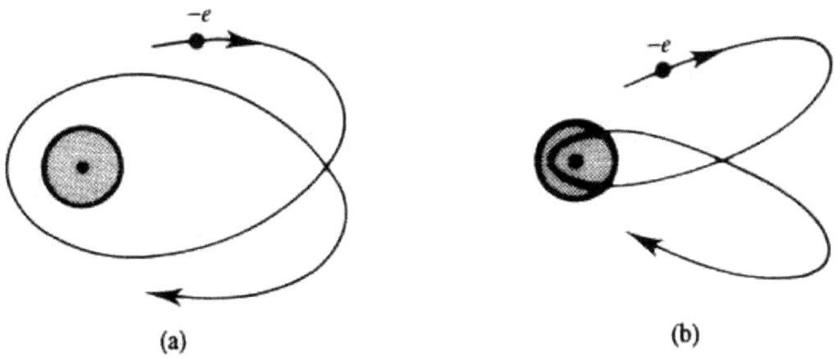

FIGURE 3.27
(a) Nonpenetrating orbit. (b) Penetrating orbit.

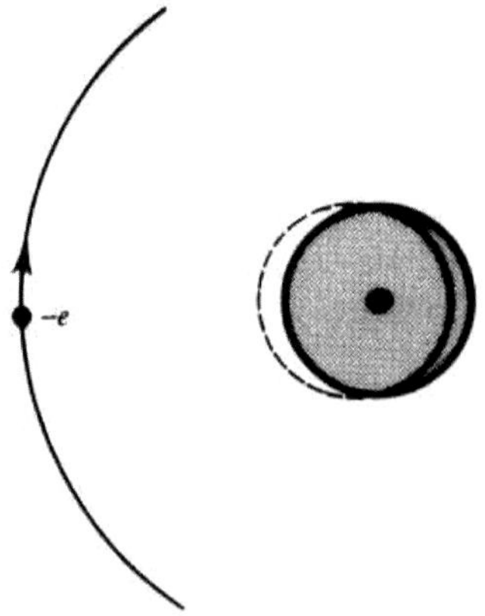

FIGURE 3.28
Schematic diagram for the polarization of the closed core

nucleus and an electrical dipole is formed. This is the polarization of the closed core. The electrical field of the electric dipole formed by this polarization acts on the valence electron again and makes it feel an additional force. This effect causes a decrease in the energy.

The effects of the polarization of the closed core and the orbit penetration nicely explain the differences between the energy levels of alkali atoms and the energy levels of the hydrogen atom.

3.6 SUMMARY

1. In order to explain the experiments on blackbody radiation, Planck introduced the hypothesis of quantization of energy exchange: $E = h\nu$. The physical meaning of Planck's constant h is the measurement of the quantization of energy, or the measurement of discrete energy.

Einstein extended Planck's hypothesis and introduced the concept of the light-quantum to explain the photoelectric effect. He proposed that the energy of a photon was $E = h\nu$. In 1917, he proposed further that the momentum of a photon was $P = h\nu/c$ so as to connect the quantities which characterize a particle (energy and momentum) with those which characterize a wave (wavelength and frequency). The bridge between them is Planck's constant h.

2. At the end of the nineteenth century, physicists began to open the door of the atom. They found the charge e and the mass m of an electron. But from these two constants alone, one could determine neither the dimension of the atomic system nor its energy. Dimension and energy are always the two basic quantities which characterize any level of the structure of matter. Still a constant was absent, it was Planck's constant.

Bohr combined the quantum idea h with e and m and obtained the dimension which characterized the atomic system, $r_1 = \hbar^2/me^2 = 0.529\,\text{Å}$, and the energy, $E = \frac{1}{2}m(\alpha c)^2 = 13.6\,\text{eV}$. Note that the product $\alpha c = e^2/\hbar$ does not include c.

3. To search for the inner link between objects which superficially are totally different is always an attractive theme in natural science. Bohr ingeniously combined the Rutherford model of the atom and Planck and Einstein's quantization of light, which were not taken very seriously at that time, with light spectra experiments which were superficially utterly irrelevant. In doing so, he explained the nearly 30-year-old puzzle of the Balmer–Rydberg formula

$$\tilde{\nu} = \frac{1}{\lambda} = R\left(\frac{1}{n^2} - \frac{1}{n'^2}\right)$$

and calculated the Rydberg constant for the first time.

The theory of Bohr was supported not only by experiments on the spectra of light, but also by the Franck–Hertz experiment, which was totally independent of the experiments on spectra. Thus the concept of quantum states was established by experimental evidence.

However, as the word "model" implies, the Bohr model had a series of difficulties which were impossible to overcome. It was exactly these difficulties that brought about a greater revolution in physics (see the following chapters, especially Chapter 7).

PROBLEMS

3.1. The work function for cesium is 1.9 eV.

(*a*) Determine the threshold frequency and threshold wavelength of the photoelectric effect of cesium.

(*b*) If one wants to obtain a photoelectron with energy of 1.5 eV, what wavelength of light is required?

3.2. For hydrogen, an ionized helium ion He^+, and a doubly ionized lithium ion Li^{2+}, calculate:

(*a*) The radii of the first and second Bohr orbits and the velocities of electrons in these orbits.

(*b*) The binding energy of an electron in their ground states.

(*c*) The first excitation energy and the wavelength of their resonance lines.

3.3. What minimum kinetic energy must an electron have in order to allow an inelastic collision between the electron and a lithium ion Li^{2+} in its ground state to take place?

3.4. A head-on, inelastic collision takes place between a moving proton and a static hydrogen atom in its ground state. What is the minimum velocity of the proton to cause the hydrogen atom to emit photons?

3.5. (*a*) In the case of thermal equilibrium, the distribution of the atoms in different energy states is given by the Boltzmann distribution, namely, the number of atoms in an excited state with energy of E_n is

$$N_n = N_1 \frac{g_n}{g_1} e^{-(E_n - E_1)/kT},$$

where N_1 is the number of atoms in the state with energy E_1, k is the Boltzmann constant, and g_n and g_1 are the statistical weights (determined by how many different ways one can put the electrons in each of the two states with energies

E_n and E_1) of the corresponding states. For hydrogen atoms at a pressure of 1 atm and a temperature of 20°C, how large must the container be to let one atom be in the first excited state? Take the statistical weights of the hydrogen atoms in the ground state and in the first excited state to be $g_1 = 2$ and $g_2 = 8$, respectively. Remember from thermodynamics $PV = \gamma RT$ where γ = number of atoms present/Avogadro's number $= N/N_A$.

(*b*) Let electrons collide with hydrogen gas at room temperature. In order to observe the H_α line, what is the minimum kinetic energy of the electrons?

3.6. In the range of wavelengths from 950 Å to 1250 Å, what spectral lines are included in the absorption spectrum of a hydrogen atom?

3.7. What is the hydrogenlike ion whose difference of wavelength between the principal line of the Balmer series and the principal line of Lyman series is 1337 Å?

3.8. The photon emitted by a transition in ionized helium He^+ from its first excited state to its ground state can ionize a hydrogen atom in its ground state and make it emit an electron. Determine the velocity of the electron.

3.9. An electron–positron pair is a bound system composed of a positron and an electron. Determine:

(*a*) The distance between the positron and the electron when the pair is in its ground state.

(*b*) Ionization potential and the first excitation potential.

(*c*) The wavelength of the resonance line.

3.10. A μ^- is an elementary particle just like the electron except its mass is 207 times as heavy as that of an electron. When the velocity of a μ^- is reasonably low, it will be captured by a proton and form a μ^- atom. Calculate:

(*a*) The first Bohr radius of a μ^- atom.

(*b*) The lowest energy of a μ^- atom.

(*c*) The shortest wavelength of the Lyman series of a μ^- atom.

3.11. The ratio of the Rydberg constant of hydrogen to that of heavy hydrogen is 0.999 728 and the ratio of their nuclear masses is $m_H/m_D = 0.500\,20$. Calculate the ratio of the mass of a proton to the mass of an electron.

3.12. A photon is emitted when a stationary hydrogen atom jumps from its first excited state to the ground state.

(*a*) What is the recoil velocity acquired by hydrogen?

(*b*) Estimate the ratio of the recoil energy of a hydrogen atom to the energy of the emitted photon.

3.13. The ground state of sodium is 3S. How many different spectral lines can be emitted when a sodium atom jumps from its 4P excited state to lower energy levels (do not consider the fine structure.)

3.14. The wavelength of the resonance line of the optical spectrum of sodium is $\lambda = 5893$ Å and the wavelength of the series limit of the subordinate series is $\lambda_\infty = 4086$ Å. Calculate:

(*a*) The energies corresponding to 3S and 3P.

(*b*) The ionization potential and the first excitation potential of sodium.

3.15. Show that Planck's formula for the energy distribution reduces to Wien's in the limit of high frequency and to the Rayleigh–Jeans distribution in the limit of low frequency.

3.16. Consider a lithium atom as a hydrogenlike atom. Determine the ionization energy of a 2S electron and explain qualitatively the reason for the difference between this value and the experimental value of 5.39 eV.

3.17. The well-known sodium yellow lines consist of 5890 Å and 5896 Å lines. Try to calculate the gap between these two lines at the limit of the sharp series, in the units of both cm^{-1} and eV.

3.18. An electron in a hydrogen atom is in the excited state of $n = 4$, $l = 1$. If only electric ($\Delta l = \pm 1$) dipole transitions are considered, diagram the possible transitions from this excited state to lower states. Give the n and l values of the lower states to which transitions occur. Why can electric dipole radiation not carry off zero angular momentum?

3.19. An energy of 7.61 eV is required to separate the atoms in a CN molecule. What is the longest wavelength of light radiation we could use to dissolve such a molecule? Where is this wavelength in the electromagnetic spectrum?

3.20. In the human body there are many C—C (carbon–carbon) bonds. To break such a bond requires 2.8 eV. What is the shortest wavelength of light to be used to break such a bond?

3.21. Assume a hydrogen atom in its ground state absorbed a 12.75 eV photon.

(*a*) After absorption, in which excited state will it be?

(*b*) Show in an energy diagram the possible transitions from that excited state. What is the shortest wavelength among the transitions?

CHAPTER 4

FINE STRUCTURE IN ATOMIC SPECTRA: ELECTRON SPIN

Have we heard the last word about spin? I do not believe so.

C. N. Yang (1985)

Bohr's atomic theory was presented in Chapter 3. In that theory, only the dominant interaction (force) in the atom, i.e., the interaction between the nucleus and the electrons by the Coulomb force, was included. The calculated energies based on this force are in good agreement with the experimental values for hydrogen and the manifestations of the energy differences—the spectral lines (Balmer series, etc.)—seem to be reasonably interpreted. However, on close inspection the observed spectral lines actually have fine structure. For example, the H_α line in the Balmer series is not a single line but consists of seven lines that are very close together. The yellow D line in sodium is perhaps the most famous example of a doublet structure under higher resolution. These doublets and multiple lines mean that we have to take into account some other forces (interactions) as the causes of these energy variations. From the viewpoint of classical mechanics, one could expect that even for a system as simple as the hydrogen atom there can be a magnetic force (interaction) in addition to a Coulomb force (interaction) between the electrons and the nucleus, since the electrons are moving around the nucleus.

In this chapter, we first introduce the concept of the magnetic moment caused by the electron orbital motion in the atom, and generalize the quantization to three dimensions (Section 4.1). Then we introduce the interaction of the atomic system with an external magnetic field (Sections 4.2, 4.5), and also the interaction initiated by the internal magnetic field of an atomic system (Section 4.4). This latter interaction points out that spatial quantization is a description based on reality and the experimental facts cannot be explained by only electron orbital motion. We have to introduce the hypothesis of electron spin (Sect. 4.3). The magnetic interaction that involves electron spin is the dominant factor in producing the fine structure in atoms. More important, the concept of electron spin is one of the most significant concepts in microscopic physics. Its profound significance reaches far beyond the scope of the atom (Section 4.5). Finally, the spectral structure in the hydrogen atom will be summarized and discussed in Section 4.6.

4.1 MAGNETIC MOMENT PRODUCED BY THE ELECTRON ORBITAL MOTION IN AN ATOM

4.1.1 Classical Expression

It is well known in classical electromagnetism that the magnetic moment μ associated with a small electric current may be expressed as (see Fig. 4.1(a))

$$\boldsymbol{\mu} = \frac{i}{c} S\boldsymbol{n}_0$$

where i is the electric current in the sense of positive charge flow, c is the speed of light ($c = 1$ in this equation in "reduced" units based on the SI system of units), S is the area enclosed by the current circuit, and $\boldsymbol{n}_0$ is a unit vector normal to the plane of the circuit.

Accordingly, there must be a magnetic moment produced by the electrons moving with velocity v around the nucleus in the atom, as shown in Fig. 4.1(b). If

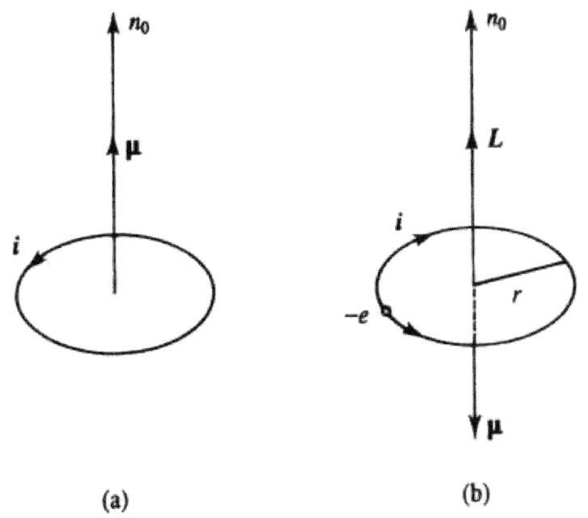

FIGURE 4.1
Illustration of the magnetic moment of a small electric circuit.

the circular frequency of the electron revolving around the nucleus is ν and the orbital radius is r, then the magnetic moment is

$$\begin{aligned}\boldsymbol{\mu} &= \frac{i}{c}\boldsymbol{S} = \frac{-e\nu}{c}\pi r^2 \boldsymbol{n}_0 = -\frac{e}{c}\frac{\mathrm{v}}{2\pi r}\pi r^2 \boldsymbol{n}_0 \\ &= -\frac{e}{2m_e c} m_e \mathrm{v} r \boldsymbol{n}_0 = -\frac{e}{2m_e c}\boldsymbol{L}\end{aligned}$$

where $\boldsymbol{L} = m_e \mathrm{v} r \boldsymbol{n}_0$ is the orbital angular momentum. Note that, because of the negative charge on the electron, $\boldsymbol{L}$ and $\boldsymbol{\mu}$ are oppositely directed.

Here the electron orbit is assumed to be circular, but the result is the same for an arbitrary closed orbit. Let

$$\gamma \equiv \frac{e}{2m_e c}$$

where m_e is the electron mass. Then we have

$$\boldsymbol{\mu} = -\gamma \boldsymbol{L} \tag{4.1}$$

This is the relation between the magnetic moment, $\boldsymbol{\mu}$, of an electron revolving around the nucleus and the angular momentum, $\boldsymbol{L}$, of the electron. It may be seen that they are in opposite directions. The direction of $\boldsymbol{L}$ is determined by the direction of rotation of the particle. However, the direction of the magnetic moment is defined by the direction of the current flow of a positive charge in the right-hand screw sense. Since the electron is negatively charged, its motion is like a positive charge flowing in the opposite direction, so $\boldsymbol{\mu}$ has an opposite sign to $\boldsymbol{L}$. The coefficient in Eq. (4.1) is called the gyromagnetic ratio.

It is also known in electromagnetism that a magnetic field exerts a force perpendicular to the direction of a current, and a magnetic moment in a homogeneous external magnetic field does not feel a force but a torque. The force and torque change sign (reverse direction) for negative charges compared to positive charges since the current $\boldsymbol{I}$ and magnetic moment $\boldsymbol{\mu}$ change sign with a change in the sign of the charge. The torque is given by

$$\boldsymbol{\tau} = \boldsymbol{\mu} \times \boldsymbol{B}$$

where $\boldsymbol{B}$ is the magnetic induction (also called magnetic flux density) or loosely called the magnetic field. The presence of a torque will cause a change of angular momentum (see Appendix 4A), i.e.,

$$\frac{d\boldsymbol{L}}{dt} = \boldsymbol{\tau} = \boldsymbol{\mu} \times \boldsymbol{B} \tag{4.2}$$

From Eqs. (4.1) and (4.2) for electrons, we have

$$\frac{d\boldsymbol{\mu}}{dt} = -\gamma \boldsymbol{\mu} \times \boldsymbol{B}$$

which we can rewrite (reversing the terms in the cross product introduces a sign change) as

$$\frac{d\boldsymbol{\mu}}{dt} = \boldsymbol{\omega} \times \boldsymbol{\mu}, \text{ where } \boldsymbol{\omega} \equiv \gamma \boldsymbol{B} \tag{4.3}$$

This is the angular velocity formula for the Larmor precession. It means that in a homogeneous external magnetic field $\boldsymbol{B}$, a magnetic moment rotating with high speed will not line up with the field direction. Instead, it will precess about $\boldsymbol{B}$ with a certain angular velocity $\boldsymbol{\omega}$. Figure 4.2 is an illustration of the precession of an atomic magnetic moment placed in a magnetic field. It is seen that the magnetic moment $\boldsymbol{\mu}$ precesses about the direction of $\boldsymbol{B}$ with a precession angular velocity $\boldsymbol{\omega}$, and the direction of $\boldsymbol{\omega}$ is parallel to $\boldsymbol{B}$.

In order to gain further understanding of $\boldsymbol{\omega}$, we analyze the precession of the vector $\boldsymbol{\mu}$. Figure 4.2b is a small sector taken from the precession plane perpendicular to $\boldsymbol{B}$. The sector radius used is just the perpendicular distance from $\boldsymbol{\mu}$ to $\boldsymbol{B}$. Obviously then

$$d\mu = \mu \sin\theta \, d\phi$$

$$\frac{d\mu}{dt} = \mu \sin\theta \frac{d\phi}{dt} = \mu \sin\theta \, \omega$$

or

$$\frac{d\boldsymbol{\mu}}{dt} = \boldsymbol{\omega} \times \boldsymbol{\mu}$$

So $\omega = d\phi/dt$ stands for the rate of change of the angle ϕ with time t. That is why we call ω the angular velocity.

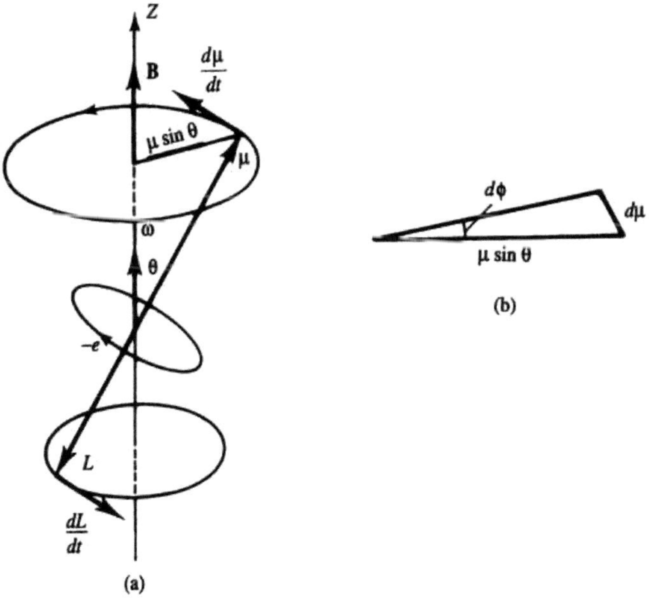

(a) (b)

FIGURE 4.2
The magnetic moment precesses about a magnetic field.

4.1.2 The Quantization Condition

In the description of an elliptical orbit in an atom, we mentioned two quantization conditions, in which the principal quantum number n is related to the energy of the atomic system and the angular quantum number n_ϕ is related to the shape of the orbit (as discussed in Section 3.5). We did not consider the orientation of the orbital plane because it would not be possible, and not necessary, to discuss the orientation of an orbital plane without a reference axis. But it becomes meaningful to do so as soon as there exists a magnetic field, since the direction of the magnetic field $\boldsymbol{B}$ may then be considered as a reference direction.

The orbital angular momentum $\boldsymbol{L} = \boldsymbol{P}_\phi$ is perpendicular to the orbital plane. The angle α of L with respect to the direction of the magnetic field (set at z) then determines the direction of the orbital plane, as shown in Fig. 4.3.

By analogy with the quantization condition of angular momentum introduced by Bohr and Sommerfeld (see Eq. 3.4),

$$L = P_\phi = n_\phi \hbar, \qquad n_\phi = 1, 2, 3, \ldots \tag{4.4}$$

We may introduce the third quantization condition as

$$L_z = P_\Psi = n_\Psi \hbar \tag{4.5}$$

For the moment, Eq. (4.5) is only an assumption. Whether it is correct or not must be judged by experiment. The first justification comes from the Stern–Gerlach experiment presented in the next section. We shall derive the quantization condition from a more fundamental level in quantum mechanics (see Chapter 7). Since

$$\cos\alpha = \frac{P_\Psi}{P_\phi} = \frac{L_z}{L} = \frac{n_\Psi}{n_\phi}$$

we have

$$-1 \leq \frac{n_\Psi}{n_\phi} \leq 1$$

Then

$$n_\Psi = n_\phi, n_\phi - 1, \ldots - n_\phi$$

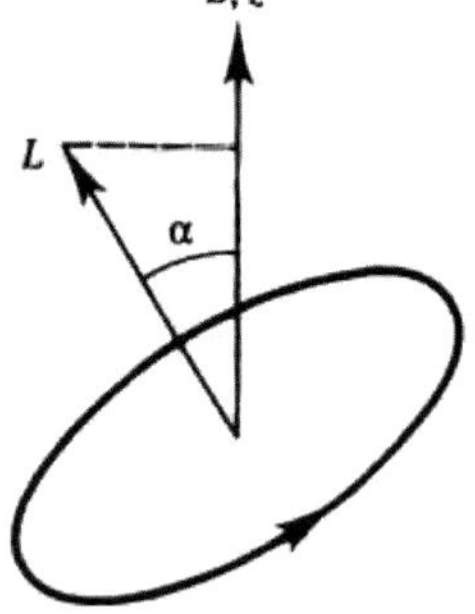

FIGURE 4.3
Orientation of the orbital angular momentum with respect to the z axis.

For convenience, we change notation and rewrite Eqs. (4.4) and (4.5) as

$$L = l\hbar, \qquad l = 1, 2, 3, \ldots \tag{4.6}$$

$$L_z = m\hbar, \qquad m = l, l-1, \ldots, -l \tag{4.7}$$

For a fixed value of l, the number of values of m is $2l + 1$.

In the following, we shall make two fundamental changes in Eq. (4.6):

$$L = \sqrt{l(l+1)}\,\hbar, = 0, 1, 2, \ldots \tag{4.8}$$

One of these shifts the initial value of l from 1 to 0, and the other changes l in the expression to $\sqrt{l(l+1)}$. You might say that these are only slight modifications, but in fact, they are quite fundamental. They refer to the essence of quantum mechanics. Why? You will find the answer in Chapter 7.

Equations (4.8) and (4.7) are the quantization conditions of angular momentum and its z component. An illustrative vector model of these is shown in Fig. 4.4. The remarkable feature of the illustration is that the vector L will never be completely along the z direction, which comes as a result of the change of Eq. (4.6) to Eq. (4.8)!

We may substitute Eq. (4.8) into the expression for the magnetic moment μ:

$$\mu = -\gamma L = -\sqrt{l(l+1)}\,\hbar\gamma = -\sqrt{l(l+1)}\,\hbar\,\frac{e}{2m_e c}$$

and substitute Eq. (4.7) into the expression for the z component of the magnetic moment μ_z:

$$\mu_z = -\gamma L_z = -\gamma m\hbar = -\frac{e\hbar}{2m_e c}\,m$$

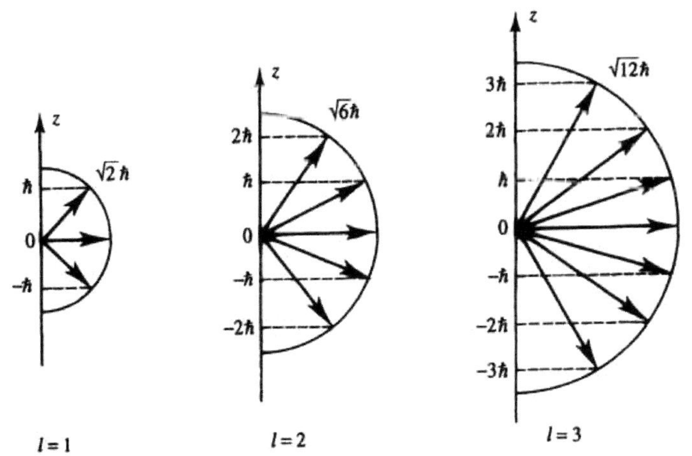

FIGURE 4.4
Illustration of the orbital angular momentum and its components (allowed projections onto the z axis). The length of the vector is $\sqrt{l(l+1)}\hbar$.

Rewrite them as

$$\mu = -\sqrt{l(l+1)}\,\mu_B \tag{4.9}$$

$$\mu_z = -m\mu_B \tag{4.10}$$

where

$$\begin{aligned}\mu_B &= \frac{e\hbar}{2m_e c} = 9.2741 \times 10^{-24}\,\mathrm{J/T} = 9.2741 \times 10^{-21}\,\mathrm{erg/G} \\ &= 0.5788 \times 10^{-4}\,\mathrm{eV/T}\end{aligned} \tag{4.11}$$

is called the Bohr magneton (here T stands for Tesla). It is the natural unit of the orbital magnetic moment and a very important constant in atomic physics. It may also be rewritten as

$$\mu_B = \frac{1}{2}\frac{e^2}{\hbar c}\frac{\hbar^2}{m_e e^2}\,e = \frac{1}{2}\alpha(ea_1) \tag{4.12}$$

where α is the fine structure constant ($\sim 1/137$), and $a_1 = \hbar^2/m_e e^2$ is the first Bohr radius. Clearly, ea_1 is the measure of the atomic electric dipole moment, while μ_B measures the atomic magnetic dipole moment. Since the latter is equal to $(\frac{1}{2})\alpha$ times the former, the magnetic interaction is, at least, two orders of magnitude less than the electric interaction.

4.2 THE STERN–GERLACH EXPERIMENT

We have seen from the above discussion that the quantized quantities include not only the size and shape of an electron orbit in an atom, the angular momentum of the electron motion, and the internal energy of atom, but also the orientation of the angular momentum in an external field. The last effect is called *spatial quantization*. It is a further unexpected manifestation of quantum mechanics that not only is the magnitude of angular momentum quantized, but also its direction in space as well.

The spatial quantization of angular momentum of atoms in an external field was first observed directly in 1921 by O. Stern and W. Gerlach. It was one of the most important experiments in atomic physics.[1] The experimental arrangement is shown in Fig. 4.5. Silver atoms were used in the original Stern–Gerlach experiment. Hydrogen atoms were used in 1927, and the results were the same. Actually, hydrogen is more difficult to use in a Stern–Gerlach experiment than silver, because hydrogen normally forms molecules (H_2) which have zero net spin. The atoms (for example, hydrogen atoms) in the container O are heated until they are vaporized. At thermal equilibrium, the atomic velocity v obeys the relation

$$\tfrac{1}{2}mv^2 = \tfrac{3}{2}kT \tag{4.13}$$

[1] W. Gerlach and O. Stern, *Z. für Physik* **9** (1922) 349.

where k is the Boltzmann constant (8.617×10^{-5} eV/K). A simple estimation shows that at a temperature $T = 10^5$ K, the energy of an atom is the order of 13 eV, which is similar to the first excitation energy in hydrogen of 10.2 eV. However, in general, the temperature within a container is far below 10^5 K so that all the hydrogen atoms will be in the ground state.

Of the hydrogen atoms which escape from the container O through a small hole, those moving along the horizontal line (x direction) joining the two slits S_1 and S_2, with the velocity v defined by Eq. (4.13), are thus selected. To the right of the slit S_2, there is a magnetic field. Since an object with a magnetic moment μ would only feel a torque to rotate its direction in a homogeneous magnetic field, the magnetic field has to be inhomogeneous in order to exert a force on an atomic beam which enters the magnetic field. Whether a magnetic field is considered homogeneous or inhomogeneous depends on whether its variation produces an observable effect. Here, the atomic beam is the object we are talking about, so the magnetic field has to be inhomogeneous on the scale of an angstrom. That is the difficult part of the experiment. One of the key features of the work of Stern and Gerlach was to make their magnet with a very sharp edge on one of the pole tips in order to create a very inhomogeneous magnetic field over a very small distance (see Fig. 4.5(b)). For such a magnet, we have

$$\frac{\partial B_z}{\partial x} = \frac{\partial B_z}{\partial y} = 0$$

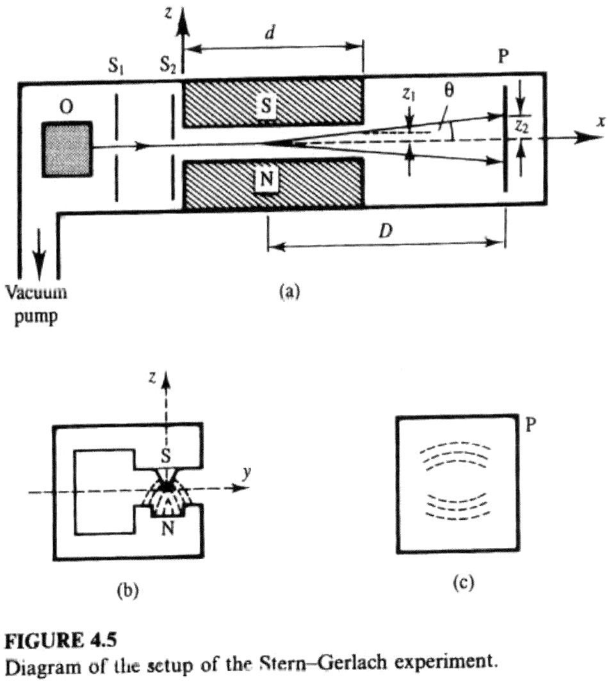

FIGURE 4.5
Diagram of the setup of the Stern–Gerlach experiment.

so there would be a force only in the z direction (see Appendix 4A):

$$F_z = \mu_z \frac{\partial B_z}{\partial z} \tag{4.14}$$

Thus an atomic beam moving initially with velocity v parallel to the horizontal direction (x) and subjected to the influence of a constant force F_z in the vertical direction (z), would follow a parabolic trajectory in the magnetic field region. This is like the motion of a projectile with a horizontal initial velocity and subject to gravity. Accordingly, the equations of the trajectories are

$$\begin{aligned} x &= \mathrm{v}t \\ z_1 &= \frac{1}{2} a_z t^2 = \frac{1}{2} \frac{F_z}{m} t^2 \end{aligned} \tag{4.15}$$

When the atomic beam reaches the exit after passing through a magnetic field region of length d, it has been deflected from the x axis by a distance of z_1, and by an angle θ, where

$$\theta = \tan^{-1}\left(\frac{dz}{dx}\right) = \tan^{-1}\left(\frac{F_z t}{m\mathrm{v}}\right) = \tan^{-1}\left(\frac{F_z d}{m\mathrm{v}^2}\right)$$

It then continues in a straight line until it hits the screen P. At this collecting point, the beam is displaced by the distance z_2 from the x axis. It may easily be shown that

$$z_2 = \mu_z \frac{\partial B_z}{\partial z} \frac{dD}{3kT} = \mu_z \frac{\partial B_z}{\partial z} \frac{dD}{m\mathrm{v}^2} \tag{4.16}$$

where D is the distance between the screen P and the middle of the magnetic field region (Fig. 4.5), and

$$\mu_z = \mu \cos \beta \tag{4.17}$$

as shown in Fig. 4.6. Note that μ_z and z_2 will not be quantized if only μ is quantized in magnitude ($\mu = -\sqrt{l(l+1)}\,\mu_B$) while $\cos \beta$ can take any arbitrary value as expected classically. So, spatial (direction in space) quantization of μ_z is a necessary condition for making the values of z_2 discrete; that is, the projection of $\boldsymbol{\mu}$ on the z direction must be quantized to have only certain discrete values as well. Conversely, whether the experimental values of z_2 take on a series of discrete values produces a test for the spatial quantization of $\boldsymbol{\mu}$. The results of the Stern–Gerlach experiment show that a hydrogen atom has only two orientations of its magnetic moment in a magnetic field. This result proves decisively that the spatial orientation of the angular momentum (and magnetic moment) of an atom in a magnetic field is quantized and not continuous (see Fig. 4.7).

The Stern–Gerlach experiment is the most direct evidence of the phenomenon of spatial quantization, and the first experiment to measure the ground-state characteristics of an atom. It was the Stern–Gerlach experiment that initiated a new field of atomic and molecular beam experiments.

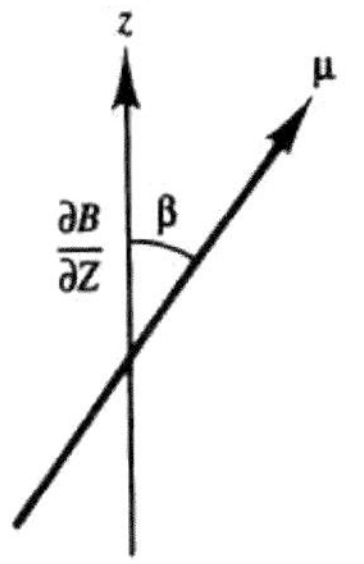

FIGURE 4.6
μ makes an angle of β with respect to the direction of z.

However, it should be pointed out that although this experiment verified the spatial quantization of an atom in a magnetic field, the experimental fact that a hydrogen or silver atom has only two orientations in a magnetic field could not be explained by the then known theory of spatial quantization. According to the theory, for a fixed value of l, the orientation values of m are $2l + 1$. Since l is an integer, $2l + 1$ must be odd. In later experimental observations, we do find examples of odd numbers of orientation, for instance, five orientations were observed for an oxygen atom in its ground state. However, only one orientation was observed for the atoms of zinc, cadmium, mercury, and tin (their beams did not split). These latter atoms passed through the system in one beam with or without the magnetic field turned on in the system. However, for the hydrogen atom and for the atoms of lithium, sodium, potassium, copper, silver, and gold, two orientations were observed. These observations mean that our description of atoms is incomplete for the moment.

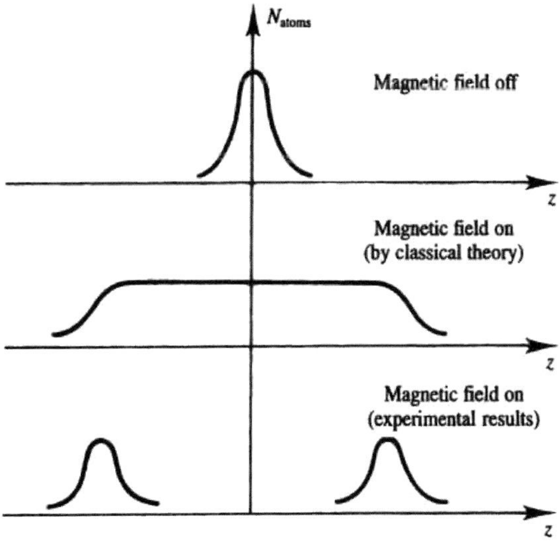

FIGURE 4.7
Results of a Stern–Gerlach experiment with hydrogen atoms. N is the number of atoms hitting the screen.

4.3 THE HYPOTHESIS OF ELECTRON SPIN

4.3.1 The Hypothesis of Electron Spin

The fact that an atomic beam splits into an even number of separate beams in some Stern–Gerlach experiments indicates that in those cases the angular momentum should be a half-integer in order to make $2l + 1$ even. But it is impossible for the orbital angular momentum to be a half-integer with integer quantized orbits.

In 1925, two 25-year-old Dutch students, G. E. Uhlenbeck and S. A. Goudsmit, on the basis of various experimental data, made a bold suggestion[2,3] that the electron is not a point charge and, in addition to its orbital angular momentum about the nucleus, an electron rotates like a top. They called this new intrinsic electron motion "spin", and the spin angular momentum is

$$L_s = \sqrt{s(s+1)}\,\hbar, \qquad s = \tfrac{1}{2} \tag{4.18}$$

If spin is subject to spatial quantization, it can have only two components in the z direction:

$$L_{s,z} = \pm\tfrac{1}{2}\hbar$$

In other words, the z-component of the spin quantum number has only two values, $\pm\frac{1}{2}$:

$$L_{s,z} = m_s\hbar, \qquad m_s = \pm\tfrac{1}{2} \tag{4.19}$$

Actually, unknown to Uhlenbeck and Goudsmit at the time, A. H. Compton had four years earlier introduced in a paper in the *Journal of the Franklin Institute* the idea of electron spin to explain the magnetic properties of the electron. Compton deduced from experiments on magnetism that the electron had a spin and that its magnitude had to be one-half the value of one unit of orbital angular momentum. This prior introduction of electron spin by Compton was acknowledged by Uhlenbeck and Goudsmit in their second paper. Nevertheless, they were the first to introduce the concept of spin to explain the spectra of atoms.

It may seem there is nothing new in proposing the idea that the electron could revolve (spin) on its axis like a top; so does the Earth, while it moves round the Sun. However, in classical physics a planet can have any arbitrary spinning motion, and no two would likely be the same. So, to propose that every electron has exactly the same spin angular momentum and that its z component can take only two values is completely unacceptable in classical physics. The electron spin is an intrinsic property of what it means to be an electron. It is even more puzzling to note that if the electron is considered as a small ball of 10^{-14} cm in radius

---

[2] The original papers are G. E. Uhlenbeck and S. Goudsmit, *Naturwissenschaften* **13** (1925) 953; *Nature* **117** (1926) 264.

[3] For a review article see S. Goudsmit, *Physics Today* **14** (1961, June) 18.

(according to recent experimental evidence, the dimensions of an electron lie far below 10^{-14} cm), with charge $-e$, and revolving on its axis like a top, it may then be shown that the tangential linear velocity at the surface, for an electron with a spin angular momentum of $\frac{1}{2}\hbar$, will be much greater than the speed of light! (You may calculate this yourself.)

Because of all these conceptual difficulties, the suggestion proposed by Uhlenbeck and Goudsmit met opposition at the very beginning from many physicists, including W. Pauli who was already famous by then. Pauli had proposed at the beginning of 1925 that in addition to the three known quantum numbers (n, l, m), a fourth was needed to give the electron a complete description. Moreover, he thought that this quantum number should be "double-valued and impossible to describe classically." Hence, he said, "I strongly doubted the correctness of this idea (spin) because of its classical mechanical character" [see Pauli's Nobel lecture[4] delivered on December 13, 1946 (the date of the actual award of the Nobel Prize to Pauli was 1945)]. The pressure was so great that Uhlenbeck and Goudsmit wanted to withdraw the paper they had submitted. However, it was too late to do so, because their adviser P. Ehrenfest had already sent the paper for publication. He said: "You are both young enough to allow yourselves some foolishness! Be a little crazy." Subsequent developments proved that the concept of electron spin is indeed a clear break with classical physics and is one of the most important concepts in microscopic physics, as will be discussed later.

4.3.2 Landé *g*-Factor

One of the consequences of an electron having spin is that there must exist a magnetic moment in connection with this spin. In analogy with Eqs. (4.9) and (4.10), according to Eqs. (4.18) and (4.19), we expect

$$\mu_s = -\sqrt{s(s+1)}\,\mu_B = -\tfrac{1}{2}\sqrt{3}\,\mu_B \tag{4.20}$$

$$\mu_{s,z} = -m_s\mu_B = \mp\tfrac{1}{2}\mu_B$$

However, these results are not consistent with experiments (see Sections 4.4 and 4.5). In order to have agreement with the experimental results, Uhlenbeck and Goudsmit had to assume that the magnetic moment of an electron produced by its spin is equal to one Bohr magneton, a value twice as large as the value expected based on a spin of $\frac{1}{2}\hbar$. The same conclusion also had been reached by Compton in 1921 that the electron spin was twice as effective in producing a magnetic field as is orbital motion in order to explain the magnetic properties of the electron.

$$\mu_s = -\sqrt{3}\,\mu_B, \qquad \mu_{s,z} = \mp\mu_B \tag{4.21}$$

[4] W. Pauli, *Nobel Lectures, Physics, 1942–1962*, Elsevier, Amsterdam (1964), p. 43.

That the direction of the magnetic moment is opposite to the direction of the spin has been supported by a wide variety of experiments. Electron spin and its associated magnetic moment were soon derived by Dirac's relativistic quantum mechanics. This means that the relations between the electron magnetic moment and its angular momentum based on orbital motion alone, Eqs. (4.9) and (4.10), are not complete. However, we may define a g-factor such that any angular momentum and its corresponding magnetic moment, and their projections on the z direction as well, may be expressed as

$$\mu_j = -\sqrt{j(j+1)}\, g_j \mu_B, \qquad \mu_{j,z} = -m_j g_j \mu_B \tag{4.22}$$

For the case of orbital angular momentum alone, $j = l$, then

$$g_l = 1 \tag{4.23}$$

and

$$\mu_l = -\sqrt{l(l+1)}\, \mu_B, \qquad \mu_{l,z} = -m_l \mu_B \tag{4.24}$$

which return to Eqs. (4.9) and (4.10). We may note that they were derived in terms of the concept of a classical orbit, in addition to the quantization condition.

For the case of spin angular momentum alone, $j = s$, then

$$g_s = 2 \tag{4.25}$$

which leads to Eq. (4.20). For the moment, it is only an assumption. The factor in Eq. (4.22) is called the Landé g-factor, or the g-factor for short. It may be expressed as

$$g = \frac{\mu \text{ measured in units of } \mu_B}{\text{the } z \text{ projection of the angular momentum, in units of } \hbar} \tag{4.26}$$

which is an important physical quantity reflecting the internal motion in matter. It should be noted that the g-factor of the electron is written, in some reference books, as $g_l = -1$, $g_s = -2$, which means that they have included the negative sign of Eq. (4.22) in the g-factor itself; in that case, the negative sign in Eq. (4.22) should be changed to positive.

4.3.3 Expression of the *g*-Factor for a Single Electron

We considered above the g-factors corresponding to the orbital and spin angular momentum of an electron, respectively. Now, we shall do it in a unified way. An electron in an atom has, generally, both orbital and spin angular momentum. The respective magnetic moments should be summed to form the total magnetic moment of the electron, as shown in Fig. 4.8.

The total magnetic moment of an electron may be calculated by using a vector model. Since $g_s/g_l = 2$, the total magnetic moment $\boldsymbol{\mu}$ resulting from combining $\boldsymbol{\mu}_s$ and $\boldsymbol{\mu}_l$ will not line up in the direction of the total angular momentum $\boldsymbol{J}$.

But $\boldsymbol{L}$ and $\boldsymbol{S}$ are precessing about $\boldsymbol{J}$, so $\boldsymbol{\mu}_l$, $\boldsymbol{\mu}_s$, and $\boldsymbol{\mu}$ will all precess about the direction of $\boldsymbol{J}$.

As seen in the figure, $\boldsymbol{\mu}$ is not a quantity with definite direction, but it may be decomposed into two components: one along the direction of $\boldsymbol{J}$ with a definite direction and magnitude is called $\boldsymbol{\mu}_j$; the other perpendicular to $\boldsymbol{J}$, which has no effect on the total on the average, because of its rotational motion around $\boldsymbol{J}$. So it is $\boldsymbol{\mu}_j$ that should be taken into account, and we shall call it the total magnetic moment of an electron.

It is also seen in the figure that $\boldsymbol{\mu}_j$ may be calculated by just adding up the components of $\boldsymbol{\mu}_l$ and $\boldsymbol{\mu}_s$ in the direction of $\boldsymbol{J}$. So

$$\mu_j = \mu_l \cos(l,j) + \mu_s \cos(s,j)$$

where (l,j) and (s,j) denote, respectively, the angles between $\boldsymbol{\mu}_l$ and $\boldsymbol{\mu}_j$, and between $\boldsymbol{\mu}_s$ and $\boldsymbol{\mu}_j$. Substituting μ_s and μ_l from Eqs. (4.20) and (4.24), and using the cosine law of trigonometry, we have that

$$\mu_j = (-g_l\hat{l}\mu_B)\frac{\hat{j}^2 + \hat{l}^2 - \hat{s}^2}{2\hat{j}\hat{l}} + (-g_s\hat{s}\mu_B)\frac{\hat{j}^2 + \hat{s}^2 - \hat{l}^2}{2\hat{j}\hat{s}} \tag{4.27a}$$

According to Eq. (4.22), μ_j is equal to $-\hat{j}g_j\mu_B$, where $\hat{j}$ is the abbreviation for $\sqrt{j(j+1)}$, etc. Here, j is the quantum number of the angular momentum $\boldsymbol{J}$. Since $\boldsymbol{J} = \boldsymbol{S} + \boldsymbol{L}$, and the quantum number corresponding to $\boldsymbol{S}$ is $s = \frac{1}{2}$, and it has only two orientations relative to L, j can take only two values, $j = l + \frac{1}{2}$ or $l - \frac{1}{2}$. If $l = 0$, the total angular momentum will be the spin angular momentum, and it has only one value $j = \frac{1}{2}$. (For the resultant of two arbitrary angular momenta, see the next chapter.) Therefore, combining Eqs. (4.22) and (4.27a),

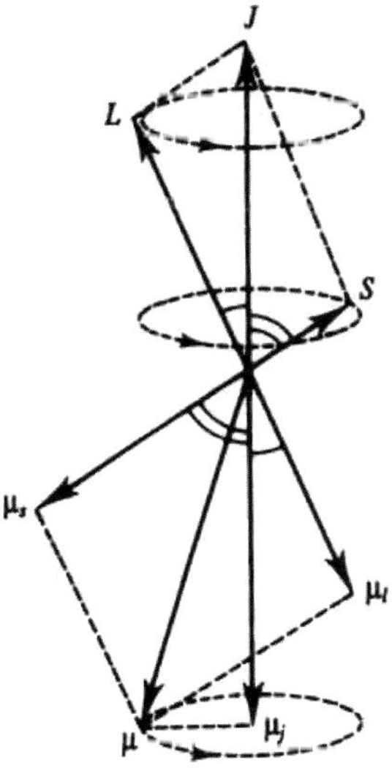

FIGURE 4.8
Correlation of the magnetic moments and the angular momenta of the electron.

$$\begin{aligned} g_j &= g_l \frac{\hat{j}^2 + \hat{l}^2 - \hat{s}^2}{2\hat{j}^2} + g_s \frac{\hat{j}^2 + \hat{s}^2 - \hat{l}^2}{2\hat{j}^2} \\ &= \frac{g_l + g_s}{2} + \left(\frac{g_l - g_s}{2}\right)\left(\frac{\hat{l}^2 - \hat{s}^2}{\hat{j}^2}\right) \end{aligned} \tag{4.27b}$$

Substituting the numerical value of the g-factors, g_l and g_s, of the electron from Eqs. (4.23) and (4.25), we get

$$g_j = \frac{3}{2} + \frac{1}{2}\left(\frac{\hat{s}^2 - \hat{l}^2}{\hat{j}^2}\right) \tag{4.28}$$

which is of great importance in the explanation of the experiments which will be discussed later. We see from Eq. (4.27b) that the inequality of g_s and g_l plays an important role here. Suppose $g_s = g_l = 1$; then we would always have $g_j = 1$. As an example, consider a state with $s = \frac{1}{2} = j$, $l = 1$ and a state with $s = \frac{1}{2}$, $l = 1$, $j = \frac{3}{2}$. Using the notation ${}^{2s+1}l_j$, where s is spin, l is the orbital angular momentum, and j is the total angular momentum, these states are written as ${}^2P_{1/2}$ and ${}^2P_{3/2}$.

$$\text{For } {}^2P_{1/2}, \quad g_j = \frac{3}{2} + \frac{1}{2}\left[\frac{\frac{1}{2}(\frac{1}{2}+1) - 1(1+1)}{\frac{1}{2}(\frac{1}{2}+1)}\right] = \frac{2}{3},$$

$$\text{For } {}^2P_{3/2}, \quad g_j = \frac{3}{2} + \frac{1}{2}\left[\frac{(-\frac{5}{4})}{\frac{3}{2}(\frac{3}{2}+1)}\right] = \frac{4}{3}$$

$$\mu_j = -\sqrt{j(j+1)}\, g_j \mu_B = -\sqrt{\tfrac{1}{2}(\tfrac{1}{2}+1)}\,\frac{2}{3}\,\mu_B = \frac{-\mu_B}{\sqrt{3}}$$

$$\mu_j = -\sqrt{\tfrac{3}{2}(\tfrac{3}{2}+1)}\,\frac{4}{3}\,\mu_B = -2\sqrt{\tfrac{5}{3}}\,\mu_B$$

We must emphasize that two assumptions were implicit in the process of the derivation of Eqs. (4.27b) and (4.28). One is that $\boldsymbol{S}$ and $\boldsymbol{L}$ are supposed to be coupled into $\boldsymbol{J}$. However, this will not be possible when there is a very strong external magnetic field, because such a field would cause $\boldsymbol{S}$ and $\boldsymbol{L}$ to precess independently about the field direction. Equation (4.28) will be correct only if the external magnetic field is not sufficiently strong as to break the $\boldsymbol{S}$–$\boldsymbol{L}$ coupling, so that $\boldsymbol{S}$ and $\boldsymbol{L}$ will be coupled into $\boldsymbol{J}$, and $\boldsymbol{J}$ will precess about the external magnetic field.

The other assumption is that we considered only a single electron. This would seem to be a very severe restriction which would limit the usage of Eq. (4.28) tremendously. But in fact, this is not the case. Indeed, for an atom, we should add up the contributions of all the electrons of the atom. But for most atoms with odd atomic numbers, the angular momenta of all the even electrons will cancel each other in pairs. Thus, the final contribution comes only from the single electron left. This cancellation of both the spin and orbital angular momenta by electrons pairing off with oppositely directed s and l is related to the Pauli Exclusion Principle presented later in Chapter 5 and to the antisymmetrization of the wavefunction of the electrons as discussed in the chapter on quantum mechancis. The Pauli Principle says that each electron must have a different set of quantum numbers. For example, when $l = 0$ (s state electron), $j = \frac{1}{2} = s$, there are two orientations of s allowed. So a $l = 0, j = \frac{1}{2}$ state can have two electrons, one electron with spin up and a second electron must have spin down so as not to violate the Pauli Principle. Thus two electrons fill the $n = 1$, $l = 0$ shell of atoms. Such a filled or closed shell is what makes helium with two electrons an inert gas. In a filled shell not only do the spins pair off in opposite directions to cancel, but for each l, its Z component (m_l) direction is such that for each positive m_l there is a negative m_l so their orbital angular momenta cancel as well. More detailed explanations follow in Chapters 5, 7, and 8. For all the single electrons outside of closed shell systems, Eq. (4.28) would be valid. For other kinds of atoms with two or more electrons outside closed shells, more than one electron will contribute to the total angular momentum of the atom and to the total magnetic moment of the atom as well. But even for these kinds of atoms, in most cases,[5] we may still use Eq. (4.28), as long as we change s and l into S and L, and recognize them, respectively, as the quantum numbers that correspond to the total spin and orbital angular momentum resultant from the electrons which contribute to them. Also, we must change j into J, which is the quantum number corresponding to the total angular momentum resultant from the S–L coupling, namely,

$$g_J = \frac{3}{2} + \frac{1}{2}\left(\frac{\hat{S}^2 - \hat{L}^2}{\hat{J}^2}\right) \tag{4.29}$$

For the moment we do not consider the contribution of the nucleus. Since the mass of the nucleus is greater than the mass of an electron by at least three orders of magnitude, and the magnetic moment is inversely proportional to the mass, the nuclear magnetic moment is at least three orders of magnitude less than the electronic magnetic moment. The contribution of the nuclear magnetic moment to the atom constitutes the so-called hyperfine interaction and is discussed in Chapter 13.

---

[5] These are the cases in which the so-called Russell–Saunders coupling dominates. As will be discussed in more detail in Chapter 5, the individual spins and orbital angular momenta of each electron outside of a closed shell can be vectorally combined (added) to form a total angular momentum, J, for all the electrons. There are two ways to couple these momenta to form a total J. Actually, Russell–Saunders coupling, which is also called L–S coupling, is correct for almost all of the ground states of atoms in which all the spins and orbital angular momenta, respectively, add (are "coupled") to form the total spin and orbital angular momentum, S and L; and then, S and L add up to form the total angular momentum J:

$$(s_1 s_2 s_3 \ldots)(l_1 l_2 l_3 \ldots) = (S, L) = J$$

The second approach is called j–j coupling, where the spin and orbital angular momenta of each electron couple (are added together) to form j_i for that electron, and the j_i are added to form J:

$$(s_1 l_1)(s_2 l_2), \ldots = (j_1 j_2 \ldots) = J$$

The atoms of $_1$H, $_3$Li, $_{11}$Na, $_{19}$K, $_{29}$Cu, $_{47}$Ag, $_{79}$Au, etc. (the subscript on the left stands for the atomic number) are examples of single-electron systems, whose ground state is $^2S_{1/2}$. (Recall the notation is $^{2s+1}l_j$ where $l = 0, 1, 2$ are labeled S, P, D.) Here S designates that the orbital angular momentum of the single electron is zero, hence j has only one value, $j = \frac{1}{2}$, which is denoted by the subscript on the right. The superscript on the left gives the value of $2s + 1$. For a single electron s is always equal to $\frac{1}{2}$, so we have $2s + 1 = 2$, which denotes a doublet. For the $^2S_{1/2}$ state, Eq. (4.28) gives $g_j = 2$, since $j = \frac{1}{2}$, so $m_j = \pm\frac{1}{2}$, and $m_j g_j = \pm 1$, see the first line in Table 4.1. For the P state with $l = 1, j = \frac{1}{2}, \frac{3}{2}$, there are two atomic states, $^2P_{1/2}$ and $^2P_{3/2}$, with $m_j = \pm\frac{1}{2}$ and $m_j = \pm\frac{3}{2}, \pm\frac{1}{2}$, respectively. In the single-electron system of $_{81}$Tl, the atomic ground state is $^2P_{1/2}$. Similarly, we have $^2D_{3/2}$ and $^2D_{5/2}$, with $m_j = \pm\frac{3}{2}, \pm\frac{1}{2}$ and $m_j = \pm\frac{5}{2}, \pm\frac{3}{2}, \pm\frac{1}{2}$, respectively. The corresponding g-factors may be calculated from Eq. (4.28). They are listed in Table 4.1 together with the values of $m_j g_j$.

4.3.4 Interpretation of the Stern–Gerlach Experiment

We introduced the Stern–Gerlach experiment in the last section, but only the orbital motion of the electron in an atom was considered so that we could not explain the even-splitting phenomena of a hydrogen atom in an inhomogeneous magnetic field. Now, we include spin in our considerations, where the total magnetic moment of the atom is the resultant of two parts, the electron orbital magnetic moment and its spin magnetic moment. Then, substituting μ_z from Eq. (4.22) into Eq. (4.16), we get

$$\begin{aligned} z_2 &= \mu_z \frac{\partial B_z}{\partial z} \frac{dD}{3kT} \\ &= -m_j g_j \mu_B \frac{\partial B_z}{\partial z} \frac{dD}{3kT} \end{aligned} \tag{4.30}$$

Since $m_j = j, j-1, \ldots, -j$, the number of values of m_j is $2j + 1$. So the number of discrete values of z_2 that appear on the photographic film will be $2j + 1$ black lines, which denote the $2j + 1$ directions of spatial orientation. From the number

TABLE 4.1
The g_j-factors and the $m_j g_j$ values of some doublets

Atomic state	g_j	$m_j g_j$
$^2S_{1/2}$	2	±1
$^2P_{1/2}$	2/3	$\pm1/3$
$^2P_{3/2}$	4/3	$\pm2/3, \pm2$
$^2D_{3/2}$	4/5	$\pm2/5, \pm6/5$
$^2D_{5/2}$	6/5	$\pm3/5, \pm9/5, \pm3$

of black lines, we may calculate the value of j and then the values of m_j. From the displaced distance of a black line to the middle line, z_2, or from the interval distance between two neighboring black lines, we may calculate the value of $m_j g_j$. Then we can combine these two to give the g-factor, which is one of the experimental methods of measuring the g-factor. Now we are ready to interpret the results of the Stern–Gerlach experiment for a hydrogen atom. As we mentioned, hydrogen atoms emitted from a high-temperature container are produced in the ground state. Hence, $n = 1$, $n_\phi = 1$, so, $l = 0, j = 0 + s = \frac{1}{2}$, and $m_j = \pm\frac{1}{2}$. Then we have from Eq. (4.28) that $g_j = 2$, and then $m_j g_j = \pm 1$, so finally,

$$z_2 = \pm\mu_B \frac{\partial B_z}{\partial z} \frac{dD}{3kT}$$

Substituting the parameters used in the experiment:

$$\frac{\partial B_z}{\partial z} = 10\,\mathrm{T/m} \text{ (here the unit of } B \text{ is Tesla)}, \; d = 1\,\mathrm{m}, \; D = 2\,\mathrm{m}, T = 400\,\mathrm{K}$$

and using the constants

$$k = 8.617 \times 10^{-5}\,\mathrm{eV/K}$$
$$\mu_B = 0.5788 \times 10^{-4}\,\mathrm{eV/T}$$

we may readily obtain

$$\begin{aligned} z_2 &= \pm 0.5780 \times 10^{-4}\,\frac{\mathrm{eV}}{\mathrm{T}} \times \frac{10\mathrm{T}}{\mathrm{m}} \times \frac{1\,\mathrm{m} \times 2\,\mathrm{m}}{3 \times 8.617 \times 10^{-5}\,\mathrm{eV/K} \times 400\,\mathrm{K}} \\ &= \pm 1.12\,\mathrm{cm} \end{aligned}$$

which means that an atomic beam of hydrogen in its ground state would be split into two beams under its interaction in an inhomogeneous magnetic field. The displaced distance of each to the middle line is the same, 1.12 cm, in good agreement with the experimental result. Experimental results for other atoms are listed in Table 4.2 (the implicit significance of each term in the table is left for the reader). The patterns in Table 4.2 are as follows: For the first two rows the beams with zero total angular momenta go through undeflected along the center line. In the third row for $J = \frac{1}{2}$ there are two $m_j = \pm\frac{1}{2}$, so one sees two beams to either side of the center line. In summary, the Stern–Gerlach experiment proves:

1. That spatial quantization of angular momentum is an experimental fact.
2. The correctness of the hypothesis of electron spin, $s = \frac{1}{2}$.
3. The correctness of the values of the spin magnetic moment of the electron, $\mu_{s,z} = \pm\mu_B$, $g_s = 2$;
4. The correctness of the use of the initial value of 0 for l, and the replacement of l by $\sqrt{l(l+1)}$ (Eq. 4.8).

TABLE 4.2
Results of Stern–Gerlach experiments

<table>
<tr><th>Atom</th><th>Ground state</th><th>g</th><th>mg</th><th>Pattern</th></tr>
<tr><td>Zn, Cd, Hg, Pd</td><td>$^1S_0$</td><td>–</td><td>0</td><td>|</td></tr>
<tr><td>Sn, Pb</td><td>$^3P_0$</td><td>–</td><td>0</td><td>|</td></tr>
<tr><td>H, Li, Na, K, Cu, Ag, Au</td><td>$^2S_{1/2}$</td><td>2</td><td>± 1</td><td>| |</td></tr>
<tr><td>Tl</td><td>$^2P_{1/2}$</td><td>2/3</td><td>$\pm 1/3$</td><td>| |</td></tr>
<tr><td rowspan="3">O</td><td>$^3P_2$</td><td>3/2</td><td>$\pm 3, \pm 3/2, 0$</td><td>| | | | |</td></tr>
<tr><td>$^3P_1$</td><td>3/2</td><td>$\pm 3/2, 0$</td><td>| | |</td></tr>
<tr><td>$^3P_0$</td><td>–</td><td>0</td><td>|</td></tr>
</table>

4.4 DOUBLET LINES OF ALKALI METALS

4.4.1 Fine Structure in the Spectral Lines of Alkali Metals: Qualitative Considerations

As we indicated in the last chapter, there are four main series of lines in the atomic spectra of alkali metals (such as lithium):

1. A "principal" series of lines, corresponding to the nP $\rightarrow$ 2S transitions
2. A "sharp" series of lines, corresponding to the nS $\rightarrow$ 2P transitions
3. A "diffuse" series of lines, corresponding to the nD $\rightarrow$ 2P transitions
4. A "fundamental" series of lines, corresponding to the nF $\rightarrow$ 3D transitions

The frequencies (or wavenumbers) of the spectral lines in each series may be expressed as the difference of two terms: one is variable, and the other is fixed, corresponding to the initial states and to the final state of the transitions, respectively. These alkali spectral series are found to have a doublet structure when they are observed carefully with a high-resolution spectrometer. For example, the doublet structures of the lines of the principal series and of the sharp series are shown in Fig. 4.9. We note that the doublet splitting (the wavenumber difference of the splitting) decreases as the wavenumber increases for the principal series of lines, but remains unchanged for the sharp series of lines.

The fact that a spectral line splits into two means that at least one of the two energy levels, which corresponds to the initial or to the final state of the transition, has been split into two. But which one? Remember that the fixed term in the formula corresponds to the final state. So, if it is the final state that is split, the magnitude of the doublet splitting of the spectral lines will be the same, which is what we observe in the lines of the sharp series. On the other hand, if the initial state splits, the magnitude of the doublet splitting will be different from line to line, in accordance with the splitting of the level from which the transition initiates, as occurs for the lines of the principal series. Since the principal series of lines corresponds to the nP → 2S transitions, and the sharp series of lines corresponds to the nS → 2P transitions, we may deduce that it is the P states, rather than the S states, that split. In other words, the $l = 1$ level splits, while the $l = 0$ level does not.

The experimental discovery of the doublet structure of the lines of the alkali metals was one of the primary reasons for Uhlenbeck and Goudsmit to propose the hypotheses of electron spin and for Pauli earlier to propose a new electron quantum number with only two values. In addition to the orbital angular momentum l, if there does exist a spin angular momentum s, and if s has only two orientations, then the $l = 1$ state will split into two states, $^2P_{3/2}$ and $^2P_{1/2}$, with $j = l \pm \frac{1}{2} = \frac{3}{2}$ and $\frac{1}{2}$; while the $l = 0$ state will remain unchanged ($^2S_{1/2}$), with $j = \frac{1}{2}$. But why do these two states, $^2P_{3/2}$ and $^2P_{1/2}$, have different energies? Also, what is the energy splitting between them? Those questions and others will be discussed next.

4.4.2 Spin–Orbit Interaction: Quantitative Consideration of Fine Structure

So far we have considered only the Coulomb force interaction between the nucleus and the electrons. Indeed, it is the dominant interaction which determines

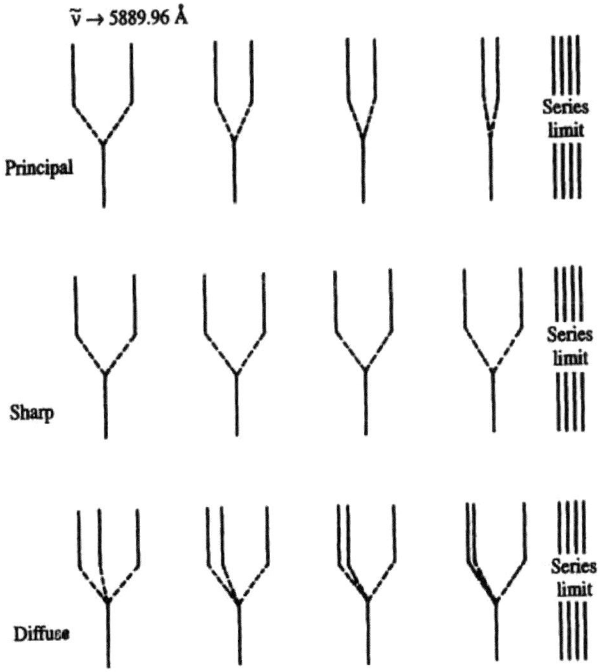

FIGURE 4.9
Doublet spectra of alkali metals.

the main character of the spectral series. But there must be a magnetic field produced by the moving charge going around the nucleus periodically. The magnetic interaction between this field and the electron spin magnetic moment will produce a fine structure in the spectral series. Let us now analyze this interaction.

In the rest frame of the atomic nucleus, the electrons are moving around the nucleus. In the rest frame of the electron the nuclear charge Ze is moving around the electron, as shown in Fig. 4.10. The electric current produced by the nuclear charge Ze is

$$i = Ze\nu = \frac{Ze\mathrm{v}}{2\pi r}$$

The orbit is considered to be circular for simplicity, but it may easily be proven that the following conclusion remains correct for any orbit of arbitrary shape. In the formula, ν is the circular frequency of motion, v is the linear velocity, and r the circular radius.

The magnetic induction B produced by the current i at the center, the location of the electron, will be

$$B = \frac{2\pi i}{cr} = \frac{Ze\mathrm{v}}{cr^2}$$

(This should be multiplied by the factor k/c to convert to the SI system of units, where c is the speed of light, and k is the proportionality constant in Coulomb's law, $k = 1/4\pi\epsilon_0$ in the SI system, see Chapter 2). Expressed in the vector form,

$$\boldsymbol{B} = \frac{Ze}{cr^3}(-\mathbf{v}) \times \boldsymbol{r} = \frac{Zec}{E_0 r^3}\boldsymbol{L} \tag{4.31}$$

where

$$\boldsymbol{L} = m_e \boldsymbol{r} \times \mathbf{v}$$

is the orbital angular momentum of the electron, and

$$E_0 = m_e c^2$$

is the rest energy of the electron.

Since the electron has a magnetic moment associated with its spin, which is called the intrinsic magnetic moment, μ_s, its potential energy in a magnetic field is of the form

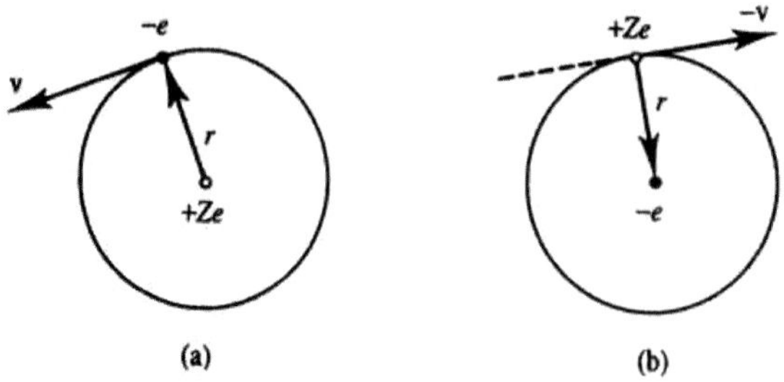

FIGURE 4.10
Hydrogenlike atom. (a) Observed in the rest frame of the nucleus. (b) Observed in the rest frame of the electron.

$$U = -\boldsymbol{\mu}_s \cdot \boldsymbol{B}$$

Using

$$\mu_s = -\sqrt{s(s+1)}\, g_s \mu_B, \quad S = \sqrt{s(s+1)}\, \hbar$$
$$\boldsymbol{\mu}_s = -g_s \frac{\mu_B}{\hbar} \boldsymbol{S}$$

and Eq. (4.31), we have

$$U = \frac{Z g_s \mu_B e c}{E_0 \hbar r^3} \boldsymbol{S} \cdot \boldsymbol{L} \tag{4.32}$$

This is the expression for the electron potential energy in the rest frame of the electron. In fact, we are more interested in having it in the rest frame of the nucleus. These two frames seem to be symmetric, but they are not, because there is a relativistic time difference between them.[6] This effect introduces a correction of a factor of $\frac{1}{2}$, so

$$U = \frac{1}{2} \frac{Z g_s \mu_B e c}{E_0 \hbar r^3} \boldsymbol{S} \cdot \boldsymbol{L} \tag{4.33}$$

The factor $\frac{1}{2}$ in Eq. (4.33) was derived by L. H. Thomas.[7] Only after the correct result was given by Thomas did Pauli write to Bohr saying that he had been convinced of the concept of electron spin. For $g_s = 2$, $\mu_B = (\hbar e/2m_e c)$, and $E_0 = m_e c^2$, the interaction energy in the rest frame of the nucleus is

$$U = \frac{Z e^2}{2 m_e^2 c^2 r^3} \boldsymbol{S} \cdot \boldsymbol{L} \tag{4.34}$$

In the SI system of units, Eq. 4.34 should be multiplied by a factor $k = 1/4\pi\epsilon_0$. We may note that the potential energy U depends only on the relative orientation of $\boldsymbol{S}$ and $\boldsymbol{L}$, but not on the direction of $\boldsymbol{S}$ or $\boldsymbol{L}$ with respect to any reference axis in space. Since U is proportional to the combination of $\boldsymbol{S}$ and $\boldsymbol{L}$, $\boldsymbol{S} \cdot \boldsymbol{L}$, it is called the "spin–orbit coupling" term. It is the additional energy produced by the interaction of the magnetic field produced by the orbital motion and the spin magnetic moment of the electron. We shall make an order of magnitude estimation before going further. For the second energy level of hydrogen ($Z = 1$), $r = 2^2 a_1$ (a_1 is the first Bohr radius), $|\boldsymbol{S}| = |\boldsymbol{L}| \simeq \hbar$, and $E_0 = m_e c^2$, so

$$U = \frac{(e^2)(\hbar c)^2}{2E_0^2 (4a_1)^3} = \frac{(14.4\,\text{eV} \cdot \text{Å})(1970\,\text{eV} \cdot \text{Å})^2}{2(0.511 \times 10^6\,\text{eV})^2 (4 \times 0.529\,\text{Å})^3} = 1.13 \times 10^{-5}\,\text{eV} \tag{4.35}$$

which is just the order of the splitting observed in experiments. Note that for the first energy level $l = 0$, and so $U = 0$.

[6] See R. M. Eisberg, *Fundamentals of Modern Physics*, Wiley, New York (1963) p. 140.

[7] L. H. Thomas, *Nature* **117** (1926) 514; *Phil. Mag*. 3 (1927) 1.

Now we do a quantitative calculation. It is quite convenient to write that

$$J = S + L$$
$$J^2 = S^2 + L^2 + 2S \cdot L$$

then

$$\begin{aligned} S \cdot L &= \tfrac{1}{2}(J^2 - S^2 - L^2) \\ &= \tfrac{1}{2}[j(j+1) - s(s+1) - l(l+1)]\hbar^2 \end{aligned} \tag{4.36}$$

For $s = \frac{1}{2}$, $l = 1$, $j = \frac{3}{2}$, $S \cdot L = \frac{1}{2}[\frac{3}{2}(\frac{3}{2}+1) - \frac{1}{2}(\frac{1}{2}+1) - (1+1)]\hbar^2 = \frac{1}{2}\hbar^2$.
For the doublet energy levels of $j = l \pm \frac{1}{2}$,

$$S \cdot L = \begin{cases} \frac{1}{2} l\hbar^2 & \text{for } j = l + \frac{1}{2} \\ -\frac{1}{2}(l+1)\hbar^2 & \text{for } j = l - \frac{1}{2} \end{cases} \tag{4.37}$$

In order to make a comparison with experiments, the average value of $1/r^3$ in Eq. (4.33) should be used. Hence the distribution of the electrons in an atom is needed, but this may only be done by quantum mechanics as in Chapter 7. Here, the result of the calculation will be given:

$$\overline{\left(\frac{1}{r^3}\right)} = \frac{Z^3}{n^3 l(l+\frac{1}{2})(l+1)a_1^3} \tag{4.38}$$

If we use Bohr's atomic theory, then $r = a_1 n^2/Z$, and

$$\overline{\left(\frac{1}{r^3}\right)} = \frac{Z^3}{n^6 a_1^3} \tag{4.39}$$

The precise result, Eq. (4.38), will coincide with Eq. (4.39) only in the case that l is very large so that $n \approx l \approx (l+1)$.

Substituting Eqs. (4.38) and (4.37) into Eq. (4.34), we obtain the spin–orbit coupling term:

$$U = \frac{(\alpha Z)^4 E_0}{4n^3} \frac{(j(j+1) - l(l+1) - \frac{3}{4})}{l(l+\frac{1}{2})(l+1)}, \neq 0 \tag{4.40}$$

where $E_0 = mc^2$ and $\alpha = e^2/\hbar c \simeq 1/137$ is the fine structure constant (note that the first Bohr radius $a_1 = \hbar/\alpha m_e c$). Note that l in Eq. (4.40) cannot be zero, otherwise, from a mathematical point of view, if both the denominator and the numerator in a fraction are zero, the result is indefinite. From the physical point of view, $l = 0$ would definitely lead to $L \cdot S = 0$. It may be pointed out that Eq. (4.40) is the same in different systems of units, as the k factor of the original formula in the SI system of units would have merged into α at this step.

For the energy difference between the two levels $j = l \pm \frac{1}{2}$, we may use Eq. (4.37), or make direct use of Eq. (4.40), and obtain

$$U = \frac{(\alpha Z)^4 E_0}{2n^3(2l+1)(l+1)}; \qquad j = l + \tfrac{1}{2}, l \neq 0$$
$$U = -\frac{(\alpha Z)^4 E_0}{2n^3 l(2l+1)}; \qquad j = l - \tfrac{1}{2}, l \neq 0 \tag{4.41}$$

The energy difference $U(j = l + \frac{1}{2}) - U(j = l - \frac{1}{2})$ is

$$\Delta U = \frac{(\alpha Z)^4}{2n^3 l(l+1)} E_0 \tag{4.42}$$

In the energy spectrum of a single-electron atom, the Coulomb interaction dominates and gives the gross structure of the spectrum, with the order of magnitude of $\alpha^2 E_0$ (where $\alpha^2 E_0/2 = \alpha^2 m_e c^2/2 \simeq 13.6\,\text{eV}$). The spin–orbit interaction gives the energy difference of the fine structure, with the order of magnitude of $\alpha^4 E_0$. That is why we call α the fine structure constant. The spin–orbit interaction is the largest relativistic effect and the main contribution to the fine structure. Note that as a result of the spin–orbit coupling the single-electron energy states with higher j are higher in energy and so less tightly bound (see Fig. 4.11).

For the splitting of the 2P state into $P_{3/2}$, $P_{1/2}$ states in the hydrogen atom, we may calculate from Eq. (4.42) that the energy difference between these two states is

$$\Delta U = \frac{0.511 \times 10^6\,\text{eV}}{2 \times 2^3 \times 1 \times 2 \times (137)^4} = 4.53 \times 10^{-5}\,\text{eV} \tag{4.43}$$

This is a precise calculation which is in agreement with the experimental results. This also shows that the gross estimation given by Eq. (4.35) is basically correct. It is not only convenient but also very important to make an order of magnitude estimation in checking the correctness of a physical idea. It may also be seen from Eq. (4.42) that the doublet splitting will increase rapidly as Z increases and will decrease rapidly as the principal quantum number n increases, and it will decrease as the orbital angular momentum quantum number increases. These results are all found to be in agreement with the experimental facts. For example, the splitting of the 2P level in the hydrogen atom has been calculated in Eq. (4.43) to be very

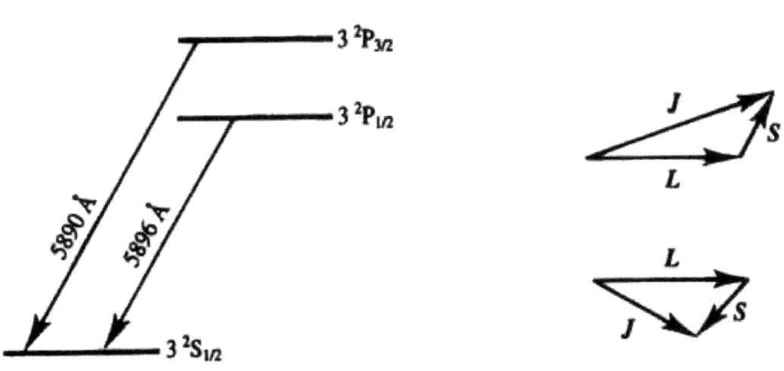

FIGURE 4.11
The yellow D lines of sodium.

small. It is observable only with a very high-resolution spectrometer. But for the famous yellow lines of the doublet in sodium, $\Delta U = 2.1 \times 10^{-3}$ eV, and the corresponding 3P splitting is 6 Å, which is quite easily observed (see Fig. 4.11).

However, calculation of the splitting of the 3P level in sodium is a little difficult. Since the atomic nucleus in sodium is shielded in part by the ten inner electrons, the outer single electron will be subjected to a potential identical to that created by the effective charge $Z_{\text{eff}}e$, instead of the nuclear charge Ze. Substituting Z_{eff} into Eq. (4.42) to replace Z, and using the experimentally measured value $\Delta U = 2.1 \times 10^{-3}$ eV, we may calculate from Eq. (4.42) ($n = 3$, $l = 1$):

$$Z_{\text{eff}} = 3.5$$

The splitting of a number of levels in sodium is shown in Fig. 4.12.

4.4.3 Note: Estimation of the Internal Magnetic Field of an Atom

Starting from

$$U = -\boldsymbol{\mu}_s \cdot \boldsymbol{B} = -\mu_z B$$

we obtain from Eq. (4.21)

$$U = \pm \mu_B B$$

where we choose the z axis along the direction of B, and assume a single electron. The energy difference is

$$\Delta U = 2\mu_B B$$

which may also be expressed for photons with $E = hc/\lambda$ as

$$\Delta E = \frac{hc\,\Delta\lambda}{\lambda^2} \tag{4.44}$$

Equating these gives

$$B = \frac{hc\,\Delta\lambda}{2\lambda^2 \mu_B}$$

For the 5890 Å line in sodium, $\Delta\lambda = 6$ Å, and the magnetic field at the electron may be estimated as

$$B = \frac{12.4\,\text{Å} \cdot \text{keV} \times 6\,\text{Å}}{2 \times (5890\,\text{Å})^2 \times (0.5788 \times 10^{-4}\,\text{eV/T})} \simeq 20\,\text{T}$$

which means that the internal magnetic field of the atom is very strong.

It should be noted that the ordinate in the energy level diagram, for example in Fig. 4.12, is energy. The difference in energy between the two energy levels, say $3\,^2D_{3/2}$ and $3\,^2D_{5/2}$, is small, or equivalently, the difference in wavenumber is small; but this does not mean, in general, that the difference in their wavelengths ($\Delta\lambda$) is small. The $\Delta\lambda$ is meaningless unless the two levels are related to another

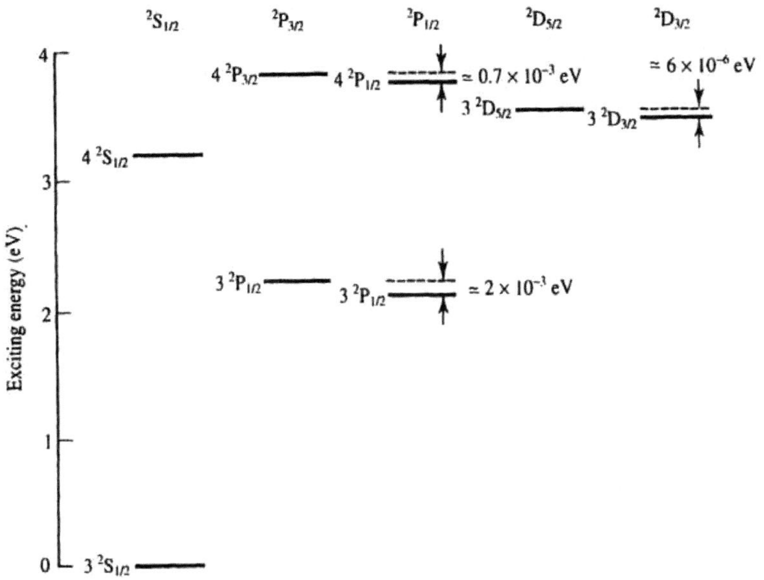

FIGURE 4.12
Energy level splittings in sodium (not to scale).

energy level. For example, the difference in wavelength is 6 Å for the yellow lines of the doublet in sodium, which means that the transitions of $3\,^2P_{3/2}$–$3\,^2S_{1/2}$ and $3\,^2P_{1/2}$–$3\,^2S_{1/2}$ differ by 6 Å in wavelength (see Fig. 4.11). For the transition between $3\,^2P_{3/2}$ and $3\,^2P_{1/2}$, the wavelength of the radiation will be 5.8×10^6 Å! The relations and the conceptual differences among the quantities ΔE, $\Delta\lambda$, and λ, may be understood from Eq. (4.44).

4.5 THE ZEEMAN EFFECT

4.5.1 The Normal Zeeman Effect

In 1896, P. Zeeman discovered that the spectral lines emitted from a light source were broadened when the source was placed in a magnetic field. Further investigation indicated that each line had been split into several lines and that it was not really "broadening."

The splitting of the spectral lines reflects a splitting of an energy state into several states with different energies. In order to understand the splitting of a line in a magnetic field, we need to know the interaction of the light source with the magnetic field.

A system with a magnetic moment of $\boldsymbol{\mu}$ placed in an external magnetic field would have a potential energy

$$U = -\boldsymbol{\mu} \cdot \boldsymbol{B} = -\mu_z B \tag{4.45}$$

where the z axis is chosen along the direction of $\boldsymbol{B}$. The main contribution to the atomic magnetic moment comes from the electrons. According to Eq. (4.22),

$$\mu_z = -mg\mu_B$$

where m is the quantum number of the z component of the angular momentum, and the subscript j has been omitted for simplicity. Substituting it into Eq. (4.45), we get

$$U = mg\mu_B B \tag{4.46}$$

Now let us consider the optical transition between two atomic levels whose energies are E_2 and E_1, respectively ($E_2 > E_1$). Suppose that there exists no external magnetic field. Then the transition energy will be

$$h\nu = E_2 - E_1$$

However, if there is a magnetic field B applied, we know from Eq. (4.46) that the energy of these two levels will then be

$$\begin{aligned} E_2' &= E_2 + m_2 g_2 \mu_B B \\ E_1' &= E_1 + m_1 g_1 \mu_B B \end{aligned} \tag{4.47}$$

respectively. Obviously, every energy level splits and each one will split into m (i.e., $2j+1$) levels. But only the difference may be observed, namely,

$$h\nu' = E_2' - E_1' = (E_2 - E_1) + (m_2 g_2 - m_2 g_1)\mu_B B$$

or

$$h\nu' = h\nu + (m_2 g_2 - m_1 g_1)\mu_B B \tag{4.48}$$

If we have a system whose spin is zero, $g_2 = g_1 = 1$, then

$$h\nu' = h\nu + (m_2 - m_1)\mu_B B \tag{4.49}$$

According to the selection rule,

$$\Delta m = m_2 - m_1 = 0, \pm 1 \tag{4.50}$$

only three values of $h\nu$ will remain. In quantum mechanics, the probability for a transition to occur between states 1 and 2 vanishes except for certain values of the quantum numbers that describe states 1 and 2. A selection rule like Eq. 4.50 gives the allowed values of the change in the m quantum numbers for which the transition probability is not zero. The origin of this selection rule is given in Chapter 7. Basically, since photons are emitted as the result of a transition, and the intrinsic (spin) angular momentum of a photon is $1\hbar$, the value of Δm (the change in the z component of angular momentum) cannot exceed one unit in first order. In other words, there will only be three spectral lines with

$$h\nu' = h\nu + \begin{pmatrix} \mu_B B \\ 0 \\ -\mu_B B \end{pmatrix} \tag{4.51}$$

This means that a spectral line will split into three lines with equal spacing in an external magnetic field. This result is in good agreement with the spectra observed in some experiments, hence the effect is named the normal Zeeman effect. The splitting of the 6438.47 Å line of a cadmium atom in an external magnetic field may be taken as an example of this kind, as shown in Fig. 4.13. Contributing to the magnetic moment of the cadmium atom are two electrons with their spin orientations opposite to each other, which makes the total spin $S = 0\,(2S+1 = 1$, a singlet), and the normal Zeeman effect is thus observed. In fact, we shall have the normal Zeeman effect only for those atoms in which the number of the electrons is even and a singlet state is formed.

It may be seen from Fig. 4.13 that the spectral line of 6438.47 Å in cadmium comes from the transition $^1D_2 \rightarrow {}^1P_1$. Of the nine transitions, there are only three energy differences. Hence only three spectral lines are observed, and each contains three transitions of different kinds. The middle line keeps the original energy,

$$h\nu' = h\nu_0 + \begin{pmatrix} \dfrac{e\hbar}{2m_e c} B \\ 0 \\ -\dfrac{e\hbar}{2m_e c} B \end{pmatrix} \tag{4.52}$$

or, expressed in frequency,

$$\nu' = \nu_0 + \begin{pmatrix} \dfrac{e}{4\pi m_e c} B \\ 0 \\ -\dfrac{e}{4\pi m_e c} B \end{pmatrix} \tag{4.53}$$

This shows that this additional frequency $eB/4\pi m_e c$ is generated only by the applied external magnetic field. From Eq. (4.53), we discover with intuition that there might not be a quantum effect. Indeed, this formula may be derived without the quantum theory. Classical theory may be used to derive the normal Zeeman effect, but it provides only three frequencies. So the quantization condition is needed in order to give the experimental results a complete description. Relying upon the classical point of view, Lorentz calculated the normal Zeeman effect; for this reason $eB/4\pi m_e c$ is sometimes called the Lorentz unit. We should note that the expression for μ_B is $\mu_B = e\hbar/2m_e$ in the SI system of units; correspondingly, the Lorentz unit (or the Larmor frequency, as it is called) would be $eB/4\pi m_e$. The frequency interval between each neighboring line is just equal to 1 in the Lorentz unit. The physical meaning of the unit is the Larmor frequency for a classical atomic system in the case when there exists no spin. Now we return to the classical expressions given in Section 4.1:

$$\boldsymbol{\mu} = -\gamma \boldsymbol{L}, \qquad \frac{d\boldsymbol{\mu}}{dt} = \boldsymbol{\omega} \times \boldsymbol{\mu}, \qquad \boldsymbol{\omega} = \gamma \boldsymbol{B} = \frac{e}{2m_e c} \boldsymbol{B}$$

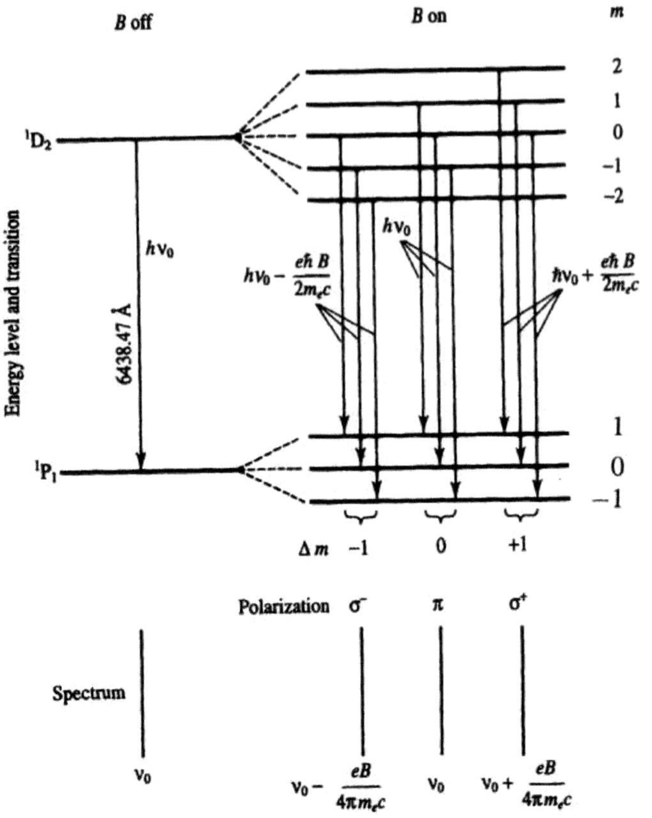

FIGURE 4.13
Zeeman effect of the spectral line $^1D_2 \rightarrow {^1P_1}$ for cadmium.

where ω is the circular frequency of Larmor precession. The Larmor frequency would be

$$\nu_L = \frac{\omega}{2\pi} = \frac{e}{4\pi m_e c} B \tag{4.54}$$

To obtain an order of magnitude, we calculate

$$\frac{\delta\nu_L}{\delta B} = \frac{e}{4\pi m_e c} = 14\,\frac{\text{GHz}}{\text{T}} \tag{4.55}$$

in which 1 GHz $= 10^9$ Hz and T is tesla. It may be readily seen that the splitting caused by an external magnetic field of 1 T would be 14×10^9 Hz.

The Zeeman effect may be used to deduce the charge-to-mass ratio of the electron, e/m_e. For example, if a spectral line of wavelength $\lambda = 6000$ Å splits because of the normal Zeeman effect in a magnetic field of $B = 1.2$ T, the difference in wavelength of the splitting will be $\Delta\lambda = 0.2013$ Å. The energy levels of the splitting are equally spaced in the normal Zeeman effect, namely,

$$\Delta E = \mu_B B \tag{4.56}$$

while $E = hc/\lambda$, so we have

$$|\Delta E| = \frac{hc}{\lambda^2}\Delta\lambda \tag{4.57}$$

Then μ_B may be calculated knowing B, λ, and $\Delta\lambda$. Combining Eqs. (4.56) and (4.57),

$$\mu_B = \frac{\Delta E}{B} = \frac{hc\,\Delta\lambda}{\lambda^2 B}$$

Meanwhile, we also know that

$$\mu_B = \frac{e\hbar}{2m_e c} = \frac{hc\,\Delta\lambda}{\lambda^2 B}$$

So

$$\frac{e}{m_e} = \frac{4\pi c^2\,\Delta\lambda}{\lambda^2 B}$$

Hence, putting in the measured values and c gives the ratio e/m_e. The value thus calculated fits well with the result that Thomson obtained in 1897. This agreement means that the derivation of the normal Zeeman effect so far has been based on correct assumptions.

4.5.2 The Polarization Character of the Zeeman Spectrum

The polarization character of the three spectral lines is indicated in Fig. 4.13 and is illustrated, more clearly, in Fig. 4.14. In order to understand the dependence of the polarization of the spectral lines in the Zeeman effect on the Δm values and on the direction of observation, the correlation of the polarization with the direction of the angular momentum in electromagnetism will be reviewed first.

For an electromagnetic wave propagating along the z direction, its electric field vector must lie in the xy plane (characteristic of a transverse wave) and may be resolved into two components E_x and E_y:

$$E_x = A\cos\omega t, \qquad E_y = B\cos(\omega t - \alpha)$$

The angle $\alpha = 0, \pi$ is for linear polarization, when the electric vector oscillates periodically in a fixed direction. If $\alpha = \pm\pi/2$ and $A = B$, the resultant electric vector has a constant magnitude but a periodically changing direction such that the end of the electric field vector travels around a circle. This corresponds to circular polarization. When we look at the light as it comes straight towards us along the z axis, if the end of the electric vector goes around in a clockwise direction, we call it right-hand (circular) polarization (σ^-) (Fig. 4.15a); if it goes around in a counterclockwise direction, we call it left-hand (circular) polarization (σ^+) (Fig. 4.15b). The fact that circular polarized light has an angular

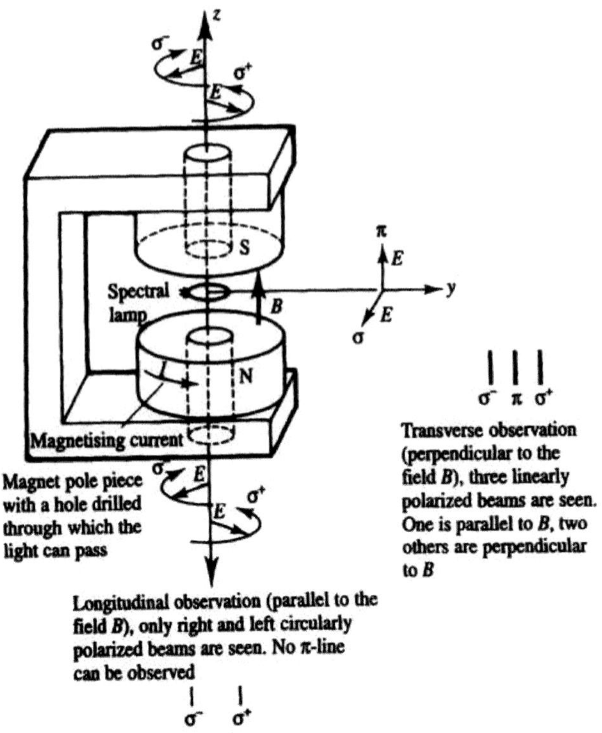

FIGURE 4.14
Observation of the polarization characteristics in the Zeeman effect.

momentum was observed by R. A. Beth in 1936. Since the direction of the angular momentum of light depends on the rotational direction of the electric vector in the right-hand screw sense, the direction of the angular momentum is opposite to the propagation direction for right-hand polarization (Fig. 4.15a), and is in the same direction as the propagation direction for left-hand polarization (Fig. 4.15b). We should note that different conventions are used in different books on the definition of right- and left-hand polarization and the direction of the angular momentum of light.

Now return to the Zeeman effect. For $\Delta m = m_2 - m_1 = 1$, the angular momentum of the atom along the direction of the magnetic field (z component) decreases by $1\hbar$. If we take the atom and the photon emitted as a unit, the angular momentum must be conserved for this unit. So the photon emitted must have an angular momentum of $1\hbar$ in the direction of the magnetic field. From Fig. 4.15(b) we know that we would have σ^+ polarization. For $\Delta m = m_2 - m_1 = -1$, the angular momentum of the atom in the direction of the magnetic field increases by $1\hbar$. Then the photon emitted must have an angular momentum in the direction opposite to that of the magnetic field, and hence σ^- polarization would be observed (Fig. 4.16). For these two spectral lines, the electric vector lies in the xy plane. When we look in the direction perpendicular to the magnetic field, say

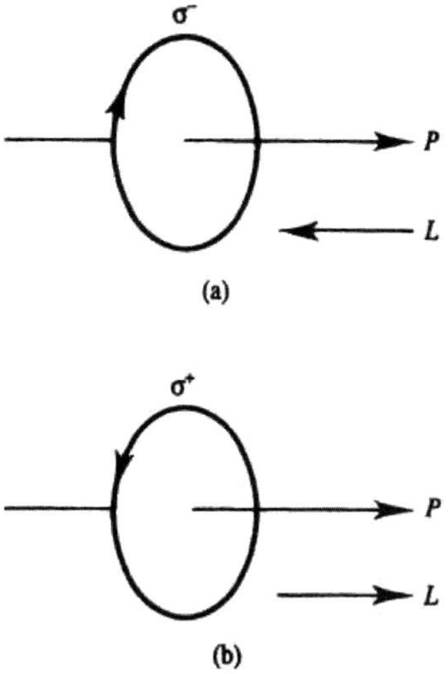

FIGURE 4.15
Definitions of the polarization and the direction of angular momentum.

in the x direction, then only waves with component E_y or E_z may be observed (Remember the transverse wave character of an electromagnetic wave. The electric vector can never be in the x direction for a light wave propagating along the x direction). Hence two linearly polarized lines (σ^+, σ^-) perpendicular to B and one (π) parallel to B (as discussed next) would be observed (Fig. 4.14).

For $\Delta m = m_2 - m_1 = 0$, the angular momentum of the atom in the direction of the magnetic field remains unchanged, but a photon has an intrinsic angular momentum $\hbar$. In order to keep the angular momentum conservation law for an emitting atom, the angular momentum of the photon emitted must be perpendicular to the magnetic field. Then, the electric vector corresponding to the light must lie in the yz plane (taking x as the direction of the angular momentum of the light), and may have the components E_y and E_z. But in fact all the photons whose direction of angular momentum lies in the xy plane satisfy the condition $\Delta m = 0$, which makes the average value of the component E_y zero. Thus, neither the component E_y nor the component E_z (because of the transverse wave character) can be observed in the direction of the magnetic field (the z direction). So that there appear no spectral lines corresponding to $\Delta m = 0$. On the other hand, only the component E_z can be observed in the direction perpendicular to that of the magnetic field (the x direction). These are the linearly polarized lines (called π lines) with electric vector parallel to the magnetic field B (Fig. 4.14).

4.5.3 The Anomalous Zeeman Effect

Not long after Zeeman discovered the splitting of the spectral line into three components in a magnetic field in 1896, H. A. Lorentz worked out the theoretical explanation. The phenomenon was later named the normal Zeeman effect. However, in December 1897, T. Preston reported that more often the number of components was not three, and the frequency intervals between components

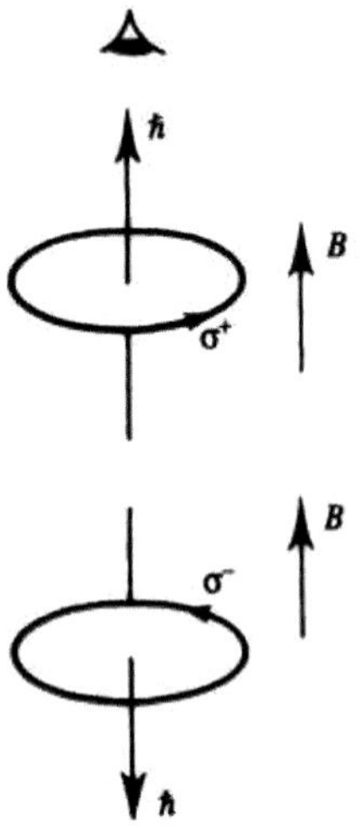

FIGURE 4.16
Definitions of the orientation of $\sigma^{\pm}$.

were not equal to one another (Fig. 4.17). In the following 28 years, many physicists tried, but none succeeded, in giving a reasonable explanation of the observed phenomenon, which was then named the "anomalous" Zeeman effect. In 1920, Sommerfeld listed the anomalous Zeeman effect as one of "the open questions in atomic physics." Pauli recalled: "One of my colleagues asked me kindly, when he saw me sitting on a park bench in Copenhagen looking dejected, what was making me so unhappy. I then answered impatiently how could one avoid despondency if one thinks of the anomalous Zeeman effect?" The difficulties of the problem may thus be seen. The anomalous Zeeman effect is another reason why Uhlenbeck and Goudsmit proposed the hypothesis of electron spin. Indeed, the puzzle of the anomalous Zeeman effect is readily solved on the basis of the hypothesis of electron spin. The reality of the spin hypothesis is thus decisively proven.

Now let us analyze the Zeeman splitting of the doublet in the principal line series of sodium. For a single-electron system like sodium, the main contribution to the magnetic moment of the atom comes from the single electron. The famous double yellow lines of sodium are from the transitions ${}^2P_{1/2,3/2} \to {}^2S_{1/2}$. The energies corresponding to the splitting lines are defined by Eq. (4.48) (where the ${}^2P_{1/2,3/2}$ g-factors were worked out earlier in section 4.3 as examples of Eq. (4.28) and the mg values are given in Table 4.1):

$$h\nu' = h\nu + (m_2g_2 - m_1g_1)\mu_B B$$

or

$$\nu' = \nu + (m_2g_2 - m_1g_1)\mathcal{L} \tag{4.58}$$

or

$$\tilde{\nu}' = \tilde{\nu} + (m_2g_2 - m_1g_1)\tilde{\mathcal{L}}$$

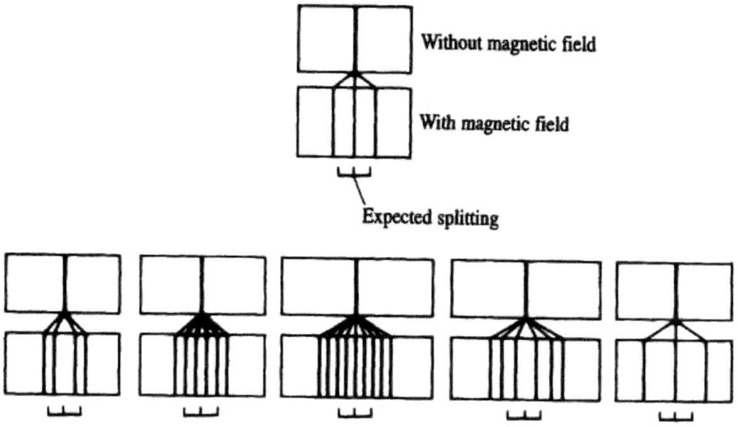

FIGURE 4.17
Experimental results of the Zeeman effect. Shown on the bottom are five examples of different experimentally observed splittings in a magnetic field of the single lines above them in the absence of a magnetic field.

where

$$\mathcal{L} = \mu_B \frac{B}{h} = \frac{eB}{4\pi m_e c} \simeq \frac{14.0 \times 10^9\,\text{Hz}}{\text{T}} B, \qquad \tilde{\mathcal{L}} = \frac{\mathcal{L}}{c} = 46.7\,\text{m}^{-1} \cdot \text{T}^{-1} \cdot B$$

is called the Lorentz unit, which is equal to the Larmor frequency in Eq. (4.54) and $\tilde{\nu} = \nu/c = 1/\lambda$ and $\tilde{\mathcal{L}} = \mathcal{L}/c$.

The related atomic states, the corresponding g-factors, and the values of mg, have been calculated in Section 4.3, and listed in Table 4.1, and are given again in Fig. 4.18. According to the selection rules for the transitions ($\Delta m = 0, \pm 1$) and Eq. (4.58), the spectral line of 5896 Å in the sodium D lines splits into four components as shown in Fig. 4.18. For example, the smallest $\nu' = \nu + (-\frac{1}{3} - 1)\mathcal{L} = \nu - \frac{4}{3}\mathcal{L}$. The frequency difference of the two neighboring spectral lines on each side of the original line is $\frac{2}{3}\mathcal{L}$, while the frequency difference of the two π lines in the middle is $\frac{4}{3}\mathcal{L}$. The 5890 Å line splits into six components, where the frequency differences of two neighboring lines are all $\frac{2}{3}\mathcal{L}$. The unshifted frequency lines before their interaction with a magnetic field of 3 T also are shown in Fig. 4.19. It may be seen that the frequency intervals between the components caused by the external magnetic field are still much less than the interval between the lines D_1 and D_2 (electron spin–orbit interaction), even when the magnetic field is as strong as 3 T. Then

$$\tilde{\mathcal{L}} = \frac{14.0 \times 10^9\,\text{Hz} \times 3\,\text{T}}{3 \times 10^{10}\,(\text{cm/s}) \cdot \text{T}} = 1.40\,\text{cm}^{-1} \text{ and } \frac{2}{3}\tilde{\mathcal{L}} = 0.93\,\text{cm}^{-1}$$

The polarization of the light is similar to that of the normal Zeeman effect, namely, $\Delta m = \pm 1$ corresponds to σ polarization, and $\Delta m = 0$ corresponds to π polarization. Observation along the direction of the magnetic field gives only σ

polarization, which is circularly polarized and observation in a direction perpendicular to that of the magnetic field gives σ and π polarization, which are both linearly polarized.

4.5.4 Supplementary Note 1: Grotrian Diagrams

It is usually inconvenient to represent on a normal energy level diagram (say, Fig. 4.18) the transitions which obey the selection rule between Zeeman sublevels and their polarizations. For this reason diagrams designed by W. Grotrian, so-called Grotrian diagrams, were introduced.

The transition between the two energy levels E_2 and E_1, with the total angular momentum $j = \frac{5}{2}$ and $j = \frac{3}{2}$, respectively, will be taken as an example. For $j = \frac{5}{2}$, there are $2j + 1 = 6$ values of m (i.e., 6 sublevels). They are denoted by points, equally spaced, on a horizontal line representing the upper energy level (see Fig. 4.20a). Similarly, the four values of m associated with $j = \frac{3}{2}$ are denoted by the equally spaced points on another horizontal line representing the lower

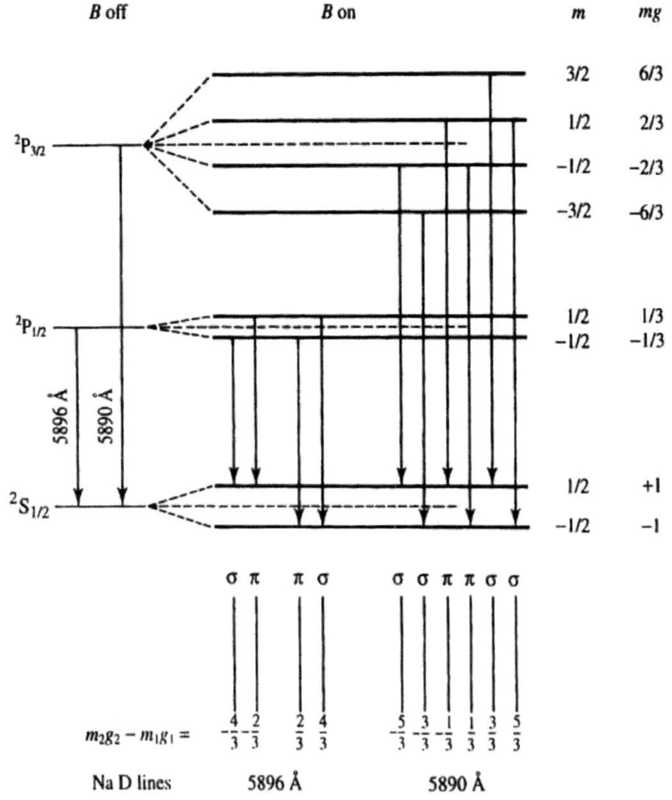

FIGURE 4.18
Zeeman effect of the D lines in sodium.

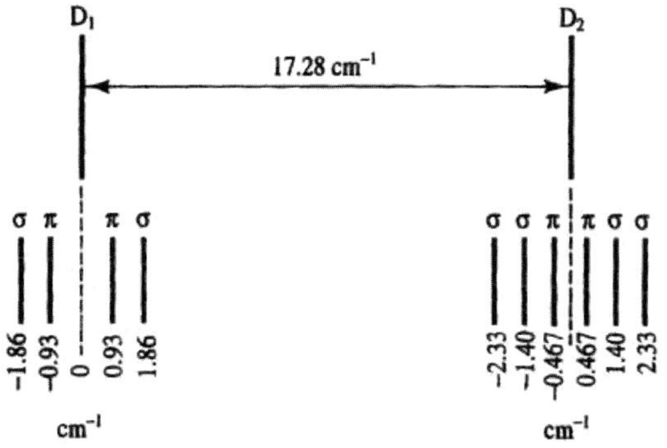

FIGURE 4.19
Zeeman splitting of the D lines in sodium in a magnetic field of 3 T. Here $\Delta\tilde{\nu} = 1/\lambda_1 - 1/\lambda_2 = 17.28\,\mathrm{cm}^{-1}$.

level, with equal values of m situated on the same vertical line. Then the transitions $\Delta m = 0$ (π polarization) are designated by vertical lines; the transitions $\Delta m = 1$ (σ^+ polarization) are designated by lines slanting towards the left; and the transitions $\Delta m = -1$ (σ^- polarization) by lines slanting towards the right. The Grotrian diagram contains no information about the energies of the transitions. It is a quick way to tell only the number and polarization of the lines and the changes in their m values. The energies of each transition are different and are determined by $(m_2 g_2 - m_1 g_1)$; see Fig. 4.18. Transitions which would be designated by lines which are not parallel to these three lines (Fig. 4.20b) are forbidden.

4.5.5 Supplementary Note 2: Paschen–Back Effect

A weak-magnetic-field assumption is implied in the interpretation of the anomalous Zeeman effect by using Eq. (4.48) or Eq. (4.46), such that the magnetic field applied is not strong enough to decouple the spin–orbit coupling. In that case, the vector representing the spin momentum and the vector representing the orbital momentum precess rapidly around their resultant J while J precesses slowly around the external magnetic field (Fig. 4.21(a)). So Eq. (4.48) is meaningful and the anomalous Zeeman effect occurs. In the presence of a strong magnetic field (Fig. 4.21(b)), the spin angular momentum and the orbital angular momentum precess individually around the external field, and are no longer coupled to form J. Then Eq. (4.48) is meaningless. Indeed, the anomalous Zeeman effect is not observed in a strong magnetic field and a different effect discovered by F. Paschen and E. Back, called the Paschen–Back effect, is observed instead. The strength of the magnetic field to decouple the spin and orbital vectors is discussed in the following section in order to give the problem a more quantitative description when Fig. 4.21(a) is replaced by 4.21(b). We must also see what must be substituted for Eq. (4.48).

The precession frequency of the angular momentum J around the external magnetic field may be estimated from the Larmor frequency. From Eqs. (4.54) and (4.55), we have

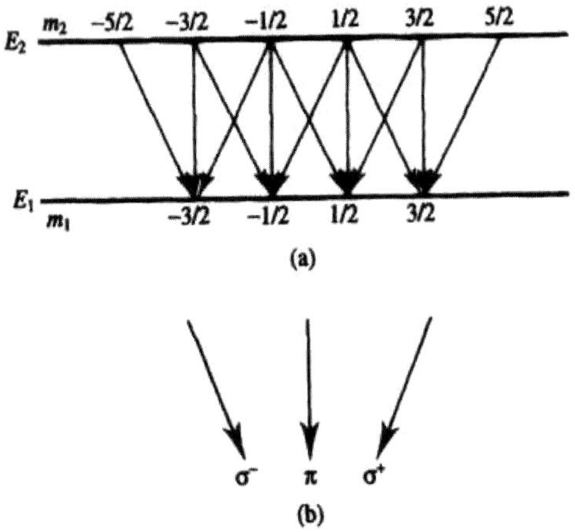

FIGURE 4.20
Grotrian diagram.

$$\nu_L = 14\,\text{GHz}\times B(\text{T}) = 1.4\times 10^6\, B(\text{G})\cdot \text{s}^{-1} \tag{4.59}$$

On the other hand, the precession frequency of the spin and the orbital angular momentum around J depends on the spin–orbit interaction energy, and may be estimated from Eq. (4.42) or Eq. (4.44). For example, for the D lines of the doublet in sodium where $\Delta\lambda = 6\,\text{Å}$, we may calculate from Eq. (4.44) that $\Delta\nu = 5\times 10^{11}$ Hz. When the external field is, generally, not larger than several tesla, $\Delta\nu$ is much larger than ν_L, as may be seen from Eq. (4.59), and the weak-field approximation is valid. It is correct to use Fig. 4.21(a) and the anomalous Zeeman effect will be observed (see Fig. 4.20).

For some value of $\boldsymbol{B}$, when the external field is sufficiently strong as to make ν_L exceed $\Delta\nu$, or to be of the same order of magnitude as $\Delta\nu$, Fig. 4.21(a) will be replaced by 4.21(b); then, the formula corresponding to Eq. (4.45) would be

$$U = -\boldsymbol{\mu}\cdot\boldsymbol{B} = \frac{e}{2m_e c}(g_s\boldsymbol{S} + g_l\boldsymbol{L})\cdot\boldsymbol{B} \tag{4.60}$$

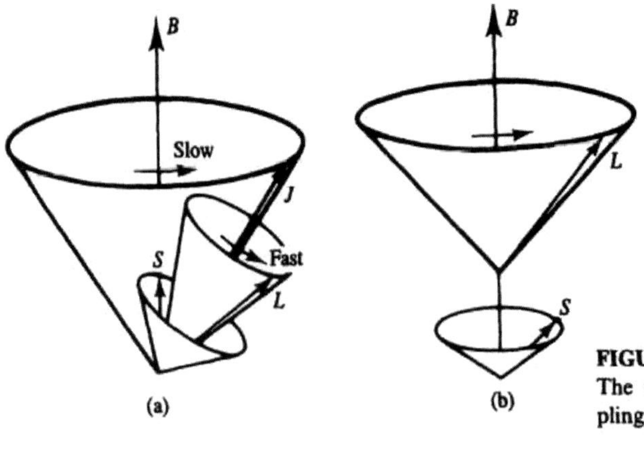

FIGURE 4.21
The vector model of spin–orbit coupling.

We may note that $\boldsymbol{\mu} = (e/2m_e c)(g_s \boldsymbol{S} + g_l \boldsymbol{L})$ is a precise expression for the magnetic moment of an electron while $\boldsymbol{\mu} = -(e/2m_e c) g_j \boldsymbol{J}$ is the average magnetic moment under the weak-field condition. See Section 4.3 and Fig. 4.8 for a better understanding. From (4.60) we have

$$U = \frac{eB}{2m_e c}(2S_z + L_z) = \frac{e\hbar B}{2m_e c}(2m_s + m_l) = \mu_B B(2m_s + m_l) \tag{4.61}$$

which is the expression for the potential energy in the presence of a strong magnetic field. The result based on this is called the Paschen–Back effect. From the selection rules

$$\Delta m_s = 0, \qquad \Delta m_l = 0, \pm 1 \tag{4.62}$$

we may readily discover the results for the transitions as ΔU approaches the normal Zeeman effect. In order to explain the Paschen–Back effect, we compare the splitting transitions between the single-electron states 3P and 3S in the weak-field case (the anomalous Zeeman effect, Fig. 4.22) and in the strong-field case (the Paschen–Back effect, Fig. 4.23).[8]

The 3S state with $m_s = \pm\frac{1}{2}$, $m_l = 0$, Eq. (4.61) is split into two components, $2m_s + m_l = \pm 1$. The 3P state with $m_s = \pm\frac{1}{2}$, $m_l = 0, \pm 1$, should have $2 \times 3 = 6$ components, but the U values corresponding to $m_s = \frac{1}{2}$, $m_l = -1$, $2m_s + m_l = 1 - 1 = 0$, and $m_s = -\frac{1}{2}$, $m_l = 1$, $2m_s + m_l = -1 + 1 = 0$ are equal to one another. Their degeneracy remains unlifted, so that 3P is split into only five components. From the selection rules, Eq. (4.62), six transitions between the two states are allowed, but there are only three different energy values, as seen in Fig. 4.23, since the shifts are all integer multiples of $\mu_B B$. Hence, only three spectral lines are observed and one of them coincides with the unshifted frequency component. In other words, under the interaction of a strong magnetic field, the anomalous Zeeman effect will be replaced by the Paschen–Back effect, which looks like the normal Zeeman effect. The energies of these three transitions where the unshifted energy is $h\nu$ are $h\nu$ and $h\nu \pm \mu_B B$. For a 10 T field, the energy shift is $0.5788 \times 10^{-4}\,\text{eV} \cdot \text{T}^{-1} \times 10\,\text{T} = 5.8 \times 10^{-4}\,\text{eV}$. Note that for sodium the unshifted energy refers to the energy difference between the 3P and 3S states without the spin–orbit coupling splitting.

Because of spin–orbit coupling, the three energy levels with $m_l \neq 0$ in Fig. 4.23 would shift either upwards ($\boldsymbol{S}$ and $\boldsymbol{L}$ in parallel approximately) or downwards ($\boldsymbol{S}$ and $\boldsymbol{L}$ in antiparallel approximately) from the original sites (as indicated by the dashed lines) according to Eq. (4.41). Obviously, two of the three spectral lines split, after taking the spin–orbit coupling into consideration. So five lines would be seen in experimental observations with a high-resolution spectrometer. The double-layer structure of the 3P state disappears completely in a strong magnetic field. By referring to Fig. 4.23, and Eqs. (4.61) and (4.41), readers are invited to plot an energy level transition diagram of the doublet in sodium for the Paschen-

[8] K. W. Ford, *Classical and Modern Physics*, Wiley, New York (1972).

Back effect with the intervals between the spectral lines quantitatively labeled analogously to Fig. 4.13.

However, we must point out that it is meaningful only in a relative sense whether a magnetic field is "strong" or "weak." As we have seen in Fig. 4.19, a magnetic field of $B = 3\,\mathrm{T}$ is called weak because in this case the Zeeman splitting of the doublet in sodium is much smaller than the interval of the doublet itself. But the same magnetic field might be strong in other cases. For example, for the first spectral line in the principal series of lithium, the fine structure splitting from spin–orbit coupling is $0.333\,\mathrm{cm}^{-1}$ (the wavelengths of the doublet are 6707.85 Å and 6708.00 Å), while the splitting caused by a magnetic field of $B = 3\,\mathrm{T}$ is $\sim 1.4\,\mathrm{cm}^{-1}$. Since the latter is much larger than the interval of the doublet, this is a strong magnetic field and the Paschen–Back effect will be expected.

The quantum numbers characterizing an electronic state are n, l, j, and m_j in the weak-magnetic-field case and n, l, m_l, and m_s in the strong-magnetic-field case

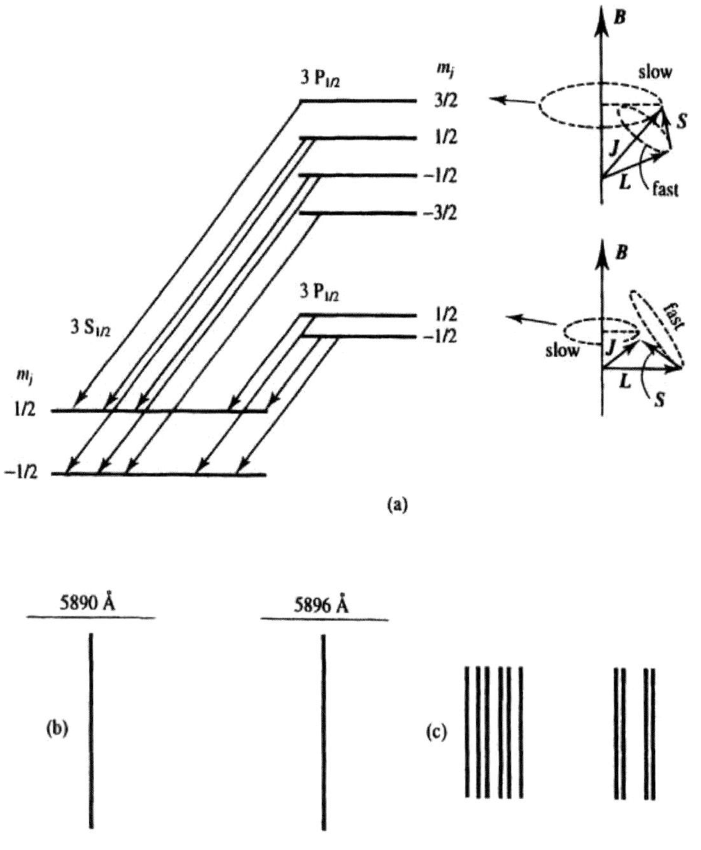

FIGURE 4.22
The energy levels and transitions of single electrons between the 3P and 3S states of sodium in a weak magnetic field, Zeeman effect. (a) Energy levels and transitions in a weak field; (b) D lines of sodium (no magnetic field); (c) Zeeman splitting of the D lines, Eq. (4.58).

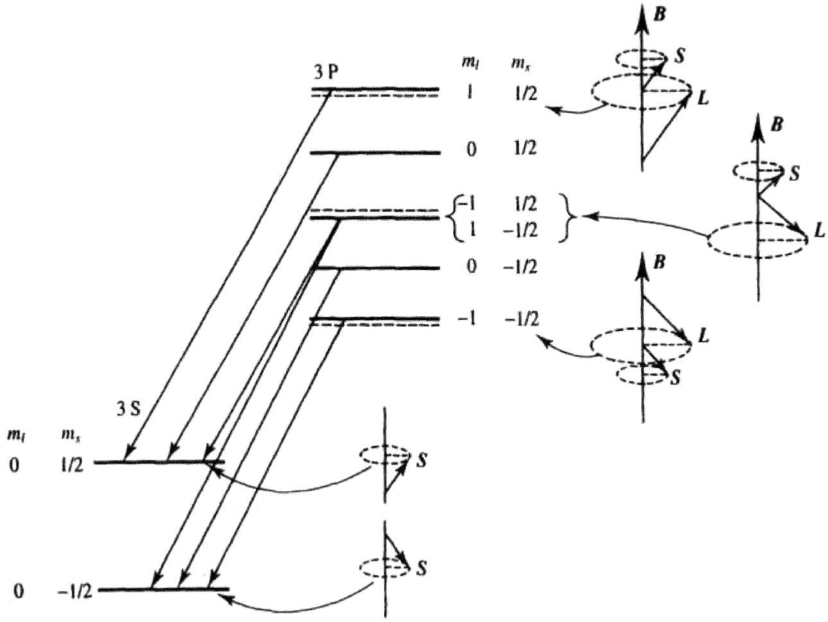

FIGURE 4.23
Energy levels and transitions between the single-electron states 3S and 3P in a strong magnetic field, Paschen–Back effect, (Eq. 4.61). The dashed lines include a spin–orbit coupling term as discussed in the text.

(see Figs. 4.22 and 4.23). These are the methods commonly used to characterize an electronic state. As we pointed out in section 4.3.1, shortly before Uhlenbeck and Goudsmit introduced electron spin into optical spectra, Pauli had suggested, on the basis of a careful inspection of the existing experimental results, that in addition to the three quantum numbers n, l, and m_l, a fourth, which took only two values, was needed in order to give the electron a complete description. But the theoretical interpretation of describing an electron state with four quantum numbers in a four-dimensional space came out only when Dirac established the foundations of relativistic quantum mechanics.

4.5.6 Conclusions

We have presented three important experimental data from three different areas. These are the Stern–Gerlach results, the doublet spectra of alkali metals, and the Zeeman effect, which have clearly shown the correctness of the hypothesis of electron spin.

The electron spin is, in fact, not like the "spin" of a gyroscope which can take any value. It is better to call it "intrinsic angular momentum." It is completely a part of the intrinsic nature (or makeup) of the electron and has nothing to do with its states of motion. One cannot stop or change the intrinsic angular momentum of a particle as the spin of a top can be changed. It has properties similar to angular momentum but it cannot be described in any classical language. It has no correspondence in classical physics. After the founding of relativistic quantum mechanics, the electron spin was no longer an assumption but a theoretical prediction. The g-factor of the electron given by the Dirac theory is 2, which coincides with the suggested value of Compton and of Uhlenbeck and Goudsmit. All seemed settled. Indeed, everyone was satisfied for about twenty years after the appearance of the Dirac theory until P. Kusch and H. M. Foley carefully measured the g-factor of the electron in 1947. With the then-new microwave method, they found that g deviated very slightly from the value of 2 (the experimental precision was five parts in ten thousand).[9–11] They had discovered the anomalous magnetic moment of the electron. The experimental result of Kusch and Foley was soon explained by J. Schwinger. In Schwinger's theory, the electron is not an isolated charged particle, because the electromagnetic field of the electron should have an effect on the electron itself. This kind of interaction is called the self-energy. The Dirac theory is incapable of calculating the self-energy. Quantum electrodynamics has to be used instead. The theoretical precision obtained by Schwinger was even greater than the experimental precision of Kusch and Foley. Since then, the precision of the g-factor of the electron has been gradually improved as a result of keen competition between theorists and experimentalists. During the process, the theoretical concepts have deepened and the experimental techniques have improved, and the winner is, of course, nothing else but physics!

The recent experimental value for the g-factor of the electron is

$$g = -2.002\,319\,304\,3662 \pm 0.000\,000\,000\,001\,5$$

Defining

$$a \equiv \frac{|g| - 2}{2}$$

we have

$$a_{\text{exp}} = 0.001\,159\,652\,1831(8)$$

while the value obtained from recent theoretical calculations is

$$a_{\text{th}} = 0.001\,159\,652\,302(112)$$

Here a_{th} may be calculated from the formula

$$a_{\text{th}} = 0.5(\alpha/\pi) - 0.328\,479(\alpha/\pi)^2 + 1.29(\alpha/\pi)^3$$

---

[9] Original papers: P. Kusch and H. M. Foley, *Phys. Rev.* **72** (1947) 1256; **74** (1948) 250.

[10] Review article: P. Kusch, *Physics Today* **19** (Feb. 1966) 23.

[11] For a popular description of the experimental measurement, see: P. Ekstrom and D. Wineland, *Science* **12** (1980) 44.

where α is the fine structure constant. Since the precision of the value of a is very high, the formula mentioned above may be used to deduce α; the disadvantage of doing so lies in the fact that the precision of α thus determined would then depend on the correctness of quantum electrodynamics. A precise, newly-found method for determining the value of α is found in papers of Klitzing, Dorda and Pepper[12] and Laughlin.[13] The difference between these is something like counting the population of the world with the precision of one person. This precision is unprecedented in natural science.

However, "Have we heard the last word about the spin?" C. N. Yang answered this question by saying "I do not believe so." It has been over 60 years since the hypothesis of spin was proposed. Yang has asked whether we are prepared to answer questions such as "Has spin a structure? Are there existing electrons of different kinds?"[14] So far, we have discovered only one kind of electron, and it has no structure down to 10^{-18} m.

4.6 SUMMARY OF THE HYDROGEN ENERGY SPECTRUM

Let us review the history of the study of the hydrogen spectrum.

Using the quantization condition of circular orbits (one dimension) and taking into account only the static electric interaction of the electrons and the nucleus, Bohr introduced an energy level diagram of the hydrogen atom (Fig. 4.24) and explained the Balmer spectral series in hydrogen in 1913. The spectral term, T (a notation from the study of spectral series where $(1/\lambda) = T(n) - T(n')$, see Eq. (3.6)) is given by the Bohr theory as

$$T = -\frac{E}{hc} = \frac{R}{n^2} \tag{4.63}$$

This was then confirmed by experimental observations such that the theoretical value of the Rydberg constant given by Bohr was in good agreement with the experimental value measured.

In 1916, on the basis of the Bohr theory, Sommerfeld considered elliptic orbits (two dimensions) and the relativistic effect of electron motion, and calculated the spectral term as

$$T = \frac{R}{n^2} + \frac{R\alpha^2}{n^4}\left(\frac{n}{k} - \frac{3}{4}\right) \tag{4.64}$$

which means that an energy level is determined by two quantum numbers (n, k) (see Fig. 4.25). The energy difference of the splitting was also calculated and was found to be in good agreement with precise measurements.

---

[12] K. V. Klitzing, G. Dorda and M. Pepper, *Phys. Rev. Lett.* **45** (1980) 494.

[13] R. B. Laughlin, *Phys. Rev.* **B23** (1981) 5632.

[14] C. N. Yang, *J. Phys. Soc. Japan* **55** (Supplement) (1986) p. 53.

In 1926, W. Heisenberg derived the spectral term by quantum mechanics as

$$T = \frac{R}{n^2} + \frac{R\alpha^2}{n^4}\left(\frac{n}{l+\frac{1}{2}} - \frac{3}{4}\right) = \frac{R}{n^2} + \Delta T_r \tag{4.65}$$

in which $k(1,2,\ldots,n)$ had been replaced by $l(0,1,2,\ldots,n-1)$. But people were surprised to learn that the calculated splitting of the spectral line, as shown in Fig. 4.26, was not in agreement with experimental measurements. So what was the problem? The problem was electron spin as introduced into atomic spectra by Uhlenbeck and Goudsmit.

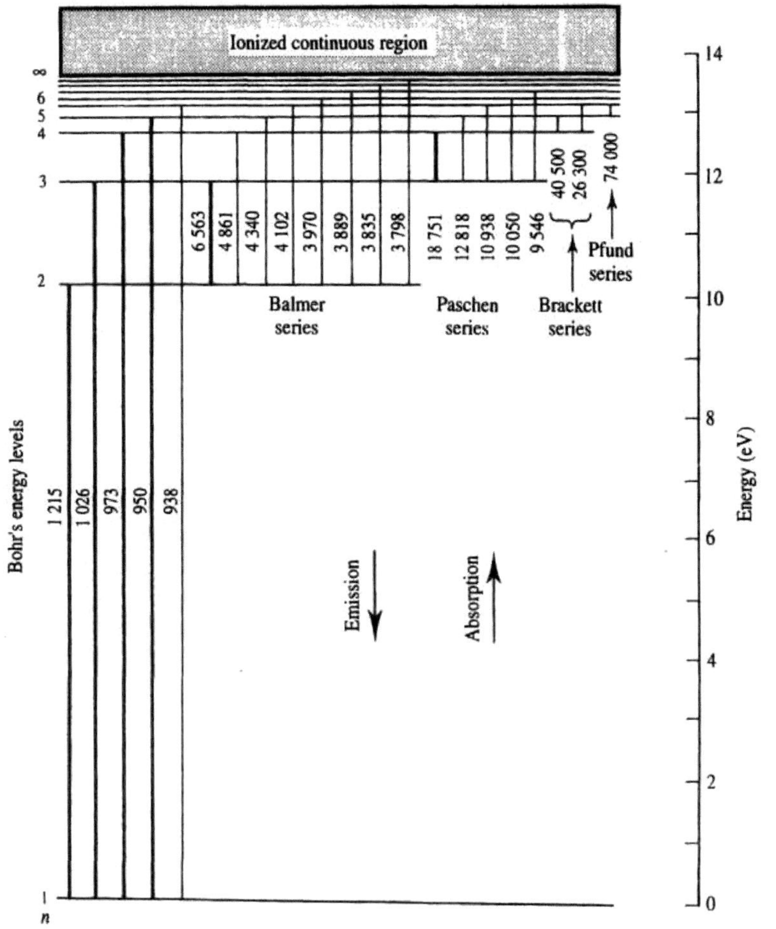

FIGURE 4.24
Energy level diagram of hydrogen given by Bohr.

In 1928, P. A. M. Dirac included electron spin in his relativistic quantum mechanics and calculated the additional term caused by the spin–orbit interaction:

$$\Delta T_{l,s} = \begin{cases} -\dfrac{R\alpha^2}{n^3 2\left(l+\frac{1}{2}\right)(l+1)}, & \text{for } j = l+\frac{1}{2} \\ +\dfrac{R\alpha^2}{n^3 2l\left(l+\frac{1}{2}\right)}, & \text{for } j = l-\frac{1}{2} \end{cases} \tag{4.66}$$

This is just Eq. (4.41) with the energy replaced by the spectral term from Eq. (4.63). Summing $\Delta T_{l,s}$ and ΔT_r given by Heisenberg, we have

$$\Delta T_r + \Delta T_{l,s} = \frac{R\alpha^2}{n^3}\left(\frac{1}{j+\frac{1}{2}} - \frac{3}{4n}\right)$$

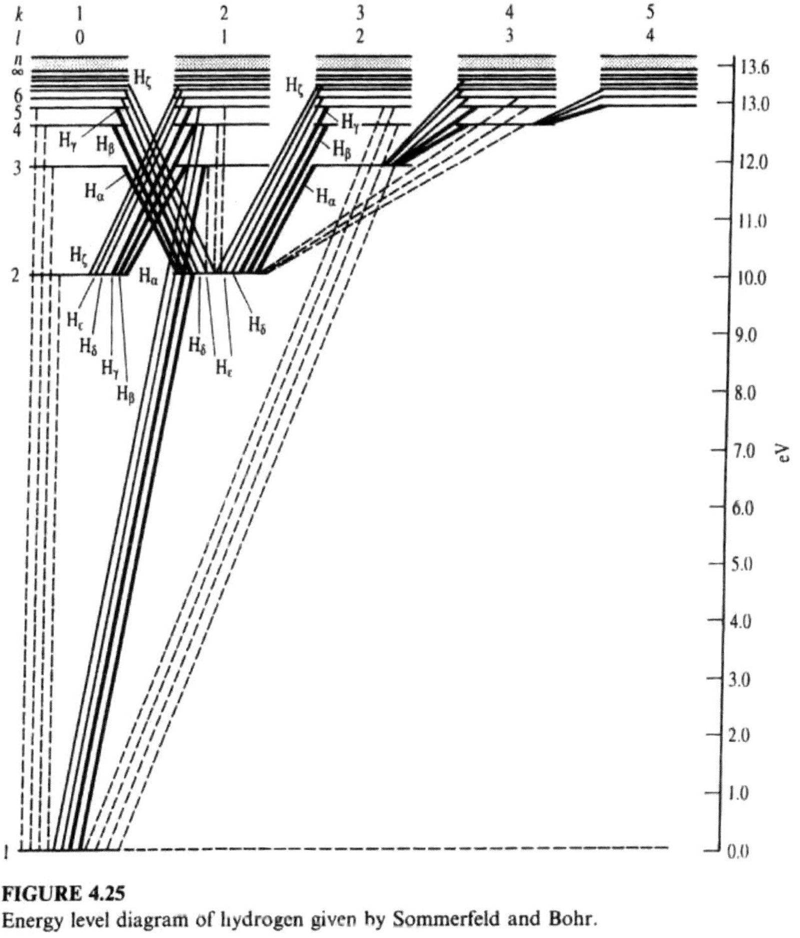

FIGURE 4.25
Energy level diagram of hydrogen given by Sommerfeld and Bohr.

Then the spectral term may be expressed as

$$T = \frac{R}{n^2} + \frac{R\alpha^2}{n^4}\left(\frac{n}{j+\frac{1}{2}} - \frac{3}{4}\right) \tag{4.67}$$

This is the summation of the term given by Bohr, and the sum of the terms of Dirac, $\Delta T_r + \Delta T_{l,s}$. The result thus obtained coincides nicely with Sommerfeld's calculation, as may be seen in Fig. 4.26 and Eq. (4.64), which means that it also fits the experimental observations well. However, their physical meanings are completely different because the expression contains a spin–orbit coupling term now. The two correct spectral terms ΔT_r and $\Delta T_{l,s}$ were not considered in the Sommerfeld theory, but some terms in them cancel each other when they are summed, which accidentally makes Sommerfeld's result fit well with the experimental results.

The Dirac theory indicates that the energy level depends only on the principal quantum number n and the total angular momentum quantum number j. It may be seen from Fig. 4.26 that the two states $^2P_{1/2}$ and $^2S_{1/2}$ with $n = 3, j = 1/2$ are degenerate, as well as the two states $^2D_{3/2}$ and $^2P_{3/2}$ with $j = 3/2$. Similarly, the two energy levels $2\,^2S_{1/2}$ and $2\,^2P_{1/2}$ of $n = 2, j = 1/2$ have no difference in energy.

However, in 1947, W. E. Lamb and his student R. C. Retherford[15] in a very precise measurement, reported a splitting interval between the two energy levels

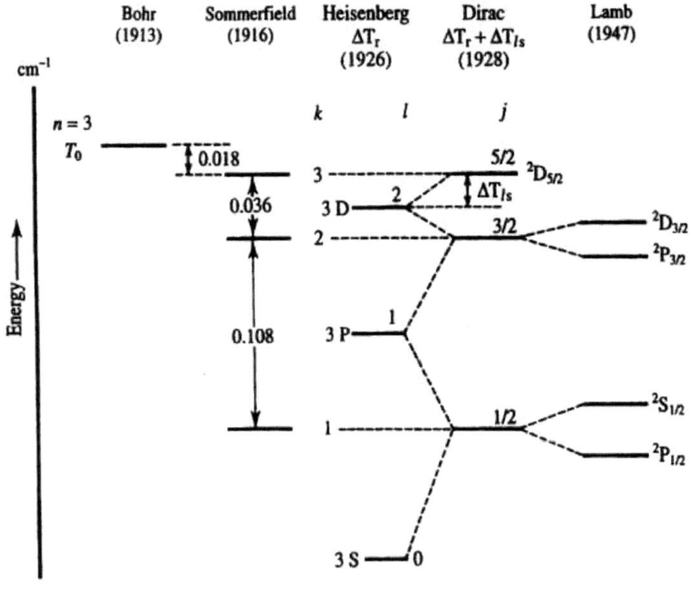

FIGURE 4.26
Evolution of the $n = 3$ level of the hydrogen atom.

[15] W. E. Lamb and R. C. Retherford, *Phys. Rev.* **72** (1947) 241.

$2\,^2S_{1/2}$ and $2\,^2P_{1/2}$, which are considered degenerate. The magnitude of the energy splitting was found to be 1057.8 MHz (4.37 μeV), see Fig. 4.27. The effect is called the Lamb shift. The shift is inversely proportional to the cube of the principal quantum number, n. It has been measured accurately so far for a few cases of $n = 2$ in H and He^+. The spectral lines of hydrogen split accordingly when this factor is taken into account. For example, the H_α line ($n=3$ to $n=2$) would be made up of seven lines[16] (see Fig. 4.27(a)), with their relative intensities shown in Fig. 4.27(b) and a high resolution spectrum with most of these lines shown in Fig. 4.27(c). The magnitude of the Lamb shift is about 1/10 of the fine structure splitting from the spin–orbit interaction for the $n = 2$ level, and it is almost negligible for the $j \neq 1/2$ levels.

The discoveries of the Lamb shift by Lamb and Retherford and of the anomalous magnetic moment of the electron by Kusch and Foley at almost the same time disclosed the limitations of Dirac's relativistic quantum mechanics. It is these two important discoveries that led to the development of quantum electrodynamics.

However, the precision of the Lamb shift, either theoretical or experimental, is still less than that of the anomalous magnetic moment of the electron. In general the theoretical and the experimental values have agreed with each other rather well since the discovery of the Lamb shift more than thirty years ago; see the review by Kugel and Murnick.[17] Not only are the experimental values and theoretical results listed there for the Lamb shift of the $n = 2$ energy level in the hydrogen atom, but the experiments and the theories of the Lamb shift in hydrogenlike ions are also reviewed. However, deviations between the experimental result and theory have appeared recently. For the $n = 2$ level in the hydrogen atom, the experimental result[18] is 1057.845(9) MHz, while recent theoretical calculations give 1057.85 (or 1057.87 depending on the choice of the proton radius) and 1057.865 (or 1057.883) MHz, respectively. Weinberg[19] notes that, given the uncertainty in the proton radius, "the agreement is excellent with the present theoretical value."

New heavy-ion accelerators will allow measurements of the Lamb shift up through hydrogenlike uranium[20,21] $_{92}U^{91+}$. In such extremely strong electric fields the very large Lamb shifts will probe quantum electrodynamics at deeper levels than ever before. Are there new effects from previously unknown phenomena that can only be seen in such ultrastrong fields? Clearly, the combination of atomic and nuclear techniques offers exciting new opportunities to delve deeper into the fundamental interactions in our universe.

---

[16] T. W. Hansch, et al., *Nature* **235** (1972) 56.

[17] H. W. Kugel and D. E. Murnick, *Rep. Progr. Phys.* **40** (1977) 297.

[18] S. R. Lundeen and F. M. Pipkin, *Phys. Rev. Lett.* **46** (1981) 232.

[19] S. Weinberg, *The Quantum Theory of Fields*, Cambridge University Press, Cambridge (1995).

[20] T. Stöhlker et al., *Phys. Rev. Lett.* **85** (2000) 3109.

[21] P. Beiersdorfer, *Rad. Phys. Chem.* **75** (2006) 1757.

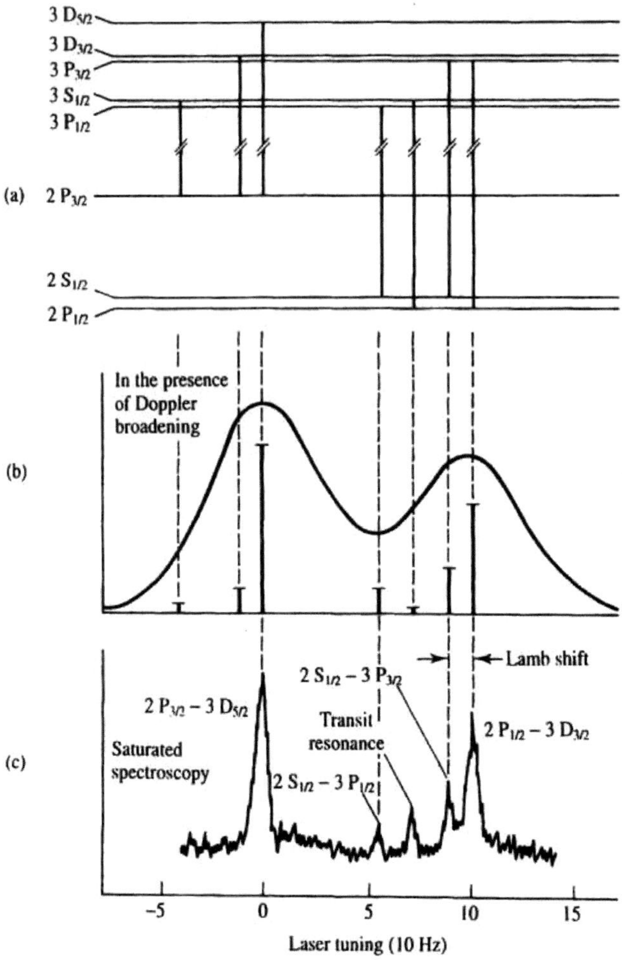

FIGURE 4.27
Fine structure of the H_α line. The seven transitions in the fine structure of the $n = 3$ to $n = 2$ transition are seen in (a). Their relative intensities are illustrated in (b) along with a poor resolution (from Doppler broadening) measurement. A high resolution measurement is shown in (c).

4.7 SUMMARY

One hypothesis. *Electron spin.* This is the most important concept introduced in this chapter. It is completely new, with no correspondence in classical physics. It has nothing to do with a particle's state of motion but is an intrinsic characteristic of the particle.

Four experiments. From different aspects, four experiments prove the existence of spin.

1. *The doublet spectra of the alkali metals*: The splitting of the spectra in the absence of an external magnetic field, because of the spin–orbit interaction of the electron in the atom. The splitting interval is given by Eq. (4.42).
2. *The Zeeman effect*: The splitting of spectral lines in the presence of a uniform external magnetic field. The splitting interval is given by Eq. (4.48), in which the g-factor is given by Eq. (4.28). These expressions are valid only when the magnetic field is weak. The Zeeman effect will be replaced by the Paschen–Back effect when the magnetic field is so strong that the magnitude of the Zeeman splitting becomes comparable to the spin–orbit interaction determined by Eq. (4.42). Then the splitting of the spectral lines will be given by Eq. (4.61).
3. *The Stern–Gerlach experiment*: The splitting of an atomic beam through its interaction with an applied inhomogeneous magnetic field. The splitting interval is determined by Eq. (4.30).
4. *Paramagnetism and ferromagnetism*: As first noted by Compton, the explanation of paramagnetism and ferromagnetism requires a magnetic moment associated with electron spin (as discussed further in Appendix 4B). Only diamagnetism can be explained without electron spin.

Four quantum numbers. n, l, m_l, and m_s or n, l, j, and m_j. Each set can give a complete description to the state of motion of an electron in an atom.

Five steps in the theoretical development of our understanding of the spectrum of the hydrogen atom. Contributions by Bohr, Sommerfeld, Heisenberg and Schrödinger, Dirac, and Schwinger.

APPENDIX 4A DIPOLE MOMENT

4A.1 ELECTRIC DIPOLE MOMENT

Shown in Fig. 4A.1 is an electric dipole placed in a homogeneous external electric field. The electric dipole moment is defined as

$$\boldsymbol{P} = q\boldsymbol{d}$$

where q is the charge on each end of the dipole, separated by the distance d.

There will be a torque acting on an electric dipole placed in a homogeneous external electric field. The torque is

$$\boldsymbol{\tau} = \boldsymbol{d} \times \boldsymbol{F} = \boldsymbol{d} \times (q\boldsymbol{E})$$

Hence we have

$$\boldsymbol{\tau} = \boldsymbol{P} \times \boldsymbol{E} \tag{4A.1}$$

Since work is done by a torque, an electric dipole will have a potential energy in a homogeneous external electrical field given by

$$U = \int_{\pi/2}^{\theta} \tau \, d\theta = -PE \cos\theta$$

The zero point of the potential energy has been defined at $\theta = 90°$ simply for convenience. Rewriting the formula given above, we have

$$U = -\boldsymbol{P} \cdot \boldsymbol{E} \tag{4A.2}$$

In fact, we may consider the two equations (4A.1) and (4A.2) as the definition of an electric dipole moment $\boldsymbol{P}$. For many systems with an unknown $q\boldsymbol{d}$, it is quite convenient to define it in such a way.

4A.2 MAGNETIC DIPOLE MOMENT

An object you may think of as an example of a magnetic dipole is a piece of permanent magnet (see Fig. 4A.2(a)). When it is put in a homogeneous external magnetic field, it will feel a torque

$$\boldsymbol{\tau} = \boldsymbol{\mu} \times \boldsymbol{B} \tag{4A.3}$$

Relating the magnet to an electric current, namely, the small coil shown in Fig. 4A.2(b), we know that it also feels a torque

$$\boldsymbol{\tau} = \frac{i}{c} \boldsymbol{S} \times \boldsymbol{B}$$

From this torque we obtain the magnetic moment of the small coil as

$$\boldsymbol{\mu} = \frac{i}{c} \boldsymbol{S}$$

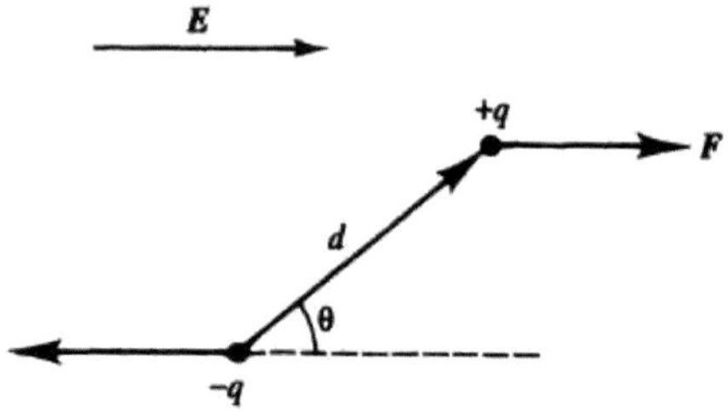

FIGURE 4A.1
Electric dipole.

Just as in the case of an electric dipole moment, the potential energy of a magnetic dipole moment placed in a homogeneous external magnetic field may be written as

$$U_B = -\boldsymbol{\mu} \cdot \boldsymbol{B} \tag{4A.4}$$

Similarly, the two equations (4A.3) and (4A.4) may be considered as the definition of a magnetic moment.

Any force may be written as minus the gradient of a potential energy, namely,

$$\boldsymbol{F} = -\Delta U = -\left(\frac{\partial U}{\partial x}\boldsymbol{i} + \frac{\partial U}{\partial y}\boldsymbol{j} + \frac{\partial U}{\partial z}\boldsymbol{k}\right)$$

Thus, the force in the z direction acting on a magnetic moment may be expressed as

$$F_z = -\frac{\partial u}{\partial Z} = \mu_x \frac{\partial B_x}{\partial Z} + \mu_y \frac{\partial B_y}{\partial Z} + \mu_z \frac{\partial B_z}{\partial Z}$$

which means that there will be a net resultant force only in an inhomogeneous magnetic field.

4A.3 FORCE AND TORQUE

We know that a force is the time rate of change of the momentum, namely,

$$\boldsymbol{F} = \frac{d}{dt}(m\boldsymbol{v})$$

while a torque will cause a change of the angular momentum, namely,

$$\boldsymbol{\tau} = \boldsymbol{r} \times \frac{d(m\boldsymbol{v})}{dt} \tag{4A.5}$$

Since

$$\boldsymbol{L} = \boldsymbol{r} \times \boldsymbol{P}$$

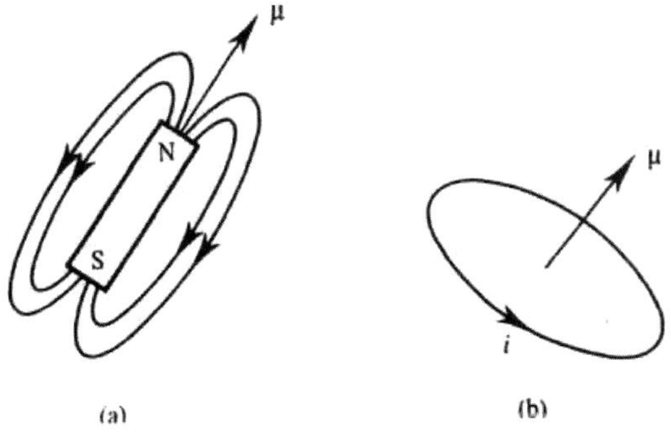

FIGURE 4A.2
Magnetic dipole.

where $\boldsymbol{P}$ is the momentum $m\mathbf{v}$ of the particle, then

$$\frac{d\boldsymbol{L}}{dt}=\frac{d}{dt}(\boldsymbol{r}\times m\mathbf{v})=\frac{d\boldsymbol{r}}{dt}\times(m\mathbf{v})+\boldsymbol{r}\times\frac{d(m\mathbf{v})}{dt}$$

Since $d\boldsymbol{r}/dt$ is in the direction of $\mathbf{v}$, the first term on the right is zero, so

$$\frac{d\boldsymbol{L}}{dt}=\boldsymbol{r}\times\frac{d(m\mathbf{v})}{dt}$$

Substituting it into Eq. (4A.5), we have

$$\boldsymbol{\tau}=\frac{d\boldsymbol{L}}{dt}$$

which means that torque is the time rate of change of the angular momentum.

APPENDIX 4B
MAGNETIC RESONANCE

4B.1 THE MAGNETISM OF MATTER

The magnetism of matter may be divided into three kinds: diamagnetism, paramagnetism, and ferromagnetism.

Suppose that the strength of the magnetic field inside a solenoid carrying an electric current is B_0 in vacuum. Then the magnetic field B will be reduced when a diamagnetic substance, like bismuth, is put into the solenoid, so $B/B_0 < 1$. On the other hand, the magnetic field will be enhanced when a paramagnetic substance, like platinum, is put into it so $B/B_0 > 1$. For a ferromagnetic substance in the solenoid, like iron, we find $B/B_0 \gg 1$.

A diamagnetic object placed in a magnetic field will repel the lines of the magnetic field, but only superconductors are completely diamagnetic and repel the lines in the way shown in Fig. 4B.1(a). A paramagnetic object placed in a magnetic field will attract the lines of the magnetic field and make the lines concentrated (see Fig. 4B.1(b)). The lines of the magnetic field will be concentrated tremendously when a ferromagnetic object is placed in the field.

What is the cause of the magnetism of matter? Consideration of this question led Compton to propose in 1921 that electrons have an intrinsic spin that gives rise to a magnetic moment.

Diamagnetism arises from electromagnetic induction in the electron orbits in an atom by an external magnetic field. It is called diamagnetism because the induced magnetic moments of the atoms are directed oppositely to the external magnetic field. Inside diamagnetic materials, the atoms have no magnetic moments, or rather, all the magnetic moments within each atom balance out so that the net moment of each atom is zero. Not only do the electrons form pairs with their spins pointing in opposite directions so that their related magnetic moments balance out, but also they pair off with their orbital angular momentum canceling so that the magnetic moments corresponding to the orbital motions of the electrons balance out exactly too, as shown in Fig. 4B.2(a). For such a system, what will happen in an external magnetic field? According to Lenz's law, turning on a magnetic field generates small extra currents inside the atom by induction. These currents are in such a direction that their magnetic fields oppose the increasing external field. For example, consider the orbital motion of the electron on the left of Fig. 4B.2. Its rotational speed will be slowed down by the induction and its magnetic moment μ will decrease. The orbital motion of the electron on the right will be increased by the induction in order to increase its magnetic moment which was originally in the direction opposite to the direction of $\boldsymbol{B}$. Hence, there will be a net magnetic moment $\mu'' - \mu'$ induced in the direction opposite to that of the external field because the two magnetic moments, which previously balanced out, no longer balance out in an external field. This is the basis of diamagnetism. We emphasize that diamagnetism is the result of a change in previously canceling moments caused by induction and not the result of previously existing permanent nonzero moments in the material itself.

The basis of paramagnetism is a different story. The prerequisite for paramagnetism is that there must exist, inside the material, atoms with permanent magnetic moments which can act independently and so tend to line up with an external magnetic field when it is applied. But what is the origin of the permanent magnetic moments?

The permanent magnetic moment is the total magnetic moment of the atom, which consists of the orbital and spin magnetic moments of the electrons and the magnetic moment of the nucleus. The contribution of the nuclear magnetic moment to paramagnetism is negligible compared with that of the electronic magnetic moments because it is three orders of magnitude smaller. For a permanent magnetic moment to exist, there must be some unpaired orbital or unpaired spin angular momenta. If the main contributor to the permanent magnetic moment were the orbital magnetic moments of the electrons then, according to Lenz's law, all materials would be diamagnetic. Lenz's law says a change in a magnetic field through a current-carrying loop induces a current to oppose the

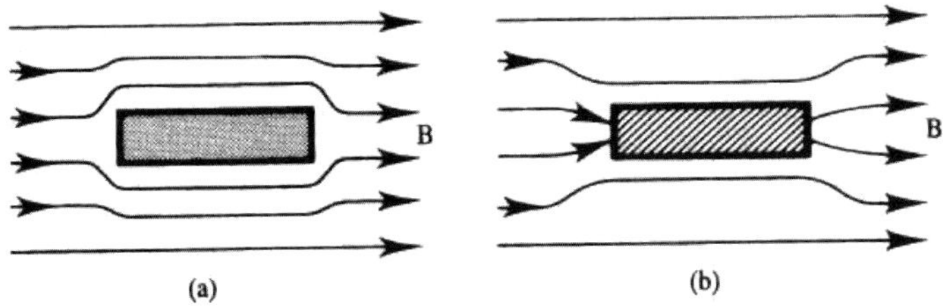

FIGURE 4B.1
(a) Diamagnetism and (b) paramagnetism.

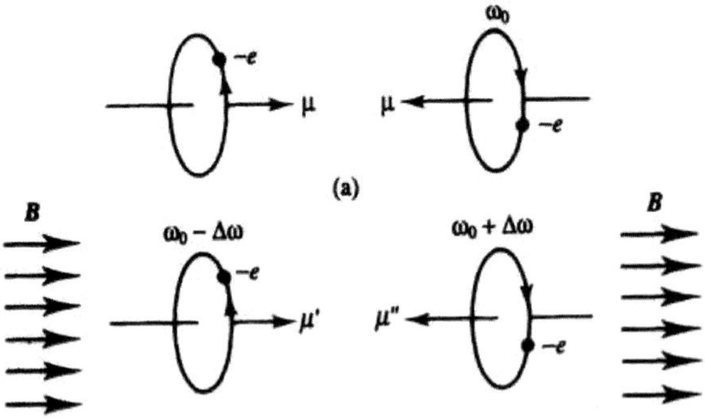

FIGURE 4B.2
Response of a diamagnetic object in an external field.

change in B. So paramagnetism comes mainly from the spin magnetic moments (the intrinsic and unchangeable magnetic moment) of the electrons. The magnitude of the spin magnetic moment cannot be changed by the electronic state of motion nor an applied external magnetic field. It is not possible to understand the paramagnetic effect from the point of view of classical physics because it is a completely quantum-mechanical phenomenon. The external magnetic field has influence only on the directions of the intrinsic (spin) moment, and in aligning them contributes to the paramagnetism.

In atoms with filled subshells, both the spin and orbital magnetic moments separately cancel in pairs. Likewise, materials where pairing of spins occurs, as in covalent crystals, are also diamagnetic. Clearly, in paramagnetic materials there must exist unpaired electrons so that the electron spins do not balance out. Hence, paramagnetic materials have atoms with their electronic subshells unfilled. Remember that, in this case, diamagnetism is present as well, but the much stronger paramagnetism dominates.

The prerequisite for a material to be ferromagnetic is also that there must exist permanent magnetic moments in the materials. The difference is that in paramagnetic materials the permanent moments are more or less isolated so that a thermal disturbance in the material destroys this alignment in the absence of an external magnetic field. In ferromagnetic materials strong coupling exists among the permanent moments, as a result of a spin exchange effect, so that the spin moments line up in the same direction over many small areas called domains. This is the so-called magnetic domain structure. Generally speaking, the directions of the magnetic moments of each domain line up in different random directions in the absence of an external magnetic field so that there is no macroscopic magnetism. But the material will be strongly magnetized in an external magnetic field because the magnetic moments of each domain have a strong tendency to line up in the direction of the field.

In summary, diamagnetism is omnipresent because it arises from Lenz's law; the spin magnetic moment is the cause of paramagnetism, which is not possible to understand in classical physics; and the magnetic domain structure is responsible for ferromagnetism.

4B.2 ELECTRON PARAMAGNETIC RESONANCE

When substances with unpaired electrons are put in an external magnetic field B, the Zeeman splitting appears because of the interaction of the spin magnetic moments of the electrons with the external field, with the energy intervals $\Delta E = g\mu_B B$. If an electromagnetic wave of frequency ν, whose energy matches the Zeeman energy inteval (i.e., $h\nu = g\mu_B B$), is applied in a direction perpendicular to that of the external field, this kind of substance will resonantly absorb energy from the electromagnetic wave. This is termed electron paramagnetic resonance (EPR), or electron spin resonance (ESR), because the paramagnetism arises from the electron spin.

Both electron paramagnetic resonance and the Zeeman effect are based on the Zeeman energy level splitting caused by the interaction of the electron magnetic moments and the external magnetic field. The Zeeman effect depends on the energy shifts in both the upper and lower states between which a valence electron makes a transition. So the transition frequencies shift only slightly because of the Zeeman level splitting in both states. However, electron paramagnetic resonance corresponds to the transitions of the unpaired electrons among the Zeeman energy levels of the ground state (see Fig. 4B.3). Moreover, electron paramagnetic resonance corresponds only to the Zeeman energy level splitting related to the electron spin magnetic moment. There is no such limitation for the Zeeman effect, where one uses thermally excited free atoms or free ions, so that the magnetic moment of the electron depends, in general, on the resultant of the electron spin and orbital magnetic moments.

The phenomenon of electron paramagnetic resonance was discovered by a Soviet scientist E. K. Tchvoilski in 1944, and the technique of electron paramagnetic resonance soon took shape. The technique is important because it is an almost unique way to detect directly unpaired electrons. Paramagnetic centers such as free radicals with unpaired electrons, the transition-metal ions, and so forth, are the active components which play important roles in the process of a chemical reaction or in the performance of materials. Studies of the structure and evolution of the paramagnetic centers by the EPR technique provide important understanding of the reaction mechanism and important clarifications of the correlation between the performance and the structure of materials. The EPR technique may be used to study photosynthesis, carcinogenic mechanisms, principles of catalysis, radiation effects, polymerization, and the intermediates of chemical exchanges and reactions. All of these are of great importance in contemporary science and technology.

For example, from the relation $h\nu = g\mu_B B$, we may measure the g-factor at resonance. If transition-metal ions are being studied, the degree of filling in the d-shell of an atom will be reflected in the value of the g-factor. Hence it is helpful to use this value to establish the valency of the ion, which is of importance to the performance of materials. For example, for a chromium oxide catalytic agent, the Cr^{5+} ions are active in polymerization of ethylene, while the Cr^{3+} ions are not.

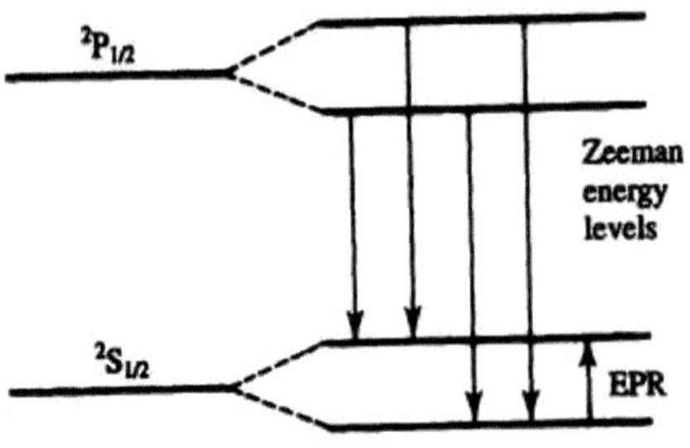

FIGURE 4B.3
Zeeman energy levels and electron paramagnetic resonance (EPR).

In addition to the g-factor, the width, line shape, and fine structure of a resonance spectral line may also be measured by the EPR technique. Each of these parameters may provide a great variety of information on the sample studied.

At the very beginning, the EPR technique was mainly used for studying free radicals, i.e., compounds containing unpaired electrons, but it soon expanded. Even for substances with no unpaired electrons, artificial methods are commonly used to initiate paramagnetic centers. Adsorption, electrolysis, pyrolysis, high-energy irradiation, oxidation–reduction, and other techniques may be used to form unpaired electrons in samples. Particularly since the spin label technique was proposed by H. McConell and co-workers in 1965, the external "paramagnetic probe" technique has been used to add a probe with unpaired electrons to molecules of a labeled substance or for diffusion into the interior of the substance. The scope of application of the EPR technique has become even wider.

4B.3 NUCLEAR MAGNETIC RESONANCE

It will be discussed in Chapter 9 that both the electrons and the nucleus possess spin. The nuclear spin gives rise to two additional phenomena. The interaction of the nuclear magnetic moment with the magnetic fields of the electron gives rise to the field of hyperfine interactions discussed in Chapter 13. The interaction of the nuclear spin magnetic moment and an external magnetic field will give rise to a "Zeeman-like" splitting of the nuclear energy levels. The magnetic resonance phenomenon corresponding to transitions among these nuclear Zeeman energy levels is called nuclear magnetic resonance (NMR). Similarly to electron magnetic resonance, the condition for nuclear magnetic resonance to occur is that $h\nu = g_N \mu_N B$, where g_N is the nuclear Landé factor and μ_N is the nuclear magneton (Section 9.3). Since μ_N is three orders of magnitude smaller than the electronic Bohr magneton μ_B, the energy intervals between the nuclear Zeeman levels are, accordingly, smaller than those of the electronic Zeeman levels. For the electron, we have

$$\frac{\nu}{B} = 14g \text{ GHz/T}$$

and for the nucleus

$$\frac{\nu}{B} = 7.6 g_N \, \text{MHz/T}$$

Taking the nucleus of hydrogen, i.e., the proton, as an example, with $g_N = 5.585\,6947$, we may calculate the values of the frequency corresponding to the different values of the magnetic field (see Table 4B.1). In the table, 1.4 T was the value of the magnetic fields used in the 1950s; 2.1 T and 2.3 T were used in the 1960s; while the value of 8.5 T was reached in the 1970s with a superconducting magnetic field. The wavelength $\lambda \simeq 5\,\text{m}$, in the meter band, is equivalent to the frequency $\nu = 6 \times 10^7\,\text{Hz}$, which corresponds to 1.4 T. In the case of electron paramagnetic resonance if $B = 1\,\text{T}$, $g \simeq 1$, we have $\nu = 1.4 \times 10^{10}\,\text{Hz}$, corresponding to $\lambda = 2\,\text{cm}$, which is in the centimeter band.

Nuclear magnetic resonance has been used widely, particularly for liquid organic compounds. Of the elements carbon, hydrogen, and oxygen contained in organic compounds, the NMR of hydrogen is the most prominent and, hence, has been most frequently studied. The abundant isotopes ^{12}C and ^{16}O of the elements carbon and oxygen, respectively, are even–even nuclei, with zero nuclear spin, so they make no contribution to the NMR. The abundance of ^{13}C is 1.11%, and of ^{17}O is only 0.038%. Their contributions to the resonance spectrum of hydrogen-containing compounds are also negligible, so hydrogen is particularly suitable for NMR analysis.

Because of the shielding of the electrons in their shells, the magnetic field acting on the nucleus is not B_{ext}, which is the external magnetic field at nuclear magnetic resonance, but $B_{\text{ext}} - \sigma B_{\text{ext}} = (1-\sigma)B_{\text{ext}}$, where σ is the coefficient of shielding. Here σB_{ext} is the amount by which the magnetic field is reduced by the effect of the inner-shell electrons in the external field B_{ext}, and so $(1-\sigma)B_{\text{ext}}$ is called the shielded magnetic field. This σ may be different for the same nucleus in different chemical environments. For example, ethylbenzene ($C_6H_5CH_2CH_3$) is composed of three functions, C_6H_5, CH_2, and CH_3. Each of them contains hydrogen, but the chemical environment of the hydrogen atoms in the different functions is different, and so is the shielding. If $B_{\text{ext.std.}}$ stands for the external magnetic field at NMR for the standard hydrogen nucleus, the external magnetic field at resonance B_{ext} for the hydrogen nucleus in different environments will deviate from $B_{\text{ext.std.}}$. We define the chemical shift as

$$\delta = \frac{B_{\text{ext.std.}} - B_{\text{ext.}}}{B_{\text{ext.std.}}} \times 10^6\,\text{ppm}$$

TABLE 4B.1

Correspondence between the NMR frequency and the external magnetic field applied

B (T)	1.4	2.1	2.3	8.5
ν (MHz)	59.6	89.4	97.9	362

which reflects the difference in electron shielding in units of ppm (parts per million). The NMR spectrum of ethylbenzene is shown in Fig. 4B.4.

The value of δ may be used as the "fingerprint" of different chemical functions. For example, for a substance of unknown chemical composition if there exists a line in the measured spectrum with $\delta = 1.22$ ppm there must be a methyl function (CH_3—) in the substance. The standard NMR spectra corresponding to various chemical functions have been established and may be used for reference in analysis. The sidebands of the spectral lines corresponding to —CH_2— (methylene) and CH_3— in Fig. 4B.4 are related to the enegy level splitting caused by the intercoupling of nuclear spin among the different chemical functions.

The development of very strong magnetic fields from superconductors and the establishment of high-resolution detectors have made it possible to observe the NMR of ^{13}C for compounds containing no hydrogen, such as CCl_4, CS_2, and others, and so to expand the scope of the applications tremendously.

In summary, for organic compounds, NMR has the advantage over EPR in the sense that it is not necessary to transform the organic compound into a free radical, so the study of the molecular structure, within the dimension of several atoms, is not limited to unpaired electrons as it is in EPR. On the other hand, the scope of the application of NMR is relatively limited as compared with EPR. The application of EPR has almost no limitation, as long as a paramagnetic center may be formed in the substance.

For the study of the interior structure of materials, many modern analytical instruments have been developed based on the absorption or emission of electromagnetic waves of different wavebands by materials and the quantum transitions between the energy levels of different types within materials. Figure 4B.5 shows the electromagnetic spectrum as function of wavelength, energy and frequency. Wavebands that cover certain regions of λ (or ν) have been given names as shown in the next to last column.

4B.4 MAGNETIC RESONANCE IMAGING IN MEDICINE

Nuclear magnetic resonance techniques now are being applied to open up remarkable new views of the interior parts and the actual dynamical working of these parts in the human body. Originally, the technique was called Nuclear Magnetic

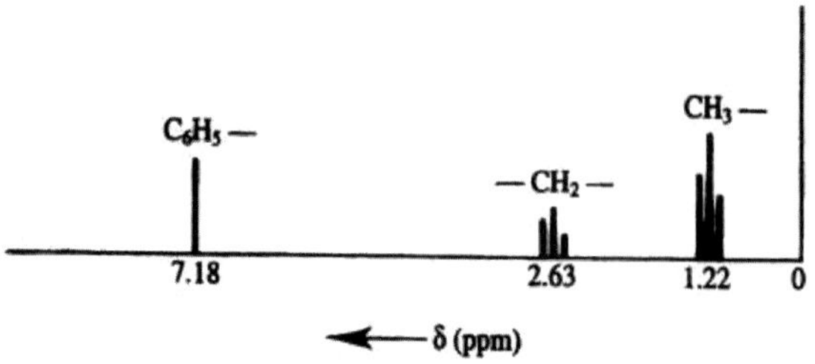

FIGURE 4B.4
NMR spectrum of ethylbenzene.

Resonance Imaging, but now simply Magnetic Resonance Imaging, MRI. It provides a powerful noninvasive technique to study the interior of the body in ways that would not have been conceived as possible even a short time ago without surgery or without using ionizing radiation such as x-rays or a radioactive dye injected into the patient. The information can often be as precise from MRI as if the doctor were directly looking at the tissue via surgery. A schematic view of a system for MRI studies is shown in Fig. 4B.6. The basic principles and operation are as follows. MRI makes use of atoms whose nuclei have a magnetic moment. As we shall see in Chapter 9, to have such a magnetic moment a nucleus must have an odd number of protons or neutrons, so the nucleus has a net spin and net magnetic moment. Such atoms include ^{1}H, ^{13}C, ^{19}F, ^{23}Na, ^{31}P and ^{39}K, which are found in biologically interesting molecules.

In the MRI system shown in Fig. 4B.6, the superconducting magnet surrounds the patient and produces a magnetic field ranging from 0.05 to 2 T ($1\,T = 10^4$ gauss, compared to the Earth's magnetic field of about 0.5 gauss). The nuclear magnetic moments of the nuclei in atoms such as ^{1}H will interact with this field to cause the moments to line up in the field. Not all will be aligned to the same degree and they will precess about the field with the Larmor frequency $\omega = \gamma B$, where γ is the gyromagnetic ratio for the nucleus and B is the external field strength. The random orientation and precession of these magnetic moments gives a net moment, M, along the direction of B.

An external magnetic field, B_1, applied perpendicular to B and rotating with the Larmor frequency will cause this moment M to change its direction. This rotation field is fortunately in the radiofrequency (RF) range. The lower frequencies (MHz) for MRI are not strongly absorbed in tissue as are the microwave frequencies for electronic Larmor frequencies, so RF radiation can be used for imaging of deep structures without tissue damage from the heating that occurs with microwaves. An RF transmitter can be applied to rotate M with respect to B; typically 90° or 180° are chosen in MRI studies for the rotation angle. When the RF field is turned off, the moment gradually returns to its original direction (spin relaxation) through thermal agitation and the electronic magnetic fields. The power of MRI is related to the fact that the relaxation rates depend on the electronic fields, which are very sensitive to subtle differences in chemical structures. Thus, differences in the relaxation rates can be used to distinguish healthy, normal functioning tissue from diseased or damaged tissue. The relaxation times are measured by a pickup coil (receiver), which gives a signal that is a function of the changing direction of M. This signal is then interpreted by the computer to give views of different areas and of different depths of the same area. But how do you view different positions, since each nucleus in a given molecule looks the same at different depths? The gradient fields shown in Fig. 4B.6 allow you to view different depths and positions. The gradient magnetic field can alter the primary field slightly as a function of distance in all three directions. Changing the primary field as a function of position changes the Larmor frequency as a function of position and so changes the signal detected.

The Vanderbilt University MRI facility with a superconducting external magnet is shown in Fig. 4B.7. Most MRI studies have been done on hydrogen. When one uses hydrogen for MRI, there is also the complication that the hydrogen may be bonded in different substances in different ways and so have different chemical environments (electronic magnetic fields). This complication is used to advantage to identify the presence and amounts of different compounds at the site of interest. More recent applications have involved other nuclei to expand the types of studies that can be made. For example, it is known that during metabo-

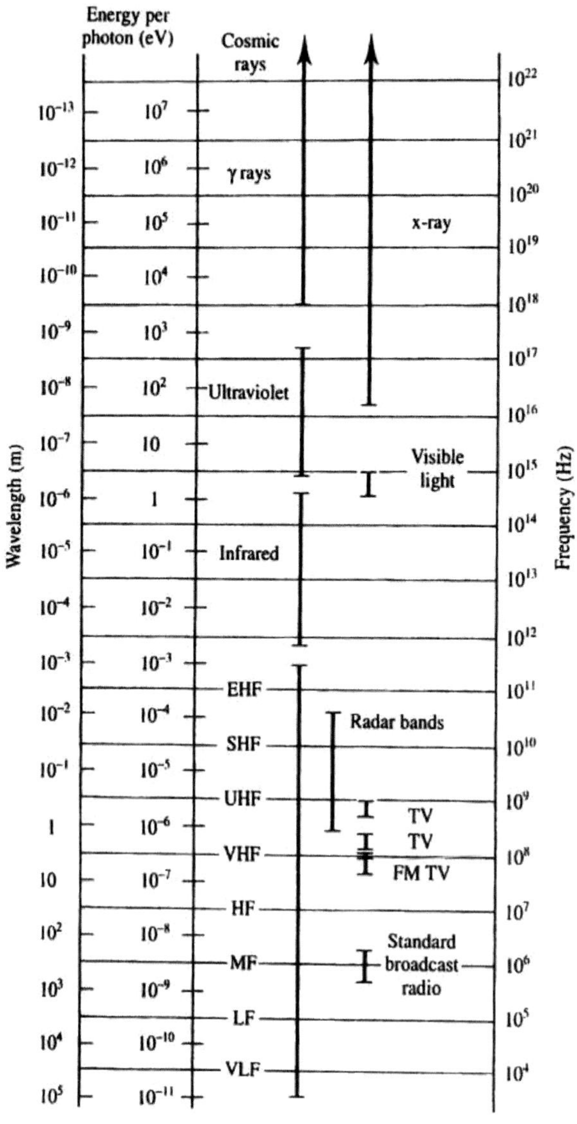

FIGURE 4B.5
The electromagnetic spectrum is identified by wavelength, energy, and frequency. The names given to certain regions of λ (or ν) are shown. Some of the terminology is for example UHF = ultra high frequency, and LF is low frequency.

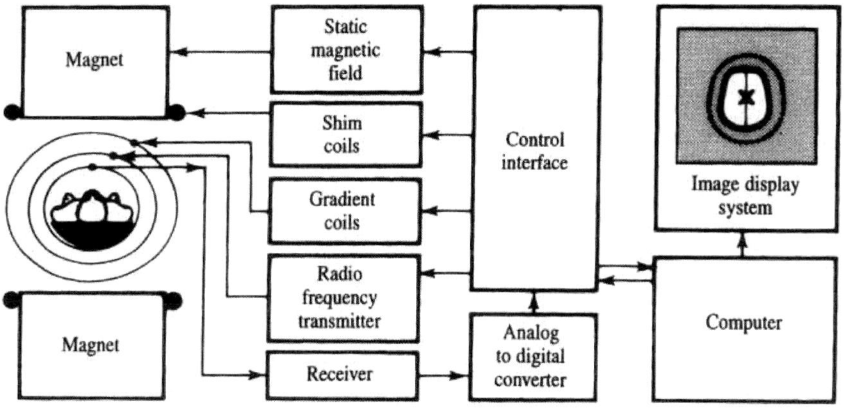

FIGURE 4B.6
Schematic of typical MRI system components. (Courtesy of Vanderbilt University Medical Center.)

lism different phosphorus compounds are used by body cells. One can determine whether the cell metabolism is normal or abnormal by measuring the relative amounts of these phosphorus compounds. In addition, much attention is given to the development of contrast agents given orally or intravenously that can enhance signals from a particular organ or region. These agents involve paramagnetic ions, paramagnetic complexes, or molecular oxygen involving ions such as gadolinium and manganese. Dynamic contrast studies may be used to study the function of an organ.

Already MRI is applicable to a wide spectrum of problems. Many tumors and other abnormalities in the brain and chest and abdominal organs can be identified, including their size and shape. Since bones do not give an MRI signal, high-quality spinal images of the brainstem and spinal cord are obtained. These can eliminate the need for myelograms in many cases. On the other hand, bone marrow does give a signal that can be used to study bone abnormality. Three different MRI scans of the head of one of the authors are shown in Figs. 4B.8 and 4B.9. In Fig. 4B.8 a sagittal (side) view of the brain is shown. The white, outer surface is the scalp tissue; the next dark area is the outer surface of the skull, the inner white area is bone marrow, and the next dark area is the inner surface of the skull. Two different top views emphasizing different features are shown in Fig. 4B.9. The detailed quality of today's MRI scans is significantly improved, especially when viewed directly on screen rather than in copies such as Figs. 4B.8 and 4B.9.

Blood flow is very important in the health of a patient. Clinical applications of MRI include measurements of blood flow in all vascular regions of the body with specific focus on neurovascular systems. MRI angiograms are routine now. In summary, MRI is now a major, new clinical diagnostic tool in medicine and is rapidly expanding into new areas.

FIGURE 4B.7
Total-body 0.5 T superconducting MRI system. (Courtesy of Vanderbilt University Medical Center.)

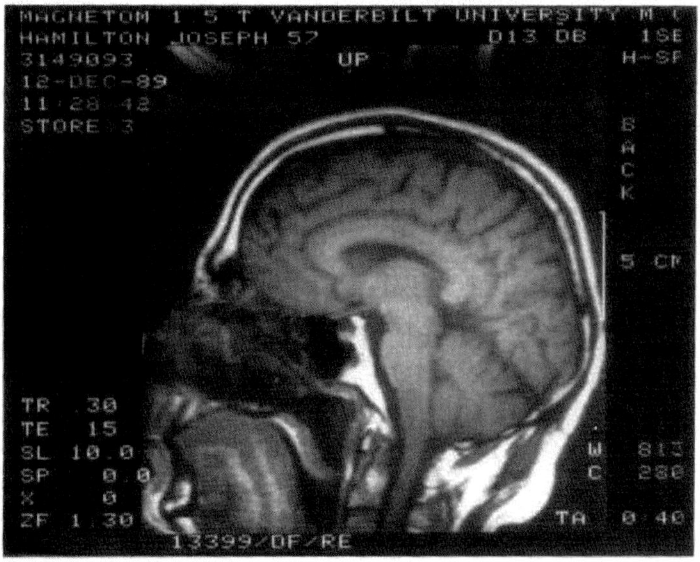

FIGURE 4B.8
Midline sagittal MRI view of a normal head (J. H. Hamilton), by using a 1.5 T superconducting system, to yield a T1-weighted high-quality anatomical image. (Courtesy of the Vanderbilt University Medical Center, Dr. Leon Partain.)

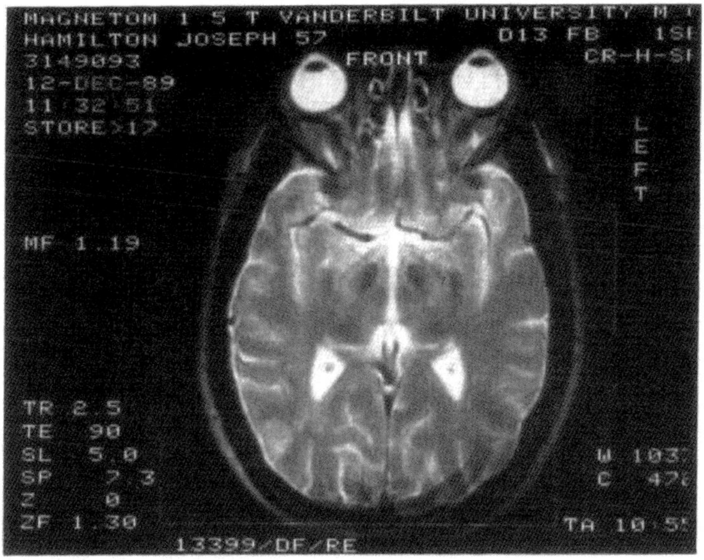

FIGURE 4B.9 (a)
Transverse MRI through the brain of a normal volunteer (J. H. Hamilton) at the level of the eyes with a 1.5 T superconducting system to yield a spin-density (proton-weighted) MRI image with somewhat different soft-tissue contrast in comparison with a T1-weighted image.

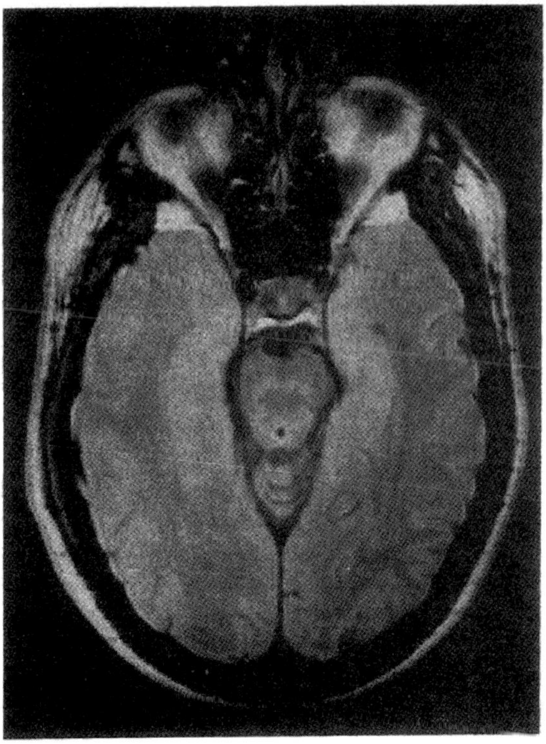

FIGURE 4B.9 (b)
Transverse MRI through the brain of the same subject taken at the same time and same level as Fig. 4B-9(a) except that the pulse system was changed to spin-echo, and the soft-tissue contrast is seen to change dramatically. (Courtesy of the Vanderbilt University Medical Center, Dr. Leon Partain.)

PROBLEMS

4.1. An electron beam goes into a uniform magnetic field of 1.2 T. What is the energy difference between the two states when the electron spin is parallel or antiparallel to the magnetic field?

4.2. An atom is in a $^2D_{3/2}$ state. What is its magnetic moment μ and the possible values of the z component of the magnetic moment μ_z?

4.3. An atom is in a $^6G_{3/2}$ state. Show that its magnetic moment is zero, and explain this fact by using the atomic vector model.

4.4. In a Stern–Gerlach experiment, a thin beam of silver atoms which are in the $^2S_{1/2}$ ground state passes through a very inhomogeneous magnetic field and reaches a screen. The length of the magnetic field region is $d = 10\,\text{cm}$. The distance between the center of the magnet and the screen is $D = 25\,\text{cm}$. The speed and mass of each Ag atom are 400 m/s and 107.87 u, respectively. The interval distance between two neighboring lines in the spectrum after the splitting is 2.0 mm. Find the value of $\partial B/\partial z$.

4.5. In a Stern–Gerlach experiment (see Fig. 4.5), $\partial B/\partial z = 5.0\,\text{T/cm}$ and the length of the magnetic field region is $d = 10\,\text{cm}$. The distance between the center of the magnet and the screen is $D = 30\,\text{cm}$. A beam of vanadium atoms in the $^4F_{3/2}$ ground state is used. Its kinetic energy is 50 meV. What is the distance between the two lines on each end?

4.6. In a Stern–Gerlach experiment, hydrogen atoms are emitted from an oven at a temperature of 400 K. They are split into two beams by the time they reach the screen. The interval distance of these two lines is 0.60 cm. If we substitute for hydrogen atoms chlorine atoms in their $^2P_{3/2}$ ground state, with the other experimental conditions remaining the same, into how many lines will the atomic beam of chlorine be split? What is the interval distance of the two neighboring lines?

4.7. The spectrum of a hydrogenlike ion exhibits the doublet structure of the principal lines in the Lyman series. Its difference in wavenumber is $29.6\,\text{cm}^{-1}$. Which hydrogenlike ion is it?

4.8. The electron in a hydrogen atom is in the 2P state. Estimate the magnetic field strength produced by the electron at the location of the proton.

4.9. Derive the normal Zeeman effect using classical theory.

4.10. A zinc atomic spectral line ($^3S_1 \rightarrow {}^3P_0$) splits because of the Zeeman effect in a magnetic field of $B = 1.00\,\text{T}$. If we look at it in the direction perpendicular to the magnetic field, how many lines will be observed? What is the wavenumber difference between the two neighboring lines? Is it a normal Zeeman effect? Draw a diagram for the transitions between the energy levels.

4.11. What is the Zeeman splitting of the D lines of the sodium doublet at a position where the magnetic field is 2.5 T.

4.12. Fine structure will be produced when an electron in a potassium atom makes a transition from the first excited state to the ground state. The corresponding wavelengths are 7664 Å and 7699 Å. If we put the atom at a position where the magnetic field is **B**, the energy levels related to these spectral lines of fine structure will make a further split.

(*a*) Give the interval distance of the splitting lines and draw a diagram of the energy levels after splitting.

(*b*) The energy difference between the highest and the lowest is ΔE_2. The initial difference is ΔE_1. If $\Delta E_2 = 1.5\Delta E_1$, what is the value of B?

4.13. If the atom is in a position where the external magnetic field $\boldsymbol{B}$ is greater than the internal magnetic field, the total spin angular momentum $\boldsymbol{S}$ and the total orbit angular momentum $\boldsymbol{L}$ will precess individually around $\boldsymbol{B}$, respectively, and $\boldsymbol{L}$ and $\boldsymbol{S}$ will no longer be coupled.

(*a*) Write the expression for the total magnetic moment $\boldsymbol{\mu}$.

(*b*) Write the expression for ΔE in such an environment.

(*c*) If the atom is sodium, calculate the energy splittings of the first excited and the ground state. Draw a diagram for the energy levels after the split, and point out which transitions are permitted according to the selection rules ($\Delta m_s = 0$, $\Delta m_l = 0, \pm 1$).

4.14. The transition 2P → 1S ($\lambda = 1210$ Å) of the hydrogen atom takes place in an external magnetic field of 4 T. If the spin–orbit interaction is neglected, what is the wavelength of this transition?

4.15. An aluminum atom has a ground state of $2P_{1/2}$. Calculate for this case the orbital angular momentum, spin, total angular momentum, and the respective magnetic moments associated with each.

4.16. If an atom has one valence electron, what are its possible g-values?

4.17. The D doublet lines of a sodium atom have wavelengths of 5890 Å and 5896 Å. What is the magnetic field induced by the electron orbital motion?

4.18. The transition $(1s3d)^1D_2 \rightarrow (1s2p)^1P_1$ in a helium atom (singlet state) produces a wavelength 6678 Å. Here 1s3d means one electron is in the $n = 1$, $l = 0$ state and the second electron in the $n = 3$, $l = 2$ state. They couple so their total spin is zero, then their total L is that of the second 3d electron and the total $J = L$, so the state is $^1D_2$ where now $^{2S+1}L_J$ is for total S, L and J. In a $B = 1.2$ T magnetic field, the Zeeman effect occurs. How many lines are there? What are their wavelengths and their polarization states for: (*a*) transverse observation? (*b*) longitudinal observation?

4.19. The transition $3\,^2P_{3/2} \rightarrow 3\,^2S_{1/2}$ in a sodium atom corresponds to a 5890 Å line. What are the variations of the wavelengths in a 1.0 T magnetic field?

4.20. In paramagnetic resonance, if the frequency of the microwave generator is 2×10^9 Hz and the magnetic field $H = 1.7 \times 10^5$ A/m, resonance is observed for Tl atoms ($^2P_{1/2}$ state). Calculate the value of the ratio e/m for the electron from this experiment.

4.21. To observe the Zeeman splitting for the D_2 line in a sodium atom ($3\,^2P_{3/2} \rightarrow 3\,^2S_{1/2}$; 5890 Å) with an optical spectrometer with resolution of $\lambda/\Delta\lambda = 105$, what is the minimum external magnetic field which must be applied?

4.22. In a magnetic field of 5.0 T, a splitting of a particular line was observed as $2\,\Delta\lambda = 0.5$ Å. What is the wavelength of this line?

4.23 Calculate the ΔU splitting in sodium using Eq. (4.42). Compare your results with the experimental results of $\Delta U = 2.1 \times 10^{-3}$ eV. Explain why these are different.

4.24. Make an energy drawing that illustrates the sharp, diffuse, and fundamental series in the alkali metals.

4.25. Explain why the maximum energy difference between the triplet transitions in the diffuse series is the same as the energy difference across the sharp series doublet transitions in the alkali metals, as seen in Fig. 4.9.

CHAPTER 5

ATOMS CONTAINING MANY ELECTRONS: THE PAULI EXCLUSION PRINCIPLE

We must expect that the eleventh electron (Na) goes into the third orbit.
Niels Bohr (1921)

No reason to expect anything; you concluded it from the spectra!!
Wolfgang Pauli (1921)

In the previous chapters, we discussed the atomic spectra of single-electron atoms, hydrogen, hydrogenlike ions, and atoms with one valence electron outside a closed shell or subshell. We also went beyond the typical energy spectra of these atoms to explain the energy fine structure of their spectra.

In this chapter we will consider an atom with two electrons (or two valence electrons) (Sections 5.1, 5.2) and introduce the Pauli exclusion principle (Section 5.3), which plays the main role in determining the placement of many electrons in an atom. From the Pauli exclusion principle, the periodic properties of the electron configurations around nuclei can be described and the physical basis for the periodic properties of the chemical elements explained (Section 5.4). The Pauli exclusion principle is the heart of this chapter. It occupies a very important place in modern physics.

5.1 THE SPECTRA AND ENERGY LEVELS OF HELIUM

Let us first discuss the simplest atom with two electrons—the helium atom. In the helium spectrum there are several spectral series, just as in the alkali spectra. However, there are two sets of helium spectra including two principal series, two primary subordinate series and two secondary subordinate series. There are obvious differences between these two sets of helium spectra. In particular, one set has only single lines, but the other set has complex structure. As described in the previous chapters, from the analysis of the spectra lines we can obtain information on the energy levels of the corresponding atom. Figure 5.1 shows the energy levels of the helium atom. It can be easily seen that there are four characteristics in the figure:

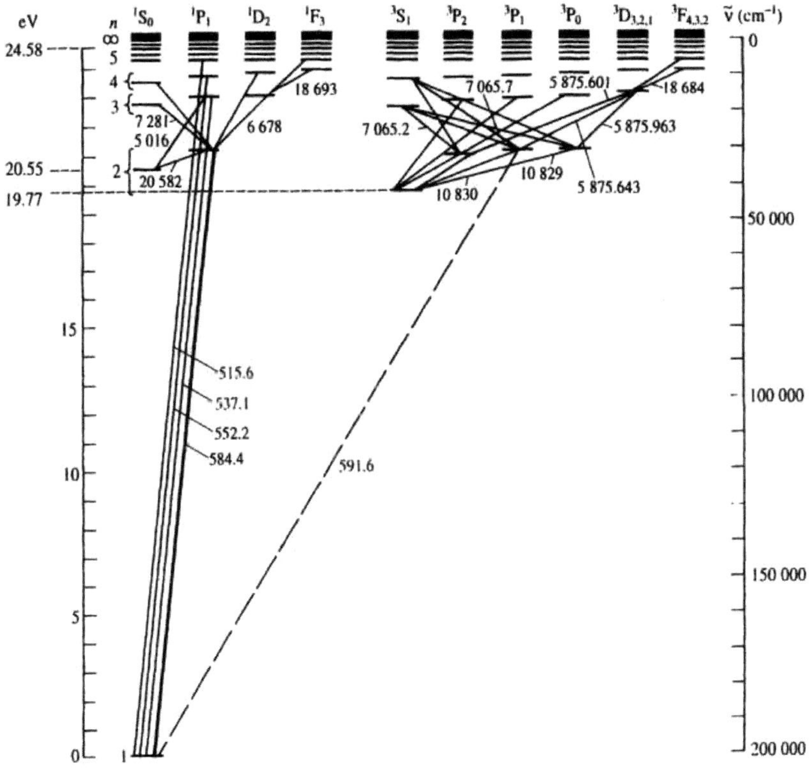

FIGURE 5.1
Energy level diagram of a helium atom

1. There exist two sets of structures and there are no transitions between the two sets. In Fig. 5.1, at first, a transition with $\lambda = 591.6\,\text{Å}$ was considered to take place between the triplet state $2\,^3P_1$ and singlet state $1\,^1S_0$ (recall that the quantum number notation is $n\,^{2S+1}L_J$), where S, L, J are sums of the individual electron spins, orbital angular momenta, and total angular momenta, but later this was shown to be from a neon contaminant. The separation into two structures leads to two independent sets of spectra. These two sets were considered to be from two kinds of helium when they were first seen. The kind with the complex spectra was called orthohelium, and the other kind with the single line spectra was called parahelium. Now it is established that there is only one kind of helium, which has two sets of energy levels.
2. There exist some metastable states. In atomic spectra, the lowest energy state is called the ground state, and all other states are called excited states. An atom in any excited state can spontaneously deexcite. But in some excited states, the atom will stay for a much longer time. Such states are called metastable states. In Fig. 5.1, the $2\,^1S_0$ and $2\,^3S_1$ are metastable states. Their long lifetimes imply that there is some selection rule which inhibits these states from decaying by spontaneous radiation.
3. The energy difference of 19.77 eV between the $1\,^1S_0$ ground state and the first excited $2\,^3S_1$ state for helium is larger than for the hydrogen atom. Also, the ionization energy (the energy to pull one electron off an atom) is the largest (24.58 eV) among all the elements.
4. In the structure of the triplet states, there is no state with $n = 1$.

These are the four characteristics of the helium spectra. They involve four physical concepts.

In the helium energy diagram shown in Fig. 5.1, the two electrons are both in the lowest state, 1s, for the ground state (recall that for a single electron the n, l, j, s are written as $n^{2s+1}l_j$ where $l = 0, 1, 2, 3$ are written s,p,d,f, and for n, l these are written nl). Excited energy levels are formed when one electron is in the 1s state, and the other is excited to one of the 2s, 2p, 3s, 3p, 3d, . . . states. This can be seen in Fig. 5.2. Of course, it is possible for the two electrons both to be in excited states. However, such states have higher energies and are populated with

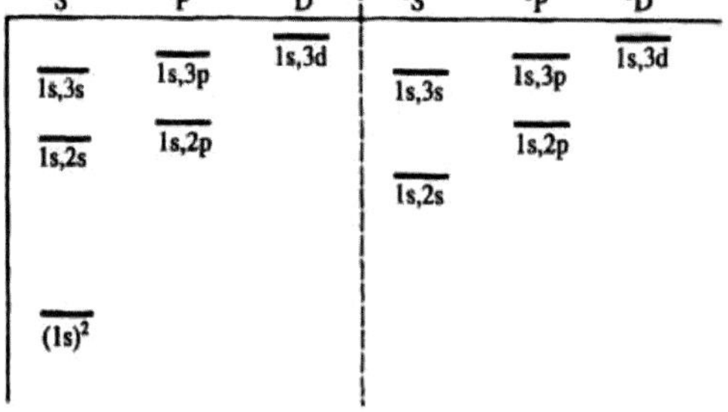

FIGURE 5.2
Energy level diagram for helium.

difficulty. For the same electron configurations, the energy level in the triplet is always lower than the corresponding one in the singlet. This can be seen for only the 1s,2s states in Figs. 5.1 and 5.2 because of the scale. The reason will be discussed in Section 5.3.

5.2 THE COUPLING OF THE TWO ELECTRONS

5.2.1 Electron Configurations

What are the electron configurations? First, consider the one electron in the hydrogen atom. For the hydrogen ground state, the electron is in the state with $n = 1, l = 0$. The notation for this electron configuration is 1s and for the ground state is $^2S_{1/2}$.

Now look at the helium atom, which has two electrons. When both electrons are in the 1s state, the configuration is expressed by 1s1s or $1s^2$. In Fig. 5.3, the possible states of the electrons are shown. In a hydrogen atom, the electron configuration in its atomic ground state is $^2S_{1/2}$. Which atomic state will be produced by a 1s1s configuration in helium will be answered in the next paragraphs.

5.2.2 *L–S* and *j–j* Couplings

Each of the two electrons in the helium atom has both orbital motion and spin. Each of these motions will produce magnetic interactions. These four motions can be described by the four quantum numbers, l_1, s_1, l_2, s_2. There are six different combinations for these four quantum numbers, which express six different types of interactions. These combinations are denoted by $G_1(s_1s_2)$, $G_2(l_1l_2)$, $G_3(l_1s_1)$, $G_4(l_2s_2)$, $G_5(l_1s_2)$, and $G_6(l_2s_1)$ where G_1 denotes the interaction between the spins of the two electrons, G_2 the interaction between the two orbital angular momenta, $G_{3,4}$, the interaction between the spin and orbital motion of the same

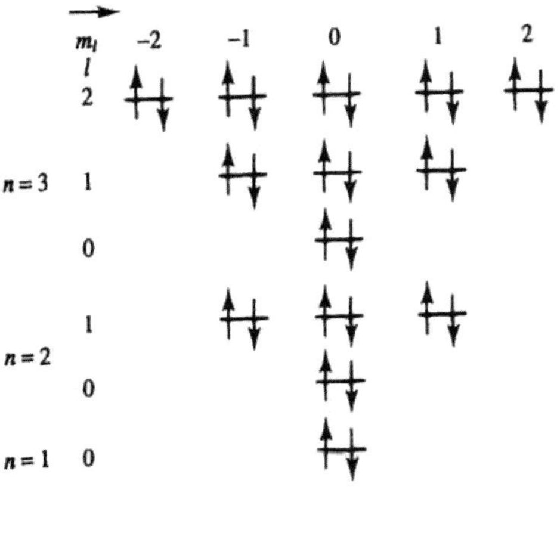

FIGURE 5.3
Configurations corresponding to different combinations of n, l, and m_l.

electron, and $G_{5,6}$, the interaction between the spin of one electron and the orbital motion of another electron. These six types of interactions have different strengths and also are different for different atoms. From physical considerations, in general, the interactions G_5 and G_6 are rather weak and can be neglected. The other four types of interactions are important.

Here we discuss two extreme cases. In one case G_1 and G_2 are dominant; that is, the spin interaction between two electrons is strong and the orbital interaction also is strong. Thus, the two spins will form a resultant total spin $\boldsymbol{s}_1 + \boldsymbol{s}_2 = \boldsymbol{S}$, and the two orbital angular momenta form a resultant total orbital angular momentum $\boldsymbol{l}_1 + \boldsymbol{l}_2 = \boldsymbol{L}$. Because the final coupling involves $\boldsymbol{L}$ and $\boldsymbol{S}$ to form $\boldsymbol{J}$, this coupling is called *L–S* coupling. In the other case G_3 and G_4 are dominant, so the spin and orbital angular momenta will first couple to their respective total angular momenta; that is, $\boldsymbol{l}_1 + \boldsymbol{s}_1 = \boldsymbol{j}_1$ and $\boldsymbol{l}_2 + \boldsymbol{s}_2 = \boldsymbol{j}_2$. Afterward these two electronic total angular momenta will combine to form the total atomic angular momentum, $\boldsymbol{j}_1 + \boldsymbol{j}_2 = \boldsymbol{J}$. Such coupling is called *j–j* coupling.

Similarly for many electrons, if the spin and orbital angular momenta are denoted by $s_1, s_2, s_3 \ldots$ and $l_1, l_2, l_3, \ldots$, *L–S* coupling can be expressed by

$$(s_1 s_2 s_3 \ldots)(l_1 l_2 l_3 \ldots) = (S, L) = J \tag{5.1}$$

and *j–j* coupling by

$$(s_1 l_1)(s_2 l_2)(s_3 l_3) \ldots = (j_1 j_2 j_3 \ldots) = J \tag{5.2}$$

It should be noted that *L–S* coupling simply implies that the interaction between the spin and orbital motion of one electron is weaker, so the main coupling comes from the interactions between the different electrons; *j–j* coupling means the spin–orbit interaction for one electron is strong, and the interactions between the different electrons are weaker.

Remember that $\boldsymbol{J} = \boldsymbol{L} + \boldsymbol{S}$ simply means the vector $\boldsymbol{J}$ is the vector sum of $\boldsymbol{L}$ and $\boldsymbol{S}$, and the triangle formed by $\boldsymbol{J}, \boldsymbol{L}, \boldsymbol{S}$ obeys the cosine law

$$J^2 = L^2 + S^2 - 2\boldsymbol{L} \cdot \boldsymbol{S} = L^2 + S^2 - 2\,|\,\boldsymbol{L}\,||\,\boldsymbol{S}\,|\,\cos\theta_{LS}$$

and its permutations of J, L, S. So for a $^3\mathrm{D}_1$ state, $J = 1$, $L = 2$, $S = 1$, the angle between J and S is

$$L^2 = J^2 + S^2 - 2\,|\,J\,|\,|\,S\,|\,\cos\theta_{JS}$$

or

$$\begin{aligned}\cos\theta_{JS} &= \frac{L(L+1) - J(J+1) - S(S+1)}{2\sqrt{J(J+1)} \times \sqrt{S(S+1)}} = \frac{6-2-2}{2\sqrt{2} \times 2} \\ &= \frac{2}{4} = 0.5 \\ \theta_{JS} &= 60^\circ\end{aligned}$$

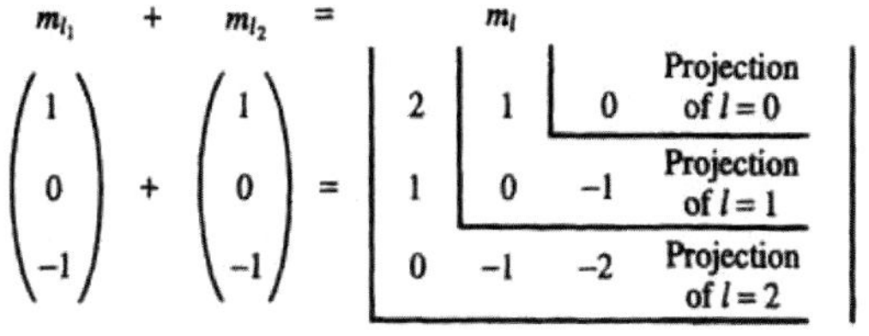

FIGURE 5.4
Coupling of two angular momenta.

5.2.3 Coupling Rule for Two Angular Moments

Let $\boldsymbol{L}_1$ and $\boldsymbol{L}_2$ be angular momenta with the quantum numbers l_1 and l_2, respectively. The values of L_1 and L_2 are

$$L_1 = \sqrt{l_1(l_1+1)}\,\hbar, \qquad L_2 = \sqrt{l_2(l_2+1)}\,\hbar \tag{5.3}$$

The summation of the two angular momenta gives $\boldsymbol{L}$,

$$\boldsymbol{L} = \boldsymbol{L}_1 + \boldsymbol{L}_2$$

where

$$L = \sqrt{l(l+1)}\,\hbar \tag{5.4}$$

and l can have only the following values from $l_1 + l_2$ to $|\, l_1 - l_2 \,|$

$$l = l_1 + l_2, l_1 + l_2 - 1, \ldots |\, l_1 - l_2 \,| \tag{5.5}$$

with an interval of 1. If $l_1 > l_2$, then there are $(2l_2 + 1)$ values. Thus, for two electrons, the total orbital angular momentum has several possible values.

Why can the values of l be taken as (5.5)? Let us consider a simple example. Take two electrons where the quantum numbers or their orbital angular momenta are $l_1 = 1$ and $l_2 = 1$, respectively. Their projections on the z direction are $m_{l_1} = 1, 0, -1$ and $m_{l_2} = 1, 0, -1$. The projection m_l of the total angular momentum is equal to the summation of m_{l_1} and m_{l_2}, as shown in Fig. 5.4. From this figure, we can see the value of l can be expressed by formula (5.5).

5.2.4 Selection Rules

In Chapters 3 and 4, it was seen that the transitions for a single-electron atom between two states involve selection rules. For an atom with several electrons, a radiative transition between two states also is limited by selection rules—in this case the selection rules for electric dipole radiation. They are given without derivation. However, note that the simplest and strongest transitions involve only one electron jumping from one state to another (two electrons jumping is a slower, higher-order process) with the emission of a photon which must carry off its intrinsic angular momentum of $1\hbar$. In higher-order processes, the photon must carry off more than one unit of angular momentum. The lowest-order radiation in quantum mechanics involves the matrix element of the electric dipole moment taken between the wavefunctions that describe the initial and final states (see

Chapter 7, Appendix C). The wavefunctions depend on quantum numbers of spin, orbital, and total angular momentum. And the matrix element of the electric dipole moment vanishes except when the initial and final quantum numbers differ by the values given below. The matrix element is a measure of the rate at which a transition occurs. Such rates can be calculated in the new quantum mechanics but not with the old Bohr quantum rules. According to the type of coupling, selection rules are divided into two types:

L–S coupling	$\Delta S = 0$
	$\Delta l = \pm 1$
	$\Delta J = 0, \pm 1$ (except no $0 \to 0$)
j–j coupling	$\Delta j = 0, \pm 1$
	$\Delta J = 0, \pm 1$ (except no $0 \to 0$)

In these selection rules, $\Delta J = 0, \pm 1$ (except $0 \to 0$) comes from $|J_f - 1| \leq J_i \leq 1 + J_f$. Also, $\Delta j = 0, \pm 1$ is easily obtained from *j–j* coupling. It will be explained in Chapter 7 why in *L–S* coupling $\Delta l \neq 0$ (this is related to the fact the photon must carry off one unit of angular momentum since its intrinsic spin is $1\hbar$, so the orbital angular momentum must change by one unit and no more to conserve angular momentum) and $\Delta S = 0$ (this is related to the Pauli Principle and the antisymmetrization of the wavefunction). Changes of S involve an oscillation of the orientation of the electron spin, angular momentum, and magnetic dipole moment. Such magnetic dipole radiation transitions have rates (probabilities of occurring) that are about 10^{-4} slower generally than electric dipole radiation.

These selection rules completely determine the spectra of the helium atom. It is just because of the rule $\Delta S = 0$ in the *L–S* coupling that no transitions can take place between the two sets of energy levels. However, the rule $\Delta S = 0$ is not strictly applicable for all atoms. This can be seen in the energy spectra of the mercury atom shown in Fig. 5.5.

In Fig. 5.5 some transitions occur between the triplet states and singlet states. This phenomenon occurs because the *L–S* coupling is not completely applicable for this case. Experimentally it is found that the $^3P_1 \to {}^1S_0$ transition with wavelength $\lambda = 2537\,\text{Å}$ is very strong. But it should be pointed out that the spectral strength observed in experiments depends not only on the corresponding transition probability but also on the number of atoms in the energy level where the transition takes place. In fact, the transition probability for $\lambda = 2537\,\text{Å}$ is relatively small.

In addition, for helium, in spite of no radiative transitions being allowed between the two sets of energy levels because of the limitation from the selection rules, helium atoms can exchange energy through collisions with each other. In collisions the radiation selection rules do not apply. Usually helium gas is a mixture of orthohelium and parahelium.

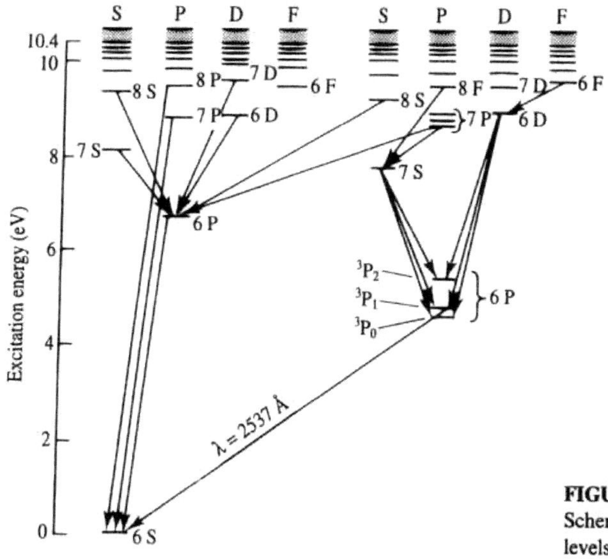

FIGURE 5.5
Schematic diagram of mercury energy levels.

5.2.5 From Electron Configurations to Atomic States

Now we consider how to form atomic states by using electron configurations. First, consider an sp configuration; that is, one electron is in an $l = 0$ state and another one is in an $l = 1$ state. The two electrons can be coupled by L–S coupling or j–j coupling.

For L–S coupling, because $s_1 = s_2 = \frac{1}{2}$, $l_1 = 0$, $l_2 = 1$, from (5.5), $S = 0, 1$ and $L = 1$. As defined previously the atomic state with $L = 1$ is called a P state. Then combine L and S to get a resultant J. From Eq. (5.5), if $S = 0$, the state with $J = L = 1$ is singlet; if $S = 1$, the states with $J = 2, 1, 0$ form a triplet. These four states can be designated by $^1P_1$, $^3P_2$, $^3P_1$, $^3P_0$. The superscript indicates the multiplicity $(2S + 1)$ and the subscript the quantum number J.

For j–j coupling, as indicated by (5.5), $j_1 = \frac{1}{2}$, $j_2 = \frac{3}{2}, \frac{1}{2}$. Thus, the resultant $J = 2, 1, 1, 0$, or $(j_1, j_2)_J = (\frac{1}{2}, \frac{3}{2})_2$, $(\frac{1}{2}, \frac{3}{2})_1$, $(\frac{1}{2}, \frac{1}{2})_1$, $(\frac{1}{2}, \frac{1}{2})_0$. Thus, J is the same for the two different couplings. We must note that the notations for atomic states are only applicable for L–S coupling. They are not applicable for j–j coupling because there are no S and L in this case.

In the above discussion of the resultant atomic states for an sp configuration, the number of states from the different couplings is the same from the geometric point of view; namely, the number of atomic states is determined completely by the electron configurations. Similarly, for an ss configuration it follows that the resultant states are $^1S_0$, $^3S_1$. The former corresponds to the two spins being antiparallel to each other, and the latter to the two spins parallel. For a pp configuration, we can have atomic states $^1S_0$, $^1P_1$, $^1D_2$, $^3S_1$, $^3P_{2,1,0}$, $^3D_{3,2,1}$.

It can be seen from these examples that for two-electron configurations the resultant states can always be divided into two kinds: one is triplet corresponding to parallel spins, and the other is singlet corresponding to antiparallel spins. This is the reason why there are two sets of transitions in the helium spectra. However, in the helium spectra, there is no $^3S_1$ state for a 1s1s configuration. Similarly, for two p electrons with the same quantum number n, only ^1S, ^1D, ^3P states occur with no ^1P, ^3S, ^3D states. Why is this? In the coupling procedures, we considered only the geometric character of the angular momenta. In order to answer this question, we must look for the physical reason. This comes from the Pauli exclusion principle.

5.3 THE PAULI EXCLUSION PRINCIPLE

5.3.1 Historical Review[1,2,3]

After he proposed a quantum theory for the hydrogen atom, Bohr sought to explain the periodic table of the elements. He realized that when an atom is in its ground state, all its electrons cannot be in the same lowest orbital according to the observed periodic features and the characteristics of the spectral lines. In particular, he pointed out that the orbitals of the helium atom are filled and thought that this filling was essentially connected to the two sets of spectral lines of the helium atom. But why is the number of electrons in one orbital limited? He guessed that only the electrons with the same quantum numbers "form harmonious interplay" with each other, so the same orbital can receive them; otherwise, there is a "disinclination to accept more electrons with other quantum numbers". But, of course, you ask why not give them all the same quantum numers?

Pauli was an outspoken critic of this farfetched explanation. When he was only 21 years old in 1921, he read Bohr's article, "The Structure Principle." When he saw "We have to expect the eleventh electron (Na) goes into the third orbit" in this article, he wrote down the comment with two exclamation points: "No reason to expect anything; you concluded it from the spectra!!"

Four years later, Pauli analyzed carefully data from atomic spectra and the Zeeman effect in a strong magnetic field and proposed his exclusion principle. It gave Bohr's interpretation of the periodic table of the elements a solid foundation. In 1940, Pauli showed that the exclusion principle is not an additional new principle for particles with half-integer spin but is a necessary result from the relativistic wave equation.

---

[1] W. Pauli, Nobel Lecture (Dec. 13, 1946), *Nobel Lectures: Physics, 1942–1962*, Elsevier, Amsterdam (1964), p. 43.

[2] W. Pauli, *Science* **103** (1946) 213.

[3] V. Weisskopf, *Physics Today* **23** (Aug. 1970) 17.

5.3.2 Exclusion Principle

Pauli[4] proposed the exclusion principle before the creation of the new quantum mechanics by Schrödinger and Heisenberg and also before he knew about electron spin. He proposed that the characterization of an electron energy state in an atom needs four quantum numbers and proposed the exclusion principle: In atoms there can be no more than one electron in each definite energy state characterized by four quantum numbers. The three known quantum numbers (n, l, m) are related to the electron moving around the nucleus; however, the fourth quantum number implies that there is a new feature of the electron. Pauli expected it to have only two values and not to be described by classical physics. After learning that Uhlenbeck and Goudsmit had proposed the assumption of electron spin, Pauli's fourth quantum number became electric spin. Compton had proposed that the electron had spin $\frac{1}{2}$ based on magnetic properties of materials (four years before Uhlenbeck and Goudsmit), but his work was evidently unknown to them. The spin quantum number m_s can have the values of $\pm\frac{1}{2}$. Four quantum numbers are needed to characterize the quantum state of an electron in an atom. The Pauli exclusion principle says: In an atom, no two electrons can have the same values for the four quantum numbers (n, l, m_l, m_s). Each electron must have a unique set of these four quantum numbers. Pauli's exclusion principle is one of the basic rules of microscopic particle motion.

This principle has an analog in classical physics, where two small balls cannot occupy the same space—namely, Newton's "impenetrability of matter." Using Pauli's exclusion principle, we can interpret the electron distribution in atoms and the periodic table of the elements. This principle can be expressed, in general, as follows:

> In a system composed of fermions (they are the microscopic particles with spin an odd multiple of $\frac{1}{2}\hbar$, such as the electron, proton, neutron, etc.), there cannot be two or more than two fermions in completely the same state, specified by the same set of quantum numbers.

What is the full physical basis of the rigorous repulsion expressed by the Pauli exclusion principle? While the Schrödinger equation can describe this mathematically, its physical basis remains an open puzzle.

5.3.3 Examples

The ground state of the helium atom. From sections 5.1 and 5.2, we know that according to the L–S coupling rule, the ground state can appear as $^1S_0$ or $^3S_1$, but that in fact only the $^1S_0$ state occurs. This is because if n, l, m_l are all the same for the two electrons, then m_s must not be the same: $m_{s_1} = m_{s_2} = \frac{1}{2}$ yielding $S = 1$ is forbidden by the exclusion principle. Therefore, no triplet $^3S_1$ can appear.

---

[4] W. Pauli, *Z für Physik* **31** (1925) 765.

In Figs. 5.1 and 5.2, we can see that the energy levels of the triplet are always lower than the energy of the corresponding singlet. For example, the 1s2s ^{3}S is lower than the 1s2s ^{1}S. So two electrons have lower energy when their spins are parallel to each other (this is also obtained from Hund's rules, see Section 5.4). Now we can use the Pauli exclusion principle to explain why two electrons with the same n, l (they are called equivalent electrons) with $l \neq 0$ prefer to have parallel spins. For equivalent electrons, if their spins are parallel (i.e., if they have the same m_s) then, from the Pauli principle, their values of m_l must be different. This corresponds to a different distribution of the direction in which each orbital angular momentum can point and likewise a difference in the spatial positions of each in orbit. The electrons prefer this difference because the two electrons repel each other. A larger spatial distance between them decreases their repulsive potential energy and makes the system more stable (lower energy). Likewise, for nonequivalent electrons, the Pauli principle makes the electrons with parallel spins possess a larger difference in spatial distribution and more stability. This will be discussed further in Chapter 7.

The helium atom has one more electron than the hydrogen atom, and hence has one more electron orbital angular momentum, one more electron spin, and corresponding magnetic moment. Thus, several types of magnetic interactions are created, but we need to point out that the electric interaction plays the main role in the dynamic properties of the helium atom. The important difference between the helium atom and the hydrogen atom comes from the electrostatic interaction between the two electrons of the helium atom. This interaction is controlled by the Pauli principle. So we can say that the helium spectrum (reflecting the dynamic features of the helium atom) is mainly determined by the Pauli exclusion principle. We give a simple explanation as follows.

In Chapter 4, we gave some estimations: The order of magnitude of the magnetic interaction is about 10^{-3} eV (for example, the separation of the Na D lines is 6 Å, corresponding to 2×10^{-3} eV), and the electrostatic interaction between two electrons separated by 0.5 Å is

$$\frac{e^2}{|\boldsymbol{r}_1 - \boldsymbol{r}_2|} \sim \frac{14.4\,\text{Å} \cdot \text{eV}}{0.5\,\text{Å}} \sim 28\,\text{eV}$$

which is much larger than the magnetic interaction. In Chapter 7, we shall see how the Pauli principle controls the size of $|\boldsymbol{r}_1 - \boldsymbol{r}_2|$. Although the electrostatic force is independent of the spin, the spin directions (parallel or antiparallel) can affect the spatial distribution of the two electrons and thereby affect the electric interaction.

The sizes of atoms. From Table 2.1 showing the radii of some atoms in Chapter 2, we see that the sizes of all atoms are nearly the same. This feature cannot be understood from classical physics and the old quantum theory. However, we can completely understand it in terms of the Pauli principle. Of course, as we shall see in Chapter 7, it is not easy to define the size of an atom. The orbits of the outer shell electrons define the size of an atom, and their orbits are very fuzzy and

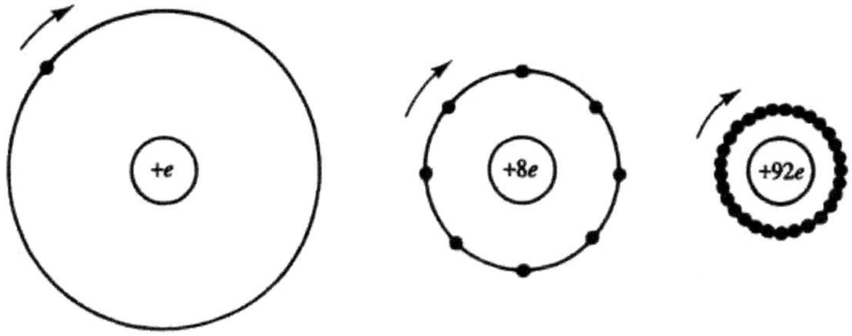

FIGURE 5.6
Bohr's orbits without the Pauli exclusion principle.

smeared out. There are different ways of determining the size of an atom. We shall discuss this more in Section 5.4.

According to Bohr's original model, the sizes of atoms would be as shown in Fig. 5.6. As the atomic number Z increases, the attractive force exerted on the electrons outside the nucleus increases and thus the distance between the electrons and the nucleus decreases. Since each electron wants to occupy the orbital that possesses the lowest energy (the $n = 1$ orbit), the attractive force exerted on each electron is the same. Therefore, with increasing Z, the radius of an atom will decrease. Of course, Bohr soon saw that the electrons were filling different orbitals. Then came the Pauli principle which does not allow all the electrons to be in the same lowest-energy orbital. So, in spite of the fact that the radius of the first orbit ($n = 1$) and each succeeding orbit becomes smaller with increasing Z, the electrons go into higher-n (larger-radius) orbits as Z increases. The net result is that the variation in the sizes of atoms with different Z is very small. There is a small periodic variation of atomic radii as explained in Section 5.4.

Electrons in metals. Metals possess an important character: In the process of heating, almost all the energy is absorbed by the nuclei. Why do the electrons not receive any heat energy? According to the Pauli exclusion principle, the arrangement of electrons in atoms is as shown in Fig. 5.7. It is very difficult to give the electrons in the lowest orbits sufficient energy to be excited, because the states near the lowest orbits are all occupied and to go to an unoccupied level requires a very large energy. A temperature of 10 000 K corresponds to only 1 eV energy, but the lattices of metals cannot survive such high temperatures. For example, aluminum melts at a few hundred degrees. Thus, only a few of the outermost electrons can receive any energy when metals are heated. However, these outermost, essentially free electrons in metals do play the important role of conducting the heat throughout the metal.

Independent nucleon motion in the nucleus. In Chapter 2, we noted that nuclei have an incredibly high density. It would seem that, in the nucleus with such high density, the nucleons would be very crowded and undergo collisions very often. Yet many nuclear experiments to be discussed in Chapter 11 show that nucleons

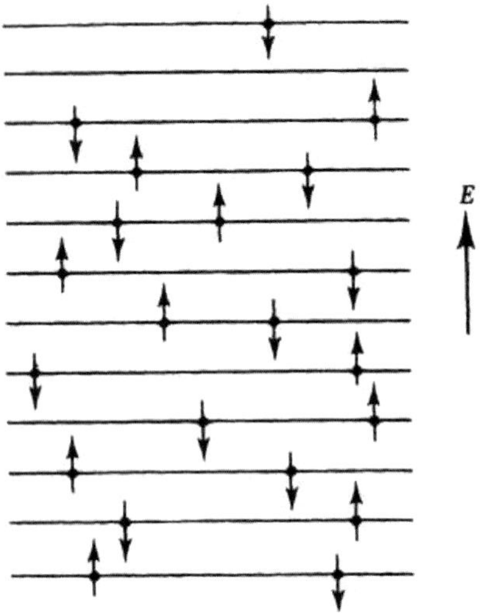

FIGURE 5.7
Filling of electron states in a metal according to Pauli exclusion principle.

move freely inside the nucleus. It seems difficult to understand how this is possible; however, it can be understood using the Pauli exclusion principle. According to that principle, because the quantized nucleon states near the ground state are occupied (filled) and the unoccupied states are far away in energy, the energy states of the nucleons cannot be changed through an energy-exchanging collision. There are no nearby empty states for the nucleons to go into, so no inelastic scattering takes place between the nucleons. Thus, the nucleons move freely.

Color quarks in nuclei. In high-energy physics quarks are considered to be the substructures of which particles such as the proton, neutron, and π meson are composed, as discussed in more detail in Chapter 14. Six types of quarks are known. Some particles are considered to be composed of three of the same type quark. Assume that the three, same-type quarks are in their ground states and they are fermions with spin $\frac{1}{2}\hbar$. For such systems with spatial quantization of spin, two quarks can have the opposite spin directions, but what is the spin direction of the third quark? It would seem that two must have the same spin direction and so appear in the same state, which violates the Pauli exclusion principle for this system. More surprising, some baryons composed of the same types of quarks have the spins of all three of their quarks pointing in the same direction so all their previously known quantum numbers are the same. Does the Pauli exclusion principle not apply to these fermions? The Pauli exclusion principle was saved by introducing a new quantum number, color, with a minimum of three different color quantum numbers required to describe the particle with three same-type quarks with aligned spins.

These five examples are given to show clearly the significance of the Pauli exclusion principle over a wide range of different physical phenomena. Imagine that there was no Pauli principle, then all the ground states of atoms would be similar, all electrons would stay at the lowest energy state, and all atoms would have the same properties. Such a world would be very dull—totally unlike our world. The variety of nature and our existence can be attributed basically to the Pauli exclusion principle.

5.3.4 The Resultant States for Equivalent Electrons

Electrons with the same two quantum numbers n and l are called equivalent electrons. Because of the effect of the Pauli exclusion principle, the atomic states formed by equivalent electrons are much fewer in number than ones formed by nonequivalent electrons. This is because many atomic states with allowed angular momenta formed by equivalent electrons are excluded by the Pauli exclusion principle. For example, two p electrons with different n can form ^{1}S, ^{1}P, ^{1}D, ^{3}S, ^{3}P, ^{3}D atomic states by L–S coupling, but for two equivalent p electrons, only ^{1}S, ^{1}D and ^{3}P can be formed. Why is this? Consider the following.

For two p electrons, with the same n, the electron configuration is np^2. By the Pauli principle, the two sets of quantum numbers (n, l, m_l, m_s) and (n, l, m_l', m_s') cannot all be the same. The m_l and m_l' have the values of $+1, 0, -1$, respectively, and m_s and m_s' have $+\frac{1}{2}$ (in simplified notation, +), $-\frac{1}{2}$ (−), respectively. If m_l and m_l' are both taken as +1, the resultant M_L is +2. Then m_s and m_s' must be different, so only one case can occur: $(1,+)(1,-)$ (the ^{1}D state) as given in the first row in Table 5.1. The ^{1}D state uses up one of the $M_S = 0$ configurations for each M_L value, 2, 1, 0, −1, −2. The $^3\mathrm{D}_1$ state would

TABLE 5.1
For an np^2 configuration, the allowed values of m_l and m_s ($\pm\frac{1}{2} = \pm$ for convenience) given as $(l_1 s_1)$ $(l_2 s_2)$ to form $M_S = m_{s_1} + m_{s_2}$ and $M_L = m_{l_1} + m_{l_2}$

M_L \ M_S	−1	0	+1
+2		$(1,+)(1,-)$	
+1	$(1,-)(0,-)$	$(1,+)(0,-)$ $(1,-)(0,+)$	$(1,+)(0,+)$
0	$(1,-)(-1,-)$	$(1,+)(-1,-)$ $(0,+)(0,-)$ $(1,-)(-1,+)$	$(1,+)(-1,+)$
−1	$(0,-)(-1,-)$	$(0,+)(-1,-)$ $(0,-)(-1,+)$	$(0,+)(-1,+)$
−2		$(-1,+)(-1,-)$	

violate the Pauli exclusion principle. The following also should be noted for two electrons. The two cases:

electron A has $m_l = 1, m_s = +\frac{1}{2}$; electron B has $m_l = 1, m_s = -\frac{1}{2}$

and

electron A has $m_l = 1, m_s = -\frac{1}{2}$; electron B has $m_l = 1, m_s = \frac{1}{2}$

are equivalent. In classical physics, two particles can always be distinguished by A and B; however, in quantum physics this is impossible for particles of the same type. Electrons are identical particles and cannot be labeled. This is one of the essential differences between classical physics and quantum physics. In Table 5.1 the ^{1}S state with $L = S = 0$ corresponds to $M_L = M_S = 0$, $m_{l_1} = m_{l_2} = 0$, $m_{s_1} = +, m_{s_2} = -$. The remaining three configurations are used up by the ^{3}P configuration which has $L = S = 1$, so $M_L = 0, \pm 1$ and $M_S = 0, \pm 1$ are needed.

If we plot Table 5.1 on the plane M_S–M_L, as in Fig. 5.8(a), each square in the figure represents different values of M_S–M_L. For example, the square in the center represents $M_S = M_L = 0$. The number in the square indicates the number of states. We can divide Fig. 5.8(a) into three figures [5.8(b), (c), (d)] and make each square stand for one state and keep the number of total states constant (to be equal to 15, see Table 5.1). Obviously, Figs. 5.8(b), (c) and (d) denote three atomic states: $L = 2$, $S = 0$ (^{1}D), $L = 1$, $S = 1$(^{3}P), $L = 0$, $S = 0$ (^{1}S) formed by an np^2 configuration. These states satisfy the Pauli principle. This method of analysis is

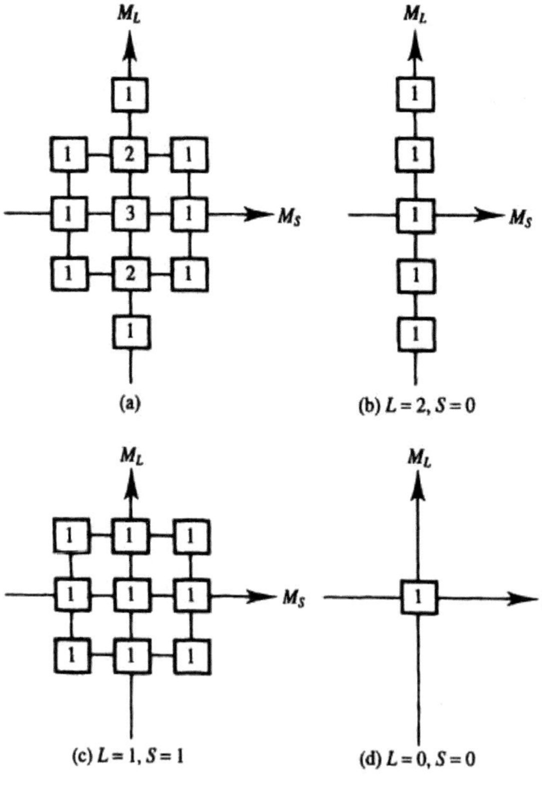

FIGURE 5.8
Diagrammatic method for determining terms of equivalent electrons (np^2 configuration).

TABLE 5.2
Terms of equivalent electrons

Electron configuration	Terms	Electron configuration	Terms
s	^2S	d, d^9	^2D
s^2	^1S	d^2, d^8	^1S, ^1D, ^1G, ^3P, ^3F
p, p^5	^2P	d^3, d^7	^2P, ^2D, ^2F, ^2G, ^2H, ^4P, ^4F
p^2, p^4	^1S, ^1D, ^3P	d^4, d^6	^1S, ^1D, ^1F, ^1G, ^1I, ^3P, ^3D, ^3F, ^3G, ^3H, ^5D
p^3	^4S, ^2P, ^2D	d^5	^2S, ^2P, ^2D, ^2F, ^2G, ^2H, ^2I, ^4P, ^4D, ^4F, ^4G, ^6S

TABLE 5.3
Terms of nonequivalent electrons

Electron configuration	Terms
ss	^1S, ^3S
sp	^1P, ^3P
sd	^1D, ^3D,
pp	^1S, ^1P, ^1D, ^3S, ^3P, ^3D
pd	^1P, ^1D, ^1F, ^3P, ^3D, ^3F
dd	^1S, ^1P, ^1D, ^1F, ^1G, ^3S, ^3P, ^3D, ^3F, ^3G

called Slater's method (first proposed by J. C. Slater[5]). Similarly, some resultant states from other equivalent electrons are listed in Table 5.2. From that table we see that p and p^5 have the same resultant states and also d^2 and d^8,.... The reason will be discussed further later. For comparison, we list some resultant states for nonequivalent electrons in Table 5.3.

5.4 THE PERIODIC TABLE OF THE ELEMENTS

5.4.1 The Periodicity of the Properties of the Elements

In 1868, L. Meyer first developed a periodic table of the elements, though he did not publish until after Dmitri Mendeleev in 1869 had proposed and published a periodic table. At that time, the table was made basically according to the order of the atomic weights. However, chemical and physical properties were important guides also. Mendeleev reversed the order of iodine ($A = 126.9$) and tellurium (127.6) from that expected from their masses because of their chemical properties.

[5] See M. Alonso and E. J. Finn, *Fundamental University Physics*, Addison-Wesley, Reading, MA (1978), vol. 3, p. 170.

In spite of the roughness of the data, the table showed a definite periodic behavior of the elements. Because only 62 elements were known at that time, the arrangement was discontinuous and some vacancies appeared. By considering this periodicity, in 1874–1875, the element scandium was discovered to be in the position between calcium and titanium. During this time, germanium and gallium were discovered to fill the two vacancies between zinc and arsenic. Although the periodic table of the elements gradually became complete, the periodicity of the properties of the elements could not be explained for fifty years.

Bohr was the first to give a physical interpretation to the periodic table. During 1916–1918, he developed the periodicity of the electronic configurations and obtained a table similar to Table 5.4 from the 1930s. According to Bohr's arrangement, the seventy-second unknown element should be similar to zirconium (Zr), while in the previous periodic table this unknown element should belong to the rare earths. In 1922 at the institute established by Bohr in the University of Copenhagen, this new element was found in zirconium ore and was named hafnium, from the Latinized form of Copenhagen, with the symbol $_{72}$Hf. Here Bohr depended on his "intuition" and the experimental data. This lack of theoretical basis was behind Pauli's early criticism of Bohr's proposals. After the Pauli exclusion principle was proposed in 1925 along with electron spin, it became clear how the periodicity of the elements comes from the periodicity of the electronic configurations and how the latter is related to the filling of the different

TABLE 5.4

Periodic table before World War II. The atomic number of the then undiscovered elements are in parentheses. (G. Seaborg, *Slack Lecture*, Vanderbilt University, 1977)

1 H																	2 He
3 Li	4 Be											5 B	6 C	7 N	8 O	9 F	10 Ne
11 Na	12 Mg											13 Al	14 Si	15 P	16 S	17 Cl	18 Ar
19 K	20 Ca	21 Sc	22 Ti	23 V	24 Cr	25 Mn	26 Fe	27 Co	28 Ni	29 Cu	30 Zn	31 Ga	32 Ge	33 As	34 Se	35 Br	36 Kr
37 Rb	38 Sr	39 Y	40 Zr	41 Nb	42 Mo	(43)	44 Ru	45 Rh	46 Pd	47 Ag	48 Cd	49 In	50 Sn	51 Sb	52 Te	53 I	54 Xe
55 Cs	56 Ba	57-71 La-Lu	72 Hf	73 Ta	74 W	75 Re	76 Os	77 Ir	78 Pt	79 Au	80 Hg	81 Tl	82 Pb	83 Bi	84 Po	(85)	86 Rn
(87)	88 Ra	89 Ac	90 Th	91 Pa	92 U	(93)	(94)	(95)	(96)	(97)	(98)	(99)	(100)				

57 La	58 Ce	59 Pr	60 Nd	(61)	62 Sm	63 Eu	64 Gd	65 Tb	66 Dy	67 Ho	68 Er	69 Tm	70 Yb	71 Lu

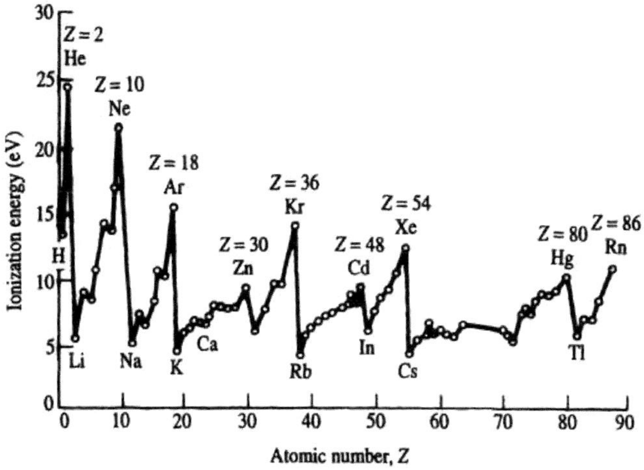

FIGURE 5.9
Ionization potentials of elements.

electron orbitals. Thus, the chemical periodicity is explained from a physical basis and chemistry became related to physics.

Figure 5.9 shows the dependence of the ionization potentials (the energy required to pull one electron off an atom) on atomic number Z. It displays obvious periodic features of the chemical properties of the elements (the values of the ionization potentials are listed in Table 5.5). In the figure, the values of Z that correspond to the peaks (high ionization potentials) were historically called magic numbers. It was because one could not understand this phenomenon at first that these Z 's were said to possess magic stability qualities. We will discuss this phenomenon in the final part of this section.

Atomic radii likewise exhibit a periodic behavior, as shown in Fig. 5.10, where the atomic radii are extracted from separation distances between atoms in crystals. This is only one way to extract the fuzzy radii of atoms. Note that atomic radii have an inverse periodic dependence with Z, so that elements with a filled electron shell have the smallest radii and elements with one electron outside a closed shell have the largest radii. This will be explained in detail by quantum mechanics (Chapter 7).

5.4.2 The Number of Electrons in the Shells

For a many-electron atom, the electron states are determined by two rules. One is the Pauli exclusion principle. The other is the minimum energy principle, which says that the system with lowest energy is most stable. The periodic table is arranged by these two rules. First, let us discuss how to group electrons into shells and how to determine the number of electrons in the shells by the first

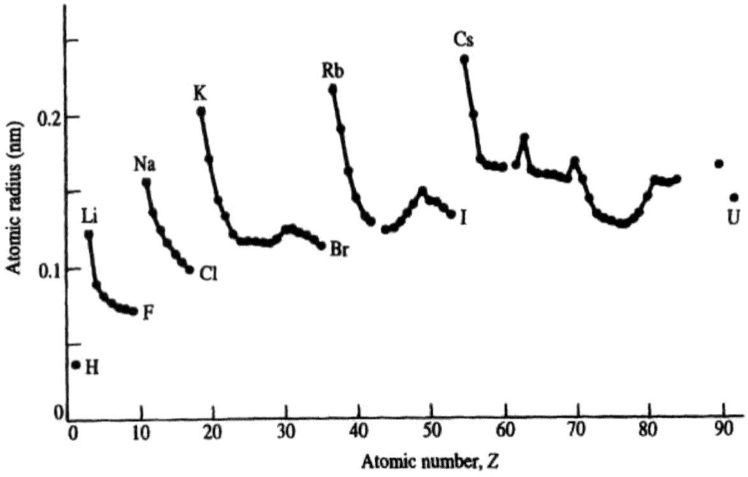

FIGURE 5.10
Trends in atomic radii. (Radii are based on ionic crystal separation.)

rule, and then consider how to determine the order of the filling of the shells by the second rule.

We know that the state of any electron in an atom is described by four quantum numbers, for example n, l, m_l, and m_s. Hence, if these four quantum numbers of an electron are given, the state of the electron is determined. For discussing atomic structure, we divide the possible atomic states into shells using the principal quantum number, n, and the angular quantum number, l. Since the energies (or energy levels) of electrons are mainly determined by the quantum number n, electrons with the same n are said to form a shell. The shells corresponding to $n = 1, 2, 3, 4, \ldots$ are named K shell, L shell, M shell, N shell, ..., respectively. In the same shell, we can have any value of l from 0 to $n-1$. Hence each shell is divided into several subshells corresponding to $l = 0, 1, 2, 3, 4, 5, \ldots$ which are denoted by the letters, s, p, d, f, g, h, and so on. Later we will see that these subshells are also further subdivided according to their different values of total angular momentum j, e.g. $P_{1/2}$, $P_{3/2}$. However, as we shall see, sometimes in the filling of shells, states with different nl can overlap in energy.

By the Pauli principle, electrons in an atom must be distributed so that each has a different set of quantum numbers. Let us see how to determine the number of electrons in a shell or subshell: For any particular value of l, there are $2l+1$ values of m_l and for each value of m_l there are two values of m_s ($m_s = +\frac{1}{2}, -\frac{1}{2}$). Hence, in a particular l subshell the maximum number of electrons allowed is

$$N_l = 2(2l+1) \tag{5.6}$$

TABLE 5.5
Atomic electron configurations, ground states, and ionization energies

Z	Symbol	Electron configuration	Ground state	Ionization potential (eV)	Radii (Å)
1	H	1s	$^2S_{1/2}$	13.598	0.53
2	He	$1s^2$	^1S	24.587	0.93
3	Li	[He]2s	$^2S_{1/2}$	5.392	1.52
4	Be	$2s^2$	$^1S_0$	9.322	1.12
5	B	$2s^22p$	$^2P_{1/2}$	8.299	0.80
6	C	$2s^22p^2$	$^3P_0$	11.260	0.77
7	N	$2s^22p^3$	$^4S_{3/2}$	14.534	0.74
8	O	$2s^22p^4$	$^3P_2$	13.618	0.74
9	F	$2s^22p^5$	$^2P_{3/2}$	17.422	0.72
10	Ne	$2s^22p^6$	$^1S_0$	21.564	1.12
11	Na	[Ne]3s	$^2S_{1/2}$	5.139	1.86
12	Mg	$3s^2$	$^1S_0$	7.646	1.60
13	Al	$3s^23p$	$^2P_{1/2}$	5.986	1.43
14	Si	$3s^23p^2$	$^3P_0$	8.151	1.17
15	P	$3s^23p^3$	$^4S_{3/2}$	10.486	1.10
16	S	$3s^23p^4$	$^3P_2$	10.360	1.06
17	Cl	$3s^23p^5$	$^2P_{3/2}$	12.967	0.97
18	Ar	$3s^23p^6$	$^1S_0$	15.759	1.54
19	K	[Ar]4s	$^2S_{1/2}$	4.341	2.31
20	Ca	$4s^2$	$^1S_0$	6.113	1.97
21	Sc	$3d4s^2$	$^2D_{3/2}$	6.54	1.60
22	Ti	$3d^24s^2$	$^3F_2$	6.82	1.46
23	V	$3d^34s^2$	$^4F_{3/2}$	6.74	1.31
24	Cr	$3d^54s$	$^7S_3$	6.766	1.25
25	Mn	$3d^54s^2$	$^6S_{5/2}$	7.435	1.29
26	Fe	$3d^64s^2$	$^5D_4$	7.870	1.26
27	Co	$3d^74s^2$	$^4F_{3/2}$	7.86	1.25
28	Ni	$3d^84s^2$	$^3F_4$	7.635	1.24
29	Cu	$3d^{10}4s$	$^2S_{1/2}$	7.726	1.28
30	Zn	$3d^{10}4s^2$	$^1S_0$	9.394	1.33
31	Ga	$3d^{10}4s^24p$	$^2P_{1/2}$	5.999	1.22
32	Ge	$3d^{10}4s^24p^2$	$^3P_0$	7.899	1.22
33	As	$3d^{10}4s^24p^3$	$^4S_{3/2}$	9.81	1.21
34	Se	$3d^{10}4s^24p^4$	$^3P_2$	9.752	1.17
35	Br	$3d^{10}4s^24p^5$	$^2P_{3/2}$	11.814	1.14
36	Kr	$3d^{10}4s^24p^6$	$^1S_0$	13.999	1.69
37	Rb	[Kr]5s	$^2S_{1/2}$	4.177	2.44
38	Sr	$5s^2$	$^1S_0$	5.695	2.15
39	Y	$4d5s^2$	$^2D_{3/2}$	6.38	1.80
40	Zr	$4d^25s^2$	$^3F_2$	6.84	1.57
41	Nb	$4d^45s$	$^6D_{1/2}$	6.88	1.41
42	Mo	$4d^55s$	$^7S_3$	7.099	1.36
43	Tc	$4d^55s^2$	$^6S_{5/2}$	7.28	1.3
44	Ru	$4d^75s$	$^5F_5$	7.37	1.33
45	Rh	$4d^85s$	$^4F_{9/2}$	7.46	1.34

TABLE 5.5 (cont.)

Z	Symbol	Electron configuration	Ground state	Ionization potential (eV)	Radii (Å)
46	Pd	$4d^{10}$	$^1S_0$	8.34	1.38
47	Ag	$4d^{10}5s$	$^2S_{1/2}$	7.576	1.44
48	Cd	$4d^{10}5s^2$	$^1S_0$	8.993	1.49
49	In	$4d^{10}5s^25p$	$^2P_{1/2}$	5.786	1.62
50	Sn	$4d^{10}5s^25p^2$	$^3P_0$	7.344	1.4
51	Sb	$4d^{10}5s^25p^3$	$^4S_{3/2}$	8.641	1.41
52	Te	$4d^{10}5s^25p^4$	$^3P_2$	9.009	1.37
53	I	$4d^{10}5s^25p^5$	$^2P_{3/2}$	10.451	1.33
54	Xe	$4d^{10}5s^25p^6$	$^1S_0$	12.130	1.9
55	Cs	[Xe]6s	$^2S_{1/2}$	3.894	2.62
56	Ba	$6s^2$	$^1S_0$	5.212	2.17
57	La	$5d6s^2$	$^2D_{3/2}$	5.577	1.88
58	Ce	$4f5d6s^2$	$^1G_4$	5.539	1.65
59	Pr	$4f^36s^2$	$^4I_{9.2}$	5.464	1.65
60	Nd	$4f^46s^2$	$^5I_4$	5.525	1.64
61	Pm	$4f^56s^2$	$^6H_{5/2}$	5.554	—
62	Sm	$4f^66s^2$	$^7F_0$	5.644	1.66
63	Eu	$4f^76s^2$	$^8S_{7/2}$	5.670	1.65
64	Gd	$4f^75d6s^2$	$^9D_2$	6.150	1.61
65	Tb	$4f^96s^2$	$^6H_{15/2}$	5.864	1.59
66	Dy	$4f^{10}6s^2$	$^5I_3$	5.939	1.59
67	Ho	$4f^{11}6s^2$	$^4I_{15/2}$	6.022	1.58
68	Er	$4f^{12}6s^2$	$^3H_6$	6.108	1.57
69	Tm	$4f^{13}6s^2$	$^2F_{7/2}$	6.184	1.56
70	Yb	$4f^{14}6s^2$	$^1S_0$	6.254	1.70
71	Lu	$4f^{14}5d6s^2$	$^2D_{3/2}$	5.426	1.56
72	Hf	$4f^{14}5d^26s^2$	$^3F_2$	6.65	1.57
73	Ta	$4f^{14}5d^36s^2$	$^4F_{3/2}$	7.89	1.43
74	W	$4f^{14}5d^46s^2$	$^5D_0$	7.918	1.37
75	Re	$4f^{14}5d^56s^2$	$^6S_{5/2}$	7.88	1.37
76	Os	$4f^{14}5d^66s^2$	$^5D_4$	8.7	1.34
77	Ir	$4f^{14}5d^76s^2$	$^4F_{9/2}$	9.1	1.35
78	Pt	$4f^{14}5d^96s^1$	$^3D_3$	9.0	1.38
79	Au	$[Xe, 4f^{14}5d^{10}]6s$	$^2S_{1/2}$	9.225	1.44
80	Hg	$6s^2$	$^1S_0$	10.437	1.52
81	Tl	$6s^26p$	$^2P_{1/2}$	6.108	1.71
82	Pb	$6s^26p^2$	$^3P_0$	7.416	1.75
83	Bi	$6s^26p^3$	$^4S_{3/2}$	7.289	1.48
84	Po	$6s^26p^4$	$^3P_2$	8.42	1.4
85	At	$6s^26p^5$	$^2P_{3/2}$		1.4
86	Rn	$6s^26p^6$	$^1S_0$	10.748	2.2
87	Fr	[Rn]7s	$^2S_{1/2}$		2.7
88	Ra	$7s^2$	$^1S_0$	5.279	2.2
89	Ac	$6d7s^2$	$^2D_{3/2}$	5.17	2.0
90	Th	$6d^27s^2$	$^3F_2$	6.08	1.65
91	Pa	$5f^26d7s^2$	$^4K_{11/2}$	5.89	—

TABLE 5.5 (cont.)

Z	Symbol	Electron configuration	Ground state	Ionization potential (eV)	Radii (Å)
92	U	$5f^36d7s^2$	$^5L_6$	6.05	1.42
93	Np	$5f^46d7s^2$	$^6L_{11/2}$	6.19	—
94	Pu	$5f^67s^2$	$^7F_0$	6.06	—
95	Am	$5f^77s^2$	$^8S_{7/2}$	5.993	—
96	Cm	$5f^76d7s^2$	$^9D_2$	6.02	—
97	Bk	$5f^97s^2$	$^6H_{15/2}$	6.23	—
98	Cf	$5f^{10}7s^2$	$^5I_8$	6.30	—
99	Es	$5f^{11}7s^2$	$^4I_{15/2}$	6.42	—
100	Fm	$5f^{12}7s^2$	$^3H_6$	6.50	—
101	Md	$5f^{13}7s^2$	$^2F_{7/2}$	6.58	—
102	No	$5f^{14}7s^2$	$^1S_0$	6.65	—
103	Lr	$6d5f^{14}7s^2$	$^2D_{5/2}$		

Source: The ionization potentials are from C. E. Moore, NSRDS-NBS34 (National Bureau of Standards, 1970); the lanthanides are from W. C. Martin, NSRDS-NBS60 (National Bureau of Standards, 1978); and the actinides are from W. C. Martin et al., *J. Phys. Chem. Ref. Data* **3** (1974) 771.

Thus we have for each l, N_l electrons in a different quantum state as shown in the table (and see Fig. 5.11):

Angular momentum l	0	1	2	3	4
Symbol	s	p	d	f	g
$N_l = 2(2l+1)$	2	6	10	14	18

For each value of n, the values of l go from 0 to $(n-1)$, so the maximum number of states with different quantum numbers in a shell of given n is

$$\begin{aligned} N_n &= \sum_{l=0}^{n-1} 2(2l+1) = 2[1+3+5+\cdots+2(n-1)+1] \\ &= 2\frac{n}{2}[1+(2n-1)] \\ &= 2n^2 \end{aligned} \tag{5.7}$$

The numbers in the last line in Fig. 5.11 are given by Eq. (5.7). Above we discussed the division of the shells and the number of electrons in a shell or subshell. But what is the order in which the electrons are filled into the shells and subshells?

	K	L		M			N			
n	1	2		3			4			
l	0 s	0 s	1 p	0 s	1 p	2 d	0 s	1 p	2 d	3 f
m_l	0	0	−1 0 1	0	−1 0 1	−2 −1 0 1 2	0	−1 0 1	−2 −1 0 1 2	−3 −2 −1 0 1 2 3
m_s	↓↑	↓↑	↓↑ ↓↑ ↓↑	↓↑	↓↑ ↓↑ ↓↑	↓↑ ↓↑ ↓↑ ↓↑ ↓↑	↓↑	↓↑ ↓↑ ↓↑	↓↑ ↓↑ ↓↑ ↓↑ ↓↑	↓↑ ↓↑ ↓↑ ↓↑ ↓↑ ↓↑ ↓↑
	K	L_1	L_2	M_1	M_2	M_3	N_1	N_2	N_3	N_4
Number of states	2	2	6	2	6	10	2	6	10	14
		8		18			32			

FIGURE 5.11
Possible states of an electron. The L_2, M_2, M_3, N_2 . . . substates are further subdivided by j in Chapter 6.

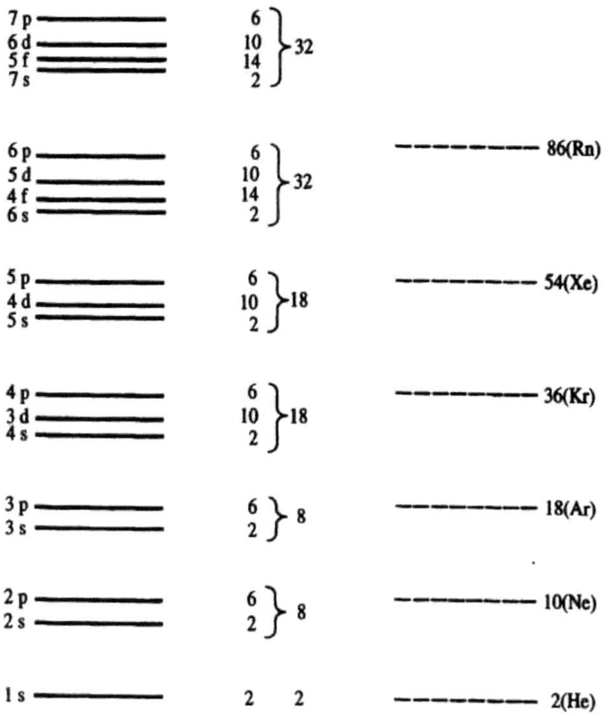

FIGURE 5.12
Electron shells of atomic levels.

5.4.3 The Energies of the Electron Configurations—The Order of the Shells

As mentioned above, the order of the filling of the shells is determined by the minimum energy principle. Based on Bohr's atomic theory, the energy increases with increasing principal quantum number n and therefore the electrons should be filled into the shells in the order of increasing n, beginning with $n = 1$. In fact, this is not always true when we look more carefully. The actual situations are shown in Table 5.5 and in Fig. 5.12. These are the orderings of the levels when the shells are being filled.

Why is the 3d level higher than 4s, so that the electrons first fill the 4s level? Let us look at Fig. 5.13. This figure shows the dependence of the electron binding energy on atomic number Z in atoms for the given electron shells. The curves in the figure come from experimental data; of course, they can be obtained from calculations, too. The energy unit in Fig. 5.13 is the Rydberg unit, Ry (1 Ry = 13.6 eV). The results in Fig. 5.13 for the 1s shell to the 3p shell are consistent with those given in Table 5.5. However, the 3d shell and 4s shell energies

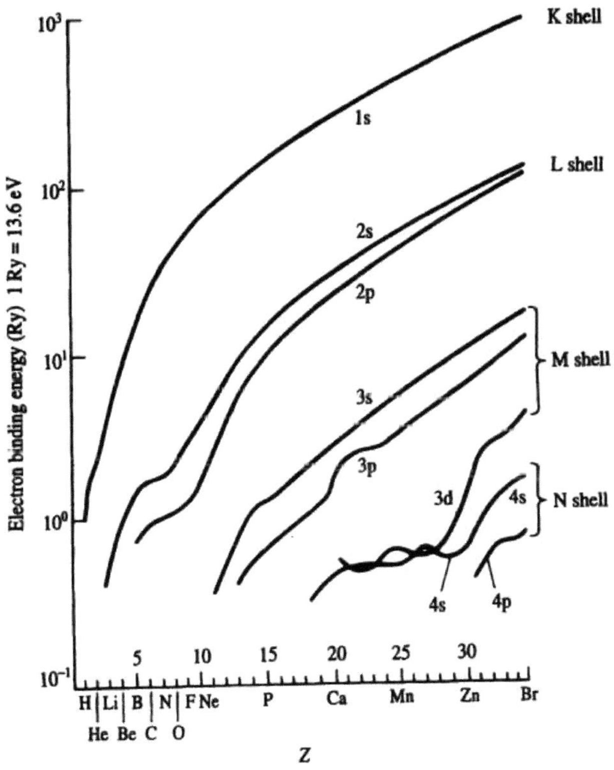

FIGURE 5.13
Electron binding energies as a function of atomic number for different shells.

are mixed between Z of about 20 to 28 even though the 4s level has a larger n than the 3d level.

It is not certain after filling the 3p subshell whether the next electron first fills the 3d or the 4s shell. For some Z values, the energy of the 4s shell is lower than that of the 3d shell, so the electrons first fill the 4s shell. However, for the heavy elements where the lower n shells are filled, the n dependence takes over again and the 3d shell is close in energy to the 3s and 3p levels, as shown in Fig. 5.14. This regular ordering by n of the energies of the levels including their subshells at higher Z is seen in the inner shell x-rays emitted by heavy elements, as described in Chapter 6. The nd and $(n+1)$s shells are mixed up only as the shells are being filled. The noble (inert) gases occur when there is a gap (large increase) in the energy between a filled shell or subshell and the next available energy state. This energy gap means that it takes considerable energy to move an electron to the next energy state and so the filled shell is stable. Also, as we find from quantum mechanics, the electron probability density for a filled shell or subshell is spherically symmetric, so there is no electric field outside the atom to produce interactions. Because of the mixing of the nd and $(n+1)$s subshells and likewise the mixing of the 4f and 6s subshells and so forth, large energy shell gaps occur after the np subshell is filled (see Fig. 5.14) and so the noble gases occur when the np subshells are filled.

The fact that the alkali metals, lithium, sodium, etc., . . . have one 3s, 4s, . . . electron whose $l = 0$ orbit extends far out from the nucleus and from the inner closed electron shells, as we shall see in Chapter 7, helps us understand the sudden large drop in binding energy of these far out electrons and the large increase in the radius of an atom with one outer ns electron. These elements

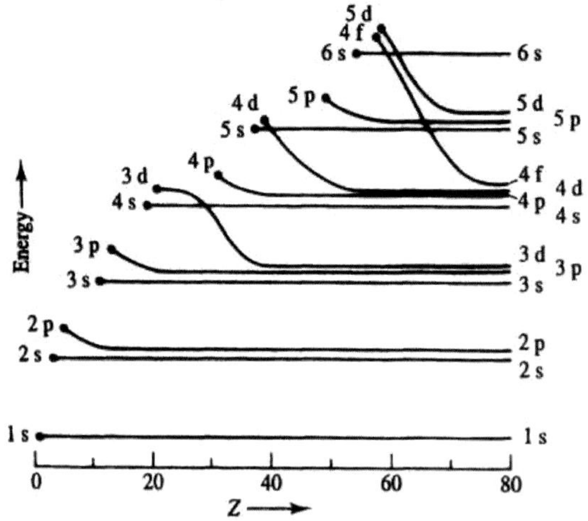

FIGURE 5.14
The ordering of the energy subshells in atoms as a function of atomic number through mercury ($Z = 80$). The curves begin on the left when a subshell begins to be occupied for a given Z. The order of the filled shells is seen on the right. The energy scale is only illustrative.

form compounds by giving up their far out electron (which is easily influenced and pulled away by other atoms). At the other end, the halogens with one vacancy in their p shell interact strongly with elements which can give up one electron to fill their p shell, so they, too, are very chemically active. Their radii are the smallest among the elements that fill a given p shell because their nuclear charge is greatest in that shell and their increased Coulomb force pull the p orbit in closest. The 4f subshell filling gives rise to the lanthanides (rare earth) elements between lanthanum and hafnium, and the 5f filling to the actinide series.

Prior to World War II, several groups were bombarding uranium with neutrons to try to make transuranic elements. In the periodic table of the elements, the order of the n = 5 and 6, l subshells was not known. In Table 5.46, elements 93 and 94 were thought to be chemically similar to rhenium and osmium. When elements 93 and 94 were discovered at the University of California, they had chemical properties more like uranium and so were thought to form a uranide series. Accordingly, elements 95 and 96 were thought to have chemical properties like those of neptunium and plutonium. But searches for new elements with such chemical properties failed. In 1944 Seaborg "conceived the idea that all known elements heavier than actinium were misplaced on the periodic table. The theory advanced was that these elements heavier than actinium might constitute a second series similar to the series of 'rare-earth' or 'lanthanide' elements."[6] His proposal proved to be correct. Thus, just as the lanthanides are very similar to each other from the filling of the 4f shell, there should be a second series, the actinide series, with chemical properties similar to actinium and to their rare-earth "sisters" from the filling of the 5f shell. Element 95 and 96 would have some chemical properties in common with europium and gadolinium, as well as with actinium. With this idea, elements 95 and 96 were quickly synthesized and chemically identified. This new chemical systematics based on the 5f filling provided the theoretical basis of the chemistry of elements up through 103 (lawrencium) and was important in their discovery. Beyond element 103, the study of the chemical properties of element 104 have established that it is similar to hafnium, the next position in the periodic table. Elements up through 111 have been firmly established and included in the 2007 periodic table issued by International Union of Pure and Applied Chemistry as shown in Table 5.6. Elements 112, 113, 114, 115, 116 and 118 have also been synthesized as discussed on Page 554. As discussed (Page 554), these very high Z elements have such large charges that they draw the electrons closer to the nucleus in relativistic orbits. It is possible that these orbital changes will alter their order with the periodic table. Thus, the study of their chemistry is very important.

Now consider the spectral lines for systems with the same number of electrons. For example, consider a potassium atom and ions with the same electron number as potassium; namely, Ca^+, Sc^{2+}, Ti^{3+}, V^{4+}, Cr^{5+}, and Mn^{6+}. These systems have 19 electrons and have similar structures that contain an atomic

[6] Glenn T. Seaborg, *Francis Slack Lectures*, J. H. Hamilton, ed., Vanderbilt University (1977).

TABLE 5.6
2007 Periodic Table of the Elements

1	2	3	4	5	6	7	8	9	10	11	12	13	14	15	16	17	18
1 **H** 1.0079																	2 **He** 4.0026
3 **Li** 6.941	4 **Be** 9.0122											5 **B** 10.811	6 **C** 12.011	7 **N** 14.007	8 **O** 15.999	9 **F** 18.998	10 **Ne** 20.180
11 **Na** 22.990	12 **Mg** 24.305											13 **Al** 26.982	14 **Si** 28.086	15 **P** 30.974	16 **S** 32.065	17 **Cl** 35.453	18 **Ar** 39.948
19 **K** 39.098	20 **Ca** 40.078	21 **Sc** 44.956	22 **Ti** 47.867	23 **V** 50.942	24 **Cr** 51.996	25 **Mn** 54.938	26 **Fe** 55.845	27 **Co** 58.933	28 **Ni** 58.693	29 **Cu** 63.546	30 **Zn** 65.38	31 **Ga** 69.723	32 **Ge** 72.64	33 **As** 74.922	34 **Se** 78.96	35 **Br** 79.904	36 **Kr** 83.798
37 **Rb** 85.468	38 **Sr** 87.62	39 **Y** 88.906	40 **Zr** 91.224	41 **Nb** 92.906	42 **Mo** 95.96	43 **Tc** [98]	44 **Ru** 101.07	45 **Rh** 102.91	46 **Pd** 106.42	47 **Ag** 107.87	48 **Cd** 112.41	49 **In** 114.82	50 **Sn** 118.71	51 **Sb** 121.76	52 **Te** 127.60	53 **I** 126.90	54 **Xe** 131.29
55 **Cs** 132.91	56 **Ba** 137.33	57-71 lanthanides	72 **Hf** 178.49	73 **Ta** 180.95	74 **W** 183.84	75 **Re** 186.21	76 **Os** 190.23	77 **Ir** 192.22	78 **Pt** 195.08	79 **Au** 196.97	80 **Hg** 200.59	81 **Tl** 204.38	82 **Pb** 207.2	83 **Bi** 208.98	84 **Po** [209]	85 **At** [210]	86 **Rn** [222]
87 **Fr** [223]	88 **Ra** [226]	89-103 actinides	104 **Rf** [261]	105 **Db** [262]	106 **Sg** [266]	107 **Bh** [264]	108 **Hs** [277]	109 **Mt** [268]	110 **Ds** [271]	111 **Rg** [272]							

57 **La** 138.91	58 **Ce** 140.12	59 **Pr** 140.91	60 **Nd** 144.24	61 **Pm** [145]	62 **Sm** 150.36	63 **Eu** 151.96	64 **Gd** 157.25	65 **Tb** 158.93	66 **Dy** 162.50	67 **Ho** 164.93	68 **Er** 167.26	69 **Tm** 168.93	70 **Yb** 173.05	71 **Lu** 174.97
89 **Ac** [227]	90 **Th** 232.04	91 **Pa** 231.04	92 **U** 238.03	93 **Np** [237]	94 **Pu** [244]	95 **Am** [243]	96 **Cm** [247]	97 **Bk** [247]	98 **Cf** [251]	99 **Es** [252]	100 **Fm** [257]	101 **Md** [258]	102 **No** [259]	103 **Lr** [262]

core composed of a nucleus and 18 electrons and one electron moving in the field of the atomic core. The difference between these systems is only in the different nuclear charges Z. For these systems, their spectral terms can be expressed by

$$T = \frac{RZ^{*2}}{n^2} \tag{5.8}$$

where Z^* is an effective charge which includes the effects from the orbital penetration and the polarization of the atomic core. The value of Z^* is between 1 and Z for a neutral atom, between 2 and Z for the singly ionized ion, between 3 and Z for the doubly ionized ion, and so forth. Because of the different Z's of the different ions, Z^* can be expressed as $Z - \sigma$. Thus (5.8) becomes

$$\left(\frac{T}{R}\right)^{1/2} = \frac{1}{n}(Z - \sigma) \tag{5.9}$$

From this formula, one can plot the dependence of $(T/R)^{1/2}$ on Z. The points with the same n fall on the same straight line and the slope of the line is $1/n$, as shown in Fig. 5.15. It is seen from the figure that whether the 3^2d or the 4^2s (here the notation is $n^{2s+1}l$ for a single-electron) spectral term is larger depends on which region of the graph one is in. According to the discussion above, we are not able to determine which shell has the lower energy when these shells are being filled. One needs a quantum-mechanical calculation. Here only a qualitative understanding of this relationship is given.

The following empirical rules can help one remember the order of filling of the electrons into the shells shown in Fig. 5.16 (except for a few cases). The filling goes from low to high $n + l$ values. The next rule is that when going to the next $n+l$ where there are several equal values of $(n + l)$, the electron is first filled into the shell with smaller n. Then fill the shell with the next higher n with the same $n + l$. When the highest n shell for a given $n + l$ is reached, one then goes to the shell with the next larger $n + l$ value and smallest n. In the squares, the a, b, c, . . . stand for the different shells. The numbers, 1, 2, 3, . . . stand for $(n + l)$. After the

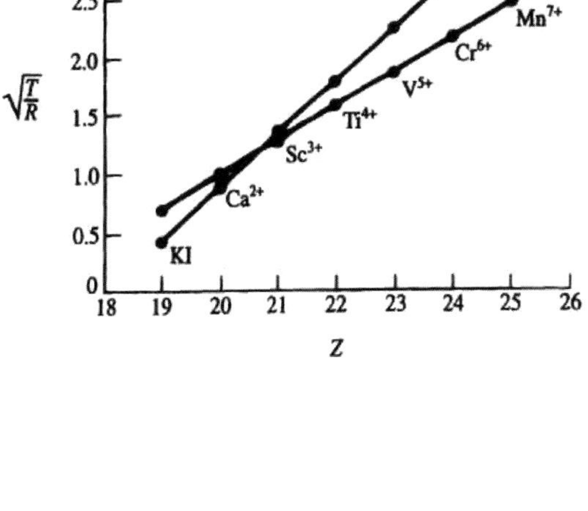

FIGURE 5.15
Moseley's diagram for isoelectronic systems.

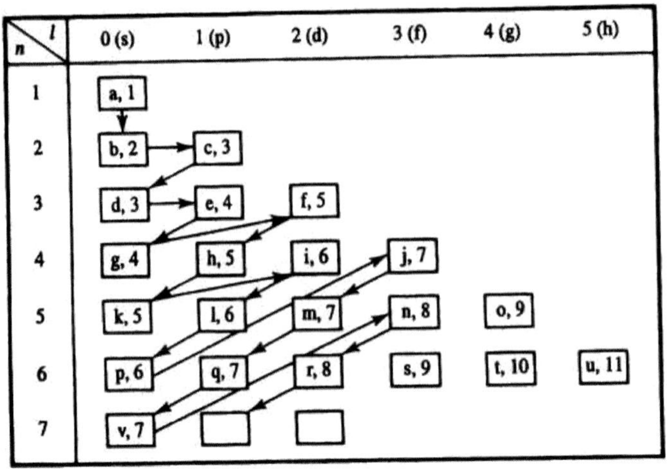

FIGURE 5.16
Order of electron filling into shells. The numbers in the boxes are $n + l$.

shell a is filled, compare shells b and c. They have different $(n + l)$ and the same n, so the b shell with small l is first filled. After b is filled, compare c and d, which have the same $(n + l)$; there the c shell, with smaller n, is first filled, and so forth.

5.4.4 Atomic Ground States

For a given electron configuration, how do you determine which state is the ground state? In Section 5.2, we showed how to find the possible resultant states from a given electron configuration. For example, in an sp configuration we have four states, $^1P_1$, $^3P_{2,1,0}$, and from np^2 (or np^4) we have five states, $^1S_0$, $^1D_2$, $^3P_{2,1,0}$. The question is what is the order in energy? Which state has the lowest energy? We need a quantum-mechanical calculation to obtain the energy values of the different states. But we can use two rules: Hund's rule to determine the order of the different atomic states and the Landé interval rule to determine the interval between the neighboring levels in the triplet states.

HUND'S RULE.[7] In 1925, F. Hund gave an empirical rule for the order in energy of the atomic states: For a group of atomic states resulting from a given electron configuration, the larger the value of S of an atomic state, the lower the energy and, for the same S, the larger the l, the lower the energy.

In 1927, he added that, for equivalent electrons: For the same l if the electron filling is less than half of the shell, the lower the J, the smaller the energy. This means that the energy level with $J = L - S$ is lowest. This is called normal

[7] N. Karayianis, Physical basis for Hund's rule. *Am. J. Phys.* **32** (1964) 216; **33** (1965) 201.

order. If the filling is more than half, then the larger the J the lower the energy ($J = L + S$). This is called reversed order.

LANDÉ INTERVAL RULE. A. Landé gave a rule for the energy interval between outer levels. For a triplet, the energy difference of the nearest levels is proportional to the larger J.

Here we give some examples to explain both Hund's and Landé's rules. For an sp outer electron configuration, one electron is in an s state, and the other in a p state. Which is the first excited state for C, Si, Ge, Sn, and Pb (the carbon family)? For L–S coupling, there are four atomic states, $^1P_1$, $^3P_2$, $^3P_1$, $^3P_0$. Using Hund's rule, the ^1P state is higher than the ^3P states and the three states of ^3P obey the normal order. The ratio of the two energy intervals from the three states is 2:1 from Landé's rule. A comparison of the experimental values shows that this is true for carbon and silicon (see Fig. 5.17). That is to say, L–S coupling is the case for carbon and silicon. There is not the same situation for tin and lead; the L–S coupling is not the only interaction for them. Hund's and Landé's rules are true only for L–S coupling.

What is the order of the energy levels for j-j coupling? Figure 5.17 shows the order of the energy levels of lead which is the standard j–j coupling result. The same is found for tin, but for germanium the data are in between j–j and L–S couplings. For all ground states and the excited states of most light elements, L–S coupling is satisfied. Pure j–j coupling is rare, and only occurs for the excited states for some heavy atoms. There the excited electrons are far from the other electrons and the interaction between these electrons is very weak, so pure j–j

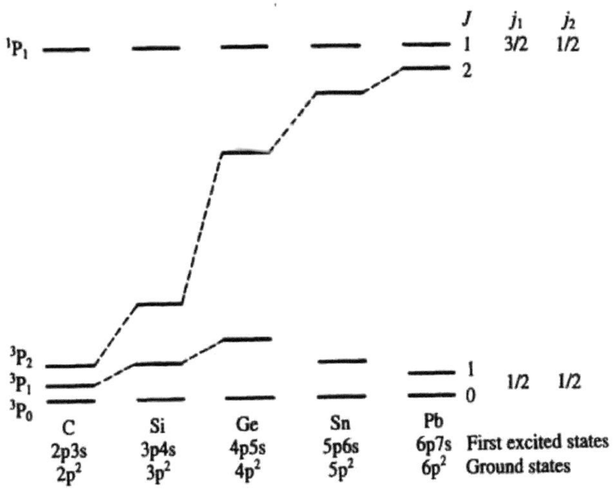

FIGURE 5.17
Observed relative positions of ^3P and ^1P terms of elements of the carbon group. (The scale is different for the various elements, but has been so chosen that each uppermost and lowest term in the diagram are the same for each element.)

coupling occurs. Further consideration of j–j coupling can be found in advanced atomic physics texts.

For the ground states of the carbon family elements, the two outermost electrons are p^2 configurations. According to Table 5.2, they can be combined to form ^1S, ^1D, ^3P. As Hund's rule points out, the lowest state is ^3P which includes $^3P_2$, $^3P_1$, and $^3P_0$. Since the number of the occupied states is less than half of the occupied number in the closed shell, this is the regular order. The energy level with the smallest J is lowest. As predicted, the ground states for C, Si, Ge, Sn, Pb are found to be $^3P_0$.

On the other hand, an oxygen atom has four p electrons in its outermost shell. So, although the resultant atomic states of a p^4 configuration are the same as those of p^2 (see Table 5.2), the number of occupied states is more than half of the occupied number in the closed shell. Thus, oxygen has the reversed order. The energy level with the largest J is lowest and the ground state of an oxygen atom is $^3P_2$ (see Table 5.5).

Now let us discuss the Landé interval rule. From the doublet observed in alkali spectra, we know that the splitting of the levels is related to the interaction between the magnetic field produced by the orbital motion and the magnetic moment of the electron spin in the absence of an external magnetic field. We can estimate the energy shift for a certain level:

$$\Delta E \sim \boldsymbol{\mu} \cdot \boldsymbol{B} \sim \hat{S}\hat{L} \cos(\boldsymbol{L}, \boldsymbol{S}) \sim \hat{S}\hat{L} \frac{\hat{J}^2 - \hat{L}^2 - \hat{S}^2}{2\hat{L}\hat{S}}$$

Since they have the same L and S, the energy difference between two levels indicated by $J+1$ and J is the same as the energy interval between the $J+1$ and J levels [recall $\hat{J}^2 = J(J+1)$]. This interval is proportional to

$$\begin{aligned}&[(J+1)(J+2) - L(L+1) - S(S+1)] - [J(J+1) - L(L+1) - S(S+1)] \\ &\quad = (J+1)(J+2) - J(J+1) \sim (J+1)\end{aligned}$$

This is Landé's rule.

We will discuss the physical basis of Hund's rule in the next paragraph. Here let us emphasize again that both Hund's and Landé's rules are true only for L–S coupling. More precisely, the notations such as ^1P, ^3P can only be used for L–S coupling. In j–j coupling there are no L and S. Sometimes they are used for simplification in j–j coupling, but only for the same j (see Fig. 5.17).

5.4.5 Explanation for the Variation of Ionization Energy

Now consider Fig. 5.18. Since the two electrons of $_2$He are in the same shell, there is nothing to shield them from the positive core. Each electron experiences a Coulomb attractive force from the nuclear charge $+2e$. So the binding energy is quite high in this case. For $_3$Li, because of the screening effect, the outermost electron feels only the effective charge $+e$ from the core, and also it is farther out

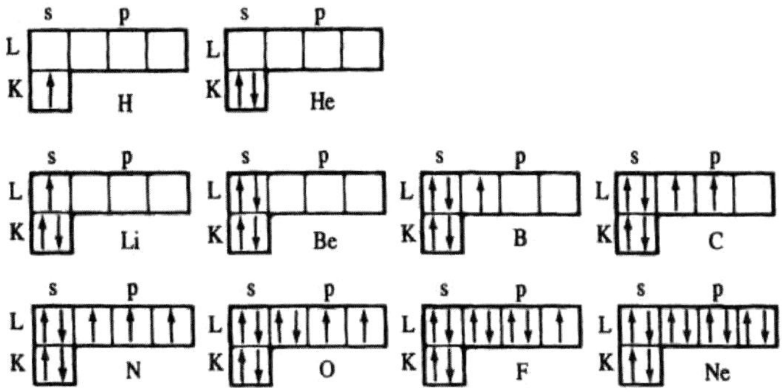

FIGURE 5.18
Electron configurations of ground states for the first ten elements.

in the 2s outer shell. So the 2s binding energy is considerably lower compared even to a hydrogen atom. But the two inner electrons see a $+3e$ core, and their binding energy is larger compared to the electrons in the $_2$He case. For $_4$Be, the two outermost electrons are in the same 2s shell. They experience the effective charge $+2e$, so their binding energy is larger than that of $_3$Li. Next is $_5$B, which has one more electron than $_4$Be. But its last electron is in the 2p shell (see Fig. 5.12) and is somewhat shielded by the two 2s electrons. So its binding energy is slightly lower than that of $_4$Be. The cores for $_6$C and $_7$N are similar to $_4$Be but with increasing Z, so their binding energies increase.

Now consider $_8$O, where the binding energy becomes smaller. The electron spin directions as the $n = 2$ shell is filled are shown in Fig. 5.18. From Eq. (5.6), the 2p shell needs six electrons to fill it. The first p electrons are parallel in spin (Hund's rule), but the Pauli exclusion principle tells us that no more than one electron can be in the same state. So the maximum-spin case is $_7$N, which has three p electrons in the same spin state. When the fourth p electron of $_8$O enters the 2p shell, it has to be antiparallel in spin with the former three, so the ground state of $_8$O is $^3P_2$ (Table 5.5). The number of electrons which have parallel spins is reduced. This means that the spatial overlaps of the electrons are increased, as discussed in Chapter 7, and the repulsive effect between the electrons is increased. This increased repulsion lowers the binding energy of $_8$O.

Last, we consider $_{11}$Na and $_{18}$Ar (Fig. 5.19). The outermost electron of $_{11}$Na is 3s, and it feels only a $+e$ core, but in the $n = 3$ shell of $_{18}$Ar each outer electron feels a $+8e$ core. So the ground state of $_{18}$Ar is much more stable than that of $_{11}$Na and its ionization energy is much higher.

5.5 SUMMARY

We have discussed the spectrum of helium and emphasized that the Pauli exclusion principle is the main reason behind the special features of its spectrum. Helium has only one more electron than hydrogen, its magnetic interaction is very small, and electric interaction is independent of spin. The marked differences

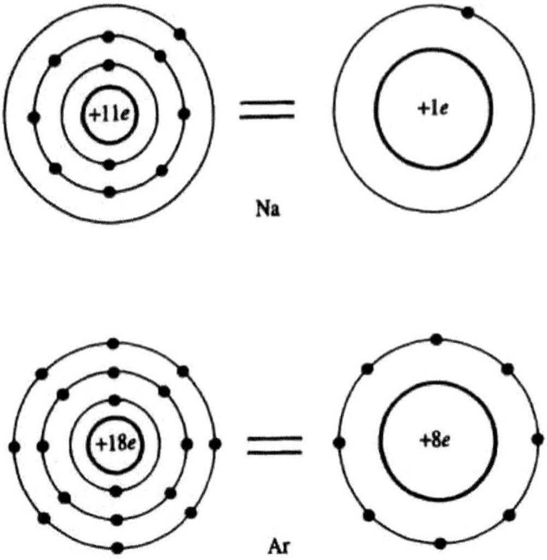

FIGURE 5.19
Orbital diagram of sodium and argon. Remember that these pictures illustrate only very roughly one electron outside of a closed shell in sodium and the closed electron shell of argon. The real electrons in these atoms have very smeared-out probability distributions which are very different from these single electrons, as discussed in Chapter 7.

in the helium spectrum cannot be explained without the pauli exclusion principle. Various periodic difference in atoms and the periodic table of the elements were shown to be related to the Pauli exclusion principle. The configurations of electrons in the ground states of atoms and the ordering of the filling of shells (for $n = 1, 2, \ldots$) and subshells (different l for given n) and their relation to the periodic table were discussed. Other examples of the Pauli exclusion principle in nuclear and particle physics were presented. The Pauli exclusion principle leads to an understanding that for electrons in atoms, the states with parallel spins must have differences in their space coordinates so they are separated in space.

The periodic table reflects the periodicity of the atomic electron configurations, and this periodicity is the result of the Pauli exclusion principle and the minimum-energy principle. Hence, the chemical properties of atoms are based on physical principles. Indeed, quantum mechanics with the Pauli exclusion principle provides an understanding of the various interactions between atoms and molecules.

PROBLEMS

5.1. The ionization energy required to pull one electron off a helium atom is 24.5 eV. If we want to ionize the two electrons one-by-one, what is the energy to be supplied?

5.2. Calculate the value of $\boldsymbol{L} \cdot \boldsymbol{S}$ of the $^4D_{3/2}$ state.

5.3. Calculate the possible values of $\boldsymbol{L} \cdot \boldsymbol{S}$ for an $S = \frac{1}{2}, L = 2$ state.

5.4. Estimate the angle between the total angular momentum and orbital angular momentum for the $^3F_2$ state.

5.5. Among hydrogen, helium, lithium, beryllium, sodium, magnesium, potassium and calcium atoms, which one shows the normal Zeeman effect? Why?

5.6. Assume there is strong spin–orbit interaction for two equivalent d electrons, and thus *j–j* coupling takes place. Calculate the possible values of the total angular momentum. If *L–S* coupling is available, what are the possible values of the total angular momentum? What is the number of the states? Is the number of the states with the same *J* the same for the two couplings?

5.7. According to the *L–S* coupling rule, what are the resultant states from the following electron configurations? Which one is lowest?
(*a*) $n\mathrm{p}^4$
(*b*) $n\mathrm{p}^5$
(*c*) $(n\mathrm{d})(n'\mathrm{d})$

5.8. The beryllium ground state has the electron configuration 2s2s. If one electron is excited to the 3p state, give the resultant atomic states by *L–S* coupling. When a transition from these states to lower states takes place, how many spectral lines can occur? Make a plot which gives the energy levels and the transitions. If the electron is excited to a 2p state instead of a 3p state, how many spectrum lines appear?

5.9. Show that an atom with a filled subshell must have a $^1S_0$ ground state.

5.10. When *L–S* coupling holds, what kinds of atomic states can be formed by an $(n\mathrm{d})^2$ configuration? Which one has the lowest energy? What is the ground state of the titanium atom?

5.11. A beam of helium ground-state atoms goes through an inhomogeneous magnetic field. How many beams will be seen at the screen? Under the same conditions, for boron atoms how many beams will be seen on the screen? Why?

5.12. Write the electron configurations for the following atoms: $_{15}\mathrm{P}$, $_{16}\mathrm{S}$, $_{17}\mathrm{Cl}$, and $_{18}\mathrm{Ar}$, and determine their ground states.

5.13. Show the atomic states by using the spectroscopic symbol for the following.
(*a*) BI (neutral boron) ground state
(*b*) NaII (Na^+) ground state
(*c*) NaIII (Na^{2+}) ground state
(*d*) NaII, first excited state

5.14. Using Hund's rule, find the ground state of an oxygen atom.

5.15. For two p equivalent electrons, what are the possible atomic states which can be constructed in *L–S* coupling? What are the possible quantum numbers of the states?

5.16. In a system with three electrons (one s, one p, and one d) what are the possible spectroscopic terms of this system in *L–S* coupling? Show the multiplicity also.

5.17. Show the atomic ground state, orbital angular momentum, spin, total angular momentum, and effective magnetic moment for aluminum and iron.

5.18. With three electrons having orbital angular momentum $l_1 = 1$, $l_2 = 2$ and $l_3 = 3$, respectively, the atom is in an S state. Determine the angle between the angular momenta of the first and the second electron.

5.19. The two electrons in a helium atom are excited to the 2p and 3d states, respectively. Show the possible values of the total angular momentum and, for each value, show the angle between l_1 and l_2.

5.20. Consider the following states for a helium atom. Show which state(s) does (do) not exist and why. Arrange them in energy order and show which is the ground state.
(a) 1s1s $^1S_0$
(b) 1s1s $^3S_1$
(c) 1s2s $^1S_0$
(d) 1s2s $^3S_1$

5.21. Determine the spectroscopic terms of the ground states in atoms of
(a) $_{71}$Lu
(b) $_{91}$Pa

5.22. Determine the number of electrons in fully occupied $n = 4$ and $n = 5$ shells.

5.23. What are all the allowed subshells for $n = 3$ in uranium? Give the quantum numbers that characterize each subshell and the number of electrons in each of these subshells.

CHAPTER 6

X-RAYS

In the field of observation, chance favors the one with a prepared mind.
Louis Pasteur

X-rays, also called Roentgen rays, were discovered in 1895 by W. K. Roentgen, who assigned this name because the true nature of the radiation was at first unknown. Later it was established that x-rays are in fact electromagnetic radiation of very short wavelength. Figure 6.1 shows their position in the range of the wavelengths of electromagnetic radiation. The range of the wavelengths of x-rays is generally from 0.01 Å to 10 Å or even longer. X-rays shorter than 1 Å are usually called hard x-rays while those longer than 1 Å are known as soft x-rays.

In this chapter, we will introduce first the discovery of x-rays, and the polarization and diffraction experiments that demonstrated the wave nature of x-rays (Section 6.1). Second, we will present the emission spectrum and the two principal mechanisms for producing x-rays (Section 6.2). Third, we will describe their Compton scattering, which shows the particle nature of x-rays (Section 6.3). Finally, we will explain the absorption of x-rays in materials (Section 6.4). In the section on ion beam analysis in Appendix I, we will present an important practical application of x-rays: proton induced x-ray emission analysis.

6.1 THE DISCOVERY OF X-RAYS AND THEIR WAVE NATURE

6.1.1 The Discovery of X-Rays

On November 8, 1895, Roentgen performed an experiment on a gas discharge in a cathode ray tube in a darkroom. The cathode ray tube is also called the Crookes

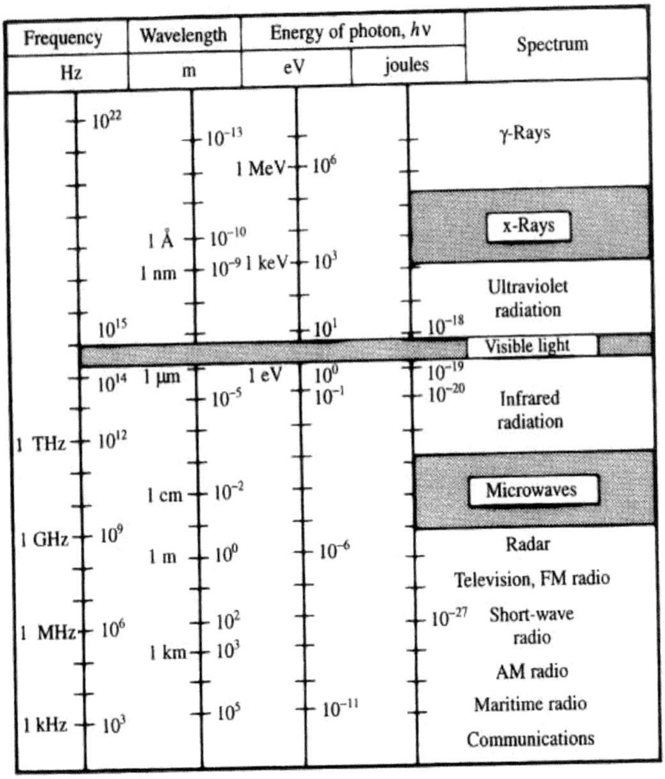

FIGURE 6.1
The electromagnetic spectrum with the different wavelength (frequency) regions labelled.

tube after the British physicist W. Crookes who developed it. In 1879 Croaker experimentally demonstrated in experiments that cathode rays were composed of charged particles, predating the later discovery of electrons. Roentgen wrapped a cathode ray tube with black paper purposely to avoid the influence of ultraviolet and visible light. He found that a fluorescent screen treated with barium platinum cyanide, $BaPt(CN)_6$, at some distance from the tube still exhibited slight fluorescence. Through repeated experiments, he made sure that the cause of the activation of the fluorescence came from the cathode ray tube, but that it was by no means the cathode rays themselves.

Over the next month or more, Roentgen made several kinds of studies on this mysterious new ray that produced the fluorescence. He found that these rays traveled straight (he observed neither reflection nor refraction), were not deflected by a magnetic field, and could go about 2 meters in air. Soon he discovered the penetrating power of the rays, which allowed photographs of the outline of a balance scale and other objects put in a closed box to be taken. He also observed the profile of the finger bone of his wife in one radiograph (see Fig. 6.2). Considering the mystery and the uncertainty of the rays he had discovered, he named them x-rays.

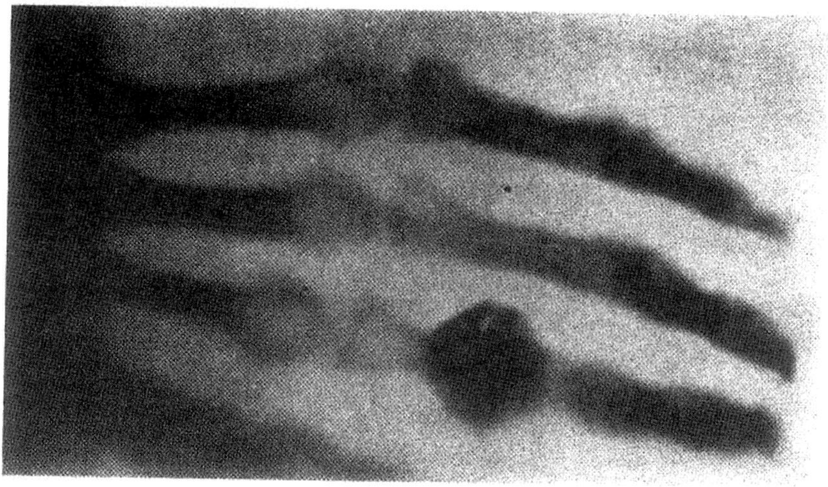

FIGURE 6.2
A radiograph of the finger bones of Mrs. Roentgen.

Roentgen read his first report "On the New Ray" at the end of 1895 and published a radiograph of the finger bone of his wife. Roentgen's discovery resulted in extremely rapid responses all over the world. Astonishingly, in 1896 there were about one thousand papers published on studies of x-rays or their application. This is truly remarkable because this was before our time when so much emphasis is placed on publication. Three months after the discovery of x-rays, a hospital in Vienna for the first time applied x-rays to take photographs in surgical therapy. Already in 1896, J. Daniels of Vanderbilt University published the first report on the harmful effects of x-rays: He discovered x-rays could cause a person's hair to fall out.

Although cathode ray tubes had been used in many laboratories for thirty years, it was Roentgen who discovered x-rays. There is some evidence that x-rays had been produced early in the nineteenth century. In 1879, Crookes often complained that film near his cathode ray tube appeared blurred with shadows. In 1890, A. W. Goodspeed and W. W. Jennings noticed that their film was especially black after they had done experiments with a cathode ray tube. These are further examples of people who did not see the new insights even when these insights were right in front of them.

Following the discovery of x-rays[1], a new period of physics began. Together with the discoveries of radioactivity and the electron in the next two years, it opened up the world of modern physics. Before these discoveries scientists thought that all the basic principles of physics were known.

[1] The original paper of Roentgen was translated and reprinted in W. K. Roentgen, *Science* **3** (1986) 227; **3** (1986) 726.

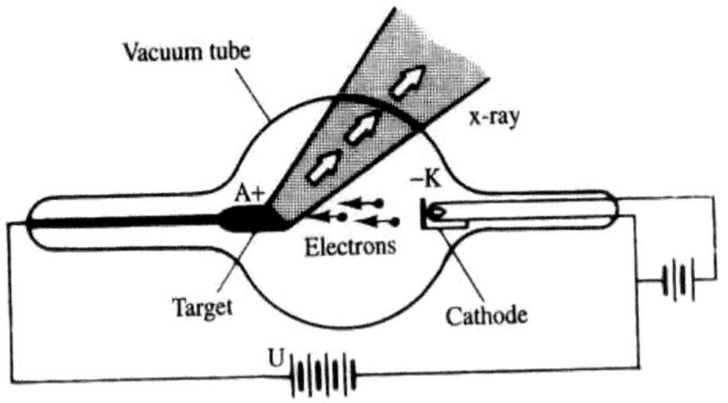

FIGURE 6.3
Schematic of an x-ray tube.

6.1.2 X-Ray Tube

There are various ways to construct tubes to produce x-rays. Figure 6.3 is a sketch of a commonly used x-ray tube which is similar to that used by Roentgen. With a pressure in the tube of 10^{-6}–10^{-8} mmHg, electrons from the heated cathode K are accelerated easily by the electric field to the anode A. The x-rays are produced when the electrons strike the anode. The anode, also called the target, is typically made of a heavy metal such as tungsten, molybdenum, or platinum, but may be made of a light metal such as chromium, iron, or copper, depending on the application. In 1895, the electrical potential difference used by Roentgen across his Crooke's tube was only several thousand volts, so the energy of the electrons was comparatively low and the rays produced were soft x-rays. Today the accelerating voltage between the cathode and the anode is generally ten thousand to one hundred thousand volts. Changing this voltage changes the energy of the electrons which strike the anode and so changes the x-ray energy (frequency).

6.1.3 The Wave Nature of X-Rays

From the equations put forward by J. C. Maxwell to describe electric and magnetic fields, he concluded that an accelerated (or decelerated) charged particle can radiate an electromagnetic wave, and this was verified experimentally by H. R. Hertz. Thus, when high-speed electrons are stopped (decelerated) by a target, they produce electromagnetic waves, and it would seem logical to think of x-rays as electromagnetic waves. But Roentgen did not observe x-ray refraction, reflection, or diffraction when using gratings which worked with visible and ultraviolet light. He therefore thought, incorrectly, that x-rays were not a form of electromagnetic radiation. It was not until 1906 that the wave nature of x-rays was for the first time experimentally confirmed by C. G. Barkla, who demonstrated the polarization of x-rays. Even then many people did not believe his conclusion.

T. Young in the eighteenth century had pointed out that the true test that a phenomenon had a wave nature was to observe a diffraction effect. In 1912, M. T. F. von Laue suggested that x-rays were electromagnetic waves with very short wavelengths, so that only the regular array of atoms in a crystal would provide grating spacings which could diffract x-rays. Von Laue's suggestions were soon verified experimentally by W. Friedrich and P. Knipping.[2] They clearly demonstrated the wave nature of x-rays and measured the wavelengths of x-rays for the first time. Next, let us look at the work of Barkla and the experiments suggested by von Laue.

6.1.4 The Polarization of X-Rays

First remember that the concept of polarization holds only for transverse waves, where the direction of the oscillating property is perpendicular to the propagation direction of the wave. If an electromagnetic wave has its electric field vector E oscillating in only one direction, that wave is said to have linear polarization (or plane polarization). In other cases, the E direction changes periodically with time. When E moves in a circle in the plane perpendicular to the wave propagation direction, the wave has circular polarization. When E moves in an ellipse in this perpendicular plane, it has elliptical polarization. If the direction of the electric vector is randomly oriented, the corresponding wave is unpolarized. Figure 6.4 is a sketch of the double-scattering equipment to observe polarization. As shown in the figure, initial unpolarized x-rays strike the first scatterer in the z direction perpendicular to the xy plane. If the x-ray is a transverse wave, at the first scatterer forced oscillation can only occur in the directions of x and y, with no oscillation in the z direction. So the waves from the first scatterer can only oscillate in the x and y directions. Thus, when observed in the direction of x, a transverse x-ray propagating in the x direction can only oscillate in the y direction because of the character of a transverse wave. One observes polarized x-rays in the x direction. When this x-ray which oscillates only in the y direction strikes a second scatterer, the x-ray from the second scatterer is polarized in the y direction. So we can observe strong x-rays in the z direction but not in the y direction. In a

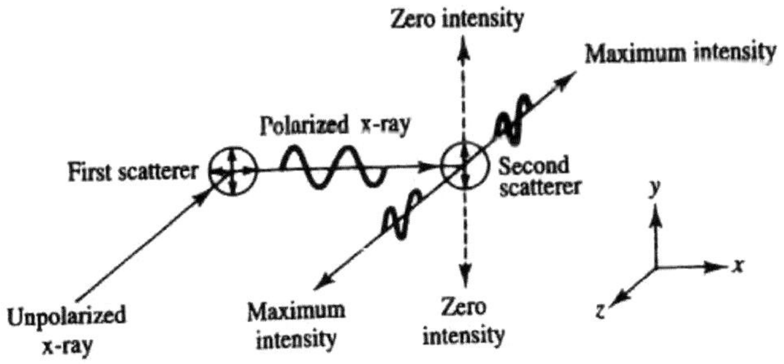

FIGURE 6.4
Schematic of a double scattering experiment.

---

[2] W. Friedrich, P. Knipping and M. Laue, *Ann. der Physik* **41** (1913) 971.

double scattering experiment, the first scatterer plays the role of the producer of polarization, and the second scatterer that of the detector of polarization. The double-scattering technique is an efficient method for measuring the polarization of a ray. Barkla carried out such an experiment to show the polarization of x-rays and for the first time confirmed the transverse wave nature of x-rays.

6.1.5 The Diffraction of X-Rays

When an electromagnetic wave passes through a slit whose size is of the same order as the wavelength or smaller, the phenomenon of diffraction appears. Since x-rays have wavelengths the order of an angstrom, it is rather difficult to make a slit of this size to observe the diffraction of x-rays. Von Laue first suggested the use of the planes of atoms in a crystal as the natural grating (set of slits) to observe the diffraction of x-rays. Figure 6.5 shows the arrangement of the chlorine ions and sodium ions in a rock salt crystal. We will see that the spacings of atoms in the planes of such a crystal grating are of the same order as x-ray wavelengths. But first look at what happens when x-rays strike crystal lattices.

As shown in Fig. 6.6(a), an incident x-ray makes an angle θ with the crystal plane. If this crystal plane is a mirror plane (that is, the incident angle is equal to the reflected angle, even though in the next part we will prove that this assumption is unnecessary), it is easy to show that the path difference between ray 1 and ray 2, $AC + CD$ is $2d\sin\theta$. When it is an integral multiple n of the wavelength (suppose that the incident x-ray is monochromatic, i.e., contains only one wavelength), we have

$$2d\sin\theta = n\lambda, \qquad n = 1, 2, \ldots \tag{6.1}$$

where d is the lattice spacing, λ is the incident x-ray wavelength, and θ is the incident angle and reflected angle. The x-rays reflected in the direction θ will have a maximum in their diffraction intensity. Formula (6.1) is called the Bragg law. When a beam of x-rays strikes a crystal, it must satisfy the Bragg law for its reflected rays to be intensified by diffraction. From this formula we know that

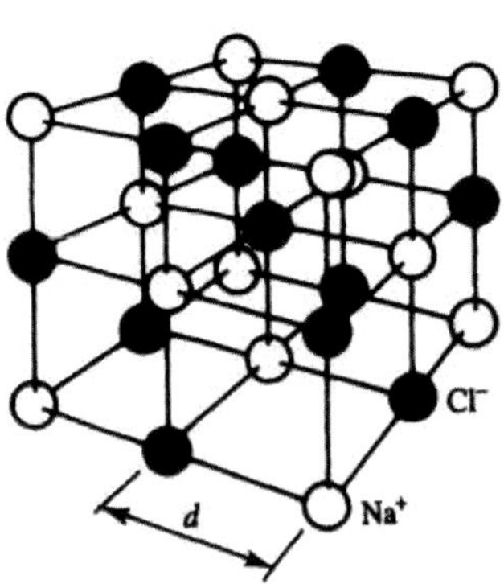

FIGURE 6.5
Arrangement of the sodium ions and chloride ions in a rock salt crystal.

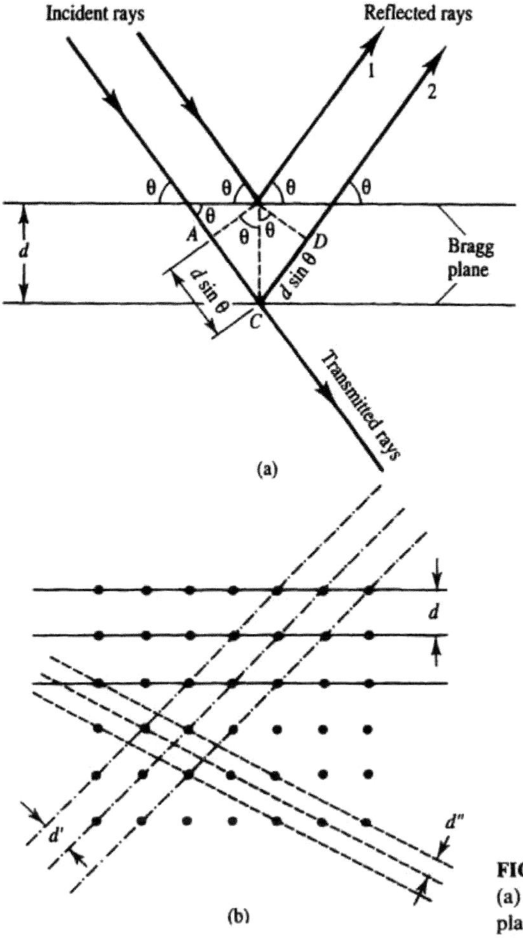

FIGURE 6.6
(a) Derivation of Bragg's formula. (b) Parallel planes of different directions in a crystal.

the wavelength must be smaller than $2d$, and, if either λ or d is known, the other can be calculated after measuring θ in a diffraction experiment.

A standard crystal with a known d can be used to measure an unknown x-ray wavelength λ, and then this x-ray can be used to measure the lattice spacing d of an unknown crystal. For a crystal of sodium chloride (Fig. 6.5), we can calculate the crystal lattice constant from Avogadro's constant: 1 g of sodium chloride contains $6 \times 10^{23}/58.5$ molecules. The density of sodium chloride is $2.163\,\mathrm{g/cm^3}$, from which we know that $1\,\mathrm{cm}^3$ of sodium chloride contains $6 \times 10^{23} \times 2.163/58.5$ molecules. However, one sodium chloride molecule is composed of two ions, so we have twice this many ions. Assume d to be the distance between two adjacent ions, then there are $1/d$ ions in one centimeter, and $(1/d)^3$ ions in a volume of $1\,\mathrm{cm}^3$. So

$$\left(\frac{1}{d}\right)^3 = 2 \times \frac{6 \times 10^{23}}{58.5} \times 2.163\,\mathrm{cm}^{-3}$$

This gives a sodium chloride crystal lattice space of $d = 2.82$ Å.

For a cubic crystal, if d is measured from a diffraction experiment, then Avogadro's constant N_A can be calculated. This is one of the experimental methods for measuring Avogadro's constant. Note that, when considering the crystal structure in Fig. 6.5, there are more than the one set of crystal planes shown in Fig. 6.6(a). Atoms in a crystal can form many sets of parallel planes in different directions, as shown in Fig. 6.6(b), with varying values of d. Also, as seen in the figure, the densities of the numbers of atoms vary on the different planes, so that the measured intensities of the reflected x-rays also vary. Now let us look at experimental results from the diffraction of x-rays in crystals.

First, we consider the Laue film method. In 1912, following Laue's suggestions, W. Friedrich and P. Knipping used continuous-wavelength x-rays in a diffraction experiment on a single crystal. A diagram of the experiment is shown in Fig. 6.7. The experimental result on a sapphire single crystal is shown in Fig. 6.8. Each dark spot in the photograph is called a Laue spot, which corresponds to one set of crystal planes. The position of the spot represents the direction of the related planes. Bragg's law gave a correct explanation of these results. It is not easy for a single-crystal to satisfy Bragg's law but, fortunately, x-rays with a continuous wavelength distribution were applied, so for the first time the beautiful pattern of crystal structure was shown.

W. H. Bragg and his son W. L. Bragg not only gave the reflection relation for x-rays but also invented the crystal reflection (diffraction) x-ray spectrometer and applied it in a series of structural analyses of crystals. Their work demonstrated that there is no NaCl in a sodium chloride crystal, only the ionized forms of Na^+ and Cl^-.

The method more commonly used than the Laue film method is the polycrystalline powder method invented by P. J. W. Debye and J. A. Scherrer.[3] Figure 6.9 is a sketch of this method and Fig. 6.10 is diffraction pattern of zirconium oxide powder. The advantage of this method is that it does not need a single crystal but rather only polycrystalline powder or metal foil. The preparation of a sample is simplified greatly, and generally monochromatic x-rays are used. Here each concentric circle on the photograph corresponds to a particular set of crystal

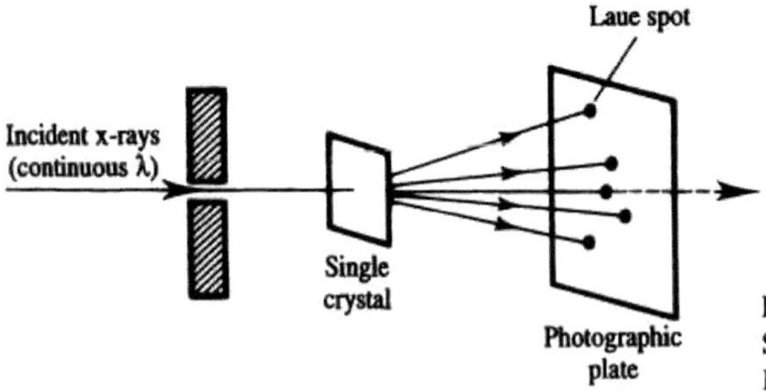

FIGURE 6.7
Schematic of the experiment in the Laue film method.

[3] P. L. W. Debye and J. A. Scherrer, *Z. für Physik* **17** (1916) 277; **18** (1917) 29; **19** (1918) 481.

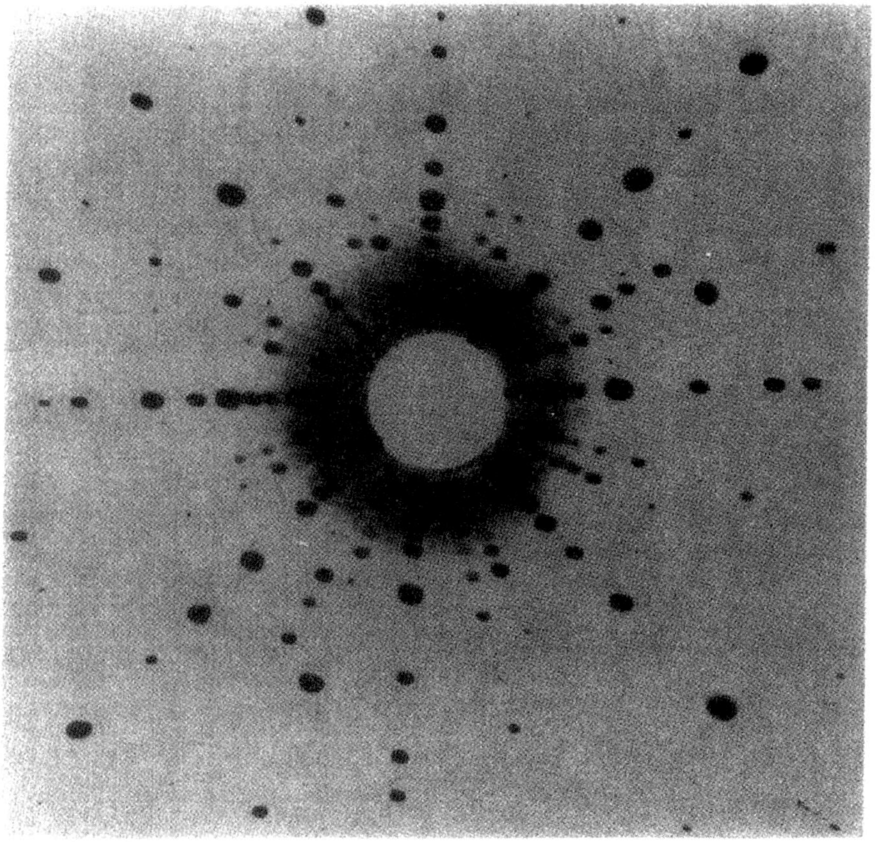

FIGURE 6.8
Single-crystal Laue film of sapphire (Al_2O_3). (Courtesy of Professor Xu Shun Sheng of Shanghai Metallurgy Institute.)

planes. The brightness of the rings reflects the density of atoms in a crystal plane as shown in Fig. 6.6(b).

When the wavelength is known and the corresponding angle of the circular ring has been measured, one can use Bragg's relation to calculate the spacing d of each crystal plane. Conversely, a known d can be used to determine the x-ray

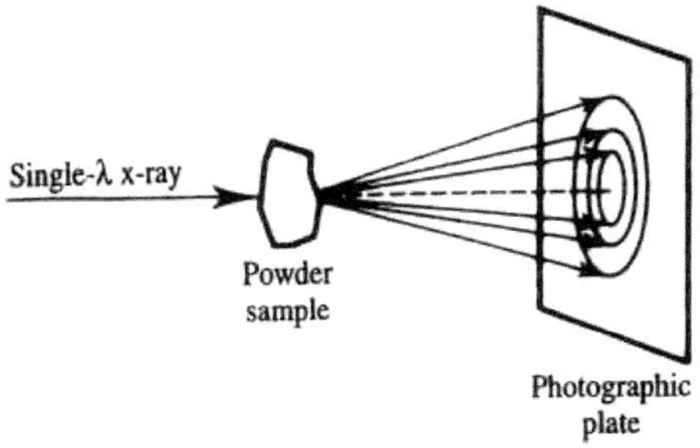

FIGURE 6.9
Schematic of the polycrystalline powder method.

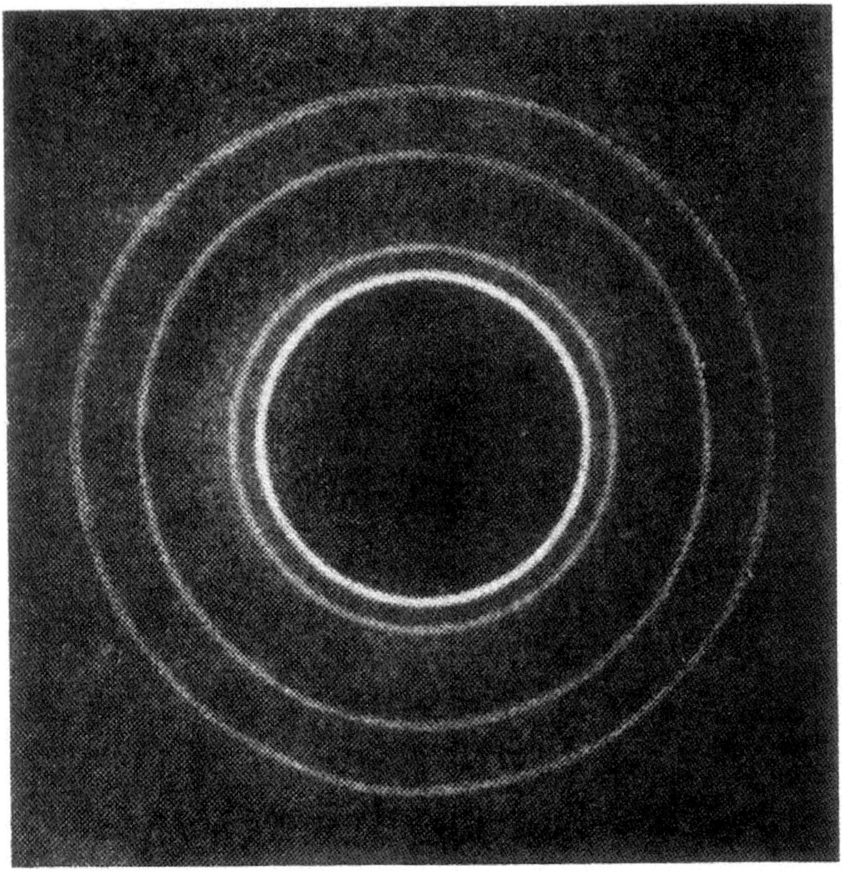

FIGURE 6.10
X-ray diffraction pattern of polycrystalline matter.

wavelength. Since it is comparatively easy to analyze, the Debye–Scherrer method is applied extremely widely in industry.

6.1.6 A Further Derivation of Bragg's Relation[4]

Now we consider a comparatively simple, two-dimensional derivation of Bragg's relation. First, let us look at Fig. 6.11(a), where θ and θ' are the incident angle and reflected angle, respectively. They are arbitrary and need not be equal to each other as required in Fig. 6.6(a). The path difference between the two rays after their reflection on the spots A and B, respectively, is

$$\overline{BD} - \overline{AE} = b\cos\theta - b\cos\theta'$$

[4] L. R. B. Elton and D. F. Jackson, *Am. J. Phys.* **34** (1966) 1036.

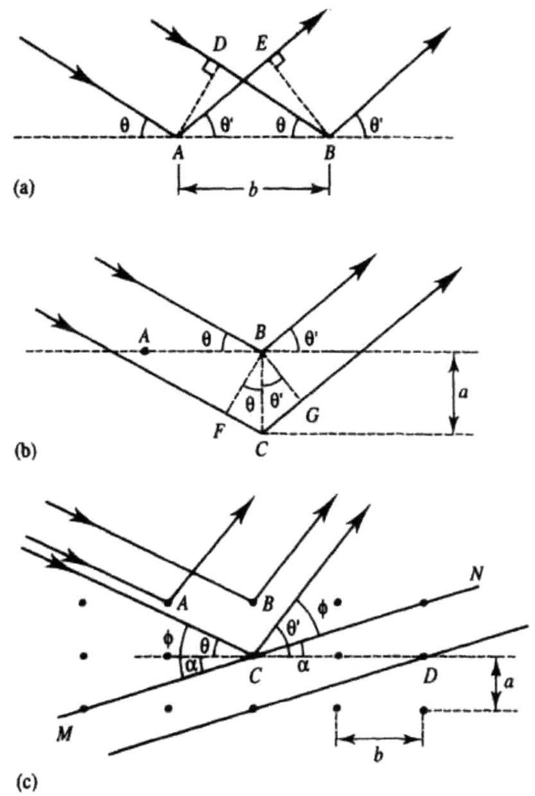

FIGURE 6.11
Schematic used in a more detailed derivation of Bragg's formula.

where b is the distance between A and B. When this is equal to an integral multiple of the wavelength, that is,

$$b(\cos\theta - \cos\theta') = k\lambda \tag{6.2}$$

the diffracted x-rays have an intensity maximum.

Next, consider the situation of an incident ray being reflected off two parallel planes. As shown in Fig. 6.11(b), the path difference between these two rays is

$$\overline{FC} + \overline{CG} = a\sin\theta + a\sin\theta'$$

Similarly, when it is an integral multiple of the wavelength, that is,

$$a(\sin\theta + \sin\theta') = l\lambda \tag{6.3}$$

we get an intensity maximum in the diffraction. In Eq. (6.2) and Eq. (6.3), k and l are positive integers and λ is the wavelength of the incident wave. When both Eq. (6.2) and Eq. (6.3) are satisfied, then the diffraction will give an intensity maximum in the direction θ'.

Now let us look at Figure 6.11(c). Select a plane MN passing through the point C as shown in the figure. It makes an angle α with the original plane.

Choose α to make sure that on the new plane MN the incident angle equals to the reflected angle, that is, choose α to satisfy

$$\theta + \alpha = \theta' - \alpha = \phi \tag{6.4}$$

Then the chosen plane MN is a mirror plane. Substituting Eq. (6.4) into Eq. (6.2), we have

$$b[\cos(\phi - \alpha) - \cos(\phi + \alpha)] = k\lambda$$

or

$$2b \sin\phi \sin\alpha = k\lambda \tag{6.5}$$

Similarly, combining Eq. (6.4) and Eq. (6.3), we get

$$2a \sin\phi \cos\alpha = l\lambda \tag{6.6}$$

To have an intensity maximum in the diffraction in all reflections, it is necessary to satisfy Eq. (6.5) and Eq. (6.6) simultaneously. Dividing Eq. (6.5) by Eq. (6.6), we obtain

$$\tan\alpha = \frac{ka}{lb} \tag{6.7}$$

Equation (6.7) tells us that the plane MN with an angle α to the crystal plane is not only a mirror plane (both incident angle and reflected angle are ϕ), but also a crystal plane (a plane regularly passing every lattice of atoms, namely, a Bragg plane (see Fig. 6.6(a)). This is because the numerator and denominator of Eq. (6.7) are integral multiples of a, b, respectively, which ensures that the plane MN is a crystal plane passing many lattices of atoms.

Suppose that n is the maximum common factor of the integers k and l, and define

$$k = nk_1, \qquad l = nl_1 \tag{6.8}$$

Then we can prove

$$\frac{b}{k_1}\sin\alpha = \frac{a}{l_1}\cos\alpha = \left(\frac{l_1^2}{a^2} + \frac{k_1^2}{b^2}\right)^{-1/2} \equiv d \tag{6.9}$$

where d is the distance between the plane MN and its adjacent plane. Thus both Eq. (6.5) and Eq. (6.6) can be converted into

$$2d \sin\phi = n\lambda \tag{6.10}$$

Thus, we have proven that when an intensity maximum of diffraction is observed in the arbitrary direction θ', then θ' definitely corresponds to a Bragg plane whose direction is determined by Eq. (6.7) and for this plane the Bragg law (6.10) holds.

6.2 MECHANISMS FOR PRODUCING X-RAYS

6.2.1 X-Ray Emission Spectra

We can obtain x-rays using the equipment shown in Fig. 6.3. Bragg's method of crystal diffraction, Fig. 6.12, can be used to measure the x-ray wavelengths and their intensities, to give an x-ray spectrum—the x-ray intensity as a function of wavelength.

Essentially, the measuring equipment is composed of three parts: the source of x-rays, the spectrometer (here it is a crystal, which corresponds to a grating or prism), and the recorder (film can be used with different sensitivity ranges that respond to different x-ray wavelengths). Modern recorders are multifunctional. Semiconductor detectors are one of the latest types of tools which function as both spectrometer and recorder.

Typical spectra are shown in Figs. 6.13(a) and (b). From these we observe that x-ray spectra are composed of two parts. One involves a continuous change of wavelength and is called the continuous spectrum; its minimum wavelength is related only to the applied voltage. The other consists of the sharp lines with discrete wavelengths which may or may not appear in the wavelength range observed. If observed, the wavelengths of these discrete peaks are determined by the target material itself. Thus, this part of the spectrum is called the characteristic spectrum and is superimposed on the continuous spectrum like a tower on a hill.

6.2.2 Continuous Spectrum—Bremsstrahlung

It is easy to understand the production of the continuous spectrum. Classical electrodynamics tells us that the acceleration (or deceleration) of charged particles produces electromagnetic radiation. When a charged particle collides with an atom (nucleus of an atom) and decelerates abruptly, the accompanying radiation is called bremsstrahlung, a German word meaning braking or stopping radiation. In an x-ray tube, when the charged electrons reach the target, their speed changes

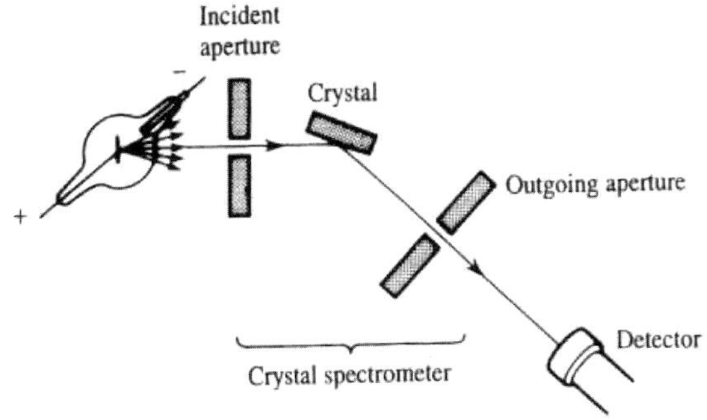

FIGURE 6.12
Schematic of the equipment to measure x-ray emission spectra.

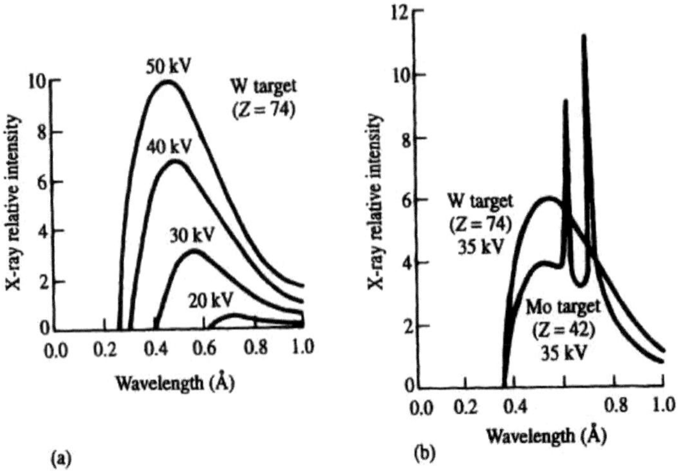

FIGURE 6.13
X-ray emission spectra. (a) Tungsten target for different applied potentials. (b) Tungsten ($Z = 74$) target and molybdenum ($Z = 42$) targets; the applied potential is fixed.

continuously under the Coulomb field of the target atoms, so the x-rays radiated have a continuous spread in wavelength.

The intensity of the bremsstrahlung is inversely proportional to the square of the mass of the incident charged particle, so a charged heavy particle such as a proton has negligible bremsstrahlung compared with that produced by an electron. The intensity of bremsstrahlung is also directly proportional to the square of the charge of the target nucleus. Since x-rays used in medicine and industry are mainly from the continuous part of the spectrum, the most commonly used anode material in x-ray tubes is tungsten, which can give off high-intensity x-rays. Tungsten is used because of its high Z, its good heat-transfer capability to dissipate the remaining energy of the electrons, and its ease in fabrication.

While the intensity of the measured continuous spectrum increases with increasing atomic number of the target, the shape of the continuous spectrum does not depend on the target material. There is a minimum wavelength $\lambda_{\min}$ (or maximum frequency $\nu_{\max}$), whose value depends only on the applied potential V, and is independent of the atomic number Z. See Fig. 6.13(a) and (b). This fact can not be explained by classical physics. The relation of the minimum wavelength $\lambda_{\min}$ of the continuous spectrum to the applied potential was first obtained from experiments by W. Duane and P. Hunt. They found

$$\lambda_{\min} = \frac{12.4}{V(\text{kV})} \quad \text{Å} \tag{6.11}$$

where V is the applied voltage in kV and $\lambda_{\min}$ is in angstroms. When an electron accelerated to a kinetic energy $T = eV$ by an electric field strikes a target, it is possible for all its energy to be converted into radiation energy. So the maximum possible energy of the emitted x-ray photon is

$$T = eV = h\nu_{\max} = \frac{hc}{\lambda_{\min}} \tag{6.12}$$

Rewriting this equation, we obtain

$$\lambda_{\min} = \frac{hc}{eV} = \frac{12.4}{V(\mathrm{kV})} \quad \text{Å} \tag{6.13}$$

For 10 000 V accelerating voltage, $\lambda_{\min} = 1.24$ Å. Thus, a purely experimental relation, which cannot be explained by classical physics (in classical electromagnetic theory the wave energy is not related to λ, so any short wavelength can be emitted), is explained beautifully by the quantum theory. Here $\lambda_{\min}$ is called the quantum limit. Its existence confirms once more the correctness of quantum theory. Equation (6.13) is very useful, for example, to obtain $\lambda_{\min}$ for a particular voltage. Alternatively, because both V and $\lambda_{\min}$ can be measured in experiments, this equation can be used to measure precisely Planck's constant h. Duane and Hunt were the first to apply this method to measure h. Their value was in good agreement with that obtained from the photoelectric effect.[5] This further demonstrates the universality of Planck's constant, which has the same value for different electromagnetic wave frequencies. This equation is the same as that given in Chapter 3 for the photoelectric effect. However, for x-rays the work for an electron to escape a metal is so small, of the order of an electron-volt, that it can be neglected here. Thus, the production of x-rays can be considered as the reverse of the photoelectric effect.

6.2.3 Characteristic Radiation—Transitions of the Inner Shell Electrons

The x-ray spectra of a molybdenum target shown in Figure 6.13(b) has a continuous x-ray spectrum produced by bremsstrahlung, which appears as the shape of a "hill," and two sharp peaks superimposed on the continuous spectrum, like two towers on the hill. The latter x-rays are the characteristic radiations of molybdenum.

The characteristic x-ray spectrum of every element has similar structure, but the energy values of the characteristic x-rays of each element are different. Characteristic x-rays can be used to identify an element as a fingerprint is used to identify a person. Characteristic x-ray spectra were discovered by C. G. Barkla in 1906. He observed that the characteristic spectra from any given element contained several series, which were labelled with the letters K,L,M .... Why were they not arranged with the letters A,B, ... instead of K,L, ...? Because Barkla at that time could not be certain whether there were still harder (shorter wavelength)

---

[5] J. A. Bearden, et al., *Phys. Rev.* **81** (1951) 70.

spectral line series, choosing letters midway between A and Z left room for both higher and lower energy series according to the order of the hardness (or penetration) of the radiation. Later, it was found that the K series lines contained $K_\alpha, K_\beta, \ldots$ and the L series contained $L_\alpha, L_\beta, L_\gamma, \ldots$.

In 1913, after measuring the spectra of 38 different elements from aluminum to gold, H. G. J. Moseley found that a diagram of the square root of the x-ray frequencies of the elements versus their atomic numbers Z gave a linear relation, as shown in Fig. 6.14. The previously unknown meaning of the integer which labeled the elements in the ordinate of the diagram was discovered by Moseley: "It is equal to the order number of the element in the periodic table." Moseley also said in the same article. "If we do not use these integers to mark the elements, or if the order is chosen incorrectly or the numbers of vacancies left for unknown elements are wrong, these rules will disappear at once ...."[6] This sentence is something of an exaggeration. In fact, in Fig. 6.14 the orders of Ho and Dy were reversed (see ref. 6). Moseley further stated, "Now Rutherford has proved that the most important constituent of an atom is its central positively charged nucleus, and Van den Brook has put forward the view that the charge carried by this nucleus is in all cases an integral multiple of the charge on the hydrogen nucleus. There is every reason to suppose that the integer which controls the x-ray spectrum is the same as the number of electrical units in the nucleus." This discovery of Moseley was an important milestone in understanding the ordering of the elements and marks the beginning of x-ray spectroscopy.

For the K_α line, Moseley found the following experimental formula:

$$\nu_{K_\alpha} = 0.248 \times 10^{16}(Z-b)^2, \qquad b \simeq 1 \tag{6.14}$$

where b is the shielding coefficient.

In the same year, Bohr published three papers setting forth the quantum theory of atoms. As soon as Moseley read Bohr's articles, he saw immediately that his experimental formula (6.14) could be derived from Bohr's theory where radiation is emitted when an electron drops from a higher to lower shell:

$$\nu_{K_\alpha} = \frac{c}{\lambda} = RcZ^2\left(\frac{1}{1^2} - \frac{1}{2^2}\right) = \tfrac{3}{4}RcZ^2 \simeq 0.246 \times 10^{16} Z^2 \tag{6.15}$$

Remember that Eq. (6.15) holds for a hydrogenlike (one electron) system. This equation is very similar to Eq. (6.14) except for a small difference in the constant and the difference of $(Z-1)^2$ in (6.14) and Z^2 in (6.15). This latter difference is because of the shielding effect of the one electron remaining in the $n = 1$ shell: when the $n = 1$ shell has a vacancy, the electrons in the $n = 2$ shell sense the attraction of positive charges of $(Z - 1)$. So Eq. 6.15 should include this effect. Then when electrons in the $n = 2$ shell make transitions to the inner shell, the radiation frequency and wavelength emitted should be

---

[6] H. G. J. Moseley, *Phil. Mag.* **26** (1913) 1024; **27** (1914) 703.

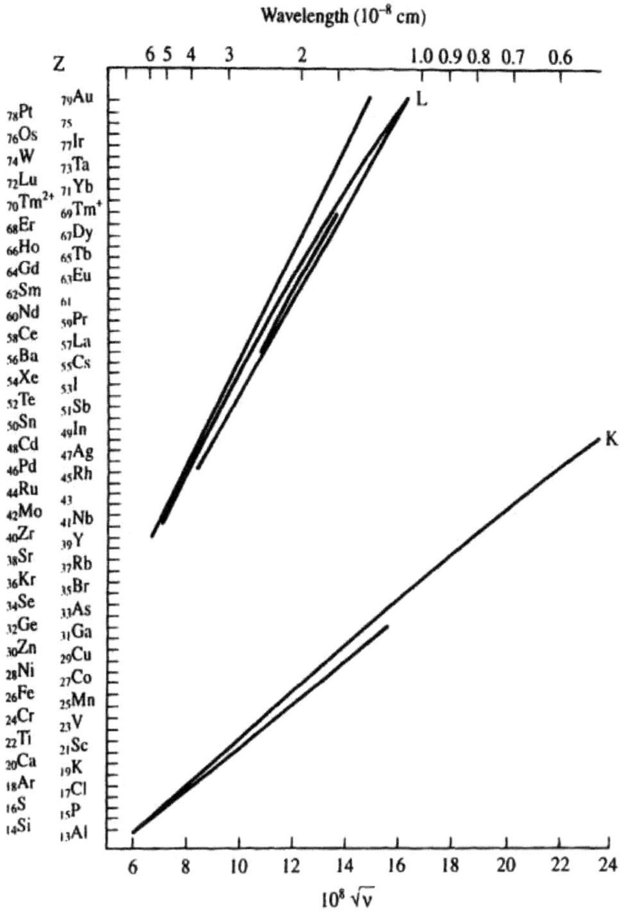

FIGURE 6.14
Moseley's plot of Z vs. $\sqrt{\nu}$.

$$
\begin{aligned}
\nu_{K_\alpha} &= 0.246 \times 10^{16}(Z-1)^2 \quad \text{Hz}, \\
\frac{1}{\lambda_{K_\alpha}} &= \tfrac{3}{4}R(Z-1)^2 \\
&= 8.23 \times 10^{-4}(Z-1)^2 \quad \text{Å}^{-1}
\end{aligned}
\tag{6.16}
$$

which can be transformed into

$$
\Delta E_{K_\alpha} = \epsilon_K - \epsilon_L = h\nu_{K_\alpha} = hRc(Z-1)^2\left(\frac{1}{1^2} - \frac{1}{2^2}\right) \simeq \tfrac{3}{4} \times 13.6 \times (Z-1)^2 \quad \text{eV} \tag{6.17}
$$

The physical meaning of this equation is very clear: the $\frac{3}{4}$ comes from the inner shell transition from $n = 2$ to $n = 1$; the 13.6 eV is the energy corresponding to the

Rydberg constant; and $(Z-1)^2$ shows that the electrons making the transition are affected by positive charges of $(Z-1)$. Now we can explain the well-known fact that the threshold energy (of electrons or photons) for production of K x-rays is larger than the energy of the K x-rays themselves. The threshold energy is the energy required to remove one electron from the $n = 1$ shell, and the energy of the K x-ray is the energy difference of the electron in the $n = 1$ and $n = 2$ shell.

Before going further, it is important to note that while Moseley's law (Eq. 6.14) or its form based on the Bohr model (Eq. 6.17) were extremely important both in terms of providing one of the first beautiful independent confirmations of Bohr's theory and of the first correct identification of the atomic numbers of several elements and the fact that element 43 is not found in nature, these equations are not exact. Equation (6.14) works well up to around $Z = 30$, but above there it begins to deviate increasingly with Z from the experimentally observed values. There is also the problem that the K_α line is in fact a doublet from the fine structure of the L shell and the separation increases with Z. For example, if we calculate the wavelength of the K_α line of zirconium $Z = 40$ using Eq. (6.16):

$$\lambda_{K_\alpha} = \frac{1}{8.23 \times 10^{-4}(Z-1)^2\,\text{Å}^{-1}} = 0.799\,\text{Å}$$

If we turn this around, then the energy of the K_α x-ray for $z = 40$ would be

$$E = \frac{hc}{\lambda_{K_\alpha}} = \frac{12.40\,\text{keV}\cdot\text{Å}}{0.799\,\text{Å}} = 15.52\,\text{keV}$$

while experimentally $E_{K_{\alpha 1}} = 15.78\,\text{keV}$ (100%) and $E_{K\alpha 2} = 15.69\,\text{keV}$ (50%), where their relative intensities are given in parentheses. If we had used Eq. (6.14), we would have come closer:

$$E_{K_{\alpha 1}} = h\nu_{K_\alpha} = h \times 0.248(Z-1)^2 \times 10^{16}\,\text{s}^{-1} = 4.136 \times 10^{-15}\,\text{eV}\cdot\text{s} \times 0.248 \times 39^2 \times 10^{16}\,\text{s}^{-1} = 15.60\,\text{keV}$$

Besides K_α x-rays, one can have K_β, K_γ where K_β is for an electron jumping from the $n = 3$ to the $n = 1$ shell, $\Delta E_{K_\beta} = \epsilon_K - \epsilon_M$ $(n = 3)$, and so forth.

In addition to the K x-rays, Moseley and others observed x-rays characteristic of transitions in which an electron jumped from higher shells to the $n = 2$ L shell, or to the $n = 3$ M shell, and so forth. Again, since the L shell is in fact three subshells and the M shell is five subshells, and so on, these spectra are even more complex under high resolution (as shown in Fig. 6.15(b)) but, of course, such resolution was not available at first. The L x-rays and M x-rays had equations with slightly different constants out front and different screening corrections, but the general form was the same (as in Eq. (6.14)) and so L x-rays were important fingerprints of elements in the early days when the K x-rays of the heaviest elements were difficult or impossible to excite or to measure because of their much higher energies.

However, let us look at a few more examples. Suppose we measure $\lambda_{K_\alpha} = 1.445\,\text{Å}$. What is the Z of the element?

$$(Z-1)^2 = \frac{1}{8.23 \times 10^{-4}\,\text{Å}^{-1} \times 1.445\,\text{Å}} = 841$$

$$Z = \sqrt{841} + 1 = 29 + 1 = 30$$

Note also that the $1/1^2$ term in Eq. (6.17) is not the energy of the K absorption edge (K binding energy), which is the energy to ionize one K shell electron, even though

$$\Delta E_{K_\alpha} = \epsilon_K - \epsilon_L$$

where ϵ_K and $\epsilon_L = hc/\lambda_{K,L}$, respectively, are the K and L shell absorption (binding) energies, respectively, since Eq. (6.15) is only for a one-electron, hydrogenlike element. Even the K electron screening in complex atoms involves the penetration of electrons from higher shells inside the K electron orbits for small parts of the time. Thus, extra screening contributes to the increasing deviation of Moseley's law for higher Z. However, if the K, L, M or other absorption edges are measured experimentally, these can be used to construct an energy level diagram. Measurement at poor resolution will produce a type of average energy for the various n(K,L,M, ...) shells. To illustrate, for $Z = 40$ one can use the first and second terms in Eq. (6.17) to obtain the energies

$$\epsilon_K = 13.6(Z-1)^2\,\text{eV} \quad \text{and} \quad \epsilon_L = (1/4)13.6(Z-1)^2\,\text{eV}$$

$$\epsilon_K = 20.7\,\text{keV} \qquad \epsilon_L = 5.17\,\text{keV}$$

to yield $E_{K_{\alpha 1}} = 15.5\,\text{keV}$ as above, which is close to experiment. For $Z = 40$, the K shell binding energy is 18.00 keV and for the $L_{I,II,III}$ subshells is 2.53, 2.31, 2.22 keV, respectively. We obtain

$$E_{K_{\alpha 1}} = 18.00 - 2.31 = 15.69\,\text{keV}$$

$$E_{K_{\alpha 2}} = 18.00 - 2.22 = 15.78\,\text{keV}$$

in agreement with experiment.

Nevertheless, because of their historical usefulness and their approximate correctness, these equations remain useful for many calculations. Of course, if the wavelengths of the radiations at which K,L, ... ejection begins to take place (the K,L, ... threshold wavelengths) are measured experimentally, then $hc/\lambda_{K,L}$ give the correct binding energies. For first-order calculations one can use the various forms of Eq. (6.17).

Moseley's experiment for the first time offered a method to measure Z precisely. Historically it was Moseley's equation that was used to determine the Z values of the elements. It was used to correct the order of $_{27}$Co and $_{28}$Ni in the periodic table, where earlier ordering by mass had put $_{27}$Co with its greater mass after $_{28}$Ni, and to show the positions of three then unknown elements $Z = 43, 61$ and 75, in the periodic table. Element $Z = 75$, rhenium, was discovered in 1925 and its x-rays fell in the proper place. The x-rays from elements 43 and 61 have not been seen among elements found in nature but these elements now are produced as radioactive elements in the laboratory. The characteristic x-ray spectra of different elements do not show the periodic changes in properties that elements do and are almost independent of the chemical states of elements. The chemical periodicity of the elements reflects the periodic changes of the electron configurations of the outer shells since the chemical properties of the elements are determined by the states of these outer shell electrons. Characteristic x-ray spectra are produced by transitions of electrons between inner shells in atoms. In recent years, very high-resolution measurements have shown, however, that even x-ray energies are related to the chemical states of the elements. But these energy shifts are several orders of magnitude less than the energies of the x-rays.[7]

Since the characteristic x-ray spectra of elements are related only to the atomic numbers, it is possible to use them as "fingerprints" in the analysis of elements. However, the production of an inner shell vacancy is first required in order to observe characteristic x-rays. When the $n = 1$, K shell is filled with two electrons, it is impossible to produce K x-rays. There must be a vacancy in the K shell to obtain K x-rays. High-energy electrons, protons, heavier ions, and x-rays can be used to eject electrons from an inner shell of an atom. Once a vacancy is produced, x-ray emission is essentially independent of the environment and is determined completely by the atomic number Z of the element.

The various methods of using x-rays as an analytical tool are classified according to the different ways of producing vacancies:

1. By an electron beam (e–x), called electron-induced x-ray emission
2. By a proton beam (p–x), called proton-induced x-ray emission (PIXE)
3. By an ion beam (i–x), called ion-induced x-ray emission
4. By x-rays (x–x), called x-ray-induced x-ray emission.

The much used proton-induced x-ray emission is described in Appendix I.

6.2.4 The Labeling of Characteristic X-Ray Radiation

Figure 6.15 is a diagram of the energy levels and x-ray transitions that can be observed. We can see that all x-rays with their final state being the $n = 1$ shell (K shell) are called K x-rays; all the x-rays with their final state being the $n = 2$ shell (L shell) are called L x-rays; and so forth (Fig. 6.15(a)). Furthermore, they are classified into K_α,K_β, ... by their different initial states. In addition to the principal quantum number n, the shells are subdivided into subshells by angular momentum, l, and total angular momentum, j, as seen in Fig. 6.15(b). The allowed (electric dipole radiation) transitions must obey the selection rules

$$\Delta l = \pm 1, \qquad \Delta j = 0, \pm 1$$

A more detailed classification of allowed x-rays is given in Table 6.1.

---

[7] P. G. Hansen, et al., *Atomic Inner-Shell Physics*, Plenum Press, New York (1983).

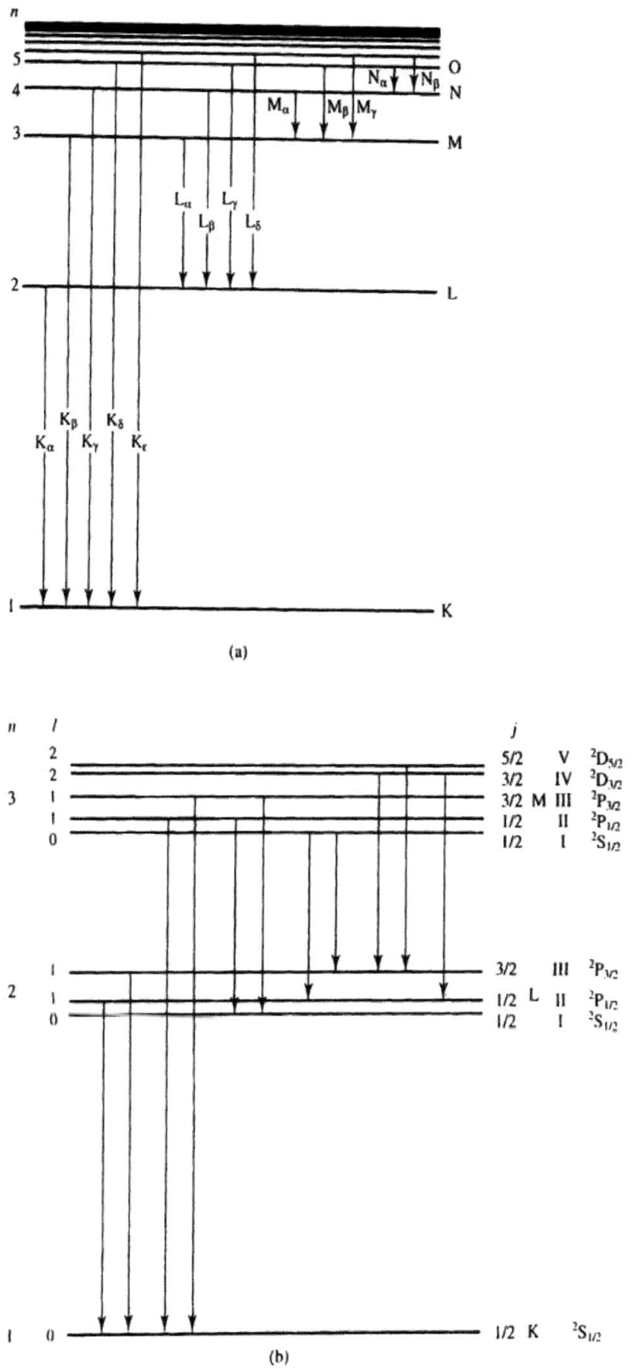

FIGURE 6.15
(a) Diagram of the atomic energy levels identified only by n and x-ray transitions between them. (b) A more complete diagram of the atomic energy levels and the allowed x-ray transitions between them.

Transitions with $\Delta l = 0, \Delta j = 0$, such as from the L_I level with $l = 0, j = \frac{1}{2}$ to the K level with $l = 0, j = \frac{1}{2}$, the $K_{\alpha 3}$ ($K - L_I$) transitions, are forbidden by the above selection rules and so occur only very rarely. They can only occur via a spin flip to give the photon its intrinsic $\hbar$ angular momentum. The $K_{\alpha 3}$ intensities vary from only 0.01 percent of $K_{\alpha 1}$ for Nd($Z = 60$) to 0.2 percent for U($Z = 92$). These forbidden transitions are also given in Appendix V.

As discussed in Section 6.2.3, the x-ray energies are fingerprints of the elements. In the laboratory elements above uranium have been produced in sufficient quantities to observe the x-rays of elements up to about $Z = 106$. Recently, heavy ions have been used to extend the study of the x-rays emitted and the binding energies of the electrons in nuclei all the way to $Z = 184$. In a heavy-ion collision where the two nuclei have energy approximately equal to the Coulomb barrier, the two relatively slow-moving heavy nuclei form for a time a nuclear molecule with $Z_c = Z_1 + Z_2$. The electrons are relativistic and so move at much higher velocities than the heavy ions. Thus, the electrons can form orbits characteristic of the combined Z_c. In Fig. 6.16 are shown, as an example, the binding energies of the 1s, 2s, $2p_{1/2}$, $2p_{3/2}$, ... electrons as calculated by W. Greiner and co-workers in Frankfurt for $Z = 178$ as a function of the distance between a lead nucleus and a curium nucleus. The calculations are in excellent agreement with recent experimental data from GSI in Germany also shown in Fig. 6.16. The K_{α} and K_{β} x-ray energies go from 300 to 800 keV. These values are much larger than those for $Z = 100$ fermium of 116–142 keV.

For this high Z, the K shell (1s) binding energy exceeds twice the rest energy of the electron (1.02 MeV) at the closest distances seen in Fig. 6.16. When the 1s

TABLE 6.1

Labeling of x-rays, where the initial electron state is given in the left column with the notation *nlj* in parentheses and the final electron state in the top row

Initial \ Final	K	L_I	L_{II}	L_{III}
L_I ($2s_{1/2}$)				
L_{II} ($2p_{1/2}$)	$K_{\alpha 2}$			
L_{III} ($2p_{3/2}$)	$K_{\alpha 1}$			
M_I ($3s_{1/2}$)			L_{η}	L_l
M_{II} ($3p_{1/2}$)	$K_{\beta 2}$	$L_{\beta 4}$		
M_{III} ($3p_{3/2}$)	$K_{\beta 1}$	$L_{\beta 3}$		
M_{IV} ($3d_{3/2}$)			$L_{\beta 1}$	$L_{\alpha 2}$
M_V ($d_{5/2}$)				$L_{\alpha 1}$
N_I ($4s_{1/2}$)			$L_{\gamma 5}$	$L_{\beta 3}$
N_{II} ($4p_{1/2}$)	$K_{\gamma 2}$	$L_{\gamma 2}$		
N_{III} ($4p_{3/2}$)	$K_{\gamma 1}$	$L_{\gamma 3}$		
N_{IV} ($4d_{3/2}$)			$L_{\gamma 1}$	$L'_{\beta 2}$
N_V ($4d_{5/2}$)				$L_{\beta 2}$
N_{VI} ($4f_{5/2}$)				
N_{VII} ($4f_{7/2}$)				

binding energy exceeds $2m_ec^2$, the spontaneous emission of positrons can occur if there is a vacancy in the K shell of the combined atom. In the Dirac theory of the electron, the infinite sea of negative energy states with $E < -2m_ec^2$ which are allowed in the theory must be filled (see Section 10.3.4). Otherwise, positive-energy electrons will fall down and fill such lower energy states. The positron is viewed as a hole in this negative energy sea. If an energy of over $2m_ec^2$ is given to a negative-energy electron, it can become a positive-energy electron, and the hole it leaves is a positron—that is the phenomenon of pair creation. In the collision of lead on curium, when there is a vacancy in the 1s shell and the distance is sufficiently close that the 1s binding exceeds $2m_ec^2$, then a negative-energy electron can fill the vacancy without any expenditure of energy and its hole, the positron, will be emitted! This spontaneous emission of positrons was predicted by Greiner and co-workers over a decade ago and has been the basis of extensive research since that time. While positrons are observed to be emitted in such collisions, the data suggest that there are additional possible processes for their formation. Now some of the positron lines are not seen in better experiments. Nevertheless, the study of such systems with extremely strong electric fields is an important area of current heavy-ion, atomic, and nuclear physics research.

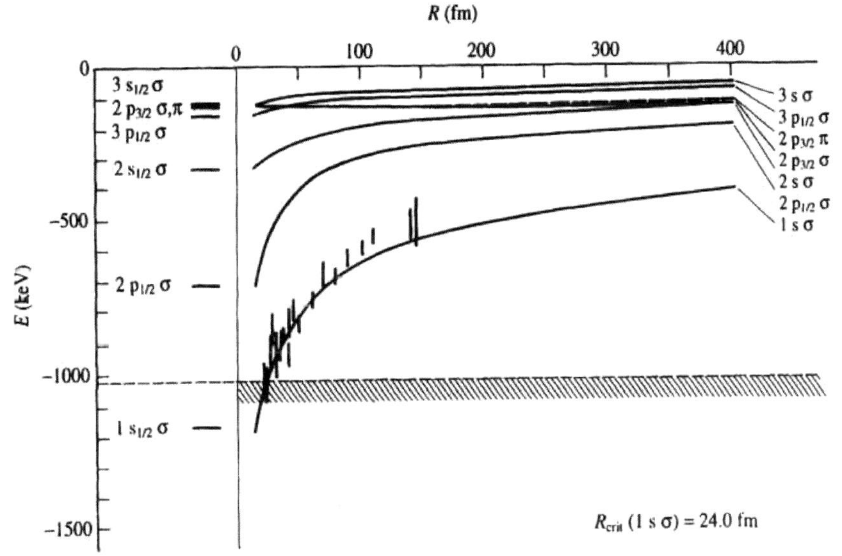

FIGURE 6.16
The curves are the energies of the electron orbitals for $Z = 178$ as calculated by W. Greiner and collaborators (Frankfurt), as a function of distance between a lead projectile nucleus and a uranium target nucleus. The experimental 1s binding energies, with error bars shown as vertical lines, were measured at GSI, Germany.

6.2.5 Auger Electrons

After a vacancy is produced in an atomic shell, x-ray emission is only one way of releasing energy. Another way, called Auger electron emission, was discovered in 1923 by P. Auger.

When one electron in the L shell drops to the K shell to fill a vacancy there, the excess energy can be given off as an x-ray (Fig. 6.17(c)), or it can be transferred to an electron in a higher shell (such as the M shell). This electron then escapes from the atom (Fig. 6.17(d)), and is called an Auger electron. Let us represent the binding energies of the electrons in the K, L and M shells by ϕ_K, ϕ_L, and ϕ_M, respectively. Then when an electron drops from the L shell to the K shell, it will release an energy of $\phi_K - \phi_L$. If this energy is given to an electron in the M shell, then $\phi_K - \phi_L - \phi_M$ is the kinetic energy of the Auger electron emitted from the M shell (Fig. 6.17(d)):

$$E_{ae} = \phi_K - \phi_L - \phi_M \tag{6.18}$$

In a more detailed treatment, Eq. (6.18) must include the various L and M subshell energies and the different screening corrections for the different subshells. Because of the many possible combinations, Auger spectra are quite complex.

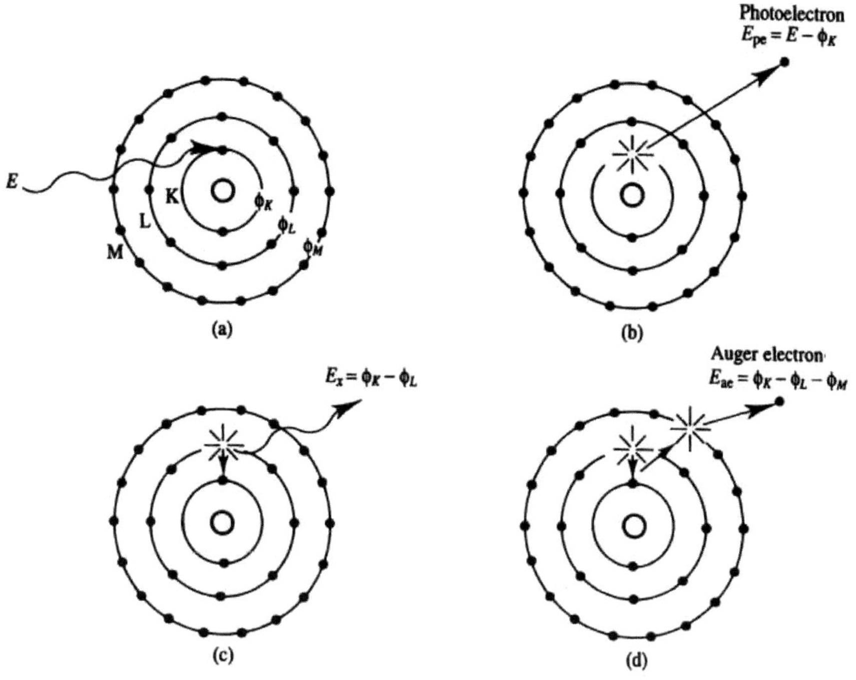

FIGURE 6.17
(a,b) Production of vacancies in electron orbitals and (c,d) the two processes which follow. Remember that these drawings are only simple illustrations of the processes that occur in electron energy shells and in no way represent the positions or motions of the electrons in the atom.

The fluorescence yield describes the competition between these processes. The fluorescence yield for the K shell is defined as

$$\omega_K = \frac{\text{number of K x-ray photons}}{\text{number of K vacancies}} \tag{6.19}$$

The fluorescence yield ω_K represents the probability of producing a K x-ray after the K shell of an atom has a vacancy, and $1 - \omega_K$ is the probability of producing an Auger electron. If $\omega_K = 90\%$, then if 100 atoms have a K vacancy, approximately 90 atoms will release K x-rays and 10 will release Auger electrons. We define ω_L, and so forth, similarly. The value of the fluorescence yield ω depends completely on the Z of the element. In general, for light elements the probability of emitting an Auger electron is large, and for heavy elements the probability of emitting x-rays is large. From Eq. (6.18) we see that the kinetic energy of an Auger electron is determined by the Z of the element, so the measurement of Auger electrons also can be used as a tool to analyze the elements in a sample. More important, the energies and intensities also depend on the details of the electron densities and screening effects, which in turn are influenced by the chemical state. In addition, because the Auger electrons have very low energies, especially those from light elements in solids, Auger electrons can only be seen from the first few atomic or molecular layers of a solid. Auger studies are one of the important techniques in probing surfaces today. Thus, Auger electron spectroscopy has now become an important tool in the study of the properties of materials.

6.2.6 Synchrotron Radiation

Synchrotrons are now an important new source of electromagnetic radiation (x-rays and γ-rays) for research. Charged particles emit radiation when accelerating. The accelerated motion may be linear or circular; however, the radiation produced by linear accelerated motion of charged particles is negligible.[8] For circular motion, the radiation intensity produced is inversely proportional to the fourth power of the mass of the charged particle. Thus, only the radiation produced by electrons which are accelerated in circular motion need be considered. The radiation produced by electrons moving in circular paths in a synchrotron accelerator is called synchrotron radiation. Energy loss by synchrotron radiation is a disadvantage in high-energy physics when high-energy electrons are used as probes. However, it is useful in other fields as a source of electromagnetic radiation because of the attractive features of the radiation.

The first feature of synchrotron radiation is that its power is very large. Using classical electrodynamics, it is easy to prove that electrons with energy E (in GeV) moving in a circle with radius of curvature R (in meters), and current intensity I (in amperes) emit a total radiation power P (in kW) of

[8] J. D. Jackson, *Classical Electrodynamics*, Wiley, New York (1975), p. 661.

$$P = \frac{88.47\, E^4 I}{R} \tag{6.20}$$

Expressed in terms of the magnetic field B (in kilogauss) which causes the electrons to move in a circle:

$$P = 2.654\, BE^4 I \tag{6.21}$$

For example, when $E = 1\,\text{GeV}$, $B = 10\,\text{kG}$ (corresponding to $R = 3.33\,\text{m}$) and $I = 0.5\,\text{A}$, $P = 13.3\,\text{kW}$!

At the present time a super-high-power x-ray tube with current of 1000 mA, high potential of 50 kV, and electron beam power of 50 kW can only produce x-rays of $P = 10\,\text{W}$ on a rotated copper target. This is over a thousand times less than the power of the synchrotron radiation from a relatively common 1 GeV synchrotron accelerator. In Germany, there is a synchrotron accelerator of 20 GeV, $I = 20\,\text{mA}$, $R = 192\,\text{m}$, which from Eq. (6.20) can reach a total power of 1500 kW!

The second characteristic of synchrotron radiation is that it gives a wide range of spectral energies. Figure 6.18 shows the different energy spectra of synchrotron radiation for 1 to 7.5 GeV electrons. The wavelengths of the x-rays vary continuously. The minimum wavelength is determined by the maximum energy of the electrons. Note the path of the electrons is only being bent, they are not being stopped, so the radiation emitted is in the low-energy x-ray region. The x-ray intensities emitted by an x-ray tube are mainly concentrated near the characteristic x-rays of the target material when the energies are sufficient to excite them, for lower-energy electrons they are near the maximum energy of the electrons (see Fig. 6.13). For instance, when using a copper target, the x-ray intensities are mainly concentrated around 1.5 Å. A rough comparison of an x-ray spectrum produced by an x-ray tube with that of synchrotron radiation is shown in Fig. 6.19.

The third characteristic of synchrotron radiation is that it is well-collimated in direction. The angular distribution of synchrotron radiation depends on the speed of the electrons. When the speed of the electrons is near the speed of light,

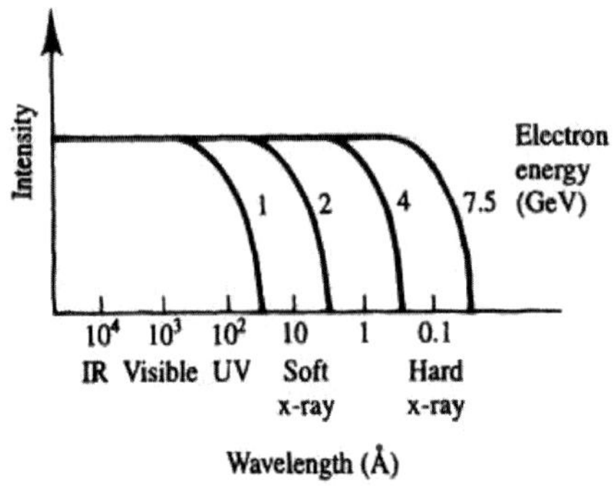

FIGURE 6.18
Energy spectra of synchrotron radiation versus intensity.

the synchrotron radiation is almost all concentrated on the line tangent to the direction of the electron motion (Fig. 6.20). For example, when the energy of an electron is 1 GeV, $\theta = 0.5$ milliradian. This collimation is as excellent as that of a laser, while the angular distribution of the x-rays from an x-ray tube is homogenous in all directions (Fig. 6.19). In addition, synchrotron radiation has a particular pulsed time structure (see Fig. 6.19), and can be used to observe instantaneously many phenomena. Also, the synchrotron radiation is a completely plane polarized wave whose polarization plane is in the plane of the circular orbit of the electrons. Common x-ray tubes possess none of these characteristics. For detailed characteristics and the applications of the synchrotron radiation, see Winick and Bienenstock[9] and Wiedemann.[9]

6.3 COMPTON SCATTERING

In 1923, A. H. Compton proved the particle nature of x-rays in an experiment which studied the scattering of x-rays by matter. In his experiment, not only the

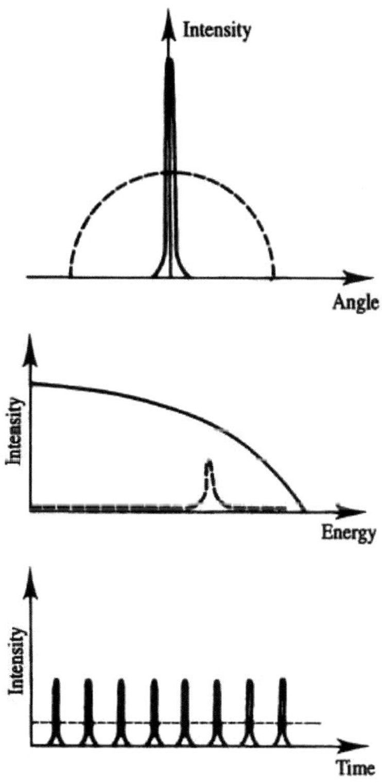

FIGURE 6.19
Comparison between the characteristics of synchrotron radiation and of the radiation from an x-ray tube, where the solid line represents the synchrotron radiation and the dashed line represents the radiation from an x-ray tube.

[9] H. Winick and A. Bienenstock, *Anna. Rev. Nucl. Part. Sci.* **28** (1978) 33; H. Wiedemann, *Synchrotron Radiation*, Springer (2002).

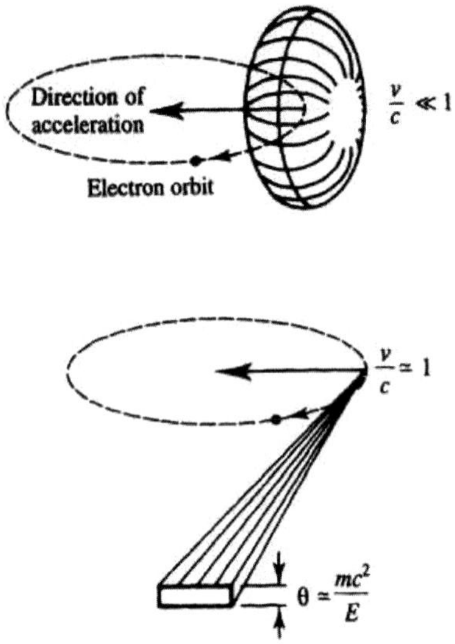

FIGURE 6.20
The directional character of synchrotron radiation.

energy but also the momentum of the x-ray photon had to be considered. Because Compton scattering showed that light traveling through space scattered like a particle with momentum and energy, this completed the proof of the quantum theory of light which had been introduced by Planck and used by Einstein to explain the photoelectric effect (which only related to the energy of the photon). It provided the first experimental evidence for Einstein's hypothesis in 1917 that a quantum of light had momentum like a particle. Many people were still suspicious of the quantum theory long after Einstein used it in 1905. But after Compton scattering had required an explanation in terms of quantum theory, objections to the quantum theory were very rare.

6.3.1 Classical Considerations

A sketch of a Compton scattering experiment is shown in Fig. 6.21. In classical electromagnetic theory, when an electromagnetic wave passes through material, the scattered radiation should have the same wavelength as the incident radiation. This is because the incident electromagnetic radiation (such as an x-ray) exerts on the electrons in the atoms of the material a periodically changing force (qE) which compels the electrons to oscillate with the same frequency as that of the incident wave. The oscillating electrons are required to emit electromagnetic waves in all directions with the same frequency as that of their oscillation (Fig. 6.22(a)). Classical electromagnetic theory has been confirmed by many macroscopic phenomena.

As a simple example, a blue coat is never seen in a mirror as red. But Compton found in his experiments on the scattering of x-rays in materials that among the scattered x-rays there were some with *increased* wavelengths in addition to those with the same wavelength as the incident x-rays. The increased wavelengths vary with the angle θ into which they are scattered. This phenomenon could not be explained by classical electromagnetic theory. However, Compton gave it a full explanation through quantum theory, and the phenomenon is called the Compton effect.[10–12]

6.3.2 The Quantum Explanation

Compton considered the observed phenomenon to be the result of the collision of an x-ray photon (particle) with free electrons. He first assumed that x-rays are composed of photons which behave as particles with both energy and momentum in passing through and colliding with matter. The relation of the wavelength λ (or frequency ν) of the x-rays to the photon energy satisfies $E_\lambda = hc/\lambda = h\nu$, and the momentum of the photon satisfies the hypothesis $p_\lambda = h/\lambda$ suggested by Einstein in 1917.

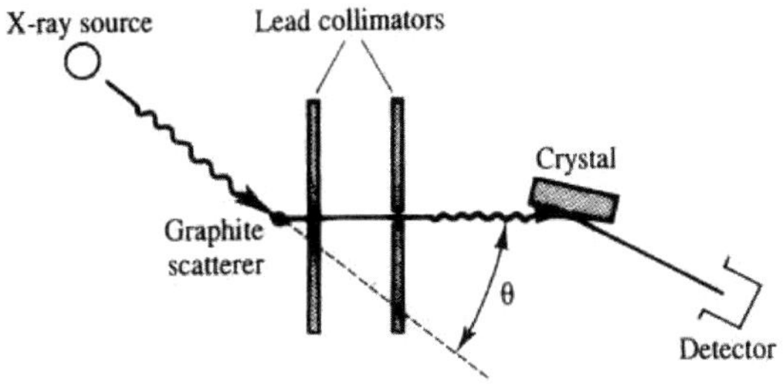

FIGURE 6.21
Schematic of a Compton scattering experiment.

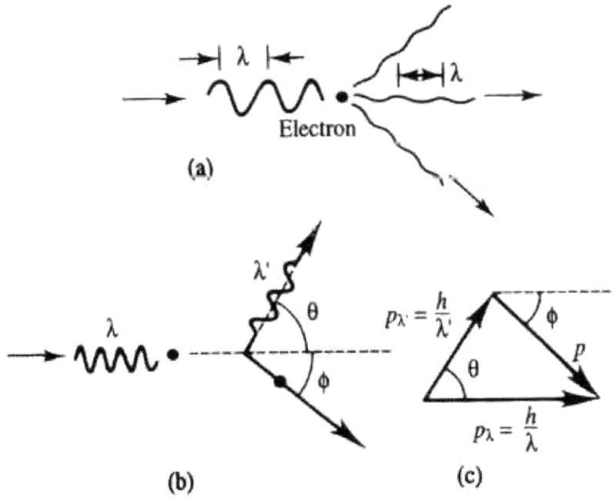

FIGURE 6.22
Radiation and electron scattering.

[10] A. H. Compton, *Phys. Rev.* **22** (1923) 409.

[11] A. H. Compton and S. K. Allison, *X-Rays in Theory and Experiment*, D. Van Nostrand, Princeton, NJ (1935).

[12] A. H. Compton, *Am. J. Phys.* **29** (1961) 817.

Assume that photons of wavelength λ collide with electrons of mass m_0 which are free and at rest. The so-called "free" and "rest" states are relative. Here they represent the situation when the binding energies of the electrons in the atom are negligible compared to the energy of the incident photon. In his experiments Compton used a molybdenum x-ray tube with a potential of 50 kV. The K_α line of molybdenum with an energy of about 20 keV is far beyond the binding energy (a few eV) of the outer shell electrons of all elements. After collision, the wavelengths of the scattered photons were observed at an angle θ with respect to the incident direction. In the collision, the electron recoils and emerges with an energy E at an angle θ with respect to the incident photon direction (see Fig. 6.22(b)). According to the conservation of energy and momentum, we find

$$\begin{aligned} h\nu + E_0 &= h\nu' + E = h\nu' + K + E_0 \\ \boldsymbol{p}_\lambda &= \boldsymbol{p}_{\lambda'} + \boldsymbol{p} \\ p_\lambda^2 + p_{\lambda'}^2 - 2p_\lambda p_{\lambda'} \cos\theta &= p^2 \\ (h\nu)^2 + (h\nu')^2 - 2h\nu\, h\nu' \cos\theta &= p^2c^2 \end{aligned} \tag{6.22}$$

where E, K and p are the total energy, kinetic energy and momentum of the recoiling electron, respectively; $E_0 = m_0c^2$ is the rest mass energy of the electron; and $p_\lambda = h/\lambda$ and $p_{\lambda'} = h/\lambda'$ are the momentum of the photon before and after collision, respectively. Since the photon is a particle moving at the speed of light, it is necessary to use the relativistic equations:

$$\begin{aligned} m &= \frac{m_0}{\sqrt{1 - v^2/c^2}} \\ E &= mc^2 \\ E^2 - p^2c^2 &= E_0^2 \end{aligned} \tag{6.23}$$

Putting these into Eq. (6.22), and rearranging, we find

$$\lambda' - \lambda = \Delta\lambda = \frac{h}{m_0c}(1 - \cos\theta) \tag{6.24}$$

This is the famous Compton scattering relation which agrees with experiments very well.

Transforming Eq. (6.24) into

$$\frac{1}{h\nu'} - \frac{1}{h\nu} = \frac{1}{m_0c^2}(1 - \cos\theta) \tag{6.25}$$

we obtain the energy equation of the scattered photon:

$$h\nu' = \frac{m_0c^2}{1 - \cos\theta + 1/\gamma}, \qquad \gamma \equiv \frac{h\nu}{m_0c^2} \tag{6.26}$$

Note that the energy of the scattered photon is a function of the energy of the incident photon as well as of θ. For $\theta = 0°$, $h\nu' = h\nu$ and the minimum $h\nu'$ occurs for $\theta = 180°$.

Substituting Eq. (6.26) into Eq. (6.22), we find the kinetic energy ($K = E_k$) of the recoiling electrons is

$$E_k = h\nu - h\nu' = h\nu \frac{\gamma(1 - \cos\theta)}{1 + \gamma(1 - \cos\theta)} \tag{6.27}$$

The electron energies are spread over a range from zero when Compton scattering occurs with $\theta = 0°$, up to a maximum value when $\theta = 180°$ and $1 - \cos\theta = 2$.

From this we can show that the maximum energy gained by a recoiling electron is

$$E_{k,\max} = h\nu \frac{2\gamma}{1 + 2\gamma} \tag{6.28}$$

In a photon detector that relies on the transfer of photon energy to an electron to determine the photon energy, such as a sodium iodide or germanium detector, $E_{k,\max}$ is called the Compton edge of the continuous distribution of absorbed photon energies which go from zero to $E_{k,\max}$. The continuous distribution is called the Compton distribution and contributes to a smooth, continuous background in a detector. Today, photon detectors are surrounded with antiCompton shields which detect the Compton-scattered photons, and these signals in the shield are used to reject the signals from the initial Compton scattering events in the detector to reduce the Compton background under the full energy peaks which correspond to photoelectric absorption.

Equations (6.24) to (6.27) are the complete set of Compton scattering relations, with the latter three derived from the initial equation (6.24). Let us do a few simple examples. Take a photon of energy about 1 MeV. For convenience let $h\nu = 1.022$ MeV or $\gamma = h\nu/m_0c^2 = 2$ and calculate $h\nu'$ for $\theta = 90°$ and 180°. Then

$$h\nu' = \frac{m_0c^2}{1 - \cos\theta + 1/\gamma} - \frac{511\,\text{keV}}{1.5 - \cos\theta}$$

$$h\nu' = \frac{511}{1.5}\,\text{keV} = 341\,\text{keV for } \theta = 90°; \qquad h\nu' = \frac{511}{2.5}\,\text{keV} = 204\,\text{keV for } \theta = 180°$$

$$E_{k,\max}(\theta = 180°) = 1.022\,\text{MeV}\ \frac{2\gamma}{1 + 2\gamma} = 1.022\,\text{MeV}\frac{4}{5} = 818\,\text{keV}$$

$$E_k(\theta = 90°) = h\nu - h\nu' = 1.022 - 0.341\,\text{MeV} = 681\,\text{keV}$$

Suppose that you determine that the maximum energy a photon can transfer to an electron is 100 keV. What is the energy of the photon?

$$E_{k,\max} = h\nu\frac{2\gamma}{1+2\gamma}, \qquad \gamma = \frac{h\nu}{511}$$

$$= \frac{h\nu}{1+1/2\gamma} = 100\,\text{keV}$$

$$h\nu = 100\,\text{keV} \times \left(1 + \frac{511}{2h\nu}\right)$$

$$h\nu^2 - 100\,h\nu - 50 \times 511 = 0$$

$$h\nu = 217.5\,\text{keV}$$

6.3.3 The Physical Meaning

COMPTON WAVELENGTH OF THE ELECTRON. The dimension of the coefficient h/m_0c in Eq. (6.24) is length. This coefficient is called the Compton wavelength of the electron. Its physical meaning is the wavelength of the corresponding photon when the energy of the incident photon equals the rest energy of the electron, that is,

$$h\nu = m_0c^2 \qquad \text{or} \qquad h\frac{c}{\lambda} = m_0c^2$$

Then

$$\lambda = \frac{hc}{m_0c^2} = \frac{12.4\,\text{Å}\cdot\text{keV}}{511\,\text{keV}} = 0.024\,26\,\text{Å} \tag{6.29}$$

and this is the Compton wavelength of the electron. From Eq. (6.24), we can see that the Compton wavelength of the electron is equal to the wavelength difference between the incident wave and scattered wave when $\theta = 90°$.

In addition to defining $\lambda_{ec} = h/m_0c$ to be the Compton wavelength of the electron, one also defines:

$$ƛ_{ec} = \frac{\hbar}{m_0c} \tag{6.30}$$

to be the reduced Compton wavelength of the electron. This can be transformed into

$$ƛ_{ec} = \frac{\hbar c}{e^2}\frac{e^2}{m_0c^2} = \frac{r_e}{\alpha} \simeq 137r_e \tag{6.31}$$

where r_e is called the classical radius of the electron and is the distance at which the repulsive potential between two electrons is equal to their rest mass energy:

$$m_0c^2 = \frac{e^2}{r_e} \tag{6.32}$$

$$r_e = \frac{e^2}{m_0c^2} \simeq 2.82\,\text{fm} \tag{6.33}$$

Equation (6.31) tells us that the reduced Compton wavelength of the electron is about 137 times the classical radius of the electron.

$\Delta\lambda$ IS DETERMINED ONLY BY θ AND IS INDEPENDENT OF λ. From Eq. (6.24) we can see that $\Delta\lambda$ is independent of the wavelength λ of the incident wave, and depends only on the value of the scattering angle θ. When $\theta = 180°$,

$$\Delta\lambda = 2\frac{h}{m_0 c} = 0.049\,\text{Å}$$

This value is the maximum shift by which the wavelength of any electromagnetic wave can be increased by Compton scattering (see Fig. 6.23).

In a practical measurement, it is the ratio $\Delta\lambda/\lambda$ that is significant. Since $\Delta\lambda$ is independent of λ only for x-rays with $\lambda \leq 1\,\text{Å}$, is it possible to make $\Delta\lambda/\lambda$ sufficiently large that it can be observed? For visible light of $\lambda \approx 5000\,\text{Å}$, $\Delta\lambda$ is the same, but even for the maximum $\Delta\lambda$ at 180°, $\Delta\lambda/\lambda = 0.049/5000 \simeq 10^{-5}$ is so small that it cannot be measured. That is why we observe the Compton effect only in x-ray scattering experiments and why, in ordinary macroscopic phenomena, classical electromagnetism explains the experiments very well.

THE RELATIONSHIP OF ΔE TO λ. Although the Compton shift in terms of wavelength is independent of the wavelength (or energy) of the incident light, the Compton shift ΔE in terms of energy depends on the wavelength (or energy) of the incident light. To explain this, we give two experimental results obtained under the same conditions. When the energy of the incident photon $E = 10\,\text{keV}$, in the direction $\theta = 90°$ the measured energy of the scattered photon $E' = 9.8\,\text{keV}$, and the relative change $\Delta E/E \approx 2\%$. When the energy of the incident photon $E = 10\,\text{MeV}$, the measured $E' = 0.49\,\text{MeV}$ at 90°, and the relative change is $\Delta E/E \approx 95\%$. This shows clearly that ΔE depends on the wavelength (or energy) of the incident photon.

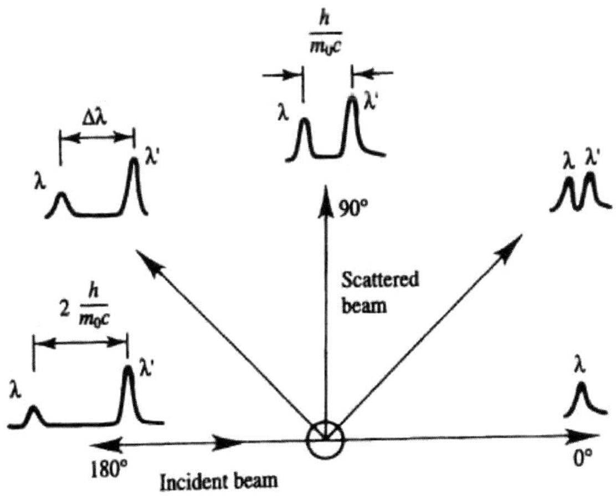

FIGURE 6.23
Change of the Compton shift with the angle θ.

Equation (6.26) tells us that the energy of a scattered photon increases as the energy of the incident photon increases. But Eq. (6.26) gives a limit to this increase except at $\theta = 0°$. For example, when $\theta = 90°$, no matter how large the energy of the incident photon, the energy of the scattered photon cannot be larger than 0.511 MeV.

COHERENT SCATTERING. At $\theta = 0$, it is obvious that $\Delta\lambda = 0$, and the incident wave is not deflected. Of course, no change of wavelength occurs there. But the situation of $\Delta\lambda = 0$ is observed not only in the direction of $\theta = 0$ but also in all directions (see Figs 6.23 to 6.25); that is, Compton scattering is always accompanied by the scattering of $\Delta\lambda = 0$, which is called coherent scattering. In turn, Compton scattering also is called noncoherent scattering. What causes the coherent scattering? When large numbers of photons strike atoms, some photons are not scattered by the outer shell electrons in the atom, which may be regarded as free electrons, but interact with the tightly bound electrons in inner shells. Since the inner shell electrons are comparatively tightly bound to the nucleus, these incident photons are scattered by the whole atom. In this situation, the electronic mass m_0 should be replaced by the atomic mass M in Eq. (6.24). Since $M \gg m_0$, $\Delta\lambda \approx 0$. So coherent scattering is the result of the interaction of incident photons with tightly bound electrons in the atom. This scattering clearly increases as the atomic number Z increases. This increased intensity for coherent scattering is confirmed by experiments (see Fig. 6.25). Essentially, coherent scattering is elastic scattering. The coherent scattering that accompanies Compton scattering gives a convenient standard spectrum line for the measurement of the Compton shift $\Delta\lambda$.

6.3.4 Compton Scattering and the Fundamental Constants

In the Compton scattering relation Eq. (6.24), Planck's constant h and the speed of light c play key roles. If $h \to 0$, or $c \to \infty$, $\Delta\lambda$ will decrease to zero, and the situation returns to classical physics. Since the Compton shift $\Delta\lambda$ is related to the three basic constants (h, c, m_0), from a measurement of $\Delta\lambda$ we may use two known constants to determine the third. Compton scattering experiments offer a method for measuring Planck's constant h independently. The value of h from Compton scattering is in good agreement with that from other methods, which again confirms the universality of Planck's constant. In fact, Compton scattering also offers us a good method of measuring the photon energy $h\nu$. A measurement of the energy of a charged particle can generally be made with high precision, so by measuring the energy of the recoiling electrons, we can determine precisely the energy of the incident photons from relation (6.27).

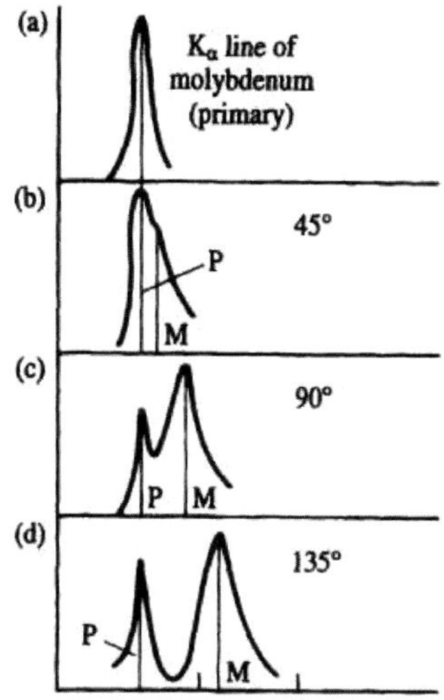

FIGURE 6.24
The scattered x-ray spectra of the molybdenum K_α line (primary x-ray) measured at different angles after being scattered by graphite. Here P is the position of the wavelength of the primary x-ray and M is the position that can be computed from Eq. (6.24). (This figure is from Compton's work, ref. 10.)

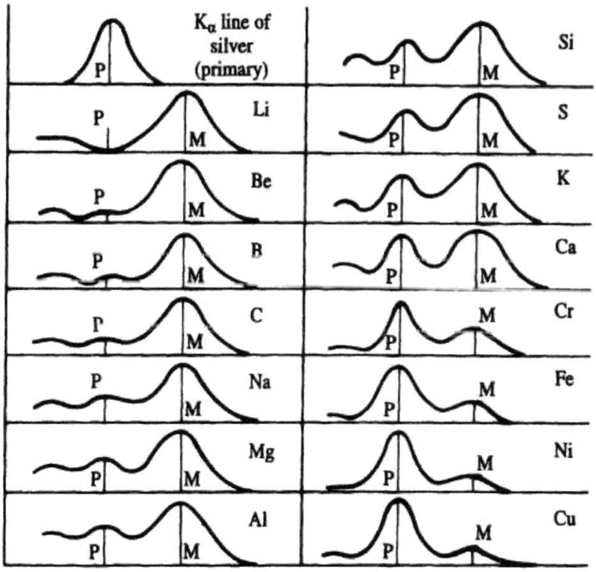

FIGURE 6.25
The x-ray energy spectra of the silver K_α line scattered by each element at a scattered angle $\theta = 120°$ (from ref. 11). This experimental result was published by Youxun Wu in 1926.

6.3.5 Note 1: Compton Profile

When deriving the Compton scattering relation, we hypothesized that the electron is not only free but also at rest. But, of course, the electron is not at rest. As shown in Fig. 6.25, the profile of the scattered x-ray spectra varies with the Z of the elements. This variation reflects the different momentum states of the motion of the electrons in the different elements.

Early research revealed that the Compton profile reflects directly the distribution of electron momenta in the outer shells of the material. In 1929, relying on this J. W. M. DuMond proved with a direct experimental method that the outer shell electrons in lithium obeyed the Fermi–Dirac distribution law instead of the Maxwell–Boltzmann distribution law. This was a most important achievement, but at that time, because of a lack of strong x-ray sources and the low sensitivity of the detectors, this kind of experiment was rather difficult. After 1965, experimental and theoretical work on the application of the Compton profile method to detection of the momentum distribution of electrons became active again. This occurred in part because strong x-ray sources and high sensitivity x-ray detectors became available to greatly increase the precision of experiments. In addition, high-efficiency computers also appeared and made the complex theoretical calculations possible. The developments of quantum chemistry and solid state physics have required experimental information about the moving states of electrons. The Compton profile has offered a convenient experimental method for detecting the distribution of electron momenta.

The Compton profile method for measuring the distribution of electron momenta has some characteristics which other methods do not have. First, it reflects only the momentum states of the outer shell electrons which are of concern in quantum chemistry and solid state physics, because the interference of the inner shell electrons is only very slight. Second, since the scattering of each electron is independent of the scattering of others, the result from an impure sample is the same as that from a pure crystal sample. This greatly simplifies the preparation of samples. And third, in general, from the profile lines one can directly gain many qualitative conclusions about some properties of electrons in molecules and solids. Because of these characteristics, the Compton profile method has been widely applied in recent years.

6.3.6 Note 2: Inverse Compton Effect

The Compton effect refers to the phenomenon whereby photons of high energy impact with electrons of low energy with the emission of a photon of lower energy, longer wavelength, and lower frequency. If the electron with which the photon impacts is of high energy, then the reverse situation occurs. Here the high-energy electron gives part of its energy to the low-energy photon, which gains energy to raise its frequency and shorten its wavelength. This phenomenon is called the inverse Compton effect. The radiation produced from it (generally in the x-ray range) is known as inverse Compton radiation. Obviously, the prerequisite for production of inverse Compton radiation is that there are both high-energy electrons and background radiation (photons). In a magnetic field, high-energy electrons may give off synchrotron radiation which itself is the background radiation field. This is sure to result in the inverse Compton effect. This kind of radiation is known as synchrotron-inverse Compton radiation. Synchrotron radiation and inverse Compton radiation are the two mechanisms by which high-energy electrons lose energy.

6.3.7 Note 3: The Photon Theory of Light

The Compton effect played an essential role in the full verification of the quantum (particle) nature of light. Planck introduced the idea that light was emitted in quanta $h\nu$ (particlelike bundles of energy) in 1900 to explain blackbody radiation. In 1905 Einstein introduced the idea that light is absorbed in quanta $h\nu$ to explain the photoelectric effect. The completion of the picture of the particle nature of light was to show that light, as it traveled through matter, could scatter and exchange energy and momentum like a particle with energy $h\nu$ and momentum $h\nu/c = h/\lambda$. Compton demonstrated this property in discovering the Compton effect. Each of these three won the Nobel Prize for their work on delineating the photon (particle) nature of light. However, the wave nature of x-rays has been confirmed by polarization and diffraction experiments. Thus, x-rays, as well as visible light, exhibit the dual nature of particles and waves.

6.4 THE ABSORPTION OF X-RAYS

When x-rays interact with material, the intensity of the x-rays is reduced as they pass through the material. This is the phenomenon of x-ray absorption. First we will consider two kinds of interactions of x-rays with matter. Then we will discuss the main subject: how the interaction of x-rays with matter leads to the absorption of x-rays. Next the important phenomena in the absorption of x-rays, absorption edges, are considered. Finally we will introduce a currently very active subject: the extended x-ray absorption fine structure.

6.4.1 Two Kinds of Interactions

Consider a well-collimated beam of monoenergetic particles passing through a slab of matter. The properties of the transmitted beam depend on the nature of the beam itself and of the material. But no matter what beam or material is used, there are always two kinds of limiting situations, two kinds of interactions, as shown in Fig. 6.26.

The first kind of interaction involves many small interactions (Fig. 6.25(a)). The interaction of heavy charged particles passing through matter is a typical example of this kind of interaction. Such particles experience many interactions in the material. Each interaction results in a small loss of energy and deflection of the direction of the particle, generally in the form of small-angle scattering. The final loss of energy and direction of deflection is the statistical summation of all the interactions. As a result, a particle beam which is monoenergetic and collimated before going into an absorbing material has its energy lowered and is no longer monoenergetic but has a spreading distribution (Fig. 6.27(a)) and an angular distribution after it passes through the absorbing material (Fig. 6.27(b)). When the absorbing material is thinner than a certain thickness, basically all particles can pass through it, and when the absorbing material is thicker than a certain limit all the energy of the particles is lost and the particles are unable to pass through it. The thickness at this limit is known as the range of the charged particles in that material. It is useful to define two ranges: an average range (where the thickness R_0 corresponds to the situation where the number of particles passing through the absorbing material is half of its original number) and the extended range R_{ext} (see Fig. 6.27(c)).

The second kind of interaction is an all-or-nothing interaction. The typical example is the photoelectric effect. In this kind of interaction, a particle either does not experience an interaction or it disappears from the beam in only one interaction (Fig. 6.26(b)). The particle which does not experience an interaction keeps the same energy and collimation after passing through the material. Let the intensity of a particle beam be I_0 rather than $N(0)$ as in Fig. 6.27. When passing through an absorbing material with a thickness of dx, if the particle beam experiences "destructive" interactions, its intensity will be reduced by dI as shown in Fig. 6.28. This quantity is obviously directly proportional to the thickness dx of the absorbing material and is also proportional to the intensity I of the beam. Let the proportionality constant be μ, then (Fig. 6.28)

$$-dI = \mu I(x)\,dx$$

After integrating,

$$I = I_0 e^{-\mu x} \qquad \text{or} \qquad \ln(I/I_o) = -\mu x \tag{6.34}$$

This is the Lambert–Beer rule, which shows that the intensity of the transmitted particles decreases exponentially with the thickness of the absorption material, where μ is called the absorption coefficient, or the $\ln(I/I_0)$ decreases linearly with x as shown in Fig. 6.28.

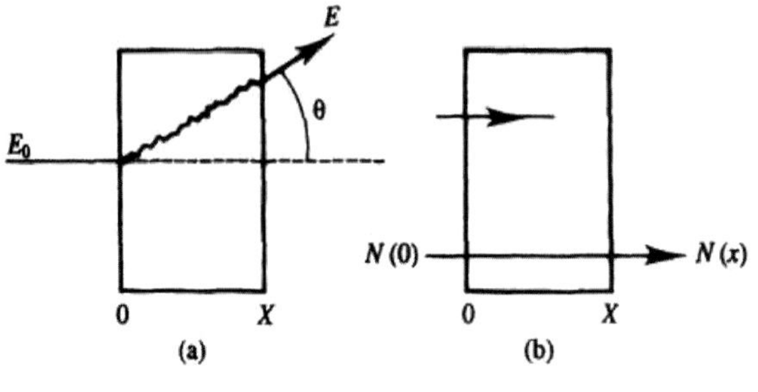

FIGURE 6.26
The two types of interactions a beam of particles can undergo in passing through matter: (a) many small interactions; (b) one interaction which removes a particle from the beam.

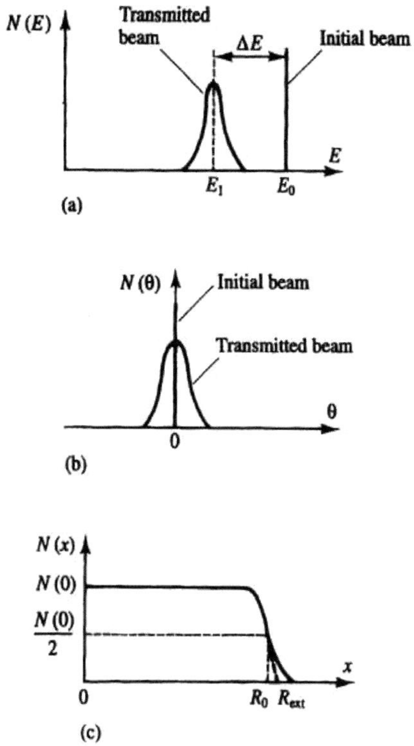

FIGURE 6.27
The interaction of heavy charged particles with matter: (a) spectrum; (b) angular distribution; (c) range.

Obviously, the concept of range has no significance in this kind of interaction. However, the value of x when $\mu x = 1$ is called the absorption length, which is the reciprocal of the absorption coefficient. It represents the thickness of the absorbing material for which the number of transmitted particles is $1/e$ (that is, 37 percent) of that of the incident particles.

If the unit of x is chosen to be centimeters (cm), then the unit of μ is cm^{-1}. This μ is called the linear absorption coefficient. We also can change the μx into $(\mu/\rho)x\rho$, where ρ is the density of the absorbing material. Here $x\rho$ represents the thickness of the absorbing material in units of mg/cm^2 and is known as the mass thickness; and μ/ρ is called the mass absorption coefficient in units of cm^2/mg. In a sense, μ/ρ is more fundamental than μ, since the value of μ/ρ does not depend on the physical state of the absorbing material (gaseous, liquid or solid). It more directly reflects the nature of the absorbing material and is more convenient to measure. Sometimes for the sake of simplicity, we use the symbol μ to represent the mass absorption coefficient. So long as one carefully notes its units, one can easily tell whether the linear absorption coefficient or the mass absorption coefficient is being used.

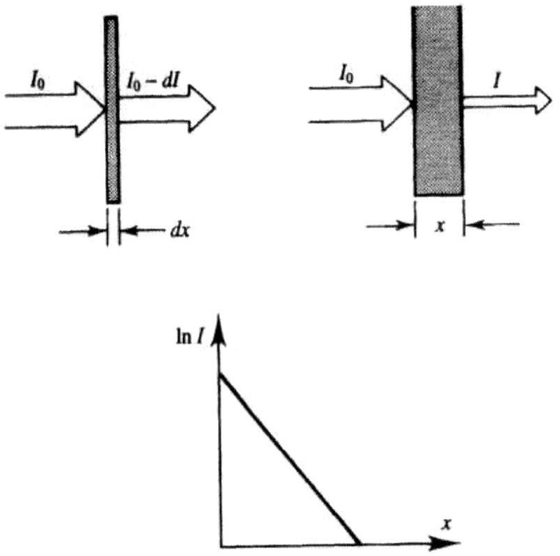

FIGURE 6.28
Characteristics of the "all-or-nothing" interaction.

6.4.2 The Interactions Between Photons and Matter

There are three main interactions between photons and matter: the photoelectric effect where the photons interact with the tightly bound electrons; Compton scattering where the photons are scattered by free electrons; and the electron–pair creation effect where, when the energy of the photons is larger than twice the rest mass of an electron (that is 1.02 MeV), photons in the atom's nuclear field are transformed into a positively and negatively charged electron pair. These are the three principal types of interactions. Other interactions are unimportant generally, but they may be important in some special situations.[13] It is easy to prove that it is impossible for free electrons to result in the photoelectric effect or in pair production. In both cases, they would violate the laws of conservation of energy and momentum. The importance of each of the three effects varies with the absorbing material as well as the energy of the photons, as shown in Fig. 6.29. Obviously, from this figure, the main interactions for x-rays are the photoelectric effect and Compton effect. For low energy and higher Z the photoelectric effect dominates; for energies of about 1–10 MeV and low Z the Compton effect dominates; and for high energies above 10 MeV and high Z pair production dominates. To illustrate the energy and Z dependence, the photoelectric cross-section (a measure of the probability of a photoelectric event occurring) for energies above the K threshold is

$$\sigma_{PE} = 4\alpha^2 \sqrt{2}\, Z^5\, \phi_0 \left(\frac{m_e c^2}{h\nu}\right)$$

[13] R. D. Evans, *The Atomic Nucleus*, McGraw-Hill, New York (1955), p. 672.

where $\alpha \simeq 1/137$ is the fine structure constant, and $\phi_0 = 6.651 \times 10^{-25}$ cm. Note that the cross-section goes down with increasing energy but increases as Z^5, a very strong Z dependence.

It is easy to see that the photoelectric effect and electron-pair production are both typical examples of the all-or-nothing interaction. In what category is the Compton effect? At first glance, it may seem to belong to the category of "many small interactions," but careful examination shows that fundamentally it also belongs to the "all-or-nothing" interaction. This is because, when the energy of a photon is not very high, the scattered photon is largely not concentrated in the forward direction and the efficient angular range of the measuring equipment is limited generally to much less than the whole solid angle of 4π. So the effect of Compton scattering is equivalent to removing the incident photons from the beam. As a result, the Compton effect can be classified approximately into the second kind of interaction. In addition, the lower-energy Compton scattered photon is very likely to be absorbed by the photoelectric effect.

Thus, the interactions between photons and matter are basically "all-or-nothing" interactions. The intensity should decrease exponentially on passing through the absorbing material.

$$I = I_0 e^{-\mu x} = I_o e^{-(\mu/\rho)x\rho} \tag{6.35}$$

where the absorption coefficient contains three parts:

$$\mu = \mu_{\text{photoelectric}} + \mu_{\text{Compton}} + \mu_{\text{pair production}} \tag{6.36}$$

Similarly, there is no concept of range here. But one can define the absorption length (also called the average free path) x_0:

$$\mu x_0 = 1, \qquad \frac{I}{I_0} = e^{-1} \tag{6.37}$$

There are also curves of half-thicknesses (mg/cm^2) of various materials and elements where, for the half thickness, $I/I_0 = \frac{1}{2}$.

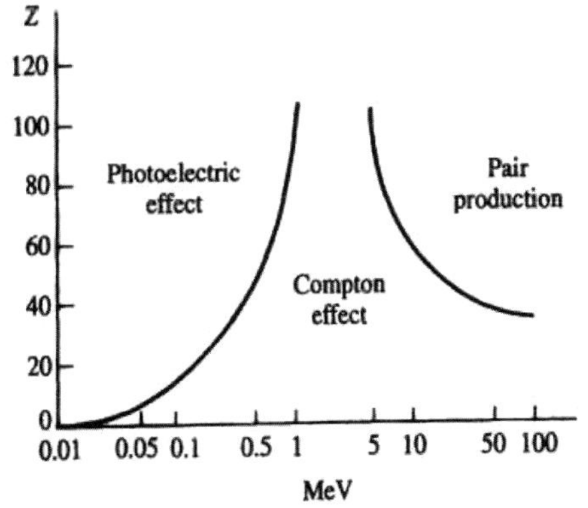

FIGURE 6.29
Relative importance of the three major interactions between photons and matter. The curves show where the interactions to either side are equal.

6.4.3 The Absorption of X-Rays

X-rays are composed of photons with energies generally speaking no more than 150 keV, the upper limit for the energies of x-rays emitted by atoms or x-ray tubes. Of course, stopping very high energy electrons can produce many high energy photons which are sometimes called x-rays and sometimes γ-rays. Here let us consider x-rays to have energies less than 150 keV. Their absorption in matter obeys Eq. (6.35). Pair production does not exist for photons below 1.02 MeV, but coherent scattering (also called Rayleigh scattering) is important in some situations. In coherent (Rayleigh) scattering the x-rays set electrons in motion with the same frequency as the x-rays, and so the electrons absorb the incident frequency and re-radiate this with the same frequency and in phase with the initial radiation, but now scattered in all directions. Thus,

$$\mu = \mu_{\text{photoelectric}} + \mu_{\text{Compton}} + \mu_{\text{coherent}}$$

For the elements carbon, aluminum, iron, and lead, the mass absorption coefficients μ/ρ and their separate components as functions of the x-ray energy are shown in Fig. 6.30. All of the μ in the figures represent mass absorption coefficients for simplicity. However, sometimes absorption coefficients in cm^{-1} are also used.

From the figures we can see that, in the low-energy region, the principal contribution is from the photoelectric effect. When the energy of the incident x-ray photons rises gradually to a certain value (which varies with the absorbing material), the contribution of the photoelectric effect is negligible and the Compton effect dominates.

To give a specific example, we list in Table 6.2 the absorption coefficient of the copper $K_{\alpha 1}$ x-ray on passing through carbon.

If the absorbing material is a homogeneous mixture of j kinds of different pure elements, then the mass absorption coefficient is

$$\mu = \sum_j w_j \mu_j \tag{6.38}$$

where μ_j represents the mass absorption coefficient of the jth element in the absorbing material. Its fractional weight in the absorption material is w_j and $\sum_j$ represents the summation over all elements. Obviously,

$$\sum_j w_j = 1 \tag{6.39}$$

As an example of Eq. (6.35), at 100 keV the linear absorption coefficient of lead is about $100\,\text{cm}^{-1}$ (check with Fig. 6.30(d)). A 0.20 cm lead foil would give an attenuation of $I/I_0 = e^{-100\,\text{cm}^{-1} \times 0.20\,\text{cm}} = e^{-20} \approx 2 \times 10^{-9}$. If we had used iron, which has a linear absorption coefficient of about $5\,\text{cm}^{-1}$ (check with Fig. 6.30(b)), what thickness would we need to produce the same reduction? This requires

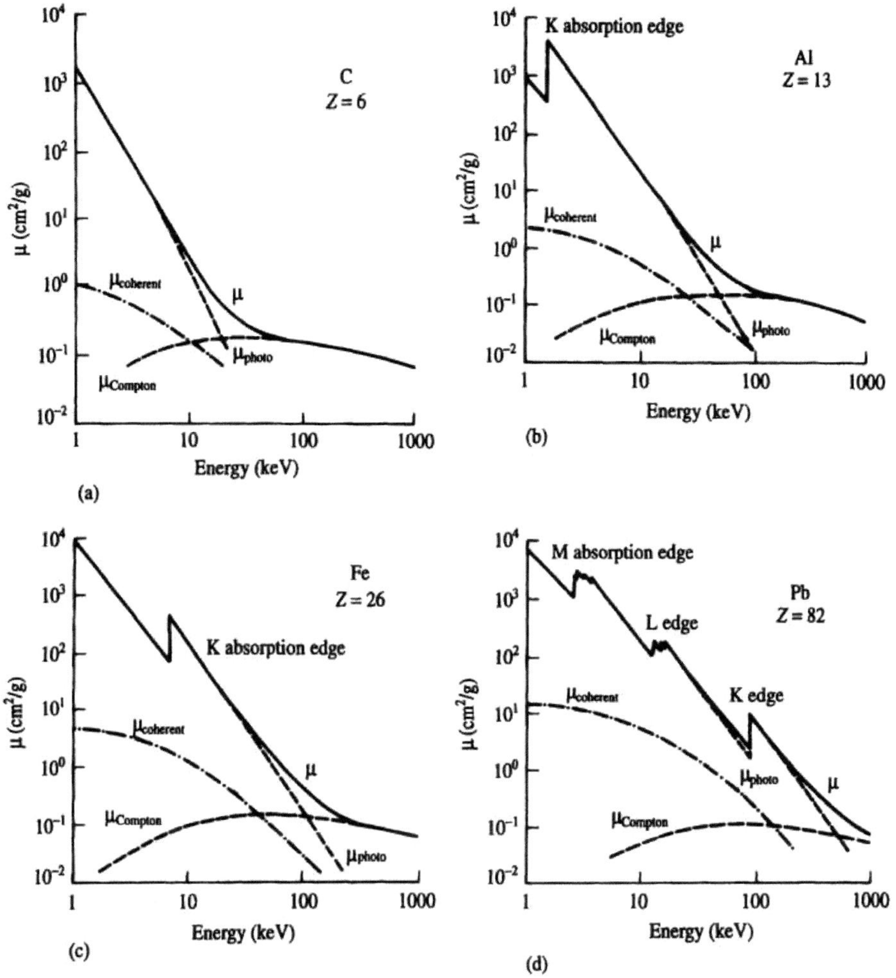

FIGURE 6.30
Relationship of the mass absorption coefficient with the energy of the incident photon for (a) carbon, (b) aluminum, (c) iron, (d) lead.

$$\mu_{Pb} t_{Pb} = \mu_{Fe} t_{Fe}$$
$$20 = 5\,\text{cm}^{-1}\, t_{Fe}$$
$$t_{Fe} = 4\,\text{cm}$$

6.4.4 The Absorption Limit

Let us examine Fig. 6.30(d) carefully. The general variation of μ with E is that the more the energy of the x-ray photon increases, the more the absorption coefficient decreases. This decrease in the absorption coefficient with increase in energy leads

TABLE 6.2
The absorption coefficient of the $K_{\alpha 1}$ x-ray (8046 eV) of copper scattered by carbon when it passes through carbon

	$\mu(cm^2/g)$	Percentage
Compton scattering	0.133	2.9
Coherent scattering	0.231	5.1
Compton + coherent	0.364	8.0
Photoelectric	4.15	92.0
TOTAL	4.51	100

to greater penetrability the higher the energy of the x-ray photon. But in the figure there are three large, sudden changes, marked K absorption edge, L absorption edge, and M absorption edge. In the L absorption edge, there are three small variations, which are called the L_I, L_{II}, L_{III} absorption edges. In the M absorption edge, there are five smaller variations called the M_I, M_{II}, M_{III}, M_{IV}, and M_V absorption edges.

Comparing with Fig. 6.15, we see at once the physical reason why the absorption edges exist. The K absorption edge represents the energy of a photon that is needed to remove one 1S electron from an atom to give what may be called a resonant absorption by the atom. A photon with energy below the K binding energy cannot remove a K electron. For energies just above the binding energy, the absorption coefficient increases abruptly. The L_I absorption edge indicates the energy of a photon which is needed to remove a 2S electron from an atom, and L_{II} and L_{III} indicate the energies for removing electrons in the $2P_{1/2}$ and $2P_{3/2}$ shells, respectively, of an atom (see Fig. 6.15(b)). We can understand the M absorption edge similarly. For aluminum and iron, only the K absorption edges are shown in Figs 6.30(b) and 6.30(c), because the absorption edges of L and M are below 1 keV. Similarly, in Fig. 6.30(a) for carbon, the very low-energy K absorption edge is not shown. The existence of absorption edges again proves clearly the reality of the shell structure of electrons in atoms.

For a quantitative understanding, Fig. 6.31 shows the K and L absorption edges of lead. The numerical values in the figure represent the energies (in keV) of the absorption edges, and the numerical values in brackets represent the corresponding wavelengths (in Å). The existence of absorption edges has important practical applications. Here we give two examples.

Example 1. In Section 6.2.3, we pointed out that, for a certain element, the threshold energy for producing K x-rays is always larger than the energy of the K x-rays of the element itself. The threshold energy for producing a K x-ray is the energy required for producing a K vacancy and this is also the energy of the K absorption edge. Thus, in the μ–E diagram of a given element, the energy position of the K x-ray is always to the left of the K absorption edge, but generally close to it, so the corresponding absorption coefficient is comparatively small.

Applying this principle to the x-rays produced by a given element, we can use a foil (called a filter) made of this element to absorb x-rays from impurities while allowing the x-rays of interest to pass through easily. Figure 6.32 gives a diagram of the transmission ratio versus energy E of x-rays produced by a molybdenum anode passing through a molybdenum foil with a thickness of 127 μm. The K absorption edge of molybdenum is 20.0 keV.

The transmission fraction I/I_0 is determined by

$$\frac{I}{I_0} = e^{-\mu(E)\rho x}$$

where $\mu(E)$ is the mass absorption coefficient of the filter (molybdenum foil), which is a function of the energy of the x-rays; ρ is the density of molybdenum; and x is the thickness of the molybdenum (127 μm). From the figure we can see that the K_α and K_β lines (17.5 and 19.6 keV, respectively) produced by molybdenum on the bremsstrahlung background spectrum (produced at the anode) are able to pass through the filter easily, but the x-rays with energy a little higher than that of the Mo K x-rays will be strongly absorbed (blocked). The "pass band" (also called the "window") of the filter is very narrow. For some trace elements whose threshold energies for production of x-rays are within the range 6–13 keV, it will be very efficient to activate them by using molybdenum x-rays that have passed through a molybdenum filter. The interference resulting from other elements with higher thresholds in the sample will be very slight.

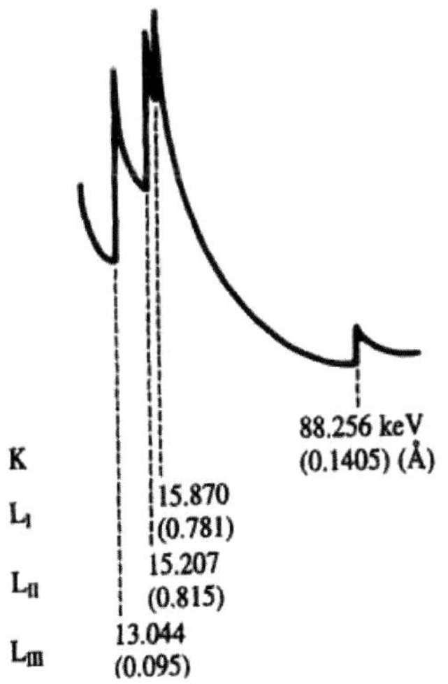

FIGURE 6.31
The K and L absorption edges of lead.

Example 2. Brass is an alloy of copper ($Z = 29$) and zinc ($Z = 30$). When x-rays strike brass, atoms of both copper and zinc can produce characteristic x-rays. The wavelengths of the characteristic K_α lines of copper and zinc, 1.539 Å and 1.434 Å, respectively, differ only slightly. When an experiment requires the analysis of both components, it is necessary to select the spectral line from copper independently of that from zinc by choosing an appropriate filter. Figure 6.33 shows the K absorption edge of nickel ($Z = 28$).

Note that here we choose the wavelengths to be the abscissa instead of the energy as used before. As shown in the figure, the absorption of edge of nickel is 1.489 Å, which is between the K_α lines of zinc and copper. The mass absorption coefficient of the K_α line of zinc in nickel is 325 cm^2/g, while that of copper is 48 cm^2/g, as shown in Fig. 6.33. This means that, when passing through a nickel foil, the K_α x-rays from zinc are absorbed much more than those from copper. If we place a nickel foil with mass thickness of 8.37 mg/cm^2 (thickness of 9.40 μm, since the density of nickel is 8.90 g/cm^3) in front of a detector, then, if the initial intensity ratio of the two kinds of K_α x-rays from copper and zinc is 1:1, after the nickel is put in place the intensity ratio will be 10:1. Let us calculate this ratio:

$$\frac{(I/I_0)_{Cu}}{(I/I_0)_{Zn}} = \frac{I_{Cu}}{I_{Zn}} = \frac{e^{-\mu_{Cu}\rho x}}{e^{-\mu_{Zn}\rho x}} = e^{-\rho x(\mu_{Cu}-\mu_{Zn})}$$

$$= e^{-8.37\times 10^{-3}(48-325)} = e^{2.32} = 10.2$$

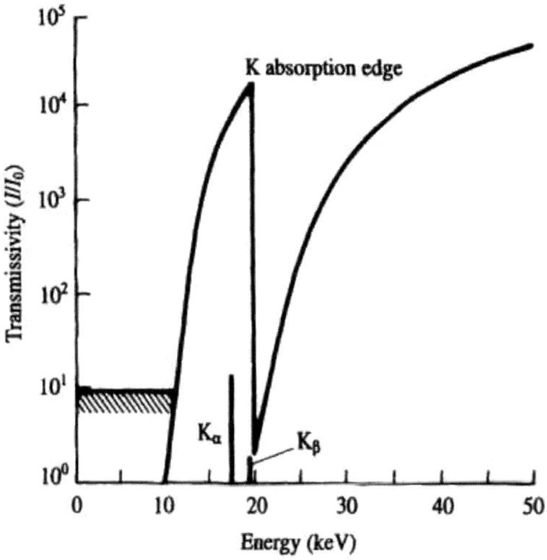

FIGURE 6.32
The relationship of the transmission ratio of a 127 μm thick molybdenum filter with x-ray energy. K_α and K_β are the characteristic K x-ray energies of molybdenum. The horizontal line represents the limits of the detector. Note that at the K absorption edge the transmission of lower-energy x-rays rises sharply, as seen in the figure, because the absorption coefficient drops sharply at the K edge to much lower absorption at lower energies (as seen in Fig. 6.30, for example).

Most of the K_α intensity from zinc is blocked and what is measured is mainly the K_α line of copper. Hence, by making use of the absorption edges, we can solve ingeniously the problem of the resolution of spectral lines and make the analysis easier.

6.4.5 Extended X-Ray Absorption Fine Structure (EXAFS)

From Fig. 6.30 we can see that, in the part above the absorption edges with higher energy, the absorption coefficient unidirectionally decreases as the energy of the photons increases. But if we observe carefully with a high-resolution spectrometer, we find that, except for a simple single-atom system (such as krypton gas), in the part just above the absorption edge with higher energy generally the absorption coefficient changes periodically as the energy of the photons increases. Figure 6.34 is a plot of the absorption coefficient of copper as a function of energy. Amplifying the part near the K absorption edge in the figure, we get Fig. 6.35 which is known as the extended x-ray absorption fine structure (EXAFS for short). The range of the extended x-ray absorption fine structure can spread over 1 keV.

The experimental observation that the absorption coefficient shows this complex structure in the higher energy part of the absorption edge was made early in the 1930s. At that time, both correct and incorrect explanations were given. But because of the inferiority of the experimental equipment, this phenomenon was not noticed widely. At the beginning of the 1970s, especially after the appearance of synchrotron radiation sources of high power, EXAFS was noticed again. Now there are not only satisfactory explanations in theory, but also important applications in biology, chemistry, solid physics, surface physics, and other fields.

How is EXAFS produced? The appearance of an absorption edge indicates that the energy of the incident photon is sufficient to remove an inner shell electron in an atom. The kinetic energy of an ejected photoelectron is $E = h\nu - E_K$ (when the electron escapes from the K shell) where E_K is the numerical value of the K absorption edge and $h\nu$ is the energy of the incident

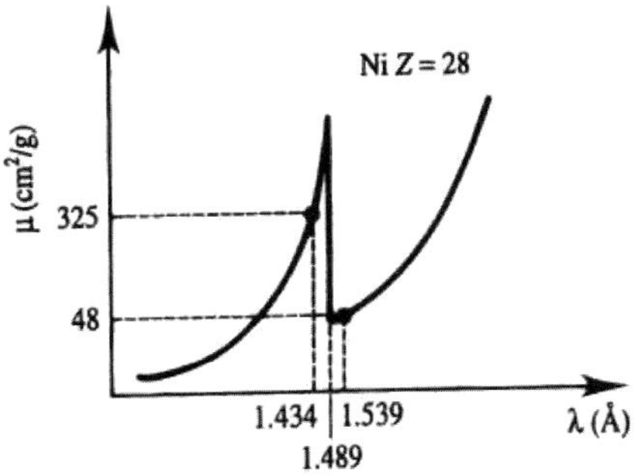

FIGURE 6.33
Different absorptions of the K_α x-rays of copper ($K_\alpha = 1.539$ Å) and zinc ($K_\alpha = 1.434$ Å) by a nickel filter.

photon. In contrast to a simple single-atom system, an atom in bulk matter is not isolated but is surrounded by other atoms as shown for a germanium crystal in Fig. 6.36. The Broglie wave (see Chapter 7) of the ejected electron can be scattered by the surrounding atoms to form an inward-directed wave which interferes with the original outward-directed wave, as illustrated in Fig. 6.36 for x-ray absorption by an electron wave scattered off atoms of germanium in a crystal. This interference can result in an increase (Fig. 6.36(a)) or a decrease (Fig. 6.36(b)) in electron intensity to produce oscillation in the absorption coefficient. Since EXAFS results from the effect of adjacent atoms, it is clearly related to the atom's environment and so we can apply EXAFS to study the environment in which an atom finds itself.

Here we give an example of applying EXAFS to studies in biology. In biology, many interesting molecules are complicated, but in many important molecular structures there are often metal centers. It is this property that makes EXAFS applicable. Complex macromolecules are difficult to make into crystals, so the x-ray diffraction method is often useless. Sometimes, even where it is possible to crystallize macromolecules, it is difficult to know whether the crystallized state is similar to the active state *in vivo*. The EXAFS method can be applied to study different situations; its objects of study may be crystals or solutions. The EXAFS method is extremely efficient in studying what atoms are located around the heavy metal center.

An important component of a nitrogen-fixing enzyme is an Mo–Fe protein whose structure is basically understood. Nitrogen is a basic element for the growth of plants, but most plants (such as rice) are unable to directly digest nitrogen from the air; they must be given nitrogen fertilizers such as ammonia. But some plants (such as groundnuts) are able to directly digest some nitrogen from the air because they have a nitrogen-fixing bacterium in their roots. The principal function is carried out by a nitrogen-fixing enzyme. Understanding of the structure of the enzyme is very important for both the artificial synthesis of nitrogen-fixing enzymes and an understanding of the nitrogen-fixation process. The Mo x-ray spectrum of a freeze-dried Mo–Fe protein shows the EXAFS spectrum clearly.

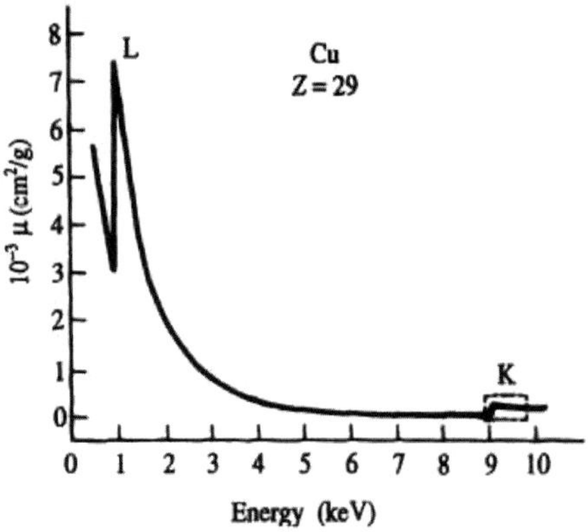

FIGURE 6.34
The K and L x-ray absorption edges of copper.

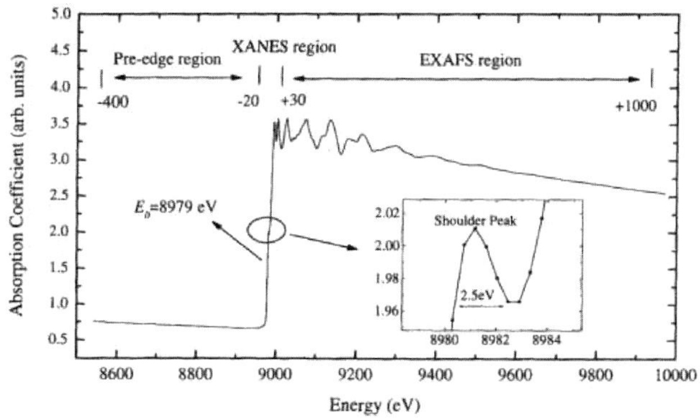

FIGURE 6.35
The fine structure of the K absorption edge of copper measured by Shanghai Synchrotron Radiation Facility in 2008.

Fitting the experimental curve to theoretical calculations, one finds the following: at a distance of 2.35 Å from molybdenum there is a sulfur shell composed of a cluster of four sulfur atoms; at a distance of 2.72 Å from these there are two or three iron atoms; and at a distance of 2.74 Å there are one or two more sulfur atoms. For the enzyme cytochrome oxidase, we know from the EXAFS spectrum that, around the copper atom there are two nitrogen atoms at a distance of 1.97 Å from the copper atom and there are two sulfur atoms, one 2.18 Å away from the copper and the other 2.82 Å from the copper.

6.4.6 Photoelectron Spectroscopy

In the course of precisely determining the binding energies of electrons in atomic shells, K. Siegbahn and his collaborators in Uppsala in the 1950s discovered the surprising fact that the binding energies of even the core electrons depend on the chemical environment. Siegbahn pioneered the field of (photo)electron spectro-

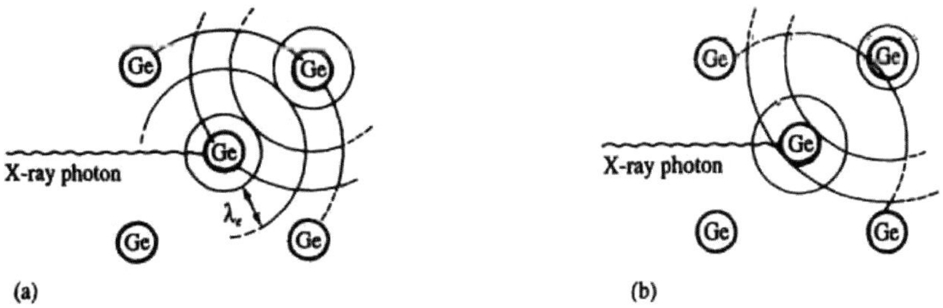

FIGURE 6.36
Production of EXAFS by germanium atoms in a crystal. (a) Increase in electron (wave) intensity caused by constructive interference of outgoing and incoming waves. (b) Decrease in electron intensity from destructive interference of outgoing and incoming waves.

scopy for chemical analysis, ESCA, for which he subsequently won the Nobel Prize in physics. The chemical shifts are small, but the high-resolution electron spectrometers that Siegbahn and his colleagues had developed for nuclear physics were capable of resolutions of better than 0.1% full width at half maximum at 1 keV electron energy. By using photons with energies just above the binding energies, the energy shifts from chemical processes are easily observed. This technique has a broad spectrum of applications as a qualitative and quantitative diagnostic tool in chemical analysis, in biological systems, in geology, in environmental studies, in surface physics research, in studies of polymers and alloys, in radiation damage studies and in a variety of industrial uses. For example, in biological systems, the total protein content in a sample can be obtained and different proteins can be distinguished. Environmental studies include measuring the different sulfur compounds in smoke particles from burning coal and studies of other air pollutants. In a different way, photoelectron spectroscopy is applied to studies of the outer shell electrons, which are important in the formation of molecular orbitals. The technique has been applied to a wide range of problems in the studies of molecules. A number of commercial companies, including DuPont, Hewlett-Packard, Perkin-Elmer, and Varian, build commercial instruments for the many different types of photoelectron spectroscopy being carried out today.

6.5 SUMMARY

6.5.1 Important Formulas

1. The minimum wavelength (quantum limit) of bremsstrahlung is

$$\lambda_{\min} = \frac{hc}{E} = \frac{12.4}{E(\text{keV})} \quad \text{Å} \tag{S.1}$$

   where E is the initial energy of the electrons as determined by the accelerating potential.

2. The approximate expression of the K_α x-ray energy is

$$E_{K_\alpha} = \tfrac{3}{4} \times 13.6 \times (Z-1)^2 \quad \text{eV} \tag{S.2}$$

   where Z is the atomic number of the corresponding element.

3. The diffraction of x-rays in a crystal, Bragg's law is according to

$$n\lambda = 2d \sin\theta \tag{S.3}$$

4. Compton scattering is expressed by

$$\Delta\lambda = \frac{h}{m_e c}(1 - \cos\theta) \tag{S.4}$$

5. The absorption of x-rays in matter follows the rule

$$I = I_0 e^{-\mu x} \tag{S.5}$$

6.5.2 Important Concepts

Bremsstrahlung; characteristic x-ray spectrum; synchrotron radiation; diffraction of x-rays; Compton scattering; the absorption of x-rays; the absorption edge; the extended absorption edge.

6.5.3 The Experimental Measurements of Some Physical Quantities

1. Applying formulas (S.1) and (S.4), we have two experimental methods to measure Planck's constant h independently.
2. Applying formula (S.3), we can measure the crystal lattice constant d, through which Avogadro's constant N_A can be calculated.
3. Applying formula (S.2), we can measure the atomic number Z.
4. Applying formula (S.3), we can measure the wavelength λ of electromagnetic radiation.
5. Applying formula (S.5), we can measure the absorption coefficient μ of electromagnetic radiation in a medium.

6.5.4 Nobel Prizes Relating to the Studies of X-Rays

1901 Wilhelm Conrad Roentgen received the first Nobel Prize in physics for the discovery of x-rays.

1914 Max Theodor Felix von Laue: Diffraction of x-rays in crystals.

1915 Sir William Henry Bragg and Sir William Lawrence Bragg: Study of crystal structure by means of x-rays.

1917 Charles Glover Barkla: Discovery of the characteristic x-rays of the elements.

1924 Karl Manne Georg Siegbahn: Discoveries in the area of x-ray spectra.

1927 Arthur Holly Compton (Fig. 6.37): Explanation of the scattering of x-rays by electrons and atoms.

1979 A. M. Cormark and C. N. Hounsfield: Nobel Prize in physiology and medicine for the invention of x-ray tomography.

1981 Kai Siegbahn: Photoelectron spectroscopy for analysis using x-rays.

2002 Riccardo Giacconi: Discovery of cosmic x-ray sources.

PROBLEMS

6.1. The minimum wavelength of the continuous x-ray spectra from an x-ray tube is 0.124 Å. What is its working potential?

6.2. Moseley's experimental method was used to measure the atomic number precisely for the first time. If the measured wavelength of a K_α x-ray of an element is 0.685 Å, what is the atomic number of the element?

6.3. The L absorption edge of a neodymium atom ($Z = 60$) is 1.9 Å. How much work is required to ionize a K electron from a neodymium atom?

6.4. Compare your calculated value from Problem 6.3 with the known K shell binding energy from Appendix V. Discuss why these are different.

6.5. Prove that for most of the elements, the intensities of the K_{α_1} x-rays are double the intensities of the K_{α_2} x-rays.

6.6. The K absorption edge of lead is 0.14 Å. The wavelengths of the spectral lines of its K series are 0.167 Å (K_α), 0.146 Å (K_β) and 0.142 Å (K_γ), respectively.

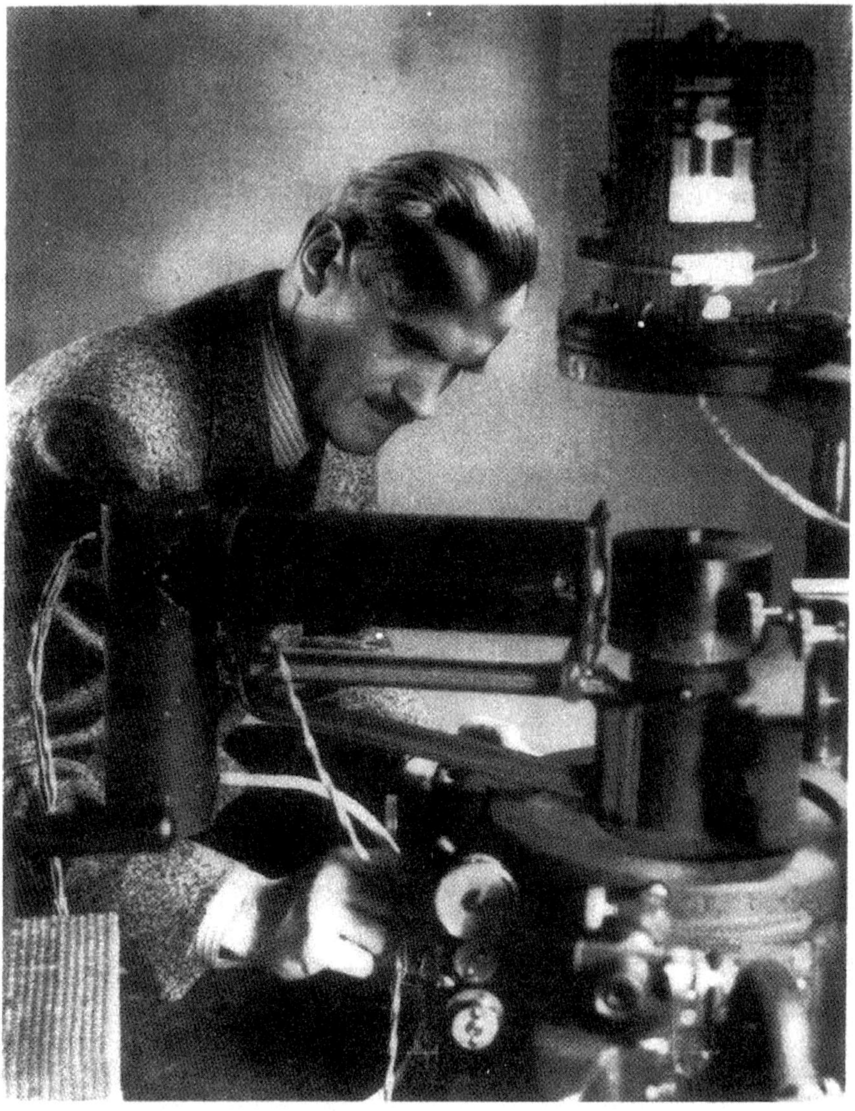

FIGURE 6.37
Professor A. H. Compton in his laboratory at the University of Chicago, 1935. (Courtesy of Professor John Compton, Vanderbilt University.)

(*a*) Draw the x-ray level diagram for lead according to these values.
(*b*) Calculate the minimum energy required to activate the L series and the wavelength of the L_α line.

6.7. A beam of monochromatic light with a wavelength of 5.4 Å is incident upon a set of crystal planes. At an angle of 120° with respect to the incident beam, a maximum first-order diffraction is observed. What is the spacing of the crystal planes?

6.8. In Compton scattering, show that $h\nu'_{\min} + E_{k,\max} = h\nu$.

6.9. In Compton scattering, if the energy of an incident photon is equal to the rest energy of an electron, what are the minimum energy of the scattered photon and the maximum momentum of the electron?

6.10. In Compton scattering, if the maximum energy which a photon could transfer to a rest electron is 10 keV, what is the energy of incident photon?

6.11. If an incident photon scatters with a proton, what is the Compton wavelength of the proton? If the energy obtained by the recoiling proton is 5.7 MeV, what is the minimum energy of the incident photon?

6.12. A scattered photon from Compton scattering interacts once with an atom. Show that when the scattered angle $\theta > 60°$, no matter how large the energy of the incident photon, the scattered photon can never produce a positive and negative electron pair.

6.13. Show that the collision of a photon with a free electron can never lead to a photoelectric effect because of the violation of conservation laws.

6.14. Show that in vacuum, it is impossible for a photon to produce a "positron–electron pair".

6.15. If the electronic state of palladium ($Z = 45$) is $1s^2 2s^2 2p^6 3s^2 3p^6 3d^{10} 4s^2 4p^6 4d^8 5s^1$
(*a*) Determine the atomic term of its ground state.
(*b*) Use the Pd K_α x-ray for Compton scattering. When the scattered angle of the photon is 60°, what is the energy of the recoiling electron (given that the shielding coefficient b of K_α is $b \simeq 0.9$)?
(*c*) In an experiment, lead with a thickness of 0.3 cm is used to shield this x-ray. If we substitute aluminum for the lead, what thickness of aluminum is needed to achieve the same shielding effect? ($\mu_{Pb} = 52.5\,\text{cm}^{-1}$, $\mu_{Al} = 0.765\,\text{cm}^{-1}$.)

6.16. You are given the wavelengths of the K_α x-rays of copper and zinc as 1.539 Å and 1.434 Å, respectively; the K absorption edge of nickel is 1.489 Å; and the mass absorption coefficients of copper and zinc in nickel are 48 and 325 cm²/g, respectively, at this wavelength. What thickness of nickel is required to increase the relative intensity ratio of the K_α-ray of copper to that of zinc by a factor of 20?

6.17. It was in Moseley's experiment that the atomic number was measured for the first time. The wavelengths of the K_α x-rays for some elements are 1.935, 1.787, 1.656, and 1.434 Å. Determine the atomic number of these elements. In this series, which element is missing? What is the wavelength of its K_α line?

6.18. An x-ray tube with aluminum as its target can produce a continuous x-ray spectrum with a short-wave limit of 5.00 Å. Is it possible to see the aluminum characteristic K x-ray in its spectrum? What is the case if its short wave limit is 10 Å?

6.19. A photon with energy of 3.00 keV collides with an electron at rest and is scattered in a direction of 45° with respect to its initial direction. What is the energy of the scattered photon and the energy, the momentum, and the direction of motion of the recoil electron?

6.20. The K absorption limit of silver is 0.485 Å, and its L_I, L_{II}, L_{III} limits are 3.240 Å, 3.510 Å, 3.690 Å. What are the wavelengths of the characteristic K_{α_1} and K_{α_2} lines of silver?

6.21. A 1 MeV photon collides with a 200 keV electron moving in the opposite direction to the photon and is scattered into 180°. What are the kinetic energies of the scattered photon and the electron after collision?

6.22. In a device, 0.45 cm thick lead is used to shield x-rays. If aluminum is used, 30.9 cm of aluminum would yield the same result. What is the absorption coefficient of aluminum if it is 52.5 cm^{-1} for lead?

6.23. The wavelengths of the K_{α_1} and K_{α_2} lines for iron x-rays are 1.9321 Å and 1.9360 Å, respectively. Calculate the splitting gap between $^2P_{3/2}$ and $^2P_{1/2}$ levels.

6.24. For gold, lead, and uranium, K x-rays appear when the voltage of an x-ray tube is 80.7, 88.3, and 115.3 kV, respectively. Determine the K absorption limits for these elements.

6.25. Use Fig. 6.32 to estimate the transmission of the Mo K_α line through a molybdenum foil. If the density of the molybdenum is 10.2 g/cm^3 and the foil thickness is 127 μm, find the absorption coefficient for the K_α line at this energy.

CHAPTER 7

INTRODUCTORY QUANTUM MECHANICS I: CONCEPTS

True genius resides in the capacity for evaluation of uncertain, hazardous, and conflicting information.

Winston Churchill

The three great discoveries at the end of the nineteenth century, namely, the discoveries of x-rays in 1895 (Chapter 6), radioactivity in 1896 (Chapter 10) and the electron in 1897 (Chapter 2) raised the curtain on the development of modern physics. Then in 1900, Planck put forward the concept of energy quantization of a radiating source to overcome the difficulties of classical physics in the explanation of blackbody radiation (Chapter 3); and in 1905, Einstein used the concept of the light-quantum to overcome the failure of classical physics to explain the experimental data on the photoelectric effect. In 1913, Bohr applied the Planck–Einstein concept of quantization to the Rutherford model of the atom (Chapter 2) and proposed the concept of quantized energy states to explain the optical spectrum of hydrogen satisfactorily (Chapter 3). With the help of the Pauli exclusion principle proposed in 1925 (Chapter 5) and the electron spin hypothesis introduced in 1921 by Compton and independently in 1925 by Uhlenbeck and Goudsmit (Chapter 4), many experimental facts such as the doublet structure of the sodium lines, the Zeeman effect, and the periodicity of the elements were explained successfully (Chapters 4 and 5).

But the quantum theory up to 1925 (called the old quantum theory) exhibited serious defects. There were shortcomings both in logic and in the treatment of practical problems (see Section 7.1). In order to establish a complete theoretical system, a new idea was needed. The idea that opened the way was the "wave–particle dualism" of matter and light (Section 7.2). In fact the wave–particle dualism of photons had been proposed by Einstein in 1905 and more clearly in 1917 (Chapter 3) and verified by Compton's work (Chapter 6). But it was not until 1924 that de Broglie extended it to particles of matter.

Following up this idea over several years of inspired work, in the years 1925–1928 Heisenberg, Born, Schrödinger, and Dirac established the new quantum mechanics which, together with relativity, formed one of the two important theoretical pillars of modern physics. They are the "revolutionary" theories of the twentieth century.

Wave–particle dualism is the most important concept in quantum mechanics. It should be emphasized that for a microbody, the "wave" or "particle" mentioned here is completely different from the waves and particles with which we work in the classical domain (Sections 7.2 and 7.4). We cannot visualize the quantum waves and particles like we do water waves and billiard balls. The innate character of quantum mechanics was reflected very clearly in the uncertainty relation proposed by Heisenberg in 1927 (Section 7.3). It is an inevitable outcome of the wave–particle dualism of a microbody.

The development of quantum mechanics includes three aspects: The first is the discovery of the important data that demanded the new concept; the second is the introduction of a series of new ideas which are different from those in classical physics; and the third is the development of mathematical physics techniques to solve practical questions.

In line with the aims of this book, we emphasize the first two aspects while introducing the third only briefly. Even for the first two aspects, some things are left for further consideration in more specialized courses.

7.1 THE DIFFICULTY OF THE BOHR THEORY

In Chapter 3, we discussed how the theory of the hydrogen atom proposed by Bohr in 1913 had many successes. The concept of quantum states proposed in the theory was verified directly by experiments. The theory explained successfully the thirty-year-old puzzle of the optical spectrum of hydrogen and made possible a theoretical calculation of the Rydberg empirical constant. It explained the basic spectrum of the characteristic x-rays quite nicely and gave insight into the periodicity of the elements using physical concepts for the first time.

However, because in Bohr's theory subatomic particles (electrons, protons, nucleus) were regarded as point masses of classical mechanics and the laws of classical mechanics were used to describe these particles, it was unavoidable that there were innate contradictions which were hard or impossible to resolve with this theory. First, it was impossible to understand conceptually why in hydrogen the electrostatic interaction between the nucleus and the electron is valid while the power of an accelerated electron to emit electromagnetic radiation disappears when the electron is in a stationary quantum state. The reasons behind the emission and absorption of radiation in the description of the process of electrons moving between two stationary states are not clear and the description of the process was very ambiguous. In order to make these contradictions more clear, we introduce the question raised by Rutherford and the criticism raised by Schrödinger.

As soon as Rutherford received Bohr's manuscript, he raised the following question:[1,2]

> How does an electron decide at what frequency it is going to vibrate when an electron jumps from one stationary state to another? It seems to me that you would assume that the electron knows beforehand where it is going to stop.

But how is that possible? Assume that the electron is in the energy state of E_1; it must absorb a photon of energy $E_2 - E_1$ to move to the energy state of E_2, while the absorption of a photon of any other energy will not cause the transition. (To simplify the discussion, we assume that there exist only two energy levels, E_1 and E_2, and also $E_2 > E_1$.) How does the electron choose the photon it needs from photons with many different energies. To select the needed photon, the electron should "know" in advance the energy level (E_2) to which it will go, as if it had been to the level before. But to "have been to the level," first of all it must absorb the needed photon, ... Thus the process degenerates into a vicious circle in logic.

Next we introduce Schrödinger's criticism, that is, the famous "damned quantum jumping."[3] According to relativity, when an electron moves from one orbit to another orbit, its velocity cannot be infinite, that is, it cannot exceed the velocity of light. Thus it takes time for the transition. Moreover, in this period of time when the electron has left the E_1 state but has not reached the E_2 state, what state is it in? This cannot be answered by the theory of "the damned jumping."

The Bohr theory (old quantum theory) not only was completely unable to answer these contradictions and difficulties in logic, it was also rebuffed by its failure to explain various practical problems. For example, this theory cannot even explain the optical spectrum of helium, the next simplest element to hydrogen. Even for hydrogen the theory could not explain the intensity of the spectral lines and its fine structure. The theory also could not describe how atoms form molecules, liquids, and solids.

In the face of these difficulties, some advocated giving up the quantum theory completely and going back to classical theory. But a host of facts increasingly revealed the importance of the quantum hypothesis. What was needed was a new idea.

---

[1]N. Bohr, *Proc. Phys. Soc.* **78** (1961) 1083.

[2]J. B. Birds, ed., *Rutherford at Manchester*, W. A. Benjamin, New York (1963), p. 127.

[3]W. Heisenberg, *Physics and Beyond*, Harper & Row, New York (1972), p. 75.

7.2 WAVE–PARTICLE DUALISM

7.2.1 The Wave and the Particle in Classical Physics

The concepts of waves and particles play very important roles in classical physics. They are the only two ways of transmitting energy; that is, the transmission of energy can always be described by waves or by particles. For example, sound causes an ear membrane to feel a vibration. This is the result of the transmission of sound energy in the form of a wave. The breaking of glass after being forcefully struck by a stone is an example of transmission of energy in the form of a particle. Experience tells us that the concept of a wave and the concept of a particle can never be used at the same time. We cannot use the concept of a wave and the concept of a particle at the same time to describe a phenomenon: This is logically impossible.

We know that an ideal particle is completely localized and in principle we can determine its mass, position, momentum, and charge with infinite accuracy. A particle can be regarded as a point mass. Although all particles in nature have a certain size, they can be treated as point mass under certain conditions. The concept of "point mass" is relative. For example, in the kinetic theory of gas molecules, although a molecule has internal structure, with a few exceptions (diffusion) it can be regarded as a point mass. Similarly, in a galaxy, stars can also be regarded as point masses. When the dimension of the particle itself can be neglected compared with the size of the system or when the inner structure of the particle is not important in the question discussed, it can be regarded as a point mass. For a point mass, as soon as its initial location and velocity have been given, in principle Newtonian mechanics can describe its future location and velocity completely.

We have understood, to some extent, waves from phenomena such as single-slit diffraction and double-slit interference. The characteristic quantities are wavelength and frequency. An ideal wave has a definite frequency and wavelength. In principle its wavelength and frequency can be determined with infinite accuracy. But the wave cannot be restrained and must extend infinitely in space.

To sum up, when saying that the location of a particle in space can be determined with infinite accuracy, we have assumed that the particle is an infinitesimal point mass; and when the wavelength and frequency of a wave are determined with infinite accuracy, the wave must extend infinitely in space.

How do we determine the wavelength of a wave? In an experiment, we can use the "beat" method. As shown in Fig. 7.1, let a wave of fixed amplitude and given frequency ν_1 (in principle this wave can be obtained from a wave generator) interfere with a wave of frequency ν_2 which is unknown. A "beat" will form (two waves of the same vibration amplitude but different frequency add to form a beat). From the existence or absence of a beat, we can determine whether there is a difference between ν_1 and ν_2. It is known from Fourier analysis that the waveform shown in Fig. 7.2 is formed by the superposition of many sine waves of different frequencies. As a simple example, take a tuning fork with a frequency

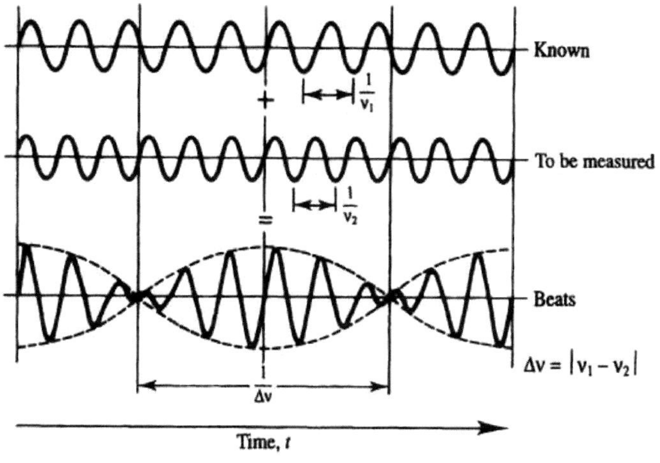

FIGURE 7.1
The formation of a beat.

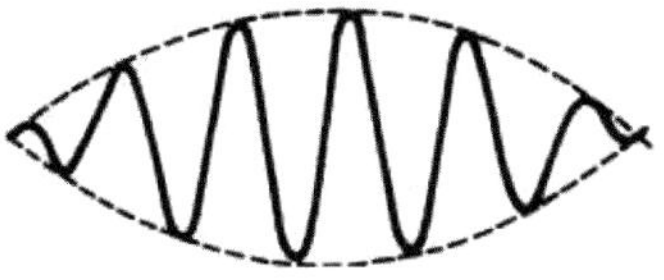

FIGURE 7.2
A beat.

of 440 Hz. If a vibrating tuning fork of similar (but not identical) frequency, f, is held near the vibrating 440 Hz one, then beats are heard. The number of beats per second is equal to the difference of their frequencies $|440 - f|$.

To determine whether there exists a beat, at least one beat must be observed. It is seen from Fig. 7.1 that the time of observing one beat is $1/\Delta\nu$; thus the time to observe at least one beat is

$$\Delta t \geq \frac{1}{\Delta\nu}, \quad \text{or} \quad \Delta t\,\Delta\nu \geq 1 \tag{7.1}$$

Assume the velocity of the wave is v; the distance the wave travels in time Δt is

$$\Delta x = \text{v}\,\Delta t$$

Substituting this into (7.1), gives

$$\frac{\Delta x}{\text{v}} \geq \frac{1}{\Delta\nu} \tag{7.2}$$

Since $\nu = \text{v}/\lambda$, thus $\Delta\nu = (\text{v}/\lambda^2)\,\Delta\lambda$; substituting this into (7.2), gives

$$\Delta x\,\Delta\lambda \geq \lambda^2 \tag{7.3}$$

Formula (7.1) indicates that in order to determine the frequency with infinite accuracy, one must spend an infinitely long time. Formula (7.3) shows that in order to determine the wavelength with infinite accuracy, one must observe over infinite space. Later we will see that the most important relation in quantum mechanics, the uncertainty relation, can be derived from these equations.

7.2.2 Wave–Particle Dualism of Light

Investigations of the nature of light have a very long history. As early as 1672, Newton thought that light was composed of particles and proposed a corpuscular theory of light. But only six years later, Huygens, a Dutch scientist, submitted a paper "Light Theory" to the Paris Institute. He regarded light as a longitudinal vibration. He derived the straight-line motion law, the reflection and refraction laws using a wave theory of light, and explained the phenomenon of double refraction. From then on, the corpuscular theory and the wave theory of light were developed and debated continuously.

Not until the beginning of the nineteenth century, after experiments that verified the interference and diffraction of light made by Fresenel, Fraunhofer and Young, was the wave theory of light accepted universally. Toward the end of the nineteenth century, Maxwell and Hertz further established that light is an electromagnetic wave. By that time, the triumph of the wave theory seemed decisive.

But as mentioned in Chapter 3, at the beginning of this century our understanding of the nature of light took a new direction. Planck introduced the idea of the emission of light in quanta. In 1905, Einstein explained the photoelectric effect using the quantum theory of light. He proposed that the energy of a photon is

$$E = h\nu \tag{7.4}$$

In 1917, he further proposed that photons not only had energy but also had momentum:

$$p = \frac{h}{\lambda}, \quad \text{or} \quad \boldsymbol{p} = \hbar \boldsymbol{k} \tag{7.5}$$

in which the wavevector $k = 2\pi/\lambda$. These two equations connect the quantities ν and $\lambda(\boldsymbol{k})$, which characterize the wave nature, with the quantities E and p, which characterize the particle nature by means of a universal constant—Planck's constant h. Light was the first entity to exhibit the contradiction of having both a particle nature and a wave nature. Equations (7.4) and (7.5) are mathematical expressions of the wave–particle dualism of light.

These characteristics of light were embodied very clearly in the Compton scattering experiments in 1923. In these experiments, the wavelengths of the x-rays were determined by a crystal spectrometer, which was based on the diffraction phenomenon of waves; but the way the scattering influenced the wavelength could only be explained by regarding the x-rays as particles. While light can exhibit a wave nature as well as a particle nature, in any particular case it can show either a wave nature or a particle nature but never the two natures at once. Further discussions on the nature of light are found in Scully and Sargent.[4]

7.2.3 The de Broglie Hypothesis

Just when many physicists were puzzled about the wave–particle dualism of light, L. de Broglie, a young French physicist who had just turned his research direction from history to physics, extended the wave–particle dualism to all matter particles. This was the new revolutionary step needed to bring forth the new quantum mechanics. In forming his ideas, de Broglie used one of the principles of the ancient Greeks, the idea of symmetry in nature.

De Broglie recalled his original thought when he received the Nobel Prize in 1929:

> Firstly the light-quantum theory cannot be regarded as satisfactory since it defines the energy of a light corpuscle by the relation $W = h\nu$ which contains a frequency ν. Now a purely corpuscular theory does not contain any element permitting the definition of a frequency. This reason alone renders it necessary in the case of light to introduce simultaneously the corpuscle concept and the concept of periodicity
>
> On the other hand the determination of the stable motions of the electrons in the atom involves whole numbers, and so far the only phenomena in which whole numbers were involved in physics were those of interference and of eigenvibrations. That suggested the idea to me that electrons themselves could not be represented as simple corpuscles either, but that a periodicity had also to be assigned to them too.

De Broglie wrote three short papers from September to October 1923, and in November 1924 he submitted his doctoral dissertation entitled "Recherches sur la Théorie des Quanta" to the School of Science of the University of Paris. In these papers he proposed the hypothesis that all matter particles have a wave–particle dualism. He thought that "any body may be accompanied by a wave and it is impossible to separate the motion of body and the propagation of wave." He gave the relation between the momentum p of a particle and the wavelength λ of its accompanying wave:

$$\lambda = \frac{h}{p} \tag{7.6}$$

This is the famous de Broglie relation and is an extension of Eq. (7.5) to particles. So by the symmetry of nature, if waves (light) can act like particles (photons), then particles can act like waves. de Broglie held that this relation was suitable for any matter particle, no matter whether its rest mass was zero or not. We can only consider it as a hypothesis and its correctness must be tested by experiments. This idea was so revolutionary that the faculty of the University of Paris at first rejected his theory and accepted his Ph.D dissertation only after inserting a disclaimer that his advisors took no responsibility for the ideas it contained.

Equation (7.6) and the mass–energy relation in relativity

$$E = mc^2 \tag{7.7}$$

---

[4]M. O. Scully and M. Sargent, *Physics Today* (March, 1972), 38.

are two of the most important relations in modern physics. The former connects the particle nature and the wave nature by means of Planck's constant (a very small quantity); and the latter connects energy and mass by means of the velocity of light (a very large quantity). It is considered a major triumph in physics to find inner connections between physical quantities which are completely different at first glance.

Equation (7.6) also gives us further understanding of the meaning of Planck's constant. When Planck introduced this constant in 1900, it represented the measurement of quantization, that is, the measurement unit of discontinuity (a discrete unit). But now after the work of Einstein and de Broglie, the concept of the wave–particle dualism of matter particles appears and Planck's constant acts as the bridge between the wave nature and the particle nature. Quantization and wave–particle dualism are the most basic concepts in quantum mechanics and the same constant h plays the crucial rule in these two concepts. This fact itself shows that there are profound internal relations between the two concepts. In any expression where Planck's constant appears, it certainly indicates the quantum mechanical characteristic of this expression.

7.2.4 Davisson–Germer Experiment[5]

In 1925 when Davisson and Germer were doing an experiment on the scattering of electrons from nickel, an accidental loss of vacuum oxidized the nickel. In order to reduce the nickel, they heated it. As a result the single-crystal structure of nickel was formed and the diffraction phenomenon of electrons from a crystal was observed for the first time. At that time they did not know of the work of de Broglie. It should be pointed out that the idea of using a crystal to do a diffraction experiment with electrons was proposed by de Broglie in 1924 in his dissertation defense. When J. B. Perrin asked him whether these waves could be experimentally verified, de Broglie replied that this should be possible by a diffraction experiment of electrons on a crystal.[6] Later, in 1927, after learning of the concept of matter waves, Davisson and Germer did the experiment more accurately. The experimental setup is shown in Fig. 7.3. The electrons emitted from the heated filament, after being accelerated by a potential difference V, were sent out of the "electron gun" with a kinetic energy of V eV and moving vertically toward a single-crystal block of nickel. The detector was placed at an angle of θ with respect to the beam direction. They recorded the intensities of the "reflected" beams at different accelerating voltages V and found that when $V = 54$ V, and $\theta = 50°$ the intensity of the reflected beam had an obvious maximum, as shown in Fig. 7.4(a). The appearance of these intense "reflected" beams can be understood by assuming that electrons have wavelengths of $\lambda = h/p$ and are reflected from a particular

---

[5] G. J. Davisson and L. H. Germer, *Phys. Rev.* **30** (1927) 705.

[6] M. Jammer, *The Conceptual Development of Quantum Mechanics*, McGraw-Hill, New York (1966), p. 247.

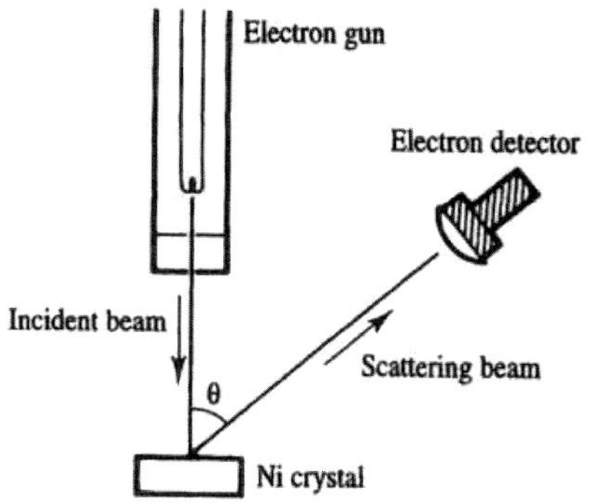

FIGURE 7.3
Schematic diagram of a diffraction experiment of electrons on a crystal.

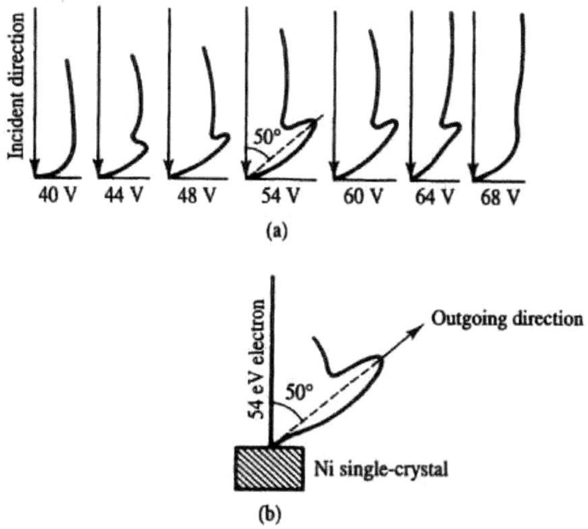

FIGURE 7.4
The results of the Davisson and Germer experiment.

family of "Bragg planes," just as occurred for x-rays described in the last chapter. The experimental results shown in Fig. 7.4 cannot be explained on the basis of the motion of particles but can be explained using interference. But according to the classical point of view, particles cannot interfere, only waves can interfere with each other. As actual examples of experimental results, see Fig. 7.5.

The scattering of electrons from a crystal is the special case of waves scattering from a crystal lattice. As shown in Fig. 7.6, the scattering plane here is a mirror plane as well as a crystal plane. Compared with Fig. 6.11 in Chapter 6, both the horizontal and vertical intervals of the crystal are a and both the incident angle and scattering angle are θ. In the figure, $\alpha = \theta/2$ and the effective interval between the planes is $d = a\sin\alpha$. Thus the condition for the appearance of an intense scattered wave beam is

$$n\lambda = 2d\cos\alpha = 2a\sin\alpha\cos\alpha = a\sin 2\alpha = a\sin\theta$$

that is,

(a)

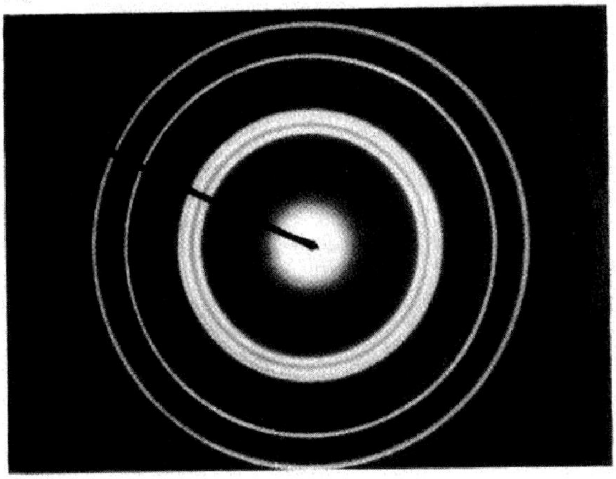

(b)

FIGURE 7.5
(a) The diffraction of electrons on an Au single crystal. (b) The diffraction pattern of electrons on an Au–V polycrystal.

$$n\lambda = a\sin\theta \tag{7.8}$$

According to de Broglie's hypothesis, the wavelength $\lambda = h/p$. When the energy is low the momentum p can be expressed classically. Thus we have

$$\lambda = \frac{h}{p} = \frac{h}{\sqrt{2mE}} = \frac{hc}{\sqrt{2mc^2E}}$$

Substituting the relevant data into this formula, we obtain the de Broglie wavelength of the electrons (nonrelativistic approximation)

$$\lambda = \frac{12400 \text{ eV} \cdot \text{Å}}{\sqrt{2 \times 0.511 \times 10^6 \text{ eV} \cdot E(\text{eV})}} = \frac{12.26}{\sqrt{E(\text{eV})}} \quad \text{Å} \tag{7.9}$$

Note that the above formula is valid only for electrons and is different from the formula for photons $\lambda = 12.4/E(\text{keV})$ given in Chapter 3. When the energy of

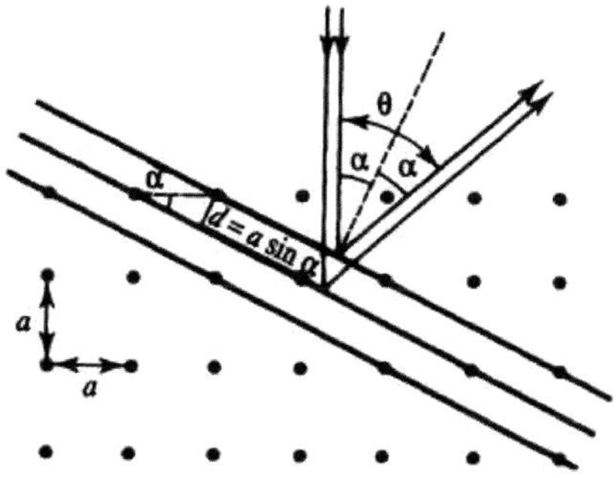

FIGURE 7.6
The enhanced "scattering" is caused by "reflection" on a family of Bragg planes with separation interval d.

the incident electrons is $E = 54\,\text{eV}$, we have $\lambda = 1.67\,\text{Å}$ from Eq. (7.9). Substituting (7.9) into (7.8) and transposing terms, we get

$$\sin\theta = \frac{n}{a}\cdot\lambda = \frac{n}{a}\cdot\frac{12.26}{\sqrt{E(\text{eV})}}\quad\text{Å} \tag{7.10}$$

For nickel, $a = 2.15\,\text{Å}$, and for an incident electron energy $E = 54\,\text{eV}$, we find

$$\sin\theta = 0.776n$$

Thus n can only be 1, that is, only one maximum exists. The maximum in the intensity of the scattered beam should be detected at the direction of $\theta = \sin^{-1} 0.776 = 50.9°$, which differs only 1° from the experimental value. Why does this difference exist? Recall that in Chapter 3 we called the creation of x-rays "the reverse of the photoelectric effect." The work function of the free electrons in a metal cannot be neglected compared to the energy of visible light, while it usually can be neglected compared to the energy of x-rays. When accurate calculations are desired, it should be taken into account. So when an electron is shot into a crystal lattice, its binding energy change increases its speed. According to Eq. (7.10), θ decreases as the energy E increases. After making a correction on the basis of these considerations, we get $\theta = 50°$, which is completely in agreement with the experimental result. This firmly proves that electrons have a wave nature and establishes the correctness of de Broglie's formula for electrons.

Since the 1930s, experiments have further shown that not only electrons but also other particles, such as neutrons, protons, nuclei, and neutral atoms exhibit a wave nature through diffraction phenomena. Their wavelengths are also determined by Eq. (7.6). Thus the correctness of de Broglie's hypothesis has been confirmed extensively. The kinetic energies of four types of particles with a de Broglie wavelength $\lambda = 1\,\text{Å}$ are given in Table 7.1. Because the energy corresponding to room temperature is

$$kT = 8.6\times 10^{-5}\,\frac{\text{eV}}{\text{K}}\times 300\,\text{K} = 0.025\,\text{eV},$$

TABLE 7.1
Kinetic energy of particles with a de Broglie wavelength $\lambda = 1\,\text{Å}$

Photon	Electron	Neutron	Helium
12.4 keV	150 eV	0.081 eV	0.02 eV

it is meaningless to discuss the de Broglie waves of atoms that are heavier than helium. This does not mean that de Broglie's relation is invalid under such conditions, but the effect cannot be seen. For example, if an object with mass of 10 μg moves at a speed of 1 cm/s, its de Broglie wavelength $\lambda = h/p = 6.6 \times 10^{-27}$ erg·s/10^{-5} g × (1 cm/s) = 6.6×10^{-22} cm. Thus while the equation is applicable, we see the de Broglie relation cannot be exhibited by macro objects: it is only seen for a microparticle.

7.2.5 de Broglie Waves and Quantum States

In Chapter 3, we pointed out that Bohr obtained the quantization condition of angular momentum $L = n\hbar$ by using the stationary-state condition, frequency condition, and correspondence principle. On the basis of angular momentum quantization, the quantization results for the first Bohr radius and the electron energy and momentum in a hydrogen atom were derived. So, assuming the quantization condition of angular momentum $L = n\hbar$, we can derive all other conclusions.

Next we will introduce the way de Broglie connected the stationary state in atoms with a standing wave and obtained the quantization condition of angular momentum very naturally. According to the de Broglie's hypothesis, the wavelength that embodies the wave nature of electrons is $\lambda = h/p = h/mv$. Now apply the de Broglie relation to an electron which is rotating around the nucleus in hydrogen. In order to let the rotation of the electron around the nucleus exist in a steady state, the wave corresponding to the electron must be a standing wave as shown in Fig. 7.7(a). That is, the phase of the wave must not change after one circle, as shown in Fig. 7.7(b), otherwise the electron wave will be destroyed. In other words, to maintain a stable constant motion of the electron, the circumference of the electron's rotation around the nucleus must be an integer times the corresponding wavelength, namely,

$$2\pi r = n\lambda = n\,\frac{h}{mv}, \quad n = 1, 2, \ldots \tag{7.11}$$

or

$$mvr = n\,\frac{h}{2\pi}$$

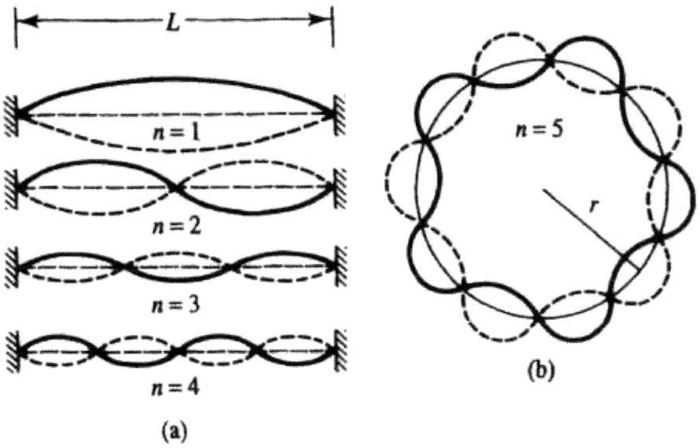

FIGURE 7.7
One-dimensional standing wave.

which is the quantization condition of angular momentum Bohr had obtained:

$$L = n\hbar, \quad n = 1, 2, \ldots \tag{7.12}$$

It is clear that a bound wave must be a standing wave and the condition of being a standing wave is the quantization condition of angular momentum. Figures 7.8(a) and (b) indicate standing waves of $n = 2$ and $n = 4$, respectively.

In order to verify what we have discussed above, we can substitute the velocity of the electron in the first Bohr orbit, $\mathrm{v} = \alpha c$, into the relation $\lambda = h/m\mathrm{v}$; so

$$\lambda = \frac{h}{m\alpha c} = \frac{h}{mc} \cdot \frac{1}{\alpha} = 2\pi \frac{\hbar}{mc} \cdot \frac{1}{\alpha}$$

in which $(\hbar/mc)(1/\alpha)$ is 137 times the reduced Compton wavelength of the electron, namely, the first Bohr radius a_1. Hence,

$$\lambda = 2\pi a_1$$

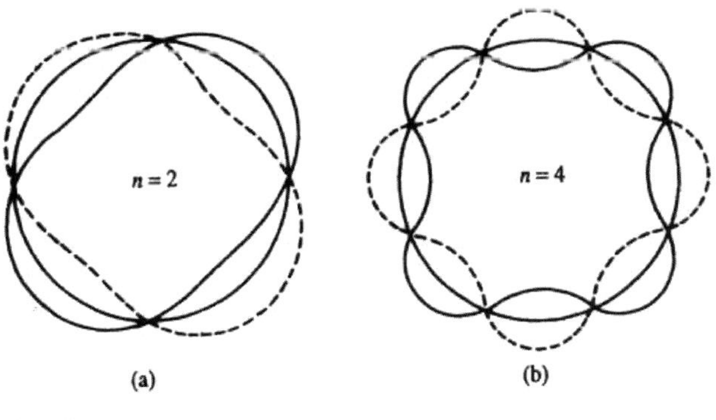

FIGURE 7.8
The standing wave diagrams of $n = 2$ (a) and $n = 4$ (b).

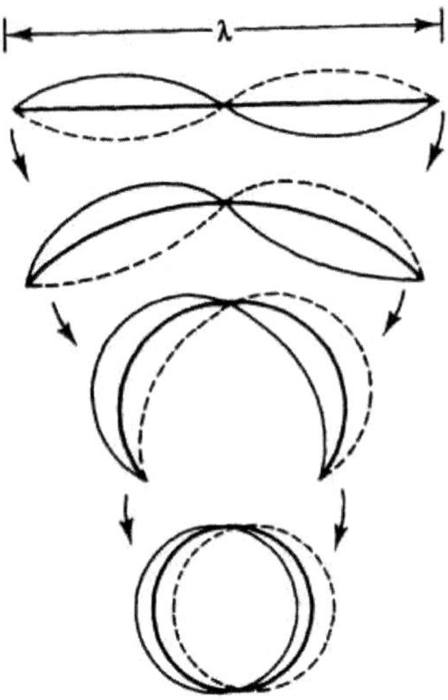

FIGURE 7.9
The standing waves and the orbits of electrons.

(see Fig. 7.9). This proves that the results obtained previously do satisfy the condition of a standing wave.

7.2.6 A Particle in a Rigid Box

Imagine a particle undergoing one-dimensional motion in a rigid box, as shown in Fig. 7.10(a). According to classical theory, we know that the kinetic energy of this particle in the box will always be $mv^2/2$ and the period of the motion will be $T = 2d/v$.

Now in the view of quantum mechanics, consider the one-dimensional motion of a microparticle in a box of width d. The de Broglie wave corresponding to the particle cannot penetrate the wall of the box, therefore the two points $x = 0$ and d will always be wave nodes, as shown in Fig. 7.10(b). The condition for this particle to exist forever in the box is that its de Broglie wave must be a standing wave and its wavelength must satisfy

$$n\frac{\lambda}{2} = d, \quad n = 1, 2, \ldots \tag{7.13}$$

That is, the box must be at least half a wavelength wide. Substituting the above formula into the de Broglie relation $p = h/\lambda$ and the nonrelativistic kinetic energy formula, $E_K = p^2/2m$, respectively, we get

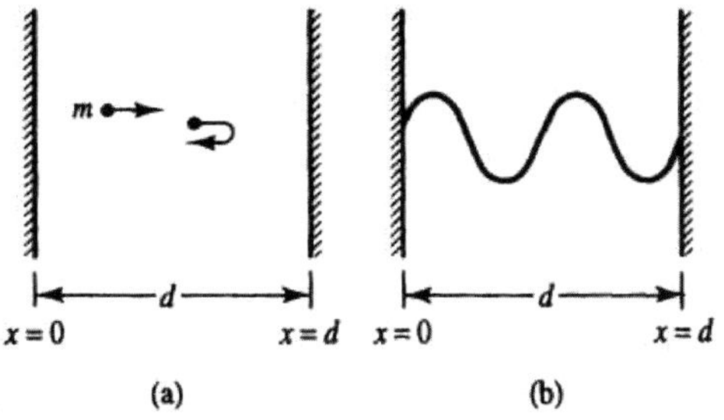

FIGURE 7.10
A standing wave in a one-dimensional rigid box.

$$p = \frac{nh}{2d} \tag{7.14}$$

$$E_K = \frac{n^2h^2}{8md^2} \tag{7.15}$$

This shows that both momentum and energy are quantized. In addition, it is clear from Eq. (7.15) that the minimum energy of a particle is $E_1(n = 1)$. This kinetic energy exists even if the temperature is absolute zero, as shown in Fig. 7.11. The dashed line in the figure indicates that the energy level $E = 0$ does not exist. It should be pointed out that although these characteristics are displayed prominently in the microcosmic domain, it is not true that they do not exist in the macrocosm. In both the macrocosm or microcosm, there is no absolute "standing still" since all atoms are in constant motion. Today this view has been verified by many observations.

The contents of the last two sections can be summed up in the statement: A bound wave yields quantization conditions. Of course, this statement itself is not just a characteristic of quantum mechanics, because there are many classical examples to illustrate it. This is easy for those who are familiar with music to understand. However, connecting a wave with a particle to obtain the quantization of the energy of the particles cannot be understood by classical physics and can only be explained by quantum mechanics.

7.2.7 A Wave and the Property of Nonlocalization

One of the characteristics of a wave is that it can extend infinitely in space. This is the property of nonlocalization of a wave. Because of this property, the dimension

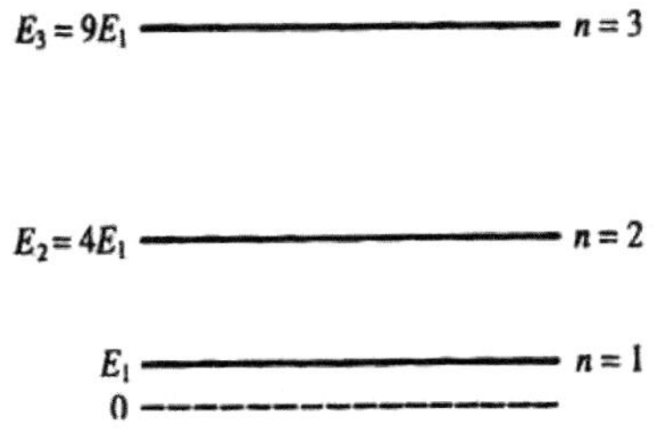

FIGURE 7.11
The energy levels in a one-dimensional rigid box.

of a box must be at least as long as half a wavelength $\lambda/2$ if we want to close a wave in a box. This proves that the shorter the wavelength, the smaller is the range required to restrain the wave, but it cannot be zero.

From the view of the de Broglie wave, electron motion in Bohr's hydrogen atom is actually the case of the de Broglie wave being closed in a Coulomb potential field. First assume that the particle (the electron in a hydrogen atom) is a simple sine wave in a box and that the boundary of the box is approximately rigid ($V = \infty$). Then add the Coulomb potential and also assume that the box is half a wavelength long, namely, that the particle in the box is in the ground state. Then we obtain the wavefunction shown in Fig. 7.12(a) [the actual wavefunction and Coulomb potential function of the hydrogen atom are shown in Fig. 7.12(b)]. Under this assumption we can obtain the kinetic energy of the particle according to Eq. (7.15):

$$E_K = \frac{h^2}{8md^2} = \frac{h^2}{8\pi^2 mr^2} = \frac{\hbar^2}{2mr^2}$$

in which the relation $d = \pi r$ has been used. This is because in a box the path of one cycle is $2d$ and the circumference is $2\pi r$, where r is the radius. Because the total energy is the sum of the kinetic energy and potential energy (now the potential energy is the Coulomb potential), we have

$$E = \frac{\hbar^2}{2mr^2} - \frac{e^2}{r} \tag{7.16}$$

From $dE/dr = 0$, we find r:

$$r_{\min} = \frac{\hbar^2}{me^2} = a_1 \tag{7.17}$$

Substituting this into Eq. (7.16) again, we obtain the ground state of the hydrogen atom

$$E = -\frac{me^4}{2\hbar^2} = -13.6\ \text{eV} \tag{7.18}$$

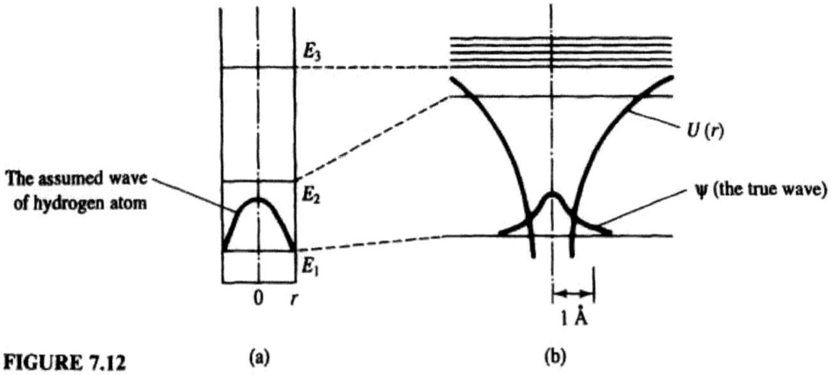

FIGURE 7.12
The energy levels and wavefunction of the hydrogen atom (simplified).

We pointed out in Chapter 2 that the most important characteristic quantities in the atomic world are the dimension a_1 and the energy E of atoms. Through the above simple calculations we obtain these two quantities that characterize the hydrogen atom and they agree with what Bohr had given.

Descriptions similar to the above three sections are found in many books on quantum mechanics. The advantage of such descriptions lies in their rather vivid illustration of the relation between the wave–particle dualism and the quantization conditions. This is why they are introduced here. But we must emphasize that all these descriptions are very approximate. The Bohr method overemphasizes the particle nature (through the concept of orbit), while the method of this section puts undue emphasis on the wave nature of matter.

7.3 UNCERTAINTY RELATIONS

The uncertainty relations, which some call, inappropriately sometimes, "expressions of being unable to accurately measure," were proposed by Heisenberg in 1927. They reflect basic laws of the motion of microparticles and have profound consequences. Among their many approximate expressions are

$$\Delta x \, \Delta p_x \geq h \tag{7.19}$$

$$\Delta t \, \Delta E \geq h \tag{7.20}$$

Expression (7.19) indicates that when a particle is limited to be in the finite range Δx in the direction of x, its momentum p_x in the x direction will have an uncertainty of Δp_x and the product of Δx and Δp_x satisfies the relation $\Delta x \, \Delta p_x \geq h$. In other words, if the location x is determined completely ($\Delta x \to 0$) then the momentum p the particle may possess will be totally undetermined ($\Delta p_x \to \infty$). When the particle is in a state where p_x is completely determined ($\Delta p_x \to 0$), we cannot locate it in the x direction, that is, the location of the particle in the x direction is totally undetermined. Note that in Eq. (7.19), Δx and Δp_x must be in the same direction; there are similar relations between Δy and Δp_y and Δz and Δp_z. But there is no similar relation between Δx and Δp_y.

The expression (7.20) indicates that if a particle can only stay in the state of energy E for a time Δt, then in this period of time the energy of the particle in that state is not exact. There is a dispersion of energy $\Delta E \geq h/\Delta t$. Only when the time period is infinite (stable state), can the energy of the state be determined completely ($\Delta E = 0$).

The descriptions of the uncertainty principle in some books are not very precise. For example, they consider that the expression (7.19) indicates that we cannot measure accurately the position x and the corresponding momentum p_x at the same time. This statement has led some people to think incorrectly that the uncertainty relation is a limitation of the measurement process. In this incorrect picture, either the particles themselves have definite momentum and location but we cannot measure them accurately at the same time, or the process of our measurement introduces an uncertainty in x and p_x. In fact the uncertainty principle reveals an important fundamental physical property of nature: a particle cannot have a definite location and a definite momentum at the same time independently of any measurement. This is a part of their nature, not of our measurement. Thus, the statement that "we cannot measure them accurately at the same time" is the inevitable outcome of the nature of reality as we know it. We must not call expressions (7.19) or (7.20) "the principle of being unable to measure accurately."

It should be pointed out that in the uncertainty relations the key quantity is again the Planck constant h. Because it is such an extremely small quantity, the uncertainty principle cannot be observed directly in the macroworld. However, it is not equal to zero and this enables the uncertainty relation to become an important law in the microworld. This is illustrated in the following two numerical examples.

First look at the hydrogen atom. From Chapter 3 we obtain the first Bohr radius of the electron $a_1 = 0.53$ Å and the first Bohr velocity $v_1 = \alpha c$. The corresponding momentum is $p_1 = mv_1 = mc\alpha$. From the viewpoint of the uncertainty relation, if the electron is in an orbit of radius a_1, its position is defined and its momentum is completely undefined. Thus the concept of "motion in an orbit" is now meaningless. If we assume that the electron is only confined to the range of a_1, that is, $\Delta x = a_1 = 0.53$ Å, then what is the $\Delta p/p$ corresponding to it? From relation (7.19), we have

$$\frac{\Delta p}{p} = \frac{h/\Delta x}{p} = \frac{hc}{mc^2\alpha\,\Delta x} = \frac{12.4\ \text{Å}\cdot\text{keV}}{511\,\text{keV}(137)^{-1}0.53\,\text{Å}} = 6.3$$

Thus it is clear that the degree of uncertainty of the momentum is so large that one can never know with any reasonable accuracy the momentum of an electron which moves in the range of a_1.

Now look at a macroscopic example. Assume that a small ball of 10 g is moving at the speed of 10 cm/s and the instantaneous position of the ball is determined reasonably accurately, for example, $\Delta x = 10^{-4}$ cm (in daily life this accuracy is very high). Then what is the uncertainty of its momentum? From relation (7.19), we have

$$\frac{\Delta p}{p} = \frac{h/\Delta x}{p} = \frac{(6.6\times 10^{-27}\,\text{g}\cdot\text{cm}^2/\text{s})/10^{-4}\,\text{cm}}{10\times 10\,\text{g}\cdot\text{cm/s}} = 6.6\times 10^{-25}$$

Thus, clearly for a macrobody the uncertainty of momentum is negligibly small and cannot be detected by any known experimental technique.

The effect of the uncertainty principle in the macrocosm is like the effect created in the microcosm when $h \to 0$. Here the correspondence principle is embodied again: Quantum physics $\to$ classical physics, when $h \to 0$.

7.3.1 Simple Derivation of the Uncertainty Principle

Here we introduce two simple derivations of the uncertainty principle. First, starting from the classical wave concept, we use the relations derived in Section 7.2.1:

$$\Delta t\, \Delta \nu \geq 1 \tag{7.21}$$

$$\Delta x\, \Delta \lambda \geq \lambda^2 \tag{7.22}$$

Equation (7.22) indicates that in order to get an isolated wave, namely, a wave packet with position well determined (small Δx) (see Fig. 7.13), we must overlap many waves, that is the smaller Δx, the larger $\Delta\lambda$. On the other hand, in order to measure the wavelength of a wave with infinite accuracy ($\Delta\lambda \to 0$), one must observe it over an infinite space ($\Delta x \to \infty$). To measure the frequency with infinite accuracy ($\Delta\nu \to 0$), the measurement time must be infinitely long.

All of these are concepts of classical physics. Now let us put in the de Broglie relation $\lambda = h/p$, that is, use the classical expressions (7.21) and (7.22) with microparticles. The new concept is created at once.

Substituting the relation $\Delta\lambda = (h/p_x^2)\,\Delta p_x$ derived from $\lambda = h/p$ into expression (7.22), we have

$$\Delta x\, \Delta\lambda = \Delta x\, \Delta p_x \frac{h}{p_x^2} \geq \lambda^2$$

or, rewriting,

$$\Delta x\, \Delta p_x \geq \frac{(\lambda p_x)^2}{h} = \frac{h^2}{h} = h$$

Thus we get the expression (7.19),

$$\Delta x\, \Delta p_x \geq h \tag{7.19}$$

Similarly, starting from expression (7.21), and substituting by using the equation $\nu = E/h$, we obtain

$$\Delta t\, \Delta E \geq h \tag{7.20}$$

Next we introduce another simple derivation of the uncertainty relation. Consider the example of single-slit diffraction in terms of a classical wave. As shown in Fig. 7.14(a), (b), and (c), when the wavelength of the incident wave is close to the width d of the slit, diffraction appears, and when $\lambda \ll d$, diffraction

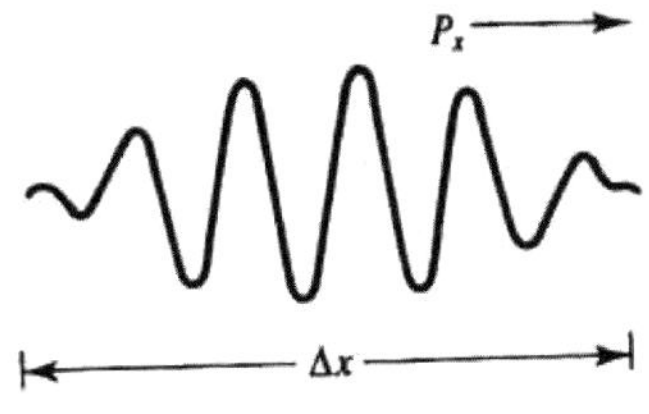

FIGURE 7.13
Wave packet.

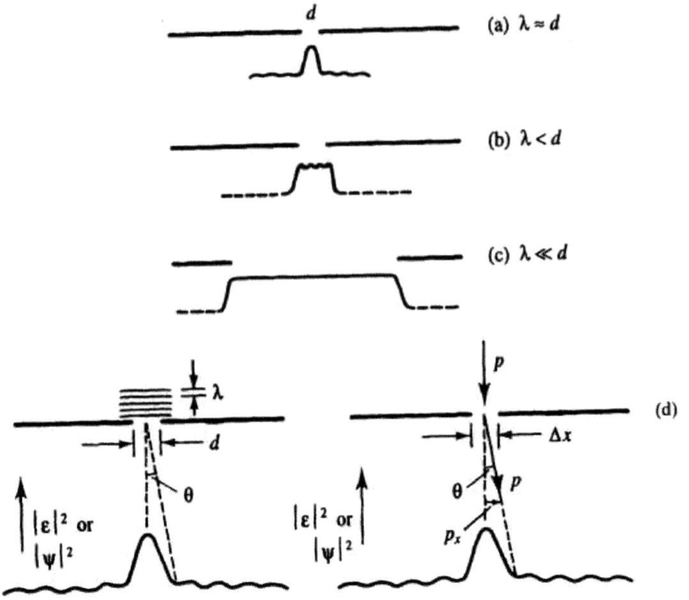

FIGURE 7.14
Single-slit diffraction.

disappears. A diffraction pattern is shown in Fig. 7.14(d) for a wave or electron with momentum p. The area of the center region (center peak) is three times as large as that of other regions. The position of the center region is determined by

$$\sin\theta = \pm\frac{\lambda}{d} \tag{7.23}$$

The minima outside the center region where the intensity is zero are determined by

$$\sin\theta = \frac{n\lambda}{d} \tag{7.24}$$

All these relations are familiar in classical physics and hold for any wave. Now consider the de Broglie wave corresponding to electrons. If the incident electron has a definite momentum p, then after passing through the slit of width d, the electron, even if only the center region is considered (75% of the electrons are included), has at least an uncertainty of momentum of $p\sin\theta$ (see Fig. 7.14(d)), that is,

$$\Delta p_x \geq p\sin\theta \tag{7.25}$$

Using Eq. (7.23) and taking $\Delta x = d$, we have

$$\Delta p_x \geq \frac{p\lambda}{\Delta x}$$

Using the de Broglie relation $p\lambda = h$ to substitute for $p\lambda$, we get Eq. (7.19):

$$\Delta p_x\,\Delta x \geq h$$

It should be pointed out that although the derivations given above reflect the nature of the uncertainty relation, they are only rough estimations. A proper derivation will give the results

$$\Delta x\, \Delta p_x \geq \frac{\hbar}{2} \tag{7.26}$$

$$\Delta t\, \Delta E \geq \frac{\hbar}{2} \tag{7.27}$$

7.3.2 Examples of the Application of the Uncertainty Relations

Example 1: The minimum average kinetic energy of bound particles. Assuming that a particle is bound in the range of r, namely, $\Delta x = r$, then according to Eq. (7.26) the momentum of the particle will have an uncertainty which is at least

$$\Delta p_x = \frac{\hbar}{2\,\Delta x} = \frac{\hbar}{2r}$$

in which the definition of Δp_x is

$$\Delta p_x = \sqrt{[(p_x - \bar{p}_x)^2]_{\text{ave}}} \tag{7.28}$$

For a particle bound in space, the average component of the momentum in any direction will inevitably be zero, that is, $\bar{p}_x = 0$, thus the relation between Δp_x and the mean-square of the momentum is

$$(\Delta p_x)^2 = (p_x^2)_{\text{ave}} \tag{7.29}$$

For three-dimensional space:

$$(p_x^2)_{\text{ave}} = \frac{1}{3}(p^2)_{\text{ave}} \tag{7.30}$$

According to these relations, we can get the minimum mean kinetic energy:

$$E = \frac{p_{\text{ave}}^2}{2m} = \frac{3\hbar^2}{8mr^2} \tag{7.31}$$

in which m is the mass of the particle. It is clear that E can never be zero. Thus we obtain the conclusion obtained in Section 7.2 again. This conclusion is obtained from the uncertainty relation and has nothing to do with what form the bounding takes. The minimum kinetic energy of a particle cannot be zero (a particle cannot fall into the bottom of a well) as long as the particle is bound in space, for example, when the particle is trapped in a potential. If the kinetic energy of a particle is zero, then the uncertainty relation demands that $\Delta x \to \infty$, so the particle cannot be bound.

Example 2: An electron cannot fall into the inside of a nucleus. Bohr's atomic theory did not give an explanation of why an electron with acceleration would not emit radiation and fall into the inside of the nucleus. The uncertainty principle shows that electrons cannot exist for extended times inside a nucleus.

As the electron approaches closer and closer to the nucleus, that is, as r becomes smaller and smaller, it will go from the atomic dimension (Å, 10^{-8} cm) to the nuclear dimension (fm, 10^{-13} cm). According to the uncertainty relation, the momentum of the electron will become more and more indefinite, or according to Eq. (7.31), the mean kinetic energy of the electron will be larger and larger. For example, when the range (distance) over which an electron can move changes from 1 Å to 3 fm, its mean kinetic energy will increase from the order of 10 eV to the order of 30 MeV. But from where can an electron get so much energy? It has no such energy source. Moreover, the energy required to pull an electron off the surface of a nucleus of $A = 1$ to 100 is only 1–10 MeV, so the electron would have too much energy to be bound inside. Thus an electron cannot be contained nor can it exist inside of a nucleus for a long time. Protons and neutrons can exist for infinite times inside the nucleus, but an electron cannot exist there for an extended time because its energy would be much too large. Thus, electrons emitted as β-rays cannot preexist inside the nucleus but must be created in the process of β decay. However, an electron in an s atomic orbit does have a finite probability of spending a brief time inside the nuclear volume. This brief penetration of electrons into the nuclear volume is allowed because the electron is not being confined to a volume the size of the nucleus.

Equation (7.31) also tells us that the mean kinetic energy of electrons in an atom and that of protons or neutrons in the nucleus differ by about 10^6. The energy of the former is the order of an electron-volt and of the latter the order of a million electron-volts. The energy difference between an atomic energy level diagram and a nuclear energy level diagram is indeed a factor of 10^6.

Example 3: The natural width of spectral lines. In an optical spectral series, if both the energy levels (states) corresponding to a certain spectral line have a definite energy, then the transition between them will certainly create a definite spectral line with a sharp energy. But as the electron jumps from one energy level down to a lower one, the electron will possess some lifetime in this upper level, so its Δt cannot be infinite. The uncertainty relation dictates that there will be a corresponding energy width ΔE of this energy level. Thus a spectral line cannot be a geometric line but must be a line with a width ΔE. This ΔE is the natural width of the optical spectral line. For example, if the lifetime of an excited state of an atom is $\Delta t = 10^{-8}$ s, then the uncertainty relation expression (7.27) gives

$$\Delta E \geq \frac{\hbar}{2\,\Delta t} = \frac{\hbar c}{2\,\Delta t c} = \frac{197 \times 10^{-15} \times 10^{6}}{2 \times 10^{-8} \times 3 \times 10^{8}}\ \text{eV} = 3.3 \times 10^{-8}\ \text{eV}$$

This is the natural width of a spectral line corresponding to an excited state and is determined by the intrinsic lifetime of the energy level. Experiments have proven definitively the existence of the natural widths of spectral lines. This relation applies to the measured natural line widths of spectral lines of light which were emitted by distant stars billions of years ago. These measurements clearly show that the uncertainty relations are not a matter of our disturbing the system by our measurement. There is no way our measurements today can disturb the distribution of photon energies in a spectral line emitted by a distant star billions of years ago. The uncertainty relations describe a basic way that nature behaves. We can, of course, increase the uncertainty in Δt, ΔE, Δx, or Δp_x by our measurements. This is included in the "greater than" symbol in these equations.

Sometimes the lifetimes of energy levels may be influenced by external conditions. For example, atoms in a gas collide with each other continually. When an atom in an excited state is impacted by other atoms, in general it will lose its excitation energy, that is, the lifetime of this excited state will be reduced. Then, according to the uncertainty principle, the effect of collisions is to broaden the width of a transition line. The width caused by this process may greatly exceed the natural line width. This is referred to as collision broadening of the natural line width. In order to reduce the line width increase produced by collisions, the light source used in optical spectra research is often in a state of low pressure; for example, the pressure is the order of 1 mmHg.

The uncertainty relation has penetrated every field of the microcosmos and examples of its application are too numerous to be mentioned one by one. But the simple examples mentioned above have demonstrated that the uncertainty relation does not "bring inaccuracy to physics" as suggested by some. Exactly the reverse, the uncertainty relation brings accuracy in the sense of predictability to the microcosmos. Now the concept of "accuracy' is based on probability rules and differs intrinsically from that in classical physics. We will elaborate further on this in the next sections.

7.3.3 Principle of Complementarity

Bohr proposed the principle of complementarity almost at the same time that Heisenberg put forward the uncertainty relation. If we consider that Heisenberg's uncertainty relations come out of the wave–particle dualism of matter mathematically, then Bohr's principle of complementarity encompasses the dualism from the point of view of philosophy. The principle of complementarity and the principle of uncertainty are the two pillars of the Copenhagen interpretation of quantum mechanics.

Bohr's principle of complementarity came out of the uncertainty relations which follow from the wave–particle dualism. Since both light and particles have the wave–particle dualism and the wave nature and the particle nature can never appear at the same time in the same measurement (see Section 7.2), the classical concepts of wave and particle are mutually exclusive when being used to describe microscopic phenomena. On the other hand, since the two characteristics of waves and particles cannot be observed to exist simultaneously, they will not conflict directly with each other in the same experiment. However, both are necessary when describing microscopic phenomena. Thus, the two descriptions give complementary views of reality; not contradictory views.

A rather general description of the principle of complementarity is: The application of certain classical concepts will inevitably exclude the application of other concepts, but the "other classical concepts" will be necessary to describe phenomena under other conditions. In order to clarify the wave–particle dualism, Bohr often illustrated it by a simple and understandable example: A silver coin has two faces but we cannot see the two faces simultaneously. To see one face excludes seeing the other. Some ideas included in Bohr's principle of complementarity had been proposed by an ancient Chinese philosopher Gong Sualong as early as 2000 years ago. He considered that by seeing a white, hard stone you can only see the white color but cannot feel its solidity; and by touching it you can only know its solidity but not its color. When visiting China in 1981, J. Wheeler noted that "in the West the idea of complementarity seems to be revolutionary, but Bohr was delighted to discover that in the East the idea of complementarity was a natural way of thinking. In order to state the complementarity principle symbolically, Bohr chose the Chinese characters 'Yin Yan' to name it."

There is a corollary to the idea of complementarity, namely, the existence of complementary variables. The Heisenberg uncertainty principles give the relations between complementary variables. You can never know exactly any two complementary variables simultaneously, and the more you know of one the less you know of the other. To know one exactly leaves the other completely undetermined. Position and momentum are complementary variables, as are energy and time. Of course, waves and particles are not complementary variables in this sense because you experience either one or the other and not both at the same time each with some uncertainty. Bohr extended his ideas into other fields. For example, thought and observation of the thought process are considered complementary variables—the more deeply you think the less you can know about or be aware of the process of thinking.

Either the uncertainty relations or the principles of complementarity will lead to the concept that "a microscopic theory is statistical", and this is completely different from the concept of "determinism" in classical physics.

In Fig. 7.15 we see P. M. Dirac and W. Heisenberg out for a stroll.

7.4 THE WAVEFUNCTION AND ITS STATISTICAL EXPLANATION

7.4.1 The Wave–Particle Dualism and the Concept of Probability

It is well known in classical mechanics that if we have equations of motion of a system which is acted on by a known force and provided the initial conditions are known, that is, the exact locations and momenta of the particles at a certain time ($t = 0$) are known, we can solve these equations and describe the locations and momenta of the particles at any time. This is "the sense of determinism" or "the strict law of causality" in classical physics. This concept has been very successful

FIGURE 7.15
P. M. Dirac and W. Heisenberg out for a stroll.

in the macrocosm, such as in the description of objects in astrophysics and in the description of the laws of motion of man-made satellites.

When changing from the macrocosmos to the microcosmos, classical physicists very naturally imitated the set of familiar and successful classical methods and expected that by observation they could determine accurately information about a certain particle, for example the location and momentum of an electron. But the view of Heisenberg and Bohr is completely different from this. In the microcosmos, we cannot determine the location and the momentum of matter or radiation simultaneously and we cannot determine them more accurately than Heisenberg's uncertainty relation has stipulated. As a result, we can only forecast the possible behaviors of particles and not their exact behaviors. Heisenberg and Bohr considered that the concept of probability was the basic concept in quantum physics and that determinism must be given up. This is the key content of the Copenhagen interpretation of quantum mechanics. This view was strongly opposed by many of the scientific leaders of that time. Next we will introduce how this view was created.

It is clear from the single-slit diffraction experiment described in the last section that because the electron's momentum has at least the uncertainty of Δp, we cannot predict accurately where the electron will appear. This uncertainty is inherent in the diffraction phenomenon and comes from the wave–particle dualism. However, there is a type of certainty in the uncertainty. For example, the probability that an electron will fall into the middle range is completely determined. From calculations, that probability is 75%.

As another example, the lifetime of a microparticle in an energy state with a width ΔE is Δt. During the period of Δt, it is completely undetermined at which moment the particle will decay (or jump to the lower energy state). But its probability has a well-defined numerical value. This type of determinism comes exactly from the uncertainty relation and the wave–particle dualism of matter.

The wave–particle dualism will certainly lead to a statistical explanation of things. The two different classical concepts, wave and particle, are connected by statistical concepts. In the case of light (radiation) it is exactly the statistical concept introduced by Einstein early in 1917, and for matter waves it is the probability explanation of de Broglie's wave proposed by Max Born in 1927.

Let us first investigate the case of radiation. Classical theory tells us that the stream of energy (energy per time per area) is proportional to the square of the electric field intensity of the wave, that is, $|E|^2$. From the particle viewpoint, it should be equal to $h\nu N$, in which N is the flux of photons, that is, the numbers of photons per unit time which strike a unit area perpendicular to their direction of motion. For example, consider a beam of very weak ultraviolet light with an intensity of $1 \times 10^{-13}\ \mathrm{W/m^2}$ which is equivalent to 10^{-8} times the light intensity of an average star on the earth's surface. From $h\nu N$, we can obtain that N is 12.5 photons/($\mathrm{cm^2 \cdot s}$). Since a photon is quantized, the non-integer here indicates that N is a mean value and the concept of probability of occurrence is included. So, N varies around 12 and its mean value is 12.5. Probability enters, so N is the most probable number of photons striking a unit area in unit time.

It is clear that the intensity of the light with a certain frequency is proportional to the number of photons and at same place the number of photons is proportional to the probability of photons appearing in this place. Further, because the intensity of light is proportional to the square of the electric field intensity of the optical wave, the probability of finding a photon in a certain place is proportional to the square of the electric field intensity of the optical wave, so

$$N \propto |E|^2$$

We know that the electric field intensity of an electromagnetic sine wave with wavelength λ and frequency ν traveling in the x direction can be written as

$$E = E_0 \sin 2\pi\left(\frac{x}{\lambda} - \nu t\right)$$

For a particle moving with constant linear momentum in the x direction, its de Broglie wave is

$$\psi = \psi_0 \sin 2\pi\left(\frac{x}{\lambda} - \nu t\right)$$

or in a more universal form:

$$\psi = \psi_0 e^{i(\boldsymbol{k}\cdot\boldsymbol{r} - \omega t)}$$

in which $|\boldsymbol{k}| = 2\pi/\lambda$, $\omega = 2\pi\nu$. Thus besides the wavelength there is an amplitude which is called the wavefunction connected with the matter wave. As Einstein explained, $|E|^2$ is "the measure of probability of photon density," and Born described $|\psi|^2$ as the probability of finding a particle in a given period of time and in a certain space interval. Born pointed out that "for a given state in space, there will be a probability determined by the de Broglie wave corresponding to the state . . .. If the wavefunction corresponding to an electron is zero at a certain point, it means that the probability of finding an electron at this point is zero".[7]

It should be emphasized that the probability explanation of the wavefunction proposed by Born is not and cannot be derived from any principle. It is one of the basic hypotheses of quantum mechanics. Although there are similarities between the wavefunction and the classical wave amplitude in form, their meanings are totally different. The classical wave amplitude can be measured, but ψ is unmeasured. In general only the probability $|\psi|^2$ is measurable. For the distribution of probability, the relative distribution is more important. It is clear that the relative distributions described by $\psi(r)$ and $C\psi(r)$ (where C is a constant) are the same, while for a classical wave doubling the amplitude quadruples the energy of the wave. So the two waves are completely different. In order to understand the difference between a quantum wave and classical wave, we will consider the double-slit interference experiment and its explanation.

7.4.2 Double-Slit Interference Experiment

In order to illustrate the characteristics of a wavefunction, let us consider the double-slit interference experiment. In optics the double-slit interference experiment was made for the first time by the British physicist Young in 1801 and was explained satisfactorily by the optical wave theory. The experiment is known as Young's experiment and was one of the most important experimental bases of optical wave theory.

We put two parallel slits, 1 and 2, in front of a light source S, where the two slits are the same distance from the light source S (see Fig. 7.16(a)). The distance between the two slits is very small, so the two slits form a pair of coherent light sources. The light emitted from slit 1 and slit 2 will superpose in space and create

[7]Max Born, *Atomic Physics*, Hafner, New York (1962).

an interference phenomenon. The curve $I_1(x)$ in Fig. 7.16(a) indicates the distribution of light intensity vs. the coordinate x recorded on the screen when only slit 1 is opened. The curve $I_2(x)$ indicates the distribution when only slit 2 is opened. Finally, curve $I_{12}(x)$ shows the double-slit interference pattern when both slit 1 and slit 2 are opened.

If S is replaced by a machine gun and bullets are shot through the two slits randomly, then according to the classical viewpoint we will get Fig. 7.16(b). The curves' meanings are similar to those of the corresponding curves in Fig. 7.16(a). The intensity distribution $n_{12}(x)$ corresponding to the two slits being opened is simply the sum of the individual intensities, that is,

$$n_{12}(x) = n_1(x) + n_2(x)$$

and no interference phenomenon exists. These are the specific expressions of the two concepts in classical physics, waves and particles, which are completely different.

What will happen if we put an electron gun at S? Electrons are emitted from S and, after passing through the two slits, they reach the screen. What is the intensity distribution of the electrons recorded by the screen? According to the classical viewpoint, we should obtain something like Fig. 7.16(b), but, in experiments, we observe that Fig. 7.16(c) is similar to 7.16(a).

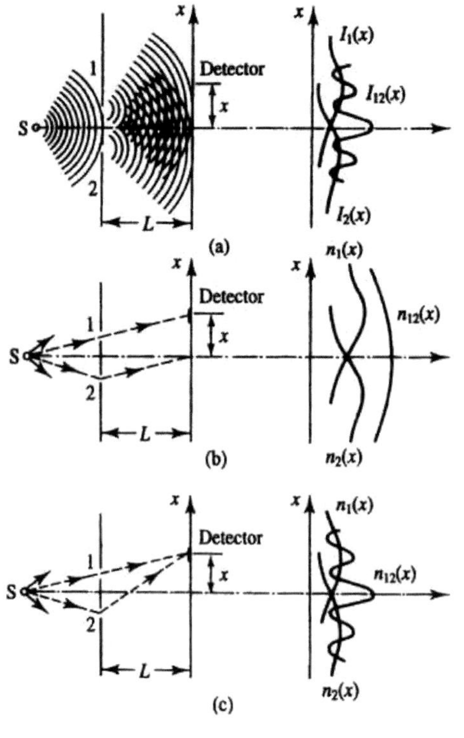

FIGURE 7.16
Schematic diagram of a double-slit interference experiment.

Experiments have found that no matter how weak the intensity of the incident light is (one photon per hour through one slit), we can still obtain the same double-slit interference pattern so long as the exposure time of the screen is long enough. In this case, from the light-quantum viewpoint, the incident light is so weak that the photons pass through the slit one by one! Meanwhile, modern experimental technology has enabled us to weaken the stream of electrons to such a degree that the time interval of emission of the electrons (or the time interval between each electron reaching the screen) is tens of thousands times longer than the time of an individual electron passing through the slit. Nevertheless, when we record the electrons on the screen, although at the beginning the distribution observed seems to be completely irregular (Fig. 7.17(a,b)), we can still obtain the distribution pattern of a double-slit interference by extending the recording time! The experimental results on electrons are shown in Fig. 7.17 as a function of time. Whether the particles we use are photons, neutrons, or protons, the results we get are similar.

All these results indicate fully that the appearance of the interference pattern reflects a common characteristic of these microparticles and the pattern is not created by the interaction of the microparticles but comes from the collective contributions of the properties of the individual particles.

For an individual microparticle (for example, an electron), we cannot forecast which slit it will pass through and at which point on the screen it will fall. Under identical experimental conditions each electron "persists in its own way," but the behavior of a large number electrons can be predicted accurately.

It should be pointed out that the results for electrons do not fit our expectations in other ways. Normally when an electron passes through slit 1, we expect that there will be no difference whether slit 2 is open or not. Similarly, when an electron passes through slit 2, we expect that there should be no difference whether slit 1 is open. In the case when slit 1 and slit 2 are opened simultaneously, the intensity of the electrons on the screen should be the sum of the two individual intensities. But this is not true! It can only indicate that somehow the two slits act simultaneously. It might seem that the electron passed through slit 1 and slit 2 at the same time. But how can an electron do that since up to now no one has discovered any structure of the electron in the range of 10^{-16} cm (it acts like a point particle) so that the radius of an electron must be less than 10^{-16} cm.

In order to "find out" how an electron passes through the two slits, we put one light source on each slit (Fig. 7.18, P_1, P_2) and a light detector (D_1, D_2) on each side of each slit. The photons emitted from the light source collide with the electrons passing through the slit and the scattered photons are recorded by the detectors. If the electron passes through the slits simultaneously, the two light detectors will give signals simultaneously (coincidence count). We control the stream of electrons so that the electrons strike the screen one by one. The light detectors are found to give only one signal each time an electron passes through and no coincidence event appears. It seems that the whole truth has come out, for we have observed the path of the electrons. But when we turn to the intensity distribution of the electrons on the screen, a result which we do not expect

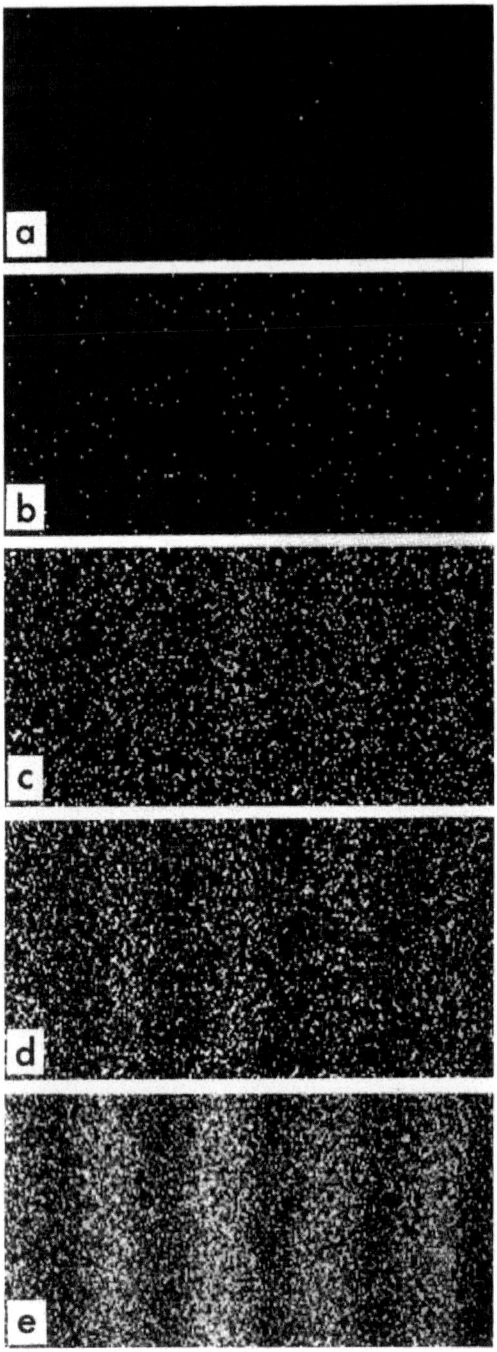

FIGURE 7.17
Build-up of the electron double-slit interference pattern. (Courtesy of A. Tonomura.)

appears. The interference pattern disappears and the measured result is the same as that of the bullets, not $I_{12}(x)$ but $I_3(x)$ in Fig. 7.18, where I_3 is the sum of the two intensities! In other words if we want to spy on the behavior of the electrons, the interference of the electrons will disappear. Turning off the light sources and repeating the experiment, we get the interference pattern again.

Someone might say that the interaction between the photons and electrons is too strong so that the interference is destroyed. How can one reduce the interaction? We cannot weaken the light intensity to reduce the number of photons, because the smaller the number of photons, the less the electrons are "being checked." Lowering the energy of the photons, that is, increasing their corresponding wavelength, increases the localization range of the photons in space (the accuracy of the localization of an electron cannot exceed this wavelength). Thus a photon with a wavelength that exceeds the slit width cannot be detected in association with a certain slit, namely, we cannot distinguish the electrons coming from the different slits. After repeated tests and careful consideration, it was concluded that in principle it is inevitable that "the observation effect causes the interference to disappear."

Finally, it should be pointed out that before 1961 both the single-slit diffraction experiment (Fig. 7.14) and the double-slit interference experiment for electrons (Fig. 7.16) were only "thought experiments"; that is, on the premise that electrons have a wave nature (the diffraction of electrons in crystals had proven this assumption) we could imagine that electrons would exhibit diffraction and interference phenomena in such circumstances. In 1961, however, C. Jonsson turned the "thought experiment" into reality. He ingeniously cut five slits on a

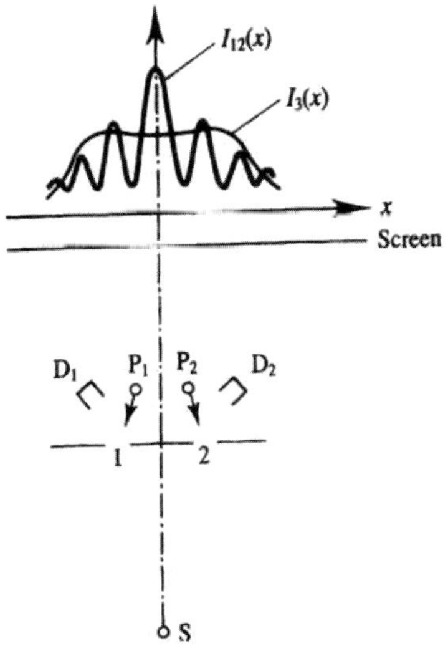

FIGURE 7.18
Using photons (from P_1, P_2) to detect the path of an electron.

[8]C. Jonsson, *Z. Physik* **161** (1961) 454.

[9]A. Tonomura, et al., *Am. J. Phys.* **57** (1989) 117.

thin metal section. Each slit was 50 μm long and 0.3 μm wide, and the interval between the two slits was 1 μm. He obtained a diffraction pattern and an interference pattern successfully for electrons. In 1989, A. Tonomura and co-workers did a beautiful two-slit electron interference experiment (see Fig. 7.17). Readers who are interested in this may consult the original papers.[8,9]

"All these experiments can absolutely not be explained by any classical method. But the kernel of quantum mechanics is exactly included in these experiments" (R. Feynman). Next we will introduce how to use the basic principle of quantum mechanics—the superposition of states principle to explain the interference experiment (for further discussion see *The Feynman Lectures*[10]).

7.4.3 Superposition of States Principle

In order to explain the double-slit experiment, we must introduce another basic principle in quantum mechanics: the principle of superposition of states. For this reason, we first repeat the basic principle mentioned in the last section: Born's statistical explanation of the wavefunction.

In the microcosmos the probability P of an event occurring is equal to the absolute square of the complex wave function:

$$P = |\psi|^2 \tag{7.32}$$

where ψ is also called the probability amplitude. For the sake of definition, we often use the following notation: If "a certain event" is generally described as "the transition from the initial state i to the final state f", the probability of this transition occurring, $w_{i\to f}$, or simply w_{if}, can be expressed in a notation introduced by Dirac as

$$w_{if} = \left|\langle f\,|i\rangle\right|^2 \tag{7.33}$$

in which $\langle f\,|i\rangle$ indicates the probability amplitude of a transition from the state i to the state f. Here $|i\rangle$ is called a ket and is analogous to the wavefunction for a state, so $|\,i\rangle$ denotes the ket vector that corresponds to the state, m, of the system. The $\langle f\,|$ is called a bra and is analogous to the complex conjugate of the wavefunction of a state. The scalar product of a bra and a ket, $\langle f\,|i\rangle$, corresponds to the integral of the product of the complex conjugate of the wavefunction for state f and the wavefunction for state i. The matrix element, $H_{fi} = \int \psi_f^* H \psi_i \, d\tau$, integrated over all space is written as $\langle f\,|H|i\rangle$.

We now list the rules that the probability amplitude $\langle f\,|i\rangle$ obeys:

Rule I. If the transition between i and f occurs in several ways (channels) which are indistinguishable in physics (see Fig. 7.19(a)) then the probability amplitude of

---

[10] R. P. Feynman, R. B. Leighton, and M. Sands, *The Feynman Lectures on Physics*, vol. 3, 2nd edition, Addison-Wesley, Reading (2005).

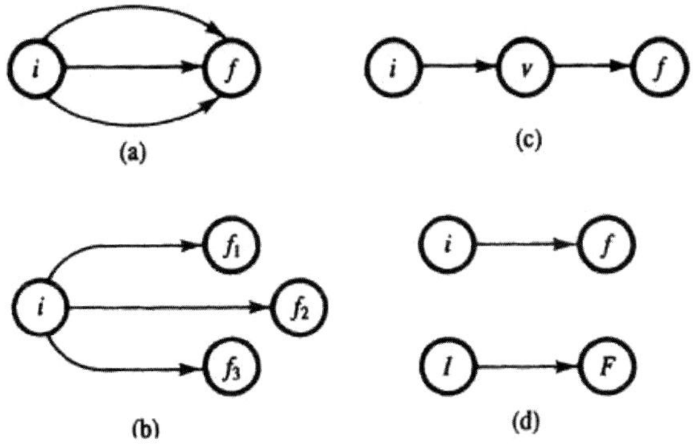

FIGURE 7.19
Illustration of the different patterns along which one can calculate the rules of probability.

the transition $i \rightarrow f$ should be the sum of the probability amplitudes of all possible transitions

$$\langle f|i\rangle = \sum_n \langle f|i\rangle_n \tag{7.34}$$

The subscript n indicates that there are n kinds of transitions.

Rule II. If there are n independent final states $f = 1, 2, 3, \ldots, n$ (see Fig. 7.19(b)), and we want to know the probability of going to an arbitrary final state, then the transition probability $|\langle f|i\rangle|^2$ to leave the state i is the sum of the probabilities of going to each of the n final states:

$$|\langle f|i\rangle|^2 = \sum_n |\langle f|i\rangle_n|^2 \tag{7.35}$$

Rule III. If the transition from state i to state f must pass through an intermediate state v (see Fig. 7.19(c)) then the total transition probability amplitude equals the product of the corresponding probability amplitudes:

$$\langle f|i\rangle = \langle f|v\rangle\langle v|i\rangle \tag{7.36}$$

Rule IV. If a system is composed of two independent microparticles and the two particles jump simultaneously (see Fig. 7.19(d)), then the transition probability amplitude of the system equals the product of the probabilities corresponding to the individual particle:

$$\langle fF|iI\rangle = \langle f|i\rangle\langle F|I\rangle \tag{7.37}$$

As a classical example, the probability of flipping coin A and it coming up heads is 50% and similarly for coin B. The probability of flipping both A and B simultaneously and both coming up heads is $0.50 \times 0.50 = 0.25$. This is as expected, since A and B being heads is but one of four possible allowed states.

It is not difficult to understand the last three rules. They are the well known addition rules of probabilities (Rule II) and product rule of probabilities for independent events (Rules III and IV). Only the first rule, which is called the superposition rule of probability amplitudes, expresses the superposition of states principle and is exactly the basis of the concepts of quantum mechanics. Feynman called it the primary principle of quantum mechanics. It is also a basic principle and cannot be derived from more fundamental concepts up to now.

7.4.4 The Explanation of the Interference Experiment

We now use the concept of probability amplitude and the rules it obeys to explain the interference experiment. Assume that an electron starts from the initial state S and passes through the wall with slit 1 and slit 2 (equivalent to the intermediate states 1 and 2) and finally is recorded by the screen (final state x) (see Figure 7.16).

Assume that only slit 1 opens; then according to Rule III we have

$$\langle x|S\rangle_1 = \langle x|1\rangle\langle 1|S\rangle$$

and the probability of the electron being recorded at x is

$$I_1(x) = |\langle x|S\rangle_1|^2 = |\langle x|1\rangle\langle 1|S\rangle|^2 \tag{7.38}$$

Similarly, when only slit 2 opens, we have

$$I_2(x) = |\langle x|S\rangle_2|^2 = |\langle x|2\rangle\langle 2|S\rangle|^2 \tag{7.39}$$

Now when both slits are open, because we cannot tell through which slit an electron passes, we must use Rule I, thus

$$\langle x|S\rangle = \langle x|1\rangle\langle 1|S\rangle + \langle x|2\rangle\langle 2|S\rangle \tag{7.40}$$

Now the transition probability will be

$$\begin{aligned} I_{12}(x) &= |\langle x|S\rangle|^2 = |\langle x|1\rangle\langle 1|S\rangle + \langle x|2\rangle\langle 2|S\rangle|^2 \\ &= I_1(x) + I_2(x) + \langle x|S\rangle_1\langle x|S\rangle_2^* + \langle x|S\rangle_1^*\langle x|S\rangle_2 \end{aligned} \tag{7.41}$$

It is clear that $I_{12}(x) \neq I_1(x) + I_2(x)$ since two more terms appear. It is exactly these two terms, the interference terms of the probability amplitudes of the two possible transitions of the electron going from the initial state to final state, that produce the interference pattern.

Let us further investigate Fig. 7.18 (or Fig. 7.20). First let us assume that the wavelength of the photons emitted from the light source P is very long so that no matter at which slit the photon is scattered by electrons it will be detected by detector D_1 or D_2; that is, the photons cannot "find out" through which slit an electron passes. At this moment, for the electrons we have two probability amplitudes

$$\langle x|1\rangle\langle 1|S\rangle = \phi_1 \quad \langle x|2\rangle\langle 2|S\rangle = \phi_2 \tag{7.42}$$

and for the photons, it is obvious from the point of view of symmetry that

$$\begin{aligned} \langle D_1|1\rangle\langle 1|P\rangle = \langle D_2|2\rangle\langle 2|P\rangle = \psi_1 \\ \langle D_2|1\rangle\langle 1|P\rangle = \langle D_1|2\rangle\langle 2|P\rangle = \psi_2 \end{aligned} \tag{7.43}$$

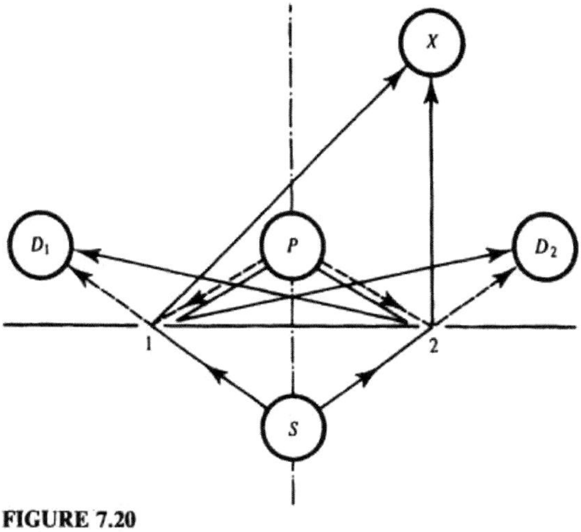

FIGURE 7.20
Diagram to illustrate the explanation of the double-slit interference experiment.

in which the first equation corresponds to the dashed line in Fig. 7.20 and the second equation is equal to the solid line starting from P in the figure.

We first calculate the probability amplitude of an electron being recorded at x and a photon being recorded at D_1 at the same time. This event includes two indistinguishable processes: the first process is where the electron passes through slit 1 and reaches x, the probability amplitude being $\langle x|1\rangle\langle 1|S\rangle = \phi_1$, meanwhile the photon reaches D_1 after being scattered by the electron near the slit 1, with a probability amplitude given by $\langle D_1|1\rangle\langle 1|P\rangle = \psi_1$. According to Rule IV, Eq. (7.37), the probability amplitude of the whole process should be

$$\langle xD_1|SP\rangle_1 = \phi_1\psi_1$$

The second possible process is that the electron passes through slit 2 and reaches x, so the probability amplitude is

$$\langle x|2\rangle\langle 2|S\rangle = \phi_2$$

Meanwhile the photon reaches D_1 after being scattered by the electron near slit 2, where its probability amplitude is

$$\langle D_1|2\rangle\langle 2|P\rangle = \psi_2$$

The probability amplitude of the whole process should be

$$\langle xD_1|SP\rangle_2 = \phi_2\psi_2$$

Because these two processes are indistinguishable, then according to Rule I, Eq. (7.34), the probability amplitude of the electron being recorded at x and the photon being recorded at D_1 at the same time is

$$\langle xD_1|SP\rangle = \phi_1\psi_1 + \phi_2\psi_2 \tag{7.44}$$

Similarly, the probability amplitude of the electron being recorded at x and the photon being recorded at D_2 at the same time is:

$$\langle xD_2|SP\rangle = \phi_1\psi_2 + \phi_2\psi_1 \quad (7.45)$$

Thus the probability of the electron being recorded at x and the photon being recorded by either detector is (Rule II, Eq. (7.35))

$$|\langle x|S\rangle|^2 = |\langle xD_1|SP\rangle|^2 + |\langle xD_2|SP\rangle|^2 \quad (7.46)$$

Substituting Eqs. (7.44), (7.45) into this equation we get

$$|\langle x|S\rangle|^2 = (|\phi_1|^2 + |\phi_2|^2)(|\psi_1|^2 + |\psi_2|^2) + (\phi_1\phi_2^* + \phi_1^*\phi_2)(\psi_1\psi_2^* + \psi_1^*\psi_2) \quad (7.47)$$

in which the second term reflects the interference effect obviously. This is the result when the path of the electron cannot be "detected" by photons.

If we shorten the wavelength of the photons so that the probability of the photons reaching D_2 after being scattered by electrons near the slit 1 is greatly reduced, that is, ψ_2 decreases, it is clear from Eq. (7.47) that the interference term thus decreases. The interference term vanishes completely when $\psi_2 = 0$. Then

$$|\langle x|S\rangle|^2 = |\psi_1|^2(|\phi_1|^2 + |\phi_2|^2) \quad (7.48)$$

and no interference pattern will be seen. So, by distinguishing the path of the electrons, we wash out the interference effect.

Equation (7.48) is the extreme case of "completely distinguishable." Another extreme case is "completely indistinguishable", where $\psi_1 = \psi_2$ and Eq. (7.47) becomes

$$|\langle x|S\rangle|^2 = 2|\psi_1|^2|\phi_1 + \phi_2|^2 \quad (7.49)$$

while Eq. (7.47) indicates the general case between the extreme cases. One can go continuously from the case of "completely distinguishable" to the case of "completely indistinguishable." With an increase in the distinguishability level of the two processes, the interference effect vanishes gradually.

It is clear from the analyses above that the explanation of the interference pattern depends on the linear summation of probability amplitudes. When both the two slits are open, even if there is only one electron, $\phi_1 + \phi_2$ is needed to describe the electron. It is true that the two slits act simultaneously. Note that we do not use the superposition principle of classical waves to explain the double-slit interference, although many books use the method of superposition of waves which is analogous to that of classical physics to explain the double-slit interference and to emphasize that the superposition of waves in classical physics is still applicable in quantum mechanics. However, the method of superposition of waves cannot explain the disappearance of the interference pattern caused by photons detecting electrons. Furthermore, that method does not provide an essential understanding of the phenomena.

Using the classical superposition principle of waves, one can actually explain the double-slit interference. Young's experiment in the early nineteenth century was explained by using this concept. In classical physics the appearance of interference phenomenon is considered as a symbol of the existence of a wave. Only when the characteristics of a wave are present, does an interference exist. Thus it is not surprising that when interference phenomenon and diffraction phenomenon appeared in experiments with microparticles the word "wave" was used to describe the microparticles. Such terms as de Broglie wave, wavefunction, wave equation, wave mechanics and so on thus appeared one after another. For historical reasons, we continue to use these terms. But it must be emphasized that a de Broglie wave is completely different from a classical wave, and we can never use the pattern of classical waves to represent microparticles. The interference and diffraction of electrons after passing through a slit have nothing to do with the pattern of a classical wave because their cause is the summation rule of probability amplitudes according to statistical rules (not the summation of probabilities). In the double-slit interference experiment, the superposition is the superposition of the two states of an electron and the interference is the interference of an electron with itself not the interference with other electrons.

Although the superposition of states in quantum mechanics and the superposition of waves in classical physics are the same in mathematical form, they are completely different in physical nature. In general the superposition of two classical waves will lead to a new wave which has new characteristics. But in the superposition of two quantum waves, the new wave (the more precise definition is the two probability amplitudes ψ_1 and ψ_2), $\psi = C_1\psi_1 + C_2\psi_2$, will not create a new state. What does it represent? If the result of measuring a certain mechanical quantity B in the case of the system being in the ψ_1 state is a definite value β_1 and the result when the system is in the ψ_2 state is another definite value β_2, then in the case of the system being in the ψ state, the result of measuring B will never be a new value different from β_1 and β_2. It is either β_1 or β_2. We cannot determine the exact one, but the probabilities of obtaining either β_1 or β_2 are completely definite, they are $|C_1|^2$ and $|C_2|^2$, respectively. The superposition of states in quantum mechanics leads to an uncertainty of the measured result in a superposition state.

Someone might raise the following objection to the above statement: Light has a wave–particle dualism and the wave nature of light is a classical concept and is common in the macrocosm. So why do we emphasize that the wave in quantum mechanics is completely different from the classical wave? This is a very good question.

First we point out that one cannot begin to talk about the concept of a classical wave when there is only one photon with the state k (wavevector), ω (angular frequency), and α (polarization state). Only when there are many photons which are in the state $k\omega\alpha$, does the phenomenon of a classical wave appear, whose characteristics correspond to those of the photon state $k\omega\alpha$. Recall, from Chapter 5, the Pauli Exclusion Principle says that two or more fermions, particles with spins $\frac{1}{2}, \frac{3}{2}, \ldots, \hbar$, cannot exist in the same quantum state. However, the other class of particles, bosons, which have integer spins, $n\hbar (n = 0, 1, 2, 3, \ldots)$, do not obey the Pauli Principle. So a boson particle can have any number of particles in the same quantum state because the Pauli Principle does not apply to them. The spin 1 photon is the unique stable boson existing in nature. This causes the "electromagnetic wave" to become a very common but also a "very specific" phenomenon. For the electron (and every fermion), a similar phenomenon will never take place and a similar "collective effect" will never occur. Classical interference (classical superposition) occurs only in an assembly of bosons.

For any microparticle the interference effect of quantum mechanics will exist. It is created by the superposition of probability amplitudes. For an assembly of bosons the classical interference still exists. It is created by the superposition of waves and coexists with the quantum interference and usually conceals the quantum interference.

The concept of wave in the concept of "the wave–particle dualism of light" is totally different from that of a light wave considered as an electromagnetic wave. A microparticle is neither a classical wave nor a classical particle, it is a special object. It has the potential power of showing characteristics analogous to the classical wave or particle in different environments. To date, although hundreds of attempts have been made, no one can describe a microparticle properly with our everyday words. As the French scientist Paul Langevin has said: We have found that in the microcosmos some concepts which were successful in the macrocosmos are inadequate. These concepts were established in order to work in the macrocosmos and were created by long contact with the world over many centuries.

7.4.5 Further Comments

In this section we have introduced two basic principles in quantum mechanics. The wave–particle dualism and the uncertainty principle combine "organically" and reflect the fundamental difference between quantum physics and classical physics. First, the difference lies in the fact that the basic law in quantum physics is a statistical law while the basic law in classical physics is determinism, strict causality. The Copenhagen School of Quantum Mechanics further states that "all laws in nature are statistical and classical causality is the extreme case of statistical laws."

The difference also lies in the fact that the statistical laws in quantum physics are totally different from the well-known statistical laws in classical physics. In classical physics "probability" is the key concept in the statistical law, while in quantum physics, the "probability amplitude" is the key concept. In classical physics the fundamental law is determinism and the statistical law is only a kind of method, a kind of tool and a kind of expedient measure of dealing with a multiparticle system. In contrast, in quantum physics, the statistical law is the fundamental law and individual particles reflect the statistical property.

It is exactly on these two different concepts of what constitute the fundamental principles that Einstein and Bohr held completely different views and launched their famous debate. Their debate, which continued for such a long time and involved such a profound question, took place between two great men who had a deep friendship. The debate between Einstein and Bohr can be divided into two stages. The Sixth Solvay conference in 1930 marked the end of the first stage. Before this conference Einstein raised a variety of criticisms countering the uncertainty relation and the principle of complementarity. He considered quantum mechanics as a theory which cannot be self-consistent and tried hard to point out the mistake of quantum mechanics in logic. At this conference, Einstein proposed the famous ideal experiment of a photon box. But he was defeated by Bohr and this forced Einstein to admit that quantum mechanics is a correct statistical theory. Then the second stage of the debate started and it continued until the death of Einstein in 1955. The focus of the debate switched to the question of the completeness of the theory. Einstein thought the statistical theory of quantum mechanics was only an expedient measure and was not the final theory. But the Copenhagen School headed by Bohr and Heisenberg thought from the very beginning that quantum mechanics was a complete theory and its mathematical base did not allow one to make further amendments.

The debate is still being carried on by some even now. Feynman wrote:[10]

> We must emphasize an important difference between classical mechanics and quantum mechanics. We discuss all along the probability of an electron reaching one place, yet even in the best experiment one cannot predict accurately what will happen. The only thing we can predict is the probability. If all these are correct, they indicate that physics has given up the idea of predicting things accurately and believes that it is impossible. The only thing we can do is to predict the probabilities of things happening. Although it does not accord with our early ideas of understanding nature and even retreats, no one can avoid it. At present we can only discuss probabilities. Although we say "at present" it is very likely to be so forever, more likely that the problem will never be solved and it is more likely that nature itself is just exactly so.

On the other hand, Dirac said in his closing speech at the conference on the development of quantum mechanics in 1972:[11]

> In my opinion it is clear that we have not obtained the basic law of quantum mechanics. The laws we are now using need to be amended greatly. Only by doing this can we get the relativistic theory. It is more likely that the amendment from the present quantum mechanics to the further relativistic quantum mechanics is as radical as the amendment from the Bohr orbit theory to the present quantum mechanics. After we have done such radical amendment, of course our concept of using statistical calculations to make theoretical physical explanations will be amended completely.

We stop the narration of the basic concepts of quantum mechanics for the time being. We hope that the discussions will inspire you to an in-depth study of quantum mechanics. In the next chapter we will introduce the Schrödinger equa-

[11] P. A. M. Dirac, *The Development of Quantum Mechanics*, Acc. Naz. Lincei, Roma (1974), p. 56.

tion. The emphasis will be on how the "present quantum mechanics" solves specific problems.

7.5 SUMMARY

1. The two important concepts in quantum mechanics: the concepts of quantization and wave–particle dualism.
2. The important relation in quantum mechanics: the uncertainty relation.
3. The basic principle in quantum mechanics: the superposition principle of states.
4. The two basic hypotheses in quantum mechanics: the statistical explanation of the wavefunction introduced in this chapter and the Schrödinger equation introduced next in Chapter 8.
5. The key constant in quantum mechanics: Planck's constant.
6. The three important experiments introduced in this chapter: the diffraction of electrons in a crystal, the single-slit diffraction, and the double-slit interference.

APPENDIX 7A THE SYMMETRY OF THE WAVEFUNCTION AND THE PAULI EXCLUSION PRINCIPLE

Consider a system which consists of two identical particles whose wavefunction is $\psi(1,2)$. Because the two particles are identical, that is, the particles cannot be distinguished from each other, then the case of particle 1 being in state a while particle 2 is in state b and the case of particle 1 being state b while particle 2 is in state a are the same. In other words, the wavefunction $\psi(2,1)$ obtained after the exchange operator P_{12} acts on $\psi(1,2)$,

$$P_{12}\psi(1,2) = \psi(2,1) \tag{7A.1}$$

is the same as the quantum state described by $\psi(1,2)$. They differ by a constant λ at most:

$$P_{12}\psi(1,2) = \lambda\psi(1,2) \tag{7A.2}$$

Using P_{12} acting on $P_{12}\psi(1,2)$ again, we have

$$P_{12}^2\psi(1,2) = \lambda P_{12}\psi(1,2) = \lambda^2\psi(1,2)$$

The state is back to the original one after being acted on twice by P_{12}. Obviously $P_{12}^2 = 1$, then

$$\lambda^2 = 1$$

and

$$\lambda = \pm 1 \tag{7A.3}$$

Thus according to formulas (7A.1) and (7A.2), we obtain

$$\psi(2,1) = \psi(1,2) \tag{7A.4}$$

and

$$\psi(2,1) = -\psi(1,2) \tag{7A.5}$$

We call the wavefunction possessing the characteristic (7A.4) a *symmetric* wavefunction, and one possessing the characteristic (7A.5) an *antisymmetric* wavefunction.

The wavefunction of a system can have two forms:

$$\psi_{\mathrm{I}} = \psi_a(1)\psi_b(2) \tag{7A.6}$$

$$\psi_{\mathrm{II}} = \psi_b(1)\psi_a(2) \tag{7A.7}$$

The probability of these two forms appearing is equivalent. It is reasonable that the system is in state ψ_{I} for the one half time and in the state of ψ_{II} for another half time. Thus the total wavefunction of the system should be the linear superposition of ψ_{I} and ψ_{II}:

$$\psi_S(1,2) = \frac{1}{\sqrt{2}}\left[\psi_a(1)\psi_b(2) + \psi_b(1)\psi_a(2)\right] \tag{7A.8}$$

$$\psi_A(1,2) = \frac{1}{\sqrt{2}}\left[\psi_a(1)\psi_b(2) - \psi_b(1)\psi_a(2)\right] \tag{7A.9}$$

which corresponds to formulas (7A.4) and (7A.5), respectively. The subscript S represents symmetry and A represents antisymmetry, and the constant $1/\sqrt{2}$ is a normalization factor.

Experiments indicate that there is a definite relation between the exchange symmetry of the wavefunction of a system with identical particles and the spin of the particles. The wavefunction of a system of identical particles whose spin is zero or an integer times $\hbar$ (such as a π meson or a photon) is symmetric and this kind of particle is called a boson. The wavefunction of a system of identical particles whose spin is an odd multiple of a half integer times $\hbar$ (such as an electron, proton, or neutron) is antisymmetric and this kind of particle is called a fermion.

For a symmetric wavefunction, formula (7A.8), the two particles can be in the same state at the same time; that is, when $a = b$, ψ_S exists. For an antisymmetric wavefunction, formula (7A.9), the two identical particles cannot be in the same state at the same time; that is, when $a = b$, $\psi_A = 0$. This is the Pauli exclusion principle: two identical fermions cannot be in the same single-particle state at the same time. Now we can express the Pauli exclusion principle as the following: For a system consisting of identical fermions, its wavefunction must be antisymmetric.

After considering the spin factor of the particle, the wavefunction of the system can be expressed as:

$$\psi(x, y, z, s_z, t) = \phi(x, y, z, t)\chi(s_z) \tag{7A.10}$$

in which $\phi(x, y, z, t)$ is the space part and $\chi(s_z)$ is the spin part. If the wavefunction ψ of the system is antisymmetric, then when ϕ is symmetric, χ is antisymmetric and when ϕ is antisymmetric, χ is symmetric.

Now let us discuss the combination of the spins of two electrons. Because the spin of an electron is $s_z = \pm\frac{1}{2}$, the total spin wavefunction may have the following forms:

$$\left\{\begin{array}{ll} & \chi_{1,1} = \alpha(1)\alpha(2) \qquad (7A.11) \\ \psi_{S,S} = & \chi_{1,0} = \frac{1}{\sqrt{2}}[\alpha(1)\beta(2) + \alpha(2)\beta(1)] \qquad (7A.12) \\ & \chi_{1,-1} = \beta(1)\beta(2) \qquad (7A.13) \\ \psi_{S,A} = & \chi_0 = \frac{1}{\sqrt{2}}[\alpha(1)\beta(2) - \alpha(2)\beta(1)] \qquad (7A.14) \end{array}\right.$$

Here α represents the state with spin up ($s_z = \frac{1}{2}$); and β represents the state with spin down ($s_z = -\frac{1}{2}$).

Figure 7A.1 shows a schematic diagram of these wavefunctions. It is clear from the figure that the z components of the spins given by formulas (7A.12) and (7A.14) are zero. But because the exchange symmetry has nothing to do with the z component (it has nothing to do with the orientation of z), different z components corresponding to the same spin must have the same symmetry. Hence formula (7A.12) is the same as formulas (7A.11) and (7A.13): all belong to $S = 1$, with formula (7A.12) being the zero component; while formula (7A.14) belongs to the zero component of $S = 0$.

It should be pointed out clearly that it is a rough approximation to use "↑" to express the z component of the spin of electrons in Fig. 7A.1(a), because s cannot be parallel with z strictly. A better picture than Fig. 7A.1(a) is Fig. 7A.1(b).

Having obtained knowledge of the spin wavefunction, we can express the total antisymmetric wavefunction as

$$\psi = \{\phi_0(r_1)\phi_{nl}(r_2) - \phi_0(r_2)\phi_{nl}(r_1)\}\begin{pmatrix}\chi_{11} \\ \chi_{10} \\ \chi_{1-1}\end{pmatrix} \tag{7A.15}$$

or

$$\psi = \{\phi_0(r_1)\phi_{nl}(r_2) + \phi_0(r_2)\phi_{nl}(r_1)\}\chi_{0,0} \tag{7A.16}$$

in which the subscript 0 of the wavefunction ϕ indicates that one particle is in the ground state and nl indicates the other particle is in a certain excited state. It is clear from (7A.15) and (7A.16) that if the spin space of the wavefunction is antisymmetric, the space part must be symmetric, and vice versa.

		S_z
$\chi_{1,1}$	$\uparrow\uparrow$	1
$\chi_{1,0}$	$\frac{1}{\sqrt{2}}(\uparrow\downarrow+\downarrow\uparrow)$	0
$\chi_{1,-1}$	$\downarrow\downarrow$	−1
χ_0	$\frac{1}{\sqrt{2}}(\uparrow\downarrow-\downarrow\uparrow)$	0

(a)

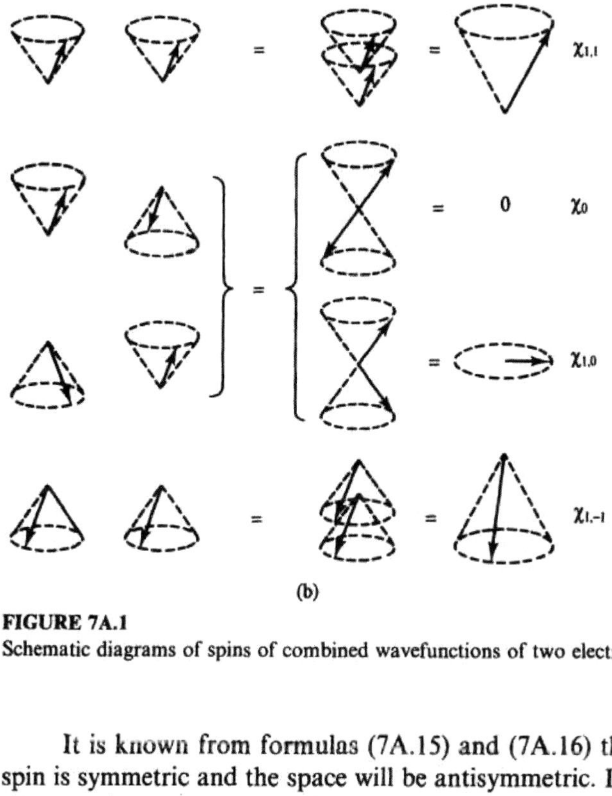

(b)

FIGURE 7A.1
Schematic diagrams of spins of combined wavefunctions of two electrons.

It is known from formulas (7A.15) and (7A.16) that $S = 1$ means that the spin is symmetric and the space will be antisymmetric. If at that moment $r_1 = r_2$, then $\psi = 0$, so r_1 cannot be equal to r_2 and r_1 in fact cannot be close to r_2. $S = 0$ means that the space part is symmetric, so that r_1 can be close to r_2. Because the electrostatic energy can be expressed as $e^2/|\boldsymbol{r}_1 - \boldsymbol{r}_2|$, as r_1 and r_2 approach each other so that the two particles are very close, the energy of repulsion is very high. Such a high energy means that the state is unstable (the two electrons repel each other). For the case where r_1 cannot be equal to r_2 so that the two particles cannot be close together, the energy is low and the state is stable. Hence we know that in the states which correspond to one electron being in the ground state and another electron being in an nl excited state, the energy of the system with $S = 1$ is the lowest, and this state is the most stable. This is exactly what we have observed in

TABLE 7A.1
The states of a system which is composed of two equivalent electrons $(p)^2$

L	D	P	S
ϕ	Even	Odd	Even
χ	Antisymmetric	Symmetric	Antisymmetric
S	0	1	0
^{2S+1}L	^1D	^3P	^1S

the optical spectrum of helium. Here we have proved Hund's rules using the Pauli principle.

It should be noted that the space wavefunction ϕ_l is proportional to the spherical harmonics $Y_{l,m}$ and the sign of $Y_{l,m}$ under inverse coordinates (its parity) is determined by $(-1)^l$ (see Chapter 8 for a discussion of $Y_{l,m}$). For a system of two particles, with total angular momentum L, the parity of $Y_{L,m}$ is proportional to $(-1)^L$. Thus for two or more fermions it is generally true that when L is an even number, the space wavefunction ϕ_L is symmetric while the spin wavefunction is antisymmetric ($S = 0$); and when L is an odd number, the space wavefunction ϕ_L is antisymmetric while the spin wavefunction is symmetric ($S = 1$). According to this rule, a system with two identical p electrons, p^2, has $L = 2, 1, 0$, so from Table 7A.1 there can only be three possible states ^1D, ^3P, and ^1S. The ^3D and ^3S states would have a symmetric total wavefunction (L even and $S = 1$), and ^1P would have a symmetric total wavefunction (L odd and $S = 0$) (the product of two antisymmetric functions is symmetric also) and so these three states are not allowed.

APPENDIX 7B
EINSTEIN'S *A* AND *B* COEFFICIENTS

A transition between two atomic energy levels generally occurs with the absorption or emission of radiation. This is the result of the interaction between the atomic system and the radiation field. A strict treatment should quantize both the atomic system and the radiation field, which is the method of quantum electrodynamics. In ordinary quantum mechanics, only the atomic system is quantized while the radiation field is treated as continuous. Quantum electrodynamics can explain the problem of spontaneous radiation while quantum mechanics can

[12]A. Einstein, *Z. Physik* **18** (1917) 121; D. ter Haar, *The Old Quantum Theory*, Pergamon Press, London, 1967.

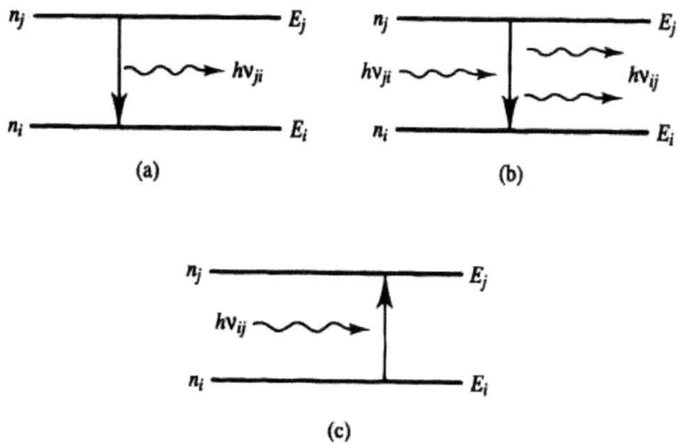

FIGURE 7B.1
Spontaneous emission (top left), stimulated emission (top right), and absorption (bottom) of electromagnetic radiation.

calculate only the absorption and the stimulated emission. Here we introduce the emission and absorption theory of radiation proposed by Einstein in 1917.[12] Using clear physical concepts, Einstein obtained the relations among the coefficients of stimulated emission and the coefficients of spontaneous emission and absorption, namely, the famous A and B coefficients. Also in his historic paper, Einstein put forward the concept of stimulated emission which provided the theoretical basis for the birth of lasers, which have had such a great influence on the world.

Now let us introduce Einstein's A and B coefficients. Consider the two energy levels of an atom E_j and $E_i(E_j > E_i)$. When a transition between the two energy levels takes place (see Fig. 7B.1), the radiative frequency of emission or absorption is determined by the quantization condition

$$h\nu_{ji} = |E_j - E_i| \tag{7B.1}$$

The transition of an atomic system from a higher energy level to a lower energy level can be divided into two categories: One is the well-known spontaneous transition (spontaneous emission) which is the transition of the system from level E_j to E_i when no external influence exists; the other is the stimulated transition (stimulated emission) introduced by Einstein, which is the transition from E_j to E_i under the external action of a radiation field. The transition of an atomic system from a lower energy level to a higher energy level takes place only when the outside can provide the energy, for example, an atomic system absorbs a photon of energy $h\nu_{ji}$ from the radiation field, an absorption transition, simply called absorption.

If at time t in a unit volume the numbers of atoms which are in the energy levels E_j and E_i are n_j and n_i, respectively, then for spontaneous emission the transition probability is obviously proportional to n_j. If we write the proportionality constant as A_{ji}, then (see Fig. 7B.1(a))

$$\frac{dn_j}{dt} = -\frac{dn_i}{dt} = -A_{ji}n_j \tag{7B.2}$$

For stimulated emission, the transition probability is not only proportional to n_j but is also proportional to the external field $u(\nu_{ji}, T)$,

$$\frac{dn_j}{dt} = -\frac{dn_i}{dt} = -B_{ji}n_j u(\nu_{ji}, T) \tag{7B.3}$$

in which $u(\nu_{ji}, T)$ represents the energy density of the radiation field per unit frequency interval. Here $u(\nu_{ji}, T)$ is not only a function of the frequency of the radiation field but also depends on the temperature T of the radiation field.

Similarly, for absorption, we have:

$$\frac{dn_j}{dt} = -\frac{dn_i}{dt} = C_{ij}n_i u(\nu_{ji}, T) \tag{7B.4}$$

The coefficients A, B, and C in the above formulas are called the spontaneous emission coefficient, stimulated emission coefficient, and absorption coefficient, respectively.

When the external field does not exist, only spontaneous radiation will take place. Then we find from formula (7B.2) that

$$n_j = n_{j0}e^{-A_{ji}t} \tag{7B.5}$$

in which n_{j0} is the number of atoms which are in the energy level E_j at the initial moment, $t = 0$. The number n_j decreases exponentially as the time t increases. The mean time τ of an atom being in the excited state can be obtained by using the formula

$$\tau = \frac{1}{n_{j0}}\int_0^\infty t|dn_j| = \int_0^\infty te^{-A_{ji}t}A_{ji}\,dt = \frac{1}{A_{ji}} \tag{7B.6}$$

Here τ is called the mean lifetime of the excited state and is also the time in which the number of atoms in the excited state decreases to $1/e$ of the original number.

When an external field exists, the three kinds of processes may take place together. In this situation:

$$-\frac{dn_j}{dt} = \frac{dn_i}{dt} = A_{ji}n_j + B_{ji}n_j u - C_{ij}n_i u = P_{\text{em}}n_j - P_{\text{ab}}n_i \tag{7B.7}$$

in which

$$P_{\text{emission}} = A_{ji} + B_{ji}u, \quad P_{\text{absorption}} = C_{ij}u \tag{7B.8}$$

represent the emission probability and absorption probability of each atom per unit time, respectively.

For a stable state, we should have

$$\frac{dn_j}{dt} = 0$$

thus

$$\frac{P_{ab}}{P_{em}} = \frac{n_j}{n_i} = \frac{C_{ij}u}{A_{ji} + B_{ji}u} \tag{7B.9}$$

or

$$C_{ij}un_i = (A_{ji} + B_{ji}u)n_j \tag{7B.10}$$

In addition, by considering the principle of thermal equilibrium and the Boltzmann law, we get

$$\frac{n_j}{n_i} = \frac{G_j}{G_i} e^{-(E_j - E_i)/kT} \tag{7B.11}$$

in which G_j (or G_i) is the degeneracy of the energy level E_j (or E_i), that is, the number of different quantum states corresponding to energy E_j (or E_i). In order to simplify the problem, we consider the case of nondegenerate states, that is, we take $G_i = G_j = 1$.

Combining formula (7B.10) with formula (7B.11), we get

$$(A_{ji} + B_{ji}u)e^{[-(E_j - E_i)/kT]} = C_{ij}u \tag{7B.12}$$

This formula should be applicable when $T \to \infty$. At that time u will be very large, $B_{ji}u \gg A_{ji}$ and the exponential part will tend to 1. Thus,

$$B_{ji} = C_{ij} \tag{7B.13}$$

and substituting this for C_{ij} in (7B.12) and solving for u gives

$$u(\nu_{ji}, T) = \frac{A_{ji}/B_{ji}}{e^{(E_j - E_i)/kT} - 1} \tag{7B.14}$$

Because Wien's thermodynamic law

$$u(\nu, T) = \nu^3 f(\nu/T) \tag{7B.15}$$

is correct for any system and certainly here as well, formula (7B.14) will lead clearly to

$$E_j - E_i \propto \nu_{ji}, \qquad \frac{A_{ji}}{B_{ji}} \propto \nu_{ji}^3 \tag{7B.16}$$

which gives the quantization condition (7B.1).

When $h\nu \ll kT$, from formula (7B.14), we can obtain

$$u(\nu_{ji}, T) = \frac{A_{ji}/B_{ji}}{h\nu/kT}$$

which must be consistent with the Rayleigh–Jeans formula

$$u(\nu_{ji}, T) = \frac{8\pi\nu_{ji}^2 kT}{c^3} \tag{7B.17}$$

Hence

$$\frac{A_{ji}}{B_{ji}} = \frac{8\pi h\nu^3}{c^3} \tag{7B.18}$$

Thus Einstein not only obtained the relation between the coefficients B and C and reduced three coefficients into two coefficients, but also obtained an expression for the ratio of A to B. Substituting it into formula (7B.14), we thus obtain Planck's famous formula:

$$u(\nu, T) = \frac{8\pi h\nu^3}{c^3} \frac{1}{e^{h\nu/kT} - 1} \tag{7B.19}$$

It is clear from the above analysis that if we omit the stimulated emission, the (-1) term in the formula for u will disappear. But this contradicts Planck's law. So in order for the quantum description of the interaction between atoms and radiation be tenable, the hypothesis of stimulated emission is necessary.

If we start from Planck's formula at the very beginning, then comparing formula (7B.14) with formula (7B.19), we can obtain the ratio of A to B (7B.18). It is clear from the formula (7B.18) that the ratio of A to B is inversely proportional to the third power of the wavelength. When we investigate long-wavelength radiation, spontaneous radiation can be neglected. When we investigate short-wavelength radiation, stimulated emission can often be omitted.

In brief, we have verified the necessity of the hypothesis of stimulated emission and derived the two relations (7B.13) and (7B.18) among the three coefficients of emission and absorption. They can be derived now by quantum electrodynamics, but Einstein had obtained these relations before the establishment of quantum mechanics in 1917.

It should be pointed out that Einstein considered an atomic system when he derived these relations, that is, an assembly of many atoms. But the coefficients A and B are the parameters which are determined by the internal structure of an individual atom and have nothing to do with an assembly of atoms and do not depend on the radiation field. In the derivation, the thermal equilibrium condition is used, but it can be proven that the relations among A, B, and C are correct whether thermal equilibrium is reached or not.

APPENDIX 7C
SELECTION RULES FOR TRANSITIONS

In quantum mechanics, we can calculate the stimulated emission coefficient and the absorption coefficent. Details of such calculations are found in quantum mechanics texts. Here we give the results,

$$B_{ji} = \frac{4\pi^2 e^2}{3\hbar^2} |\langle \boldsymbol{r}_{ji} \rangle|^2 \tag{7C.1}$$

and discuss the transition selection rules which can be extracted from this formula.

In formula (7C.1) the term

$$e\langle \boldsymbol{r}_{ji} \rangle = e\langle j|\boldsymbol{r}|i \rangle = \mathrm{e}\iiint \psi_j^* \boldsymbol{r} \psi_i \, d\tau \tag{7C.2}$$

is called the electric dipole moment matrix element. We express the quantum numbers of states i and j as $n_1 l_1 m_1$ and $n_2 l_2 m_2$, respectively, and as an example substitute from Chapter 8 the wavefunction of hydrogen into the formula (7C.2) to see under what conditions the formula is not equal to zero. Because the wavefunction can be divided into r, ϑ, ϕ separate parts, we can resolve $\boldsymbol{r}$ into three corresponding parts:

$$r_x = x = r \sin\vartheta \cos\phi = r \sin\vartheta \, \frac{e^{i\phi} + e^{-i\phi}}{2}$$

$$r_y = y = r \sin\vartheta \sin\phi = r \sin\vartheta \, \frac{e^{i\phi} - e^{-i\phi}}{2i}$$

$$r_z = z = r \cos\vartheta$$

Because the ϕ part of the formula (7C.2) is

$$\int_0^{2\pi} e^{-im_2\phi} \boldsymbol{r} e^{im_1\phi} \, d\phi$$

the x, y, z components of the integral are, respectively,

$$\int_0^{2\pi} [e^{i(m_1 - m_2 + 1)\phi} + e^{i(m_1 - m_2 - 1)\phi}] \, d\phi \tag{7C.3}$$

$$\int_0^{2\pi} [e^{i(m_1 - m_2 + 1)\phi} - e^{i(m_1 - m_2 - 1)\phi}] \, d\phi \tag{7C.4}$$

$$\int_0^{2\pi} e^{i(m_1 - m_2)\phi} \, d\phi \tag{7C.5}$$

In order for formulas (7C.3) and (7C.4) not to be zero, it is necessary that

$$m_2 - m_1 = \Delta m = \pm 1$$

In order to have formula (7C.5) not be equal to zero, it is necessary that

$$m_2 - m_1 = \Delta m = 0$$

Thus we obtain the selection rules for an electric dipole moment transition:

$$\Delta m = \pm 1, 0 \tag{7C.6}$$

For the ϑ part, we only need to use the characteristics of associated Legendre polynomials

$$\begin{aligned} \cos\vartheta P_l^m &= \frac{(l-m+1)P_{l+1}^m + (l+m)P_{l-1}^m}{2l+1} \\ \sin\vartheta P_l^m &= \frac{P_{l+1}^{m+1} - P_{l-1}^{m+1}}{2l+1} \end{aligned} \tag{7C.7}$$

and the orthogonality of P_l^m, that is,

$$\int P_{l_2}^{m_2^*} P_{l_1}^{m_1} \, d\vartheta \tag{7C.8}$$

is not equal to zero only when $l_2 = l_1$. Similarly, when we evaluate the ϑ component of Eq. (7C.2), we have an integral over ϑ of $P_{l_2}^{m_2} \cos\vartheta$ (*or* $\sin\vartheta$)$P_{l_1}^{m_1}$. From Eq. (7C.7), we see that $\cos\vartheta$ (*or* $\sin\vartheta$)$P_{l_1}^{m_1}$ involves $P_{l_1\pm1}$. So, from terms like $\int P_{l_2}P_{l_1\pm1}$ we obtain the result that l_2 must be equal to $l_1 \pm 1$, or

$$\Delta l = \pm 1 \tag{7C.9}$$

This l selection rule is related to the fact the photon has an intrinsic angular momentum of $1\hbar$ and carries off $1\hbar$. So, to conserve angular momentum for electric dipole radiation, the photon must connect states differing by $1\hbar$ ($\Delta l = \pm 1$).

Formulas (7C.6) and (7C.9) are the selection rules for electric dipole radiation. They are suitable for ultraviolet light and visible light whose wavelengths are larger than the radii of atoms. For x-rays whose wavelengths are close to the radii of atoms, higher multipole radiations with $|\Delta l| > 1$ (for example from the electric quadrupole moment) should be considered. There is also magnetic dipole radiation that is related to oscillation in the orientations of the electron spin and magnetic dipole moments.

Higher multipole radiations also can be important in highly ionized atoms which are seen in stars and heavy-ion collisions, for example.

APPENDIX 7D
THE PRINCIPLES OF LASERS

Forty years after Einstein put forward the concept of stimulated emission, the laser based on this concept was developed. Lasers not only rejuvenated old optics, but also made a profound influence on many fields of science and technology. Lasers are one of the great scientific and technological achievements of the century.

We have mentioned before that if there exists an atomic system as well as light in a cavity, the interaction between them can be described by three basic processes—spontaneous emission, stimulated absorption, and stimulated emission.

Atoms in an excited state will move from a higher energy level E_2 to a lower energy level E_1 spontaneously and emit photons. This is called spontaneous emission. For each atom in an excited state, the process of spontaneous emission is independent, hence the photons emitted are also independent of each other. Thus their emission directions and initial phases are different. So for a common light source, we can see light in all directions. We will find that a laser beam has totally different characteristics.

Assume that an atom is in a lower energy level E_1 and a photon is incident on it. If the energy $h\nu$ of the photon is exactly equal to the energy difference of a pair of energy levels of the atom $(E_2 - E_1)$, then the atom can absorb the photon and move to the higher energy level. This transition does not happen spontaneously, but happens under the stimulation of the external photon and is called stimulated absorption. The stimulated absorption energy of an atom per unit time per unit volume is

$$C_{12}u(\nu_{21})N_1h\nu_{21} \tag{7D.1}$$

in which C_{12} is the Einstein coefficient, $u(\nu_{21})$ is the energy density of the radiation field in the cavity, and N_1 is the density of atoms in the lower energy level.

Stimulated emission is the fundamental process that makes lasers possible. If initially an atom is in a higher energy level E_2 at the moment a photon is incident on it, and if the energy of the photon $h\nu_{21}$ is exactly equal to the energy difference $E_2 - E_1$ between a pair of energy levels of an atom, then this atom will be stimulated by the external photon to move from the higher energy level E_2 to the lower energy level E_1 by emitting a photon which is exactly the same as the external photon. This is the process of stimulated emission. As a result of stimulated emission, the atom decays from the higher energy level to the lower energy level and the number of photons increases from one to two. The photon created by the stimulated emission has the same frequency, emission direction, phase, and polarization state as the incident photon. Thus the stimulated emission gives rise to an amplification of the light (increase in identical photon intensity). The stimulated emission energy of an atom per unit time per unit volume is

$$B_{21}u(\nu_{21})N_2h\nu_{21} \tag{7D.2}$$

We have proven in Appendix 7B that $B_{21} = C_{12}$.

For an atomic system, if there exists a light signal of energy density $u(\nu_{21})$, there will be both stimulated absorption and stimulated emission. From the point of view of the stimulated emission, the net stimulated emission energy will be

$$B_{21}u(\nu_{21})N_2h\nu_{21} - C_{12}u(\nu_{21})N_1h\nu_{21} = (N_2 - N_1)B_{21}u(\nu_{21})h\nu_{21} \tag{7D.3}$$

Thus in order to let the stimulated emission exceed the stimulated absorption, that is, obtain output from a laser, it is necessary to have the density of atoms in the higher energy level N_2 be larger than the density of atoms in the lower energy level N_1. If we can think of a way to keep $N_2 > N_1$, then by using two mutually parallel, highly reflective mirrors at each end of the laser to make the light photons oscillate back and forth between the mirrors, coherent light can be amplified continuously in the cavity of the laser until finally an intense laser beam is formed. The photons which are emitted by stimulation have the same energy (frequency), phase, and polarization and move in the same direction. Such a light beam is said to be coherent.

In 1954. the first monochromatic amplifier was made in the microwave range. It was called Microwave Amplification by the Stimulated Emission of Radiation, or maser for short. A few years later a light-stimulated emission device in the visible light region was made and called a laser, for Light Amplification by the Stimulated Emission of Radiation. These days, no matter in which waveband, they are called by the name laser.

The key to creating a laser is to have the density of the atoms N_2 in the higher energy level larger than the density of atoms N_1 in the lower energy level. But in the general case of thermal equilibrium, the densities satisfy the Boltzmann distribution

$$\frac{N_2}{N_1} = e^{-(E_2-E_1)/kT} \tag{7D.4}$$

Because $E_2 > E_1$, then $N_2 < N_1$. Thus in thermal equilibrium, it is always true that the stimulated absorption exceeds the stimulated emission and no net stimulated emission appears. In order to create a net stimulated emission, we must destroy the thermal equilibrium and create some conditions to give $N_2 > N_1$. For a specific energy level, if the case of $N_2 > N_1$ appears, we call this "the inversion of the number of particles," or "population inversion," that is, the distribution of particles is opposite to the Boltzmann distribution.

In order to create the occurrence of nonthermal equilibrium or the appearance of an inversion of the number of particles, we must stimulate the atomic system in some way to cause the number of particles in the higher energy state to increase. In the gas laser, the method of "discharge stimulation" is used. In a solid laser or dye laser, a pulsed light source, "light stimulation," is used to provide the energy to the working material of the laser. We call the various stimulation methods by the name "pump," because one is pumping the atoms up from N_1 to N_2.

Because the particles naturally move toward lower energy states in various ways, a system which is in nonthermal equilibrium is unstable and will move toward thermal equilibrium. When the particles go from the higher energy level to the lower energy level, the number of particles N_2 of the higher energy level will decrease while the number of particles in the lower energy level will increase. In order to maintain $N_2 > N_1$ and get the laser output continuously, we must continue to "pump" the particles up to the higher energy level. The energy put in to pump the laser is always much greater than the energy output of the laser, so lasers have rather low efficiency generally.

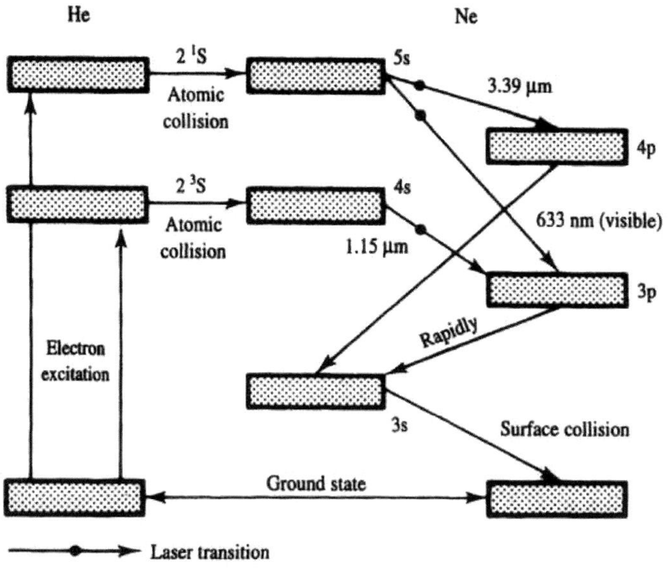

FIGURE 7D.1
The energy level diagram of a He–Ne laser system.

Now let us look at the He–Ne laser as an example to demonstrate the working principle of a laser. The ratio of helium to neon in a He–Ne laser is from 5 : 1 to 10 : 1. The actual ratio is determined by the particular structure of the cavity. Figure 7D.1 is the energy level diagram related to helium and neon. The energy levels $2\,^1S$ and $2\,^3S$ of helium are given in the notation $n^{2S+1}L$ where L and S are the total orbital angular momentum and spin of the two electrons in He. Some of the energy levels of neon are noted in small letters which represent the quantum numbers of the single electron which is in an excited state while the other nine electrons are in the ground state.

Neon can produce many laser spectral lines. Its three most intense lines are shown in Fig. 7D.1. The most commonly used line is the red light of 6328 Å (5s → 3p). The other two lines are 3.39 μm (5s → 4p) and 1.15 μm (4s → 3p). Both are in the infrared waveband. Note that the transition between the 5s and 4s level is forbidden since $\Delta l = 0$ is not allowed. The higher energy levels of the three laser lines are 5s and 4s, respectively, and the lower energy levels are 3p and 4p. We now discuss how the inversion of the number of particles of the corresponding energy levels is realized.

At room temperature, the overwhelming majority of the neon atoms are in the ground state. If an electric voltage is applied to a He–Ne laser tube and produces a gas discharge, the electrons in the tube are accelerated by the electric field and have reasonably large kinetic energy. These fast electrons undergo inelastic collisions with the atoms of He or Ne and transfer their energy into the internal energy of the atoms to excite He and Ne to their low excited states. Because the probabilities of the He atoms colliding with electrons and being excited to the $2\,^1S$ and $2\,^3S$ states are larger than those of Ne atoms and both the He states are metastable states which only slowly decay back to the $1\,^1S$ ground state, there are more atoms in these two He excited states than those of Ne. However, because the energies of the states of 5s and 4s of Ne and the states $2\,^3S$ and $2\,^1S$ of He are nearly the same, it is very easy for a "resonance transfer" of the energy to take place when the two kinds of atoms collide, that is, a He atom transfers its energy to Ne and decays to its ground state and the Ne atom is excited to 5s or 4s. To produce a laser, besides increasing the number of particles in the higher energy level, it is necessary to decrease the number in the lower energy levels. The lifetimes of the lower energy 4p and 3p levels are much shorter than those of the higher energy 5s and 4s levels. This is favorable to the inversion of the number of particles.

The stimulated atoms decay from the 5s, 4s states to the 4p, 3p states and the atoms which are in the 4p, 3p states decay to the 3s state quickly. However, the 3s state is a metastable state. If there are many particles in the 3s state, they can capture the photons which come from the spontaneous emission of $4p \to 3s$ or $3p \to 3s$ or collide with electrons and go back to the 4p or 3p states which is not advantageous to the inversion of the number of particles. To overcome this disadvantage, we can make the discharge tube rather long and thin so that the atoms will collide with the wall of the tube frequently. With the help of such collisions, the Ne atoms in the 3s state can hand over their energy to the wall through a "radiationless transition" and go back to the ground state to reduce the number of Ne atoms in the 3s state and help exhaust the 4p and 3p states. A nonradiative transition means that the decay process does not emit radiation: An atom in an excited state can transfer its energy to a surrounding atom in the form of heat or transfer it into the vibrations of molecules.

It should be pointed out that the inversion of the number of particles is only one necessary condition to produce a laser, not the sufficient condition. The inversion of the number of particles only ensures that when the light travels in the laser tube, the gain coefficient is larger than zero and amplification occurs. In order to obtain a laser, we must have the reflection rate of the light on the mirrors at each end of the laser tube be very high so as to have the light undergo repeated reflections and continuous amplification on each pass through the system. In other words, the loss of light in the tube (especially on the mirrors) must be much lower than the amplification gain. For the two mirrors, one should be nearly total reflecting and the other partially reflecting, so the laser light can be emitted out that end when it has built up sufficient intensity.

Figure 7D.2 is a schematic diagram of a He–Ne laser tube. To date, the smallest He–Ne laser tube in the world is of length 14.6 cm, diameter 2.5 cm, weight 70 g, and power 0.5 mW.

Lasers have excellent monochromaticity: For example, the frequency corresponding to the red light of 6328 Å emitted from a He–Ne laser tube is 4.74×10^{14} Hz, but its frequency width can be as narrow as 7×10^3 Hz and its relative frequency width $\Delta\nu/\nu = 1.6 \times 10^{-11}$. Thus its monochromaticity is 10 000 times better than that of the best common monochromatic light source. To date

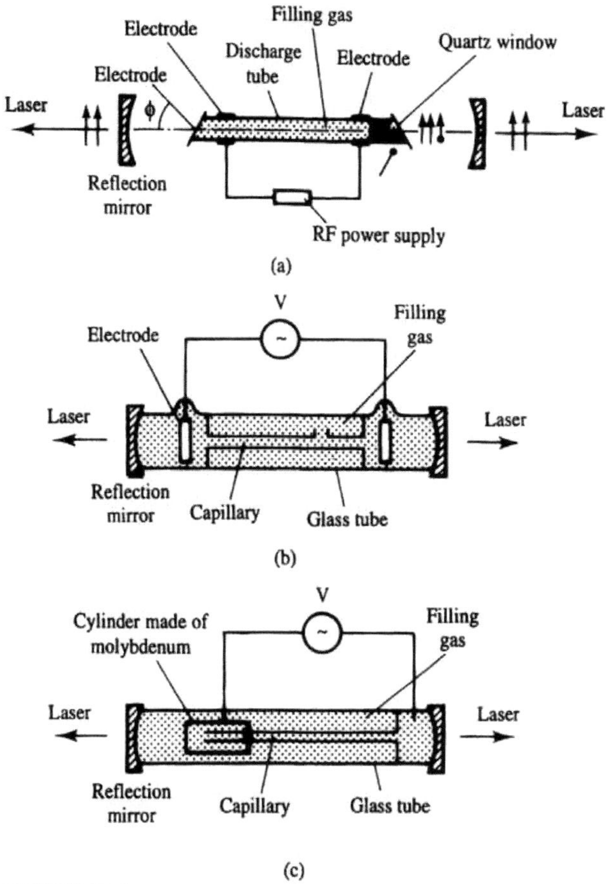

FIGURE 7D.2
Schematic diagram of three versions of a He–Ne laser. (a) produces linearly polarized light and has a small tube to contain the gas. (b) and (c) differ only in the way the voltage is applied across their capillary tubes. The capillary tubes are where the laser light is produced. The tubes are small to keep the laser light at the center of the mirrors

the smallest value of $\Delta\nu/\nu$ reached by a laser is the order of 10^{-14}. If such a laser is used to control a clock, then the error of this clock per year will be less than 1×10^{-6} s.

In addition, laser light has the excellent characteristic of very slow divergence. For example, a laser beam of only a few hundred watts of power has been sent out from the Earth to the Moon and then reflected back to the Earth. In this way, the distance between the Earth and the Moon, about 384 400 000 m, has been measured to an accuracy of 3 cm. The neodymium–YAG laser beam used has spread over only 1 mile by the time it reaches the moon. Over the last 20 years these laser signals have confirmed several predictions of general relativity and measured the Moon's recession of 1.5 inches/year from the Earth.[13]

The interaction of laser light with materials covers a wide range of phenomena, depending on laser wavelength, pulse length, and pulse energy. In many materials processing applications which depend on thermal vaporization, the critical laser parameter is average power. Continuous-wave (cw) or long-pulse (μs to ms) lasers with high average power deliver laser light on time scales which are long compared to thermal diffusion times. Continuous-wave welding and metalworking lasers, for example, have average powers ranging from hundreds of watts to tens of kilowatts. In other phenomena—such as laser fusion or laser surgery—it is critical to deliver sufficient laser energy to cause material ablation (energetic removal of material) on timescales short compared to thermal diffusion times. Infrared, visible, and ultraviolet lasers have been developed which can meet these requirements by providing pulses with energies up to a hundred kilojoules in pulse lengths of 10^{-9} to 10^{-8} s at repetition rates from 10 kHz to one or two pulses per day (laser fusion). Finally, there are many interesting phenomena which depend on pulse intensity, as measured in watts or watts·cm^{-2}. Picosecond and femtosecond lasers are of special interest because they excite the material with light pulses of duration comparable to or shorter than the vibrational period of molecules or ions in a solid lattice and so open up dynamical studies on timescales never dreamed possible. The shortest pulses produced to date[14] range down to 6.5×10^{-15} s (approximately three and a half cycles of the electric field vector!), while the highest peak intensities—generated by amplifying picosecond laser pulses at repetition rates of a few hertz—are now in the vicinity of 10^{19} W · cm^{-2}.[15]

In 1995, M. Scully and his US–German–Russian collaborators working at the National Institute of Standards and Technology in Colorado have reported evidence for lasing without population inversion (see *Physics Today*, Sept. 1995, p.19). Their method employs a destructive interference effect to prevent the more populated lower level from absorbing energy from the field and returning to the lasing level. So the upper, lasing level can have many fewer atoms than the ground state — no population inversion is required. This phenomenon, which has been sought for years, has potentially important applications in nonlinear optics.

PROBLEMS

7.1. Calculate the corresponding wavelengths of electrons whose energies are 10, 100 and 1000 eV, respectively.

7.2. Assume that the wavelengths of a photon and an electron are both 4.0 Å, and calculate:
(*a*) The ratio of the momentum of the photon to that of the electron
(*b*) The ratio of the energy of the photon to that of the electron

---

[13] D. C. Morrison, *Science* **246** (1989), 447.

[14] R. L. Fork, et al., *Optics Letters* **12** (1987) 483.

[15] M. D. Perry and G. Mourou, *Science* **264** (1994) 917.

7.3. Assume that the kinetic energy of an electron is equal to its rest mass. Then calculate:
(*a*) The velocity of the electron
(*b*) The corresponding de Broglie wavelength

7.4. A narrow beam of thermal neutrons strikes a crystal. The energy of the thermal neutrons can be obtained from the Bragg diffraction pattern. If the atomic interval of the crystal is 1.8 Å and the first-order Bragg scattering angle is 30°, calculate the energy of the thermal neutrons.

7.5. The acceleration voltage used in electron microscopes is generally very high and the velocity of the electrons after acceleration is very large, so the relativistic correction must be considered. Prove that the relation between the de Broglie wavelength of the electrons and the acceleration voltage is

$$\lambda = \frac{12.26}{\sqrt{V_r}} \text{ Å}$$

in which $V_r = V(1 + 0.978 \times 10^{-6}\,V)$ is called the relativistic correction voltage, where the unit of the acceleration voltage of the electrons is volt.

7.6. (*a*) Prove that the ratio of the Compton wavelength of a particle to its de Broglie wavelength is

$$\sqrt{\left(\frac{E}{E_0}\right)^2 - 1}$$

in which E and E_0 are the total energy of the moving particle and its rest energy, respectively.
(*b*) What is the kinetic energy of an electron when its de Broglie wavelength is equal to its Compton wavelength?

7.7. An excited state of an atom emits an optical line of wavelength 6000 Å. The accuracy of measuring the wavelength is $\Delta\lambda/\lambda = 10^{-7}$. Calculate the lifetime of this atomic state.

7.8. An electron is enclosed in a system whose dimension is 10 fm, which is the order of the dimension of an atomic nucleus. Calculate its minimum kinetic energy. What does this say about the existence of electrons in nuclei?

7.9. Use the uncertainty relation to answer Rutherford's question and Schrödinger's criticism (Section 7.1).

7.10. Use the uncertainty relation to explain the necessity of changing $L = l\hbar$ into $\sqrt{l(l+1)}\hbar$ (Chapter 4).

7.11. Demonstrate that in the microcosmos the concept of "electron orbit" has lost its meaning but on a TV screen the concept of "the trace of an electron" can still be used.

7.12. A 0.01 kg bullet has the same speed as an electron, i.e. 500 m/s with a measured accuracy of 0.01%. If the position and speed of each are measured simultaneously in the same experiment, determine the respective minimum uncertainty of the position of each.

7.13. Our world would be much more complicated if the Planck constant were 660 J · s instead of 6.6×10^{-34} J · s. In such a case, if a football player with 100 kg body weight runs at a speed of 5 m/s, what is his de Broglie wavelength? In the view of his friend moving toward him, what is the minimum uncertainty of his position.

7.14. When we calculate the de Broglie wavelength for the proton using the nonrelativistic formula, what is the energy of the proton if the calculated error reaches 5%?

7.15. For de Broglie wavelengths of the proton of 1 Å and 1 fm, what are the corresponding proton energies?

7.16. When 50 eV electrons vertically strike a MgO crystal surface, we observe a diffraction enhancement at a scattering angle 55.6°. What is the crystal constant? For 100 eV electrons, at what angle would we observe the diffraction enhancement?

7.17. What is the maximum wavelength or minimum energy of a photon beam we can use to observe an object with the size of 0.5 Å? If we use an electron or proton beam to observe it, what is the minimum energy of the electrons or protons?

7.18. Calculate the corresponding wavelengths of the following particles:
(*a*) 1 eV free electron
(*b*) 1 eV free neutron
(*c*) 1 eV particle with mass of 1 g.

7.19. Use the symmetry of the wavefunction and the Pauli Exclusion Principle to explain which states are allowed for two equivalent d electrons, a d^2 configuration. Make a table like Table 7A.1.

7.20. In the ground state of hydrogen, a bound electron has the Bohr radius a_o. Estimate the minimum kinetic energy of the electron by the uncertainty principle. What is the estimated total energy for the electron?

7.21. Calculate the minimum energy of a proton confined inside a nucleus of 10^{-14} m radius from the uncertainty principle.

7.22. Explain why, in an operating laser, you must have more atoms in the upper state which you want to stimulate to emit the laser light than in the lower state.

CHAPTER 8

INTRODUCTORY QUANTUM MECHANICS II: THE SCHRÖDINGER EQUATION

I have deep faith that the principle of the universe will be beautiful and simple.

A. Einstein

8.1 SCHRÖDINGER EQUATION

8.1.1 The Establishment of the Schrödinger Equation

When the concept of matter waves proposed by de Broglie arrived in Zürich, Schrödinger, a student of Debye, following the suggestion of his teacher, made a report on matter waves. In the report, Schrödinger demonstrated clearly how de Broglie connected waves with particles and how the quantization conditions came naturally from the connection. After the report, Debye made the comment: "If there is a wave, there should be a wave equation."[1] de Broglie had not given the wavefunction of a particle in a potential field nor the variation of the wave function with time. Schrödinger put forward a wave equation soon after his

[1] See an interesting memorial article: Felix Bloch, *Physics Today* **19** (Dec. 1976) 23.

report, but no one realized that the equation was so important. Later it became the famous Schrödinger equation of quantum mechanics. This equation like Newton's equations of motion cannot be derived from more fundamental hypotheses. It is the basic equation in quantum mechanics and its correctness can only be tested by experiments.

Let us introduce one method of deducing the Schrödinger equation. For a nonrelativistic particle of mass m and momentum p moving in a potential field $V(x)$, its energy (first consider the one-dimensional motion) can be written as

$$E = \frac{p^2}{2m} + V(x) \tag{8.1}$$

Using the de Broglie relation $E = \hbar\omega$, $p = \hbar k$ (see Section 7.2), this becomes

$$\hbar\omega = \frac{(\hbar k)^2}{2m} + V(x) \tag{8.2}$$

For a free particle described by a wave, we can write a wavefunction of the particle in the form of a plane wave by analogy with the equation for an electromagnetic wave propagating with wave vector k and angular frequency ω:

$$\Psi(x, t) = \psi_0 e^{i(kx-\omega t)} \tag{8.3}$$

Now our task is to find an equation which is both consistent with Eq. (8.2) and has the solution of Eq. (8.3) when $V(x) = 0$. It is clear from Eq. (8.3) that

$$\begin{aligned} i\hbar\frac{\partial}{\partial t}\Psi &= E\Psi \\ -i\hbar\frac{\partial}{\partial x}\Psi &= p\Psi \\ -\hbar^2\frac{\partial^2}{\partial x^2}\Psi &= p^2\Psi \end{aligned} \tag{8.4}$$

Using Eq. (8.1) and assuming $V(x) = 0$, we have

$$\left(i\hbar\frac{\partial}{\partial t} + \frac{\hbar^2}{2m}\frac{\partial^2}{\partial x^2}\right)\Psi = \left(E - \frac{p^2}{2m}\right)\Psi = 0 \tag{8.5a}$$

or

$$i\hbar\frac{\partial}{\partial t}\Psi(x, t) = -\frac{\hbar^2}{2m}\frac{\partial^2}{\partial x^2}\Psi(x, t) \tag{8.5b}$$

For the general state of a free particle which is a superposition of plane waves, we can prove easily that it still satisfies Eq. (8.5).

For the case of $V(x) = V_0$, a constant (which is still the case where no force from some type of impact or action exists), it is easy to show that Eq. (8.3) is the solution of Eq. (8.6):

$$-\frac{\hbar^2}{2m}\frac{\partial^2\Psi}{\partial x^2} + V_0\Psi = i\hbar\frac{\partial\Psi}{\partial t} \tag{8.6}$$

and is consistent with

$$\frac{(\hbar k)^2}{2m} + V_0 = \hbar\omega \tag{8.7}$$

Now we extend Eq. (8.6) to the case of a general potential field $V(x)$ and consider that the motion of its particles satisfies the equation

$$-\frac{\hbar^2}{2m}\frac{\partial^2\Psi}{\partial x^2} + V(x)\Psi = i\hbar\frac{\partial\Psi}{\partial t} \tag{8.8}$$

This is the one-dimensional Schrödinger equation. Compare it with the classical relation (8.1). It is clear that the only thing we have done in formula (8.1) is to make the following substitutions:

$$E \rightarrow i\hbar\frac{\partial}{\partial t}, \qquad p \rightarrow -i\hbar\frac{\partial}{\partial x} \tag{8.9}$$

and then act on the wavefunction Ψ.

Obviously the extension to the case of three dimensions is very easy:

$$\left[-\frac{\hbar^2}{2m}\nabla^2 + V(\boldsymbol{r})\right]\Psi(\boldsymbol{r}, t) = i\hbar\frac{\partial}{\partial t}\Psi(\boldsymbol{r}, t) \tag{8.10}$$

where ∇^2 is the Laplace operator, which in cartesian coordinates is

$$\nabla^2 \equiv \frac{\partial^2}{\partial x^2} + \frac{\partial^2}{\partial y^2} + \frac{\partial^2}{\partial z^2}$$

The solution for free particles when $V(\boldsymbol{r}) = 0$ is

$$\Psi(\boldsymbol{r}, t) = \psi_0 e^{i(\boldsymbol{k}\cdot\boldsymbol{r} - \omega t)} \tag{8.11}$$

where the classical expression corresponding to formula (8.10) is

$$E = \frac{\boldsymbol{p}^2}{2m} + V(\boldsymbol{r}) \tag{8.12}$$

Going from (8.12) to (8.10) is simply the result of making the following substitutions:

$$E \rightarrow i\hbar\frac{\partial}{\partial t}, \qquad \boldsymbol{p} \rightarrow i\hbar\nabla \tag{8.13}$$

and acting on the wavefunction.

Equation (8.10) is the general form of the famous Schrödinger equation. We emphasize again that it is the fundamental equation in quantum mechanics and it can only be considered as a hypothesis. In fact we can take the formula (8.13) and the existence of a wavefunction (8.11) together as the basic hypothesis in quantum mechanics.

We also note that Eq. (8.10) is a linear equation for Ψ, that is, if both Ψ_1 and Ψ_2 are solutions of the equation, then $C_1\Psi_1 + C_2\Psi_2$ (where C_1 and C_2 are two constants) is also a solution of the equation, which is exactly what the superposition principle of the wavefunction demands.

8.1.2 The Schrödinger Equation of a Stationary State

When the potential field $V(\boldsymbol{r})$ does not explicitly include the time t, the specific solution of Eq. (8.10) can be obtained by the method of separation of variables, that is, we can write the wavefunction as

$$\Psi(\boldsymbol{r}, t) = \psi(\boldsymbol{r})T(t) \tag{8.14}$$

Substitute Eq. (8.14) into Eq. (8.10) and divide both sides of Eq. (8.10) by $\Psi(\boldsymbol{r}, t)$ as given by (8.14). Since the time and space variables are separable now, we obtain

$$\frac{i\hbar}{T}\frac{dT}{dt} = \frac{1}{\psi}\left[-\frac{\hbar^2}{2m}\nabla^2 + V(\boldsymbol{r})\right]\psi \tag{8.15}$$

Since the left-hand side depends only on t and the right-hand side only on $\boldsymbol{r}$, the two sides are independent and each must equal the same constant. Thus

$$\frac{i\hbar}{T}\frac{dT}{dt} = E \tag{8.16}$$

$$\frac{1}{\psi}\left[-\frac{\hbar^2}{2m}\nabla^2 + V(\boldsymbol{r})\right]\psi = E \tag{8.17}$$

in which E is a separating constant and does not depend on $\boldsymbol{r}$ and t. E has the dimension of energy. The solution of Eq. (8.16) is

$$T = T_0 e^{-iEt/\hbar} \tag{8.18}$$

If the constant T_0 is included into the constant in $\psi(\boldsymbol{r})$, then

$$\Psi(\boldsymbol{r}, t) = \psi(\boldsymbol{r})e^{-iEt/\hbar} \tag{8.19}$$

The probability density is

$$\Psi^*\Psi = \psi^*\psi \tag{8.20}$$

which does not depend on the time t. We also have from Eq. (8.17):

$$\left[-\frac{\hbar^2}{2m}\nabla^2 + V(\boldsymbol{r})\right]\psi = E\psi \tag{8.21}$$

which is the time-independent Schrödinger equation.

Here we want to emphasize that we require (to ensure physical reality in our interpretation of the wavefunction) that the wavefunction used to describe the particles satisfy the following three conditions: $\psi(x)$ must be (a) single-valued, (b) finite, and (c) continuous. (a) Because the probability of a particle appearing at a certain place can only have one value, the wavefunction should be single-valued. (b) Since obviously the probability cannot be infinite, the wavefunction must be finite at every place. (c) The probability cannot have sudden discontinuities, thus the wavefunction must be continuous at every place. These three conditions are generally called the standard conditions. The standard conditions are very important when quantum mechanics is used to solve practical problems. These conditions helped determine the form of the Schrödinger equation.

8.1.3 Examples

EXAMPLE 1. : ONE-DIMENSIONAL INFINITE POTENTIAL WELL. Consider a particle moving in a one-dimensional space. The potential is zero in a certain range [from $x = 0$ to $x = d$] and outside the range the potential is infinite (Fig. 8.1); that is,

$$V(x) = \begin{cases} 0, & 0 < x < d \\ \infty, & x \geq d, x \leq 0 \end{cases} \tag{8.22}$$

Such a potential is called a one-dimensional infinite potential well. In the well, the stationary Schrödinger equation of the system is

$$\frac{\hbar^2}{2m}\frac{d^2\psi}{dx^2} + E\psi = 0 \tag{8.23}$$

If we define

$$B^2 \equiv \frac{2mE}{\hbar^2} \tag{8.24}$$

then the equation can be rewritten as

$$\frac{d^2\psi}{dx^2} = -B^2\psi \tag{8.25}$$

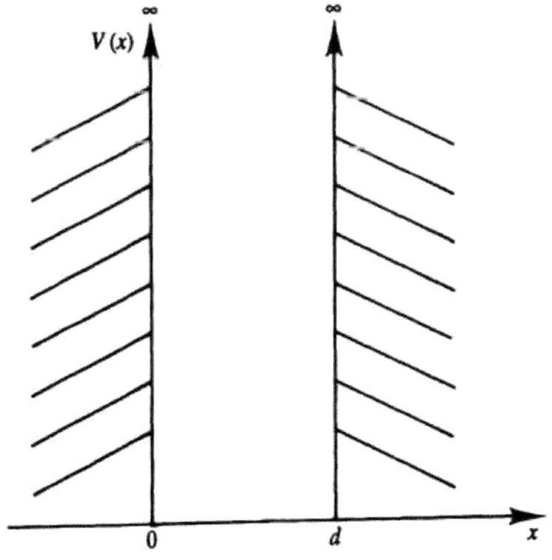

FIGURE 8.1
One-dimensional infinite square well potential.

Because $\psi(0) = 0$, the wavefunction is a sine function:

$$\psi = A \sin Bx \tag{8.26}$$

and because $\psi(d) = 0$ (since outside the well $V(x) = \infty$, the wavefunction must be zero at $x = 0$ and $x = d$). Substituting these into the formula (8.26), we can obtain the quantization conditions:

$$n\pi = Bd, \quad n = 1, 2, \ldots \tag{8.27}$$

Thus Eqs. (8.26) and (8.24) are, respectively,

$$\psi_n(x) = A \sin \frac{n\pi x}{d} \tag{8.28}$$

$$E_n = \frac{n^2 h^2}{8md^2}, \qquad n = 1, 2, \ldots \tag{8.29}$$

which is exactly the formula (7.15).

Now let us sum up what we have done. Outside the well, because $V(x) = \infty$, $\psi(x)$ can only be zero, otherwise the equation will lose its meaning. Inside the well, $V(x) = 0$, so the Schrödinger equation is Eq. (8.23). Because after double differentiation the wavefunction becomes itself again, the solution of the equation certainly has the form of an exponential function. Because the result is negative after double differentiation, the exponential power will certainly include the imaginary number i and the solution will be a sine or cosine function. The condition $\psi(0) = 0$ excludes the solution of a cosine function and $\psi(d) = 0$ leads to the quantization condition.

Still the constant A in the wavefunction (8.28) has not been determined. We can use the normalization condition to determine A. Because the probability of a particle appearing somewhere in the whole space is by definition 1, the wavefunction satisfies the normalization condition

$$\int_{-\infty}^{\infty} |\psi_n|^2 \, dx = 1$$

Substituting formula (8.28) into this and paying attention to the valid range, we obtain

$$\int_{-\infty}^{\infty} A^2 \sin^2 \frac{n\pi x}{d} \, dx = \int_0^d A^2 \sin^2 \frac{n\pi x}{d} \, dx = A^2 \frac{d}{2} = 1$$

From this we obtain the normalization coefficient $A = \sqrt{2/d}$. Substituting into formula (8.28), we obtain the normalized wavefunction

$$\psi_n = \sqrt{\frac{2}{d}} \sin \frac{n\pi x}{d} \tag{8.30}$$

In Fig. 8.2 the wavefunction for $n = 1, 2, 3$, and the corresponding probabilities, are given. It should be pointed out that when $n = 1$ the probability of a

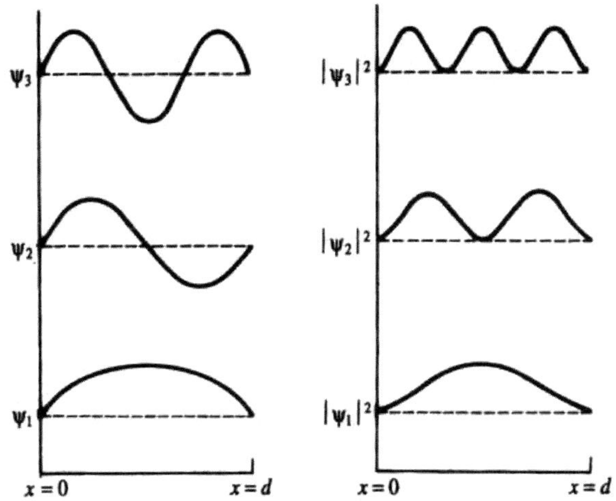

FIGURE 8.2
The wavefunctions and probabilities of particles in a one-dimensional infinite well potential.

particle appearing is $|\psi_1(x)|^2$, whose maximum is in the middle of the well; for $n = 2$, $|\psi_2(x)|^2$ is zero in the middle and has maximum on each side of the middle. So where is the particle actually? This is a question which cannot be answered in classical language. We can only say that the particle has certain probabilities of being here and there.

EXAMPLE 2: ONE-DIMENSIONAL FINITE POTENTIAL WELL. Now let us consider the motion of particles in a one-dimensional finite potential well. The one-dimensional finite potential well, as shown in Fig. 8.3, can be expressed as

$$V(x) = \begin{cases} 0, & \text{when } |x| < d/2 \\ V_d, & \text{when } |x| \geq d/2 \end{cases} \tag{8.31}$$

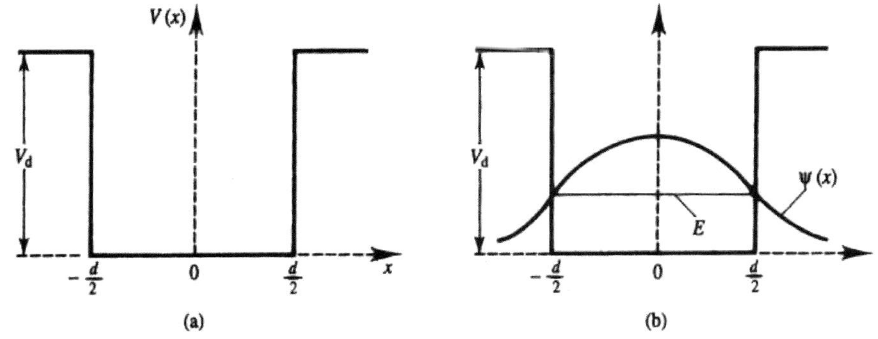

FIGURE 8.3
One-dimensional finite potential well ($E < V_d$).

The case of particles being inside the well ($|x| < d/2$) is the same as given in the last section and the wavefunction is a sine function. Outside the well ($|x| \geq d/2$), the Schrödinger equation the system satisfies is

$$\frac{d^2\psi}{dx^2} = \frac{2m(V_d - E)}{\hbar^2}\psi \equiv C^2\psi \tag{8.32}$$

where

$$C^2 \equiv \frac{2m(V_d - E)}{\hbar^2}$$

The solution of the equation is

$$\psi(x) = \begin{cases} A_+ e^{Cx}, & x < 0 \\ A_- e^{-Cx}, & x > 0 \end{cases} \tag{8.33}$$

in which both C and $A_\pm$ are constants. (Solutions with the opposite sign in the exponential are rejected because they grow without bound.) This result indicates the fundamental difference between microparticles and classical particles. In the case of $E < V_d$, according to the viewpoint of classical physics, the particle can never go outside the walls of the well, but in quantum mechanics the particle has a certain probability of appearing outside the walls of the well.

Suppose quantum mechanics was observable in our macroworld. Figure 8.4 pictures an event that then might occur in the macrocosm while friends are drinking tea in the drawing room. A car, originally parked in the next-door garage, suddenly appears in the room! Two physical concepts are expressed: A car governed by the uncertainty relation can never be stable in a garage (the garage walls are the walls of the potential well), and a body has a certain probability of penetrating the walls of a finite potential well (note that the wall of the garage cannot be infinitely hard and so it corresponds to a finite potential well). This picture is redrawn in imitation of the original by G. Gamow. The main contribution of Gamow to science is the formula (8.40) which will be given next. He also wrote many popular works to make the behavior of microparticles in modern physics understandable to nonscientists, for example, *Mr. Tompkins Meets the Atom*. An account of his scientific career can be found in his autobiography.[2]

Now let us investigate the potential well from the standpoint of the uncertainty relation. From physical concepts or from the mathematical equation, it is easy to understand that when the potential field outside the well is $V(x) = \infty$ then the wavefunction there is zero. According to the uncertainty relation, when $\psi(x) = 0$, one has $\Delta x = 0$ and this certainly leads to $\Delta p = \infty$. Thus $V(x) = \infty$, which is the infinite potential well. If $V(x)$ is finite, then $\Delta p \neq \infty$, and the uncertainty relation will lead to $\Delta x \neq 0$. The probability will certainly not be zero outside the well. This is the finite potential well.

[2] G. Gamow, *My World Line*, Viking Press, New York (1970).

FIGURE 8.4
A car rushes into the parlor.

Let us further look at Fig. 8.5. According to the classical point of view, if the particle is inside the potential well its energy E is less than the depth V of the well: $E < V$. Then the particle can only move inside the well (Fig. 8.5a). If the particle is outside the potential well where its energy is greater than the well depth, $E > V$, then the potential well has no influence on the motion of the particle (Fig. 8.5b). But according to the quantum point of view, the situation is totally different. When $E < V$, the particle still has a probability of appearing outside the well (Fig. 8.5c). When $E > V$, the kinetic energy of the particle will change at the boundary of the potential well. A change of kinetic energy is equivalent to a change of wavelength. This indicates that the particle has both reflection and penetration at the boundary of the well (Fig. 8.5d). It is like looking at a shop window, when one sees the goods as well as one's face.

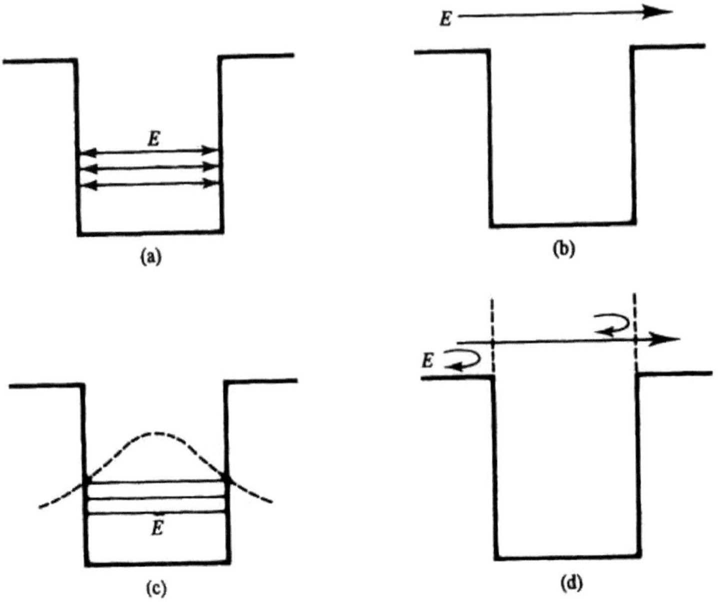

FIGURE 8.5
The classical diagrams (a, b) and quantum diagrams (c, d) of a particle inside a finite potential well (a, c) and outside the finite potential well (b, d), respectively.

EXAMPLE 3: TUNNEL EFFECT. Now consider the problem of the penetration of a square potential barrier, such as is shown in Fig. 8.6:

$$V(x) = \begin{cases} 0, & 0 > x,\ x > a \\ V_0, & 0 \le x \le a \end{cases} \tag{8.34a}$$

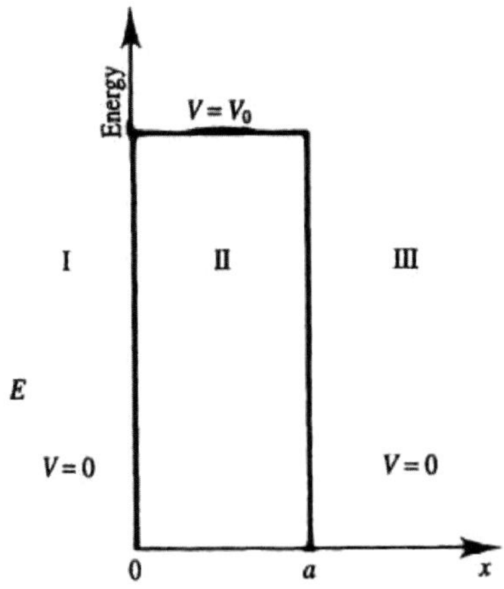

FIGURE 8.6
A square barrier potential of height V_0.

When the energy E of the incident particle is lower than V_0, from the viewpoint of classical mechanics the particle cannot enter the barrier and will be reflected completely. But quantum mechanics will give a quite different conclusion. Let us start from the one-dimensional stationary Schrödinger equation:

$$\frac{d^2\psi}{dx^2} = \frac{2m}{\hbar^2}[V(x) - E]\psi \tag{8.34b}$$

and then solve this equation in three regions:

(I) Left of origin: $$\frac{-\hbar^2}{2m}\frac{d^2\psi}{dx^2} = E\psi \tag{8.34c}$$

(II) Inside barrier: $$\frac{-\hbar^2}{2m}\frac{d^2\psi}{dx^2} + V_0\psi = E\psi \tag{8.34d}$$

(III) Right of barrier: $$\frac{-\hbar^2}{2m}\frac{d^2\psi}{dx^2} = E\psi \tag{8.34e}$$

The solutions are

$$\text{(I)}\ \psi = A_1 \exp\left(-i\frac{\sqrt{2mE}}{\hbar}x\right) + B_1 \exp\left(i\frac{\sqrt{2mE}}{\hbar}x\right) \tag{8.35a}$$

$$\text{(II)}\ \psi = A_2 \exp\left(-i\frac{\sqrt{2m(E - V_0)}}{\hbar}x\right) + B_2 \exp\left(i\frac{\sqrt{2m(E - V_0)}}{\hbar}x\right) \tag{8.35b}$$

$$\text{(III)}\ \psi = A_3 \exp\left(-i\frac{\sqrt{2mE}}{\hbar}x\right) + B_3 \exp\left(i\frac{\sqrt{2mE}}{\hbar}x\right) \tag{8.35c}$$

It is clear that the wavefunctions in regions II and III are not zero. Particles that originate in region I can pass through region II and enter region III since the wavefunction, as shown in Fig 8.7, extends into that region. Now determine the constants.

In a second-order equation we require that the function and its first derivative be continuous. To establish a boundary condition, the experiment is such that the absolute magnitude of the incoming wave is 1 (normalize incoming wave to 1):

$$\exp\left(i\frac{\sqrt{2mE}}{\hbar}x\right) \text{ is a wave going to the right:}$$

$$\exp\left(-i\frac{\sqrt{2mE}}{\hbar}x\right) \text{ is a wave going to the left}$$

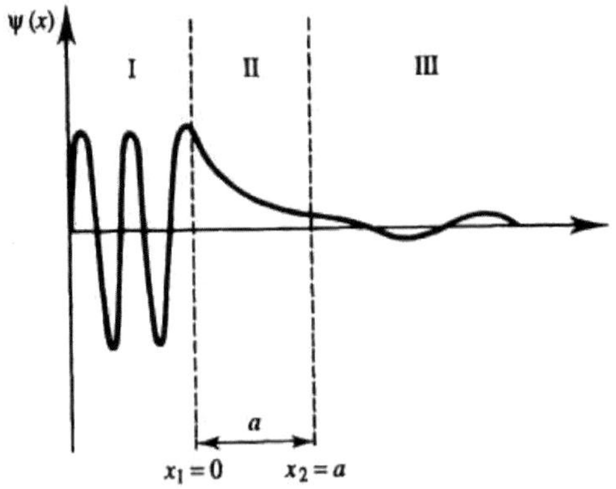

FIGURE 8.7
Penetration of a wavefunction through a square barrier of thickness $D(= a$ in Fig. 8.6), where $x_2 = a$ and $x_1 = 0$ in Fig. 8.6.

Thus, as a boundary condition, the wave going to the right has an absolute magnitude of 1, so

$$B_1 = 1$$

A_1 gives the magnitude of the reflected wave, and from the boundary condition we cannot say anything about it here. Look at the solutions at the boundaries:

At $x = 0$, from ψ we find

$$A_1 + 1 = A_2 + B_2 \tag{8.36a}$$

and from $d\psi/dx$,

$$\sqrt{\frac{E}{E - V_0}}\,(1 - A_1) = -A_2 + B_2$$

Combining these gives

$$\begin{aligned} A_2 &= \frac{1}{2}A_1\left(1 + \sqrt{\frac{E}{E - V_0}}\right) + \frac{1}{2}\left(1 - \sqrt{\frac{E}{E - V_0}}\right) \\ B_2 &= \frac{1}{2}A_1\left(1 - \sqrt{\frac{E}{E - V_0}}\right) + \frac{1}{2}\left(1 + \sqrt{\frac{E}{E - V_0}}\right) \end{aligned} \tag{8.36b}$$

At $x = a$, from ψ:

$$\begin{aligned} &A_2 \exp\left(-i\frac{\sqrt{2m(E - V_0)}a}{\hbar}\right) + B_2 \exp\left(i\frac{\sqrt{2m(E - V_0)}a}{\hbar}\right) \\ &= A_3 \exp\left(-i\frac{\sqrt{2mE}a}{\hbar}\right) + B_3 \exp\left(i\frac{\sqrt{2mE}a}{\hbar}\right) \end{aligned} \tag{8.36c}$$

from $d\psi/dx$:

$$\sqrt{\frac{E-V_0}{E}}\left[-A_2\exp\left(-i\frac{\sqrt{2m(E-V_0)}a}{\hbar}\right)+B_2\exp\left(i\frac{\sqrt{2m(E-V_0)}a}{\hbar}\right)\right]$$
$$=-A_3\exp\left(-i\frac{\sqrt{2mE}a}{\hbar}\right)+B_3\exp\left(i\frac{\sqrt{2mE}a}{\hbar}\right)$$

On the left-hand side we had particles hitting the barrier; some went in, and some were reflected. But on the right-hand side of the barrier there can only be particles going out. There can be no wave traveling to the left in this region. So,

$$A_3=0$$

is another boundary condition. The important thing is $|A_1|^2=R$, the reflection coefficient. This gives the magnitude of the wave traveling backward. $|B_3|^2=T$ is the transmission coefficient (the magnitude of the beam on the right). By manipulating the above equations and remembering that $R+T=1$, for $E>V_0$, we find

$$T=\frac{E}{E+\frac{1}{4}\frac{V_0^2}{E-V_0}\sin^2\left(\frac{a}{\hbar}\sqrt{2m(E-V_0)}\right)} \tag{8.37}$$

The interesting thing is the following: Classically when $E>V_0$ all the particles pass through the first edge, slow down inside, then leave the other side with the same velocity they had on entering. But quantum mechanics says that even with $E>V_0$ there are some reflected particles with reflection coefficient, $R=1-T$.

Now look at the case of $E<V_0$. Classical mechanics says that if $E<V_0$, all particles are reflected. Quantum mechanics says this is not so. From T, the transmission coefficient, quantum mechanics says that even when $E<V_0$ some particles are transmitted. Consider $E<V_0$, then $\sqrt{E-V_0}=-i\sqrt{V_0-E}$ and we find

$$R=\frac{1}{1+\frac{4E(V_0-E)}{V_0^2}\frac{1}{\sinh^2\left(\frac{a}{\hbar}\sqrt{2m(V_0-E)}\right)}}$$
$$T=\frac{4E(V_0-E)}{4E(V_0-E)+V_0^2\sinh^2\left(\frac{a}{\hbar}\sqrt{2m(V_0-E)}\right)} \tag{8.38}$$

Suppose $(a/\hbar)\sqrt{2m(V_0-E)}\gg 1$, then one can approximate sinh by

$$\frac{1}{2}\exp\left(\frac{a}{\hbar}\sqrt{2m(V_0-E)}\right)$$

Also, assume that $4E(V_0 - E)$ is small with respect to the exponential. Then

$$T \approx 16\frac{E}{V_0}\left(1 - \frac{E}{V_0}\right)\exp\left(-\frac{2a}{\hbar}\sqrt{2m(V_0 - E)}\right) . \qquad (8.39)$$

Even when $E < V_0$, some particles are able to get through. The number of particles that get through is a strong function of the barrier width a: the thicker the barrier, the lower the transmission. The transmission is also very sensitive to V_0 and E. Classically, conservation of energy says that the particle cannot be inside the barrier because it would have negative energy. But quantum mechanics says the particle can leak through the barrier in spite of conservation of energy.

There is an exponential drop of the wavefunction inside the barrier, but there is always a finite probability of the wavefunction on the other side. This is the *tunnel effect*. In spite of conservation of energy, particles can leak through the barrier. To ask what the particle's velocity is inside the barrier is meaningless. Its velocity is imaginary. One cannot speak of the particle being in this region, one must talk only of the wavefunction.

Gamow derived the relation for the penetration through a barrier for the first time and described it as a tunnel effect in which the particle tunnels through the barrier. He used it to explain the experimental fact that an atomic nucleus can undergo radioactive α decay. He initiated the precedent of quantum mechanics being used in the nuclear field and explained the penetration of a barrier in a way that cannot occur in classical physics. Gamow's work made a deep impression on Rutherford, who originally took a skeptical attitude to quantum mechanics.

Consider an α particle and a nucleus as an example of barrier penetration. When an α particle is far from a nucleus, the Coulomb repulsion is strong and represents a barrier to keep the α out of the nucleus (see Fig. 10.9). An α inside the nucleus is held in by the short-range, very strong force that holds the nucleus together (negative potential) and, as well, sees the positive-energy Coulomb barrier as a barrier to be penetrated. An α particle inside the nucleus has $E < V$ (Coulomb barrier), so classically it cannot get out of the nucleus. Yet quantum mechanics lets it leak out by Eq. (8.39). The α vibrates back and forth in the well with a finite probability that each time it strikes the barrier it may leak out. This gives a method to calculate probability for α decay.

Let us approximate the nuclear potential by the square potential barrier in Fig. 8.6. The α will be vibrating back and forth with some energy in this well. First, how many times per second (n), does the α strike the barrier wall of a nucleus. Take $v_\alpha \approx 10^9$ cm/s and the nuclear radius $R \approx 10^{-12}$ cm. In this case the α particle strikes the walls about 10^{21} times per second ($= n$). Let P be the probability that the α will leak through when it strikes the wall. The lifetime of the nucleus for α decay is

$$\tau = \frac{1}{nP}$$

For a large square barrier of height V,

$$P = T \approx 16\frac{E}{V}\left(1 - \frac{E}{V}\right)\exp\left(\frac{-2a}{\hbar}\sqrt{2m(V-E)}\right) \tag{8.39}$$

If

$$\frac{a}{\hbar}\sqrt{2m(V-E)} \gg 1$$

the term in front of the exponential is of the order of 1 for α decays, so the probability of penetration is

$$P = e^{-G}$$

where

$$G = \frac{2}{\hbar}a\sqrt{2m(V-E)} \tag{8.40}$$

As a rough example, take $a = 3 \times 10^{-12}$ cm, $m_\alpha = 6.4 \times 10^{-24}$ g. For ^{238}U, $V - E = 12$ MeV, $v_\alpha \approx 10^9$ cm/s, and radius $\simeq 10^{-12}$ cm:

$$n = \frac{v_\alpha}{R} \simeq 10^{21} \text{ times per second}$$

$$\frac{2a}{\hbar}\sqrt{2m(V-E)} \approx 90$$

$$\tau = \frac{1}{nP} = 10^{-21}e^{90} \approx 10^{18}\,\text{s} \approx 10^{11}\text{ years}$$

which, given our rough approximations, is close to the actual value of 4.5×10^9 years. The half-life of the nucleus is extremely sensitive to $V - E$.

One thing we have learned is that there are certain more-or-less "classic" or important types of Schrödinger equations that can be solved. In general a Schrödinger equation may not be solvable, but if one knows the important types that are solvable, then approximations can be made to get a solution of an arbitrary equation. The approximation methods used to obtain solutions are to a great extent based on the well-known solutions to the Schrödinger equation.

EXAMPLE 4: ONE-DIMENSIONAL HARMONIC OSCILLATOR POTENTIAL. In classical physics, simple harmonic motion is known to be very common in nature. A small vibration of any system can often be divided into several independent one-dimensional simple harmonic motions. The energy of one-dimensional harmonic motion can be expressed as

$$E = \frac{p^2}{2m} + \frac{kx^2}{2} \tag{8.41}$$

in which $V = kx^2/2$ and k is the elastic coefficient. In classical physics, the motion of a harmonic oscillator can be expressed as a sine function,

$$x(t) = x_0 \sin(\omega t + \delta)$$

in which the angular frequency is

$$\omega = \sqrt{\frac{k}{m}} \tag{8.42}$$

The classical energy can be expressed as

$$E = \frac{1}{2}kx_0^2 \tag{8.43}$$

where x_0 is the position of the particle when its kinetic energy is zero and is called the "turning point." Obviously E can take any value.

In quantum mechanics, for $V(x) = \frac{1}{2}kx^2$ the Schrödinger equation is

$$-\frac{\hbar^2}{2m}\frac{d^2\psi(x)}{dx^2} + \frac{1}{2}m\omega^2x^2\psi(x) = E\psi(x) \tag{8.44}$$

This is a differential equation with a nonconstant coefficient, which is different from the preceding three examples, but it still belongs to one of the small number of examples which can be solved precisely. A complete calculation is given in Appendix 8A. Here we only give the results:

$$E_n = \left(n + \frac{1}{2}\right)\hbar\omega, \qquad n = 0, 1, 2, \ldots \tag{8.45}$$

$$\psi_n(y) = \sqrt{\frac{1}{2^n n! \sqrt{n}}}\, e^{y^2/2}\, H_n(y) \tag{8.46}$$

in which $y = \sqrt{m\omega/\hbar}\, x$, and $H_n(y)$ is a Hermite polynomial. The forms of the first few Hermite polynomials are

$$\begin{aligned} H_0(y) &= 1 \\ H_1(y) &= 2y \\ H_2(y) &= 4y^2 - 2 \\ H_3(y) &= 8y^3 - 12y \\ H_4(y) &= 16y^4 - 48y^2 + 12 \\ H_5(y) &= 32y^5 - 160y^3 + 120y \end{aligned} \tag{8.47}$$

Note that the solution to the Schrödinger equation gives quantized energies for the particle as expected. It also can be proven that the transitions of the particle between different energy levels obey the selection rule $\Delta n = 1$; that is, a transition can only occur between two adjacent energy levels. It is clear from these results that the harmonic oscillator potential has three characteristics: (1) it has a zero point energy (the particle has an energy which remains when n is zero), which is not a characteristic peculiar to a harmonic oscillator potential; (2) the energy intervals between two adjacent levels are the same (equal intervals), which is the distinguishing feature of the harmonic oscillator potential; (3) the transitions can only go from one level to the next level. Combining the last two characteristics, we can make the statement: All transitions will give radiation of the same frequency and there will be only one spectral line in the energy spectrum being measured. These characteristics are often used to verify the reliability of using a harmonic oscillator as a theory to describe a system. Because both the harmonic oscillator potential and the square well potential are reasonably simple and can yield precise solutions of the Schrödinger equation, they are often used in theoretical calculations as the starting point of the first approximation. For example, for the motion of a high-speed electron in a crystal, theoretically it is first assumed that a harmonic oscillator potential acts on the electron. Then the results are compared with the radiation spectrum of the electron in the crystal obtained by experiment. If the potential that acts on the electron is strictly a harmonic oscillator potential, then in an experiment only one spectral line of a certain frequency (a certain energy) will be measured. But, as could be expected, there are differences between practical cases and the theoretical one. According to this difference, we can amend the theory. Although the harmonic oscillator potential is simple, it is important in both experimental work and theoretical work.

Note. The first characteristic of the harmonic oscillator potential mentioned above, the existence of a zero point energy, can be obtained immediately from the uncertainty relation. Assume the uncertainty of the position of a particle to be $\Delta x = a$; then the uncertainty of the momentum is at least $\Delta p = \hbar/2a$. The minimum momentum of the particle is

$$p = \Delta p = \frac{\hbar}{2a}$$

Substituting this into formula (8.41), we obtain the energy

$$E = \frac{\hbar^2}{8ma^2} + \frac{1}{2}m\omega^2 a^2 \tag{8.48}$$

which is a function of a. When $dE/da = 0$, this has a minimum value; that is, when

$$\frac{dE}{da} = -\frac{\hbar^2}{4ma^3} + m\omega^2 a = 0 \tag{8.49}$$

then

$$a^2 = \frac{\hbar}{2m\omega}$$

From formula (8.48), we have

$$E = \tfrac{1}{2}\hbar\omega \tag{8.50}$$

which corresponds to the formula (8.45) when $n = 0$ and is the zero point energy of the harmonic oscillator potential.

8.2 AVERAGE VALUE AND OPERATOR

8.2.1 Average Values

Since the fundamental laws in quantum mechanics are statistical laws and the wavefunction ψ only contains a meaning of probability, then naturally one only obtains the average value for any physical quantity to be compared with the quantity observed in experiment. We can obtain the wavefunction from the Schrödinger equation, but how do we then extract the average value after we have the wavefunction?

Let us first recall the classical case. If a uniform rod has length L and the coordinate x can be any point between 0 and L, then its average value is

$$\bar{x} = \frac{\int_0^L x\,dx}{\int_0^L dx} = \frac{L^2/2}{L} = \frac{L}{2}$$

which is the center of gravity of a uniform rod.

If the rod is not uniform and its density has a distribution $\rho(x)$, then

$$\bar{x} = \frac{\int_0^L x\rho(x)\,dx}{\int_0^L \rho(x)\,dx}$$

In general, the average value of an arbitrary function $f(x)$ in its definition zone of $x[0, L]$ is calculated as

$$\overline{f(x)} = \frac{\int_0^L f(x)P(x)\,dx}{\int_0^L P(x)\,dx} \tag{8.51}$$

in which $P(x)$ is the probability distribution of $f(x)$ in the definition zone of x and $\overline{f(x)}$ is the weighted average. In general, $P(x)$ satisfies the normalization condition

$$\int_0^L P(x)\,dx = 1$$

Now turn to the case of quantum mechanics. Obviously, $\psi^*\psi$ is equivalent to the probability distribution in x space. Then the average of multiple measurements of position (sometimes this is called the expectation value) is

$$\bar{x} = \int_{-\infty}^{\infty} \psi^*(x) x \psi(x)\, dx \tag{8.52}$$

Here $\psi(x)$ satisfies the normalization condition

$$\int_{-\infty}^{\infty} \psi^*(x)\psi(x)\, dx = 1 \tag{8.53}$$

which indicates that the probability of finding the particle somewhere in the whole space is 100 percent, which is obviously true.

The average of a measurable function $f(x)$ at any position is

$$\overline{f(x)} = \int_{-\infty}^{\infty} \psi^*(x) f(x) \psi(x)\, dx \tag{8.54}$$

The x component of momentum, p_x, is measurable. How does one calculate its average value? It seems very easy at first glance: Only the substitution of $p_x(x)$ into Eq. (8.54) is needed. But it is not so easy, because $p_x(x)$ cannot be expressed in a suitable form to go into this equation. Here $p_x(x)$ indicates a value corresponding to a particular x in the language of an "orbit" and this violates the uncertainty relation directly.

Up to now we have been considering the space in which the position x is taken as an independent variable (take the eigenfunction of x as the function vector), called the position representation. In this representation, $p_x(x)$ does not exist. Only for an existing function $f(x)$ can we use the above-mentioned method to calculate the average.

8.2.2 The Introduction of Operators

For momentum p_x, we must go over to a different representation: a momentum representation instead of a position representation. Then we can calculate its average in the momentum representation:

$$\bar{p}_x = \int_{-\infty}^{\infty} \phi^*(k) p_x \phi(k)\, dk \tag{8.55}$$

The meaning of the function $\phi(k)$ is analogous to that of $\psi(x)$: the probability of the momentum of a particle falling in the momentum interval p_x to $p_x + dp_x$ is $\phi^*(k)\phi(k)$ where k is the wave vector and has a one-to-one relation with p_x so that $p_x = \hbar k$.

The position representation can be associated with the momentum representation by Fourier analysis:

$$\phi(k) = \frac{1}{\sqrt{2\pi}} \int_{-\infty}^{\infty} \psi(x) e^{-ikx}\, dx \tag{8.56}$$

$$\psi(x) = \frac{1}{\sqrt{2\pi}} \int_{-\infty}^{\infty} \phi(k) e^{ikx}\, dx \tag{8.57}$$

The constant before the integral guarantees the normalization. Substituting Eq. (8.56) into (8.55), we have

$$\begin{aligned}\bar{p}_x &= \frac{1}{2\pi} \int_{-\infty}^{\infty} \left[\int_{-\infty}^{\infty} \psi^*(x') e^{ikx'}\, dx'\, k\hbar \int_{-\infty}^{\infty} \psi(x) e^{-ikx}\, dx \right] dk \\ &= \frac{i\hbar}{2\pi} \int_{-\infty}^{\infty} \left[\int_{-\infty}^{\infty} \psi^*(x') e^{ikx'}\, dx' \int_{-\infty}^{\infty} \psi(x) \frac{\partial}{\partial x} (e^{-ikx})\, dx \right] dk\end{aligned}$$

Taking the second integral in the brackets as the partial integral and noting that $\psi(\infty) = 0$, we have

$$\bar{p}_x = -i\hbar \int_{-\infty}^{\infty} \left\{ \frac{1}{\sqrt{2\pi}} \int_{-\infty}^{\infty} \left[\frac{1}{\sqrt{2\pi}} \int_{-\infty}^{\infty} \psi^*(x') e^{ikx'}\, dx' \right] e^{-ikx}\, dk \right\} \frac{\partial \psi(x)}{\partial x}\, dx$$

Compared with formula (8.56), we find that the integral in the square brackets is $\phi^*(k)$. Thus from formula (8.57), we know that the integral in the braces is $\psi^*(x)$. So,

$$\bar{p}_x = \int_{-\infty}^{\infty} \psi^*(x) \left(-i\hbar \frac{\partial}{\partial x} \right) \psi(x)\, dx \tag{8.58}$$

Compare this formula with $\bar{x}$, that is with formula (8.52). We find that in order to calculate $\bar{p}_x$ in the position representation, that is to calculate $\bar{p}_x$ from $\psi(x)$, we only need to change $p_x(x)$ as follows:

$$\hat{p}_x = -i\hbar \frac{\partial}{\partial x} \tag{8.59}$$

and so,

$$\bar{p}_x = \int_{-\infty}^{\infty} \psi^*(x) \hat{p}_x \psi(x)\, dx \tag{8.60}$$

Then we can calculate $\bar{p}_x$ the same way as calculating $\bar{x}$, in which $\hat{p}_x$ is called the operator of the x component of momentum. In three-dimensional space, it is obvious that the extension of formula (8.59) is

$$\hat{p} = -i\hbar\nabla \tag{8.61}$$

where

$$\nabla = \frac{\partial}{\partial x}\boldsymbol{i} + \frac{\partial}{\partial y}\boldsymbol{j} + \frac{\partial}{\partial z}\boldsymbol{k}$$

in cartesian coordinates, where $\boldsymbol{i}$, $\boldsymbol{j}$, and $\boldsymbol{k}$ are unit vectors in the x, y, and z directions.

In the position representation, the operators of all the measurable physical quantities which can be written as a function of x are equal to themselves, that is,

$$f(x) \rightarrow \hat{f}(x)$$

For example the operator of x is x.

Similarly, we have the energy operator:

$$\hat{E} = i\hbar\frac{\partial}{\partial t} \tag{8.62}$$

Formulas (8.59), (8.61), and (8.62) are the formulas (8.9) and (8.13) used in the last section. We must emphasize again that the Schrödinger equation is not derived, it is a basic hypothesis in quantum mechanics and this hypothesis is equivalent to assuming the existence of energy operators and momentum operators and at the same time the existence of a wavefunction.

Similarly we have angular momentum operators:

$$\begin{aligned}
\hat{L}_x &= \hat{y}\hat{p}_z - \hat{z}\hat{p}_y = -i\hbar\left(y\frac{\partial}{\partial z} - z\frac{\partial}{\partial y}\right) \\
\hat{L}_y &= \hat{z}\hat{p}_x - \hat{x}\hat{p}_z = -i\hbar\left(z\frac{\partial}{\partial x} - x\frac{\partial}{\partial z}\right) \\
\hat{L}_z &= \hat{x}\hat{p}_y - \hat{y}\hat{p}_x = -i\hbar\left(x\frac{\partial}{\partial y} - y\frac{\partial}{\partial x}\right)
\end{aligned} \tag{8.63}$$

These come from the relations

$$L_x = yp_z - zp_y \qquad L_y = zp_x - xp_z \qquad L_z = xp_y - yp_x$$

or

$$\boldsymbol{L} = \boldsymbol{r} \times \boldsymbol{p}$$

After putting these into spherical coordinates, Eqs. (8.63) become (prove this for yourself):

$$\hat{L}_x = i\hbar\left(\sin\phi\frac{\partial}{\partial\theta} + \cot\theta\cos\phi\frac{\partial}{\partial\phi}\right)$$
$$\hat{L}_y = -i\hbar\left(\cos\phi\frac{\partial}{\partial\theta} - \cot\theta\sin\phi\frac{\partial}{\partial\phi}\right) \tag{8.64}$$
$$\hat{L}_z = -i\hbar\frac{\partial}{\partial\phi}$$

and also

$$\hat{L}^2 = \hat{L}_x^2 + \hat{L}_y^2 + \hat{L}_z^2 = -\hbar^2\left[\frac{1}{\sin\theta}\frac{\partial}{\partial\theta}\left(\sin\theta\frac{\partial}{\partial\theta}\right) + \frac{1}{\sin^2\theta}\frac{\partial^2}{\partial\phi^2}\right] \tag{8.65}$$

From these formulas, we can easily prove an important characteristic of operators; that is that in general the product of two operators cannot be commuted. For example,

$$\hat{L}_x\hat{L}_y \neq \hat{L}_y\hat{L}_x, \text{ thus } (\hat{L}_x\hat{L}_y - \hat{L}_y\hat{L}_x) \equiv [\hat{L}_x, \hat{L}_y] \neq 0 \tag{8.66}$$

It is easily proven from Eq. (8.63) that

$$[\hat{L}_x, \hat{L}_y] = i\hbar\hat{L}_z \qquad [\hat{L}_y, \hat{L}_z] = i\hbar\hat{L}_x \qquad [\hat{L}_z, \hat{L}_x] = i\hbar\hat{L}_y \tag{8.67}$$

but

$$[\hat{L}^2, \boldsymbol{L}] = 0 \tag{8.68}$$

That is, the operator L^2 can commute with L_x, L_y, and L_z. In a course on quantum mechanics, you will learn that formulas (8.67) and (8.68) include very important physical content.

With the appropriate operator, we can express the average of any mechanical quantity A of a group of particles in three-dimensional space as

$$\bar{A} = \int \psi^* \hat{A}\psi\, d^3x \tag{8.69}$$

in which $\hat{A}$ is the operator corresponding to the mechanical quantity A.

8.2.3 Eigenvalue and Eigenequation

In classical mechanics, the sum of the kinetic energy and potential energy, $p^2/2m + V$, is called the Hamiltonian. Because the operator corresponding to $p^2/2m$ is $-\hbar^2\nabla^2/2m$, and the operator of $V(\boldsymbol{r})$ is itself,

$$H = \left[-\frac{\hbar^2}{2m}\nabla^2 + V(\boldsymbol{r})\right] \tag{8.70}$$

is called the Hamiltonian operator. In this way, the stationary Schrödinger equation can be written as:

$$H\psi = E\psi \tag{8.71}$$

In mathematics, the general definition of an operator is that it can change one function into another function after acting on this function, namely,

$$\hat{O}f = g \tag{8.72}$$

But when the difference between function f and function g is only a constant

$$\hat{O}f = \lambda f \tag{8.73}$$

the function f is called an eigenfunction. In general λ is a set of numbers called the eigenvalue spectrum. The corresponding equation is called an eigenequation. If one eigenvalue corresponds to n eigenfunctions, then the eigenfunction is n-degenerate.

Hence the stationary Schrödinger Eq. (8.71) is an eigenequation and the problem of solving the Schrödinger equation is actually the problem of obtaining the eigenfunctions and the eigenvalues.

8.3 THE SOLUTION OF THE SCHRÖDINGER EQUATION OF THE HYDROGEN ATOM

8.3.1 Schrödinger Equation in a Central Force Field

The first problem Schrödinger wanted to solve after he had established the wave equation was not the case of a square potential well or barrier mentioned in Section 8.1, but that of the hydrogen atom. The description of the hydrogen atom by the Schrödinger equation achieved great success and brought the Schrödinger equation general attention and recognition.

For a system of one electron and one atomic nucleus, the charge and mass of the atomic nucleus are Ze and M, respectively, and the charge and mass of the electron are e and m_e, respectively. Obviously, the main term in the potential energy is the electrostatic interaction between the two charges

$$V(r) = -\frac{Ze^2}{r} \tag{8.74}$$

where $V(r)$ is just a scalar quantity function of the distance r between the electron and the nucleus. Thus, just as in classical mechanics, it is a central force field. For this kind of force field, there is no component in the direction perpendicular to the radius vector. Thus the angular momentum is conserved:

$$\boldsymbol{r} \times \boldsymbol{F} = \frac{d\boldsymbol{L}}{dt} = 0 \qquad \boldsymbol{L} = \text{constant} \tag{8.75}$$

In this section we do not consider the form of $V(r)$ for the time being but investigate the Schrödinger equation in the case of a central force field and its general results. We also want to observe in this case what the conservation of angular momentum means.

The three-dimensional stationary Schrödinger equation is as Eq. (8.21), that is,

$$\left[-\frac{\hbar^2}{2m}\nabla^2 + V(\boldsymbol{r})\right]\psi = E\psi \tag{8.76}$$

Now m should be transformed into the reduced mass of the electron–nucleus system, $\mu = m_e M/(m_e + M)$. Since $V(\boldsymbol{r}) = V(r)$ is a spherically symmetric potential, it is convenient to use spherical polar coordinates instead of cartesian coordinates (see Fig. 8.8):

$$\begin{aligned}\nabla^2 &= \frac{\partial^2}{\partial x^2} + \frac{\partial^2}{\partial y^2} + \frac{\partial^2}{\partial z^2} \\ &= \frac{1}{r^2}\frac{\partial}{\partial r}\left(r^2\frac{\partial}{\partial r}\right) + \frac{1}{r^2 \sin\theta}\frac{\partial}{\partial\theta}\left(\sin\theta\frac{\partial}{\partial\theta}\right) \\ &\quad + \frac{1}{r^2\sin^2\theta}\frac{\partial^2}{\partial\theta^2}\end{aligned} \tag{8.77}$$

The position of the nucleus is taken as the origin of the coordinates. Equation (8.76) thus becomes

$$\begin{aligned}&-\frac{\hbar^2}{2\mu}\left[\frac{1}{r^2}\frac{\partial}{\partial r}r^2\frac{\partial}{\partial r}\psi\right] \\ &-\frac{\hbar^2}{2\mu r^2}\left[\frac{1}{\sin\theta}\frac{\partial}{\partial\theta}\left(\sin\theta\frac{\partial\psi}{\partial\theta}\right) + \frac{1}{\sin^2\theta}\frac{\partial^2\psi}{\partial\phi^2}\right] + V\psi = E\psi\end{aligned} \tag{8.78}$$

Compared with Eq. (8.65), we immediately see that the second term on the left-hand side of Eq. (8.78) is proportional to $\hat{L}^2$, thus

$$-\frac{\hbar^2}{2\mu}\left[\frac{1}{r^2}\frac{\partial}{\partial r}r^2\frac{\partial}{\partial r}\right]\psi + \frac{\hat{L}^2}{2\mu r^2}\psi + V(r)\psi = E\psi \tag{8.79}$$

Readers should prove that the first term on the left-hand side of formula (8.79) is proportional to the r-component of the momentum P of the electron and explain

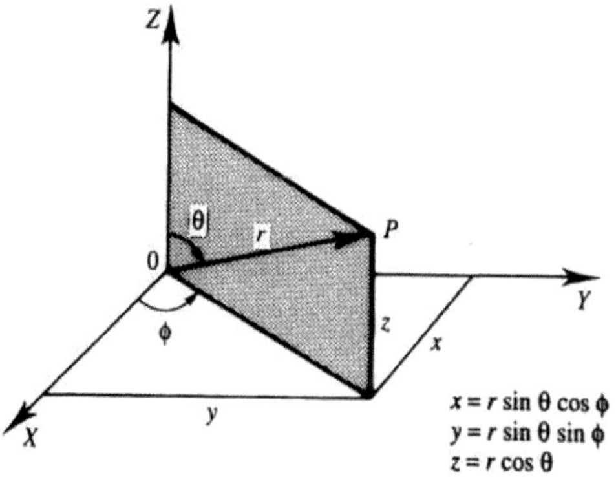

FIGURE 8.8
Spherical polar coordinates.

the meaning of formula (8.79). Because $V(r)$ is a central field, we can greatly reduce the problem by using the method of separation of variables. Let

$$\psi(r,\theta,\phi) = R(r)Y(\theta,\phi) \tag{8.80}$$

Substituting this into Eq. (8.79) and making the proper rearrangement (and dividing by ψ), we have

$$\frac{1}{\hbar^2 Y}\hat{L}^2 Y = \frac{1}{R}\frac{d}{dr}\left(r^2\frac{d}{dr}R\right) + \frac{2\mu r^2}{\hbar^2}(E - V(r)) \tag{8.81}$$

Since the left-hand side of the equation is concerned only with θ and ϕ and the right-hand side only with r, and further since r, θ, ϕ are independent variables, then the equality of the two sides certainly means that both sides are equal to the same constant. We define the constant as α and call it the separating constant. Thus we obtain a rather simple partial differential equation,

$$\hat{L}^2 Y(\theta,\phi) = \alpha\hbar^2 Y(\theta,\phi) \tag{8.82}$$

and an ordinary differential equation,

$$\left[-\frac{\hbar^2}{2\mu r^2}\frac{d}{dr}\left(r^2\frac{d}{dr}\right) + \frac{\alpha\hbar^2}{2\mu r^2} + V(r)\right]R(r) = ER(r) \tag{8.83}$$

Both of these equations are eigenequations. The second equation is the eigenequation of the energy, whose solution depends on the explicit form of the potential field $V(r)$. We will discuss it in detail in the next section. Now let us consider the first equation, which is independent of the specific form of the potential field and is the universal equation of a central force field. We can further separate the variables for Eq. (8.82) as

$$Y(\theta,\phi) = \Theta(\theta)\Phi(\phi) \tag{8.84}$$

Substituting this and the expression for $\hat{L}^2$ (8.65) into Eq. (8.82) and rearranging, we obtain

$$\frac{\sin\theta}{\Theta}\frac{d}{d\theta}\left(\sin\theta\frac{d\Theta}{d\theta}\right) + \alpha\sin^2\theta = -\frac{1}{\Phi}\frac{d^2\Phi}{d\theta^2} = \nu \tag{8.85}$$

in which we again introduce a separating constant ν. Thus we obtain two equations. For the first equation

$$\frac{d^2\Phi}{d\phi^2} + \nu\Phi = 0 \tag{8.86}$$

Its solution is easily obtained:

$$\begin{aligned}\Phi &= Ae^{i\sqrt{\nu}\phi} + Be^{-i\sqrt{\nu}\phi}, \qquad && \nu \neq 0 \\ &= C + D\phi && \nu = 0\end{aligned} \tag{8.87}$$

in which A, B, C, and D are constants. The standard condition on a wavefunction demands that the wavefunction be single-valued at any point in space, that is,

$\Phi(\phi) = \Phi(\phi + 2\pi)$. To satisfy this condition, we must have $D = 0$ when $\nu = 0$ so then Φ is a constant; and for the solution for $\nu \neq 0$, $\sqrt{\nu}$ must be an integer, expressed as m where $(\sqrt{\nu} \equiv m)$. Thus we obtain the specific solutions:

$$\Phi_m = \frac{1}{\sqrt{2\pi}} e^{im\phi}, \qquad m = 0, \pm 1, \pm 2, \ldots \tag{8.88}$$

in which the coefficient $1/\sqrt{2\pi}$ is obtained according to the normalization condition

$$\int_0^{2\pi} \Phi_m^* \Phi_m \, d\phi = 1 \tag{8.89}$$

Obviously, the wavefunction (8.88) is the eigenfunction of the operator $\hat{L}_z$:

$$\hat{L}_z \Phi_m = L_z \Phi_m$$

or (8.90)

$$-i\hbar \, d\phi \left(\frac{1}{\sqrt{2\pi}} e^{im\phi} \right) = m\hbar \left(\frac{1}{\sqrt{2\pi}} e^{im\phi} \right)$$

Thus we get Bohr's quantization condition,

$$L_z = m\hbar \tag{8.91}$$

except that now m is not the same as Bohr's n, which also appeared in the energy equation, and the series of m values starts with 0, not 1.

Now let us turn to the other equation deduced from Eq. (8.85):

$$\frac{\sin\theta}{\Phi} \frac{d}{d\theta} \left(\sin\theta \frac{d\Theta}{d\theta} \right) + \alpha \sin^2\theta = \nu = m^2 \tag{8.92}$$

In order to solve this equation, let us first make the change

$$u = \cos\theta \tag{8.93}$$

Substituting this into $\Theta(\theta)$, we obtain a new function $P(u)$:

$$\Theta(\theta) = P(u) \tag{8.94}$$

Thus Eq. (8.92) becomes

$$\frac{d}{du}\left[(1 - u^2)\frac{dP}{du}\right] + \left(\alpha - \frac{m^2}{1 - u^2}\right) P = 0 \tag{8.95}$$

The solution of Eq. (8.95) can be found in texts on the methods of mathematical physics. Here we only list the results.

Result I. Just as in the case of solutions for Φ_m (Eq. 8.88), solutions to Eq. (8.95) exist only for

$$\alpha = l(l+1), \qquad l = 0, 1, 2, \ldots$$
$$|m| \leq l, \qquad m = 0, \pm 1, \pm 2, \ldots, \pm l \tag{8.96}$$

Thus from Eq. (8.82), we obtain

$$\hat{L}^2 Y_{l,m} = L^2 Y_{l,m} = \alpha^2 \hbar^2 Y_{l,m} = l(l+1)\hbar^2 Y_{l,m} \tag{8.97}$$

Hence the eigenfunction of the operator $\hat{L}^2$ is $Y_{l,m}$ and the corresponding eigenvalues are

$$L^2 = l(l+1)\hbar^2$$

or

$$L = \sqrt{l(l+1)}\,\hbar$$

It is clear that not only the z component L_z of the total angular momentum but also the total angular momentum itself is quantized.

Result II. The solution of Eq. (8.95) is

$$P(u) = P_l^{|m|}(u) \tag{8.98}$$

in which $P_l^{|m|}(u)$ is an associate Legendre polynomial:

$$P_l^{|m|}(u) = (1-u^2)^{|m|/2} \frac{d^{\,|m|}}{du^{|m|}} P_l(u) \tag{8.99}$$

and

$$P_l(u) = \frac{1}{2^l l!} \frac{d^l}{du^l}(u^2-1)^l \tag{8.100}$$

is a Legendre polynomial.

Summing up Eqs. (8.84), (8.88), (8.94), and (8.98), we find that the eigenfunction of the operator $\hat{L}^2$ is a spherical harmonic:

$$Y_{l,m}(\theta, \phi) = N_{l,m} P_l^{|m|}(\cos\theta) e^{im\phi} \tag{8.101}$$

in which N_{lm} is a normalization constant. According to the normalization condition

$$\int_0^{2\pi}\int_0^{\pi} Y_{l,m}^*(\theta, \phi) Y_{l,m}(\theta, \phi) \sin\theta \, d\theta \, d\phi = 1 \tag{8.102}$$

we have

$$N_{l,m} = \sqrt{\frac{(l-|m|)!(2l+1)}{4\pi(l+|m|)!}} \tag{8.103}$$

As described in Chapter 3, the state of $l = 0$ is called an s state and the states of $l = 1$ and $l = 2$ are called p state and d states, respectively. The expressions for some spherical harmonics are

$$l = 0: \qquad Y_{0,0} = \frac{1}{\sqrt{4\pi}}$$

$$l = 1: \qquad Y_{1,0} = \sqrt{\frac{3}{4\pi}}\cos\theta$$

$$Y_{1,\pm1} = \sqrt{\frac{3}{8\pi}}\sin\theta\, e^{\pm i\phi}$$

$$l = 2: \qquad Y_{2,0} = \sqrt{\frac{5}{16\pi}}(3\cos^2\theta - 1)$$

$$Y_{2,\pm1} = \mp\sqrt{\frac{15}{8\pi}}\sin\theta\cos\theta\, e^{\pm i\phi}$$

$$Y_{2,\pm2} = \sqrt{\frac{15}{32\pi}}\sin^2\theta\, e^{\pm 2i\phi}$$

The θ parts of the wavefunction for different l are shown in Fig. 8.9. We find that for l reasonably large, the probability distribution of the electron when $m = 0$ concentrates near the z axis, while the probability distribution of an electron when $m = l$ concentrates on the x–y plane. This can be predicted by the correspondence principle.

Spherical harmonics, as the eigenfunctions of the orbital angular momentum (see Eq. 8.82), have a very important quality under the inversion of coordinates through the origin (called parity transformation). If

$$(r, \theta, \phi) \rightarrow (r, \pi - \theta, \phi + \pi) \tag{8.104}$$

then it can be proven (you should prove this for yourself) that

$$Y_{l,m}(\pi - \theta, \phi + \pi) = (-1)^l Y_{l,m}(\theta, \phi) \tag{8.105}$$

Thus the difference between the new spherical harmonic obtained from inversion and the old one is only in the sign, and the sign is determined by the value of l. The sign is positive when l is even, and $Y_{l,m}$ is then said to have even parity. The sign is negative when l is odd, and $Y_{l,m}$ then has odd parity. Because another part $R(r)$ of the wavefunction does not change sign when reversing the coordinates, the oddness or evenness of l determines the parity of the wavefunction.

We use an operator $\hat{P}$ to express the inversion of coordinates. If $\psi(\boldsymbol{r})$ has definite parity then

$$\hat{P}\psi(\boldsymbol{r}) = \psi(-\boldsymbol{r})$$

Obviously,

$$\hat{P}^2\psi(\boldsymbol{r}) = \hat{P}\psi(-\boldsymbol{r}) = \psi(\boldsymbol{r})$$

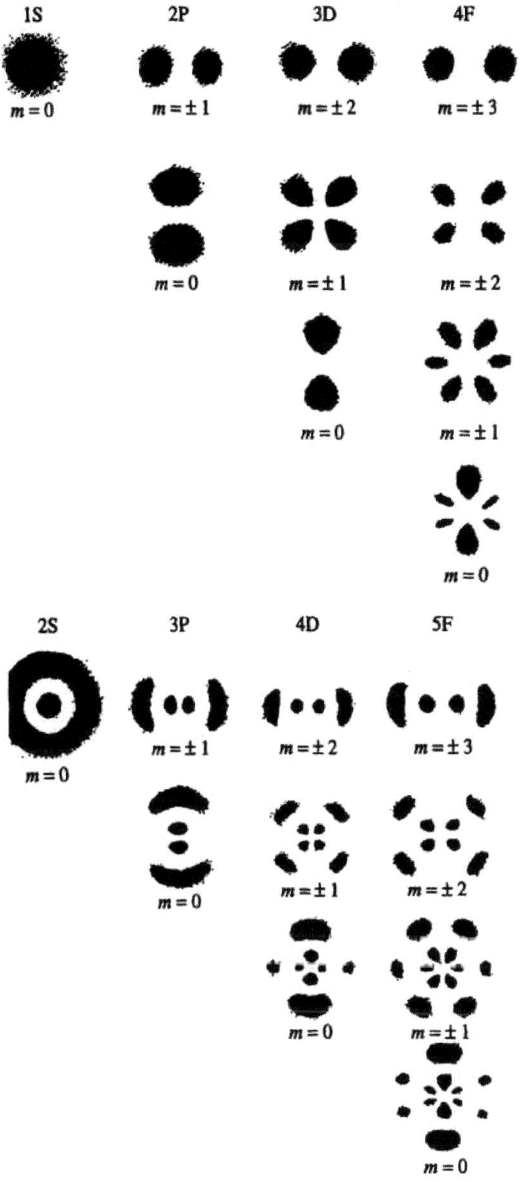

FIGURE 8.9
The patterns of the angular part of the wavefunction in a central force field for $l = 0, 1, 2, 3$ (S, p, d, f), with z components $m = 0, 1, 2, 3$, and $n = 1, 2, 3, 4, 5$.

Thus the eigenvalue of $\hat{P}$ is $P = \pm 1$ and the parity of the wavefunction ψ can only be positive or negative.

Generally speaking, when there is no external force acting on a system, the Hamiltonian will not change under the operation of the parity operator. Thus the parity of a wavefunction is a constant of the motion and will not change with time.

8.3.2 The Motion of Electrons in a Coulomb Field

Now we consider the radial equation (8.83). In order to solve this equation, we must know the explicit form of $V(r)$. We select formula (8.74), that is we consider the motion of electrons in a Coulomb field. In this case,

$$\left[-\frac{\hbar^2}{2\mu r^2}\frac{d}{dr}\left(r^2\frac{d}{dr}\right)+\frac{l(l+1)\hbar^2}{2\mu r^2}-\frac{Ze^2}{r}\right]R = ER \tag{8.106}$$

We only consider the case of electrons being in bound states, so that E is negative. When E is positive, for any value of E, Eq. (8.106) has a solution which satisfies the standard conditions of a wavefunction, so that the energy of the particle is continuous. This is equivalent to saying that the particle can leave the fixed charge and move to infinity (the atom is ionized). Note that the potential well of bound electrons includes not only the Coulomb potential but also the centrifugal potential, which is the second term on the left-hand side of equation (8.106). The two potentials can be combined into an effective potential:

$$V_{\text{eff}}(r) = \frac{l(l+1)\hbar^2}{2\mu r^2} - \frac{Ze^2}{r} \tag{8.107}$$

The form of the potential is shown in Fig. 8.10. We pointed out in Section 7.2 that the energy of the electrons bound in a potential well can only be discrete, namely, that the energy is quantized. Here we will see this more clearly.

It is not easy to solve Eq. (8.106), but we can investigate its general behavior and extreme case. This is a common method in physics and can usually give us a clear picture quickly. First we define

$$k^2 = -\frac{2\mu E}{\hbar^2} \tag{8.108}$$

Because E is negative, k is positive. Taking $l = 0$ (the ground state of angular momentum) and substituting Eq. (8.108) into (8.106) we get

$$\left[\frac{1}{r^2}\frac{d}{dr}\left(r^2\frac{d}{dr}\right)+\frac{2\mu Ze^2}{\hbar^2 r}\right]R = k^2 R$$

or

$$\frac{d^2R}{dr^2}+\left(\frac{2}{r}\frac{dR}{dr}+\frac{2\mu Ze^2}{\hbar^2 r}R\right) = k^2 R \tag{8.109}$$

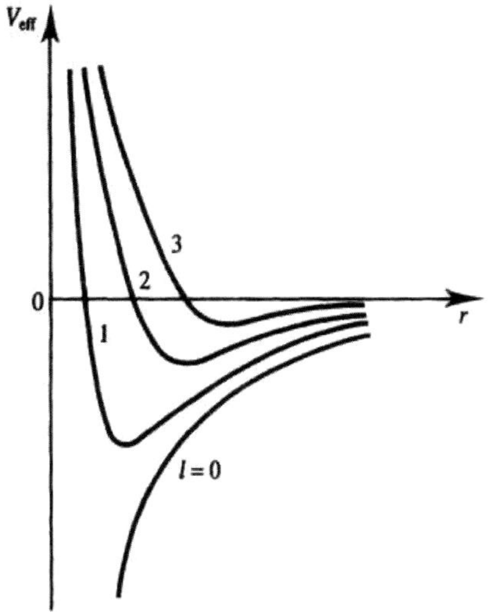

FIGURE 8.10
The effective potential of a hydrogen atom.

Now let us consider the asymptotic solution as $r \to \infty$:

$$\frac{d^2R}{dr^2} - k^2R = 0$$

Its solution is

$$R = C_1 e^{kr} + C_2 e^{-kr} \tag{8.110}$$

The wavefunction should go to zero when r is very large, thus $C_1 = 0$. In order to let $R \sim e^{-kr}$ be correct for all r, we must let the value of the term in parentheses in Eq. (8.109) be zero, that is,

$$-\frac{2kR}{r} + \frac{2\mu Ze^2 R}{\hbar^2 r} = 0$$

or

$$k = \frac{\mu Ze^2}{\hbar^2} - \frac{Z}{a_1} \tag{8.111}$$

in which a_1 is the first Bohr radius. Substituting formula (8.111) into (8.108), we obtain the ground-state energy

$$E_1 = -\frac{\hbar^2 k^2}{2\mu} = -Z^2 \frac{1}{2}\mu c^2 \left(\frac{e^2}{\hbar c}\right)^2 = -Z^2 E_{11} \tag{8.112}$$

in which

$$E_{11} = \frac{1}{2}\mu c^2 \left(\frac{e^2}{\hbar c}\right)^2 \simeq 13.6\,\text{eV} \tag{8.113}$$

is the Bohr ground-state energy.

On the basis of our experience, it is easy to guess that the eigenvalue spectrum of the energy can be expressed as

$$E_n = -\left(\frac{Z}{n}\right)^2 E_{11} \tag{8.114}$$

A careful calculation proves the correctness of this expression.

Now let us go back to Eq. (8.106) and solve for the wavefunction more rigorously. Besides the k introduced in formula (8.108), we introduce two other dimensionless parameters:

$$\gamma = \frac{\mu Z e^2}{k\hbar^2} \qquad \rho = 2kr \tag{8.115}$$

Then the radial equation becomes

$$\frac{d^2R}{d\rho^2} + \frac{2}{\rho}\frac{dR}{d\rho} + \left[\frac{\gamma}{\rho} - \frac{1}{4} - \frac{l(l+1)}{\rho^2}\right]R = 0 \tag{8.116}$$

First let us see the asymptotic solution. When ρ is very large, the terms of the denominator containing ρ can be omitted. Thus

$$\frac{d^2R}{d\rho^2} - \frac{1}{4}R = 0 \tag{8.117}$$

Similar to the solution (8.110), we have

$$R = e^{-\rho/2}, \qquad \text{where } \rho \to \infty \tag{8.118}$$

From this we can assume the solution of Eq. (8.116) is

$$R(\rho) = e^{-\rho/2}F(\rho) \tag{8.119}$$

Substituting this into (8.116), we get

$$\frac{d^2F}{d\rho^2} + \left(\frac{2}{\rho} - 1\right)\frac{dF}{d\rho} + \left[\frac{\gamma - 1}{\rho} - \frac{l(l+1)}{\rho^2}\right]F = 0 \tag{8.120}$$

This equation has a singularity at $\rho = 0$. In general, to solve this problem we use a form of power series to solve the equation

$$F(\rho) = \rho^s \sum_{j=0}^{\infty} a_j \rho^j \tag{8.121}$$

in which s is an integer to be determined. It must be larger than zero in order to insure the finiteness of the wavefunction when $r = 0$. Substituting formula (8.121) into Eq. (8.120) and rearranging, we obtain

$$\sum_{j=0}^{\infty}[(s+j)(s+j-1)+2(s+j)-l(l+1)]a_j\rho^{s+j-2} + \sum_{j=0}^{\infty}[(\gamma-1)-(s+j)]a_j\rho^{s+j-1}=0 \tag{8.122}$$

In order to make the equation tenable, the coefficients of the power terms of ρ at all levels should be zero. Taking the zero-degree term ($j=0$) and letting the coefficient of the lowest power term of ρ be zero, we get

$$s(s-1)+2s-l(l+1)=0$$

or

$$s(s+1)=l(l+1)$$

This equation has two solutions: $s=l, s=-(l+1)$. Since $l>0$ and s must be larger than zero, we can only take the solution

$$s=l$$

Thus

$$R(\rho)=e^{-\rho/2}\rho^l\sum_{j=0}^{\infty}a_j\rho^j \tag{8.123}$$

This formula will tend to infinity when $\rho\to\infty$, which does not coincide with the standard condition of a wavefunction. Hence the series can only contain limited terms; for example, let γ be the integer n:

$$(\gamma-1)-(s+j)=0 \qquad \text{or} \qquad j+l+1-n=0 \tag{8.124}$$

As a result we obtain [according to formula (8.115)]

$$n=\frac{\mu Ze^2}{k\hbar^2} \tag{8.125}$$

Hence

$$E_n=-\frac{(k\hbar)^2}{2\mu}=-\left(\frac{Z}{n}\right)^2E_\infty \tag{8.126}$$

which is exactly Eq. (8.114).

We rewrite Eq. (8.123) again:

$$R(\rho)=e^{-\rho/2}\rho^lG(\rho) \tag{8.127}$$

That is, let

$$F(\rho)=\rho^lG(\rho)$$

Equation (8.120) now becomes

$$\rho\frac{d^2G}{d\rho^2}+[2(l+1)-\rho]\frac{dG}{d\rho}+[n-(l+1)]G=0 \tag{8.128}$$

which is well known in mathematical physics as the associated Laguerre equation, and its solution is the associated Laguerre function:

$$G(\rho) = L_{n+l}^{2l+1}(\rho) \tag{8.129}$$

The normalized radial wavefunction is

$$R_{n,l}(r) = -\left[\left(\frac{2Z}{na_1}\right)^3 \frac{[n-(l+1)]!}{2n[(n+l)!]^3}\right]^{1/2} \exp\left(-\frac{Zr}{na_1}\right)\left(\frac{2Zr}{na_1}\right)^l L_{n+l}^{2l+1}\left(\frac{2Zr}{na_1}\right) \tag{8.130}$$

It is known from formula (8.124) that

$$n \geq l+1 \tag{8.131}$$

The following are expressions for some radial wavefunctions:

$$R_{1,0}(r) = 2\left(\frac{Z}{a_1}\right)^{3/2} \exp\left(-\frac{Zr}{a_1}\right)$$

$$R_{2,0}(r) = 2\left(\frac{Z}{2a_1}\right)^{3/2}\left(1-\frac{Zr}{2a_1}\right)\exp\left(-\frac{Zr}{2a_1}\right)$$

$$R_{2,1}(r) = \frac{1}{\sqrt{3}}\left(\frac{Z}{2a_1}\right)^{3/2}\left(\frac{Zr}{a_1}\right)\exp\left(-\frac{Zr}{2a_1}\right)$$

$$R_{3,0}(r) = 2\left(\frac{Z}{3a_1}\right)^{3/2}\left[1-\frac{2Zr}{3a_1}+\frac{2}{27}\left(\frac{Zr}{a_1}\right)^2\right]\exp\left(-\frac{Zr}{3a_1}\right)$$

$$R_{3,1}(r) = \frac{4\sqrt{2}}{3}\left(\frac{Z}{3a_1}\right)^{3/2}\left[\frac{Zr}{a_1}-\frac{1}{6}\left(\frac{Zr}{a_1}\right)^2\right]\exp\left(-\frac{Zr}{3a_1}\right)$$

$$R_{3,2}(r) = \frac{2\sqrt{2}}{27\sqrt{5}}\left(\frac{Z}{3a_1}\right)^{3/2}\left(\frac{Zr}{a_1}\right)^2\exp\left(-\frac{Zr}{3a_1}\right)$$

The radial probability densities calculated using these wavefunctions are shown in Fig. 8.11. The location of the vertical solid lines in the figure corresponds to $\langle r\rangle$, namely, the average value of the radial coordinates of the electrons:

$$\begin{aligned}\langle r\rangle &= \int_0^\infty\int_0^\pi\int_0^{2\pi} \psi_{n,l,m}^* r \psi_{n,l,m} r^2 \sin\theta \, dr\, d\theta\, d\phi \\ &= \frac{n^2 a_1}{Z}\left[1+\frac{1}{2}\left(1-\frac{l(l+1)}{n^2}\right)\right]\end{aligned} \tag{8.132}$$

The reader is expected to prove formula (8.132) according to the definition

$$\langle r^k\rangle = \int_0^\infty (R_{n,l})^2 r^k \cdot r^2\, dr$$

and to use the expressions for R to prove the following expressions:

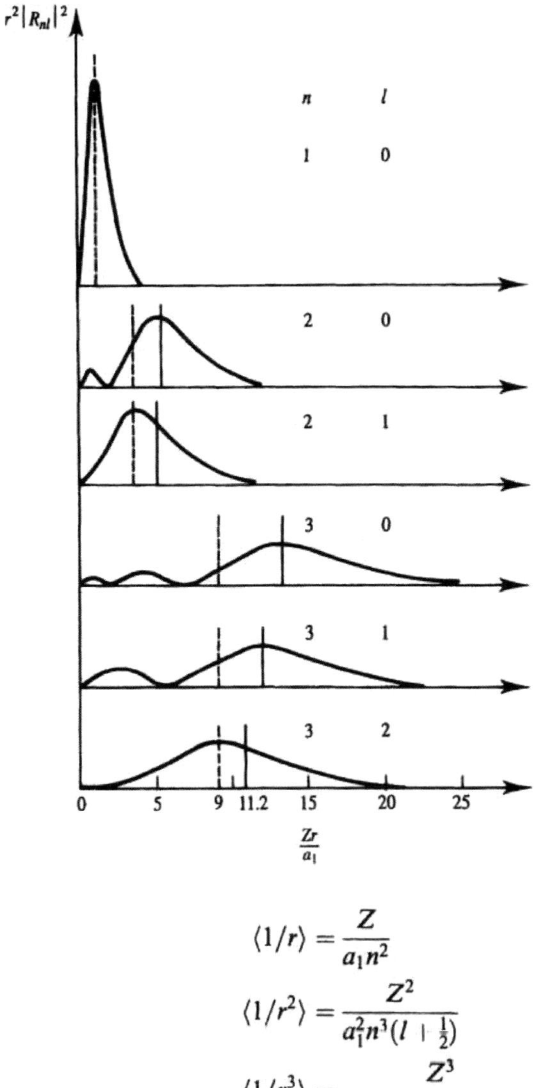

FIGURE 8.11
The electron radial probability densities in a hydrogen atom for $n = 1, 2$, and 3. The vertical solid lines represent the position of $\langle r \rangle$ and the dashed lines represent the position of a_n, where a_1 is the first Bohr radius.

$$\langle 1/r \rangle = \frac{Z}{a_1 n^2}$$

$$\langle 1/r^2 \rangle = \frac{Z^2}{a_1^2 n^3 (l + \frac{1}{2})}$$

$$\langle 1/r^3 \rangle = \frac{Z^3}{a_1^3 n^3 l(l + \frac{1}{2})(l + 1)}$$

$$\langle r^2 \rangle = (\tfrac{1}{2} Z^2) a_1^2 n^2 [5n^2 + 1 - 3l(l + 1)]$$

It is seen from the figure that the difference between $\langle r \rangle$ and the Bohr radius

$$a_n = \frac{n^2 a_1}{Z} \tag{8.133}$$

(indicated by a vertical dashed line) decreases as l increases. Even though $\langle r \rangle$ does not coincide with a_n, for various cases in Fig. 8.11, when $n = l + 1$, the maximum of the probability function (the place where the electrons most probably appear) is at a_n.

In the Bohr model, only one quantum number and two quantization conditions are involved:

$$E_n = -\left(\frac{Z}{n}\right)^2 E_\infty \quad \text{and} \quad L_z = n\hbar$$

In a wave mechanical calculation for the hydrogen atom, there are three quantum numbers and three eigenvalue equations:

$$\begin{aligned} \hat{H}\psi_{n,l,m} &= -\left(\frac{Z}{n}\right)^2 E_\infty \psi_{n,l,m} \\ \hat{L}_z\psi_{n,l,m} &= m\hbar\psi_{n,l,m} \\ \hat{L}^2\psi_{n,l,m} &= l(l+1)\hbar^2\psi_{n,l,m} \end{aligned} \tag{8.134}$$

It should be pointed out that although the wavefunction is dependent on three quantum numbers, the energy eigenvalue depends only on n (in first order, but a more exact calculation introduces weak dependencies on other quantum numbers). This indicates that the eigenfunction is degenerate; that is, many different eigenfunctions correspond to one eigenvalue. Because $n \geq l+1$, for $n = 1, 2, 3, \ldots, l = 0, 1, 2, \ldots$, for each n there will be n possible l values; and because $l \geq |m|$, for each l there will be $(2l+1)$ possible m values. Thus the degeneracy of the total wavefunction is

$$\sum_{l=0}^{n-1}(2l+1) = n^2$$

8.3.3 Comments

The wave mechanics based on the Schrödinger equation together with the matrix mechanics developed by Heisenberg somewhat earlier (1925) formed two expressions of quantum mechanics. Schrödinger proved that the two formalisms are completely equivalent. Because the Schrödinger equation is more easily understood, wave mechanics is introduced first.

The appearance of matrix mechanics and wave mechanics marked the birth of the new quantum mechanics, which is different from the old quantum hypothesis. Together with Born's statistical explanation of the wavefunction and Heisenberg's uncertainty relation, they form the system of nonrelativistic quantum mechanics. They explained not only the hydrogen atom but also the helium atom and the properties of other atoms and molecules for the first time.

However, while Schrödinger quantum mechanics was compatible with the concept of the spin of the electron, it did not predict it because it was a nonrelativistic theory. In 1928 Dirac developed a relativistic quantum mechanics and showed that an electron must have an intrinsic spin of $\frac{1}{2}$ and an intrinsic magnetic moment with $g = 2$. Although relativistic quantum mechanics has obtained amazing achievements to date, it still cannot be considered as perfect either.

APPENDIX 8A HARMONIC OSCILLATOR SOLUTION

The harmonic potential (Fig. 8A.1) is a very important type. For example, this potential is a good approximation for other potentials such as the nuclear potential. The Schrödinger equation in one dimension is

$$-\frac{\hbar^2}{2m}\frac{d^2\psi}{dx^2}+\frac{1}{2}kx^2\psi = E\psi \tag{8A.1}$$

We want the time-independent, stationary states. The first thing is to simplify constants [multiply both sides by $2(m/k\hbar^2)^{1/2}$]. Rewriting

$$\frac{-1}{\sqrt{km/\hbar^2}}\frac{d^2}{dx^2}\psi+\sqrt{\frac{km}{\hbar^2}}x^2\psi = \frac{2E}{k}\sqrt{\frac{km}{\hbar^2}}\psi \tag{8A.2}$$

Let us introduce the following dimensionless variables:

$$x^2\sqrt{\frac{km}{\hbar^2}} = y^2 \qquad \lambda = \frac{2E}{\hbar}\sqrt{\frac{m}{k}} \tag{8A.3}$$

The classical frequency is

$$\omega_c = \sqrt{\frac{k}{m}} \qquad \text{so} \qquad E = \tfrac{1}{2}\lambda\hbar\omega_c \tag{8A.4}$$

Equation 8A.2 becomes

$$\frac{d^2\psi(y)}{dy^2}+(\lambda - y^2)\psi(y) = 0 \tag{8A.5}$$

Now there is only one parameter, and the constants are removed. If a differential equation simplifies asymptotically, then it is interesting and useful to look at the asymptotic solution,

$$y \to \infty \qquad \frac{d^2\psi}{dy^2} \approx y^2\psi \tag{8A.6}$$

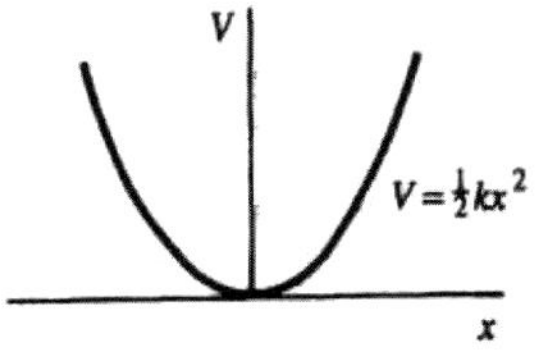

FIGURE 8A.1

An asymptotic solution is

$$\psi \approx e^{-\frac{1}{2}y^2} \tag{8A.7}$$

This equation can be multiplied by a function of y, but this function must be small in the asymptotic solution. Now check this solution (8A.7) against (8A.6) as $y \to \infty$ and keep the leading term in y:

$$\frac{d\psi}{dy} = -ye^{-\frac{1}{2}y^2} \tag{8A.8}$$

$$\frac{d^2\psi}{dy^2} = +y^2e^{-\frac{1}{2}y^2} \tag{8A.9}$$

So it checks. A more complex solution would be

$$\psi(y) = e^{-\frac{1}{2}y^2} H(y) \tag{8A.10}$$

$$\frac{d\psi}{dy} = (H' - yH)e^{-\frac{1}{2}y^2} \tag{8A.11}$$

$$\frac{d^2\psi}{dy^2} = (H'' - 2yH' - H + y^2H)e^{-\frac{1}{2}y^2} = -(\lambda - y^2)\psi(y) \tag{8A.12}$$

Here we simplify by writing

$$\frac{dH}{dy} = H' \quad \text{and} \quad \frac{d^2H}{dy^2} = H'' \tag{8A.13}$$

For the above to be true,

$$H'' - 2yH' + (\lambda - 1)H = 0 \tag{8A.14}$$

This equation can be expanded around y, because of its simple dependence on y:

$$H(y) = \sum_{n=0}^{\infty} a_n y^{n+\rho} \quad \text{where} \quad a_0 \neq 0 \tag{8A.15}$$

Now, put this series back into Eq. 8A.14:

$$\sum_{n=-2}^{\infty} y^{n+\rho}[a_{n+2}(n+\rho+2)(n+\rho+1) - 2(n+\rho)a_n + (\lambda - 1)a_n] = 0$$
$$a_{-1} = a_{-2} = 0 \tag{8A.16}$$

The coefficients in brackets must equal zero, so we have a recursion relation

$$a_{n+2} = a_n \frac{2(n+\rho) - \lambda + 1}{(n+\rho+1)(n+\rho+2)} \tag{8A.17}$$

or

$$a_{n+2}(n+\rho+1)(n+\rho+2) = a_n[2(n+\rho) - \lambda + 1]$$

$$n = -2: \qquad a_0\rho(\rho - 1) = 0 \tag{8A.18}$$

$$\rho = 0, 1 \qquad \text{since} \qquad a_0 \neq 0 \tag{8A.19}$$

$n = -1:$

$$a_1 \rho(\rho + 1) = 0 \tag{8A.20}$$

$$a_1 = 0 \qquad \text{this equation is meaningless}$$

$$a_1 \neq 0 \qquad \text{so, } \rho \text{ must equal } 0 \tag{8A.21}$$

Clearly, we get two solutions. If $a_1 = 0$, all odd terms are zero; and we get a solution with all even terms by starting with a_0:

$$\rho = 0 \text{ for even series} \qquad \rho = 1 \text{ for odd series}$$

For $\rho = 0$,

$$\frac{a_{n+2}}{a_n} = \frac{2n - \lambda + 1}{(n+2)(n+1)} = \frac{2}{n} \frac{1 - \dfrac{\lambda - 1}{2n}}{\left(1 + \dfrac{2}{n}\right)\left(1 + \dfrac{1}{n}\right)} > \frac{2c}{n} \tag{8A.22}$$

where c can be as close to 1 as we choose by choosing n_0 and $n > n_0$. Consider the series:

$$e^{y^2} = \sum \frac{y^{2k}}{k!} = \sum b_n y^{2k} \tag{8A.23}$$

The coefficient b_n of y^n is $1/k!$ where $2k = n$. So

$$b_n = \frac{1}{\left(\frac{n}{2}\right)!}$$

Then

$$\frac{b_{n+2}}{b_n} = \frac{(n/2)!}{[(n+2)/2]!} = \frac{1}{(n+2)/2} \sim \frac{2}{n} \qquad \text{for large } n \tag{8A.24}$$

Our harmonic oscillator (HO) series has the same asymptotic recursion relation as this, so

$$\sum a_n y^n > \sum e^{y^2} \tag{8A.25}$$

Therefore, as it stands, our HO series is greater than a series that is known to diverge. Hence, our HO series must break off so that the wavefunction will not be infinite at infinity.

When the coefficient of $a_{n+2} = 0$, the series breaks off. From Eq. (8A.17) (remember here $\rho = 0$) the criterion for $a_{n+2} = 0$ is for some value of the integer n, for say the lth term in the series:

$$\lambda = 2n + 1 \tag{8A.26}$$

Thus from

$$\lambda = 2n + 1 = \frac{2E}{\hbar\omega_c} \tag{8A.27}$$

we find the eigenvalues of the harmonic oscillator to be

$$E_n = \hbar\omega_c(n + \tfrac{1}{2}) \tag{8A.28}$$

This is almost the expression Planck got, $E_n = n\hbar\omega_c$. The $\frac{1}{2}$ was missing in the old theory. This is characteristic of the old theory; it always gave *about* the right answer.

This additional $\frac{1}{2}$ factor was verified even before wave mechanics derived it: it is called the "zero point energy."

As k (the force constant) goes to zero, the potential flattens out, and the levels move closer together. So, $k = 0$ gives a continuum of allowed energy solutions. The levels are not quite the same as discrete levels.

We still have not decided what ρ to take. We can add or not add the odd terms: for $\rho = 0$, the odd terms drop out. One can thus get an odd (sum of odd terms only) and an even (sum of even terms) solution. This is characteristic of many differential equations and Hamiltonians:

$$\mathcal{H} = -\frac{\hbar^2}{2m}\frac{d^2}{dx^2} + \frac{1}{2}kx^2 \tag{8A.29}$$

The first term is always even. The operator is always even, but the potential may or may not be. Here the potential is even also; so if we find a solution, $\psi(x)$, then $\psi(-x)$ is a solution also. Likewise,

$$\psi(x) + \psi(-x) \qquad \text{and} \qquad \psi(x) - \psi(-x) \tag{8A.30}$$

are both solutions. The first is an even solution, and the second an odd solution. In Schrödinger equations where the potential is an even function, there always exist both an odd and an even solution. The behavior of ψ with the exchange of x and $-x$ is called *parity*:

$$\text{Parity is even } (+1) \text{ if } \psi(x) = \psi(-x) \tag{8A.31}$$

$$\text{Parity is odd } (-1) \text{ if } \psi(x) = -\psi(-x) \tag{8A.32}$$

One speaks of the potential as being of either even or odd parity. So the first solution in Eq. (8A.30) has even parity and the second solution has odd parity. H_n is the solution with highest order term y^n. H_n is a polynomial in y, and the parity of H gives the parity of ψ. The n indicates the parity of the solution. These functions are called Hermite polynomials. They obey the recursion relation

$$H_{n+1} = \left(2y - \frac{d}{dy}\right)H_n(y) \tag{8A.33}$$

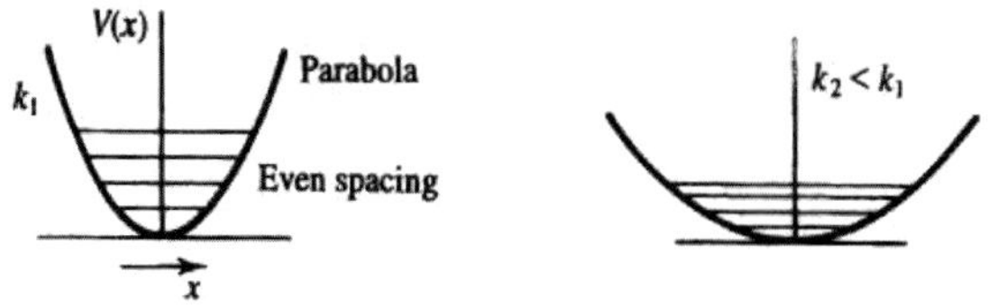

FIGURE 8A.2
Spectrum of HO oscillation energies—note that the levels are evenly spread.

After normalizing the wavefunction through $\int \psi^2 \, d\tau = 1$, one finds the wavefunction for the harmonic oscillator as

$$\psi_n(x) = \sqrt[4]{\frac{m\omega_c}{\pi\hbar}} \frac{1}{\sqrt{2^n n!}} \exp\left(-\frac{1}{2}\frac{\omega_c m x^2}{\hbar}\right) H_n\left(\sqrt{\frac{m\omega_c}{\hbar}}x\right) \tag{8A.34}$$

Note here that the equation is written in terms of x. If one writes $\psi_n(y)$ as in Eq. (8.46), the normalization constant is different, integrating over y there and x here to produce the normalization constant.

PROBLEMS

8.1. Given the wavefunction of a particle

$$\psi = N \exp\left\{-\frac{|x|}{2a} - \frac{|y|}{2b} - \frac{|z|}{2c}\right\}$$

calculate:

(*a*) The normalization constant N

(*b*) The probability of the x coordinate of the particle being in the range of 0 to a

(*c*) The probabilities of the y coordinate and z coordinates being in the range of $-b \to +b$ and $-c \to +c$, respectively

8.2. A system is composed of a proton and an electron. Assuming that its normalized space wavefunction is $\psi(x_1, y_1, z_1; x_2, y_2, z_2)$, in which the subscripts 1 and 2 represent the proton and electron, respectively, express:

(*a*) The probability density of finding the proton at the point (1,0,0) while the electron is at the point (0, 1,1)

(*b*) The probability density of finding the electron at (0, 0, 0) while the proton can be anywhere

(*c*) The probability of finding that both the proton and the electron are in a sphere of radius 1 with its center at the origin of the coordinates.

8.3. For a particle moving in a one-dimensional infinite potential well of width a, calculate the averages $\bar{x}$ and $(x - \bar{x})^2$ in an arbitrary eigenstate ψ_n and prove that these results coincide with the classical results when $n \to \infty$.

8.4. Calculate the positions where the radial electron charge densities of the 1s state and the 2p state of a hydrogen atom are maximum, respectively.

8.5. Assume that a hydrogen atom is in the state

$$\psi(r, \theta, \phi) = \frac{1}{\sqrt{\pi a_1^3}} e^{-r/a_1}$$

in which a_1 is the first Bohr radius. Calculate the average value of the potential

$$U(r) = -\frac{e^2}{r}$$

8.6. Prove the following commutation relations:

$$[x, p_x] = i\hbar$$
$$[x, p_y] = 0$$
$$[x, L_x] = 0$$
$$[x, L_y] = i\hbar z$$
$$[p_x, L_x] = 0$$
$$[p_x, L_y] = i\hbar p_z$$

8.7. Assume that a particle of mass m is moving in a square well potential which can be expressed as

$$V(x) = \begin{cases} \infty, & x < 0 \\ 0, & 0 \le x \le a \\ V_0, & x > a \end{cases}$$

For the bound state condition where $E < V$:

(*a*) Calculate the expression for the energy levels of the particle

(*b*) Prove that the condition for at least one bound state to exist in the well is that the well depth V_0 and the well width a satisfy

$$V_0 a^2 \ge \frac{\hbar^2}{32m}$$

8.8. In the macroworld, what is the probability of a car appearing in the parlor (Fig. 8.4)?

8.9. Imagine a microparticle of mass m undergoing one-dimensional motion in a rigid box of width d. Prove that the momentum and energy of this particle are quantized by finding their allowed values.

8.10. A particle is in its ground state in an infinite square well of width d. Calculate the probability of finding the particle within $\Delta x = 0.01d$ at:
(a) $x = \frac{1}{2}d$ (b) $x = \frac{1}{4}d$ (c) $x = d$

8.11. Calculate the kinetic energy E_n of a particle moving in a one-dimensional harmonic potential $V(x) = \frac{1}{2}kx^2$.

8.12. If a 1 eV electron hits a rectangular potential barrier which has a height of 2 eV, and if the transmission probability is 10^{-3}, what is the width of the barrier?

8.13. Schrödinger's equation can be written in the system of spherical polar coordinates r, θ, ϕ, where the ϕ part is

$$\frac{d^2\Phi(\phi)}{d\phi^2} = -m^2\Phi(\phi)$$

Prove that its solution is

$$\Phi(\phi) = \frac{1}{\sqrt{2\pi}} e^{im\phi}$$

and show that m must be an integer, if the solution is physically meaningful.

8.14. For a one-dimensional harmonic oscillator, the energy operator H_0 can be written as (cf. Eq. 8.44)

$$H_0 = \frac{-\hbar^2}{2m}\frac{d^2}{dx^2} + \frac{1}{2}m\omega^2 x^2$$

If the moving particle has charge e, and a constant electrical field E is applied along the moving direction, what is the energy operator H_0 in such a case? What is the eigenvalue? Write the eigenfunction as well as compare it with Eq. 8.46.

8.15. Find the reflection coefficient at the barrier wall ($x = 0$) for a particle (with energy $E > 0$) moving from negative to positive values of x toward the step barrier where:

$$V(x) = \begin{cases} -V_0, & x < 0 \\ 0, & x > 0 \end{cases}$$

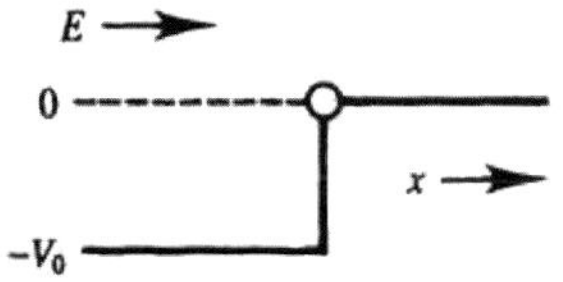

8.16. For a microscopic particle, its Hamiltonian operator can be written as

$$\hat{H} = -\frac{\hbar^2}{2m}\nabla^2$$

At $t = 0$, the particle is in the state $\psi(\boldsymbol{r}, 0) = A\sin^2(\boldsymbol{k}\cdot\boldsymbol{r})$. Find its state at any time $\psi(\boldsymbol{r}, t)$.

8.17. Calculate $[\hat{x}, \hat{p}_x^2]$ and $[\hat{x}, \hat{p}_x^3]$ by using the operator formula

$$[\hat{A}, \hat{B}\hat{C}] = [\hat{A}, \hat{B}]\hat{C} + \hat{B}[\hat{A}, \hat{C}]$$

8.18. Verify the commutation relations, Eqs. (8.67) and (8.68); that is, $[\hat{L}_x, \hat{L}_y] = i\hbar L_z$, and so forth.

8.19. Verify the anticommutation relation

$$\boldsymbol{l} \times \boldsymbol{r} + \boldsymbol{r} \times \boldsymbol{l} = 2i\hbar\boldsymbol{r}$$

8.20. If a particle of mass m is in the state $\psi(x) = A[\sin^2 kx + \frac{1}{2}\cos kx]$, calculate its average momentum and kinetic energy.

8.21. The definition of the fluctuation Δx is

$$\Delta x = \sqrt{\overline{x^2} - \bar{x}^2}$$

Find Δx for the hydrogen atom ground state.

CHAPTER 9

BASIC CONCEPTS OF NUCLEAR PHYSICS

> Nuclear physics is on the threshold of new and exciting developments stemming both from new understandings of the nuclear many-body system and new higher energy probes—electrons, kaons and heavy ions.
>
> D. Allan Bromley (1988)

9.1 THE NUCLEUS OF THE ATOM

9.1.1 Introduction

Since 1911 when Rutherford first advanced his nuclear model of atoms, the atom has been divided into two parts: a positively charged nucleus and its surrounding electrons. Some take Rutherford's discovery as the start of nuclear physics. The properties of the electrons are the subject of atomic physics, and the properties of nuclei are the subject of nuclear physics. As described in Chapter 13, there are interactions between these two regimes that impact on both areas. On the other hand, one can make the claim that nuclear physics began with the discovery of radioactivity in 1896 by Becquerel. The α-, β- and γ-rays were indeed the first ambassadors to us from the unknown world of the nucleus. Without these ambassadors from naturally occurring radioactive substances in nature, the nucleus may have remained unknown to us. Indeed, as we shall see, had the α decay energies of the long-lived uranium and thorium isotopes been even a few hundred keV higher than they are, the half-lives for their decays would have been so short that they would not have been present in the Earth's crust for Becquerel to discover radioactivity, for Rutherford and his colleagues to use their energetic α-particles to discover the atomic nucleus, and for Hahn and Strassman to discover induced fission of uranium, which led to nuclear energy. This is but one example of how even small, seemingly inconsequential changes in certain physical properties or certain physical constants would have radically changed our world.

In 1895 scientists thought that all the forces of nature were known—the gravitational and electromagnetic forces. When Rutherford discovered that all the positive charge and nearly all the mass of an atom are contained in the extremely small nucleus of the atom, it was recognized immediately that there had to be a new force in nature and that it was both the strongest and the shortest range force in nature in comparison to the known forces. Strongest because it overcomes the Coulomb repulsion of the positive charges in the nucleus, and extremely short in its range (distance over which it can act) because it was not observed at distances greater than about 10^{-14} m. The discovery of this new force was a major milestone in science. In another twenty years the beta radioactive decay of nuclei was to provide the crucial evidence for still another force in nature, the weak nuclear force.

Nuclear physics, in its quest to understand the basic constituents of matter, spawned its companion discipline of elementary particle (high-energy) physics and developed experimental techniques and theoretical methods that play important roles in atomic, molecular, and condensed matter physics as well as in astrophysics, cosmology, and many other basic and applied fields. Today, nuclear physics extends from studies of the basic properties of nuclei including their masses, sizes, shapes, and the individual and collective motions of nucleons inside the nucleus to their relation to the underlying strong nuclear force, to the connection between the meson–nucleon and the quark–gluon descriptions of strongly interacting systems, and to tests of the electroweak interaction and the so-called standard model of particle physics, as illustrated in Fig. 9.1. The nucleus itself is a unique laboratory. As many have emphasized, "One of the central motivations for probing all aspects of nuclear structure is that the atomic nucleus is the major, if not the only, testing ground in an important intermediate realm of quantum-mechanical many-body problems, namely systems incorporating an intermediate number of particles: too few to be treated by statistical methods but too many to be treated easily one by one."[1]

The study of radioactive decays has been extended very far from near stable ones in nature out to exotic new nuclei that exhibit new motions, structures, and decay modes. Reactions between nuclei can probe the nucleus under new extremes of temperature up to ten million times hotter than the surface of the Sun for the whole nucleus and much hotter in spots within the nucleus. While nuclei are normally incompressible, ultrarelativistic heavy-ion collisions will probe nuclei at up to five to ten times normal nuclear densities—the densities reached otherwise only in neutron stars and supernovas and at temperatures even higher than in stars. Under these conditions the nucleons may undergo a phase change to a new

[1] J. H. Hamilton and J. Maruhn, *Scientific American* **254** (July 1989) 80.

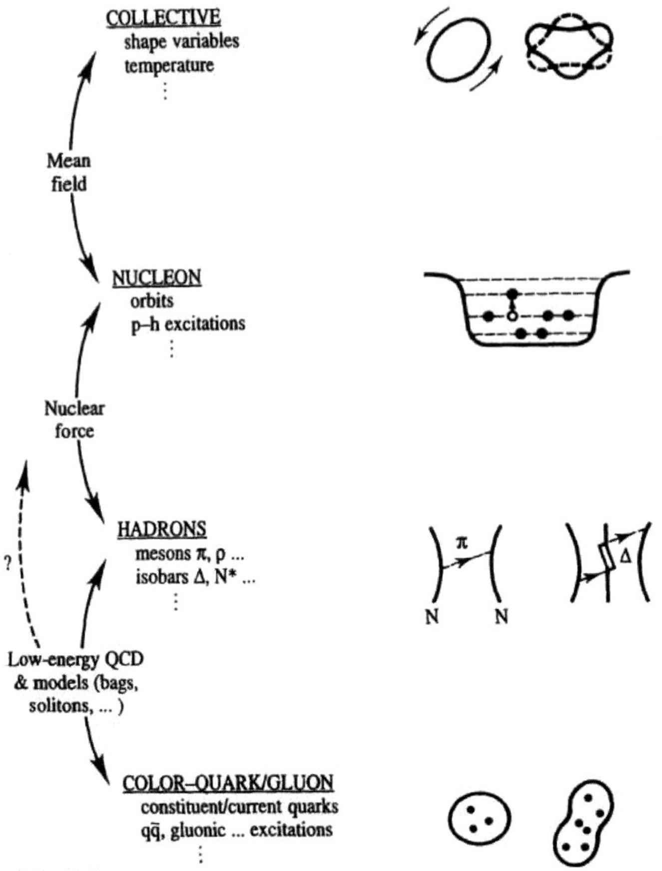

FIGURE 9.1
Illustration of nuclear properties and the models which are used to describe them, with increasing resolution from the top to the bottom. At the most basic level the nucleus consists of quarks and gluons.

state of matter, a quark–gluon plasma such as occurred at the very beginning of time following the big bang that formed our universe. Such research brings nuclear and particle physics together again. Nuclear phenomena today range from the shortest distances we can test ($< 10^{-16}$ m) to those of the most spectacular astrophysical events 10^{21} kilometers away and in times from $< 10^{-21}$ s to greater than 10^{22} years, or 10^{32} years including the proton decay. Finally, nuclear physics has made important contributions to many areas of research outside the natural sciences. The applications of nuclear techniques have transformed our world in many ways, from life-saving developments in nuclear medicine to important new long-range sources of energy to replace our rapidly vanishing fossil supplies, to giving us enormous potential for self-destruction with nuclear weapons. Nuclear physics has given new challenges to the human intellect to view our world, from our fundamental understanding of the basic forces which govern the world to fundamental questions of peace or self-destruction unparalleled in human history.

In this chapter, we shall introduce the basic concepts of nuclear physics, including the nucleus itself and the properties of nuclear ground states. Radioactive decay laws are presented in Chapter 10. The nuclear force and nuclear models are considered in Chapter 11 and nuclear reactions and nuclear energy in Chapter 12. Many of the important milestones in nuclear physics are noted in the Introduction.

9.1.2 Atomic Center: The Nucleus

The radius of a typical nucleus ($\sim 10^{-14}$–10^{-15} m) is a ten-thousandth to a hundred-thousandth (10^{-4}–10^{-5}) of that of an atom, but its mass is greater than 99% of the mass of the whole atom. The density of nuclear matter is beyond our imagination, about $10^{17}\,\mathrm{kg/m^3}$ or $10^{14}\,\mathrm{g/cm^3}$. This is 100 million million times the density of water, $1\,\mathrm{g/cm^3}$. In addition, it contains all the positive charge of the atom. Basically, the physical, chemical, and light radiation properties of the elements are related to their electrons, which are extremely far away from the nuclei of the atoms. On the average, the electrons are in orbits with radii the order of 10^{-10} m. In everyday life, we experience many phenomena related to atomic physics, but nuclear phenomena are very seldom seen directly. However, since the detonation of nuclear ("atomic") bombs in 1945, the importance of the nucleus of the atom has been well known.

9.1.3 Constituents of the Nucleus

Before the neutron was discovered, only the electron and proton were known as "basic" particles. With only these two particles, it became more and more difficult as time went on to explain the constitution of the nucleus. For example, the mass of the ^{4}He nucleus is four times that of the proton, and its charge is +2e. If it consists of protons and electrons, it must have 4 protons and 2 electrons. But the dimension of the nucleus is about 5 fm. If an electron is inside of the nucleus, its de Broglie wavelength must be less than $2d \simeq 10\,\mathrm{fm}$. Thus, its momentum is

$$p = \frac{h}{\lambda} = \frac{hc}{\lambda c} \geq \frac{1240\,\mathrm{fm \cdot MeV}}{10\,\mathrm{fm}\,c} = 124\,\frac{\mathrm{MeV}}{c}$$

Assuming that it is nonrelativistic, the electron velocity

$$\mathrm{v} = \frac{p}{m} = \frac{pc^2}{mc^2} = \frac{124c\,\mathrm{MeV}}{0.511\,\mathrm{MeV}} \simeq 240c$$

which is 240 times the velocity of light. So the assumption is wrong. Then use the relativistic equation:

$$E^2 = (pc)^2 + (mc^2)^2$$

Since $pc = 124\,\text{MeV} \gg mc^2 = 0.511\,\text{MeV}$, $E \simeq pc = 124\,\text{MeV}$. But there is no experimental evidence that an electron inside of a nucleus can have such a high energy! For example, the maximum electron energies in beta decay go from only 18 keV to a few MeV. There is no force that can hold 124 MeV electrons inside a nucleus.

If a nucleus consisted of protons and electrons, we also could not explain the nuclear spin. For example, the nuclear mass of ^{14}N is about 14 times that of the proton and its charge is $7e$. According to the hypothesis of protons–electrons only, there must be 14 protons and 7 electrons each with spin $\frac{1}{2}$ inside of a ^{14}N nucleus. If the total number of particles in its nucleus is 21, its spin must be a half integer. But the ^{14}N spin is one unit of angular momentum from experiment. Similar arguments hold for other nuclei, such as ^{2}H. The question of just what particles were in the nucleus had become a major problem by the time Chadwick discovered the neutron. After the discovery of the neutron, Heisenberg suggested that the nucleus consists of protons and neutrons. The history of the discovery of the neutron is described in Chapter 12.

The masses of the neutron and proton are almost the same:

$$m_n = 1.008\,665\,\text{u} \qquad m_p = 1.007\,277\,\text{u}$$

where u is the atomic mass unit and both have spin $\frac{1}{2}$. The atomic mass unit is based on the mass of the isotope ^{12}C being 12.00 exactly. Heisenberg called the constituents of the nucleus (neutrons and protons) nucleons. A neutron and proton are different states of the nucleon.

We use $^{A}_{Z}X_N$ to represent a given atomic nucleus. Here N and Z are the number of neutrons and protons in a nucleus, respectively, A is the nucleon number, $A = N + Z$, and X is the symbol for the element which is determined by Z, for example, $^{4}_{2}He_2$, $^{14}_{7}N_7$, and $^{16}_{8}O_8$. Nuclei with the same Z and different N and A are called isotopes of the element. In nature, three stable isotopes of oxygen occur, $^{16}_{8}O_8$, $^{17}_{8}O_9$, and $^{18}_{8}O_{10}$. As long as X is the same, we can write simply ^{A}X to express a particular isotope of an element because the general physical and chemical properties are the same, for example, ^{235}U and ^{238}U. However, their nuclear properties are quite different. Note A also indicates the approximate mass of an atom in atomic mass units, and Z is called the atomic number of an element.

9.1.4 Chart of the Nuclides

The periodic table of the elements orders them by their chemical properties. For nuclei there are also regularities in nuclear shapes and radioactive decay properties and magic closed shell nuclei analogous to the noble gases. Some of these regularities can be illustrated in a chart of the nuclides (Fig. 9.2), but now the regularities are a function of the atomic number Z and the neutron number N. A chart of the nuclides is a two-dimensional plot of N and Z. Figure 9.3 is an enlarged picture of a small part of the chart of the nuclides for better illustration. Each isotope is represented by a square. The abundances of the stable isotopes are given and the decay modes and half-lives of the radioactive ones are given.

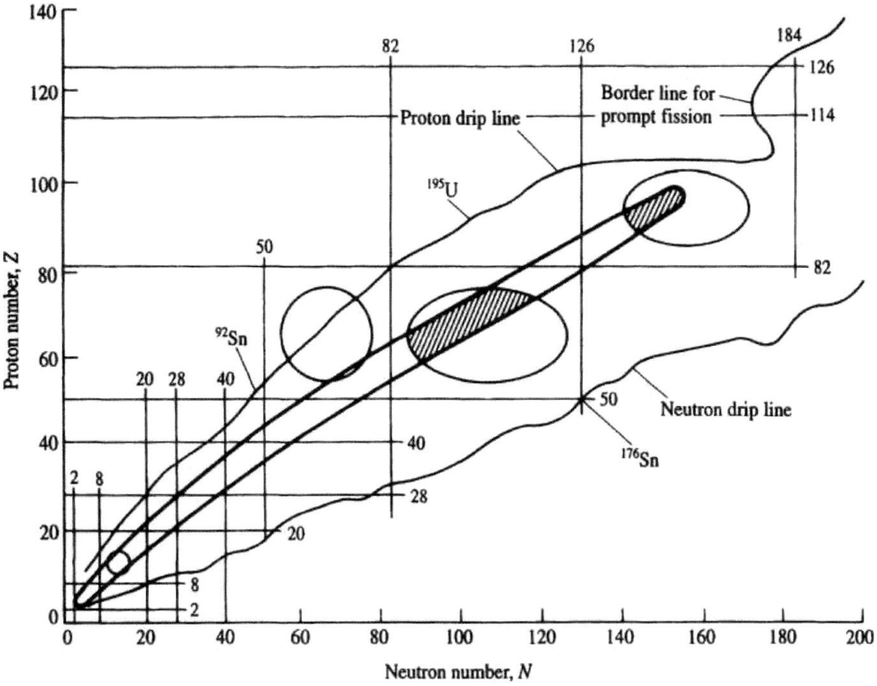

FIGURE 9.2
Chart of the nuclides. The spherical shell model magic (closed shell) numbers, which give extra stability (binding) to the nucleus (see Chapter 11) are indicated by vertical and horizontal lines. The long narrow enclosed region in the middle of the figure encompasses the stable and radioactive nuclei which are found in nature. Nuclei in the regions enclosed by the circles and ellipses and well removed from the spherical magic numbers are deformed in their ground states. The proton drip line and the neutron drip line are indicated. This figure is based on the liquid drop model which leads to the proton drip line being too low in Z above $N = 130$.

There are more than 325 nuclides found in nature. Of these, 263 are stable and the remaining are radioactive. Some 1900 so-called "artificial" radioactive nuclides have been produced in the laboratory. The term "artificial" is misleading since these nuclides are as real as the stable ones; the term signifies nuclides which are not found in nature but can be made in the laboratory. A better terminology is to call these induced or artificially produced radioactivities. The total number of nuclides which have been identified is of the order of 3000 out of potentially 6000-10,000 that theoretically can exist.

If we examine the chart of the nuclides, we find that almost all the stable nuclides fall along a smooth curve or near it. This region is called the valley of stability, or more specifically the valley of beta stability where decay does not occur. For light nuclides, this curve overlaps the line $N = Z$. But as N and Z increase, the stability line is inclined to the direction of $N > Z$. Above the stability line, one has a neutron-deficient region where there are too few neutrons in these nuclei for them to be stable against radioactive decay. Below that line there is a

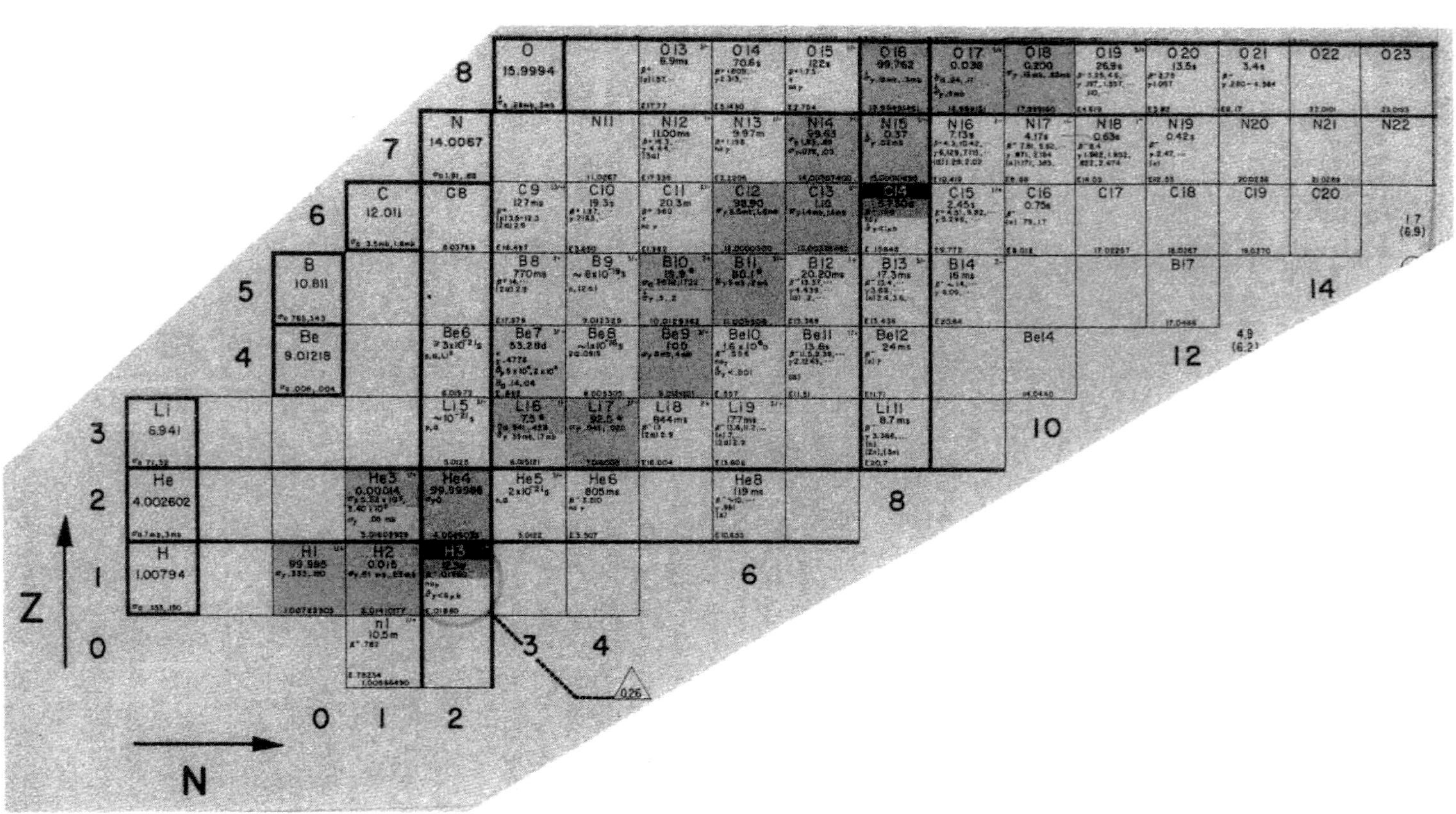

FIGURE 9.3
Partial chart of the nuclides prepared by Knolls Atomic Power Laboratory, 13th Ed., (1983) to better illustrate the information presented in such plots.

neutron-rich region where there are too few protons for these nuclei to be stable against radioactive decay. On either side of the line there are limits called the proton and neutron drip lines, where the nuclei can no longer bind another proton or neutron, respectively, for even any short length of time. If such a system is made, the unbound proton or neutron dips out of the nucleus. As one moves away from stability to neutron-rich or neutron-deficient nuclides, the half-lives of these nuclides become increasingly shorter and go essentially to zero at the two drip lines (shown in Fig. 9.2). Between the two drip lines there are 6000-7000 nuclides. For proton-rich nuclei the above statement is not strictly accurate. The Coulomb barrier acts to hold the unbound proton in the nucleus for times long compared to gamma emission ($\leq 10^{-12}$ s). It is thus possible to observe gamma emission in competition with proton emission for many nuclei produced in a heavy ion reaction with 10 to 15 neutrons less than the proton drip line! Such exotic species beyond the proton drip line cannot be produced at present, but new radioactive ion beam accelerators will be able to probe this entirely new class of nuclear matter where the Coulomb force is no longer a small perturbation on the nuclear mean field force but becomes comparable to the mean field. Such studies are not possible for neutron-rich nuclei.

In 1900, there were only two basic forces known: the gravitational force and the electromagnetic force. The discovery of the existence of a tiny inner core in every atom that contains all the positive charge immediately led to the realization that there is an additional force in nature. Since the nuclear charge has as a building block the proton with one positive charge unit, uranium with $Z = 92$ has at least 92 protons in its nucleus. By Coulomb's force law these 92 positive charges should repel each other, and since they are contained in a nucleus with a radius $< 10^{-14}$ m, the Coulomb repulsive force is tremendous. A proton sitting just outside a typical nucleus would feel a repulsive force, with a repulsive potential energy of the order of 100 MeV. The fact that nuclei exist is evidence that there is another force and it is by far the strongest force in nature since it can override the tremendously strong repulsive Coulomb electrical force. Furthermore, since Rutherford scattering follows the Coulomb force law down to very small distances between an α-particle and a nucleus, $\sim 10^{-14}$ m, the force must be very, very short range ($R_{\text{Nucl.F.}} \lesssim 10^{-14}$ m), in stark contrast to the very long-range electrical force that extends more than 10^5 times farther out to hold the electrons in orbit about the nucleus. Of course, the Coulomb force between the protons does lessen the effect of the attractive nuclear force. In the region of light nuclei, the Coulomb repulsion effect is not strong and the nuclei are more stable when $N = Z$ because of the Pauli exclusion principle. Since neutrons and protons are fermions with spin $\frac{1}{2}$, one can put only two neutrons and two protons in each energy state.

However, as Z increases, the situation changes. The Coulomb interaction is long range and is proportional to $Z(Z - 1)$, since each proton can act on all the other protons in a nucleus. As we shall see, the nuclear force is so short-range that one nucleon can interact with only a few neighboring nucleons and so this interaction is proportional to A (see Sections 9.3 and 11.2). With increasing Z, the

Coulomb force increases faster than the nuclear force. To have a stable nucleus at higher Z one must add more neutrons to give extra bonding from the nuclear force to cancel out the repulsive action of the Coulomb force. So as $Z(A)$ increases, the neutron number of the stable nuclides increases more rapidly than the number of protons. Thus, the line of stability bends down in Fig. 9.2. Beyond $Z = 83$, bismuth, adding more neutrons does not give sufficient attraction to overcome the Coulomb repulsive force to form stable nuclei, so nuclei above $Z = 83$ are radioactive. The observed half-lives for their radioactive decay get shorter and shorter as Z continues to increase through $Z = 112$. For example, the $T_{1/2}$ of isotope of $Z = 105$ (2-34 s), 107 (10 ms-1 s), 109 (2-42 ms), 111 (1.6 ms). These were discovered at the heavy-ion laboratory in Darmstadt, Germany. The discovery of elements 114, 115, 116, 118 with $T_{1/2}$ ranges of 114 (0.3-21 s), 115 (32-87 ms) and 116 (6-53 ms) were made at the Flerov Laboratory in Dubna, Russia. However, note these neutron rich nuclei with $Z = 114$-118 have longer half lives than the nuclei with $Z = 109$-112 presumably because they are closer to $N = 184$.

At some limit as Z increases, even very short-lived radioactive nuclides cannot exist, and the chart of the nuclides ends. But where is that end? For over three decades it has been proposed that in the region of $Z \sim 114$ to 126, and $N = 184$, there should be an island of perhaps stable or at least long-lived elements[2] known as superheavy elements because of gaps in the proton and neutron single particle energies for $Z = 114$ and 126 and $N = 184$ that give them added stability (see shell model, Chapter 11). Scientists have eagerly sought to discover them. Many different target and projectile combinations have been tried and the Russians have looked for them in rocks from the ocean floor and molten salts erupting from deep in the Earth's crust. Unfortunately, we have not been able to make elements from $Z = 114$-126 (which are magic shell numbers for protons) with their neutron number $N = 184$ which is a magic number in the shell model for neutrons to give special stability to these nuclei (see magic numbers, Fig. 9.2 and Chapter 11. However, one of the motivations to build an accelerator which can provide beams of radioactive neutron-rich nuclei is to study the region around $Z = 114$-126, $N = 184$. Beams of stable isotopes found in nature have too few extra neutrons to reach $N = 184$ for even $Z =$ 114. Thus, research in this field continues.

9.2 NUCLEAR GROUND STATE PROPERTIES

9.2.1 Nuclear Shapes and Densities

Despite its very tiny size we have learned much about the shapes and sizes of atomic nuclei. Atomic nuclei are not expected to have precisely defined shapes and sizes since the particles that make up the nucleus follow "smeared-out"

[2] P. Armbruster, et al., *Scientific American* (May 1989) 66.

probability distributions as seen in quantum mechanics. An early model of nuclei was that of a liquid water drop with a spherical ground-state shape. Indeed, such a model is useful in describing certain behaviors of nuclear matter. However, we now know that nuclei can take on a variety of different shapes in addition to spherical, as shown in Fig. 9.4, from axially symmetric prolate shapes (like an American football or more like a rugby football) to axially symmetric oblate shapes (like a thick pancake or discus) to axially symmetric but reflection-asymmetric "pear" (Western pears, not round Korean pears) shapes and on to triaxial shapes with no symmetry. As we shall see, the long-accepted idea of each nucleus having just one of these shapes now has been radically changed, so that a given

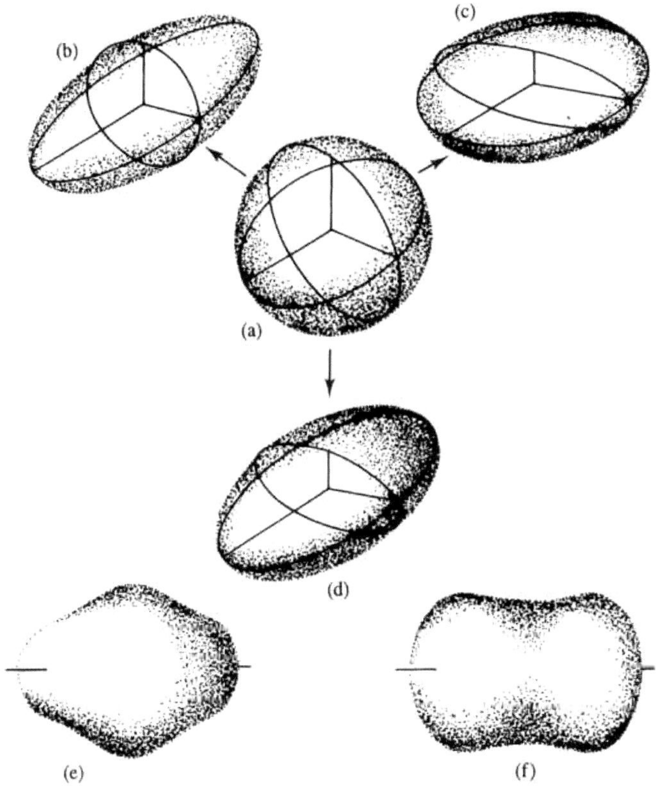

FIGURE 9.4
The range of shapes of atomic nuclei is much broader than was once thought. For spheroids (top) each shape is described by the lengths of three mutually perpendicular axes. In a spherical nucleus (a) the axes are all equal. In a prolate spheroid, or football-shaped nucleus (b), one axis is longer than the other two, which are equal. In an oblate spheroid, or discus-shaped nucleus (c), one axis is shorter than the other two, which are equal. In a triaxial nucleus (d), the axes are all unequal. Some nuclei manifest additional, higher-order deformations, which make their shape more complex. Examples include ground states of ^{234}U (e) and ^{180}Hf (f), which have large positive and large negative hexadecapole moments.

nucleus may have one set of energy levels with one shape and another set with a totally different shape which overlaps in energy the first set. Such nuclear shape coexistence was a major discovery in studies of nuclei far from stability.[1,3] Such changes of shape in one nucleus are found to be an important feature of nuclei throughout the periodic table. Finally, even higher-order shape variations, such as bulges or depressions in the middle of the football shapes (see Fig. 9.4) are being observed. We will look at such detailed views of the nuclear shapes in Chapter 11. First, let us get a general view of nuclear sizes and densities.

One can probe the nuclear size and shape by studying the scattering of charged particles—electrons, protons and alphas—off nuclei. With protons and alphas, one can see at what distance deviations from the Rutherford scattering formula occur to signify that nuclear interactions are beginning. The interference of the Coulomb and nuclear interactions is sensitive to the fine details of nuclear shape such as the hexadecapole moments illustrated in Fig. 9.4. Electrons, because they do not experience the nuclear force, can probe both the shape and the nuclear density as a function of distance from the center of the nucleus. Of course, what you probe in such experiments is the charge density not the nuclear mass density since neutrons with zero charge do not experience the Coulomb force.

What happens to the nuclear density as you add more nucleons? You might expect that increasing the nucleon number would increase the attractive force that holds the nucleus together as each added nucleon interacts with all the other nucleons. In such a case, the density would increase with A and decrease with distance from the center of the nucleus. Surprisingly, as shown in Fig. 9.5, experiments have revealed that both such expectations are untrue. Nuclear charge densities are remarkably constant as a function of distance and, more surprisingly, remarkably constant within 10 percent with a change in A from 10 to 250. In Fig. 9.5 are shown nuclear charge densities measured by scattering charged particles off nuclei. It is assumed that the neutron density matches the charge density, and there is some evidence for this assumption, so the total density would look as in Fig. 9.5. Experiments to measure the neutron density are very hard, so the neutron density is an open question. There is evidence that very neutron-rich exotic nuclei far from stability (e.g., ${}^{11}_{3}\mathrm{Li}_8$) have neutron halos on their outer surface, so that their neutron density does not follow the proton density. The constancy of the nuclear density with distance and mass number gives us important clues into the nature of the nucleon force—that it is very short-range and that it saturates, as discussed in more detail later. Finally, we should note that one of the fascinating areas of current research is the measurement of small variations in the nuclear density by electron scattering to provide more detailed tests of our understanding of nuclear forces.

---

[3] J. H. Hamilton, in *Treatise on Heavy Ion Science*, vol. 8, *Nuclei Far From Stability*, Allan Bromley, ed., Plenum Press, New York (1989), pp. 2–98.

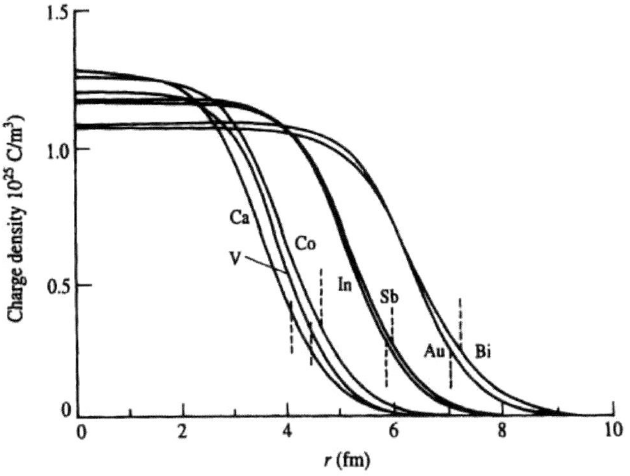

FIGURE 9.5
Charge density as a function of radius for several nuclei according to electron scattering experiments by R. Hofstadter [*Annu. Rev. Nucl. Sci.* 7 (1957) 231]. For each nucleus, the radius $R = R_0A^{1/3}$ is marked by a dashed line.

The fact that the nuclear density is constant, independent of A, says that the number of protons and neutrons per unit volume is approximately constant for all A. Thus,

$$\frac{A}{\frac{4}{3}\pi R^3} \simeq \text{constant}$$

$$R^3 \propto A, \qquad \text{and} \qquad R \propto A^{1/3}$$

While nuclei do not have precise radii but rather smeared-out probability distributions for the nucleons, and nonspherical nuclei can only be described by root-mean-square radii, we need some approximation to consider such problems as determining when two nuclei touch. As a first-order approximation we can define the nuclear radius as

$$R = R_0A^{1/3} \tag{9.1}$$

where R_0 is a proportionality constant to be determined by experiment. The values of R_0 range from 1.0×10^{-15} m to 4.5×10^{-15} m as determined in different ways. For most applications a value of 1.2×10^{-15} m is used, where 10^{-15} m is 1 femtometer (fm) but is more often written as 1 fermi, in honor of Enrico Fermi.

The radius of ^{27}Al from Eq. (9.1) is

$$R = 1.2 \times 10^{-15}\,\text{m} \times 27^{1/3} = 3.6 \times 10^{-15}\,\text{m}$$

and of ^{216}Ra is

$$R = 1.2 \times 10^{-15}\,\text{m} \times 216^{1/3} = 7.2 \times 10^{-15}\,\text{m}$$

So, with 8 times as many nucleons, ^{216}Rn has a radius only twice as great as ^{27}Al. Note that Eq. (9.1) assumes that all nuclei are spherical though, as we shall see, they are not. In cases of deformation the mean-square radius will be important.

The density of nuclear matter is given by

$$\rho = \frac{m}{V} = \frac{A \cdot 1\,\text{u}}{\frac{4}{3}\pi R^3} = \frac{A \cdot 1\,\text{u}}{\frac{4}{3}\pi R_0^3 A} = \frac{1\,\text{u}}{\frac{4}{3}\pi R_0^3} = 2 \times 10^{17}\,\text{kg/m}^3 = 2 \times 10^{14}\,\text{g/cm}^3$$

Recall that the density of water is 1 g/cm^3 and that of lead is 11 g/cm^3, and the average density of the Earth about 5.5 g/cm^3. If a small sugar cube of 1 cm^3 were made of nuclear matter, it would have a mass

$$m = \rho V = 2 \times 10^{14}\,\text{g/cm}^3 \times 1\,\text{cm}^3 = 2 \times 10^{14}\,\text{g} = 2 \times 10^{11}\,\text{kg}$$
$$\simeq 200 \text{ million tons!}$$

From Fig. 9.5 one sees the nuclear charge density is not constant but depends weakly on Z and/or A. This dependence can be written as

$$\rho(r) \sim \frac{Z}{A}\,\rho_{\text{mass}}(r)$$

where ρ_{mass} is the mass density. One also sees in Fig. 9.5 that the charge density goes gradually to zero. The charge density can be parametrized as

$$\rho(r) = \frac{\rho(0)}{1 + e^{(r-a)/b}} \tag{9.2}$$

where $a = 1.07A^{1/3} \times 10^{-15}\,\text{m}$ and $b = 0.55 \times 10^{-15}\,\text{m}$. Note that a is the nuclear radius where the charge density is half the maximum and $2b$ is a measure of the thickness of the nuclear surface since most of the change in the nuclear density from $\rho(0)$ to $\rho =$ zero occurs over this distance. At a distance $r = \frac{1}{2}R_0A^{1/3}$, for ^{209}Bi,

$$\frac{r-a}{b} = \frac{0.6A^{1/3} - 1.07A^{1/3}}{0.55} = -5.07 \qquad \text{and} \qquad e^{-5.07} = 0.0063.$$

So, $\rho(r) \simeq \rho(0)$ as shown in Fig. 9.5. The fact that nuclear charge density is nearly constant from ^{40}Ca to ^{209}Bi in Fig. 9.5 is evidence that the nuclear force saturates, so that each nucleon can interact with only a few neighbors and not with all the nucleons (see Chapter 11). It is like being in a room with 100 people where everyone is talking (interacting); you can talk (interact) with a few nearby people but not with all 100.

9.2.2 Nuclear Masses: "$1 + 1 \neq 2$"

Since the nucleus consists of neutrons and protons, one might think that the nuclear mass is equal to the sum of its constituent neutron and proton masses. But that is not the case. Look at the simplest nuclear example, deuterium, an isotope of hydrogen. In seawater there are about 150 deuterium atoms per million hydrogen atoms. The deuterium ($^2_1H_1$) nucleus (called the deuteron) consists of a neutron and a proton.

Neutron mass	$m_n = 1.008\,665\,\text{u}$
Proton mass	$m_p = 1.007\,276\,\text{u}$
Sum of both	$m_n + m_p = 2.015\,941\,\text{u}$

But the $^2_1H_1$ mass $m_d = 2.013\ 553\ \text{u} \neq m_n + m_p$. The difference between the two figures is

$$m_p + m_n - m_d = 0.002\ 388\ \text{u} = 2.224\ \text{MeV}$$

where 1 u = 931.5 MeV.

When a neutron and a proton combine to form a deuteron, part (2.224 MeV) of their rest mass energy is released. This energy is called the nuclear binding energy. This is the energy that must be supplied to separate the neutron and proton into two free particles. For example, when a γ-ray with 2.225 MeV is absorbed by a deuterium nucleus, the deuterium nucleus becomes a free proton and a free neutron.

While binding energy occurs in molecular and atomic physics, these binding energies are significantly larger in nuclear physics and still more so in high-energy physics. For example, when two hydrogen atoms combine to form a hydrogen molecule, about 4 eV of binding energy is released. The energy which corresponds to the static mass of a hydrogen atom (proton + electron) is about 938.3 MeV + 0.511 MeV $\approx$ 1000 MeV. The ratio of the molecular binding energy to the mass of one atom is

$$\frac{4\,\text{eV}}{1000\,\text{MeV}} = 4 \times 10^{-9}$$

Thus, molecular binding is so small that it can be ignored in nuclear problems.

When an electron and a proton combine to form a hydrogen atom, the electron's atomic binding energy of 13.6 eV is released. The ratio of the electron's binding energy to its mass is

$$\frac{13.6\,\text{eV}}{511\,\text{keV}} \simeq 3 \times 10^{-5}$$

But when a neutron and a proton combine to form a deuterium nucleus, this ratio is

$$\frac{2.224\ \text{MeV}}{938\ \text{MeV}} \simeq 2 \times 10^{-3} (0.2\%)$$

We shall see that in high-energy physics (see Chapter 14) this ratio approaches 1 and can even be greater than 1. Such conditions provide new challenges to our understanding of the structure of matter.

Finally, note because the neutron is heavier than the proton, a free neutron can beta decay into a proton, electron, and antineutrino (see Section

10.3.3). However, inside a nucleus the binding energy given up to form the nucleus from free protons and free neutrons can keep a neutron inside the nucleus from beta decaying. Here, for example, if the neutron in ${}^2_1H_1$ were to undergo beta decay, it would form ${}^2_2He + e^- + \bar{\nu}$ ($\bar{\nu}$ for antineutrino). The ${}^2He$ is not bound by the nucleon force sufficiently and breaks up into two protons. But since $m_d < 2m_p$ without adding m_e, the decay of ${}^2_1H_1$ into ${}^2_2He + m_e + \bar{\nu}$ cannot occur. On the other hand, ${}^3_1H_2$ can beta decay to stable ${}^3_2He_1 + e^- + \bar{\nu}$. Here one can say one of the neutrons inside the nucleus can have sufficient mass to decay into a proton plus electron and antineutrino. The nuclear binding energy determines which combinations of N and Z are stable to beta decay and provides the explanation for Fig. 9.2. When there are too many neutrons, the masses of the initial and final nucleus (as determined by their binding energies) are such that a neutron can change into a proton in beta decay and, if there are too many protons to have a stable nucleus, the proton can change into a neutron in beta decay (Section 10.3.4). Thus, the binding energies of the neutrons and protons play a crucial role in determining which nuclei are stable and which are radioactive.

9.2.3 Binding Energy

For a nucleus of mass m and total nuclear binding energy B:

$$m = Zm_p + Nm_n - \frac{B}{c^2}$$

In general, it is easier to measure and to use the atomic mass M instead of the nuclear mass m, so we rewrite it as:

$$M = Zm_p + Nm_n + Zm_e - \frac{B}{c^2}$$

or

$$B = (Zm_p + Nm_n + Zm_e - M)c^2 = [ZM({}^1H) + Nm_n - M({}^A_ZX_N)]c^2 \qquad (9.3)$$

This is a general expression for the total nuclear binding energy of a nucleus, where $M({}^1H)$ and $M({}^AX)$ are the atomic masses of the atoms of ${}^1H$ and isotope ${}^AX$. Here we ignore the very small electron binding energies which are included in the atomic masses in the first expression since the nuclear masses range from 10^9 to 10^{11} eV, the total electron masses from 10^6–10^8 eV, and the total electron binding energies from only 10 to 10^5 eV.

In Appendix II the mass defects, $[M(X) - A]c^2 = \Delta$, are given. Note that Δ is often a negative number. You can find atomic masses given in other texts or handbooks. From Appendix II, $M(A_X) = A + \Delta(A_X)$. For example,

$$M({}^4He) = 4 + \frac{2.425\ \text{MeV}}{931.5\ \text{MeV/u}}$$
$$= 4.002603\text{u}$$

To calculate the total binding energy of ${}^4_2\mathrm{He}_2$:

$$\begin{aligned} B(^4\mathrm{He}) &= 2M(^1\mathrm{H}) + 2m_n - M(^4\mathrm{He}) = 2\Delta(\mathrm{H}) + 2\Delta(\mathrm{n}) - \Delta(^4\mathrm{He}) \\ &= 2(7.289\ \mathrm{MeV})(^1\mathrm{H}) + 2(8.071\ \mathrm{MeV}) - 2.425\ \mathrm{MeV} = 28.295\ \mathrm{MeV} \end{aligned}$$

Using Appendix II, we also can calculate the energy to pull one neutron out of ^{4}He to form ^{3}He and a free neutron, which is given by

$$\begin{aligned} M({}^3_2\mathrm{H}_1) + \mathrm{m}_n - M({}^4_2\mathrm{H}_2) &= (A+\Delta)_{^3\mathrm{He}} + (A+\Delta)_n - (A+\Delta)_{^4\mathrm{He}} \\ &= \Delta(^3\mathrm{He}) + \Delta(\mathrm{n}) - \Delta(^4\mathrm{He}) \\ &= 14.931 + 8.071 - 2.425 = 20.578\ \mathrm{MeV} \end{aligned}$$

The binding energy per nucleon, *B/A*, is another important quantity. This also is called the average binding energy or binding energy ratio. For example, in deuterium the binding energy ratio is $2.224/2 \approx 1.1$ MeV and in ^{4}He it is 28.295/4 ≈ 7 MeV. From experimental atomic masses and Eq. (9.3), we can get the binding energies of the nuclides. Figure 9.6 is a curve of the binding energy ratio vs. mass number. This curve occupies an important place in nuclear physics, as we shall see, because it provides insight into the nature of the force that holds the nucleons together in the nucleus as well as insights into nuclear fission and fusion, and α and cluster decays.

From Fig. 9.6, we see that the binding energy (BE) ratio curve has a maximum around $A = 56$: the B/A ratios of nuclides around $A \approx 56$ are greater than those of lighter or heavier nuclides. At low mass there is some oscillation, after which the B/A ratio goes smoothly to a maximum of about 8.8 MeV/A around $A \approx 56$, and after that falls off gradually.

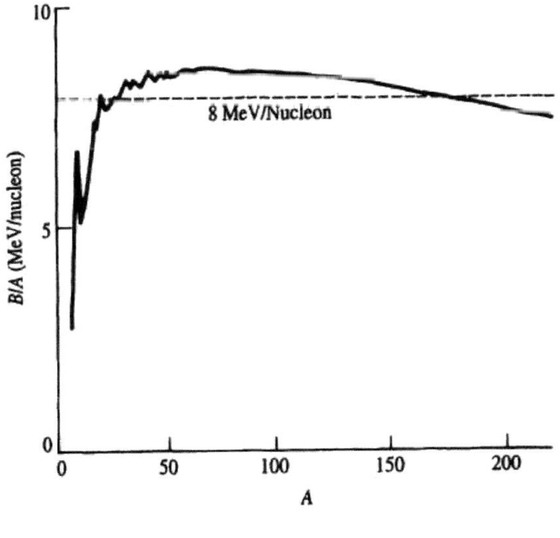

FIGURE 9.6
Average nuclear binding energy per nucleon as a function of *A*.

From the figure one can also see that if two light nuclei with small B/A ratios are combined to form a nucleus with a larger B/A, that is, nuclei that are loosely bound (small B/A) form a more tightly bound (larger B/A) nucleus, then energy is released. This is the process of nuclear fusion. Take as an example, the fusion of $^{12}C(B/A = 7.68\,MeV/A)$ with $^{16}O(B/A = 7.98\,MeV/A)$ to form $^{28}Si(B/A = 8.44\,MeV/A)$:

$$\begin{aligned} BE(^{12}C) &= 7.68 \times 12 = 92.16\,MeV \\ BE(^{16}O) &= 7.98 \times 16 = \underline{127.68\,MeV} \\ \text{Sum} & \qquad\qquad\quad\;\; 219.84\,MeV \\ BE(^{28}Si) &= 8.44 \times 28 = 236.32\,MeV \end{aligned}$$

The energy released in this case is 16.48 MeV. At the other end, if a heavy nucleus like ^{235}U splits (fission) into two lighter ones with larger B/A ratios, then again energy is given off. Nuclear reactors and the first type of nuclear bombs get their energy from the fission of heavy nuclei. Hydrogen bombs and the Sun and stars get their energy by the fusion of hydrogen and other light nuclei into heavier ones. The so-called nuclear energy is the energy that is released when the nuclear binding energy is increased and the total nuclear mass decreases, as expressed in the mass–energy equation of Einstein, $E = mc^2$. The remarkable feature of the data is that from $A = 12$ to 240 the B/A curve is constant to within about 20 percent. We shall see that this tells us something very important about the nature of the nuclear force that holds the nucleus together.

9.2.4 Semi-empirical Mass Formula

The curve of the nuclear binding energy per nucleon came from experimental results. So far, we cannot deduce from first principles a nuclear mass formula or a nuclear binding energy formula which would allow us to calculate all the nuclear masses (or binding energies) precisely.

In 1935, C. F. von Weizsäcker gave a semi-empirical nuclear mass formula. There are three main terms in that formula for the nuclear binding energy: the volume, surface and Coulomb terms

$$B = B_V - B_S - B_C \tag{9.4}$$

From Fig. 9.6, we can see that the binding energy ratio, B/A, is approximately constant, except for light nuclei. So, B is proportional to mass number A. The constancy of B/A comes about because the nuclear force saturates. The fact the nuclear force saturates (see Chapter 11 also) means that each nucleon can only interact with a few neighboring nuclei—not with all A nucleons inside the nucleus. If it interacted with the other $A - 1$ nucleons, then the binding energy would depend on the number of nucleon pairs, which is $A(A - 1)$. Then B would be proportional to A^2.

B_V, the first term in Eq. (9.4), is the volume energy, which is the main term of the binding energy. From experiments, we know that the radius of a nucleus is approximately

$$R = R_0 A^{1/3} \tag{9.1}$$

where R_0 is a constant, $R_0 \simeq 1.2\,\text{fm}$. So the nuclear volume is approximately proportional to the mass number A. Thus, as for a water drop, the nuclear binding energy is nearly proportional to the nuclear volume. The volume energy is

$$B_V = a_V A \tag{9.5}$$

where a_V is a constant. We would expect that the dominant term B_V/A would be approximately constant since $B/A \simeq$ constant.

The surface energy contribution is negative because the particles on the surface have less binding, since there are no particles outside of them as occurs for all the others inside the nucleus. Since the surface area is proportional to $R^2 \sim (A^{1/3})^2$,

$$B_S = a_S A^{2/3} \tag{9.6}$$

where a_S is a constant.

In Eq. (9.4) the third term is the Coulomb energy, which is also negative because of the Coulomb repulsion of the positively charged protons by each other. Suppose a nucleus is a sphere and the charge is uniformly distributed. How do we calculate the Coulomb energy?

Imagine that the nuclear charge (Ze) is moving in from infinitely far away and draw different spheres about some center. From r to $r + dr$, the charge is $dq = 4\pi r^2\, dr \cdot \rho$ (where ρ is the charge density). Inside the sphere the charge is $\frac{4}{3}\pi r^3 \rho$. So when dq is moved from infinitely far away (where the electrical potential is zero) to a point on the sphere of radius r, the work done is

$$dW = \frac{\frac{4}{3}\pi r^3 \rho \times 4\pi r^2\, dr\, \rho}{r}$$

where $\rho = Ze/\frac{4}{3}\pi R^3$. Thus, to constitute a sphere with the radius R and with charge Q, the work, or Coulomb energy is

$$B_C = \int_0^R dW = \int_0^R \frac{(4\pi)^2}{3} r^4 \rho^2\, dr = \frac{3}{5}\frac{(Ze)^2}{R}$$

When we do this, we suppose that the nuclear electric charge is continuously brought together. But in fact, inside nuclei, charge comes in a discrete unit, namely the charge of the proton. The protons exist already, so it is not necessary to bring charge together to form every proton. The work is $\frac{3}{5}e^2/R$ for producing one proton, so we must subtract $\frac{3}{5}(e^2/R)Z$ from B_C, so

$$B_C = \frac{3}{5}\frac{(Ze)^2}{R} - \frac{3}{5}\frac{e^2}{R}Z = \frac{3}{5}Z(Z-1)\frac{e^2}{R} \tag{9.7}$$

This is the Coulomb energy.

We also can deduce it in another way. A proton can interact with the other $(Z-1)$ protons. So, there are $\frac{1}{2}Z(Z-1)$ pairs. In a nucleus with radius R, the Coulomb energy between every pair is $\frac{6}{5}e^2/R$. So, the total Coulomb energy is $\frac{3}{5}Z(Z-1)e^2/R$. Substituting Eqs. (9.4) to (9.6) into (9.3), we have

$$B = a_V A - a_S A^{2/3} - \frac{3}{5}Z(Z-1)\frac{e^2}{R} \tag{9.8}$$

Figure 9.7 shows the sum of these three terms. We adjust the constants a_V and a_S so that the curve obtained from Eq. (9.7) agrees with the experimental one. In fact, except for some details, the two are roughly the same. This agreement shows that it is reasonable to consider the nucleus as a liquid drop.

9.2.5 A More Complete Mass Formula

We can improve considerably on Eq. (9.7). In fact, even the semi-empirical mass formula which was published in 1935 by Weizsäcker is more complex than (9.8). Since then, various refined semi-empirical mass formulas have been developed. One of the more complex recent ones is that given by Möller and Nix.[4] The standards by which to judge a semi-empirical mass formula include: (a) the explicit physical principles used, (b) the use of as few adjustable parameters as possible, (c) the calculations should be in better agreement with experiment, and (d) the formula should be able to explain other nuclear properties which are related to the nuclear mass. In their most recent calculations, Möller et al.[4] calculated the ground-state masses of 8979 nuclei for ^{16}O to $A=339$ by use of an improved macroscopic–microscopic model. The calculated masses of 1654 known nuclides

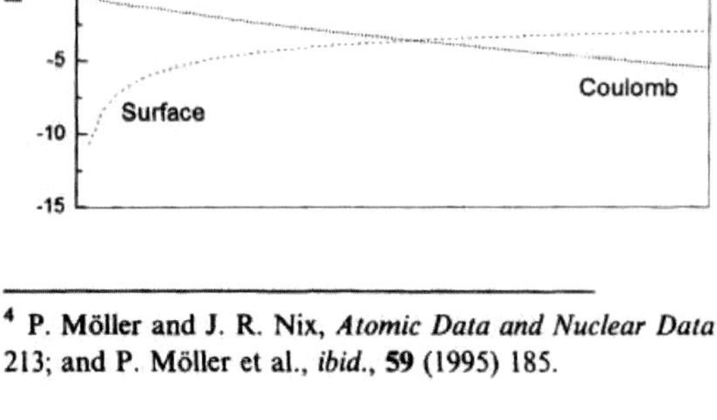

FIGURE 9.7
The primary contributions to the total binding energy, namely the volume, Coulomb, and surface contributions, are shown

[4] P. Möller and J. R. Nix, *Atomic Data and Nuclear Data Tables* **26** (1981) 165; *ibid.*, **39** (1988) 213; and P. Möller et al., *ibid.*, **59** (1995) 185.

by using nine adjustable parameters gave a rms deviation of 0.669 MeV. The rms deviation basically is

$$\sqrt{\sum_i^n \frac{(\chi_{\text{calc}} - \chi_{\text{exp}})_i^2}{n}}$$

where χ_{exp} is the experimental value, χ_{calc} is the calculated one, and n is the number of nuclides. One unexpected prediction of their 1981 calculations[4] was a new region of deformed nuclei centered around $N = Z = 38$ with the largest ground-state deformations known for nuclei above $A \simeq 30$, as shown in Fig. 11.23. In an unusual congruence of theory and experiment, this unexpected deformed new region was simultaneously and quite independently discovered by the Vanderbilt group and their collaborators the same year, as discussed in Chapter 11. Since modern science began to be developed based on the experimental approach a few centuries ago, there have been three elements in a good science: (1) collecting reliable data by experiments, (2) correlating these data into rules and laws, and (3) the use of these rules and laws to predict new phenomena. The ability of a theory to predict previously unknown phenomena is considered an important triumph for a theory.

A more complete binding energy formula (still not as complex as in reference 4), which is still referred to as a Weizsäcker mass–binding energy formula, is

$$B = a_V A - a_S A^{2/3} - a_C Z^2 A^{-1/3} - a_{\text{sym}}(Z - N)^2 A^{-1} + B_P + B_{\text{shell}} \tag{9.9}$$

The first three terms have been discussed. The coefficients of the first two terms come from experiment and the third can be calculated (Eq. 9.6). One set is

$$a_V = 15.8\,\text{MeV} \qquad a_S = 18.3\,\text{MeV} \qquad a_C = 0.72\,\text{MeV} \tag{9.10}$$

The fourth term is a symmetry energy term. When the neutron number is equal to the proton number in a nucleus, this term is zero. Otherwise, because of the Pauli exclusion principle, when the number of one kind of nucleon is greater than that of the other, the excess of the one kind must be in a higher energy state, so the nucleus has less binding energy. The symmetry term is derived in Chapter 11. The coefficient can be calculated from Eqs. (11.19) and (11.20), but the result is not accurate. It is more reliable to extract it from fits to experimental data. When the values in Eqs. (9.10) are used in such a fit, one obtains

$$a_{\text{sym}} = 23.2\,\text{MeV} \tag{9.11}$$

The fifth term is the pairing energy. In nature we find that even–even nuclei with even Z and even N numbers are the most stable, and odd–odd nuclei the most unstable (odd numbers of protons and odd numbers of neutrons). For example, of the stable nuclides in nature, 166 are even–even nuclides and only 9 are odd–odd. There is a gain in binding when the protons and neutrons can form pairs of protons and pairs of neutrons. These experimental facts lead to the following for B_P:

$$B_P = \begin{cases} a_P A^{-1/2} & \text{even–even nucleus} \\ 0 & \text{odd } A \\ -a_P A^{-1/2} & \text{odd–odd nucleus} \end{cases} \tag{9.12}$$

where $a_P = 11.2\,\text{MeV}$.

In Eq. (9.9), the final term is to correct for shell effects associated with the spherical shell model magic numbers (2, 8, 20,...). Further information on the correction for shell structure effects is found in reference 4. The shell model is discussed in Chapter 11. Note that the values of the coefficients a_V, a_S, a_C, and a_P given in Eqs. (9.10), (9.11), and (9.12) are from a recent fit to a particular set of data. There are other values based on different sets of data. In some earlier formulations $A^{-3/4}$ is used with a different a_P in Eq. (9.12).

9.3 NUCLEAR GROUND STATE SPINS AND MOMENTS

9.3.1 Nuclear Spin

In 1924 Pauli made the suggestion that the nucleus as a whole has spin—angular momentum. But it was only after Chadwick discovered the neutron in 1932 that the origin of nuclear spin could be understood. From experiments, we find that the neutron and proton are both fermions which have the same intrinsic spin as an electron ($\frac{1}{2}\hbar$). Since the nucleus consists of neutrons and protons, its spin should be the sum of the spins of the neutrons and protons. However, as we shall see, just as for electrons, the nucleons can have both intrinsic spin and orbital angular momentum which combine to a total spin (angular momentum) for the nucleus.

From experiments it is known that in their ground states: (a) the spins of all even–even nuclei (that is, their N number and Z number are even) are zero; (b) the spins of all even–odd (even numbers of Z or N and odd numbers of N or Z) nuclei are half integer multiples of $\hbar$; and (c) the spins of all odd–odd nuclei are integer multiples of $\hbar$. Obviously, some strong mechanism for combining spins is operating to produce these total spins. Of course, in an excited state, the situation is different; for example, the spins of excited states of even–even nuclei can be nonzero. Spins of excited states are discussed further in Chapter 11.

9.3.2 Magnetic Dipole Moment

In Chapter 4 the electron magnetic dipole moment is given as

$$\boldsymbol{\mu}_e = -\frac{e\hbar}{2m_e c}(g_{e,l}\boldsymbol{L} + g_{e,s}\boldsymbol{S}) \tag{9.13}$$

For convenience, we have separated $\hbar$ from the angular momentum, so

$$|\boldsymbol{L}|^2 = l(l+1) \qquad |\boldsymbol{S}|^2 = s(s+1) \tag{9.14}$$

For an electron, $g_{e,l} = 1$, but only for $g_{e,s} = 2$ do we obtain results which are consistent with experiments. From the Dirac equation, we find that because the electron spin is $\frac{1}{2}\hbar$, we can have $g_{e,s} = 2$ if the electron is a point charge. While in modern electron theory $g_{e,s}$ has some small corrections, we still use in general

$$\boldsymbol{\mu}_e = -\frac{e\hbar}{2m_e c}(\boldsymbol{L} + 2\boldsymbol{S}) = -(\boldsymbol{L} + 2\boldsymbol{S})\,\mu_B \tag{9.15}$$

where

$$\mu_B \equiv \frac{e\hbar}{2m_e c} = 0.5788 \times 10^{-4}\,\text{eV/T} \tag{9.16}$$

is called the Bohr magneton.

About 1930, scientists knew that the proton is a fermion with $\frac{1}{2}\hbar$ spin just like that of an electron and thought of it, too, as a point charge. The only difference is its mass. So, $g_{p,s} = 2$ was the prediction of Dirac's theory. Then

$$\boldsymbol{\mu}_p = \frac{e\hbar}{2m_p c}(\boldsymbol{L} + g_{p,s}\boldsymbol{S}) \tag{9.17}$$

with $g_{p,s} = 2$; and

$$\mu_N = \frac{e\hbar}{2m_p c} = 3.152 \times 10^{-8}\,\text{eV/T} \tag{9.18}$$

which is called a nuclear Bohr magneton. As the proton mass is 1836 times that of an electron, the nuclear magneton is 1836 times smaller than the Bohr magneton. In addition, Eqs. (9.17) and (9.15) have different signs. Stern found experimentally that

$$g_{p,s} = 5.6 \tag{9.19}$$

The best modern value is 5.58 (see Appendix I). This is quite different from the initial theoretical prediction!

What about the neutron? As the neutron does not carry electric charge, one would expect $g_{n,l} = 0$ and $g_{n,s} = 0$. But from experiments,

$$\boldsymbol{\mu}_n = \frac{e\hbar}{2m_n c} g_{n,s}\boldsymbol{S} \qquad g_{n,s} = -3.82 \tag{9.20}$$

As the neutron does not carry charge, the magnetic moment associated with any orbital angular momentum is zero. But that related to the spin need not be zero if the neutron is not a simple point particle. While the neutron as a total body does not carry charge, its magnetic moment tells us there is some charge distribution inside the neutron. The sign of the neutron spin magnetic moment is the same as that of the electron. So, like an electron, its spin direction is inverted with respect to its magnetic moment. From the proton and neutron magnetic moments, we can see that both of them are not point particles. They must have some internal structure. A good theory of the proton and neutron structures must explain the experimental results of their magnetic moments.

Knowing the neutron and proton magnetic moment values, we want to know the magnetic moment of the whole nucleus. First, look at the deuteron. If the deuteron is in its ground state, from Eqs. (9.17), (9.19), (9.20) we obtain (remember that the projection of $\boldsymbol{S}$ on the z axis is $\frac{1}{2}$)

$$\mu_p = 2.79\mu_N \qquad \mu_n = -1.91\mu_N$$

The more accurate results given in Appendix III are

$$\mu_P = 2.792\,847\,356\mu_N \qquad \mu_n = -1.913\,042\,73\mu_N \tag{9.21}$$

Their sum is $0.879\,804\,63\mu_N$, but the experimental value for the deuteron magnetic moment is $0.857\,438\,23\mu_N$ which is significantly different. This difference shows that in addition to the spin motion of the nucleons, the orbital motion of the nucleons contributes to the magnetic moment. The deuteron ground state is not completely an orbital spin-zero, S state. If it were a pure S state where the n and p spins are aligned to give spin $j = 1$, then its μ would be the sum of the terms in Eq. (9.21). It is necessary, therefore, to include the effects of the orbital motion of the nucleons inside the nucleus in order to calculate the nuclear magnetic moment correctly. In a nucleus of 100 to 200 nucleons, this might seem to be too complex. As we shall see, however, pairing effects greatly simplify the situation. The nuclear magnetic moment is an important test of any model of the structure of nuclei. Indeed, the recent combinations of low-temperature nuclear orientation and nuclear magnetic resonance facilities coupled on-line with isotope separators and accelerators have opened up very precise measurements of magnetic moments to provide detailed tests of nuclear models even in nuclei far from stability (reference 3). Moreover, the interaction of the nuclear magnetic moment and the nuclear quadrupole moment (discussed next) with the atomic electrons gives rise to the field of hyperfine interactions as discussed in Chapter 13.

9.3.3 The Electric Quadrupole Moment

As we know, at a point in space a distance r from a point charge e, the electric potential is

$$\phi = \frac{e}{r}$$

If a body has a charge density ρ, at a point a distance r outside of it we obtain a similar result:

$$\phi = \frac{1}{r}\int \rho\, dv \tag{9.22}$$

where $\int \rho\, dv$ is a volume integration over the region where ρ exists and represents the total charge of the system. When r is very large compared to the size of the charged system, this is the full potential and reduces to e/r again. At closer distances to the charged system ($r \to$ size of the system), other electric moments can be experienced. For example, if the system consists of a pair of positive and negative charges, $e^{\pm}$, separated by a small distance, d, Eq. (9.22) will not give the whole picture. Because the total charge is zero, the potential will be zero when r is very much larger than d. However, such a system has an electric dipole moment, ed, that produces a potential as r approaches close to d. For a group of point charges ϵ_i, the z projection of the electric dipole moment is $\sum_i \epsilon_i z_i$. For a body with arbitrary charge distribution, the potential at a distance r produced by a dipole moment is:

$$\phi = \frac{1}{r^2}\int \rho z \, dv \tag{9.23}$$

Now look at a system such as shown in Fig. 9.8. There, both the total charge and dipole moment are zero. There can be no ϕ from Eqs. (9.22) or (9.23). But this charge has an electric quadrupole moment $\sum_i \epsilon_i z_i^2$ that produces a potential. The z axis is shown in Fig. 9.8. In general, the potential at a position r produced by a quadrupole charge distribution is

$$\phi = \frac{1}{r^3}\int \rho(3z^2 - r^2)\, dv \tag{9.24}$$

The potential from an arbitrary charge configuration is

$$\phi = \frac{1}{r}\int \rho \, dv + \frac{1}{r^2}\int \rho z \, dv + \frac{1}{r^3}\int \rho(3z^2 - r^2)\, dv + \ldots \tag{9.25}$$

It has been shown by theory and experiment that the nuclear electric dipole moment is zero. The nuclear electric quadrupole moment is given by

$$Q = \frac{\phi}{e} = \sum_p (3z_p^2 - r_p^2) \tag{9.26}$$

The subscript p expresses the fact that only the protons p contribute. So to sum over p, we take a z axis and make the nuclear spin projection along the z axis a maximum, that is, $m_I = I$. The unit of Q is a barn (b) where 1 b = $10^{-24}\,\text{cm}^2 = 10^{-28}\,\text{m}^2$.

If a nucleus is an axially symmetric ellipsoid with the length of the semi-major axis of symmetry given by c, and the length of the other equal semi-minor axis by a, we obtain

$$Q = \tfrac{2}{5} Z(c^2 - a^2) \tag{9.27}$$

In a spherical nucleus $c = a$, so $Q = 0$; in a prolate nucleus, $Q > 0$; and in an oblate nucleus, $Q < 0$ (see Fig. 9.9). Some nuclear moments are given in Table 9.1.

FIGURE 9.8
Example of a charge distribution with no electric dipole moment but an electric quadrupole moment.

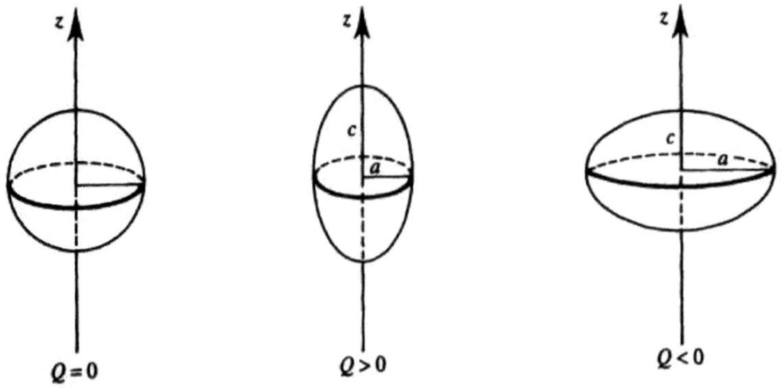

FIGURE 9.9
The relation between nuclear shape and quadrupole moment.

TABLE 9.1
Some experimental values of the nuclear spins, magnetic dipole moments and electric quadrupole moments for several nuclei

Nucleus	Spin ($\hbar$)	Magnetic dipole moment (μ_N)	Electric quadrupole moment (b)
n	$\frac{1}{2}$	−1.9131	0
^{1}H	$\frac{1}{2}$	2.7928	0
^{2}H	1	0.8574	0.002 82
^{3}H	$\frac{1}{2}$	2.9789	0
^{3}He	$\frac{1}{2}$	−2.1275	0
^{4}He	0	0	0
^{7}Li	$\frac{3}{2}$	3.2563	−0.045
^{12}C	0	0	0
^{13}C	$\frac{1}{2}$	0.7024	0
^{176}Lu	7	3.1800	8.000
^{235}U	$\frac{7}{2}$	−0.35	4.1
^{238}U	0	0	0
^{241}Pu	$\frac{5}{2}$	−0.730	5.600

From Table 9.1 we can see that the quadrupole moment of the deuteron is not zero:

$$Q_d = 0.002\,82\,\text{b} = 0.282\,\text{fm}^2$$

This result also supports the conclusion that the deuteron ground state is not completely an S ($L = 0$) state. An S state has spherical symmetry and $Q = 0$. So, the deuteron has a nonspherical shape. Calculations give the measured electric quadrupole moment when the ground state is taken as a mixture with 96 percent an $L = 0$ state and 4 percent an $L = 2$ state. Since $J = 1$, these states ($^{2S+1}L_J$) are

$^3S_1$ and $^3D_1$. In the latter state L and S are antiparallel to give $J = 1$. This mixture also gives the correct magnetic dipole moment while a pure $^3S_1$ configuration will not, as noted in the last section. The very large quadrupole moments for nuclei like ^{176}Lu and ^{241}Pu as given in Table 9.1 were the first crucial data that implied that these nuclei and their neighbors had significant deviations from a spherical shape. Their positive signs indicated they had prolate shapes. Indeed, the sign and size of the quadrupole moment are very important parameters to be measured for nuclei because of their close relation to the nuclear shape. These values are very important tests of the predictions of nuclear models. From Table 9.1 we also can see that when $I = 0$ or $\frac{1}{2}$, $Q = 0$. This can be derived in quantum mechanics.

Nuclei may also have higher multipole moments that are associated with more exotic shapes than axially symmetric ellipsoids. Measurements of the hexadecapole moments and even higher moments in very heavy nuclei are active areas of research that probe important details of the structure of nuclei. Positive hexadecapole moments lead to a bulge around the middle of a prolate deformed nucleus, seen in the depiction of the nucleus of ^{234}U in Fig. 9.4 as observed in the scattering of α-particles by ^{234}U. A negative hexadecapole moment produces a depression in the middle of a prolate deformed nucleus to yield a "peanut" shape as reported in ^{180}Hf, also shown in Fig. 9.4.

Currently a "hot" topic in nuclear structure in both theory and experiment is the investigation of which nuclei have ground and/or excited states that have stable reflection-asymmetric shapes in addition to the long-known octupole vibrational shapes? Such shapes are characterized by $l = 3$ octupole deformation. There is evidence for stable octupole deformation in which the nucleus is shaped like a "pear" (as illustrated in Fig. 11.12 in Chapter 11 for an excited octupole vibrational band). Such a shape does not have symmetry of reflection about its center of mass. An even more interesting question in current research is whether the theoretically predicted octupole deformations with "banana" shapes occur in nuclei.

9.4 SUMMARY

1. The nucleus is the level of matter intermediate between the atom and the substructure of protons and neutrons. The nuclear and atomic (orbital electron) domains can interact via the nuclear electric and magnetic moments. There is also a mingling of nuclear physics and particle physics through the strong force. The nucleus is fundamentally important because it is a unique many-body, quantum-mechanical laboratory with too few particles to be treated by statistical methods but too many to be treated easily one-by-one. Nuclear physics is not only a major area of basic research but also has important applications in a wide spectrum of other fields of basic research and societal problems including nuclear astrophysics, nuclear solid-state physics, nuclear chemistry, nuclear biology, geology, nuclear energy, nuclear medicine (for both diagnostic and

treatment), art, history, anthropology, and a broad range of industrial problems.

2. Basically the nucleus consists of neutrons (of number N) and protons (of number Z), both of which are called nucleons which are held together by the strong force. The nuclear density is nearly constant for all nuclei from ^{12}C to ^{238}U and the nuclear radius is a constant times $A^{1/3}(A = N + Z)$. The average binding energy per nucleon is also nearly constant for all nuclei. In a semi-empirical mass formula there are three main contributions to the nuclear binding energy. The largest depends on the nuclear volume, which is proportional to A; the next depends on the nuclear surface and it lessens the binding; there is also a Coulomb energy term which becomes increasingly important as Z increases; and finally there are other terms which depend on the nuclear symmetry between N and Z, on the pairing of neutrons together and of protons together, and on the magic shell gap corrections.

3. Because of the Pauli principle and the fact the mass difference between the neutron and proton is small (so the nucleon binding energy offsets the mass difference between the two), inside a stable nucleus the neutron cannot undergo beta decay like the free neutron does. For light nuclei, the neutron number is about equal to the proton number. Because of the long-range action of the Coulomb force, for nuclei stable to beta decay the ratio of neutron number to proton number (N/Z) increases as the mass number A increases to offset the increased Coulomb force. There are about 325 nuclides found in nature and an additional 1900 radioactive nuclides have been produced in the laboratory on the proton-rich and neutron-rich sides of the stability line. Potentially, another 4000 should be observable between the proton and neutron drip lines. One of the major frontiers in nuclear physics today is to extend our knowledge to nuclei far from the β-stability line on the proton- and neutron-rich sides and to new superheavy elements with $Z > 110$.

4. The nucleus, like the atom, has a total spin and magnetic moment from the sum of the individual angular momenta and magnetic moments of the nucleons. The nucleus may also have electric quadrupole and higher-order moments which are closely related to the nuclear shape.

PROBLEMS

9.1. Calculate the volume occupied by a wooden table top with a mass of 50 kg (110 lb) if all the atoms could be collapsed to have a density equal that of nuclear matter. Compare this volume with the volume of the table if the density of its wood is 0.95 g/cm^3.

9.2. Calculate the volume and radius of the Earth if all its matter were condensed to a density equal that of nuclear matter. The mean radius of the Earth is 6.38×10^6 m. The mass of the earth is 6×10^{24} kg.

9.3. Explain why a free neutron is unstable (radioactive), but in first order the proton is not.

9.4. Explain why a neutron inside a helium nucleus or that of another stable element is stable and does not decay.

9.5. What does the shape of the B/A curve (Fig. 9.6) tell you about how elements below iron and above iron were produced in nucleosynthesis.

9.6. Calculate the binding energy and the binding energy per nucleon of ^{6}Li, ^{19}F, ^{40}Ca, ^{56}Fe, and ^{192}Pt.

9.7. Calculate the energy released if ^{238}U were to spontaneously fission into $^{93}_{36}$Kr and $^{142}_{56}$Ba with the emission of three neutrons.

9.8. If there is energy to be gained by ^{238}U splitting into two lighter fragments with higher B/A values, why has all the ^{238}U not done so?

9.9. If the nuclear surface thickness, $2b$, were twice as great as we observe it to be (if $b \sim 1.0F$) what would this do to the binding energies of nuclei? Discuss the physics behind your answer.

9.10. Explain why the symmetry term and the Coulomb term in the mass formula are negative.

9.11. From Table 9.1 what can you say about the nuclear shapes of ^{2}H and ^{7}Li?

9.12. Since ^{2}H has only one proton and one neutron, what does its having a quadrupole moment tell you about what is happening to the proton and neutron?

9.13. In Table 9.1 both ^{176}Lu and ^{241}Pu have large positive quadrupole moments. How can they have such different magnetic dipole moments?

9.14. Calculate the binding energy per nucleon for ^{16}O, ^{17}O, and ^{18}O.

9.15. Compare the binding energy of the extra neutron when added to ^{15}O, ^{16}O, and ^{17}O. Is the binding energy gained when adding a neutron to ^{15}O the same as the energy needed to extract one neutron from ^{16}O? What do these data tell you about the stability of ^{16}O?

9.16. Compare the extra binding energy when one proton is added to ^{39}K and to ^{40}Ca.

9.17. Compare the one-neutron and two-neutron binding energies (the energy to extract one and two neutrons) for ^{120}Sn. Why is the one-neutron binding energy not half the two-neutron binding energy?

9.18. Compare the one- and two-proton binding energies for ^{48}Ca and discuss how these compare with what you would expect.

9.19. Calculate the energy to be gained if three helium nuclei were fused to form ^{12}C.

9.20. Locate in a periodic chart of the elements the nine odd–odd nuclei which are found in nature.

9.21. Explain in simple terms why ^{7}Li can have spin $\frac{3}{2}$, but ^{13}C has spin $\frac{1}{2}$ as given in Table 9.1.

9.22. Explain in simple terms why even-A ^{176}Lu has spin 7, but even-A ^{238}U has spin zero in Table 9.1.

9.23. Show that a in Eq. (9.2) is the nuclear radius at the point where the charge density is one-half its maximum value.

9.24. Use Eq. (9.2) to calculate the ratio of the charge density of ^{209}Bi to its maximum value when $r = R_0 A^{1/3}$ for $R_0 = 1.2\,\text{fm}$. Compare this with Fig. 9.5.

9.25. Note that ^{48}Ca has 20 percent more nucleons than ^{40}Ca. Compare the nuclear radius of stable ^{48}Ca and ^{40}Ca. How would ^{48}Ca look in Fig. 9.5?

CHAPTER 10

RADIOACTIVE DECAY

Nothing in the world can take the place of persistence. Talent will not; nothing is more common than unsuccessful men with talent. Genius will not; unrewarded genius is almost a proverb. Education will not; the world is full of educated failures. Persistence and determination alone are omnipotent.

Calvin Coolidge

10.1 RADIOACTIVE DECAY LAWS

In 1896, Becquerel discovered uranium radioactivity and with the Curies opened the door to the nucleus of the atom. Soon after, three types of decay were established. The α-, β-, and γ-rays emitted by uranium were the first ambassadors to our world to tell us about their world. Radioactive decay phenomena continue to provide us with many important messages about the internal workings of the nuclear world and many important applications. The separate α and β decay modes were identified by Rutherford in 1899, followed by the identification of γ decay. In discovering more than 2500 new radioactive nuclides, many new types of decay modes have been found with several new modes discovered since 1983. Additional types of radioactivity have been theoretically predicted but have yet to be observed. These include one- and two-neutron radioactivity as well as other heavy cluster radioactivities.

The types of radioactive decays which have been discovered are grouped into the following:

1. *α Decay*. Emission of a doubly ionized ^{4}He nucleus with two units of positive charge.
2. *β Decay*
   - β^- Decay: emission of an electron and antineutrino.
   - β^+ Decay: emission of a positron and neutrino.
   - Electron capture (EC): The nucleus captures one electron which is in an atomic shell outside the nucleus.
   - Double β decay: Emission of two electrons (or positrons) and two antineutrinos (neutrinos) at the same time. There may also occur double β decay by the emission of two electrons without the emission of two neutrinos.

     β^-, β^+ and EC are collectively called β decay.
3. *γ Decay*. Emission of electromagnetic radiation with energies in the range of a few keV up to a few MeV.
   - Internal conversion (IC): The nucleus gives its excitation energy to an atomic electron, and the electron is ejected from the atom.
   - Internal pair creation: The nucleus gives its excitation energy to create an e^-–e^+ pair which carries away the remaining energy.
   - Internal conversion and γ decay are competing processes at all energies. Pair creation competes with them when the available energy is greater than $2m_ec^2(\sim 1.02\,\text{MeV})$.
4. *Spontaneous fission* (SF). The nucleus spontaneously breaks into two nuclei with about the same mass (or on rare occasions three with one low Z) with 0-10 neutrons emitted. This is discussed together with α decay.
   - Heavy cluster radioactivities: Light nuclei like ^{14}C and ^{24}Ne are emitted in a highly asymmetric spontaneous fission but without neutron emission.
5. *More rare decay types*:
   - p Radioactivity: Emission of a proton from a ground or isomeric state.
   - 2p radioactivity; emission of 2 protons from a ground or excited state.
   - β-Delayed p, 2p and 3p emission: After β^+ decay if the nucleus is left in sufficiently highly excited state it can emit one or two protons.
   - β-Delayed n emission: After β^- decay, the highly excited nucleus emits one, two, or more neutrons.
   - β-Delayed d, t or α: After β^+ decay, the highly excited nucleus emits a deuteron (d = ^{2}H$^+$), triton (t = ^{3}H$^+$), or α.
   - β-Delayed spontaneous fission: After β^-, β^+ or EC, the highly excited nucleus fissions into two near equal mass isotopes.

Radioactive decays obey the principle that nature seeks its lowest energy state. Radioactive decay occurs when the products of the decay have less mass than the mass of the initial radioactive nucleus and its electrons. Such decays obey the following conservation rules (laws): (a) conservation of energy; (b) conservation of linear momentum; (c) conservation of angular momentum; (d) conservation of electric charge; (e) conservation of nucleon number, $A = N + Z$; (f) conservation of lepton number (number of light particles). Earlier there was an additional conservation law called parity conservation, but that was shown not to hold in beta decay. Can lepton number conservation be violated? Can the free proton decay and violate nucleon number conservation? These are fundamental questions of major interest today.

10.1.1 The Exponential Decay Law

The nucleus is a quantum system, so a nuclear decay is a spontaneous quantum transformation. Nuclear decays obey quantum statistics. The exact time when a given radioactive nucleus will decay cannot be predicted, but for a large number of radioactive nuclei, their statistical decay law is well known—the exponential decay law.

The number of nuclei which decay in the time dt is given by the decrease in the number of nuclei present, dN. This number is proportional to the number of nuclei present N and the time dt:

$$-dN = \lambda N\,dt \tag{10.1}$$

where λ is a proportionality constant and the negative sign indicates that dN is a reduction in the number of nuclei. After integration, we find

$$N(t) = N_0 e^{-\lambda t} \tag{10.2}$$

where N_0 is the number of nuclei at $t = 0$. This is the exponential decay law.

Rewriting Eq. (10.1) gives

$$\lambda = \frac{-dN/dt}{N}$$

Here the numerator represents the number of nuclear decays in unit time, and the denominator represents the total number of nuclei. Thus, λ represents the probability of decay of one nucleus in unit time and is called the decay constant. The numerical value of λ expresses the statistical probability of decay of each radioactive atom in a group of identical atoms, per unit time. For example, if $\lambda = 0.01\ \mathrm{s}^{-1}$ for a certain isotope, then each of its atoms has a 0.01 (1 percent) chance of decaying in 1 s and a 0.99 (99%) chance of not decaying in any 1 s interval. When any given nucleus will decay is not predicted, but through the specific decay constant its probability of decaying at any moment is predicted.

In many cases a radioactive nucleus has two or more independent and alternative modes of decay. For example, ^{238}U can decay either by α-emission or by spontaneous fission. The isotope ^{64}Cu has probabilities for decay in any of three competing independent ways: β^- emission, β^+ emission, or electron capture. When two or more independent modes of decay are possible, the nuclide is said to exhibit dual decay. The competing modes of decay of any nuclide have independent partial decay constants given by the probabilities $\lambda_1, \lambda_2, \lambda_3, \ldots$ per second, and the total probability of decay is represented by the total decay constant λ, defined as

$$\lambda = \lambda_1 + \lambda_2 + \lambda_3 + \ldots$$

The branching ratios are

$$\frac{\lambda_1}{\lambda} = f_1, \qquad \frac{\lambda_2}{\lambda} = f_2, \qquad \ldots$$

10.1.2 The Half-Life

The half-life for the decay of a nucleus, denoted by $T_{1/2}$, is the time during which half the original number of nuclei decay. When $t = T_{1/2}$, $N = N_0/2$. From (10.2) we have

$$\begin{aligned} \frac{N_0}{2} &= N_0 e^{-\lambda T_{1/2}} \\ T_{1/2} &= \frac{\ln 2}{\lambda} = \frac{0.693}{\lambda} \end{aligned} \tag{10.3}$$

Both $T_{1/2}$ and λ are characteristic constants for a given radioactive isotope. The larger λ is, the shorter $T_{1/2}$ is. Just as there are partial λs in dual decay, there is a partial half-life for each of the dual decay modes:

$$(T_{1/2})_1 = \frac{0.693}{\lambda_1} = \frac{0.693}{f_1\lambda} = \frac{0.693}{f_1 0.693/T_{1/2}} = \frac{T_{1/2}}{f_1}$$

For example, for ^{64}Cu, EC is 41 percent, β^+ is 19 percent, β^- is 40 percent, and $T_{1/2} = 12.7\,\text{h}$. So,

$$f_{EC} = \frac{\lambda_{EC}}{\lambda} = 0.41 \qquad \text{and} \qquad (T_{1/2})_{EC} = \frac{12.7\,\text{h}}{0.41} = 31.0\,\text{h}$$

This would be the half-life of ^{64}Cu, if its β^- and β^+ branches did not exist.

As an example of the half-life, consider the decay of ^{13}N with $T_{1/2} = 9.96$ minutes. After about 10 minutes, the number of ^{13}N nuclei is reduced to one-half its initial number. After another 10 minutes, the half which had not decayed in the first 10 minutes does not all decay, but that half is reduced by one-half again, so one-fourth of the original number remains. This is illustrated in Fig. 10.1, which gives the ratio of $N(t)/N_0$ as a function of time expressed in units of an arbitrary $T_{1/2}$.

10.1.3 The Mean Life

In any population there is the concept of the mean or average life span of the people. Similarly, nuclei have a type of average lifetime called the mean life of a nuclide. (Sometimes this is simply called the lifetime.) When $t = 0$, the number of radioactive nuclides is N_0, and at a time t later, $N(t) = N_0 e^{-\lambda t}$. So, during $t \to t + dt$, the number of nuclei which decay is

$$-dN = \lambda N\, dt$$

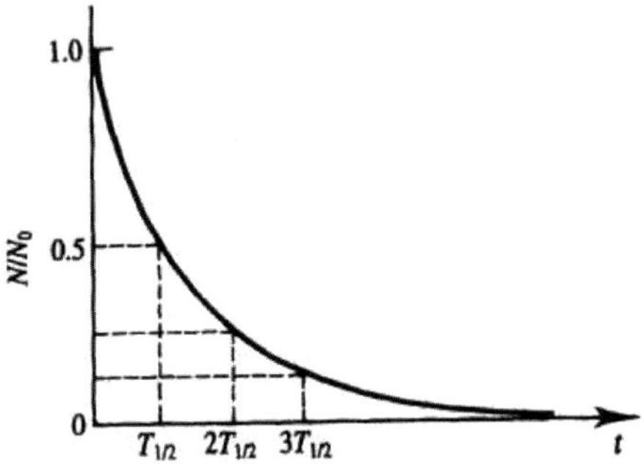

FIGURE 10.1
An illustration of the fraction of radioactive atoms remaining as a function of time in units of an arbitrary half-life $T_{1/2}$.

Each of these nuclei lived for a time t before it decayed in the time dt. The total lifetime of this part of the total N_0 which decayed is

$$\lambda N t \, dt$$

Some nuclei will decay at $t \approx 0$ and some will decay only as $t \to \infty$. The total time over which all the nuclei have lived is the sum over the time of existence of each atom. For a large number of atoms this sum can be approximated as an integral:

$$\int_0^\infty \lambda N t \, dt$$

The mean life ("average" lifetime) is this total time divided by the total number of atoms.

$$\tau = \frac{\int_0^\infty \lambda N(t) t \, dt}{N_0} = \frac{1}{\lambda} = \frac{T_{1/2}}{\ln 2} = 1.44 T_{1/2} \tag{10.4}$$

Note that the mean life is the reciprocal of the decay constant. It is longer than a half-life by the factor 1.44. The mean life (average lifetime) is the time it takes for N to decay to N_0/e, that is, the time for the nuclei to decay to $1/e$ of their original number.

Substituting (10.4) into (10.2), we obtain

$$N = N_0 e^{-1} \simeq 0.37 N_0$$

So after one mean life of a radioactive isotope, the number of nuclei left is 37 percent of the original number.

10.1.4 λ Is a Characteristic Parameter of a Radioactive Nuclide

The decay constant λ, half-life $T_{1/2}$, and mean life τ are related by Eq. (10.4). Each of them has a unique value that characterizes the radioactive decay of each nuclide. Since every radioactive nuclide has its own characteristic λ and no two are exactly the same, λ (or $T_{1/2}$ or τ) is a very important "fingerprint" of a radioactive nucleus.

We can establish the presence of a particular nuclide by measuring its λ. In practice, there are some situations where two nuclides have very similar half-lives, so an accurate measurement is required to distinguish them.

For example, to analyze the carbon impurity in monocrystalline silicon, we can bombard the silicon with protons from an accelerator. If there is carbon in the silicon, ^{12}C can absorb a proton and become ^{13}N. The ^{13}N is a β^+ radioactive isotope. But in natural silicon there is 4.7 percent ^{29}Si, which also can absorb a proton and become ^{30}P. The ^{30}P also is a β^+ radioactive isotope. However, ^{13}N and ^{30}P have different half-lives, 9.96 minutes and 2.5 minutes, respectively. It is easy to distinguish them from a half-life measurement, so we can use this technique to measure the amount of carbon impurity (see Appendix I).

A large number of experiments have shown that the decay constant λ is essentially unchanged by external conditions. Over variations in sample temperature from 24 to 1500 K, of pressure from 0 to 2000 atmospheres or of magnetic field from 0 to 83 000 Gauss, no noticeable changes have been observed in λ. This is to be expected since the nucleus is buried very deep inside the atom and is shielded by the orbiting electrons. In some special situations small variations in λ have been observed. These are all decays that involve the atomic electrons in their decay. For example, the radioactive nuclide ^7Be decays by electron capture (i.e., it captures an atomic electron in the K or L shell). One might expect its decay constant to be related to the state of the outer electrons. When ^7Be is in the metallic state, its decay constant is 0.13 percent larger than in some chemical combinations. When the pressure is increased to 270 kbar (1 atmosphere = 1.013 bar), the ^7BeO volume is reduced 10 percent and λ is increased 0.6 percent. Such small variations in the electron capture and internal conversion of isomers have no important influence.

However, the situation will be quite different in stellar interiors and other astronomical environments where matter is essentially fully ionized. Under terrestrial conditions certain radioactive nuclei decay by the capture of a bound atomic electron. At the high temperatures found in stellar interiors where the atoms are completely ionized, such nuclei can only decay via electron capture from the continuum. In high-energy nuclear cosmic rays not only are there no bound electrons to be captured, there also are extremely few electrons around to be captured from the continuum. Thus, dramatic increases in the half-lives may occur for such cosmic ray nuclei. For example, the half-life of ^7Be is approximately 70 days in the Sun, instead of 53 days as observed on Earth. The half-life of ^{54}Mn is estimated to be $(1\text{–}2) \times 10^6$ years in cosmic rays, compared to 312 days on Earth.

10.1.5 Activity

Let us define A to be the number $(-dN/dt)$ of radioactive nuclei decaying in unit time. Then

$$A(t) \equiv -\frac{dN}{dt} = \lambda N(t) = \lambda N_0 e^{-\lambda t} = A_0 e^{-\lambda t} \tag{10.5}$$

The activity, A, also obeys the exponential law. The parameter which measures the activity (the number of disintegrations per second) is $A(t)$ where $A = \lambda N$ is the product of λ and N. In the case of dual decay, the partial activities $\lambda_1 N$, $\lambda_2 N, \ldots$ also decrease as $e^{-\lambda t}$ not as $e^{-\lambda_1 t}, \ldots$ because

$$\frac{\lambda_1 N}{\lambda_1 N_0} = \frac{N}{N_0} = e^{-\lambda t}$$

Note that the decrease of each partial activity with time is related to the depletion of the total supply of atoms N, and this depletion depends on the combined action of all the competing decay modes. The activity depends on $\lambda(T_{1/2})$ that is, for a given number of atoms of a substance, the activity increases as λ increases. For example, the activity of one gram of ^{235}U is much greater than that of one gram of ^{238}U because the half-life of ^{235}U is much shorter than that of ^{238}U.

A commonly used unit of activity is the curie (Ci), though in practice sub-multiples are used:

1 Ci = 3.7×10^{10} disintegrations/s
1 mCi = 3.7×10^{7} disintegrations/s
1 μCi = 3.7×10^{4} disintegrations/s

The activity of 1 g of ^{226}Ra is about 1 Ci, and the earlier definition of 1 Ci was the number of disintegrations in 1 g of radium in 1 s. The SI unit of activity is the becquerel (Bq):

1 Bq = 1 disintegration/s
1 Ci = 3.7×10^{10} Bq

For example, in natural potassium there is 0.012 percent abundance of ^{40}K, which is radioactive. It is present in various glasses and in our bodies and it is the principal source of radioactivity in all human beings. Its half-life is 1.3×10^{9} years: Its λ is extremely small, so its activity is correspondingly weak. Each person has about 0.1 μCi (3.7×10^{3} Bq) of radioactive ^{40}K in their body; this level of ^{40}K is too weak to significantly affect our health.

There are also units that indicate the effect on matter of absorbing nuclear radiation. The curie and becquerel are activity units which describe properties of the radioactive isotope itself. The roentgen (abbreviation Ren or R), rad and gray are units which measure the effects that the radiation produces on materials. These effects are determined not only by the intensity (activity) of the radioactive material, but also by the character of the radiation (α, β, γ, n, etc.) and the properties of the material which receives the radiation.

1 Roentgen (1 R) is the radiation intensity which produces 2.58×10^{-4} C of charge in 1 kg of air.

1 rad is an absorbed dose of 0.01 J of radiation energy in 1 kg of radiated material (originally defined as 100 erg/g).

1 gray (1 Gy) is an absorbed dose of 1 J of radiation energy in 1 kg of radiated material (1 Gy = 100 rad). The gray is the SI unit of absorbed radiation dose.

There are also units that measure biological effects. The relative biological effectiveness (RBE) measures the biological effectiveness of a particular type of ionizing radiation in producing a specific biological effect relative to x-rays of about 200 keV. The rem is one unit of RBE compared to 1 rad of x-radiation. However, the absorption of equal amounts of energy per unit mass for different irradiation conditions does not insure the same biological effect on a living system. Differences of up to a factor of 10 in the biological effect can occur depending on whether heavy ions or electrons are the source of the energy deposition. Ionizing radiation can alter the chemical makeup of biological molecules to produce biological damage. The linear energy transfer, L, which is essentially dE/dX, the rate of energy loss with distance along the path, determines the amount and permanency of biological damage. So, even when the total energy deposited per unit mass is the same, radiations such as heavy charged particles with large L values give rise to much greater biological effects than radiations such as electrons that have low L values. Thus, the concept of dose equivalent was introduced to better measure the likely biological effect of a particular radiation exposure. A dose equivalent unit is defined as the amount of any type of radiation which, when absorbed in a biological system, leads to the same biological effect as one unit of absorbed dose delivered by low-L radiation. The dose equivalent, H, is found by multiplying the product of the absorbed dose, D, and the quality factor, Q, that characterizes the specific radiation:

$$H = DQ$$

where Q increases as the linear energy transfer, L, increases. If D is in rads, then H is in rem (roentgen equivalent man). More recently, new units have been introduced. When D is expressed in grays, the dose equivalent is called the sievert (SV), where 1 SV = 100 rem. Fast electrons, including most β-rays, have sufficiently low L value that they have $Q \simeq 1$ in all cases. Since the energy of x-rays and γ-rays is transferred to create fast electrons, they likewise have $Q \simeq 1$. Heavy charged particles have much higher L values and larger Q, so their dose equivalent is larger than the absorbed dose. At the energies typical for α-particles, $Q = 20$. Thus, we can understand why α-emitting sources trapped in the lungs can be so dangerous in producing lung cancer. Neutrons transfer most of their energy to form heavy charged particles, so $Q_n \gg 1$ and varies with neutron energy. For an absorbed dose of 3 Gy delivered by α-particles with $Q = 20$, one finds $H = 3 \times 20 = 60$ SV. Radiation exposure limits for radiation workers are given in units of dose equivalent.

10.1.6 Measurement of a Long Half-Life

The half-life is a fingerprint of a radioactive nuclide. Since determining the half-life is an important method of identifying a radioactive nuclide, let us look at how this may be done. A simple method is to measure the radioactive intensity at some moment and then measure the time during which the activity reduces to one half. For greater accuracy one repeats the measurement over several half-lives. The half-life can be extracted from a weighted least-squares fit of a log plot of the activity vs. time. However, this method cannot be used for radioactivities with very long lifetimes. For example, the ^{238}U half-life is 4.5×10^9 years. We cannot wait until it has decayed to one-half or even to any reasonable fraction of the initial number. For radioactivities with extremely short lifetimes, again this method is not easily applied. Here we discuss only the measurement of a very long half-life.

For uranium, we could measure the activity A of a sample and calculate N from its mass and Avogadro's number. Then from $A = \lambda N$, we can obtain λ. For example, we will find the activity of 1 mg of ^{238}U to be $A = 740\,\alpha$-particles/min $= 12.3\ \alpha$-particles/s; and 238 g of uranium contains 6.0×10^{23} atoms. So,

$$\lambda = \frac{A}{N} = \frac{12.3\,\mathrm{d/s}}{6 \times 10^{20}/238} = 4.88 \times 10^{-18}\ \mathrm{s}^{-1}$$

and

$$T_{1/2} = \frac{0.693}{\lambda} = \frac{0.693}{4.88 \times 10^{-18}\,\mathrm{s}^{-1}} = \frac{1.42 \times 10^{17}\,\mathrm{s}}{3.154 \times 10^{7}\,\mathrm{s/yr}} = 4.5 \times 10^{9}\ \text{years}$$

The longer the half-life, the weaker the activity. Remember that the statistical error in a certain number of counts is given by the square root of the number of counts. If the total number of counts is 900 in some reasonable length of time, the statistical error in the half-life is about 3 percent; for 10^4 counts, the error is reduced to 1 percent. In order to get sufficient counts to obtain good accuracy we must increase N.

Suppose I had 0.1 μg of ^{252}Cf, which has a half-life of 2.65 years. What is the disintegration rate per second of this sample and the activity in microcuries?

$$\frac{dN}{dT} = \lambda N = \frac{0.693}{T_{1/2}} \times N = \frac{0.693}{2.65\,\mathrm{y} \times 3.15 \times 10^{7}\,\mathrm{s/y}} \times \frac{6.02 \times 10^{23}\ \text{atoms}}{252\,\mathrm{g}} \times 10^{-7}\,\mathrm{g}$$

$$= 2.0 \times 10^{6}\,\mathrm{d/s} = \frac{2.0 \times 10^{6}\,\mathrm{d/s}}{3.7 \times 10^{4}\,\mathrm{d/s/\mu Ci}} = 54\,\mu\mathrm{Ci}$$

Since ^{252}Cf undergoes both α decay (96.9 percent of the time) and spontaneous fission (SF) (3.1 percent), how many spontaneous fissions occur per second?

$$\text{SF rate} = 2.0 \times 10^{6}\,\mathrm{d/s} \times 0.031 = 6.2 \times 10^{4}\ \text{fission/s}$$

10.1.7 Simple Cascade Decay

Many radioactive nuclides do not become stable nuclei through just one decay. Their daughter nuclei are also radioactive. The daughters continue to decay until they arrive at a stable nuclide. This is a cascade decay. There are three cascade decay chains found in nature and a fourth in the series whose initial parent ^{237}Np is too short-lived for this chain to be found in nature. These four series are ordered by their mass number A, beginning with the thorium series with $A = 4n$ (n an integer), followed by $A = 4n + 1$, $4n + 2$ and $4n + 3$ as shown in Figs. 10.2–10.5. Note that each of the naturally occurring cascades begins with an isotope whose half-life is on the order of a billion years or more—a time which is comparable to the age of the Earth. We can only follow the decay of a radioactive isotope for about ten half-lives, so to be observed in nature an isotope must have a half-life no shorter than one-tenth the age of the Earth whose age is $\approx 5 \times 10^9$ years or be continuously produced by the decay of such a long-lived isotope or by some nuclear reaction that continues to occur in nature.

In Table 10.1 are given the decay properties of the well-established cases of long-lived radioactivities found in the Earth's surface. Geological age measurements are based on studies of these long-lived isotopes, especially ^{40}K, ^{87}Rb, ^{232}Th, ^{235}U and ^{238}U. For example, today uranium is a mixture of 99.3 percent abundant ^{238}U ($T_{1/2} = 4.5 \times 10^9$ y) and 0.72 percent ^{235}U ($T_{1/2} = 0.70 \times 10^9$ y). Based on our current understandings of heavy element production in the earlier stars that produced the material for our Earth, these should have been present in roughly equal amounts in our Earth when it was formed as well as other uranium isotopes like ^{236}U. The difference in the abundances of ^{235}U and ^{238}U from when the Earth was formed until now indicates they have been decaying for about five

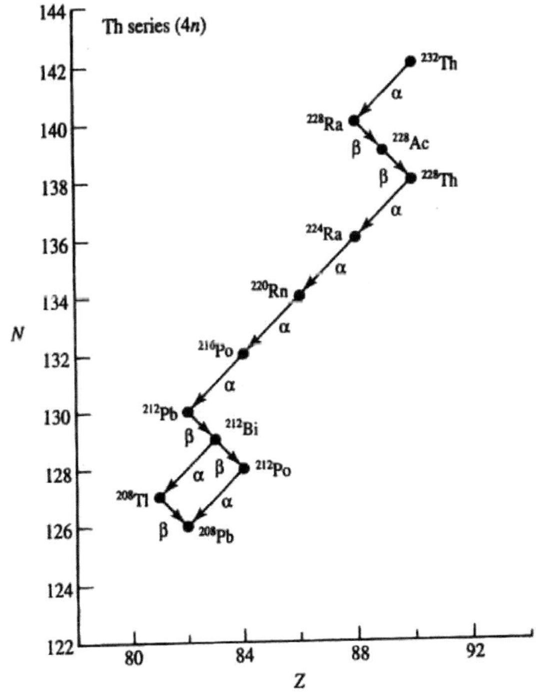

FIGURE 10.2
The first decay cascade with $A = 4n$, the thorium chain which is found in nature.

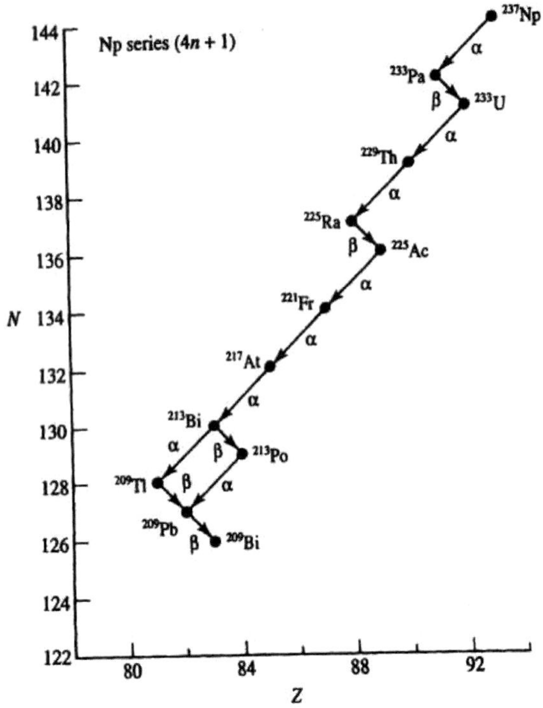

FIGURE 10.3
The second decay cascade with $A = 4n + 1$, the neptunium chain is not found in nature, but some of its decay products are found, so it was present in the early formation of the Earth.

billion years, which in turn is the age of the Earth. Assume $N_0(238) = N_0(235)$ when the Earth was formed (actually $N_0(235) \simeq 0.29N_0(238)$ is thought to be correct). Now,

$$\frac{N_{235}}{N_{238}} = \frac{N_0\, e^{-\lambda_{235}t}}{N_0\, e^{-\lambda_{238}t}} = e^{(\lambda_{238}-\lambda_{235})t} = \exp\left[0.693\left(\frac{1}{4.5 \times 10^9} - \frac{1}{0.7 \times 10^9}\right)t\right]$$

$$= e^{-0.836t} = 0.0072$$

$$t = 5.9 \times 10^9 \text{ years}$$

There is geophysical evidence that ^{236}U was present in the Earth based on observation of products formed in its decay. But even if equally present at the Earth's formation, ^{236}U with $T_{1/2} = 2 \times 10^7$ y would have effectively decayed away to where it would be undetectable in a few billion years since we can only detect a radioactive element for about 10 half-lives. Its absence today again suggests a few billion years age for our Earth. Short-lived activities present in the Earth include ^{14}C ($T_{1/2} = 5600$ y), because it is produced in our atmosphere by cosmic rays in the reaction $^{14}\text{N} + \text{n} \to (^{15}\text{N})^* \to {}^{14}\text{C} + \text{p}$. Carbon-14 activity is useful for dating objects which have lived during the last $\sim 30{,}000$ y.

To understand a cascade decay, consider a simple two-step decay:

$$\text{A} \to \text{B} \to \text{C}$$

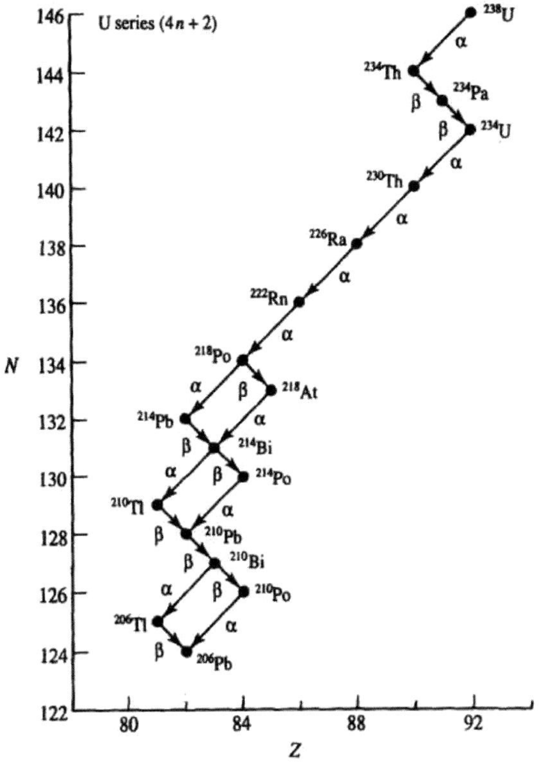

FIGURE 10.4
The third decay cascade with $A = 4n + 2$, the uranium chain which is found in nature.

Decay of A obviously obeys the exponential decay law. But what about B? On the one hand it is lost by decay but on the other it is produced by decays from A. So the number of B nuclei changing in a unit time is

$$\frac{dN_B}{dt} = \lambda_A N_A - \lambda_B N_B \tag{10.6}$$

Here N_A is given by

$$N_A = N_{A_0} e^{-\lambda_A t}$$

where N_{A_0} is the number of atoms of A at $t = 0$. If $N_{B_0} = 0$ at $t = 0$, the solution of (10.6) is

$$N_B = N_{A_0} \frac{\lambda_A}{\lambda_B - \lambda_A} (e^{-\lambda_A t} - e^{-\lambda_B t}) \tag{10.7}$$

Thus, the number of daughter atoms present at a given time depends not only on its own decay constant but also on its parent's (A's) decay constant. Equation (10.7) is different from the simple exponential law. If there were N_{B_0} initial atoms present, too, then one would need to add the term $N_{B_0} e^{-\lambda_B t}$ to Eq. (10.7).

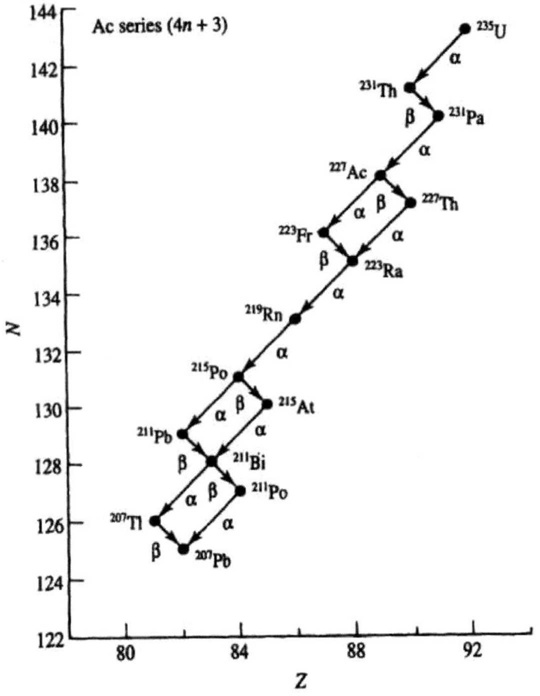

FIGURE 10.5
The fourth decay cascade with $A = 4n + 3$, the actinium chain which is found in nature.

TABLE 10.1
Parent radioactive nuclides found in nature because of their very long half-lives

Atomic number, Z	Mass number, A	Percentage abundance in nature	Half-life (years)	Radioactive decays observed
^{19}K	40	0.0117	1.3×10^9	β^-, EC, β^+
^{37}Rb	87	27.83	4.8×10^{10}	β^-
^{48}Cd	113	12.2	9×10^{15}	β^-
^{49}In	115	95.77	4.4×10^{14}	β^-
^{52}Te	123	0.91	1.3×10^{13}	EC
^{52}Te	130	34.49	2×10^{21}	Growth of $^{130}_{54}$Xe[a]
^{57}La	138	0.089	1.3×10^{11}	β^-, EC, γ
^{60}Nd	144	23.8	2.1×10^{15}	α
^{62}Sm	147	15.07	1.1×10^{11}	α
^{62}Sm	148	11.3	7.1×10^{15}	α
^{64}Gd	152	0.20	1.1×10^{14}	α
^{71}Lu	176	2.6	3.6×10^{10}	β^-, γ
^{72}Hf	174	0.16	2.0×10^{15}	α
^{75}Re	187	62.6	5.1×10^{10}	β^-
^{76}Os	186	1.6	2.0×10^{15}	α
^{78}Pt	190	0.013	6×10^{11}	α
^{90}Th	232	100	1.4×10^{10}	α, γ
^{92}U	235	0.715	7.0×10^8	α, γ
^{92}U	238	99.28	4.5×10^9	α, γ

[a]Indirect evidence for double β decay.

Consider the special situation where the half-life of the daughter nuclei is less than that of its parent. From (10.7), when $t \gg 1/\lambda_B$, we have

$$N_B \simeq N_{A_0} \frac{\lambda_A}{\lambda_B - \lambda_A} e^{-\lambda_A t} \tag{10.8}$$

At such time, the daughter nuclei decay according to the parent's half-life. This condition is called transient equilibrium. This is an important result which has practical applications. For example, it can give a way to transport a short half-life daughter nuclide for use elsewhere. In hospitals the radioactive nuclide ^{113}In, which has a half-life of only 99.5 minutes, is used often in diagnostic procedures. If we produce ^{113}In in a reactor and transport it to a hospital, little is left. For example, transportation alone from Beijing to Shanghai, or even from Oak Ridge National Laboratory to nearby Nashville or Atlanta takes about 3 hours or more. Only one-fourth of the original ^{113}In would be left even if it were in a chemical form suitable for use. Preparation for use adds to the time and the amount of decay.

We also can obtain ^{113}In from the decay of ^{113}Sn. Because $^{113}\text{Sn} \rightarrow {}^{113}\text{In}$ has a half-time of 115 days, Eq. (10.8) shows that, after some time, ^{113}In decays with the ^{113}Sn half-life. In addition, when $t \gg 1/\lambda_B$, if $\lambda_A \ll \lambda_B$, as in the case of ^{113}Sn–^{113}In, λ_A can be dropped and Eq. (10.8) gives

$$\lambda_A N_A \simeq \lambda_B N_B \tag{10.9}$$

In this case the parent's activity is equal to the daughter's. They are in equilibrium. This special case, $\lambda_A \ll \lambda_B$ when Eq. (10.9) holds, is called secular equilibrium. The number of daughter nuclei decaying is equal to the number being produced from its parent's decays. A hospital wishing to use ^{113}In in diagnostic studies orders a mixed ^{113}Sn–^{113}In radioactive source. The ^{113}In can be separated from ^{113}Sn by a chemical method in which a liquid passing through the material that holds the mixed source carries the ^{113}In out and leaves the ^{113}Sn behind. After the ^{113}In is separated out, the ^{113}Sn decay continues to produce ^{113}In. After some time, a new ^{113}In sample can be separated again. This situation is like obtaining milk from a cow, so ^{113}Sn is sometimes called a "cow" or a "generator" from which ^{113}In can be milked regularly.

To see how often a generator can be milked, let us calculate the number of atoms of ^{113}In 6 hours after milking the ^{113}Sn generator and see how this number compares with the number at equilibrium where $\lambda_B N_B = \lambda_A N_A$:

$$\begin{aligned} N_B &\approx N_A \frac{\lambda_A}{\lambda_B - \lambda_A} e^{-\lambda_A t} \simeq N_A \frac{\lambda_A}{\lambda_B} e^{-\lambda_A t} \\ &= N_A \frac{\lambda_A}{\lambda_B} \exp\left(\frac{-0.693 \times 6}{115 \times 24}\right) = 0.998 \frac{\lambda_A N_A}{\lambda_B} \end{aligned}$$

So $\lambda_B N_B = 0.998\, \lambda_A N_A$ 6 hours after the ^{113}In was removed.

Equation (10.9) also provides a method of measuring the short half-life in such cases:

$$T_{1/2_B} = \frac{N_B}{N_A} \times T_{1/2_A} \tag{10.10}$$

By measuring $T_{1/2A}$ (the long lifetime), N_A, and N_B, we can calculate the short lifetime.

10.1.8 Isotope Production

In addition to the more than 60 radioactive nuclides that are found in nature (see Table 10.1 and Figs. 10.2–10.5), more than 2000 others have been produced by different methods in the laboratory. These are often called *artificial* radioactive nuclides in reference to the fact they are produced in the laboratory and do not naturally occur in nature. Induced radioactivities is a better name. They are produced in reactors or accelerators through nuclear reactions. While such an isotope is being produced, it is also decaying simultaneously. The rate of change of N is

$$\frac{dN}{dt} = P - \lambda N \tag{10.11}$$

where P is the nuclide "productivity" rate (number produced per unit time) and N is the nuclide number at the beginning of the production time. In certain nuclear reactions, P is a constant. Rewriting the above equation, one has

$$\frac{dN}{dt} + \lambda N = P$$

This is a nonhomogeneous differential equation whose solution is

$$N = \frac{P}{\lambda}(1 - e^{-\lambda t})$$

The activity is

$$A = \lambda N = P(1 - e^{-\lambda t}) = P(1 - 2^{-t/T_{1/2}}) \tag{10.12}$$

We can see that after the time $t = T_{1/2}$, A is one-half of P. After $2T_{1/2}$, $A = P(1 - 2^{-2})$ is 75 percent of P (see Table 10.2 or Fig. 10.6). No matter how long the production time is, the maximum activity A is the production rate P.

In a practical situation, it is not economically sensible to extend the production for a time longer than three to four half-lives. From Fig. 10.6 and Table 10.2 we can see that the activity of a nuclide being produced does not increase linearly

TABLE 10.2

$t/T_{1/2}$	A/P
1	0.5
2	0.75
3	0.875
4	0.9375
5	0.9688

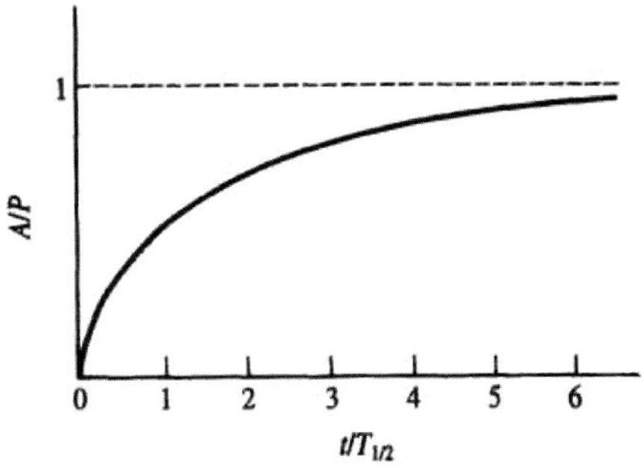

FIGURE 10.6
An isotope production diagram where the ratio of activity to production is plotted versus time in units of half-life.

with the time. When $t \geq 3T_{1/2}$, the activity increases slowly. When the time is over $5T_{1/2}$, the activity is basically unchanging.

10.2 ALPHA, PROTON, HEAVY CLUSTER AND SPONTANEOUS FISSION DECAYS

10.2.1 The Necessary Condition for α Decay

Alpha decay can be written as

$$^{A}_{Z}X_{N} \rightarrow ^{A-4}_{Z-2}Y_{N-2} + \alpha(^{4}_{2}He_{2}) \tag{10.13}$$

Note the conservation of charge and nucleon number. Before decaying, the parent nucleus X is at rest. From energy conservation, we have

$$m_X c^2 = m_Y c^2 + m_\alpha c^2 + E_\alpha + E_r \tag{10.14}$$

where, m_X, m_Y, and m_α are the parent nucleus, daughter nucleus, and α particle rest masses, respectively, and E_α and E_r are the kinetic energies of the α and recoiling daughter. We change the nuclear mass into atomic masses:

$$m_X = M_X - Zm_e$$
$$m_Y = M_Y - (Z-2)m_e$$
$$m_\alpha = M_{He} - 2m_e$$

Here we neglect the binding energies between the electrons and nucleus and M_X, M_Y, and M_{He} are the X, Y, and helium atomic masses, respectively, and m_e is the electron mass. The α decay energy $E_0 = E_\alpha + E_r$ is

$$E_0 = [M_X - (M_Y + M_{He})]c^2 \tag{10.15}$$

Obviously, if α decay occurs, $E_0 > 0$ and

$$M_X(Z, A) > M_Y(Z-2, A-4) + M_{He} \tag{10.16}$$

Thus, for a nuclide to undergo α decay, the parent mass (before decay) must be greater than the sum of the masses of the daughter and the helium atom produced in the decay. For example,

$$^{210}_{84}Po \rightarrow ^{206}_{82}Pb + \alpha$$

Here:

$$M(^{210}\text{Po}) = 209.9829\,\text{u}$$
$$M(^{206}\text{Pb}) = 205.9744\,\text{u}$$
$$M(^{4}\text{He}) = 4.0026\,\text{u}$$

So Eq. (10.16) is satisfied. The ^{210}Po may undergo α decay, which is observed in experiments:

$$E_0 = (209.9829 - 205.9744 - 4.0026\,\text{u})c^2 = 0.0059\,\text{u} \times 931.5\,\text{MeV/u} = 5.5\,\text{MeV}$$

Similarly, one can show that α decay cannot occur in ^{64}Cu, for example, because Eq. (10.16) is not satisfied.

10.2.2 α Decay Energies and Energy Level Diagrams

The α decay energy Q_0 is an important parameter. According to Eq. (10.15), we can get the decay energy Q_0 from the atomic masses. But in some cases, for example, a newly discovered nuclide, the nuclear and atomic masses before decay are not known. We can get $Q_0 = E_\alpha + E_r$ by measuring the kinetic energies, E_α and E_r, then use Eq. (10.15) to get the mass of the unknown nuclide. If M_Y is also unknown, then Q_0 tells us only about the mass difference between the parent and daughter, but even this is an important parameter for comparison with theoretical calculations of masses. Since the daughter nuclear mass is large compared to that of the α, the recoil kinetic energy E_r is very small and difficult to measure. However, we can deduce a relation between E_r and E_α from momentum conservation, so that a measurement of E_α yields Q_0.

If the parent nucleus is at rest, the initial momentum is zero. Then

$$m_Y \mathrm{v}_Y = m_\alpha \mathrm{v}_\alpha$$

The daughter nucleus recoil energy is

$$E_r = \frac{1}{2} m_Y \mathrm{v}_Y^2 = \frac{1}{2} m_\alpha \mathrm{v}_\alpha^2 \frac{m_\alpha}{m_Y} = \frac{m_\alpha}{m_Y} E_\alpha$$

So,

$$Q_0 = E_\alpha + E_r = \left(1 + \frac{m_\alpha}{m_Y}\right) E_\alpha \simeq \left(1 + \frac{4}{A-4}\right) E_\alpha = \frac{A}{A-4} E_\alpha \qquad (10.17)$$

For ^{210}Po,

$$Q_0 = \frac{209.9829}{205.9744} E_\alpha \approx \frac{210}{206} E_\alpha$$

Thus, we can use the nuclear mass number ratio instead of the nuclear masses because the error introduced is very small (the order of 1 part in 10^4 or 1 keV out of 10 MeV). Equation (10.17) is important because it shows that the decay energy can be obtained from a measurement of the α kinetic energy, E_α.

Experimentally α-particle kinetic energies can today be measured with high precision using solid-state detectors; earlier, magnetic spectrometers were used. In many α decays, α-particles (α-groups) with several different energies are observed. For example, when ^{212}Bi(ThC) decays to ^{208}Tl there are six different α-particles (α-groups) observed, as shown in Fig. 10.7. From the figure we can obtain each group's α-particle kinetic energy, E_α, and then using Eq. (10.17) calculate the Q_0 as listed in Table 10.3. From Fig. 10.7 we can see that an α-particle energy spectrum has discrete energy lines and not a continuous energy distribution. The different decay energies tell us about the discrete energy states in the daughter nucleus, since α decay takes place between two discrete, fixed nuclear energy states. We can draw a nuclear energy-level diagram for the daughter from such data. When ^{212}Bi releases α_0 (the α with the maximum kinetic

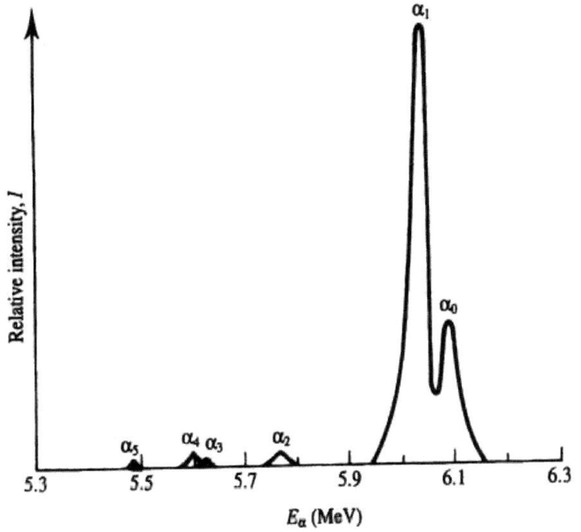

FIGURE 10.7
The α energy spectrum in the decay of ^{212}Bi.

TABLE 10.3
The α-particle energies and decay energies of ^{212}Bi(ThC)

Group	E_α	Q_0
α_0	6.090	6.207
α_1	6.051	6.167
α_2	5.768	5.879
α_3	5.626	5.734
α_4	5.607	5.715
α_5	5.481	5.586

energy), it decays to the ground state of ^{208}Tl, with a maximum decay energy, Q_0, denoted by E_{00}. Here the second zero subscript indicates that the energy corresponds to decay to the ^{208}Tl ground state. When ^{212}Bi emits α_1, it decays into an excited state of ^{208}Tl, which we would assume is the first excited state with the decay energy E_{01}. After this decay, the nucleus in this excited state can decay by γ-ray emission and return to the ground state. The γ-ray energy is the energy difference between the excited and ground states and also is equal to the difference of the corresponding α decay energies. Thus, we should see γ-rays with energies.

$$E_{00} - E_{01} = 0.040\,\text{MeV} = E_{\gamma_1}$$
$$E_{00} - E_{02} = 0.328\,\text{MeV} = E_{\gamma_2}$$
$$E_{00} - E_{03} = 0.473\,\text{MeV} = E_{\gamma_3}$$
$$E_{00} - E_{04} = 0.493\,\text{MeV} = E_{\gamma_4}$$
$$E_{00} - E_{05} = 0.621\,\text{MeV} = E_{\gamma_5}$$

As shown in Fig. 10.8, the above five γ-rays are also observed in the ^{212}Bi decay. That these γ-rays follow after the five different α-groups $(\alpha_1, \alpha_2, \ldots)$ as expected can be established by an α–γ coincidence experiment. That they do follow the α decay was first established by studying the energies of the competing internal-

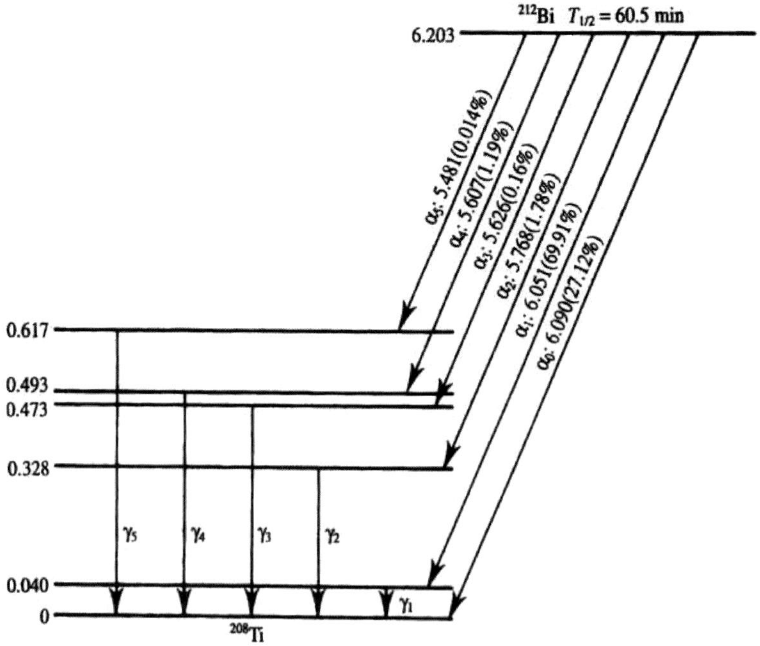

FIGURE 10.8
The ^{212}Bi α decay and ^{208}Tl energy level diagram. The energies are in MeV.

conversion electrons whose energies reflected the K, L,... shell binding energies in the daughter. For example, the energy of the K internal conversion electrons for the E_{00}–E_{02} transition is $E_{e_{K2}} = E_{\gamma_2} - E_{B_K}$. No γ-ray follows the α_0 decay here, but it is possible that the α_0 decay is not between two ground states, and then even α_0 can be followed by a γ-ray. In principle, one could also see γ-rays corresponding to transitions from one of the higher excited states to a lower excited state such as $E_{02} - E_{03} = 0.145\,\text{keV}$. Whether such decay branches occur depends on the spins, parities, and nuclear configurations of the states, as will be discussed later.

10.2.3 α Decay Mechanism and Half-Life

The α-particle emitted in α decay comes from inside the nucleus where the α-particle is held in by the attractive nuclear force (the negative potential). Once outside the nucleus, the α-particle is repelled by the Coulomb force barrier. Thus, a potential is formed as shown in Fig. 10.9. From classical physics, an α-particle whose energy is below the potential barrier cannot escape from inside the nucleus nor can it come into the nucleus from the outside. Outside the α would undergo Rutherford scattering by the Coulomb potential barrier.

Let us estimate the height of the potential barrier, which is the Coulomb potential energy when the α-particle and parent nucleus are just touching, so that the distance of separation of the two charges is about the sum of the nuclear radius and α-particle radius:

$$E_B = \frac{2Ze^2}{R} = \frac{2Z \times 1.44\,\text{MeV} \cdot \text{fm}}{1.2(A^{1/3} + A_\alpha^{1/3})\,\text{fm}} = 2.4\frac{Z}{(A^{1/3} + A_\alpha^{1/3})}\,\text{MeV} \qquad (10.18)$$

Here the nuclear radius formula (9.3) with $r_0 = 1.2\,\text{fm}$ is used and Z, A, and A_α are the nuclear charge, nuclear mass number, and α-particle mass number,

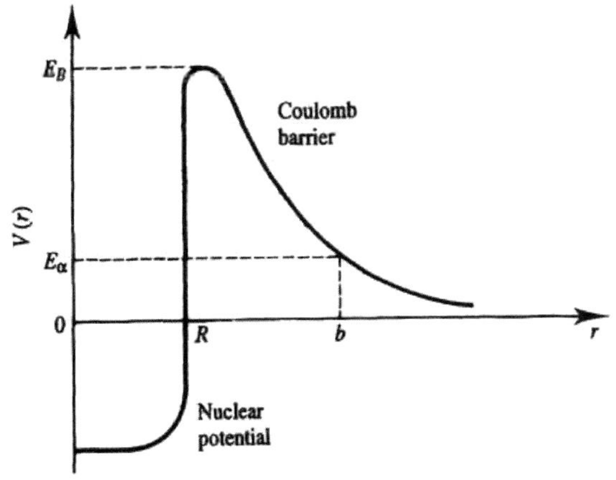

FIGURE 10.9
The nuclear potential barrier as seen by an α particle inside a nucleus.

respectively. We have also used $e^2 = 1.44\,\text{fm}\cdot\text{MeV}$, so the unit of E is MeV. For example, for $^{212}_{84}\text{Po}$, we obtain

$$E_B = \frac{2.4 \times 84\,\text{MeV}}{212^{1/3} + 4^{1/3}} \simeq 26\,\text{MeV}$$

But when ^{212}Po undergoes α decay, the kinetic energy of the α-particle is 8.78 MeV. This energy is far below the potential barrier height. In classical physics, an α-particle cannot escape to outside of the nucleus. However, in quantum theory, there is a probability of penetrating the potential barrier (see Section 8.1). We can use Eq. (8.40) to estimate the probability of an α-particle getting out of a nucleus. We rewrite (8.40):

$$\begin{aligned} P &= e^{-G} \\ G &= \frac{2}{\hbar} D\sqrt{2m(V-E)} \qquad D = x_2 - x_1 \end{aligned} \tag{10.19}$$

This is fitted to a rectangular well potential, but any arbitrary shape potential barrier can be cut apart into many small rectangular potential barriers. Thus,

$$G = 2\sqrt{\frac{2m}{\hbar^2}} \int_R^b \sqrt{V-E}\, dr \tag{10.20}$$

For the Coulomb potential barrier (Fig. 10.9), the integration has an analytical solution. Since

$$b = \frac{Z_1 Z_2 e^2}{E} \tag{10.21}$$

we have

$$\begin{aligned} \int_R^b \left[\frac{Z_1 Z_2 e^2}{r} - E\right]^{1/2} dr &= \sqrt{Z_1 Z_2 e^2} \int_R^b \left(\frac{1}{r} - \frac{1}{b}\right)^{1/2} dr \\ &= b\left(\frac{Z_1 Z_2 e^2}{b}\right)^{1/2} \left[\cos^{-1}\sqrt{\frac{R}{b}} - \sqrt{\frac{R}{b}\left(1 - \frac{R}{b}\right)}\right] \\ &= \sqrt{Z_1 Z_2 e^2 b}\; F\left(\frac{R}{b}\right) \end{aligned} \tag{10.22}$$

When $E \ll E_B$, $b/R \gg 1$, in a first approximation, we find

$$F\left(\frac{R}{b}\right) \simeq \frac{\pi}{2} - 2\sqrt{\frac{R}{b}}$$

So we have

$$G \simeq 2\sqrt{\frac{2m}{\hbar^2}}\, \frac{Z_1 Z_2 e^2}{\sqrt{E}} \left(\frac{\pi}{2} - 2\sqrt{\frac{R}{b}}\right) \tag{10.23}$$

For an α-particle, $Z_1 = 2$, $mc^2 = 3750\,\text{MeV}$, $Z_2 = Z$ is the daughter's nuclear charge number, and $e^2/\hbar c = 1/137$; then

$$G \simeq \frac{4Z}{\sqrt{E_\alpha}} - 3\sqrt{ZR} \tag{10.24}$$

where E_α is the kinetic energy of the emitted α-particle in MeV and $R \simeq 1.2(A^{1/3} + A_\alpha^{1/3})$ fm. Now we can obtain $P(e^{-G})$, the probability of an α penetrating the potential barrier. However, we also must know how many times the α-particle strikes the potential barrier in a second (the collision frequency, n). If we know n, we can find the α-particle penetration probability nP per second, which is the α decay probability λ. This number is the ratio of the α-particle velocity to the distance across the nucleus (twice the parent radius R_p) (the inverse of the time it takes an α to cross the nucleus)

$$n = \frac{\text{v}}{2R_p} \tag{10.25}$$

If an α-particle's kinetic energy is E_k (MeV) inside the nucleus, its velocity is

$$\text{v} = \left(\frac{2E_k}{m_\alpha}\right)^{1/2} = c\left(\frac{2E_k}{m_\alpha c^2}\right)^{1/2} = c\left(\frac{2E_k}{3750}\right)^{1/2} \simeq \sqrt{E_k}\,6.9 \times 10^6\ \text{m/s}$$

Thus,

$$n \simeq (2.88 \times 10^{21}) A_p^{-1/3} E_k^{1/2}\ \text{s}^{-1}$$

We have used $R_p = r_0 A_p^{1/3}$ for the parent radius, where A_p is the parent nuclear mass number.

Now we can estimate the α decay mean life:

$$\tau = \frac{1}{\lambda} = \frac{1}{nP} \simeq (3.5 \times 10^{-22}) A_p^{1/3} \frac{1}{\sqrt{E_k}} e^{(4Z/\sqrt{E_\alpha} - 3\sqrt{ZR})} \tag{10.26}$$

So, we can obtain the following relation between τ and E_α:

$$\ln \tau = AE_\alpha^{-1/2} + B \tag{10.27}$$

where the τ is in seconds, E_α is in MeV, and A and B are constants which are different for different parent nuclei. This provides an understanding of the empirical Geiger and Nuttall law which relates the decay constant λ and the range of an α-particle ($R \sim E_\alpha : \log_{10} \lambda = a + b \log_{10} R_\alpha$). This empirical law was derived from experiments in 1911 long before the new quantum mechanics.

Because the above is a rough estimation, we do not expect the calculated result to be very quantitatively consistent with the experimental data. But the relation between τ and the α-particle kinetic energy E_α (10.27) is consistent with experiments, as seen in Fig. 10.10. From this we can understand why, when E_α varies by only a factor of 2, τ varies by 10^{20} (see Table 10.4)! The agreement of theory and experiment shows the correctness of using the potential barrier penetration to explain α decay. Exact calculations give results which are

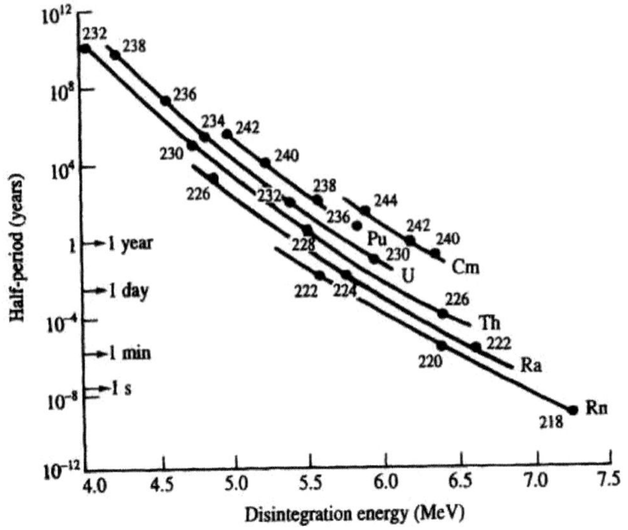

FIGURE 10.10
A plot of the logarithm of $T_{1/2}$ for α decay as a function of the α energy to illustrate their nearly linear dependence. The Z dependence is also seen.

TABLE 10.4
Experimental energies and half-lives for the ground states of several α-emitting nuclides and the α energies for their most intense α group whose decay percentages are given. These groups are not always the most energetic group

	E_α(MeV)	Decay percentage	$T_{1/2}$
^{212}Po	8.78	100	0.30 μs
^{217}Rn	7.74	100	0.54 ms
^{216}Po	6.78	100	0.15 s
^{209}At	5.65	4.1	5.4 h
^{228}Th	5.42	73	1.91 y
^{226}Ra	4.78	94	1600 y
^{235}U	4.40	55	7.04×10^8 y
^{232}Th	4.01	77	1.41×10^{10} y

very consistent with experiments. Calculating α decay half-lives was the first successful application of the new quantum mechanics inside nuclei. As another example of the very delicate balance we observe in nature, if the α decay energies of the first member of each of the decay chains which are found in nature had been only 0.1–0.5 MeV greater, their half-lives would have been too short for them to still be present in the Earth, which was formed a few billion years ago, as can be seen from Eq. 10.27 and Fig. 10.10. In that case there would have been no naturally occurring α-emitters and, without them, the fact atoms have a nucleus may have remained hidden to us.

10.2.4 Some Important Historical Events in α Studies

Following their discovery, α-particles from radioactive decay played crucial roles in the history of nuclear physics, including the following:

1. In 1903, Rutherford proved that an α-particle is a doubly ionized ^{4}He atom with two units of positive charge and four mass units.
2. In 1911, Rutherford developed an atomic model with a nucleus based on α-particle scattering.
3. In 1919, Rutherford achieved the first nuclear reaction and nuclear transmutation using α-particles.
4. In 1928, Gamow explained α decay using the new quantum mechanics.
5. In 1932, Chadwick discovered the neutron using α-particles.
6. In 1934, Joliot-Curies identified the first radioactive nuclide produced in a nuclear reaction using α-particles.

10.2.5 Proton Radioactivity and Proton Decay

The term "proton decay" may be used with two meanings: One is that a nucleus emits a proton spontaneously just as in α and β decays where a nucleus emits an α- and β-particle. The other is that the proton itself may be unstable and so can undergo decay. In order to distinguish these two, the former is called proton radioactivity and the latter the proton decay.

A simple calculation for proton radioactivity in the same way as for α decay indicates that proton radioactivity should not occur. However, in nuclei very far from stability the conditions do become favorable for such decay. The first example of proton radioactivity was the long-lived excited state (called an isomeric state as discussed under γ decay) of ^{53m}Co, which emits protons with energy 1.59 MeV and $T_{1/2} = 17$ s. Here, $M(^{53m}_{27}\text{Co}) > M(^{52}_{26}\text{Fe}) + M(\text{p})$.

As one moves farther and farther from stability on the proton-rich side, the probability that a nucleus can emit a proton increases substantially. Ground-state proton radioactivity was discovered in 1983 in $^{151}_{71}\text{Lu}_{80}$, which emits a proton with 1.23 MeV energy with a half-life of about 85 ms. Note that in nature the stable isotope of lutetium is $^{175}_{71}\text{Lu}_{104}$, so the new one with 24 neutrons less is very far off stability! The $^{151}_{71}\text{Lu}_{80}$ was produced by using the fusion reaction of ^{58}Ni + ^{96}Ru at the UNILAC Heavy Ion Accelerator in Darmstadt, Germany. A second example quickly followed, $^{147}\text{Tm}_{78}$ (the stable isotope is ^{169}Tm, so the new one has 22 neutrons less), which was produced by the fusion reaction of ^{58}Ni + ^{92}Mo at UNILAC and emits a 1.05 MeV proton with a half-life of about 0.42 s. Several other examples are now known in nuclei very far off stability. This is but one example of the new physics seen in nuclei far from stability that could not be observed in nuclei near stability.

Goldanskii in Russia has pointed out that some nuclei which are stable to one-proton radioactivity could emit two protons. Such a process is of much interest. Goldanskii has noted that the two-proton barrier penetration is analogous to the tunnelling of electrons between metals in superconducting and normal states and so may be called a nuclear Josephson effect. In 2002, two-proton radioactivity was discovered in the decay of ^{45}Fe with a half life of 2.6 ms and 70% decay branch.

The study of proton radioactivities is becoming a more important area of research as new accelerator facilities allow one to make new neutron-deficient nuclei still farther from stability. As more and more neutrons are removed from the nucleus of an element of given Z, one finally reaches a point where one proton becomes unbound. This point is called the proton drip line, and the last bound nucleus is $^A_Z X_{N_0}$ where N_0 is the neutron number for the lighest-mass, bound isotope of the element. Nuclei with neutron number less than this should not be observable. However, even though the nuclear force no longer binds the last proton, there is still the Coulomb barrier, which can hold the unbound proton inside a nucleus after it is formed for a time which is long compared to the time for γ-decay. It has been shown that one can study the γ-decay and thus the nuclear structure of nuclei beyond the proton drip line. Such nuclei lie in a totally new region of nuclear matter where the Coulomb force is comparable to the nuclear mean field. This opens up for study a new form of nuclear matter which may be quite different from any we have observed so far. This is discussed in more detail in Chapter 11.

In 1983, a massive experimental effort to search for the decay of the proton itself found that the lifetime for the decay of the proton is greater than 6.5×10^{31} years. With 8000 tons of pure water and 200 photomultipliers as the detector, data were continuously recorded for 130 days. No case of proton decay was observed in this large volume of water. This will be discussed in more detail din Chapter 14. The current limit is greater than 2×10^{32} years.

10.2.6 Spontaneous Fission and Cluster Radioactivities

In 1940, within a year after the discovery of neutron-induced fission of uranium, Flerov and Petrzhak in Russia discovered that ^{238}U has a very small branch for radioactive decay by spontaneous fission (SF) in which the nucleus without external influence splits apart into two nearly equal parts. Just as in the case of induced fission discussed later in the chapter on nuclear reactions, spontaneous fission has a very broad mass distribution. In most cases this involves two broad mass distributions peaked around A = 95 and A = 140 for uranium, while in some heavier nuclei the mass distribution peak is for symmetric fission. In general, SF is a process with a long half-life, and since α decay completes with it, SF often is only a weak decay branch. For elements with atomic number lower than thorium ($Z = 90$), the fission potential barriers are so high that no spontaneous fission has been found. The SF half-life for ^{230}Th is $T_{1/2} \geq 1.5 \times 10^{17}$ years, and for ^{235}U and ^{238}U the SF half-lives are 3×10^{17} years and 8.2×10^{15} years, respectively, so their spontaneous fission branching is very weak.

But as new higher-Z nuclides were discovered, spontaneous fission was found to be more and more important. In many elements, SF quickly becomes the main decay mode. For example, for ^{252}Cf ($T_{1/2} = 2.6$ y) SF is 3.09 percent, but for ^{254}Cf ($T_{1/2} = 60.5$ d), it is 99.69 percent, and α decay is only 0.31 percent. For $^{259}_{101}$Md ($T_{1/2} = 1.6$ h) SF is 100 percent. Spontaneously fissile nuclides have many very useful applications. For example, although SF is only 3 percent in ^{252}Cf, very small quantities of it make very efficient compact energy sources for many applications. Also, ^{252}Cf emits neutrons in spontaneous fission and so it is a useful and very compact neutron source that eliminates the need for an accelerator or reactor for many applications.

With new, large γ-ray detector arrays, one can now measure the coincidences between prompt γ-rays emitted by the two fission products. On the order of 100 diffferent isotopes are formed, and each emits 10 to 130 and more prompt γ-rays followed by the β and γ decays of the products, so the spectra are extremely complex. However, in the triple-γ-coincidence mode with such arrays, where three or more γ-rays are observed in coincidence, it is possible to gate on one γ-ray in each partner and look at the γ-rays in coincidence with both. This double gating eliminates the presence of γ-rays from several isotopes and γ-rays following beta decay when a single gate is used.[1,2] This unique gating on both partners was also used to determine the neutron multiplicities (number of neutrons emitted when different correlated pairs are formed) directly for the first time. In the Mo/Ba pairs in SF of ^{252}Cf, as shown in Table 10.5, from zero up to 9 neutron emission have been observed[1,2] (the number of neutrons is the difference in the sum of the masses ($A_1 + A_2$) of the two fragments and the $A = 252$, for example, ^{104}Mo–^{142}Ba pair has $252 - 104 - 142 = 6$ neutrons emitted). Such data open up a new era in understanding the fission process. Studies of the prompt γ-rays in SF are important in elucidating the structure of neutron-rich nuclei which cannot be populated in fusion reactions. We will return to this process in Chapter 11.

In 1962, another Russian scientist discovered an isomer of ^{242}Am which decays by SF with $T_{1/2}$ of only 14 ms. This is 21 orders of magnitude less than the half-life for SF of the ^{242}Am ground state. Similar SF isomers were found in other nuclei in this region. After a few years, it was found that this isomer is quite different from previously known isomers (Section 10.4). The previously known isomers decay to their ground states through γ-ray or internal conversion emission. In those cases, because the isomer's spin is very different from that of its

---

[1] J. H. Hamilton et al., in *Progress in Particle and Nuclear Physics*, Vol. 35, Pergamon Press, Oxford (1995), p. 635.

[2] C. Goodin et al., *Phys. Rev.* **C74** (2006), p. 017309.

TABLE 10.5

Relative yields of the correlated fragment pairs $Y(A_L, A_H)$ for $Z_L/Z_H = 42/56$ in the spontaneous fission of ^{252}Cf. The matrix is normalized to 100 SF events. The results here are from high statistical data obtained with 102 Ge detectors in Gammasphere[2] (described in Chapter 11).

	^{102}Mo	^{103}Mo	^{104}Mo	^{105}Mo	^{106}Mo	^{107}Mo	^{108}Mo	^{109}Mo
^{138}Ba				0.4 (2)	0.35(11)		0.8(4)	
^{139}Ba			0.41(8)	0.36(12)	2.3(2)	2.7(6)	4.8(8)	2.5(1)
^{140}Ba		0.4(1)	0.19(12)	2.9(9)	12(1)	11(1)	22(5)	2.5(2)
^{141}Ba		1.2(9)	1.8(8)	13(1)	37(1)	26(3)	39(3)	3.2(3)
^{142}Ba	0.007(3)	6(1)	27(1)	46(1)	92(2)	41(2)	51(3)	5.5(2)
^{143}Ba	4(1)	19(1)	60(1)	93(3)	111(2)	32(3)	59(9)	2.8(8)
^{144}Ba	10.0(3)	51(1)	102(4)	99(2)	72(2)	10(1)		
^{145}Ba	21(1)	53(2)	79(2)	46(2)	22(1)			
^{146}Ba	17(1)	30.6(6)	39(3)	11(1)	4.5(2)			
^{147}Ba	4.3(2)	14(1)	5.1(4)	2.1(2)				
^{148}Ba	2.1(2)	1.8(2)	0.5(2)					

ground state, its γ-ray transition probability is very small and its lifetime is very long. But the difference between a SF isomer and its ground state is not spin, it is a large difference in shape. Actinide nuclei have ground states which are prolate deformed ellipsoids with a long axis about 25 percent greater than the short one, $\beta_2 \simeq 0.25$ (see Chapter 11). But these SF isomers have a long axis that is twice as long as the short axis (that is an axis ratio of 2:1), $\beta_2 \sim 0.6$. Since they are so elongated, it is easier for them to undergo spontaneous fission—their barrier to fission is much lower. These isomers are called fission isomers or shape isomers. The symbol, f, is put after the mass number in the upper-left side of the nuclide symbol, for example ^{242f}Am, to distinguish it from an ordinary isomer, for example ^{113m}In discussed earlier. In 1994 groups in Oak Ridge, Warsaw, and Frankfurt predicted a third minimum in several actinide nuclei at larger deformation. The potential energy surface for ^{252}Cf shown in Fig. 10.11a has a third such minimum at $\beta_2 \simeq 0.9$ and $\beta_3 \simeq 0.7$, almost as deep as the ground state (see Fig. 10.11b) for a hyperdeformed (HD) shape with 3:1 axis ratio. Note that in this minimum the shape is highly reflection-asymmetric about its center. Possible indirect evidence for this HD minimum has been found,[1] but its direct observation is a new challenge.

Originally the theories of α decay and spontaneous fission were developed independently even though both are, in a sense, forms of potential barrier penetration. In the traditional approach, one started with α-particle preformation followed by barrier particle penetration. Spontaneous fission in the liquid drop model, however, was related to collective deformation, oscillations and the fission barrier.

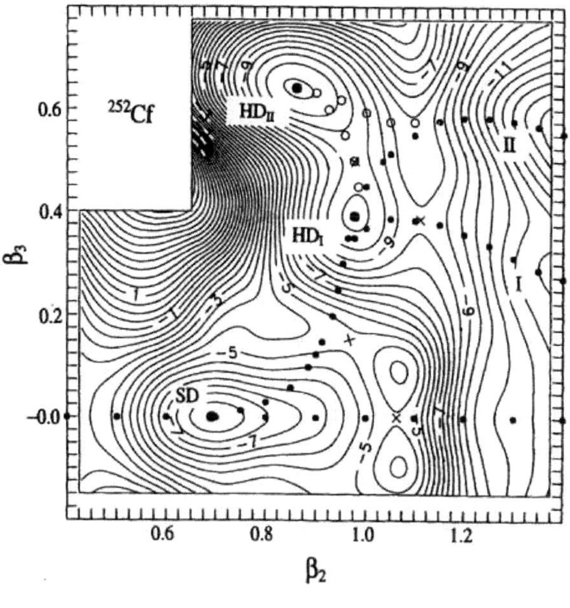

FIGURE 10.11 (a)
Potential energy surface for ^{252}Cf for $0.4 \leq \beta_2 \leq 1.4$. The lines indicate the paths to scission.

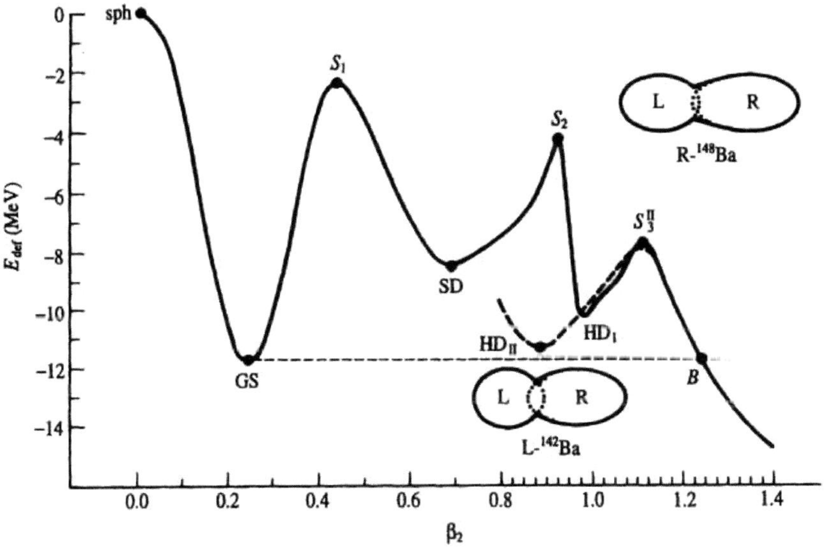

FIGURE 10.11 (b)
The static path II for ^{252}Cf from the HD minimum at $\beta_2 \sim 0.9$ and $\beta_3 \sim 0.7$. The lower shape is that calculated for the HD minimum and the upper shape for the path just beyond point B in the figure. [From J. H. Hamilton et al., in *Progress in Particle and Nuclear Physics*, vol. 35, Pergamon Press, Oxford, (1995), p. 625.]

In 1980, A. Sandulescu, D. Poenaru and W. Greiner[3] sought to understand these processes within a unified approach by using fragmentation theory and the two-centered shell model developed in Frankfurt. An unexpected consequence of their work was the prediction of a "new type of decay of heavy nuclei intermediate between fission and α decay." They predicted that heavy nuclei from radium through uranium should have observable decay branches via heavy clusters such as ^{14}C in radium nuclei to ^{28}Mg in uranium and plutonium. These predicted new radioactivities were such that the other product in the decay was $^{208}_{82}\mathrm{Pb}_{126}$ or an isotope only one or two nucleons removed from it. These decay modes were predicted to have half-lives in observable ranges because of the very strong binding energy of double magic $^{208}_{82}\mathrm{Pb}_{126}$ (see Section 11.2.2 on nuclear shell model). If one looks at the curve of binding energy per nucleon (Fig. 9.4), the fact the curve peaks around ^{56}Fe indicates that there is energy to be gained by all nuclei above about $^{90}_{40}\mathrm{Zr}_{50}$ undergoing fission. So why are there stable nuclei heavier than zirconium? The Coulomb barrier to fission is sufficiently high and the energy released in the process sufficiently low that the half-lives for most of these to fission is much greater than 10^{30} years. Such long half-life decays are stable for all practical purposes.

In 1984 Rose and Jones (see ref. 3) observed the predicted ^{14}C radioactivity of ^{223}Ra ($T_{1/2} = 11.4$ d) in secular equilibrium with ^{227}Ac ($T_{1/2} = 22$ y) with a branching ratio of $(8.5 \pm 2.5) \times 10^{-10}$ compared to α decay. This case was quickly confirmed by other laboratories and several other ^{14}C radioactivities are now known. A year later, the ^{24}Ne decay of ^{232}U (breakup into ^{24}Ne and ^{208}Pb) and ^{231}Pa with branching ratios of $(1\text{–}4) \times 10^{-12}$ were reported. Greiner and co-workers[3] have given a review of the experimental and theoretical work on these important new modes of nuclear decay. More recently, ^{26}Ne ^{28,30}Mg and ^{34}Si emissions have been reported and there has been a suggestion of ^{48}Ca emission, where again in each case the accompanying partner is ^{208}Pb or a nearby isotope. There are numbers of new challenges from ^{5}He to Sn radioactivities yet to be observed. There have also been predictions by the Frankfurt group of cluster radioactivities for much lighter nuclei associated with other closed shells. Heavy cluster emission induced by the Coulomb field in a heavy-ion collision has been proposed as a promising new way to study these processes.

The zero-neutron emission in spontaneous fission can be understood as a new form of cluster radioactivity in which the very large decay energy (recall that up to 10 neutrons can be emitted with average binding energy of about 7 MeV) appears as kinetic energy of the fragments since the internal excitation energies (excited level energies) of the fragments are below the one-neutron binding energy of 4.8–7 MeV.[4] For the first time in cluster radioactivity, fine structure in both partners is observed in this work; for example, several excited states were observed in both partners. These Mo/Ba, Zr/Ce zero-neutron emissions may be related to the

[3] W. Greiner et al., in *Treatise on Heavy Ion Science*, D. Allan Bromley, ed., Plenum Press, New York, p. 641; and *Scientific American* (March, 1990), p. 58.

[4] J. H. Hamilton et al., *J. Phys. G: Nucl. Part. Phys. Lett.* **20** (1994) L85.

double closed shell at $^{132}_{50}Sn_{82}$. These results complete the full mass range of cluster decays from ^4He, through ^{14}C, ^{26}Ne, ^{34}Si up to $^{104}_{42}Mo$, $^{148}_{56}Ba$ and $^{104}_{40}Zr$, $^{148}_{58}Ce$.

10.3 BETA DECAY

In nuclear β decay, the nuclear charge number changes but the nucleon number is not changed. Until 1934 only β^- (negatively charged electron) decay was known. Then I. Curie and F. Joliot (wife and husband) produced and identified new radioactivities in the laboratory that were not found in nature and that decayed by emission of a β^+ (a positively charged electron). Their discovery of new laboratory-produced radioactivities won them the Nobel Prize. Later, orbital electron capture (EC), which competes with β^+ decay, was discovered by L. Alvarez in 1937.

10.3.1 The Puzzle of β Decay

Four years after Becquerel discovered radioactivity, one of the rays was found to be negatively charged and was called beta (β). These β^- rays were subsequently shown to be negatively charged electrons. From measurements over several decades, it was well established that the energy spectrum of the β rays is continuous. The emitted electron energies vary continuously between zero and a maximum value, $E_{\beta m}$, which is different for each β-decaying nucleus. Figure 10.12 is the ^{210}Bi β^- energy spectrum, which is strikingly different from the α-particle discrete energy spectrum with only a few specific energies as seen in Fig. 10.7. Very early, there was also the puzzle of discrete energy lines as in α spectra which were found to sit on top of the continuous beta spectrum in some decays (as shown later in Fig. 10.19). The discrete β-ray lines, as they were called at first, were subsequently found to be atomic electrons ejected from the atom in the decay of an excited nuclear state and were not from inside the nucleus as were the β-rays (see internal conversion processes, Section 10.4). This difference in the energy spectra of β^--rays and α-rays was quite disturbing. At first it was thought that the light-mass β-rays were simply losing energy in coming out of the sample. Careful

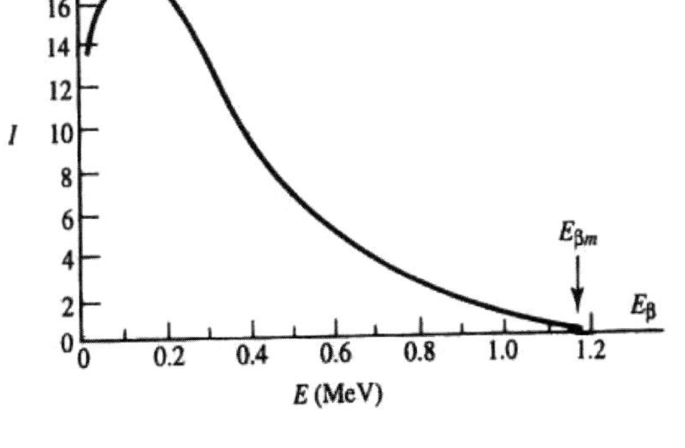

FIGURE 10.12
The ^{210}Bi β spectrum as a function of energy.

measurements of the energy heating of the sample did not reveal that the missing energy was being absorbed in the sample. With the development of quantum mechanics two major puzzles emerged:

1. The nucleus is a quantum system and its energy is discrete. A nuclear decay is a transition between two different nuclear states, each with a fixed energy. So the energy released in radioactive decay must be discrete. The energies of α decays agreed with this understanding. How can the β^- energy spectrum be continuous if energy is conserved?
2. Since the uncertainty relation does not allow electrons to exist inside the nucleus, what is the origin of the electrons in β^- decay?

10.3.2 The Neutrino Hypothesis

In response to the first puzzle, N. Bohr proposed that if the experiments say so, then indeed the principle of energy conservation did not hold in β decay. Other laws had been overthrown, why not this one? He even used this to explain energy production in stars (before fusion energy production).

A solution to this puzzle was proposed by Pauli. In 1930 he suggested that if one assumes that there is a light neutral particle emitted together with every electron in β decay, and that the sum of the energies of the "neutron" and electron is a constant, then the continuous spectrum of the electron in β decay could be explained. Pauli called the particle the "neutron", the neutral one. After the discovery of what we now call the neutron by Chadwick in 1932, Fermi re-named the Pauli particle the neutrino, which in Italian means little neutron or little neutral one. In other words, the decay energy must be distributed among the electron, e, and the neutrino ν, and the daughter nucleus:

$$E_0 = E_e + E_\nu + E_r$$

An arbitrary distribution does not violate momentum conservation. Since the electron rest mass is nearly 2000 times less than the nuclear mass, the recoil kinetic energy of the daughter $E_r \approx 0$. The decay energy E_0 essentially is distributed between the electron and the neutrino. When the neutrino energy $E_\nu \approx 0$, then $E_e \approx E_{\beta m} = E_0$; the electron carries off the maximum energy (Fig. 10.12, the end-point value). When $E_\nu \approx E_0$, then $E_e \approx 0$, so the electron can take any energy value from 0 to $E_{\beta m}$.

In order to have charge and angular momentum conservation before and after β decay, the neutrino charge must be zero and its spin must be $\frac{1}{2}\hbar$. Without the neutrino, angular momentum as well as energy is also not conserved in the β decay process. Later experimental data showed that the neutrino is also necessary for linear momentum conservation because the electron is not emitted at 180° to the recoil direction. Experimental data indicate $E_{\beta m} \approx E_0$; this implies the neutrino rest mass is nearly zero. The first accurate limit on the neutrino rest mass came from the early work of L. M. Langer and co-workers at Indiana on the shape of the beta spectrum of ^3H in the 1950s, in which they set an upper limit of $m_\nu c^2 < 500$ eV for the β-decay neutrino. Studies of ^3H at Mainz have now put an upper limit on $m_\nu c^2$ of 2.3 eV. Theoretical estimates put the mass as finite but < 1 eV. Whether the neutrino rest mass is zero or finite is a very important question that can critically determine the answers to what form grand unified theories may take and perhaps whether the universe is open or closed (whether it will continue to expand or will eventually stop expanding and collapse back).

The neutrino hypothesis of Pauli in 1930 was a bold action. At that time, the only known "basic" particles were the electron and the proton. The neutrino would be the third particle. But since the neutrino carries no electric charge, is essentially massless, and interacts extremely weakly with matter, it is very difficult to observe it directly. Not until 1956, 26 years after Pauli's hypothesis, was direct evidence for neutrinos found in experiments which observed their interaction in matter.

After Pauli had proposed the neutrino hypothesis, many people did not believe in it. Fermi not only accepted it but solved the second puzzle of β decay. In 1934 Fermi developed the weak interaction theory of β decay, which is one of the most inspired and long-standing theories in modern physics. Fermi proposed that just as a photon is produced by the nucleus when it made a transition from an excited state to another state, the electron and the neutrino are produced in the β decay process itself. He pointed out that the essence of β^- decay is that a neutron is changed into a proton, and the essence of β^+ and EC is that a proton is changed into a neutron. The proton and the neutron are viewed as two different states of the nucleon. The conversion of a neutron into a proton corresponds to a transition from one quantum state to another in which the electron and the neutrino, which were not inside the nucleus before decaying, are created and emitted. This is analogous to the creation and emission of a photon when an atom makes a transition between two different atomic states. The photon was not inside the atom before the transition. Just as the electromagnetic interaction produces the photon, there must be a new interaction that produces the electron and the neutrino. This new interaction is called the weak nuclear interaction. As the name suggests, it is much weaker than the strong nuclear interaction that holds the nucleus together. The strong nuclear interaction is discussed in some detail in Chapter 11, and both the weak and strong nuclear forces are considered again in Chapter 14.

The Fermi theory of β decay explained the continuous distribution of energies of the β^- and β^+ for what are called allowed transitions where the beta particle and neutrino, both with spin $\frac{1}{2}$, can carry off zero or one unit of angular momentum when their spins are antiparallel and parallel, respectively, and the nuclear states have the same spin or differ by one unit. About four years after Fermi proposed his theory, E. J. Konopinski and G. Uhlenbeck developed the theory of forbidden β decay in which the parities of the two nuclear states are different and/or the nuclear spin change between the parent and daughter is greater than 1. It was another 11 years before L. M. Langer and co-workers at Indiana confirmed the β energy distribution predicted by the Konopinski–Uhlenbeck theory of forbidden β decay. Since the electron and neutrino both have spin $\frac{1}{2}$, they can easily carry off 0 or 1 unit of angular momentum. When the parities of the two nuclear states are both even (their wavefunctions are equal under reflection—see parity conservation later in this chapter), then the angular momentum (nuclear spin) change between the two states must be zero or $1\hbar$ for an allowed transition. For the same spin change but a change in parity ($\Delta\pi = -1$), the transition is slowed down, or forbidden. For $\Delta I = 0, 1, 2, \Delta\pi = -1$, the β transitions are once (first) forbidden. Carrying off more angular momentum also slows down the β decay because β particles must be formed farther out in the nucleus to carry off the extra angular momentum, e.g. $\Delta I \geq 2$. For $\Delta\pi = +$ (no parity change) and $\Delta I = 2, 3$, one has a twice (second) forbidden decay and the lifetime for decay is again increased. This can be extended to third and higher forbidden decays. A more complete description of the theory of beta decay is found in Appendix 10A. Brief discussions of what happens and the conditions for beta decay to occur follow here.

10.3.3 β^- Decay

In general β^- decay is written as

$$ {}^{A}_{Z}\mathrm{X} \rightarrow {}^{\;\;A}_{Z+1}\mathrm{Y} + \mathrm{e}^- + \bar{\nu}_e \qquad \text{or} \qquad \mathrm{n} \rightarrow \mathrm{p} + \mathrm{e}^- + \bar{\nu}_e \tag{10.28}$$

where electric charge, mass, energy, momentum, and nucleon number A are conserved. We will come back to why an antineutrino, written as $\bar{\nu}_e$, is emitted in β^- decay because of lepton number conservation (see Chapter 14). Since there are no leptons on the left of Eq. (10.28), and the e^- on the right is a lepton, the neutrino on the right must be a lepton antiparticle to conserve lepton number.

As in α decay, the β^- decay energy E_0 can be expressed in terms of the nuclear masses (m_X, m_Y) or the atomic masses (M_X, M_Y, where the electron binding energies can be neglected and the extra electron mass in the middle of Eq. (10.29) is included in M_Y which has $Z+1$ electrons):

$$E_0 = [m_X - (m_Y + m_e)]c^2 = [M_X - M_Y]c^2 \tag{10.29}$$

This is the energy difference between the two states in the parent and daughter nuclei. So, the condition for β^- decay to occur is

$$M_X(Z, A) > M_Y(Z+1, A) \tag{10.30}$$

For example, the ^{3}H (tritium is another name for ^{3}H) β^- decay is written as

$$ {}^{3}_{1}\mathrm{H}_2 \rightarrow {}^{3}_{2}\mathrm{He}_1 + \mathrm{e}^- + \bar{\nu}_e \tag{10.31}$$

The atomic masses of ^{3}H and ^{3}He are 3.0160495 u and 3.0160291 u, respectively, so Eq. (10.30) is satisfied and β^- decay (Eq. 10.31) occurs. Here, E_0 = $(0.0000204\ \mathrm{u})c^2 \simeq 2\times10^{-5}\ \mathrm{u}\times 931.5\ \mathrm{MeV/u} = 18.6\ \mathrm{keV}$. The ^{3}H decay diagram is shown in Fig. 10.13a. Note that since the mass of the neutron is greater than the sum of the masses of the proton and electron, free neutrons undergo β decay. The half-life for the free neutron decay is 10.6 minutes.

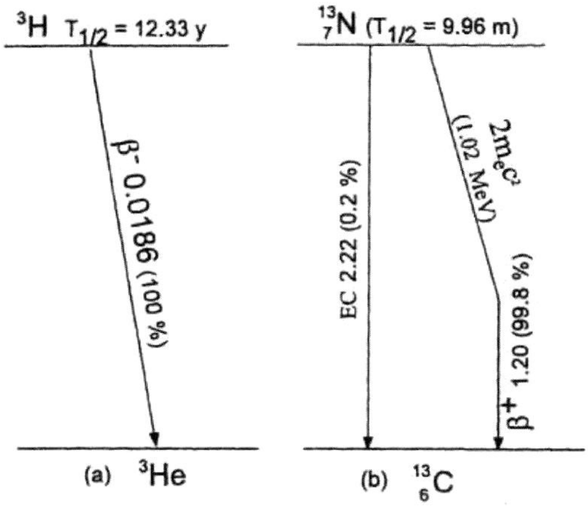

FIGURE 10.13
(a) The β^- decay scheme of ^{3}H. The energy is in MeV.
(b) The $^{13}_{7}$N decay scheme.

In Fig. 10.13a, we put the smaller Z nuclides on the left, larger Z on the right; β^- decay is expressed using the arrow from top-left to bottom-right. The number 0.0186 expresses the maximum kinetic energy of the β particle (e^-) in MeV. This is the decay energy, since ^{3}H β^--decays only into the ^{3}He ground state, as shown by the 100% decay probability in the diagram. The number 12.33 years is the half-life of ^{3}H.

Tritium is an important material that is used to produce thermonuclear reactions. It is one of the fuels in fusion reactors and hydrogen bombs. But because it is radioactive with a half-life of 12.33 years and is a gas that easily enters the body, it could pose a health risk if large quantities were released into the atmosphere in an accident in a fusion reactor. On the other hand, ^{3}H has many important uses as a tracer in medical and industrial research. When ^{3}H replaces stable hydrogen in a molecule of biological or industrial interest, the presence of this molecule can be traced through a system by observing the decay of ^{3}H.

In Fig. 10.14 are plotted the masses of the A = 75 odd-A and A = 76 even–even and odd–odd nuclei. One sees that to the left in the figures, the masses of the lighter nuclei are greater for Z than $Z + 1$ and so β^- decay to higher-Z nuclei is observed. When the masses of the higher-Z nuclei are such that $M(Z + 1) > M(Z)$ then β^+ decay (discussed next) occurs. One easily sees from this figure that nuclei which have fewer neutrons than stable nuclei undergo β^+-decay back to stability and nuclei which have more neutrons than the stable isotopes (neutron-rich) β^--decay back to stability.

10.3.4 β^+ Decay

Before discussing β^+ decay we should discuss the discovery of the positron by C. D. Anderson in 1932, since this was before β^+decay was discovered by F. and I. Joliot-Curie in 1934, see Section 12.1.2. The β^+ particle is a positron, which is the

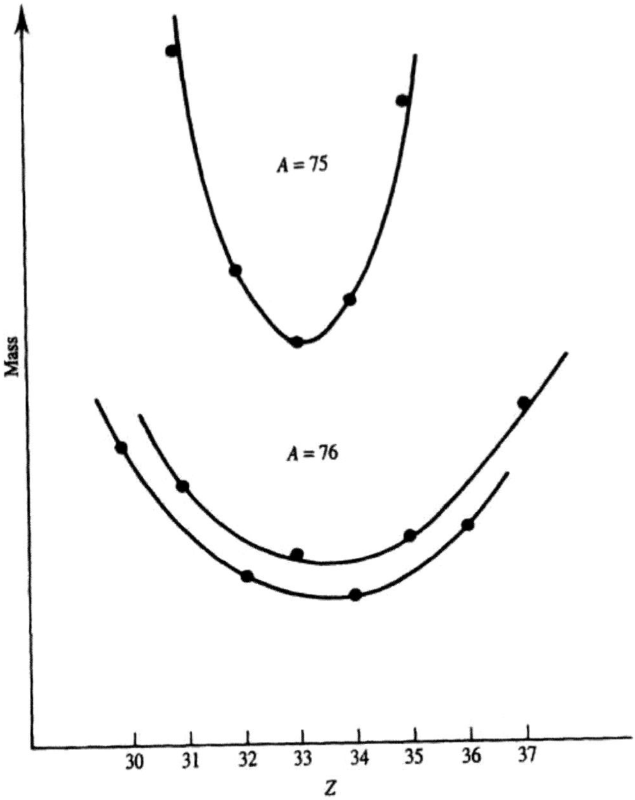

FIGURE 10.14
The masses of nuclei with $A = 75$ and 76 as a function of Z. The nuclei with the lowest masses (greatest binding) are stable. Those to the left of center will undergo β^- decay and those to the right will undergo β^+ decay.

antiparticle of the electron and so has opposite electrical properties. P. A. M. Dirac in 1928 worked out the relativistic wave equation for the electron in the new quantum mechanics. Relativistically the electron total energy is $E = T + m_ec^2$, the sum of its rest mass energy and relativistic kinetic energy (Chapter 1). In his solution he found that the energies of the electron could be either positive or negative. Normally, in classical physics, one would have thrown away the negative energies as having no physical significance, but quantum mechanics provided a totally new way of understanding the world. So, Dirac said these negative energy states should be as good as the positive energy states. The positive energy states of a free electron began at an energy of $m_ec^2 = 511$ keV with a continuum of positive kinetic energies above that. The negative energy states of a free particle began at $-m_ec^2$ and continued toward negative infinity. Now, if there exist in nature such negative energy states, Dirac argued they must all be filled. Otherwise, since nature seeks the lowest possible energy state, electrons in our world would jump to these lower negative energy states and such jumps are not observed.

So, the world is filled with an infinite sea of negative energy state electrons which we cannot see because there are no unoccupied near-by energy states for them to go to in a collision. However, if in a collision one could give at least $2m_ec^2$ to one of the negative energy electrons, then it could be promoted to a positive energy electron. The hole left behind in the infinite sea would act like a particle with the same mass but opposite electrical charge to the electron. Since the only positive charged particle known was the proton, Dirac tried to make the proton this particle but could not. There must be another positive-charged particle with the rest mass of an electron.

Anderson and co-workers were studying the absorption in matter of γ rays emitted in radioactive decay. When γ rays with energy greater than $2m_ec^2 = 1.022\,\text{MeV}$ were absorbed in lead, they found on some occasions the γ rays did not undergo the photoelectric or Compton scattering processes as described earlier (Chapter 6) but produced a pair of particles. One of these particles was an electron, and the other had the same mass but opposite electric charge to the electron, e^+; the positron was discovered. In the Dirac picture, the γ ray gives its energy to a negative energy electron and promotes it to a positive energy electron, leaving behind a hole which acts as a positron. This new, totally unexpected particle fit the equally unexpected prediction of the Dirac theory and was a tremendous triumph for the new quantum mechanics. Why had this not been observed before? Recall from Chapter 6, the probability for such pair production, as it is called, depends strongly on Z as well as γ-ray energy. Secondly, γ rays in vacuum cannot convert their energy to a $e^+\!-e^-$ pair and conserve linear and angular momentum. The conversion must take place in the field of a nucleus to conserve angular momentum.

In the Dirac picture, since the positron is a hole in the negative energy sea when it loses its kinetic energy and comes to rest in matter, a positive energy electron in the matter on coming in contact with it will fall down to the negative energy state and fill the hole. The e^+ and e^- annihilate each other. Their rest mass energy of $2m_ec^2$ must go into some other form. In $e^+\!-e^-$ annihilation, two γ rays of $511\,\text{keV} = m_ec^2$ each are given off. Why two? If the $e^+\!-e^-$ annihilate at rest, their total linear and angular momenta are zero. One photon has angular momentum of $1\hbar$ and linear momentum $p = h\nu/c$. So, one-photon emission is forbidden unless the electron is tightly bound to the whole nucleus to conserve momentum. The simplest allowed process that conserves both these angular momenta is two-photon emission, but only if the two photons are emitted at an angle of 180° (back-to-back) to each other so their net linear and angular momenta are zero. This sharp 180° angle between the two photons is what makes positron emission tomography possible (Section 12.5). The 511 keV photons emitted in the $e^+\!-e^-$ annihilation are called annihilation radiation. It is possible to have three photons if the annihilation occurs while the positron is moving, but this process occurs only rarely.

So, the Dirac theory predicted the existence of a new particle, the positron, and the process of pair production and pair annihilation. In time, the e^+, came to be thought of as the antiparticle of the electron. It was not until after World War II that scientists began to seriously consider if there could be other particle–

antiparticle pairs. Do the proton and neutron have antiparticles? This is considered in Chapter 14. It was even longer before one thought about the proposed neutrino having an antiparticle and the concept of lepton conservation (Chapter 14). Now, whether or not you like the Dirac conceptual picture of a world filled with negative energy state electrons (and seas of other such negative energy state particles, also), this conceptual physical picture provides an excellent basis for describing the solutions of the relativistic wave equation found by Dirac.

In general, β^+ decay can be written:

$$^A_ZX \rightarrow {}_{Z-1}^{\ A}Y + e^+ + \nu_e \qquad \text{or} \qquad p \rightarrow n + e^+ + \nu_e \tag{10.32}$$

The e^+ is an antiparticle lepton, so the neutrino must be a particle lepton to conserve lepton number. The decay energy in terms of nuclear and atomic masses, respectively, is

$$E_0 = [m_X - (m_Y + m_e)]c^2 = [M_X - M_Y - 2m_e]c^2 \tag{10.33}$$

Note the difference between Eq. (10.33) and Eq. (10.29) for β^- decay. The β^+ decay energy as expressed in terms of atomic masses on the right is equal to the difference of the rest energies of the parent atom and daughter atom less the rest energy of two electrons. The two electron rest energies occur because the daughter atomic mass, M_Y, has one less electron than M_X rather than one more electron as in β^- decay. Also, E_0 is equal to the maximum kinetic energy of the emitted positron. The condition for β^+ decay is

$$M_X(Z, A) > M_Y(Z-1, A) + 2m_e \tag{10.34}$$

For example,

$$^{13}_{\ 7}N_6 \rightarrow {}^{13}_{\ 6}C_7 + e^+ + \nu_e \tag{10.35}$$

The masses of ^{13}N and ^{13}C are 13.005 739 u and 13.003 354 u, respectively, so β^+ decay is allowed with a decay energy

$$\begin{aligned} E_0 &= (13.005\,738 - 13.003\,355\ \text{u})c^2 - 2m_ec^2 \\ &= 0.002\,383\ \text{u} \times 931.5\ \text{MeV/u} - 1.022\ \text{MeV} = 2.220 - 1.022\ \text{MeV} = 1.20\ \text{MeV} \end{aligned}$$

Note that since the mass of the proton is less than the mass of the neutron, a free proton cannot undergo β^+ decay to a neutron. A proton can β^+-decay into a neutron only inside a nucleus where a difference in the nuclear binding energy between the parent and daughter makes up the mass difference between the proton and neutron. Recall that ^{13}N is produced when ^{12}C absorbs a proton. From the measurement of the ^{13}N radioactivity, the amount of ^{12}C in a sample can be deduced (see Appendix I). Its decay diagram is shown in Fig. 10.13b.

As in the case of α decay, measurements of the maximum β^+ and β^- energies provide data to determine the mass differences between nuclei. Indeed, for neutron-rich nuclei far from stability, β^- endpoint energies are the only way to get information on nuclear masses. In such cases one must take care to ensure that one has observed the total decay energy by determining whether the β decay feeds the ground state or an excited state. As seen by comparing Figs. 10.12 and 10.15a,

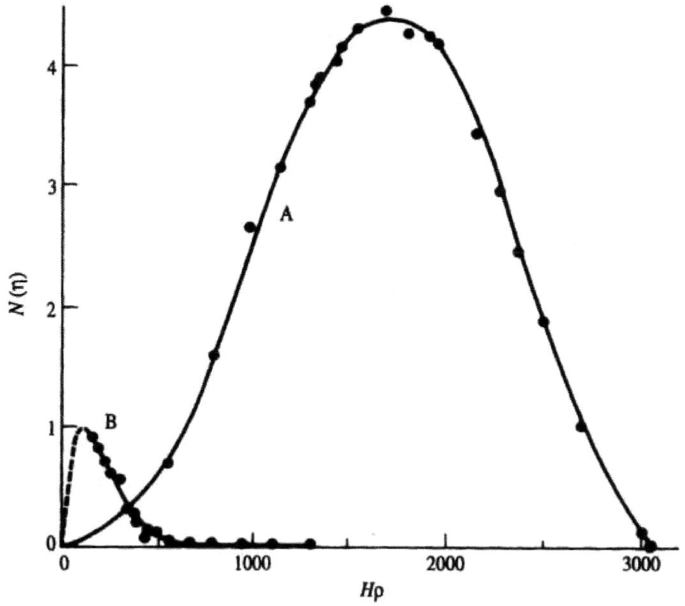

FIGURE 10.15 (a)
The β^+ spectrum, $N(\eta)$ (number of beta particles in each linear momentum interval where η is the momentum), as a function (curve A) of the β^+ momentum or magnetic field strength times the radius of the electron orbit in the field in the decay of ^{22}Na. Also shown (curve B) is a low-energy negative electron distribution from atomic electrons which are shaken off in the rearrangement of the electrons to the new nuclear charge. [J. H. Hamilton, et al., *Phys. Rev.* **112** (1958) 2010.]

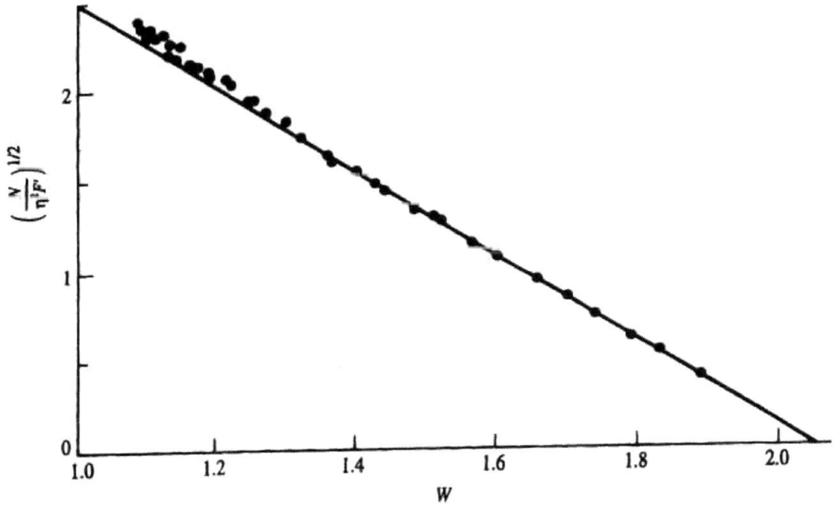

FIGURE 10.15 (b)
A Fermi–Kurie plot of the β^+ spectrum of ^{22}Na.

there is also a difference in the distribution of β^- and β^+ electrons. In Fig. 10.15 the β^+ spectrum from the decay of ^{22}Na is peaked more toward the higher energies than is the β^- spectrum in Fig. 10.12 (there is a difference in the momentum and energy distribution in the two figures, but this does not alter the conclusion). This difference arises because the β^- electrons with negative charge are attracted by the positively charged nucleus and so have somewhat less energy, while the positively charged β^+ are repelled by the nucleus and gain some energy after leaving the nucleus. To extract the β end-point energies, Kurie proposed what is called a Kurie or Fermi–Kurie plot where the β^- and β^+ distributions are corrected by a Coulomb force correction factor, F', as shown in Fig. 10.15b. For an allowed decay the Fermi–Kurie plot is a linear function of the total β energy. An FK plot generally is not a linear function for forbidden decays. The maximum energy is easily obtained from the extrapolation of the straight line in Fig. 10.15b. Also, in Fig. 10.15a one sees a low-energy negative electron distribution. When the nuclear charge suddenly changes by one unit, the atomic electrons must readjust their binding energies. In this rearrangement the atomic electrons can also take on some of the β decay energy as well as rearrangement energy and be ejected from the atom. These so-called "shake-off" electrons are the low-energy negative electrons in Fig. 10.15a.

10.3.5 Orbital Electron Capture (EC)

In the Dirac theory of the electron, the positron predicted by the theory can be pictured as a vacancy in the negative energy sea of electrons. Thus, β^+ decay can be pictured as a proton capturing a negative energy sea electron with the hole that is left going off as a positron. If a proton in a nucleus can capture a negative energy electron, then it should also be able to capture positive energy electrons. Indeed the process does occur where the parent nucleus captures an orbital electron from an atomic shell. In this process, one of the protons in the parent nucleus captures an orbital electron to become a neutron, and only a neutrino is emitted. This process is called electron capture. Recall that the K shell electrons are nearest to the nucleus and indeed have some probability for being inside the nucleus, so K electron capture occurs most easily but captures from L and higher shells also occur with decreasing probabilities.

Orbital electron capture can be written

$$ {}^{A}_{Z}\mathrm{X}_N + \mathrm{e}_i^- \rightarrow {}_{Z-1}^{\;\;A}\mathrm{Y}_{N+1} + \nu_e \qquad \mathrm{p} + \mathrm{e}^- \rightarrow \mathrm{n} + \nu_e \tag{10.36} $$

Here, there is one lepton particle on each side of the equation, the e^- and ν_e respectively, so lepton number is conserved along with nucleon number. So, the decay energy of the electron capture from the ith shell is

$$ E_{0i} = [m_X + m_e - m_Y]c^2 - W_i \tag{10.37} $$

where W_i is the binding energy of the ith shell electrons in the parent atom. Since the binding energy differences of the electrons in the atoms can be neglected, Eq. (10.37) can be converted into the atomic masses to yield

$$E_{0i} = [M_X - M_Y]c^2 - W_i \tag{10.38}$$

That is, the decay energy for ith shell electron capture is equal to the difference in the rest energies of the parent and daughter atoms less the ith shell electron binding energy. The condition for electron capture to occur is

$$M_X(Z, A) - M_Y(Z - 1, A) > \frac{W_i}{c^2} \tag{10.39}$$

Thus, in two neighboring isobars, only when the atomic mass difference between the parent and the daughter is greater than the corresponding mass equivalence of the ith shell electron binding energy of the parent can the ith shell orbital electron capture decay occur. Note for the ^{13}N case calculated earlier, $M_X - M_Y = 2.220$ MeV $= E_0(\text{EC})$ compared to $E_0(\beta^+) = 1.200$ MeV.

The K-shell electrons are nearest to the nucleus, so the K-capture probability is the largest. In quantum mechanics the $n = 1$ (K shell) wavefunctions are larger near the nucleus, and for a given n the s ($l = 0$) electron wavefunctions are larger in the vicinity of the nucleus. But when $W_K/c^2 > (M_X - M_Y) > W_L/c^2$, K-capture does not occur, but L-capture does. In that case, the L-capture probability is the maximum. An example of such a case is the ^{205}Po decay.

Because $2m_ec^2 \gg W_i$, a nucleus which can undergo β^+ decay always has electron capture decay competing with the β^+ decay. However, when $W_i < (M_X - M_Y)c^2 < 2m_ec^2$, only electron capture decay occurs, with no β^+ decay. An example is ^{55}Fe; its atomic mass energy is only 0.232 MeV greater than ^{55}Mn, so β^+ decay cannot occur, but EC can. Its half-life for EC is 2.7 years. The energy released (0.231 MeV) is distributed between the daughter nucleus and the neutrino, but essentially all of the energy goes to the neutrino. But it is very difficult to measure either the neutrino energy or the nuclear recoil energy. However, after the orbital electron is captured, there is a vacancy in the ith shell of the daughter nucleus. Then a characteristic x-ray or Auger electron (see Chapter 6) is emitted. It is easy to measure the ^{55}Mn K_α x-ray with energy 5.9 keV to confirm that an EC decay has occurred.

In $^{64}_{28}Ni$, $^{64}_{29}Cu$, and $^{64}_{30}Zn$ the mass differences are such that β^-, β^+, and EC can all occur for ^{64}Cu (see Fig. 10.16). The masses are like those of ^{76}Ge, ^{76}As, and ^{76}Se in Fig. 10.14. There is a 38.7 percent probability for β^- decay to occur, 0.48 percent probability for K-capture decay to a ^{64}Ni excited state, a 43.2 percent probability for the K-capture to the ^{64}Ni ground state, and a 17.6 percent probability for decay to the ^{64}Ni ground state by β^+ decay. This unusual situation of competing β^- and β^+ decay can occur only in the decay of a few odd-Z–odd-N nuclei near stability, like ^{40}K (Table 10.1), ^{64}Cu or ^{76}As (see Fig. 10.14) where nucleon pairing effects are favorable to lower the masses of the even–even nuclei with Z to either side of an odd–odd nucleus. In such cases the extra pairing energy makes both even–even nuclei stable (they cannot undergo β or α decay).

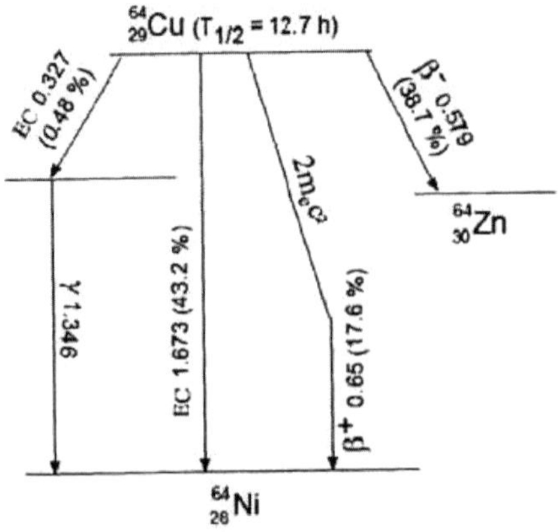

FIGURE 10.16
The decay scheme of ^{64}Cu, which can decay by β^-, β^+, or EC.

10.3.6 The Other Decay Modes Associated with β Decay

INVERSE β DECAY (NEUTRINO ABSORPTION). Theoretically, the process in which a proton absorbs an antineutrino and a neutron absorbs a neutrino can occur:

$$\begin{aligned} \bar{\nu} + \mathrm{p} &\rightarrow \mathrm{n} + \mathrm{e}^+ \\ \nu + \mathrm{n} &\rightarrow \mathrm{p} + \mathrm{e}^- \end{aligned} \tag{10.40}$$

These processes are, in a sense, the inverse of β decay. They are like electron capture except there are no ν and $\bar{\nu}$ around for the neutron and proton to absorb. As in β^+/EC decay and β^- decay, these processes convert a proton into a neutron and a neutron into a proton, respectively, through the weak interaction. However, note that in Eq. (10.40) the neutrino absorbed is the opposite in type to the one that is emitted in β^+ and β^- decay. So, the process $\mathrm{p} \rightarrow \mathrm{n} + \mathrm{e}^+ + \nu$ can be thought of as the absorption by the proton of a negative energy electron whose hole is the positron and by particle–antiparticle symmetry the absorption of a negative energy antineutrino whose hole is the neutrino. Thus, the proton can absorb an antineutrino but not a neutrino. Electron capture is then the proton absorbing a positive energy electron and a negative energy antineutrino. Similarly the neutron decay involves the absorption of the negative energy antineutrino whose hole is the ν. So, just as the proton absorbs the opposite type of electron, e^-, in electron capture to the e^+ it emits in β^+ decay, the proton and neutron absorb the opposite type of neutrino to the one they emit in β decay, provided, of course, that the neutrino and antineutrino are not identical particles. Technically, these (Eq. 10.40) are observed only as induced reactions by bombarding nuclei with ν or $\bar{\nu}$ and are not observed as decay modes since there are no ν or $\bar{\nu}$ surrounding nuclei to be captured like the atomic e^- to produce electron capture.

In 1956 C. L. Cowan and F. Reines provided the first direct evidence for the existence of the antineutrino by observing the first process of Eq. (10.40). Their experiments also showed that indeed neutrinos interact with matter very, very weakly. This very weak interaction is expected because they have no electric charge with which to interact and do not interact by the strong nuclear force, but only by the weak nuclear force. The experiment was done near a nuclear reactor where there was a high flux of antineutrinos from β^- decay of the fission products. The antineutrino was absorbed by a proton in the H_2O target, and the neutron and e^+ were detected by the γ-rays following neutron capture in cadmium and e^+ annihilation. A neutrino attempting to pass through the Earth along its diameter has only one chance in 10^{12} of being captured and can easily pass through the Earth without interacting. This ultraweak interaction is most fortunate, since the Earth is subject to a flux of neutrinos of about 6.5×10^{10} per cm^2 per second from the Sun. Such a flux of neutrons, electrons, or gamma rays would be lethal to all life.

Beginning in 1955, Davis and collaborators studied the reaction $\nu + {}^{37}_{17}Cl \rightarrow {}^{37}_{18}A + e^-$ (which is the inverse of ${}^{37}_{18}A + e^- \rightarrow {}^{37}_{17}Cl + \nu$) with neutrinos from the Sun and the same reaction with $\bar{\nu}$ from a reactor to see whether ν and $\bar{\nu}$ were distinguishable particles. The reaction was observed with ν but not $\bar{\nu}$. These and other data clearly support the fact that the ν and $\bar{\nu}$ are not identical. Indeed, in the Dirac theory of spin-$\frac{1}{2}$ particles, the neutrino must have a nonidentical antiparticle. However, even though the positron predicted by Dirac as the antiparticle of the electron was confirmed in 1932, the idea that there are two neutrinos (ν and $\bar{\nu}$) in β decay, which also was embedded in the Dirac theory, was not seriously considered until the 1950s when the search for the antiproton and the nature of antiparticles became important general questions. Textbooks before about 1955 will show only one type of neutrino emitted in both β^- and β^+ decay. Subsequent work by Lederman and co-workers at Brookhaven in 1962 demonstrated that the neutrinos produced in the decay of pi-mesons are associated with mu-mesons but not with electrons. Later, a third neutrino associated with the tau lepton was discovered. Thus there are separate ν_e and $\bar{\nu}_e$, ν_μ and $\bar{\nu}_\mu$ and ν_τ and $\bar{\nu}_\tau$ as discussed in Chapter 14. However, the nature of all these ν and $\bar{\nu}$ remains a very important question as we shall see in the discussion of double β decay. We complete our discussion of neutrino physics there and in Chapter 14.

DOUBLE β DECAY AND THE NATURE OF THE NEUTRINO. Because of the pairing energy discussed earlier, even–even nuclei are considerably more tightly bound than their odd–odd neighbors. There are a number of even–even nuclei which cannot undergo single β decay from mass–energy considerations but which can undergo double β decay as seen in Fig. 10.13 for ^{76}Ge. Double β decay is a two-step weak interaction which can be written as

$$
\begin{aligned}
{}^{A}_{Z}X_N &\rightarrow {}^{A}_{Z+2}Y_{N-2} + 2e^- + 2\bar{\nu}_e \\
{}^{A}_{Z}X_N &\rightarrow {}^{A}_{Z-2}Y_{N+2} + 2e^+ + 2\nu_e
\end{aligned}
\qquad (10.41)
$$

Since it is a two-step weak interaction, candidates for double β decay have half-lives the order of 10^{20} years and longer. Thus, the process is very difficult to observe. In the mass formula (Chapter 9) the symmetry energy term, which is proportional to $(N - Z)^2$, is important in double β decay because it allows an even-even nucleus, for example with $Z + 2$ and $N - 2$, to be more tightly bound than a nucleus with Z and N, or vice-versa. This can be seen in Fig. 10.14, where the mass of $^{76}_{32}$Ge is lower than that of $^{76}_{33}$As but greater than $Z + 2$ ^{76}Se, so double β decay is allowed. Since heavier stable nuclei have a neutron excess, there are more possible double β^- decaying isotopes than double β^+ decaying ones. The latest results on $2\,\nu$ double beta decay are given in Table 10.6a (taken from F. T. Avignone III, S.R. Elliott and J. Engel, Rev. Mod. Phys., **80**, 481 (2008). As seen there, 2ν double beta decay is now well established. When both 2ν and $0\,\nu$ double beta decay can occur in the same nucleus, it is important to have as accurate as possible $2\,\nu$ decay results.

TABLE 10.6a
A list of the values of $T^{2\nu}_{1/2}$ for various isotopes (see discussion Avignone, *et al.*).

Isotope	$T_{1/2}{}^{2\nu}$ (yr)	Isotope	$T_{1/2}{}^{2\nu}$ (yr)
^{48}Ca	$(4.2_{-1.0}^{+2.1}) \times 10^{19}$	^{128}Te	$(2.5 \pm 0.3) \times 10^{24}$
^{76}Ge	$(1.5 \pm 0.1) \times 10^{21}$	^{130}Ba EC-EC(2ν)	$(2.2 \pm 0.5) \times 10^{21}$
^{82}Se	$(0.92 \pm 0.07) \times 10^{20}$	^{130}Te	$(0.9 \pm 0.1) \times 10^{21}$
^{96}Zr	$(2.0 \pm 0.3) \times 10^{19}$	^{150}Nd	$(7.8 \pm 0.7) \times 10^{18}$
^{100}Mo	$(7.1 \pm 0.4) \times 10^{18}$	^{238}U	$(2.0 \pm 0.6) \times 10^{21}$
^{116}Cd	$(3.0 \pm 0.2) \times 10^{19}$		

Double beta decay is very important because it probes the current standard model in which the spin-1/2 neutrinos are Dirac fermions and so ν and $\bar{\nu}$ are not identical. There is also the possibility that neutrinos are Majorana fermions such that a neutrino is identical to its antineutrino — the neutrino is its own antiparticle. In this case, in double β^- decay, the antineutrino emitted in the first β^- decay is absorbed as a neutrino with the emission of the second e^-. This neutrinonless double beta decay requires $\nu_e \equiv \bar{\nu}_e$. The neutronless double beta decay would violate lepton number conservation (see Chapter 14) in contradiction to the standard model and strongly support grand unified models which favor Majorana neutrinos. Nonconservation of lepton number is tied up with the possible existence of a neutrino rest mass. A Majorana neutrino must have a rest mass. β-decay experiments have put an upper limit of 2.3 eV/c^2 on the ν_e mass. Current searches for both two-neutrino and neutrinoless double beta decay, as illustrated in Fig. 10.16b, are important areas of current research because they put significant constraints on theoretical developments and the neutrino rest mass.

Several groups have studied the process of neutrinoless double beta decay. As noted above, F. T. Avigone III et al. have recently summarized the known results and discussed future plans. Table 10.6b gives the lower limits on the half-life and upper limits on the Majorana neutrino rest mass from neutrinoless double beta decay. The limits on the Majorana neutrino rest mass from neutrinoless double beta decay are even lower than the neutrino rest mass extracted from single beta decay. These data

TABLE 10.6b

A list of recent $\beta\beta(0\nu)$ experiments and their 90% confidence level (except as noted) limits on $T_{1/2}^{0\nu}$. The electron neutrino mass $<m_{\beta\beta}>$ limits are those quoted by authors.

Isotope	$T_{1/2}^{0\nu}$ (yr)	$<m_{\beta\beta}>$ (eV)
^{48}Ca	$>5.8 \times 10^{22}$	<3.5-22
^{76}Ge	$(2.23_{-0.31}^{+0.44}) \times 10^{25}$ (1σ)	0.32±0.03
^{76}Ge	$>1.57 \times 10^{25}$	<0.33-1.35
^{82}Se	$>2.1\times 10^{23}$	<1.2-3.2
^{100}Mo	$>5.8 \times 10^{23}$	<0.6-2.7
^{116}Cd	$>1.7\times 10^{23}$	<1.7
^{128}Te	$>7.7 \times 10^{24}$	<1.1-1.5
^{130}Te	$>3.0\times 10^{24}$	<0.41-0.98
^{136}Xe	$>4.5 \times 10^{23}$	<0.8-5.6
^{150}Nd	$>3.6\times 10^{21}$	

constrain the mass of the Majorana neutrino to be less than 2 eV and likely less than 1 eV. This mass is an important number in constraining grand unified theories. A. Heidelberg-Moscow Collaboration claimed to observe evidence for a $\beta\beta(0\nu)$ decay in ^{76}Ge. The data yield a neutrino mass of 0.32 ± 0.03 eV. It is not definite that these results are valid. This work has produced a number of critical replies. Additional experiments are needed to confirm or refute these results.

Double beta decay studies involve the extremely long-lived radioactive decays of a variety of isotopes. Very special techniques had to be developed to search for such decays, and very large quantities of these isotopes are needed to search for these rare decay modes. The importance of observing neutrinoless double-beta decay is seen by the large number of different experiments under development or proposed as given in Table 10.7 (Avignone, et al. give the references). Note the sizes of the isotopes to be used in each experiment. The large number of different experiments underscores the importance of these long-term searches. Avignone, Elliott and Engel emphasized this in their concluding remarks: "The answer to the question "Is the neutrino its own antiparticle?" is critical for theories of particle mass. It is also needed to help uncover the reasons matter dominates over antimatter in our Universe. The neutrino mass will not only tell us about how the high scale at which the standard model breaks down (e.g., through the see-saw), but also will have implications for the large-scale structure of the Universe. Finally, lepton-number violation is significant in its own right. $\beta\beta(0\nu)$ is the only practical method for investigating the particle-antiparticle question. And if the neutrino is its own antiparticle, $\beta\beta(0\nu)$ will have the best sensitivity to neutrino mass of any laboratory technique."

What the correct standard model is for the particles of nature and what a grand unified theory will be like are among the most important questions in particle physics today. Studies of processes that involve neutrinos will provide critical answers to these questions and to a fundamental understanding of the nature of neutrinos. Moreover, these answers will have far-reaching astrophysical implications as discussed next.

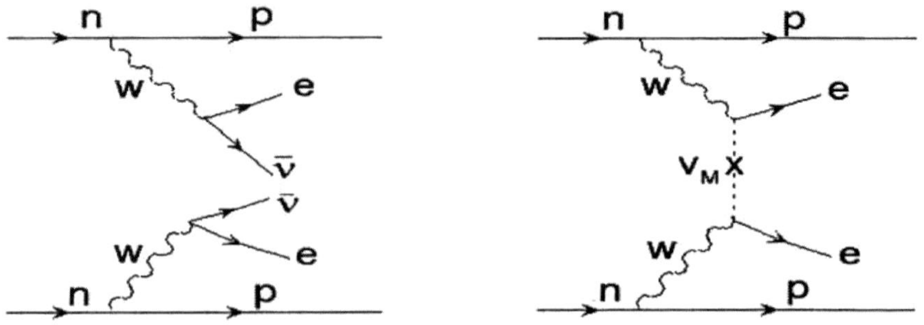

FIGURE 10.16b

Feynman diagrams for two neutron and neutrinoless double beta decay. (See Chapter 14 for further discussion).

TABLE 10.7

Experiments for neutrinoless double β-decay (References see Avignone et al.).

Experiment	Isotope	Mass	Technique	Present Status
CANDLES	^{48}Ca	few tons	CaF_2 scint. crystals	Prototype
CARVEL	^{48}Ca	1 ton	$CaWO_4$ scint. crystals	Development
COBRA	^{116}Cd	418 kg	CZT semicond. det.	Prototype
CUORICINO	^{130}Te	40.7 kg	TeO_2 bolometers	Running
CUORE	^{130}Te	741 kg	TeO_2 bolometers	Proposal
DCBA	^{150}Nd	20 kg	^{enr}Nd foils and tracking	Development
EXO-200	^{136}Xe	200 kg	Liq. ^{enr}Xe TPC/scint.	Construction
EXO	^{136}Xe	1-10 ton	Liq. ^{enr}Xe TPC/scint	Proposal
GEM	^{76}Ge	1 ton	^{enr}Ge det. in liq. nitrogen	Inactive
GENIUS	^{76}Ge	1 ton	^{enr}Ge det. in liq. nitrogen	Inactive
GERDA	^{76}Ge	$\approx$ 35 kg	^{enr}Ge semicond. det.	Construction
GSO	^{160}Gd	2 ton	Gd_2SiO_5: Ce crys. scint. in liq. scint.	Development
MAJORANA	^{76}Ge	120 kg	^{enr}Ge semicond. det.	Proposal
MOON	^{100}Mo	1 ton	^{enr}Mo foils/scint.	Proposal
SNO++	^{150}Nd	10 ton	Nd loaded liq. scint.	Proposal
SuperNEMO	^{82}Se	100 kg	^{enr}Se foils/tracking	Proposal
Xe	^{136}Xe	1.56 ton	^{enr}Xe in liq. scint.	Development
XMASS	^{136}Xe	10 ton	Liquid Xe	Prototype
HPXe	^{136}Xe	tons	High pressure Xe gas	Development

In the early 1990's, several new neutrino detection facilities came into operation; the Baksan Valley Laboratory in Russia, a second in the Gran Sasso Laboratory in the tunnel in the Alphs and the Russian-American Gallium Experiment, SAGE. Their data, in turn, inspired much larger new facilities such as the Super-Kamiokande with a 50 kiloton water Cherenkov detector in Japan, and the Sudbury Neutrino Observatory (SNO) in a Canadian mine. Then laboratories were built to detect neutrinos produced in reactors (for example, in Japan), and accelerators (Los Alamos and Fermi Lab) which were well removed from the detectors. Today there are many deep, underground laboratories in operation or nearing so, all over the world. Even larger, 100-1000 kiloton detector experiments are being proposed or are under development such as LAGUNA in Europe, Hyper K in Japan, and the Deep Underground Science and Engineering Laboratory (DUSEL) in the US, which still needs final congressional approval. There is not space to discuss all the individual discoveries being made or the planned experiments. Excellent summaries of the current detectors and their results are found in "Neutrino Oscillations Present Status and Future Plans," by J.A. Thomas and P.L. Vahle, World Scientific, 2008. A listing and brief descriptions of all the deep underground laboratories are found in Nuclear Physics News, **18**, Issue 4, October-December 2008.

Here we will summarize the results, present knowledge as of 2008, and future questions and challenges to be addressed. The experimental data are most simply described by introducing the phenomena of neutrino oscillations, flavor transition, $\nu_e \to \nu_\mu$ $\nu_\mu \to \nu_\tau$, etc. Such oscillations imply neutrinos have mass. Mikheyev-Smirnov-Wolfenstein (MSW) proposed an effect in which electron neutrinos interact with electrons in their passage out of the Sun (or through the Earth) and are converted to muon neutrinos. This flavor oscillation requires a sufficient mass difference and mixing between the flavor eigenstates. The nonzero neutrino masses required by MSW are less than 1 eV. Neutrino flavor changing interactions in matter should give rise to a day-night effect in detectors but such an effect is still an open question. Neutrino flavor oscillations also can occur in vacuum.

With accelerators and reactors, atmospheric oscillations have been observed. This is a very active area of research. The distance between the production site and detection site is an important parameter. Detection of longer distances are under development, even to having the source and detectors in a different country – Japan and Korea, as discussed in the above reference. The standard neutrino scenario, as summarized by Smirnov (as X is: 0810.2668v1)[hep-ph] October 2008 (and in *Neutrino Oscillations* above), is as follows. Neutrino interactions are described by the standard electroweak model; there are only three types of light neutrinos (three flavor and three mass states); there are no sterile neutrinos. Neutrinos have mass ($\leq$ 1 eV); neutrinos mix with two large and one small to zero mixing angles and the observed masses and mixing angles have a pure vacuum origin.

Let us briefly discuss each of the above. Gage interactions of neutrinos are well known. However, we have no information on Yukawa coupling with the Higgs Boson (Chapter 14) and these could be important for leptogenesis (the ν–$\bar{\nu}$ asymmetry in nature). There are open questions about neutrino interactions with nucleons and nuclei and the strong interaction.

Oscillation experiments are sensitive only to the differences in masses of the neutrinos. The masses of the heaviest neutrinos should be less than 1 eV but the mass of the lightest is unknown. However, there can be very massive neutrinos or even families of such with GeV and above masses. Also, there can be sterile neutrinos which do not interact with anything else in the standard model and so are very difficult to detect. The values of the mixing angles are of much interest, especially θ_{13}, because it impacts CP violation for neutrinos.

Some brief mention of some current neutrino programs, their success and questions they leave unanswered, are in order. The Superkamiokande experiment in Japan observed the solar neutrino oscillations, a ν_μ/ν_e suppression in the atmosphere and $\nu_\mu \to \nu_\tau$ oscillation. In 2009, it will again be a detector for an oscillation experiment based on a neutrino beam. Hyper-K is under consideration with 100-1000 kilotons of water. SNO which stopped taking data in 2006 gave clear evidence for solar ν oscillations also to help solve the missing solar ν problem. No day-night effect was observed.

The Kam LAND reactor experiment with a 175 km baseline between reactor production of $\bar{\nu}_e$ and their detection observed $\bar{\nu}_e$ disappearance and made the most precise Δm_{12}^2 measurement. By purifying its liquid, scintillation target and reducing internal backgrounds, it is preparing for direct observation of low-energy, ^{7}Be solar neutrinos to hopefully make the first direct test of the MSW mechanism of solar ν oscillation in matter. The Main Injection Neutrino Oscillation Search (MINOS) is 735 km baseline experiment with ν produced at Fermi Lab. A deficit of ν_μ charge current events was observed. In 2008, MINOS completed its second year of operation. Further research will seek evidence for the sub-dominate $\nu_\mu \to \nu_e$ mode and for $\nu_\mu \to \bar{\nu}_\mu$ oscillations.

The LSND experiment with the 800 MeV proton accelerator at Los Alamos National Laboratory studied neutrinos from $\pi^+ \to \mu^+ \, \nu_\mu$ and $\mu^+ \to e^+\bar{\nu}_\mu\nu_e$ decays. Evidence for $\bar{\nu}_\mu \to \bar{\nu}_e$ oscillations was observed. This observation is known as the LSND anomaly because it does not fit with the picture of neutrinos described above. It will require new sterile neutrinos. At Fermi Lab, the Mini Booster Neutrino Experiment (Mini BOONE) was built to test the LSND evidence for neutrino oscillation. It did so by searching for ν_e appearance in a ν_μ beam. No significant excess of ν_e events above background was observed. These results are consistent with no oscillation within a two neutrino appearance-only oscillation model. The LSND-Mini BOONE results require very speculative schemes for both to be right – sterile ν and extra dimensions, CPT violation and sterile ν or 3 sterile ν's, etc. The T2K experiment is a future, long baseline experiment, designed to precisely measure ν_μ disappearance and to look for ν_e appearance in a ν_μ beam. It will use an accelerator being built in Tokai and the Super-Kamiokande detector. Double Chooz, under construction in 2009, will have two liquid scintillation detectors at 270 m and 1 km from the CHOOZ reactor in France. It seeks to make a precise measurement of the

θ_{13} mixing angles. The Daya Bay reactor experiment in China, under development, likewise seeks to make a precise measurement of θ_{13}. These experiments should end around 2013. Then the next phase will involve development of order of magnitude larger mass detectors with further background reduction.

All the large new facilities are being built to address important unsolved problems in understanding neutrinos. Oscillation and other experiments are expected to address the following unresolved problems.

- Which is the lightest neutrino, or stated another way, what is the hierarchy of mass is $m_3^2 >$ or $< m_1^2$?
- What is the size of the mixing angle θ_{13}?
- Is there CP violation?
- Is there a Day-Night or MSW matter effect?

Other questions that may not be answered by oscillation experiments.

- What is the mass of the lightest ν?
- Are neutrinos Majorana or Dirac particles?
- Are there more than three neutrinos and are there massive neutrinos?
- Are there sterile neutrinos?
- Do exotic neutrino interactions exist?
- Better understanding of leptogenesis – the ν-$\bar{\nu}$ asymmetry in nature?
- Astrophysical issues and neutrino astronomy.
- We still have not identified the underlying physics of neutrinos.
- Is there new physics beyond the standard model?

In his conclusions, Smirnov said, "neutrino physics is in the transition phase." Significant progress in terms of the standard model scenario has been made. Tests of this scenario and searches for physics beyond this model are the objectives for the future. There is still no clear fundamental theory nor understanding of the origin of neutrino mass and mixing. The latter problem is the driving force for future experiments.

β-DELAYED NEUTRON EMISSION. As one goes to nuclei very far from the stable nuclei found in nature, the energies for β^- and β^+ decay go up to 10 MeV and more. In such cases, following β^- decay to an excited state in the daughter, the excited state energy may be so high that there is a probability that a neutron is promptly emitted rather than a γ-ray. This neutron is called a β-delayed neutron because it is emitted with the half-life of the β^- decay which it follows immediately (within $\leq 10^{-10}$ s).

Figure 10.17 is an important example of such a decay. The ^{87}Br is a product of the fission of excited ^{236}U formed when ^{235}U absorbs a neutron. It is a

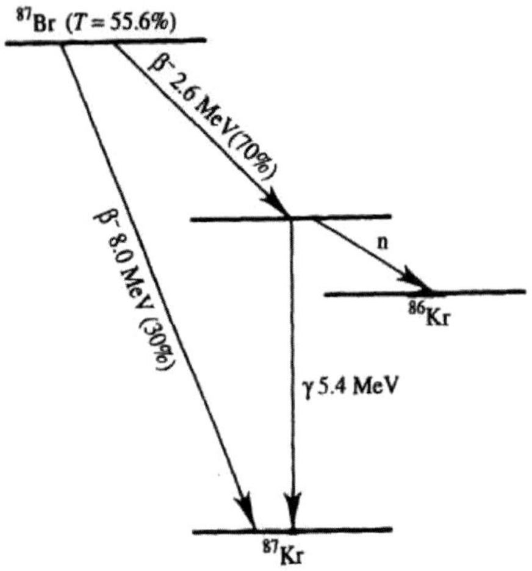

FIGURE 10.17
The decay scheme of ^{87}Br, which undergoes β-delayed neutron emission.

β^- radioactive nuclide with a half-life of 55.6 s. It decays to an excited state of ^{87}Kr with 70 percent probability. This state has sufficient excitation that it can essentially instantaneously emit a neutron to become ^{86}Kr. The neutron emission thus follows the half-life of the ^{87}Kr parent. As we shall see in Chapter 12, this process is critical in the working of nuclear reactors. The process of beta-delayed neutron emission was identified by E. Booth and J. Dunning (Columbia University) and F. Slack (Vanderbilt University) in one of the early 1939 classic experiments on nuclear fission shortly after fission was discovered by O. Hahn and F. Strassman and before there was a voluntary ban on the publication of results related to fission to keep information from the Germans. There are number of β^-delayed neutron emitters found in fission products and in other neutron-rich areas of the periodic table.

Since 1979, β^-delayed two neutron and three-neutron emissions have been observed. The β^- decay of ^{11}Li to ^{11}Be with a half-life of 8.5 ms populates an 8.84 MeV excited state in ^{11}Be which is observed to decay by the emission of two neutrons to ^{9}Be and of three neutrons to ^{8}Be. The ^{11}Li nucleus is very interesting because there is experimental evidence that it has a wide outer shell of pure neutron nuclear matter. It looks like a ^{9}Li core with a two-neutron halo outside with the radius of ^{11}Li, comparable to that of ^{208}Pb! This can offer exciting opportunities to study such unusual pure neutron matter which otherwise is thought to occur only in stars composed of neutrons. One of the new frontiers in nuclear physics in the twenty-first century will involve the building of accelerators to produce beams of radioactive nuclides for further research. An intense beam of ^{11}Li could provide exciting new opportunities to probe new forms of nuclear matter. Other unique new research would be to make still more exotic nuclei farther from stability and to measure the nuclear reaction rates of highly unstable nuclei which are involved in nucleosynthesis in stars where such reactions occur. Such reaction rates are essential to help unravel how heavy elements are produced in stars.

β-DELAYED PROTON, TWO-PROTON, THREE-PROTON, DEUTERON, TRITON, AND α-PARTICLE EMISSIONS. In very proton-rich nuclei when β^+ decay occurs, the nucleus may be left in an excited state of sufficiently high energy that this state can decay by the emission of a proton, two protons, three protons, a deuteron (^{2}H) ^{3}H, or ^{4}He rather than emit γ-rays. Again, the proton, two-proton, three-proton, ^{2}H, ^{3}H, or ^{4}He decays from the excited state are prompt so they follow the half-life of the β^+ emitting parent. Each of these β-delayed decays has been observed. An example of β^+-delayed proton and α emission is the β^+ decay of ^{114}Cs to an excited state of ^{114}Xe which then emits a proton and becomes ^{113}I. The ^{114}Xe excited state also can emit an α-particle and become ^{110}Te with a certain probability. The nucleus of ^{114}Cs has 19 fewer neutrons than its stable isotope.

As another example, ^{35}Ca β^+ decays to an excited state in ^{35}K which then emits a proton to become ^{34}Ar. Note that natural calcium has stable isotopes from ^{40}Ca to ^{48}Ca, so ^{35}Ca has 5–13 fewer neutrons than stable calcium nuclei. In the decay of a light nucleus like ^{35}Ca, many different proton groups are observed to depopulate different excited states of the daughter. Their intensities are a measure of the β population of these excited states and are the only way of obtaining accurate information on β decays to high-lying excited states for testing nuclear models. These are additional examples of why the study of nuclides far from β stability is a major frontier in nuclear physics today. Various unexpected new properties can be observed in exotic nuclides which probe nuclear matter under extreme conditions to test and challenge our understanding of the nucleus.

β-DELAYED SPONTANEOUS FISSION. Again, after β decay of a nucleus far off stability, the daughter may be in a state of sufficiently high energy that it can undergo immediately spontaneous fission into two light fragments plus one or more neutrons. First β^-- and β^+-delayed spontaneous fission were observed, and in 1989 electron-capture-delayed spontaneous fission was unambiguously identified. Examples of these processes are:

$$^{246}_{99}\text{Es} \rightarrow \beta^+ + \nu + \left(^{246f}_{98}\text{Cf}\right)^* \tag{10.42a}$$

promptly followed by

$$\left(^{246f}_{98}\text{Cf}\right)^* \rightarrow {}^{138}_{54}\text{Xe} + {}^{107}_{44}\text{Ru} + \text{n} \tag{10.42b}$$

and

$$^{236}_{91}\text{Pa} \rightarrow \beta^- + \bar{\nu} + \left(^{236f}_{92}\text{U}\right)^* \tag{10.43a}$$

promptly followed by

$$\left(^{236f}_{92}\text{U}\right)^* \rightarrow {}^{139}_{53}\text{I} + {}^{94}_{39}\text{Y} + 3\text{n} \tag{10.43b}$$

These reactions give only one possible example out of the broad distribution of fission products that can be found in any spontaneous fission. The f in the superscript indicates a spontaneous fission decay mode. Recently, in Russia, β-delayed spontaneous fission has been observed even in much lighter nuclei, very far off stability, such as $^{180}_{80}Hg$.

CONCLUDING REMARKS ON α AND β DECAYS. Originally, known α decays were concentrated only in the heavy nuclei, but now are observed extensively in rare-earth nuclei and more recently even in the region around $A = 100$. However, β decays occur across the whole periodic chart. In the chart of the nuclides as one goes away from the valley of β-stability (the line of stable nuclei with no β decay), the neutron-deficient isotopes decay back to the stability line by β^+ and EC decay while the neutron-rich nuclides decay back to the stability line by β^- decay (Fig. 10.18).

The experimental and theoretical study of β decay and the weak interaction have been and continue to be an exciting fountain of knowledge in physics. The Fermi theory of β decay was a milestone. In 1956, T. D. Lee and C. N. Yang made the astonishing suggestion that parity is not conserved in the weak interaction, and in the next year this was proven in the β decay of ^{60}Co by C. S. Wu and co-workers. In 1967, building on the work of S. Glashow, S. Weinberg and A. Salam proposed a unified theory of the weak interaction and the electromagnetic interaction. In 1983, at the world's largest particle accelerator in CERN, Switzerland, the $W^{\pm}$ and the Z^0, the particles which mediate the weak interaction, were found.

Studies of the rest mass and the nature of the neutrino are important areas of research today. New neutrino detectors are giving us a new window into our universe. An early example was the detection of a neutrino burst from the supernova in 1987. Then, of course, much of the information we have about nuclear masses, nuclear energy levels, and the structure of nuclei have come from studies of β-decaying nuclei.

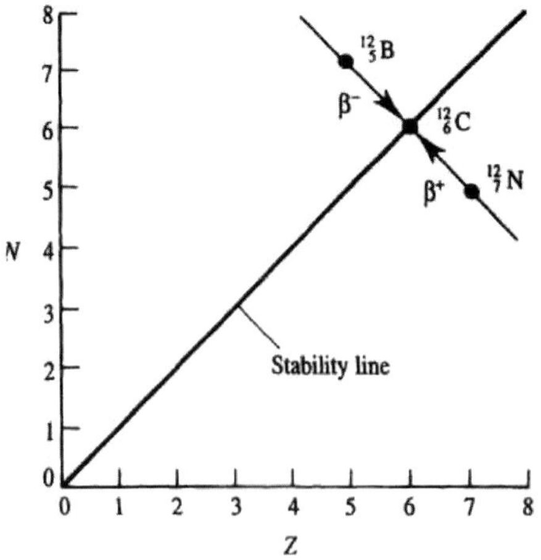

FIGURE 10.18
Illustration of β^+ and β^- decay back to stability.

10.4 GAMMA DECAY, INTERNAL CONVERSION AND PAIR PRODUCTION

10.4.1 Gamma Decay

As described in more detail in the next chapter, nuclei as well as atoms can exist only in certain quantized energy states. Just as a transition between two atomic energy levels in an atom may lead to the emission of a photon, so may a transition between two energy levels in a nucleus lead to photon emission, but the energy range of the photons is quite different. The photon energies in a nuclear transition are between a few keV and several MeV, while the photon energies in atomic transitions go from about one eV for transitions involving valence electrons up to 100 keV for transitions involving the innermost electrons of heavy elements. Measurements of γ-ray energies are one of the major ways of gaining information about the energy states in a nucleus. The patterns of the γ-ray energies alone can tell us the structure of a nucleus, as we shall see. There is also a difference in the types of electromagnetic radiation emitted by atoms and nuclei. Recall that in atoms there is a selection rule on the change in the total angular momentum between the two atomic states of $\Delta j = 0, \pm 1$ for dipole radiation, where the photon carries off only its intrinsic angular momentum of $1\hbar$. There are essentially always atomic states available in an atom to fulfill this condition, so higher multipole radiations where $2\hbar$ and more angular momentum must be carried off are not observed. In nuclei the lowest-order electromagnetic radiation is also electric dipole radiation, with lifetimes of the order of less than 10^{-12} s. However, where there are essentially always two atomic states that can be connected by electric dipole radiation, the changes in spin and parity between two nuclear states often do not allow the emission of electric dipole radiation and so higher-order processes commonly occur in nuclear transitions.

In γ decay one must have conservation of mass–energy, linear momentum, and angular momentum. If E_i is the initial total energy of a nucleus and E_f the final total energy, then

$$E_i = E_f + E_\gamma + T_R \qquad \text{and} \qquad 0 = \boldsymbol{p}_\gamma + \boldsymbol{p}_R \tag{10.44}$$

where T_R and $\boldsymbol{p}_R$ are the kinetic energy ($p_R^2/2M$ nonrelativistically) and linear momentum of the nucleus. Since

$$E_\gamma = p_\gamma c = p_R c \tag{10.45}$$

then

$$T_R = \frac{p_R^2}{2M} = \frac{E_\gamma^2}{2Mc^2} \tag{10.46}$$

Since the rest energy of one nucleon is 938 MeV, $2Mc^2 \sim 2A \times 10^3$ MeV. Since $E_\gamma \lesssim 5$ MeV,

$$T_R \leq \frac{25}{2A \times 10^3}\ \text{MeV}$$

which is $< 10^{-3}$ for $A = 10$ and $< 10^{-4}$ for $A = 100$ and so can generally be neglected. The recoil energy does play a part in the Mössbauer effect discussed later in this chapter.

In Chapter 9 we described how an arbitrary charge distribution can be described in terms of static multipole moments: electric monopole, electric dipole, electric quadupole moments, and so forth. The electric fields associated with these vary as $1/r^2$ (from the total charge or monopole moment), $1/r^3$ for the dipole moment, $1/r^4$ for the quadrupole moment, and so forth. Magnetic fields are similarly described, except that all magnetic fields that have been observed arise from electric currents and no magnetic monopoles have been observed. So the lowest-order magnetic field is that associated with the magnetic dipole moment associated with an electric current as discussed in Section 4.1. There are similar higher-order moments of the current distributions as well—magnetic quadrupole, and so forth.

Classically when a charge or current distribution varies with time, a radiation field is produced. If the distributions vary sinusoidally with a frequency ω, a radiation field characteristic of the distribution is produced. For example, if one has an electric dipole located on the Z axis ($+q$ and $-q$ above and below the origin separated by a distance z) with electric dipole moment of qz that oscillates in time so the instantaneous dipole moment is $p(t) = qz \cos \omega t$, then electric dipole radiation is produced. If one has a time-varying magnetic dipole moment $\mu(t) = iA \cos \omega t$, this produces magnetic dipole radiation.

Classically, for dipole radiation the power radiated through a small area A varies as $\sin^2 \theta$, where θ is the angle between the z axis and the direction of A. An electric or magnetic dipole does not radiate energy along its axis. This must also be true in quantum mechanics by the correspondence principle. Quadrupole radiation and other higher multipoles will have different angular distributions. A measurement of the directional distribution of the radiation can determine the multipole character of the radiation. It will not determine the parity of the field. Electric and magnetic fields of the same multipole have opposite parities. Electric dipole radiation has odd parity and magnetic dipole radiation has even parity.

The total radiated power W over a sphere (total energy emitted per unit time) for an electric dipole with dipole moment amplitude p_0 is

$$W = \frac{C}{12\pi\epsilon_0} \frac{p_0^2}{\lambda^4} = \frac{1}{12\pi\epsilon_0} \frac{\omega^4}{c^3} p_0^2 \tag{10.47}$$

and for a magnetic dipole with magnetic dipole moment amplitude μ is

$$W = \frac{c\mu_0}{12\pi\lambda^4} \mu^2 = \frac{1}{12\pi\epsilon_0} \frac{\omega^4}{c^5} \mu^2 \tag{10.48}$$

[See D. Carson and P. Lorrain, *Electromagnetic Fields and Waves*, Freeman, San Francisco (1962).]

Similar expressions exist for higher multipoles. One defines the multipole order by 2^L where $L = 1$ is dipole and $L = 2$ is quadrupole, $L = 3$ is octupole, and so forth. As we shall see L is the angular momentum carried off by the radiation. In the case of γ-ray emission in a transition between two nuclear states, L is related to the angular momentum of the two states as shown below. The radiation fields also have a parity given by $(-1)^L$ for electric and $(-1)^{L+1}$ for magnetic multipoles. The electric and magnetic multipoles of the same order have opposite parity. The final note is that classically the power radiation for any multipole moment depends on the square of that moment. To go to quantum mechanics, one replaces multipole moments by multipole operators, and the decay probability is governed by the square of the matrix element of the multipole operator. So the total transition probability, T, neglecting polarization of the radiation and including the possibility of both electric and magnetic multipoles contributing, is

$$T(j_i \to j_f) = \frac{8\pi}{2j_i+1} \sum_{L,\pi} \frac{\omega^{2L+1}}{[(2L+1)!!]^2} \frac{L+1}{L} |\langle j_f \| \mathcal{M}(\pi L) \| j_i \rangle|^2 \tag{10.49}$$

where π stands for electric or magnetic. Since the approximation $\omega R \ll L$ may be used for nuclear γ-rays, the multipole operators $\mathcal{M}$ are given by

$$\begin{aligned} \mathcal{M}(ML, m) &= \frac{-i}{(L+1)} \int \boldsymbol{j}(\boldsymbol{r}')\boldsymbol{L}(r')^L Y_{Lm}(\hat{r})\, dV' \\ \mathcal{M}(EL, m) &= \int \rho(\boldsymbol{r}')(r')^L Y_{Lm}(\hat{\boldsymbol{r}}')\, dV' \end{aligned} \tag{10.50}$$

Note that the multipole operator is the integral over $\boldsymbol{j}(\boldsymbol{r}')$ the nuclear current density and the vector potential $\boldsymbol{A}^{(\pi)}$ of the radiated field. (In Appendix 10A it is noted this form was used by Fermi to generate the form of the β decay matrix elements.) Here the potential has been transformed into its radial dependence and spherical harmonics, Y_{Lm}, and for $\mathcal{M}(EL, m)$ further transformed into the charge density, which can be expanded into multipoles. For details, see K. Alder and R. M. Steffan, *The Electromagnetic Interaction in Nuclear Spectroscopy*, W. D. Hamilton, ed., North Holland, Amsterdam (1975).

The selection rules for a transition from a state of total angular momenta, j_i and m_i to a state j_f m_f are that $|j_i - j_f| \le L \le j_i + j_f$ and $m_f = m + m_i$. So the photon can carry off any angular momentum between the sum and difference of the angular momenta of the two states. These rules are a consequence of the rotational properties of T and the states $|jm\rangle$. One can also define a reduced transition probability with the energy factor ω^{2L+1} taken out.

$$B(\pi L, j_i \to j_f) = \frac{|\langle j_f \| \mathcal{M}(\pi L) \| j_i \rangle|^2}{(2j_i+1)} \tag{10.51}$$

This depends only on the structure of the system and not the transition energy. It is important in lifetime measurements and Coulomb excitation.

One can calculate the transition probability for a single proton making a transition from one shell model state to another. These are called the Weisskopf single-particle estimates. These single-particle estimates for the decay constants and lifetimes expected for the transition of a single nucleon from one energy state to a lower energy state by the emission of different multipole radiations as a function of decay energy are given in Table 10.8 and Fig. 10.19, respectively.

The multipolarities are related to the angular momentum carried off by the radiation. For example, since the photon can carry off any angular momentum between the sum, $j_i + j_f$ and difference, $|j_i - j_f|$ of the angular momenta of the two states, if $j_i = j_f = 2\hbar$, then $L = 0, 1, 2, 3, 4$ are allowed. If the parities of the two states are both even then E2, E4, M1, and M3 are allowed. However, since the photon has spin of $1\hbar$, $L = 1$ dipole radiations have the shortest lifetimes. The lifetimes increase as L increases. This is seen in the rapid decrease in the constant in front of the transition probabilities in Table 10.8. For example from Table 10.8 for a 1-MeV γ-ray, $T_{sp}(\text{M1})/T_{sp}(\text{E2}) \sim 10^5$ and $T_{sp}(\text{M1})/T_{sp}(\text{M3}) \sim 10^{11}$. So one would expect to see only M1 radiation even though M3, E2, and E4 are allowed by the spin and parity selection rules. There is also a strong dependence on the energy, $(E)^{2L+1}$ for each multipole L so that, as the transition energy increases, the lifetime decreases rapidly. As we shall see, there are nuclear structure effects which can slow down or speed up these single-particle (SP) lifetimes by factors up to 10^5 and more. This can lead to different competition between different multipolarities so that a $2^+ \rightarrow 2^+$ transition may be E2 instead of M1. A comparison of experimental and theoretical SP lifetimes can provide important insight into the structure of a nucleus. Such comparisons were among the important data that established the nuclear shell model of Mayer and Jensen and the collective model of Bohr and Mottelson, as discussed in the section on nuclear models in Chapter 11. Thus, measurements of the lifetimes of nuclear states continues to be an important area of nuclear structure research.

Excited energy states in nuclei were first observed through their population in α or β decays, and their decay by γ-ray and internal conversion electron emission. L. Meitner first established that the γ-ray and internal conversion electrons were emitted by the daughter following α and β decay. This method of studying nuclear energy levels continues to offer unique opportunities to probe the levels of new nuclei far from stability. Beginning in the late 1940s,

TABLE 10.8

Single-particle estimates for the transition probability of a single proton to emit a photon ($R_0 = 1.2 \times 10^{-15} A^{1/3}$ m; $\mu_P = 2.79$; E_γ in MeV)

Multipolarity	T_{sp} (s^{-1})
E1	$1.0 \times 10^{14} A^{2/3} E_\gamma^3$
M1	$2.9 \times 10^{13} E_\gamma^3$
E2	$7.4 \times 10^7 A^{4/3} E_\gamma^5$
M2	$8.4 \times 10^7 A^{2/3} E_\gamma^5$
E3	$3.4 \times 10^1 A^2 E_\gamma^7$
M3	$8.7 \times 10^1 A^{4/3} E_\gamma^7$
E4	$1.1 \times 10^{-5} A^{8/3} E_\gamma^9$
M4	$4.8 \times 10^{-5} A^2 E_\gamma^9$
E5	$2.5 \times 10^{-12} A^{10/3} E_\gamma^{11}$
M5	$1.7 \times 10^{-11} A^{8/3} E_\gamma^{11}$

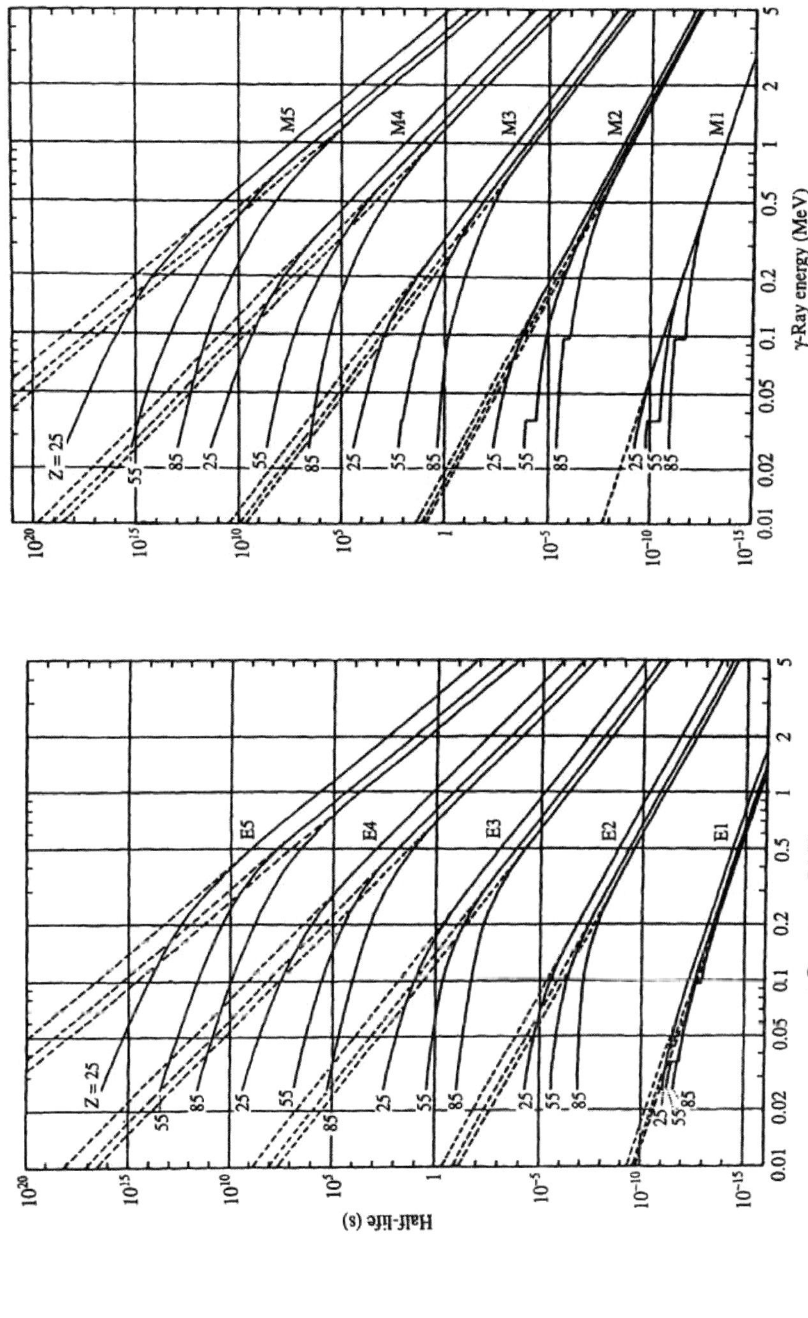

FIGURE 10.19
The Weisskopf single-particle lifetimes for different multipole γ-radiations. The solid curves indicate the correction to the level lifetime for internal conversion.

accelerated charged particles such as the proton and α began to be used extensively to excite the energy levels of stable nuclei by the Coulomb force or by inelastic scattering. Now the other major technique is to use a heavy-ion nuclear reaction to populate the excited states including those of nuclei far from stability. States up to very high angular momentum, more than $60\hbar$, have been observed.

An excited nuclear state generally promptly ($< 10^{-10}$ s) decays to a lower-energy excited state or the ground state with the emission of electromagnetic radiation. This radiation is called a γ-ray, and the transition is called a γ-ray transition or γ decay. However, γ decay may be retarded to yield lifetimes up to months and years. As an example, the ^{60}Co decay scheme is shown in Fig. 10.20. We can see that in β^- decay ^{60}Co goes to a 2.50-MeV excited state in ^{60}Ni with a half-life of 5.27 years. The maximum energy of the β-particle is 0.309 MeV. The lifetimes of the excited states in ^{60}Ni are very short ($< 10^{-11}$ s) for decay through the emission of 1.17-MeV and 1.33-MeV γ-rays in cascade to the ground state. So, when one ^{60}Co nucleus emits one β^--particle, two photons follow immediately. The spins (in units of $\hbar$) and parities (even +, odd −) of the three states are 4^+, 2^+, 0^+, so each transition is a $\Delta I = 2$ no-parity-change, electric quadrupole transition. From Fig. 10.19 we can see that the single-particle lifetimes for 1.2-MeV E2 γ-rays are less than 10^{-11} s. The real lifetimes for the ^{60}Co gamma rays are shorter than this because of collective effects. A 1-curie ^{60}Co source has 3.7×10^{10} decays per second. This includes 3.7×10^{10} β-particles per second and $2 \times 3.7 \times 10^{10}$ γ photons per second. The number of radiations emitted is triple the number of decays. In general, the number of nuclear decays is considerably less than the number of radiations (γ-rays) emitted in a given time. In medicine ^{60}Co has been and still is used in many places to irradiate tumors. Unfortunately, the intensity of the activity is reduced by one-half every 5.27 years. The ^{60}Co decay has a simple spectrum of two γ-rays. Nuclei far from stability may emit hundreds of γ-rays, as seen in Fig. 10.21.

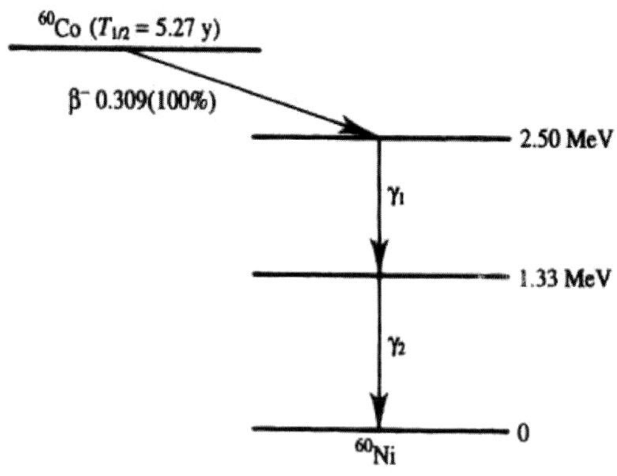

FIGURE 10.20
The ^{60}Co decay scheme.

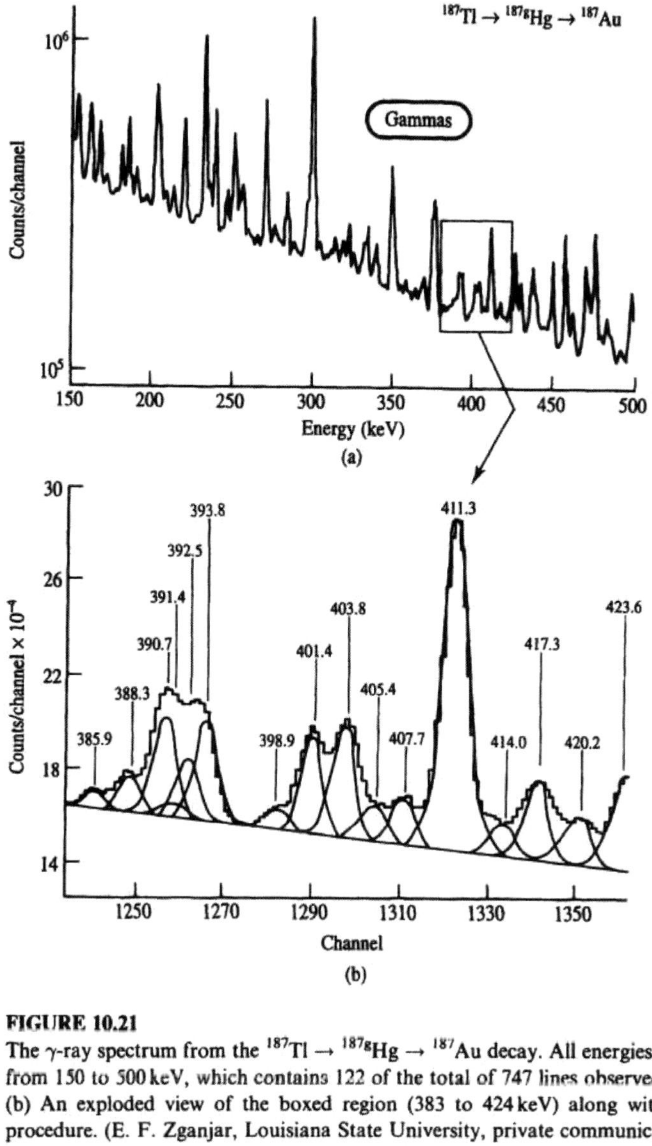

FIGURE 10.21
The γ-ray spectrum from the $^{187}Tl \rightarrow {}^{187g}Hg \rightarrow {}^{187}Au$ decay. All energies are in keV. (a) The portion from 150 to 500 keV, which contains 122 of the total of 747 lines observed between 36 and 2555 keV. (b) An exploded view of the boxed region (383 to 424 keV) along with the results of the fitting procedure. (E. F. Zganjar, Louisiana State University, private communication.)

10.4.2 Internal Conversion

In some cases, when the nucleus goes from an excited state to a lower-energy state, photon emission does not occur. Instead the nucleus gives its excess energy to one of the atomic electrons outside the nucleus. This electron then is thrown out of the atom. This is called the internal conversion (IC) process. The ejected atomic electron is called an internal conversion electron.

If the γ photon energy is

$$E_\gamma = E_u - E_l \tag{10.52}$$

where the E_u and E_l are the energies of the upper and lower energy levels (here we neglect the nuclear recoil when releasing the photon), the internal conversion electron energy is

$$E_e = E_u - E_l - W_i \tag{10.53}$$

where W_i is the i-shell electron binding energy (as given in Appendix IV). Obviously, the energy spectrum of internal conversion is discrete, but also complex as shown in Fig. 10.22. These discrete internal conversion lines are the puzzling "discrete energy β-rays" which were observed to sit on top of the continuous energy β-ray spectrum in many of the first studies of radioactivity. The continuous β-rays contribute to the background seen under the sharp lines in Fig. 10.22. They have the same nature (a negative electron) but have a different origin from the negative β decay radiation. While C. D. Ellis and co-workers introduced the name internal conversion electrons in 1924, this name did not gain full acceptance until H. Bethe in his classic article on nuclear physics in 1937 proposed that α, β, and γ be reserved for particles coming from the nucleus and that the ejected atomic electron be called an internal conversion electron. The discovery that the W_i in Eq. (10.53) is characteristic not of the parent but of the daughter formed by the β decay of the parent proved that the discrete lines were not from the same process as β decay.

When a nucleus makes a transition from one energy level to another, the probabilities of emitting a photon and an internal conversion electron are functions of the changes in spin and parity between the two nuclear energy levels, the energy difference, the angular momentum carried away by the radiation, and the Z of the nucleus. The total "internal conversion coefficient," α, for a given transition is defined as the ratio of the internal conversion electron emission probability from all atomic shells, N_e, to the photon emission probability, N_γ. Since the electron can be ejected from any shell which contains electrons, K, L, M, ..., each shell and each subshell has its own conversion coefficient:

$$\begin{aligned} &\alpha \equiv \frac{N_e}{N_\gamma} \\ &\alpha_K = \frac{N_{e,K}}{N_\gamma} \qquad \alpha_L = \frac{N_{e,L}}{N_\gamma} \qquad \alpha_{L_I,L_{II},L_{III}} = \frac{N_{e,L_I,L_{II},L_{III}}}{N_\gamma} \end{aligned} \tag{10.54}$$

and so forth.

At first it was thought that internal conversion is a process during which an excited state emits a photon and then this photon gives its energy to an orbital electron (i.e., a photoelectric effect), but that viewpoint is wrong. The probability of first producing a γ-ray and then producing the photoelectric effect is much smaller than what is observed. In internal conversion, the nuclear energy is transferred directly to the atomic electron through the Coulomb field which couples them.

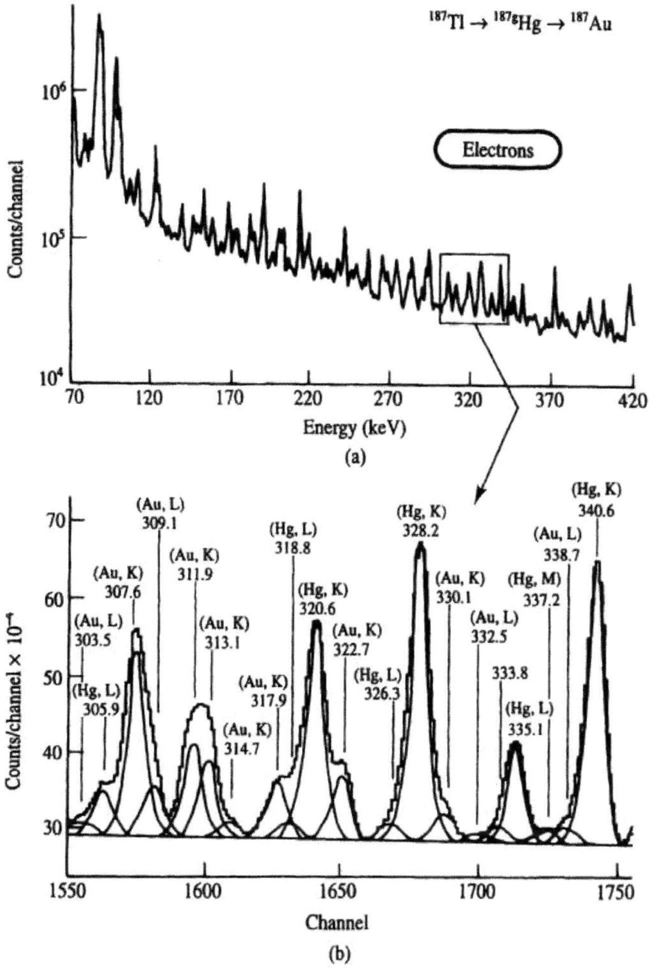

FIGURE 10.22
The conversion-electron spectrum from the $^{187}Tl \rightarrow {}^{187g}Hg \rightarrow {}^{187}Au$ decay. All energies are in keV. (a) The portion from 70 to 420 keV, which contains 152 lines. The energy range spanned by part (a) exactly matches the 150 to 500 keV range of Fig. 10.21(a) if you subtract 80 keV. (b) An exploded view of the boxed region (303 to 344 keV) along with the results of the fitting procedure. The K-shell binding energy is 80.7 keV in Au and 83.1 keV in Hg. Thus the regions in parts (b) of the two figures enable one to make a 1:1 comparison for conversion in the K-shell. The L and M subshell lines from lower energy transitions are also indicated. The line at 330.1 keV corresponds to K-conversion of a 410.8 keV transition in Au. Clearly one could never get the γ-ray intensity of such a transition nor its multipolarity from a single spectrum shown in Fig. 10.21. Coincidence data were used to extract intensities of the 410.8 keV γ-ray in Au and the 411.3 keV γ-ray in Hg. In addition to "cleaning up" such complicated spectra as those displayed in Fig. 10.21 and here, coincidence techniques developed at UNISOR enable one to extract quite accurate internal-conversion coefficients for transitions which have intensities even as low as 0.05 percent of the total decay intensity. (Louisiana State University.)

Internal conversion electrons, like γ-rays, are characterized by the angular momentum, l, carried off and the parity change between the two states, as discussed in the previous section. The conversion electron and conversion coefficients are characterized by the same multipole order (E1, M1, E2, ...) as the γ-rays associated with the transition. Some typical conversion coefficients are shown in Fig. 10.23.[5] One notes, as expected, the strong dependence on the atomic shell. For example, the K electrons, which are closest to the nucleus, have the largest probability of being emitted. Also, α_K exhibits a rapid increase with Z because the Coulomb interaction between the electrons and the nucleus increases and this force is how the energy is transferred. The internal conversion process is most important for low-energy, high-Z, and high-order multipole transitions. The first detailed calculations of conversion coefficients were done in the 1950s by M. Rose (Oak Ridge) for a point nucleus. Soon after his calculations were begun, L. Sliv and I. M. Band in Russia showed that the finite size of the nucleus must be included in the calculations.

When the spin and parity of both nuclear levels are 0^+, it is impossible for a nucleus to emit a photon which has intrinsic spin of $1\hbar$. However, in such a case internal conversion electrons are observed. In fact, observing internal conversion electrons, but no photons, is one of the experimental methods for discovery of excited 0^+ states. For example, both authors have used this technique for many years to discover excited 0^+ states in nuclei across the periodic table. As an illustration, Fig. 10.24 shows the strong conversion lines and no γ-rays for a 522-keV transition in ^{186}Hg. These established the first excited 0^+ level in ^{186}Hg. These internal conversion electrons carry off no angular momentum and are called electric monopole (E0) radiation ($l = 0$). E0 conversion electrons are emitted because the electrons have a probability of being inside the nuclear volume where they can interact directly with the nucleus. As we shall see, because their emission arises through their penetration into the nuclear volume, E0 transitions provide unique probes of changes in shape between two nuclear energy levels. Such studies at the University Isotope Separator in Oak Ridge (UNISOR) were used to establish the coexistence of quite different shapes for different energy levels in the same nucleus, as discussed in the next chapter.

10.4.3 Internal Pair Creation

In 1937, R. J. Oppenheimer and co-workers, while interpreting the discovery of positrons and $e^{\pm}$ pair creation by γ-rays, predicted that when the energy difference between two nuclear states exceeds $2m_ec^2$ (1.02 MeV), it becomes possible for the excited nuclear level to give up its energy through the creation of an electron–positron pair. This process, known as internal pair creation, competes with γ and internal conversion emission above 1.02 MeV. It becomes more important as the

[5] J. D. Cole, W. Lourens, J. H. Hamilton, and B. van Nooijen, *Graphical Representation of K-Shell and Total Internal Conversion Coefficients from Z = 30–104*, Delft University Press (1984).

transition energy increases and can dominate a decay for energies of a few MeV. The probability for internal conversion electron emission decreases rapidly with energy and is negligibly small above 2 MeV where internal pair production becomes important. The $e^- - e^+$ pair is created in the Coulomb field of the nucleus. Internal pair creation becomes the dominant mode for high-energy (few MeV) $0^+ \rightarrow 0^+$ transitions. New internal pair spectrometers have been built recently at UNISOR and elsewhere to search for high-energy 0^+ states which could be the first states of bands of levels built on superdeformed and other exotic shapes.

10.4.4 Isomeric Transitions

As noted above, the lifetime of a nucleus in an excited state is typically very short, less than 10^{-12} s. In some cases, however, the lifetime of the excited nuclear state is very much longer, even days to years. Excited states with long lifetimes are called "isomers" or metastable states. A small m is placed between the atomic mass number and the chemical symbol to identify such a state. For example, ^{113m}In is an isomer of ^{113}In. Its half-life is 99.5 minutes for decay to the ground state of ^{113}In with a 65 percent probability of emitting a γ-ray and 35 percent probability of emitting an internal conversion electron (see Fig. 10.25). As shown in Fig. 10.25, one source of ^{113m}In is ^{113}Sn, which decays to ^{113m}In with a 97.8 percent EC probability or decays to a higher excited state with 2.2 percent probability, and this state promptly emits a 255-keV γ-ray to go to ^{113m}In.

There is no definite value of the half-life above which one can call a state isomeric. Earlier, when the lower limit for the measurement of a half-life was of the order of 10^{-9} s, this value was used by some to define isomers as those states which were longer-lived. However, the symbol m is not generally used until the half-life is as long as 10^{-6} s or even 10^{-3} s.

There is some difference in the properties of the isomeric state and ground state which hinders the decay. Most often this difference is a large change in spin between the states, a parity change, and a small energy difference, as can be seen from the lifetimes in Fig. 10.19. Sometimes it is related to large differences in their shapes. Large spin changes and low decay energy can sufficiently hinder decay to the ground states so that the level undergoes β or α decay rather than internal conversion and γ decay. As discussed earlier, very large differences in the nuclear shape in very heavy actinide elements can lead to spontaneous fission of the isomer (fission isomer) rather than emission of a photon or conversion electron. Recall from the discussion of lasers that lasers require a metastable (isomeric) atomic state into which the electrons can be pumped and from which they can be induced to emit laser light. The energies of the photons in a laser beam are typically of the order of a few eV. One increases the laser power by increasing the number of photons. Much effort is being directed today toward developing a γ-ray laser (a graser) in which the photon energy could be 10^3 to

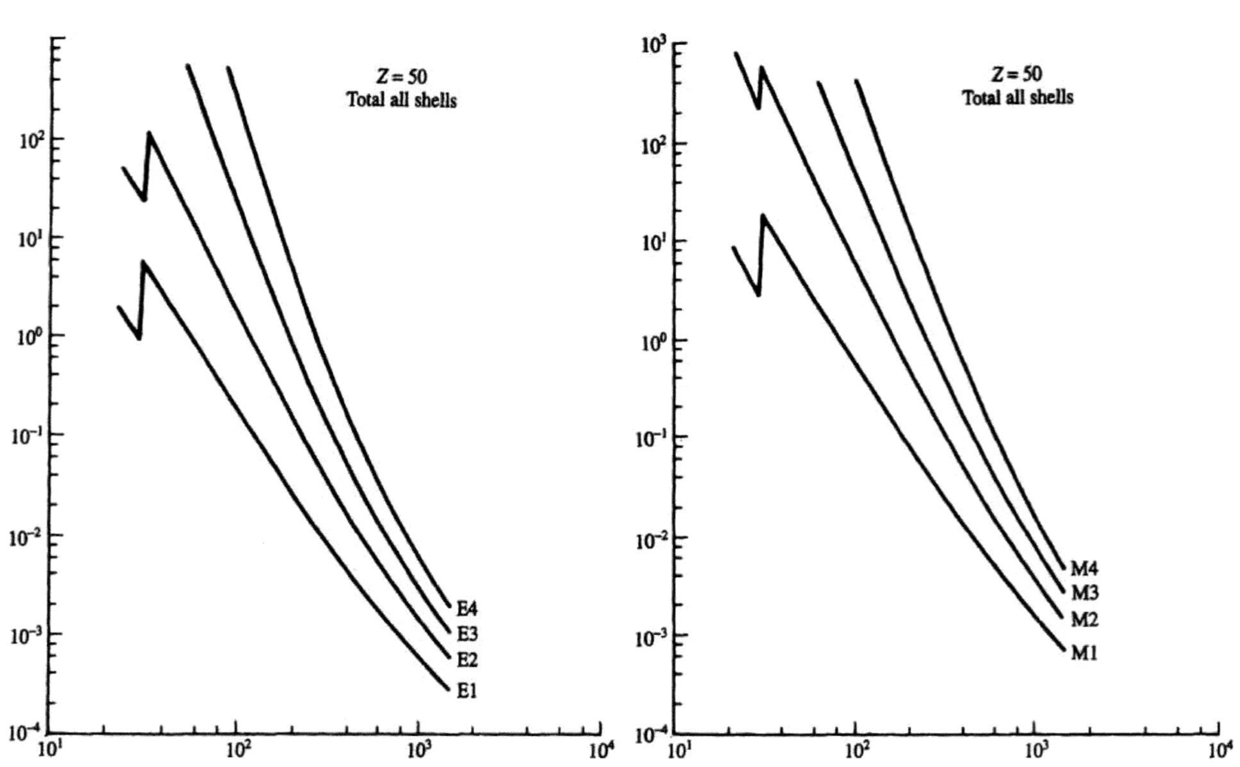

(a)

FIGURE 10.23
Selected examples of internal conversion coefficients. [J. D. Cole, W. Lourens, J. H. Hamilton, and B. van Nooijen, *Graphical Representation of K-Shell and Total Internal Conversion Coefficients from* Z *= 30–104*, Delft University Press (1984).] The energies are in keV.

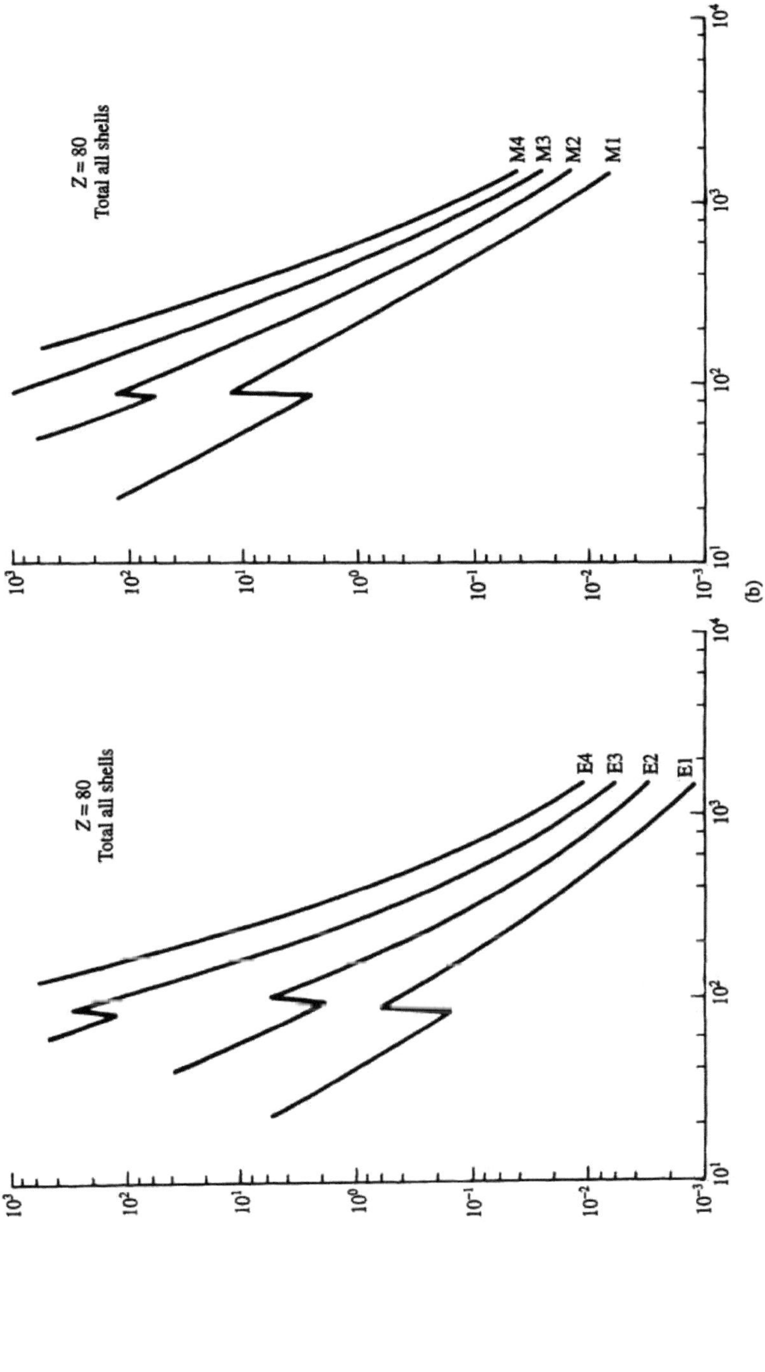

FIGURE 10.23 *(continued)*

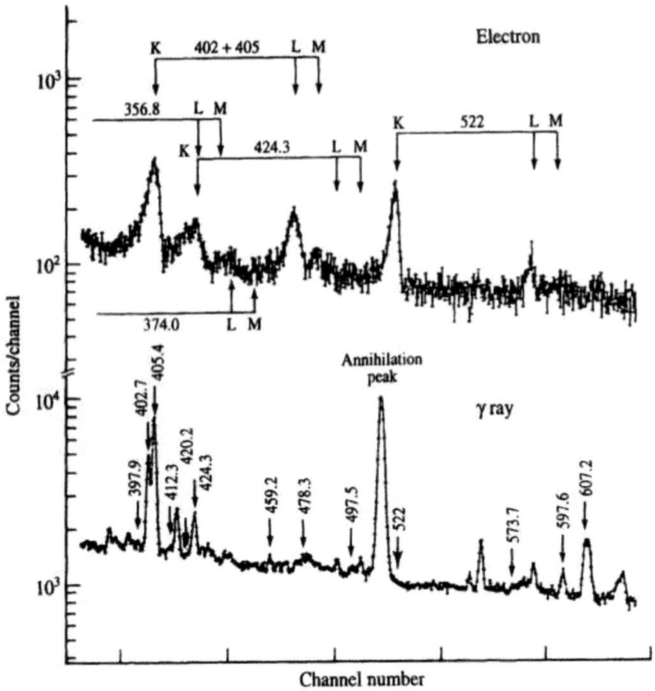

FIGURE 10.24
A portion of the conversion electron (upper) and γ-ray (lower) spectra of the decay of ^{186}Tl to ^{186}Hg. [J. D. Cole et al., *Phys. Rev.* **C16** (1977) 2010.]

10^6 times greater (keV to MeV). The power in such a device would be staggering. Construction of such a device requires the identification of a nuclear metastable (isomeric) state which can be pumped and stimulated to emit γ-ray energy laser light.

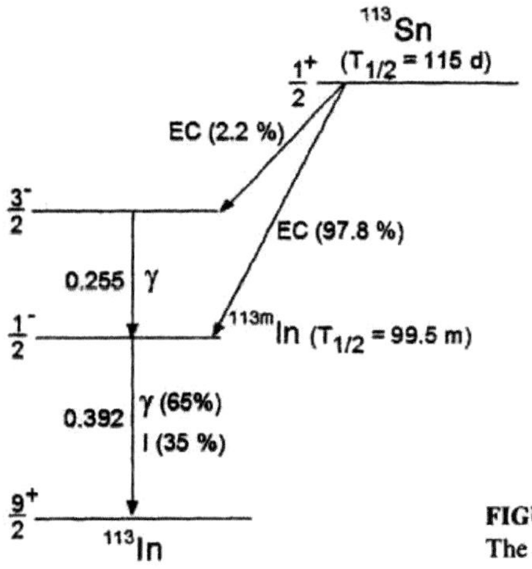

FIGURE 10.25
The decay scheme of the isomer ^{113m}In.

10.4.5 Mössbauer Effect

The Mössbauer effect, or recoilless γ resonance absorption was found by R. J. Mössbauer in 1958 and won him the Nobel Prize. In formula (10.52), the nuclear recoil energy, E_R, was neglected. The equation should be written

$$E_0 = E_u - E_l = E_\gamma + E_R = h\nu + E_R \tag{10.55}$$

Now let us estimate the size of E_R.

Suppose the nucleus prior to emission is not moving. From momentum conservation, the nuclear momentum after emitting the photon should be equal to that of the emitted photon:

$$p = m\text{v} = \frac{h\nu}{c}$$

So

$$E_R = \frac{1}{2}m\text{v}^2 = \frac{p^2}{2m} = \frac{(h\nu)^2}{2mc^2} \tag{10.56}$$

For example, the first excited state of ^{57}Fe goes to its ground state with the emission of a 14.4-keV photon (see Fig. 10.26). Then

$$E_R = \frac{(14.4)^2(\text{keV})^2}{2 \times 57\,\text{u} \times 931\,\text{MeV/u}} = 2 \times 10^{-3}\,\text{eV} \tag{10.57}$$

Compared to E_γ of 14.4 keV, this is very small so, in general, (10.52) is a good approximation. But E_R is large compared to its energy level width. According to the uncertainty relation, any excited state with a lifetime τ must have some energy level width $\Gamma \sim \Delta E$ as given by the uncertainty relation. For example, the 14.4-keV energy level in ^{57}Fe* has a half-life of 9.8×10^{-8} s, and an energy level width of

$$\Gamma = \frac{\hbar}{\tau} = \frac{\hbar c}{\tau c} = 4.7 \times 10^{-9}\,\text{eV} \tag{10.58}$$

This is much less than E_R. How, then, does E_R influence what can happen?

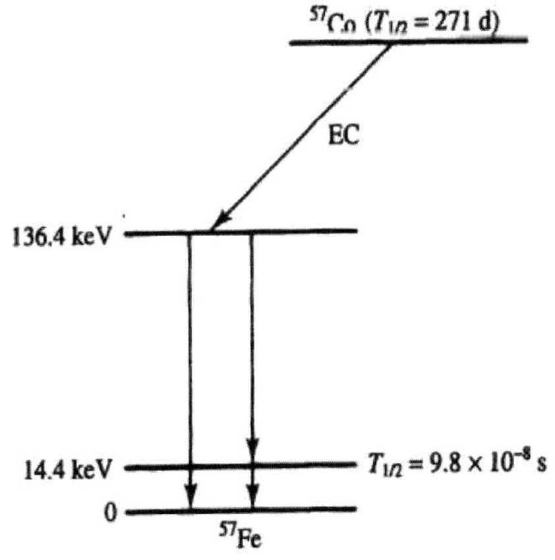

FIGURE 10.26
The decay scheme of ^{57}Co.

First, look at the atomic situation. A sodium D line is produced by an excited state of a sodium atom going back to the ground state. If we use the sodium D line to excite the ground state of a sodium atom, a resonance absorption occurs in which the photon energy and energy level difference are the same. In this case the energy level width is $\Gamma \sim 10^{-8}$ eV, and the recoil energy $E_R \sim 10^{-11}$ eV which is negligible. Here the emission and absorption energies have a spread of 10^{-8} eV but a shift of only 10^{-11} eV, so they almost completely overlap. Consequently, it is very easy to observe resonance absorption in atoms. But in the nucleus the situation is different because the transition energies are so much greater. After emitting a γ-ray, the nucleus recoils and the emitted γ-ray energy is reduced to

$$E_{\gamma_e} = E_0 - E_R \tag{10.59}$$

When absorption of a photon occurs, a nucleus also has a recoil energy. So, for resonance absorption to occur, the γ-ray energy must be larger than the level energy:

$$E_{\gamma_a} = E_0 + E_R \tag{10.60}$$

where the E_0 is determined by formula (10.55). The difference between the emitting and absorbing energies is $2E_R$ (see Fig. 10.27).

Obviously, only where the emission spectrum and the absorption spectrum overlap each other (in the shaded part of the figure), can resonance γ-ray absorption occur. Resonance absorption can occur when $E_R < \Gamma$. When $E_R \gg \Gamma$ [for example in ^{57}Fe, formulas (10.57) and (10.58)], there is no overlap between the emission and absorption spectra and so resonance absorption cannot occur.

Mössbauer discovered in studying the resonance absorption of the 129-keV γ-ray of ^{191}Ir that the nuclear recoil is essentially eliminated by putting the nuclei into a crystal lattice where the ^{191}Ir atom is bound to the crystal. Then, the recoil will be by the whole crystal to which the nucleus and atom are bound instead of the individual nucleus. In the equation for E_R, the m is the crystal mass, which may be 10^{20} or more larger than that of an atom [see formula (10.56)]. So E_R is then essentially zero and the processes of emission and absorption are essentially recoilless. This is the Mössbauer effect.

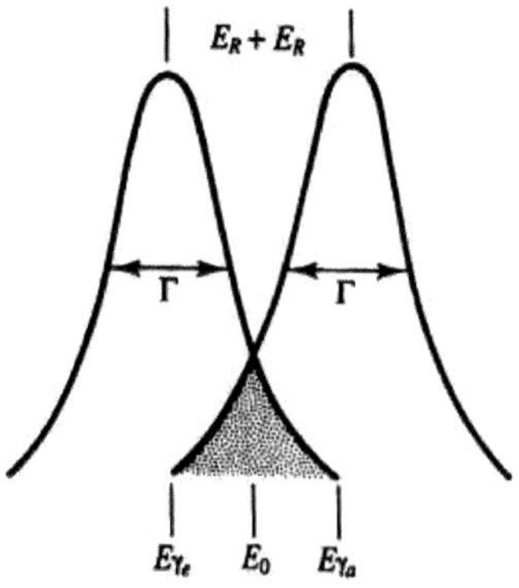

FIGURE 10.27
A γ-ray emission spectrum and absorption spectrum, each with an energy width Γ.

In the case of recoilless γ-ray emission, for the ^{57}Fe 14.4-keV γ-ray, $\Gamma/E_0 \approx 3 \times 10^{-13}$. Such a perturbation is too small to be observed. The Mössbauer effect is used in various precision measurements of frequency differences. For example, the Mössbauer effect has been used to measure the gravitational red-shift of light from a star.

Suppose a luminous star with radius R emits a photon with energy $h\nu$. Then, far away from the star, the photon energy is not $h\nu$, because to get there the photon has to overcome the star's gravitational potential and so loses a part of its energy according to Einstein's general theory of relativity. The gravitational potential is

$$V = -G\frac{Mm}{R} \tag{10.61}$$

where M is the star mass, G is the gravitational constant, and m is the mass of the photon ($m = h\nu/c^2$). So, far from the luminous star, the photon energy is

$$E = h\nu - GM\left(\frac{h\nu}{c^2}\right)\frac{1}{R} = h\nu\left(1 - \frac{GM}{Rc^2}\right) \tag{10.62}$$

Thus, the frequency is lower, and the wavelength longer: λ is shifted toward the red wavelengths in the spectrum. This is called the gravitational red-shift. The red-shift from a star can be of the order of one part in 10^{-6}. However, this can easily be masked by various Doppler shifts from motions including rotation, thermal vibrations, and convection currents. Although the gravitational red-shift could be calculated, for years it was very difficult to measure. Using the Mössbauer effect, it was measured precisely for the shift in the gravitational field of the Earth. When the distance of the photon from a massive object is different, the gravitational potential is different and the frequency is different. For a difference in distance of 22.6 m at the surface of the Earth, the expected frequency change is 2.46×10^{-15}. Remarkably, Pound and Rebka[6] measured this extremely small change in 1960 using the Mössbauer effect on the 14.4-keV transition in ^{57}Fe.

Currently 46 different elements have been used to observe the Mössbauer effect. These have involved 91 different isotopes and 112 different transitions. The most used isotopes are ^{57}Fe, ^{119}Sn, and ^{151}Eu. The highest resolution achieved is 10^{-16} for a ^{67}Zn compound to 10^{-13} for ^{57}Fe.

10.4.6 Brief Summary of the Properties of α, β, and γ Radiations in Radioactive Decays and Their Interactions in Matter

A comparison of several properties of α, β, and γ decay of nuclei and of the radiations are given in Table 10.9.

---

[6] R. V. Pound and G. A. Rebka, *Phys. Rev. Lett.* **4** (1960) 337.

TABLE 10.9
Properties of α, β, and γ decays

	α	β	γ
	Helium nucleus **^{4}He**	**Electron** e^-, e^+	**Electromagnetic radiation (photon)**
$T_{1/2}$			
minimum	$\sim 10^{-7}$ s	$\sim 10^{-2}$ s	$\sim 10^{-17}$ s
maximum	$\sim 10^{15}$ y	$\sim 10^{15}$ y	≥ 30 y
E_0			
minimum	~ 4 MeV	~ 10 keV	~ 10 keV
maximum	~ 10 MeV	~ 10 MeV	~ 10 MeV
Mechanism	Potential barrier penetration	Weak interaction	Electromagnetic interaction
Energy distribution of products	Monoenergetic	β^+, β^-: continuous EC: monoenergetic	Monoenergetic
Decay energy	$M_X - (M_Y + M_{He})$	β^-: $M_X - M_Y$ β^+: $M_X - (M_Y + 2m_e)$ EC: $M_X - (M_Y + W_i/c^2)$	$E_u - E_l$
Penetration through matter (stopping distance)	Few cm of air	Few mm of aluminum	Few cm and more of lead

Alphas and other heavy charged particles in passing through matter interact with the material primarily through the attractive Coulomb force between their positive charge and the negative charge on the orbital electrons in the atom of the material. Rutherford scattering of α-particles and other heavy particles on nuclei (Chapter 2) occurs very very rarely. As the heavy particle approaches the first atom on the surface, the long range Coulomb force is felt by the atomic electrons. Each electron in the vicinity of the particle feels a force that varies inversely with the square of the separation distance. There will be an energy transfer, which may raise the electron to a higher orbit (excitation) or remove it from the atom (ionization). The energy transferred comes from the heavy particle and so its energy is reduced. The maximum energy that a charged particle of mass M and kinetic energy E (nonrelativistic energies) can transfer to an electron of mass m_e in a single collision is $4Em_e/M$. For an α-particle, this is Em_e/M_{nucleus} or about 1/8000 of E. Since the maximum energy transferred for an α-particle is so small, the α-particle must undergo thousands of such collisions to give up all its energy. Of course any heavy particle interacts with many electrons at any time, with the result that it continuously loses energy until it stops. Since heavy particles are so massive compared to electrons, heavy particles are not deflected but move in straight lines, while the electrons go in many different directions. Some of the electrons receive sufficient energy to interact with electrons in other atoms and produce further ionization. Such electrons with sufficient energies to ionize atoms are called delta-(δ-) rays. When one looks in detail at the path of an α-particle, one can see the paths of the δ-rays to either side of the α-particle's path. The ion pairs formed by a heavy particle are the basis of heavy charged-particle detectors. The total energy of the particle can be related to the total number of electrons or ion pairs formed.

The linear stopping power S (also called specific energy loss) for charged particles in matter is defined as the differential energy loss by the particle per differential path length

$$S = -\frac{dE}{dx} \tag{10.63}$$

The specific energy loss is given by the Bethe formula:

$$-\frac{dE}{dx} = \frac{4\pi e^4 z^2}{m_e \mathrm{v}^2} NZC(\mathrm{v}) \tag{10.64}$$

where

$$C(\mathrm{v}) = \ln\left(\frac{2m_e \mathrm{v}^2}{I}\right) - \ln\left(1 - \frac{\mathrm{v}^2}{c^2}\right) - \frac{\mathrm{v}^2}{c^2} \tag{10.65}$$

Here z and v are the atomic number and velocity of the primary heavy charged particle; m_e and e are the electron rest mass and charge; Z and N are the atomic number and number of atoms per unit volume of the atoms in the stopping material; and I is the average excitation and ionization potential of the absorbing material (generally treated as a parameter to be determined experimentally for each element). If the material is a compound, then Z, N, and I for each element in the compound must be used in Eq. (10.64) and the loss for each element summed to get its total. For nonrelativistic charged particles ($\mathrm{v} \ll c$) only the first term in C is important.

The linear stopping power increases rapidly with atomic number of the heavy charged particle for the same velocity; only the z dependence varies when comparing different particles. However, remember that nonrelativistically $\mathrm{v}^2 = 2E/m$. Thus, for the same particle energy, the linear stopping power increases directly with mass (which can be approximated by A for relative comparison).

Because of the $Z^2 M_\alpha$ dependence of the linear stopping power, α-particles with even several MeV of energy lose their energy very quickly, so they can penetrate only a few centimeters of air or a few sheets of paper.

For a given particle, the linear stopping power increases as $1/E$ as shown in Fig. 10.28. Eventually the particles are moving so slowly that they begin to pick up or lose one or more electrons (depending on z) and so lose energy at a lower rate. Finally the energy loss goes to zero when a heavy particle picks up enough electrons to become neutral and has so little energy it can no longer lose an electron in a collision. As can be seen from Fig. 10.28, much of a heavy charged particle's energy loss is at the end of its range just before it stops. This fact is important in using heavy charged particles for cancer treatment.

Beta particles also lose energy to electrons, but because they have the same mass their path is highly irregular, changing directions sharply and sometimes even being scattered back out of the material. Since a β-particle and e^- are identical, after a collision one cannot know which is the β and which is the e^-.

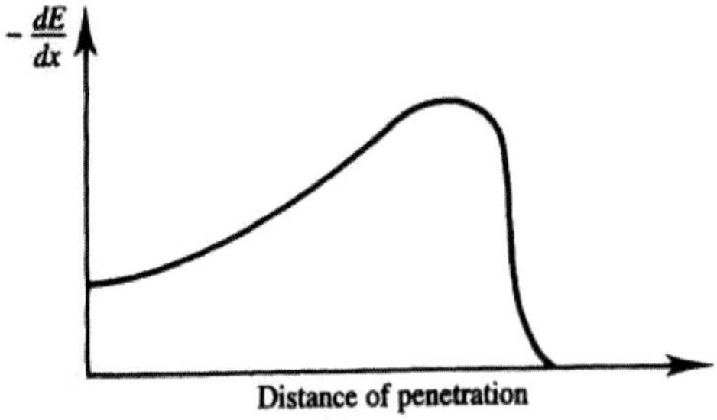

FIGURE 10.28
The average specific energy loss along the α tracks for a parallel beam. The α energy decreases with distance of penetration.

Thus there is no simple relationship between the range and energy of a β-particle as there is for an α-particle. Also, as we shall see, because of their smaller charge and much smaller mass, β-particles lose energy much more slowly than α-particles and so can penetrate a few millimeters of aluminum.

Gamma rays pass straight through matter until they undergo a single event in which through the photoelectric effect, Compton effect, or pair production they give up part or all of their energy. Thus, a 1-MeV γ-ray can penetrate a centimeter or more of lead. These interactions and their relative probabilities in different-Z materials were discussed in Chapter 6.

10.5 SUMMARY

Radioactivity is the spontaneous slow decay of a nucleus into one or more other elements, generally with the emission of particles and/or photons. Decay can occur when the initial mass is greater than that of the products after the decay. Radioactivity decays all follow an experimental decay law, $N = N_0 e^{-\lambda t}$ where $\lambda = (\ln 2)/T_{1/2}$ has a characteristic value for each decay. An unstable nuclide with too many neutrons will undergo β^- decay, where a neutron in the nucleus decays into a proton, electron, and antineutrino, and an unstable nuclide with too many protons will undergo β^+ decay or EC, where a proton decays into a neutron, positron and neutrino or neutron and neutrino, respectively; in heavier nuclei, α decay and, very far from stability, n and p decay, respectively, occur. A nuclide in an excited state will undergo γ decay or internal conversion, and pair production when the decay energy is above $2m_ec^2$. In still higher energy states populated by β decay, particle emission can compete with γ decay (β-delayed particle emission).

Very heavy unstable nuclides can undergo spontaneous fission, including highly asymmetric fission involving a heavy cluster with $A = 14$–48 and cluster decays like ^{146}Ba–^{106}Mo. Proton, α, heavy cluster, and spontaneous fission decays involve barrier penetration.

Beta decay involves a new interaction—the weak nuclear interaction introduced by Fermi. Its force range is shorter than that of hadronic interactions by a factor of about 10^{-2}. Initially to conserve energy, a new particle the neutrino was introduced by Pauli and it was incorporated into Fermi's theory. Neutrinos are now observed to interact with nuclei via the weak interaction in what is called inverse beta decay. Beta decays are classified by the change in spin and change in parity that occur. Double beta decay can also occur when single β decay is not energically possible.

Gamma decay, internal conversion, and pair production involve the electromagnetic interaction. These decays are classified by their electric (E) and magnetic (M) character and determined by the parity changes associated with the angular momentum carried away by the photon, electron, or $e^{\pm}$ pair, e.g., E0, E1, E2, ..., M1, M2, M3 .... Each of these multipole radiations has a characteristic lifetime for a single particle making the transition but these can be significantly altered by nuclear structure effects.

Studies of neutrinos are of major interest because they provide important insights into a wide range of critical problems from how energy is produced in stars to the standard model of particle physics and grand unified theories.

All known decay processes obey the conservation of energy, linear momentum, angular momentum, electric charge, nucleon number, and lepton number. However, parity conservation is observed to be violated in the weak interaction. Whether nucleon number and lepton number conservation are ever violated is a question of major interest now.

APPENDIX 10A
THEORY OF BETA DECAY[7,8]

The distinguishing characteristic of β decay is its energy or momentum spectrum (Figs. 10.11 and 10.14). The shape of the spectrum is related to the sharing of energy between the electron and neutrino and the relative masses of the electron and neutrino. If their masses were equal, we would expect a symmetric sharing of the energy, so the electron distribution would be symmetric about a point midway between zero and the decay energy. The fact the average electron energy is shifting toward low energies tells us the neutrino has very low mass. The shape of the electron distribution of ^{3}H has been used over many years to set lower and lower limits for m_ν. The present limit is about 2 eV. The basic energy distribution can be understood from simple statistical theory. Assume that all states of motion are equally probable. Classical statistics of motion are described by having a point in phase space where we have $\boldsymbol{r}_e$, $\boldsymbol{r}_\nu$, $\boldsymbol{p}_e$, and $\boldsymbol{q}_\nu$ (the position and momentum of e and

[7] One of us (J.H.H.) is deeply indebted to E. J. Konopinski from whom he learned the theory of β decay in two special series of lectures, of which the second in the summer of 1958 at Indiana University forms the basis of this section, and to L. M. Langer from whom he learned the techniques and importance of measuring beta spectra to test the theory.

[8] For more details of β decay, see E. J. Konopinski, *The Theory of Beta Radioactivity*, Oxford University Press, Oxford (1966).

ν), a 12-dimensional phase space. Now e and ν can be put into any position but p and q are restricted.

The decay energy

$$W_0 = W + K \tag{10A.1}$$

the relativistic energies of the e and ν.

$$W^2 = c^2p^2 + m^2c^4 = p^2 + 1 \tag{10A.2}$$

in units where $c = m_e = \hbar = 1$

$$K^2 = q^2 + \nu^2 \tag{10A.3}$$

where we know the rest mass ν^2 of the neutrino can be neglected.

Every point in phase space is equally probable except for the restriction on p and q. Since we are not interested in the direction of emission,

$$d\lambda \sim d\mathbf{p}\, d\mathbf{r}_e\, d\mathbf{q}\, d\mathbf{r}_\nu \tag{10A.4}$$

If V is the total volume into which e and ν are dumped and direction is neglected, so that the angular intergrations go through, then the number of states available to an electron with p in dp is

$$4\pi p^2\, dp\, V \tag{10A.5}$$

and the number of states available to a neutrino with q in dq is

$$4\pi q^2\, dq\, V \tag{10A.6}$$

so

$$\frac{d\lambda}{\lambda} = \frac{p^2\, dp\, q^2\, dq}{\int p^2\, dp \int q^2\, dq} \tag{10A.7}$$

Since $K\, dk = q\, dq$ and $W\, dW = p\, dp$, (10A.8)

$$\frac{d\lambda}{\lambda} = \frac{pW\, dW\, qK\, dK}{\int_1^{W_0} pW\, dW \int qK\, dK} \tag{10A.9}$$

But since $W_0 = W + K$ for a fixed W, $dK = \delta W_0 \to 0$ and so the integration over dK is a vanishing element and cancels in the numerator and denominator (the ratio is finite).

Let

$$f_0 = \int^{W_0} pWqK\, dW \tag{10A.10}$$

then the fraction of the decays in which the energy is in the range dW is

$$\frac{d\lambda}{\lambda} = \frac{pWqK\, dW}{f_0} \tag{10A.11}$$

or

$$\frac{d\lambda}{\lambda\, dW} = \frac{pWqK}{f_0} = W(W^2 - 1)^{1/2}(W_0 - W)^2 \tag{10A.12}$$

This is the energy spectrum when the neutrino mass is negligible.

This is still not quite correct. So far we have assumed that the density function is $|\psi_e|^2 = |\psi_0|^2$ where $|\psi_0|^2$ is the density function when the electron is at infinity where there is no Coulomb effect. However, we know that as $r \to 0$ the β^- is held into the nucleus by the positive charge and so has an enhanced probability of being near the nucleus while the β^+ is pushed out by the positive charge. So there must be some correction to account for the Coulomb attraction on the β^-, which leads to an enhancement of the number of low-energy β^-, and a pushing out to yield a decrease in the number of low-energy β^+.

The assumption is introduced that various electron waves are excited in proportion to their density at the nucleus. Then, we assume that

$$d\lambda \sim F(Z, W) = |\psi_e(0)/\psi_0|^2 \tag{10A.13}$$

This Coulomb correction factor $F(Z, W)$ is called the Fermi function. For Z not too large,

$$F(Z, W) = \frac{2\pi\nu}{1 - e^{-2\pi\nu}} \tag{10A.14}$$

where $\nu = \pm Ze^2/h\text{v}$ for $e^{\mp}$, where v is the electron velocity and Z is the charge of the nucleus after β decay.

Substituting for $K = q = W_0 - W$ and $p = (W^2 - 1)^{1/2}$ (recall $mc^2 = 1$):

$$\frac{d\lambda}{\lambda\, dW} = W(W^2 - 1)^{1/2}(W_0 - W)^2 \frac{F(Z, W)}{f(Z, W)} \tag{10A.15}$$

where

$$f(Z, W) = \int_1^{W_0} W(W^2 - 1)^{1/2}(W_0 - W)^2 F(Z, W)\, dW \tag{10A.16}$$

Call $n(W)$ the number of electrons observed with energy W. Then

$$[(n(W)/pWF)]^{1/2} \sim W_0 - W \tag{10A.17}$$

if $n(W)$ is proportional to the statistical spectrum. A plot of the left side of Eq. (10A.17) as a function of W, as shown in Fig. 10.14b, is called a Kurie (or Fermi–Kurie) plot.

If only the nuclear charge and the energy released influence the decay rates and their influence is the same as for the spectrum, then

$$\begin{aligned} d\lambda &= CW(W^2 - 1)^{1/2}(W_0 - W)^2 F(Z, W)\, dW \\ \lambda &= \frac{1}{\tau} = \frac{\ln 2}{T_{1/2}} = Cf(Z, W) \end{aligned} \tag{10A.18}$$

where $f(Z, W)$ is given by Eq. (10A.16). Let $T_{1/2} = t$. Then (10A.19)

$$ft = \frac{\ln 2}{C} \tag{10A.20}$$

The ft values are called comparative half-lives. They should be constant at least for all similar decays. For example, if the spectrum is not statistical, then this could change ft. These comparative half-lives have characteristic values for different types of decays, as we shall see.

Beta decay takes place between two states of spin and parity $I_i\pi_i$ and $I_f\pi_f$. The electron and neutrino each have intrinsic spin of $\frac{1}{2}\hbar$. For the simplest case, called allowed decays, to conserve angular momentum $I_i - I_f = S_e + S_\nu = S$, where S is the sum of the intrinsic spin of the electron and neutrino. The sum S can have the value of either 0 or 1, corresponding to the electron and neutrino coming out with antiparallel or parallel spins, respectively. $S = 0$ is a singlet state and $S = 1$ is a triplet state since there are three possible orientations of S. The simplest decays, called allowed decays, are ones where the parity does not change for the two states. For $\Delta I = |I_i - I_f| = 0$, $S = 0$ and for no parity change (written $\Delta\pi = 0$ meaning no change) the allowed β decay obeys the "Fermi-selection rule." For $S = 1$, $I_i = I_f + 1$ (so $\Delta I = 0$ or 1) and $\Delta\pi = 0$ the decay obeys Gamow–Teller (GT) selection rules, but there can be no GT decay between two spin-zero states ($I_i = I_f = 0$). Fermi's name is associated with the first case because in his initial theory of β decay these were the only unhindered types of decay. Later George Gamow and Edward Teller pointed out that the triplet decays may be equally unhindered. A $0^+ \to 0^+$ β decay is pure Fermi and a $1^+ \to 0^+$ is pure GT. For other cases the ratio of Fermi to GT decays is an interesting number. Each decay is given a coupling constant, C_F and C_{GT}, that is a measure of the strength of that type of decay, so that C_F^2 and C_{GT}^2 measure the intensities with which the β interaction gives allowed singlet and triplet decays.

In addition to the two types of allowed decays, there are also forbidden decays which are hindered (slower decay rates) compared to allowed decays. The hindrances are related to changes in the spins and parities between the initial and final states. For the same $\Delta I = 0, 1$, if the parity is different for the initial and final state $\Delta\pi = 1$ (yes), the transition is hindered and such transitions are called once- or first-forbidden β decays. Once-forbidden decays can also occur for $\Delta I = 2$ and $\Delta\pi = 1$ and are called unique once-forbidden decays. Several transition matrix elements can contribute to $\Delta I = 0, 1$, $\Delta\pi = 1$ β decays, but only one to the $\Delta I = 2, \Delta\pi = 1$ decays, so the spectrum has a unique shape factor (energy dependence) in addition to $(W_0 - W)$, and the Kurie plot is not a straight line. Konopinski and Uhlenbeck worked out the theory of forbidden β decay soon after Fermi's theory of allowed decay (see ref. 8) and L. M. Langer and H. Price (*Phys. Rev.* **75**, (1949) 1109) discovered the first spectrum to exhibit the once-forbidden, unique shape to confirm the theory of forbidden decay. The $\Delta I = 2$ once-forbidden decays are hindered more than the $\Delta I = 0, 1$ once-forbidden decays because of the extra angular momentum carried off. There are higher-order degrees of forbiddenness associated with larger changes in ΔI of 3, 4, 5 units. We can understand why carrying off 2, 3, and more units of angular momentum increasingly retards the beta decay (giving longer half-life) since the electron and neutrino must be created farther out from the center of the nucleus to carry off orbital as well as their spin angular momentum. The I^π selection rules for the various allowed and forbidden decays are given in Table 10A.1. Since the forbidden decays are retarded, we expect them to have larger ft values.

There is also a class of decays known as mirror decays, for example, where an odd proton in a particular shell model state changes into the same orbital odd neutron and vice versa. These should be super-allowed (fast) with smaller ft values than normal allowed decays. On the other hand, there can be $\Delta l \neq 1$ (l is the orbital angular momentum) allowed decays with $\Delta I = 0, 1$, the so-called "l-forbidden" allowed decays. These are expected to be hindered with respect to normal allowed decays and so have larger ft values than normal allowed decays. For forbidden decays it is generally more useful to give $\log_{10} ft$ rather than the ft values since the different forbidden decays have much larger ft values. In Tables 10A.2 and 10A.3 are found examples of ft and $\log ft$ values for different types of decays (from ref. 8).

"Mirror nuclei'" are the simplest and fastest (lowest ft value) of all β decayers. These nuclei have one extra nucleon outside an inert core which has equal numbers of neutrons and protons. The extra nucleon (or nucleon hole) is what is transformed in the β decay process. Some examples of mirror nuclei are given in Table 10A.2. Their ft values vary from 1137 s for ^{3}H to 5680 s for ^{35}A, while their half-lives vary from $T_{1/2} = 12.4$ y for ^{3}H to $T_{1/2} = 0.6$ s for ^{43}Ti, a variation of a factor of 6×10^8. The ft values, which vary less than a factor of 6, are remarkably constant in light of the variation in $T_{1/2}$. Relative to $T_{1/2}$ the ft values are constant to one part in 10^8! "Mirror" decays involve the fact that nuclear forces (see Chapter 11) are charge independent. This means that the interaction between two protons is the same as between two neutrons or the same as between a

TABLE 10A.1
Classifications of transitions[a]

Degree of forbiddenness	$\Delta I^{\pi_i \pi_f}$	J	β-Moments
Allowed	0^+	0	$C_V\langle 1 \rangle$
	$0^+, 1^+$ (No $0 \to 0$)	1	$C_A\langle \sigma \rangle$
Once-forbidden	0^-	0	$\left\{ \begin{array}{l} C_A\langle i\boldsymbol{\sigma} \cdot \hat{\boldsymbol{r}} \rangle \\ C_A\langle \gamma_5 \rangle \end{array} \right.$
	$0^-, 1^-$ (No $0 \to 0$)	1	$\left\{ \begin{array}{l} C_V\langle i\hat{\boldsymbol{r}} \rangle, \quad C_A\langle (\boldsymbol{\sigma} \times \hat{\boldsymbol{r}}) \rangle \\ C_V\langle \alpha \rangle \end{array} \right.$
	2^- (1F unique)	2	$C_A\langle \boldsymbol{\sigma} \cdot \boldsymbol{T}_2^1 \rangle$
Twice-forbidden	2^+	2	$\left\{ \begin{array}{l} C_V\langle Y_2 \rangle, \quad C_A\langle \boldsymbol{\sigma} \cdot \boldsymbol{T}_2^2 \rangle \\ C_V\langle \boldsymbol{\sigma} \cdot \boldsymbol{T}_2^1 \rangle \end{array} \right.$
	3^+ (2F unique)	3	$C_A\langle \boldsymbol{\sigma} \cdot \boldsymbol{T}_3^2 \rangle$
n-times forbidden	$n^{(-)^n}$	n	$\left\{ \begin{array}{l} C_V\langle Y_n \rangle, \quad C_A\langle \boldsymbol{\sigma} \cdot \boldsymbol{T}_n^n \rangle \\ C_V\langle \boldsymbol{\alpha} \cdot \boldsymbol{T}_n^{n-1} \rangle \end{array} \right.$
	$(n+1)^{(-)^n}$	$n+1$	$C_A\langle \boldsymbol{\sigma} \cdot \boldsymbol{T}_{n+1}^n \rangle$

[a] Note $\pi_i \pi_f = +$ means $\pi_i = \pi_f$.

TABLE 10A.2
Experimental *ft* values for a few mirror transitions

Parent	Daughter	*ft* exp. (s)	Transition
${}_0^1n_1$	${}_1^1H_0$	1180 ± 35	$s_{1/2} \rightarrow s_{1/2}$
${}_1^3H_2$	${}_2^3He_1$	1137 ± 20	$s_{1/2} \rightarrow s_{1/2}$
${}_4^7Be_3$	${}_3^7Li_4$	2300 ± 78	$p_{3/2} \rightarrow p_{3/2}$
${}_8^{15}O_7$	${}_7^{15}N_8$	4475 ± 30	$p_{1/2} \rightarrow p_{1/2}$
${}_{10}^{19}Ne_9$	${}_9^{19}F_{10}$	1900 ± 100	$s_{1/2} \rightarrow s_{1/2}$
${}_{18}^{35}A_{17}$	${}_{17}^{35}Cl_{18}$	5680 ± 400	$d_{3/2} \rightarrow d_{3/2}$

From ref. 8.

TABLE 10A.3
Ranges of experimental $\log_{10} ft$ values and typical average values

Type β transition	Range $\log_{10} ft$	Typical $\log_{10} ft$
Mirror allowed	3–4	3.3
Allowed	4.0–5.8	5
l-Forbidden allowed	5.0–8	6.6
Once-forbidden	5.1–8.9	6–7
Once-forbidden unique	7.1–9.9	8.4
Twice-forbidden	10.9–13.5	12–13
Twice-forbidden unique	11.5–13.7	12.5–13.5
3F	17–19	
4F	18–24	

neutron and a proton. The relatively minor electromagnetic interaction and the neutron–proton mass difference, of course, alter this independence.

In β decay a neutron is transformed into a proton or vice versa. For charge-independent forces one might think there would be a final configuration identical to the initial one. When there are unequal numbers of neutrons and protons, the Pauli exclusion principle comes into play so that the transformed nucleon cannot assume exactly the same motions with respect to the other nucleons as the original nucleon had. So considerable rearrangement in structure must occur for $Z \neq N$. "Mirror" nuclei are those which have equal numbers of neutrons and protons in both cores and one transforming nucleon outside the core. In this case their energies will differ only by the electrostatic repulsion of the Z proton pairs less the mass difference of the neutron and proton. This energy difference is the energy available for β decay of the mirror nuclei. For example, ${}_1^3H_2$ is heavier by one neutron–proton mass difference than ${}_2^3He$, and so β decays to ${}^3He$, but the mass difference is reduced by the electrostatic repulsion of the two protons in ${}^3He$. For ${}_8^{15}O_7$ the electrostatic repulsion of its many pairs of protons (calculate the numbers of pairs) compared to that of ${}^{15}N$ easily is greater than the extra neutron–proton mass difference of ${}_7^{15}N_8$ and so ${}^{15}O$ decays to ${}^{15}N$.

The next simplest decays involve nuclei which have equal number of nucleons $N = Z = n$ (where n is even) in a core and two nucleons—two protons, a neutron–proton, or two neutrons—outside this core. This is because the Pauli principle allows two like nucleons in an antiparallel spin state to have the same spatial distributions. So $A = 4n + 2$ are a second class of favored transitions. For example, one can have an "isobaric triad" formed by three isotopes with the same $N = Z = n$ core with one having two protons, one a neutron and a proton, and one two neutrons outside the core. An example is the pure Fermi $0^+–0^+$ β decays of ${}^{38}_{20}Ca_{18} \to {}^{38}_{19}K_{19} \to {}^{38}_{18}A_{20}$; the latter $ft = 3140 \pm 400\,s$. Others include ${}^{6}_{2}He_4 \to {}^{6}_{3}Li_3$, with the lowest known $ft_{exp} = 808 \pm 325\,s$; ${}^{14}_{6}C_8 \to {}^{14}_{7}N_7 \leftarrow {}^{14}_{8}O_6$ with $ft_{exp} \sim 10^9$ and $10^{7.6}$, respectively. The last three are spin 0 to spin 1 pure GT decays.

The charge independence of the nucleon force is also related to the nucleon isospin and isospin selection rules. If the nuclear force is really charge independent, then nucleon states are not only eigenstates of the energy and total angular momentum but also of total isospin:

$$T = \sum_{a=1}^{A} \tfrac{1}{2}\tau^a \tag{10A.21}$$

where protons have $T_Z = \frac{1}{2}(\tau = 1)$ and neutrons $T_Z = -\frac{1}{2}(\tau = -1)$.

States with different T values are expected to have different energies. It is found experimentally that states with the lowest T tend to have the lowest energies. The isospin-symmetry-breaking electrostatic interaction will become non-negligible in heavier nuclei where states have a mixture of different T values.

We also have

$$T_Z = \tfrac{1}{2}(Z - N) \qquad \text{where} \qquad T_Z = \pm T, \pm(T-1), \ldots \tag{10A.22}$$

with the lowest energy state expected to have $T = \frac{1}{2}|Z - N|$.

For charge-independent nuclear forces, the electrostatic repulsion would make the $T_Z = -T$ state, the substate with fewest protons, the most stable. For pure Fermi decays, one now has the additional selection rules $\Delta T = 0$ and $\Delta T_Z = \pm 1$ when T is a good quantum number, that is when the charge-dependent electromagnetic forces are negligible. For mirror nuclei, each one has $T_Z = \frac{1}{2}$ or $-\frac{1}{2}$ and so $T = \frac{1}{2}$ for their ground states.

The members of a triad can form a $T = 1$ isomultiplet with $T_Z = 0, \pm 1$. One may argue that the $N = Z$ member of the triad with $T_Z = 0$ should have a $T = 0$ ground state and not be a member of a $T = 1$ isomultiplet. In such a case there should be a low-lying excited state with $T = 1, T_Z = 0$. While the $0 \to 0$ decays of ${}^{10}C$ and ${}^{14}O$ go to second and first excited states of ${}^{10}_{5}B_5$ and ${}^{14}_{7}N_7$, in agreement with this expectation, the zero-spin ground states of ${}^{34}Cl$ and ${}^{42}Sc$ are $T = 1, T_Z = 0$ and undergo ground-state $0 \to 0$ decay. Studies of 0–0 β transitions, particularly as a function of Z, offer the opportunity to test the charge independence of nucleon forces and isospin mixing. New radioactive ion beam accelerators will open up the opportunity to study new far-off-stability 0–0 transitions to make such tests.

Now look at how Fermi constructed his theory of β decay. For a formal theory of β decay, one needs to know how to describe spin-$\frac{1}{2}$ particles—Dirac particles, or fermions.

Such particles have four-component wavefunctions to describe them, so the Schrödinger equation becomes four equations. Such four-component functions need to be represented in terms of some basic set of wavefunctions and couplings:

$$i\hbar\frac{\partial}{\partial t}\begin{pmatrix}\psi_1\\ \psi_2\\ \psi_3\\ \psi_4\end{pmatrix} = \begin{pmatrix}H_{11} & \cdot & \cdot & H_{14}\\ \cdot & \cdot & \cdot & \cdot\\ \cdot & \cdot & \cdot & \cdot\\ H_{41} & \cdot & \cdot & H_{44}\end{pmatrix}\begin{pmatrix}\psi_1\\ \psi_2\\ \psi_3\\ \psi_4\end{pmatrix} \tag{10A.23}$$

There are 16 basic matrices which can be used to express H_{ij}. To get the Hamiltonian to describe Dirac particles, one picks a set. One can find in ref. 8 the theory of how to describe Dirac particles, which we must skip over here. Fermi constructed his original theory of β decay in analogy with γ-radiation and the interaction of a field with a charged particle. The perturbation Hamilton responsible for γ decay is

$$H_\gamma = ie\,\beta\gamma_\mu A_\mu \tag{10A.24}$$

where A_μ $(\boldsymbol{A}, i\phi)$ is the four-vector potential; β, γ_μ are basic matrices; and e is the charge on electron (summed over $\mu = 1$–4 understood). The coupling energy density of γ-radiation is $\psi^\dagger H_\gamma \psi$.

The energy density in a volume or radiated into a volume is

$$\int \psi^\dagger H_\gamma \psi\, dV = ie\int (\psi^\dagger\, \beta\, \gamma_\mu\, \psi) A_\mu\, dV \tag{10A.25}$$

where A_μ is a function of space and time, and β and γ_μ are only commands which tell how to multiply $\psi^\dagger$ and ψ. So

$$\psi^\dagger\, H_\gamma\, \psi = h_\gamma = ie\, j_\mu\, A_\mu \tag{10A.26}$$

where j_μ is a four-vector current density $(-i\rho\mathbf{v}/c,\ \rho) = \psi^\dagger\, \beta\, \gamma_\mu\, \psi$.

By analogy, Fermi assumed that the rate at which a nucleus will β-radiate will be proportional to the current associated with the change of a neutron into a proton or a proton into a neutron in β decay. So

$$h_\beta \sim g\, j_\mu(\mathrm{n} \to \mathrm{p}) \tag{10A.27}$$

where there must be some constant in nature like charge e to give the strength of the β interaction. He called this constant g and one finds the strength of g from the rate of β decay.

How do you formulate a four-vector current associated with the β decay (n → p) transformation? Now instead of a change in energy states as in electromagnetic radiation, we have a change of character n → p. Fermi said

$$\begin{aligned} j_\mu(\mathrm{n} \to \mathrm{p}) &= \bar{\psi}_p \gamma_\mu \psi_n \\ j_\mu(\mathrm{p} \to \mathrm{n}) &= \bar{\psi}_n \gamma_\mu \psi_p \end{aligned} \tag{10A.28}$$

where $\bar{\psi} = \psi^\dagger \beta$. What about the electron and neutrino? What is the vector potential for β decay, A_μ^β? Fermi said

$$\begin{aligned} A_\mu^\beta &\sim \bar{\psi}_e \gamma_\mu \psi_\nu (\mathrm{n} \to \mathrm{p}) \\ &\sim \bar{\psi}_\nu \gamma_\mu \psi_e (\mathrm{p} \to \mathrm{n}) \end{aligned} \tag{10A.29}$$

So the coupling energy density is

$$h_\beta = g(\bar{\psi}_p \gamma_\mu \psi_n)(\bar{\psi}_e \gamma_\mu \psi_{\bar{\nu}}) + g(\bar{\psi}_n \gamma_\mu \psi_p)(\bar{\psi}_\nu \gamma_\mu \psi_e) \tag{10A.30}$$

where the second term is the complex conjugate of the first and is necessary for relativistic invariance.

Thus the essential feature of the Fermi theory is that the coupling energy density is proportional to each of the fields which enter into the interaction,

$$h_\beta \sim \psi_n \psi_p \psi_e \psi_{\bar{\nu}} \tag{10A.31}$$

where this is a contact interaction so each wavefunction acts at the same point. There is no action at a distance.

However, this is not general. There are other ways of forming these four-vector products. There are 16 products of $\psi_n \psi_p$ and 16 products of $\psi_{\bar{\nu}} \psi_e$, or 256 products for both. We reduce the number of these products by requiring that the products np and $\bar{\nu}e$ be of the same form so that their product is a scalar.

One set of sixteen basic 4×4 matrices is

$$\begin{aligned} &1 = \begin{pmatrix} 1 & 0 \\ 0 & 1 \end{pmatrix} \quad \rho_1 = \begin{pmatrix} 0 & 1 \\ 1 & 0 \end{pmatrix} \quad \rho_2 = \begin{pmatrix} 0 & -i \\ i & 0 \end{pmatrix} \quad \rho_3 = \begin{pmatrix} 1 & 0 \\ 0 & -1 \end{pmatrix} \\ &\boldsymbol{\sigma} = \begin{pmatrix} \boldsymbol{\sigma} & 0 \\ 0 & \boldsymbol{\sigma} \end{pmatrix} \quad \rho_1 \boldsymbol{\sigma} \quad \rho_2 \boldsymbol{\sigma} \quad \rho_3 \boldsymbol{\sigma} \end{aligned} \tag{10A.32}$$

where each nonzero component is a 2×2 matrix and the 2×2 $\boldsymbol{\sigma}$ are

$$\sigma_1 = \begin{pmatrix} 0 & 1 \\ 1 & 0 \end{pmatrix} \quad \sigma_2 = \begin{pmatrix} 0 & -i \\ i & 0 \end{pmatrix} \quad \sigma_3 = \begin{pmatrix} 1 & 0 \\ 0 & -1 \end{pmatrix} \tag{10A.33}$$

However, with these 16 we can lose track of the covariance. It is better to define the following, where we will know the type of product formed—scalar, vector or whatever:

$\gamma_\mu\ (-i\boldsymbol{\alpha}\beta, \beta) = \Gamma_V$	4 matrices	Vector
$\gamma_\mu^2 = 1 = \Gamma_S$	1 matrix	Scalar
$\gamma_\mu \gamma_\nu = \Gamma_T\ (\mu \neq \nu)$	6 matrices	Tensor
$\gamma_\mu \gamma_\nu \gamma_\rho = \Gamma_A / i\ (\mu \neq \nu \neq \rho)$	4 matrices	Axial vector
$\gamma_1 \gamma_2 \gamma_3 \gamma_4 = \gamma_5 = \Gamma_P$	1 matrix	Pseudoscalar

The golden rule for decay is

$$d\lambda = \frac{2\pi}{\hbar} \rho_E |\langle f\, | H_\beta | i \rangle|^2 \tag{10A.34}$$

where ρ_E is the density of states per unit energy and the last term is the square of the transition matrix element where

$$\langle f\,|H_\beta|\,i\,\rangle = \int \psi_f^* H_\beta \psi_i \equiv \sum \int h_\beta\, dV \tag{10A.35}$$

where H_β is the perturbation that changes states ψ_i into ψ_f, so the integral gives the projection of $H_\beta\psi_i$ onto ψ_f, and the integration is over all variables and in the last term is summed over internal variables as well. This equation gives us the rate at which the desired final state is being produced by the transition.

Earlier we wrote

$$h_\beta = \psi^\dagger H_\beta \psi. \tag{10A.36}$$

The general form of h_β is

$$h_\beta = g \sum_x C_x(\bar{\psi}_p \Gamma_x \psi_n)(\bar{\psi}_e \Gamma_x \psi_{\bar{\nu}}) + \text{ complex conjugate} \tag{10A.37}$$

Remember to form a scalar h_β, the p–n and e–ν couplings must both be scalar or both be vector or both be tensor, or both be axial vector, or both be pseudoscalar, so one has only 16 ways of coupling. In giving Eq. (10A.37), Fermi selected out of the five possible couplings the vector coupling γ_μ and, as we shall see, considered only the leading term.

This equation is still not the most general one since it does not include parity nonconservation in β decay. Remember that parity nonconservation in β decay came in the late 1950s. With parity nonconservation included,

$$h_\beta = g \sum_x (\bar{\psi}_p \Gamma_x \psi_n)[\bar{\psi}_e \Gamma_X (C_x + C_x' \gamma_5)\psi_{\bar{\nu}}] + \text{c.c.} \tag{10A.38}$$

where X stands for the scalar, vector, tensor, axial vector, and pseudovector couplings that give the 16 ways of coupling the wavefunctions. The γ_μ's provide one way of writing these 16 couplings.

If we use the experimentally determined fact that the e^- emitted in β decay is left-handed (its spin is opposite to the direction of its linear momentum), then $C_{VA} = C_{VA}'$ and $C_{STP} = -C_{STP}'$. So

$$h_\beta = \underset{X=VA}{g}\; C_X(\bar{\psi}_p \Gamma_x \psi_n)[\bar{\psi}_e \Gamma_X (1+\gamma_5)\psi_{\bar{\nu}}] + \text{c.c.} \tag{10A.39}$$

or

$$h_\beta = g \sum_{X=STP} C_X(\bar{\psi}_p \Gamma_x \psi_n)[\psi_e \Gamma_X (1-\gamma_5)\psi_{\bar{\nu}}] + \text{c.c.} \tag{10A.40}$$

where $(1+\gamma_5)\psi_{\bar{\nu}}$ holds for left-handed neutrinos and $(1-\gamma_5)\psi_{\bar{\nu}}$ holds for right-handed neutrinos.

M. Goldhaber, L. Grodzins and A. Sunyar carried out a classic experiment to show that ν is left-handed, so we can drop the STP couplings. Thus, β decay involves only the vector and axial vector couplings and is called the V–A interaction.

In his original formulation, Fermi chose the product $\bar{\psi}_p \gamma_\mu \psi_n$ because they form four components of a four-vector and he wanted such a covariant four-vector. Remember the γ_μ's are only matrix commands that tell you how to multiply two spinors together. There are different choices one can make for γ_μ. One is

$$\gamma_\mu(\boldsymbol{\gamma}, \gamma_4) \equiv (-i\beta\, \boldsymbol{\alpha}, \beta) \tag{10A.41}$$

where $\gamma_{1,2,3} = -i\beta\alpha_{1,2,3}$ and $\gamma_4 = \beta$, $\boldsymbol{\alpha} = \rho_1 \boldsymbol{\sigma}$, $\beta = \rho_3$ and $\beta^2 = 1$. This choice has the advantage of separating the two large components of the wavefunction from the two small components which go to zero in the nonrelativistic limit. Using these γ_μ will lead to two terms in C_V. Remember that $\bar{\psi} = \psi^\dagger \beta$. Then the fourth component with γ_4 takes the form

$$h_\beta \sim g C_V(\psi_p^\dagger \psi_n)\{\psi_e^\dagger(1+\gamma_5)\psi_{\bar{\nu}}\} \tag{10A.42}$$

In the language of creation operators where τ_+ operating on a neutron changes it into a proton, so $\psi^\dagger \tau_+ \psi \equiv \psi_p^\dagger \psi_n$, and dropping $\psi^\dagger\psi$ in h_β to get H_β. With H_β the first-order terms in C_V and C_A become

$$\int \psi_f^* H_\beta \psi_i = \sqrt{2}\, g\, C_V \int \psi_f^* \sum_{a=1}^{A} \tau_+^a \psi_i \cdot \mathcal{L}'(\boldsymbol{r}^a) + \sqrt{2}\, g\, C_A \int \psi_f^* \sum_{a=1}^{A} \tau_+^a \boldsymbol{\sigma}^a \psi_i \cdot \mathcal{L}'(\boldsymbol{r}^a) + \text{c.c.} \tag{10A.43}$$

where

$$\int \psi_f^* \sum_{a=1}^{A} \tau_+^a \psi_i \equiv \int 1 = \langle 1 \rangle \tag{10A.44}$$

and $\mathcal{L}'$ is the electron-neutrino term, and

$$\beta\gamma_5\boldsymbol{\gamma} = i\boldsymbol{\sigma} \qquad \int \psi_f^* \sum_{a=1}^{A} \tau_+^a \boldsymbol{\sigma}^a \psi_i \equiv \int \boldsymbol{\sigma} = \langle \boldsymbol{\sigma} \rangle \tag{10A.45}$$

There are selection rules for $\langle 1 \rangle$ of $\Delta I = 0$ and $\Delta\pi = 0$ and for $\langle \boldsymbol{\sigma} \rangle \Delta I = 0, \pm 1, \Delta\pi = 0$.

The remaining three components in C_V are $\beta(-i\beta\boldsymbol{\alpha}) = -i\boldsymbol{\alpha}$, so there is a term in h_β,

$$-g\, C_V(\psi_p^\dagger \boldsymbol{\alpha} \psi_n)\{\psi_e^\dagger \boldsymbol{\alpha}(1+\gamma_5)\}\, \psi_{\bar{\nu}} \tag{10A.46}$$

where $(\psi_p^\dagger \boldsymbol{\alpha} \psi_n) \equiv \langle \boldsymbol{\alpha} \rangle$, and in C_A we left out

$$g\, C_A(\psi_p^\dagger \gamma_5 \psi_n)[\psi_e^\dagger \gamma_5(1+\gamma_5)\, \psi_{\bar{\nu}}] \tag{10A.47}$$

where $(\psi_p^\dagger \gamma_5 \psi_n) \equiv \langle \gamma_5 \rangle$.

These terms have selection rules $\Delta I = 0, \pm 1$ but $\Delta\pi =$ yes. So they describe first-forbidden decays.

In allowed decays one also approximates the $\mathcal{L}'(\boldsymbol{r}^a)$ by its value at $r = 0$ since the e and ν lepton wavelengths are very long compared to the nuclear size and so change little over the nuclear radius. So $\mathcal{L}'$ can be taken outside the integral. For $\Delta I \geq 2$, where the e and ν must be formed away from the nuclear center, this $r = 0$ approximation is not good and the $\Delta I \geq 2$ transitions came from the second-order terms in $\mathcal{L}'$.

Now we note that the singlet Fermi transitions correspond to the matrix element $\langle 1 \rangle$ and the GT decays to $\langle \boldsymbol{\sigma} \rangle$. By analogy with electromagnetic radiation, $\langle 1 \rangle$ and $\langle \boldsymbol{\sigma} \rangle$ may be called the nuclear moments of beta radiation.

Why do we say $\langle 1 \rangle$ has no spatial dependence? All the operator does is tell you the ath nucleon in ψ_i should be a proton in the future, so we have

$$\sum_a \int \psi_f^* \psi_i \tag{10A.48}$$

The wavefunctions are orthogonal to one another, so if they have different dynamical properties this integral is zero. Thus $\langle 1 \rangle = 0$ unless $\boldsymbol{I}_f = \boldsymbol{I}_i$, $\Delta I = 0$, the Fermi selection rule. One can show $\langle \boldsymbol{\sigma} \rangle = 0$ unless $\boldsymbol{I}_f = \boldsymbol{I}_f + \boldsymbol{1}$, or $\Delta I = 0, \pm 1$ with no $I = 0$–0.

For the neutron decay,

$$|\langle 1 \rangle|^2 = 1 \qquad \text{but} \qquad \langle \boldsymbol{\sigma} \rangle^2 = |\langle \sigma_x \rangle|^2 + |\langle \sigma_y \rangle|^2 + |\langle \sigma_z \rangle|^2 = 3 \tag{10A.49}$$

So

$$((ft)_{\text{neutron}})^{-1} \sim (C_F)^2 + 3(G_{GT})^2 \tag{10A.50}$$

Finally,

$$\frac{d\lambda}{dp} = \frac{1}{(2\pi)^5} p^2 q^2 \, d\Omega_e \, d\Omega_{\bar{\nu}} \, V^2 \times 2 \frac{g^2}{V^2} |C_V \langle 1 \rangle \mathcal{L}(0) + C_A \langle \boldsymbol{\sigma} \rangle \mathcal{L}'(0)|^2 \tag{10A.51}$$

where for $r = 0$,

$$\mathcal{L} = u^\dagger (1 + \gamma_5) \frac{w}{\sqrt{2}} \qquad \text{and} \qquad \mathcal{L}' = \frac{u^\dagger \boldsymbol{\sigma} (1 + \gamma_5) w}{\sqrt{2}} \tag{10A.52}$$

For e and ν in different momentum states p and q,

$$\psi_e = \frac{u e^{i\boldsymbol{p}\cdot\boldsymbol{r}}}{\sqrt{V}} \qquad \text{and} \qquad \psi_{\bar{\nu}} = \frac{w e^{-i\boldsymbol{q}\cdot\boldsymbol{r}}}{\sqrt{V}} \tag{10A.53}$$

where one wants one electron in volume V and the $1/V^2$ cancels with V^2 in ρ_E. Note that u and w express the dependence on inner variables. All experiments follow from this equation. For example, for unpolarized nuclei this equation becomes the sum of the squares of four terms instead of the square of the sum of four terms.

PROBLEMS

10.1. Verify that the activity of 1 g of ^{238}U is 0.33 μCi. The ^{238}U half-life is 4.5×10^9 years, and 1 mg ^{238}U emits 740 α-particles per minute.

10.2. While they are living, organic bodies continuously take in both ^{14}C and ^{12}C, so their ratio of ^{14}C to ^{12}C is about constant, 1.3×10^{-12}. After a body dies, no more ^{14}C is taken in, and the ^{14}C present at death continuously decreases from radioactive decay. By measuring the decay rate per gram of a substance, we can calculate when it died. If the measured β decay rate for 100 g of carbon from a skeleton is 300 decays/minute, how long has the skeleton been dead?

10.3. What is the decay probability on the fifth day of a radioactive element with a mean life of 10 days?

10.4. Determine whether ^{218}Ra and ^{226}Ra can undergo α decay. If they can, what are the maximum possible energies of α-particles they emit at rest? If they cannot, what would you expect their decay modes to be?

10.5. Complete the following by giving all N, Z, and A numbers and the other particles or elements involved in these decay and inverse decay processes:

(*a*) ${}_{11}Na_{11} \rightarrow \beta^+ +$

(*b*) ${}^{184}_{80}Hg + e^- \rightarrow$

(*c*) $\nu + {}^{71}_{31}Ga \rightarrow$

(*d*) ${}^{230}_{90}Th \rightarrow {}_{88}Ra +$

Discuss all the conservation laws you used to arrive at your answers.

10.6. When ^{47}V undergoes β^+ decay, the maximum β^+ energy is 1.89 MeV. Calculate the neutrino energy in the K capture process in ^{47}V to the same level in ^{47}Ti to which the β^+ goes. Calculate the maximum β^+ energy for decay between the ground states of ^{47}V and ^{47}Ti from their masses. What does a comparison of this value with the measured energy tell you about the 1.89-MeV β^+ group?

10.7. Since the α particle can escape out of a nucleus through the barrier penetration principle, why is proton radioactivity (proton emission) so very rare? By analogy with α decay, write the conditions for a nucleus to emit protons spontaneously and show whether ^{235}U and ^{200}Hg could be expected to undergo proton radioactivity.

10.8. If you have the same number of ^{238}U and ^{235}U atoms, say 238 and 235 grams of each respectively, what are the activities (disintegration/second) of these two? Express this activity in curies.

10.9. Derive an equation between Z and A above the β-stable line nuclei by using the nuclear semi-empirical binding energy formula to obtain the nuclear mass and minimizing this equation with respect to changes in Z.

10.10. From the data given below, make a plot and determine the half-life of the isotope. The numbers of counts per second observed at six second intervals starting at 0, 6 s, 12 s, ... were 3310, 2960, 2370, 1995, 1800, 1620, 1440, 1310, 1220, and 1005, respectively. The error is the square root of the counts.

10.11. Following β decay of the parent, you observe the following γ-rays in the daughter nucleus with the energies, relative γ-ray intensities, and multipolarities as shown. If the daughter nucleus has $Z = 50$ and $N = 70$ and a ground state spin and parity of 0^+, what are the energies of the excited levels in the daughter nucleus and what are their allowed spins and parities?

(*a*) 834 keV, 100%, pure E2

(*b*) 1.464 MeV, 4%, pure E2

(*c*) 2.515 MeV, 0.25%, pure E3

(*d*) 1.051 MeV, 7%, pure E1

(*e*) 630 keV, 25%, M1 + E2

(*f*) 894 keV, 10%, pure E2

(*g*) 1.681 MeV, 1.0%, pure E1

10.12. Use Fig. 10.19 to estimate the single-particle lifetimes for γ-rays with the energies and multipolarities given in Problem 10.11. Also, from that figure estimate whether internal conversion would be important, that is, would compete successfully with γ-ray emission for some sizable fraction of the decay of the excited state.

10.13. Suppose you measure the intensities of the electrons and γ-rays depopulating an excited state of $^{184}_{80}$Hg at 367 keV to its ground state. You find for every 1000 γ-rays there are about 55 electrons from all shells. What is the total internal conversion coefficient for this transition and what would you predict the multipolarity of the transition to be based on Fig. 10.23? What would be the spin and parity of the excited state (remember that the spin and parity of the ground state of an even–even nucleus are 0^+)?

10.14. By using Fig. 10.23, compare the total internal conversion coefficient for a 500-keV transition in Sn and in Hg. Roughly, what is the physical basis for this difference?

10.15. What type of decay process is electric monopole radiation? What is the angular momentum carried off by such radiation and what is the parity change between the states? What is different about this type of radiation compared to other electric multipole radiations?

10.16. When an excited nuclear state has a half-life (or mean life) of seconds to minutes to decay, it is called an isomeric state. What are some properties that an excited state must have compared to the ground state of that nucleus in order for the excited state to be isomeric (i.e., to live so long)? Can you give a semi-classical argument to justify why you would expect some of these differences to retard a transition and make it isomeric?

10.17. Calculate the recoil energy of the 129-keV γ-ray emitted by ^{191}Ir if the ^{191}Ir atom is free to move. Compare this with the energy spread expected on the basis of the energy uncertainty for the 129-keV excited state which has a mean life of 0.11×10^{-9} s. What did Mössbauer discover that eliminated this recoil?

10.18. A radioactive source of an element with atomic number Z is observed to undergo β^- decay, γ decay and internal conversion emission. How could you use a measurement of the energy of the γ-ray and the K internal conversion electrons associated with the decay of a particular nuclear level to show that γ-ray emission and K internal conversion emission come after and not before the β decay?

10.19. From the mass data in Appendix II, make a plot of the masses for the $A = 199$ and 200 chains analogous to Fig. 10.14. Discuss all the possible decays you would expect to observe.

10.20. From the mass data in Appendix II, make a plot of the masses for the $A = 99$ and 100 chains analogous to Fig. 10.14. Discuss all the possible decays you would expect to observe.

10.21. If you start with 0.1 g of pure ^{238}U, what will be the approximate activity of ^{234}Th after 100 days? Compare this with the activity of ^{238}U.

10.22. Nuclei that are neutron-rich compared to stable nuclei undergo what type of β decay to go back towards stability? A few nuclei are found to decay by both β^- and β^+ (positron) emission. How is this possible? Explain what the physical characteristics of the parent and daughter must be for this to occur.

10.23. Some theories indicate that the proton can decay. If its mean lifetime is about 1.2×10^{32} years, how much water must be used if we want to measure the proton decay (for example, measuring one decay event per month)?

10.24. On the basis of the difference in their penetrating power and ionizing power, discuss the relative danger of α, β, and γ rays in (a) killing cells and (b) inducing cell mutation (e.g., cancer), if the source is outside the body.

10.25. Answer Question 10.24 if the source is inside the body, say, in the lungs.

10.26. You have 60 g sample of ^{242}Pu which has a half-life of 3.79×10^5 years. It has a spontaneous fission half-life of 6.8×10^{10} years. What is the spontaneous fission branching ratio? Calculate the number of spontaneous fissions per second.

10.27. Complete the following by giving all N, Z, and A numbers and other particles or elements involved in these decays. Discuss all the conservation laws you used to arrive at your answer.

(*a*) $^{223}\text{Ra} \rightarrow {}^{209}\text{Pb}+$

(*b*) $^{252}\text{Cf} \rightarrow \text{Zr}_{62} + 4\text{n} +$

(*c*) $^{252}\text{Cf} \rightarrow {}^{104}\text{Mo} + {}^{142}\text{Xe} + 2\text{n} +$

(*d*) $^{242}\text{Pu} \rightarrow {}^{132}\text{Sn} + \text{Ru}_{66} +$

CHAPTER 11

NUCLEAR FORCES AND NUCLEAR MODELS

Nuclear models are at the heart of nuclear physics; they connect our thinking about the structure of a new microcosmos (the nucleus) with reality, and through that stimulate our imagination.

W. Greiner

11.1 NUCLEAR FORCES

11.1.1 General Characteristics

Before we knew about nuclei, only two kinds of interactive forces were known; the gravitational force and the electromagnetic force. The electromagnetic force between two nuclear particles is many factors of 10 stronger than the gravitational force between them. When the very tiny nucleus of the atom was discovered with essentially all the mass and all the positive charge of the atom, it became apparent that there had to be a new force in nature—the nuclear force—to hold the nucleus together. The repulsive Coulomb force between positive charges at distances less than the nuclear radius ($\sim 10^{-14}$ m) would be extremely large. There had to be a nuclear force then, which is attractive and much stronger than the electromagnetic force to hold the nucleus together. It is the enormous strength of the nuclear force that binds the neutrons and protons in a nucleus so tightly that the nuclear density is $\sim 10^{14}$ g/cm^3!

From the time of the discovery of the nucleus and the inferred nuclear force in 1911 until the discovery of the neutron in 1932, the composition of the nucleus was a real problem. One had only two particles, ionized hydrogen (the proton) and the electron. Thus, to get 238 nuclear mass units for ^{238}U with $Z = 92$, one had to postulate that its nucleus had 238 protons and 146 electrons. However, this picture had many problems, such as how to put electrons in the new wave picture into a volume the size of a nucleus and how to account for the rotational (spin) properties of nuclei as discussed earlier. With the discovery of the uncharged neutron (as discussed in Chapter 12), this problem was solved and scientists could begin to explore the nuclear force between nucleons (the name for both protons and neutrons).

We have since obtained much knowledge about the nuclear force; however, we still cannot write it down in any mathematical form in the way we can for the gravitational and electromagnetic forces. However, as we shall see, considerable progress has been made toward developing realistic nuclear force models which can even be used to calculate the properties of nuclei. Here we introduce the basic experimentally determined properties of the nuclear force.

1. Very short range. The nuclear force has a short range. This short range accounts for the fact that this strong force was not observed until nuclei were discovered. The nuclear force is felt only over distances the size of a nucleus, $\leq 10^{-14}$ m. Outside this distance, a low-energy α-particle feels only the long-range Coulomb force in scattering off a nucleus.

2. Saturation. If the nuclear force of one nucleon could act on every other nucleon inside the nucleus like the long-range $1/R^2$ Coulomb force, the nuclear binding energy would be proportional to the number of nuclear pairs $A(A-1)$, that is, proportional to A^2. But, from experiment, the total binding energy, B, is proportional to A, that is, proportional to the nuclear volume (since $A \sim R^3$). Recall that in Fig. 9.2, B/A is approximately constant. This is very much like a liquid. For example, the energy which is needed to boil two liters is twice that for one liter. So, the nuclear force acts between only a few neighboring nucleons and not between all the nucleons in a nucleus, just as occurs for the force between molecules in a liquid. Stated another way, the nuclear force has the property of saturation. One proton or neutron can interact through the nuclear force with only a few neighbors and then it is saturated. Other evidence for the saturation of the nuclear force is the essential constancy of the nuclear density from the center of a nucleus to its edge and the essential constancy of the nuclear density as a function of mass from $A = 10$ to 250. If the nuclear force of one nucleon acted on all the others, then the densities should be greater near the center and should increase with increasing A. Note that you can have a short range without saturation over that range, but saturation implies short range.

3. Very strong interaction. The nuclear force is a very strong interaction. In fact, the nuclear force is about 100 times greater than the Coulomb force inside the nucleus. This has to be so or the long-range Coulomb repulsive force would split nuclei apart and neither nuclei nor our universe would exist. However, the fact that the long-range, repulsive Coulomb force does not saturate but exerts itself between every proton in the nucleus leads to observable effects as Z increases. This is the origin of the bending of the stability line from $N = Z$ toward $N > Z$ for heavier nuclei and produces the very slow decrease in B/A above $A \sim 60$.

4. Charge independence. Heisenberg proposed in 1932 that the nuclear force between a proton and a proton, F_{pp}, is equal to that between a neutron and a neutron, F_{nn}, and between a proton and a neutron, F_{np}:

$$F_{pp} = F_{nn} = F_{np}$$

The fact that $F_{pp} = F_{nn}$ we can call the charge symmetry of the nuclear force, and that $F_{nn} = F_{np}$ we call the charge independence of this force. The first experimental evidence came in 1937. Later, from 1946 to 1955, more precise experiments were done. Now the reliability that $F_{pp} = F_{nn}$ is more than 99 percent and that $F_{pp} = F_{nn} = F_{np}$ is more than 98 percent. However, studies of possible small deviations from 100 percent in both cases are topics of much current research interest.

5. Strong repulsive core. Although the attractive nuclear force is very strong, the nucleons do not get infinitely close. Nuclei have finite sizes. So, there must also exist a very strong repulsive force between two nucleons when they get too close together, to override the strong attractive force that pulls them together.

6. Spin dependence. The nuclear force between two nucleons depends on their relative spin directions. The deuteron ($^2_1H_1$) provides one piece of evidence for this. The proton and neutron can combine in an $L = 0$ state so that the total spin and angular momentum $S = J = 1$ with spins parallel or $S = J = 0$ with spins antiparallel, to form a $^3S_1$ (triplet) state and a $^1S_0$ (singlet) state. If the nuclear force is spin-independent, these two states should have essentially the same energy. However, the observed spin of the deuteron ground state is 1, and the $S = 0$ state is observed not to be bound. So, when the proton and neutron spins are parallel, the nuclear force is stronger and binds the neutron and proton to form ^2H. Thus, the nuclear force must have a spin dependence. Likewise, in n–p scattering, when the n–p form an $S = 1$ state the cross-section is quite different in magnitude from that when $S = 0$.

Figure 11.1a shows the potential between a proton and a proton derived from p–p scattering, where the positive hump is the Coulomb barrier. We can also get Fig. 11.1b from p–n scattering experiments. We cannot study V_{nn} by n–n

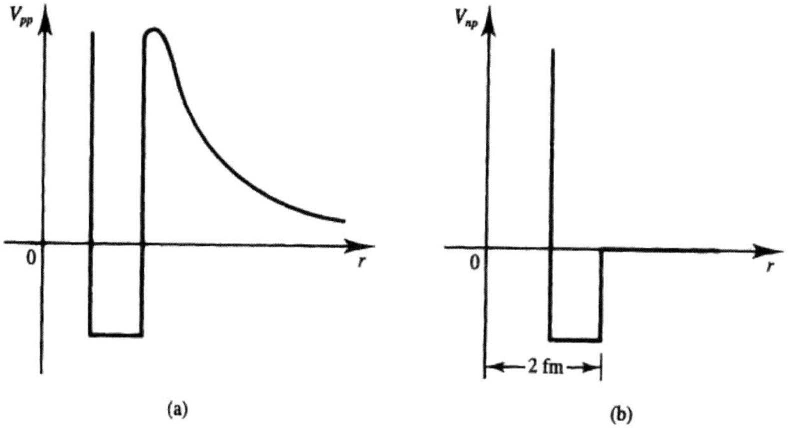

FIGURE 11.1
(a) The potential between two protons; the positive hump on the right side is the Coulomb barrier. (b) The potential between a neutron and a proton; this latter has no Coulomb repulsion. In both figures, the steep rise at smaller r is the effect of the very strong repulsive core.

scattering experiments, because we cannot make pure neutron targets. But from various indirect experiments, it is found that V_{nn} is the same as Fig. 11.1b.

From such research we have obtained the following knowledge about the nuclear force. When the distance between two nucleons is about 0.8–2.0 fm, the nuclear force is attractive. At less than 0.8 fm, it is repulsive. At more than 10 fm, it disappears. Our knowledge of the nuclear force is very good for $r > 2.0$ fm. But for $r \sim 0.8$–2.0 fm, it is known only partly and for $r < 0.8$ fm it is known only poorly.

11.1.2 Meson Theory of the Nuclear Force

It is well known how two particles with electric charge interact: two charges with the same sign repel and two charges with opposite signs attract. At first it was strange that although two bodies with charge did not touch each other, they interacted with each other. This led to the concept of electric field. Similarly, one could ask about the nucleon force field. However, we have developed an alternate approach for charges in which the electromagnetic interaction produces an exchange force between particles with charge in which they exchange "imaginary photons." For example, in Fig. 11.2 is shown an interaction between two electrons. One electron enters diagonally from left to right and the other from right to left. The horizontal axis is the distance x and the vertical axis is the time t. At the point A, the left electron emits a virtual photon and changes its direction of motion from the recoil. The right electron absorbs this imaginary photon and recoils to change its direction. The total effect is that electron "a" came from the left to the right and electron "b" came from the right toward the left and, after

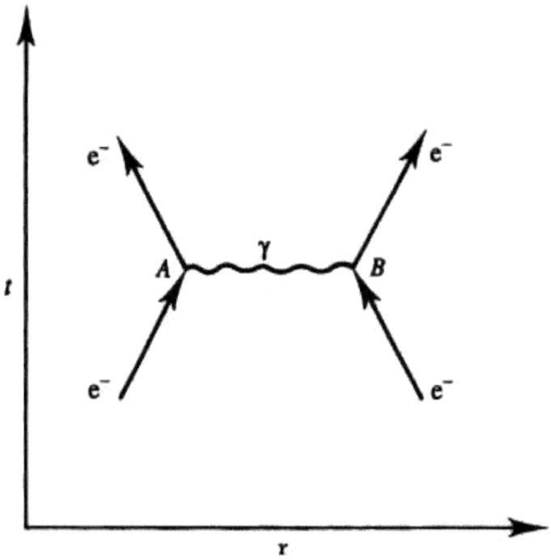

FIGURE 11.2
The interaction between two electrons.

interacting, "a" moves from right to left and "b" from left to right. Between charged particles, such virtual (imaginary) photons are exchanged continuously.

In 1935 a Japanese scientist, H. Yukawa, put forward the meson theory of the nuclear force. He thought that the nuclear force is a kind of exchange force. The interaction between two nucleons occurs through exchanging some medium (particle). He also estimated the mass of the medium (the exchanged particle) according to the range of the nuclear force. The method is as follows: An imaginary particle released by nucleon (a) is absorbed by nucleon (b) after it passes through a distance Δx (see Fig. 11.3). The time (the imaginary particle lifetime) is Δt. Even if the imaginary particle moved with the speed of light, the distance Δx is no more than $c\,\Delta t$. From the uncertainty relation, we can get the maximum energy transformation during the time Δt:

$$\Delta E = \frac{\hbar}{\Delta t} = \frac{\hbar}{\Delta x/c} = \frac{\hbar c}{\Delta x} \tag{11.1}$$

If this energy were all transformed to the imaginary particle's rest mass, its rest mass m would be

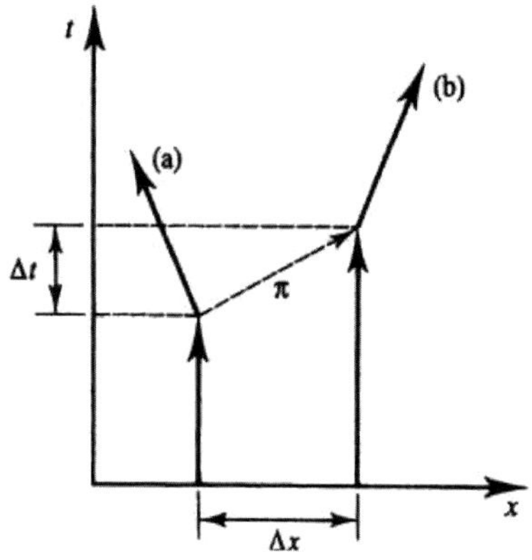

FIGURE 11.3
The interaction between two nucleons represented by a and b.

$$\Delta E = \frac{\hbar c}{\Delta x} = mc^2 \qquad m = \frac{\hbar}{c\,\Delta x} \tag{11.2}$$

If this were real energy, something would be wrong. It violates energy conservation. So, we cannot observe an exchange particle releasing such energy in an experiment. However, for an "imaginary" particle, it is allowed by the uncertainty relation that there exists the energy ΔE which is not conserved during the time Δt. For the electromagnetic interaction, because the force range is indefinitely large, the mass of the exchange particle must be zero. The photon is the transferring medium (exchange particle).

Now consider the nuclear force, for which $\Delta x \sim 2.0\,\text{fm}$. The exchange particle has a rest mass energy

$$mc^2 = \frac{\hbar c}{\Delta x} \simeq \frac{197\ \text{fm}\cdot\text{MeV}}{2.0\ \text{fm}} \simeq 100\ \text{MeV} \tag{11.3}$$

The mass of this particle is about 200 times the electron rest mass, but 10 times smaller than the masses of the proton and neutron. Yukawa called the new exchange particle the "meson." When Yukawa gave his theory, this particle was not known. Immediately, scientists began to search for the meson. During 1936–1937 the μ meson with mass 207 times that of the electron was found, which satisfied the rough mass condition. This was seemingly a great triumph for theory. But soon the μ meson was found to interact only very weakly with nucleons. Since it does not join in the strong interaction nuclear force, it is not the meson which Yukawa had predicted. In 1947 the π meson, which does mediate the strong interaction between nucleons, was found. There are in fact three π mesons, π^+, π^-, and π^0. The m_{π^+} and m_{π^-} masses are each 273.3 times that of an electron, while m_{π^0} is 264 times that of an electron. The π^+, π^-, and π^0 exchanges are shown in Fig. 11.4.

Diagrams as in Fig. 11.4, which represent interactions, are called Feynman diagrams. The imaginary (exchange) particle theory has been verified in many experiments and is a great success for nuclear theory. However, ultimately the

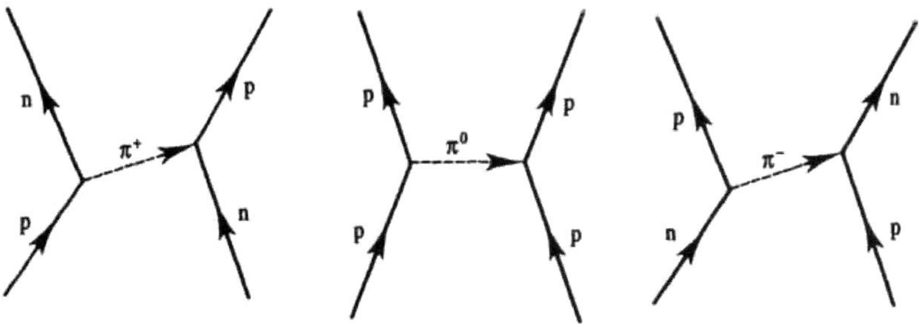

FIGURE 11.4
The π meson (exchange particle) mediates the nuclear force as illustrated for the $\pi^{\pm}$ and π^0 exchanges between nucleons.

origin of the strong force in nature is the color force mediated by the gluon, as we shall see in Chapter 14.

11.2 NUCLEAR MODELS

In a basic description of nuclei, we want to know how the nucleons inside the nucleus move and behave in terms of the nuclear force. While great strides are being made, this is a difficult problem that so far remains unsolved. We know that protons and neutrons, like electrons and atoms, obey the laws of quantum mechanics and electrodynamics. The problem is the force that holds the nucleus together. In atoms, the interaction force is the Coulomb force whose properties are well known. The Coulomb interaction between the electrons and the nucleus plays the key role in the motion, so the problem can be solved in a straightforward manner.

But in the nucleus, the main interaction force is the nuclear force. While we know much about the nature of this force as discussed in the last section, we cannot write it in closed form as we can the Coulomb force. Even if we understood the nuclear force thoroughly, we would still meet another thorny problem—the many-body problem. In atoms, the interaction of the electrons on each other is at most a small perturbation. However, it is the interaction among the many nucleons in the nucleus that holds the nucleus together. So, it is not possible to get a solution like that for the two-body Coulomb problem between an electron and a nucleus in an atom. One may think that a statistical approach would work, but the number of nucleons is not sufficiently high for use of statistical methods. These problems are also the reason the nucleus is so interesting and important in a fundamental sense. It is a unique many-body quantum system, with too many particles to treat the individual interactions and too few particles to use statistical methods. As a result of these difficulties, different nuclear models have been proposed in order to describe the motion of the nucleons inside the nucleus and the structure of nuclei in some approximate way. Sometimes a special model may be used to explain only one property. However, theoretical developments and new, large computers have opened up the possibility of doing more microscopic nuclear model calculations that include features of the nuclear force. In such cases one seeks to derive nuclear properties from the fundamental nucleon–nucleon interaction. Note that in addition to the force between two nucleons, there is an additional contribution from three-body correlations (interactions) that must be included when treating nuclei with many nucleons. The interplay between microscopic and macroscopic nuclear models and new experimental results is at the forefront in nuclear physics today.

Many nuclear models have been put forward since 1932. Earlier we mentioned the "liquid drop model," which is a collective model in which the nucleons move collectively. The opposite extreme is the independent particle model, such as the "Fermi gas model" in which the nucleons move independently. We shall introduce this model briefly. Then we shall consider the two most successful nuclear models: the spherical nuclear shell model and the nuclear collective model. The former was proposed independently in 1949 by M. G. Mayer and by J. H. D. Jensen and co-workers, and won them the Nobel Prize in Physics in 1963. The latter was developed in 1952 by Å. Bohr and B. Mottelson following an earlier suggestion by J. Rainwater and won them the Nobel Prize in Physics in 1975. The fact that nuclei can be divided into these two broad categories can be seen by looking at Fig. 11.5a,b. In Fig. 11.5a the first excited 2^+ energies of even–even nuclei are given, and in 11.5b the transition probabilities for the electric quadrupole transitions between the first 2^+ and the 0^+ ground states, B(E2: $0^+ \rightarrow 2^+$) upward, are given. In certain regions (category 1) the 2^+ energy is large (0.5–4 MeV) and the transition probability very small (nearly vanishing) (generally, the left side of each figure and around $A = 120$ and 200 where there are peaks in Fig. 11.15a and valleys in 11.15b), around the spherical magic numbers, as we shall see; while in other regions (category 2) such as the rare earths and actinides, the energies are very low (≤ 100 keV) and the transition probabilities very high, suggesting some collective enhancement.

A critical feature of the interaction of nucleons in the nucleus that is important in both the shell model and collective model is nucleon pairing. While the nucleon–nucleon interaction is spin-dependent, so that the proton and neutron when combining to form ^{2}H have their spins aligned to form an $S = J = 1$ state, there is a strong pairing force inside the nucleus so that each proton pairs off with another proton and each neutron with another neutron in a spin-zero ($S = 0$) state. Thus, the spin of an even Z–even N nucleus is always zero. M. Mayer made

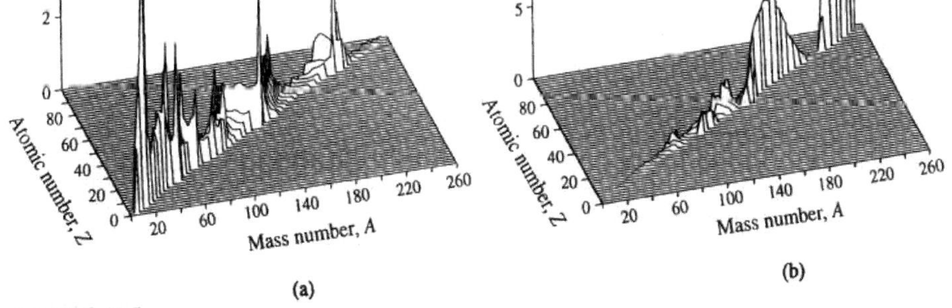

FIGURE 11.5
(a) Energies of the first excited 2^+ states in even–even nuclei as a function of mass number. These energies are not scaled by their moment of inertia. For example, a 200 keV 2^+ state in ^{74}Kr would scale to 28 keV for comparison with a nucleus with $A = 240$. (b) The electric quadrupole transition probability from the ground state to the first excited 2^+ state as a function of A. [S. Raman et al., *Phys. Rev.* **37** (1988) 805.]

use of this in her shell model to explain why odd-A nuclei would have the spin of the last unpaired particle. The Bardeen–Cooper–Schrieffer (BCS) theory of superconductivity provided a major breakthrough in understanding pairing in nuclei. The physical idea at the heart of superconductivity is that two electrons of opposite spin ($S = 0$) are paired within a material to form a type of bound state. Such a *Cooper pair* behaves like a boson. A. Bohr, B. R. Mottelson, and D. Pines, and then S. Belyaev, introduced BCS theory into nuclear physics. In even–even nuclei, there is an energy gap arising from the pairing of nucleons, just as there is in a superconductor, but now an energy gap for protons and another one for neutrons. Below the pairing gap, only collective excitation of rotations and vibrations can occur in even–even nuclei. Exciting a single particle from one orbital to another requires energy to break a pair. After considering the shell and collective models, we will consider the extensions of these models and the need for more microscopic descriptions.

11.2.1 Fermi Gas Model

This model is the original independent-particle model. It treats nucleons as a molecular gas without interaction inside a potential well. Because nucleons are fermions, nuclei can be taken as a Fermi gas. The factor that determines the nucleon motion is the Pauli exclusion principle. The nucleons can move independently through the nucleus without interacting because in the ground state all the allowed energy states are filled, according to the Pauli principle, to a certain level (Fermi level) and the states above are not energetically accessible to the nucleons below. To illustrate, suppose two protons collide. For one proton to go to a higher energy state above the Fermi level would mean the other proton would have to go to a lower energy level to conserve energy. But all the lower energy states are filled, so the two particles cannot interact and exchange energy.

Since neutrons and protons have different charges, the shapes and depths of the potentials that bind them are not the same, as seen in Fig. 11.6, where B' is the experimental binding energy (the energy to pull one particle out of the nucleus) and E_c is the Coulomb energy, which is given by Eq. (9.7). The bottom of the proton well is higher than that of the neutron well by the amount E_c. The upper part of the proton well has a Coulomb barrier which acts to keep charged particles outside the nucleus from entering and those inside from leaving. If a proton coming from outside wants to enter a nucleus, its energy must exceed this barrier. At energies just below the barrier height, there is the quantum "tunnel effect," which gives some probability of penetration.

In the potential well, there are some discrete energy levels. When the nucleus is in its ground state, the nucleons are all in the lowest energy states available to them according to the Pauli principle. Each energy state in the neutron well can contain two neutrons, and in the proton well two protons, where one has its spin up and the other has spin down. When a nucleus is in its ground state, the highest filled nucleon energy level is called the Fermi energy level, E_F ($E_{F,n}$ or $E_{F,p}$).

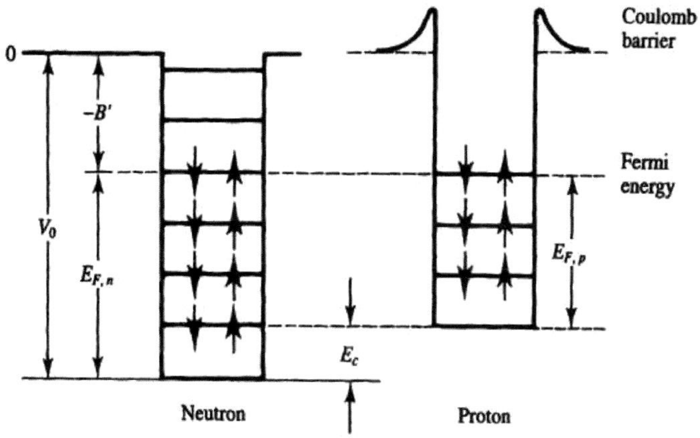

FIGURE 11.6
The neutron and proton square well potentials.

From Section 7.5 (Eq. 7.29) the energy levels of a particle in a one-dimensional square well are

$$E_n = \frac{n^2 h^2}{8md^2}, \qquad n = 1, 2, 3, \ldots \tag{11.4}$$

where m is the particle mass and d the potential well width. We can expand this to three dimensions by considering a cubic body potential well with volume d^3. We have

$$\begin{aligned} E_{n_1 n_2 n_3} &= \frac{h^2}{8md^2}\left(n_1^2 + n_2^2 + n_3^2\right) \\ n_1 &= 1, 2, 3, \ldots \\ n_2 &= 1, 2, 3, \ldots \\ n_3 &= 1, 2, 3, \ldots \end{aligned} \tag{11.5}$$

In contrast to the one-dimensional case, the energy degeneracy increases. There is one ground state $(n_1, n_2, n_3) = (1, 1, 1)$, but there are three first excited states, (2, 1, 1), (1, 2, 1), (1, 1, 2), with the same energy. The degeneracy increases with increasing energy. For example, (1, 2, 6), (1, 6, 2), (2, 6, 1), (2, 1, 6), (6, 1, 2), (6, 2, 1), (3, 4, 4), (4, 3, 4), and (4, 4, 3) all have the same energy.

We must know how many states are at the E_F energy level and below the E_F energy level in order to get the Fermi energy. One needs to know how many (n_1, n_2, n_3) combinations satisfy the condition

$$n_1^2 + n_2^2 + n_3^2 \leq \frac{8mE_F d^2}{h^2} \tag{11.6}$$

If we define

$$R^2 \equiv \frac{8mE_F d^2}{h^2} \tag{11.7}$$

then (11.6) becomes

$$n_1^2 + n_2^2 + n_3^2 \leq R^2 \tag{11.8}$$

If we take n_1, n_2, n_3 as the three axes of the rectangular coordinates, then R is the radius of the corresponding spherical coordinate. The states with the same energy are on the same spherical surface with radius R. Every (n_1, n_2, n_3) corresponds to a crystal lattice. The higher the energy is, the larger the spherical surface, the larger the crystal lattice, and the higher the degeneracy is. How many groups of (n_1, n_2, n_3) are in accord with Eq. (11.6)? This question is equivalent to asking how many crystal lattices are on the spherical surface with radius R. For all positive n_1, n_2, n_3, this is 1/8 of the volume of the sphere. That is,

$$\frac{1}{8}\frac{4\pi}{3}R^3 = \frac{\pi}{6}\left(\frac{8mE_F d^2}{h^2}\right)^{3/2} \tag{11.9}$$

Now, first consider the neutrons. There are two neutrons in each energy state. From the Pauli principle, in a volume d^3, the neutron number is

$$N = \frac{\pi}{3}\left(\frac{8m_n E_{F,n} d^2}{h^2}\right)^{3/2} \tag{11.10}$$

The nuclear volume can be written

$$d^3 = \frac{4\pi}{3} R_0^3 = \frac{4\pi}{3} r_0^3 A \tag{11.11}$$

So we can obtain the neutron maximum kinetic energy, that is, the Fermi energy:

$$E_{F,n} = \frac{\hbar^2}{2m_n r_0^2}\left(\frac{9\pi N}{4A}\right)^{2/3} \tag{11.12}$$

In the same way, we can obtain the proton maximum kinetic energy:

$$E_{F,p} = \frac{\hbar^2}{2m_p r_0^2}\left(\frac{9\pi Z}{4A}\right)^{2/3} \tag{11.13}$$

where r_0 is the nuclear radius constant (1.20 fm).

Correspondingly, we can write the neutron and proton maximum momentum:

$$\begin{aligned} p_n &= \frac{\hbar}{r_0}\left(\frac{9\pi N}{4A}\right)^{1/3} \\ p_z &= \frac{\hbar}{r_0}\left(\frac{9\pi Z}{4A}\right)^{1/3} \end{aligned} \tag{11.14}$$

To obtain the above equations, we have used the nonrelativistic relation between energy and momentum:

$$E_F = \frac{p_F^2}{2m} \tag{11.15}$$

According to this, we are able to calculate the average kinetic energy:

$$\langle E \rangle = \frac{\int_0^{p_F} E\, d^3p}{\int_0^{p_F} d^3p} = \frac{3}{5}\left(\frac{p_F^2}{2m}\right) \tag{11.16}$$

and the total kinetic energy of a nucleus:

$$\begin{aligned} \langle E(Z,N) \rangle &= N\langle E_N \rangle + Z\langle E_Z \rangle = \frac{3}{10m}\left(Np_n^2 + Zp_z^2\right) \\ &= \frac{3}{10m}\frac{\hbar^2}{r_0^2}\left(\frac{9\pi}{4}\right)^{2/3}\left(\frac{N^{5/3} + Z^{5/3}}{A^{2/3}}\right) \end{aligned} \tag{11.17}$$

Here we take both the neutron and proton masses as m approximately and also take both their potential well widths as the same. Of course, from the beginning, we made the supposition that both the protons and neutrons are respectively in independent motion. From (11.17) we can conclude what we mentioned in Section 9.1, namely, that when $Z = N$, $\langle E(Z,N) \rangle$ is a minimum. In order to study the behavior near the minimum value, we take $Z - N = \delta$, and let $Z + N = A$ be a definite value. We have

$$N = \frac{1}{2}A\left(1 - \frac{\delta}{A}\right) \qquad Z = \frac{1}{2}A\left(1 + \frac{\delta}{A}\right)$$

If $\delta/A \ll 1$, we can use the binomial theorem:

$$(1 + x)^n = 1 + nx + \frac{n(n-1)}{2}x^2 + \dots$$

Then, when $N - Z$, expression (11.17) becomes

$$\langle E(Z,N) \rangle = \frac{3}{10m}\frac{\hbar^2}{r_0^2}\left(\frac{9\pi}{8}\right)^{2/3}\left(A + \frac{5}{9}\frac{(Z-N)^2}{A} + \dots\right) \tag{11.18}$$

The first term is proportional to A, which contributes to the volume energy. The second term can be written as

$$a_{\text{sym}}\frac{(Z-N)^2}{A} \tag{11.19}$$

where

$$a_{\text{sym}} = \frac{1}{6}\left(\frac{9\pi}{8}\right)^{2/3}\frac{\hbar^2}{mr_0^2} \tag{11.20}$$

Equation (11.19) is what we used earlier in Section 9.2. From Eq. (11.18) we can see very clearly that for nuclei with the same A, when $Z = N$, the nuclear energy is a minimum, that is, the $Z = N$ nucleus is the most stable one for a given A.

11.2.2 Nuclear Shell Model

INTRODUCTION. The exact nuclear many-body Hamiltonian has the following structure:

$$H(1,\ldots,A) = \sum_{i=1}^{A}\left(-\frac{\hbar^2}{2m}\nabla_i^2\right) + \sum_{i<j}^{A}(V_{\text{Coul}}(i,j) + V_{\text{nucl}}(i,j)) \tag{11.21}$$

where we have used the shorthand notation

$$(i) = (\boldsymbol{r}_i, \boldsymbol{s}_i, \boldsymbol{t}_i). \tag{11.22}$$

The first term in Eq. (11.21) represents the familiar quantum-mechanical expression of the kinetic energy operator for the ith nucleon, while the second and third terms describe the two-body Coulomb and nuclear interactions between nucleons i and j. The latter depend not only on the spatial positions $(\boldsymbol{r}_i, \boldsymbol{r}_j)$ of the two nucleons but also on their spin vectors $(\boldsymbol{s}_i, \boldsymbol{s}_j)$ and isospin vectors $(\boldsymbol{t}_i, \boldsymbol{t}_j)$. The above Hamiltonian is invariant under a number of symmetry operations; that is, the Hamiltonian commutes with the corresponding symmetry operator S:

$$[H, S] = 0 \tag{11.23}$$

This implies that we can always find nuclear many-body wavefunctions which are simultaneous eigenfunctions of H and S. Examples of the particular symmetry operator S are the center-of-mass momentum vector $\boldsymbol{P}$, the total angular momentum operators $\boldsymbol{J}^2, J_z$ of the nuclear many-body system, the particle number operator N, and the parity operator Π.

It should not be too surprising that these symmetries will, in general, no longer hold if one approximates the exact many-body Hamiltonian by a sum of one-body operators, as one does in the phenomenological shell models (e.g., Nilsson model) or even in the far more sophisticated self-consistent nuclear mean-field theories (so-called Hartree–Fock approximation)

$$H_{\text{shell}}(1,\ldots,A) = \sum_{i=1}^{A}\left[-\frac{\hbar^2}{2m}\nabla_i^2 + V_{\text{Coul}}(i) + V_{\text{nucl}}(i)\right] \tag{11.24}$$

As we can see from Eq. (11.24), the shell model Hamiltonian H_{shell} has a much simpler structure than the exact Hamiltonian, Eq. (11.21); this is because the Coulomb and nuclear *mean-field potentials* $V(i)$ depend only on the coordinate of *one* nucleon at a time. It turns out that, because of this approximation, most of the symmetry properties obeyed by the exact Hamiltonian are violated by the shell model Hamiltonian H_{shell}. In particular, the translational invariance is always violated in the shell models, because the origin ($\boldsymbol{r}_i = \boldsymbol{0}$) of the phenomenological mean-field potential is fixed at a certain point in space, in contradiction to the translational invariance of space. (Note that this is not true for the exact Hamiltonian (11.21), which depends on the distance vectors $(\boldsymbol{r}_i - \boldsymbol{r}_j)$ between any two nucleons.) This symmetry violation in the nuclear shell models causes serious consequences, such as the appearance of spurious states in the nuclear excitation spectrum. If one uses shell model wavefunctions, one therefore needs to correct for these deficiencies. In general, this can be done by applying certain quantum-mechanical *projection operators* to the nuclear many-body wavefunction. The mathematical structure of these projection operators is very complicated and is far beyond the scope of this book.

We mention, however, that the most sophisticated nuclear many-body theories currently available, the so-called cranked Hartree–Fock–Bogoliubov theories (which contain both a self-consistent mean field and a pairing field), violate not only the translational invariance of space but also the rotational invariance of space; furthermore, the pairing field breaks the invariance of the Hamiltonian with respect to the particle number operator. Therefore, besides the above- mentioned center-of-mass projection, one needs to perform additional total angular momentum projections and particle number projections.

EXPERIMENTAL EVIDENCE AND EARLY ATTEMPTS. The atomic shell structure is essential in explaining the chemical periodicity of the elements. In the periodic table the appearance of an inert gas indicates that a particular electron shell or subshell is closed, to give this atom its special stable character. When the atomic number Z is equal to 2 (He), 10 (Ne), 18 (A), 36 (Kr), 54 (Xe),..., the chemical element is most stable and does not interact with other atoms to form molecules. These atomic magic numbers are explained by the filling of shells and subshells in the atomic shell model and the appearance of gaps in the energy spectrum of the shells (significant jumps in the electron energy in going from one shell to the next).

After 1932, experimental data on nuclei repeatedly revealed that there exists a series of magic numbers of protons and neutrons that give special stability to those nuclei which have such a Z and/or N number. It was found that when the proton number or the neutron number is equal to one of

$$2, 8, 20, 28, (40), 50, 82, 126$$

the nucleus is particularly stable and has a spherical shape. The evidence for the number 40 is weaker, which is why it is shown in parentheses. The experimental evidence is discussed below. These numbers are called magic for nuclei because of the stability of such nuclei. Although the nuclear magic numbers are different from those in the atom, the presence of such nuclear stability suggests some type of shell structure and shell closure as found in atoms. But the first nuclear shell models did not predict the correct magic numbers. Why was this?

Some felt the failure was fundamental in that the nuclear shell model lacked a physical basis as was found in atoms. The reason that a shell structure exists in atoms is that there one has a relatively fixed central body, the nucleus. All the electrons independently move around the central Coulomb force supplied by the nucleus. With this force we solve the Schrödinger equation, add the Pauli exclusion principle, and obtain the atomic shell model (structure). The atomic shell model beautifully explains the periodic table (see Chapter 5). But there is no such central force and physical principle for the nucleus. All nucleons are equal. It is a "democracy" with very strong interactions between the nucleons. How can a nucleon move "independently"?

Then it was proposed that nucleons inside the nucleus are in independent motion in some average potential well (approximated by a square well potential, harmonic oscillator potential, or some other), where the force is the average force generated by the action of all the nucleons on each other. With this assumption, the Schrödinger equation was solved for various potentials, but these did not give magic numbers consistent with the experiments. The left and right sides of Fig. 11.7 show the energy levels of a three-dimension harmonic oscillator and an infinite square well, respectively. These two potentials are limiting cases. The actual situation may be between the two of them, so that the energy levels would be as in the middle of Fig. 11.7. But no matter what the potential shape was, the predicted magic numbers did not agree with experiment.

Another reason why the nuclear shell model was not taken seriously in the early years was the success of the collective "liquid drop model," which is contrary to the individual particle motion of the shell model. The liquid drop model was not only able to explain why the nuclear binding energy is proportional to the nucleon number A, but was also used by Bohr and Wheeler to calculate nuclear reaction cross-sections and to explain the exciting new phenomenon of nuclear fission.

However, facts that supported a nuclear shell model accumulated continuously. For example:

1. Beginning with H, Ca ($Z = 20$) is the first element with more than four stable isotopes. Its six isotopes span an unusually wide range, from ${}^{40}_{20}\mathrm{Ca}_{20}$ to ${}^{48}_{20}\mathrm{Ca}_{28}$ which have N either much smaller or much larger than N of nearby stable nuclei, indicating that Z and N of 20 and 28 have special stability.
2. The element Sn ($Z = 50$) has 10 stable isotopes which is more than any other element. The element ${}^{209}_{83}\mathrm{Bi}_{126}$ is the last stable isotope.
3. For $N = 20$ there are five stable isotones (isotopes with the same N and different Z), while $N = 19$ and 21 have no stable isotones. The $N = 50$ and $N = 82$ isotones have even more stable elements, 6 and 7, respectively.
4. In the even $Z\,(Z > 28)$ nuclides, no isotope has an abundance of more than 60 percent, with three exceptions: ${}^{88}\mathrm{Sr}_{50}$ 82.56 percent, ${}^{138}\mathrm{Ba}_{82}$ 71.66 percent, ${}^{140}\mathrm{Ce}_{82}$ 88.48 percent. Clearly the nuclei with neutron numbers 50 or 82 are more stable.
5. A large binding energy is released when addition of one proton or neutron yields one of the magic numbers, but there is only a small energy release when a proton or neutron is added to a nucleus with a magic Z or N. Conversely, the separation energy for one proton or one neutron is very large for a

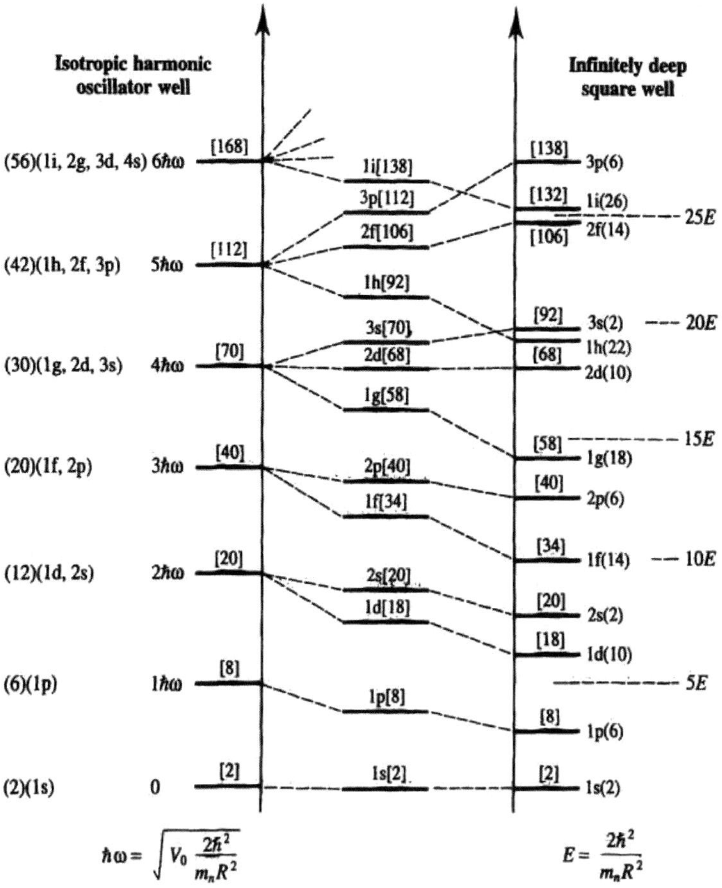

FIGURE 11.7
Energy levels for a harmonic oscillator potential well on the left and a square potential well on the right. In neither case do the predicted magic numbers (associated with shell gaps in the energy) as shown inside the brackets (e.g., for the harmonic oscillator 2, 8, 20, 40, 70, ...) agree with the experimental ones. (Remember the harmonic ground state energy is $\frac{1}{2}\hbar$, not 0.)

nucleus with Z or N magic but is very small for a nucleus with one more proton or neutron than for a magic Z or N.

6. The probability of nuclei with neutron numbers 20, 28, 50, 82, or 126 capturing a neutron is much lower than that of their neighboring nuclei.
7. The energy of the first excited state of ^{208}Pb is by far the largest of the Pb isotopes (see Fig. 11.8). This unusually high first excited state energy is also true for $^4_2He_2$, $^{16}_8O_8$, $^{40}_{20}Ca_{20}$, $^{48}_{20}Ca_{28}$, and $^{90}_{40}Zr_{50}$, as seen in Fig. 11.5a.

NUCLEAR SPIN–ORBIT COUPLING—THE MAYER–JENSEN SHELL MODEL. As the number of experimental facts that indicated a nuclear shell structure continued to increase, the shell model was considered again. Clearly a new

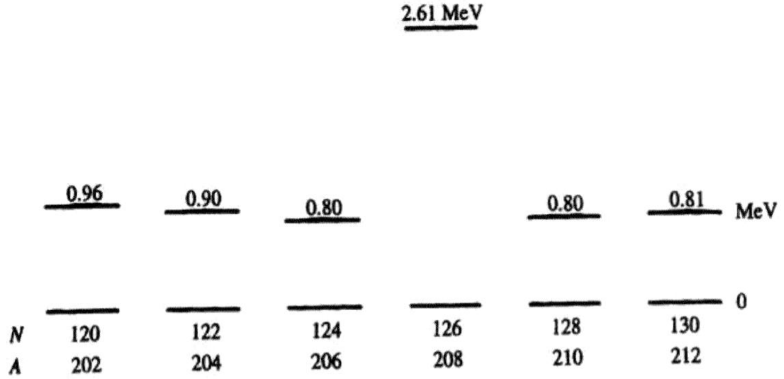

FIGURE 11.8
The first excited states of the even–even lead isotopes from ^{202}Pb to ^{212}Pb.

approach was needed. In 1949 M. Mayer in the United States and independently H. Jensen and co-workers in Germany proposed the new idea that led to the development of a nuclear shell model for spherical nuclei that successfully predicted the nuclear magic numbers. Their key step was the addition of a nuclear spin–orbit coupling force that was both much stronger and had the opposite sign to the spin–orbit coupling force in atoms. This strong coupling of the spin and orbital motions of the nucleons in the nucleus with the opposite sign is essential to give energy splittings (shell gaps) in agreement with the experimental spherical nuclear magic numbers (see Fig. 11.9).

The force which gives rise to the coupling of the spin and orbital motions of the nucleons in the nucleus has a different origin from the coupling of the spin and orbital motion of electrons in atoms. The spin–orbit coupling in nuclei is related to the force between nucleons and is not electromagnetic in origin as is the spin–orbit coupling in atoms (recall for atomic electrons, $\Delta E_{ls} \sim \boldsymbol{B}_l \cdot \boldsymbol{\mu}_s$). However, just as in atoms, the nuclear spin–orbit coupling splits a nuclear energy level which is labeled by the orbital angular momentum l into two levels $j = l \pm s$ (except for $l = 0$), but with the difference that the $j = l - \frac{1}{2}$ lies higher in energy (less binding), and $j = l + \frac{1}{2}$ lies lower in energy.

The order of the level which lies lowest is inverted from that in an atom (see Chapter 4) because of the change of sign of the spin–orbit coupling in nuclei compared to atoms. The splitting magnitude also is much larger because of the much larger strength of the nuclear spin–orbit coupling, and it increases with increasing l. In quantum mechanics, one has a closed shell when one has put all the particles allowed into a given quantum state and there is a sizeable gap in energy to the next allowed energy state of the particle. Thus, it is the relative concentration of the energy levels that determines the magic numbers; that is, gaps in the individual proton and neutron energy levels occur at the magic numbers. From Fig. 11.9, we can see that it is the strong splitting of the high-l orbitals by the spin–orbit coupling, pushing up the levels with $j = l - \frac{1}{2}$ and pushing down levels with $j = l + \frac{1}{2}$, that results in relative concentration of the energy levels and yields gaps at the numbers indicated by experiment. For example, it is the pushing

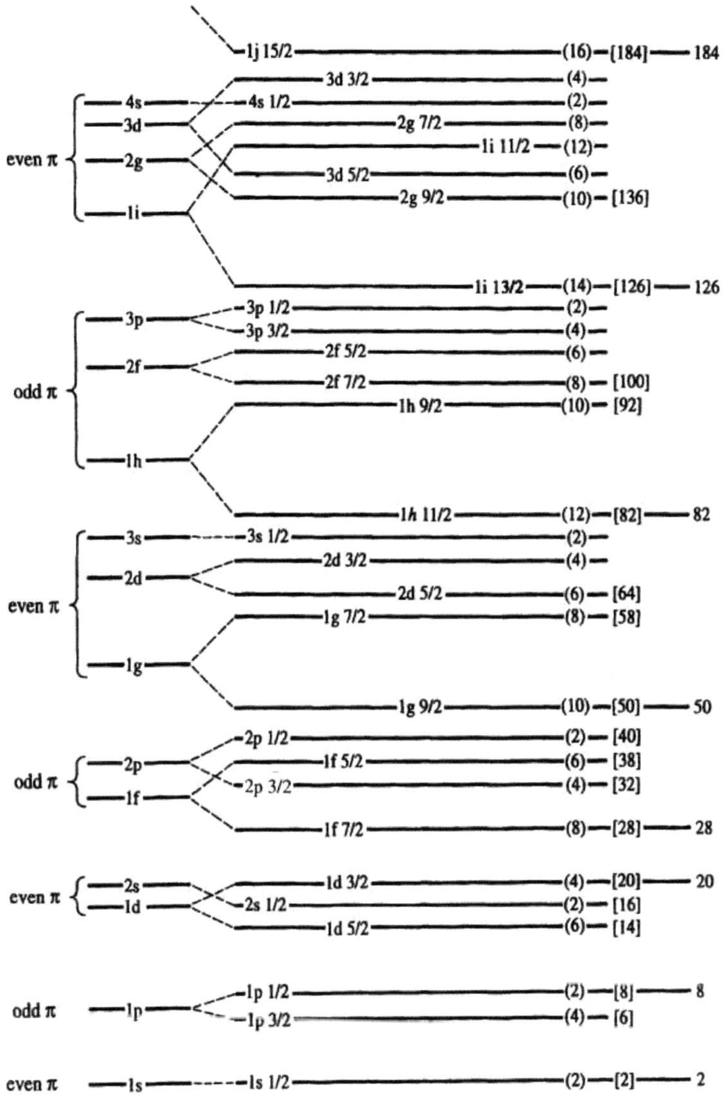

FIGURE 11.9
The strong nuclear spin–orbit coupling, with opposite sign from that in atoms, results in splittings of the energy levels on the left (levels in the middle of Fig. 11.7) to form new energy levels with gaps in their energies at the magic numbers which are found in experiments, as shown on the right. The exact relative positions of the different states that fall between two shell gaps depend on A and are somewhat different for proton and neutron for a given A. For example, the energies of the $2p_{3/2}$, $1f_{5/2}$, $2p_{1/2}$ orbitals vary slighty with A for neutrons, so the $2p_{3/2}$ is lower than $1f_{5/2}$ for $A \lesssim 110$ and the $2p_{1/2}$ is lower than $1f_{5/2}$ for $A \lesssim 60$.

up of the $g_{7/2}$ and pushing down of the $g_{9/2}$ orbitals that leads to the spherical shell gap (magic nucleon number) of N and Z of 50. Just as in atoms, the levels are identified by nl_j, but in nuclei the n quantum number is not the dominant factor in the energy as it is in atoms and the spin–orbit splitting is much larger in nuclei.

The starting point of the nuclear shell model is the premise that the nucleons move independently in an average potential field. So the nucleons move in a single-particle potential $V(i)$ which depends on the nucleons' spatial, spin, and charge (isospin) coordinates. To this one must add an effective residual interaction v_{ij} between the nucleons, where this contribution is treated in first- or second-order perturbation theory. This means that the effective residual interactions are weak and can be treated as perturbations. However, there is the problem that this approach does not lead to a wavefunction which separates into a product of wavefunctions describing the center-of-mass motion and internal motions. We can introduce a potential $V(R)$ that acts on the center of mass only and add $\sum_{ij} v_{ij}$ for the nucleon interactions. If we choose $V(R)$ attractive and sufficiently strong to give bound states of the center mass, then the wavefunction that describes the A nucleons is a localized wavefunction which can be approximated by a combination of single-particle wavefunctions related to the same point in space. We then have the nuclear wavefunction expressed in terms of independent particles moving in a single-particle potential. This is the essence of the shell model. The $V(R)$ is arbitrary and the above holds for many different attractive potentials, such as the harmonic oscillator or infinite square well.

But how can the nucleons in the nucleus move freely on fixed orbits without bouncing off each other and exchanging energy? The Pauli exclusion principle is the key to the answer. The nucleons are in motion in a mean potential field (called a self-consistent field) generated by all of them. The lowest energy levels are all occupied. So a nucleon in one of these cannot undergo a collision and exchange energy with another nucleon in which each would move to a nearby level because they are all filled. To go to a higher, unfilled shell still requires the other nucleon to go to a lower-energy filled shell. Occasionally, collisions do occur which result in the excitation of one of the nucleons into an energy level above the Fermi level. However, since this violates energy conservation by an amount ΔE^*, it can only occur for a short time according to the uncertainty principle where $\Delta t = \hbar/\Delta E^*$. Then it must return to the original state. Thus, nucleons move freely and independently in the nucleus, keeping their particular energy level.

The great success of the shell model is attributed to this fact that nucleons can scatter out of their orbitals for only very short times (within the limits of the uncertainty principle) because of energy conservation and the Pauli principle. However, even "very short time" means that nucleons do not occupy their mean field orbits all the time. The deviation from full occupancy thus provides a quantitative measure of the inadequacy of the mean field description. But can we measure such deviations experimentally? By 1995, experiments could provide evidence that some orbitals are occupied with only 70 percent probability.

Thus, 45 years after the introduction of the nuclear shell model, we have experimental results establishing some limits on a mean field description. Now there is a focus on multiparticle correlations as a new way of going beyond the mean field theory. Studies of multinucleon correlations has been an emphasis of the continuous electron beam accelerator at Newport News, Virginia.

SUCCESSES OF THE MAYER–JENSEN SHELL MODEL. In the first extreme single-particle shell model with spin–orbit coupling, the nucleons moved in a mean potential and the properties of the nucleus were those of the single unpaired nucleon, all the paired nucleons contributing nothing. This model had some success but too many failures. The single-particle shell model then evolved so that the nucleus consisted of a closed-shell core, where the orbitals for N and Z were filled, and the interactions of the "valence" nucleons outside the core were included to give rise to the nuclear properties. In contrast, an individual (or independent) particle model has all the nucleons in the nucleus taken into account.

One of the major early successes of the nuclear shell model was the prediction of the spins and parities of most nuclear ground states rather well. For nuclei with one nucleon outside of a closed shell or one nucleon vacancy (hole) in a closed shell which also acts like a particle, the nuclear ground-state spin and parity are determined by the extra nucleon (or nucleon hole) because the nucleons inside of the closed shell have zero angular momentum just as in atoms (see Chapter 5). For example, ${}^{17}_{8}O_9$ and ${}^{15}_{8}O_7$ have respectively one nucleon outside of and one hole in the closed $p_{1/2}$ shell ($N = 8$). The "extra" nucleon in ${}^{17}O$ is in the $d_{5/2}$ shell, which is the next available level (see Fig. 11.9) and the hole in ${}^{15}O$ is in the $p_{1/2}$ shell. The notation here is the same as for atoms, l_j, where $l = 0, 1, 2, 3, \ldots$ are labeled s,p,d,f,.... For ${}^{17}O$ and ${}^{15}O$, the angular momentum must be $\frac{5}{2}$ (in units of $\hbar$) and $\frac{1}{2}$, respectively. Their parity is determined by $(-1)^l$, so $l = 2$ is positive (even) and $l = 1$ is negative (odd). These $\frac{5}{2}^+$ and $\frac{1}{2}^-$ I^π shell model assignments are consistent with experiments. A few other examples are given in Table 11.1.

When there are two nucleons outside of a closed shell, according to the

TABLE 11.1
Shell model predictions of l_j and experimental ground-state spins and parities of some nuclides. A −1 indicates a hole state

Nuclides	Shell model predictions	Experimental results, spin-parity
${}^{17}_{8}O_9$	$d_{5/2}$	$\frac{5}{2}^+$
${}^{17}_{9}F_8$	$d_{5/2}$	$\frac{5}{2}^+$
${}^{39}_{19}K_{20}$	$d^{-1}_{3/2}$	$\frac{3}{2}^+$
${}^{123}_{51}Sb_{72}$	$g_{7/2}$	$\frac{7}{2}^+$
${}^{209}_{82}Pb_{127}$	$g_{9/2}$	$\frac{9}{2}^+$
${}^{209}_{83}Bi_{126}$	$h_{9/2}$	$\frac{9}{2}^-$

angular momentum coupling rule in Chapter 5 the composite total angular momentum may have many values. For example, if there are two neutrons in a $1g_{7/2}$ shell, the values of the composite angular momenta may be 0, 1, 2, ..., 7; but the Pauli principle reduces the allowed values to 0, 2, 4, 6. Inside the nucleus, the situation is simpler yet. From experiments we know that all even–even nuclei have zero total spin. How can this be if the nucleons have spin $\frac{1}{2}$ and various orbital angular momenta? To explain this and other phenomena, a strong pairing interaction between nucleons was introduced. Thus, protons and neutrons form pairs with both their spins and orbital angular momenta opposed, so that $J = 0$ for the pairs. So, all even–even nuclei have ground-state spin zero and parity positive. As noted earlier, the BCS theory of superconductivity was subsequently introduced into nuclear physics to explain nucleon pairing. This introduction gave rise to pairing gaps in nuclei for both protons and neutrons just as for electron pairs in superconductors.

Hence, the nucleons inside of closed shells do not make any contribution to the angular momentum, nor do even numbers of neutrons or protons outside closed shells contribute to nuclear ground states. This indicates that, in addition to the average interaction from the net potential and the spin–orbit force, there is a strong pairing interaction which strongly favors the pairing of protons and neutrons in $J = 0$ states. This is a residual nuclear interaction that is not described by the spherically symmetric net potential of the shell model or by the spin–orbit interaction. The large energy gain from pairing is why nuclei strongly prefer to have even Z and even N and is the basis of the pairing term in the semi-empirical mass formula.

The nuclear ground state spin and parity in odd–even nuclei are determined by the unpaired particle, since all pairs couple to $J = 0$. For example, ${}^{93}_{41}Nb_{52}$, its two neutrons which are in the $1g_{7/2}$ state outside of the closed shell 50 have a composite J of zero. Its nuclear spin is determined by its one proton in the $1g_{9/2}$ orbital outside of the at least semi-magic closed shell of 40 protons. Thus, the shell model predicts that the ${}^{93}Nb$ ground state has spin and parity $\frac{9}{2}^{+}$, in agreement with experiment.

The nuclear spins of odd–odd nuclei are determined by the coupling between the final unpaired odd neutron and odd proton. Because the proton and neutron intrinsic spins are $\frac{1}{2}$ (so their total $s = 0$ or 1) and the orbital angular momentum is always an integer, the coupling results must be an integer. Only nine odd–odd nuclei are found in nature, again strong evidence for the importance of the nuclear pairing force. Four of them are stable, and their ground-state spin is 1 or 3. The other five are unstable nuclides whose spins are 2, 4, 5, 6, or 7. A great many short-lived radioactive odd–odd nuclei have been produced in the laboratory and without exception their spins also are integers.

The allowed quantum energy states in the spherical shell model are quite different for even–even and odd-A nuclei. In even–even nuclei, the pairing force causes all the protons and neutrons to form pairs with zero angular momentum. The energy needed to break a pair of particles in order to promote a particle to the next highest orbital is large, so the low-lying energy states of near-spherical even–

even nuclei with N and/or Z near the magic numbers are generated by collective vibrations of the nucleus as a whole. (Collective motion is treated in the next section.) However, when both N and Z are magic numbers, as in $^{16}_{8}O_{8}$, $^{40}_{20}Ca_{20}$ and $^{208}_{82}Pb_{126}$, these nuclei have such strong spherical structures that the vibrational modes are very high in energy, and indeed the lowest states are formed by the breaking of a pair with the promotion of one particle to a higher orbital, leaving a hole, to form a particle–hole excitation, or through the promotion of a pair (or pairs) of particles to a higher orbital (a two-particle–two-hole state, 4p–4h, ...). The energies of the first excited states in spherical double magic nuclei are exceptionally large, and the pattern of energies and spins and parities is quite different from any other nuclei. For example, the first and second excited states for $^{16}_{8}O_{8}$ are 0^+ 6.05 MeV and 3^- 6.13 MeV; for $^{40}_{20}Ca_{20}$, 0^+ 3.35 MeV and 3^- 3.73 MeV; and for $^{208}_{82}Pb_{126}$, 3^- 2.61 MeV and 5^- 3.20 MeV. The marked difference between the levels of spherical double magic $^{208}_{82}Pb_{126}$ and $^{132}_{50}Sn_{82}$ shown in Fig. 11.10a and those of other nuclei can be seen by comparing them with those of a typical near-spherical, even–even vibrational nucleus shown later in Fig. 11.15 and of even–even deformed nuclei shown in Fig. 11.12a,b. In Fig. 11.10b are shown the levels in double magic $^{40}_{20}Ca_{20}$. However, here one sees additional states that are not predicted by the spherical shell model.

In contrast, in odd-A nuclei, energy levels can be produced by the promotion of the odd particle to the higher orbitals. One can also break a pair and form a multiparticle or multiparticle–hole configuration in odd-A nuclei, too. For example, three neutrons outside a closed shell could form a three-particle state. In general, breaking of pairs and promotion of particles to higher-energy orbitals leads to a higher excitation energy for such a state. However, when the nucleus becomes deformed, the energy of an orbital can be significantly lowered or raised as a function of the size of the deformation, to open up different orbital occupations at low energy as we shall see later.

A major success of the nuclear shell model was the prediction of the islands of nuclear isomerism in regions around closed shells. Consider Fig. 11.9. When the odd N or Z is around 38–52, the $p_{1/2}$ and $g_{9/2}$ orbitals are close and the $g_{7/2}$ is not too far away. Similarly around 82, one has the $s_{1/2}$, $d_{3/2}$, and $h_{11/2}$ orbitals close in energy. In these regions the odd particle may have one of these orbitals for the ground state and the first excited state of the nucleus may be the other orbital. Since they are so different in spin, their γ-ray decay is $E3(g_{7/2} \rightarrow p_{1/2})$ or $M4(h_{11/2} \rightarrow d_{3/2})$ which have long lifetimes for decay. Thus, the excited state is isomeric. Examples of the many isomers found around these closed shells include radioactive $^{91}_{42}Mo_{49}$ with a $\frac{9}{2}^+$ ground state and 653 keV, $\frac{1}{2}^-$ 1.1 min isomeric state; stable $^{107}_{47}Ag_{60}$ with a $\frac{1}{2}^-$ ground state and 93 keV, $\frac{7}{2}^+$ 44 s isomeric state; and stable $^{137}_{56}Ba_{81}$ with a $\frac{3}{2}^+$ ground state and a 662 keV, $\frac{11}{2}^-$, 2.6 min isomeric state. Remember that the isomeric state can be formed either by promoting the odd particle to a higher orbital or by breaking a pair in a lower orbital, with one particle forming a pair with the particle in the original orbital and leaving the other particle in the next lower orbital in Fig. 11.9.

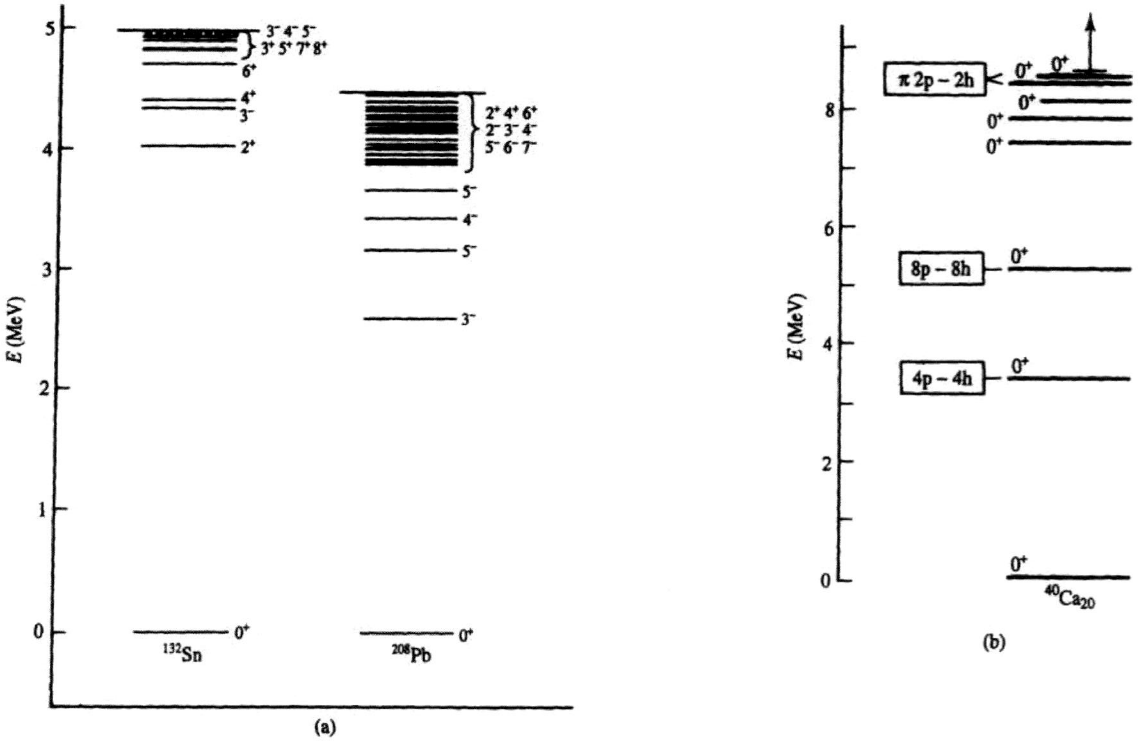

FIGURE 11.10
(a) Energy levels in the more recently discovered far-off-stability radioactive nucleus $^{132}_{50}Sn_{82}$ and the well-known spherical double magic stable nucleus $^{208}_{82}Pb_{126}$ found in nature are remarkably similar, establishing that both are spherical double magic nuclei. (b) Energy levels in spherical double magic $^{40}_{20}Ca_{20}$. In addition to the spherical shell model states expected in this double magic nucleus, one also sees other states that cannot be explained in that model. These states arise from the promotion of pairs of particles from the filled shells below the shell gap to intruder orbitals from above the gap. The intruder orbitals drop in energy with nuclear deformation. These 2-particle–2-hole, 4p–4h, and so forth, states as shown have significant deformation.

Another triumph of the nuclear shell model was the correct prediction of the sign and roughly the magnitude of ground-state magnetic moments of nuclei. For one nucleon in a shell model orbital, the magnetic moment, μ_{sp}, is

$$\mu_{sp} = j\left(g_l \pm (g_s - g_l)\frac{1}{2l+1}\right)$$

for $j = l \pm \frac{1}{2}$, and the orbital (g_l) and spin (g_s) g-factors are $g_l = 1, 0$ and $g_s = 5.58$, -3.82 for protons and neutrons, respectively. For example, for $^{17}O(d_{5/2})$,

$$\mu_{sp} = \frac{5}{2}\left(0 + (-3.82 - 0)\frac{1}{2 \times 2 + 1}\right) = \frac{-5}{10} \times 3.82 = -1.91$$

to be compared with the observed value of -1.89. For $^{17}F(d_{5/2})$, $\mu_{sp} = 4.79$ and $\mu_{ob} = 4.72$. The agreement is not as good quantitatively in heavy nuclei because of the polarization effect of the closed shells: $^{209}Bi(h_{9/2})$ $\mu_{sp} = 2.62$, $\mu_{ob} = 4.08$.

The shell model predictions of the quadrupole moments, Q, are generally in agreement with experiment near the spherical magic numbers; in particular, one can understand the sign change in Q in going from an odd-Z nucleus just below a magic number to one just above as seen in Fig. 11.11 for 8, 20, 28, 40, 50, 82, and 126. For a single proton in an orbital nl_j,

$$Q_{sp} = \frac{-(2j-1)}{2(j+1)}\langle j|r^2|j\rangle$$

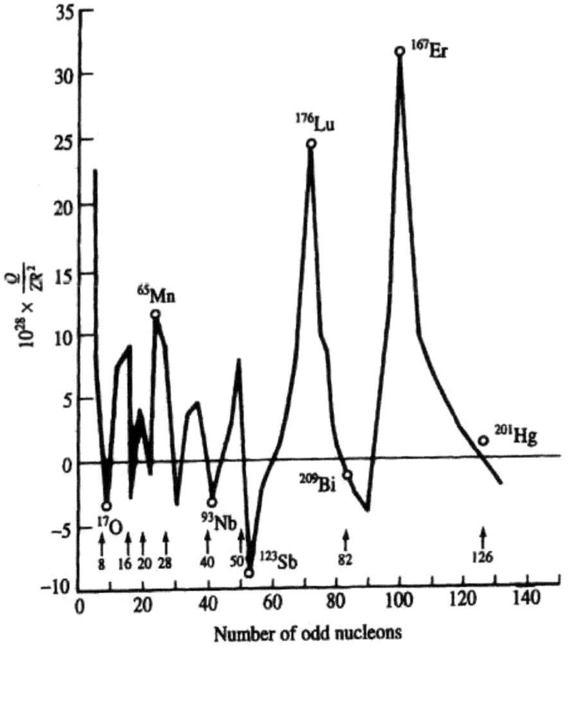

FIGURE 11.11
Reduced nuclear quadrupole moments as a function of the odd Z or odd N number of nucleons. The quantity Q/Zr^2 gives a measure of the nuclear deformation independent of the size of the nucleus.

TABLE 11.2
Experimental values of the ground-state quadrupole moments of a few selected nuclei and values calculated in a spherical shell model

Nuclides	Character	j	$Q_{exp}(\mathrm{fm}^2)$	$Q_{shell}(\mathrm{fm}^2)$	Q_{exp}/Q_{shell}
${}^{17}_{8}O_9$	Double magic + 1 neutron	$\frac{5}{2}$	−2.6	−0.1	26
${}^{39}_{19}K_{20}$	Double magic + 1 p (hole)	$\frac{3}{2}$	5.5	5	≈ 1
${}^{175}_{71}Lu_{104}$	Between two shells	$\frac{7}{2}$	560	−25	−22
${}^{209}_{83}Bi_{126}$	Double magic + 1 p	$\frac{9}{2}$	−35	−30	≈ 1

Q_{sp} depends on the average value of r^2, where for $j \gg 1$, $Q_{sp} \simeq \langle r^2 \rangle$, so the charge distribution is completely concentrated in the equatorial plane. For a proton hole in a closed shell, the quadrupole moment changes sign to $-Q_{sp}$. A neutron carries no charge, so a single-neutron configuration has no quadrupole moment. A hole in an outer shell orbital of a spherical nucleus (unfilled hole around the middle) makes the nucleus look prolate ($+Q$), and an extra particle outside a spherical core makes a nucleus look somewhat pancakelike, i.e., oblate ($-Q$). However, the shell model predictions are often drastically different from those found in experiments in other regions (see Table 11.2). The quadrupole moments are much larger than predicted in nuclei where both Z and N are far from any spherical shell model magic number, as occurs in rare-earth nuclei ($50 < Z < 82, 82 < N < 126$) and actinide nuclei ($82 < Z < 126, 126 < N$). In Fig. 11.11 one sees the large quadrupole moments for these nuclei. Also, the shell model transition rates for electromagnetic decays between nuclear energy levels are slower than the experimental ones in certain types of cases and much faster than predicted in others. Moreover, in Fig. 11.10b, even in spherical double magic ${}^{40}_{20}Ca_{20}$, additional states not predicted in the spherical shell model are seen. These failures of the shell model brought about nuclear collective models. However, even without the above limitations the spherical shell model quickly reached computational limitations. In the shell model one starts with a nuclear mean field and adds the effects of the residual interactions among the valence nucleons outside the closed shells. Earlier shell models had serious computational difficulties beyond the sd shell. In a nucleus like ${}^{164}Er$, one has the order of 10^{32} configurations!

11.2.3 The Collective Model

ROTATIONAL MOTION IN DEFORMED NUCLEI. As noted above, when N and Z are well removed from spherical magic numbers, for example, in the rare-earth and actinide nuclei, the quadrupole moments calculated using the spherical shell model are much less than the experimental ones. How are these large quadrupole moments explained?

In Section 9.3 we introduced the concept of the electric quadrupole moment. The quadrupole moment of a charge system is a measure of the deviation of its charge distribution from a spherical shape. If a nucleus has a large quadrupole moment, its shape deviates significantly from spherical. J. Rainwater in 1950 pointed out that if a nuclide has a large quadrupole moment, its closed core is never spherical as expected in the shell model. It is deformed by the valence nucleons. Since most of the nucleons in the nucleus are in the core, the core has most of the charge, so even a very small deformation produces a large quadrupole moment. For example, for ^{17}O, if we assume that it is an ellipsoid with equal short axes and the difference between its long and short axes is only 7 percent, the calculated quadrupole moment is consistent with experiments.

The nuclear deformation not only has an effect on the quadrupole moment but also gives rise to new nuclear degrees of freedom and motion. These new motions clearly alter the nuclear energy level diagram. For example, in quantum mechanics, there is no rotational motion in spherical nuclei, but there is in deformed nuclei. Why can rotational motion occur in deformed nuclei? In spherical nuclei, any axis which goes through the center is a symmetry axis. When a spherical body turns angularly through an angle, ϕ, this cannot change the wavefunction. That is

$$\frac{\partial \psi}{\partial \phi} = 0$$

If we take this axis as the z axis, from Eq. (8.64), the angular momentum z component is zero:

$$\hat{L}_z = -i\hbar \frac{\partial}{\partial \phi} = 0$$

There is no way to distinguish the fact that the nucleus has rotated since there is no spatial direction (axis) from which to measure the rotation. So there is no observable collective rotational motion for a spherical system in quantum mechanics.

If a nucleus is a symmetrical ellipsoid with a permanent deformation (shaped somewhat like a rugby football), then again there is no way to distinguish a different rotational position about the long z symmetry axis. So since there is no way to observe it, there is no collective rotation about this axis. Intrinsic or single-particle angular momentum can point along the symmetry axis, but not rotational angular momentum. However, collective rotations around the x or y axes are allowed, because one can measure the rotation with respect to the direction of the symmetry axis. This is analogous to saying that nuclei cannot undergo rotational motion around the long axis like the rotation of a forward pass in American football, but can tumble end-over-end like a football when it is place-kicked. To illustrate, assume that the total nuclear angular momentum is zero (an even–even nucleus). Consider rotation around the x axis where the nuclear rotation angular momentum is $\boldsymbol{R}$. Then the rotation energy is

$$E_{\text{rot}} = \frac{R^2}{2\Im} \tag{11.25}$$

where $\Im$ is the rotational moment of inertia around the x axis. From quantum mechanics, we can write the Schrödinger equation

$$\frac{\hat{R}^2}{2\mathfrak{I}}\psi = E\psi \tag{11.26}$$

The operator $\hat{R}^2$ is given in Eq. (8.65). Its eigenvalue and eigenfunction come from the following equation [see Eq. (8.97)]:

$$\hat{R}^2 Y_{I,M} = I(I+1)\hbar^2 Y_{I,M}, \qquad I = 0, 1, 2, \ldots \tag{11.27}$$

A zero-spin nucleus is invariant to an x–y plane reflection and the parity of the harmonic $Y_{I,M}$ with odd I changes its sign under reflection, so it cannot be taken as an eigenfunction. The allowed I are only even. From Eqs. (11.26) and (11.27) we have

$$E_I = \frac{\hbar^2}{2\mathfrak{I}} I(I+1), \qquad I = 0, 2, 4, \ldots \tag{11.28}$$

This is the nuclear rotational energy formula that Bohr and Mottelson proposed in 1953. It is the same formula that was known to describe a "dumbbell" molecule with two connected atoms rotating about an axis perpendicular to the line joining the two atoms. Soon such nuclear rotational spectra were observed in rare-earth and actinide nuclei. Figure 11.12a is an early example noted by Bohr and Mottelson. More generally one has

$$E_I(\text{rotation}) = AI(I+1) + BI^2(I+1)^2 + CI^3(I+1)^3 + \ldots \tag{11.29}$$

where A is the intrinsic matrix element and the higher-order corrections $B, C, \ldots$ are inertial parameters that characterize the nonadiabaticity of the rotation. Interactions between the rotational motions and vibrational motions (to be discussed) can give rise to B, for example.

The experimental first excited state energy (93 keV) in ^{180}Hf was used to determine the parameter $\hbar^2/2\mathfrak{I}$:

$$93.3\ \text{keV} = \frac{\hbar^2}{2\mathfrak{I}} \times 2 \times 3 \qquad \text{so} \qquad \frac{\hbar^2}{2\mathfrak{I}} = 15.55\ \text{keV}$$

With this parameter and Eq. (11.28) the 4^+, 6^+, 8^+ energies were calculated as shown in Fig. 11.12a (in parenthesis) e.g., $E_4 = 15.55\ \text{keV} \times 4 \times 5 = 311$ keV. The calculated values are in reasonable agreement with the experimental ones, but there are still some deviations. These deviations have essentially been removed (as shown in brackets) in subsequent, more detailed calculations in which the interaction of the nuclear rotational motion with the nuclear vibrational motion is taken into account to give higher-order corrections to Eq. (11.28).

Figure 11.12b is an example of more recent studies of a well-deformed even–even nucleus and Fig. 11.12c of an odd-A nucleus. Two additional bands of levels labeled $K^\pi = 0^+$ (β vibration) and $K^\pi = 2^+$ (γ vibration) are seen in Fig. 11.12b along with negative parity bands of levels. These are rotational bands built on vibrational excitations, as discussed next. You will also see in Fig. 11.12b,c a new quantum number, K, introduced to describe a state in a deformed nucleus. The K quantum number is the intrinsic angular momentum of a state along its symmetry axis. An intrinsic level of spin $I = K$ can have a rotational band built on it, where the rotational angular momentum must be along an axis perpendicular to the symmetry axis. As shown in Fig. 11.13, the intrinsic angular momentum can be that of a single particle, j, as shown or from the intrinsic structure with $K^\pi = 0^\pm, 1^\pm, 2^\pm, \ldots$ in an even–even nucleus. Combining the intrinsic angular momentum (K) and rotational angular momentum (R) gives the total angular momentum of a state, where we find $I^\pi = 0^+, 2^+, 4^+ \ldots$ or $1^-, 3^-, 5^- \ldots$ for $K = 0$, and $I = K, K+1, K+2, \ldots$ when $K \neq 0$. The energies of the levels in the rotational band are given by

$$W_I = W_0(K) + \frac{\hbar^2}{2\mathfrak{I}}[I(I+1) + a(-1)^{I+1/2}(I + \tfrac{1}{2}\delta_{K,1/2}] \tag{11.30}$$

which is based on the energy relationship of rotational levels in a linear molecule. Here, $W_0(K)$ depends on the intrinsic structure of the band; the term in brackets gives the energies of the rotational states built on this intrinsic state, and δ is a delta function, which is 1 if $K = \frac{1}{2}$ and 0 if $K \neq \frac{1}{2}$. The $I(I+1)$ term occurs because the rotational motion is strongly coupled only to total angular momentum states for which the orbital angular momentum is not zero. For $K = \frac{1}{2}$, only the spin angular momentum is not zero, and a decoupling of the rotational motion occurs where a, the decoupling parameter, measures the degree of decoupling. One can still use this equation to extract $\mathfrak{I}$ from the differences in energies of two levels and to predict the energies of the higher members. The ground states of even–even nuclei have $K = 0$. For $K = \frac{1}{2}$, $(-1)^{I+1/2}$ creates a signature term that shifts the energies of the $\frac{1}{2}, \frac{5}{2}, \ldots$ states compared to the $\frac{3}{2}, \frac{7}{2} \ldots$ members as seen in Fig. 11.12c. The magnitude and direction of the signature splitting depends on the size and magnitude of a. For $K \geq \frac{3}{2}$, smaller signature splittings also are observed from K mixing from Coriolis effects.

The shape of a deformed nucleus can be described by the length of the radius vector pointing from the origin to the surface with an expansion in terms of spherical harmonics:

$$R(\theta', \phi') = R_0[1 + \sum_{\lambda=0}^{\infty} \sum_{\mu=-\lambda}^{\lambda} \alpha_{\lambda\mu} Y_{\lambda\mu}(\theta', \phi')] \tag{11.31}$$

where R_0 is the radius of a sphere with the same volume ($r_0 A^{1/3}$) and θ' and ϕ' are the polar angles with respect to an arbitrary coordinate set. The constant α_{00} describes changes in the nuclear volume; since this is considered constant (the incompressibility of nuclear matter), $\alpha_{00} = 0$. Since $\lambda = 1$ describes a translation of the system as a whole, $\alpha_{1\mu} = 0$. The lowest order $\lambda = 2$ is quadrupole deformation. There is also evidence for higher-order deformations: $\lambda = 3$, octupole; $\lambda = 4$, hexadecapole; and higher ones. If one transforms the system to the body-fixed system whose three axes 1,2,3 coincide with the principal axes of the nuclear mass distribution via a matrix $D^\lambda_{\mu\mu}$ that describes the rotation of the three Euler angles, one has

$$a_{\lambda\mu} = \sum_{\mu'} \alpha_{\lambda\mu'} D^\lambda_{\mu\mu'} \tag{11.32}$$

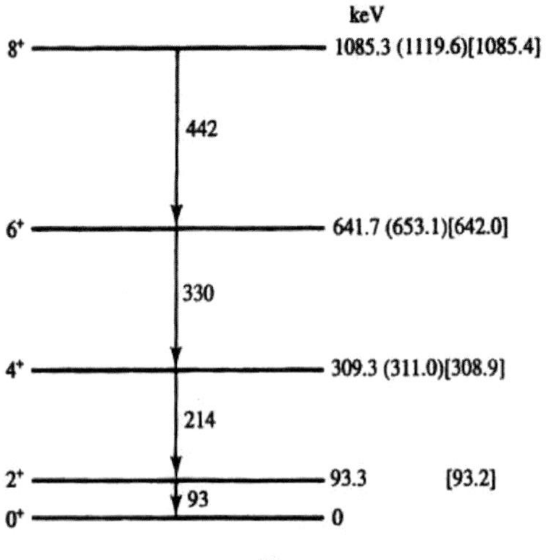

(a)

FIGURE 11.12
(a) An early example of a band of rotational energy levels observed in ^{180}Hf. The energies in parentheses are based on $\hbar^2/2\mathfrak{I}$ as determined from the 2^+ energy and Eq. (11.28). The energies in brackets include the effects of rotation–vibration interaction. (b) The low-energy rotational and vibrational band structures with K^π shown for ^{232}Th as determined in several Coulomb excitation measurements at GSI in Germany. The $K = 4^+$ band is proposed to be the two-phonon γ-band. The K^π values for each rotational band are given.

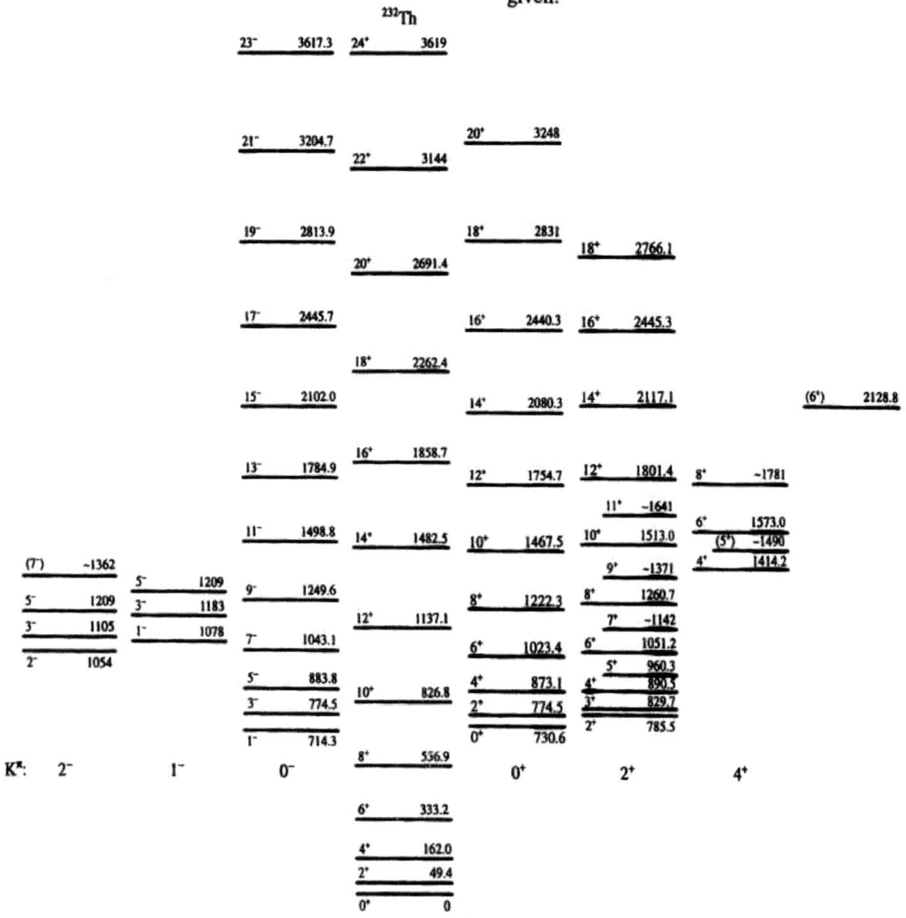

(b)

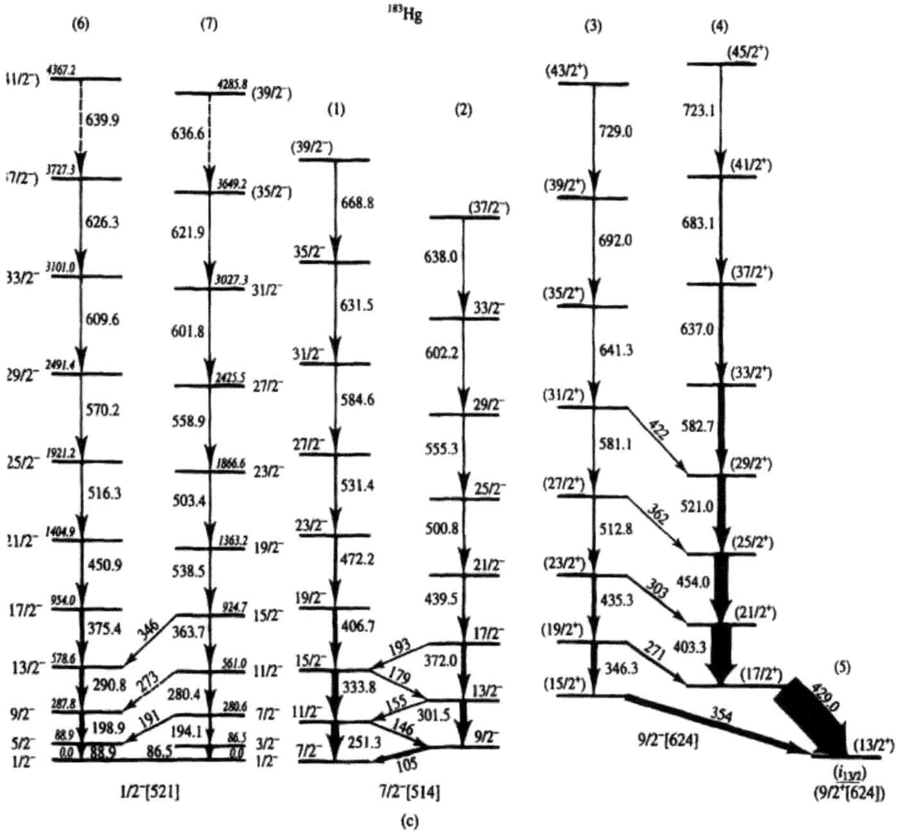

FIGURE 11.12
(c) Rotational bands built on the Nilsson single-particle orbitals, which are given below each band. Note the signature splitting in the $\Omega = K = \frac{1}{2}$ band on the left (K is the spin of the band head, e.g., $\frac{1}{2}, \frac{7}{2}$ for the four bands on the left) (J. K. Deng, Ph.D. thesis, Vanderbilt University, 1993).

So for quadrupole deformation, the radius of the surface is defined in terms of the angles θ and ϕ in the body-fixed system:

$$R(\theta, \phi) = R_0[1 + \sum_{\mu} a_{2\mu} Y_{2\mu}(\theta, \phi)] \tag{11.33}$$

Since the body-fixed axes are the principal axes of the nucleus, the five $a_{2\mu}$ can be expressed in terms of three Euler angles and a_{20} and $a_{22} = a_{2\,-2}$ with $a_{21} = a_{2\,-1} = 0$.

The physical description of the system is more conveniently expressed in terms of the variables β and γ where

$$\begin{aligned} a_{20} &= \beta_2 \cos\gamma \\ a_{22} &= \frac{1}{\sqrt{2}} \beta_2 \sin\gamma \end{aligned} \tag{11.34}$$

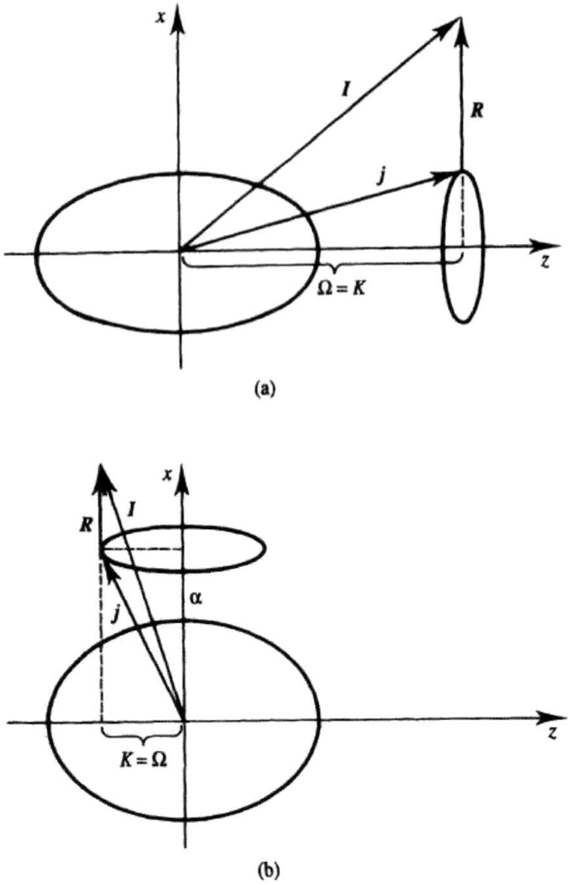

FIGURE 11.13
As shown, K is the projection or component of the intrinsic angular momentum of a state on the symmetry axis of a deformed nucleus. Here the intrinsic angular momentum is that of a single particle with angular momentum j and its projection K on the symmetry axis for a single particle is also called Ω. R is the rotational angular momentum, which is perpendicular to the symmetry axis. The total angular momentum I is the vector sum of these angular momenta. (a) An orbital with large Ω; (b) an orbital with low Ω.

and

$$\sum_{\mu} |a_{2\mu}|^2 = \beta_2^2 \tag{11.35}$$

Thus, β_2 is the measure of the total deformation of the nucleus.

$$R(\theta, \phi) = R_0\left[1 + \beta_2 \cos\gamma\ Y_{20}(\theta, \phi) + \frac{1}{\sqrt{2}}\beta_2 \sin\gamma\ \{Y_{22}(\theta, \phi) + Y_{2-2}(\theta, \phi)\}\right] \tag{11.36}$$

The values $\gamma = 0^\circ$ and $\pm 120^\circ$ correspond to prolate ellipsoids where axis 3 is the symmetry axis and 1, 2 are equal to give a shape like an American or rugby football, with the symmetry axis being the long axis; $\gamma = \pm 60^\circ$ and 180° correspond to oblate ellipsoids which have shapes like a discus or pancake, with the symmetry axis being the shortest axis; and γ not a multiple of 60° corresponds to a triaxial shape with no symmetry axis. For our purposes we need consider only γ values from 0 to 60°.

The differences of the three axes from the radius of a sphere with the same volume in the body-fixed frame are

$$\delta R_\kappa = R_0\sqrt{\frac{5}{4\pi}}\,\beta_2\cos\left(\gamma - \frac{2\pi}{3}\kappa\right) = R_\kappa - R_0 \tag{11.37}$$

where $\kappa = 1, 2, 3$.

For $\gamma = 0$ one finds for R_1, R_2 and R_3 that, from Eq. (11.37),

$$\begin{aligned} R_1 = R_2 &= R_0\left(1 - \frac{1}{2}\sqrt{\frac{5}{4\pi}}\,\beta_2\right) \\ R_3 &= R_0\left(1 + \sqrt{\frac{5}{4\pi}}\,\beta_2\right) \end{aligned} \tag{11.38}$$

It can be shown that

$$\beta_2 \sim \frac{R_3 - R_1}{(R_3 + R_1)/2} \tag{11.39}$$

so $\beta_2 > 0$ is a prolate ellipsoid ($\gamma = 0°$) and $\beta_2 < 0$ is an oblate ellipsoid ($\gamma = 60°$). For the rare-earth and actinide nuclei, β_2 ranges from 0.22 to 0.26 for prolate shapes.

One can plot the nuclear potential energy surface for a given nucleus with Z, N in the β–γ plane, as shown in Fig. 11.14, where β is measured by the distance out from the origin (sphere) and γ is the angle between the $\gamma = 0$ prolate and

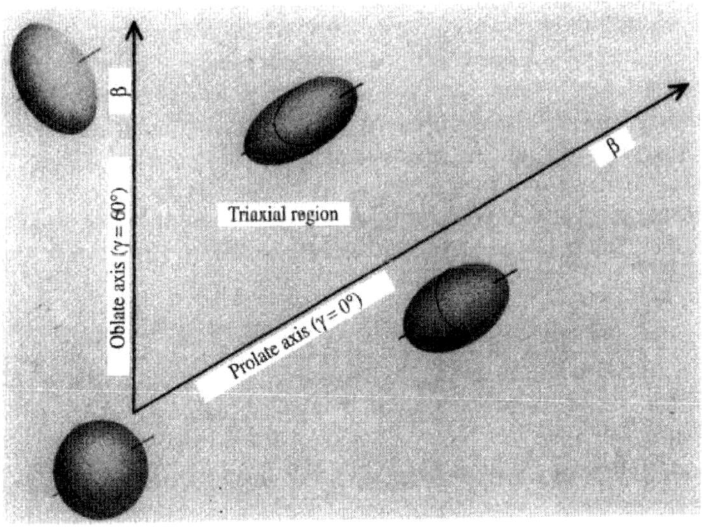

FIGURE 11.14
Various nuclear shapes are shown in the β–γ plane between $\gamma = 0$ on the prolate shape axis and $\gamma = 60°$ oblate shape axis. In between, nuclei are spherical. At the origin a nucleus is spherical.

$\gamma = 60^\circ$ oblate axis. The location of a minimum in the potential in this plane indicates the shape of the nucleus. The ground state of a nucleus occurs when its potential energy is a minimum. The existence of more than one minimum at different deformations in the potential (see Fig. 11.26) can lead to shape coexistence and isomers as discussed later.

Soon after Bohr and Mottelson introduced their collective model, S. G. Nilsson introduced the next step by calculating the proton and neutron single-particle energies in a deformed potential rather than the spherical potential used in the spherical shell model. The spherical single-particle levels then split and change energies rapidly as the deformation (β_2) changes, as illustrated in Fig. 11.24 in the next section. The single-particle energy levels are now labeled by a new set of asymptotic quantum numbers $[Nn_z\Lambda]\Omega^\pi$, where N is the principal quantum number, n_z is the number of nodes in the oscillations along the symmetry axis, Λ the component of orbital angular momentum along the symmetry axis, Ω is the projection of the single-particle angular momentum j on the symmetry axis (see Fig. 11.13), and π is the parity. The low-Ω orbitals are down-sloping with prolate deformation (see Fig. 11.23b) and the high-Ω orbitals are down-sloping for oblate deformation. Examples of rotational bands in odd-A nuclei are seen in Fig. 11.12c.

Lest there be some confusion, remember that β is used to identify three different phenomena discussed here and in the next section: (1) β decay, the process of the radioactive decay of a nucleus; (2) β vibration, where a prolate deformed nucleus vibrates in and out along its long symmetry axis; and (3) nuclear deformations ($\beta_2, \beta_4, \beta_6$), which are measures of the differences in deformation of the nuclear charge from a sphere.

Finally, note in Fig. 11.5b that the γ-ray reduced transition probabilities for electric quadrupole radiation, $B(E2; 0^+ \to 2^+)$, show strong enhancements, 50–200 times the single-particle estimates, in the regions where the quadrupole moments are large and E_{2^+} is small. These strong enhancements likewise indicate that strong collective effects are occurring, such as collective rotation. Thus, strong collective enhancements of the electromagnetic transition probabilities are another signature of large deformation. The superdeformed nuclei with much larger deformation ($\beta_2 \sim 0.6$) than rare-earth and actinide nuclei ($\beta_2 \sim 0.25$) have γ-ray transition probabilities on the order of 1000 times the single-particle estimates.

VIBRATIONAL MOTION IN SPHERICAL AND DEFORMED NUCLEI. Collective motion appears not only in deformed nuclei but also in spherical nuclei. In 1877, Rayleigh studied the vibrations of a liquid with electric charge. In 1936 N. Bohr pointed out that a particle system which is gathered together by an attraction can undergo collective vibration. About 1953, A. Bohr and B. Mottelson studied in detail the collective vibrations of both spherical and deformed nuclei. Figure 11.15 shows the theoretical prediction of the pure spherical harmonic vibrational spectrum of a spherical even–even nucleus. In a pure spherical harmonic vibrator, the energy levels are generated by one-phonon vibration

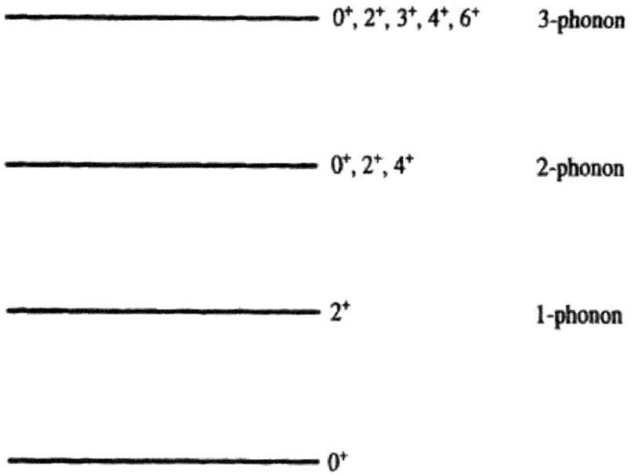

FIGURE 11.15
A theoretical vibrational spectrum for an even–even, spherical nucleus. The two-phonon energy is twice the one-phonon energy, and so forth. Anharmonicity will split the degenerate energies of two-phonon, three-phonon and higher phonon levels.

($\hbar\omega$), two-phonon vibration (at twice the energy of the one phonon, $2\hbar\omega$), three-phonon vibration (at $3\hbar\omega$), and so forth. The spins and parities of the energy levels of an even–even nucleus allowed by quantum mechanics, for example, are 2^+ (one quadrupole phonon), a degenerate $0^+, 2^+, 4^+$ triplet (two quadrupole phonon) and a degenerate $0^+, 2^+, 3^+, 4^+, 6^+$ quintet (three quadrupole phonon). The one-phonon energies of spherical and near-spherical nuclei are in the range of 0.6–1.0 MeV. These energies are much lower than the energy needed to break a pair of particles and to promote one to a higher orbital. Of course, no nucleus is like a pure harmonic vibrator. Real nuclei include additional terms in the nuclear potential (force) which lead them to be described as anharmonic vibrators. The anharmonic term removes the degeneracies of the two-phonon triplet, three-phonon quintet and higher phonons found for a pure harmonic vibrator (or oscillator). Thus, the appearance of all three members of the two-phonon triplet and the magnitude of the splitting between them are measures of the deviation of a nucleus from a pure harmonic vibrator. Many examples of two-phonon triplets with very small to large ($\sim$ 200 keV) splittings have been observed, along with some higher phonon members. Figure 11.16a gives the energy levels of ^{102}Ru which are in good agreement with the theoretical prediction for a vibrational nucleus as shown in Fig. 11.15. The levels built on the 0^+ excited state at 854 keV in ^{98}Zr (Fig. 11.16) also exhibit a pattern that is characteristic of one-, two-, and three-phonon vibrational states. However, subsequently excited states in ^{98}Zr have been investigated in the spontaneous fission of ^{252}Cf, and now a band of levels is observed to 16^+ built on the 854 keV 0^+ state and they look like a weakly deformed band (Fig. 11.16b). Such data underscore the difficulty of interpreting the structure of a nucleus from what seems like rather extensive data. Various approaches to include anharmonicities in the pure vibrator model have been developed. One of the more successful is called the variable moment of inertia model, where the energy levels of a nucleus are fitted by allowing the moment of inertia of a nucleus to be a function of two parameters.

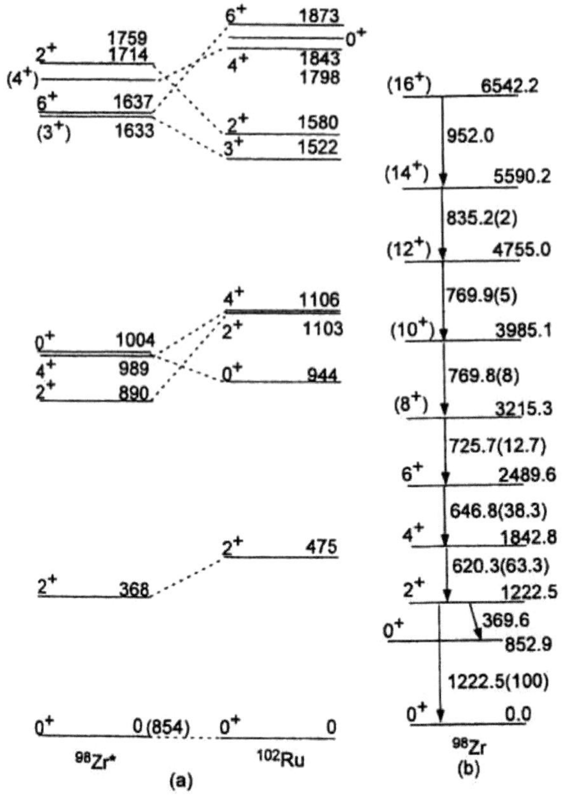

FIGURE 11.16
(a) One-, two- and three-phonon levels in ^{102}Ru report built on an excited 0^+ state in ^{98}Zr. [Fig. from R.A. Meyer et al., *Phys. Lett.* **B177**, 271 (1986)].
(b) High spin states in ^{98}Zr from spontaneous fission of ^{252}Cf. [J.H. Hamilton et al., *Prog. in Part. and Nuclear Physics* **35**, 635 (1995)].

As well as near-spherical nuclei, deformed nuclei can also vibrate. At low energies in deformed nuclei, three different types of vibrational states are observed: a quadrupole one-phonon β vibration, a quadrupole one-phonon γ vibration, and an octupole one-phonon vibration. As discussed, the shape of a deformed nucleus can be described by two variables, β and γ.

In Fig. 11.17 one can see that in a β vibration the nuclear shape is pulsating in and out along the long axis while maintaining a circular cross-section perpendicular to the long axis, thus changing the nuclear mean square radius (since such a vibration changes the deformation, β_2, it is called a β vibration). In a γ vibration, the short sides of the nucleus are moving in and out to generate a small deviation from spherical symmetry perpendicular to the long axis (since such a vibration slightly disturbs the symmetry when $\gamma = 0$ with little change in the mean square radius, it is called a γ vibration). In the octupole vibration, a nucleus takes on a reflection-asymmetric "pear" shape in lowest order and vibrates back and forth in a shape like that of an American or European pear, as shown (not Korean pears, which are round!). The one-phonon β, γ, and octupole vibrational energies are typically 0.5–1.5 MeV in deformed nuclei and so are much higher than the energies of the first few rotational states.

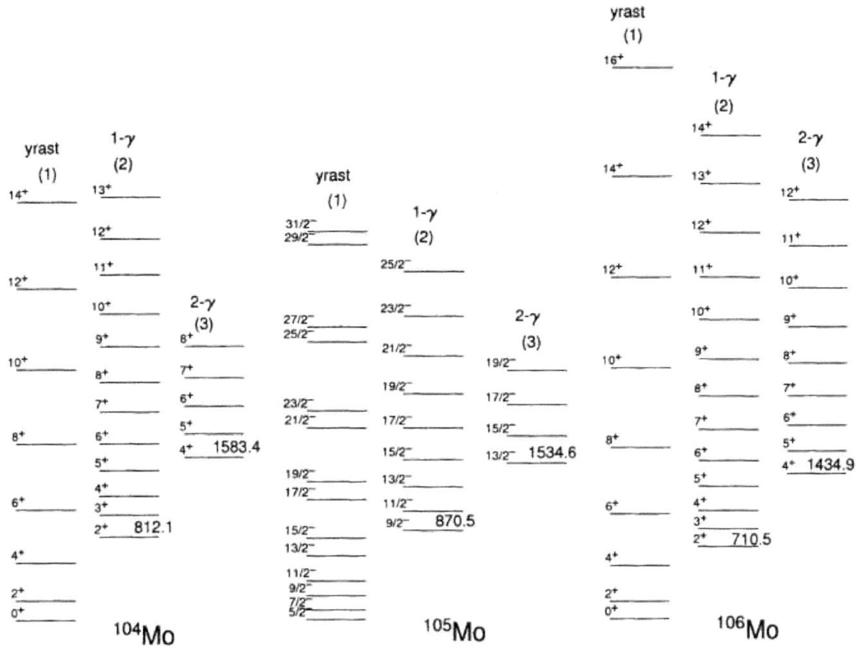

FIGURE 11.16(c)
The yrast ground band, one γ phonon and two γ phonon bands in ^{104,105,106}Mo (Tsinghua-Vanderbilt collaboration). Note the striking similiarities in energy of the one and two γ phonon bands in ^{105}Mo formed from coupling the odd neutron to ^{104}Mo, compared to these bands in ^{104}Mo.

Once a nucleus experiences such a vibration, it can have a rotational band built on each vibration with the allowed spins and parities as shown. Note the close analogy with molecular spectra, where a dumbbell molecule has both vibrational states and rotational bands built on these vibrational states. However, the dumbbell molecule has only the β-type vibration in which the atoms vibrate in and out along the line connecting the atoms. The discovery of such rotational bands built on the ground and β, γ, and octupole vibrations, as predicted by Bohr and Mottelson, provided beautiful confirmation of the theory. A comparison of the theoretical predictions in Fig. 11.17 with some recent data in Fig. 11.12b demonstrates the detailed agreement found between theory and experiment in strongly deformed nuclei. There is still controversy over whether two β-phonon and two-octupole-phonon vibrations in deformed nuclei exist or not. Two γ-phonon vibrational bands are seen, for example, in ^{232}Th (Fig. 11.12b), neutron-rich $^{104-106}$Mo,

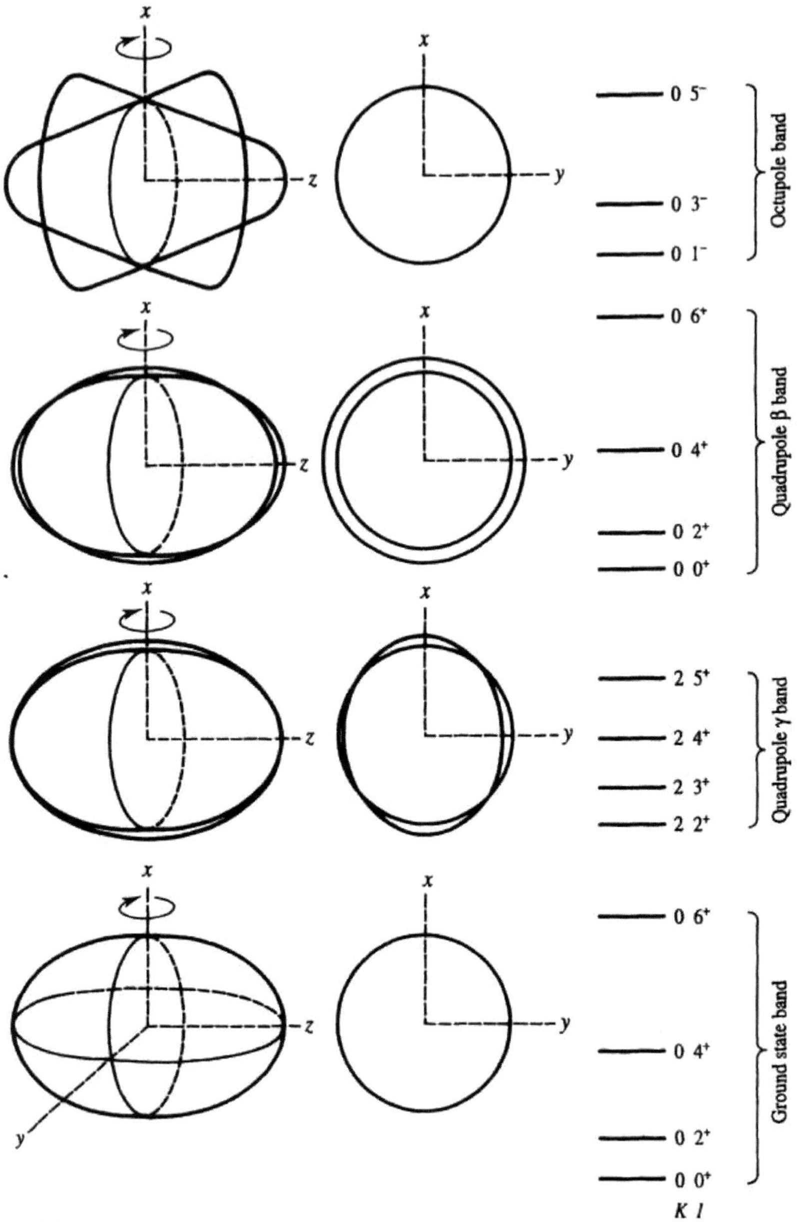

FIGURE 11.17
Illustration of the rotational bands built on the ground state and β, γ, and octupole vibrational states in a well-deformed nucleus.

Fig. 11.16c) ^{112}Ru, and for the first time in an odd-Z nucleus, ^{103}Nb with a possible 3γ phonon band. See J. Wang, *Phys. Lett. B* **675** (2009) 420.

Then came the discovery that the precise energies of the levels and the intensities of the connecting γ-rays are more complex than first thought, with interactions between the rotational and vibration motions producing perturbations just as in molecules and even perturbations produced by interactions between the β and γ vibrations. Nevertheless, the detailed successes of the Bohr–Mottelson model are quite striking. However, more complete data for these β and γ bands do not agree even with rotational–vibrational model predictions which include β–γ interactions, demonstrating the need for a more microscopic description of the β vibrational bands.[1,2] The anomalies in the branching ratios and M1 and E2 radiation mixing ratios in transitions from the β bands to the ground-state rotational bands were the first major failure of the predictions of the Bohr–Mottelson model as noted by Mottelson at the 1967 Nuclear Physics Conference in Tokyo, and called for a more microscopic description.[1,2]

At the same time, the long-standing prediction of the Bohr–Mottelson model that there should be large E0 admixtures in $\Delta I = 0$ transitions from the β band to ground band but not from the γ band to ground band was confirmed in deformed rare-earth nuclei.[3] Remember that the E0 transitions involve only internal conversion electrons and arise from the penetration of the atomic electrons into the nuclear volume. Thus, they are sensitive to changes in the root mean square radius of a nucleus. The large E0 transitions found in the $\Delta I = 0$ transitions out of the β bands in ^{154}Gd confirmed[3] that the β vibrations are along the symmetry axis and change the nuclear radius as predicted by the Bohr–Mottelson model. Large E0 admixtures are not found for transitions from the γ bands as predicted in the model. These data were an important first direct confirmation of the nature of the β and γ vibrational shape changes predicted in the Bohr–Mottelson model.

11.3 TOWARD A UNIFIED MODEL DESCRIPTION OF NUCLEI

11.3.1 Nuclear Shape Coexistence in the Region below $Z = 82$

By 1970 nuclei were pictured as having one of three shapes—spherical to near-spherical for double magic nuclei and those around the spherical magic closed

---

[1] J. H. Hamilton, *Radioactivity in Nuclear Spectroscopy*, J. H. Hamilton and J. C. Manthuruthil, eds., Gordon and Breach, New York (1972), p. 935.

[2] J. H. Hamilton et al., *Phys. Rev. Lett.* **22** (1969) 65; *ibid.* **23** (1969) 1178; *ibid.* **25** (1970) 946.

[3] J. H. Hamilton et al., *Proc. Int. Nuclear Physics Conf.*, ed. R. L. Becker et al., Academic Press, New York (1967), p. 919.

shells; well-deformed prolate shapes far from the closed shells; and a soft, not well-defined shape in the transition region between these two shapes.[4] Each nucleus had one fixed shape that it maintained in time, and its energy levels were characteristic of that shape. The different patterns of the energy levels associated with different even–even nuclear types (shapes), spherical double magic, near-spherical, soft transitional, and well-deformed nuclei, are illustrated in Fig. 11.18. Previously, even–even nuclei exhibited one of these patterns, and the pattern revealed its fixed shape and structure. The patterns shown there are seen for

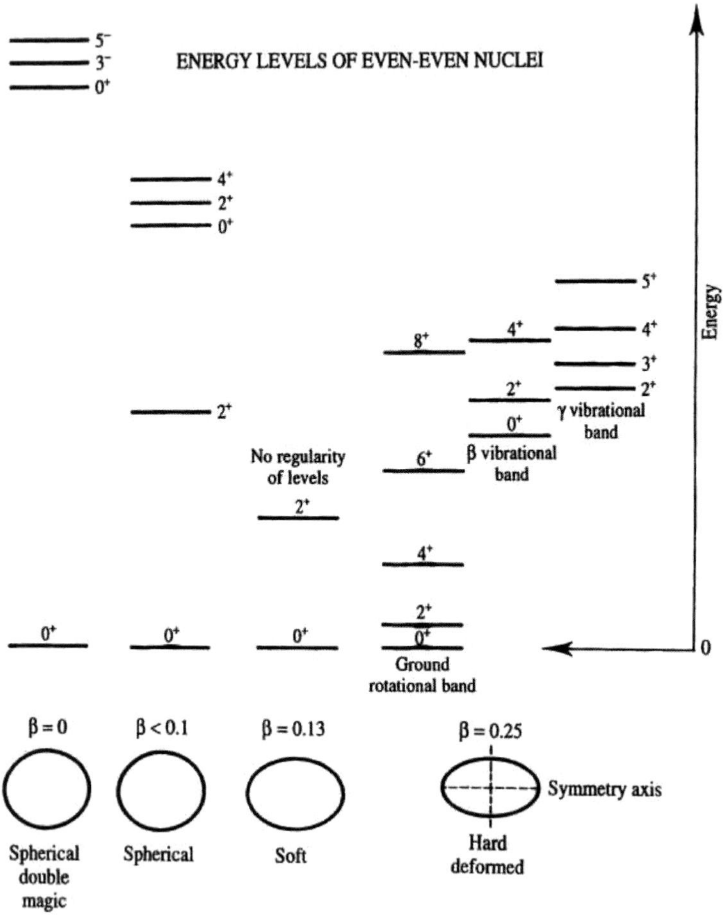

FIGURE 11.18
The patterns of the low-lying energy levels of even–even nuclei are illustrated going from left to right for (a) spherical double magic nuclei, (b) spherical and near-spherical vibrational nuclei, (c) soft transitional nuclei, and (d) well-deformed, prolate nuclei.

[4] M. Baranger and R. A. Sorenson, *Scientific American* (Aug. 1969), p. 58.

double magic nuclei as illustrated in Fig. 11.10; for vibrational even–even spherical and near-spherical nuclei in Fig. 11.16; and for the rotational bands built on the ground state and the β and γ vibrational band levels in a well-deformed nucleus in Fig. 11.12. The only regularity in even–even transitional nuclei is a first excited 2^+ state with an energy between those of vibrational and rotational nuclei. Much research has been devoted to developing a unified model that can describe all nuclei instead of having to use different models for different regions of N and Z. S. G. Nilsson took a major step in this direction when he calculated single-particle energies in a deformed potential to give single-particle energies as a function of deformation. One would also like to see such a model that included realistic nuclear forces to describe the nucleon–nucleon interactions.

Studies of nuclei which are very far from the valley of β stability are revealing an unexpected richness and diversity of nuclear shapes, structures, and decay modes which are transforming and challenging our understanding of the nucleus.[5,6] Discoveries of new decay modes such as proton radioactivity of nuclear ground states, β-delayed two-neutron and β-delayed three-neutron decay, and β-delayed two-proton decay were discussed in the last chapter. Here we consider the discoveries of new motions and structures in nuclei such as the coexistence of multiple shapes with different deformation in one nucleus, unexpected new regions of very strong deformation, the importance of reinforcing shell gaps for both protons and neutrons on the nuclear shape, and the new shell gaps that stabilize a nucleus for a deformed shape ("deformed" magic numbers). The coexistence of states built on both spherical and well-deformed shapes in the same nucleus provides a new testing ground for unified models that can describe both types of structures. The importance of the discoveries and the broad ranges of unexplored nuclear terrain still further from stability have made studies of nuclei far from stability a major frontier in nuclear science. At the same time, the field and the techniques which have been developed to explore exotic short-lived decays have been used to discover new elements and have opened up new research in atomic, particle, and solid-state physics, astrophysics, and beyond to applied areas. By "far from stability," we here mean nuclei with many (5–25) fewer or more neutrons than in stable nuclei of a given Z or with high angular momentum of up to 50–60$\hbar$ and more. Selected examples are described in this section to illustrate the richness and diversity of the shapes and structures which now are found to exist in even one nucleus. Equally as important, they illustrate how sometimes experimental data have challenged our theoretical understanding even to calling for new theoretical approaches while at other times theorists have predicted new phenomena that are so challenging to experimentalists as to require the development of new detector facilities or new accelerators.

In Chapter 13 we discuss how, as one removes neutrons from a stable nucleus, there is an isotope shift in the optical spectral energies for the same

[5] J. H. Hamilton and J. Maruhn, *Scientific American* (July 1986), p. 80.

[6] J. H. Hamilton, *Treatise on Heavy Ion Science*, vol. 8, D. A. Bromley, ed., Plenum Press, New York (1989), p. 2.

electron transition in different isotopes of an element from the change in the nuclear mean radius. In 1972, using optical pumping techniques, the Mainz group of E. Otten discovered that as neutrons were removed from mercury nuclei a sudden large change occurred in the isotope shift between $^{187}_{80}Hg$ and ^{185}Hg (see Fig. 11.19) in the opposite direction from that expected. As the neutron number decreases in mercury nuclei, the energy of the optical transition being observed decreased in a regular fashion as expected for smoothly decreasing nuclear radii until ^{185}Hg, where the energy shifts to the same energy as in ^{196}Hg. One interpretation was that this represented a sudden large change in ground-state mean-square radius (change in deformation) from a spherical shape in ^{187}Hg to a well-deformed shape in ^{185}Hg in order for ^{185}Hg to have a mean-square radius (isotope shift) as big as ^{196}Hg, which has 11 more neutrons. This was surprising for these $Z = 80$ nuclei because they are very close to the strong spherical magic shell gap at $Z = 82$. However, ^{184}Hg and ^{186}Hg were found to be near-spherical in their ground states but to shift to well-deformed shapes above their second excited state on the basis of the energies of the γ-rays between the first five or six states populated in nuclear reactions observed at Lawrence Berkeley and Chalk River Nuclear Laboratories in 1973.

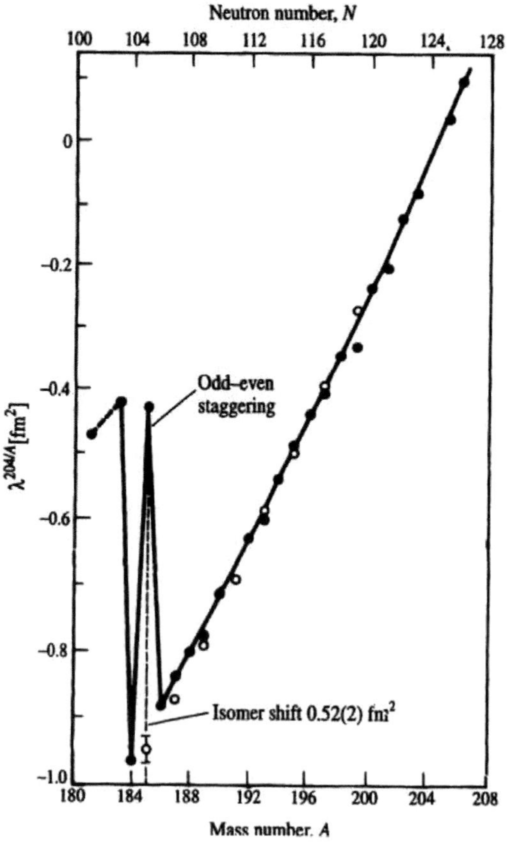

FIGURE 11.19
The isotope shift of an atomic transition in the mercury isotopes relative to ^{204}Hg as a function of A. In the odd-A nuclei, the shift for both the ground state and a nearby isomeric state are shown. The shifts shown can be taken as a measure of the changes of the charge (nuclear) radii as A changes.

Then came the discovery in 1974–76 that the nuclei of ^{72}Se and of ^{184}Hg, ^{186}Hg, and ^{188}Hg had separate, full bands of levels that are characteristic of both near-spherical and well-deformed shapes, as shown in Fig. 11.20 for the mercury nuclei (see footnotes 5,6). As can be seen by comparing the experimental data in Fig. 11.20 with the spectra characteristic of different shapes in Fig. 11.18, states built on deformed shapes are observed to appear suddenly in $^{184,186,188}Hg$ (and over ten years later in $^{180,190}Hg$) overlapping in energy and so coexisting with states built on the very stable, near-spherical ground states as observed from ^{198}Hg to ^{180}Hg. The first coexisting bands of levels in ^{184}Hg to ^{188}Hg were identified from the decay of the new far-off-stability Tl nuclei with $A = 184, 186, 188$ (15–21 neutrons less than stable ^{203}Tl) at UNISOR. The discovery and investigation of the properties of far-off-stability Tl nuclei like ^{184}Tl to ^{190}Tl and ^{183}Hg to ^{190}Hg required the development of new, specialized facilities. In a typical heavy-ion fusion or spallation reaction (described in Chapter 12), many different radioactive isotopes are produced.

To isolate one isotope or one mass chain, e.g. ^{188}Ti, ^{188}Hg, ^{188}Au, required development of new facilities in 1970's where magnetic isotope separators were placed on-line with particle accelerators so that the decay of new short-lived products ($T_{1/2}$ down to 0.1 s and less) could be studied. Major cooperative facilities were developed at the Center for European Nuclear Research (CERN, Geneva) using high-energy protons to produce spallation reactions, at the Holifield Heavy Ion Research facility (ORNL) and subsequently at GSI, Germany for heavy-ion reactions and at nuclear reactors to study fission products in Sweden, Germany and

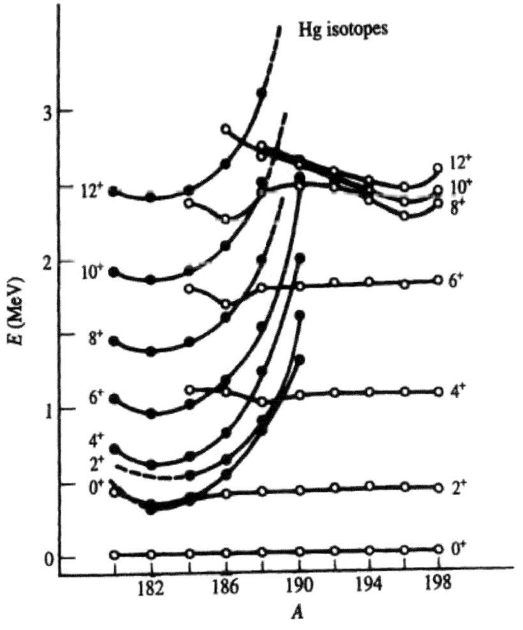

FIGURE 11.20
The regularly spaced vibrational-like levels of the near-spherical ground states of the $Z = 80$ mercury isotopes, which are close to the spherical magic number $Z = 82$, from ^{180}Hg to ^{198}Hg (solid lines). These levels are seen to coexist with a series of levels (open circles and dashed lines) with energy spacing characteristic of a well-deformed nucleus, $\beta_2 \sim 0.25$, in far-from-stability ^{180}Hg to ^{190}Hg. [J. K. Deng, Ph.D. Thesis 1993, Vanderbilt University; and ref. 6.]

Brookhaven National Laboratory. As an example, the University On-line Isotope Separator at Oak Ridge (UNISOR), is shown in Fig. 11.21. Shown in the figure is the focal plane of the magnetic isotope separator where the different mass-separated isotopes are focused following their production in a heavy-ion collision, extraction from the target in the separator ion source, and separation by mass/charge ratio in a 90° deflection electromagnet that extends out of the shielding wall at the top center.

As shown in Fig. 11.20, the bands built on the different shapes overlap in energy but maintain their quite different shapes—nuclear shape coexistence. Earlier, a different type of shape coexistence was observed in the fission isomers (Chapter 10) where the ground state and isomer had very different shapes ($\beta_2 \sim$ 0.25 and 0.6 respectively) but the energy levels associated with each shape are very widely separated in energy so they do not overlap in energy. So the idea that there could be a second minimum in the nuclear potential energy surface at a different deformation was known from the actinide fission isomers. However, theoretically, it was not expected that two such minima in the potential energy could coexist with such small energy differences (500–800 keV in $^{184-188}$Hg) and still have a barrier between the two minima so that there would be separate bands of states in each potential well (minimum). Thus, the proposed new form of shape coexistence discovered in ^{72}Se and ^{184}Hg to ^{188}Hg met resistance at first. As the new form of shape coexistence gained acceptance, it was then thought to be only an oddity in a few nuclei far off stability. Now we are finding such shape coexistence in nuclei throughout the periodic table, even in nuclei with spherical magic numbers such as tin ($Z = 50$) and lead ($Z = 82$). Over the decade after the initial work down to ^{185}Hg, the isotope shift studies were extended to ^{181}Hg, as shown in Fig. 11.19, and to the isomers down to ^{185}Hg. Note that while the isomers and ground states of ^{187}Hg to ^{199}Hg have the same near-spherical shape, in ^{185}Hg they have two quite different shapes, as observed in ^{184}Hg to ^{188}Hg. Since 1990, shape coexistence has been seen in ^{176,178,180}Hg and ^{190}Hg. Theoretically, such shape coexistence was predicted not to exist in the latter case and its discovery led to more refined theoretical calculations.

Even more complex multiple shape coexisting structures now are being seen. For example, four different shape coexisting bands are found in ^{185}Au and ^{187}Au, as shown in Fig. 11.22. This figure also illustrates another model approach to understanding the modes of excitation in nuclei. Earlier it was pointed out that in odd-A nuclei you can expect to have energy levels generated by the promotion of the odd particle to different, allowed energy orbitals. In addition, these single-particle states may couple to the even–even core to give bands of levels from the rotation or vibration of the core. Such situations are treated by particle–core coupling models. In ^{185}Au and ^{187}Au one can have particle states built on even–even ^{184}Pt and ^{186}Pt cores or hole states built on ^{186}Hg and ^{188}Hg cores. Shape coexistence is observed in all four of these core nuclei, with the ground state Pt cores being more deformed than the excited bands and vice versa in the Hg cores. Here, too, large E0 transitions are observed in transitions between the bands with different shapes. As seen in Fig. 11.22, bands built on the $h_{9/2}$ particle

FIGURE 11.21
Near the center the focal plane box (with three circular openings on top) of the UNISOR electromagnetic isotope separator which is on-line with the Holifield Heavy Ion Accelerator is shown along with three beam lines for the study of three different mass-separated isotopes. The left and center beam lines are for electron and γ-ray measurements and the right line is for on-line laser spectroscopy of short-lived nuclei. The separator magnet extends through the shielding walls at the top to the target area where the isotopes are produced in heavy-ion reactions. The central line subsequently had a low-temperature (10 mK) nuclear orientation refrigerator coupled on-line to the mass separator for studies of oriented nuclei.

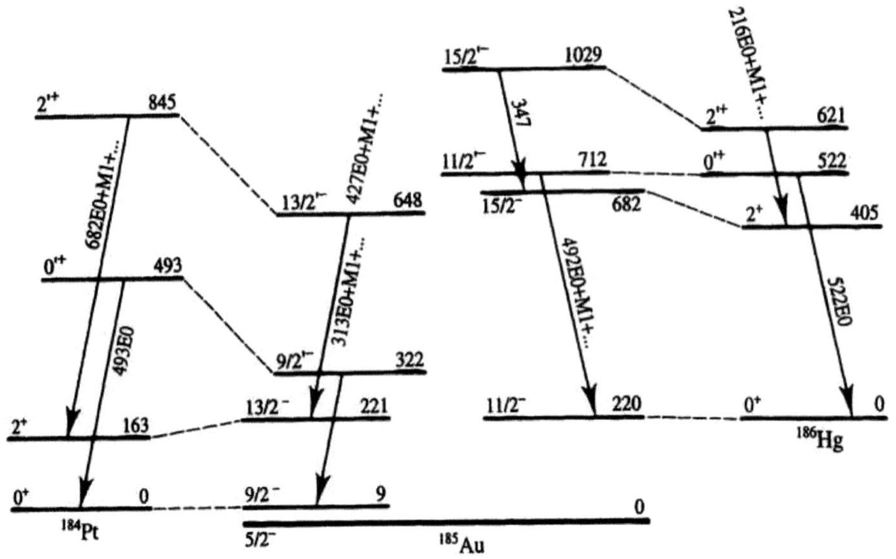

FIGURE 11.22
Portions of the two $h_{9/2}$ particle-state bands in ^{185}Hg formed from coupling to the two different shaped cores in ^{184}Pt and the two $h_{11/2}$ hole-state bands formed from coupling to the two different shaped cores in ^{186}Hg, compared to the ground-band and excited-band members in ^{184}Pt and ^{186}Hg. The excitation energies are in keV, and on the right are the spins and parities. The transitions are labeled by the energies (in keV) and multipole character of the emitted γ-rays (taken from the work of E. F. Zganjar, Louisiana State University and J. L. Wood, Georgia Institute of Technology with the isotope separator at Oak Ridge National Laboratory). Note the relative energy spacing of the 0^+ and 2^+ members of a band are an indication of their relative deformation (Eq. 11.24), so the smaller the gap the larger the deformation.

level coupled to two different shapes in the Pt cores, and those built on the $h_{11/2}$ hole state coupled to the two different shapes in the Hg cores are observed. Particle–core coupling models probe both the structure of the cores as well as the single-particle states. These shape coexisting nuclei allow us to probe how a particle or hole state couples to quite different deformations in a nucleus with the same N and Z. Thus, shape coexisting nuclei are an important new testing ground for the development of a unified nuclear model that works for all classes of nuclei. Further information can be found elsewhere.[5,6]

11.3.2 New Islands of Superdeformation Associated with New Deformed Magic Numbers

Following upon the discovery of shape coexistence in $^{72}_{34}\mathrm{Se}_{38}$ and $^{74}_{34}\mathrm{Se}_{40}$, new regions with the largest ground-state deformation known above $A = 30$ were discovered (see refs. 5,6). In an unusual congruence of theory and experiment, the 1981 calculations of ground state masses (Chapter 9) and quadrupole deformations β_2 for over 4000 nuclei by Möller and Nix and the experimental work on

${}^{74,76}_{36}Kr_{38,40}$ by a Vanderbilt–ORNL collaboration simultaneously and independently revealed this new region with the largest ground-state deformation known. The results of the 1995 Möller et al. calculations of β_2 are shown in Fig. 11.23. There one can easily see the spherical shell model magic numbers in lightest gray ($\beta_2 \leq 0.05$) and centered around 38–40 a black region signifying $\beta_2 > 0.4$ (earlier we used β for quadrupole deformation but now we need to be specific and use β_2). Indeed, centered around $N = Z = 38$ and around $Z = 38$, $N = 60$, 62 are two new islands of superdeformation with quadrupole deformation $\beta_2 \simeq 0.40$–0.45. These nuclei have a long-to-short axis ratio of 3:2. The term superdeformation was first used to describe the very large deformation of ^{100}Sr discovered by the on-line isotope separator group at CERN, Geneva. Recall that the rare-earth and actinide nuclei have ground states with $\beta_2 \sim 0.25$. However, the first examples of superdeformation were the actinide fission isomers with $\beta_2 \sim 0.6$ as first interpreted by Strutinski in 1967. A new, unexpected interpretation was given for the new islands of superdeformation (see refs. 5,6). Look at the calculations of the single-particle levels as a function of deformation shown in Fig. 11.24. The spherical shell gaps (spherical magic numbers of 28 and 50 and the weaker spherical subshell gap at 40) are clearly seen for $\epsilon_2 = 0$ ($\epsilon_2 \sim \beta_2$). Note, however, that there are numerous other shell gaps at various prolate and oblate deformations. Do they have any significance?

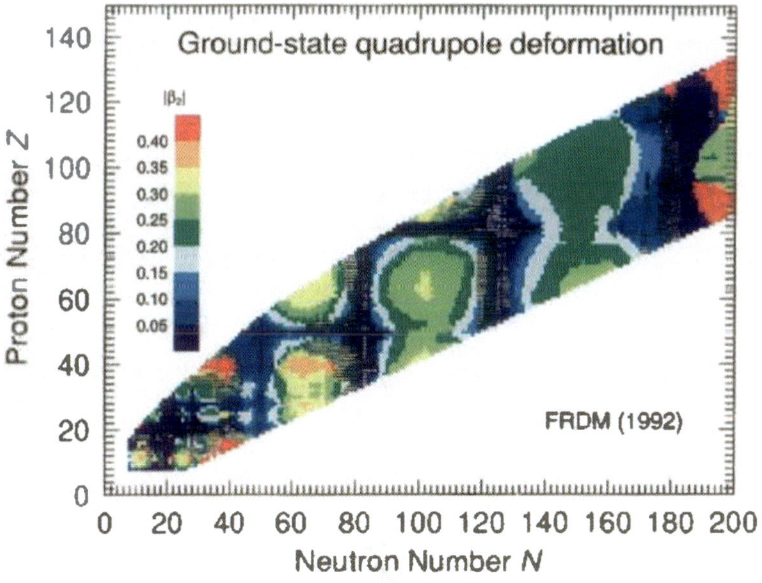

FIGURE 11.23
Calculated ground-state quadrupole deformation, β_2, for 8979 nuclei by P. Möller et al., *Atomic and Nuclear Data Tables* 59 (1995) 185.

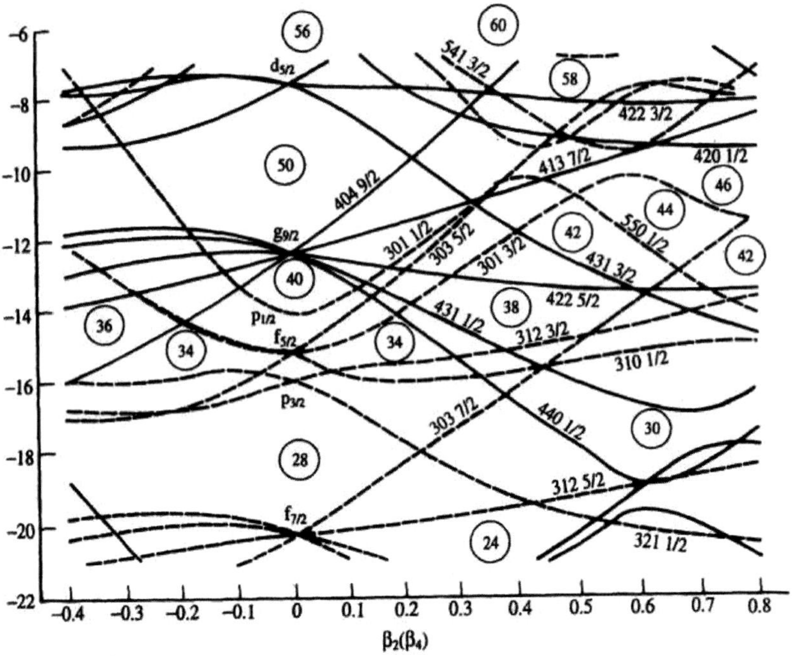

FIGURE 11.24
Predicted single-particle (e_ν) Nilsson configurations as a function of β_2 for light nuclei (β_4 is minimized for each value of β_2). For $N \approx Z$ this figure can be applied to both single-neutron and single-proton configurations. Particle numbers corresponding to gaps in the spectrum of single-particle states (shown inside circles) are associated with the variety of exotic shapes predicted for light nuclei with $N \approx Z$, as discussed in the text and in Table 11.3. The spherical orbitals, e.g. $f_{7/2}$, $p_{3/2}$ ... are shown at zero deformation and the Nilsson asymptotic quantum numbers $[Nn_z\Lambda]\,\Omega$ for the orbitals with deformation. [Adapted from W. Nazarewicz et al., *Nucl. Phys.* **A435** (1985) 397.]

On the basis of an analysis of all the data in the regions of these two islands, it was pointed out (see ref. 6) that these shell gaps at different deformations can be just as important as the spherical shell gaps but they become important only when both N and Z have shell gaps for the same deformation so that the protons and neutrons reinforce each other's push toward that deformation. Indeed, the same proton–neutron reinforcing shell gap effect holds for spherical nuclei, too (ref. 6). Thus, ${}^{68}_{28}Ni_{40}$ and ${}^{90}_{40}Zr_{50}$ have energy levels characteristic of a spherical double magic nucleus because the weaker spherical subshell gap at 40 is reinforced by a strong spherical magic number for shell gaps for the N or Z partner in each case, $Z = 28$ and $N = 50$, respectively. This same reinforcement of spherical subshell–shell gaps at $Z = 64$, $N = 82$ makes ${}^{146}_{64}Gd_{82}$ also a new, spherical double magic nucleus (see ref. 6). Recall that the nuclear quadrupole moments also show a change in sign at 40 (Fig. 11.11) that is characteristic of a spherical magic number closed shell. It was these and other such data that led to 40 being considered at least semi-magic for a spherical shape. However, there is a switch in importance from the 40 spherical gap to the $Z = 38$ and $N = 38$ shell gaps at $\beta_2 \sim 0.4$ when both N and Z approach 38 to reinforce each other and drive these nuclei to superdeformation, and similarly when $Z = 38$ and $N = 60, 62$. Thus, $N = Z = 38$ and $N = 60$, 62 are "deformed" magic numbers where N and Z reinforce each other to stabilize a nucleus at a particular deformation just as the spherical shell gaps stabilize a nucleus for a spherical shape. The influence of these new "deformed" shell gaps on the nuclear shape rapidly vanishes as either nucleon number (N or Z) moves away from where the two gaps reinforce each other. Structures change dramatically for changes in N or Z of 2 or 4 away from 38 and 60, 62 (see Fig. 11.26b discussed later). These shell gaps and the size of the deformation have been related to an increased importance of the proton–neutron interaction contribution in addition to that of the nuclear mean field and perhaps to a reduction of the pairing force between two nucleons. Thus, these studies give new insight into the importance of various components of the nuclear force.

The expansion of the shell model to include "deformed" magic numbers through a reinforcement of shell gaps is a significant extension of the spherical nuclear shell model. The calculations of the single-particle energies have been extended to even larger deformation as shown in Fig. 11.24. As one looks at Fig. 11.23, one is struck by the presence of many shell gaps similar to those at 38 and 60, 62. Now the calculations reveal new shell gaps for superdeformation with $\beta_2 \sim 0.6$ and the new hyperdeformed shapes where $\beta_2 \sim 0.8$–1 for $Z = N = 42, 44, 46$ (the latter is even beyond the range of Fig. 11.23) as predicted first by Dudek (Strasbourg) and colleagues and by Nazarewicz et al. (Oak Ridge) in several mass regions. Are they all equally as important? For example, are there oblate nuclei with equally large deformation, $\beta_2 = -0.4$, as theory predicts for $N = Z = 36$? Are there hyperdeformed nuclei with $\beta_2 \sim 1.0$? The determination of the importance of these other gaps is a clear challenge for experimentalists. Based on the initial observations that such deformed shell gaps are important only when both N and Z reinforce each other for the same deformation, many of the new deformed gaps will be difficult to observe. However, in 1994 evidence for the reinforcement of the $Z = 38$ and $N = 44$ shell gaps at superdeformation with $\beta_2 \sim 0.5$ was found at high spins in ^{82}Sr and neighbors. Table 11.3 gives the first "deformed" double magic N and Z made by their reinforcement of shell gaps at the same deformation. Deformed double magic nuclei like $^{152}_{66}Dy_{86}$ and $^{192}_{80}Hg_{112}$ achieve their double magic character at high spins – not in their ground states and similarly in other subsequently discovered nuclei like ^{60}Zn.

The study of reinforcing shell gaps at still larger deformation, for $N = Z = 42$ and 44, will require a new generation of detection devices and accelerators capable of accelerating radioactive ion beams to produce such exotic nuclei still farther from stability. Some of the exotic nuclear shapes that have been theoretically predicted along the $N = Z$ line between 30 and 50 are shown in Fig. 11.25. Moreover, along the $N = Z$ line, the protons and neutrons fill the same orbitals to enhance the proton and neutron overlaps (e.g., in quantum mechanics this means their wavefunctions overlap so that the interaction between the protons and neutrons is maximal). In this case the proton–neutron interaction

TABLE 11.3

First regions of Z and N where superdeformation was discovered. The underlying physics involves the reinforcement of proton and neutrons shell gaps for the same superdeformation to stabilize the superdeformation. The reinforcing shell gaps responsible for the superdeformation are shown.[6]

Mass region	Year of discovery	States	Reinforcing deformed shell gaps
	Nuclei with 3:2 axis ratios, $\beta_2 \sim 0.4$–0.5		
$A \sim 100$	1979	Ground states	$Z = 38$, $N = 60, 62$
$A \sim 76$	1981	Ground states	$Z = 38$, $N = 38$
$A \sim 82$	1994	High-spin states	$Z = 38$, $N = 44$
$A \sim 130$	1987	High-spin states	$Z = 58$, $N = 72, 74$
$A \sim 190$	1989	High-spin states	$Z = 80$, $N = 112, 116$
	Nuclei with 2:1 axis ratios, $\beta_2 \sim 0.6$		
Actinides	1967	Fission isomers	$Z = 94$, $N = 144, 148$
$A \sim 150$	1986	High-spin states	$Z = 64, 66$, $N = 84, 86$

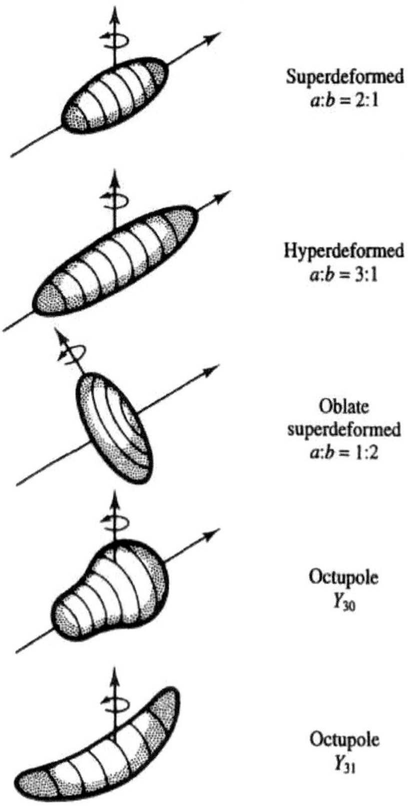

FIGURE 11.25
Examples of different exotic nuclear shapes which are predicted in various theories to occur along the $N = Z$ line.

enhances the nuclear collectivity. Also, significant isospin breaking can occur as a result of the exchange symmetry between protons and neutrons filling the same orbitals. This is a consequence of the charge independence of nuclear forces (see discussion in Appendix 10A).

An initiative to combine a new generation recoil mass spectrometer and a radioactive ion beam, RIB, facility was completed at Oak Ridge National Laboratory in the late 1990's. Starting with radioactive ion beams in which the ions already have a deficiency (excess) of neutrons leads to much larger cross-sections for the production of neutron-deficient (rich) nuclei farther off stability. Such facilities opened up studies of previously inaccessible exotic nuclei far from stability to out beyond the proton drip line, as discussed in the next section. These nuclei not only test the spherical shell model far from stability, for example around ^{78}Ni and ^{100}Sn, they also make possible improved measurement of beta delayed neutron emission probabilities which are important in reactor design (see Chapter 12). New generation radioactive ion beam facilities will extend the range of nuclei which can be studied.

From studies of nuclei far from stability, extensive systematic data now are being obtained over long N chains for a given Z to yield new vistas of the changing nuclear landscape. These data often show rapid changes in structure that provide theorists with unprecedented information to develop and refine their models of the structure of nuclei. As an example, consider the first 2^+ energies in the even–even strontium isotopes in Fig. 11.26a. Recall the inverse relation between E_{2^+} and deformation discussed earlier. They have been traced from one region of superdeformation ($\beta_2 \sim 0.45$) at $N = 38$ across to $N = 50$, an essentially spherical double magic nucleus ($\beta_2 \approx 0$), to a plateau of near-spherical nuclei for $N = 52$–58 ($\beta_2 \sim 0.15$), to a sudden, dramatic change to superdeformed shapes ($\beta_2 \sim 0.45$) again at $N = 60, 62$.

Now look at Fig. 11.26b. The nuclei $^{90}_{40}\mathrm{Zr}_{50}$ and $^{96}_{40}\mathrm{Zr}_{56}$ have large 2_2^+ energies characteristic of spherical double magic nuclei. There is only a slightly less sharp change in structure between $N = 58$ and 60 for Zr nuclei than for the Sr nuclei. However, by Mo at $Z = 42$, the sharp transition in shape between $N = 58$ and 60 is disappearing. The $N = 60$, 62 shell gaps must be reinforced by $Z = 38$, 40 to give superdeformation. From this figure and the theoretical predictions seen in Fig. 11.24, one can begin to see why the mass 70 and 100 regions are major testing grounds for further development of nuclear models.

Nuclear models must be able to trace over long sequences of nuclides both the smooth and the rapid changes in structures and to treat the multiple nuclear shapes seen in one to several nuclei in the chain. The generalized collective model of W. Greiner's group in Frankfurt is based on a description of collective states as quadrupole surface excitations of the nucleus to give rise to an average potential in which the nucleons move. Another approach is to calculate potential-energy surfaces in what is called the cranked shell model where the effects of deformation and rotation are included. The interacting boson model (IBA) developed by A. Arima and F. Iachello with its various mathematical symmetries is based on

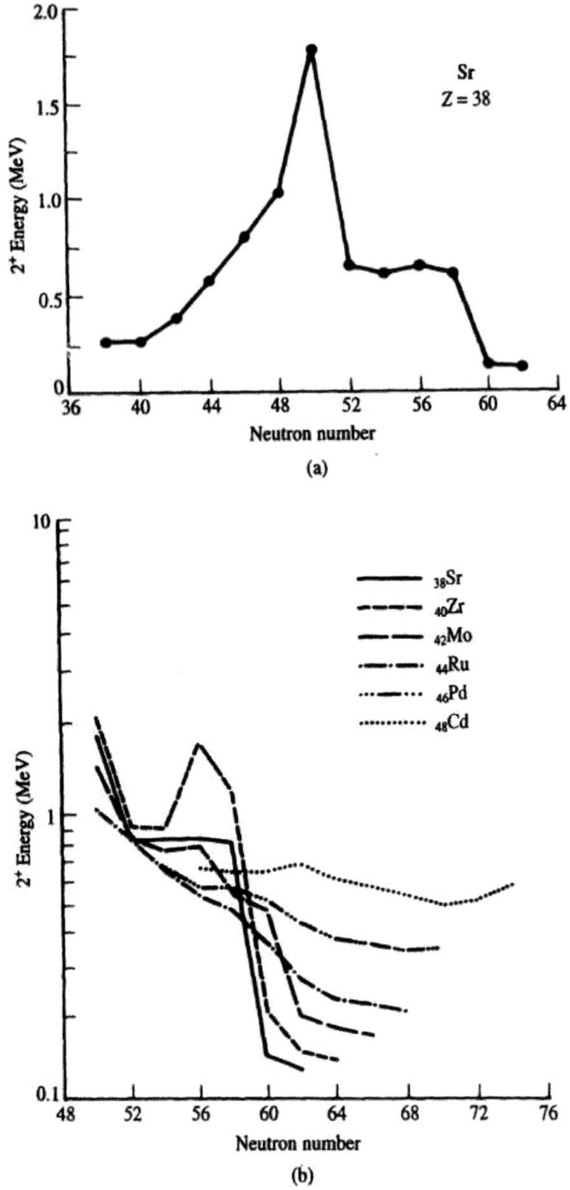

FIGURE 11.26
(a) The first 2^+ energies in the even–even strontium nuclei from $N = 38$ to 62 (more than 50 percent change in N). Recall the inverse correlation between this energy and nuclear deformation. (b) The first 2^+ energies in the even–even Sr ($Z = 38$), Zr (40), Mo (42), Ru (44), Pd (46), and Cd (48) nuclei from spherical magic $N = 50$ to 62.

treating pairs of protons and neutrons as bosons with spins of 0 or 2. The generalized collective model and IBA approaches are equivalent descriptions, the beauty of the generalized collective model is related to its vivid physical pictures of the nuclear potential energy surfaces (and nuclear shapes) with one and more minima in the potential to describe the changing and multiple structures, as illustrated in Fig. 11.27; and the beauty of IBA is related to its mathematical simplicity and

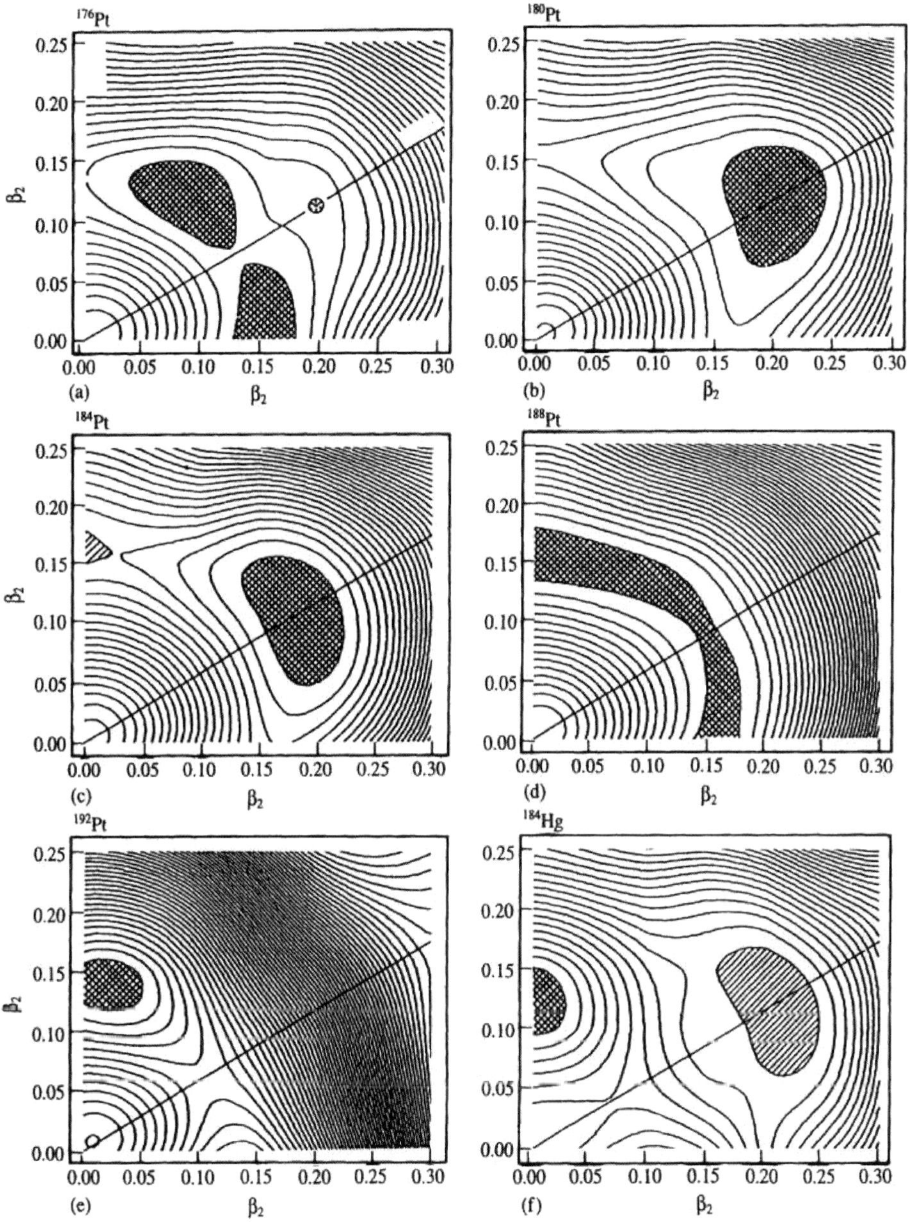

FIGURE 11.27 (a)

Potential energy surface calculations for ^{184}Hg and ^{176}Pt, ^{180}Pt, ^{184}Pt, ^{188}Pt, and ^{192}Pt as a function of deformation β_2. The lines are equipotential contours. The deformation for pure oblate shapes ($\gamma = 60°$) is plotted along the vertical axis and the pure prolate shapes ($\gamma = 0°$) along the diagonal line. The crosshatched regions are the lowest minima and the diagonal lines the nearby shape coexisting minimum. [R. Bengtsson et al., *J. Phys. G Nucl. Phys.* **12** (1986) L223.]

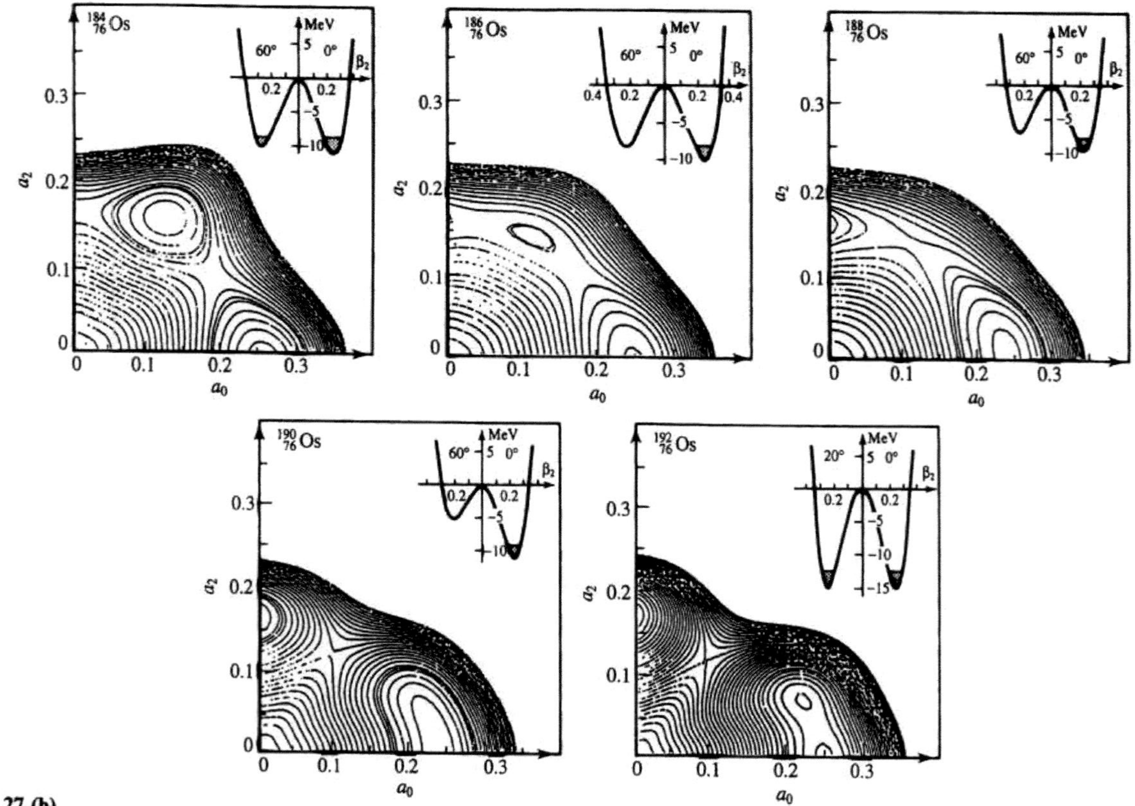

FIGURE 11.27 (b)

Potential energy surface calculated in the generalized collective model for ^{186}Os, ^{188}Os, ^{190}Os by the Frankfurt group. The insets show the depths of the potential for the values of γ as given (Courtesy W. Greiner.)

elegance. Of course, nucleons are not bosons but fermions. The mathematical symmetry approach has been employed in other ways that include symplectic group structure and a fermion dynamical-symmetry model in which the protons and neutrons are treated as fermions.

One can easily see how the nuclear landscape (shape) changes as a function of neutron number for a given Z in Fig. 11.27. In Fig. 11.27a, the nuclear potential energy surfaces in the β–γ plane for ^{184}Hg and ^{176}Pt, ^{180}Pt, ^{184}Pt, ^{188}Pt, ^{192}Pt are shown. In the β–γ plane, the distance from the origin along any line is a measure of β_2 while the angle between the x-axis ($x = 0°$) and the line drawn through a minimum in the β–γ plane measures the value of γ. Remember, $\gamma = 0$ is prolate and $\gamma = 60°$ is oblate and for $0 < \gamma < 60°$ a nucleus is triaxial. Note in ^{184}Hg the potential energy has two minima: the deepest along the oblate axis with $\beta_2 \sim 0.12$ gives rise to a near-spherical ground state, and an excited-state second minimum at $\beta_2 \sim 0.22$ on the prolate axis gives rise to excited states in this second well with a deformed shape coexisting with the near-spherical shape. In the platinum nuclei, ^{192}Pt has only a near-spherical oblate ground-state minimum, while ^{188}Pt is a more triaxial γ-soft (there is no barrier in the γ plane, so γ has no definite value) transitional nucleus, but ^{184}Pt has its deepest minimum for $\beta_2 \sim 0.2$, prolate deformed and a second minimum for a coexisting less-deformed, oblate shape (the reverse of ^{184}Hg), while ^{180}Pt has only a prolate minimum and ^{176}Pt has a triaxial ground state with a prolate shape coexisting excited band. In a similar fashion, the potential energy surfaces for a series of osmium isotopes as calculated in the generalized collective model by the Frankfurt group are shown in Fig. 11.27b. All these nuclei exhibit a γ-softness (very low barrier in the direction of $\gamma \neq 0$) in the triaxial shape direction with deeper prolate minima in ^{186}Os, ^{188}Os, ^{190}Os.

More detailed microscopic models continue to be developed by different groups. A. Faessler and his group at Tübingen have had considerable success in carrying out very large-scale computer calculations which include realistic nucleon–nucleon interactions derived from the Bonn one-boson exchange potential, neutron–proton interactions, and correlations among the different configurations in what is called the Hartree–Fock–Bogoliubov approach.

11.3.3 Nuclei at High Angular Momentum

Another major area of current research is the study of nuclei at high angular momentum. The existence of different shapes and structures in nuclei has been extensively enlarged through studies of nuclei at higher and higher angular momenta. In the early 1970s, in Stockholm, it was discovered that the moments of inertia of the ground-state rotational bands in deformed rare-earth nuclei underwent a sudden large change above a certain angular momentum around 12–14$\hbar$, a bending back of the moment of inertia as a function of rotational frequency ($\hbar\omega \sim E\gamma/2$) (Fig. 11.28b). This effect was interpreted in terms of the promotion of a pair of particles to a high-j (total angular momentum), low-Ω (projection of j on the symmetry axis) orbital where the angular momentum of the pair was aligned with that of the rotating core to give rise to a new 2-particle rotation-aligned rotational band with much larger moment of inertia. The physical picture is the following: a high-j requires a large radius orbit and its alignment requires the plane of the orbit to contain the long axis of the nucleus. Such orbits give rise to a significant increase in the moment of inertia of the nucleus. Such a crossing band is seen in the data for the band built on the ground state of ^{164}Er in Fig. 11.28a. Note the change above 14^+ from the regular increasing energy of the γ-rays between the levels of the rotational band built on the ground state as expected from $E = (\hbar^2/2\mathfrak{I})I(I+1)$ (Eq. 11.28) to E_γ (16–14) $\simeq$ E_γ (18–16). This gives rise to the bending back in the moment of inertia shown in Fig. 11.28b. The band that begins at the (12^+) 2519.3 keV level becomes the yrast band by the 16^+ level. An yrast level is the lowest energy level for a given spin. By spin 16,

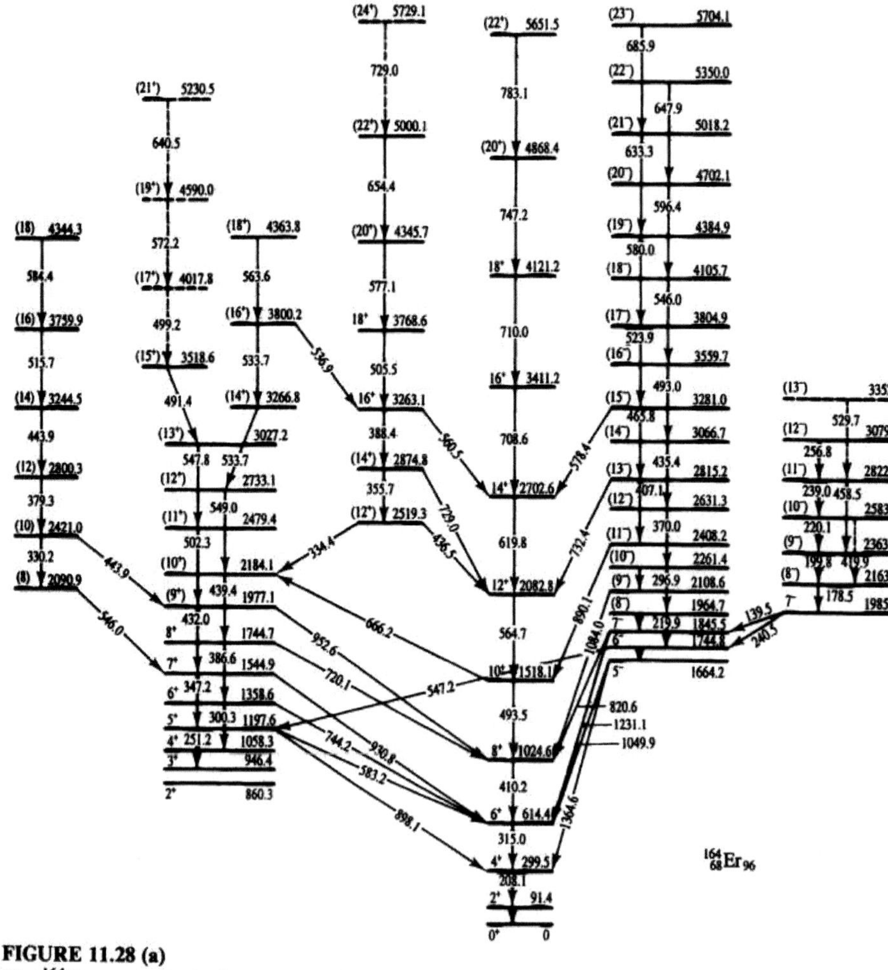

FIGURE 11.28 (a)
The ^{164}Er energy levels. [From N. R. Johnson and co-workers at Oak Ridge and Vanderbilt.]

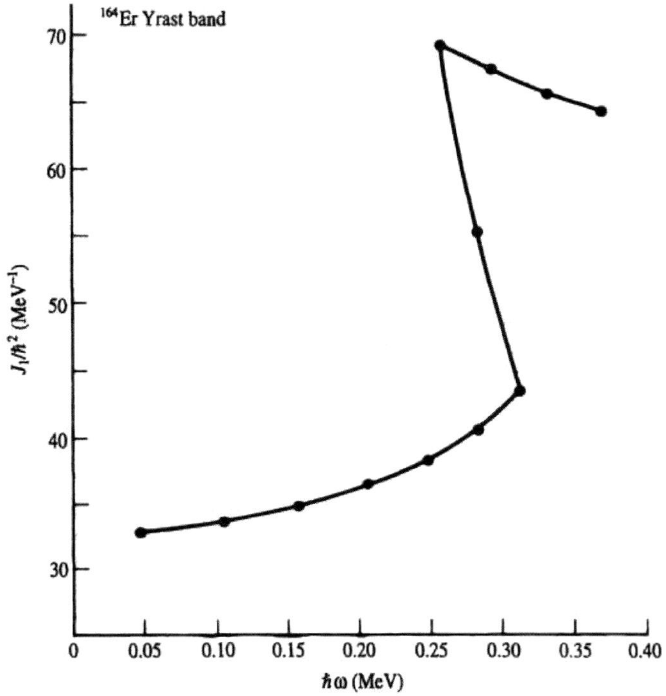

FIGURE 11.28(b)
The bending back of the moment of inertia, $J_1/\hbar^2 = (2I-1)/E_\gamma$, as a function of rotational motion $\hbar\omega \sim E_\gamma/2$ for ^{164}Er.

rotation of the ground state is not the most energy efficient way for ^{164}Er to carry angular momentum and a new way described above becomes more efficient.

The study of how a nucleus most efficiently carries angular momentum and how its structure and shape change with increasing angular momentum has been an important frontier in nuclear structure physics since the above discovery in Stockholm. New surprises continue to be made. Near-spherical nuclei are found to go over to well-deformed structures and well-deformed prolate structures change how they carry angular momentum from collective rotational motion to the summed angular momentum of increasing numbers of pairs of rotation-aligned particles, as the nuclear angular momentum increases. Many different shapes and structures including those with collective structures and those with multiple single-particle structures are found to coexist at higher angular momenta.

Another unexpected development was the discovery by a Liverpool group that at angular momentum of 30–60$\hbar$ some nuclei suddenly took on a different superdeformed shape, $\beta_2 \sim 0.6$, where the ratio of the long to the short axis is 2:1. Superdeformed nuclei with $\beta_2 \sim 0.45$ and 3:2 axis ratios were subsequently found at high spins in the cerium region around $A = 130$, light mercury and lead nuclei with $A = 190$–196, and nuclei with $A = 60$–82. Superdeformed shapes have very large moments of inertia and very small increase in transition energy between the superdeformed levels as the angular momentum increases, as illustrated in Fig. 11.29. Some nuclei are found to have several superdeformed bands. In some cases these high-energy SD bands are excitations of a lower SD band, while others may be excited bands built on a core in a neighboring nucleus. Again, the physics behind these new superdeformed states at high spins in nuclei around $A = 80$, 130, 150, and 190 is related to the reinforcing shell gap phenomena discussed earlier. In these new regions both N and Z have shell gaps at $\beta_2 \sim 0.6$ or 0.45 to form new islands of high-spin superdeformation[7] as was found earlier for the new islands of ground-state superdeformation around $Z = N = 38$ and $Z = 38$, $N = 60$, 62. Table 11.3 gives the regions where superdeformed shapes have been observed, and the new deformed magic numbers for Z and N which reinforce each other at the same superdeformation. With these new deformed magic numbers, ${}^{76}_{38}Sr_{38}$, ${}^{98}_{38}Sr_{60}$, and ${}^{100}_{38}Sr_{62}$ in their ground states, and ${}^{152}_{66}Dy_{86}$ and ${}^{192}_{80}Hg_{112}$ in excited bands are superdeformed double magic nuclei analogous to the spherical double magic nuclei like ${}^{4}_{2}He_2$, ${}^{16}_{8}O_8$, ${}^{40}_{20}Ca_{20}$, and ${}^{208}_{82}Pb_{126}$.

Theoretical calculations attempting to understand superdeformation found a second minimum at $\beta_2 \sim 0.6$ as observed experimentally. However, more surprisingly, it was predicted that at higher angular momentum there should be levels characterized by hyperdeformed shapes with $\beta_2 \sim 0.8$–1.2, a 3:1 axis ratio! An

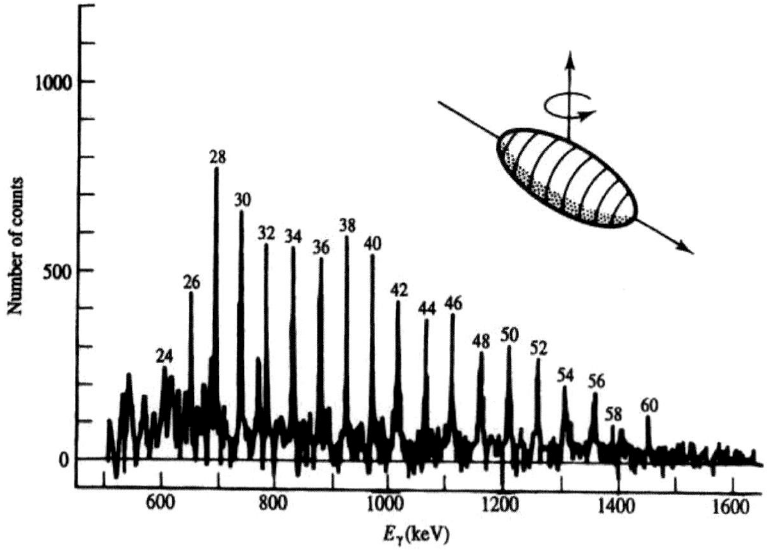

FIGURE 11.29
A coincidence γ-ray spectrum of the superdeformed band in ^{152}Dy. The close, equally spaced lines are thought to de-excite the band from spin $60\hbar$ to spin $24\hbar$ (From P. Twin and co-workers, Liverpool-Daresbury Nuclear Laboratory.)

---

[7] R. V. F. Janssens and T. L. Khoo, *Annu. Rev. Nucl. Part. Sci.* **41** (1991) 321.

illustration of the initial work of Dudek at Oak Ridge and now Strasbourg is shown in Fig. 11.30. There one sees minima in the potential energy surface for normal, super-, and hyperdeformed shapes. Predictions of hyperdeformation were extended by Dudek, Nazarewicz, and others to several regions of the periodic

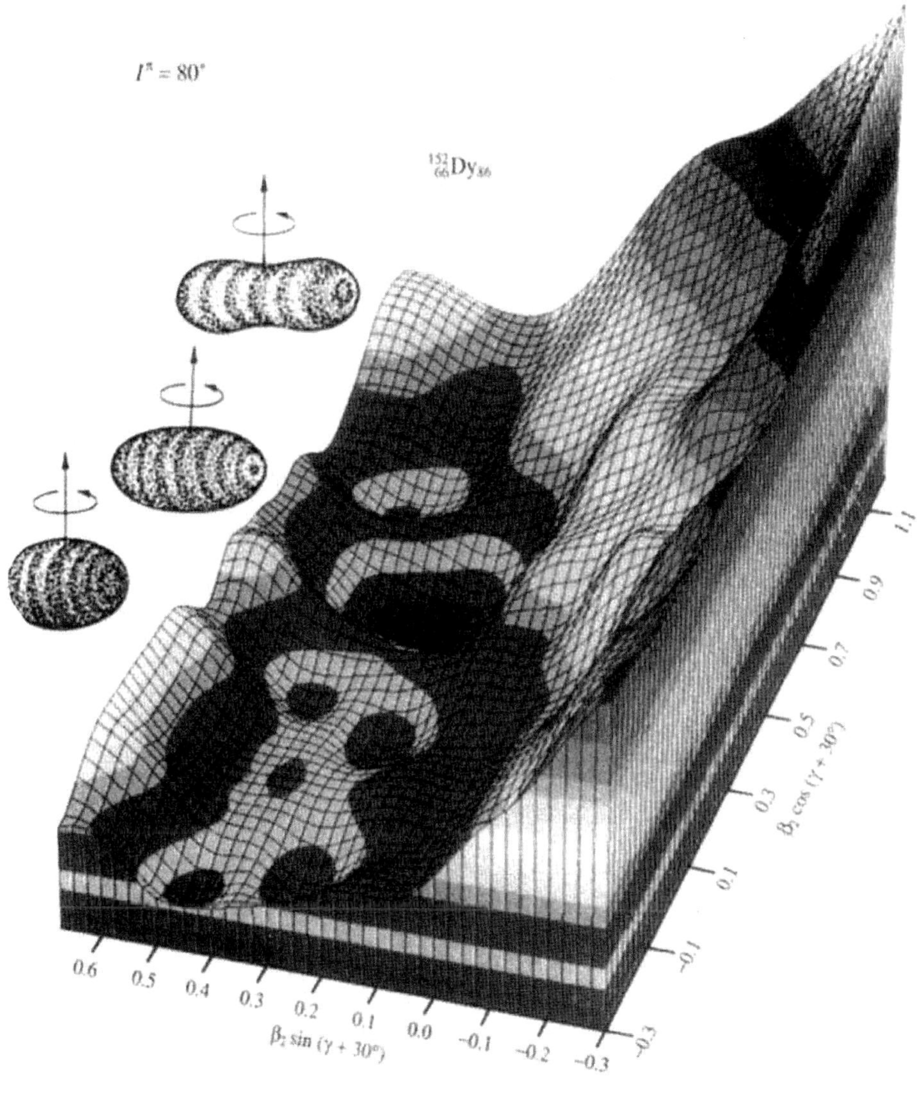

FIGURE 11.30
A potential energy surface calculation from Dudek (Oak Ridge–Strasbourg) for ^{152}Dy. The surface has a lowest minimum at a small oblate deformation for its ground state, then at medium angular momentum $\sim 30\hbar$ becomes well-deformed prolate and on to a superdeformed $\beta \sim 0.6$ shape at higher spin. Evidence for all these shapes has been found. Finally, there is a new hyperdeformed minimum $\beta \sim 1.0$ predicted at still higher spin.

table, for example, in $N = Z$ nuclei (see Fig. 11.26b), and represent a major challenge for experimentalists. The observation of hyperdeformation can provide critical tests of our understanding of how nuclei carry angular momentum at different shapes and probe at new extremes the theoretical models used to describe such motion.

However, the superdeformed bands are typically populated in a heavy-ion reaction only 1–5 percent of the time, and the hyperdeformed levels are expected to be considerably less populated. Such very weak structures have not been seen by large arrays of 100 Ge detectors even when they had shields to suppress (reduce) the background events that arise when a γ-ray Compton scatters out of a Ge detector and leaves a broad, flat distribution of energies rather than giving up all its energy to create an event where the full γ-ray energy is absorbed to form a well-defined photo peak. From an unexpected direction there came strong evidence in early 1995 for hyperdeformation—namely, from the study of spontaneous fission of ^{252}Cf where following scission $^{146}_{50}Ba_{90}$ is sometimes created with a hyperdeformed shape with a long-to-short axis ratio of 3:1. This discovery is supported by the calculations of Dudek and co-workers, which have shell gaps for $Z = 56$ and $N = 88$, 90 at $\beta_2 \sim 1.2$–1.4. This may be another example of reinforcing shell gaps. New γ-ray tracking detector arrays under development will make searches for hyper deformation a rich field for investigation.

Another surprise was the discovery in the 1990s that many of the superdeformed bands had energies and moments of inertia, as illustrated in Fig. 11.31, which were identical to those of another superdeformed band either in the same or neighboring nucleus. Then, it was discovered that such identical bands existed in normal deformed rare-earth odd-A and neighboring even–even nuclei, and more recently, in nuclei in the $A = 70$ range. The first results were all for proton-rich

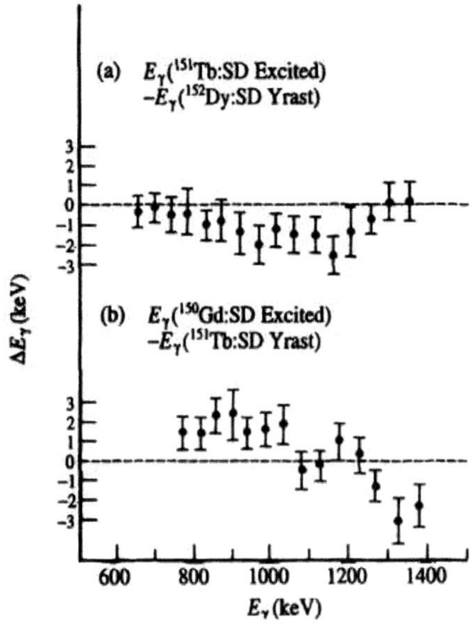

FIGURE 11.31

The differences in the γ-ray energies between the bands in (a) ^{151}Tb (excited superdeformed) and ^{152}Dy (yrast superdeformed) and (b) ^{150}Gd (excited superdeformed) and ^{151}Tb (yrast superdeformed). For the ^{151}Tb–^{152}Dy pair the difference is, on average, less than 1 keV while for the ^{150}Gd–^{151}Tb pair the difference is slightly larger. [T. Byrszi et al., *Phys. Rev. Lett.* **64** (1990) 1650, reproduced with permission.]

nuclei, but more recently several such cases have been observed in neutron-rich nuclei such as ^{98}Sr and ^{100}Sr, (the energy differences of the transitions out of the known first five states are 2.8 ± 2.0, 1.8 ± 0.3, -2.0 ± 0.3, -2.1 ± 0.3, and -2.2 ± 0.3 keV[8]) and yrast and octupole bands in ^{144}Ba and ^{146}Ba. How can two bands have identical transition energies and moments of inertia? Each nucleus has a different number of nucleons and, moreover, it was long ago shown that, as a result of pairing, the moments of inertia of odd-A nuclei should be 20–25 percent *greater* than those of their even–even neighbors. So identical bands, especially in such profusion in many parts of the periodic table from $A \simeq 70$ to 200, are a major, unexpected surprise. Such identical bands present another challenge for theoretical models.

11.3.4 New Facilities to Meet New Challenges

To illustrate the way in which a field advances through new instrumentation, the development of new facilities to probe the structure of nuclei under greater extremes of angular momentum and further from stability and to open new area of nuclear astrophysics are described briefly. By the late 1990's, new much larger γ-ray detector facilities became available in the United States and Europe. Projects in Europe include EUROBALL, EUROGAM, GASP and NORDBALL. In the United States, Gammasphere has 102 Compton-suppressed germanium detectors with larger volumes than previous detectors. By close packing, the geranium fills about 50 percent of the total 4π solid angle. This system enhanced the ability to see weak processes by a factor of 100 to 1000 over previous facilities. This is an incredible gain in resolving power. By comparison, the Hubble Orbiting Space Telescope gives only a factor of 10 higher resolving power than previously existed. A view of Gammasphere is shown in Figure 11.32.

However, even when Gammasphere was only 40 percent completed, the call by the U.S. Federal Funding Agencies in early 1995 for a new five-year long-range plan helped inspire ideas for a next generation of γ-ray detectors. Already the Europeans had proposed a new cluster germanium array with a factor of 5 better resolving power than Gammasphere to detect weak processes. The new idea proposed in the United States is for a segmented array of small-sized, close-packed pure germanium covering 4π geometry with no Compton suppression shields. The germanium material would be segmented into such small volumes that one could add the signals produced in the small neighboring detectors by a single γ-ray to eliminate the Compton-scattered events. This device would improve the resolving power to detect weak processed by over a factor of 100 to Gammasphere.

[8] J. H. Hamilton et al., *Prog. Part. Nucl. Phys.* **35** (1995) 635.

FIGURE 11.32
A view of the Gammasphere detector system with 102 large-volume germanium detectors with bismuth germinate shields to detect and reject Compton scattering events in the germanium detectors.

This detector, Greta, is being built at Lawrence Berkeley National Lab. The future of nuclear structure and other processes that can be studied through γ-ray detection will be strongly influenced by this and similar devices. Before going on, it is important to stress that many areas of research today depend on facilities which require numbers of years to reach completion from conception of the idea to obtaining funding, to designing, and to building. The idea for an "ultimate" detector illustrates how scientists must always be thinking of new experimental ways to attack problems — sometimes even before research has fully begun with a new facility coming on line, like Gammasphere, if one is to carry out the best science ten or more years later.

In addition to γ-ray detector devices, to study exotic nuclei produced far off stability with very low (μbarn and lower) cross-sections one must have higher sensitivity to select the products of interest. For example to select exotic nuclei produced in very weak ($\leq$ 10 μb) reaction channels at a given mass and to have sufficient intensity to carry out γ–γ coincidence studies in coincidence with a given mass reaction product called forth the development of a new generation recoil mass spectrometer (RMS), to separate by mass and Z the products in heavy-ion reactions in a Vanderbilt-Oak Ridge collaboration. It was designed to use an inverse reaction in which a heavy projectile like tin or nickel is used to bombard a light target like carbon. The next generation RMS has a momentum achromat to provide an immediate focus before the first electrostatic deflector where the primary beam can be rejected to better than 1 in 10^{13}. This allows studies of inverse reactions where the beam and recoil products have almost the same energy. This RMS can be used with γ-ray and particle detector arrays around the target or the mass-separated fragments at the focal plan. At the focal plane of the RMS one can study proton and alpha radioactivities and β-delayed proton and neutron decays. The lightest known α-decay chain ^{110}Xe $\rightarrow$ ^{106}Te $\rightarrow$ ^{102}Sn was discovered in 2006 with the ORNL RMS 25 years after the ^{112}Xe chain was discovered at GSI.

Finally, there is the problem of the Doppler shift of the γ-rays in the 90° detectors. In an inverse reaction, this could degrade the resolution by a factor of 3–5. A new high-efficiency detector design has been developed in which one has four smaller detectors in one cryostat, so each has a small solid angle but the sum of their signals gives high efficiency. The Doppler shift of γ-rays detected at an angle of θ to the beam from recoiling compound nuclei is given to first order in v/c by

$$\Delta E = 1 + \beta E \cos\theta \tag{11.40}$$

where β = v/c and c is the light speed, and E is the intrinsic γ energy emitted from rest nuclei. The velocity of the compound nuclei can be calculated with the help of the conservation of linear momentum. Note that at 90° the detector's finite solid angle can lead to either a positive or negative contribution to the energy for different events, degrading the energy resolution. One can use thin targets so that the γ-rays in the forward and backward direction have the same shift. The new proposed "ultimate" detector with its small crystals would be a significant improvement.

As noted earlier when discussing extending our knowledge along the N = Z line, radioactive ion beam (RIB) facilities provide the best means for producing new, exotic nuclei out to the proton and neutron drip lines and beyond the proton drip line. There are two basic ways to produce radioactive ion beams; one is fragmentation of high-energy relativisitic heavy-ion beams on a relatively thin target and separation of the high-energy recoil products in a fragment mass analyzer; or use of an intense proton or deuteron beam to fission uraninum producing radioactive products which are extracted from a thick target, separated by mass and isobar (same mass, different Z) with a magnetic isotope separator on-line and then re-accelerated. The advantage of the former is there are no hold-up times or chemistry involved, but with the disadvantage of very high energies and poor momentum resolution. These problems can be overcome with a cooler ring to slow the recoils down and then re-accelerate them,

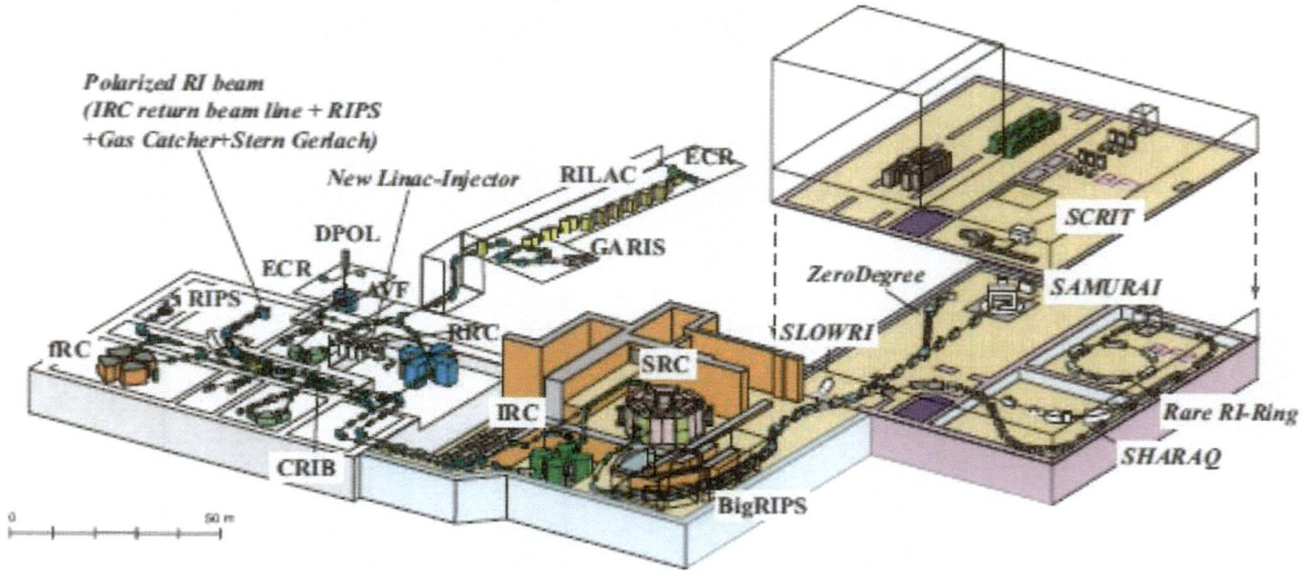

FIGURE 11.33
A schematic view of the RIKEN Radioactive Ion Beam Factory. The accelerator complex consists of three cyclotrons to give energies up to 345 MeV/nucleon. A linear accelerator and ring cyclotron are used as injectors. RIBs are produced via projectile fragmentation of heavy ions or in-flight fission of uranium ions by superconductng fragment separator, BigRIPS which consists of 14 superconducting quadrupole triplets and 6 room temperature dipole magnets.

but the advantage of short lifetimes is lost. The advantage of the latter is variable energy and high beam quality but with restrictions on the half-lives that can be accelerated related to the hold-up times of the radioactive ions in the target-ion source of the mass separator. One can use fission of relativistic beams or neutron induced fission to produce neutron-rich radioactive ion beams.

Both approaches have been developed. Low energy radioactive ion beam facilities were developed for example at Leuven, Belgium, TRIUMF, Canada, and University of Notre Dame, US, particularly to study reactions of astrophysical interest of light nuclei. Exotic radioactive beams such as ^{11}Li were produced at RIKEN in Japan and GANIL in France via fragmentation. At GSI, beams up to 2 GeV/A were used to produce proton and neutron rich nuclei at relativistic energies via fragmentation or Coulomb fission. A similar facility has been developed at Michigan State University. Relativistic radioactive beams of individual exotic products can be separated in a fragment mass analyzer. These radioactive beams also can be injected into a cooler ring to slow down their energies and then the beams can be reaccelerated. Exciting results are being obtained (see proceedings of the *Radioactive Ion Beam Conference* in 2009 held at Michigan State University and the *Fifth International Conference on Exotic Nuclei and Atomic Masses* at Poland in 2008, see Eur. Phys. Journal A, 2009).

The first low-energy radioactive ion beam facility in the United States became active in Oak Ridge in 1996. It can accelerate radioactive beams with half-lives down to a few minutes. A cyclotron is used to produce the radioactive isotopes which are then mass- and isobar- separated, and then accelerated with a 25-MeV electrostatic accelerator. As another example of the important physics being done at this RIB facility, β-delayed neutron probabilities for nuclei near double magic ^{78}Ni were measured. These probabilities for $^{76\text{-}78}$Cu and ^{83}Ga were found to be two to five times higher than previously reported. The large discrepancies with previous results indicate the necessity of radioactive beam purification for these difficult measurements. These βn emissions not only shed light on the evolution of nuclear structures in neutron rich nuclei, they impact our understanding of rapid-neutron capture sites of nucleosynthesis and the design of fission power reactors. (J. Winger et al., Phys. Rev. Lett. **102**, 142502 (2009)).

Upgrades of RIB facilities have been made in Canada, Germany, France, Russia and the USA. Major efforts are being made worldwide to build new billion dollar RIB facilities to attack the problems described here. Japan initiated the first such major new facility which became operational in 2007 as shown in Fig. 11.33. It includes three cyclotrons and a massive spectrometer to separate high energy fragmentation products. At GSI in Germany, another major new RIB facility is under way. In fall 2008, the US announced its decision to build a major new RIB facility at Michigan State University. The design of this facility is based on fragmentation technique as shown in Fig. 11.34.

Let us illustrate some of the novel physics that can be addressed. Of major importance can be the use of RIBs to produce and probe totally new states of nuclear matter. Proton-rich nuclei beyond the proton drip line cannot exist. However, the Coulomb barrier in heavy nuclei is sufficiently high that the unbound proton can be held inside the nucleus for times sufficiently long that γ-ray decay of the excited

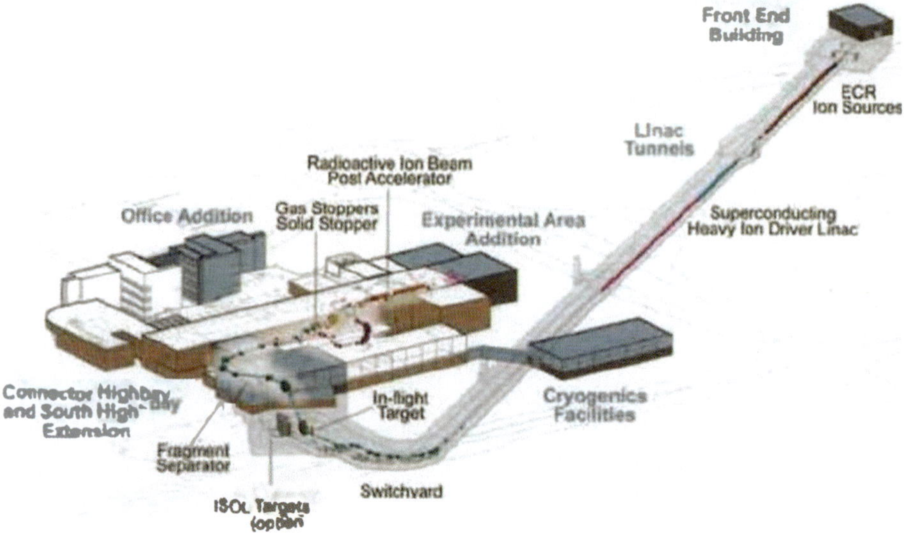

FIGURE 11.34 A new facility for Radioactive Ion Beam, FRIB, has been sited at Michigan State University. As proposed, it has a superconducting heavy ion linear accelerator driver and an inflight fragment separator for fast beam which can be stopped and reaccelerated.

levels can occur before the proton drips off. For example, it may be possible to observe in a RIB reaction the γ-rays in ^{177}Bi, which has 15 fewer neutrons than the predicted proton drip line. Such exotic nuclei can comprise a completely new form of nuclear matter in which the Coulomb force is no longer a small perturbation on the nuclear mean field but can be comparable to or greater than it. Our present understandings of nuclear matter are based on a mean-field approach as the dominant factor in the behavior of nuclear matter. How will nuclear matter behave and what will the structure of nuclei look like in this new region? Thus, with RIBs we should be able to probe fundamental properties of the nuclear force under completely new conditions.

As outlined in Table 11.4, such RIBs open up a variety of exciting studies from new states of nuclear matter to new nuclear structures and reactions to broad areas of nuclear astrophysics. In nuclear astrophysics one can measure, for the first time, unknown reaction rates, unknown nuclear shapes, masses, and isomers, all of which are critical to the understanding of inhomogeneous big bang nucleosynthesis, break-out reactions from the CNO energy cycle, and Ne–Na–Mg and Mg-Al-Si minicycle reactions. Recent calculations indicate that inhomogeneous big band (IBB) nucleosynthesis not only produces very light nuclei, $A \leq 9$ (not strongly produced in other nucleosynthetic processes) but also produces much heavier nuclei.

One of the important early contributions of the RIB facilities to astrophysics will be studies of nuclei in the rp (rapid proton capture)-process path of nucleosynthesis. Thermonuclear runaway can occur in material accreting on the surface of a white dwarf in interacting, close binary stellar systems leading to emission of jets, see Fig. 11.35.

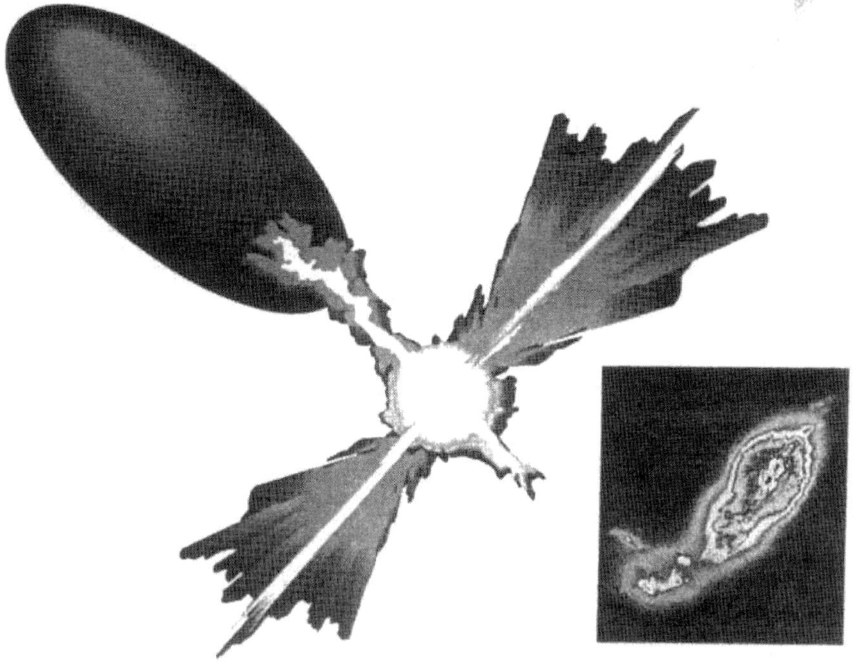

FIGURE 11.35
Graphical conceptions of the cataclysmic binary star system R Aquarii in which matter extracted from a red giant accumulates in an accretion disk around a white-dwarf companion until it detonates in a thermonuclear explosion with a long narrow jet. In a Hubble Space Telescope enhanced photograph sensitive to the light of doubly-ionized oxygen, the jets corresponding to an explosion thought to have occurred in the late 1970s, can be traced over 4×10^{11} km.

TABLE 11.4
Interesting scientific questions that can be addressed with RIB facilities

Nearly self-conjugate nuclei ($N \approx Z$)	*General nuclear structure problems (everywhere)*
New doubly-magic ^{100}Sn	Proton drip line (proton and delayed-proton radioactivities)
Exotic nuclear shapes	New regions of deformation
Superallowed β decay	Long isotopic and isotonic chains (pn interaction)
Tests of conserved vector current in standard model	proton transfer reactions
	Subbarrier transfer
Proton-rich heavy nuclei	neutron rich nuclei
Coulomb redistribution	
"Schizophrenic nuclei"	*Astrophysics*
Octupole deformations	rp-Process nucleosynthesis
"Coriolis correlations"	CNO cycle and breakout
	Na–Ne–Mg and Mg–Al–Si minicycles
	p Process nucleosynthesis
	Inhomogeneous big bang nucleosynthesis
	Neutrino-induced nucleosynthesis

Such an outburst, observed as a nova, or if very energetic as an x-ray burst, has all the ingredients (hydrogen-rich accreted material from the red giant, light heavy ions from the white dwarf, and high temperatures) to ignite an explosive hydrogen-burning process to drive the rp-process. If the temperatures and densities are sufficiently high, such an rp-process may proceed all the way to mass 100. Both reaction rates and specific nuclear properties, such as masses, binding energies, quantum numbers of energy levels in the Gamow window, level densities, and the existence and lifetimes of isomeric states, are needed to understand nucleosynthesis in the rp-process path. Nuclei not accessible with stable beams and stable targets can be studied with RIB facilities. Then for the first time, one will have many of the necessary reaction rates and nuclear structures experimentally measured to test theoretical calculations of nucleosynthesis.

Present and future radioactive ion beam facilities will accelerate neutron-rich nuclei. Very little is known about neutron-rich nuclei in comparison to the wealth of information on proton-rich nuclei, which can be produced in fusion reactions. Neutron-rich RIBs would open up opportunities to make and study an enormous number of new, unknown neutron-rich nuclei which sit in different regions of the single-particle spectra from proton-rich nuclei. Moreover, it would open up further extensions of the periodic table to still higher-Z superheavy elements. By using cold fusion reactions on stable targets, e.g. $^{70}Zn + ^{208}Pb \rightarrow ^{278}112$ the group at GSI, Germany, first identified new elements with Z = 108-112. The half lives and cross sections dropped so fast that using such reactions to produce higher Z nuclei seemed not possible. Then a Russian-Livermore collaboration used a neutron-rich ^{48}Ca beam in hot fusion reactions on radioactive uranium, plutonium, curium and californium to discover elements Z = 114, 115, 116 and 118. Their production cross sections and lifetimes were larger and longer than expected because the extra neutrons brought the produced nuclei nearer the N = 184 spherical magic number as shown in Fig. 11.36. However, beams of neutron-rich nuclei offer the only possibility to reach very neutron-rich superheavy nuclei such as Z = 114-120 with N = 184. The predicted spherical neutron shell gap around N = 184 could give sufficient stability that such nuclei could have half-lives much longer than those observed for even recently discovered Z = 114-118. Let us illustrate why beams of neutron-rich nuclei are needed. The most neutron-rich heavy elements are ^{208}Pb and ^{209}Bi with N = 126. If one wanted to add 32 to 34 protons to ^{208}Pb to form element Z = 114 or 116, the heaviest Z = 34 is ^{80}Se. The combined N of 126 + 46 = 172.

Another very interesting question opened up by the discoveries of new elements 114-118 is where do these fit in the periodic chart of the elements (see Chapter 5, page 210). Is element 118 a nobel gas as the normal ordering would indicate? In such high Z nuclei, the electrons are more strongly bound and their relativistic effects shrink their orbits. How such orbital shrinking effects their chemical properties is not known. Thus there are both important nuclear structure questions and chemistry questions.

Then there is the whole field of the synthsis of heavy elements in stars along paths of rapid and slow neutron capture in neutron-rich nuclei. These are the routes which make most of the heavy elements we have on Earth, and very little is known

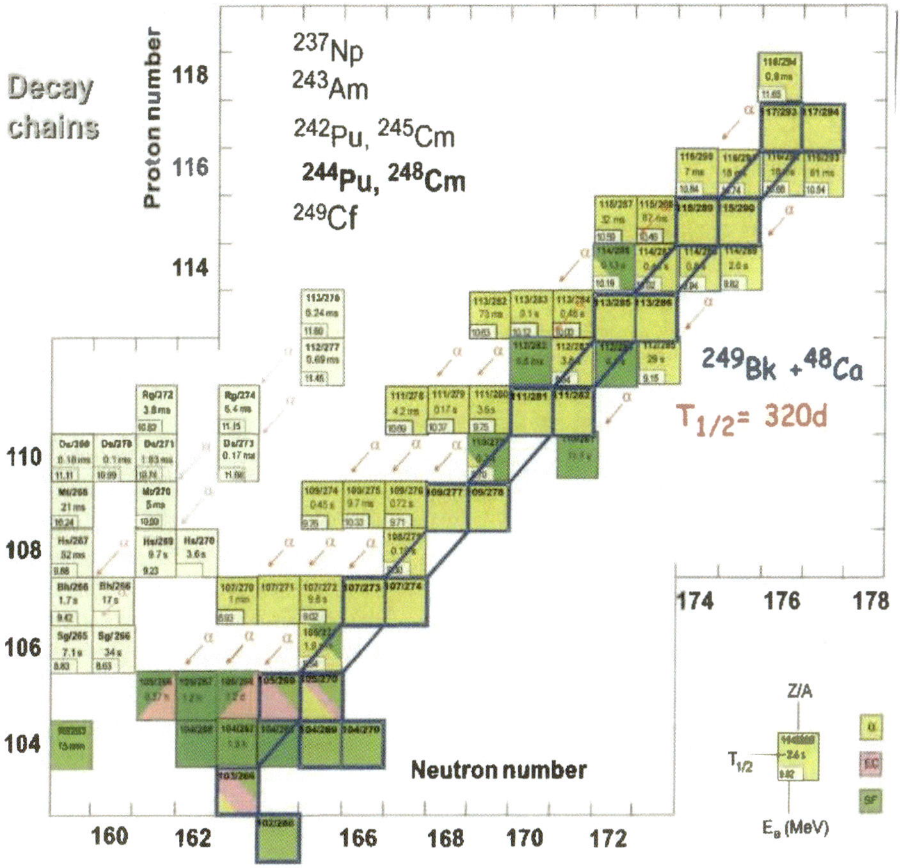

FIGURE 11.36
New superheavy elements discovered in GSI, Z = 108-112 on left, and discovered in Dubna 114-118 on right. A new experiment to discover element 117 in the $^{249}Bk + ^{48}Ca$ reaction by a Vanderbilt-Dubna-ORNL-LLNL collaboration is also shown.

about the structure and the reaction rates of these nuclei to test theoretical models of nucleosynthesis along these paths.

In summary, new experimental facilities including new radioactive ion beam facilities, γ-ray arrays like Greta, other detector systems and digital electronics will give unprecedented views of the structure of nuclei under new extreme conditions, including weak, presently inaccessible reaction channels, very weak decay paths at high spin, and other exotic decay processes and provide unique opportunities to probe experimentally for the first time many astrophysical phenomena, and provide energetic radioactive ions that can be implanted deep into materials for the first time for solid-state physics research.

Here the latest nuclear research developments in the understanding of the structures of nuclei have been traced to give a flavor of the current interplay of theory and experiment, of how they push and pull on each other, as one takes the lead and then the other. This research also illustrates the continuing challenges to develop significantly better experimental facilities to probe new theoretical predictions, to explore unknown regions where new phenomena may be lurking, and to come forward with new theoretical insights to challenge experimentalists.

Similar tracings and interplays could be presented in other equally important, major areas of nuclear physics if space allowed. In Chapter 14 major efforts uniting both nuclear and particle physics are discussed. For example, the ultra relativistic heavy-ion collider, RHIC, at Brookhaven National Laboratory collides 100 GeV/A gold ions ($100 \times 197 = 19.7$ TeV gold) on 100 GeV/A gold ions and the Large Hadron Collider, LHC, at CERN will collide 2.7 TeV/A lead ($2.7 \times 208 = 562$ TeV) on 2.7 TeV/A lead (see also Chapter 12). In such collisions one can create a new state of matter — the quark–gluon plasma. Such a plasma is thought to have been the first form of matter in the earliest universe following the big bang that led to the universe we see today. Other equally important questions about strange matter and antimatter may be probed. The 6-GeV Continuous Electron Beam Acclerator Facility at Newport News probes such questions as the quark structure of the proton and the role of multinucleon correlations. All these facilities require the development of very large, new detector facilities. Clearly, there are many frontiers to be conquered that will challenge our experimental technology and theoretical understanding as we seek to understand nuclear matter from its collective and single-particle motions right on through to the quark–gluon constituents of the nucleons as discussed in Chapter 14.

11.4 SUMMARY

1. The nucleons inside the nucleus are held together by the strong force. The force between two nucleons is a factor of 100 stronger than the Coulomb force. The strong force between nucleons is very short range ($\leq 10^{-14}$ m) and saturates, so the nucleons can interact only with a few nearby nucleons. Thus, the average binding energy per nucleon (B/A), except for the light nuclei, is nearly constant (~ 8 MeV), and the total nuclear binding energy is proportional to the nuclear mass number. Also, it follows that a nucleus acts in some ways like a liquid drop. The nuclear volume is proportional to the nuclear mass number, and the nuclear density is nearly constant (independent of A). Nuclei are almost incompressible, so there is a repulsive hard core in the force between nucleons. In addition, the strong force between nucleons is charge independent. The force between nucleons can be treated as the exchange of a virtual π meson between two nucleons. However, the ultimate origin of the strong force in nuclei is the true strong color force mediated by the gluon as discussed in Chapter 14.

2. It is still not possible to write the nuclear force in closed form as we can the Coulomb force. Even if it were known, we do not know how to solve the many-body problem between the nucleons. In atoms the interactions between the electrons are very small, and one has basically only a two-body electron–nucleus problem. Thus, nuclear models which incorporate various features of the nucleon–nucleon force have been developed to give us insight into the shapes and structure of nuclei. These models fall into two broad categories: (a) independent-particle models, and (b) collective models.

3. The Fermi gas and spherical shell models are independent-particle models in which the nucleons move independently on orbits inside the nucleus under an attractive mean field generated by the action of all the nucleons on each other. The nuclear shell model makes use of the fact that the nuclear force depends on how the spin and orbital motions of the nucleons are coupled to explain the nuclear spherical magic numbers for Z and N at which nuclei are particularly stable. However, the origin of the nuclear spin–orbit coupling is different from that in atoms. It is related to the nuclear force and is not a magnetic interaction as in atoms. Moreover, it is much stronger and has the opposite sign from that in atoms. The Mayer–Jensen shell model predicts the spherical magic numbers of 2, 8, 20, 28,..., which give special spherical stability to nuclei with N and/or Z having a magic number. The spins, parities, and magnetic moments of nuclei near these spherical closed shell magic numbers are also correctly predicted. There is also a nuclear pairing force that couples two protons and two neutrons in pairs so that the total angular momentum of each pair is zero.

4. When nuclei are far away from these spherical closed shells, they are observed to have considerable deformation from a spherical shape as in the rare-earth and actinide nuclei. These deformed nuclei can undergo a new mode of excitation, namely, rotational states, which quantum mechanics shows are not accessible to spherical nuclei. Both spherical and deformed nuclei, however, can have vibrational modes of excitation, which are more extensive in deformed nuclei. Thus, the patterns of the energy levels in spherical and deformed nuclei are quite different.

5. Recent research has shown a significantly greater diversity and interplay of these two cases of spherical and deformed shapes. The coexistence of both these nuclear shapes is now seen throughout the periodic table, especially in nuclei under extreme conditions far from stability in their neutron number for a given Z or in their rotational motion at high spins. The spherical shell model with its spherical magic numbers has been significantly extended to include new deformed magic numbers for nuclear ground states and for bands of high-spin excited states associated with superdeformed shapes. However, these new deformed magic numbers become important only when the nuclei have both protons and neutrons at or very near (only a few particles away) deformed magic numbers (shell gaps) associated with the same deformation so that they reinforce each other. New superdeformed shapes are associated with new deformed double magic nuclei, such as for $Z = 38 = N$; $Z = 38$, $N = 60$, 62; $Z = 66$,

$N = 86$; and $Z = 80$, $N = 112$. In neutron rich nuclei, the disappearance of some spherical magic number like 20 and new magic numbers are found. These new data have called for new, more microscopic descriptions based on more realistic nucleon-nucleon interactions to explain the new vistas of changing nuclear shapes and structures.

6. New experimental facilities are being developed worldwide to open up presently inaccessible areas of nuclear research. For example, radioactive ion beam accelerators will open up presently inaccessible studies of (a) nuclear matter under new extremes never seen before including the structures and shapes of new exotic proton and neutron rich nuclei; (b) proton radioactivities and β-delayed particle emission; (c) many areas of nuclear astrophysics including inhomogeneous big bang nucleosynthesis, CNO cycle breakout, rapid proton and neutron capture and slow neutron capture; (d) super heavy elements. New mass- and Z-selection devices and large γ-ray detection systems also are being developed.

7. The 6-GeV Continuous Electron Beam Accelerator Facility under upgrade to 12 GeV and the ultrarelativistic heavy ion colliders, RHIC at Brookhaven National Laboratory and LHC at CERN provide new opportunities to probe the substructure of nucleons. Fundamental questions include what the interior of a nucleon is really like and the deconfinement of quarks and gluons into a new state of matter — the quark-gluon plasma (as further discussed in Chapters 12 and 14).

PROBLEMS

11.1. Justify the short-range character of the nuclear force from the fact that the nuclear density is constant or nearly so.

11.2. Show that in light nuclei one should find $N = Z$ on the basis of charge symmetry ($F_{pp} = F_{nn}$).

11.3. Calculate the coefficient of the symmetry energy using Eqs. (11.19) and (11.20) and compare it with the experimental value (Section 9.11). Why is the calculated coefficient so different from experiments?

11.4. On the basis of the spherical nuclear shell model (Fig. 11.9), determine the expected spins and parities of the ground states of ${}^{39}_{20}Ca_{19}$, ${}^{41}_{20}Ca_{21}$, ${}^{119}_{50}Sn$, ${}^{121}_{51}Sb$, and ${}^{207}_{81}Tl$. What would you predict for the spin and parity of the first excited state in each of these nuclei? Compare your ground-state predictions with the experimental results in Appendix II. Your excited state predictions can be compared with the data in: www.nndc.bnl.gov.

11.5. Determine the spins and parities of the following nuclei according to the spherical shell model: ${}^{9}Be$, ${}^{27}Si$, ${}^{37}Cl$, ${}^{73}Ge$, and ${}^{92}Mo$.

11.6. Compare the total binding energies of ${}^{13}C$ and ${}^{13}N$. Since both nuclei have the same number of nucleons and the nuclear force is said to be charge independent, explain why these nuclei have different binding energies.

11.7. From the data in Fig. 11.12b calculate the moment of inertia from the 2^+ energy of 49.4 keV. Then calculate the energies of the 4^+, 6^+, and 8^+ levels and compare with experiment. Show that the theoretical $4^+/2^+$ energy ratio is 3.3 in the rotational model. How does the experimental value in ^{232}Th compare with this ratio?

11.8. For the 774.5 keV 2^+ level in ^{232}Th (Fig. 11.13), calculate the moment of inertia and use it to predict the energies of the 4^+, 6^+, and 8^+ members of this band. How does the moment of inertia and the comparison of calculated and experimental energies compare with the results of Problem 11.7?

11.9. The first 2^+ level in ^{180}Hf is at 93 keV (Fig. 11.12a). From Fig. 10.19 estimate the single-particle lifetime of this transition. How would you expect this single-particle lifetime to compare with the experimental value? Discuss in general terms the physical reasons of your comparison of these two lifetimes.

11.10. Compute the moment of inertia for each transition from the yrast states (the states) with the lowest energy for each spin, $2^+, 4^+, 6^+, 8^+, \ldots$) up to 20^+ in Fig. 11.28a and plot as a function of rotational frequency $\hbar\omega = \frac{1}{2}E_\gamma$ (the transition energy). What can you infer about the structure or shape of this nucleus as a function of rotational frequency?

11.11. What is the energy needed to take one neutron or one proton out from the nucleus of ^{14}N? Explain the reason for the difference.

11.12. Explain why you can expect large electric monopole radiation in transitions between β bands and ground bands when the spins of the initial and final states are the same but not in transitions between γ bands and ground bands when the spins are the same.

11.13. From a chart of the nuclides or Appendix II, determine the number of even-Z-even-N nuclei, even-Z–odd-N nuclei, odd-Z–even-N nuclei, and odd-Z–odd-N nuclei that are stable in nature. What can you conclude about the importance of the pairing force between protons and between neutrons?

11.14. How is it possible for the so-called spin–orbit force to have a different sign in nuclei from the spin–orbit force for the electrons in the atomic shells?

11.15. Suppose you discovered that the first excited states in three new nuclei (call them A, B, and C) very far from stability but in different regions of the periodic table between $A = 50$ and 250 had energies of 3.3 MeV, 800 keV and 80 keV, respectively. What can you infer about the structure and shape of nuclei A, B and C, and why?

11.16. What is meant by the term islands of isomerism? What is the origin of such islands and where are they expected to be found?

11.17. For years $Z = 40$ was considered a reasonably good spherical magic number associated with the gap in the spherical single-particle energies for protons and neutrons at 40. Now we know that 40 is spherical magic (gives extra stability to a spherical nuclear shape) only under certain conditions. What are those conditions? What do we know about ^{80}Zr, which has 40 protons and 40 neutrons?

11.18. What is meant by nuclear shape coexistence? Discuss one example of experimental evidence for nuclear shape coexistence.

11.19. What does the term single-particle estimate for the lifetime of a nuclear level mean? How can a nuclear level have a lifetime 100 times shorter than its single-particle estimate?

11.20. Show that superdeformed nuclei with $\beta \sim 0.4$ and 0.6 and hyperdeformed nuclei with deformation $\beta \sim 1.0$ have long-to-short axis ratios of approximately 3:2, 2:1 and 3:1, respectively.

11.21. Use the chart of the nuclides to select a neutron-rich nucleus for acceleration to interact with a heavy stable target around lead so as to produce element $Z = 114$ with $N = 184$.

CHAPTER

12

NUCLEAR INTERACTIONS AND REACTIONS

We cannot control atomic energy to any extent which would be of any value commercially, and I believe we are not likely ever to be able to do so.

E. Rutherford (1933)

12.1 INTRODUCTION TO NUCLEAR INTERACTIONS AND REACTIONS

12.1.1 Overview

As discussed in Chapter 10, studies of the radiations emitted in the spontaneous decays of nuclei are one of the major ways of gaining insight into the behavior of nuclear matter. In this chapter, alternate ways of sending in particles to probe nuclei by nuclear reactions and electromagnetic interactions are discussed. Such reactions probe both the structure of nuclei and the reaction mechanisms between nucleons and between nuclei. There are many different classes of particle-nucleus and nucleus-nucleus reactions; they are divided by mass and energy, beginning with beams of the lightest particles—photons, neutrinos, and electrons—and going all the way up to beams of uranium on targets from ${}^{1}_{1}H$ to ${}^{249}_{98}Cf$ and on to colliding beams of ${}^{208}Pb$ on ${}^{208}Pb$. It is not possible to cover all the types of reactions and interactions here. Only a few highlights and general discussions are given.

Electron accelerators with energies up to a few hundred MeV continue to be a major source of precision information on nuclear charge distributions and densities and to provide other nuclear structure information. The 6-GeV electron accelerator, the Continuous Electron Beam Accelerator Facility in Virginia, is opening new ways to probe nuclear matter including its quark structure. The electron-proton collider facility in Germany provided new insights into the quark structure of the proton before closing. The work done with both these accelerators is discussed in Chapter 14.

The strong nuclear-interacting probes can be divided into several groups. The first group includes light ions, neutrons, protons, deuterons, ^3H, ^3He and α-particles, which have been the most used probes historically. The second group, called heavy ions, includes all particles heavier than the α-particle. Heavy-ion beams provide unique opportunities for transfer of large angular momentum and mass to create nuclei under new extreme conditions of temperature, of angular momentum, of high Z (beyond the known elements) and with Z/N far from stability, and for the study of exotic new phenomena such as nuclear shock waves, quantum electrodynamics of strong fields, and the quark–gluon plasma. Another category comprises mesons, positive and negative pions and kaons. The mesons mediate the force between nucleons in the nucleus and so provide important insight into this force. In addition, by double charge exchange, π^+ going to π^- or vice versa in the reaction, new nuclei farther from stability can be probed. Finally, antiproton probes are now being used to study the nucleon–nucleon interaction.

The energies of the bombarding particles span a very wide range from below 1 eV up to a few hundred GeV. Reactions in which the total bombarding energy is under 100 MeV or under 5–6 MeV per nucleon for heavier nuclei are called low-energy heavy-ion reactions; from 100 MeV to about 1 GeV they are called intermediate-energy reactions; and above 1 GeV they are called high-energy reactions. The motivation for such studies covers a wide range, from studies of the structure of nuclei and the reaction mechanism, to the production of radioactive nuclei for basic and applied research, and on to energy sources and the creation of new forms of matter in which the nucleus dissolves into its constituents of quarks and gluons. At the highest energies the fields of elementary particle (high-energy) and nuclear physics are coming together again, as discussed in Section 14.4.3. One of the important questions in nuclear physics today is at what energy and in what way the quark structure of the nucleon manifests itself in nuclear reactions. Such questions are considered in more advanced texts.

Another frontier in studies of nuclear reactions is nuclear astrophysics, where it is important to measure many of the nuclear reaction rates of far-off-stability, short-half-life radioactive nuclei with protons, neutrons, and perhaps α-particles. Such reactions are involved in the production of the heavy elements present in the universe, in particular as found in the Earth. The evolution of stars and the production of heavy elements in stars are considered briefly in the last section. In addition, radioactive ion beams open up the production of observable quantities of presently inaccessible nuclei still further from stability for the study of their exotic decay modes and nuclear structure. Suffice it here to say that opportunities are expanding for accelerating radioactive nuclei to probe such questions. Major new radioactive ion beam facilities to study such questions are under development in Europe, the United States, and Asia. In addition, major new efforts to study x-ray and γ-ray sources in our solar system and in the universe are underway. From 1991-99, the Compton Gamma Ray Observatory in orbit about the Earth, provided unique insights into structure and evolution of the universe.

Nuclear reactions is a broad subject which is divided into different areas. A basic area is reaction kinematics: the application of the general laws of energy and momentum conservation, both nonrelativistically and relativistically. This does not involve any interaction mechanism among the particles, but considers only whether the reaction could occur. A second area is reaction dynamics, where the different interaction mechanisms among particles are considered. Interactions range from electromagnetic interactions, including Coulomb excitation and photon–nuclear reactions, to mesonic charge exchange. The two extremes of reactions are represented by compound fusion reactions, where the two nuclei fuse and equilibrate their energy, and direct nuclear reactions between only two or a few nucleons; there is a wide range in between, from preequilibrium and partial fusion, to high energies where, for different energies and conditions, one may have proton-induced spallation of many different particle combinations, deep inelastic transfer of a few nucleons, fragmentation of the projectile into a variety of different nuclei (all three of which produce exotic nuclei far off stability), and nuclear shock waves. Nuclear reaction dynamics are an important probe of the states of motion of the nucleons inside the nucleus. Nuclear reactions are used to study the structure, motions, and decay properties of nuclei up to extreme conditions of high spin, high temperature or being far from stability, and are used to produce tailor-made radioactive nuclei for specific applied purposes from energy production to use in medical and industrial diagnostic procedures, and other uses. The study of the structure of nuclei, as discussed in Chapter 11, involves the identification of the quantized energy levels in the nucleus and their properties—spin, parity, electromagnetic decay strengths, for example. These can be probed by studying the energies and angular distribution of scattered particles, or through studies of the γ-rays, conversion electrons, and electron–positron pairs emitted in the decay of the states formed in the nuclear reactions.

First, we look at several nuclear reactions which have played important roles in the development of nuclear physics in order to give some knowledge of nuclear reactions. Then we develop the nuclear reaction kinematics—the Q value equation with examples of its application. Next, we introduce the important concept of reaction cross-section, which measures the nuclear reaction probability. Finally, we briefly discuss the theory of the compound nucleus reaction mechanism.

12.1.2 Several Famous Early Nuclear Reactions

THE FIRST ARTIFICIAL NUCLEAR REACTION. In 1919, Rutherford used 7.68 MeV α-particles which were emitted by ^{212}Po as projectiles to bombard nitrogen gas. He found that more energetic light particles were produced, which turned out to be protons. The reaction was

$$^{4}_{2}\alpha_2 + ^{14}_{7}N_7 \rightarrow ^{1}_{1}p + ^{17}_{8}O_9 \tag{12.1}$$

The α-particle fuses with the ^{14}N nucleus and this breaks up into ^{17}O and a proton. Such a reaction in shortened version is written as: ^{14}N(α, p)^{17}O. This was the first time in history that anyone had realized the dream of the alchemists of transforming one element into another, of "turning things to gold." Of course, it was unknown then that nuclear reactions turning hydrogen into helium are going on at a tremendous rate to produce the Sun's energy and that of all stars. Closer to us, cosmic-ray neutrons are continuously producing radioactive ^{14}C from ^{14}N. This production allows us to use ^{14}C to date the time when previously living systems died.

The setup for Rutherford's experiment is shown in Fig. 12.1. The distance between the α source and the fluorescent screen was 28 cm. From estimates of the energy loss, no α-particle could reach the fluorescent screen. Indeed, when the system was filled with CO_2 gas, no flash could be seen in the microscope. But when it was filled with N_2 gas, flashes were observed. After careful analyses, Rutherford decided that the particles reaching the screen were protons. Later, using a cloud chamber, the α-particle traces, the very short-range ^{17}O traces, and the long-range proton traces were observed and recorded. (Range is the distance the particle travels in the gas of the chamber and is related to the particle's kinetic energy.)

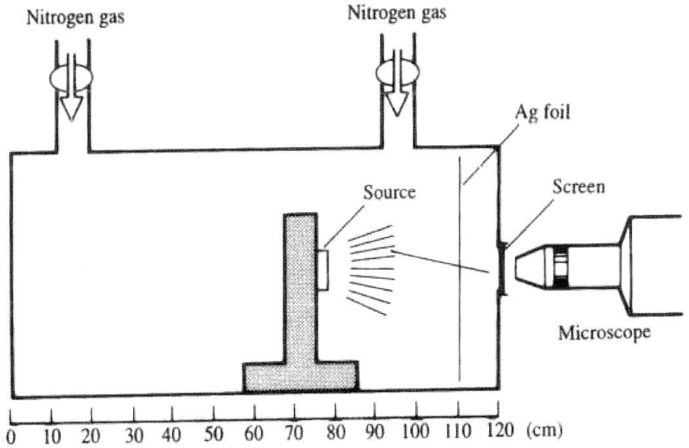

FIGURE 12.1
The experimental setup for the first observed transmutation of elements by a nuclear reaction.

THE FIRST NUCLEAR REACTION PRODUCED WITH AN ACCELERATOR. In 1932, J. D. Cockcroft and E. T. S. Walton invented the Cockcroft–Walton accelerator, which produced up to 500 000 volts on a terminal. This could accelerate protons to 500 keV to produce the following reaction:

$$\begin{aligned} &{}^1_1\mathrm{p} + {}^7_3\mathrm{Li}_4 \rightarrow {}^4_2\alpha + {}^4_2\alpha \\ &{}^7\mathrm{Li}(\mathrm{p}, \alpha)^4\mathrm{He} \end{aligned} \qquad (12.2)$$

Each α-particle had a kinetic energy of 8.9 MeV. Remarkably, by putting in an energy of only 0.5 MeV, an output energy of 17.8 MeV was achieved. This is an example of a reaction that releases energy. They won the Nobel Prize in Physics for their development and use of this new accelerator. Other accelerators quickly came into use: the Cyclotron developed by E. O. Lawrence (California), and the Van de Graaff electrostatic accelerator by R. G. Van de Graaff (Alabama–MIT).

THE FIRST NUCLEAR REACTION TO PRODUCE NEW RADIOACTIVE NUCLEI. In 1934 Frédéric and Irène Joliot-Curie produced and identified the first new radioactive nuclide produced in a nuclear reaction by bombarding aluminum with α-particles:

$$\begin{aligned} &\alpha + {}^{27}\mathrm{Al} \rightarrow \mathrm{n} + {}^{30}\mathrm{P} \\ &{}^{27}\mathrm{Al}(\alpha, \mathrm{n})^{30}\mathrm{P} \end{aligned} \qquad (12.3)$$

The product ^{30}P undergoes a new type of β decay, β^+ radioactivity, with a half-life of 2.6 minutes:

$$^{30}\mathrm{P} \rightarrow {}^{30}\mathrm{Si} + \beta^+ + \nu \qquad (12.4)$$

The ^{30}P is called an artificial radioactive nucleus in many books. As noted earlier, this name is somewhat misleading. It signifies only that the activity is produced in the laboratory and is not found to occur naturally on Earth. It is just as real as those found terrestrially in nature. In fact, many such activities are found in supernovas, so they are produced in nature but their half-lives are too short for them to still be present in the Earth's crust.

THE REACTION THAT BROUGHT ABOUT THE DISCOVERY OF THE NEUTRON. This is the reaction

$${}^4_2\alpha + {}^9_4\mathrm{Be} \rightarrow {}^1\mathrm{n}_1 + {}^{12}_6\mathrm{C} \qquad (12.5)$$

This reaction was observed by W. Bothe and H. Becker in 1930, but they thought that the neutral entity was a γ-ray. Later, the Joliot-Curies repeated this experiment and stopped the reaction products in wax. They found that protons with an energy of about 6 MeV were ejected from the wax. They explained this as a type of "Compton effect" produced by a γ-ray. It is easy to estimate (the formula is in Chapter 6) that if a 6 MeV proton is to be produced, the γ-ray energy has to be at least 60 MeV. But it is impossible to produce such a high-energy γ-ray from an α-particle bombarding ^{9}Be to produce ^{12}C. (See Problem 12.1.)

In 1932, J. Chadwick repeated this experiment. He not only let the neutral reaction product bombard hydrogen, but also had it bombard helium and nitrogen. Then, comparing the recoils of the hydrogen, helium, and nitrogen, he concluded that what was produced was a neutral particle whose mass was almost equal to the proton's mass. He called it the neutron.

In these examples, the following conservation laws were followed: electric charge conservation, nucleon number conservation, and angular momentum conservation in addition to total energy and linear momentum conservation, which are the basis of the Q value equation that follows.

12.2 REACTION KINEMATICS

12.2.1 The Q Value Equation

In general, a nuclear reaction can be expressed as

$$i + \mathrm{T} \rightarrow l + \mathrm{R} \tag{12.6}$$

or

$$\mathrm{T}(i, l)\mathrm{R}$$

where i is the incoming particle, T is the target, l is the outgoing, generally light-mass, particle, and R is the residual (product) nucleus. Their static masses and kinetic energies are M_i, M_T, M_l, M_R; and K_i, K_T, K_l, K_R, respectively. No matter what happens in the reaction, applying energy conservation always yields

$$M_ic^2 + K_i + M_Tc^2 + K_T = M_lc^2 + K_l + M_Rc^2 + K_R \tag{12.7}$$

The reaction energy Q is given by

$$\begin{aligned} Q &\equiv [(M_i + M_T) - (M_l + M_R)]c^2 \\ &= (K_l + K_R) - (K_i + K_T) \end{aligned} \tag{12.8}$$

This is the definition of the reaction Q value, which is the difference between the energy of the total masses before the reaction and the energy of the total masses after the reaction. Using Eq. (9.3), we can express the Q value in terms of the nuclear binding energies or, using Appendix II, in terms of the mass defect Δ:

$$\begin{aligned} Q &= (B_l + B_R) - (B_i + B_T) \\ &= (\Delta_i + \Delta_T) - (\Delta_l + \Delta_R) \end{aligned} \tag{12.9}$$

If $Q > 0$, the initial masses are greater than the final masses (binding energies of the final masses are greater than the binding energies of the initial masses; the greater binding makes the final masses less) so the reaction gives off energy. It is called an exoergic (exothermic) reaction. If $Q < 0$, the reaction requires kinetic energy to make up the deficit in masses and is an endoergic (endothermic) reaction.

For example, in the reaction Eq. (12.2), ^{7}Li(p, α)^{4}He, $M_i = 1.007\,825$ u, $M_T = 7.016\,004$ u, and $M_l = M_R = 4.002\,603$ u. From (12.8), where $Q/c^2 =$

$M_i + M_T - (M_l + M_R)$, we can calculate $Q/c^2 = 0.018\,623$ u, or $Q = 17.35$ MeV; this is an exoergic reaction that gives off energy. Or we can write $Q = \Delta_H + \Delta_{Li} - 2\Delta_{He} = 7.289 + 14.908 - 2 \times 2.425$ MeV $= 17.35$ MeV. As an example of an endoergic reaction that requires energy, let us calculate Q for the reaction ${}^{76}_{34}Se(\alpha, 2n){}^{78}_{36}Kr$:

$$Q = (\Delta_\alpha + \Delta_{Se}) - (2\Delta_n + \Delta_{Kr}) = 2.425 - 75.252 - (2 \times 8.071 - 74.160)$$
$$= -14.809\,\text{MeV}$$

We can use momentum conservation to write another equation for the Q value. According to momentum conservation, the incoming particle's momentum should be equal to the vector sum of the outgoing particle's momentum and that of the residual nucleus. As shown in Fig. 12.2 for a fixed target,

$$\boldsymbol{p}_i = \boldsymbol{p}_l + \boldsymbol{p}_R$$

or

$$p_R^2 = p_i^2 + p_l^2 - 2p_i p_l \cos\theta \qquad (12.10)$$

Using the classical relation $p^2 = 2MK$, this can be written as

$$M_R K_R = M_i K_i + M_l K_l - 2\sqrt{M_i M_l K_i K_l}\cos\theta \qquad (12.11)$$

Merging this with (12.8), by substituting for K_R, we have

$$Q = \left(1 + \frac{M_l}{M_R}\right)K_l - \left(1 - \frac{M_i}{M_R}\right)K_i - \frac{2\sqrt{M_i M_l K_i K_l}}{M_R}\cos\theta \qquad (12.12)$$

This is the nuclear reaction Q equation.

For a given nuclear reaction, if the masses both before and after the reaction are known, from Eq. (12.8), we can calculate the reaction energy. The masses of stable projectiles and targets are known. If one of the other nuclear masses M_l or M_R is unknown, then, by measuring the kinetic energy K_i and K_l of the incoming particle and outgoing particle and the angle θ between them, we can calculate the Q value from Eq. (12.12) in terms of the unknown M_l or M_R. This Q (in terms of M_l or M_R) can be used in Eq. (12.8) to calculate the mass of the unknown nucleus. Many nuclear masses have been determined in this way.

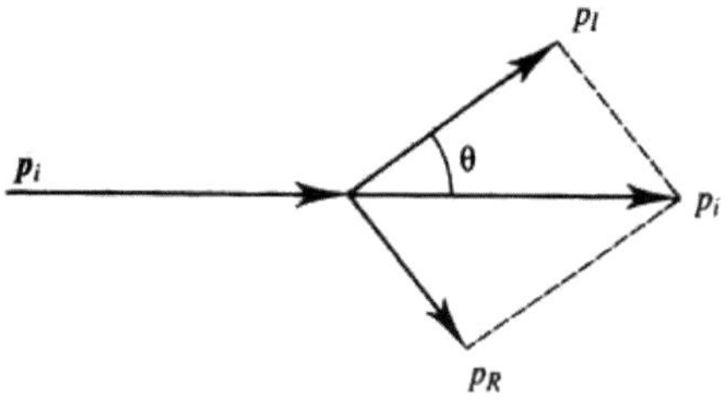

FIGURE 12.2
Momentum conservation in a nuclear reaction.

The following very useful expression which involves the Q value is obtained by putting Eq. (12.12) in the form of a quadratic equation ($ax^2 + bx + c$, where $x = \sqrt{K_l}$) and solving it:

$$K_l(\theta) = (u \pm \sqrt{u^2 + w})^2 \tag{12.13}$$

where

$$\begin{aligned} u &\equiv \frac{\sqrt{M_i M_l K_i}}{M_l + M_R} \cos\theta \\ w &\equiv \frac{M_R Q + K_i(M_R - M_i)}{M_l + M_R} \end{aligned} \tag{12.14}$$

Note that the kinetic energy of the outgoing particle is a function of the outgoing angle θ. There is a different kinetic energy at different angles. For a particular nuclear reaction, as long as we know the kinetic energy of the incoming particle K_i and the masses, we can calculate $K_l(\theta)$. This expression is not only suitable for an outgoing light-mass particle, but also applies for the residual nucleus if we only exchange R and l.

In order to have the radical sign in Eq. (12.13) make sense, we must have

$$u^2 + w \geq 0$$

An exoergic reaction with $w > 0$ must have $u^2 + w > 0$. For an endoergic reaction, if $w < 0$ then the incoming energy K_i must at least be greater than some value so that $u^2 + w \geq 0$ for the reaction to occur. The minimum K_i is determined by the following equation:

$$\frac{M_i M_l K_i}{(M_l + M_R)^2} \cos^2\theta + \frac{M_R Q + K_i(M_R - M_i)}{M_l + M_R} = 0 \tag{12.15}$$

When $\theta = 0$, K_i is a minimum. This is the threshold of the endoergic reaction:

$$K_{\text{th}} = -Q \frac{M_l + M_R}{M_l + M_R - M_i} \tag{12.16}$$

Rewrite Eq. (12.8) as

$$M_i + M_T = M_l + M_R + \frac{Q}{c^2} \tag{12.17}$$

For the case $M_T \gg Q/c^2$, the threshold equation (12.16) can be rewritten:

$$K_{\text{th}} = -Q \frac{M_T + M_i}{M_T} \tag{12.18}$$

Consider the K_{th} for ^{76}Se(α, 2n)^{78}Kr. Note that for the masses we can use the A values as a reasonable approximation,

$$K_{\text{th}} = -(-14.83)\,\text{MeV} \frac{76 + 4}{76} = 15.61\,\text{MeV}$$

For elastic scattering (in the strict sense, elastic scattering does not belong in the nuclear reaction category since there is no excitation of the nucleus, only recoil kinetic energy is given to the target), $Q = 0$, $M_i = M_l = M_1$, $M_T = M_R = M_2$. Then,

$$K_l(\theta) = \left\{ \frac{M_1}{M_1 + M_2} \cos\theta \pm \left[\left(\frac{M_1 \cos\theta}{M_1 + M_2} \right)^2 + \frac{M_2 - M_1}{M_2 + M_1} \right]^{1/2} \right\}^2 K_i \qquad (12.19)$$

In general, in the above equations, if there are masses in both the numerator and denominator, then the error introduced by using the nuclear mass number A instead of M is not greater than 1 percent. One can also have inelastic scattering in which some of the energy of the projectile goes into internal excitation energy of the nucleus, leaving the target nucleus in an excited state which will then decay by electromagnetic processes (γ-emission, internal conversion, and pair decay).

12.2.2 Coulomb Barrier and Coulomb Excitation

There is another important parameter in the collision of two positively charged ions—the Coulomb barrier. For two ions with charge Z_1 and Z_2 and mass numbers A_1 and A_2 to come sufficiently close to touch (their centers are separated by the sum of their radii) and so be in the range of the nuclear force between their nucleons, the kinetic energy of the projectiles must equal or exceed the repulsive Coulomb barrier between them when they touch. This barrier is

$$V = \frac{1.44 Z_i Z_T \,\text{MeV}\cdot\text{fm}}{R_i + R_T} = \frac{1.44 Z_i Z_T \,\text{MeV}\cdot\text{fm}}{R_0\left(A_i^{1/3} + A_T^{1/3}\right)} = \frac{1.2 Z_i Z_T \,\text{MeV}}{A_i^{1/3} + A_T^{1/3}} \qquad (12.20)$$

where $R_0 = 1.2\,\text{fm}$. The Coulomb barrier for bombarding ${}^{60}_{28}\text{Ni}$ with ${}^{16}_{8}\text{O}$ is

$$V = \frac{1.2 \times 8 \times 28\,\text{MeV}}{16^{1/3} + 60^{1/3}} = \frac{268}{6.43} = 41.7\,\text{MeV}$$

For the ${}^{76}\text{Se}(\alpha, 2n)$ reaction,

$$V = \frac{1.2 \times 2 \times 34\,\text{MeV}}{4^{1/3} + 76^{1/3}} = \frac{81.6\,\text{MeV}}{5.82} = 14.0\,\text{MeV}$$

In order to conserve linear momentum as well as energy, since the incoming projectile has momentum, in a compound reaction where the two nuclei fuse into one, the outgoing nucleus must have linear momentum. Thus, in order to conserve linear momentum the projectile kinetic energy K_i must exceed V so the real Coulomb barrier threshold kinetic energy K_{CBT} (the subscript CBT indicates Coulomb barrier threshold):

$$K_{CBT} = V\left(1+\frac{m_i}{m_T}\right) = \frac{1.2Z_iZ_T}{A_i^{1/3}+A_T^{1/3}}\left(1+\frac{m_i}{m_T}\right)$$
$$= \frac{1.2Z_iZ_T}{A_i^{1/3}+A_i^{1/3}}\left(1+\frac{A_i}{A_T}\right) \tag{12.21}$$

For the reaction ^{16}O projectile on ^{60}Ni,

$$K_{CBT} = 41.7 \times \left(1+\frac{16}{60}\right)\,\text{MeV} = 41.7 \times 1.267\,\text{MeV} = 52.8\,\text{MeV}$$

For $^{76}Se(\alpha, 2n)$,

$$K_{CBT} = 14.0 \times \left(1+\frac{4}{76}\right)$$
$$= 14.0 \times 1.053 = 14.7\,\text{MeV}$$

This kinetic energy must be exceeded in any reaction between two charged particles, including endoergic as well as exoergic ones. For an endoergic reaction Eq. (12.16) also holds, and the kinetic energy must satisfy this equation as well as Eq. (12.21). Note that for the $^{76}Se(\alpha, 2n)$ reaction $K_{th} = 15.6\,\text{MeV}$ is greater than $K_{CBT} = 14.7\,\text{MeV}$, so K_{th} is the important threshold. When the Coulomb barrier exceeds the K_{th} threshold energy in an endoergic reaction, the excess energy goes into internal energy of the nucleus. This can leave the nucleus in a highly excited state.

In a heavy ion collision, this energy can be up to a factor of 10 and more different for the production of the same compound nucleus, depending on which nucleus is the projectile and which is the target. In the case above of ^{16}O on a ^{60}Ni target, $K_{CBT} = 1.267V$ where V is the Coulomb barrier (Eq. 12.20), but for ^{60}Ni in ^{16}O it is $4.75V$. This difference increases with the mass differences. Consider the reaction of ^{12}C and ^{120}Sn. If ^{12}C is the projectile $A_i/A_T = 1/10$ and $K_{CBT} = 1.1V$, but if ^{120}Sn is the projectile $A_i/A_T = 10$ and $K_{CBT} = 11V$. However, most of this energy goes into kinetic energy of the recoil in order to conserve momentum, not internal energy of the nucleus. Why would one want to use ^{120}Sn as a projectile? In certain experiments this so-called inverse reaction has advantages which can be the difference between doing and not doing an experiment. As examples, if one wants to measure the lifetime of a nuclear state by observing the Doppler shift of the γ-rays emitted, then the velocity of the recoil and Doppler shifts are much greater for a heavy projectile on a light target. If one wants to identify the masses of the recoil fragments by sending them through a recoil mass spectrometer which has a relatively small geometric entrance solid angle of, for example, 10 msr in a good system, the inverse reaction can kinematically focus nearly all the recoils into this small geometric solid angle, vastly improving the collection efficiency and reducing a prohibitive 60-day experiment to 3–6 days.

12.2.3 Examples of the Application of the Q Value Equation

IDENTIFY THE TARGET NUCLEUS. The target in a nuclear reaction is often not pure. Oxygen and carbon often appear in targets. For example, in studying the reaction of deuterium and samarium, SmF or SmO may be sprayed on a carbon foil as a target. How, then, can we distinguish the samarium reaction products from the oxygen, carbon, and fluorine reaction products?

For simplicity, consider only elastic scattering, where kinetic energy is conserved. Any nuclear reaction always has elastic scattering associated with it. Sometimes we can use the elastically scattered particle energy as the standard energy to correct other particle energies. Consider how to use the Q value equation to distinguish elastic scattering of deuterons off ^{12}C from that off ^{144}Sm. Using Eq. (12.19), we can calculate the scattered deuteron energy K_l. For the ^{12}C(d, d) reaction with $K_i = 2.5\,\text{MeV}$ we find

$$K_l = \left\{\frac{2\cos\theta}{2+12} \pm \left[\left(\frac{2\cos\theta}{14}\right)^2 + \frac{10}{14}\right]^{1/2}\right\}^2 K_i = 0.956K_i \qquad (\text{for } \theta = 30°)$$

θ	K_l/K_i	K_l (MeV)
30°	0.956	2.39
150°	0.533	1.33

We can see that when the scattering angle varies from 30° to 150°, the K_l value differs by 1.06 MeV.

For the ^{144}Sm(d, d) reaction with $K_i = 2.5\,\text{MeV}$ we find:

θ	K_l/K_i	K_l (MeV)
30°	0.998	2.49
150°	0.949	2.37

Here K_l has only a 0.12 MeV difference between the two angles. This is much less than for carbon. So, by measuring at two scattering angles, we can determine the kind of nucleus off which the outgoing particle scattered.

REDUCING THE KINETIC BROADENING. In a typical reaction, several different outgoing particles are produced. In order to distinguish them, we must keep the energy width of the outgoing particles (ΔK_l) as small as possible so that they do not overlap each other. There are several contributions to ΔK_l, for example, the energy uncertainty of the incoming particles, the target thickness, and other factors. One of the important sources is the "kinetic broadening,"

$\Delta K_l/\Delta\theta$, related to the finite size of the detector solid angle. From Eq. (12.12), we can calculate this contribution to be

$$\frac{\Delta K_l}{\Delta\theta}=\frac{2\sqrt{A_iA_lK_iK_l}\,\sin\theta}{\sqrt{\frac{A_iA_lK_i}{K_l}}\cos\theta-(A_l+A_R)} \tag{12.22}$$

Here the nuclear mass number is used instead of the masses, but this introduces a negligible error. From this equation, we see that if we want to reduce ΔK_l, we must use a small $\Delta\theta$ for the detector. But this also reduces the number of outgoing particles detected, so it will take a longer time to obtain the same statistical error. The important thing Eq. (12.22) tells us is that for a fixed $\Delta\theta$, ΔK_l is related to θ. When $\theta=\pi/2$, ΔK_l is maximum, so we must as far as possible avoid using 90°. When $\theta=0$, the energy resolution is best. Sometimes, in order to obtain a constant ΔK_l, we must use a different acceptance angle $\Delta\theta$ at each scattering angle. Here we have given only two examples of the importance of the Q value equation in actual experiments. Further details can be found elsewhere.[1]

12.2.4 Reaction Cross-Section

One can treat reactions by assuming that inside the target every nucleus occupies an effective area of interaction, σ. When an incoming particle strikes the target, the projectile must hit this effective area for a reaction to occur with a nucleus. Then, in a target with thickness t, area A, and N nuclei per unit volume, the effective area for a reaction to occur is $N(tA)\sigma$. The reaction probability of the incoming particles hitting a nucleus in a target with area A is this effective area divided by the total area: $N(At)\sigma/A$. This probability must be equal to n_r/n_i, where n_i and n_r are the numbers of incoming and outgoing (reaction) particles (either total numbers or per unit time), respectively. So,

$$\frac{N(At)\sigma}{A}=\frac{n_r}{n_i} \tag{12.23}$$

and

$$\sigma=\frac{n_r}{n_iNt}$$
$$=\frac{\text{number of outgoing particles (the number of reaction particles)}}{\text{number of incoming particles}\times\text{number of target nuclei per unit area}} \tag{12.24}$$

This is the definition of a nuclear reaction cross-section. If all the different outgoing particles are counted, Eq. (12.24) expresses the probability of one incoming particle reacting with one target nucleus (the total cross-section). If

[1] For example, R. D. Evans, *The Atomic Nucleus*, McGraw-Hill, New York (1955).

only one type of outgoing particle is counted, it gives the reaction cross-section for that particular reaction.

Note that σ has the dimension of an area, and its unit is a barn (b) where

$$1\ \text{b} = 10^{-24}\ \text{cm}^2$$
$$1\ \text{mb} = 10^{-27}\ \text{cm}^2$$

A typical nuclear radius is 6 fm. Its classical geometric cross-section is $\pi R^2 = 1.1 \times 10^{-24}\ \text{cm}^2 = 1.1\ \text{b}$. So, the reaction cross-section unit and the nuclear geometric size are of the same order of magnitude.

It is important to understand that the cross-section (probability) for a particular reaction to occur is not related to the geometric size of either the target or projectile, but their nuclear properties. This can be seen by looking at the cross-sections for several different nuclei to absorb thermal neutrons and emit γ-rays, (n, γ) reactions as shown in Table 12.1.

From the above data one sees that the neutron capture cross-section for a nucleus can be a thousand times less than the geometrical area of the nucleus to one hundred thousand times larger than the geometrical area. Obviously, these cross-sections tell us something quite interesting about the structure of these nuclei. In particular, note that $^{16}_{8}O_8$ and $^{208}_{82}Pb_{126}$ are spherical double magic nuclei (as discussed in the last chapter). Their spherical double magic nature makes them very stable—very unwilling to give up a particle (large binding energy) or to absorb a particle such as a neutron.

Now look at the cross-section from the standpoint of the production of an isotope. Consider the reaction ^{27}Al(n, γ)^{28}Al. For a reaction cross-section $\sigma = 2$ mb, a target thickness $t = 0.2$ mm, and an incoming neutron flux of 10^{10}/(cm$^2 \cdot$ s) let us calculate the number of γ-rays emitted.

TABLE 12.1

Thermal neutron capture cross-sections for several elements

Nucleus	Thermal neutron capture cross-section
$^{16}_{8}O$	0.18 mb
$^{64}_{30}Zn$	0.46 b
$^{112}_{48}Cd$	2 b
$^{113}_{48}Cd$	2×10^4 b
$^{133}_{55}Cs$	27 b
$^{157}_{64}Gd$	2.5×10^5 b
$^{158}_{64}Gd$	2.4 b
$^{198}_{79}Au$	2.6×10^4 b
$^{204}_{80}Hg$	0.4 b
$^{208}_{82}Pb$	0.5 mb

The number of target nuclei per unit volume N of aluminum with a density 2.7 g/cm^3 and Avogadro's number of atoms in 27 g is

$$N = \frac{\rho}{A} A_0 = \frac{2.7}{27} \times 6.02 \times 10^{23} = 6.02 \times 10^{22}\,\text{nuclei/cm}^3$$

So, from (12.24) we can find

$$n_r = \sigma N t n_i = 2 \times 10^{-27}\,\text{cm}^2 \times 6.02 \times 10^{22}/\text{cm}^3$$
$$\times\, 2 \times 10^{-2}\,\text{cm} \times 10^{10}/(\text{cm}^2 \cdot \text{s}) = 2.4 \times 10^4\,\text{photons}/(\text{cm}^2 \cdot \text{s})$$

Thus, for every 1 million neutrons hitting the target, only 2.4 γ-rays are emitted (only 2.4 reactions occur to make ^{28}Al). In this case the neutrons are very "poor shots." Although some nuclear reactions release much energy [for example, in Eq. (12.2) the Q value is 17.35 MeV], the reaction probabilities are so very small that the losses outweigh the gains. No wonder Rutherford suggested that "anyone who believes that energy can be obtained from nuclei is in a dream." But, at the same time, he recognized that the discovery of the use of atomic energy would open a new chapter in human development.

12.3 COULOMB EXCITATION, COMPOUND NUCLEUS REACTIONS, AND OTHER REACTIONS

12.3.1 Coulomb Excitation

Rutherford discovered the nucleus by bombarding atoms with low-energy α-particles and observing their scattering. When the α energies become sufficiently high to overcome the Coulomb barrier and enter the nucleus, nuclear reactions occur. When the energy of the projectile is well below the Coulomb barrier for a target, the two nuclei do not come close enough for the nuclear force to be experienced. The projectile then undergoes scattering by the Coulomb force between the two nuclei. This force can transfer energy from the projectile into internal motion of the nucleus via the Coulomb field. One can excite the quantized energy states of one or both nuclei by the Coulomb field. This process is called Coulomb excitation.

Coulomb excitation continues to be one of the important ways of probing the structure of nuclei. More important than learning about the energies of excited states in a nucleus is the fact that the probability of Coulomb-exciting a level depends on the electromagnetic transition probability between the two states. Only the electric multipoles will be excited, since the magnetic interaction is the order of v/c smaller than the electric interaction. In even–even nuclei, ground states are 0^+, and the lowest excited states are typically 2^+, 4^+, and 3^-. These can be excited directly by electric quadrupole (E2), octupole (E3), and hexadecapole (E4) transitions. The cross-section for Coulomb excitation is directly related to $B(\text{E}\lambda)$, the transition probability for Eλ radiation, so a measurement of the cross-section yields $B(\text{E}\lambda)$. The transition probability is, in turn, closely related to the shape (deformation) of the nucleus. A measurement of the cross-section can easily identify collective effects which strongly enhance B(E2). Transition probabilities between collective nuclear states can be factors of ten to several hundred times the single-particle transition probabilities, for example, between the rotational states of deformed nuclei. Thus, Coulomb excitation is a very useful process for studying collective states in nuclei.

The $B(E\lambda)$ depend on the square of the transition matrix elements, $M(E\lambda)$, which in turn are related to the charge distributions, that is, the nuclear quadrupole moment and the deformation. From such measurements one can extract the electric quadrupole deformation β_2, the hexadecapole deformation β_4, and in the very heavy actinide elements even the higher-order deformation β_6.

In Coulomb excitation one can measure the energies and intensities of the γ-rays which de-excite the levels, or of the scattered projectiles such as protons and α-particles, with magnetic spectrographs to determine the energies of the nuclear levels and the transition probabilities between them. An example of an α-particle spectrum following Coulomb excitation is shown in Fig. 12.3. The B(E2) and B(E3) transition probabilities extracted from these data are 200(2), 3.9(1), and 11.9(13) times larger than the single-particle estimates for the 75-keV 2_g^+, 992-keV 2_γ^+, and 1289-keV 3^- states, respectively. These data show the very strong collective enhancement expected (see Chapter 9) for the first 2^+ member of the ground rotational band in the well-deformed ^{160}Gd nucleus. The B(E4) was also extracted

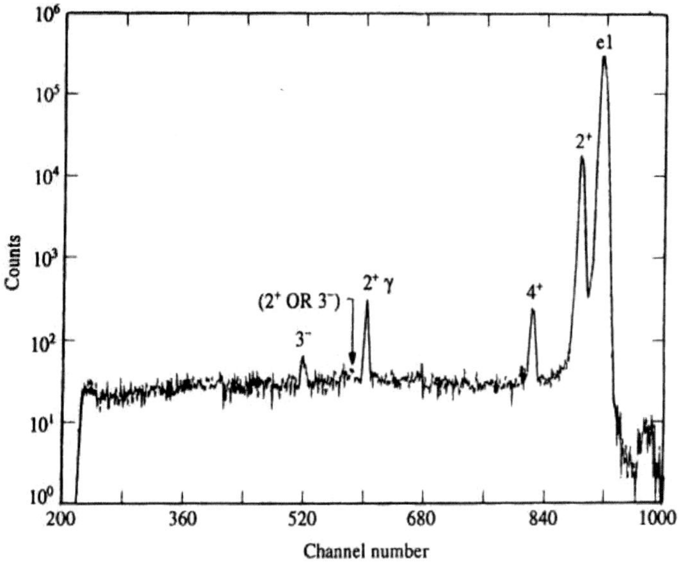

FIGURE 12.3
Spectrum of elastically (from the ground state) and inelastically scattered 15-MeV α-particles from ^{160}Gd at the Oak Ridge National Laboratory EN tandem. The inelastic scattering to the first 2^+ and 4^+ states of the ground rotational band and to the 2^+ level of the γ-vibration band and the 3^- octupole band level are clearly seen. [From R. M. Ronningen et al., *Phys. Rev.* **C16** (1977) 2208.]

from the direct excitation of the 4^+ state. The β_2 and β_4 deformations were obtained from these results.

Initially protons and α-particles were used to excite only the first one or two levels in nuclei. The large electric charge of heavier ions from $^{16}_{8}O$ up to $^{208}_{82}Pb$ opened up the study of higher-energy states via multiple Coulomb excitation. The ^{208}Pb nucleus is ideal because it is spherical double magic with no low-lying collective rotations. Its 3^- first excited state is at 2.6 MeV and is essentially not Coulomb excited. While much work was done with beams of krypton and xenon, the UNILAC accelerator at GSI in Germany was unique in the world with the only available beams of ^{208}Pb for over a decade beginning around 1975. An example of the Coulomb excitation of ^{248}Cm is shown in Fig. 12.4. This experiment illustrates the international flavor and cooperation in science today. The ^{248}Cm target was prepared at the unique Transuranic Laboratory in Oak Ridge and loaned to Vanderbilt University. It was then shipped to GSI to make use of the ^{208}Pb beam there. Because ^{248}Cm is a very well-deformed nucleus with a large quadrupole moment, Coulomb excitation with lead ions opened up a good test of the rotational model to much higher spins than ever before. Collective states up to 28^+ and tentatively 30^+ were Coulomb-excited. The $B(E2)$ values followed the rigid rotor values for a deformed nucleus very well up to 28^+. Beams of ^{208}Pb are now available at various other laboratories. The fact ^{248}Cm is radioactive required precautions to be taken to handle the target because of the health hazards and because any flaking of the target could contaminate the accelerator. Since extensive searches were being made for new super-heavy elements of $Z = 108$-126 by their α decay, any such contamination could washout their identification.

Nearly all Coulomb excitation studies have been done on stable and a few very long-lived radioactive targets. However, RIB facilities have opened up Coulomb excitation of short-lived radioactive nuclides. At HRIBF, the first Coulomb excitation of radioactive $^{132\text{-}136}Te$ and $^{126\text{-}134}Sn$ found unexpected variations of B(E2) with A. The technique of intermediate-energy Coulomb excitation of fast rare isotope beams was developed at Michigan State and opened important studies in the region of A = 30-50. New insights were found into the breakdown of known and the appearance of new spherical shell gaps and the so-called "Island of Inversion" where deformed intruder configurations are formed by promoting two protons across the N = 20 shell gap. These examples illustrate the importance of building new two-stage radioactive ion beam accelerators in which the first stage produces the radioactive ions and the second stage accelerates them for experiments.

12.3.2 Compound Nuclear Reactions

Nuclear reaction cross-sections are very important data. Experimental measurements of reaction cross-sections have been and continue to be important in both basic and applied nuclear physics. The calculations of cross-sections from a theoretical point of view have been and continue to be an important challenge for

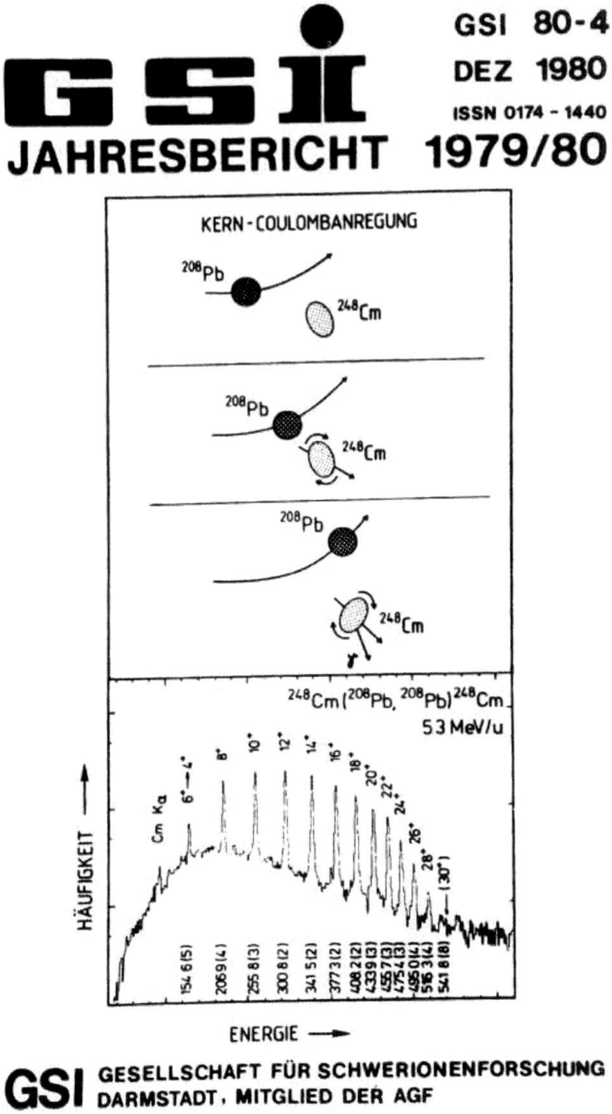

FIGURE 12.4
The front cover of the 1979/80 Annual Report (*Jahresbericht*) of Gesellschaft für Schwerionenforschung, GSI (Organization for Heavy Ion Research), Darmstadt, Germany, highlights one of the over 150 experiments reported in that year. The nuclear Coulomb excitation (Kern-Coulombanregung) of the well-deformed $^{248}_{96}$Cm nucleus by the spherical, double magic $^{208}_{82}$Pb nucleus with an energy of about 5 MeV/*A* is illustrated. The large charges of both nuclei Coulomb-excited ^{248}Cm into rotational motion but did not excite ^{208}Pb. The allowed rotational states of the ground state of ^{248}Cm were Coulomb-excited to a record $30\hbar$ in a Vanderbilt-GSI collaboration as seen in the spectrum of γ-rays de-exciting ^{248}Cm following its excitation. This spectrum required sophisticated detection of both the lead and curium ions in coincidence with the γ-rays and complex analysis procedures to correct for the large Doppler shifts of the γ-rays from the recoiling heavy ions. (Courtesy of GSI.)

nuclear theory. Like the nuclear structure problem, we cannot find the reaction probabilities by solving the Schrödinger equation. We can only infer the reaction mechanism by developing nuclear reaction models from the experimental data, and then check the model against experiments to correct it. Various models have been proposed. Such models are developed for a particular type of reaction. Here we briefly introduce the compound nucleus reaction model.

The compound nuclear model was proposed by N. Bohr in 1936. In this model, the incoming particle and target nucleus fuse together to form a compound nucleus in an excited state. This compound nucleus exists for a time sufficiently long that the incoming particle or nucleons in the projectile and the nucleons inside the target nucleus are mixed together completely. After some time (long on a nuclear scale of 10^{-21} s), some particle or particle group has a probability of obtaining sufficient energy to escape from the compound nucleus. This is the compound nucleus decay. Thus, the compound nucleus model divides into two steps: One is compound nucleus formation, in which the incident nucleons lose all their identity; the other step is compound nucleus decay. These two are totally independent processes. The decay pattern is not related to how the compound nucleus was formed, so the reaction cross-section can be written as the product of two terms:

$$\sigma = \sigma_{\text{formation}} \cdot \sigma_{\text{decay}} \tag{12.25}$$

This idea has been proven in many nuclear reactions. For example, $\alpha + {}^{60}\text{Ni}$ forms ${}^{64}\text{Zn}^*$, as does $\text{p} + {}^{63}\text{Cu}$ along with numbers of other projectile–target combinations. If we choose the proton and α energies, K_p and K_α, such that

$$K_p + B_p = K_\alpha + B_\alpha \tag{12.26}$$

where B_p and B_α are the binding energies of the p and α to the target nucleus, respectively, then forming the compound nuclei in both reactions leaves them in the same excited state. How the compound nucleus decays is not related to its formation pattern. It could decay through various channels such as (each arrow in the following points to a different channel)

$$\begin{aligned} &\nearrow \text{n} + {}^{63}\text{Zn} \\ {}^{64}\text{Zn}^* &\rightarrow \text{p} + \text{n} + {}^{62}\text{Cu} \\ &\searrow 2\text{n} + {}^{62}\text{Zn} \end{aligned} \tag{12.27}$$

and other channels depending on the energy available. From Eq. (12.25) and the fact σ_{decay} is not related to the formation, we have the following relations:

$$\begin{aligned} \frac{\sigma[{}^{63}\text{Cu}(\text{p},\text{n}){}^{63}\text{Zn}]}{\sigma[{}^{60}\text{Ni}(\alpha,\text{n}){}^{63}\text{Zn}]} &= \frac{\sigma_{\text{formation}}(\text{p} + {}^{63}\text{Cu})}{\sigma_{\text{formation}}(\alpha + {}^{60}\text{Ni})} = \frac{\sigma[{}^{63}\text{Cu}(\text{p},\text{pn}){}^{62}\text{Cu}]}{\sigma[{}^{60}\text{Ni}(\alpha,\text{pn}){}^{62}\text{Cu}]} \\ &= \frac{\sigma[{}^{63}\text{Cu}(\text{p},2\text{n}){}^{62}\text{Zn}]}{\sigma[{}^{60}\text{Ni}(\alpha,2\text{n}){}^{62}\text{Zn}]} \end{aligned} \tag{12.28}$$

Experiments have verified these relations. The relative cross-sections for the production of various far-off-stability nuclei in the mass 70 region are shown in Fig. 12.5. Note how, as the energy increases, the cross-sections for two particles out decrease and the cross-sections for three to nine particles out successively increase and become important. Theoretical calculations of these cross-sections at a particular energy are given in Table 12.2.

One success of the compound nucleus model was to explain a resonance reaction which had been difficult to explain earlier. A resonance reaction occurs when the sum of the incoming particle's energy and the energy from fusion of the projectile with the target nucleus just equals the excitation energy of a particular level. Then the reaction cross-section is a maximum.

It is also possible to fuse two nuclei at energies just below the Coulomb barrier through quantum-mechanical effects. This is called subbarrier fusion and is an important area of current research.

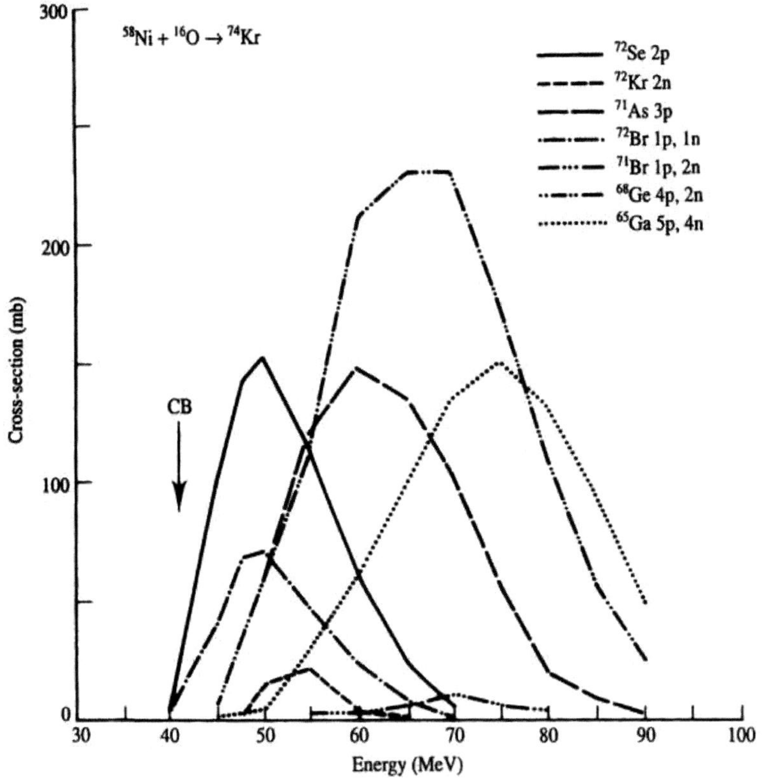

FIGURE 12.5
Cross-sections for various reaction channels when ^{58}Ni is bombarded with ^{16}O with the energy given on the horizontal axes were calculated to estimate the yields for the study of different nuclei in the region around $A = 70$.

TABLE 12.2

Calculated reaction products and cross-sections of 48 MeV ^{16}O on ^{58}Ni

Reaction observed	Product	Percentage	Cross-section, σ(mb)
1p	^{73}Br	0.3	1.02
2n	^{72}Kr	0.1	0.34
1p, 1n	^{72}Br	19.7	67.24
2p	^{72}Se	42.6	145.40
2p, 1n	^{71}Se	3.25	11.09
3p	^{71}As	10.5	35.84
2p, 2n	^{70}Se	0.050	0.17
2p, 3n	^{69}Se	0.850	2.90
3p, 2n	^{69}As	14.2	48.47
3p, 3n	^{68}As	0.1	0.34
4p, 2n	^{68}Ge	6.3	21.50
4p, 4n	^{66}Ge	1.5	5.12
5p, 4n	^{65}Ga	0.550	1.88

Another important parameter in a compound fusion reaction is the maximum angular momentum which can be brought into the compound nucleus. This is particularly important when one seeks to study nuclear states with very large angular momentum of 30–60$\hbar$ and higher. Such high-angular-momentum states can occur at relatively low excitation energies through rotational motion of the nucleus, especially if the nuclear moment of inertia is very large as it is for super-deformed ($\beta_2 = 0.4$–0.6) and hyperdeformed ($\beta_2 = 0.8$–1.0) nuclei. Under the assumption that all the energy goes into rotational motion, the maximum angular momentum in units of $\hbar$ for projectile A_p on a target A_t is

$$l_{\max} = 0.32(A_p^{1/3} + A_t^{1/3})[\mu(E - V)]^{1/2} \tag{12.29}$$

where μ is the reduced mass $m_p m_t/(m_p + m_t)$, E is the energy in the laboratory system in MeV, and V(MeV) is the Coulomb barrier. This is a semi-empirical equation that gives reasonable agreement with experiments. For example, calculate the maximum angular momenta brought into the ^{68}Ge compound nucleus in the reaction $^{58}Ni(^{12}C,2p)^{68}Ge$ at a beam energy of 39 MeV:

$$l_{\max} = 0.32 \times [12^{1/3} + 58^{1/3}]\left[\frac{12 \times 58}{70}\left(39 - \frac{1.2 \times 6 \times 12}{12^{1/3} + 58^{1/3}}\right)\right]^{1/2}$$

$$\simeq 15\hbar$$

12.3.3 Other Types of Reactions

Some reactions cannot be explained using the compound nucleus model. For example, stripping reactions like $^{27}Al(d, p)^{28}Al$. Here, the deuteron interacts with the target nucleus and is stripped of a neutron, with the proton continuing on. In a pick-up reaction like $^{13}C(^3He, \alpha)^{12}C$, the incoming ^3He interacts with the target nucleus to pick up a neutron and continue onward as ^4He. The products of these two types of reactions are concentrated primarily in the forward direction: The angular distribution of the particles coming out is peaked in the forward direction. These are types of what are called direct reactions because effectively a single nucleon in the projectile or target is interacting directly. This forward peaking does not exist in a compound nuclear reaction, where the incident particle history is wiped out following its fusion to form the compound system.

Capture reactions in which a nucleus captures a proton or neutron or even an α-particle and emits one or more γ-rays are another important class of reactions. In Table 12.1 we gave some cross-sections for different nuclei to capture thermal (few eV energy) neutrons, for example, the $^{16}O(n, \gamma)^{17}O$ reaction. At much higher temperatures to overcome the Coulomb barriers, $^{17}F(p, \gamma)^{18}Ne$ or $^{15}O(\alpha, \gamma)^{19}Ne$ reactions can occur. These capture reactions played a major role in the creation of our world as we know it. As discussed in Chapter 11, the heavy elements present in our world were made by proton and neutron capture reactions in supernova explosions, for example. The rapid proton capture process in proton-rich nuclei is thought to have extended up to $A \simeq 80$ while the slow neutron and rapid neutron capture processes produced elements up through uranium. The nuclei involved in these proton and neutron capture reactions that produced heavy elements are radioactive, so the cross-sections for these reactions are unknown. The lack of knowledge of these capture cross-sections is a severe limitation on the development of theoretical models of nucleosynthesis. Of course, equally as important in nucleosynthesis models is knowledge of the half-lives, decay modes, shapes, isomeric states, masses, and other properties of these radioactive nuclei as discussed in Chapter 11. A major driving force behind the development of radioactive ion beam facilities is to measure proton and neutron capture cross-sections for many radioactive nuclei which are needed in nucleosynthesis calculations.

In heavy elements such as uranium, the capture of even a few eV neutrons can raise the internal energy of the product so that it undergoes fission rather than γ decay. In Section 12.4.4 we discuss the fission of uranium. For eV neutron capture, ^{235}U will undergo fission (n, f) but $^{238}U(n, \gamma)^{239}U$. Moreover, these cross-sections have opposite energy dependencies with increasing neutron energies—(n, f) goes down and (n, γ) goes up. In addition to neutron capture-induced fission, proton capture in heavy elements around uranium can induce fission, e.g., $^{238}U(p, f)$. Heavier ion reactions can also induce fission. In the collision of two heavy nuclei, the Coulomb field of the two nuclei can induce fission. At GSI, Coulomb fission of 750 MeV/A ^{238}U on light (Be) and heavy (Pb) targets has been studied. Many new neutron-rich, radioactive nuclei have been identified in this way, as reported at the ENAM 2004 Conference. This is a powerful new technique that opens up much new research on very neutron-rich nuclei. For example the new, very neutron-rich spherical double magic nucleus $^{78}_{28}Ni_{50}$ was discovered in such a ^{238}U reaction at GSI.

Indeed, nuclear reaction phenomena are very diverse and rich in content, especially in the heavy-ion reactions which have been developed in recent years. The energies of nuclear reactions now range from low energies to ultrarelativistic energies for heavy ions. Let us briefly review the diversity of what happens as the energy of a projectile increases. Heavy-ion compound nuclear reactions can be used to make exotic nuclei far from stability. For example, the reactions ^{181}Ta(^{16}O, xn) and ^{180}W(^{14}N, xn) ^{184,186,188}Tl ($x = 9$–11) were used to first identify $^{184}_{81}$Tl, $^{186}_{81}$Tl, and $^{188}_{81}$Tl (discussed in Chapter 11). These isotopes have 15–21 fewer neutrons than the stable thallium isotopes of $^{203}_{81}$Tl and $^{205}_{81}$Tl. Energetic ($\sim$ 600 MeV) proton-induced spallation reactions in which multineutron–multiproton (p, xn,yp) reactions produce a wide range of proton-rich nuclei is another major way used to produce nuclei far from stability. Radioactive ion-beam accelerators will extend the range of far-off stability nuclei which can be produced by starting with a neutron-deficient or neutron-rich beam.

Exotic transfer reactions with light ions have been observed such as (^{4}He, ^{11}C). Multinucleon transfer reactions in what are called heavy-ion deep inelastic collisions at 8–12 MeV/A were first used in Russia to make exotic neutron-rich heavy nuclei like ^{15}B, ^{17}C, ^{18}N by bombarding ^{232}Th with 131-MeV ^{16}O. In the last several years studies of such reactions have been extended across the periodic table to make new neutron-rich nuclei all the way to the actinide region. As one example, the isotopes $^{60}_{24}$Cr and $^{68}_{28}$Ni were seen in the reaction of $^{76}_{32}$Ge at 11.5 MeV/A on a tungsten target. At these same energies other types of reactions are observed, such as partial fusion reactions: for example, 167 MeV ^{14}N(11.9 MeV/A) on ^{154}Sm with the emission of very energetic p, d, t in the forward direction followed by the fusion of the remaining part of the projectile. More exotic reactions have been observed where, in bombarding 300-MeV ^{32}S(9.4 MeV/A) and 600-MeV ^{58}Ni(10.3 MeV/A) tantalum, protons with up to 100 MeV energy and α-particles with up to 150–200 MeV total energy have been observed. These protons and α-particles have 10–100 times the incident energy per nucleon and represent 30 percent to over 50 percent of the total energy of the incident projectiles! Obviously, some strong collective effect is occurring to give rise to such a high fraction of the incident energy in a projectile with 32–58 nucleons being given to one-particle (p) or a four-particle cluster (α).

As the energy of the projectile increases to 40–200 MeV/A, a new phenomenon occurs in which the projectile fragments into many different nuclear species to make both very neutron-rich and proton-rich nuclei. For example, $^{22}_{6}$C, $^{23}_{7}$N, $^{29}_{10}$Ne, and $^{30}_{10}$Ne were discovered from the fragmentation of 44-MeV/A ^{40}A on a tantalum target and $^{51}_{28}$Ni, $^{52}_{28}$Ni, $^{50}_{27}$Co, $^{51}_{27}$Co, $^{52}_{27}$Co, and $^{48}_{28}$Fe from 55-MeV/A ^{58}Ni on nickel. You may want to compare the neutron numbers of these nuclei with those of the stable isotopes of these elements. In 1994, the long-sought $^{100}_{50}$Sn was identified in the fragmentation of relativistic 1.1 GeV/A ^{127}Xe and of 63 MeV/A ^{112}Sn. Other exotic reactions include charge exchange with mesons, for example, $^{9}_{4}$Be$_{5}$(π^-, π^+)$^{9}_{2}$He$_7$ or transfers $^{10}_{4}$Be$_6$($^{14}_{6}$C$_8$, $^{14}_{8}$O$_6$)$^{10}_{2}$He$_8$. The recently discovered spherical double magic $^{10}_{2}$He$_8$ has 2^+ and 3^- levels at 4.3 and 7.9 MeV to support its double magic character. Here ^{10}He has 4 times the number of neutrons of stable ^{4}He.

The intensities of the products found in these fragmentation reactions and others are such that two-stage accelerators are being used in which the first accelerator produces radioactive nuclei and the second accelerator accelerates these radioactive nuclei to higher energies for further studies. Such a facility can make even more exotic nuclei farther from stability. In addition, whole new areas can be opened up. For example, studies of radioactive beams of ^{11}Li indicate that it has an outer shell of neutrons surrounding an inner core of protons and neutrons. The radius of the outer neutron shell is much greater than the radius of its inner core which is ^{9}Li. Beams of ^{11}Li give us the possibility of studying pure neutron matter which otherwise is available in our universe only in neutron stars. Radioactive ion beam studies with present facilities are already providing important new capabilities (note the use of ^{19}Ne in medical research discussed in the section 12.5). New facilities designed specifically for radioactive ion beam research will provide exciting, new opportunities as also discussed later.

As heavy ion energies increase to 1 GeV/A and beyond, new regions are reached where one sees other completely new phenomena. W. Greiner and the Frankfurt School predicted nuclear shock waves can occur in very relativistic heavy ion collision. The discovery of these shock waves in nuclear matter opened up new ways of heating up and compressing nuclear matter to explore the nuclear equation of state under high temperature and high density.

The next level of matter to be probed is the quark-gluon structure of the nucleon. In the standard model nucleons are composed of three quarks held together by gluons (see Chapter 14). Electrons which are pointlike (structureless), can probe the nucleus with unique precision. Moreover, they interact with matter by the best understood of the fundamental forces in nature-the Coulomb force. The 6 GeV Continuous Electron Beam Accelerator Facility, CEBAF, at the Jefferson Lab. in Virginia, probes at a new depth the structure of nuclei and the nucleons in nuclei, as discussed in Chapter 14.

As also discussed in more detail in Chapter 14, quarks come in six types or flavors, u, d, s, c, b, t. Neutrons and protons contain only u and d quarks. The lightest hyperon (hypernucleon) with another combination is the lamda (Λ) with uds-quark combination. Because the strong and electromagnetic interactions do not allow decay where a flavor change occurs (s to u or d), the Λ decays via the weak interaction. "Flavor physics in nuclei" is the study of nuclear many-body systems that include one or more heavy quarks (s, c, b, t).

Coupled with the goal of understanding the quark structure of nucleons in nuclear matter is the goal of nuclear physicists to probe the equation of state of nuclear matter as a function of temperature and density from normal nuclear matter all the way up to the formation of the theoretically predicted quark-gluon plasma. The states of nuclear matter which can be formed at different temperatures and densities are illustrated in Fig. 12.6.

Nuclear matter is normally incompressible, but at low temperature densities of 5-10 times normal are found in supernova explosions and neutron stars. At low matter density and very high temperature, one comes to the region that characterized the universe in its very earliest times following the big bang.

Between these limits at high temperature and density one has the unbound quark-gluon plasma. The quark-gluon plasma, in which nucleons evaporate into their constituents, is a type of "holy grail" of both nuclear and particle physicists today. It previously existed only in the very earliest time following the big bang before the universe cooled down sufficiently to allow quarks to form hadrons. The creation and study of this new fifth state of matter provides fundamental tests of our understanding of the structure of matter and provides crucial input into the development of any grand unified theory that encompasses all the known forces.

Ultrarelativistic heavy-ion accelerators have begun to probe a fuller range of the temperature and density seen in Fig. 12.6. The first such facility, which began operations in 2001, is the Relativistic Heavy Ion Collider (RHIC) located at the Brookhaven National Laboratory, as shown in Fig. 12.7.

The first acceleration stage at RHIC is the original MP Tandem Van De Graaf. These heavy ion beams gain more energy in a new booster synchrotron such that they can be efficiently bunched and accelerated by the venerable AGS high-energy physics facility. Two Nobel prizes in physics were won for research done at the AGS. Upon exiting the AGS the heavy ion beam is split into two oppositely directed beam lines which provide the final synchrotron acceleration to 100 GeV/nucleon, meaning that the beams are traveling at 99.99% the speed of light. The beam lines converge at six selected intersection regions about a 5-kilometer circumference where four sepa-

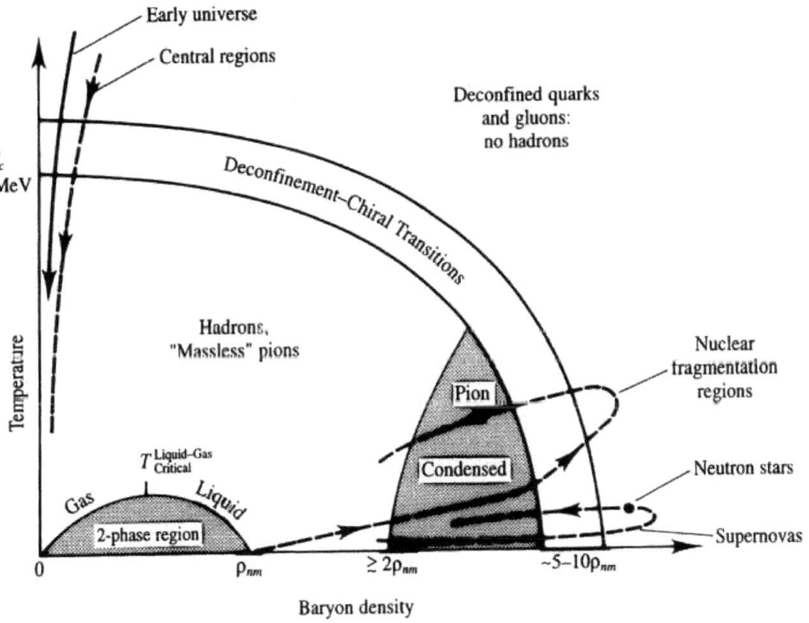

FIGURE 12.6
The expected phases of nuclear matter at different temperatures and nucleon (baryon) densities. As temperature T and density ρ increase, a hadron (strong interaction particle) gas phase is reached, going through a liquid–gas phase transition, and at still high T and ρ there is a transition to the region of deconfined quarks and gluons—the quark–gluon plasma.

FIGURE 12.7: A photograph of the complex of accelerator stages which comprise the Relativistic Heavy Ion Collider, RHIC, constructed at the Brookhaven National Laboratory. Heavy ion beams are first accelerated in the Tandem Van De Graaf (lower right), and then gain more energy successively in the Booster, AGS, and the final 5-kilometer synchrotron rings.

rate experiments (BRAHMS, PHENIX, PHOBOS, and STAR) began taking data in 2001. The journal Nuclear Physics devoted a special two-issue volume (NP A757, 1-2, August 8, 2005) for the four experiments assessing the physics results of the first three years of their RHIC operations. These experiments were unanimous in concluding that a new state of matter had been formed whose properties could not be understood with conventional hadronic physics models. Since then, the experimental evidence for a quark-gluon plasma, QGP state produced in heavy ion collisions at RHIC has become even more compelling. Current interpretations of the data suggest that the QGP state at RHIC behaves like a strongly coupled system, more like a liquid than like a gas of non-interacting constituents. In this respect, the QGP state at RHIC has much in common with strongly coupled states such as liquid crystals or superconductors, although the temperatures at RHIC are about 10^{12} K as compared with the 10-300 K typical of the condensed matter domain.

Given the exciting physics frontier opened by the experiments at RHIC, a new program of relativistic heavy ion research is being planned to take place at the Large Hadron Collider, LHC, which has been constructed in the LEP tunnel (see Chapter 14) at CERN. The LHC is designed to collide 7 TeV protons on 7 TeV protons, and likewise collide ^{208}Pb beams of 2.75 TeV each. Thus the lead collisions at 5.5 GeV per nucleon pair will be at almost 30 times the energy of the heavy ion collisions at RHIC. At such energies the QGP state will be produced at much higher temperatures and will be expected to be larger and live longer than the QGP state at

RHIC. The experimental signatures of the QGP as pioneered by the RHIC experiments should show up even more dramatically at the LHC.

The first ^{208}Pb beam collisions in the LHC are presently anticipated to take place in the summer of 2010. Two separate, international experimental collaborations are preparing to take heavy ion data: the ALICE (*A Large Ion Collider Experiment*) detector collaboration and the CMS (*Compact Muon Solenoid*) detector collaboration. The LHC detectors will ship their raw data to a multitude of institutions around the world on a daily basis for immediate processing. For this purpose the WLCG (Worldwide LHC Computing Grid) was developed, to be the new distributed paradigm for handling the immense amounts of data that these experiments will be generating annually.

12.4 FISSION AND FUSION: ATOMIC ENERGY UTILIZATION

12.4.1 The Fission Discovery

After the discovery of the neutron, many scientists began to study nuclear reactions using neutrons, which easily penetrate a nucleus because there is no Coulomb repulsion. One particular area of research was the production of new elements beyond the heaviest known, uranium. In 1939, O. Hahn and F. Strassman in Germany discovered that when neutrons bombarded uranium, some medium-heavy nuclear products such as Ba ($Z = 56$) were formed. L. Meitner explained this as follows: After absorbing a neutron, the excited uranium nucleus divides into two fragments with about equal masses. Hahn and Strassman had discovered induced nuclear fission, the splitting apart of a heavy nucleus into two nearly equal lighter fragments. Fermi and co-workers in similar experiments had observed radioactivities from uranium fission earlier, but they did not recognize that these new activities had been produced by this important new phenomenon. Only a year later, in 1940, G. N. Flerov and K. A. Petrzhak, in Russia, discovered spontaneous fission, in which certain heavy elements have a probability of decay by spontaneously splitting apart into two similar mass fragments instead of undergoing α or β decay. In 1947, the Chinese physicists Qian Sanqiang and He Zehui discovered fission into three fragments where one of them is generally the α-particle. But the probability for this is very small.

The discovery of uranium fission was first confirmed on January 25, 1939, by a group at Columbia University that included E. Fermi and F. Slack (Vanderbilt University) and soon thereafter around the world. However, caution immediately set in because the process seemed to involve so few events. In *Newsweek* on February 13, 1939, we read that "the hope of running machines with forces released from atoms has little immediate chance of realization." The magazine's report concluded: "In his radio program last week Fred Allen interviewed a fictitious atom smasher, one Professor Gaffney Fubb. Queried by Allen as to the practical possibilities of his laboratory work, Professor Fubb answered: 'Well, someone may come in some day and want half an atom.' " The caution arose because the fission was being studied primarily with fast neutrons, and the probability of fast neutron absorption by uranium was very low. However, it soon became known that the rare isotope ^{235}U had a high probability of fissioning when it absorbed a slow (thermal energy) neutron.

There are many different channels that lead to two fragments in the fission process. For example, ^{235}U plus a neutron forms ^{236}U* which may break up into ^{144}Cs and ^{91}Rb plus a neutron, into ^{140}Xe and ^{94}Sr plus two neutrons, into ^{136}I and ^{99}Y plus a neutron, or into many other such combinations of neutron-rich medium-mass nuclei. The mass distribution of the nuclides produced in uranium fission is shown in Fig. 12.8. Let us illustrate these three possibilities and their subsequent decays:

1.

$$n + {}^{235}U \rightarrow {}^{236}_{92}U^* \rightarrow {}^{144}_{55}Cs + {}^{91}_{37}Rb + 1n$$

$${}^{144}_{55}Cs \xrightarrow{\beta^-} ({}^{144}_{56}Ba)^* \xrightarrow{\gamma} {}^{144}_{56}Ba \xrightarrow{\beta^-} {}^{144}_{57}La \xrightarrow{\beta^-} {}^{144}_{58}Ce \xrightarrow{\beta^-} {}^{144}_{59}Pr \xrightarrow{\beta^-} {}^{144}_{60}Nd \text{ (stable)}$$

$$n \searrow$$

$${}^{143}_{56}Ba \xrightarrow{\beta^-} {}^{143}_{57}La \xrightarrow{\beta^-} {}^{143}_{58}Ce \xrightarrow{\beta^-} {}^{143}_{59}Pr \xrightarrow{\beta^-} {}^{143}_{60}Nd \text{ (stable)}$$

$${}^{91}_{37}Rb \xrightarrow{\beta^-} {}^{91}_{38}Sr \xrightarrow{\beta^-} {}^{91}_{39}Y \xrightarrow{\beta^-} {}^{91}_{40}Zr \text{ (stable)}$$

2.

$$n + {}^{235}U \rightarrow {}^{236}_{92}U^* \rightarrow {}^{140}_{54}Xe + {}^{94}_{38}Sr + 2n$$

$${}^{140}_{54}Xe \xrightarrow{\beta^-} {}^{140}_{55}Cs \xrightarrow{\beta^-} {}^{140}_{56}Ba \xrightarrow{\beta^-} {}^{140}_{57}La \xrightarrow{\beta^-} {}^{140}_{58}Ce \text{ (stable)}$$

$${}^{95}_{38}Sr \xrightarrow{\beta^-} {}^{94}_{39}Y \xrightarrow{\beta^-} {}^{94}_{40}Zr \text{ (stable)}$$

3.

$$n + {}^{235}U \rightarrow {}^{236}_{92}U^* \rightarrow {}^{136}_{53}I + {}^{99}_{39}Y + n$$

$${}^{136}_{53}I \xrightarrow{\beta^-} {}^{136}Xe \text{ (stable)}$$

$${}^{99}_{39}Y \xrightarrow{\beta^-} ({}^{99}_{40}Zr)^* \xrightarrow{\gamma} {}^{99}_{40}Zr \xrightarrow{\beta^-} {}^{99}_{41}Nb \xrightarrow{\beta^-} {}^{99}_{42}Mo \xrightarrow{\beta^-} {}^{99}_{43}Tc \xrightarrow{\beta^-} {}^{99}_{44}Ru \text{ (stable)}$$

$$n \searrow$$

$${}^{98}_{40}Zr \xrightarrow{\beta^-} {}^{98}_{41}Nb \xrightarrow{\beta^-} {}^{98}_{42}Mo \text{ (stable)}$$

Note in examples (1) and (3) that ^{144}Cs and ^{99}Y are β-delayed neutron emitters. The neutron emissions from the excited states of ^{144}Ba and ^{99}Zr follow the half-lives for the β^- decay of their parents since they are emitted promptly, less than 10^{-10} s after the β decay. These neutrons are in addition to the prompt neutrons emitted in the fission itself. The fission fragments must be neutron-rich nuclides because ^{236}U has an even larger neutron excess over its Z than medium-weight nuclei, so the products are not able to decay by β^+ or EC. It must be pointed out that not only neutrons can induce fission in a heavy nucleus; other particles such as protons, deuterons, α-particles, photons, and heavy ions can also induce fission. However, neutron-induced fission has a special importance.

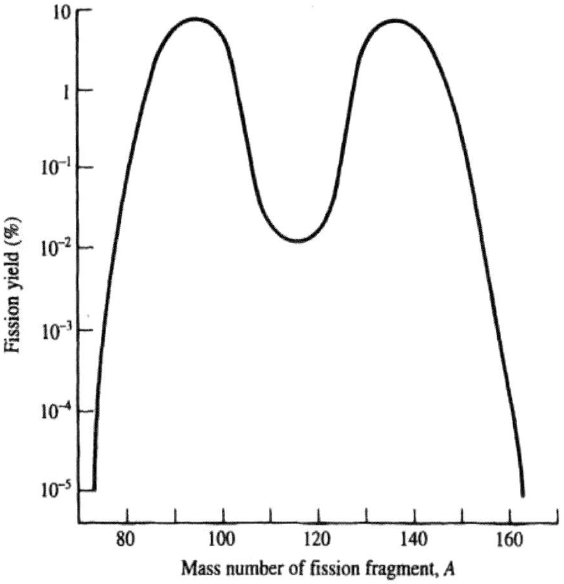

FIGURE 12.8
The mass distribution of the products formed in the neutron-induced fission of ^{235}U.

The discovery of neutron-induced fission received great attention not only because energy was released during the fission process but, more importantly, because every fission is accompanied by the emission of 2–3 neutrons. The average number of neutrons emitted in the neutron-induced fission of ^{235}U is 2.5. These neutrons can induce fission in other ^{235}U nuclei, and so on, to form a self-sustaining chain reaction to continuously build up energy release! It took less than four years from the discovery of fission to the production of a chain reaction and only another two years to the tremendous energy release in atomic bombs. The speed of transfer from science to technological exploitation on the scale required in this case is truly remarkable.

12.4.2 Fission Mechanism

After the discovery of fission, Bohr and Wheeler immediately explained the fission process using the nuclear liquid drop model and the compound nucleus reaction mechanism. After a neutron is captured, a compound nucleus is formed. The compound nucleus is in an excited state. This excitation will produce collective oscillations and changes in shape just as when a liquid drop fuses with another. At this time, there is competition between the surface tension, which strives to restore the nucleus to a more spherical shape, and the Coulomb repulsion force, which seeks to push the charges farther apart to elongate the nucleus further and finally make it break into two parts (see Fig. 12.9). The other important factors are: (a) the height of the barrier against fission which holds the nucleus together, typically 6 MeV or so in a heavy nucleus, and (b) the excitation energy of the excited nucleus, which raises the nucleus toward the top of the barrier.

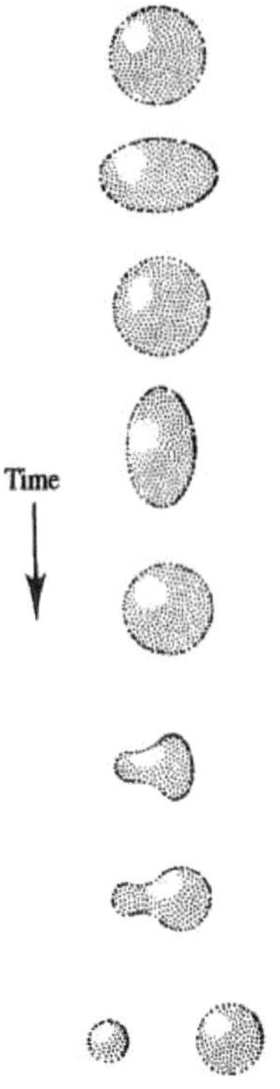

FIGURE 12.9
Nuclear deformation in the fission process.

Thus, whether fission occurs depends on the excitation energy of the compound nucleus and the ratio of the Coulomb energy E_c to the surface energy E_s. First we look at the ratio:

$$\chi \sim \frac{E_c}{E_s} \sim \frac{Z^2/A^{1/3}}{A^{2/3}} = \frac{Z^2}{A} \tag{12.30}$$

where χ is called the fissionable ratio. It is proportional to Z^2/A. The larger Z^2/A is, the larger the fission probability. When $Z^2/A > 50$, the Coulomb force is so large that the nucleus cannot exist.

The two nuclei ^{235}U and ^{239}Pu have very favorable properties for producing a neutron-induced fission chain reaction and so are the most used isotopes. Their Z^2/A values are

$$\begin{aligned} &\mathrm{n} + {}^{235}\mathrm{U} = {}^{236}\mathrm{U} \qquad Z^2/A = 35.9 \\ &\mathrm{n} + {}^{239}\mathrm{Pu} = {}^{240}\mathrm{Pu} \qquad Z^2/A = 36.8 \end{aligned} \tag{12.31}$$

The only nuclide found in nature which can be induced to fission by thermal energy (eV range) neutrons is ^{235}U. But ^{235}U is only 0.72 percent abundant in natural uranium. The 99.27 per cent abundant ^{238}U in natural uranium only undergoes fission by absorbing a fast neutron (at least 1 MeV energy). Why is there such a large difference in neutron energy required? The Z^2/A of ^{238}U is a little less, but the more important reason is the difference of the excitation energy of the compound nucleus. ^{235}U is an odd-A nucleus. The odd neutron can form a neutron pair with the new neutron and the extra binding energy from fusing the pair goes into excitation of ^{236}U. But ^{238}U is an even–even nucleus, and the addition of an extra neutron involves less binding energy since there is no pairing energy for this neutron. Thus, the neutron and ^{235}U combine very tightly to give off a binding energy of 6.43 MeV to form ^{236}U in high excited states. The ^{236}U fission potential barrier is only 5.3 MeV, and it is easy for fission to occur. In contrast, an extra neutron and ^{238}U combine loosely with a binding energy of 4.85 MeV, while the ^{239}U fission potential barrier is 5.45 MeV. So in general ^{239}U undergoes γ and β^- decay. With neutrons of 1 MeV kinetic energy, this extra energy can raise the energy of the excited ^{239}U to higher than the barrier to fission. However, there are no easy sources of 1-MeV neutrons for practical applications and, moreover, such fast neutrons can easily escape the uranium. The same arguments explain why ^{239}Pu is an excellent fissionable nuclide. Indeed, ^{239}Pu is better than ^{235}U because on the average it gives off more neutrons to sustain the chain reaction.

Although we cannot use ^{238}U for thermal neutron-induced fission, it can be used to produce ^{239}Pu by thermal neutron capture and β^- decay. The ^{238}U is used to "breed" the fissionable material ^{239}Pu as follows:

$$\begin{aligned} &\mathrm{n} + {}^{238}\mathrm{U} \rightarrow {}^{239}\mathrm{U} + \gamma \\ &{}^{239}\mathrm{U} \rightarrow {}^{239}\mathrm{Np} + \mathrm{e}^- + \bar{\nu}_\mathrm{e}(T_{1/2} = 24\ \mathrm{min}) \\ &{}^{239}\mathrm{Np} \rightarrow {}^{239}\mathrm{Pu} + \mathrm{e}^- + \bar{\nu}_\mathrm{e}(T_{1/2} = 2.35\ \mathrm{days}) \end{aligned} \tag{12.32}$$

This is one example of obtaining fissionable material in a breeder reactor that breeds its own fuel. When fission was found in 1939, there was no enriched ^{235}U material, and ^{239}Pu was an "unknown" element in the periodic table. In 1945, the first atomic bomb was exploded in America. The two atomic bombs which were exploded over Hiroshima and Nagasaki used ^{235}U and ^{239}Pu as fuel, respectively. There were tremendous technical problems to be overcome in both cases: (a) to physically separate ^{235}U from ^{238}U to obtain sizable quantities of sufficiently pure ^{235}U, and (b) to chemically separate the new element ^{239}Pu which had an unknown chemistry and was found in the uranium fuel rods along with many other highly radioactive products from ^{235}U fission. Francis Slack from Vanderbilt led the team that developed the barriers which were used in the enormous Oak Ridge plants to separate ^{235}U from ^{238}U by gaseous diffusion, while Glenn Seaborg, who discovered plutonium, worked out the chemistry of plutonium and the techniques for its chemical separation from other highly radioactive isotopes. In 1964, China exploded its first atomic bomb. At the first, Western scientists thought that it was a ^{239}Pu bomb. Quite surprisingly, it was a uranium bomb. China too had mastered the sophisticated uranium separation technology.

12.4.3 Spontaneous Fission (SF)

In high-Z nuclei beginning around uranium, Flerov and Petrzhak in Russia had already discovered in 1940 that a nucleus can undergo spontaneous fission into two intermediate-weight nuclei. We now understand this as the same type of process as α decay and the recently discovered heavy cluster radioactivities. All are forms of potential barrier penetration. This process was discussed in Chapter 10 along with α decay and the new heavy cluster radioactivities, including zero-neutron cold spontaneous fission.

12.4.4 Fission Energy and Its Applications

From the binding energy diagram (Fig. 9.6) we can see that when a heavy nucleus breaks into two intermediate-weight nuclei, the average binding energy per nucleon (B/A) will increase by about 1 MeV; that is, on the average every nucleon will give up 1 MeV energy. The exact value will depend on the actual situation but, on the average, when ^{235}U absorbs a thermal (very low energy) neutron and fissions, the energy released is about 200 MeV. The energy released goes into kinetic energy of the two fragments, the kinetic energy of the neutrons released, and the kinetic energy of the β decay products. For example, the energy is distributed in the following way in the induced fission of ^{235}U:

Kinetic energy of the fragments	170 MeV
Kinetic energy of the neutrons emitted	5 MeV
Kinetic energy of the β^- particles and γ-rays emitted following fission	15 MeV
Energy in the $\bar{\nu}$ emitted with the β^-	10 MeV

Neglecting the neutrino energy and some of the γ-ray energy which escapes, there is still about 185 MeV energy available for use from each fission event.

One uranium fission can provide 185 MeV of energy. This is almost 100,000,000 times greater than the energy available in the binding of two atoms. This value is also over 10 times greater than the energy released in an exoergic reaction, for example, even for $^7Li(p, \alpha)^4He$ which has an unusually large $Q = 17.35$ MeV because of the two double magic α- particles formed. However, the important thing in the fission of ^{235}U by neutron absorption is that the fission releases 2.5 neutrons on the average. If one of these neutrons can react with another ^{235}U nucleus, a self-sustaining chain reaction can occur to continuously release energy. The chain reaction is what makes it possible to develop large-scale use of the energy locked in the nucleus of the atom.

To create a self-sustaining reaction even with pure ^{235}U, its volume must be sufficiently large that the neutrons cannot easily escape from the surface of the uranium. Only when its volume is greater than some "critical volume", can a chain reaction happen. In an atomic bomb, material enriched to 90 percent ^{235}U is divided into two parts. Each part does not have a critical volume. An explosion drives the two parts together to form a critical volume in which a chain reaction can occur.

There is another important factor in producing a chain reaction in uranium: the energy of the neutrons. Although ^{235}U is only 1/140 abundant in natural uranium, the cross-section for the capture of a thermal (few eV) neutron by ^{235}U is 200 times greater than the ^{238}U neutron capture γ-emission cross-section, so a chain reaction seems possible with natural uranium. But the neutrons emitted in the fission process are not thermal neutrons: They have a distribution in energy that peaks near 1 MeV. As the neutron energy increases, the ^{235}U neutron-induced fission cross-section decreases, and the $^{238}U(n, \gamma)$ cross-section increases. Hence, for fast (MeV) neutrons the ^{238}U absorption probability is large and $^{235}U(n, f)$ cross-section is a few orders of magnitude less than for thermal neutron capture, so that a chain reaction can never happen. When the neutron energy is greater than 1 MeV, ^{238}U also has the possibility to fission, but the probability of an inelastic scattering is 10 times greater than that of fission.

The key problem is to slow down the neutrons to thermal energies. When a fast neutron collides with ^{238}U, elastic scattering can occur. Can this slow the neutron down? Indeed it can, but the mass difference between a neutron and ^{238}U is so large that in each collision the neutron loses only a little energy. To reduce the neutron energy from 1.0 MeV to thermal energies would require at least 2000 collisions. In so many collisions, the probability of absorption by ^{238}U is too large. Once absorbed by ^{238}U, it cannot release a neutron and produce more fission. A light element must be used to slow down the neutrons. Hydrogen would seem to be the best, since after only 18 collisions with hydrogen a 1-MeV neutron would have a thermal energy. But the hydrogen (n, γ) cross-section is quite large, so hydrogen is also not a good candidate. The most common materials used to moderate (slow down) the neutron energies are heavy water (with deuterium) and graphite (carbon). In 1942, the first nuclear reactor used natural uranium as a fuel and graphite as the moderator material. In a reactor the chain reaction is regulated so that energy is continuously produced without an explosion, as shown in Fig. 12.10.

The other key part of the reactor is the control rod which regulates the number of neutrons available to produce fission. The control rods must contain

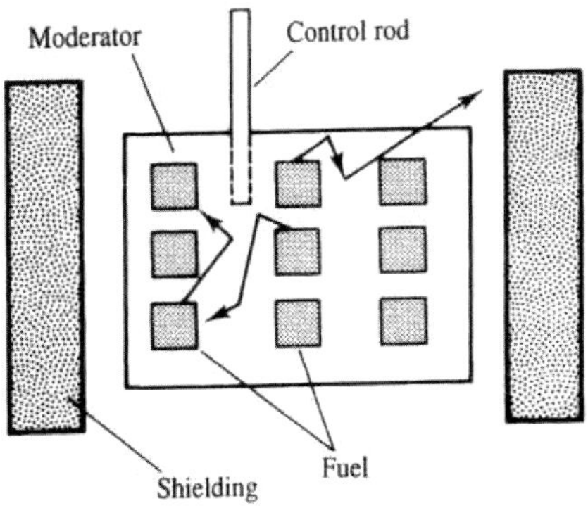

FIGURE 12.10
Schematic diagram of a nuclear fission reactor.

elements which have very high cross-sections for the absorption of neutrons, such as cadmium, gadolinium (see Table 12.1), and boron. By pulling a control rod in and out of the uranium core, the number of neutrons and number of fission events can be controlled. However, a chain reaction is very fast, a thousand generations of neutron absorptions can occur in 1 second. In so short a time, how can we control the chain reaction? The solution lies in the beta-delayed neutrons, which were first identified by Booth, Dunning, and Slack shortly after fission was discovered. As shown earlier, some of the fission products after β decay are in such highly excited states that neutron emission competes with γ emission following β decay. Beta decay of the fission products occurs a few seconds to a few minutes after the fission and the β-delayed neutrons follow immediately. Thus, the neutron emission follows the β decay half-life. These β-delayed neutrons give enough time to control the reaction rate.

Under controlled conditions, the energy produced in nuclear reactors can be used to generate electricity. Until nuclear fusion reactors become practical, fission reactors are an important source of available energy to replace our rapidly depleting supplies of coal and oil. In 2008 there were 438 nuclear power stations and 44 under construction in 30 countries, producing 14 percent of the world's electricity. There are over 100 nuclear reactor powered naval vessels. Nuclear reactors are also important sources of neutrons for basic and applied research in nuclear and solid-state physics and other areas, and are used to produce radioactive nuclides that are used in medicine and industry. The 1994 Novel Prize in Physics was won for the development of neutron scattering as a tool to probe the structure of matter. The Oak Ridge Graphite Reactor (Fig. 12.11), the first working reactor, was designed by Fermi (Fig. 12.12) to produce plutonium for one of the first two atomic bombs. In 1946, Shull and Wollan initiated a program in neutron scattering that led to the 1994 Novel Prize for Shull. Neutrons have the important property of possessing a magnetic moment, so they can probe properties of materials that cannot be studied by x-ray crystallography. Major, new neutron science facilities are being developed or have become operational such as the Spallation Neutron Source at Oak Ridge National Laboratory.

12.4.5 Fusion of Light Nuclei

The fission of heavy elements is one way to obtain energy from the nucleus. But looking at the binding energy per nucleon curve for light nuclei, one can see there is also energy to be gained from the binding energy by the fusing together of two light nuclei, like hydrogen, to form a heavier nucleus. For example,

$$\begin{aligned}
{}_1^2\mathrm{d} + {}_1^2\mathrm{d} &\rightarrow {}_2^3\mathrm{He} + \mathrm{n} + 3.25\,\mathrm{MeV} \\
{}_1^2\mathrm{d} + {}_1^2\mathrm{d} &\rightarrow {}_1^3\mathrm{H} + \mathrm{p} + 4.0\,\mathrm{MeV} \\
{}_1^2\mathrm{d} + {}_1^3\mathrm{H} &\rightarrow {}_2^4\mathrm{He} + \mathrm{n} + 17.6\,\mathrm{MeV} \\
{}_1^2\mathrm{d} + {}_2^3\mathrm{He} &\rightarrow {}_2^4\mathrm{He} + \mathrm{p} + 18.3\,\mathrm{MeV}
\end{aligned} \tag{12.33}$$

FIGURE 12.11
Workers use a long rod to push uranium slugs into one of the openings in the concrete face to the tubes which hold the uranium inside the Graphite Reactor at Oak Ridge, the world's first working reactor. It also produced the first radioisotopes for use in medicine in 1946. The openings to other tubes include spaces for reactor control rods and for neutron activation of materials. The graphite moderator material which slows down the neutrons emitted in fission to thermal speeds occupies the space between the tubes which run through the core. [*Oak Ridge National Laboratory Review*, **25** (3–4) (1992).]

FIGURE 12.12
Enrico Fermi directed design of the Chicago Pile 1 and the Graphite Reactor at Oak Ridge shown in Fig. 12.11. [*Oak Ridge National Laboratory Review*, **25** (3–4) (1992).]

The total energy given off in the reactions in Eq. (12.33) is

$$6\text{d} \rightarrow 2\,{}^4\text{He} + 2\text{p} + 2\text{n} + 43.15\,\text{MeV} \tag{12.34}$$

In the energy released, the contribution of every initial nucleon is 3.6 MeV which is about four times the energy per nucleon in the neutron-induced fission of ^{235}U. Not only is the energy released per nucleon very large, but the supply of deuterium in ocean water is practically limitless. To put this in perspective, the fusion of one ounce of deuterium is equal to burning 78 000 gallons of gasoline and the d + ^{3}H fusion yield is over five times greater. How do we capture this enormous source of energy? There is an important difference between fission and fusion: Thermal neutrons are easily absorbed by ^{235}U to induce fission with the emission of additional neutrons to produce a self-sustaining chain reaction. However, the deuteron has a positive charge so two deuterons do not fuse together at room temperature because of their repulsive Coulomb force.

In order to fuse together through the short-range nuclear force, the deuterons must overcome the long-range repulsive Coulomb force. The distance between two nucleons must be less than 10 fm for the nuclear force to take effect. The height of the repulsive Coulomb potential barrier between two deuterons at 10 fm from Eq. (12.20) is then

$$E_c = \frac{e^2}{r} = \frac{1.44\,\text{fm} \cdot \text{MeV}}{10\,\text{fm}} = 144\,\text{keV} \tag{12.35}$$

Two deuterons can fuse together if each deuteron has 72 keV kinetic energy. If this kinetic energy is in thermal energy ($\frac{3}{2}kT$), it corresponds to $kT = 48$ keV and a temperature $T = 5.6 \times 10^8$ K. But consider the following two factors: The particle has some probability to penetrate the potential barrier and its kinetic energy has some distribution. The kinetic energies of some particles are greater than the average kinetic energy ($\frac{3}{2}kT$). By including these factors, the fusion temperature can be reduced to 10 keV ($T \sim 10^8$ K). This still is an extremely high temperature. At that temperature, all atoms are fully ionized and matter is in its fourth state—plasma. Such high temperatures are realized in the Sun and other stars whose source of energy is the fusion of light elements into heavier ones.

If we want to realize self-sustaining fusion and to obtain energy from it, even such a high temperature alone is not sufficient. In addition to heating the plasma, there are other conditions. The density of the plasma must be sufficiently large, and the temperature and the density must be maintained for a sufficiently long time. In 1957, J. D. Lawson gave the following two conditions as necessary for the deuterium + tritium reaction to occur:

$$n\tau = 10^{14}\,\text{s/cm}^3 \qquad T = 10\,\text{keV} \tag{12.36}$$

where n is the density, τ is the confinement time and T the temperature in keV ($10\,\text{keV} \simeq 10^8$ K). These are the necessary conditions to realize a self-sustaining fusion reaction with a net energy output. Unfortunately, it is very difficult to sustain a plasma of such a high temperature for any time. It is likewise very difficult to find a "container" that not only can stand up to the 10^8 K temperature but also will not conduct the heat away and will not reduce the temperature of the plasma through collisions with the container or desorption of material from the container walls.

12.4.6 Solar Energy—Gravitational Confinement Fusion

In the cosmos the main source of energy is nuclear fusion. The Sun and other stars shine continuously as the result of nuclear fusion. Inside the Sun, there are two principal reactions:

1. The carbon cycle, which Bethe proposed in 1938:

$$\begin{aligned} &p + {}^{12}C \rightarrow {}^{13}N \\ &{}^{13}N \rightarrow {}^{13}C + e^+ + \nu \\ &p + {}^{13}C \rightarrow {}^{14}N + \gamma \\ &p + {}^{14}N \rightarrow {}^{15}O + \gamma \\ &{}^{15}O \rightarrow {}^{15}N + e^+ + \nu \\ &p + {}^{15}N \rightarrow {}^{12}C + \alpha + \gamma \end{aligned} \tag{12.37}$$

as shown in Fig. 12.13. In this cycle, the carbon nucleus is a catalyst whose number neither increases nor decreases. The total result is

$$4p \rightarrow \alpha + 2e^+ + 2\nu + 26.7\,\text{MeV} \tag{12.38}$$

2. The proton–proton cycle, which can be expressed as

$$\begin{aligned} &p + p \rightarrow d + e^+ + \nu \\ &p + d \rightarrow {}^3He + \gamma \\ &{}^3He + {}^3He \rightarrow \alpha + 2p \end{aligned} \tag{12.39}$$

as shown in Fig. 12.14. The net effect again is $4p \rightarrow \alpha + 2e^+ + 2\nu + 26.7\,\text{MeV}$.

Which reaction is more important depends on the temperature of the plasma. When the temperature is less than 1.8×10^7 K, the proton–proton cycle dominates. In the center of the Sun the temperature is only 1.5×10^7 K,

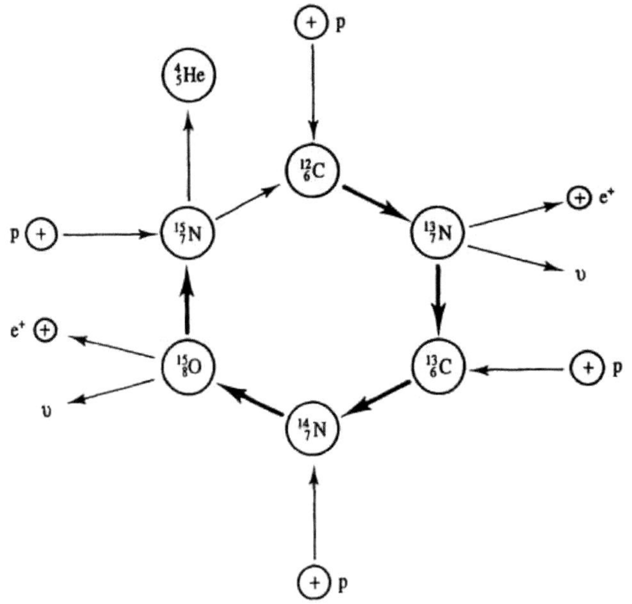

FIGURE 12.13
The carbon cycle for the fusion of hydrogen.

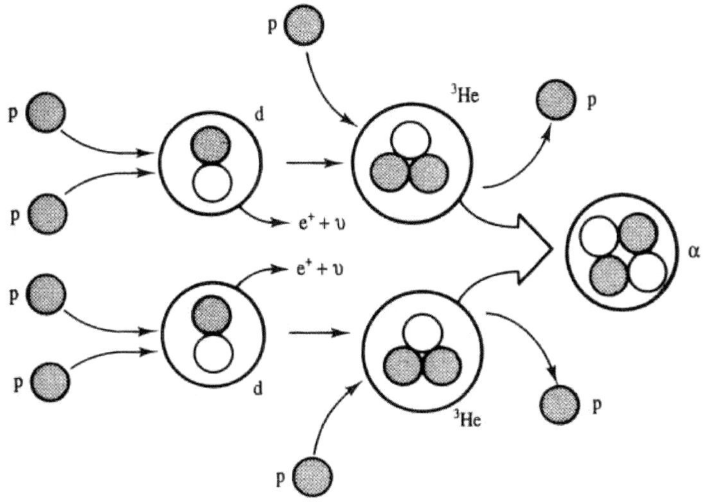

FIGURE 12.14
The proton–proton fusion cycle in which protons combine to form helium (Eq. 12.39).

so the proton–proton cycle provides 96% of the total energy production. A higher temperature is needed to fuse p + ^{12}C since its Coulomb barrier is nearly four times higher than for p + p. In many younger stars, however, the carbon cycle is more important. In both cases, the final result is that four protons fuse and 26.7 MeV of energy is released. Each proton contributes 6.7 MeV, so 6.7/938 ($\sim$1/140) of the proton rest mass is turned into energy. The 6.7 MeV per proton is eight times greater than the average energy contribution per nucleon in ^{235}U fission (200 MeV/236) and 10^8 times greater than the energy released in chemical processes.

In the Sun, 5×10^{16} kg of hydrogen are "burned" (turned into α-particles) per day. This is equivalent to 3.5×10^{14} kg per day of mass turning into energy (the difference in mass of the α-particles produced and the original hydrogen). The energy released corresponds to 9×10^{10} hydrogen bombs exploding, if each one is in the megaton class. While 5×10^{16} kg of hydrogen per day seems an extremely large amount to burn, in fact, compared to the Sun's total mass of about 2×10^{30} kg, which is 3.3×10^5 times the mass of the Earth, it is quite small. The gravitational attraction produced by the Sun's enormous mass is what confines the plasma together at such a high temperature (10^7 K) to produce the thermonuclear reactions. But even at such temperatures, the energies of the particles are still well below the height of the repulsive Coulomb barrier (see the section above). So fusion still involves penetration of the Coulomb potential barrier. The proton–proton cycle is very slow. The primary reason is that in the p + p reaction to form a deuteron, one of the protons must undergo β^+ decay, which is a weak nuclear process. The probability for this to occur is very small, so the p + p reaction cross-section is only 10^{-23} b. It is therefore extremely rare in a p-p collision for two protons to fuse. In the Sun, the time to go through one carbon cycle is about 6×10^6 years. The time for one proton–proton cycle is about 3×10^9 years. The slow reaction rate ensures that the Sun's mass does not change noticeably even after a few hundred million years. Nevertheless, the energy produced is remarkable because of the extremely large number of protons and the large number of collisions. Every second the energy that the Sun beams to the Earth is only 5 parts out of 10^{12} of its total energy produced, but what reaches Earth is 10^5 times more than all the energy used by all the people on it.

In summary, the Sun, the fusion reactor which nature provides for us, produces fusion reactions at a very slow rate. Because of its enormous mass, its plasma at 6000 K temperature in its outside layer and 1.5×10^7 K in its center, is confined in a "big container" with 7×10^8 meter radius. This reaction rate is too slow for us to use to build such a fusion reactor on the Earth. Only stars, with their giant gravitational attraction, can contain such a hot plasma for the long times required.

We must have other methods to produce artificial fusion energy on Earth. In 1946 a noted physicist who had worked on the development of the atomic bomb said that the United States could produce a fusion reactor in a year or two, just as we had produced a fission reactor and atomic bombs, if we invested similar money and manpower into the effort. All the principles were known. Unfortunately, he did not realize that the technology was not ripe for such development of a fusion power reactor. It was possible to develop a thermonuclear (hydrogen) bomb on a rather short timescale, however. It is important to recognize in science that there can be an enormous difference between knowing the principles of how to do something and having the technology to exploit the principles. As the history of fusion reactor research has shown, the expenditure of vast sums of money and manpower are no guarantee of quick success, or even long-term success, although we are now nearing the achievement of a controlled fusion reactor.

12.4.7 The Hydrogen Bomb—Inertial Confinement Fusion

We need to find a reaction which has a larger cross-section at a lower temperature. If the temperature is not to be too high, the Coulomb barrier must be low. It is lowest for two hydrogen isotopes.

From Eq. (12.33) we can see that a reaction with large cross-section and large energy release is that of d($^2_1$H) + t($^3_1$H):

$$^2_1\text{H} + {}^3_1\text{H} \rightarrow \alpha + \text{n} + 17.58\,\text{MeV} \qquad (12.40)$$

At the same deuteron energy, the cross-section for the d + t reaction is 100 times greater than that for d + d.

In natural hydrogen, deuterium (d) is only 0.015 percent abundant. In about 7000 hydrogen atoms, there is one deuterium atom. But there is so much hydrogen that there are tremendous amounts of deuterium readily available. However, tritium does not exist in nature, but it can be produced through the following reaction:

$$n + {}^{6}\mathrm{Li} \rightarrow \alpha + {}^{3}\mathrm{H} + 4.9\,\mathrm{MeV} \quad (12.41)$$

So, ^{6}Li and ^{2}H can be used to make a thermonuclear (hydrogen) bomb. The steps in the process are as follows. First a common explosive is used to bring together two pieces of a separated fissionable material (^{235}U or ^{239}Pu) to give a critical mass. Then a fission chain reaction occurs, releasing an enormous amount of energy and producing high temperature and pressure. At the same time, a vast number of neutrons are emitted. These react with ^{6}Li to produce ^{3}H (Eq. 12.41). Finally, the ^{2}H and ^{3}H under the high temperature and pressure from the fission explosion undergo a fusion reaction. Note that the 14-MeV neutrons produced in the ^{2}H + ^{3}H reaction can make ^{238}U fission (and remember that ^{238}U is much cheaper than using ^{235}U or ^{239}Pu as the fissionable material). So, mixing ^{238}U with the ^{6}Li and ^{2}H lowers the cost. The whole fission–fusion process is finished in a moment. A pure fission atomic bomb is equivalent to a few hundred thousand tons of TNT, but a hydrogen bomb (with fission plus fusion) can be equivalent to up to ten million tons of TNT. The hydrogen bomb is an uncontrollable, thermonuclear reaction and up to now is the only method of obtaining fusion energy on Earth on a large scale. It must use fission to ignite it.

The energy distribution in a typical fission-fusion bomb is as follows:

Exploded quake and shock wave	50%	Residual radiation	10%
Thermal radiation	35%	Early radiation	5%

A pure fusion reaction does not produce residual radiation (radioactive products), but instead has much more early radiation, especially neutron radiation. If the ratio of the fusion energy release to the fission release is increased, the number of neutrons will be increased and the energy in the quake and shock wave and thermal radiation will be reduced. This is the principle of the neutron bomb, which also is called an "enhanced radiation weapon." The neutron bomb was developed as a tactical weapon in which the neutrons can kill the enemy soldiers with significantly less damage to buildings and equipment and with smaller amounts of radioactive products. It has not been possible to make a pure fusion bomb that is entirely "clean" of radioactive products because the high temperatures for fusion require the use of fission to ignite the fusion.

A hydrogen bomb represents a type of inertial confinement. For many years, scientists have sought ways to produce controllable fusion by inertial confinement. One such way is laser inertial confinement: A deuterium–tritium gas mixture is placed inside a small ball of 400 μm diameter at a pressure of 30–100 atmospheres. Powerful lasers (of the order of 10^{12} watts) irradiate the small ball simultaneously from all directions. In a deuterium–tritium (DT) mixture with a density a thousand times that of liquid DT. The number density is 3×10^{25} nuclei/cm^{2}. Then a confinement time of only 30×10^{-12}s satisfies the Lawson criteria. When the temperature is 10^{8} K, the fusion reaction results.

Pulsed power machines magnetically accelerate wire arrays or solid liners to convert electrical energy to kinetic energy of an implosion for inertial fusion. These devices also can accelerate electron beams, ions beams, and imploding plasmas for many applications. Power concentration experiments in the 1980s and 1990s showed that electron beams have too long a range and preheat the fuel. Research on the production and focusing of ion beams was interrupted by a major breakthrough in imploding plasmas in 1995 when it was found that increasing the number of wires in a wire array from 20 to 200 dramatically increased the output x-ray power from wire array z pinches. Although this discovery came just after the decision to invest in lasers for the National Ignition Facility, it enabled the Particle Beam Fusion Accelerator II (PBFA II) at Sandia National Laboratories to exceed its design goals by a factor of two and produce 230 TW and 1.6 MJ of x-rays for a wide variety of applications in high energy density physics including fusion. In 1997, PBFA II was permanently converted to the Z Machine shown in Fig. 12.15 to optimize it for driving imploding plasmas. In addition to records in x-ray drive, experiments on Z have set records in the number of DD neutrons produced with x-ray driven inertial fusion in the laboratory (2×10^{10} neutrons), in the velocity of a solid flyer plate for shock experiments (30 km/s), in isentropic compression of matter to high pressures in the laboratory (3.25 megabars in copper), and in x-ray line radiation (450 kilojoules at 1 keV, 100 kilojoules at 5 keV, and 18 kilojoules at 8 keV.)

The knowledge gained in high energy density plasma experiments support three approaches for fusion energy. In the Double Ended Hohlraum, current is passed through wire arrays that are above and below the implosion capsule. The resulting magnetic field drives the wire plasma to the axis, where the kinetic energy of the implosion is converted to x-rays, which flow through the hohlraum to uniformly deposit energy in the central capsule. In the Dynamic Hohlraum, the single wire array is imploded onto a low-density foam where the kinetic energy is converted to x-rays that flow through the foam to ablatively drive the central capsule. The development of an improved theory of electrical resistivity for high energy density materials on Z has stimulated research on directly driving the pusher with the magnetic pressure from current passing through it—avoiding the x-ray production, transport, and ablative acceleration processes altogether. Simulations of Quasi Spherical Direct Drive indicate that the efficiency of coupling the electrical energy to energy in the fusion fuel is about 10 times that of x-ray drive, substantially reducing the cost of the facility required for fusion. Research is underway to determine the best option for a future high yield facility. The low cost and high efficiency of pulsed power accelerators meet the requirements for fusion energy but the stand off is a major limit on the duty factor of the accelerator and power production chambers. More work is needed on the technology to make the projected cost of electricity economically advantageous compared to currently available sources.

12.4.8 Controllable Fusion Reactor – Magnetic Confinement

The history of the magnetic confinement of fusion reactions covers more than 50 years, and the first efforts to obtain controllable fusion involved this process. It is still the most promising technique. In a magnetic confinement experiment, the plasma is kept inside a high vacuum container. The plasma is prevented from coming into contact with the sides of the container by the presence of a magnetic field. Ions in the plasma may move along the magnetic field lines, but those moving perpendicular to the field lines are deflected by the action of the Lorentz force. The plasma also is heated by an electromagnetic field.

FIGURE 12.15: Z Machine at Sandia National Laboratories during a shot. This facility is the world's largest driver for the pulsed power approach to inertial confinement fusion. (Courtesy of J. Pace VanDevender)

Unfortunately, the intensity of available magnetic fields is only the order of 10^5 Gauss (10 teslas), so a high temperature plasma with magnetic confinement is difficult to achieve. In inertial confinement one tries to increase the ion density n to get the necessary conditions (12.36), while in magnetic confinement one tries to increase the confinement time in order to achieve fusion.

There are many types of magnetic confinement devices. The most promising is the Tokamak, which was invented in the Soviet Union in 1950 and translated means toroidal magnetic chamber (as illustrated in Fig. 12.16). Research with Tokamaks has been carried out in many countries including besides Russia, China, Japan and the

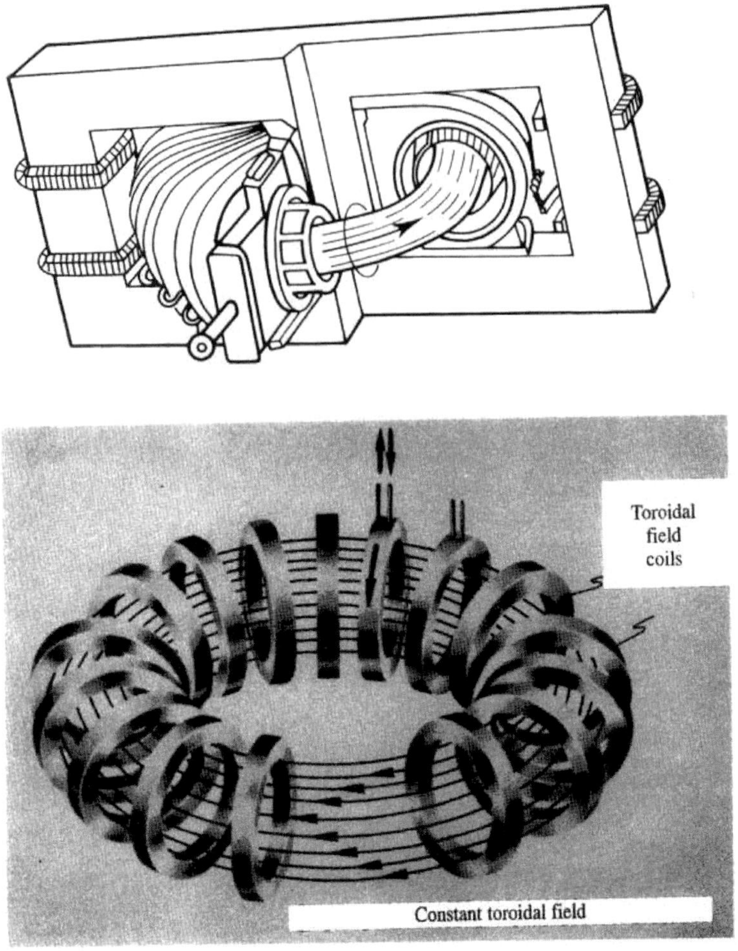

FIGURE 12.16
A Tokamak (circulator arrangement).

United States. The ring coils of the circulator's main machine can produce a ring-magnetic field of a few times 10^4 Gauss along the axis line of the ring-tube. The iron core transformer induces the plasma electrical current. The ring plasma current forms a secondary coil of the transformer which has only one loop. Since the inducing plasma electrical current produces the ohmic heating, this field also is called the heating field.

In the United States, the Princeton Tokamak Fusion Test Reactor (TFTR) (shown in Figure 12.17) continuously ran from 1982 to 1997. In this arrangement the large radius of the vacuum chamber is 2.5 m and the small one is 0.85 m. In the world's first experiments utilizing a 50/50 mix of deuterium and tritium, a world record temperature of 510 million °C was achieved. At the center of the TFTR the

FIGURE 12.17
A close-up photograph of the Princeton Tokamak Fusion Test Reactor. (Courtesy Princeton University.)

plasma pressure has reached six atmospheres which is comparable to that needed for a commercial fusion reactor.

Following the 1985 US-Soviet Summit Meeting, a joint, multinational program to build the world's first fusion engineering test reactor, called the International Thermonuclear Experimental Reactor (ITER) was established. It was thought completion would be around the year 2000, and that a fusion power demonstration plant could be in operation within 25 years. However, reaching a multinational consensus on this project proved more difficult than anticipated. The ITER now has been sited in France with 2018 expected completion. The overall objective of ITER is to demonstrate the scientific and technological feasibility of fusion energy for peaceful purposes. ITER would accomplish this by demonstrating controlled ignition and extended burn of deuterium-tritium plasma. ITER represents the necessary and essential step towards practical demonstration of controlled fusion power. The importance of the INTER mission is evidenced by participation of all major industrial nations of the world, including the United States, China, Europe, India, Japan, Korea, and Russia. Conservative estimates of world population and energy demand growth show that we can expect serious energy shortfalls within 50 to 75 years. In addition, problems of acid rain, air pollution, and the potential of global warming will necessitate an eventual shift to non-fossil fuel energy production. During the twenty-first century, a significant fraction of the world's energy must come from non-fossil fuels. Fusion energy is clearly a major candidate to supply our long-range energy needs toward the end of the century.

In 1989, researchers at the University of Utah reported that they had observed the fusion of deuterium at room temperature. If true, this would have been a fantastic development, but one immediately asks how it is possible given what we have said about the temperatures required to fuse deuterium with deuterium or tritium. In principle, it could be possible if there is a chemical process whereby two deuterium nuclei can be brought sufficiently close together through the structure of the material that the two nuclei could tunnel their Coulomb barrier at room temperature—cold fusion. This was the process that was reported to have been observed. If true, the financial profits from the discovery would be astronomically large. Unfortunately, efforts to reproduce this reported cold fusion have not been successful. One must be especially careful to be objective when there is great fame and/or fortune to be made from one's discovery.

There is another type of cold fusion in which a negative muon stopped in a mixture of deuterium and tritium can induce the D-T fusion. In this case the muon replaces the electron in an atom which then has a radius which is smaller than the normal hydrogen radius by the ratio of the electron/muon mass; this brings the atoms very much closer together to allow tunneling. A single muon can initiate up to 150 fusion reactions under certain conditions. This is called muon-catalyzed fusion. Unfortunately, here too the energy production is much too low for energy use on a large scale.

12.5 SOME SELECTED APPLICATIONS OF NUCLEAR PHYSICS

Nuclear physics has made major impacts in a remarkably broad spectrum of basic and applied fields, including critical societal problems. Energy from the nucleus of the atom, as just discussed, has radically altered our world. The development of nuclear energy as a replacement for our rapidly diminishing fossil fuels appears crucial for the long-term future of mankind, while nuclear weapons threaten our very existence. Here we describe a few examples to illustrate the far-ranging breadth of the applications of nuclear techniques today. A number of applications of nuclear technologies have been discussed in other sections of the text, and others are considered in some detail in Appendix I. The remark of Philip Handler, former President of the U.S. National Academy of Sciences and a distinguished biochemist, could be echoed in many fields today as well as then. He said, in the late 1980s, "Biological sciences have been moved forward fifty years in the past five through the adoption of many of the techniques of nuclear physics."

Materials science is central in the development of new technologies. The use of nuclear techniques has opened up the development of critical new materials tailored to meet specific applications. Ion implantation techniques have been crucial in developing new materials for microelectronics and computer chips. The scattering of thermal-energy neutrons from research reactors or pulsed neutron sources from accelerators have opened up studies of the structural order of materials. Very "cold" neutrons (with much lower than thermal energies) are offering unique opportunities to probe with resolution over 10 000 times better than with a light microscope everything from the properties of crystals and glasses to production of new synthetic materials and on to the localization of protein molecules in the field of microbiology. The new spallation accelerator for neutron science at Oak Ridge National Laboratory is opening up new research in material science with additional impacts on electronics, transportation, environment, biology and medicine. It also includes new opportunities in basic research, such as testing time reversal and parity conservation. Several facilities for neutron science in Europe have been upgraded, and new ones proposed there, too.

Neutron scattering offers unique opportunities to study samples containing hydrogen and magnetic order in materials. For example, since uncharged neutrons easily penetrate deep into matter and since they strongly scatter off hydrogen nuclei, they have become a crucial diagnostic tool for studies of protons and other biomolecules on to detecting and locating oil leaks deep inside complex engines, even during operation. They have been used to locate water condensation in honeycomb aluminum structures used in the aircraft industry; such condensation produces aluminum oxide which has much less structural strength than aluminum. Another recent application includes understanding and improving the new high-temperature superconductors. For example, neutron irradiation of the high-temperature superconductor $YBa_2Cu_3O_7$ increased its current-carrying capacity by a factor of 100 at 77K. Neutron capture or inelastic scattering have been employed to provide high sensitive detection of elements which present major risks to society. For example, one can detect the sulfur content of coal on the conveyor belt using neutrons, and can modify high-sulfur-content coal by mixing with low-sulfur-content coal to meet sulfur content standards. This same technique is now being used to detect the nitrogen and explosive materials in airline luggage for the prevention of airline sabotage and to detect nuclear material in cargo being shipped.

At Oak Ridge National Laboratory the heavy-ion accelerators were used to simulate neutron damage of materials used in reactors at fluxes that would have taken many years to duplicate in a reactor. This research enabled the laboratory to develop a new alloy that resists the swelling and cracking over time that was occurring in metal pipes used for cooling and other critical purposes in nuclear reactors. The expected cost savings from the use of this new neutron-damage-resistant alloy in nuclear reactors will be tens of billions of dollars.

Another area is the time evolution of the depolarization of the spins of polarized positive or negative muons implanted in a sample—muon spin relaxation. This technique has been used in numerous areas with important recent applications in characterizing the properties of high-temperature superconductors, including superconducting electron carrier densities and penetration depths of magnetic fields. Positron annihilation in materials probes other key properties of materials.

In January 1990 the space shuttle *Columbia* chased and caught a 30-ton probe that had been orbiting at 17 400 miles/hour for six years and brought it back to Earth. Materials on board had been exposed to cosmic rays for nearly six years to see what radioactivities would be induced in them. From these data one can estimate problems from such induced radioactivities in future space stations.

Another very important study of nuclear processes is in the problem of system failure in microelectronics systems, including microcomputer chips. It is now well established that the error rate in computers in Denver is considerably greater than at sea level, presumably because of the much larger cosmic ray flux. The problem of single-event failure induced in a system from cosmic ray and background radiations is becoming more and more important as microelectronic chips are made smaller and smaller. Problems of the failure rate of computers and other microelectronic devices in the radiation environment of space must be studied and solved as we move toward greater exploration of space and the development of space stations.

People around the world watched in August 1989 in awe as we saw on TV the incredible photographs of Neptune sent back by the *Voyager* spacecraft. These pictures were only possible because *Voyager* employed radioisotope thermoelectric generators RTG's as power sources. These RTG's use the energy from the radioactive decay of ^{238}Pu to generate electricity. These power sources exceeded both their predicted performance and predicted lifetime. *Gailileo*, the first nuclear-powered spacecraft since *Voyager* in 1977, was launched in October 1989 to begin its journey to explore Jupiter and its moons. RTG's have powered 24 NASA missions including the *Cassini* mission to Saturn and others, whose extreme distances from the sun, for example, require RTG's.

Nuclear astrophysics is a major frontier in understanding the evolution of our universe. The birth and death of a star are related to the nuclear processes that occur beginning with hydrogen fusion, helium fusion, on to neutron capture and nuclear reactions of short-lived radioactive isotopes in nova and supernova explosions. Studies of the reaction cross-sections and decay properties of radioactive nuclei are providing critical new data to improve our understandings of stellar evolution and the production of heavy elements in the universe. The picture we have of the formation of the universe, our cosmology, is that the universe as we see it began 13.7 ± 0.2 billion years ago when all the matter and energy was concentrated at incredibly high densities and temperature in a small volume or singularity.

This erupted with a big bang in which matter and energy were thrown out to produce the expanding universe we see now. There were various important stages in this expansion, as illustrated in Fig. 12.18. Before the first 10^{-43} s, all the known forces were united into one force and the laws of physics are thought to have been different, to allow for an inflationary universe that expanded at a more rapid rate than our present laws would allow. Then at 10^{-43}–10^{-35} s, the gravitational force separated from the strong, weak, and electromagnetic forces. Soon after, the strong force separated and after about 10^{-10} s, the electromagnetic and weak forces became separate forces. From 10^{-35} s to 10^{-10} s, the world was filled with quarks,

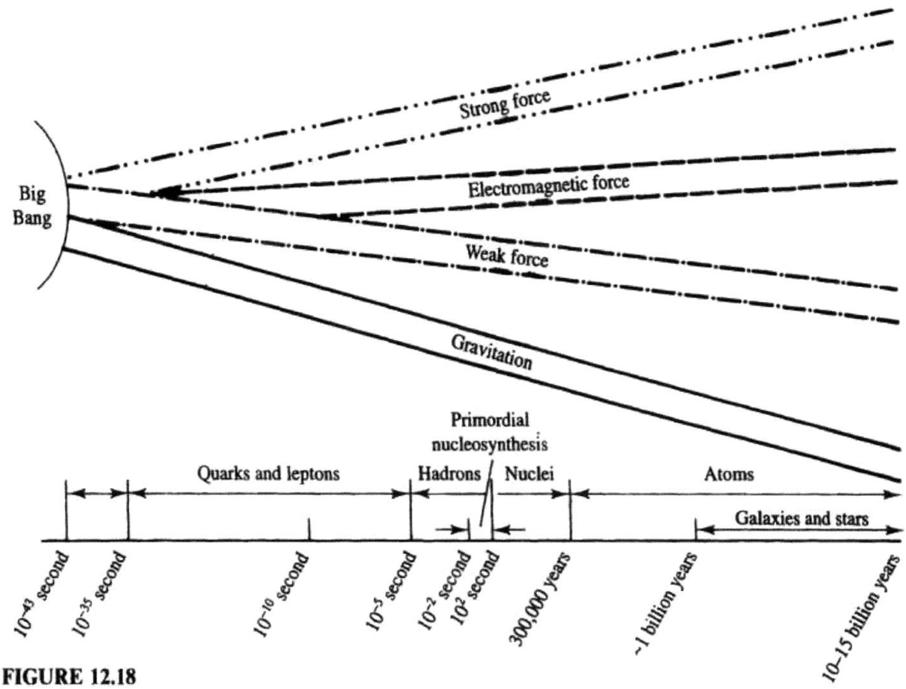

FIGURE 12.18
The expansion and cooling of the universe, showing when the single original force split into new forces and when particles, nuclei, atoms, and stars emerged.

leptons and photons. The temperatures were too hot for quarks to stick together. About 10^{-5} s the temperature dropped to less than 10^{13} K, so that quarks formed hadrons, protons, neutrons, and mesons.

In the early universe, after protons and neutrons condensed out, their ratio was frozen at about 6/1 at about 1 s. Primordial nucleosynthesis occurred between 10^{-2} and 10^{2} s. However, even at 1 s nuclei with $A \geq 2$ had very, very small abundances. The first nucleus to form would be $\mathrm{n} + \mathrm{p} \rightarrow \mathrm{d} + \gamma$ (2.225 MeV). From $t = 1$ min to 2–3 min, the abundances of deuterium, ^{3}H, and ^{3}He became large and ^{4}He was quickly formed. After sufficient deuterium had formed, the other reactions included $\mathrm{d} + \mathrm{p} \rightarrow {}^3\mathrm{He} + \gamma$, $\mathrm{d} + \mathrm{n} \rightarrow {}^3\mathrm{H} + \gamma$, and later $\mathrm{d} + \mathrm{d} \rightarrow {}^3\mathrm{H} + \mathrm{p}$ or $\rightarrow {}^3\mathrm{He} + \mathrm{n}$, and quickly all the neutrons were tied up in ^{4}He in reactions like ${}^3\mathrm{H} + \mathrm{p} \rightarrow {}^4\mathrm{H} + \gamma$, ${}^3\mathrm{He} + \mathrm{n} \rightarrow {}^4\mathrm{He} + \gamma$, ${}^3\mathrm{He} + \mathrm{d} \rightarrow {}^4\mathrm{He} + \mathrm{p}$, and ${}^3\mathrm{He} + \mathrm{d} \rightarrow {}^4\mathrm{He} + \mathrm{n}$. Since there are no stable $A = 5$ nuclei and $^8\mathrm{Be}(\alpha + \alpha)$ is unstable, ^{4}He is the end-product. The relative primordial abundance of ^{4}He by weight is about 0.24, which is what is observed today in a variety of systems. Nuclei continued to be the primary form of matter to about 300,000 years. About this time the temperature dropped below 10^4 K ($\sim$ 1 eV), so electrons and nuclei began to combine to form atoms. Note 1 eV is the order of the outer-shell electron binding energies in atoms. For a more complete description of the Big Bang see *The Early Universe* by E. W. Kolb and M. S. Turner (Addison-Wesley, New York, 1990).

One of the major evidences for the big bang was the discovery by Arno Penzias and Robert Wilson of a cosmic background microwave radiation. The radiation corresponds to black-body radiation with a temperature of 2.7K, which is consistent with the temperature of a gas that has been cooling since the big bang. This won them the 1978 Nobel Prize in Physics.

Charged particle reactions involving protons and α particles are the primary processes by which elements up to $A = 60$ were formed in stars. When a star is first formed, hydrogen fusion (p + p) is the primary reaction. As a star's hydrogen is used up, the star undergoes gravitational collapse that leads to high temperature, so ^{4}He + ^{4}He can occur, but ^{8}Be breaks up into 2α in a time of about 10^{-16} s. It was realized by Fred Hoyle that for the small concentration of ^{8}Be to fuse with ^{4}He to form ^{12}C, there had to be an excited resonance state in ^{12}C with a much larger cross-section. The needed resonance state was found at 7.65 MeV in ^{12}C. From ^{12}C one can form $^{12}\text{C} + {}^4\text{He} \rightarrow {}^{16}\text{O} + \gamma$, $^{16}\text{O} + {}^4\text{He} \rightarrow {}^{20}\text{Ne} + \gamma$, but the Coulomb barrier becomes too large by ^{24}Mg to continue α capture. With ^{12}C present, one also has the carbon cycle, converting four protons to ^{4}He as well. After ^{4}He is used up, gravitational collapse occurs again, and reactions like $^{12}\text{C} + {}^{12}\text{C} \rightarrow {}^{20}\text{Ne} + {}^4\text{He}$ and $^{16}\text{O} + {}^{16}\text{O} \rightarrow {}^{28}\text{Si} + {}^4\text{He}$ can occur. Reactions such as (p, γ) and (α, n) also occur with lower probability.

Earlier we discussed the hot CNO cycle for energy production. Other reactions can occur in this cycle to change the energy production, as shown in Fig. 12.19. However, there are α-particle break-out reactions not shown in that figure. For example, at temperatures near 0.5×10^9 K the β decay of ^{14}O can be bypassed by $^{14}\text{O}(\alpha, \text{p})^{17}\text{F}(\text{p}, \gamma)^{18}\text{Ne}\ (\beta)^{18}\text{F}(\text{p}, \alpha)\ ^{15}\text{O}$, which would give a 60 percent increase in energy production. At $T > 0.5 \times 10^9$ K, reactions like $^{15}\text{O}(\alpha, \gamma)^{19}\text{Ne}$ and $^{18}\text{F}(\text{p}, \gamma)^{19}\text{Ne}$ lead to ^{19}Ne (p, γ), and the rapid proton capture process (rp-process) with a hundredfold increase in the power of the hot CNO cycle. These unknown α and p capture rates on these radioactive nuclei are critical for understanding whether such processes occur. The Radioactive Ion Beam (RIB) facilities around the world are giving RIBs of ^{14}O, ^{15}O, ^{17}F, and ^{18}F with which to study these reaction cross-sections.

Theoretically, the breakout reaction $^{15}\text{O}(\alpha,\gamma)\ ^{19}\text{Ne}(\text{p},\gamma)\ ^{20}\text{Na}$ can lead to rp-processes all the way to mass 100. Such rp-processes can occur in a white dwarf accreting hydrogen from a red giant binary star or similarly for a neutron star. The rp-process path is determined by the proton capture cross-sections and the β decay rates. With new radioactive beam facilities, one can measure many of the proton capture rates and determine the nuclear shapes, masses, and isomers that can influence the relative p-capture/β decay rates.

Above ^{56}Fe no energy is to be gained by fusion of lighter elements since the binding energy per nucleon curve starts to decrease after that (see Fig. 9.6); neutron capture processes are involved in heavier element production. New insights into the roles of the rapid neutron capture process (r-process) and the slow neutron capture process (s-process) are coming from studies of nuclei far from stabililty. The major new RIKEN RIB factory shown earlier in Fig. 11.33 began operation in 2007. Major new RIB facilities are underway in Europe, see the Facility

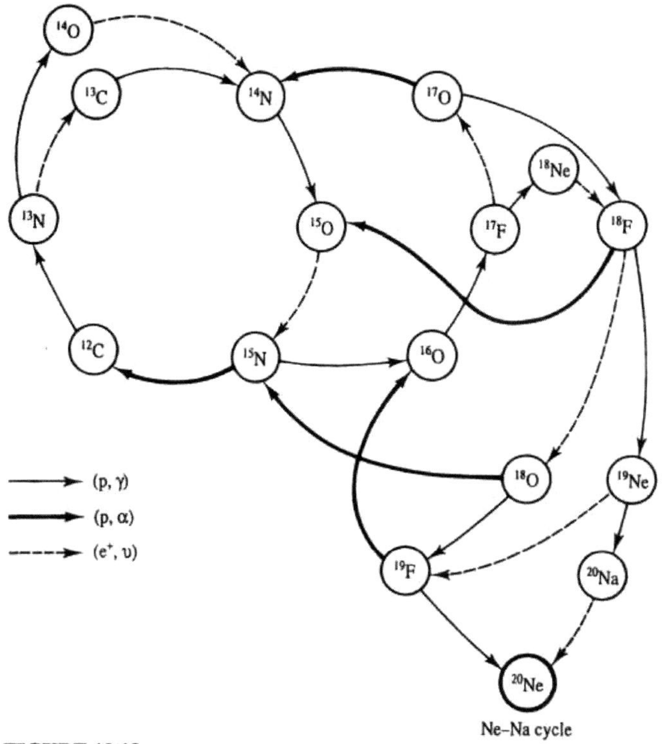

FIGURE 12.19
Schematic of hydrogen burning in the "hot" CNO cycle. [From C.E. Rolfs and W.S. Rodney, Couldrons in the Cosmos, University of Chicago Press, Chicago (1980)]. Two of the four exotic nuclei (^{13}N, ^{17}F, ^{18}F, and ^{19}Ne) for which proton-induced reactions are proposed, ^{17}F and ^{18}F, have been studied at the Oak Ridge RIB facility. Likewise, α-particle-induced "breakout" on ^{14}O and ^{15}O also can be studied.

for Antiproton and Ion Research (FAIR) in Fig. 14.5 and in the Untied Sates in Fig. 11.34. Clearly RIB facilities will open many new windows to view our universe.

Let us illustrate the rapid and slow neutron-capture process. The most abundant stable isotope found near the chain of fusion reactions is ^{56}Fe. If there is a large flux of neutrons, then

$$^{56}\text{Fe} + \text{n} \rightarrow {}^{57}\text{Fe} + \gamma$$
$$^{57}\text{Fe} + \text{n} \rightarrow {}^{58}\text{Fe} + \gamma$$
$$^{58}\text{Fe} + \text{n} \rightarrow {}^{59}\text{Fe}\ (T_{1/2} = 45\,\text{days}) + \gamma$$

What happens next depends on the neutron flux intensity. If this flux is too low, then ^{59}Fe β-decays to ^{59}Co, which can capture a neutron to form ^{60}Co, etc. If the flux is sufficiently high, then $^{59}\text{Fe} + \text{n} \rightarrow {}^{60}\text{Fe}(T_{1/2} = 3 \times 10^5\,\text{y})$, which can capture a neutron to $^{61}\text{Fe}(T_{1/2} = 6\,\text{min})$, then undergo n capture to ^{62}Fe(68 s) and beyond. This continues until a β-decay half-life is reached which is shorter than the mean-life for n capture. Then β^- decay to a higher Z occurs, and n capture occurs in that element until β^- decay occurs before n capture. When the time scale is long so that there is time for all intervening β decays to occur, it is called the s- (slow) process. When the process proceeds so that there is time for only the most short-lived β decays, it is called the r- (rapid) process.

The neutrons may be produced during helium burning by reactions like $^{13}C(\alpha, n)^{16}O$, but at red-giant temperatures the neutron density is only of the order of $10^{14}/m^3$. Such densities give rise to the s-process. The r-process requires many orders of magnitude higher fluxes. Such fluxes are believed to occur in supernova explosions or in neutron stars. Theoretical explanations of supernova are active areas of research. Nevertheless, there is clear experimental evidence for the r-process. The s- and r- processes produced all the elements up through uranium found in the Earth; our planet's heavy elements are the remnants of earlier stars. Elements above $A > 209$ are produced through the r- but not the s-process. One can date the age of the solar system by using decay processes such as $^{40}K \to {}^{40}A$, $^{87}Rb \to {}^{87}Sr$ and $^{238}U \to {}^{206}Pb$ or the $^{238}U(T_{1/2} = 4.5 \times 10^9\,y)$ to $^{235}U(T_{1/2} = 0.7 \times 10^9\,y)$ ratio to yield an age of about 4.5×10^9 y. The present ratio of $^{235}U/^{238}U \simeq 0.0072$ and at the formation of our solar system is estimated to have been 0.29 following their earlier synthesis and decays to the time of the formation of the solar system.

When supernova 1987 was observed, a University of Florida–Goddard Space Flight Laboratory team quickly adapted for balloon flight their newly designed BGO Compton-shielded germanium γ-ray detector. With the help of the U.S. Air Force and the NSF ground station in Antarctica, they launched the first balloon flight off Antarctica, as shown in Fig. 12.20, and observed γ-rays from supernova 1987. They and other groups observed γ-rays from the decay of short-lived $^{56}Co(T^{1/2} = 78\,\text{days})$ to confirm for the first time the expectations of heavier element production in supernova explosions.

Many different types of gamma-ray emission processes go on in the universe. Another way which nuclear physics is illuminating astronomy and our understanding of the universe was a US space probe called the Compton Gamma Ray Observatory (CGRO) which took data from 1991 to 1999 to measure gamma rays of extraterrestrial origin. The observatory is illustrated in Fig. 12.21. The CGRO had four different γ-ray detectors. We will consider two of them. The Burst and Transient Source Experiment (BATSE) observed 2704 GRBs and found two types. Short bursts which last from 0.002 to 2 seconds and long burst from 2 s to several minutes. While there are only theories about short bursts, long bursts are believed to originate at the furthest edges of the observable universe, billions of light years away. They briefly shine hundreds of times brighter than a supernova and a million trillion times brighter than our sun. The first x-ray after glow of a GRB was seen by the Dutch-Italian satellite Beppo SAP which had both γ-ray and an x-ray detectors. There are now several γ and x-ray detector satellites, NASA's High Energy Transits Expolorer, the Rossi X-ray Timing Explorer and the Europe Space Agency, Integral. A worldwide network called the Gamma-ray Burst Coordinate Network coordinates

FIGURE 12.20 (a)
The balloon launch in Antarctica of the BGO shielded germanium detector by a University of Florida–Goddard Space Flight–U.S. Air Force team.

satellite and ground based after glow observatories. There is clear evidence that GRBs are linked to supernova. The favored model for GRB's involves the collapse of the core of an especially energetic supernova dubbed a hypernova. Another exciting new discovery was made by EGERT on CGRO, a new class of objects known as blazars. There objects are quasars at cosmological distances that emit the majority of their electro-magnetic energy in 30 MeV to 30 GeV γ-rays.

Geological and archeological applications continue to provide unique answers to important questions, many with broad historical and cultural

FIGURE 12.20 (b)
The detector is housed in the gondola (as shown) which is attached to the balloon. (Courtesy of A. C. Rester, University of Florida.)

importance. We have already mentioned the use of uranium and other very long-lived isotopes to date the age of the Earth and the ages of rocks, and the use of ^{14}C, which is continuously produced in the atmosphere, to date the time of death of earlier living systems. At death the $^{14}C/^{12}C$ ratio decreases with the half-life of ^{14}C, so this ratio measures the time in which atmospheric carbon ceased to be taken into a sample. Earlier ^{14}C dating involved difficult measurements of the low-energy β-rays from ^{14}C. In recent years the field of accelerator mass spectrometry (AMS) has opened up remarkable new sensitivity for ^{14}C dating. Three accelerator laboratories recently dated the "Shroud of Turin" that had been thought to be the burial cloth of Jesus at A.D. 1325 with an uncertainty of about 30 years. Similar AMS studies of wood found in an iron-smelting pit in Newfoundland was dated to A.D. 997 with an uncertainty of 13 years. This provides evidence that Scandinavians were in North America long before Columbus. Nuclear techniques to unravel the composition of pottery and thus to locate its origins have given new insights on the travels and trade of ancient peoples. An Ion Beam Analysis Laboratory with a $2 \times 2\,\mathrm{MV}$ tandem accelerator

FIGURE 12.21
An artist's drawing of the Gamma-Ray Observatory placed in orbit in 1991. The solar power panels are shown on each side. Several of the different γ-ray detector systems for the different experiments to be carried out are shown in the drawing. (Courtesy U.S. National Aeronautic and Space Administration, Huntsville.)

has been built underground at the Louvre Museum in Paris to open up new insights into art and history.

Another particularly fascinating application concerns the question of what brought on the extinction of the dinosaurs. There is considerable agreement now that about 65 million years ago the Earth was struck by a sizable piece of debris such as a meteoroid, asteroid, or comet that enveloped the Earth in a dense cloud of dust from its break-up. This blocked the sunlight which, in turn, killed the vegetation. Other important ingredients were the shock heating of the atmosphere and probably nearly pure nitric acid rain. This led to the extinction of dinosaurs and indeed of most living species through starvation. Meteorites have concentrations of iridium that are ten thousand times larger than material in the crust of the Earth. Using neutron activation analysis, a marked enhancement of iridium at over 70 locations around the world at depths that correspond to the time of extinction of the dinosaur has been observed with a sensitivity of a few parts per trillion.[2] This discovery has brought on new searches for iridium enhancements at times associated with other sudden disappearances of different species.

Nuclear medicine is one of the brightest showplaces of the applications of nuclear physics. It is estimated that currently over 10 million Americans receive nuclear medical treatments each year, and this number will continue to grow.

---

[2] L. Alvarez, *Physics Today* (July 1987), p. 24.

Long before the development of nuclear reactors in World War II and the exploitation of particle accelerators, radioactive radium was one of the few weapons in the medical arsenal for killing cancer cells. Beginning in the 1950s, there has been an explosion in the applications of nuclear techniques with important new developments continuing to be opened up every few years.

Radioactive-isotope-tagged compounds continue to play a significant role in diagnostic medicine. More recently, there have been developed new ways to study the interior of the human body in greater detail than ever before. One major new technique is magnetic resonance imaging (MRI), in which magnetic fields are used to provide unprecedented views of the interior of the body without invasive surgery or the injection of radioactive isotopes. This new technique was discussed earlier in Chapter 4. A second technique, positron emission tomography, PET, is now a very powerful means to study the dynamic workings of the interior of the body. These techniques are revolutionizing diagnostic medicine. They can give dynamical views of the interior of different parts of the body that can eliminate much exploratory surgery. MRI and PET are complementary techniques. Each has its own unique capabilities to probe different features of the body. In PET a cyclotron accelerator is used to produce short-lived β^+-decaying isotopes like $^{11}C(T_{1/2} = 20.4\,min)$, $^{15}O(T_{1/2} = 2.0\,min)$ and $^{18}F(T_{1/2} = 110\,min)$. An ^{11}C-, ^{15}O-, or ^{18}F-tagged source is sent to the area of interest. The two photons emitted 180° with respect to each other when the e^+ from their β decay annihilates with an e^- are used to locate the tagged molecules and so determine the actual functioning of the area of interest. A photograph of one of the compact cyclotrons is shown in Fig. 12.22. A patient is shown in the position for a brain scan in Fig. 12.23. The location of the radioactive sample is determined by observing in a circular array of detectors the two annihilation photons in any pair of detectors located 180° apart. Studies can be done, for example, by tagging glucose (the energy fuel for the brain and body) with a radioisotope and by observing how and where the fuel is being used in the organ or area of interest. For example, one can see as a function of width and depth the actual dynamic working of the left side of the brain when the subject hears speech and the right side of the brain where music is heard. Examples of the use of PET to study particular problems of the brain and heart are shown in Figs. 12.24 and 12.25.

Figure 12.24 illustrates a myocardial tissue viability study consisting of two separate scans of the same slice through the heart. The left scan was taken 4 minutes after the injection of 20 mCi of ^{13}N-labeled ammonia. Ammonia interacts, to first order, as a microsphere, in that after it enters the tissue it remains there. Therefore, the amount of activity in a given region is proportional to the amount of blood flowing to that tissue. (The results of these studies are presented in color to indicate the different intensities; as reproduced here the different colors come out as different shades of gray.) Referring to Fig. 12.24, the lighter regions have higher flow than the dark regions. Note the left side and top of the heart muscle have sharply decreased or no blood flow. The scan on the right was taken 40 minutes after a 10-mCi injection of ^{18}F-labeled 2-fluoro-2-deoxy-D-glucose

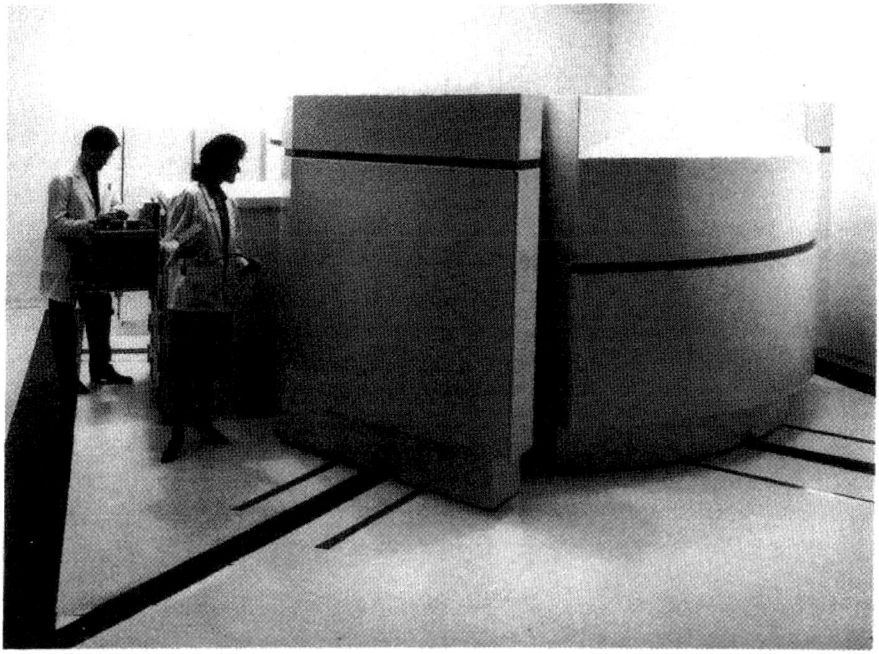

FIGURE 12.22
A compact cyclotron like the one recently installed at Vanderbilt University Medical Center. The actual cyclotron magnet and inner workings are inside the shielding walls seen in the photograph. After production, the short-lived isotopes can be transferred by pneumatic tubes to a patient in a nearby room. (Courtesy of Siemens Gammasonics, Inc.)

(FDG). This tracer is taken up by a particular tissue in proportion to how much glucose it is using. The purpose of this study was to determine how much tissue was infarcted (with no chance of recovery), how much was ischemic (could recover with increased blood perfusion), and how much was normal. Again, the left side and top regions are dark to indicate no metabolic activity—it is infarcted. This person was judged to be a good candidate for a heart transplant. Following the heart transplant, a slice was taken through the heart that was removed at the place corresponding to the two upper PET scans. One sees clearly that the upper and left parts of the heart are no longer useful heart muscle as seen in the bottom and right side but have become gristle that cannot function to pump blood from the heart.

In Fig. 12.25, the upper view is a top view of the brain in an FDG study of a patient with Huntington's disease. The intensity scale is that the degree of blackness is proportional to glucose metabolism (the darker areas have the greater metabolic activity). This study is normal in all areas except the basal ganglia (the two small light-gray areas to the left and right of center in the upper right center of the image). Normally, they would have the same metabolic activity as the structure below or to either side (caudate nucleus), but not in this case. This is

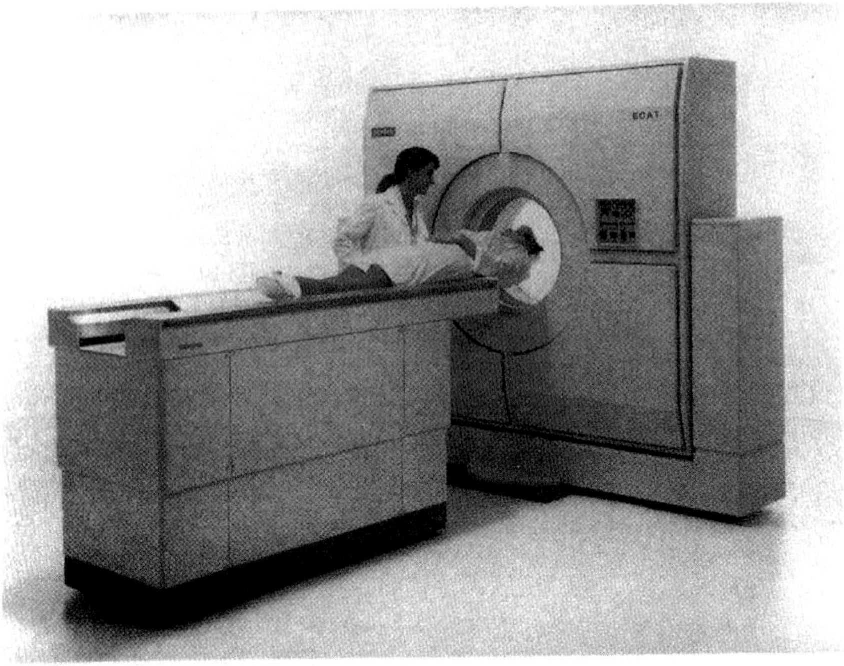

FIGURE 12.23
A patient is shown having a brain scan. The bismuth germinate γ-ray detectors form a complete circle inside the ring around the patient's head. (Courtesy of Siemens Gammasonics, Inc.)

clearly seen by comparing the upper view with that of a normal patient shown below where two very dark half-moon-shaped areas are seen to the right and left of the center. The basal ganglia are responsible for some of the motor functions of the body, and decreased metabolic activity in these structures is consistent with the symptoms of Huntington's disease. The MRI scan of this patient was completely normal.

Dramatic improvements in cancer therapy from the early days of radium continue to be made. The next stage was the use of strong γ-ray sources such as a 1-Ci source of ^{60}Co and electron linear accelerators—the latter continue as the present major facility for radiotherapy. However, facilities for neutron, proton, pion, and heavy-ion irradiations are being employed because of their significant advantages in different situations over x-rays and electrons. The point is to kill the cancer cells. The primary issue is how to deposit the energy precisely at the location of the tumor. Variations in tissue densities in different patients also create uncertainties. Lawrence Berkeley Laboratory has pioneered a technique in which ^{19}Ne ions produced in a fragmentation reaction are deposited in the patient. The PET technique is used to locate precisely where a small dose of ^{19}Ne is stopped and to adjust it to the right position before the major dose is given. This can be critical when the tumor is near other vital organs or tissues. At Michigan State University a new, small (25 tons vs. several hundred tons)

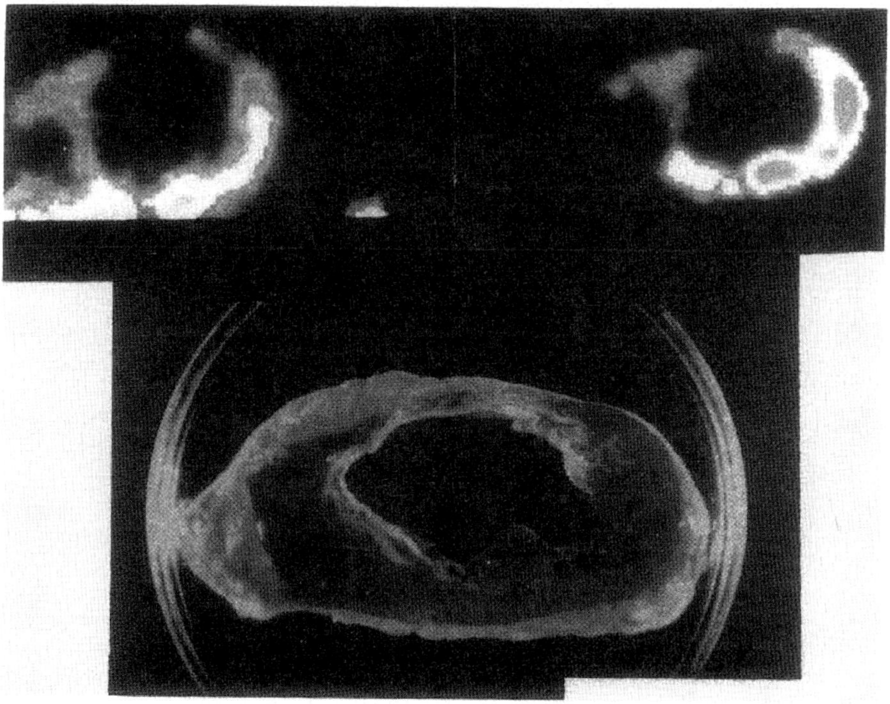

FIGURE 12.24
PET images of the heart under two different conditions as described in the text. Based on these pictures, the patient had a heart transplant. The lower picture is a photograph of a slice through the heart that had been removed at the same place on the upper two scans. (Courtesy of Dr. D. Delbeke, Vanderbilt University Medical Center.)

superconducting cyclotron has been developed. Such a cyclotron, which can be positioned so that it can rotate around a patient on a 14-foot diameter became operational at the Harper-Grace Hospital in Detroit in the early 1990s.

Exciting new developments are being made in functional MRI and functional PET studies. Functional studies allow studies of how various parts of the body actually are functioning, not simply is there a tumor or other damage present. For example, these allow us to see which parts of the brain are involved in a wide range of activities: thinking (as illustrated in Figure 12.26), seeing, hearing responses to music and voice occur on opposite sides of the brain and all the emotions from stimuli that evolve fear, hunger, sadness, happiness, etc. Not only can they tell whether a patient's brain is functioning normal but how the brain changes with age and other factors. By knowing what parts of the brain are involved in what processes, one can design better drugs to target the area of concern. These are a major breakthrough in psychology and medicine.

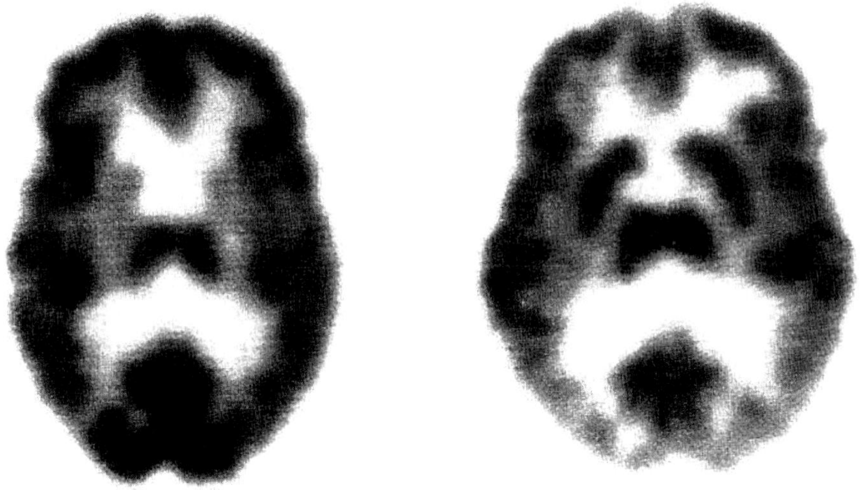

FIGURE 12.25. PET images of the brains of a patient with Huntington's disease (left and of a normal patient (right) as described in the text. (Courtesy of Dr. D. Delbeke, Vanderbilt University Medical Center)

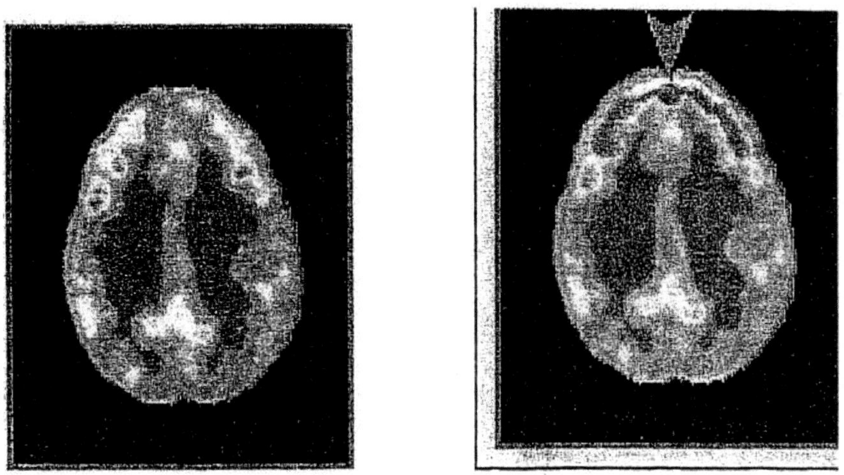

FIGURE 12.26. PET scan of a brain at rest on the left and a brain performing a thinking task on the right. Note the arrow in the scan on the right notes the area of the brain that is activated by thinking. Courtesy of Dr. R. Kessler, Vanderbilt University Medical Center.

12.6 SUMMARY

1. In the cosmos, most of the mass of stars and planets is in nuclear matter and the energy source of the stars comes from nuclear reactions. Most of the mass of the universe resides in the dark matter whose nature is not known. The strong and weak nuclear interactions and the electromagnetic and gravitational interactions together govern the history of the cosmos, including the formation and evolution of the stars and planets, and the elements of which they are composed. Thus, nuclear physics plays a central role in our understanding of the universe.

2. A nuclear reaction occurs when two particles or nuclei interact via the strong nuclear force. A nuclear reaction involves both kinematics and dynamics: the Q value equation expresses the kinematics and the reaction cross-section expresses the dynamics. Two important classes of nuclear reactions that were considered are fusion and direct reactions. Nuclei can also interact via the electromagnetic Coulomb force with other nuclei in Coulomb excitation or with electrons.

3. Two reactions which have important applications are neutron-induced fission and the fusion of light nuclei. These are the two important ways to extract energy from the nucleus of the atom. Nuclear reactors provide major applications of nuclear fission for energy production, research, production of radioisotopes for medicine and industry, and atomic weapons. Nuclear fusion in stars is the major source of energy in the universe. We are nearing the development of controlled nuclear fusion reaction.

4. New directions in nuclear reaction research include studies of a rich variety of phenomena. Heavy-ion reactions go from compound fusion to deep inelastic (multinucleon transfer) to multifragmentation and Coulomb fission of the projectile to make exotic nuclei out to the proton and neutron drip lines. New radioactive ion-beam facilities will extend the range of nuclear reactions which can be studied, including ones of astrophysical interest and of the production of exotic nuclei still farther from stability. At high energies, nuclear shock waves are observed as one begins to compress nuclear matter at high temperatures and probe the nuclear equation of state. New ultrarelativistic heavy-ion of colliders can reach densities several times normal nuclear densities and new temperatures at which the nucleons make a phase transition to a new state of matter, the quark-gluon plasma or liquid. Intense beams of high energy electrons will probe the quark-gluon structure of the nucleons and test our current standard model of quantum chromodynamics (QCD) in new ways.

5. As discussed in this chapter and elsewhere in the text, the techniques and principles of nuclear physics have a wide range of significant applications to a broad spectrum of major societal problems as well as in other areas of basic research.

PROBLEMS

12.1. Show that bombarding ^{9}Be with α-particles of less than 10 MeV cannot produce ^{12}C and photons with energy on the order of 60 MeV.

12.2. Calculate the Q values for the following reactions, where the first term is the projectile.
(*a*) $\alpha + {}^{14}_{7}\mathrm{N} \to {}^{17}_{8}\mathrm{O} + \mathrm{p}$ (*b*) $\mathrm{p} + {}^{9}\mathrm{Be} \to {}^{6}\mathrm{Li} + \alpha$
(c) ${}^{16}_{8}\mathrm{O} + {}^{58}_{28}\mathrm{Ni} \to {}^{72}_{34}\mathrm{Se} + \mathrm{p} + \mathrm{n}$ (*d*) ${}^{44}_{20}\mathrm{Ca} + {}^{28}_{14}\mathrm{Si} \to {}^{70}_{32}\mathrm{Ge} + 2\mathrm{p}$
The atomic masses are given in Appendix II.

12.3. Calculate the threshold energy for each reaction in Problem 10.22.

12.4. A proton beam is used to bombard a fixed tritium target to produce the reaction $\mathrm{p} + {}^3\mathrm{H} \to \mathrm{n} + {}^3\mathrm{He}$. How much energy must the proton have for the reaction to occur? If the incoming proton energy is 3.00 MeV, and the angle between the outgoing neutron and incoming proton is 90°, what are the energies of the outgoing neutron and ^{3}He, respectively?

12.5. Calculate the Coulomb barrier for ${}^{16}_{8}\mathrm{O} + {}^{56}_{26}\mathrm{Fe}$. What kinetic energy is required to overcome this barrier (a) if ^{16}O is the projectile, (b) if ^{56}Fe is the projectile? If the decay of a compound nucleus is independent of its mode of formation, what are the advantages and disadvantages of using ^{16}O or ^{56}Fe as the projectile?

12.6. Calculate the energy released by the thermal neutron-induced fission of ${}^{235}_{92}\mathrm{U}$ into ${}^{140}_{54}\mathrm{Xe}$ and ${}^{94}_{38}\mathrm{Sr}$. Compare with the energy released for the neutron-induced fission of ${}^{239}_{94}\mathrm{Pu}$ into ${}^{142}_{56}\mathrm{Ba}$ and ${}^{96}_{38}\mathrm{Sr}$ or ${}^{142}_{56}\mathrm{Ba}$ and ${}^{95}_{38}\mathrm{Sr}$.

12.7. (*a*) A neutron with energy E_0 collides with a static carbon nucleus. Show that after N collisions, the neutron energy is approximately equal to $(0.72)^N E_0$.
(*b*) A thermal neutron can cause ^{235}U to fission easily, but the energies of the neutrons produced in the fission are rather high (MeV). If graphite is used as the moderator in the reactor, how many collisions are needed for a fast neutron with energy 2.0 MeV to be slowed down to a thermal neutron with an energy 0.025 eV?

12.8. The resonance reaction of the light nucleus ^{19}F under proton bombardment often is used to energy calibrate accelerators. For example:

Proton energy (keV)	Reaction	Width (keV)
224.4	^{19}F(p, γ)	1.0
340.4	^{19}F(p, $\alpha\gamma$)	4.5
873.5	^{19}F(p, $\alpha\gamma$)	5.2
935.3	^{19}F(p, $\alpha\gamma$)	8.0
1085.0	^{19}F(p, $\alpha\gamma$)	4.0

(*a*) Determine the energies of a few excited energy levels of ^{20}Ne.
(*b*) Calculate the mean lifetime of the compound nucleus, ^{20}Ne.

12.9. If the power produced is 10^9 watt using the $^2\mathrm{H} + {}^3\mathrm{H}$ reaction, calculate how much tritium is used in one year. If using coal, how much coal would be used in one year to produce this power? (Coal provides 3.3×10^7 J/kg.)

12.10. How many neutrons are absorbed after thermal neutrons pass through 1.0 mm of iron? The iron thermal neutron capture cross-section is 2.5 b.

12.11. In the reaction ^{7}Li(p,d)^{6}Li, the binding energies are known to be B (^{7}Li) = 39.246 MeV; B(^{6}Li) = 31.995 MeV; B(d) = 2.225 MeV. What is the Q value of the reaction? What is the threshold energy E_{th}? Explain why E_{th} is bigger than Q (absolute value).

12.12. What is the energy needed from outside to make it possible for a ^{12}C nucleus to split into three ^{4}He nuclei?

12.13. If the total mass of the universe is of the order of 10^{54} g and it is compressed to the density of nuclear matter (2×10^{14} g/cm^3), what is the radius of the universe at that time? Compare this value with the radius of the Sun (7×10^8 m).

12.14. It is known that in a natural uranium mine at the present time the ratio of the number of atoms, $N(^{206}\text{Pb})/N(^{238}\text{U}) = 1/2.785$, and the half-life for ^{238}U decaying to ^{234}Th is $T = 4.5 \times 10^9$ years. After decaying to ^{234}Th, it continuously decays to various nuclei whose lifetimes are all much less than 10^9 years, with the final decay product being ^{206}Pb. Determine the age of the uranium mine assuming no ^{206}Pb in the initial uranium ore (approximately the age of the Earth).

12.15. If the relative abundance of ^{235}U and ^{238}U was 1:2 when the Earth was formed, what is the age of the Earth based on their abundance today and their lifetimes?

12.16. What are β-delayed neutrons? Why are they critical to producing usable energy from fission?

12.17. Calculate the Coulomb barrier for ^{60}Ni on ^{120}Sn. Could these two nuclei fuse at energies below their barrier energy? What quantum-mechanical phenomena can you use to justify your answer?

12.18. What distinguishes a direct reaction with α-particle projectiles from a fusion reaction with α-particles on the same target?

12.19. Calculate the energy for a deuteron to undergo a stripping reaction in which a neutron is transferred to a ^{60}Ni nucleus. Compare this energy to that in the same reaction on ^{120}Sn.

12.20. Calculate the energy for a deuteron to undergo a pick-up reaction in which a neutron is stripped off a ^{60}Ni nucleus.

12.21. Why is the Coulomb excitation of nuclear energy levels such a useful tool to probe the structure of certain types of nuclei? What information is gained from such studies?

12.22. The production rate for ^{56}Mn is 5.00×10^8/s if it is produced by bombarding deuterium on ^{55}Mn. Determine the activity of ^{56}Mn after 10 hours of bombardment. The half-life of ^{56}Mn is 2.579 h.

12.23. What is the maximum angular momentum which can be transferred to the compound nucleus in the reaction ^{34}S + ^{154}Gd at a ^{34}S energy of 200 MeV?

CHAPTER 13

HYPERFINE INTERACTIONS

Imagination is more important than knowledge.

A. Einstein

13.1 INTRODUCTION

In Chapter 3, we considered only the electrostatic interaction between the electrons and the nucleus in looking at the gross structure of atomic spectra. In Chapter 4, the concept of electron spin was introduced in order to explain the fine structure of spectra. But we still considered the nucleus as a massive point charge whose main contribution is through its electric charge (Ze) and did not consider any effect from the fact it has a spread-out, rotating charge distribution.

However, we learned in Chapter 9 that a nucleus has a finite size and in some cases has a nonspherical shape rather than being a point mass. It has a charge distribution (which can include an electric quadrupole moment), an angular momentum I, and a magnetic moment μ as well. All of these properties will have effects on the motion of the electrons in an atom and cause further splitting of the atomic spectral lines. This kind of splitting is much smaller than the fine structure splitting, so it is designated by the name of hyperfine structure. Hyperfine interactions occur between the electrons and nuclear moments.

The orders of magnitude of the gross structure (electrostatic Coulomb interaction), fine structure, and hyperfine structure are given in Table 13.1 for comparison.[1] The splitting of the energy levels are expressed in three different systems of units: energy in eV, wave number in cm^{-1}, and frequency in s^{-1}.

TABLE 13.1
Orders of magnitudes of the energies in three units for the central Coulomb, fine structure, and hyperfine structure interactions in atoms

Interaction	$1/\lambda$, cm^{-1}	eV	ν, s^{-1}
Central Coulomb	30 000	4	10^{15}
Fine structure	1–1000	10^{-4}–10^{-1}	3×10^{10}–3×10^{13}
Hyperfine structure	10^{-3}–1	10^{-7}–10^{-4}	3×10^{7}–3×10^{10}

As can be seen in Table 13.1, the hyperfine splitting is three orders of magnitude smaller than the fine structure. With the development of very high-resolution optical spectroscopic techniques (the highest resolution now is better than 10^{-9}) and Mössbauer spectroscopy (see Section 10.4.5), studies of hyperfine interactions are now a major area of research. The field of hyperfine interactions gradually has become an independent branch of physics, with its own special journal, *Hyperfine Interactions*, and international conferences held every three years.

The hyperfine structure of spectral lines was first observed experimentally by A. Michelson (1891), and C. Fabry and A. Perot (1897). The theoretical explanation was given first in 1924 by Pauli, even before Uhlenbeck and Goudsmit had postulated electron spin.[2] Pauli suggested that a nucleus has a total angular momentum I and magnetic moment μ which are responsible for the hyperfine structure of spectral lines.

By symmetry arguments of parity and time-reversal, the only multipole (2^k-pole) hyperfine interactions which do not vanish are the magnetic moments for odd k and the electric moments for even k. Then $k = 0$ corresponds to the electric monopole interaction, i.e., the nucleus is considered as a charged symmetric sphere of finite size; $k = 1$ corresponds to the magnetic dipole; $k = 2$ to the electric quadrupole; $k = 3$ to the magnetic octupole, and $k = 4$ to the electric hexadecapole, and so forth.

In this chapter we shall confine ourselves to the two lowest orders of interaction: that of a nuclear magnetic dipole moment interaction with an electronic magnetic field (Section 13.2), and that of a nuclear electric quadrupole moment with an electronic electric field gradient (Section 13.3). These two interactions turn out to be of the same order of magnitude for non-s electrons, as we shall see, and they are about 10^8 times larger than the octupole and hexadecapole interactions.

Because of their different masses and nuclear charge distributions, different isotopes of an element also have very small shifts of spectral lines, which are called isotope shifts. Their order of magnitude is in the same range as the hyperfine (hf) splitting, and hence will be discussed briefly in this chapter too.

[1]G. K. Woodgate, *Elementary Atomic Structure*, McGraw-Hill, New York (1970).
[2]S. Goudsmit, "Pauli and Nuclear Spin," *Physics Today* **14** (June 1961) 18.

13.2 MAGNETIC DIPOLE HYPERFINE INTERACTION

13.2.1 General Expression

Suppose the nuclear magnetic moment is $\boldsymbol{\mu}_I$ and the magnetic field at the nucleus, from the motion of the electrons, is $\boldsymbol{B}_{el}$. The Hamiltonian of the magnetic dipole interaction is

$$\mathcal{H}_m = -\boldsymbol{\mu}_I \cdot \boldsymbol{B}_{el} \tag{13.1}$$

We also have

$$\boldsymbol{\mu}_I = g_I \mu_N \boldsymbol{I} \tag{13.2}$$

where g_I is the effective g-factor of the nucleus; μ_N is the nuclear magneton, which is smaller than the Bohr magneton by a factor of about 1/1840 (according to Eq. (9.18)), so the interaction (13.1) is 10^{-3} times smaller than the fine structure interaction between the electron spin and orbital motion as shown in Table 13.1.

The magnetic field $\boldsymbol{B}_{el}$ produced by the electrons is proportional to their total angular momentum $\boldsymbol{J}$, so Eq. (13.1) can be written as

$$\mathcal{H}_m = A\boldsymbol{I} \cdot \boldsymbol{J} \tag{13.3}$$

This is the most general expression for the magnetic hyperfine interaction. In Eq. (13.3), A is called the magnetic hyperfine interaction constant that manifests the degree of hyperfine splitting of the energy levels. In general, A is determined experimentally or evaluated by theory. Although A has a similar name to the fine structure constant ($\alpha = 1/137$), the two have quite different meanings: α is an important universal constant while A is only a parameter or a constant for a given state of an atom.

13.2.2 Magnetic Dipole Hyperfine Interaction of a Single-Electron Atom

First consider a single-electron (hydrogen or hydrogenlike) atom for which the electronic orbital angular momentum $l \neq 0$. Then, semiclassically the magnetic field at the nucleus consists of two parts: that arising from the orbital motion of the electron of charge $-e$, velocity $\mathbf{v}$, and coordinate $\boldsymbol{r}$ about the nucleus as origin, and that arising from the spin magnetism $\boldsymbol{\mu}_s$ of the electron at a distance r from the nucleus:[3]

$$\boldsymbol{B}_{el} = \frac{(-e\mathbf{v}) \times (-\boldsymbol{r})}{cr^3} - \frac{1}{r^3}\left[\boldsymbol{\mu}_s - \frac{3(\boldsymbol{\mu}_s \cdot \boldsymbol{r})\boldsymbol{r}}{r^2}\right], \qquad r \neq 0 \tag{13.4}$$

where $-\boldsymbol{r}$ is the coordinate of the nucleus with respect to the electron. With $\boldsymbol{\mu}_s = -g_s\mu_B\boldsymbol{s} = 2\mu_B\boldsymbol{s}$ (for $g_s = 2$ exactly, where μ_B is the Bohr magneton) and $-e\boldsymbol{r} \times \mathbf{v}/c = -2\mu_B\boldsymbol{l}$, Eq. (13.4) can be rewritten as

[3] J. D. Jackson, *Classical Electrodynamics*, Wiley, New York (1976), p. 187.

$$B_{el} = -2\frac{\mu_B}{r^3}\left[l - s + \frac{3(s \cdot r)r}{r^2}\right], \qquad l \neq 0 \tag{13.5}$$

and the Hamiltonian of Eq. (13.1) is

$$\mathscr{H}_m = \left(2\mu_B\frac{\mu_I}{I}\right)\frac{I \cdot N}{r^3} \tag{13.6}$$

where

$$N = l - s + \frac{3(s \cdot r)r}{r^2} \tag{13.7}$$

Now we have the expression for the energy (Hamiltonian) operator. If the wavefunction of the electron is known, the hyperfine splitting of the energy levels can then be calculated by quantum mechanics (Section 8.2).

Instead of a detailed calculation, we will write down the results for the magnetic hyperfine splitting of the energy levels as follows:

$$\Delta E = \frac{a_j}{2}\{F(F+1) - J(J+1) - I(I+1)\} \tag{13.8}$$

which is the mean (expectation) value $\langle\mathscr{H}_m\rangle$ of the operator $\mathscr{H}_m$ (see Section 8.2). Here F is the quantum number of the total angular momentum $\boldsymbol{F}$ for an atomic system, and

$$F = I + J \tag{13.9}$$

is a conservative quantity (the total angular momentum of an isolated atom should be conserved in the absence of an external field). The $\boldsymbol{I}$ and $\boldsymbol{J}$ are the total angular momenta of the nucleus and the electrons, respectively. They are coupled into $\boldsymbol{F}$ through the hyperfine interaction in the sense that $\boldsymbol{I}$ and $\boldsymbol{J}$ precess about $\boldsymbol{F}$ with their magnitudes unchanged as shown in Fig. 13.1. From Eq. (13.3) it is easy to understand the origin of the terms in the brackets of Eq. (13.8):

$$\begin{aligned} F^2 &= I^2 + J^2 + 2I \cdot J \\ I \cdot J &= \tfrac{1}{2}\{F(F+1) - I(I+1) - J(J+1)\} \end{aligned} \tag{13.10}$$

where j should take the place of J in the single-electron case.

The expression for a_j in Eq. (13.8) is

$$a_j = \left(2\mu_B\frac{\mu_I}{I}\right)\left\langle\frac{1}{r^3}\right\rangle\frac{l(l+1)}{j(j+1)}, \qquad l \neq 0 \tag{13.11}$$

which is proportional to the mean value $\langle 1/r^3\rangle$ or $1/r^3$, where r is the coordinate of the electron. Thus the closer to the nucleus the electron is, the stronger is the hyperfine interaction. When the wavefunction is known, one can calculate $\langle 1/r^3\rangle$.

For s electrons, $l = 0$,

$$a_s = \left(2\mu_B\frac{\mu_i}{I}\right)\frac{8\pi}{3}|\psi(0)|^2, \quad l \neq 0 \tag{13.12}$$

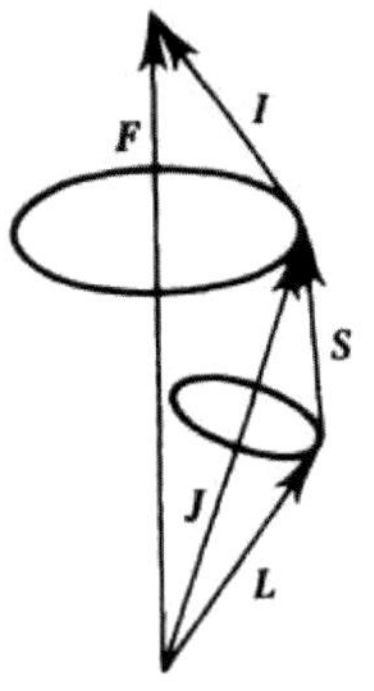

FIGURE 13.1
The vector model of the hyperfine interaction representing $\boldsymbol{J} + \boldsymbol{I} = \boldsymbol{F}$ with precession.

which is proportional to the probability density $|\psi(0)|^2$ of s electrons at the nucleus.

13.2.3 Magnetic Hyperfine Structure of a Hydrogen Atom

For hydrogenlike ions with nuclear charge Ze, we have

$$|\psi(0)|^2 = \frac{Z^3}{\pi a_1^3 n^3} \tag{13.13}$$

hence

$$a_s = \left(2\mu_B \frac{\mu_I}{I}\right) \frac{8}{3} \frac{Z^3}{a_1^3 n^3}, \quad l \neq 0 \tag{13.14}$$

For $l \neq 0$,

$$\left\langle \frac{1}{r^3} \right\rangle = \frac{Z^3}{a_1^3 n^3 (l + \frac{1}{2}) l (l+1)} \tag{13.15}$$

Hence

$$a_j = \left(2\mu_B \frac{\mu_I}{I}\right) \left(\frac{Z}{a_1 n}\right)^3 \frac{1}{(l + \frac{1}{2}) j (j+1)}, \quad l \neq 0 \tag{13.16}$$

It turns out that Eq. (13.16) reduces to Eq. (13.14) for $l = 0, j = \frac{1}{2}$. So, a_s seems to be a specific example of a_j. But this is a mere coincidence for hydrogenlike atoms. Generally, the expression for a_j cannot be used for $l = 0$.

Now we consider the hyperfine structure of the $1s\,^2S_{1/2}$ electron ground state ($l = 0, j = \frac{1}{2}$) of hydrogen. The nuclear spin (proton spin) is $I = \frac{1}{2}$. As seen from Eq. (13.9), the ground level splits into two hyperfine levels $F = I \pm J = \frac{1}{2} \pm \frac{1}{2}$, $F = 0$ and $F = 1$, separated by an energy interval where

$$\Delta[\Delta E(F = 1) - \Delta E(F = 0)] = h\,\Delta\nu = a_s \tag{13.17}$$

The $\Delta E(F=0)$ and $E(F=1)$ are the energy shifts of the $j=\frac{1}{2}$ level, respectively, given by Eq. (13.8). Substitution of the proton magnetic moment $\mu_I = 2.792\,77\mu_N$ for μ_I in Eq. (13.14) and constants μ_B, μ_N, and a_1 in Appendix III, leads to a value

$$\Delta\nu = 1.421\,120\,293\,\text{GHz} \tag{13.18}$$

Another way to calculate this is to change (13.14) to

$$a_s = \frac{2}{3}\,2g_p\alpha^4\left(\frac{m_e}{M_p}\right)(m_ec^2)\,\frac{Z^3}{n^3} \tag{13.19}$$

where $g_p = 5.585\,54$ is the g-factor of the proton, $\alpha = 1/137$, $m_e/M_p = 1/1836$, $m_ec^2 = 0.511\,\text{MeV}$, $Z=1$, $n=1$, hence $\Delta\nu = a_s/h \cong 1.42\,\text{GHz}$. Converting this into wavelength ($\lambda = c/\Delta\nu$) gives 21 cm (the well-known hyperfine wavelength of the hydrogen atom; remember this is not a difference of two wavelengths). The experimental value obtained from an extremely precise measurement is

$$\Delta\nu_{\text{exp}} = 1.420\,405\,751\,768(2)\,\text{GHz} \tag{13.20}$$

with an experimental error of only ± 0.002 Hz! The difference between the experimental value and the calculated one is about 1 part in 1400, which is much larger than the expected uncertainty. Obviously, there are some refined corrections to be made. For example, the reduced mass instead of the electron mass should be used, and because of the motion of the nucleus and the finite size of the nucleus, the realistic $|\psi(0)|^2$ is a little less than that from Eq. (13.13). However, after making these corrections, the problem was still not solved until the anomalous magnetic moment of the electron was introduced. When the value 2 in Eq. (13.14) (see also Section 4.5) was replaced by

$$g_s = 2 \times 1.001\,159\,652\,193$$

excellent agreement with the experimental result was achieved.[4] The hyperfine splitting frequency of the ground level of hydrogen, Eq. (13.20), has been used as a time standard, a so-called atomic clock. N. F. Ramsey won the 1989 Nobel Prize in Physics for his development of the atomic clock, which revolutionized the accuracy of time measurements.

13.2.4 Magnetic Hyperfine Interaction for Many-Electron Atoms

For many-electron atoms, we can write down the Hamiltonian of the magnetic hyperfine interaction as

[4] S. B. Crampton, D. Kleppner, and N. F. Ramsey, *Phys. Rev. Lett.* **11** (1963) 338.

$$\mathscr{H}_m = \left(2\mu_B \frac{\mu_I}{I}\right) \boldsymbol{I} \cdot \left\{ \sum_i \left[\frac{\boldsymbol{l}_i}{r_i^3} - \frac{1}{r_i^3}\left(\boldsymbol{s}_i - \frac{3(\boldsymbol{s}_i \cdot \boldsymbol{r}_i)\boldsymbol{r}_i}{r_i^2} \right) \right] + \sum_j \left(\frac{8\pi}{3} \right) |\psi_j(0)|^2 \boldsymbol{s}_j \right\} \tag{13.21}$$

where the first term is the summation over all the electrons with $l \neq 0$ (with subscript i), and the second term is over the electrons with $l = 0$ (s electrons, with subscript j).

Because of the spherical symmetry of the closed shells in atoms, in Eq. (13.21) we need only to sum over the valence electrons. In some cases even for valence electrons the effects caused by their angular momenta may offset each other. For example, a pair of ns^2 valence electrons forming the $^1S_0$ state would not produce any hyperfine splitting.

In any case, the total Hamiltonian of the magnetic hyperfine interaction can always be written in a form like Eq. (13.3), that is,

$$\mathscr{H} = A(J)\boldsymbol{I} \cdot \boldsymbol{J}$$

where $\boldsymbol{J}$ is the total angular momentum of all the electrons or of the valence electrons in an atom, and $A(J)$ is proportional to μ_I/I and B_{el}. Hence we can calculate the mean value of $\mathscr{H}_m$ by quantum mechanics. The general expression for the energy splitting related to the hyperfine interaction is

$$\Delta E = \tfrac{1}{2}A(J)\{F(F+1) - J(J+1) - I(I+1)\} \tag{13.22}$$

which determines the main properties of the hyperfine structure of the spectral lines:

1. Hyperfine levels are labeled by the total angular momentum quantum number F of an atomic system including the nucleus and the electrons. Since $\boldsymbol{F} = \boldsymbol{I} + \boldsymbol{J}$, for a fixed I and J, the possible values of F are

$$(I+J),\ (I+J)-1,\ \ldots,|I-J|-1,\ |I-J|$$

Thus, F has $(2I+1)$ values for $I < J$ or $(2J+1)$ values for $J \leq I$. In other words, the number of F values is determined by the smaller of I and J. For example, what is the nuclear spin of an odd-A nucleus whose ground state is $^5H_7$ and for which 10 lines are seen in the hyperfine structure? If $J \leq I$ then there will be $2J + 1 = 15$ lines, which is greater than the 10 seen, so $I < J$ and there are $2I + 1 = 10$ lines and $I = \frac{9}{2}$.

2. For a fixed F, the quantum number M_F of its components has $(2F+1)$ values. An F level is $(2F+1)$-fold degenerate in the absence of an external field on the atomic system.

3. The relative intensities of the hyperfine spectral lines obey a sum rule; that is, the ratios of the intensities of the lines, labeled by different F, are dependent upon their statistical weights (the number of M_F substates, i.e., $2F+1$). For example, in Fig. 13.2 the ratio of the line intensities of transitions to $F = 7$

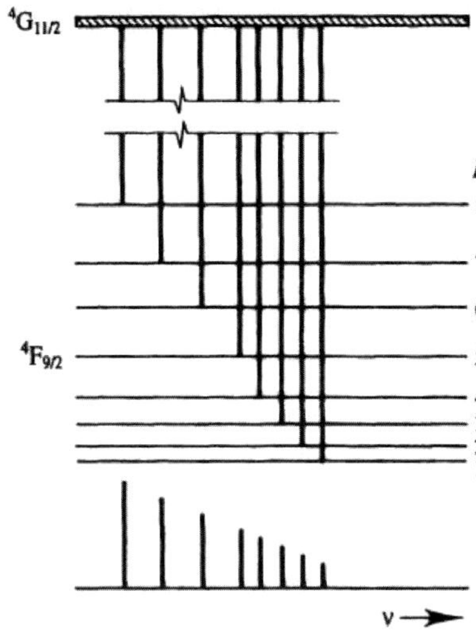

FIGURE 13.2
The hyperfine structure of the transition $5d^26s\,^4F_{9/2} \rightarrow 5d^26p\,^4G_{11/2}$ in ^{139}La.

and to $F = 1$ is $(2 \times 7 + 1)/(2 \times 1 + 1) = 5$. (Here we assume that the $^4G_{11/2}$ level has a negligible hyperfine structure, see Problem 13.6.)

4. $A(J)$ is a measure of the hyperfine splitting and there is an interval rule,

$$\Delta E(F+1) - \Delta E(F) = A(J)(F+1) \quad (13.23)$$

which is given directly from Eq. (13.22). From $F_{\min} = |I - J|$ to $F_{\max} = (I + J)$ the intervals between adjacent levels increase by a value of $A(J)$ every step.

5. The selection rule for electric dipole transitions between hyperfine levels is

$$\Delta F = 0, \pm 1, \qquad \Delta m = 0, \pm 1 \quad (13.24)$$

and the transition for $F = 0 \rightarrow F = 0$ is forbidden.

A typical example is shown in Fig. 13.2. The structure of the hyperfine splitting of the transition $5d^2\,6s\,^4F_{9/2} \rightarrow 5d^2\,6p\,^4G_{11/2}$ (the latter was unresolved) fundamentally obeys the above rule.

In Fig. 13.3 we can see the further splitting of the D-lines of sodium ($\lambda = 5890$ Å and 5896 Å) because of the hyperfine interaction. Shown in Fig. 13.4 are the experimental hyperfine spectral lines obtained by laser spectroscopy, which are in good agreement with the theoretical predictions for the transitions from the $^2P_{3/2}$ sublevel of $F = 2$ to the $^2S_{1/2}$ sublevel of $F = 2$.

13.2.5 Determination of the Nuclear Spin from Magnetic Hyperfine Structure

Since the hyperfine structure of spectral lines is closely related to the nuclear spin of a luminescent atom, we can determine the nuclear spin from the experimental

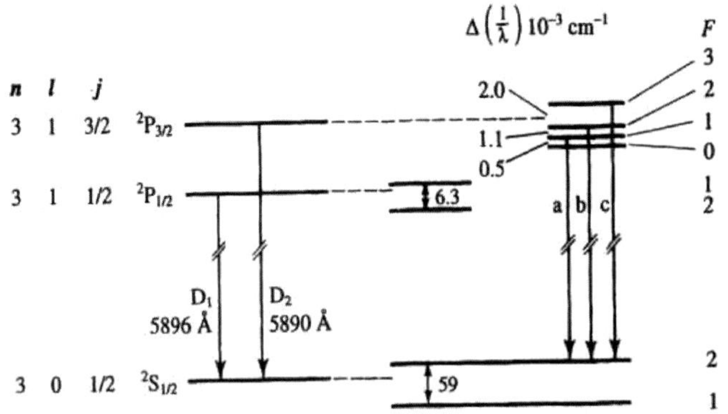

FIGURE 13.3
The fine and hyperfine structure of the atomic levels of ^{23}Na (nuclear spin $I = \frac{3}{2}$). The spins of the electron states are on the left and of the total atoms on the right.

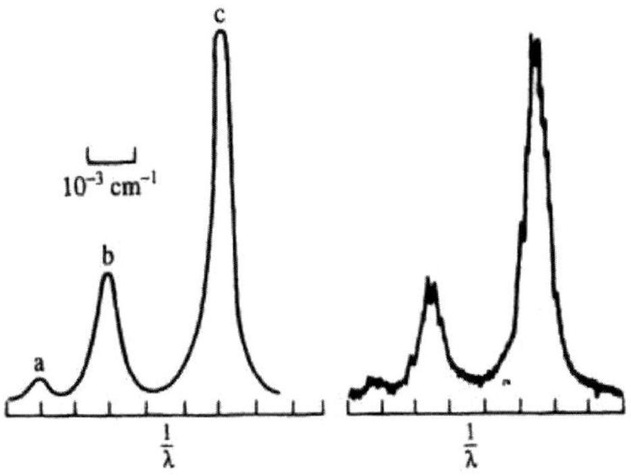

FIGURE 13.4
A part of the spectrum of the hyperfine structure of the D_2 lines in ^{23}Na (the left-hand curve is calculated and the experimental result is shown on the right).

data on hyperfine splitting. Three methods can be used to determine I. In order to make the interpretation simpler, we assume that there is a transition in which the initial level has negligible hyperfine structure for its large electron angular momentum and the final level with a known J has a larger hyperfine structure.

1. From the number of spectral lines of the hyperfine splitting: If the number of different F is less than $(2J + 1)$, then $I < J$, and $(2I + 1) = F$. So we can know the nuclear spin $I = (F - 1)/2$. If $I \geq J$, the number of splitting lines is equal to $(2J + 1)$; we cannot then determine the value of I by this method.
2. From the relative intensities of the hyperfine lines: Since the relative line intensities in the hyperfine structure are proportional to the statistical weight $(2F + 1)$ values, which go from $[2(I + J) + 1]$ to $(2|I - J| + 1)$, that is, to the specific distribution of the line intensities, one can determine the value of I by an accurate measurement of the intensities of the hyperfine spectral lines.

3. From the interval rule: From Eq. (13.22), the relative intervals are proportional to $F = I + J$, $I + J - 1, \ldots, |J - I|$. So if J is known, the intervals measured by experiment lead to a determination of the I value.

Now let us illustrate the ways of determining a nuclear spin using as an example the hyperfine structure and the spectral lines of the transition $5d^2 6s\,^4F_{9/2} \rightarrow 5d^2 6p\,^6G_{11/2}$ at 6250 Å in ^{139}La (as shown in Fig. 13.2).

The magnetic dipole interaction is the largest for configurations with an unpaired s electron which gives rise to a contact interaction. Thus, neglecting the contribution of the $5d^2$ electrons, which are paired and are also much further out from the nucleus, we assume that the $5d^2 6s$ configuration has a reasonably large hyperfine structure and the $5d^2 6p$ has only a very small hyperfine structure, since the 6p electron is further from the nucleus. Therefore, the spectral lines reflect the structure of the splitting level. In the experiment at least seven but fewer than nine components could be distinguished, so the number is seven or eight. The nuclear spin of an odd–even nucleus has to be a half integer. Since the number of the components is less than $2J + 1 = 10$, so I is less than J $(= 9/2)$. Only $2I + 1 = 8$ is allowed, to yield $I = \frac{7}{2}$ since $2I + 1 = 7$ would yield $I = 3$, which is excluded.

Also, the relative intervals of the hyperfine lines were measured from $F = 8 \rightarrow F = 2$. The experimental intervals were 111, 96.3, 82.6, 67.9, 57.4 and 42.7 in units of $10^{-3}\,\mathrm{cm}^{-1}$, which must follow $(I + \frac{9}{2})$, $(I + \frac{7}{2})$, $(I + \frac{5}{2})$, $(I + \frac{3}{2})$, $(I + \frac{1}{2})$. These data are consistent only with a value $I = \frac{7}{2}$. Let us illustrate this. For $I = \frac{7}{2}$, the intervals should follow 8, 7, 6, 5, 4, 3. Normalizing 8 to 111, the others would be 97.1, 83.2, 69.4, 55.5 and 41.6 in agreement with the data. If $I = \frac{5}{2}$, the intervals follow 7, 6, 5, 4, 3, 2, and normalizing 7 to 111 yields 95.1, 79.3, 63.4, 47.6, 31.7, which are clearly in disagreement with the data.

Note that one can also determine the nuclear magnetic moment μ_I from the magnetic hyperfine structure if one knows B_{el}. However, we shall not discuss this problem here.

13.2.6 Nuclear Level Splitting by the Magnetic Hyperfine Interaction

The magnetic hyperfine interaction between the nuclear magnetic moment μ_I and the magnetic field related to J produces not only an atomic level splitting, as we have discussed, but also a nuclear level splitting.

Since the nuclear wavefunction is confined to a region much, much smaller than an atom, the magnetic field intensity from the motion of the atomic electrons is nearly constant at the nucleus. Thus, the hyperfine splitting of the nuclear levels can be considered as a Zeeman effect as if the nucleus were in a fixed magnetic field B_{el}.

The Hamiltonian of the magnetic hyperfine interaction is

$$\mathscr{H} = \boldsymbol{\mu}_I \cdot \boldsymbol{B}_{el} \tag{13.25}$$

and

$$\boldsymbol{\mu}_I = g_I \mu_N \boldsymbol{I} \tag{13.26}$$

Then

$$\mathcal{H} = -g_I \mu_N \boldsymbol{I} \cdot \boldsymbol{B}_{el} \tag{13.27}$$

We select the direction of $\boldsymbol{B}_{el}$ as the z axis of the nucleus. The projection of the nuclear angular momentum $\boldsymbol{I}$ on the z axis is $I_z = \boldsymbol{I} \cdot \boldsymbol{B}_{el} = M_I B_{el}$ and the energy level shifts from the nuclear Zeeman effect are

$$\Delta E = -g_I \mu_N B_{el} M_I \tag{13.28}$$

where B_{el} for a given atomic state is a constant, g_I is the nuclear g-factor, and μ_N is the nuclear magneton.

Now look at an example to illustrate this effect. Figure 13.5(a) shows the transition between the ground state and the first excited state of the ^{57}Fe nucleus. The angular momenta of the ground state and the first excited state are $I = \frac{1}{2}$ and $I = \frac{3}{2}$, respectively, with magnetic moments of $\mu_{1/2} = 0.090\,24\mu_N$ and $\mu_{3/2} = -0.1547\mu_N$. The hyperfine splitting of the nuclear levels is shown in Fig. 13.5(b). Since the direction of the magnetic moment of the first excited level is opposite to its angular momentum for the ^{57}Fe nucleus, the components for $M_I > 0$ have a positive energy shift (move up to higher energies), while in the ground state, where I and μ_I are in the same direction, the shift for $M_I > 0$ is negative (they move down to lower energies). For an iron atom, the internal magnetic field produced by the electrons at the nucleus is $B_{el} = 33\,\text{T}$. In general, this is much stronger than the fields one can obtain in a laboratory. Even so, the ΔE from Eq. (13.28) is still only of the order of 10^{-5} eV. When compared to the excitation energy (14.4 keV for ^{57}Fe), $\Delta E/E \sim 10^{-9}$ is far smaller than the resolution of the best nuclear radiation detectors commonly used. However, it can be easily resolved by Mössbauer spectroscopy (Section 10.4). The experimental method is briefly described as follows.

When the γ-rays emitted by radioactive ^{57}Co, whose decay scheme is shown in Fig. 10.25, pass through an iron crystal which acts as an absorber without a recoil, resonant γ-ray absorption occurs, and the intensity of the γ-rays transmitted through the absorber is a minimum. Since the nuclei are strongly bound in the crystal lattice, the whole crystal must recoil and, because of the enormous

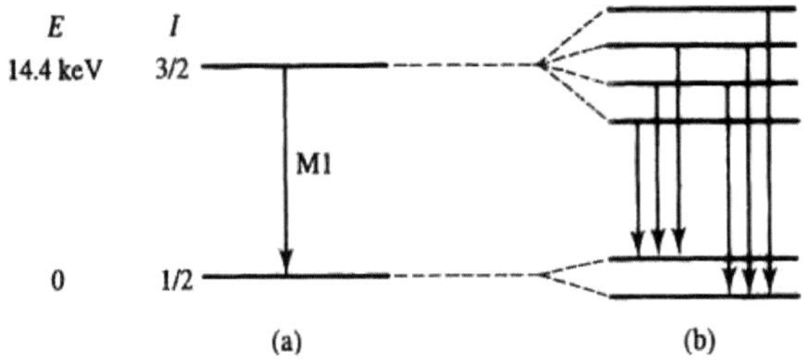

FIGURE 13.5
The hyperfine splitting of the ^{57}Fe nuclear levels. The 14.4 keV transition is magnetic dipole (M1) radiation.

mass of the whole crystal, it recoils with essentially zero velocity. How do we know we "have reached the minimum"? This can be checked by changing the γ-ray energy slightly so the resonance absorption does not occur. The way to change the energy slightly is to let the source and absorber be in relative motion with a velocity v. Then the effective value of the γ-ray energy "seen" by the absorber differs from the real energy by a small Doppler shift energy:

$$\Delta E_D = \frac{v}{c} E_0 \tag{13.29}$$

When E_0 is equal to 14.4 keV, v = 1 mm/s corresponds to $\Delta E = 4.8 \times 10^{-8}$ eV. In this way the Mössbauer absorption spectrum of ^{57}Fe was measured and is shown in Fig. 13.6, where two abscissas are used. The lower is the energy ΔE_D and the upper represents the velocity instead of the energy from Eq. (13.29). The latter is often used as the typical scale in Mössbauer spectroscopy.

In order to keep the emission and absorption spectra identical (the necessary condition for resonant absorption), the nuclei in the source and the absorber should have the same hyperfine structure, that is, they be under the identical B_{el} or in the "same electronic or atomic environment." However, in general this is not the case. For instance, the ^{57}Fe nucleus in the excited level following the decay of ^{57}Co is in a ^{57}Co atomic environment. On the other hand, the ^{57}Fe nucleus of the absorber is in an iron atomic environment. Thus their internal magnetic fields are different. To let them have the same environment experimentally, the radioactive source of ^{57}Co can be electroplated onto the surface of a metal iron foil and then thermally diffused at high temperature into the metal substrate. The Mössbauer effect may then occur.

We can use stainless steel as a substrate of the ^{57}Co source to make the Mössbauer spectra simple and clear. Because stainless steel has no effective internal magnetic field at room temperature, the ^{57}Fe emission spectrum has no hyperfine structure but rather exhibits a single 14.4 keV line. However, the absorption

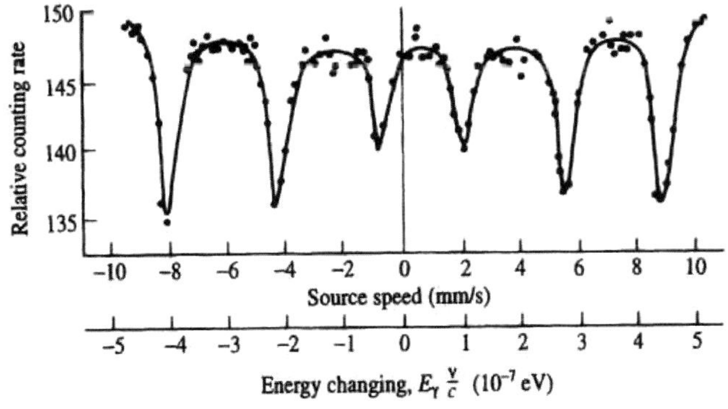

FIGURE 13.6
Mössbauer spectrum showing the hyperfine splitting of ^{57}Fe.

spectrum has a hyperfine structure. Therefore, the Mössbauer spectrum manifests a hyperfine structure pattern for the ^{57}Fe 14.4 keV transition. The spectrum obtained by this method is shown in Fig. 13.6, where the six γ absorption lines with different energies correspond to the six transition lines in Fig. 13.5 (note the selection rule $\Delta M = 0, \pm 1$). The first observations of the magnetic hyperfine structure of a nuclear level by the Mössbauer effect were made in 1959.[5]

Remember that Eq. (13.28) also holds for an external magnetic field B acting on the nucleus. For example in a $I \neq 0$ state $\Delta E = M_I g_I \mu_N B$ and since $\Delta M_I = \pm 1$, the energy differences between two M states is $\Delta(\Delta E) = g_I \mu_N B = h\nu$, where ν is the resonance frequency and the nuclear magnetic moment is

$$\mu_I = \sqrt{I(I+1)} g_I \mu_B$$

The precessional frequency of a magnetic moment about the z axis defined by the field is $\omega = (M_I B)/\hbar$. When only the spin motion of a particle is considered, then, for example, $\omega = g_{e,s}(\mu_B/\hbar)B$ for electrons and $\omega = g_{p,s}(\mu_N/\hbar)B$ for protons; also $g_{e,s} = 2$ and $g_{p,s} = 5.58$. For example, for electrons in a field of 1 T,

$$\omega_e = 2 \times \frac{57.88 \times 10^{-6}\,\text{eVT}^{-1}}{6.58 \times 10^{-16}\,\text{eV} \cdot \text{s}} \cdot 1\,\text{T} = 17.6 \times 10^{10}\,\text{s}^{-1}$$

13.3 ELECTRIC QUADRUPOLE HYPERFINE INTERACTION

13.3.1 General Expression

When the nuclear charge distribution departs from spherical symmetry, the nucleus has an electric quadrupole moment (Section 9.3). The interaction of a nuclear electric quadrupole moment with the gradient of the electric field produced by the electrons will cause an additional energy of hyperfine interaction.

Let the direction of the electron angular momentum J be along the z axis. The gradient of the electric field from the electrons whose wavefunction is cylindrically symmetric is

$$\phi_z = -\frac{\partial E_z}{\partial Z} = \frac{\partial^2 V_e}{\partial Z^2} \tag{13.30}$$

where V_e is the electric potential produced by the electrons. A quantum mechanical calculation indicates that the energy shift of the electric quadrupole hyperfine interaction is

$$\Delta E_Q = \frac{B}{4} \frac{\frac{3}{2}K(K+1) - 2I(I+1)J(J+1)}{I(2I-1)J(2J-1)} \tag{13.31}$$

where $K = F(F+1) - J(J+1) - I(I+1)$ and

[5]P. V. Pound and G. A. Rebka, *Phys. Rev. Lett.* **3** (1959) 554.

$$B = eQ\left\langle \frac{\partial^2 V_e}{\partial Z^2} \right\rangle \tag{13.32}$$

Here $\langle \partial^2 V_e/\partial Z^2 \rangle$ is the mean value of the gradient of the electric field and is dependent on the electron wavefunction, and Q the quadrupole moment. B is called the electric quadrupole hyperfine interaction constant.

We should note that the quadrupole interaction vanishes in the following cases:

1. For an atom in an S state, because the total electronic orbital angular momentum $L = 0$ and the electron charge distribution is spherically symmetrical, $\langle \partial^2 V_e/\partial Z^2 \rangle$ at the nucleus is equal to zero.
2. When the nuclear spin $I = 0$ or $I = \frac{1}{2}$, the nuclear quadrupole moment $Q = 0$.
3. When the total angular momentum of the electrons $J = 0$ or $J = \frac{1}{2}$, then the electron wavefunction is spherically symmetrical, and the gradient of the electric field is zero at the nucleus.

There is only a magnetic hyperfine interaction for the above three cases, so the hyperfine splitting is relatively simple. In general cases, however, the magnetic dipole and electric quadrupole interactions between the electrons and the nucleus in an atomic system exist simultaneously. The total experimental effect is a summation of the two parts. Combining Eqs. (13.22) and (13.31), we have the total energy shift

$$\Delta E = A\,\frac{K}{2} + B\,\frac{\frac{3}{2}K(K+1) - 2I(I+1)J(J+1)}{4I(2I-1)J(2J-1)} \tag{13.33}$$

This is the most general expression for the hyperfine energy shift of the atomic levels.

13.3.2 An Example of a Quadrupole Hyperfine Shift of Atomic Energy Levels

Because of the second term (if $B \neq 0$) in Eq. (13.33), the structure of the hyperfine splitting level no longer obeys the interval rule (13.23). The quadrupole interaction gives rise to a small departure from the interval rule because its dependence on F is different from that of the magnetic dipole interaction. When B/A is sufficiently large, even the order of the F-levels may be different from that when $B/A = 0$.

Let us consider the line 4031 Å in ^{55}Mn as an example of a small departure from the interval rule. Precise optical measurements on this line have been made and, from the results, Q of the ^{55}Mn nucleus has been evaluated. The transition is $3d^5\,4s^2\,{}^6S_{5/2} \rightarrow 3d^5\,4s4p\,{}^6P_{7/2}$ (see Fig. 13.7). For ^{55}Mn, $I = \frac{5}{2}$. The S-term has no quadrupole interaction because it is spherically symmetrical, and indeed the magnetic dipole interaction vanishes also because of the symmetry of the half-filled d^5 shell. The $^6P_{7/2}$ level, on the other hand, has both magnetic dipole

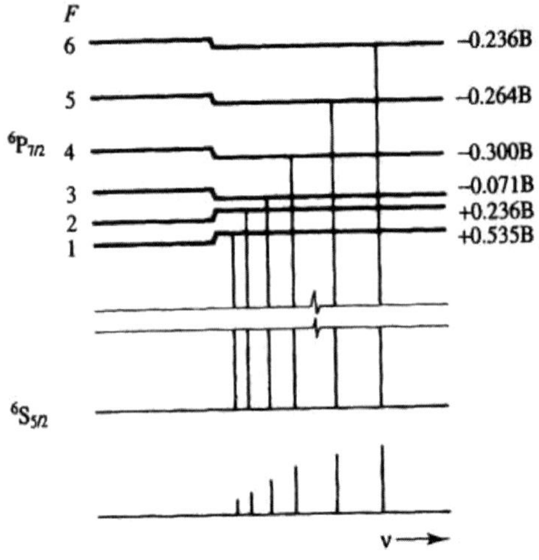

FIGURE 13.7
The hyperfine structure of the $^6P_{7/2}$ level in ^{55}Mn.

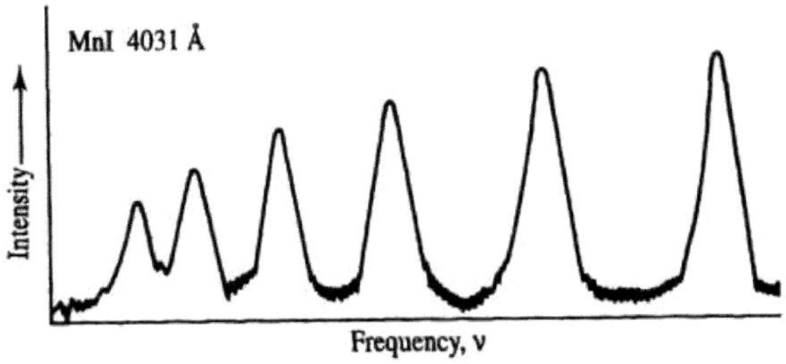

FIGURE 13.8
The hyperfine structure of the 4031 Å line from the transition $^6P_{7/2} \rightarrow {}^6S_{5/2}$ in ^{55}Mn.

and electric quadrupole hyperfine structure. The experimental spectrum is shown in Fig. 13.8. The intervals between the F-levels, extracted from the data in Fig. 13.8, are given in the second column of Table 13.2, and with $A(^6P_{7/2}) = (14.29 \pm 0.05) \times 10^{-3}\,\text{cm}^{-1}$ the expected AF values obeying the interval rule are given in the third column. The departures from the interval

TABLE 13.2
Hyperfine splitting of ^{55}Mn

$F \rightarrow F-1$	Experimental data for intervals ($10^{-3}\,\text{cm}^{-1}$)	Calculated from Eq. (13.23) for $A(^6P_{7/2})F = 14.29$ ($10^{-3}\,\text{cm}^{-1}$)	Difference ($10^{-3}\,\text{cm}^{-1}$)
$6 \rightarrow 5$	86.65 ± 0.11	85.74	+0.91
$5 \rightarrow 4$	72.55 ± 0.22	71.45	+0.10
$4 \rightarrow 3$	56.53 ± 0.20	57.16	−0.63
$3 \rightarrow 2$	41.96 ± 0.30n	42.87	−0.91

rule are shown in column four, and these are a measure of $B(^6P_{7/2})$. From these data, the value $B(^6P_{7/2}) = (2.2 \pm 0.4) \times 10^{-3}\,\mathrm{cm}^{-1}$ is obtained. Since $B \ll A$, the electric quadrupole interaction is much, much less than the intervals. The value of Q for the ^{55}Mn nucleus is about 0.35 barns.

13.3.3 Nuclear Energy Level Shift by Quadrupole Hyperfine Interaction

Similarly to the magnetic dipole interaction, the electric quadrupole hyperfine interaction also produces a nuclear energy level shift:

$$\Delta E_Q = \frac{1}{4}\,eQ\left\langle\frac{\partial^2 V_e}{\partial Z^2}\right\rangle \frac{3M_I^2 - I(I+1)}{I(2I-1)} \tag{13.34}$$

where M_I is the projection of the nuclear angular momentum I on the $\partial^2 V_e/\partial Z^2$ direction (z axis) as shown in Fig. 13.9.

In actual cases, the quadrupole interaction is present together with the magnetic interaction. The hyperfine structure of the nuclear energy levels of ^{57}Fe with both interactions is shown in Fig. 13.10. According to Eq. (13.34) the quadrupole shifts are $+B/4$ for $M_I = \pm\frac{3}{2}$ and $-B/4$ for $M_I = \pm\frac{1}{2}$.

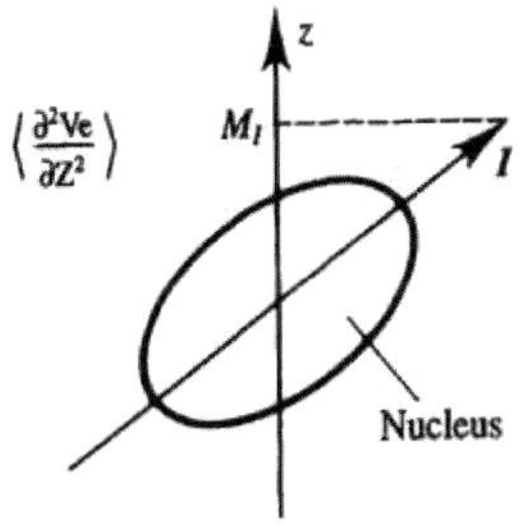

FIGURE 13.9
The projection of the nuclear angular momentum I on an electric field gradient direction.

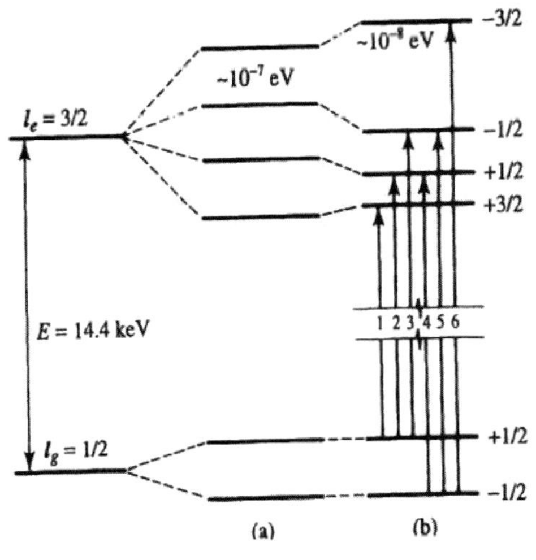

FIGURE 13.10
Six transition lines of ^{57}Fe corresponding to the 14.4 keV resonance. (a) The splitting from a pure magnetic dipole interaction. (b) A weaker quadrupole interaction is added to (a) as a small correction.

13.4 ISOTOPE SHIFT

In this section we discuss the influence of the finite mass and volume of a nucleus on its atomic energy levels.

13.4.1 Isotope Shift from the Mass Effect

In Chapter 3, we have shown that for an atom whose nucleus has a finite mass M the Rydberg constant is

$$R_A = R_\infty \frac{1}{1+\dfrac{m}{M}} \tag{13.35}$$

where m is the electronic mass and R_∞ is the universal Rydberg constant, which corresponds to the case of infinite nuclear mass. The quantized energy expression for hydrogenlike atoms (Sections 3.2, 3.3) is

$$E_n = -\frac{R_A hcZ^2}{n^2} \tag{13.36}$$

There is a frequency difference for two isotopes whose nuclear masses, M and $(M+\delta M)$, differ by δM: $\delta\nu = h\nu' - h\nu$ (where $h\nu = E_{n^*} - E_n$ with $n^* > n$). The fractional frequency shift can be easily calculated:

$$\frac{\delta\nu}{\nu} = \frac{(E'_{n^*} - E'_n) - (E_{n^*} - E_n)}{E_{n^*} - E_n} \tag{13.37}$$

$$= \frac{R_A(M+\delta M) - R_A(M)}{R_A(M)} = \left[\frac{1}{1+\dfrac{m}{M+\delta M}} - \frac{1}{1+\dfrac{m}{M}}\right] \Bigg/ \frac{1}{1+\dfrac{m}{M}} \tag{13.38}$$

since $E' \simeq R_A(M+\delta M)$ and $E \simeq R_A(M)$ and the $[1/n^{*2} - 1/n^2]$ factors and other constants cancel out. Then,

$$\frac{\delta\nu}{\nu} = \frac{m\delta M}{M(M+\delta M+m)} \simeq \frac{m\delta M}{M(M+\delta M)} \tag{13.39}$$

where the electron mass m is neglected in the last expression. Hence, for the same optical line, the atoms of an isotope with greater mass emit photons with higher energy, larger wavenumber, and shorter wavelength.

Although we discussed only the case of hydrogenlike atoms, for many-electron atoms the relation between spectral line shift and nuclear mass is nearly the same as Eq. (13.39). This phenomenon has been observed and is called the isotope mass shift effect of spectral lines.

Because of the proportionality to $1/M$, the mass effect is very small for heavy elements. For instance, for two neighboring isotopes with $A \sim 100$,

$$\delta\nu/\nu = \frac{\frac{1}{1840} \times 1}{100 \times 101}$$

which is only about 5×10^{-8}. Hydrogen isotopes produce the greatest mass shift effect. Between hydrogen and deuterium, $\Delta\nu/\nu \sim 2.7 \times 10^{-4}$ which may be observed in an ordinary optical spectrometer. In just this way, H. Urey discovered the existence of the deuterium isotope of hydrogen in 1932 (Section 3.3).

13.4.2 Isotope Shift from the Volume Effect

We know that the electrostatic potential energy between the electron and a nuclear charge with a distribution (the nucleus has a particular size and shape) is different from that where the nucleus is assumed to be a point charge. Since the charge density of an s electron does not vanish in the region of the nucleus (i.e., $|\psi(0)|^2 \neq 0$), when an s electron is inside the nucleus the potential $V(r)$ is no longer $-Ze^2/r$ as it is when we regarded the nucleus as a point charge.

For simplicity we assume the nuclear charge distribution is uniform and spherical with radius R. By classical electromagnetic theory (Gauss' law) the potential energy of an electron at a distance r from the center of the nucleus is

$$V(r)\begin{cases} V_0(r) = -\dfrac{Ze^2}{r}, & r > R \\ \dfrac{Ze^2}{R}\left(-\dfrac{3}{2}+\dfrac{r^2}{2R^2}\right), & 0 \leq r \leq R \end{cases} \tag{13.40}$$

Thus the volume effect gives rise to a significant modification of the electrostatic interaction energy between the nucleus and the electron. The Hamiltonian is

$$\mathscr{H}_s = V(r) - V_0(r) \tag{13.41}$$

By quantum mechanics the correction value of the energy difference is

$$\begin{aligned} \Delta E &= \int_0^\infty \psi^*[V(r) - V_0(r)]\psi \cdot 4\pi r^2\, dr \\ &= |\psi(0)|^2 \int_0^R [V(r) - V_0(r)]4\pi r^2\, dr \end{aligned} \tag{13.42}$$

where, in the last step, we assume the electronic wavefunction in the region of the nucleus is constant. Since the nuclear radius is more than four orders of magnitude smaller than the atomic radius, this assumption is reasonable.

Using Eq. (13.40) we have

$$\Delta E = \frac{2\pi}{5}|\psi(0)|^2 Z^2 e^2 R^2 \tag{13.43}$$

So far we have considered only one electron outside the nucleus. For many-electron atoms the expression (13.43) should be summed over all the s electrons.

We emphasize that ΔE only represents a difference between the potential energy appropriate to an extended nucleus and that appropriate to a nucleus regarded as a point charge. For different isotopes of an element, because of the difference of neutron number in their nuclei, the energy shift related to the varying R is

$$\delta(\Delta E) = \frac{4\pi}{5}|\psi(0)|^2 Ze^2 R^2 \frac{\delta R}{R} \tag{13.44}$$

This indicates:

1. For the same element, an atomic energy level of an isotope with a larger nuclear radius (or larger neutron number) is relatively higher than that of an isotope with smaller radius. This is in agreement with the experimental facts.

   As only an S-term has an isotope shift, the shift direction of the spectral line depends upon the direction of the transition, from an S-term to a non-S-term or the opposite. When an S-term is the final level, the spectral line has a red-shift and, conversely, the line has a blue-shift for an S-term as the initial level.
2. From Section 9.2 the nuclear radius $R \sim A^{1/3}$, thus

$$\delta(\Delta E) \propto \frac{\delta R}{R} \propto \frac{\delta A}{A} \tag{13.45}$$

   For the isotopes of even mass numbers, $\delta A = 2$, the interval of the energy level shifts should be equal.
3. The isotope shift $\delta(\Delta E)$ is proportional to the s electron density at the nucleus $|\psi(0)|^2 \propto Z^3$. Hence the volume shift effect grows as the atomic number increases, and so it is particularly obvious for heavy elements. Since the mass shift decreases approximately with $1/M^2$, it is negligible for heavy elements. Both theory and experiments indicate that the volume isotope shift lies within the same order of magnitude as the hyperfine splitting (Table 13.1).

As an example, the energy levels and approximate 2537 Å lines of the $6\,^3P_1 \to 6\,^1S_0$ transitions in the even mercury isotopes are shown in Fig. 13.11, where the isotope shifts of the energy levels are measured relative to the transition energy in ^{202}Hg where the shift is equal to zero. We can see that the $6\,^1S_0$ energy levels shift to higher energies as the mass of the isotope increases. In addition, the energy level differences between neighboring even isotopes $\delta(\Delta E)(\delta M = 2)$ are approximately equal. All these facts are in agreement with the expectations of theory. Note that the shift direction of the energy levels related to the volume effect is opposite to that from the mass effect. The isotope shift has been traced in the light mercury isotopes down to ^{183}Hg.

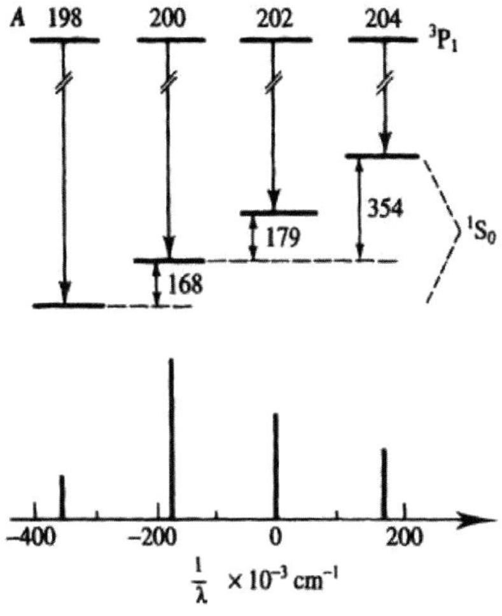

FIGURE 13.11
Atomic energy level shifts of mercury even isotopes.

13.4.3 Isomer Shift of the Atomic Energy Levels

In the above discussion, a change in nuclear volume with change in neutron number for a spherical charge distribution gives rise to a shift in the atomic energy levels through the electrostatic interaction. There can also be a change in the electrostatic interaction between an excited state (isomer) and ground state of the same nuclide (with equal proton number and neutron number) when they have different charge distributions (different shapes, such as spherical and deformed). In such a case there is an atomic energy level shift related to the change in mean square radius between the isomer and its ground state.

13.4.4 Examples of Isotope and Isomer Shifts of Atomic Energy Levels

Let us look at an example of studies of the isotope shift and isomer shift of atomic spectral lines that have provided crucial insights in the study of nuclear shape changes. In Section 11.3 we discussed evidence for nuclear shape coexistence where in the same isotope there are bands of overlapping energy levels which are built on quite different shapes. One of the first two areas where such nuclear shape coexistence was found was in the light-mass mercury isotopes very far from the beta-stable isotopes. The first clue to this came from isotope shift studies, which saw a sudden large change in the atomic transition frequency in going from ^{187}Hg to ^{185}Hg, as seen in Fig. 11.19. Note that in extending the data of Fig. 11.18 to nuclei very far from stable mercury nuclei, there is a gradual shrinking of the nuclear size (isotope shift in frequency) as one pulls out neutrons. But suddenly in going from 187 to 185 there is a drastic change in the isotope shift so that ^{185}Hg looks like ^{197}Hg. One interpretation was that there was a sudden onset of large nuclear deformation and ^{185}Hg is a well-deformed nucleus in its ground state with much greater mean square radius. But, of course, there are several competing effects and the sudden onset was only one possible explanation. This sudden onset of large deformation was subsequently established from studies of the nuclear energy levels, as discussed in Chapter 11.[6] Thus the ordinate in Fig. 11.18 now is known to be proportional to the mean square nuclear radius. These Hg nuclei from $A = 190$ down to the lightest known $A = 176$ all have separate bands of energy levels built on different, coexisting nuclear shapes. The even–even nuclei all have ground states with small deformations, $|\beta| \simeq 0.12$ (a near-spherical shape) and excited levels built on this shape, and low-energy, excited bands which overlap in energy with the ground band levels but built on deformed shapes with $\beta \simeq 0.25$ where the band heads are short-lived, shape-isomeric states.

In the odd-A nuclei the odd neutron has a polarizing effect in ^{185}Hg, to couple more easily to the low-lying deformed core in ^{184}Hg, so the ground state of ^{185}Hg has significant deformation ($\beta \simeq 0.25$). However, as seen earlier in Fig. 11.18, there is a near-spherical isomeric state in ^{185}Hg with $|\beta| \simeq 0.12$ that is the ground state of ^{187}Hg and the heavier mercury isotopes. Note that in ^{187}Hg to ^{199}Hg their isomeric and ground states both have the same small near-spherical deformation. Both the isotope and isomer shift effects have been observed for all these nuclei as shown in Fig. 11.18. Note also that there is an oscillation in the ground-state deformation between odd- and even-A as N decreases below 186. The near-spherical ground states for $A \geq 187$ continue as the ground states of the even–even nuclei below this, but become the isomeric states for the lighter odd-A nuclei.

Such isotope shifts have been used recently to demonstrate sudden changes in deformation with changes in neutron number in several mass regions. The classic sudden onset of deformation between $N = 88(\beta_2 \sim 0.10)$ and $90(\beta_2 \sim 0.25)$ was mapped and found to occur only when $Z = 64 \pm 2$ with an unexpected disappearance of this sudden onset for these same neutron numbers when Z is above or below the above limits. This was traced to the importance of reinforcement of proton and neutron shell gaps at $Z = 64$ (see ref. 6). The sudden onset of deformation in the new island of deformation centered around $N = Z = 38$ observed in studies of the nuclear energy levels was confirmed by isotope shift data. (See ref. 6 for details of the new studies.) Such isotope shift studies continue to be a very important source of information on changes in nuclear shapes with neutron number and between isomers and ground states.

[6]See J. H. Hamilton, "Structure of Nuclei Far from Stability", in *Treatise on Heavy Ion Science*, A. Bromley, ed., vol. 8, Plenum Press, New York (1989), p.1.

13.4.5 The Isomer Shift of the Nuclear Energy Levels

As noted above, a difference in nuclear volumes causes a shift of the atomic energy levels similar to the magnetic dipole and electric quadrupole hyperfine interaction. It is also possible for a ground and isomeric state of the same nucleus (i.e., same neutron and proton number) to have a different nuclear charge distribution and, thus, different nuclear size between the excited state (e) and ground state (g), that is, $\langle r^2 \rangle_e$ and $\langle r^2 \rangle_g$. Therefore, the electric interaction between nuclei and electrons outside is different for the different nuclear states. Thus, because of the electric interaction, the energy is shifted from E_g to E_g' and from E_e to E_e'. Usually, the shift $\Delta = (E_e - E_g) - (E_e' - E_g')$ is too small to be measured, but the Mössbauer effect offers a possibility. Since the shift Δ depends upon the chemical environment of the nuclei, it is different for the source (s) and for the absorber (a) in a Mössbauer measurement. Thus, we have $\delta = \Delta_a - \Delta_s$, which is called an isomer shift or chemical shift that is different from the δ in NMR in Appendix 4A. There it is related to a magnetic effect, but here it is related to an electrical effect. Indeed, we could have called Δ an isomer shift, but for practical reasons, we prefer to use δ to obtain the information about the s electron density for different compounds.

The isomer shift can be written as

$$\delta = \frac{2\pi}{3} Ze^2 \langle r^2 \rangle_g \{ |\psi(0)|_a^2 - |\psi(0)|_s^2 \} \tag{13.46}$$

where $\psi(0)$ is the s electron wave function at the nuclear site.

13.4.6 The Hyperfine Structure Anomaly of Isotopes: Bohr–Weisskopf Effect[7]

From Eq. (13.16) or Eq. (13.21), we might expect that for two isotopes whose atoms are in the identical state the ratio of their hyperfine constants A_1 and A_2 will be equal to the ratio μ_I/I, that is,

$$\frac{A_1}{A_2} = \frac{(\mu_I/I)_1}{(\mu_I/I)_2} \tag{13.47}$$

where the nuclear charge, the atomic energy levels, and the angular quantum numbers for the two isotopes are identical and only μ_I and I are different.

However, A. Bohr and V. Weisskopf found

$$\frac{A_1}{A_2} = (1 + \Delta) \frac{(\mu_I/I)_1}{(\mu_I/I)_2} \tag{13.48}$$

[7] A. Bohr and V. F. Weisskopf, *Phys. Rev.* 77 (1950) 95.

where Δ is called the hyperfine structure anomaly for the isotopes 1 and 2. This effect is the result of the finite size of the nucleus: the nuclear magnetic dipole moment must be regarded as distributed over the volume of the nucleus, so the interaction $-\boldsymbol{\mu}_I \cdot \boldsymbol{B}_{el}$ must be averaged over the nuclear volume. The differential effect between isotopes of different nuclear size can be significant for electrons which penetrate into the nucleus (s and $p_{1/2}$ electrons). This leads to anomalies of the order of $\Delta \sim 1$ percent. Obviously, the effect is only revealed when very accurate measurements of μ_I and of A are made independently. Nevertheless, a considerable amount of work on the hyperfine structure anomaly has been done and has given further information about nuclear structure.

13.4.7 Isotopic Separation by Lasers

We have learned that the atomic levels of the isotopes belonging to the same element have only a very small "hyperfine" difference. However, because of their very precisely adjustable wavelengths, high-precision tunable lasers can selectively excite and ionize practically any single quantum state of an atom or a molecule of different isotopes, and thus have been used to separate isotopes. The fundamental principle of isotopic separation by laser radiation is briefly described as follows (for simplicity we assume there are only two isotopes in a certain natural element).

In Fig. 13.12 we show the ground state (g), excited state (e), and ionization state (i) of the atoms or molecules. The solid lines represent the atomic or molecular levels of the isotope I that we want to separate from the other isotope II whose levels are represented by the dashed lines.

Suppose we irradiate the natural element by a laser with frequency that is equal to the resonant frequency of the transition from g_{I} to e_{I} for the isotope I. Since the tunable dye laser has an extremely narrow line width and good selectivity, it can produce the resonant excitation to pump the atoms of the isotope I from the g_{I} state to the e_{I} state. But the atoms of the isotope II are not excited and are still in the ground state. At the same time the atoms are irradiated by a second laser whose photon energy $h\nu_2$ is not sufficient to cause any resonant excitation of the atoms of the isotope II in the ground state, but is sufficient to ionize the atoms of isotope I in its excited state, that is, $h\nu_2 > (E_i - E_e)$. Then the atoms (ions) of

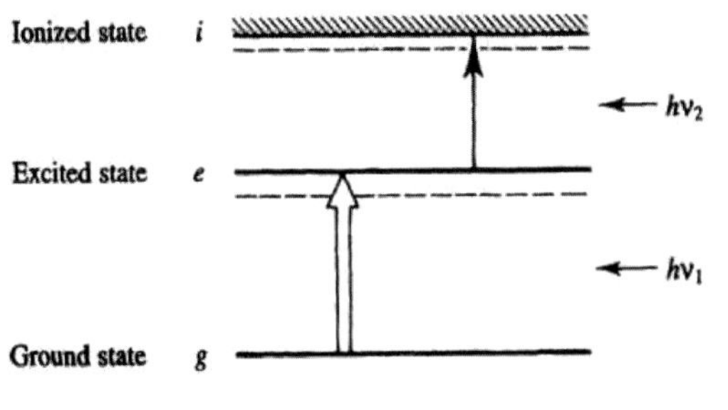

FIGURE 13.12
Diagram for the principle of isotopic separation by lasers.

the isotope I may be separated from those of the element II by an external electric field.

In recent years many laboratories around the world have been devoted to the study of isotope separation by lasers. In particular, the separation of ^{1}H and ^{2}D, ^{6}Li and ^{7}Li, and ^{235}U and ^{238}U are significant.

13.4.8 Study of the Isotope Shift by Muonic Atoms

If an atom contains a negative muon μ^- (see Chapter 14) instead of an electron, it is called a muonic atom. The muon is 207 times heavier than an electron but has the same charge as an electron. Muons can be easily produced by bombarding matter with energetic protons (> 440 MeV) in intermediate- or high-energy accelerators. They can be captured into outer atomic orbits by nuclei. In making transitions from outer orbits to inner ones, the muons radiate x-rays whose energy region is the order of MeV. Since the negative muons behave like heavy electrons, we may simply apply the results of the Bohr model.

From Eq. (13.39), the isotope shift $\delta\nu/\nu$ from the mass effect is proportional to m_μ/M^2, where $m_\mu = 207m_e$ is the mass of the muon. This means that the isotope shift by the mass effect is 200 times greater for muonic atoms than for electronic atoms.

In Eq. (8.133) if we replace the electron mass by the muon mass, we have the orbital radii of the muonic atoms,

$$a_{\mu n} = \frac{\hbar^2}{Ze^2 m_\mu} n^2 \tag{13.49}$$

where $a_{\mu n}$ is 200 times smaller than the radius of the corresponding electronic orbits because of the ratio of the electron mass to the muon mass. So the muon is much closer to the nucleus than the electron. Thus, the probability density for finding the muon inside the nucleus, $|\psi_\mu(0)|^2$, is much greater than for the electron. From Eqs. (13.44) and (13.13) we have

$$\delta(\Delta E) \propto |\psi(0)|^2 \propto \frac{1}{a_1^3} \propto m^3 \tag{13.50}$$

and

$$\frac{\delta_\mu(\Delta E)}{\delta_e(\Delta E)} \propto \frac{|\psi_\mu(0)|^2}{|\psi_e(0)|^2} \propto \left(\frac{m_\mu}{m_e}\right)^3 \sim (207)^3 \sim 10^7 \tag{13.51}$$

The isotope shift from the volume effect is seven orders of magnitude greater for muonic atoms than for electronic atoms. Thus, muonic atoms are very sensitive probes for studying the nuclear charge distribution and even the details of the nuclear potential.

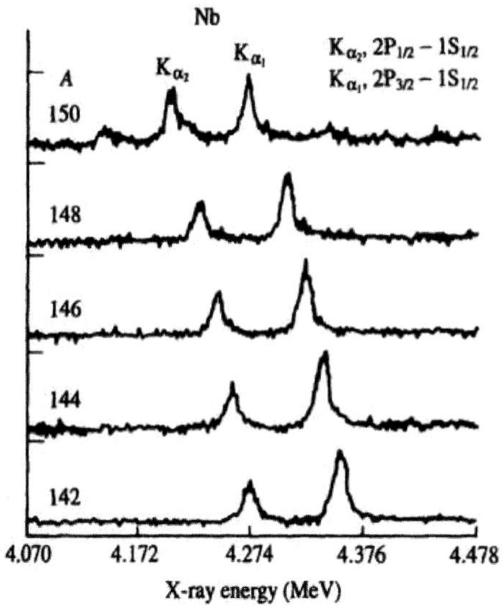

FIGURE 13.13
X-ray energy shifts of the muonic atoms of niobium isotopes.

Figure 13.13 shows the isotope shifts in the spectra of K_{α_1} and K_{α_2} x-rays produced from the transitions $2P_{3/2} \rightarrow 1S_{1/2}$ and $2P_{1/2} \rightarrow 1S_{1/2}$ for the muonic atoms of niobium isotopes. From the figure we can see that when the nucleon number of Nb increases by the addition of one neutron, the energy shifts of the x-rays are the order of

$$10\,\text{keV} \sim 10^5\,\text{cm}^{-1} \qquad \text{to be compared with} \qquad (10 \times 10^{-3}\,\text{cm}^{-1})$$

where the factor in parentheses is a typical isotope shift for electronic atoms. Thus, the shift of muonic atoms is about 10^7 greater than that of electronic atoms, in good agreement with theoretical calculations.

13.5 SUMMARY

The study of hyperfine interactions not only contains a wealth of significant information itself but has also provided the main experimental technique that has opened up the study of many new phenomena in physics, chemistry, and biology.

The development of hyperfine interaction research has been propelled by its many current applications in a variety of research areas, including the techniques of the Mössbauer effect, perturbed angular correction, nuclear magnetic resonance, laser spectroscopy, atomic and nuclear polarization, muonic atoms, and ionic, atomic, and molecular beam methods.

Hyperfine interactions bridge the fields of solid state, atomic, and nuclear physics with broad applications in other fields including chemistry and medicine.

PROBLEMS

13.1. From the hyperfine structure of the ^{23}Na atomic spectrum, it was found that the D_1 line (5896 Å) splits into two lines with an interval 0.023 Å, which primarily is from the $3\,^2S_{1/2}$ level splitting (the $3\,^2P_{1/2}$ splitting lines were unresolved). What is the order of magnitude of the energy difference between the two sublevels of $3\,^2S_{1/2}$?

13.2. Each hyperfine level in a state of ^{42}K atoms splits into five components in a strong magnetic field. What is the nuclear spin of ^{42}K?

13.3. The hyperfine structure of the ^{235}U atomic optical spectrum consists of eight lines. What is the nuclear spin of ^{235}U? The ground state of a ^{235}U atom is $^5L_6$.

13.4. When ^{25}Mg atoms in their ground state ($^1S_0$) are examined by the NMR in a magnetic field $B = 0.540$ T, a resonance frequence $\nu_0 = 1.4$ MHz is measured. The ^{25}Mg nuclei have spin $I = \frac{5}{2}$. Calculate the g-factor and the magnetic moment of the nucleus.

13.5. Calculate the precession frequencies of the electrons, protons, and neutrons in a magnetic field $B = 0.450$ T.

13.6. Consider the hyperfine structure of the transition $5d^26s\,^4F_{9/2} \rightarrow 5d^26p\,^4G_{11/2}$ in ^{139}La (Fig. 13.2).

(*a*) Why might you suspect that the magnetic dipole hyperfine structure of the $^4F_{9/2}$ level is considerably larger than that of the $^4G_{11/2}$ level?

(*b*) Would you expect an electric quadrupole interaction to give a striking departure from the interval rule in the $^4F_{9/2}$ level?

13.7. Show that the $2S_{1/2}$ level of hydrogen has a hyperfine splitting of 177.56 MHz.

13.8. Calculate the hyperfine splitting of a hydrogenlike atom of ^{39}K ($I = \frac{3}{2}$ nuclear spin) in its $1S_{1/2}$ ground state. Its g-factor is $g_K = 0.39$.

13.9. Give a diagram of the hyperfine splitting for the deuterium $1S_{1/2}$ level and calculate the splitting in the unit of eV. Deuterium has $I = 1$ and $g_D = 0.857$.

13.10. There is a well-known hyperfine transition of 21 cm in hydrogen. What is the corresponding number in deuterium?

13.11 The ground-state spectroscopic term of a ^{59}Co atom is $^4F_{9/2}$, determined by the outer electrons outside the nucleus. Its hyperfine structure splits into eight components. What is the nuclear spin of ^{59}Co?

2. Experiments show the electric quadrupole moment of ^{165}Ho is 300 fm^2. If $R_0 = 1.2A^{1/3}$ fm, what is the deviation of its nuclear shape from a spherical shape; for example, what is $\delta R_0/R_0$? Here R_0 is the radius of a sphere which has the same nuclear volume as the spheroid and δR_0 is the difference of b, the symmetric axis of the spheroid, from R_0.

13.13. Atomic beams of ^{9}Be($^3S_0$) and ^{13}C($^3P_0$) pass an ultra-nonuniform strong magnetic field (a Stern–Gerlach experiment), and their beams split into four and two lines, respectively. Determine the nuclear spins of ^{9}Be and ^{13}C.

13.14. Show the difference of the hyperfine structures for the cases of strong and weak external magnetic fields. Take the ground state of hydrogen $^2S_{1/2}$ as an example and make the diagram.

13.15. The ground state of ^{17}O is $1s^22s^22p^4\,^3P$. Give its energy level diagram with fine and hyperfine structure. Note that the nuclear spin of ^{17}O is $I = \frac{5}{2}$.

13.16. Calculate the variation of the frequency of the spectroscopic lines related to the mass effect of isotopes for: (*a*) ^{7}Li and ^{6}Li and (*b*) ^{22}Ne and ^{20}Ne.

13.17. If the hyperfine splitting is known, it is possible to estimate the magnetic field which the electrons feel from the relation

$$\Delta E \sim g_j \mu_B B_{\text{eff}} m_j$$

Make a rough estimate of the magnetic field for the hydrogen ground state.

13.18. Calculate the hyperfine splitting ratio of the $2\,^2P_{1/2}$ and $1\,^2S_{1/2}$ levels of a hydrogen atom and give the splitting gap for the 2p state in units of eV.

13.19. If the energy spacing of the nuclear energy levels remained constant for a sequence of nuclei with constant Z and decreasing neutron numbers, what would you say about the structure of these nuclei (see Chapter 11)? If the measurements of the mean square charge radii using the hyperfine interaction indicated that a significant increase in the RMS radii occurred at certain neutron numbers for constant Z as N decreased, what would you conclude was occurring as a function of neutron number? If both of the above effects were observed, how would you go about explaining these conflicting data, for example, what general guidelines would you use? (Such a situation actually occurs in the light thallium isotopes: see J. A. Bounds, et al., *Phys. Rev. Lett.* **55** (1985) 2269.)

CHAPTER 14

HIGH-ENERGY PHYSICS

One thing I have learned in a long life: that all our science, measured against reality, is primitive and childlike—and yet it is the most precious thing we have.

A. Einstein

14.1 TOWARD A DEEPER STRATUM

14.1.1 Dimension and Energy

High-energy physics is one of the frontiers in contemporary physics. It seeks to explore a deeper stratum of the structure of matter than the nucleus. In the nineteenth century, atoms were thought to be indivisible. Then came the discoveries of the electron and radioactive decays. So, atoms could be broken up into bits and pieces. With the discovery of the neutron in 1932, it was thought that we had all the particles needed to make atoms and all matter, namely, protons, neutrons, and electrons. The proton and neutron were considered indivisible, and hence "elementary," just as atoms were thought of as indivisible at the end of the last century. High-energy physics is also called elementary particle physics, because it seeks to discover the fundamental, "elementary," building blocks of all matter. This search has pushed physics towards unfolding a deeper stratum of matter than the nucleus of the atom. Already in 1932, a new particle, the positron, theoretically predicted by Dirac to be the antiparticle of the electron was discovered, and another new particle, the neutrino, had been predicted by Pauli. However, evidence that the neutron and proton are not elementary particles but have internal structure was crucial in the push to seek a deeper stratum.

The name elementary particle physics may seem to be the proper title of this field. However, as prophetically pointed out by ancient Chinese philosophers, nothing is immutable. Today's "elementary" particles may not be elementary tomorrow. The correctness of the idea that nothing is immutable has been driven home to us time and again in the development of physics. Our latest ideas may likewise turn out to be primitive and childlike.

In dissecting the structure of matter into smaller pieces, we are in need of a finer knife. The energy needed to separate an atom is about 10 eV. Then the energy increases to the order of MeV to separate particles from a nucleus. To separate out deeper strata of matter requires from tens of MeV to tens of GeV on to tens of TeV. The smaller the dimension of the observed object, the shorter the wavelength of the "light" wave needed in the observation. Thus, the energy needed becomes greater and greater the smaller the object. That is why the search for the basic particles of matter is called high-energy physics. These two terminologies are unified by the de Broglie relation in quantum mechanics.

14.1.2 High-energy Particle Sources

How "high" is the energy needed? The answer to this question has been ever increasing. The Bevatron at Berkeley was built to accelerate protons to the then record of 6.2 GeV to search for antiprotons. The discovery of this new particle called for still higher-energy machines using protons of about 30 GeV at Brookhaven National Laboratory and the Central European Nuclear Research Center (CERN). Still these energies did not probe deep enough. Let us survey the world's largest accelerators available for producing high-energy particles.

The proton synchrotron at Fermi National Accelerator Laboratory (FNAL) (see Fig. 14.1) had the highest energy particle beam in early 2009. The proton orbit is circular with a diameter of about 2.2 km. The final proton energy was upgraded to reach 2000 GeV(2 TeV) in the center of mass by the use of superconducting magnets. The Large Hadron Collider (LHC) at CERN is to turn on in 2009 with an expected proton energy of 14 TeV COM. The 3 km long electron linear accelerator at the Stanford Linear Accelerator Center (SLAC) delivered electrons of 50 GeV until it was shut down.

In the above summaries, it is the kinetic energies of the proton or electron that are given. When particles of these kinetic energies collide with a stationary target, only part of their energy, the so-called "energy of the center of mass system," is effective as a probe of new substructures. The remainder of the energy will be transformed into kinetic energy of the particles emitted from the target. This is analogous to the case where a moving car collides with a stationary one: only part of the kinetic energy will do damage to the car being hit, because part of the energy must be transferred as kinetic energy to conserve momentum in addition to total energy. However, in a head-on collision between two cars of equal but oppositely directed momenta (zero net momentum) and the same total kinetic energy as the first case, the damage produced will be many times greater.

FIGURE 14.1
Exterior of Fermi National Accelerator Laboratory (its synchrotron has a diameter of 2.2 km). (Courtesy of Fermi National Accelerator Laboratory.)

When an electron of 22 GeV laboratory energy (E_L) collides with a stationary proton, only 30% of this energy can be used to probe its substructure. The effective energy (E_{CM}) is 7 GeV, with the remaining 15 GeV going into the motion of the proton. The energy in the center of mass system, E_{CM}, is related to that in the laboratory, E_L, by the expression

$$E_{CM} \simeq \sqrt{2m_T E_L}$$

where m_T is the mass of the target nucleus in units of energy. Hence, the greater the E_L, the less the effectiveness ($E_{CM}/E_L \sim \sqrt{2m_T/E_L}$). For example, if a proton at energy of 400 GeV (E_L) collides with a proton at rest, then $E_{CM} = \sqrt{2 \times 0.938 \times 400}$ GeV $= 27.4$ GeV. The effectiveness, $E_{CM}/E_L = 27.4/400$, is only 7%. If $E_L = 1000$ GeV, then $E_{CM} = 43$ GeV, and the effectiveness drops to 4%.

The Q value of a reaction as defined earlier is simply the difference in rest mass energies of the initial and all the final particles. For relativistic particle collisions, the threshold kinetic energy has the form

$$K_{\text{th}} = \frac{\text{(total mass of all the particles on both sides of the equation)}}{2 \times \text{mass of the target}} \text{X(-Q)}$$

where $Q = (\sum \text{ initial mass} - \sum \text{ final mass})c^2$ in MeV. One can use this to calculate the threshold energy for the production of a particle of a particular mass. For example, suppose you wanted to produce a proton–antiproton pair by bombarding protons on protons: $\mathrm{p + p \to p + p + p + \bar{p}}$. (Note the conservation of particle number, where the antiparticle counts as -1.) Then

$$K_{\text{th}} = 6 \times m_p c^2 \times (-(2-4)) m_p c^2 / 2 m_p c^2 = 6 \times m_p c^2 = 6 \times 938 \text{ MeV} = 5.6 \text{ GeV}$$

The Bevatron at Berkeley was designed to give 6.2 GeV protons, just above this threshold, in order to be able to produce antiprotons. Thus, they were able to discover the antiproton.

In order to increase the effective energy, particle colliders were developed. For example, in a head-on collision between an electron and a positron at equal energies of 18 GeV, since their total momentum is zero, the effective energy will be 36 GeV. High energy colliders in operation in 2009 include:

KEKB (Japan)	electron-positron collider	8 + 3.5 GeV
NEPD-2000 (Russia)	electron-positron collider	1 + 1 GeV
BEPC# (China)	electron-positron collider	1.9 + 1.9 GeV
Tevatron (USA)	proton-proton collider	1 + 1 TeV
RHIC (USA)	proton-proton collider	0.1 + 0.1 TeV
	Au-Au collider	0.1/A + 0.1/A TeV/A
LHC (CERN)	proton-proton collider	7 + 7 TeV
	Pb + Pb collider	2.7/A + 2.7/A TeV/A

The former SPS, p-$\bar{\text{p}}$ collider and LEP e^--e^+ collider are indicated in Fig. 14.2. The LEP was converted to a Large Hadron Collider (LHC). Note LHC crosses the border between France and Switzerland four times in the 16.6 mile (27 km) circumference. The SPS had nearly seven times the energy but was much smaller than LEP. The orbit of LEP had to be much larger than that of SPS to reduce the acceleration during bending so as to reduce for the electrons the synchrotron radiation which has an inverse mass dependence. On an orbit the size of SPS, the electrons would quickly lose too much energy to synchrotron radiation. The LHC at CERN has the highest energy with 14 GeV center of mass energy for protons. To illustrate the enormous detector sizes now required at such energies, the PHENIX detector at RHIC is shown in Fig. 14.3 and the L3 detector for LEP is shown in Fig. 14.4. As another example of facilities, Fig. 14.5 shows the current accelerators and the new antiproton and RIB facilities under development at GSI to probe problems in both particle and nuclear structure physics.

Figure 14.2
An aerial view of the CERN accelerators. The initial PS and ISP accelerator and storage ring are the small circles on the left. The 400 GeV proton collider is SPS (solid line), and the 60 GeV on 60 GeV e^+-e^- collider LEP (dashed line) had a 27 km circumference. The LEP is now converted to LHC. (Photo provided by the Photographic Services of CERN).

The advantage of a collider is its high effective energy. The effective energy of the earlier 270 + 270 GeV hadron collider at CERN was 540 GeV. The massive W and Z intermediate bosons were discovered in 1983 at this collider. Now it has been upgraded to 400 + 400 GeV. To reach 540 GeV of effective energy with an accelerator with a stationary proton target would require a particle energy of 1.6×10^5 GeV. With the same proportions as the proton synchrotron at FNAL, which was 400 GeV and 2.2 km in diameter, the diameter of an accelerator of 1.6×10^5 GeV would be nearly equal to that of the Earth at the equator!

On the other hand, a collider has the disadvantage of low probability of two particles hitting. Nevertheless, the energies required to probe deeper into the structure of matter shifted the frontier of accelerator-based high energy physics to colliders. Upgrades to improve the luminosity of colliders are always on going. Most colliders involved equal-mass particles and antiparticles in the collision. A unique collider was HERA, where electrons collided with protons with a total energy of 850 GeV but it was closed in 2007.

Before accelerators reached energies sufficient to probe the structure of matter below that of the nucleus, many new "elementary" particles were discovered in cosmic rays in the late 1940s—the golden age of cosmic ray studies. Detectors ranged from simple nuclear emulsions to triggered cloud chambers with magnetic deflection and photographic recording. Cosmic ray stations were

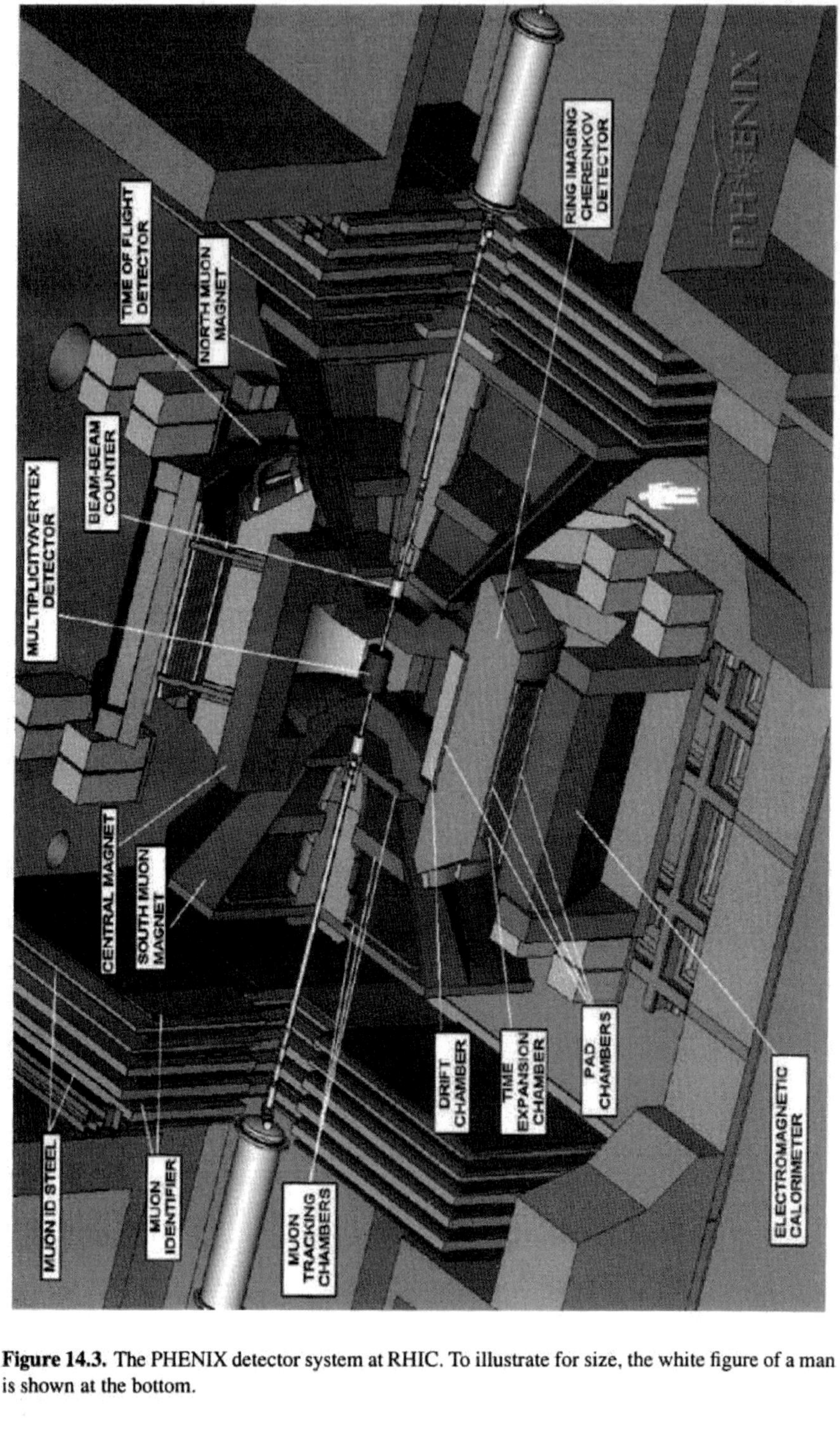

Figure 14.3. The PHENIX detector system at RHIC. To illustrate for size, the white figure of a man is shown at the bottom.

FIGURE 14.4
The giant L3 detector developed by Ting and collaborators for e^+-e^- collision studies at a total energy of approximately 91 GeV at LEP. The photograph shows the octagonal magnet and coils and the large support tube which houses most of the detector. (Courtesy of S. C. C. Ting, MIT.)

built as high up as possible, and more recently an array of detectors in coincidence to record showers from primary events up to 10^{14} GeV was constructed in Tibet. High-altitude balloons carry emulsions still higher. Cosmic ray studies continue to have a special place because the energies of man-made accelerators are still far below those of "natural" sources of high-energy particles. Particles with energies up to 10^{13} GeV have been found in cosmic rays. So, while cosmic ray intensities are low and drop quickly ($I \sim E^{-2.6}$) as the energy increases, nonetheless cosmic rays are still a useful tool for high-energy physics studies.

How to increase the usable energy of particles obtained from high-energy accelerators is an unending quest that continues to push out the frontiers of

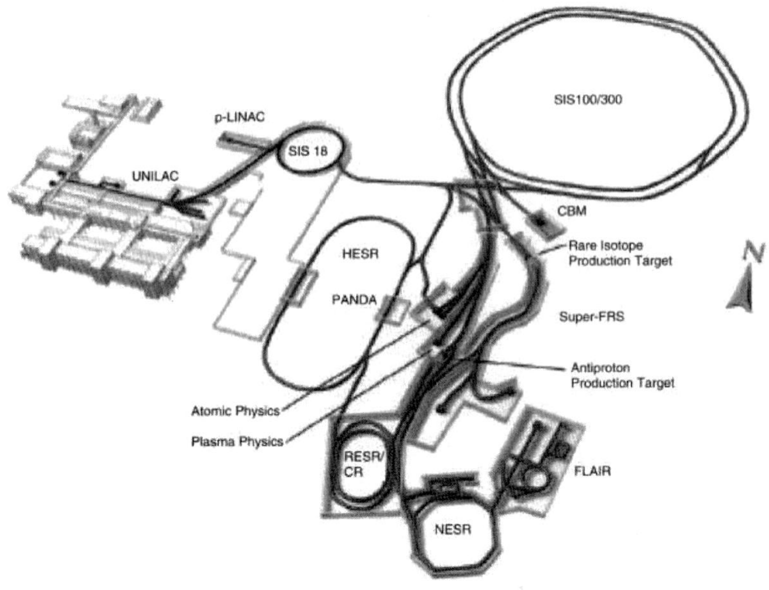

Figure 14.5. Layout of the existing (UNILAC, SIS18) and the planned FAIR facility: the superconducting synchrotrons SIS100 and SIS 300, the collector ring CR, the accumulator ring RESR, the new experimental storage ring NESR, the super fragment separator SuperFRS, the proton linac, the antiproton production target, and the high energy storage ring HESR. Also shown are the experimental stations for plasma physics, nuclear collisions (CBM), radioactive ion beams (Super-FRS), atomic physics experiments, and low-energy antiproton and atomic physics (FLAIR).

technology. At present all colliders are circular or near so. But charged particles moving in a circle have the disadvantage of losing energy by emission of electromagnetic radiation from the acceleration to bend their path; the higher the energy for a given radius of the accelerator, the greater the radiation energy loss. An alternate approach is a linear collider. A collider of this type was completed in the United States, the Standford Linear Collider (SLC) at SLAC. It had 50 GeV e^+ on 50 GeV e^-. The energy was selected so that the effective energy was above the Z^0 mass to allow study of its properties. The first Z^0s were produced at SLC for research in 1989. In competition with SLC to study the Z^0 was the 60 GeV e^+ on 60 GeV e^- LEP collider at CERN where, since the LEP turned on also in 1989, even larger numbers of Z^0s were produced. Research at these colliders has been terminated. In their place, a much large International Linear Collider has been proposed to explore new frontiers.

14.1.3 Characteristics of High-energy Physics

1. One might think that the explosion of an atomic bomb may be tied to high-energy physics, because it emits a tremendous amount of energy. But, as indicated in Chapter 12, all the microscopic processes involved in an atomic or hydrogen bomb belong to the low-energy regime. The energies of the neutrons that induce nuclear fission are only about 0.025 eV. The fission energy per nucleon in the splitting of ^{235}U is less than 1 MeV, and even in a fusion process the contribution per nucleon is no more than 6–7 MeV (1 MeV is equivalent to 1.6×10^{-13} joule; and 4.18 J of energy is needed to increase the temperature of 1 g of water by 1° C!). It is the enormous size of Avogadro's constant, which links the macro- and microscopic sides of the process, that makes nuclear energy so tremendously powerful.

The energy involved in the microscopic processes of high-energy physics research belongs to the GeV to TeV regimes. There is still no way to achieve such energies macroscopically. Hence, high-energy physics is purely basic research which has to first order no connection with the exploration of energy resources. On the other hand, if one could make macroscopic quantities of antiprotons and store them, their annihilation with protons would produce tremendous energy. Although such a possibility has been proposed, it seems very unlikely and is outside the field of high-energy physics itself.

2. The defining characteristic of high energy physics is $\Delta/mc^2 \lesssim 1$, where Δ denotes the binding energy, and mc^2 the rest energy of the system. For example, for a molecule consisting of atoms, $\Delta \sim 4$ eV, $mc^2 \sim$ few GeV, and $\Delta/mc^2 \sim 10^{-9}$; in an atom consisting of electrons and a nucleus, an outer electron has $\Delta \sim 10$ eV, $mc^2 \sim 0.511$ MeV (the rest energy of electron), so the ratio $\Delta/mc^2 \sim 10^{-5}$; even in a nucleus formed by nucleons, $\Delta \sim 8$ MeV/u, $mc^2 \sim$ GeV, $\Delta/mc^2 \sim 10^{-2}$. Only in the high-energy physics regime do we find the ratio $\Delta/mc^2 \lesssim 1$.

The paradox of $1 + 1 \neq 2$ already was found in the nuclear regime where $\Delta/mc^2 \sim 10^{-2}$ (Section 9.2). So we expect the challenge of $\Delta/mc^2 \lesssim 1$ to be so great as to change our ideas of the structure of matter. Indeed, E. Fermi and C. N. Yang proposed in 1949 a constituent model of the following elementary particles:

$$\mathrm{p} + \bar{\mathrm{n}} = \pi^+ \qquad \left.\begin{matrix} \mathrm{p} + \bar{\mathrm{p}} \\ \mathrm{n} + \bar{\mathrm{n}} \end{matrix}\right\} = \pi^0$$
$$\mathrm{n} + \bar{\mathrm{p}} = \pi^-$$

Since the masses of $\pi^\pm$ mesons are 139.6 MeV, while the sum of the masses of a proton and antineutron $M(\mathrm{p}) + M(\bar{\mathrm{n}}) = 1877.8$ MeV, the mass difference of the constituents is great. The binding energy of 1736.2 MeV would be equal to 92.1% of the rest energy, so $\Delta/mc^2 \sim 1$. Although the model proposed by Fermi and Yang was not accepted subsequently, the idea of "a π^+ consisting of a p and $\bar{\mathrm{n}}$" is obviously different from that of "an atom consisting of a nucleus and electrons" or "a nucleus consisting of neutrons and protons."

3. The production and annihilation of a particle and its antiparticle are commonly seen in high-energy physics. Thus, the year of the prediction or discovery of the first antiparticle (e^+) may be taken as the commencement of high-energy physics.

14.1.4 Historical Review

1928 P. Dirac developed the relativistic theory of the electron and predicted the existence of the positron (e^+).

1930 W. Pauli proposed the hypothesis of the neutrino.

1932 C. D. Anderson discovered the positron in cosmic rays.

1932 J. Chadwick discovered the neutron.

1934 E. Fermi proposed the theory of annihilation and production of particle–antiparticle pairs and the weak interaction theory of beta decay.

1935 H. Yukawa proposed a meson theory of the nuclear force, and predicted the existence of mesons.

1936 C. D. Anderson, S. H. Neddermeyer, and co-workers discovered the muon in cosmic rays.

1947 C. F. Powell and co-workers discovered the π meson in cosmic rays.

1955 E. Segre, O. Chamberlain, and co-workers discovered the antiproton with a 6.2 GeV proton accelerator.

1956 F. Reines and C. L. Cowan directly observed the antineutrino interacting in matter for the first time. B. Cork and co-workers experimentally proved the existence of the antineutron; and T. D. Lee and C. N. Yang proposed nonconservation of parity in the weak interaction.

1963 M. Gell-Mann and G. Zweig independently proposed a quark model for the structure of hadrons. Some Chinese scientists proposed, at the same time, the stratum model.

1967 S. Weinberg and A. Salam proposed a unified theory of the weak and the electromagnetic interactions based on the work of S. L. Glashow, and predicted the existence of intermediate bosons, the W and Z.

1970 S. L. Glashow and co-workers predicted the existence of a fourth quark (the charm quark).

1974 S. C. C. Ting and co-workers, and B. Richter and co-workers independently discovered the J/ψ particle, which is considered to be composed of charm and anticharm quarks.

1975 M. Perl and co-workers discovered the heavy τ lepton.

1978 L. Lederman and co-workers discovered the Υ, which is composed of a $b\bar{b}$ pair of quarks.

1983 The intermediate bosons, the $W^{\pm}$ and Z^0 were discovered by the experimental group at CERN headed by C. Rubbia.

1995 Definitive evidence for the top quark was reported at Fermilab

2001 Direct observation of the tau neutrinos at Fermilab

The photograph in Fig. 14.6 includes a number of the Nobel laureates mentioned above.

FIGURE 14.6
Eight scientists who won the Nobel Prize. Left to right: E. Segre, C. N. Yang, O. Chamberlain, T. D. Lee, E. McMillan, C. D. Anderson, I. I. Rabi and W. Heisenberg. (Courtesy of E. Segre, Berkeley, CA.)

14.2 PARTICLE FAMILIES AND INTERACTIONS

14.2.1 Particle Families

HISTORY. The particle spectra shown in Figs. 14.7 to 14.10 trace the history of the discovery of "elementary" particles starting with the particles known in 1930. However, not all of the particles known today are listed in the last graph. Only those with a lifetime greater than 10^{-20} s are listed, since otherwise the total number of particles would exceed eight hundred. It should be noted that the timescale for a particle with speed approaching the speed of light to traverse a distance of 1 fm is of the order of magnitude of 10^{-23} s. Hence, in high-energy physics a particle with a lifetime greater than 10^{-20} s is considered to be long-lived.

When the antiparticles of those shown are added, there are 45 particles in Fig. 14.10. Excepting the exchange particle of the electromagnetic interaction, the photon (γ), and that of the gravitational interaction, the not-yet-discovered graviton (g), the remaining particles can be divided into two categories: leptons and hadrons.

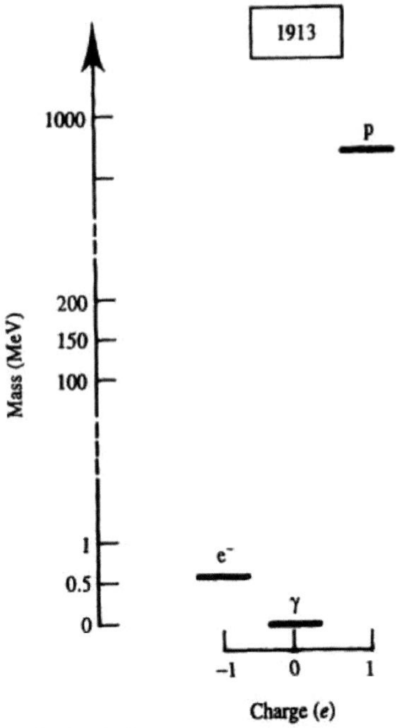

FIGURE 14.7
Elementary particles known in 1913.

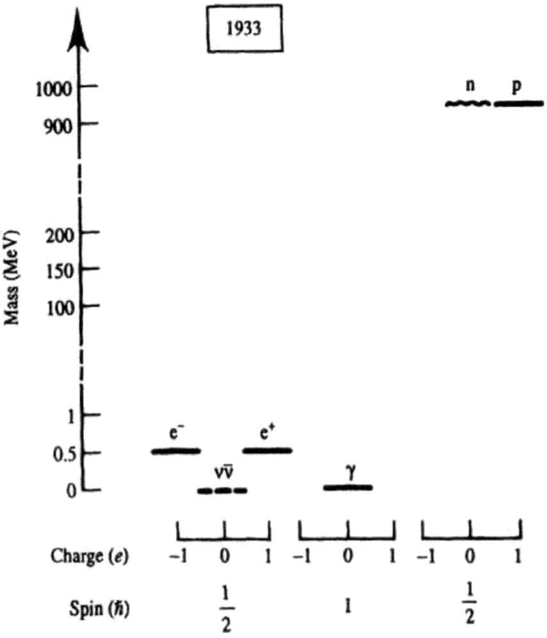

FIGURE 14.8
Elementary particles known in 1933. The wavy line under the neutron indicates that it is unstable and decays, and the dashed line under the neutrinos indicates that these had only been theoretically predicted. The same wavy and dashed line notations will be used in the subsequent figures.

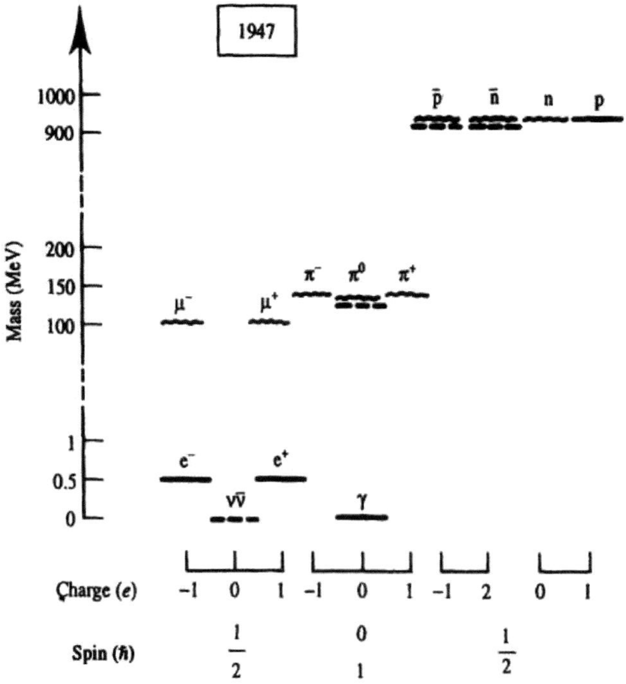

FIGURE 14.9
Elementary particles known in 1947.

LEPTONS. Originally the name lepton was used for the lightest-mass particles. But since the more recently discovered, quite massive τ particle is grouped as a lepton, obviously light mass is not the defining characteristic. To be precise, a lepton is defined as a fermion that does not participate in any strong interactions. No internal structure has yet been found for a lepton, so lack of internal structure may be counted as another characteristic of a lepton. Ten leptons in Fig. 14.10 are established including the e^-, ν_e, ν_μ, μ^-, τ^- and their antiparticles. Based on symmetry arguments, there must exist ν_τ, $\bar{\nu}_\tau$ neutrinos associated with the τ lepton, but they have not been found yet.

The muon was discovered in 1936 right after the prediction of the existence of a meson with similar mass by Yukawa. Accordingly, it was then named the meson, because its mass was intermediate between that of an electron and that of a proton. But this particle was found later not to participate in strong interactions, hence it could not be the particle predicted by Yukawa in his meson exchange theory of the strong nuclear force. The most essential characteristic of a Yukawa meson is its participation in strong interactions rather than its mass. So now the particle discovered in 1936 is called the muon instead of the μ meson.

A muon has almost the same properties as an electron, except that it is about 200 times heavier. Sometimes it is called a heavy electron. While they have different lifetimes (2 μs for the muon, which is a tremendously long lifetime

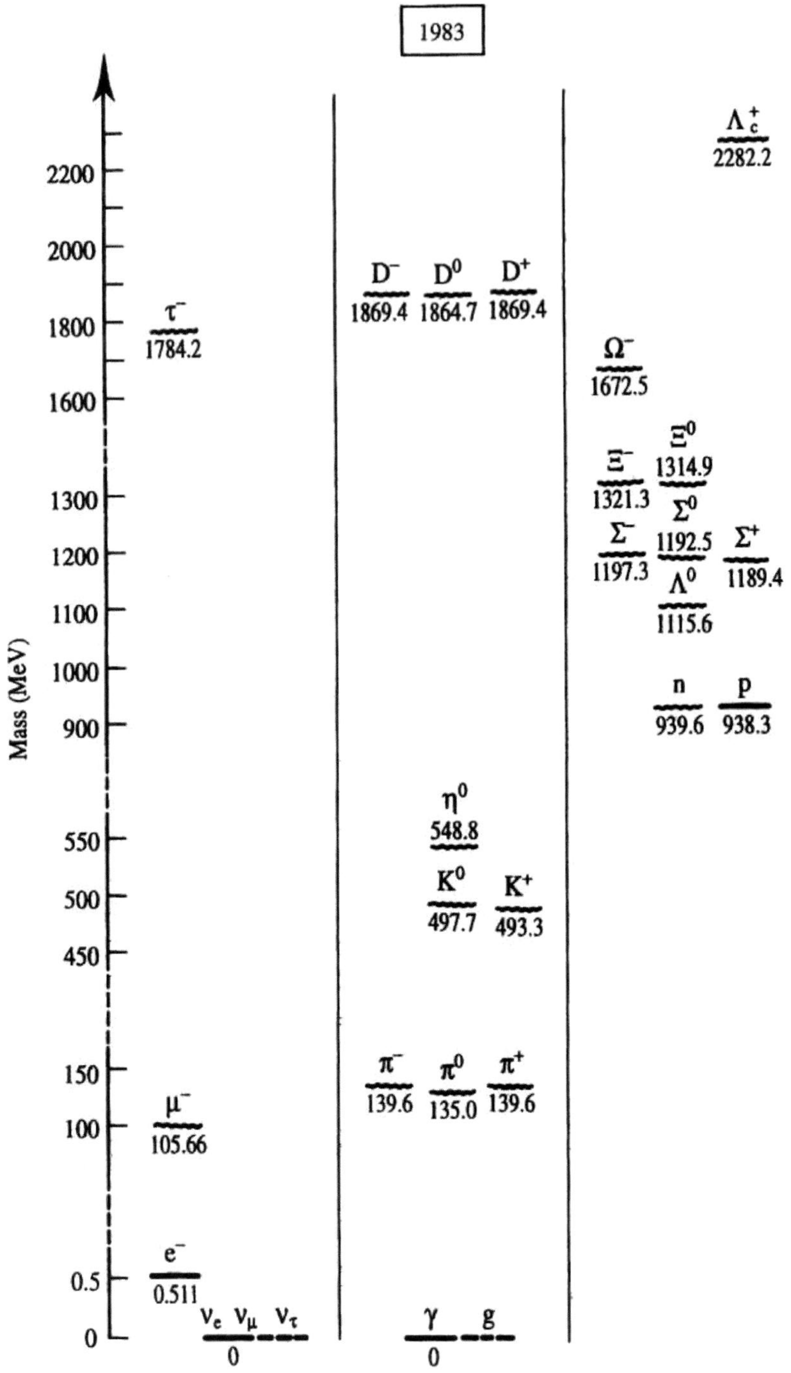

FIGURE 14.10
Elementary particles known in 1983 with lifetimes greater than 10^{-20} s.

by particle physics standards, and infinite for the electron), this difference is not considered important, because it originates from their mass differences. (Think through why this is so. Compare with the neutron to proton decay.) At present, the muon remains a very mysterious particle. We do not know why it is here or what special role it plays. As I. Rabi put it, "Who ordered it?" However, the tauon, or superheavy electron, is understood even less. Even the most sophisticated experimental techniques have revealed no indication that a lepton has any internal structure down to 10^{-16} cm (10^{-3} fm). So the electron, the first elementary particle discovered, remains an elementary particle.

HADRONS. All the particles that participate in the strong interaction are called hadrons. All hadrons have an internal structure that is composed of quarks (stratons) (see the next section). Hadrons can be divided into two groups: mesons and baryons. The former are bosons (integral spin particles), and the latter fermions (half-odd-integer spin).

Mesons are shown in the middle of Fig. 14.10. Also listed for convenience are two exchange particles, the γ and g, since both of these are also bosons. The $\pi^{\pm}$ are the antiparticles of each other, and the antiparticle of the π^0 is itself. Hence π mesons include three types of particles. The K^0 and K^+ mesons have their corresponding antiparticles, and hence include four types of particles. The K^0 meson is the only one listed in Table 14.1 that has two different lifetimes, one decaying relatively fast with a mean lifetime of 0.89×10^{-10} s (denoted as K_s^0), the other decaying relatively slowly with a mean lifetime of 5.2×10^{-8} s (denoted as K_L^0). The antiparticle of the η^0 meson is itself. The D meson was not discovered until 1976. It is the first meson discovered with a nonzero charm number (see Table 14.1). The meaning of charm number, baryon number, lepton number, and others will be discussed later. The D meson is similar to the π meson, that is, the $D^{\pm}$ are the antiparticles of each other, and D^0 is its own antiparticle. So the D includes three types of particles. There are nine types of mesons in the graph. By taking into account the antiparticles of the K mesons, the number increases to eleven. With the photon, whose antiparticle is itself, there are 12 types of bosons.

Baryons, which are fermions, are listed to the right in Fig. 14.10. Except for the nucleons—the neutron and proton the others are called hyperons. There are 10 types of baryons in the figure, and these double to 20 particles with their antiparticles included.

Disregarding the ν_τ and g (which are not yet discovered), there are 42 particles and their antiparticles included in Fig. 14.10. The properties of these particles are given in Table 14.1.[1] Some of the symbols given there have been explained above while others will be discussed later.

While particles and their antiparticles have the same mass, the mass difference of other particle pairs is different from pair to pair, as seen in the masses in

[1] For a more detailed table of elementary particles, see Particle Physics Booklet, 2008, available at: PDG@LBL.gov

TABLE 14.1

"Elementary" particles[a]

Young man, if I could remember the names of these particles, I would have been a botanist—Enrico Fermi

Family	Particle	Mass (MeV)	Charge Q	Spin parity J^π	Isospin T	T_Z	Baryon no. B	Lepton no. L_e	L_μ	L_τ	Strangeness S	Charm $\mathscr{C}$	Lifetime (s)	Main decay	Antiparticle
Photon	γ	$0(< 5 \times 10^{-22})$	0	1^-	0,1	0	0	0	0	0			Stable		γ
Leptons	ν_e	$0(< 1 \times 10^{-5})$	0	$\frac{1}{2}$			0	+1	0	0			Stable		$\bar{\nu}_e$
	ν_μ	$0(< 0.52)$	0	$\frac{1}{2}$			0	0	+1	0			Stable		$\bar{\nu}_\mu$
	e^-	0.511	−1	$\frac{1}{2}$			0	+1	0	0			Stable		e^+
	μ^-	105.66	−1	$\frac{1}{2}$			0	0	+1	0			2.2×10^{-6}	$\mu^- \to e^- + \nu_\mu + \bar{\nu}_e$	μ^+
	τ^-	1784.2	−1	$\frac{1}{2}$			0	0	0	+1			3.2×10^{-13}	$\tau^- \to \mu^- + \bar{\nu}_\mu + \nu_\tau$	τ^+
														$\tau^- \to e^- + \bar{\nu}_e + \nu_\tau$	
Mesons	$\pi^\pm$	139.6	±1	0^-	1	±1	0	0	0	0	0	0	2.6×10^{-8}	$\pi^+ \to \mu^+ + \nu_\mu$	$\pi^\mp$
														$\pi^- \to \mu^- + \overline{\nu_\mu}$	
	π^0	135.0	0	0^-		0	0	0	0	0	0	0	0.83×10^{-16}	$\pi^0 \to \gamma + \gamma$	π^0
	η^0	548.8	0	0^-	0	0	0	0	0	0	0	0	2×10^{-19}	$\eta^0 \to 3\pi^0$	η^0
														$\eta^0 \to \gamma + \gamma$	
														$\eta^0 \to \pi^+ + \pi^- + \pi^0$	
	K^+	493.7	+1	0^-	$\frac{1}{2}$	$+\frac{1}{2}$	0	0	0	0	+1	0	1.24×10^{-8}	$K^+ \to \mu^+ + \nu_\mu$	K^-
	K_S^0	497.7	0	0^-		$-\frac{1}{2}$	0	0	0	0	+1	0	0.89×10^{-10}	$K_s^0 \to \pi^+ + \pi^-$	$\bar{K}_s^0$
	K_L^0												5.18×10^{-8}	$K_L^0 \to \pi^0 + \pi^0 + \pi^0$	$\bar{K}_L^0$
	D^+	1869.4	+1	0^-	$\frac{1}{2}$	$+\frac{1}{2}$	0	0	0	0	0	+1	9.1×10^{-13}	$D^+ \to K^- + \pi^+ + \pi^+$	D^-
	D^0	1864.7	0	0^-		$-\frac{1}{2}$	0	0	0	0	0	+1	4.8×10^{-13}	$D^0 \to K^- + \pi^+ + \pi^0$	D^0

TABLE 14.1 (cont.)

Family	Particle	Mass (MeV)	Charge Q	Spin parity I^{π}	Isospin T	Isospin T_Z	Baryon no. B	Lepton no. L_e	Lepton no. L_{μ}	Lepton no. L_{τ}	Strangeness S	Charm $\mathscr{C}$	Lifetime (s)	Main decay	Antiparticle
Baryons	p	938.3	+1	$\frac{1}{2}^+$	$\frac{1}{2}$	$+\frac{1}{2}$	+1	0	0	0	0	0	Stable ($>10^{32}$y)		$\bar{\text{p}}$
	n	939.6	0	$\frac{1}{2}^+$		$-\frac{1}{2}$	+1	0	0	0	0	0	917	$\text{n} \to \text{p} + \text{e}^- + \bar{\nu}_e$	$\bar{\text{n}}$
	Λ^0	1115.6	0	$\frac{1}{2}^+$	0	0	+1	0	0	0	−1	0	2.6×10^{-10}	$\Lambda^0 \to \text{p} + \pi^-$ $\Lambda^0 \to \text{n} + \pi^0$	$\bar{\Lambda}^0$
	Σ^+	1189.4	+1	$\frac{1}{2}^+$	1	+1	+1	0	0	0	−1	0	0.8×10^{-10}	$\Sigma^+ \to \text{p} + \pi^0$ $\Sigma^+ \to \text{n} + \pi^+$	$\bar{\Sigma}^-$
	Σ^0	1192.5	0	$\frac{1}{2}^+$		0	+1	0	0	0	−1	0	5.8×10^{-20}	$\Sigma^0 \to \Lambda^0 + \gamma$	$\bar{\Sigma}^0$
	Σ^-	1197.3	−1	$\frac{1}{2}^+$		−1	+1	0	0	0	−1	0	1.48×10^{-10}	$\Sigma^- \to \text{n} + \pi^-$	$\bar{\Sigma}^+$
	Ξ^0	1314.9	0	$\frac{1}{2}^+$	$\frac{1}{2}$	$+\frac{1}{2}$	+1	0	0	0	−2	0	2.9×10^{-10}	$\Xi^0 \to \Lambda^0 + \pi^0$	$\bar{\Xi}^0$
	Ξ^-	1321.3	−1	$\frac{1}{2}^+$		$-\frac{1}{2}$	+1	0	0	0	−2	0	1.64×10^{-10}	$\Xi^- \to \Lambda^0 + \pi^-$	$\bar{\Xi}^+$
	Ω^-	1672.5	−1	$\frac{3}{2}^+$	0	0	+1	0	0	0	−3	0	0.82×10^{-10}	$\Omega^- \to \Lambda^0 + \text{K}^-$ $\Omega^- \to \Xi^- + \pi^0$ $\Omega^- \to \Xi^0 + \pi^-$	$\bar{\Omega}^+$
	Λ_c^+	2285.1	+1	$\frac{1}{2}^+$	0	0	+1	0	0	0	0	1	1.1×10^{-13}	$\Lambda_c^+ \to \Lambda^0 + \pi^+ + \pi^+ + \pi^-$ $\Lambda_c^+ \to \text{p} + \text{K}^- + \pi^+$	$\bar{\Lambda}_c^-$

[a]Note the latest notation for antiparticles is to take the charge outside the bar over the particle, so the antiparticle of the Σ^+, earlier written as $\overline{\Sigma^+}$ with the bar over the particle and its charge, is now written $\bar{\Sigma}^-$ with the charge changed and outside the bar (see G. P. Yost et al., *Phys. Lett.* **204** (1988) 1 or *Phys. Rev.* **D50** (1994) 1173).

Fig. 14.10. The masses of the proton and neutron are only slightly different. This may be ascribed to the difference in their electromagnetic properties. As indicated in Chapter 9, the proton and neutron can be treated as different states of one particle, the nucleon. In analogy to the fact that a particle of spin $\frac{1}{2}\hbar$ has two states (two components on the z axis), the concept of isospin was introduced. One defines the isospin of a nucleon as $T = \frac{1}{2}$. The nucleon has $2T + 1 = 2$ components, a proton with spin up ($T_z = \frac{1}{2}$) and a neutron with spin down ($T_z = -\frac{1}{2}$). (Earlier, the definition of T_z was just the opposite in nuclear physics where a neutron was assigned $T_z = \frac{1}{2}$, and a proton $T_z = -\frac{1}{2}$. Now nuclear physicists use the same sign convention as high-energy physicists, where it is based on electric charge.) In particle physics the Z component of a multiplet member is given by $T_z = Q - \bar{Q}$, where Q is the particle charge in units of electron charge, and $\bar{Q}$ is the average charge of the multiplet computed by adding all the charges and dividing by the number of particles. An alternative definition is $Q = T_z + (B + S)/2$, where B is baryon number and S the strangeness quantum number. Isospin is not a real spin but a way of mathematically counting particle states. Each multiplet has an isospin quantum number, T, which characterizes a vector $\boldsymbol{T}$ in isospin space with the proper T_z components. The isospin states of a particle with $T = 1$, which has $2T + 1 = 3$ components, are $T_z = 1, 0, -1$; an example is the Σ^+, Σ^0, and Σ^- triplet ($T_z(\Sigma^+) = 1 - 0/3 = 1$). The slight mass differences between them also originate from the differences in their electromagnetic properties. From Fig. 14.10 one sees that the Ξ particle is an isospin doublet ($T = \frac{1}{2}$), the Λ^0, Ω^-, and Λ_c^+ are isospin singlets ($T = 0$), the π meson and D meson are isospin triplets ($T = 1$), the K meson has $T = \frac{1}{2}$, and the η meson has $T = 0$. Conservation of T_z in decay and reaction processes is equivalent to conservation of electric charge and so is nothing new. Conservation of T_z may be violated only in the weak interaction. In decays and reactions, T is conserved in strong interaction processes with rare exceptions. This is equivalent to saying that the energies from the interaction (force) between nucleons depends only on T and not T_z. Here we learn something new about the nucleon interaction. Conservation of T may be violated in electromagnetic and weak processes (see Table 14.3).

The electron mass (0.511 MeV) can be completely ascribed to its electromagnetic self-energy. Mass differences caused by different electromagnetic properties are generally of the order of 1 MeV, while those caused by strong interactions are of the order of 100 MeV (recall how the mass of a π meson was predicted). In Fig. 14.10, the masses of the strong interacting particles go from 135 MeV (minimum) to 2282 MeV (maximum). The muon is a very strange particle in this respect, since it has a mass close to that of the π meson, but does not participate in the strong interaction. What is the origin of its mass? These problems are among the open questions in contemporary high-energy physics. The LHC and Tevatron both seek to discover the predicted Higgs particle which is important to understanding masses as discussed elsewhere.

14.2.2 Interactions

Listed above are 42 types of particles with lifetimes greater than 10^{-20} s. Among them, there might exist $42 \times 41/2! = 861$ pairs of interactions. The number of pairs will even reach 100 000 if we take the particles of shorter lifetimes into account. One of the wonders of nature and one of the great achievements of physics is that all of these pairs of interactions may be classified into only three distinct types: the gravitational, electroweak (with the unification of the electromagnetic and weak), and strong interactions. However, we will discuss the electromagnetic and weak interactions separately. All physical phenomena can be described in terms of one or the other of these interactions. Comparisons of these are shown in Table 14.2.

1. The response times to these different-strength interactions are different. There is a characteristic time for each interaction. Hadrons with lifetimes less than 10^{-20} s, which are not listed in Table 14.1, decay in terms of the strong interaction; for example,

$$\rho^+ \to \pi^+ + \pi^0 \qquad \text{with} \qquad \tau = 4 \times 10^{-23}\ \text{s}$$

2. Not only charged particles, but also neutral ones, such as the π^0, η^0 and Σ^0, may decay in terms of the electromagnetic interaction via intermediate charged states of their quark constituents:

$$\pi^0 \to \gamma + \gamma$$
$$\eta^0 \to \gamma + \gamma$$
$$\Sigma^0 \to \Lambda^0 + \gamma$$

A photon participates uniquely in the electromagnetic interaction so that a process is electromagnetic in nature at the appearance of photons. The characteristic times of the π^0, η^0, and Σ^0 decays are 0.8×10^{-16}, 8×10^{-19}, and 5.8×10^{-20} s, respectively, which are within the range of the electromagnetic interaction.

3. Neutrinos participate only in the weak interaction. For example,

$$\pi^+ \to \mu^+ + \nu_\mu, \qquad \pi^- \to \mu^- + \bar{\nu}_\mu$$

with the characteristic time of 2.60×10^{-8} s, which is in the range of the weak interaction ($\pi^\pm$ are the antiparticle of each other, so they have the same lifetime). Also,

$$K^+ \to \mu^+ + \nu_\mu$$

TABLE 14.2

The comparison of the basic interactions (basic forces) observed in nature (courtesy of *Contemporary Physics Education Project*, Lawrence Berkeley National Laboratory)

	Interaction				
	Gravitational	Weak	Electromagnetic	Strong	
Property		(Electroweak)		Fundamental	Residual
Acts on	Mass–energy	Flavor	Electric charge	Color charge	[a]
Particles experiencing	All	Quarks, leptons	Electrically charged	Quarks, gluons	Hadrons
Particle mediating	Graviton (not yet observed)	W^+ W^- Z^0	γ	Gluons	Mesons
Strength (relative to electromagnetic) for two u quarks at:					
10^{-18} m	10^{-41}	0.8	1	25	Not applicable to quarks
3×10^{-17} m	10^{-41}	10^{-4}	1	60	
for two protons in nucleus	10^{-36}	10^{-7}	1	Not applicable to hadrons	20

[a] The strong binding of the color-neutral protons and neutrons to form nuclei is due to residual strong interactions between their color-charged constituents (quarks). It can be viewed as an exchange of mesons between hadrons.

where the appearance of the ν_μ indicates that this is a process caused by the weak interaction. Indeed, the lifetime is about 10^{-8} s. But, a K meson can also participate in the following processes:

$$\pi^- + p \rightarrow K^0 + \Lambda^0$$

$$K^+ + p \rightarrow \Sigma^+ + \pi^+$$

with reaction times of 10^{-23} s, which are typical of a strong interaction process. This means that a K meson may be involved in either the strong interaction in a reaction or in the weak interaction in a decay. Nucleons have similar properties: they are involved in the strong interaction in the nucleus and nuclear reactions, but in the weak interaction during beta decays.

One of the greatest achievements of physics in the nineteenth century was the unification of electricity and magnetism. The contributions to modern culture of this unification, for example, in electric power generation and electric motors, transformed society. In recent years, starting from experimental facts, a unified theory of the weak and electromagnetic interactions was postulated and proven by a series of new experiments. Among them, the discovery of the intermediate bosons, the $W^\pm$ and Z^0, in 1983, as predicted by the unified theory, is considered an epochal event. The $W^\pm$ and Z^0 particles were discovered at the proton–antiproton collider of 270 + 270 GeV at CERN.[2] The $W^\pm$ mediates the charge-changing weak interactions like the known β^- neutron decay or β^+ decay of a proton in a nucleus. On the other hand, the Z^0 mediates neutral current (non-charge-changing) weak interaction processes and was not expected since no neutral current processes were known. In 1973, such a neutral current process was found at CERN in a triumph of the electroweak theory. The theoretical predictions of the masses of the $W^\pm$ and Z^0 are, respectively,

$$m_{W^\pm} = 83.375 \pm 0.015 \text{ GeV} \quad \text{and} \quad m_{Z^0} = 91.1874 \pm 0.0021 \text{ GeV}$$

Which are in excellent agreement with the 2008 experimental results (see ref. 1),

$$m_{W^\pm} = 80.398 \pm 0.025 \text{ GeV} \quad \text{and} \quad m_{Z^0} = 91.1876 \pm 0.0021 \text{ GeV}$$

another great triumph of the electroweak theory.

Note in Table 14.2, that the weak interaction and electromagnetic interaction are essentially of the same strength at a distance of 10^{-18} m. The weak charge, g_W, which expresses the strength of the weak interaction, is comparable to the electric charge that expresses the strength of the electromagnetic interaction. The weak interaction is not intrinsically weak. It is the very large masses of the $W^\pm$ and Z^0 that make the range of the interaction very short so that it occurs rarely to make the weak interaction appear weak. In Fig. 14.11 is shown the decay of the neutron in terms of its quark structrure and W^-. By the uncertainty principle, a virtual particle of 80 GeV cannot go far. For $\Delta E \Delta t = \hbar$, 80 GeV $\Delta t = \hbar$ leads to $\Delta t \sim 10^{-26}$ s. At a velocity c, it reaches only to $\sim 10^{-18}$ m. The width of the Z is used to set a limit on the number of neutrino families.

[2]G. Arnison, et al., *Phys. Lett.* **122B** (1983) 95; **126B** (1983) 398; *CERN Courier* **23** (1983) 355. The latest masses are found in *Phys. Rev.* **D50** (1994) 1191.

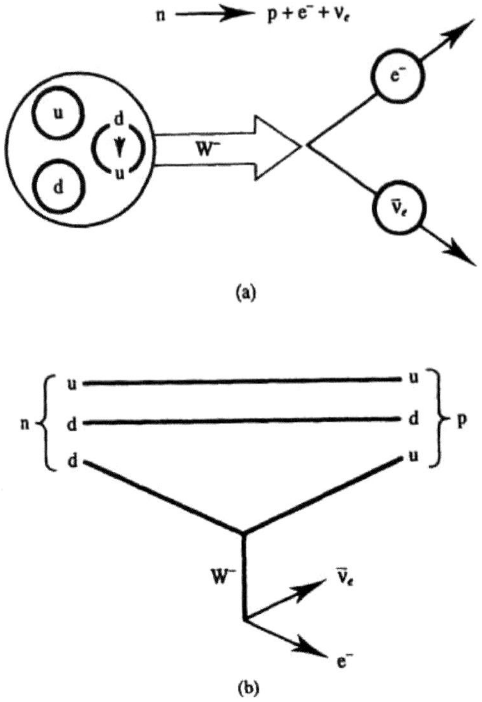

FIGURE 14.11
(a) Neutron decay via a virtual W^- boson into a proton, electron, and neutrino. Inside the neutron on the left, one down quark changes into an up quark to turn the neutron into a proton via a virtual W^- that mediates the weak interaction decay into the e^- and $\bar{\nu}_e$. (b) Neutron decay in terms of the quark structure.

One of the major challenges in high-energy physics is the possible unification of the strong, weak, and electromagnetic interactions simultaneously. Some theories have predicted that the strengths of these interactions are almost the same magnitude at an energy of 10^{14} GeV. An even greater goal in Grand Unified Theories is to include gravity with them.

14.2.3 Resonances

The first resonance was discovered by Fermi and co-workers in 1952 in the π^+ meson bombardment of protons. The experimental excitation curve (collision cross-section versus incident energy) of $\pi^+ + \mathrm{p}$ is shown as the solid curve in Fig. 14.12, while the dashed curve represents that for a π^- as incident particle. The peaks which appear on both curves are called resonances. Such measurements may be called the Frank and Hertz experiments of high-energy physics. Not only $\pi^+ + \mathrm{p}$ and $\pi^- + \mathrm{p}$, but also the bombardments of π^+, π^0, or π^- on the neutron produce such resonances. If these resonances are related to p or n, resonances related to the π mesons might be produced as well.

We shall take the first $\pi^+ + \mathrm{p}$ resonance as an example. The center-of-mass energy of this resonance is 1232 MeV, so it is denoted as $\Delta(1232)$, where $1232\,\mathrm{MeV}/c^2$ is the mass of the resonance. The spin of this resonance is measured to be $\frac{3}{2}$, its baryon number is 1, and $T = \frac{3}{2}$. However, its mass is not definite, but $1232 \pm 60\,\mathrm{MeV}/c^2$. Its full width at half maximum is about 115 MeV, from which

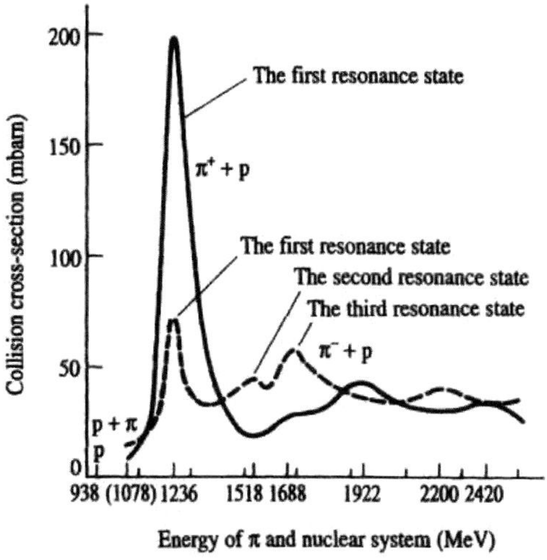

FIGURE 14.12
Resonances seen in the collision cross-section between a π^+ meson and proton and a π^- meson and proton.

a lifetime of 6×10^{-24} s can be calculated. Such a lifetime is typical of a resonance. Traveling at the velocity of light, the distance such an object can travel in 10^{-23} s is the order of 10^{-15} m. This distance is smaller than a nucleus and is about the diameter of a proton! (The relativistic time dilation will stretch out the lifetime and permit it to travel further, but this is only a factor of 10–100 at the most.) The electric charge of the $\Delta(1232)$ can be 2, 1, 0, −1.

Is a resonance a particle? In the Frank and Hertz experiment in atomic physics, the electron bombardment of mercury atoms yielded a series of lines, resonances, which were the excited states of mercury atoms (compare Fig. 3.18 and Fig. 14.12). The energy difference of these mercury excited states from the ground state is of the order of 10 eV, which is negligible compared to the rest energy of a mercury atom, so these resonances would not be mistaken for new types of mercury atoms. In nuclear physics, protons can be used to bombard and to excite a target nucleus into its excited states. For example, excited states of a compound nucleus can be observed by proton bombardment:

$$p + {}^{27}\text{Al} \rightarrow ({}^{28}\text{Si})^* \rightarrow {}^{28}\text{Si} + \gamma$$

where $({}^{28}\text{Si})^*$ are excited states of a known nucleus, ^{28}Si. However, the $\Delta(1232)^{++}$ (+ + means it has two positive charges, twice the charge of the proton) resonance produced in the reaction

$$\pi^+ + p \rightarrow \Delta(1232)^{++} + \pi^0$$

had never been seen before. Its mass is greater than the sum of the masses of the π and p! Such a system is called a resonance particle or simply a resonance. Because of its very short lifetime (and distance traveled before decay), one only sees such resonance particles indirectly from their decay modes. Many Δ resonances going up to 3200 MeV/c^2 and spin $\frac{11}{2}$ have been observed to date.

There are also similar states for mesons such as the ρ as observed in the following reaction:

$$\pi^- + \text{p} \rightarrow \rho^0 + \text{n} \rightarrow \pi^+ + \pi^- + \text{n}$$

The ρ is a member of an important class called vector mesons with spin 1. The ρ rest mass is $770 \pm 77\,\text{MeV}/c^2$. Its full width is similar to that of $\Delta(1232)$, so its lifetime is also $\sim 10^{-23}$ s.

More than 300 different baryon and meson resonances like those discussed above have been discovered. If such resonances were included in the table of elementary particles, the total number of particles would exceed 400! In ancient Greek thought, all matter was composed of four types of elements: water, fire, earth, and air. The ancient Greeks placed a high value on this idea because of its simplicity and elegance in describing nature. Today physicists continue to place high value on simplicity and elegance in formulating an understanding of nature. We seek to describe things, no matter how complicated they are, within a simple picture or framework. Obviously 400 "elementary" particles would not be expected to be elementary. Thus, a deeper substructure was sought to explain the particles.

14.3 CONSERVATION RULES

14.3.1 Baryon Number, Lepton Number, and Isospin Conservation

All the known conservation laws, conservations of mass and energy, angular momentum, linear momentum, and charge, remain valid in high-energy physics. No violations of these conservation laws have been found in any events. However, some new conservation rules have been discovered.

Let us look at the process

$$\text{p} \rightarrow \text{e}^+ + \gamma \tag{14.1}$$

It has never been observed to occur, although it violates none of the above known conservation laws. Something must be preventing it from occurring.

Let us introduce the baryon number B, such that all baryons are assigned $B = 1$ and antibaryons $B = -1$. Meson and leptons have $B = 0$ (see Table 14.1). In all observed processes, the algebraic sum of the baryon numbers on each side of an arrow is conserved, even when the number of particles is not conserved. For example, in the cases

$$\begin{aligned} \Lambda^0 &\rightarrow \text{p} + \pi^- \\ \Lambda^0 &\rightarrow \text{n} + \pi^0 \end{aligned} \tag{14.2}$$

which are observed experimentally, the baryon number is conserved. The baryon number is not conserved in the process of Eq. (14.1), so it can be understood as violating the baryon conservation law. The creation of a quark-gluon plasma also violates this law.

The processes

$$\begin{aligned} \nu_\mu + \mathrm{n} &\rightarrow \mathrm{e}^- + \mathrm{p} \\ \bar{\nu}_\mu + \mathrm{p} &\rightarrow \mathrm{e}^+ + \mathrm{n} \end{aligned} \tag{14.3}$$

have never been observed, although they too violate no known conservation laws, including the new conservation of baryon number. In order to understand why these and other reactions including leptons do not occur, two types of lepton numbers, L_e and L_μ, are introduced,

$$L_e = 1, \text{ for } \mathrm{e}^-, \nu_\mathrm{e}; \qquad L_e = -1, \text{ for } \mathrm{e}^+, \bar{\nu}_e$$

$$\mathrm{L}_\mu = 1, \text{ for } \mu^-, \nu_\mu; \qquad L_\mu = -1, \text{ for } \mu^+, \bar{\nu}_\mu$$

along with conservation laws that the algebraic sum of L_e and L_μ on each side must be conserved independently in any process. Each of the processes in Eq. (14.3) violates the conservation of both L_e and L_μ lepton numbers, so that they cannot happen. Looking again at Eq. (14.1), we see the process violates both the conservation of baryon number and that of lepton number. Similarly, while the mass of the τ lepton allows

$$\tau^+ \rightarrow \mathrm{p} + \gamma$$

conservation of baryon and lepton number conservation does not. Instead, one finds (remember for leptons that negatively charged particles (e^-, μ^-, τ^-) are the particles and the positive charged ones the antiparticles)

$$\begin{aligned} \tau^+ &\rightarrow \mathrm{e}^+ + \nu_\mathrm{e} + \bar{\nu}_\tau \\ \tau^+ &\rightarrow \mu^+ + \nu_\mu + \bar{\nu}_\tau \\ \tau^+ &\rightarrow \pi^+ + \bar{\nu}_\tau \end{aligned}$$

The processes

$$\begin{aligned} \nu_\mu + \mathrm{n} &\rightarrow \mu^- + \mathrm{p} \\ \bar{\nu}_\mu + \mathrm{p} &\rightarrow \mu^+ + \mathrm{n} \end{aligned} \tag{14.4}$$

are observed, in which B, L_e, and L_μ are all conserved. No violations of the conservation of baryon number and lepton number have been observed in any reactions or decays. However, as discussed elsewhere, neutrino oscillation violates lepton number conservation. By analogy, one would assume that there is also a similar lepton conservation number L_τ for the τ lepton.

There are also similar conservation laws for isospin and the z component of isospin. Both are conserved in the strong with rare exceptions and the latter in electromagnetic interactions but both can be violated in the weak interaction with ΔT and $\Delta T_Z = \frac{1}{2}$ for nonleptonic processes.

14.3.2 Meson Number Conservation

There are no conservation laws for numbers of mesons of any type. Mesons are bosons and the Pauli principle does not apply to bosons, so they can be produced in any number in nuclear and particle reactions as long as energy, momentum, and electric charge are conserved. So

$$p + p \to p + p + \pi^+ + \pi^- + \pi^0 + \pi^0$$

can occur, as well as with additional π mesons, if there is sufficient energy and if all other conservation laws are obeyed.

14.3.3 Strangeness Conservation

The K^0 and Λ^0 were both new particles which were discovered in cosmic rays simultaneously with the discovery of the π meson in 1947. What was observed first were new vee-shaped tracks where $\pi^- + p$ (strong interaction) $\to V^0 \to \pi^- + p$ (weak interaction). The lifetime for strong interactions is the order of 10^{-23} s but the V^0 had a lifetime of 10^{-10} s so it decays via the weak interaction. The fact that the decay rate of the V^0 (now called Λ^0) is 10^{-13} of the production rate was why the particles whose decay gave the vee-shaped tracks were called "strange" particles.

Several years later, the process

$$\pi^- + p \to K^0 + \Lambda^0 \tag{14.5}$$

was demonstrated in an accelerator laboratory, as shown in Fig. 14.13. Note only the Λ^0 gives a vee-shaped track in the picture. The K and Λ can be produced in different ways, for example

$$p + p \to \Lambda^0 + K^0 + p + \pi^+ \tag{14.6}$$

In all cases these two particles are observed to be created simultaneously, so their production is called "associated production." Note that baryon number is conserved in this reaction but not meson number. However, there is no restriction on their decay mode by the weak interaction. The processes

$$\begin{aligned} K^0 &\to \pi^+ + \pi^- \\ \Lambda^0 &\to \pi^- + p \end{aligned} \tag{14.7}$$

belong to the weak interaction regime and so have a rather long lifetime characteristic of the weak interaction. Each particle can decay separately. So while they must be produced by the strong interaction, they may decay by the weak interaction. This is very strange because the production and decay of particles should be closely correlated according to quantum mechanics. The associated production and separate decay of the K^0 and Λ^0 led Nishijima and Gell-Mann to introduce a new quantum number, strangeness, in 1953.

Nishijima and Gell-Mann independently proposed that in addition to the mass, charge, spin, isospin, baryon number, and lepton numbers, an elementary particle may have a new quantum number S called strangeness. This quantum number must be conserved in a strong interaction, but strangeness conservation could be violated in processes governed by the weak interaction.

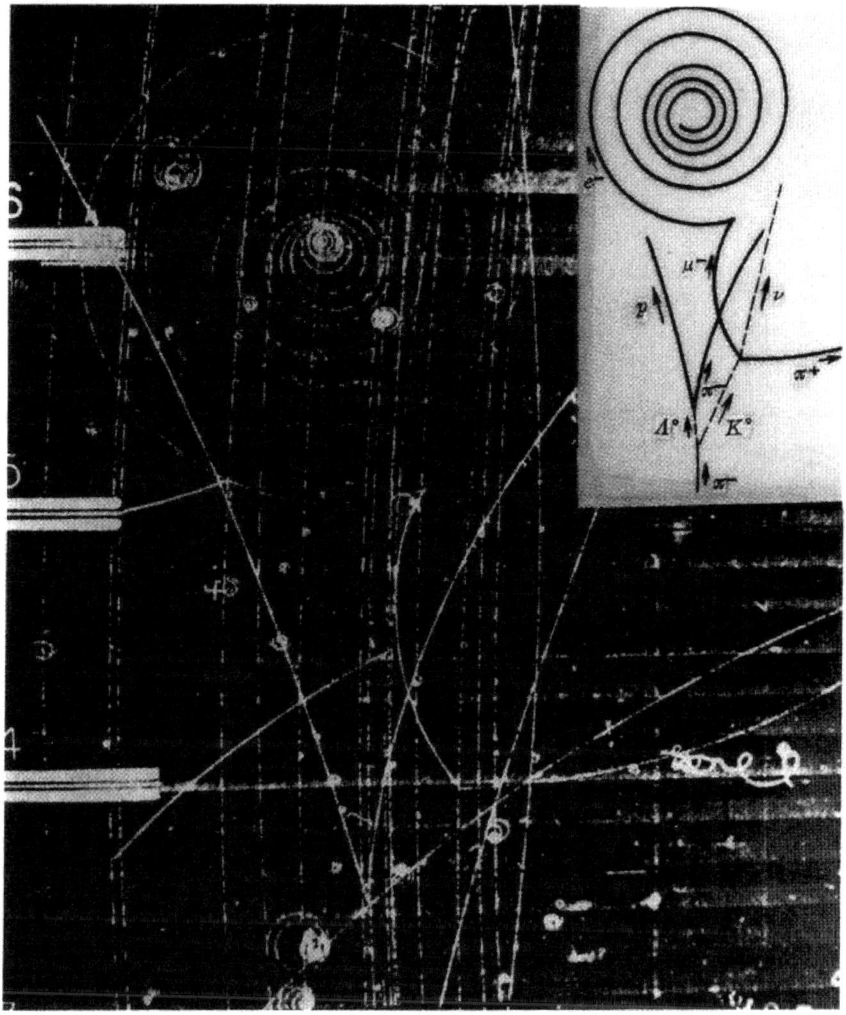

FIGURE 14.13
The production and decay of the K^0 and Λ^0 in a laboratory experiment of a π^- beam striking protons. An interpretation of the tracks in the photograph is given in the upper right-hand corner. The π^- track stops with no track indicating particles coming out. Then at some distance away (following the decay of the neutral Λ^0 and K^0 which produce no tracks by ionization) a vee-shaped track and two other tracks appear.

All the particles discovered previously were assigned a strangeness $S = 0$, since they had exhibited no evidence for it. However, the K^0 was assigned $S = 1$, and the Λ^0 was assigned $S = -1$. So, to conserve strangeness in their production in a strong interaction, they must be produced in pairs. For example,

$$\begin{aligned} &\pi^- + p \to K^0 + \Lambda^0 \\ S:\ &0 + 0 = 1 - 1 \end{aligned} \tag{14.8}$$

$$\begin{aligned} &p + p \to K^0 + \Lambda^0 + p + \pi^+ \\ S:\ &0 + 0 = 1 - 1 + 0 + 0 \end{aligned} \tag{14.9}$$

But in the decay process,

$$\begin{aligned} &K^0 \to \pi^+ + \pi^- \\ S:\ &1 \neq 0 + 0 \end{aligned} \tag{14.10}$$

strangeness is not conserved. Hence, it must be a weak interaction process with a long decay timescale, and it is.

In the process

$$p + \pi^- \to n + K^+ + K^- \tag{14.11}$$

the conservation rule for strangeness in a strong interaction led to the assigning of strangeness for K^+ of $S(K^+) = +1$, and the K^- was assigned $S(K^-) = -1$. Similarly, the process

$$p + \pi^- \to \Sigma^- + K^+ \tag{14.12}$$

led to $S(\Sigma^-) = -1$. And

$$p + K^- \to \Sigma^+ + \pi^- \tag{14.13}$$

led to $S(\Sigma^+) = -1$. Since the Σ^- and Σ^+ have the same S, and also the same baryon number B, as can be deduced from the conservation of baryon number, they cannot be the antiparticles of each other. From Table 14.1 one could find the S value for each particle. In summary, strangeness is a new quantum number whose conservation holds in the strong interaction but is violated in the weak interaction where $\Delta S = 1$ is allowed. Note that the introduction of the strangeness quantum number does not tell us about some fundamental property of a particle, like mass and spin do. Assigning strangeness is a way of keeping track of what production and decay modes are allowed. As we shall see, strangeness is associated with one of the six quarks. We will consider associated production in terms of the quark model in Section 14.4.

The four quantum numbers charge Q, isospin T_z, strangeness S, and baryon number B are found not to be independent but related by the expression

$$Q = T_z + \frac{S + B}{2} \tag{14.14}$$

Hence, the fourth quantum number can be obtained from the other three.

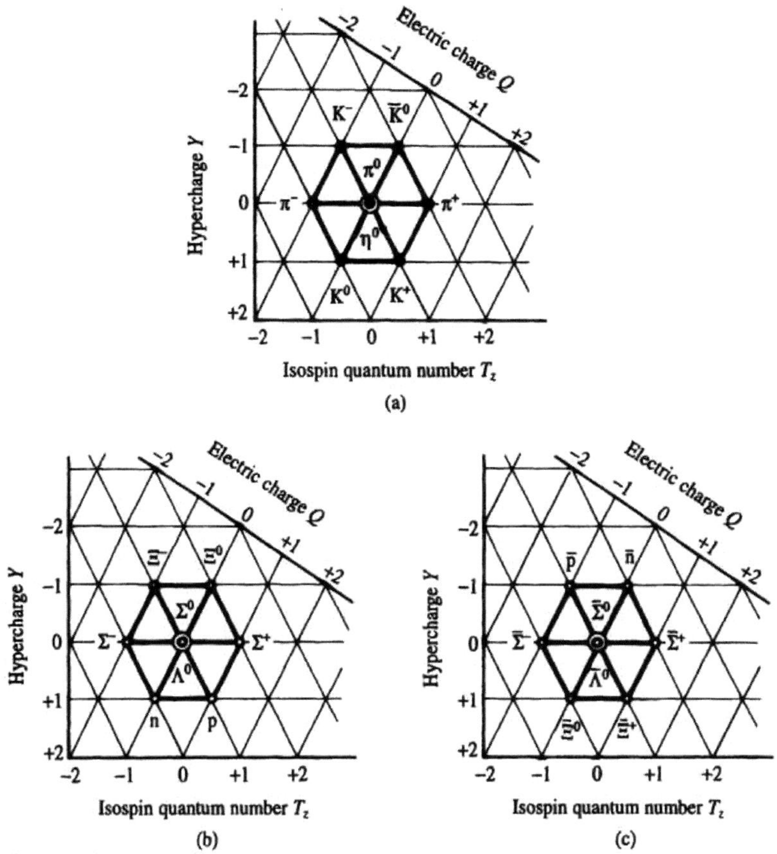

FIGURE 14.14
The symmetry of hadrons.

If a new quantum number Y, called hypercharge, is defined as

$$Y = B + S \tag{14.15}$$

then we have

$$Q = T_z + \frac{Y}{2} \tag{14.16}$$

Shown in Fig. 14.14 are T_Z, Y, Q plots for mesons, baryons, and anti-baryons, respectively. The theoretical explanation for the symmetrical patterns shown in the plots has been worked out already.

14.3.4 Violation of Parity and Time Reversal and CPT Conservation

The parity of a microscopic particle is a physical quantity that characterizes its motion. If a wavefunction $\psi(x)$ is used to describe the state of motion of a microscopic particle, then under the inversion of coordinates through the origin we will have

$$\psi(\boldsymbol{x}) \to \psi(-\boldsymbol{x}) \tag{14.17}$$

This process, called the parity operation, can be expressed by using a parity operator P:

$$P\psi(\boldsymbol{x}) = \psi(-\boldsymbol{x}) \tag{14.18}$$

Clearly, another parity operation will return the wavefunction to the original wavefunction:

$$P^2\psi(\boldsymbol{x}) = \psi(\boldsymbol{x}) \tag{14.19}$$

Hence, P has two eigenvalues only:

$$P = \pm 1 \tag{14.20}$$

The parity of ψ must be even or odd. States belonging to the eigenvalue $+1$, that is, $\psi(\boldsymbol{x}) = \psi(-\boldsymbol{x})$, are referred to as even parity states and those belonging to the value -1, $\psi(\boldsymbol{x}) = -\psi(-\boldsymbol{x})$, are odd parity states.

The parity principle claims that the parity of a wavefunction is a constant of motion for an isolated system. So a wavefunction of definite parity (even or odd) keeps that parity (even or odd) for all time. Physically, the parity principle can be stated in the following form: any process which occurs in nature can also occur as it is seen reflected in a mirror. In other words, nature is mirror-symmetric. The mirror image of any object is also a possible object in nature. The motion of any object as seen in a mirror is also a motion which would be permitted by the laws of nature.

The parity principle was found to be correct in many experiments after the introduction of the concept of parity. Parity conservation was considered to be a fundamental principle of nature. Then came the τ–θ paradox. It arose because the so-called τ and θ particles had all the same properties except for their decay modes:

$$\tau \to \pi + \pi + \pi \qquad \text{but} \qquad \theta \to \pi + \pi$$

Since the intrinsic parity of a meson is odd (see Table 14.1), the above 3π and 2π systems would have different parities. If parity were conserved in both processes, then the τ and θ must be different particles. But how could these two different particles have exactly the same mass, lifetime, and other properties? Physicists were deeply puzzled about the contradiction. In 1956 T. D. Lee and C. N. Yang in seeking to resolve the "τ–θ paradox" suggested that the weak interactions did not conserve parity, so that then the τ and θ could be the same particle. This was an astonishing proposal—a flat contradiction to the plausible existing knowledge!

At the suggestion of Lee and Yang, C. S. Wu and co-workers employed a sample of polarized ^{60}Co and measured the angular distribution of the electrons (β rays) emitted. First let us analyze the decay of ^{60}Co in view of the parity principle. Note that $^{60}_{27}$Co is an odd–odd nucleus, with a spin of $I = 5\hbar$, and decays in the following mode ($T_{1/2} = 5.3$ years):

$$^{60}\text{Co} \rightarrow {}^{60}\text{Ni} + e^- + \bar{\nu}_e$$

What will the angular distribution of the electrons emitted be? For an unpolarized ^{60}Co source, which is a collection of ^{60}Co nuclei with their spin directions randomly distributed, the observed angular distribution of the electrons would be isotropic, no matter whether the angular distribution of the electrons emitted from ^{60}Co is isotropic or not. Now what happens when a polarized instead of an unpolarized ^{60}Co source is used? The ^{60}Co nuclei can be polarized (nuclear spins all aligned in the same direction) by holding them at a low temperature of 0.01K inside a solenoid which produces an external magnetic field of several hundred Gauss to align the spins—a low-temperature nuclear orientation device. At this temperature a high proportion of ^{60}Co nuclei whose spins are aligned along the magnetic field direction are frozen in that orientation. What would the angular distribution of the electrons emitted from these polarized nuclei be?

Suppose that the directions of the emitted electrons concentrated more or less in the spin direction of the nucleus (see Fig. 14.15(a)). Shown in Figs. 14.15(b) and (c) are its images reflected in mirrors parallel and normal to the nuclear spin, respectively. Note that the direction of the nuclear spin and its rotation are defined according to the right-hand-screw rule so that the directions of the emitted electrons change (do not change), while the direction of the nuclear spin remains unchanged (changed) in Figs. 14.15(c), (b), respectively, as compared to Fig. 14.15(a). In each case, the emission of the electrons turns out to be concentrated in the opposite direction to the nuclear spin in a reflecting mirror. If there had been no violation of parity conservation, both cases would have been

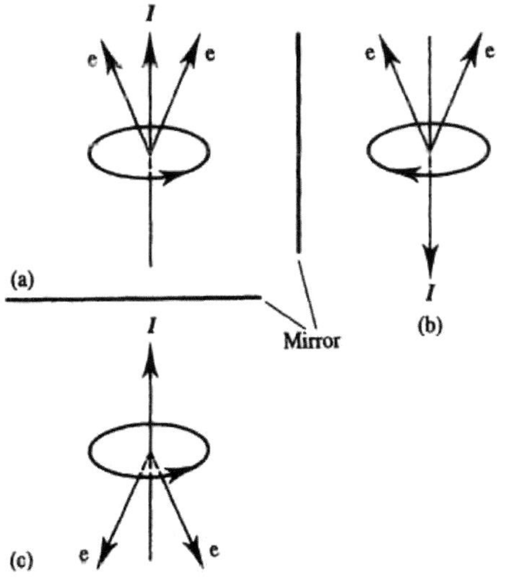

FIGURE 14.15
Image of nonisotropic emission of electrons in β-decay. (b) and (c) are the images of (a) reflected in mirrors parallel and normal to the nuclear spin, respectively.

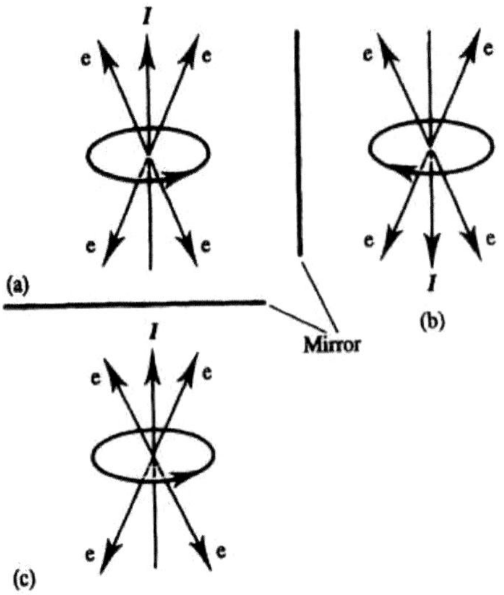

FIGURE 14.16
Images of isotropic emission of electrons in β-decay.

realistic. Therefore, it is a demand of the parity principle that the angular distribution of the emitted electrons must be isotropic, as shown in Fig. 14.16. The principle of parity conservation was considered so fundamental that, in general, it was expected that the experimental test of Lee and Yang's proposal of nonconservation in the weak interaction would prove their idea wrong. As a graduate student one of us (J.H.H.) remembers E. K. Konopinski reading at a seminar his letter from the foremost theoretical authority of the day, Pauli, who soundly proclaimed that the experiments would prove the ideas of parity nonconservation wrong. Pauli wrote, "I do not believe in a left-handed God."

In the ^{60}Co experiment,[3,4] the relative intensity of electrons with an emission angle (which is the angle between emitted electrons and the nuclear spin) greater than 90° was found to be 40% higher than that with the emission angle less than 90°. In other words, the event shown in Figs. 14.15(b) or (c) was preferred by nature and not Fig. 14.16. Thus, parity nonconservation in the weak interaction was then proven! The failure of such an important, long-held principle was truly shocking. Moreover, there is no precedent in the history of physics where the failure of a well-established principle showed up with such large effects in the first experiment. Usually the first doubts are based on small deviations which hardly exceed the limits of error, and only after the passing of time and the application of great effort by many people is the effect substantiated. Figure 14.17 is a photograph of C. S. Wu and W. Pauli.

Violations of the parity principle in the weak interaction were confirmed quickly in many experiments. What was happening? The reason for the violation

[3]For the original paper, see C. S. Wu, et al., *Phys. Rev.* **105** (1957) 1413.

[4]For a review article, see C. S. Wu, *Alpha-, Beta, and Gamma-Ray Spectroscopy*, K. Siegbahn, ed., North-Holland Co. (1965), p. 1415.

FIGURE 14.17
Photograph of W. Pauli and C. S. Wu. (Courtesy E. Segre, Berkeley, CA.)

lies in the fact that the neutrino is a particle of maximum parity violation, as proposed independently by Yang and Lee, by Salam, and by Landau in 1957. The spin of a neutrino points opposite to its direction of motion for all time, the left-handed neutrino as it is called. On the contrary, the antineutrino, the spin of which points in its direction of motion, is right-handed. The mirror image of a left-handed neutrino is a right-handed neutrino, while the image of a right-handed antineutrino is a left-handed antineutrino. Neither image exists in nature, that is, neither the right-handed neutrino nor the left-handed antineutrino ever exists. So maximum parity violation happens in the case of neutrinos. This type of weak interaction is called the two-component neutrino theory.

The decay of ^{60}Co (spin $I = 5\hbar$) leads to an excited state of ^{60}Ni with $I = 4\hbar$, where part of the ^{60}Co spin angular momentum is taken away by the $\bar{\nu}$ and e^-. Hence, the latter two particles must have the same directions of spin angular momentum as the nucleus of ^{60}Co. Note that the $\bar{\nu}$ is right-handed—the spin of $\bar{\nu}$ coincides with its direction of motion—while the e^- has a preferential direction of emission (movement) opposite to the spin of ^{60}Co, so that the e^- can only be left-handed (see Fig. 14.18). This was demonstrated in subsequent experiments.[5]

Since all the particles and their antiparticles have just the opposite helicity, and similarly in a β^+ decay, the spin of the parent and daughter nuclei and of e^+

[5]The experimental design for testing whether the electrons emitted from ^{60}Co are left-handed or right-handed is very interesting. See H. Frauenfelder, et al., *Phys. Rev.* **106** (1957) 386.

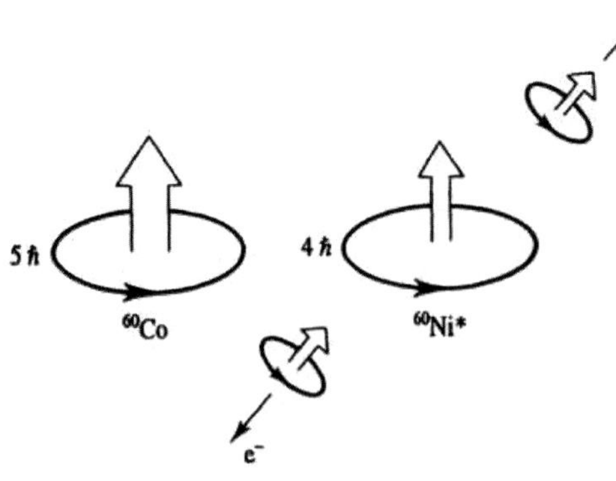

FIGURE 14.18
The spin directions and motion directions in a β-decay of ^{60}Co.

and ν line up in the same direction, the preferential directions of movement of the e^+ and ν must be opposite to those appearing in a β^- decay. Indeed, in the decay of the ^{58}Co nucleus, the e^+ was measured to be emitted preferentially in parallel to the spin of ^{58}Co; that is, many more positrons (e^+) with the emission angle less than 90° were found than those with the angle greater than 90° relative to the nuclear spin direction.

Since the helicity of a neutrino (or antineutrino) is fixed (longitudinal polarization), its spatial inversion does not exist, and hence the concept of parity has no meaning for a neutrino. Particles of this kind have to travel at the speed of light. Otherwise, the helicity of the particle would be seen to change at the moment one passed it, if one was traveling in a spaceship with a speed greater than that of the particle. Stated another way, if v_ν = constant, but v_R for a rocket is changing, then you see a flip of the helicity when $v_\nu = v_R$.

This character of the neutrino implies that the conservation laws may be preserved if there exists a mirror which acts not only as a spatial inversion operation (P operation) but also to change a particle into its antiparticle, called the charge conjugation operation, or C operation. For example, under the CP joint operation, a left-handed neutrino becomes a right-handed antineutrino which does exist in nature. Some people were unhappy about the violation of P symmetry until they found that the CP symmetry was still preserved. They expected that the CP invariance might be used as a criterion in a broader sense.

However, in 1964, CP violation was found in the K meson. It is a very strange particle! This time it was the decay of the K^0 meson which exhibits a CP violating decay at a level of 0.16%. More recently violation of CP invariance has been seen in b quark baryons at a larger level of violation. Through the CPT theorem (according to relativistic quantum field theory the product CPT is always conserved) a CP violation indicates a T (time inversion operation) violation also occurs. Time reversal invariance means that physical processes are independent of whether the process goes forward or backward in time. The T operator reverses the direction of time. So, the analysis of the K^0 decay and b quark establish T violation, too. No evidence of the violation of CP or T have been found in any other systems. How strange that nature, through these decay modes, can distinguish the direction of the flow of time on this microscopic level. Time reversal invariance is being investigated in all interactions.

Within the Standard Model of particle physics, CP symmetry is broken by complex phases in the Yukawa couplings (that is, the couplings of the Higgs scalar to quarks). The consistent Standard Model description of CP violation in terms of Yukawa couplings, known as the Kobayashi-Maskawa (KM) mechanism, agrees with all measurements to 2009. However, within experimental accuracy and theoretical uncertainties, additional subdominant sources of CP violation from new physics are allowed. In addition, CP violation is a necessary condition for baryogenesis, the process of dynamically generating the matter-antimatter asymmetry of the Universe. Despite the phenomenological success of the KM mechanism, it fails by several orders of magnitude to account for the observed asymmetry. This failure strongly suggests that there are additional sources of CP violation beyond the KM mechanism. The possibility of new sources of CP violation motivates major experimental efforts to find deviations from the predictions of the KM mechanism. Future searches for CP violation in neutrino oscillations could yield new insight beyond the Standard Model. The present measurements of CP asymmetries provide some of the strongest constraints on the weak couplings of quarks. Future measurements of CP violation in K, D, B, and Bs meson decays will provide additional constraints on the flavor parameters of the Standard Model and can probe new physics.

14.3.5 Charm, Bottom and Top Conservation

Hadrons and mesons are composite particles composed of six different types of quarks. Strangeness conservation is related to strange quarks. Similar conservation laws for charm, bottom, and top are associated with the charm, bottom, and top quarks. Strangeness, charm, bottom and top are each conserved in the strong and electromagnetic interaction but can change by one unit in the weak interaction. So when one adds the number of s, c, b, t quarks and subtracts their antiquarks on each side of a decay or reaction equation, then S, $\mathscr{C}$, $\mathscr{B}$, and $\mathscr{T}$ are constant in the strong and electromagnetic interaction, but in the weak interaction the quark flavor can change by one. The preferred change sequence is t → b → c → s. One has a new equation,

$$Q = Tz + (B + S + \mathscr{C} + \mathscr{B} + \mathscr{T})/2 \tag{14.21}$$

14.3.6 Summary: Conservation Laws and Interactions

The conservation laws, in addition to those of energy, electric charge, linear momentum and angular momentum are summarized in Table 14.3. The following are examples of the conservation laws.

Case 1:		$\Lambda^0 \to p + \pi^-$
	S:	$-1 \to 0 + 0$
	Baryon:	$1 \to 1 + 0$

TABLE 14.3
Summary of conservation laws in addition to those of energy, electric charge, and linear and angular momentum, which are always conserved
("Yes" means conserved; "No" means not conserved)

Quantity conserved	Strong	Electromagnetic	Weak
Electron lepton number	Yes	Yes	Yes
Muon lepton number	Yes	Yes	Yes
Tauon lepton number	Yes	Yes	Yes
Baryon number	Yes	Yes	Yes
Isospin magnitude	No	No	No ($\Delta T = \frac{1}{2}$ for nonleptonic)
Isospin z component	Yes	Yes	No ($\Delta T_Z = \frac{1}{2}$ for nonleptonic)
Strangeness	Yes	Yes	No ($\Delta S = 1$)
Parity	Yes	Yes	No
Charge conjugation	Yes	Yes	No
Time reversal (or CP)	Yes	Yes	Yes (but 10^{-3} violation in K^0 decay)
Charm	Yes	Yes	No ($\Delta\mathscr{C} = 1$)
Bottom (beauty)	Yes	Yes	No ($\Delta\mathscr{B} = 1$)
Top (truth)	Yes	Yes	No ($\Delta\mathscr{T} = 1$)

This must be a weak interaction decay because of the nonconservation of strangeness.

Case 2:

$$\bar{\Lambda}^0 \to p + \pi^-$$
$$S:\quad 1 \to 0 + 0 \qquad \Delta S = -1$$

This is also a weak interaction decay for the same reason.

Case 3:

$$\pi^- \to \mu^- + \bar{\nu}_\mu$$
$$S:\quad 0 \to 0 + 0 \qquad \Delta S = 0$$

Strangeness is conserved in this case, but the μ^- and ν participate only in the weak interaction, so that it is also a process governed by the weak interaction. The Λ^0 particle and $\bar{\Lambda}^0$ its antiparticle have the same lifetime of 2.5×10^{-10} s, and the lifetime of the π^- is 2.6×10^{-8} s, so they all belong to the weak interaction regime.

In the weak interaction, strangeness can change by at most one unit, so, $\Delta S = 0, \pm 1$ and no more.

Case 4:

$$\Omega^- \to \quad p + \pi^- + K^-$$
$$S:\quad -3 \quad \to \quad 0 + 0 \; -1 \qquad \Delta S = 2$$

This decay cannot happen because of the S selection rule mentioned above. It can only be

$$\begin{array}{lll} & \Omega^- \to \Lambda^0 + K^- & \\ S: & -3 \to -1 + -1 & \Delta S = 1 \\ & \Lambda^0 \to p + \pi^- & \\ S: & -1 \to 0 + 0 & \Delta S = 1 \end{array}$$

Both belong to the weak interaction regime.

Case 5:

$$\begin{array}{ll} & \Lambda^0 \to n + \gamma \\ S: & -1 \to 0 + 0 \end{array}$$

This is forbidden because a process with a γ participating must be related to the electromagnetic interaction and strangeness must be conserved there. Besides, the isospin components (T_Z) of Λ^0, n, and γ are 0, $-\frac{1}{2}$, and 0, respectively, so $\Delta T_Z \neq 0$, which is also forbidden in the electromagnetic interactions.

Case 6:

$$\pi^+ + p \to \pi^+ + \pi^0 + p$$

This is a strong interaction process with no violations of conservation of baryon number, isospin T, the third component of isospin T_Z, or other conservation rules.

Case 7:

$$\begin{array}{lll} & D^+ \to K^- + \pi^+ + \pi^+ & \\ S: & 0 \to -1 + 0 + 0 & \Delta S = -1 \\ \mathscr{C}: & 1 \to 0 + 0 + 0 & \Delta \mathscr{C} = -1 \end{array}$$

This is called a favored decay of the D^+. It is a weak interaction process that violates both strangeness conservation and charm conservation. There is also an unfavored (slow) decay of the D^+:

$$\begin{array}{lll} & D^+ \to \pi^+ + \pi^+ + \pi^- & \\ S: & 0 \to 0 + 0 + 0 & \Delta S = 0 \\ \mathscr{C}: & 1 \to 0 + 0 + 0 & \Delta \mathscr{C} = -1 \end{array}$$

Again, this is a weak interaction with strangeness conserved, but charm not conserved.

14.4 THE QUARK MODEL

14.4.1 Internal Structure of Hadrons

So far, no internal structure has been found for any lepton down to 10^{-18} m, hence they can still be treated as point particles. Data on the g-factor of the electron suggest that the electron radius may be the order of 10^{-20} cm (10^{-4} of the accepted upper limit), as discussed by one of the 1989 Nobel Laureates.[6] However, this is not the case for hadrons.

In 1933, Stern measured the magnetic moment of the proton to be $2.79\mu_N$ and in 1940 F. Bloch measured the neutron magnetic moment to be $-1.91\mu_N$, respectively. These deviate significantly from the predictions of Dirac's theory. How could the neutron, which carries no charge, have a nonzero magnetic moment? Its magnetic moment indicates that a neutron has an internal structure.

In 1956, R. Hofstadter and co-workers found,[7] by bombarding protons with high-energy electrons, that the proton had a charge distribution with a radius of about 0.7 fm. Shortly afterwards, positive and negative charges were found inside the neutral neutron. The radius of its charge distribution is 0.8 fm. The proton and neutron charge distributions are shown in Fig. 14.19. The neutron charge distribution with an inner positive charge and outer layer of negative charge is consistent with its negative magnetic moment. So neither the proton nor the neutron are point particles. The next evidence that hadrons had an internal structure came from elastic scattering of neutrinos by hadrons. The measured cross-section for such neutrino scattering increases linearly with neutrino energy as predicted for the elastic scattering of a pointlike neutrino by pointlike particles.

The electromagnetic interaction is relatively well known. The electron, a point charge without internal structure, does not participate in the strong interaction and can be an ideal tool to probe the internal structure of hadrons. Earlier the problem was that in the years around 1960 the available energies of electrons was not sufficiently high (the probe was not sufficiently fine) to be used to detect any fine structure inside a proton.

In the years around 1970, the proton was bombarded by GeV electrons, and "the shadow of Rutherford scattering" was seen (similar to the large back-angle scattering of the α particle by the tiny nucleus). At very large scattering angles, the inelastic scattering cross section for electrons was measured to be 40 times greater than for a proton ball with its charge uniformly distributed. It was as if the electrons were hitting very small, hard objects. Thus, the proton is composed of a number of "hard cores." This discovery won the Nobel Prize for J. Friedman, H. Kendall, and R. Taylor.

---

[6]H. Dehmelt, *Science* **247** (1990) 539.

[7]See R. Hofstadter, *Science* **136** (1962) 1013.

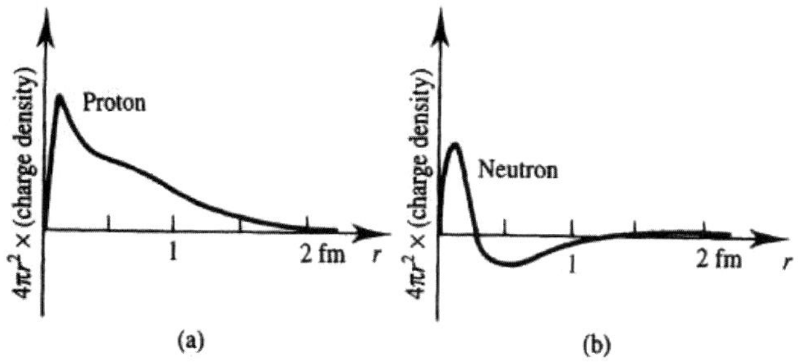

FIGURE 14.19
Charge distributions for the proton and neutron as determined from electron scattering.

14.4.2 Quark Model

The quark model of hadron structure was proposed independently by Gell-Mann and Zweig in 1964. A similar "straton" model, was proposed by several Chinese physicists at almost the same time. It is the quark (or straton) model's assertion that all the baryons (antibaryons) are composed of three kinds of quarks (antiquarks), and all the mesons are composed of a quark and an antiquark of different kinds.

Originally there were proposed three quarks, an up quark (u), a down quark (d), and a strange quark (s). Then the charm quark (c) was added to the family. Next, the bottom (or beauty) quark (b) was discovered and the top (truth) quark (t) was proposed to complete the theory. The top quark was first identified at FermiLab in 1994 and definitively confirmed in early 1995 by subsequent work there. The different types are also called different flavors of quarks. Their characteristics are listed in Table 14.4.

There are only 10 possible combinations of three (uds) quarks, as shown in Table 14.5, and 9 possible quark–antiquark combinations shown in Table 14.6.

TABLE 14.4
Properties of the up, down, strange, charmed, bottom, and top quarks. All are spin-$\frac{1}{2}$ particles

	Charge	Isospin		Baryon no.	Strangeness	Charm	Bottom (beauty)	Top (truth)
Quark flavor	Q	T	T_Z	B	S	$\mathscr{C}$	$\mathscr{B}$	$\mathscr{T}$
d	$-\frac{1}{3}$	$\frac{1}{2}$	$-\frac{1}{2}$	$\frac{1}{3}$	0	0	0	0
u	$\frac{2}{3}$	$\frac{1}{2}$	$\frac{1}{2}$	$\frac{1}{3}$	0	0	0	0
s	$-\frac{1}{3}$	0	0	$\frac{1}{3}$	−1	0	0	0
c	$\frac{2}{3}$	0	0	$\frac{1}{3}$	0	1	0	0
b	$-\frac{1}{3}$	0	0	$\frac{1}{3}$	0	0	−1	0
t	$\frac{2}{3}$	0	0	$\frac{1}{3}$	0	0	0	1

TABLE 14.5
Three-quark combinations

Combination	Charge	Spin	Baryon no.	Strangeness
uuu	+2	$\frac{3}{2}$	1	0
uud	+1	$\frac{1}{2}, \frac{3}{2}$	1	0
udd	0	$\frac{1}{2}, \frac{3}{2}$	1	0
ddd	−1	$\frac{3}{2}$	1	0
dds	−1	$\frac{1}{2}, \frac{3}{2}$	1	−1
dss	−1	$\frac{1}{2}, \frac{3}{2}$	1	−2
sss	−1	$\frac{3}{2}$	1	−3
uus	+1	$\frac{1}{2}, \frac{3}{2}$	1	−1
uss	0	$\frac{1}{2}, \frac{3}{2}$	1	−2
uds	0	$\frac{1}{2}, \frac{3}{2}$	1	−1

TABLE 14.6
Quark–antiquark combinations

Combination	Charge	Spin	Baryon no.	Strangeness
$u\bar{u}$	0	0,1	0	0
$u\bar{d}$	+1	0,1	0	0
$u\bar{s}$	+1	0,1	0	+1
$d\bar{u}$	−1	0,1	0	0
$d\bar{d}$	0	0,1	0	0
$d\bar{s}$	0	0,1	0	+1
$s\bar{u}$	−1	0,1	0	−1
$s\bar{d}$	0	0,1	0	−1
$s\bar{s}$	0	0,1	0	0

The quark compositions of baryons and mesons are shown in Figs. 14.20 to 14.22. The figures illustrate the underlying mathematical symmetry of the quark model. One sees easily how the first three quarks complete the symmetry of the particles shown. The quark model is not merely a way of ordering the particles in elegant geometrical figures such as shown. It has been strikingly successful in predicting the rest masses, decay modes, lifetimes, and other properties of these particles. In Fig. 14.20 the Λ^0 and Σ^0 are both shown as uds combinations. How can they be different particles? The Λ^0 is an isospin singlet, $T = 0$, and the Σ^0 is the $T_Z = 0$ member of the $T = 1$ isospin triplet with Σ^- having $T_Z = -1$ and Σ^+ having $T_Z = +1$.

It can be easily demonstrated that there are no violations of any known quantum rules for hadrons in Figs. 14.20 to 14.22 except for the spin $\frac{3}{2}$ for the Ω^-, Δ^- and Δ^{++} particles, which are composed of three quarks of the same kind, sss, ddd, and uuu, respectively, with their spins all aligned. These particles cannot have all the same quantum numbers and still have the Pauli exclusion principle

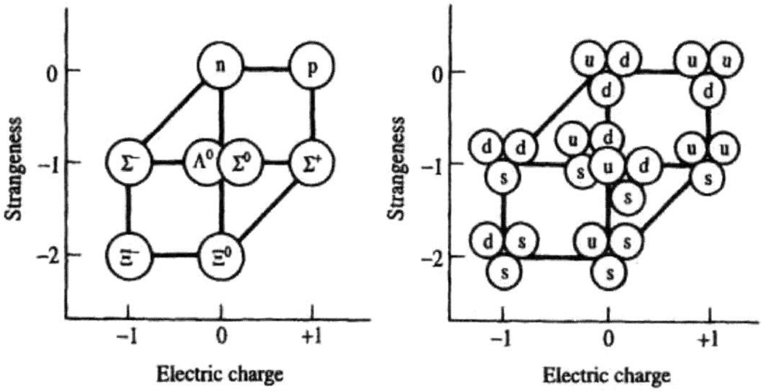

FIGURE 14.20
(a) The spin-$\frac{1}{2}$ baryons as a function of strangeness and electric charge. (b) The spin-$\frac{1}{2}$ three-quark combinations as a function of strangeness and electric charge. A comparison of the two figures yields the quark structures of these baryons.

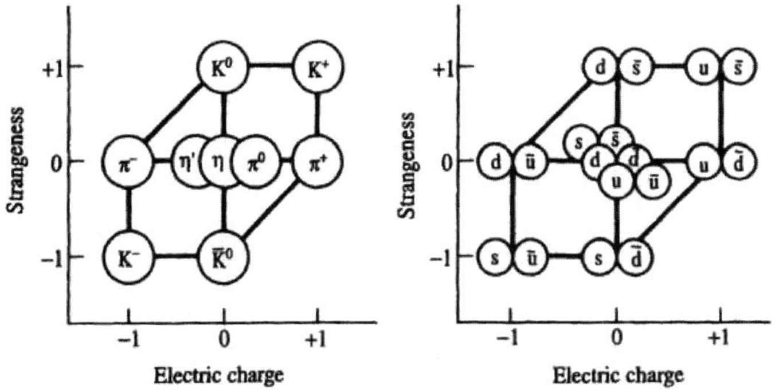

FIGURE 14.21
(a) The spin-0 mesons as a function of strangeness and electric charge. (b) The spin-0 quark–antiquark combinations as a function of strangeness and electric charge. A comparison of the two figures yields the quark structures of these mesons.

satisfied because quarks are fermions. In order to resolve this contradiction, O. Greenberg proposed in 1964 that each quark could have a new quantum number of three types—three different colors. Here color is not used in the literal sense of the word but denotes the fact that each quark must have at least three different new quantum numbers for the Pauli principle to hold. Since hadrons do not exhibit color, the three quarks must combine to be colorless. Color was chosen as the analogy because three colors can combine to be colorless (red, yellow and blue pigments mix to give colorless black; or the primary light colors, red, green and blue when mixed give white light, so some use r, y, b and others r, g, b for the three colors of quarks). Mesons also do not exhibit color, but in that case the color of one quark is cancelled by the anticolor of the antiquark. Note that the quark–antiquark structure of the meson is determined by the requirements of

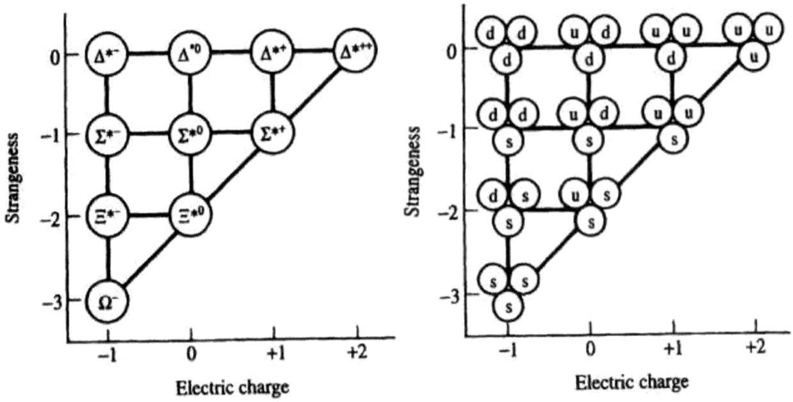

FIGURE 14.22
(a) The spin-$\frac{3}{2}$ baryons as a function of strangeness and charge. (b) The spin-$\frac{3}{2}$ three-quark combinations as a function of strangeness and electric charge. A comparison of the two figures yields the quark structures of these baryons.

their having spin = 0 and baryon number $B = 0$. The problem of the Pauli principle seemed solved but at the cost of tripling the numbers of quarks.

In 1974, S. Ting and co-workers and B. Richter and co-workers discovered independently a new particle,[8] which they named the J and ψ particle, respectively, or the J/ψ (gipsy) particle, as it is called now. The particle has a mass three times the proton mass, appears in the form of a resonance shown in Fig. 14.23, but has a lifetime more than 100 times longer than that of ordinary hadron resonances. It was the narrowness of energy spread of the massive (3097 MeV/c^2) resonance that indicated it was not decaying into hadrons by the expected strong interaction. It is a peculiar particle; as Ting said "It is just like a new species of mankind with an expected lifetime of ten thousand years." It was soon shown by theoretical calculations that the J/ψ particle is composed of a charm quark and an anticharm quark ($c\bar{c}$), called charmonium in analogy with e^+e^- positronium.

Ting and Richter shared the 1976 Nobel Prize in Physics for their discovery. The charm quark does not exist in ordinary hadrons. It had been theoretically predicted by S. L. Glashow and co-workers in 1970. The J/ψ particle was the first experimental evidence for the existence of another new quark, the charm quark. It provides additional strong evidence for the reality of the quark model. Charm quarks and ordinary quarks are different in nature. It is very difficult for one to be transformed into the other, so the lifetime of the J/ψ particle is tremendously long.

The D mesons were found in 1976. They are the first mesons reported to include charm quarks, with the following compositions: $D^0(c\bar{u})$, $D^+(c\bar{d})$, and $D^-(\bar{c}d)$. Mesons with a c quark have $\mathscr{C} = 1$ and those with a $\bar{c}$ have $\mathscr{C} = -1$.

[8]S. C. Ting, *Science* **196** (1977) 1167; B. Richter, *Science* **196** (1977) 1286.

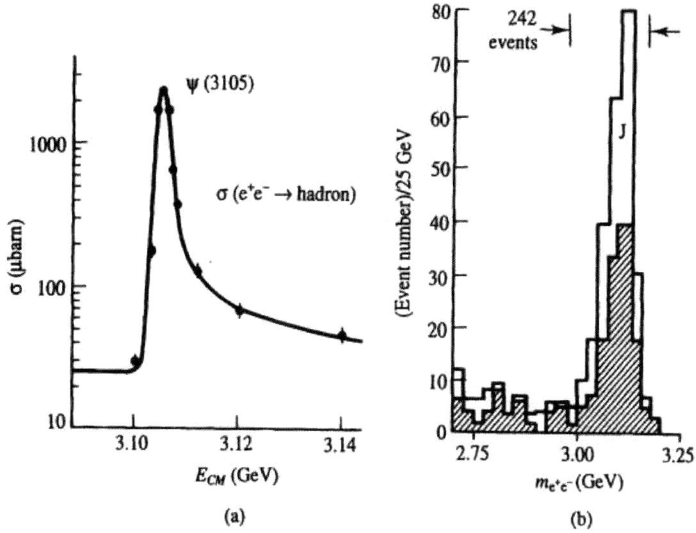

FIGURE 14.23
Resonances which yielded the discovery of the J/ψ particle.

One could make Fig. 14.21 a three-dimensional plot with that figure being the $\mathscr{C} = 0$ plane and the D^0 and D^+ lying above on the $\mathscr{C} = +1$ plane and their antiparticles along with the D_s^+ meson on the $\mathscr{C} = -1$ plane below.

The history of the discovery of particles with strangeness was just the opposite to that of those with charm. The sequence of discovery was the K meson ($S \neq 0$), then the ϕ meson composed of $s\bar{s}(S = 0)$. With charm, the particle $J/\psi(c\bar{c})$ ($\mathscr{C} = 0$) was found first, then the D meson ($\mathscr{C} \neq 0$). In another reversal, strangeness was introduced to describe existing experimental observations, while charm was introduced theoretically before the corresponding particles were discovered.

Besides the u, d, s, and c quarks, from symmetry considerations, the existence of another type of quark, the bottom (beauty) quark (b) requires still another one, the top (truth) quark (t). The upsilon (Υ) particle, with a mass of 9.5 GeV, was discovered in 1977. It is considered to be composed of $b\bar{b}$. Three excited states of the Υ, the Υ', Υ'' and Υ''' have been seen, and all but the Υ''' have narrow width. At the Cornell e^+-e^- collider, the broad Υ''' resonance has been seen to decay into a $B\bar{B}$ pair of mesons where the B^+ is $b\bar{u}$ and B^0 is $\bar{b}d$, and a spectroscopy of the b quark is underway. A high intensity "b factory" is now under development in the United States to probe this quark further. The first evidence for the t quark was reported at FermiLab in 1994, and subsequently confirmed there in 1995.

Let us look at some examples of quark structure. Consider the proton (uud) and neutron (udd), which form an isospin doublet. The u and d quarks form an isospin doublet which has the same mass and interaction apart from electromagnetic effects. Since changing u and d makes no difference in the strong interaction, the proton and neutron have the same strong interaction. The understanding of isospin is the same as understanding why there are two quarks (u,d) which are different only in their electromagnetic properties. This is another of the unanswered challenges.

The reason for the associated production of the Λ^0 and K^0 is easily shown from their quark structure. Remember, quarks and antiquarks can annihilate just as other particles and antiparticles, and energy can form quark–antiquark pairs. Look at

$$\begin{aligned} \pi^- + p &\rightarrow \Lambda^0 + K^0 \\ d\bar{u} + uud &\rightarrow uds + d\bar{s} \\ S{:}\quad 0 + 0 &\rightarrow -1 + 1 \end{aligned}$$

Here the $\bar{u}$ and one u quark annihilate each other, and the energy goes to create an s and $\bar{s}$ quark. Strangeness is conserved in this strong interaction. The K^0 decay is

$$\begin{aligned} K^0 &\rightarrow \pi^- + \pi^+ \\ d\bar{s} &\rightarrow d\bar{u} + u\bar{d} \\ S{:}\quad 1 &\rightarrow 0 + 0 \end{aligned}$$

Here the $\bar{s}$ decays to a $\bar{d}$ with the creation of a $u\bar{u}$ pair from the decay energy. Strangeness is not conserved as a strange antiquark becomes a nonstrange $\bar{d}$. This must go by the slow, weak interaction which does not conserve strangeness. This can be illustrated as

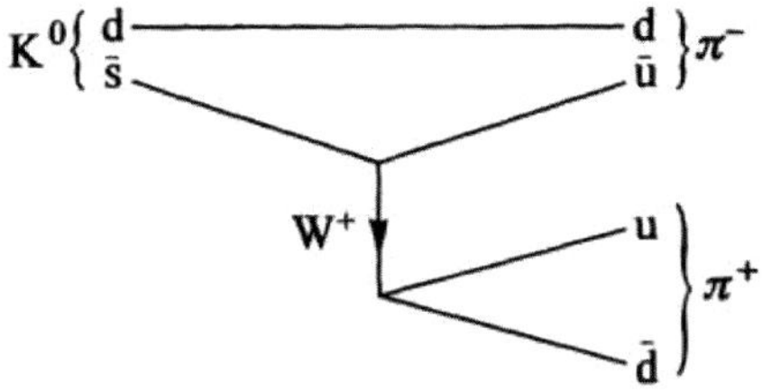

The production and decay of the J/ψ can be written as

$$\begin{aligned} e^+ + e^- \rightarrow \gamma \rightarrow J/\psi &\rightarrow \pi^+ + \pi^0 + \pi^- \\ \rightarrow c\bar{c} &\rightarrow u\bar{d} + d\bar{d} + d\bar{u} \end{aligned}$$

Here the $c\bar{c}$ pair annihilation creates a $u\bar{u}$ pair and two $d\bar{d}$ pairs. This process is forbidden (hindered) because of the difficulty of going from $c\bar{c}$ to $u\bar{u}$. The $c\bar{c}$ also can annihilate into a photon which can produce a $\mu^+ + \mu^-$ pair by the electromagnetic interaction. Another example is the decay of the $\eta_c(c\bar{c})$, as shown in Fig. 14.24, where an $s\bar{s}$ pair along with a $u\bar{u}$ and $d\bar{d}$ pair are favored. Why does J/ψ not decay first into a charmed meson pair such as $D^0\bar{D}^0$, one with a c and the other with $\bar{c}$? This is not possible because these least massive charmed mesons have masses above the mass of the 3097 MeV/c^2 J/ψ. The second excited state of the J/

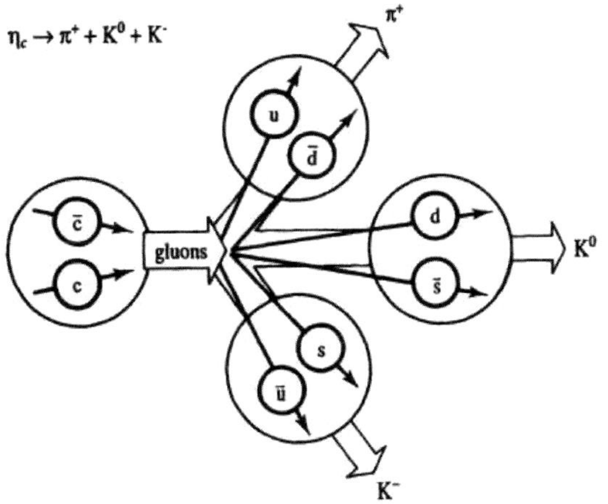

FIGURE 14.24
The decay of $\eta_c(c\bar{c})$ into three mesons, including two with strangeness, via the gluon field.

ψ, ψ'' (3767), has sufficient mass to decay into the $D\bar{D}$, so the ψ'' resonance has a large mass width. There are many charmed particles which are possible, but besides the D^0 and $D^{\pm}$, only the $D_s^{\pm}(c\bar{s}, \bar{c}s)$ and Λ_c (udc) have been observed. Note that the D_s is composed of a strange and a charm quark.

In reactions and decays of elementary particles, the following rules hold:

(a) Strong interactions cannot change the flavors (types) of quarks, e.g.,

$$\pi^- + p \to \Lambda^0 + K^0$$
$$d\bar{u} + uud \to uds + d\bar{s}$$

On the left the total number of quarks is 2d and one u quark ($u\bar{u}$ cancel), and on the right 2d and 1u ($s\bar{s}$ cancel).

(b) Weak interactions can change the flavors of quarks by emitting or absorbing a $W^{\pm}$ boson, e.g.,

$$\begin{array}{ccc} n & \to & p + e^- + \nu_e \\ (udd) & & (uud) \end{array}$$

$d \to u + W^-$ and W^- decays into $e^- + \bar{\nu}_e$. Other examples are

$$\bar{u} \to \bar{d} + W^-,\ \bar{s} \to \bar{u} + W^+$$

(c) Weak neutral current processes cannot change quark flavor, e.g.,

$$c \to u + Z^0 \text{ is not allowed,}$$

$$\nu_\mu + p \to \nu_\mu + p \text{ scattering goes via } Z^0.$$

(d) Quark–antiquark pairs can be created (if the energy is available) and annihilated just as e^-–e^+ creation and annihilation occurs.

Ever since the quark model was proposed in 1964, extensive searches have been made for evidence of the existence of quarks as free particles. As yet, there has been no decisive evidence for the existence of free quarks. On the other hand, there have been many good agreements between the deductions of the quark model and various experimental data to strongly support the existence of quarks. Various models of "quark confinement" have been proposed to explain the failure to observe free quarks. Perhaps quarks are very massive, with such a large binding energy that present accelerators do not have sufficient energy to unbind them. Perhaps nature does not allow free quarks to exist. Our understandings of the problem are still at the very beginning.

How are quarks held together to form hadrons and mesons? They are held together by what in fact is the real strong interaction, whose propagator is the new particle called the gluon. There is indirect evidence for the existence of gluons, but again no free gluon or glue-ball has been observed. One such evidence is in e^+-e^- collisions, where quark–antiquark pairs are formed and then decay into two jets of particles. At higher energies, three jets are seen. Since only $q\bar{q}$ are produced (not 3 quarks), one of them must radiate a gluon that gives rise to the third jet. As one looks at the role of the gluons in the binding of quarks, one sees that gluons must also come with color.

The color quantum number, it turns out, is not simply a way of counting particle states allowed by the Pauli principle but is the basis of the binding of quarks. *The color quantum number is related to the true strong interaction in the same way as electric charge is related to the electromagnetic interaction.* The electromagnetic interaction involves the exchange of photons which are emitted and absorbed by electric charge. The true strong interaction involves the exchange of gluons which are emitted and absorbed by *color charge*. Now we must distinguish the true strong (color) interaction from what is generally called the strong (or strong nuclear) interaction as observed between hadrons. The true strong (color) interaction is the interaction which binds quarks into mesons and hadrons. The meson exchange theory of the strong nuclear force is but a residual manifestation of the true strong interaction mediated by the gluon.

We now speak of the nuclear binding as a residual strong interaction, as shown in Table 14.2 earlier. Table 14.2 is from the 1988 chart prepared by the Fundamental Particles and Interactions Chart Committee as an outgrowth of the Conference on Teaching of Modern Physics. The chart says of the "Residual Strong Interaction—The strong binding of the color-neutral protons and neutrons to form nuclei is due to residual strong interactions between their color-charged constituents. It is similar to the residual electrical interaction which binds electrically neutral atoms to form molecules. It can be viewed as the exchange of mesons between the hadrons." For example, the Van der Waal's force that binds molecules is not a fundamental force but a residual interaction related to electric charge.

Lepton–nucleon scattering data have indicated the presence of objects which have no weak or electromagnetic interactions in the nucleon. The gluon has no weak or electric charge and so does not participate in these interactions, but it does have color charge. Note there is an important difference between gluons and photons. Gluons carry color charge but photons do not carry electric charge. Likewise, all leptons, photons, and the $W^{\pm}$ and Z^0 bosons do not have color charge and so are not involved in the strong interaction.

We return to the question of why we do not see free quarks or color particles, i.e. quark confinement.[9] As we know, there is a fine structure constant $\alpha = e^2/\hbar c$, which presents the basic strength of electromagnetic interaction between electric charges. In strong interaction there is an analogous coupling constant , which describes the strength of interaction between color charges. D. J. Gross[10] proposed an asymptotic freedom theory, which tells that the coupling constant is predicted to vanish as the distance shrinks to zero and the quarks almost free. As quarks start to separate, the coupling constant becomes very large for large distance so that the attractive force holding the quarks together increases rapidly, which leads to quark confinement. For his theory, D. J. Gross shared 2004 Nobel prize in physics with H. P. Politzer and F. Wilczek.[11]

Now look at the binding of the quarks by the gluon. It turns out that one must have eight fields (with eight possible combinations of color charge), and hence eight colored gluons to accomplish the binding of three colors of quarks. Each gluon has a color, red (r), green (g), blue (b) and anticolor $\bar{r}$, $\bar{g}$, $\bar{b}$. Six of the combinations are easy to see, $r\bar{g}$, $r\bar{b}$, $g\bar{r}$, $g\bar{b}$, $b\bar{r}$, and $b\bar{g}$, with the other two being linear combinations of $r\bar{r}$, $b\bar{b}$ and $g\bar{g}$. The two with color are $(r\bar{r} - g\bar{g})/\sqrt{2}$ and $(r\bar{r} + g\bar{g} - 2b\bar{b})/\sqrt{6}$. How do these colored gluons bind quarks? Take a colorless hadron with rgb quarks. The r quark can change to a g quark by emitting an $r\bar{g}$ gluon and the g quark can absorb the $r\bar{g}$ gluon to become an r quark and the system remains rgb, colorless. Here we have sketched only in broad strokes the basic ideas in the quark–gluon picture with color charge in what is called the "standard model" of quantum chromodynamics (QCD) which has many analogies with as well as significant differences from quantum electrodynamics (QED).

Probing the limits of the standard model is a major thrust of particle physics today. New surprises are already coming. At the electron–proton collider HERA (Hadron–Electron Ring Accelerator) near Hamburg, experiments are beginning to probe the inner sanctum of the proton (see *Science* **264** (1994) 1843). Quarks are so much smaller than the proton that the proton could be mostly empty space with three quarks bouncing around in the vast space inside the proton. The recent HERA experiments indicate that, on the contrary, the interior is filled with activity, with virtual quarks coming and going. At a given time the evidence indicates the presence of several virtual quarks and many gluons. There seems to be a surprisingly large admixture of polarized strange quarks. This raises a whole new set of questions. Finally, there is evidence for a new, mystery particle. Perhaps it is a temporary lump of gluon, but it may be something else. The

---

[9] See, for example, E. M. Henley and A. Garcia, *SubAtomic Physics*, World Scientific (2007).

[10] D. J. Gross, *Phys. Rev. Lett.* **32** (1974) 1071.

[11] Gosta Ekspong, *Nobel Lectures in Physics*, World Scientific (2008).

internal structure of the proton can also be probed with the new continuous electron beam accelerator facility at CEBAF in Virginia.

There are other fascinating questions about the internal structure of the proton and neutron. We have argued that the spins of hadrons come from summing the spins of the quarks. Thus, particles like the spin-$\frac{3}{2}$ Ω^- and Δ^- were assumed to be composed of three spin-$\frac{1}{2}$ quarks all with their spins aligned. This would violate the Pauli Exclusion Principle, so each quark was assigned a new quantum number, color, which comes in three types so that each quark has a different color quantum number. Recently, evidence has come from CERN, and from the Stanford Linear Accelerator Laboratory, for a "spin crisis" where perhaps only 20–30% of the nucleon's spin comes from its three main quarks.[12] The evidence comes from very high-momentum transfer experiments which probe the extremities (tail) of the nucleon wavefunction.

At low energies the spin is accounted for by its quark constituents. The presence of virtual quarks inside the proton could provide some of the missing part of the proton's spin and part by quark orbital motion and spins of gluons. Scientists hope that the unique facilities at HERA, where beams of polarized (where all their spins are aligned in one direction) electrons or positrons will enable them to probe this question. Recall that the force between two nucleons that binds them is spin-dependent; that is, it depends on whether the two nucleon spins are parallel or antiparallel. However, the answer may still not be forthcoming since the Hermes experiment at HERA cannot reveal whether gluons carry spin (see *Science* **267** (1995) 1767). Clearly, all these new data present exciting challenges to our fundamental understandings of the way nature behaves at its basic level, including an understanding of the internal structure of nucleons.

14.4.3 New Synthesis of Particle and Nuclear Physics

New accelerators such as CEBAF, the Relativistic Heavy Ion Collider (RHIC) at Brookhaven National Laboratory, and the Large Hadron Collider (LHC) at CERN, are bringing together again the separate fields of high-energy particle physics and nuclear physics. CEBAF will deliver 10^{15} electrons per second at 4 GeV. Quantum chromodynamics (QCD) provides a fundamental understanding of the "strong interaction" between quarks and gluons. However, QCD has primarily been tested only in the very high-energy regime where the interaction becomes weak. CEBAF will probe the transition regime between the high-energy region and the strongly interacting region where our understanding of the underlying physics is very rudimentary. High-energy physics is concerned with the ultimate laws of nature, while CEBAF seeks to understand how the real world works. Research will include such questions as the strange-quark content of the

---

[12] R. L. Jaffe, *Physics Today* **48** (1995) 24.

proton, gluonic degrees of freedom, multinucleon correlations in nuclei, the neutron electric form factor, and hypernucleon physics.

A major motivation to build RHIC and LHC is to see whether one can create free quarks and gluons in the new state of matter called the quark–gluon plasma. W. Greiner and co-workers in Frankfurt first predicted the existence of nuclear shock waves in nuclear matter in relativistic heavy-ion collisions. They pioneered the understanding of how such shock waves could provide the key mechanism for compressing and heating nuclear matter to the high densities and high temperatures required to explore the nuclear equation of state, and to produce a quark–gluon plasma which presumably existed in the very early universe following the Big Bang before hadrons were formed. RHIC is expected to have 100 GeV/u ^{197}Au (100×197 GeV $= 20$ TeV) colliding with 100 GeV/u ^{197}Au (40 TeV total collision energy). LHC, being built in the LEP tunnel, will have 7 TeV protons colliding with 7 TeV protons with unsurpassed intensitites and 2.7 TeV/u ^{208}Pb ($2.7 \times 208 = 562$ TeV) with 562 TeV Pb (1125 TeV total collision energy). RHIC is expected to be completed in 1999 and LHC by 2005. These facilities likewise will unite nuclear and particle physicists again in research.

At the much higher energies, densities, and temperatures that can be achieved with RHIC and LHC, one expects to probe the standard model of QCD to new limits and to help lead in the development of a Grand Unified Theory of all forces. Are the quarks the fundamental binding blocks or is there a still lower structure or stratum? Are there more than six basic flavors of quarks? These are but some of the exciting questions to be considered in the future.

Greiner in Frankfurt has pointed out, in a provocative paper,[13] the extension of the periodic system beyond super-heavy, high-Z nuclei, high-spin nuclei, isospin and high densities and temperatures as discussed earlier, to unexplored regions of strange matter and antimatter which can be probed in relativistic heavy-ion collisions. This is illustrated in Fig. 14.25 where, in addition to the normal Z–A (or N) plot, one has anti-Z and anti-A axes along with a strangeness and antistrangeness axis.

First look at antimatter production. Theoretically one expects for antimatter that the $\bar{Z}$(anti Z) – $\bar{N}$(anti N) chart will exhibit shell structure like the chart found for normal matter, Fig. 9.2. But can we make antinuclei? In ultra-relativistic collisions where nuclear shock waves can compress nuclear matter, there are two production processes. First, in such very high-energy nucleus–nucleus collisions the many antiprotons and antineutrons produced could coalesce step by step to build up antinucleon clusters. Such processes can occur, but the large amount of normal matter available for annihilation will be a serious limiting factor. Greiner and his colleagues have pointed out there is also a collaborative antimatter production mechanism. It is related to the spontaneous positron emission and vacuum decay process in QED of very strong fields. In meson

[13] W. Greiner, "On the Extension of the Period System into the Sectors of Strangeness and Antimatter," *Int. J. Mod. Phys. E* **5** (1996) 1.

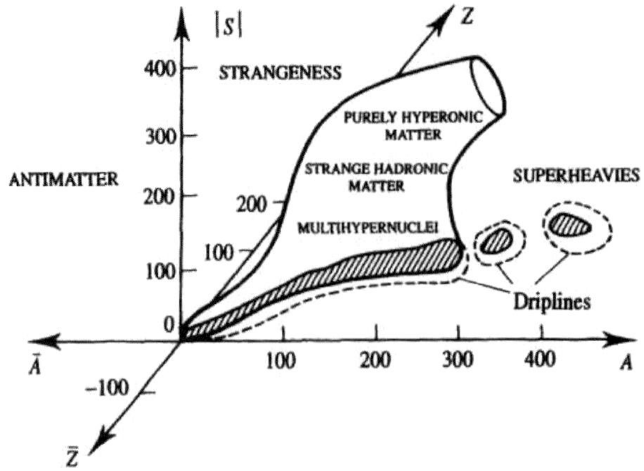

FIGURE 14.25
Schematic view of the standard periodic system in the Z–A plane and its extension into the sectors of antimatter ($\bar{Z}, \bar{A}$) and strangeness (S). (From Greiner.)

field theory, there is a difference in the nuclear shell potential for nucleons and antinucleons. The net result is that, in such heavy-ion collisions, clusters of nucleons in the negative energy continuum could be spontaneously emitted, leaving behind clusters of holes—clusters of antinucleons in bound states. This collective creation process of antimatter clusters has a high potential for giving rise to higher production probabilities of antimatter. It seems that one should be able to produce light antinuclei such as $\bar{\mathrm{d}}$, $\overline{\mathrm{He}}$, $\bar{\mathrm{C}}$ and $\bar{\mathrm{O}}$. While first-order calculations indicate similar closed-shell structures, do our models really work under such extreme conditions, are the binding energies and masses the same, and at what level of sophistication will differences show up? How would antimatter annihilation occur—collectively or step by step, $\mathrm{p\bar{p}}$, $\mathrm{n\bar{n}}$ . . . ? Then consider the questions of antiatoms, that is, antinuclei with positrons in orbits. Both experimentally and theoretically, the equations are major challenges.

Now look at strange matter, as shown in Fig. 14.26. One must now extend the Bethe–Weizäcker mass formula into this new region, where we replace one of the nucleon quarks by a strange quark. Of course, we can ask the same question about nuclei and nuclear matter with charm, bottom, or top quarks. Such questions will arise in the future, but we have had evidence for strange nuclear matter since 1953, long before quarks and strangeness were ever proposed. Strange matter has been produced by the exchange of a strange quark with a normal nucleon quark, with the nucleon bound inside a nucleus. For example:

$$
\begin{array}{llll}
\mathrm{K^-} + \mathrm{n} \rightarrow & \Lambda^0 + & \pi^- \\
(\mathrm{s\bar{u}}) & (\mathrm{udd}) & (\mathrm{uds}) & (\mathrm{\bar{u}d}) \\
\mathrm{K^-} + \mathrm{p} \rightarrow & \Sigma^+ + & \pi^- \\
(\mathrm{s\bar{u}}) & (\mathrm{uud}) & (\mathrm{uus}) & (\mathrm{\bar{u}d})
\end{array}
$$

Nuclei containing one Λ^0 or Σ or Ξ in place of a proton or neutron are called hypernuclei. By 1991, nuclei with one Λ^0 include ${}^{3,4}_{\Lambda}\mathrm{H}$, ${}^{4,5,6,7,8}_{\Lambda}\mathrm{He}$, ${}^{6,7,8,9}_{\Lambda}\mathrm{Li}$, ${}^{7,8,9,10}_{\Lambda}\mathrm{Be}$, ${}^{9,10,11,12}_{\Lambda}\mathrm{B}$, ${}^{12,13,14}_{\Lambda}\mathrm{C}$, ${}^{14,15}_{\Lambda}\mathrm{N}$, ${}^{16,18}_{\Lambda}\mathrm{O}$, ${}^{27}_{\Lambda}\mathrm{Al}$, ${}^{28}_{\Lambda}\mathrm{Si}$, ${}^{32}_{\Lambda}\mathrm{S}$, ${}^{40}_{\Lambda}\mathrm{Ca}$, ${}^{51}_{\Lambda}\mathrm{V}$, ${}^{56}_{\Lambda}\mathrm{Fe}$, ${}^{89}_{\Lambda}\mathrm{Y}$,

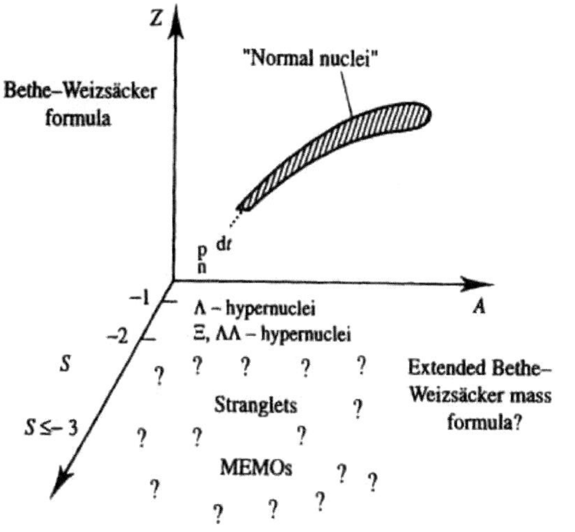

FIGURE 14.26
Schematic view of the extension of the periodic system (proton Z and nuclear number A plane) into the direction of strangeness S. (From Greiner.)

${}^{90}_{\Lambda}Zr$, ${}^{209}_{\Lambda}Bi$; and those with one Σ or Ξ are ${}^{6,7}_{\Sigma}H$, ${}^{4}_{\Sigma}He$, ${}^{9,12}_{\Sigma}Be$, ${}^{12,16}_{\Sigma}C$, ${}^{8}_{\Xi}He$, ${}^{11}_{\Xi}B$, ${}^{13,15}_{\Xi}C$, ${}^{17}_{\Xi}O$, ${}^{28}_{\Xi}Al$, and ${}^{29,30}_{\Xi}Mg$. In addition double Λ^0 hypernuclei have been reported: ${}^{6}_{\Lambda\Lambda}He$, ${}^{10}_{\Lambda\Lambda}Be$, and probably ${}^{13}_{\Lambda\Lambda}B$. The Λ or Σ have single particle levels inside the nucleus just as the nucleons do and measurements of the π^- energies give the single particle energies. The lifetimes of such systems are typically 10^{-10} s as a result of the weak decay, for example, $\Lambda \rightarrow p + \pi^-$. It might also be possible to form strange matter from a quark–gluon plasma where droplets of u, d, and s quarks are found. Such objects are called stranglets. The binding energies of the Λ in light nuclei have been calculated in a spherical relativistic mean field model and found to agree quite well with experiments.

In addition to hypernuclei, there are 13 double configurations with strangeness which are metastable, such as $\Sigma^-\Sigma^-$, $\Lambda^0\Lambda^0$, $\Sigma^+\Sigma^+$ (strangeness 2); $\Xi^-\Sigma^-$, $\Xi^-\Lambda^0$ (strangeness 3), up to $\Omega^-\Omega^-$ (strangeness 6). There are also metastable triplets, five with one or two nucleons such as Λnp, $\Xi^-\Sigma^-$n, or $\Xi^0\Lambda^0$p, and two without normal nucleons, $\Xi^-\Xi^0\Lambda^0$ and $\Omega^-\Xi^-\Xi^0$. In relativistic heavy-ion colissions the following configurations are potentially metastable (1) several Λs, (2) several Λs or Σs with neutrons, or (3) several Ξs with Λs and neutrons. Such systems are called MEMOs—metastable exotic multistrange objects.

Very high-energy heavy-ion collisions offer the only way at present to make multiple strange nuclei. Calculations predict, for example, that for Pb + Pb at 160 GeV/u some 60 Λ and 40 Σ will be produced, while at 20 000 GeV/u there will be 100–500 Λ and 400 Σ. Clearly these new forms of matter will provide challenges both experimentally and theoretically well into the future.

14.4.4 Grand Unified Theories

The successes of the unification of the weak and electromagnetic interactions have been so remarkable that physicists are much encouraged toward the development of a Grand Unified Theory, GUT, to include all the interactions. The first step would be to unify the strong interaction with the electromagnetic and weak interactions and finally to unify these with gravity.

In the electroweak theory, there is a spontaneous symmetry breaking that leads to the differences in the electromagnetic and weak interactions. If the underlying symmetry of the theory were broken, these would be the same interaction. Indeed, at some sufficiently high energy this symmetry should apply. High energy means short distance, short de Broglie wavelength, $\lambda = \hbar c/E$ or $\Delta x \sim \hbar/\Delta p_x$. At a distance of 10^{-18} cm the weak and electromagnetic interactions between two quarks are essentially equal, as shown in Table 14.1.

One can ask at what energy the weak and electromagnetic interactions will approach the strong interaction. Indications are that at an energy of about 2×10^{14} GeV these three interactions come together. From the de Broglie wavelength one can find the distance at which this occurs:

$$\frac{\lambda}{2\pi} = \frac{\hbar c}{E} = \frac{0.2\,\text{GeV} - F}{2 \times 10^{14}\,\text{GeV}} \sim 10^{-30}\,\text{m}.$$

Mathematically, one says that the U(1) symmetry of the electromagnetic interaction, with SU(2) of the weak interaction and SU(3) symmetry of the strong color interaction, are the result of a symmetry breaking of a grand unified interaction. Much effort has been directed to obtain a new gauge symmetry that incorporates the U(1), SU(2) and SU(3) symmetries. The U(1), SU(2) and SU(3) coupling constants are not really constants but vary with energy scale.

Various proposals for a Grand Unified Theory have been made. Each has its own particular difficulty. The simplest proposal is the SU(5) theory introduced by Georgi and Glashow. It has many attractive features but has difficulties, including its prediction of the partial (for only the following decay mode) lifetime of $4.5 \times 10^{29\pm1.7}$ years for the proton decay, $p \rightarrow e^+ + \pi^0$. The current lower limit for this lifetime is the order of 10^{32} years.

The prediction of proton decay requires further discussion. In particular, we see that the reaction $p \rightarrow e^+ + \pi^0$, while conserving electric charge, does not conserve baryon number or lepton number! Previously we spoke as if baryon number and lepton number were absolutely conserved quantities. However, any absolute conservation law must be correlated with an exact invariance principle and symmetry. Charge consevation is related to gauge invariance and the massless field of the photon. There is no known gauge invariance and massless field that can be related to lepton and baryon number conservation. Thus, lepton and baryon number conservation laws are probably only approximate conservation laws. They are nearly exact because of the very large energy of unification. Unification of leptons and quarks requires that leptons and baryons not be conserved.

While the energy for Grand Unification of the basic interactions to occur is outside the present limits of our laboratories, in the very early universe ($\sim 10^{-40}$ s after the Big Bang that started our present universe) the conditions were such that all the interactions were unified. After that time the symmetry (unifying gauge invariance) was broken and our three (four) basic interactions began to appear.

The detection of proton decay with a lifetime greater than 10^{30} years requires very special sensitivity. The upper limit of 10^{32} years for p decay into $e^+ + \pi^0$ was first obtained by using 8000 tons of very highly purified water detector that was located in a deep salt mine to eliminate cosmic ray background. The decay of the proton is a critical datum for Grand Unified Theories. Much larger, more sophisticate, even deeper underground locations are being developed to search for the proton decay, study neutrinos, search for dark matter and other exotic phenomena as discussed in section 10.3.

Other attempts at Grand Unified Theories go under such names as supersymmetry and supergravity. Particle physics becomes closely connected with both nuclear physics, astrophysics, and cosmology at this level, as we seek to understand the structure of our universe from its smallest to its largest dimensions. Finally there are major new frontiers opening up in particle-astrophysics through the discoveries of dark matter and dark energy.

There is strong evidence that the matter which makes up stable nuclei and atoms is not, in fact, the dominant form of matter in the universe. The cleanest evidence for this comes from observations of the rotation of galaxies. Consider a galaxy of mass M. If the motion of a gas cloud or a rare star, lying far outside the luminous part of the galaxy, is measured, then we would expect the velocity v and the distance form the center of the galaxy, R, to be related by

$$\frac{v^2}{R} = \frac{GM}{R^2}$$

so that v should scale as $1/\sqrt{R}$. The crucial assumption here is that the mass of the galaxy is all contained in the part that is producing light, so that outside of this region, the mass of the galaxy can be treated as a single point mass in the center. This scaling of velocity with distance is what is observed, for example, in planets orbiting the sun. In galaxies, however, the velocity outside the luminous region does *not* scale as $1/\sqrt{R}$; instead, the velocity is roughly constant at large distances from the center of the galaxy. This sort of behavior is known as a "flat rotation curve". In order to produce such a scaling of velocity with distance, additional matter must be present in the region outside of the visible part of the galaxy. This additional matter is known as dark matter, since it produces no visible light. Further evidence for dark matter can be seen in the motions of galaxies within galaxy clusters: the amount of matter needed to bind galaxy clusters together is much larger than what is observed in the visible galaxies.

These observations do not require the additional dark matter to be anything exotic. They are consistent with the dark matter being in the form of baryons, i.e., the same protons and neutrons that provide most of the mass in ordinary matter. However, other observations rule out this possibility. Precision measurements of the cosmic microwave background, along with the predictions of element production in the early universe, provide a reliable estimate of the total density of baryonic matter in the universe. This baryon density falls well below the total density of matter in the universe. Hence, we are forced to conclude that the bulk of the dark matter is not ordinary baryonic matter, but something more exotic.

The exact nature of dark matter is not well-understood at the present time, but the most likely possibility is some as-yet undiscovered stable elementary particle. This particle feels the force of gravity, which allows it to cluster gravitationally, but its interactions with ordinary matter are extremely weak, which explains why it is "dark".

In addition to dark matter, the universe appears to contain an even more exotic component, called dark energy. Evidence for dark energy emerged in the late 1990s, from two groups examining the expansion of the universe at early times. The rate at which the universe expands is directly related to the total density of mass-energy in the universe. By observing very distant supernovae, billions of light-years away, these two groups were able to measure this expansion rate (and thus, the density) both at the present time, and at much earlier times. Their observations imply that there must be an additional component to the universe, beyond both ordinary baryonic matter and dark matter, which has been dubbed dark energy. In order to satisfy these supernova observations, the dark energy must have some very exotic properties. Unlike both baryonic matter and dark matter, it does not cluster gravitationally; it appears to be smoothly distributed throughout the universe. The density of the dark energy does not decrease as the universe expands; instead, it appears to be roughly constant. Finally, the pressure of the dark energy must be negative.

All of these properties can be accommodated by modifying Einstein's theory of general relativity to add what is called a "cosmological constant" to the equations, or by adding a more exotic dark energy component. At this point, the correct explanation is far from certain.

What is well-understood is the relative contribution of baryonic matter, dark matter, and dark energy to the total mass-energy density of the universe. Current observations imply that the universe is roughly 5% baryonic matter, 20% dark matter, and 75% dark energy. Thus, the more exotic components of the universe are also the most abundant. The known elementary particles are given in Tables 14.4 and 14.7.

Table 14.7 Elementary particles in a new stratum.[a] The mediators of the interactions are given first, followed by the leptons with lepton number one. The quarks are given earlier in Table 14.4.

Particle	Mass (MeV/c^2)	Charge Q	Spin-parity J^π	Mean life
γ	$< 3 \times 10^{-33}$	0	1^-	Stable
G	0	0	1^-	
$W^\pm$	80398(25)	±1	1	$3.108(62) \times 10^{-25}$s
Z^0	91187.6(2.1)	0	1	$2.638(3) \times 10^{-25}$s
g	0	0	2	
e^-	0.510998910(13)	−1	1/2	Stable
ν_e	$< 2.6 \times 10^{-6}$	0	1/2	
μ^-	105.658367(4)	−1	1/2	$2.197019(21) \times 10^{-6}$s
ν_μ	< 0.19	0	1/2	
τ^-	1776.84(17)	−1	1/2	$290.6(1.0) \times 10^{-15}$s
ν_τ	< 18.2	0	1/2	

[a] See ref. 1.

The antiparticles are not listed in the table. Every antiparticle and its corresponding Particle possess identical mass, spin and lifetime, but opposite value for electric charge and lepton number.

14.4.5 Summary and Perspectives

1. Now we have six types of leptons; The e, μ, τ, and their corresponding neutrinos ν_e, ν_μ, ν_τ, and their antiparticles. So there are twelve leptons altogether. Important questions to be answered include (see Chapter 10 for a more detailed list of questions):

Are there more leptons including sterile and very massive (> 1 GeV) ones?
What are the ν rest masses and what are their mixing angles?
Are neutrinos a Dirac or a Majorana particle?
Does CP violation occur for leptons?
Are the lepton number conservation laws absolute or only good approximations?

2. There are six quark flavors in the Standard Model: These are the u, d, s, c, b, t quarks, and their antiparticles. But each has three colors, so that there are 36 quarks altogether. The quarks are held together in mesons and hadrons by gluons of which there are eight different colors.

Are all of these elementary particles?
What is really going on inside the proton and neutron?
Why are free quarks and free gluons so hard to find and what is the nature of a quark-gluon liquid?
Can we find the Higgs particle and are there more Higgs including very massive ones?
How are the leptons related to quarks?
Can they be unified on the basis of a deeper stratum?
Is baryon number conservation absolute or only a good approximation?
Quarks were introduced to bring simplicity and order to the multiplicity of particles. But now we need over 50 particles to explain all the data, and maybe more will be needed in the future.
Is there a deeper, simpler level that will order our new multiplicity of particles?

3. We have the following basic interactions:

The gravitational interaction (its mediator g has not been found)
The electroweak interaction
The electromagnetic interaction (photon mediator)
The weak interaction ($W^\pm$ and Z^0 mediators)
The strong interaction (gluon)
Residual strong interaction between hadrons (π mediator)
Fundamental strong interaction between quarks and gluons (gluon mediator)

The unified theory of the weak interaction and the electromagnetic interaction has been a great success. Now we have proposals of Grand Unified Theories which would include the strong interaction together with the electromagnetic and weak interactions and finally gravity. The proton decay, neutrino rest mass, and violation of lepton and baryon conservation are critical issues in the development of Grand Unified Theories.

Other important questions are bound up with the questions of the existence or nonexistence of a deeper stratum or strata both for leptons and quarks. Is there a more fundamental layer and fundamental interaction deeper than that of color charge? What is the real nature of color charge?

A fascinating journey of discovery is in prospect for those whose curiosity, imagination, and sense of adventure compel them to sign on. There will always be interplay between theory and experiment. Sometimes interest in theory may dominate. However, experiments are at the heart of physics, so we hope those entering into physics will be challenged to take up the call and apply all their imagination and ingenuity to design, build, and use new and more powerful facilities to probe the mysteries of the structure of our universe.

PROBLEMS

14.1. Which of the following processes can happen? What is the governing interaction of the process, if it does happen? What is violated if it does not happen? In the latter case, how would you change the left or right side of the arrow to make it correct?

(*a*) $e^- \to \nu_e + \gamma$	(*g*) $\pi^- + p \to \Sigma^+ + K^-$
(*b*) $p + \bar{p} \to n + \bar{\Sigma}^0 + \bar{K}^0$	(*h*) $K^- + p \to \Lambda^0 + K^0$
(*c*) $\mu^- \to e^- + \gamma$	(*i*) $p + p \to p + n + \pi^+ + \pi^+ + \pi^-$
(*d*) $\Sigma^0 \to \Lambda^0 + \gamma$	(*j*) $\mu^+ \to \pi^+ + \nu_\mu$
(*e*) $\Sigma^+ \to p + \gamma$	(*k*) $\tau^- \to \mu^- + \nu_\mu + \nu_\tau$
(*f*) $\Xi^- \to \Lambda^0 + \pi^-$	(*l*) $n + \nu_e \to p + \mu^-$

14.2. Show why the $W^\pm$ and Z^0 could not be discovered at the 500 GeV proton fixed-target accelerator at the Fermi National Laboratory.

14.3. Calculate the lifetime of J/ψ resonance if the width of the resonance is only 0.06 MeV/c^2. What would the width have been if it had a lifetime expected for decay by the strong interaction into hadrons?

14.4. Use the quark numbers to calculate the charge, baryon number, and strangeness for the K^0 in Fig. 14.21, Σ^- in Fig. 14.20, the Ω^- and Σ^{*+} in Fig. 14.22.

14.5. Only the (upsilon) Υ''' is observed to decay into a $B\bar{B}$ pair. If the B meson has a mass of 5.27 GeV/c^2, what can you infer about the masses of the Υ', Υ'' and Υ'''? Check your results against their known masses.

14.6. For a colorless meson with $r\bar{r}$, give an example of the color of a gluon that will bind these two quarks and show how the binding occurs through the gluon exchange. What color is present in the new colorless meson after the gluon exchange?

14.7. Calculate the threshold for the production of K^- and K^+ in the reaction

$$p + p \to p + p + K^- + K^+$$

Explain how the conservation rules are followed in this reaction.

14.8. Analyze the following reactions in terms of their quark structures:
(*a*) $\gamma + p + \pi^+ + n$
(*b*) $p + p \to p + \Lambda^0 + K^0 + \pi^+$
(*c*) $p + p \to p + p + p + \bar{p}$
(*d*) $K^- + p \to K^+ + K^0 + \Omega^-$
(*e*) $\pi^- + p \to \Lambda^0 + K^0$

14.9. Analyze the following decays in terms of their quark structures:
(*a*) $\Sigma^- \to \pi^- + n$
(*b*) $\Lambda^0 \to p + \pi^-$
(*c*) $K^0 \to \pi^+ + \pi^-$
(*d*) $K^+ \to \mu^+ + \nu_\mu$
(*e*) $\Delta^{*+} \to p + \pi^0$

14.10. Discuss the favored and unfavored decays of the D^+ (Case 7 in Section 14.3.6) in terms of the quark structure of the particle. What does the differences in these two decays tell you about the progression of quark flavor changes?

14.11. What was the solar neutrino problem?

14.12. What is meant by neutrino oscillation and how does this relate to the solar neutrino problem?

14.13. Give at least one possible decay mode of the following:
(*a*) $\bar{p}$ (*b*) Ω^+ (*c*) $\bar{K}^0$

14.14. What is the structure of the J/ψ particle? What is so unusual about this particle that its discovery is called the J/ψ or October revolution?

14.15. Discuss briefly the difference between the fundamental strong interaction mediated by the gluon and the residual strong interaction between hadrons mediated by the π meson.

14.16. Describe at least one piece of experimental evidence for why quarks must come in at least three "colors."

14.17. What is meant by saying quarks in nucleons are "colorless," and why do we say this?

14.18. What is "strange" about the strange quark? Why was strangeness proposed as a quantum number?

14.19. What is a resonance? Should such an entity be called a particle? Why or why not?

14.20. Give a possible decay mode of $\bar{n}$.

14.21. If an $\bar{n}$ annihilates instead of decaying, describe all the characteristics of the annihilation process, including how the various conservation laws involved are obeyed.

14.22. Calculate the threshold for the production of two π^+ and two π^- in the reaction

$$p + p \rightarrow p + p + \pi^+ + \pi^+ + \pi^- + \pi^-$$

Explain why there are two protons on the left and right, but four π mesons on the right and none on the left.

14.23. Discuss why electric charge conservation can be considered an absolute conservation law but baryon and lepton conservation laws cannot be considered absolute (at least at present).

APPENDIX I

ION BEAM ANALYSIS

Ion beam analysis is based on the interaction between an ion beam and materials. It can be used to analyze the elemental compositions in both quality and quantity and the structures of materials. It has been developed into an effective analytical tool in the last two decades. Since the first international conference on ion beam analysis held in 1973, a series of conferences have been held regularly every two years and the contents have been expanded to cover many fields (proceedings of these conferences can be found in *Nuclear Instruments and Methods B*). Ion beam analysis involves three major analytical methods: scattering, especially Rutherford backscattering; ion induced x-ray emission; and nuclear reactions.

In this appendix we briefly introduce the three analytical methods and show some results on ion beam analysis obtained at the Nuclear Physics Laboratory at Fudan University to give some practical examples of the applications of atomic and nuclear physics.

I.1 BACKSCATTERING TECHNIQUE

I.1.1 Fundamental Principles[1]

The ions from an ion source in an accelerator are accelerated to a certain energy, selected by an analyzing magnet, focused, and collimated. The monochromatic ion beam is then incident on a sample in a vacuum target chamber. An elastic

[1] W. K. Chu, J. W. Mayer, and M. A. Nicolet, *Backscattering Spectroscopy*, Academic Press, New York (1976).

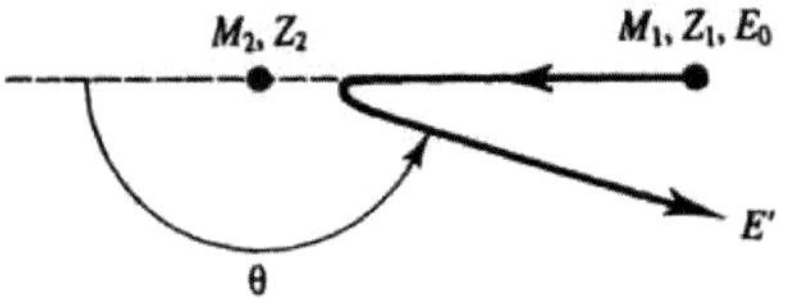

FIGURE I.1
Diagram of Rutherford backscattering.

collision will take place between the incident ions and the target (see Fig. I.1) if the energy of the ions is below the threshold for nuclear reactions (non-Coulomb interaction). The ion energy measured at a scattering angle θ is E'. In elastic scattering, according to the conservation laws of energy and momentum, E' can be expressed as

$$E' = K(\theta)E_0 \tag{I.1}$$

and

$$K(\theta) = \left(\frac{M_1 \cos\theta + \sqrt{M_2^2 - M_1^2 \sin^2\theta}}{M_1 + M_2} \right)^2 \tag{I.2}$$

where E_0 is the incident ion energy, M_1, Z_1 and M_2, Z_2 are the masses and atomic numbers of the incident ions and the target nuclei, respectively, and K is defined as the kinematic factor.

For example, if the projectiles are 2 MeV ^{4}He ions ($M_1 = 4$, $Z_1 = 2$), the E' for various target nuclei, scattered at $\theta = 180°$, are calculated and listed in Table I.1. The spectra corresponding to these nuclei and energies E' are shown in Fig. I.2. As seen in the figure, the peaks of various elements are separately located on the energy coordinate, and the corresponding peak area for each element [i.e., the integrated counts of $N(E')$] is proportional to the concentration of that element in the target and to the Rutherford scattering cross-section [see Eq. (2C.6), Chapter 2] for the case of a thin film target in which the energy loss of the particles is negligible.

For thick targets, the projectiles passing through the target will lose energy in collisions with the electrons in the target on both incoming and outgoing paths in addition to the energy loss in elastic collisions, reducing the energy of the ions from E_0 to E''. The energy loss per unit path length, related to ionization and

TABLE I.1
The backscattering energies of 2 MeV ^{4}He ions for carbon, silicon, and copper

Target	M_2	E'(MeV)	Target	M_2	E'(MeV)
C	12	0.50	Mo	96	1.69
Si	28	1.12	Pd	106	1.72
Cu	63	1.55	Au	197	1.84

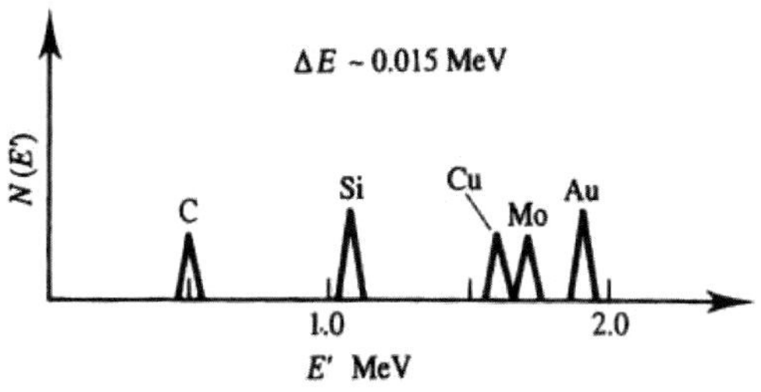

FIGURE I.2
Backscattering spectrum of 2 MeV ^{4}He ions from a thin target with various elements.

excitation, is often expressed as dE/dx and is called the stopping power. Because of this, the energy of the ions backscattered from atoms of the same kind is related to the depth at which the collisions take place (Fig. I.3). Therefore, from the analysis of the backscattering spectra, the depth profiles of atoms in the target can be obtained. When the projectiles are normally incident on a target, the energy difference ΔE between the particles scattered from the atoms on the first layer and that at a given depth t is

$$\Delta E = E' - E'' = KE_0 - E'' = [S]t \tag{I.3}$$

where for small thickness t (energy loss between the front and back is neglected)

$$[S] = \left(\frac{dE}{dx}\bigg|_{E_0} K + \frac{dE}{dx}\bigg|_{KE_0} \frac{1}{|\cos\theta|} \right) \tag{I.4}$$

is called the backscattering energy loss factor, where $dE/dx|_{E_0}$ is the energy loss for a particle with energy E_0 on the way in before scattering and $dE/dx|_{KE_0}$ is the energy loss on the way out with energy KE_0 after scattering at a depth t. We can see from Eq. (I.3) that there is a linear dependence between ΔE and t; hence the energy scale in the scattering spectra can easily be transformed into a depth scale of the target.

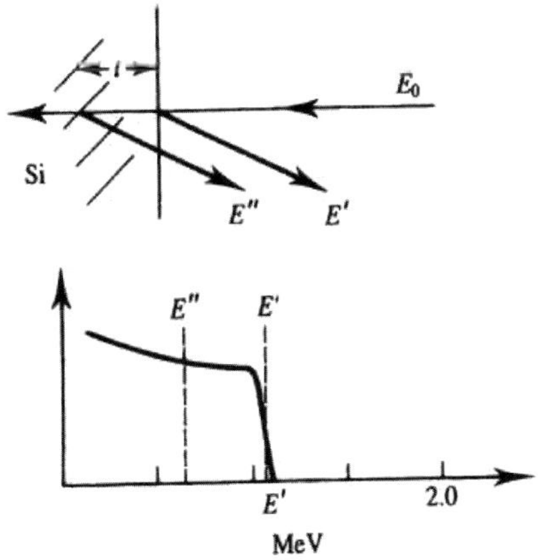

FIGURE I.3
Backscattering spectrum of 2 MeV ^{4}He ions from a thick silicon target.

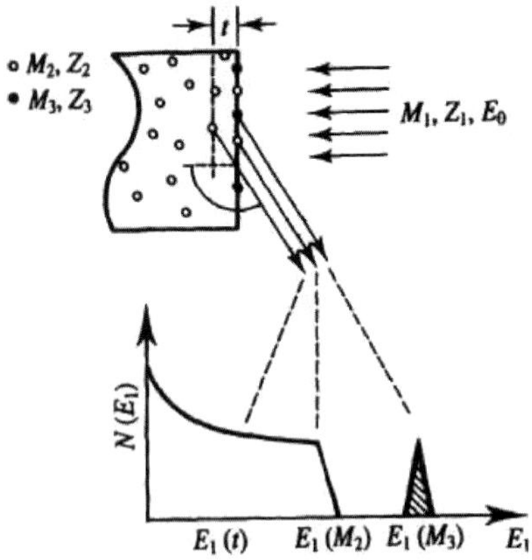

FIGURE I.4
Backscattering spectrum from a thin film of a heavy impurity on a thick substrate.

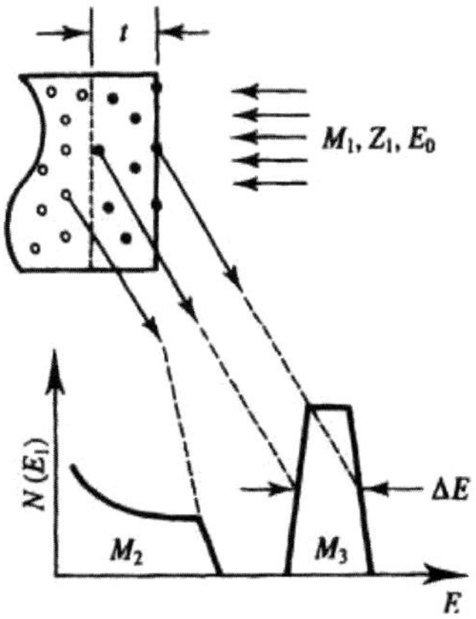

FIGURE I.5
Backscattering spectrum from a thick film of a heavy element on a thick substrate.

For a thick substrate with a thin film of impurity element (M_3, Z_3) on it, the backscattering spectrum measured is as shown in Fig. I.4 (assuming $Z_3 > Z_2$), and the spectrum will approach that in Fig. I.5 as the distribution of the impurity element reaches a certain thickness.

I.1.2 Experimental Examples

Measurement of the thickness of thin solid films. The thicknesses of thin metal films of Au, Ta, Co, and so forth, as well as thin films of silicon nitride and silicon dioxide on Si substrates can easily be measured by the backscattering method. The α backscattering spectrum measured by a surface barrier detector, conventionally used for charged particle detection, is shown in Fig. I.6, where the peaks of Si and Au are indicated. From the width of the Au peak, the thickness of the Au layer was calculated to be 141 Å($\pm 5\%$). It is very difficult to measure the thicknesses of thin films which are opaque or have thicknesses

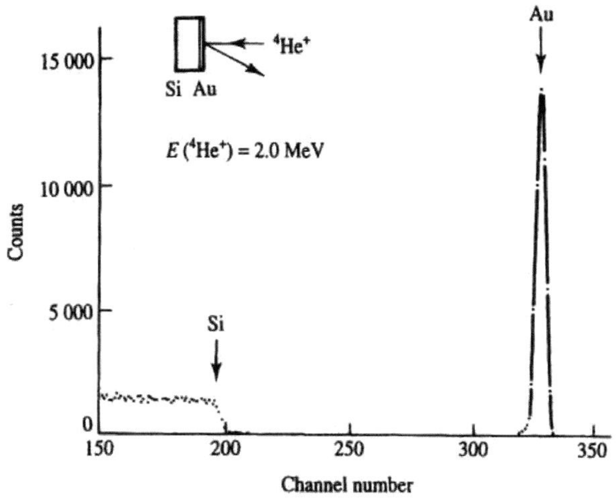

FIGURE I.6
Backscattering spectrum recorded in a surface barrier detector. The horizontal coordinate represents MCA channel number, which is proportional to the energy of the backscattering ions.

below 100 Å by ordinary optical methods. However, there is no such limitation for the backscattering method.

Determination of the stoichiometry of compounds and the relative elemental composition of a mixture of materials. The backscattering spectrum of a magnetic bubble material grown by epitaxy technology on a gadolinium–gallium target substrate is shown in Fig. I.7. Since it contains five elements, Fe, Ga, Y, Sm, and O, this thin film generally has the similar stoichiometry as the garnet, X_8O_{12}, where X denotes other elements except oxygen and the subscripts indicate the relative number of atoms. From the heights of the steps (see Fig. I.7) for each of the metal elements in the spectrum, combined with the scattering cross-sections and the energy loss factor correction, the composition of the film can be determined. In this case, the measured and the expected compositions are $Y_{2.32}\,Sm_{0.38}\,Ga_{1.2}\,Fe_{3.8}\,O_{12}$ and $Y_{2.6}\,Sm_{0.4}\,Ga_{1.2}\,Fe_{3.8}\,O_{12}$, respectively.

The composition of a magnetic bubble film influences its characteristics. One may regulate the content of these films by making composition measurements until an optimized elemental ratio is reached.

Measurement of the depth profile of chlorine in silicon dioxide. In the thermo-oxidation technology of certain semiconductor devices, chlorine doping is very effective in improving the characteristics of the devices. Measurements of the depth profiles of Cl in SiO_2 are helpful in understanding the behavior of the doped Cl ions. Figure I.8 shows a typical $^4He^+$ backscattering spectrum of this kind. Backscattering peaks of the Si substrate, Si and O in the surface layer of silicon dioxide, and Cl are indicated with arrows in the figure. The relative natural abundances of ^{37}Cl and ^{35}Cl are 24.23% and 75.77%, respectively, and their backscattering peaks are clearly separated from each other.

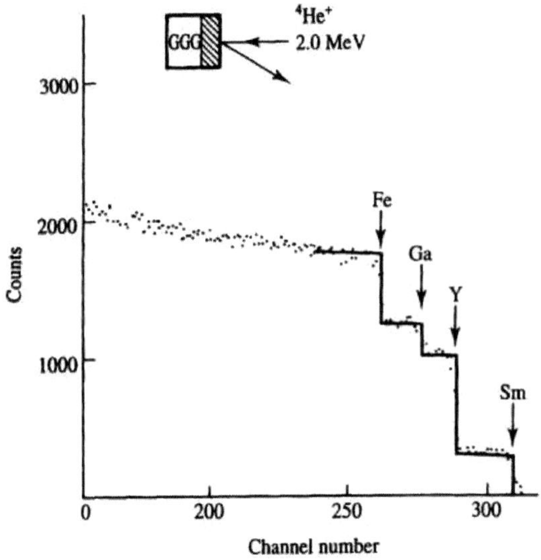

FIGURE I.7
Backscattering spectrum of a magnetic bubble material (GGG - gadolinium–gallium garnet substrate).

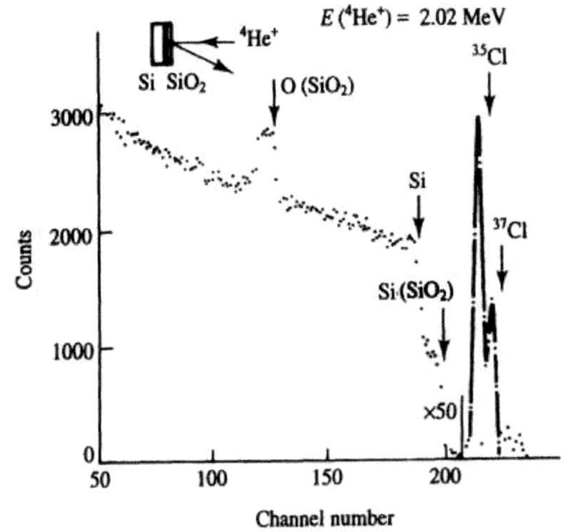

FIGURE I.8
Backscattering spectrum of a silicon sample doped with chlorine.

The corresponding depth of a Cl peak, the depth profile (width) of Cl, the concentration of Cl at the peak position, and the total amounts of Cl per unit area of samples can be determined from the backscattering spectrum.

The depth corresponding to the Cl peak is calculated as follows. First, the energy shift of the Cl peak, $\Delta E_{Cl} = K_{Cl}E_0 - E_{Cl}$, is obtained from the energy $K_{Cl}E_0$ of the Cl ions backscattered at the surface (in this case there is no Cl on the surface so no $K_{Cl}E_0$ peak on the far right in the figure) and the energy E_{Cl} corresponding to the Cl peak in the spectrum. Second, the depth position corresponding to the Cl peak can be calculated from the backscattering energy loss factor [S] of Cl in SiO, $t = \Delta E_{Cl}/[S]_{Cl}^{SiO_2}$. In the case shown, the experimental results indicate the depth position of the Cl is at the interface between the SiO layer and the Si substrate, which is in agreement with the conclusion obtained by other methods.

The concentration of Cl at the peak position is calculated from

$$N_{Cl} = \frac{H_{Cl}}{H_{Si}^{SiO_2}} \frac{\sigma_{Si}}{\sigma_{Cl}} \frac{[S]_{Cl}^{SiO_2}}{[S]_{Si}^{SiO_2}} N_{Si}^{SiO_2}$$

where H_{Cl} and $H_{Si}^{SiO_2}$ are the heights of the Cl peak and the surface peak of Si in SiO_2, respectively; σ_{Si} and σ_{Cl} are the scattering cross-sections of He^+ by Si and Cl; $[S]_{Cl}^{SiO_2}$ and $[S]_{Si}^{SiO_2}$ are the backscattering energy loss factors from Cl and Si in SiO_2; and $N_{Si}^{SiO_2}$ is the atomic density of Si in SiO_2. In the calculations there are two corrections to be included; one is the isotopic abundance ratio of the Cl element, and the other is the dependence of the cross-sections σ_{Cl} on energy. The calculated results are shown in Fig. I.9 as a depth profiling of Cl in the sample.

These are practical examples of the application of Rutherford backscattering. Although the history of Rutherford scattering extends over 70 years, because of limitations in detection techniques, backscattering did not become an analytical technique until 1967. In that year Turkevich put an α-radioactive source and detectors together in Surveyor 5, which landed on the Moon. The elemental compositions of the lunar soil were successfully analyzed by the system. The analytical results were in essential agreement with those obtained by chemical analysis made two years later from soil samples brought back from the Moon. The additional reason for the backscattering becoming so noteworthy is that the technique, combined with channeling effects discovered in the 1960s, provides a means of determining the lattice site location of impurities and lattice disorders. In recent years, backscattering and channeling techniques have been used to make detailed monolayer analyses at the surfaces and the interfaces of solids in combination with ultrahigh-vacuum chambers and other tools of surface analysis.[2]

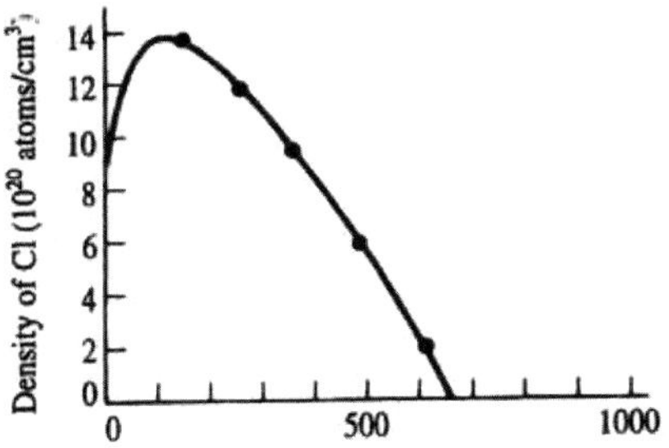

FIGURE I.9
Depth profiling of Cl in SiO_2.

[2] L. C. Feldman, *Nucl. Instrum. Meth.* **191** (1981) 211.

I.2 PROTON INDUCED X-RAY EMISSION: PIXE ANALYSIS

I.2.1 Fundamental Principles

When energetic protons impinge on a sample, there are certain probabilities of ionizations which eject inner-shell electrons out of the atoms and create inner vacancies. When an atomic inner vacancy is filled by an outer-shell electron, two processes can happen; one is the emission of a characteristic x-ray, and the other is Auger electron emission (see Chapter 6). If the production of inner vacancies is assumed to be in the K shell, the relative probability of these two processes,

$$\omega_K = \frac{\text{Number of K x-ray photons}}{\text{Number of K shell vacancies}}$$

is called the fluorescence yield of the K shell of a given element. The characteristic x-ray, Auger electron, and fluorescence yields all are unique functions of the element in which the vacancy occurs. Thus they are fingerprints of the elements. If the primary vacancies of the elements in a sample are induced by protons, the concentrations of the different elements can be determined by the energies and the number of x-rays emitted. This is called the proton induced x-ray emission (PIXE) analytical method.

The advantage of using protons to induce the x-rays rather than electrons is that protons have much less Bremsstrahlung (see Chapter 6; the intensity of Bremsstrahlung is inversely proportional to the square of the mass of the incident charged particles). The detection limit would be 10^2–10^4 times lower using electrons. In addition, the analysis of samples can be performed in the atmosphere or in a nitrogen environment by extracting the proton beam from a thin foil at the end of a vacuum chamber. This cannot be achieved with an electron beam. Such an extracted beam is particularly effective for the analysis of precious or very large antiquities and even live biological samples. Nonvacuum analysis can also be done by x-ray excitation, but it is quite difficult to focus and collimate the x-rays and the source intensity and analytical sensitivity are lower unless synchrotron radiation is used.

Figure I.10 shows a typical experimental setup for nonvacuum analysis by PIXE.

I.3 PRACTICAL EXAMPLES OF THE APPLICATION OF PIXE

I.3.1 Analysis of a Famous Sword[3]

A sword owned by Yue King of ancient China, Goujian, was unearthed from the No. 1 tomb at Wang Hill in Jiang Ling county of Hubei province in 1965.

[3] J. X. Chen, et al., *Nucl. Instrum. Meth.* **168** (1980) 437.

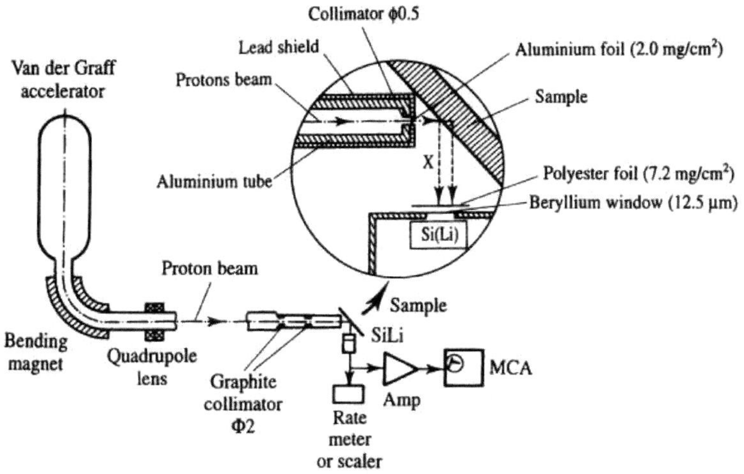

FIGURE I.10
Diagram of an experimental arrangement for nonvacuum analysis by PIXE.

Another unearthed at the same time was an auxiliary sword with the same patterns on its surface as the first one, but without any inscription on it. Although they were made in the Spring and Autumn Period in Chinese history (770–476 B.C.) and had been buried for more than 2500 years, the surfaces of the swords still have a metallic lustre and surprisingly sharp edges! These two are the most precious swords in the ancient sword museum of China and are well known around the world. If one wants to analyze such priceless relics it is first of all important to ensure that the analytical method to be used is nondestructive of the samples. The second consideration is the dimensions of the samples, which often are quite large (these swords are 64.1 cm in length). Finally, a detailed analysis of the different parts of a sample may be required. Based on these considerations, nonvacuum analysis by PIXE is the most effective method at present.

With the support of the museum of Hubei province and the National Antique Bureau of China, the two swords were successfully analyzed at Fudan University, in cooperation with the Shanghai Nuclear Institute and the Beijing Iron and Steel College. At the same time, a sword named Zhougou, and an arrowhead made in the Qin Dynasty of China (221–207 B.C.) were also analyzed in this way. Figure I.11 shows the PIXE spectra of different parts of the Goujian sword.

As seen in Fig. I.11, the composition of the sword is essentially copper and tin, as well as a small amount of iron and lead. The iron is an absolute impurity and its presence is an essential cause of rusting. The results show that the iron concentration of the sword is much lower than that in copper mined in China, and a sulphurating treatment helpful for anticorrosion was used at the surface of the sword. These results provide important insights into the level of metallurgical technology in ancient China.

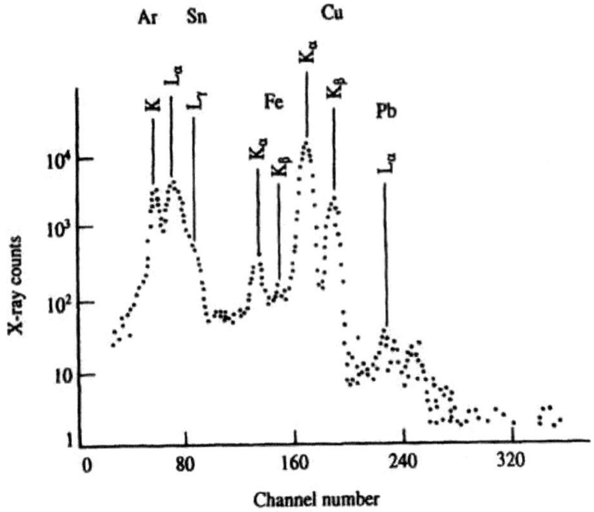

(a)

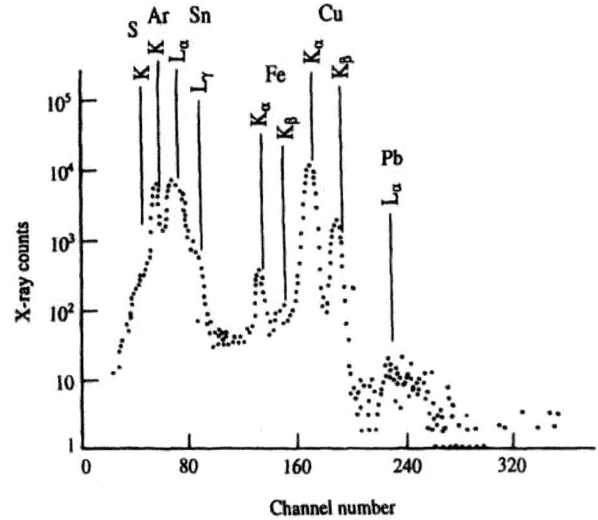

(b)

FIGURE I.11
(a) PIXE spectrum of the different parts of the Goujian sword in the yellow pattern. (b) PIXE spectrum of the different parts of the Goujian sword in the black pattern.

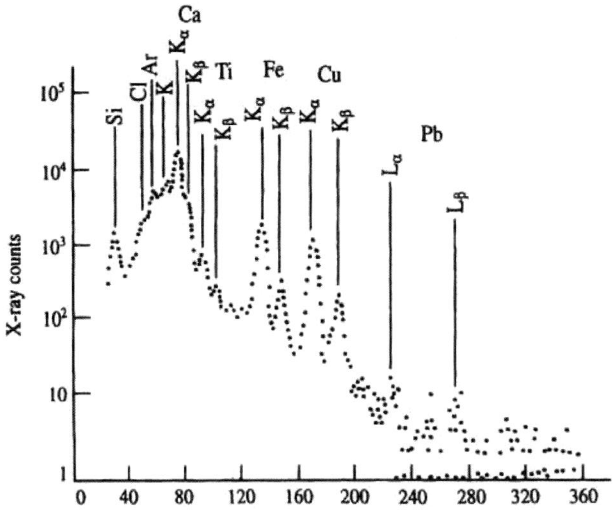

(c)

FIGURE I.11 (cont.)
(c) PIXE spectrum of the different parts of the Goujian sword in the glass decoration on the handle.

The results in Fig. I.11(c) show that the glass on the sword handle is a type of K–Ca glass; it is the oldest so far discovered in China.

The Ar peak in Fig. I.11 comes from an argon trace in the atmosphere. If the analysis is done in a helium environment, the Ar interference peak can be eliminated.

I.3.2 Analysis of Human Hair

The trace elements is human hair are usually a reflection of the situation in the human body. The study of the correlation between trace elements in the human body and diseases has become an important field in modern biological medicine.

Figure I.12 shows the PIXE spectrum of a hair sample taken from a person working in a factory using GaAs. Atomic absorption spectroscopy (AAS) was previously used as a routine method for hair analysis. However, a considerable amount of hair is required for each test and the samples have to be destroyed (i.e., the analysis is destructive), and only a few elements can be determined in each analysis. Using the PIXE method, only a few strands of hair are required and they can be used repeatedly (i.e., the analysis is nondestructive). The PIXE spectrum shown in Fig. I.12 was obtained by bombarding a hair sample for 30 minutes with 2 MeV protons at 20 nA beam current. The inset table in the figure gives a comparison between AAS and PIXE. As another example, it has been observed that

Sample	AAS (p.p.m.)		PIXE (p.p.m.)
	Zn	As	As
1	112.9	13.5	12.4 ± 1.4
2	122.8	4	5.7 ± 1.5
3	148.7	7	11.0 ± 2.6

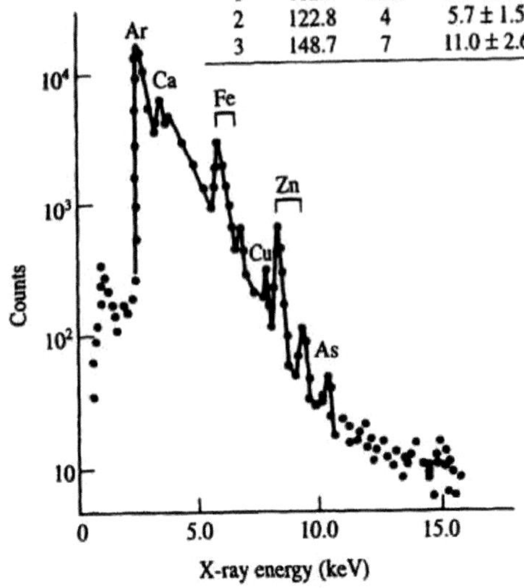

FIGURE I.12
PIXE spectrum of human hair contaminated with arsenic.

the Cu concentration is significantly lower in hair from children who are deficient in intelligence compared to normal children.[4]

I.3.3 Analysis of Air Pollution

It is very important in environmental protection to monitor air pollution frequently in cities. PIXE plays a significant role in this case, too. Figure I.13 shows a comparison between PIXE spectra of aerosols sampled from Lhasa in Tibet and from Shanghai, China. Obviously, the air is much cleaner in Lhasa than in Shanghai.

PIXE has been applied in many ways and recently developed towards microbeam methods (scanning proton microprobe or proton microscopy).[5]

I.4 NUCLEAR REACTION ANALYSIS

By utilizing a specific nuclear reaction induced by ion beam and measuring the reaction products, the species of the nuclei and their amounts in the target can be determined.

Example 1: Analysis of carbon. An excess concentration of carbon in semiconductors will severely influence the properties of electronic devices. PIXE is not very suitable for the analysis of carbon because of its small excitation cross-sec-

[4] J. X. Chen et al., *Nucl. Instrum. Meth.* **181** (1981) 269.

[5] See, for example, L. Zhong et al., *Nucl. Instr. Meth.* **B260** (2007) 283.

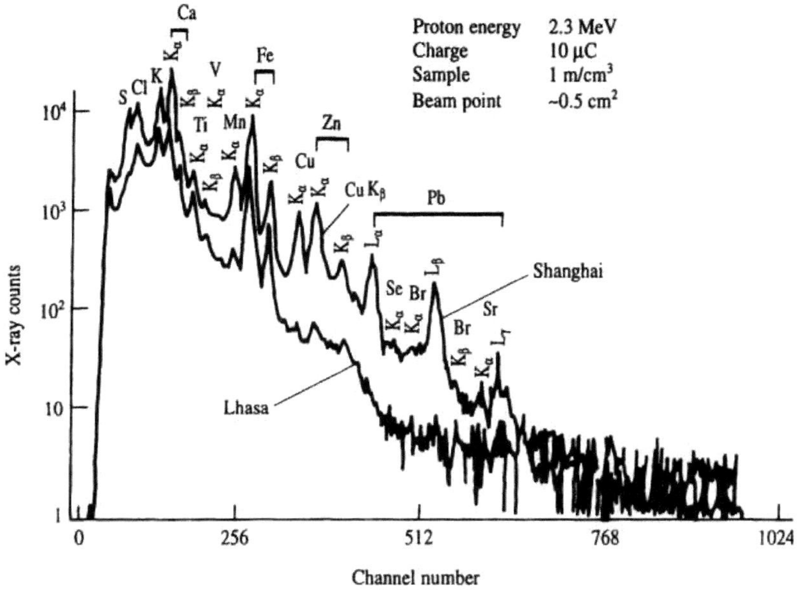

FIGURE I.13
PIXE analysis of an aerosol sample.

tion, lower x-ray energy, and difficulty in detection. It is also very difficult to analyze carbon with the Rutherford backscattering technique because, on the one hand, the Rutherford scattering cross-section is proportional to the square of the atomic number (for example, for 2 MeV incident α particles, the scattering cross-section of C is only 15% of that of Si), and on the other hand, the scattering peak of carbon in a backscattering spectrum will be difficult to see on the continuum background from the silicon substrate. Hence, unless the concentration of the C atoms compared to the Si atoms in the sample is more than 20%, the accuracy needed for carbon analysis cannot be obtained by backscattering of 2 MeV α particles.

If a nuclear resonant scattering of carbon ($^{12}C(\alpha, \alpha)^{12}C$ reaction) induced by α particles at 4.26 MeV is utilized, the cross-section of the resonance is 200 times greater than Rutherford scattering, so that a sensitivity of 0.1% can be achieved for the analysis of carbon in a silicon sample.

If we measure the β^+ radioactivity of ^{13}N (lifetime $T_{1/2} = 10\,\text{min}$) produced by the $^{12}C(d,n)^{13}N$ reaction, the trace of carbon in silicon samples also can be determined. In this case, the number of C atoms per unit volume in silicon samples can be calculated absolutely from the expression

$$n = \frac{N \ln 2}{I\epsilon T_{1/2}(1 - e^{-\lambda t_1})e^{-\lambda t_2}(1 - e^{-\lambda t_3}) \int_{E_{th}}^{E_0} \frac{\sigma(E)}{(-dE/dx)} dE}$$

where t_1 is the irradiation time, t_2 is the time interval between the end of irradiation and the start of a measurement, and t_3 is the duration of the measurement of the radioactivity of which the total count is N; I is the intensity of the beam current; ϵ is the efficiency of the detector; $\sigma(E)$ is the cross-section of the $^{12}C(d,n)^{13}N$ reaction; and $-dE/dx$ is the stopping power of a deuteron beam in silicon samples. Both σ and dE/dx vary with the deuteron energy. In addition, E_0 is the bombarding energy and $E_{th} = 0.329$ MeV is the threshold of the reaction.

The method involving measurement of the radioactivity mentioned above is called charged particle activation analysis. (If neutrons are used to induce nuclear reactions, it is called neutron activation analysis, which is not in the category of ion beam analysis). The total content, rather than the depth profile, of carbon is determined by this method.

To determine the depth profile of carbon in a sample, the $^{12}C(d,p)^{13}C$ reaction may be used and the energy spectrum of the emitted protons measured. However, the $^{12}C(^3He,p)^{14}N$ reaction at 2.42 MeV has recently been suggested as a better alternative. With this the depth resolution analysis of carbon in silicon samples is 109 Å, which is three times better than that obtained from the (d,p) reaction.

Example 2: Analysis of oxygen. Figure I.14 shows a charged particle spectrum obtained from the bombardment of an Si_3N_4 film (thickness 1370 Å) with 830 keV deuterons. There are peaks from reactions with C and N in addition to O. The relative concentration of O to N in the film can easily be deduced, with

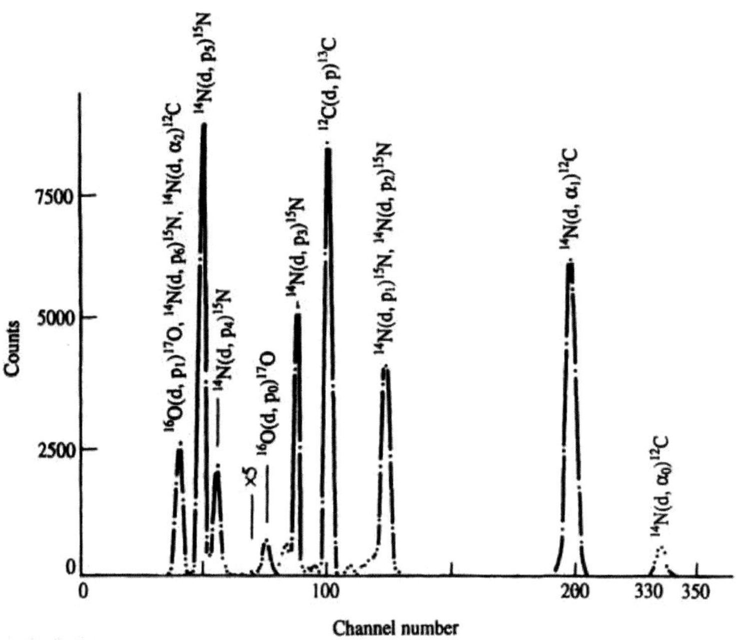

FIGURE I.14
Charged particle spectrum obtained from the bombardment of a 1370 Å Si_3N_4 film with 830 keV deuterons.

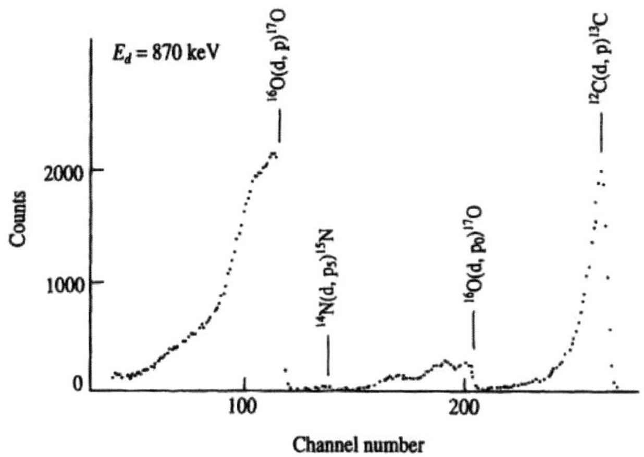

FIGURE I.15
Proton spectrum from the bombardment of an ancient arrowhead with 870 keV deuterons.

an analytical sensitivity of about 0.3% relative to the number of N atoms in the film.

Figure I.15 shows the energy spectrum of the particles emitted from the bombardment of a metal arrowhead (thick sample), made in the Qin Dynasty of China, using 870 keV deuterons.[6] In the spectrum we can see the peaks from the

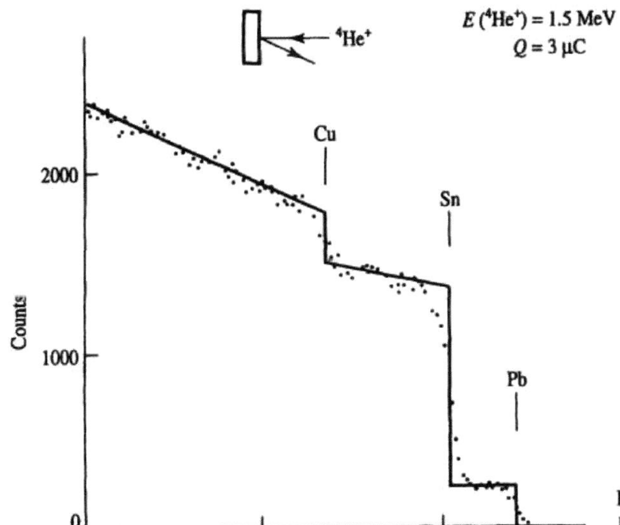

FIGURE I.16
Rutherford backscattering spectrum from the surface of the arrowhead.

[6] H. S. Chen, et al., *Nucl. Instrum. Meth.* **191** (1981) 391.

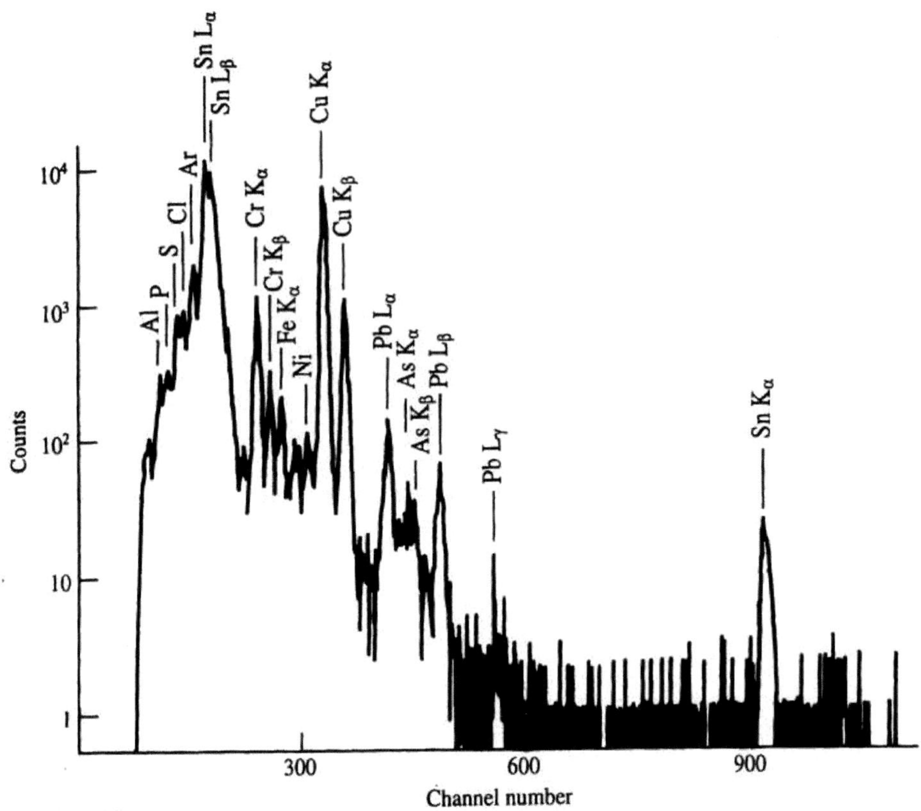

FIGURE I.17
PIXE spectrum from the arrowhead.

reactions of O, C, and N with deuterons. In Fig. I.14, p_1, p_2, etc., represent the protons with various energies corresponding to the different energy levels of the nuclei such as the ^{15}N formed in the reaction $^{14}N(d,p)^{15}N$. For comparison, the backscattering and PIXE spectra from the same arrowhead are shown in Fig. I.16 and Fig. I.17, respectively. Although the backscattering only provides information on the main elements in the sample, its calculation is very simple and the quantitative results are easily extracted. PIXE has a higher sensitivity [even trace elements (for example, chromium) can be clearly displayed in the spectrum] but for a thick target the calculation of the PIXE analysis is quite complicated. However, we may take the quantitative results for copper and tin obtained from the backscattering spectrum as an internal standard of the concentrations and then calculate the relative concentration of other elements. Not only backscattering but also the PIXE spectra provide little information on C, N, and O. Combining the results from the nuclear reaction, backscattering, and PIXE we can conclude that the surface of the arrowhead was treated by a chrominizing process and that there was a fine and tightly bound layer which consisted of CuO, SnO, and Cr_2O_3 at the surface. These make the arrowhead rust-free.

The use of the $^{16}O(^3He, \alpha)^{15}O$ reaction at 2.42 MeV in combination with the $^{12}C(^3He,p)^{14}N$ reaction has been suggested for the simultaneous analysis of carbon and oxygen. A depth resolution of 98 Å has been reached.

Example 3: Analysis of hydrogen. The analysis of hydrogen in materials is of significance because the amount of hydrogen will determine or influence the physical and chemical properties of various materials to a great extent. For instance, amorphous silicon doped with hydrogen becomes a semiconductor, which makes it possible to fabricate cheaper solar cells on a large scale. In the petroleum industry, catalysis and the poisoning of catalysts in thermosplitting and reformation reactions are closely related to the behavior of hydrogen. In superconductor studies, a significant problem is how to obtain the factor that determines or limits the phase transition point of superconductors. Most of these studies concentrated on Nb_3Ge because it had one of the early high-phase transitions points ($\approx$23.2 K) prior to 1986. Lanford, et al.[7] discovered that increasing the hydrogen content in Nb_3Ge reduced the phase transition point tremendously. Also, the strength of glass is closely related to the behavior of hydrogen. These are but a few important examples.

The analysis of hydrogen is rather difficult. Backscattering and PIXE techniques are unsuitable for analysis of hydrogen because of its small atomic number and mass. However, hydrogen concentration and its depth profile in materials can be effectively determined by certain specific resonant nuclear reactions, such as $^1H(^{15}N,\alpha\gamma)^{12}C$; $^1H(^{19}F,\alpha)^{16}O$; and $^1H(^{35}Cl,p)^{35}Cl$, and others. Using the $^1H(^{15}N,\alpha\gamma)^{12}C$ reaction with a width of 4 keV at the resonant energy of 6.385 MeV, depth resolutions of 40 Å and 18 Å have been reached for the analysis of hydrogen in glass and copper materials, respectively. Leich, et al.[8] used the $^1H(^{19}F, \alpha\gamma)^{16}O$ reaction to analyze the depth profiles of hydrogen in lunar soil samples. The depth resolution was estimated to be 50 Å and the analyzed depth was 3 μm.

Another effective method of hydrogen analysis is elastic recoil detection (ERD), which can be performed by a lower-energy helium beam from a small accelerator. The experimental conditions are quite simple, and the target chamber and detection system are nearly the same as used in backscattering.

Figure I.18 is a schematic diagram of a typical setup for an ERD experiment. When the Coulomb interaction occurs between the incident helium ions and the nuclei of hydrogen in the targets, the scattered protons will recoil in the forward direction. If the energy spectrum of the recoiling protons is measured, information on the concentration and depth profile of hydrogen in the target can be obtained. Because of its energy loss on the incoming path in the target material, the incident energy E_0 of ^4He ions will be reduced to E_1 before collision with the

[7] W. A. Lanford, et al., *Nucl. Instrum. Meth.* **149** (1978) 1.

[8] D. A. Leich, et al., *Earth and Planetary Sci. Lett.* **19** (1973) 305.

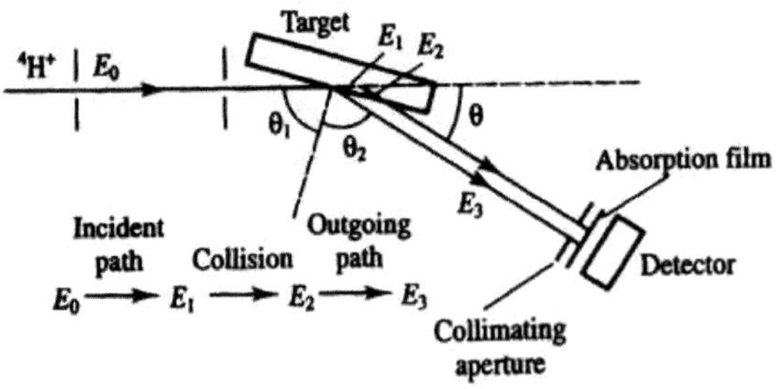

FIGURE I.18
Diagram of the experimental setup for hydrogen depth profile measured by ERD. Note in the figure the ray, including the proton direction that makes an angle θ_2 with the normal, appears to come from the surface of the crystal. This was done to illustrate the angle θ_2 and is not a real path. The He^+ ions actually enter the surface and lose energy before scattering off the protons which come out, as seen in the upper ray with energy E_3.

protons. After the collision, the energy transferred to the protons is E_2, and the detected energy of the recoil protons after losing energy on the outgoing path is E_3. Here θ_1 and θ_2 are the angles made by the normal to the surface and the

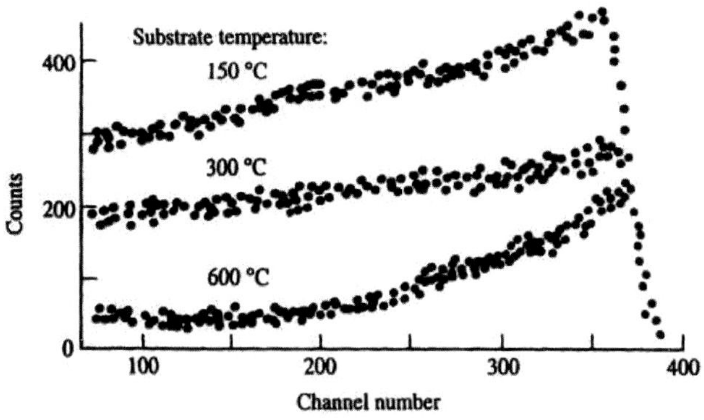

FIGURE I.19
ERD spectra of amorphous silicon films: $E_{He} = 2.1$ MeV, $Q = 30\,\mu$C.

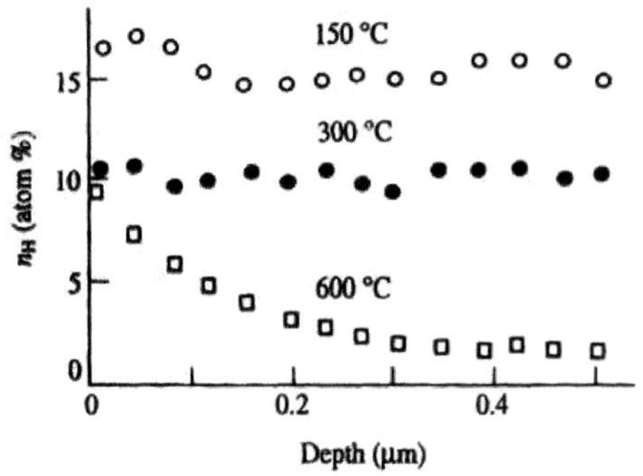

FIGURE I.20
Depth profiles of hydrogen in amorphous silicon films

direction of the incident beam and the recoil protons, respectively. In this experiment $\theta_1 = \theta_2 = 75°$ were chosen and the recoil angle was 30°.

Figure I.19 shows the ERD spectra of amorphous silicon films fabricated by plasma deposition. The hydrogen depth profiles deduced from the ERD spectra are shown in Fig. I.20.

The hydrogen concentration in plasma-deposited silicon nitride films is closely related to the corrosion resistance of the films. The corrosion resistance behavior of silicon nitride films is tested using a corrosive solution of BOF (HF 300 ml, NH_4F 600 g, H_2O 1000 g) and the hydrogen concentrations of the films are analyzed by ERD. The results show that if the hydrogen concentration changes by a factor of 2, the corrosion rate will vary by a factor of 20.[9] These data indicate how great the influence of hydrogen concentration can be on the properties of materials.

Ion beam analysis is one of the newest analytical methods and is still developing. New applications are being developed each year.

[9] H. Chen, et al., *Proc. 6th Int. Conf. on Ion Beam Analysis*, Arizona State University, Th. 47, also *Atomic and Nuclear Physics* (Chinese) **5**, 203 (1973).

APPENDIX II

TABLE OF ISOTOPES

Table II.1 lists all the known isotopes and all the known isomers (excited nuclear states) with a lifetime of 1 s or longer. The table is reproduced from the *Nuclear Wallet Cards*, J. K. Tuli, National Nuclear Data Center, Brookhaven National Laboratory (2005) and are used with permission.

The first column of each page lists the atomic number Z, the chemical symbol, and the mass number A for each isotope or isomer (the letter m after the mass number indicates an isomer).

The second column gives the spin and the intrinsic parity (+ or −) of the nucleus.

The third column gives the difference, Δ (mass excess = M − A), between the atomic mass of the isotope and A × (1 u), expressed in MeV where $\Delta(^{12}\mathrm{C}) = 0$ by definition. Since 1 MeV is equivalent to $1.073\,535 \times 10^{-3}$ u, the atomic mass is

$$M = A \times (1\,\mathrm{u}) + \Delta \times 1.073\,535 \times 10^{-3}\,\mathrm{u}$$

For example, for ^{16}O, the value of Δ is −4.737 MeV, so that

$$M = 16 \times (1\,\mathrm{u}) - 4.373 \times 1.073\,53 \times 10^{-3}\,\mathrm{u} = 15.994\,914\,66\,\mathrm{u}$$

The fourth column gives the abundance for naturally occurring isotopes or the half-life for artificially produced isotopes. The italic number on the right gives the uncertainty in the last decimal; for example, for ^{16}O the abundance is 99.762% and its uncertainty is ±0.015%.

The fifth column gives the decay modes. Most of the symbols in this column are self-explanatory; ϵ stands for electron capture and/or β^+ decay; IT for isomeric transition from one excited state to another; SF for spontaneous fission; and combined symbols, such as ϵp, for an initial decay followed by a beta delayed, secondary decay (here a proton).

TABLE II.1

Known isotopes and known isomers with lifetime ≥ 1 s

Nuclide Z	El	A	Jπ	Δ (MeV)	T½, Γ, or Abundance	Decay Mode
0	n	1	1/2+	8.071	10.24 m *2*	β–
1	H	1	1/2+	7.289	**99.985% *1***	
		2	1+	13.136	**0.015% *1***	
		3	1/2+	14.950	12.32 y *2*	β–
		4	2–	25.9	4.6 MeV *9*	n
		5		32.9	5.7 MeV *21*	n
		6	(2–)	41.9	1.6 MeV *4*	n
		7		49s	29×10^{-23} y *7*	
2	He	3	1/2+	14.931	**0.000137% *3***	
		4	0+	2.425	**99.999863% *3***	
		5	3/2–	11.39	0.60 MeV *2*	α, n
		6	0+	17.595	806.7 ms *15*	β–
		7	(3/2)–	26.10	150 keV *20*	n
		8	0+	31.598	119.0 ms *15*	β–, β–n 16%
		9	(1/2–)	40.94	65 keV *37*	n
		10	0+	48.81	0.17 MeV *11*	2n?
3	Li	3		29s	unstable	p?
		4	2–	25.3	6.03 MeV	p
		5	3/2–	11.68	≈1.5 MeV	α, p
		6	1+	14.087	**7.59% *4***	
		7	3/2–	14.908	**92.41% *4***	
		8	2+	20.947	838 ms *6*	β–, β–α
		9	3/2–	24.954	178.3 ms *4*	β–, β–n 50.8%
		10	(1–,2–)	33.05	1.2 MeV *3*	n
		11	3/2–	40.80	8.59 ms *14*	β–, β–nα 0.027%, β–n
		12		50.1s	<10 ns	n?
4	Be	5	(1/2+)	38s	?	p
		6	0+	18.375	92 keV *6*	p, α
		7	3/2–	15.770	53.22 d *6*	ε
		8	0+	4.942	6.8 eV *17*	α
		9	3/2–	11.348	**100%**	
		10	0+	12.607	1.51×10^{6} y *6*	β–
		11	1/2+	20.174	13.81 s *8*	β–, β–α 3.1%
		12	0+	25.08	21.49 ms *3*	β–, β–n ≤ 1%
		13	(1/2–)	33.25	2.7×10^{-21} s *18*	n
		14	0+	40.0	4.84 ms *10*	β–, β–n 94%, β–2n 6%
		15		49.8s	<200 ns	n?
		16	0+	57.7s	<200 ns	2n?
5	B	6		43.6s	unstable	2p?
		7	(3/2–)	27.87	1.4 MeV *2*	p, α
		8	2+	22.921	770 ms *3*	ε, εα
		9	3/2–	12.416	0.54 keV *21*	p,
		10	3+	12.051	**19.8% *3***	
		11	3/2–	8.668	**80.2% *3***	
		12	1+	13.369	20.20 ms *2*	β–, β–3α 1.58%
		13	3/2–	16.562	17.33 ms *17*	β–
		14	2–	23.66	12.5 ms *5*	β–, β–n 6.04%
		15		28.97	9.93 ms *7*	β–, β–n 93.6%, β–2n 0.4%
5	B	16	0–	37.08	<190 ps	n
		17	(3/2–)	43.8	5.08 ms *5*	β–, β–n 63%, β–2n 11%, β–3n 3.5%, β–4n 0.4%
		18	(4–)	52.3s	<26 ns	n?
		19	(3/2–)	59.4s	2.92 ms *13*	β–, β–n 72%, β–2n 16%
6	C	8	0+	35.09	230 keV *50*	p, α
		9	(3/2–)	28.910	126.5 ms *9*	ε, εp 83%, εα 17%
		10	0+	15.699	19.26 s *5*	ε
		11	3/2–	10.650	20.334 m *24*	ε
		12	0+	0.000	**98.89% *1***	
		13	1/2–	3.125	**1.11% *1***	
		14	0+	3.020	5700 y *30*	β–
		15	1/2+	9.873	2.449 s *5*	β–
		16	0+	13.694	0.747 s *8*	β–, β–n 99%
		17		21.04	193 ms *13*	β–, β–n 32%
		18	0+	24.93	92 ms *2*	β–, β–n 31.5%
		19		32.42	49 ms *4*	β–n 61%, β–
		20	0+	37.6	14 ms *+6–5*	β–, β–n 72%
		21	(1/2+)	46.0s	<30 ns	n?
		22	0+	53.3s	6.1 ms *+14–12*	β–, β–n 61%, β–2n > 0%
7	N	10	(1–)	38.8	20×10^{-23} y *14*	p?
		11m	1/2+	24.62	1.58 MeV *+75–52*	p
		12	1+	17.338	11.000 ms *16*	ε
		13	1/2–	5.345	9.965 m *4*	ε
		14	1+	2.863	**99.634% *20***	
		15	1/2–	0.101	**0.366% *20***	
		16	2–	5.684	7.13 s *2*	β–, β–α 1.2×10^{-3}%
		17	1/2–	7.87	4.173 s *4*	β–, β–n 95.1%
		18	1–	13.11	624 ms *12*	β–, β–n 14.3%, β–α 12.2%
		19		15.86	271 ms *8*	β–, β–n 54.6%
		20		21.77	130 ms *7*	β–, β–n 57%
		21	(1/2–)	25.25	85 ms *7*	β–, β–n 81%
		22		32.0	18 ms *4*	β–, β–n 36%, β–2n < 13%
		23		38.4s	14.1 ms *+12–15*	β–, β–n 42%, β–2n 8%
		24		47.5s	<52 ns	n?
		25		56.5s	<260 ns	n?
8	O	12	0+	32.05	0.40 MeV *25*	p
		13	(3/2–)	23.112	8.58 ms *5*	ε, εp = 100%
		14	0+	8.007	70.641 s *20*	ε
		15	1/2–	2.856	122.24 s *16*	ε
		16	0+	–4.737	**99.762% *16***	
		17	5/2+	–0.809	**0.038% *1***	
		18	0+	–0.781	**0.200% *14***	
		19	5/2+	3.335	26.88 s *5*	β–
		20	0+	3.797	13.51 s *5*	β–
8	O	21	(5/2+)	8.06	3.42 s *10*	β–
		22	0+	9.28	2.25 s *15*	β–, β–n < 22%
		23		14.6	82 ms *37*	β–, β–n 31%
		24	0+	19.1	65 ms *5*	β–, β–n 18%
		25	(3/2+)	27.4s	<50 ns	n
		26	0+	35.7s	<40 ns	n
		27		45.0s	<260 ns	n?
		28	0+	53.8s	<100 ns	n?
9	F	14	(2–)	32.7s	?	p
		15	(1/2+)	16.8	1.0 MeV *2*	p
		16	0–	10.680	40 keV *20*	p
		17	5/2+	1.952	64.49 s *16*	ε
		18	1+	0.874	1.8291 h *4*	ε
		19	1/2+	–1.487	**100%**	
		20	2+	–0.017	11.07 s *6*	β–
		21	5/2+	–0.048	4.158 s *20*	β–
		22	(4+)	2.79	4.23 s *4*	β–, β–n < 11%
		23	(3/2,5/2)+	3.33	2.23 s *14*	β–
		24	(1,2,3)+	7.56	400 ms *50*	β–, β–n < 5.9%
		25	(5/2+)	11.27	50 ms *6*	β–, β–n 14%
		26	1+	18.3	9.6 ms *8*	β–, β–n 11%
		27	(5/2+)	24.9	5.0 ms *2*	β–, β–n 77%
		28		33.2s	<40 ns	n
		29	(5/2+)	40.3s	2.5 ms *4*	β–, β–n, β–
		30		48.9s	<260 ns	n?
		31		56.3s	>260 ns	β–?, β–n?
10	Ne	16	0+	24.00	122 keV *37*	p
		17	1/2–	16.46	109.2 ms *6*	ε, εp = 100%, εα
		18	0+	5.317	1672 ms *8*	ε
		19	1/2+	1.751	17.22 s *2*	ε
		20	0+	–7.042	**90.48% *3***	
		21	3/2+	–5.732	**0.27% *1***	
		22	0+	–8.025	**9.25% *3***	
		23	5/2+	–5.154	37.24 s *12*	β–
		24	0+	–5.951	3.38 m *2*	β–
		25	(3/2)+	–2.11	602 ms *8*	β–
		26	0+	0.43	192 ms *6*	β–, β–n < 0.2%
		27	(3/2+)	7.1	32 ms *2*	β–, β–n 2%
		28	0+	11.2	19 ms *3*	β–, β–n 16%
		29	(3/2+)	18.1	15.6 ms *5*	β–, β–n 17%, β–2n < 2.9%
		30	0+	23.1	5.8 ms *2*	β–, β–n < 26%
		31		30.8s	3.4 ms *8*	β–
		32	0+	37.3s	3.5 ms *9*	β–
		33		46.0s	<260 ns	n?
		34	0+	53.1s	>1.5 μs	β–?, β–n?
11	Na	18	(1–)	24.19	1.3×10^{-21} s *4*	p?, ε?
		19	(5/2+)	12.93	<40 ns	p
		20	2+	6.848	447.9 ms *23*	ε, εα 20.05%
		21	3/2+	–2.184	22.49 s *4*	ε
		22	3+	–5.182	2.6027 y *10*	ε

TABLE II.1 (cont.)

Nuclide Z El	A	Jπ	Δ (MeV)	T½, Γ, or Abundance	Decay Mode
11 Na	23	3/2+	−9.530	**100%**	
	24	4+	−8.418	14.951 h *3*	β−
	25	5/2+	−9.358	59.1 s *6*	β−
	26	3+	−6.862	1.077 s *5*	β−
	27	5/2+	−5.517	301 ms *6*	β−, β−n 0.13%
	28	1+	−0.99	30.5 ms *4*	β−, β−n 0.58%
	29	3/2+	2.66	44.9 ms *12*	β−, β−n 21.5%
	30	2+	8.36	48 ms *2*	β−, β−n 30%, β− 1.17%, β−α 5.5×10^{-5}%
	31	3/2+	12.7	17.0 ms *4*	β−, β−n 37%, β− 0.9%
	32		19.1	13.2 ms *4*	β−, β−n 24%, β− 8%
	33		24.9	8.1 ms *4*	β−, β−n 47%, β−2n 13%
	34		32.8s	5.5 ms *10*	β−, β−n≤100%, β−
	35		39.6s	1.5 ms *5*	β−, β−n
	36		48.0s	<260 ns	n?
	37		55.3s	>1.5 μs	β−?, β−n?
12 Mg	19		33.0	?	2p?
	20	0+	17.57	90.8 ms *24*	ε, εp≈27%
	21	5/2+	10.91	122 ms *3*	ε, εp 32.6%, εα<0.5%
	22	0+	−0.397	3.8755 s *12*	ε
	23	3/2+	−5.474	11.317 s *11*	ε
	24	0+	−13.934	**78.99% 4**	
	25	5/2+	−13.193	**10.00% 1**	
	26	0+	−16.215	**11.01% 3**	
	27	1/2+	−14.587	9.458 m *12*	β−
	28	0+	−15.019	20.915 h *9*	β−
	29	3/2+	−10.62	1.30 s *12*	β−
	30	0+	−8.911	335 ms *17*	β−
	31		−3.22	230 ms *20*	β−, β−n 1.7%
	32	0+	−0.95	86 ms *5*	β−, β−n 5.5%
	33		4.89	90.5 ms *16*	β−, β−n 17%
	34	0+	8.8	20 ms *10*	β−, β−n
	35	(7/2−)	16.2s	70 ms *40*	β−, β−n 52%
	36	0+	21.4s	3.9 ms *13*	β−
	37	(7/2−)	29.2s	>260 ns	β−, β−n
	38	0+	35.0s	>260 ns	β−?
	39		43.6s	>260 ns	n?
	40	0+	50.2s	1 ms *sy*	β−?, β−n?
13 Al	21	(5/2+)	26.1s	<35 ns	p
	22	(3)+	18.18s	59 ms *3*	ε, εp≈60%, ε 0.9%, εα 0.31%
	23	3/2+	6.77	0.47 s *3*	ε, εp≈1.1%
	24	4+	−0.057	2.053 s *4*	ε, εα 0.04%, εp 1.6×10^{-3}%
	24m	1+	0.369	131.3 ms *25*	IT 82%, ε 18%, εα 0.03%
	25	5/2+	−8.916	7.183 s *12*	ε
	26	5+	−12.210	7.17×10^{5} y *24*	ε
13 Al	26m	0+	−11.982	6.3452 s *19*	ε
	27	5/2+	−17.197	**100%**	
	28	3+	−16.850	2.2414 m *12*	β−
	29	5/2+	−18.215	6.56 m *6*	β−
	30	3+	−15.87	3.60 s *6*	β−
	31	(3/2,5/2)+	−14.95	644 ms *25*	β−
	32	1+	−11.06	33 ms *4*	β−
	33	(5/2+)	−8.53	41.7 ms *2*	β−, β−n 8.5%
	34		−2.9	42 ms *6*	β−, β−n 27%
	35		−0.1	38.6 ms *4*	β−, β−n 41%
	36		5.8	90 ms *40*	β−, β−n<31%
	37		9.9	10.7 ms *13*	β−
	38		16.1	7.6 ms *6*	β−
	39	(3/2+)	21	7.6 ms *16*	β−
	40		29.3s	>260 ns	β−, β−n
	41		35.7s	>260 ns	β−
	42		43.7s	1 ms *sy*	β−?, β−n?
14 Si	22	0+	32.2s	29 ms *2*	ε, εp 32%
	23		23.8s	42.3 ms *4*	ε, εp≈73%, ε2p<4%
	24	0+	10.75	140 ms *8*	ε, εp 38%
	25	5/2+	3.82	220 ms *3*	ε, εp
	26	0+	−7.145	2.234 s *13*	ε
	27	5/2+	−12.384	4.16 s *2*	ε
	28	0+	−21.493	**92.230% 19**	
	29	1/2+	−21.895	**4.683% 8**	
	30	0+	−24.433	**3.087% 5**	
	31	3/2+	−22.949	157.3 m *3*	β−
	32	0+	−24.081	132 y *13*	β−
	33	(3/2+)	−20.49	6.18 s *18*	β−
	34	0+	−19.96	2.77 s *20*	β−
	35		−14.36	0.78 s *12*	β−
	36	0+	−12.5	0.45 s *6*	β−, β−n<10%
	37	(7/2−)	−6.6	90 ms *60*	β−, β−n 17%
	38	0+	−4.1	>1 μs	β−, β−n
	39	(7/2−)	1.9	47.5 ms *20*	β−
	40	0+	5.5	33.0 ms *10*	β−, β−n
	41		14	20.0 ms *25*	
	42	0+	18.4s	13 ms *4*	β−
	43		26.7s	>260 ns	β−?, β−n?
	44	0+	32.8s	10 ms *sy*	β−?, β−n?
15 P	24	(1+)	32.0s	?	p?, ε?
	25	(1/2+)	18.9s	<30 ns	p
	26	(3+)	11.0s	43.7 ms *6*	ε, εp
	27	1/2+	−0.72	260 ms *80*	ε, εp 0.07%
	28	3+	−7.159	270.3 ms *5*	ε, εp 1.3×10^{-3}%, εα 8.6×10^{-4}%
	29	1/2+	−16.953	4.142 s *15*	ε
	30	1+	−20.201	2.498 m *4*	ε
	31	1/2+	−24.441	**100%**	
	32	1+	−24.305	14.262 d *14*	β−
	33	1/2+	−26.337	25.34 d *12*	β−
	34	1+	−24.558	12.43 s *8*	β−
15 P	35	1/2+	−24.858	47.3 s *7*	β−
	36	4−	−20.25	5.6 s *3*	β−
	37		−18.99	2.31 s *13*	β−
	38		−14.8	0.64 s *14*	β−, β−n<10%
	39		−12.9	0.25 s *8*	β−, β−n 26%
	40	(2−,3−)	−8.1	125 ms *25*	β−, β−n 15.8%
	41		−5.3	100 ms *5*	β−, β−n 30%
	42		0.9	48.5 ms *15*	β−, β−n 50%
	43		5.8	36.5 ms *15*	β−, β−n
	44		12.1s	18.5 ms *25*	β−
	45		17.9s	>200 ns	β−?
	46		25.5s	>200 ns	β−
16 S	26	0+	26.0s	≈10 ms	2p?
	27	(5/2+)	17.5s	15.5 ms *15*	ε, εp 2.3%, ε2p 1.1%
	28	0+	4.1	125 ms *10*	ε, εp 21%
	29	5/2+	−3.16	187 ms *4*	ε, εp 47%
	30	0+	−14.063	1.178 s *5*	ε
	31	1/2+	−19.045	2.572 s *13*	ε
	32	0+	−26.016	**95.02% 9**	
	33	3/2+	−26.586	**0.75% 1**	
	34	0+	−29.932	**4.21% 8**	
	35	3/2+	−28.846	87.51 d *12*	β−
	36	0+	−30.664	**0.02% 1**	
	37	7/2−	−26.896	5.05 m *2*	β−
	38	0+	−26.861	170.3 m *7*	β−
	39	(3/2,5/2,7/2)−	−23.16	11.5 s *5*	β−
	40	0+	−22.9	8.8 s *22*	β−
	41	(7/2−)	−19.0	1.99 s *5*	β−, β−n
	42	0+	−17.7	1.013 s *15*	β−
	43		−12.0	0.28 s *3*	β−, β−n 40%
	44	0+	−9.1	100 ms *1*	β−, β−n 18%
	45		−3	68 ms *2*	β−, β−n 54%
	46	0+	0.7s	50 ms *8*	β−
	47		8.0s	>200 ns	β−?
	48	0+	13.2s	≥200 ns	β−
	49		22.0s	<200 ns	n
17 Cl	28	(1+)	26.6s	?	p?
	29	(3/2+)	13.1s	<20 ns	p
	30	(3+)	4.4s	<30 ns	p
	31		−7.07	150 ms *25*	ε, εp 0.7%
	32	1+	−13.330	298 ms *1*	ε, εα 0.05%, εp 0.03%
	33	3/2+	−21.003	2.511 s *3*	ε
	34	0+	−24.440	1.5264 s *14*	ε
	34m	3+	−24.293	32.00 m *4*	ε 55.4%, IT 44.6%
	35	3/2+	−29.014	**75.77% 4**	
	36	2+	−29.522	3.01×10^{5} y *2*	β− 98.1%, ε 1.9%
	37	3/2+	−31.761	**24.23% 4**	
	38	2−	−29.798	37.24 m *5*	β−
	38m	5−	−29.127	715 ms *3*	IT
	39	3/2+	−29.800	55.6 m *2*	β−

TABLE II.1 (cont.)

Nuclide Z El	A	Jπ	Δ (MeV)	T½, Γ, or Abundance	Decay Mode
17 Cl	40	2–	-27.56	1.35 m 2	β–
	41	(1/2+,3/2+)	-27.31	38.4 s 8	β–
	42		-24.9	6.8 s 3	β–
	43		-24.2	3.07 s 7	β–
	44		-20.2	0.56 s 11	β–, β–n<8%
	45		-18.4	400 ms 43	β–, β–n 24%
	46		-14.7	232 ms 2	β–, β–n 60%
	47		-10.5s	101 ms 6	β–, β–n>0%
	48		-4.7s	≥200 ns	β–
	49		0.3s	≥170 ns	
	50		7.3s	20 ms sy	β–?
	51	(3/2+)	13.5s	>200 ns	β–
18 Ar	30	0+	20.1s	<20 ns	p?
	31	5/2+	11.3s	15.1 ms 13	ε, εp 69%, ε 7.6%
	32	0+	-2.200	98 ms 2	ε, εp 43%
	33	1/2+	-9.384	173.0 ms 20	ε, εp 38.7%
	34	0+	-18.377	844.5 ms 34	ε
	35	3/2+	-23.047	1.775 s 4	ε
	36	0+	-30.232	**0.3365% 30**	
	37	3/2+	-30.948	34.95 d 4	ε
	38	0+	-34.715	**0.0632% 5**	
	39	7/2–	-33.242	269 y 3	β–
	40	0+	-35.040	**99.6003% 30**	
	41	7/2–	-33.068	109.61 m 4	β–
	42	0+	-34.423	32.9 y 11	β–
	43	(5/2–)	-32.010	5.37 m 6	β–
	44	0+	-32.673	11.87 m 5	β–
	45		-29.771	21.48 s 15	β–
	46	0+	-29.72	8.4 s 6	β–
	47	(3/2–)	-25.9	1.23 s 3	β–, β–n<0.002%
	48	0+	-23.7s	0.48 s 40	β–
	49		-18.1s	≥170 ns	
	50	0+	-14.5s	≥170 ns	
	51		-7.8s	>200 ns	β–?
	52	0+	-3.0s	10 ms	β–
	53	(5/2–)	4.6s	3 ms sy	β–, β–n
19 K	32		20.4s	?	p?
	33	(3/2+)	6.8s	<25 ns	p
	34	(1+)	-1.5s	<25 ns	p
	35	3/2+	-11.17	178 ms 8	ε, εp 0.37%
	36	2+	-17.426	342 ms 2	ε, εp 0.05%, εα 3.4×10^{-3}%
	37	3/2+	-24.800	1.226 s 7	ε
	38	3+	-28.801	7.636 m 18	ε
	38m	0+	-28.670	924.2 ms 3	ε
	39	3/2+	-33.807	**93.2581% 44**	
	40	4–	-33.535	1.248×10^{9} y 3 **0.0117% 1**	β– 89.28%, ε 10.72%
	41	3/2+	-35.559	**6.7302% 44**	
	42	2–	-35.022	12.321 h 25	β–
	43	3/2+	-36.593	22.3 h 1	β–
	44	2–	-35.81	22.13 m 19	β–

Nuclide Z El	A	Jπ	Δ (MeV)	T½, Γ, or Abundance	Decay Mode
19 K	45	3/2+	-36.61	17.3 m 6	β–
	46	(2–)	-35.42	105 s 10	β–
	47	1/2+	-35.696	17.50 s 24	β–
	48	(2–)	-32.12	6.8 s 2	β–, β–n 1.14%
	49	(3/2+)	-30.32	1.26 s 5	β–, β–n 86%
	50	(0–,1,2–)	-25.4	472 ms 4	β–, β–n 29%
	51	(1/2+,3/2+)	-22.0s	365 ms 5	β–, β–n 47%
	52	(2–)	-16.2s	105 ms 5	β–, β–n=64%, β–
	53	(3/2+)	-12.0s	30 ms 5	β–, β–n=67%, β–2n=17%
	54		-5.4s	10 ms 5	β–, β–n>0%
	55		-0.3s	3 ms sy	β–?, β–n?
20 Ca	34	0+	13.2s	<35 ns	p
	35		4.6s	25.7 ms 2	ε, εp 95.7%, ε2p 4.2%
	36	0+	-6.44	102 ms 2	ε, εp 57%
	37	3/2+	-13.16	181.1 ms 10	ε, εp 82.1%
	38	0+	-22.059	440 ms 8	ε
	39	3/2+	-27.274	859.6 ms 14	ε
	40	0+	-34.846	$>3.0\times10^{21}$ y **96.94% 16**	2ε
	41	7/2–	-35.138	1.02×10^{5} y 7	ε
	42	0+	-38.547	**0.647% 23**	
	43	7/2–	-38.409	**0.135% 10**	
	44	0+	-41.468	**2.09% 11**	
	45	7/2–	-40.812	162.61 d 9	β–
	46	0+	-43.135	$>0.28\times10^{16}$ y **0.004% 3**	2β–
	47	7/2–	-42.340	4.536 d 3	β–
	48	0+	-44.214	2.3×10^{19} y +12–6 **0.187% 21**	2β– 84%, β– <25%
	49	3/2–	-41.289	8.718 m 6	β–
	50	0+	-39.571	13.9 s 6	β–
	51	(3/2–)	-35.86	10.0 s 8	β–, β–n
	52	0+	-32.5	4.6 s 3	β–, β–n≤2%
	53	(3/2–,5/2–)	-27.9s	90 ms 15	β–, β–n>30%
	54	0+	-23.9s	>300 ns	β–
	55		-18.1s	>300 ns	β–?
	56	0+	-13.4s	10 ms sy	β–?
	57		-7s	5 ms sy	β–?, β–n?
21 Sc	36		13.9s	?	p?
	37		2.8s	?	p?
	38	(2–)	-4.9s	<300 ns	p
	39	(7/2–)	-14.17	<300 ns	p
	40	4–	-20.523	182.3 ms 7	ε, εp 0.44%, εα 0.02%
	41	7/2–	-28.642	596.3 ms 17	ε
	42	0+	-32.121	681.3 ms 7	ε
	42m	(7)+	-31.505	61.7 s 4	ε
	43	7/2–	-36.188	3.891 h 12	ε
	44	2+	-37.816	3.97 h 4	ε
	44m	6+	-37.545	58.61 h 10	IT 98.8%, ε 1.2%

Nuclide Z El	A	Jπ	Δ (MeV)	T½, Γ, or Abundance	Decay Mode
21 Sc	45	7/2–	-41.068	**100%**	
	45m	3/2+	-41.056	318 ms 7	IT
	46	4+	-41.757	83.79 d 4	β–
	46m	1–	-41.615	18.75 s 4	IT
	47	7/2–	-44.332	3.3492 d 6	β–
	48	6+	-44.496	43.67 h 9	β–
	49	7/2–	-46.552	57.2 m 2	β–
	50	5+	-44.54	102.5 s 5	β–
	50m	(2,3)+	-44.28	0.35 s 4	IT>97.5%, β–<2.5%
	51	(7/2)–	-43.22	12.4 s 1	β–
	52	3(+)	-40.4	8.2 s 2	β–
	53	(7/2–)	-37.6s	>3 s	β–, β–n
	54	(3,4+)	-34.2	0.36 s 6	β–
	55	(7/2–)	-29.6	0.115 s 15	β–, β–n
	56	(1+)	-25.3s	35 ms 5	β–, β–n
	56	(6+,7+)	-25.3s	60 ms 7	β–, β–n
	57		-20.7s	13 ms 4	β–
	57	(7/2–)	-20.7s	13 ms 4	β–n
	58	(3+)	-15.2s	12 ms 5	β–
	59		-10.0s	10 ms sy	β–?, β–n?
	60		-4.0s	3 ms sy	β–
22 Ti	38	0+	9.1s	<120 ns	2p?
	39	(3/2+)	1.5s	31 ms +6–4	ε, εp 14%
	40	0+	-8.9	53.3 ms 15	ε, εp
	41	3/2+	-15.7s	80.4 ms 9	ε, εp=100%
	42	0+	-25.122	199 ms 6	ε
	43	7/2–	-29.321	509 ms 5	ε
	44	0+	-37.549	60.0 y 11	ε
	45	7/2–	-39.006	184.8 m 5	ε
	46	0+	-44.123	**8.25% 3**	
	47	5/2–	-44.932	**7.44% 2**	
	48	0+	-48.488	**73.72% 3**	
	49	7/2–	-48.559	**5.41% 2**	
	50	0+	-51.427	**5.18% 2**	
	51	3/2–	-49.728	5.76 m 1	β–
	52	0+	-49.465	1.7 m 1	β–
	53	(3/2)–	-46.8	32.7 s 9	β–
	54	0+	-45.6	1.5 s 4	β–
	55	(3/2–)	-41.7	1.3 s 1	β–
	56	0+	-38.9	200 ms 5	β–, β–n
	57		-33.5	60 ms 16	β–, β–n
	58	0+	-30.8s	59 ms 9	β–
	59	(5/2–)	-25.2s	30 ms 3	β–
	60	0+	-21.6s	22 ms 2	β–
	61		-15.6s	>300 ns	β–?
	62	0+	-11.7s	10 ms sy	β–?
	63		-5.2s	3 ms sy	β–?, β–n?
23 V	40		10.3s	?	p?
	41		-0.2s	?	p?
	42		-8.2s	<55 ns	p
	43		-18.0s	>800 ms	ε
	44	(2+)	-24.1	111 ms 7	ε, εα

TABLE II.1 (cont.)

Nuclide Z El	A	Jπ	Δ (MeV)	T½, Γ, or Abundance	Decay Mode
23 V	44m	(6+)	-24.1	150 ms *3*	ε
	45	7/2−	-31.88	547 ms *6*	ε
	46	0+	-37.073	422.50 ms *11*	ε
	47	3/2−	-42.002	32.6 m *3*	ε
	48	4+	-44.475	15.9735 d *25*	ε
	49	7/2−	-47.957	329 d *3*	ε
	50	6+	-49.222	1.4×10^{17} y *4*	ε 83%,
				0.250% *2*	β− 17%
	51	7/2−	-52.201	**99.750% *2***	
	52	3+	-51.441	3.743 m *5*	β−
	53	7/2−	-51.849	1.60 m *4*	β−
	54	3+	-49.89	49.8 s *5*	β−
	55	(7/2−)	-49.2	6.54 s *15*	β−
	56	(1+)	-46.1	216 ms *4*	β−
	56	1+	-46.1	216 ms *4*	β−n 0.03%
	57	(3/2−)	-44.2	0.35 s *1*	β−, β−n 0.04%
	58	(1+)	-40.2	185 ms *10*	β−
	59	(5/2−,3/2−)	-37.1	75 ms *7*	β−
	60		-32.6	68 ms *5*	β−
	60m		-32.6	122 ms *18*	β−, β−ε
	61	(3/2−)	-29.4s	47 ms *1*	β−
	62		-24.4s	33.5 ms *2*	β−
	63	(7/2−)	-20.9s	17 ms *3*	β−
	64		-15.4s	>150 ns	β−
	65		-11.3s	10 ms *sy*	β−?, β−n?
24 Cr	42	0+	6.0s	13 ms *+4−2*	ε
	43	(3/2+)	-2.1s	21.6 ms *7*	ε, εp 23%, ε 6%
	44	0+	-13.46s	53 ms *+4−3*	ε, εp>7%
	45		-19.0	50 ms *6*	ε, εp>27%
	46	0+	-29.47	0.26 s *6*	ε
	47	3/2−	-34.56	500 ms *15*	ε
	48	0+	-42.819	21.56 h *3*	ε
	49	5/2−	-45.331	42.3 m *1*	ε
	50	0+	-50.259	$>1.3\times10^{18}$ y	2ε
				4.345% *13*	
	51	7/2−	-51.449	27.7025 d *24*	ε
	52	0+	-55.417	**83.789% *18***	
	53	3/2−	-55.285	**9.501% *17***	
	54	0+	-56.932	**2.365% *7***	
	55	3/2−	-55.107	3.497 m *3*	β−
	56	0+	-55.281	5.94 m *10*	β−
	57	3/2−,5/2−,7/2−	-52.524	21.1 s *10*	β−
	58	0+	-51.8	7.0 s *3*	β−
	59	(1/2−)	-47.9	0.46 s *5*	β−
	60	0+	-46.5	0.57 s *6*	β−
	61		-42.2	0.27 s *2*	β−
	62	0+	-40.4	209 ms *12*	β−, β−n
	63	(1/2−)	-35.5s	129 ms *2*	β−, β−n
	64	0+	-33.2s	43 ms	β−
	65	(1/2−)	-27.8s	27 ms *3*	β−, β−n?
	66	0+	-24.8s	10 ms *6*	β−
	67		-19.0s	≈50 ms	β−?
25 Mn	44	(2−)	6.4s	<105 ns	ε, p
	45	(7/2−)	-5.1s	<70 ns	p
	46	[4+]	-12.4s	34 ms *+5−4*	ε, εp 22%
	47		-22.3s	100 ms *50*	ε, εp>3.4%
	48	4+	-29.3	158.1 ms *22*	ε, εp 0.28%,
					εα<6.0×10^{-4}%
	49	5/2−	-37.62	382 ms *7*	ε
	50	0+	-42.627	283.29 ms *8*	ε
	50m	5+	-42.398	1.75 m *3*	ε
	51	5/2−	-48.241	46.2 m *1*	ε
	52	6+	-50.705	5.591 d *3*	ε
	52m	2+	-50.328	21.1 m *2*	ε 98.25%, IT 1.75%
	53	7/2−	-54.688	3.74×10^{6} y *4*	ε
	54	3+	-55.555	312.12 d *6*	ε, β−<2.9×10^{-4}%
	55	5/2−	-57.711	**100%**	
	56	3+	-56.910	2.5789 h *1*	β−
	57	5/2−	-57.487	85.4 s *18*	β−
	58	1+	-55.91	3.0 s *1*	β−
	58m	(4)+	-55.83	65.2 s *5*	β−≈80%, IT≈20%
	59	(5/2)−	-55.48	4.59 s *5*	β−
	60	0+	-53.18	51 s *6*	β−
	60m	3+	-52.91	1.77 s *2*	β− 88.5%, IT 11.5%
	61	(5/2)−	-51.6	0.67 s *4*	β−
	62	1+	-48.0	92 ms *13*	β−
	62	(3+,4+)	-48.0	671 ms *5*	β−, β−n
	63	(5/2−)	-46.4	0.29 s *2*	β−
	64		-42.6	89 ms *4*	β−, β−n 1.42%
	65		-40.7	92 ms *1*	β−, β−n 6.92%
	66		-36.3s	64 ms *2*	β−, β−n 10.88%
	67	(5/2−)	-33.4s	47 ms *4*	β−, β−n
	69	5/2−	-25.3s	14 ms *4*	β−
26 Fe	45	(3/2+)	13.6s	3.8 ms *+20−8*	2p
	46	0+	0.8s	12 ms *+4−3*	
	47		-6.6s	21.8 ms *7*	ε, εp
	48	0+	-18.16s	44 ms *7*	ε, εp>3.6%
	49	(7/2−)	-24.6s	70 ms *3*	ε, εp≥52%
	50	0+	-34.48	155 ms *11*	ε, εp≈0%
	51	5/2−	-40.22	305 ms *5*	ε
	52	0+	-48.332	8.275 h *8*	ε
	52m	(12+)	-41.512	45.9 s *6*	ε
	53	7/2−	-50.945	8.51 m *2*	ε
	53m	19/2−	-47.905	2.526 m *24*	IT
	54	0+	-56.252	**5.845% *35***	
	55	3/2−	-57.479	2.737 y *11*	ε
	56	0+	-60.605	**91.754% *36***	
	57	1/2−	-60.180	**2.119% *10***	
	58	0+	-62.153	**0.282% *4***	
	59	3/2−	-60.663	44.495 d *9*	β−
	60	0+	-61.412	1.5×10^{6} y *3*	β−
	61	3/2−,5/2−	-58.92	5.98 m *6*	β−
	62	0+	-58.90	68 s *2*	β−
	63	(5/2)−	-55.5	6.1 s *6*	β−
26 Fe	64	0+	-54.8	2.0 s *2*	β−
	65		-50.9	1.3 s *3*	β−
	66	0+	-49.6	0.44 s *6*	β−
	67		-45.7	0.47 s *5*	β−, β−n 1.13%
	68	0+	-43.1	187 ms *6*	β−
	69	1/2−	-38.4s	109 ms *9*	β−
	70	0+	-35.9s	94 ms *17*	β−
	71	(7/2+)	-31.0s	>150 ns	β−
	72	0+	-28.3s	>150 ns	β−
27 Co	47		10.7s	?	p?
	49		-9.6s	<35 ns	ε, p
	50	(6+)	-17.2s	44 ms *4*	ε, εp>54%
	51	(7/2−)	-27.3s	>200 ns	ε
	52	(6+)	-33.92s	115 ms *23*	ε
	53	(7/2−)	-42.64	240 ms *9*	ε
	53m	(19/2−)	-39.45	247 ms *12*	ε≈98.5%, p≈1.5%
	54	0+	-48.009	193.28 ms *7*	ε
	54m	(7)+	-47.812	1.48 m *2*	ε
	55	7/2−	-54.028	17.53 h *3*	ε
	56	4+	-56.039	77.233 d *27*	ε
	57	7/2−	-59.344	271.74 d *6*	ε
	58	2+	-59.846	70.86 d *6*	ε
	58m	5+	-59.821	9.04 h *11*	IT
	59	7/2−	-62.228	**100%**	
	60	5+	-61.649	1925.28 d *14*	β−
	60m	2+	-61.590	10.467 m *6*	IT 99.76%, β− 0.24%
	61	7/2−	-62.898	1.650 h *5*	β−
	62	2+	-61.43	1.50 m *4*	β−
	62m	5+	-61.41	13.91 m *5*	β−>99%, IT<1%
	63	7/2−	-61.84	27.4 s *5*	β−
	64	1+	-59.79	0.30 s *3*	β−
	65	(7/2)−	-59.17	1.20 s *6*	β−
	66	(3+)	-56.1	0.18 s *1*	β−
	67	(7/2−)	-55.1	0.425 s *20*	β−
	68	(7−)	-51.4	0.199 s *21*	β−
	68m	(3+)	-51.4	1.6 s *3*	β−
	69	7/2−	-50.0	0.22 s *2*	β−
	70	(6−)	-45.6	119 ms *6*	β−
	70m	(3+)	-45.6	0.50 s *18*	β−
	71		-43.9	79 ms *5*	β−, β−n 2.61%
	72	(6−,7−)	-39.3s	62 ms *3*	β−, β−n 4.8%
	73		-37.0s	41 ms *4*	β−
	74	0+	-32.2s	>150 ns	β−
	75	(7/2−)	-29.5s	>150 ns	β−
28 Ni	48	0+	18.4s	>0.5 μs	ε
	49		9.0s	12 ms *+5−3*	ε, εp?
	50	0+	-3.8s	12 ms *3*	εp 70%, ε
	51	(7/2−)	-11.4s	>200 ns	ε
	52	0+	-22.65s	38 ms *5*	ε, εp 17%
	53	(7/2−)	-29.4s	45 ms *15*	ε, εp≈45%
	54	0+	-39.21	104 ms *7*	ε
	55	7/2−	-45.34	202 ms *3*	ε

TABLE II.1 (cont.)

Nuclide Z El	A	Jπ	Δ (MeV)	T½, Γ, or Abundance	Decay Mode
28 Ni	56	0+	-53.90	6.075 d 10	ε
	57	3/2-	-56.082	35.60 h 6	ε
	58	0+	-60.228	**68.077% 9**	
	59	3/2-	-61.156	7.6×10^4 y 5	ε
	60	0+	-64.472	**26.223% 8**	
	61	3/2-	-64.221	**1.140% 1**	
	62	0+	-66.746	**3.634% 2**	
	63	1/2-	-65.513	100.1 y 20	β-
	64	0+	-67.099	**0.926% 1**	
	65	5/2-	-65.126	2.5172 h 3	β-
	66	0+	-66.006	54.6 h 3	β-
	67	(1/2)-	-63.743	21 s 1	β-
	68	0+	-63.464	29 s 2	β-
	69	9/2+	-59.979	11.4 s 3	β-
	69m	1/2-	-59.658	3.5 s 5	β-
	70	0+	-59.1	6.0 s 3	β-
	71		-55.2	2.56 s 3	β-
	72	0+	-53.9	1.57 s 5	β-, β-n
	73	(9/2+)	-49.9s	0.84 s 3	β-
	74	0+	-48.4s	0.68 s 18	β-, β-n
	75	(7/2+)	-43.9s	0.6 s 2	β-, β-n 8.43%
	76	0+	-41.6s	0.24 s +55-24	β-, β-n
	77		-36.7s	>150 ns	β-?
	78	0+	-34s	>150 ns	β-
29 Cu	52	(3+)	-2.6s	?	p
	53	(3/2-)	-13.5s	<300 ns	ε, p
	54	(3+)	-21.7s	<75 ns	p
	55	3/2-	-31.6s	>200 ns	ε
	56	4+	-38.6s	94 ms 3	ε
	57	3/2-	-47.31	196.3 ms 7	ε
	58	1+	-51.662	3.204 s 7	ε
	59	3/2-	-56.357	81.5 s 5	ε
	60	2+	-58.344	23.7 m 4	ε
	61	3/2-	-61.984	3.333 h 5	ε
	62	1+	-62.798	9.67 m 3	ε
	63	3/2-	-65.579	**69.17% 3**	
	64	1+	-65.424	12.700 h 2	ε 61%, β- 39%
	65	3/2-	-67.264	**30.83% 3**	
	66	1+	-66.258	5.120 m 14	β-
	67	3/2-	-67.319	61.83 h 12	β-
	68	1+	-65.567	31.1 s 15	β-
	68m	(6-)	-64.845	3.75 m 5	IT 84%, β- 16%
	69	3/2-	-65.736	2.85 m 15	β-
	70	(6-)	-62.976	44.5 s 2	β-
	70m	(3-)	-62.875	33 s 2	β- 52%, IT 48%
	70m	1+	-62.734	6.6 s 2	β- 93.2%, IT 6.8%
	71	(3/2-)	-62.711	19.5 s 16	β-
	72	(1+)	-59.783	6.6 s 1	β-
	73	(3/2-)	-58.987	4.2 s 3	β-
	74	(1+,3+)	-56.006	1.594 s 10	β-
	75	(3/2-)	-54.1	1.224 s 3	β-, β-n 3.5%
	76m		-50.976	0.641 s 6	β-, β-n 3%
29 Cu	76m		-50.976	1.27 s 30	β-
	77		-48.6s	0.469 s 8	β-
	78		-44.7s	342 ms 11	β-
	79		-42.3s	188 ms 25	β-, β-n 55%
	80		-36.4s	>300 ns	β-
30 Zn	54	0+	-6.6s	?	2p?
	55		-14.9s	>0.5 μs	ε, p
	56	0+	-25.7s	>0.5 μs	ε, p
	57	(7/2-)	-32.8s	38 ms 4	ε, εp≥65%
	58	0+	-42.30	84 ms 9	ε
	59	3/2-	-47.26	182.0 ms 18	ε, εp 0.1%
	60	0+	-54.19	2.38 m 5	ε
	61	3/2-	-56.35	89.1 s 2	ε
	61m	1/2-	-56.26	<430 ms	IT
	61m	3/2-	-55.93	0.14 s 7	IT
	61m	5/2-	-55.59	<0.13 s	IT
	62	0+	-61.17	9.186 h 13	ε
	63	3/2-	-62.213	38.47 m 5	ε
	64	0+	-66.004	$>2.8\times10^{16}$ y **48.63% 60**	2ε
	65	5/2-	-65.912	243.66 d 9	ε
	66	0+	-68.899	**27.90% 27**	
	67	5/2-	-67.880	**4.10% 13**	
	68	0+	-70.007	**18.75% 51**	
	69	1/2-	-68.418	56.4 m 9	β-
	69m	9/2+	-67.979	13.76 h 2	IT 99.97%, β- 0.03%
	70	0+	-69.565	$>1.3\times10^{16}$ y **0.62% 3**	2β-
	71	1/2-	-67.33	2.45 m 10	β-
	71m	9/2+	-67.17	3.96 h 5	β-, IT≤0.05%
	72	0+	-68.131	46.5 h 1	β-
	73	(1/2)-	-65.41	23.5 s 10	β-
	73m		-65.41	5.8 s 8	IT, β-
	74	0+	-65.71	95.6 s 12	β-
	75	(7/2+)	-62.47	10.2 s 2	β-
	76	0+	-62.14	5.7 s 3	β-
	77	(7/2+)	-58.7	2.08 s 5	β-
	77m	(1/2-)	-57.9	1.05 s 10	IT>50%, β-<50%
	78	0+	-57.34	1.47 s 15	β-
	79	(9/2+)	-53.4s	0.995 s 19	β-, β-n 1.3%
	80	0+	-51.8	0.54 s 2	β-, β-n 1%
	81		-46.1s	0.29 s 5	β-, β-n 7.5%
	82	0+	-42.5s	>150 ns	β-
	83	(5/2+)	-36.3s	>150 ns	β-
31 Ga	56		-4.7s	?	p?
	57		-15.9s	?	p?
	58		-24.0s	?	p?
	59		-34.1s	?	p?
	60	(2+)	-40.0s	70 ms 13	ε 98.4%, εp 1.6%, εα<0.02%
	61	3/2-	-47.09	168 ms 3	ε
	62	0+	-52.00	116.18 ms 4	ε
31 Ga	63	(3/2-)	-56.547	32.4 s 5	ε
	64	0+	-58.834	2.627 m 12	ε
	65	3/2-	-62.657	15.2 m 2	ε
	66	0+	-63.724	9.49 h 7	ε
	67	3/2-	-66.880	3.2623 d 15	ε
	68	1+	-67.086	67.71 m 9	ε
	69	3/2-	-69.328	**60.108% 9**	
	70	1+	-68.910	21.14 m 3	β- 99.59%, ε 0.41%
	71	3/2-	-70.140	**39.892% 9**	
	72	3-	-68.589	14.095 h 3	β-
	73	3/2-	-69.699	4.86 h 3	β-
	74	(3-)	-68.050	8.12 m 12	β-
	74m	(0)	-67.990	9.5 s 10	IT 75%, β-<50%
	75	(3/2)-	-68.465	126 s 2	β-
	76	(2+,3+)	-66.297	32.6 s 6	β-
	77	(3/2-)	-65.992	13.2 s 2	β-
	78	(3+)	-63.707	5.09 s 5	β-
	79	(3/2-)	-62.51	2.847 s 3	β-, β-n 0.09%
	80	(3)	-59.1	1.676 s 14	β-, β-n 0.86%
	81	(5/2-)	-58.0	1.217 s 5	β-, β-n 11.9%
	82	(1,2,3)	-53.1s	0.599 s 2	β-, β-n 19.8%
	83		-49.4s	0.308 s 1	β-, β-n 37%
	84		-44.1s	0.085 s 10	β-, β-n 70%
	85	(3/2-)	-40.1s	>150 ns	β-
	86		-34.4s	>150 ns	β-
32 Ge	58	0+	-8.4s	?	2p?
	59		-17.0s	?	2p?
	60	0+	-27.8s	≈30 ms	ε?, 2p?
	61	(3/2-)	-33.7s	39 ms 12	ε, εp≈80%
	62	0+	-42.2s	129 ns 35	ε
	63	(3/2-)	-46.9s	142 ms 8	ε
	64	0+	-54.35	63.7 s 25	ε
	65	(3/2)-	-56.4	30.9 s 5	ε
	66	0+	-61.62	2.26 h 5	ε
	67	1/2-	-62.658	18.9 m 3	ε
	68	0+	-66.980	270.95 d 16	ε
	69	5/2-	-67.101	39.05 h 10	ε
	70	0+	-70.563	**20.37% 18**	
	71	1/2-	-69.908	11.43 d 3	ε
	72	0+	-72.586	**27.31% 26**	
	73	9/2+	-71.298	**7.76% 8**	
	73m	1/2-	-71.231	0.499 s 11	IT
	74	0+	-73.422	**36.73% 15**	
	75	1/2-	-71.856	82.78 m 4	β-
	75m	7/2+	-71.717	47.7 s 5	IT 99.97%, β- 0.03%
	76	0+	-73.213	1.2×10^{25} y 14 **7.83% 7**	2β-
	77	7/2+	-71.214	11.30 h 1	β-
	77m	1/2-	-71.054	52.9 s 6	β- 81%, IT 19%
	78	0+	-71.862	88.0 m 10	β-
	79	(1/2)-	-69.49	18.98 s 3	β-
	79m	(7/2+)	-69.30	39.0 s 10	β- 96%, IT 4%

TABLE II.1 (cont.)

Nuclide Z El	A	Jπ	Δ (MeV)	T½, Γ, or Abundance	Decay Mode
32 Ge	80	0+	-69.52	29.5 s 4	β–
	81	(9/2+)	-66.3	7.6 s 6	β–
	81m	(1/2+)	-65.6	7.6 s 6	β–
	82	0+	-65.6	4.55 s 5	β–
	83	(5/2+)	-60.9s	1.85 s 6	β–
	84	0+	-58.2s	0.947 s 11	β–, β–n 10.8%
	85		-53.1s	535 ms 47	β–, β–n 14%
	86	0+	-49.8s	>150 ns	β–
	87	(5/2+)	-44.2s	≈0.14 s	β–, β–n
	88	0+	-40.1s	≥300 ns	β–?
	89		-33.7s	>150 ns	β–
33 As	60		-6.4s	?	p?
	61		-18.1s	?	p?
	62		-25.0s	?	p
	63	(3/2–)	-33.8s	?	p
	64		-39.5s	18 ms +43–7	ε?
	65		-47.0s	128 ms 16	ε
	66		-51.5	95.79 ms 22	ε
	67	(5/2–)	-56.6	42.5 s 12	ε
	68	3+	-58.90	151.6 s 8	ε
	69	5/2–	-63.09	15.2 m 2	ε
	70	4+	-64.34	52.6 m 3	ε
	71	5/2–	-67.894	65.28 h 15	ε
	72	2–	-68.230	26.0 h 1	ε
	73	3/2–	-70.957	80.30 d 6	ε
	74	2–	-70.860	17.77 d 2	ε 66%, β– 34%
	75	3/2–	-73.032	**100%**	
	76	2–	-72.289	1.0942 d 7	β–
	77	3/2–	-73.917	38.83 h 5	β–
	78	2–	-72.817	90.7 m 2	β–
	79	3/2–	-73.636	9.01 m 15	β–
	80	1+	-72.16	15.2 s 2	β–
	81	3/2–	-72.533	33.3 s 8	β–
	82	(1+)	-70.3	19.1 s 5	β–
	82m	(5–)	-70.3	13.6 s 4	β–
	83	(5/2–,3/2–)	-69.9	13.4 s 3	β–
	84	(3–)	-66.1s	3.24 s 26	β–, β–n 0.28%
	85	(3/2–)	-63.3s	2.021 s 10	β–, β–n 59.4%
	86		-59.2s	0.945 s 8	β–, β–n 33%
	87	(3/2–)	-56.0s	0.56 s 8	β–, β–n 15.4%
	88		-51.3s	≥300 ns	β–?, β–n?
	89		-47.1s	≥300 ns	β–?
	90		-41.5s	>150 ns	β–?
	91		-36.9s	>150 ns	β–
	92		-30.9s	>300 ns	β–
34 Se	65		-32.9s	<50 ms	ε
	66	0+	-41.7s	33 ms 12	ε
	67		-46.5s	133 ms 11	ε, εp 0.5%
	68	0+	-54.21	35.5 s 7	ε
	69	(1/2–,3/2–)	-56.30	27.4 s 2	ε, εp 0.05%
	70	0+	-62.05	41.1 m 3	ε
	71	5/2–	-63.12	4.74 m 5	ε
34 Se	72	0+	-67.89	8.40 d 8	ε
	73	9/2+	-68.22	7.15 h 8	ε
	73m	3/2–	-68.19	39.8 m 13	IT 72.6%, ε 27.4%
	74	0+	-72.213	**0.89% 4**	
	75	5/2+	-72.169	119.779 d 4	ε
	76	0+	-75.252	**9.37% 29**	
	77	1/2–	-74.600	**7.63% 16**	
	77m	7/2+	-74.438	17.36 s 5	IT
	78	0+	-77.026	**23.77% 28**	
	79	7/2+	-75.918	2.95×10^{5} y 38	β–
	79m	1/2–	-75.822	3.92 m 1	IT 99.94%, β– 0.06%
	80	0+	-77.760	**49.61% 41**	
	81	1/2–	-76.390	18.45 m 12	β–
	81m	7/2+	-76.286	57.28 m 2	IT 99.95%, β– 0.05%
	82	0+	-77.594	9.1×10^{19} y 9 **8.73% 22**	2β–
	83	9/2+	-75.341	22.3 m 3	β–
	83m	1/2–	-75.112	70.1 s 4	β–
	84	0+	-75.95	3.10 m 10	β–
	85	(5/2+)	-72.43	31.7 s 9	β–
	86	0+	-70.54	15.3 s 9	β–
	87	(5/2+)	-66.58	5.50 s 12	β–, β–n 0.2%
	88	0+	-63.88	1.53 s 6	β–, β–n 0.99%
	89	(5/2+)	-59.2s	0.41 s 4	β–, β–n 7.8%
	90	0+	-55.9s	>150 ns	β–?
	91		-50.3s	0.27 s 5	β–, β–n 21%
	92	0+	-46.6s	>300 ns	β–
	93	(1/2+)	-40.7s	>150 ns	β–
	94	0+	-36.8s	>150 ns	β–
35 Br	67		-32.8s	?	p?
	68		-38.6s	<1.2 μs	p?
	69		-46.5s	<24 ns	p
	70	0+	-51.4s	79.1 ms 8	ε
	70m	9+	-49.1s	2.2 s 2	ε
	71	(5/2)–	-57.1	21.4 s 6	ε
	72	1+	-59.02	78.6 s 24	ε
	72m	1–	-58.91	10.6 s 3	IT≈100%, ε
	73	1/2–	-63.63	3.4 m 2	ε
	74	(0–)	-65.31	25.4 m 3	ε
	74m	4(+)	-65.29	46 m 2	ε
	75	3/2–	-69.14	96.7 m 13	ε
	76	1–	-70.289	16.2 h 2	ε
	76m	(4)+	-70.186	1.31 s 2	IT>99.4%, ε<0.6%
	77	3/2–	-73.235	57.036 h 6	ε
	77m	9/2+	-73.129	4.28 m 10	IT
	78	1+	-73.452	6.46 m 4	ε≥99.99%, β–≤0.01%
	79	3/2–	-76.068	**50.69% 7**	
	79m	9/2+	-75.861	4.86 s 4	IT
	80	1+	-75.890	17.68 m 2	β– 91.7%, ε 8.3%
	80m	5–	-75.804	4.4205 h 8	IT
	81	3/2–	-77.975	**49.31% 7**	
35 Br	82	5–	-77.496	35.282 h 7	β–
	82m	2–	-77.451	6.13 m 5	IT 97.6%, β– 2.4%
	83	3/2–	-79.009	2.40 h 2	β–
	84	2–	-77.80	31.80 m 8	β–
	84m	6–	-77.48	6.0 m 2	β–
	85	3/2–	-78.61	2.90 m 6	β–
	86	(2–)	-75.64	55.1 s 4	β–
	87	3/2–	-73.86	55.65 s 13	β–, β–n 2.6%
	88	(2–)	-70.73	16.29 s 6	β–, β–n 6.58%
	89	(3/2–,5/2–)	-68.57	4.40 s 3	β–, β–n 13.8%
	90		-64.62	1.91 s 1	β–, β–n 25.2%
	91		-61.51	0.541 s 5	β–, β–n 20%
	92	(2–)	-56.58	0.343 s 15	β–, β–n 33.1%
	93	(5/2–)	-53.0s	102 ms 10	β–, β–n 68%
	94		-47.8s	70 ms 20	β–, β–n 70%
	95	(3/2–)	-43.9s	>150 ns	β–
	96		-38.6s	>150 ns	β–
	97	(3/2–)	-34.7s	>150 ns	β–
36 Kr	69		-32.4s	32 ms 10	ε
	70	0+	-41.7s	52 ms 17	ε, εp≤1.3%
	71	(5/2)–	-46.9	100 ms 3	ε, εp 5.2%
	72	0+	-53.941	17.1 s 2	ε
	73	3/2–	-56.552	27.3 s 10	ε, εp 0.25%
	74	0+	-62.332	11.50 m 11	ε
	75	5/2+	-64.324	4.29 m 17	ε
	76	0+	-69.014	14.8 h 1	ε
	77	5/2+	-70.169	74.4 m 6	ε
	78	0+	-74.180	≥2.3×10^{20} y **0.35% 1**	2ε
	79	1/2–	-74.443	35.04 h 10	ε
	79m	7/2+	-74.313	50 s 3	IT
	80	0+	-77.893	**2.28% 6**	
	81	7/2+	-77.694	2.29×10^{5} y 11	ε
	81m	1/2–	-77.503	13.10 s 3	IT, ε 2.5×10^{-3}%
	82	0+	-80.590	**11.58% 14**	
	83	9/2+	-79.982	**11.49% 6**	
	83m	1/2–	-79.940	1.83 h 2	IT
	84	0+	-82.431	**57.00% 4**	
	85	9/2+	-81.480	3916.8 d 25	β–
	85m	1/2–	-81.175	4.480 h 8	β– 78.6%, IT 21.4%
	86	0+	-83.266	**17.30% 22**	
	87	5/2+	-80.709	76.3 m 5	β–
	88	0+	-79.69	2.84 h 3	β–
	89	3/2(+)	-76.73	3.15 m 4	β–
	90	0+	-74.97	32.32 s 9	β–
	91	5/2(+)	-71.31	8.57 s 4	β–
	92	0+	-68.79	1.840 s 8	β–, β–n 0.03%
	93	1/2+	-64.0	1.286 s 10	β–, β–n 1.95%
	94	0+	-61.1s	212 ms 5	β–, β–n 1.26%
	95	1/2	-56.0s	114 ms 3	β–, β–n 2.87%
	96	0+	-53.0s	80 ms 8	β–, β–n 3.8%
	97		-47.9s	63 ms 4	β–, β–n 8.2%

TABLE II.1 (cont.)

Nuclide Z El	A	Jπ	Δ (MeV)	T½, Γ, or Abundance	Decay Mode
36 Kr	98	0+	−44.8s	46 ms 8	β−, β−n 7%
	99	(3/2+)	−39.5s	40 ms 11	β−, β−n 11%
	100	0+	−36.2s	>150 ns	β−
37 Rb	71		−32.3s	?	p?
	72	(3+)	−38.1s	<1.2 μs	p
	73		−46.1s	>30 ns	ε, p>0%
	74	(0+)	−51.917	64.9 ms 5	ε
	75	(3/2−)	−57.222	19.0 s 12	ε
	76	1(−)	−60.480	36.5 s 6	ε, εα 3.8×10^{-7}%
	77	3/2−	−64.825	3.77 m 4	ε
	78	0(+)	−66.936	17.66 m 8	ε
	78m	4(−)	−66.833	5.74 m 5	ε 90%, IT 10%
	79	5/2+	−70.803	22.9 m 5	ε
	80	1+	−72.173	33.4 s 7	ε
	81	3/2−	−75.455	4.570 h 4	ε
	81m	9/2+	−75.368	30.5 m 3	IT 97.6%, ε 2.4%
	82	1+	−76.188	1.273 m 2	ε
	82m	5−	−76.119	6.472 h 5	ε, IT<0.33%
	83	5/2−	−79.075	86.2 d 1	ε
	84	2−	−79.750	33.1 d 1	ε 96.2%, β− 3.8%
	84m	6−	−79.286	20.26 m 4	IT
	85	5/2−	−82.167	**72.17% 2**	
	86	2−	−82.747	18.642 d 18	β− 99.99%, ε 5.2×10^{-3}%
	86m	6−	−82.191	1.017 m 3	IT, β−<0.3%
	87	3/2−	−84.598	4.97×10^{10} y 3 **27.83% 2**	β−
	88	2−	−82.609	17.773 m 11	β−
	89	3/2−	−81.713	15.15 m 12	β−
	90	0−	−79.362	158 s 5	β−
	90m	3−	−79.255	258 s 4	β− 97.4%, IT 2.6%
	91	3/2(−)	−77.745	58.4 s 4	β−
	92	0−	−74.772	4.492 s 20	β−, β−n 0.01%
	93	5/2−	−72.618	5.84 s 2	β−, β−n 1.39%
	94	3(−)	−68.553	2.702 s 5	β−, β−n 10.01%
	95	5/2−	−65.85	377.5 ms 8	β−, β−n 8.73%
	96	2+	−61.22	202.8 ms 33	β−, β−n 14%
	97	3/2+	−58.36	169.9 ms 7	β−, β−n 25.1%
	98	(0,1)	−54.22	114 ms 5	β−, β−n 13.8%, β− 0.05%
	99	(5/2+)	−50.9	50.3 ms 7	β−, β−n 15.9%
	100		−46.7s	51 ms 8	β−, β−n 6%, β−2n 0.16%
	101	(3/2+)	−43.6	32 ms 5	β−, β−n 28%
38 Sr	73		−31.7s	>25 ms	ε, εp>0%
	74	0+	−40.7s	>1.2 μs	ε
	75	(3/2−)	−46.6	88 ms 3	ε, εp 5.2%
	76	0+	−54.24	7.89 s 7	ε, εp 0.34%
	77	5/2+	−57.804	9.0 s 2	ε, εp<0.25%
	78	0+	−63.174	2.5 m 3	ε
	79	3/2(−)	−65.477	2.25 m 10	ε
	80	0+	−70.308	106.3 m 15	ε
38 Sr	81	1/2−	−71.528	22.3 m 4	ε
	82	0+	−76.008	25.55 d 15	ε
	83	7/2+	−76.80	32.41 h 3	ε
	83m	1/2−	−76.54	4.95 s 12	IT
	84	0+	−80.644	**0.56% 1**	
	85	9/2+	−81.103	64.84 d 2	ε
	85m	1/2−	−80.864	67.63 m 4	IT 86.6%, ε 13.4%
	86	0+	−84.524	**9.86% 1**	
	87	9/2+	−84.880	**7.00% 1**	
	87m	1/2−	−84.492	2.815 h 12	IT 99.7%, ε 0.3%
	88	0+	−87.922	**82.58% 1**	
	89	5/2+	−86.209	50.57 d 3	β−
	90	0+	−85.942	28.90 y 3	β−
	91	5/2+	−83.645	9.63 h 5	β−
	92	0+	−82.868	2.66 h 4	β−
	93	5/2+	−80.085	7.423 m 24	β−
	94	0+	−78.840	75.3 s 2	β−
	95	1/2+	−75.117	23.90 s 14	β−
	96	0+	−72.94	1.07 s 1	β−
	97	1/2+	−68.79	429 ms 5	β−, β−n≤0.05%
	98	0+	−66.65	0.653 s 2	β−, β−n 0.25%
	99	3/2+	−62.19	0.269 s 1	β−, β−n 0.1%
	100	0+	−60.2	202 ms 3	β−, β−n 0.78%
	101	(5/2−)	−55.4	118 ms 3	β−, β−n 2.37%
	102	0+	−53.1	69 ms 6	β−, β−n 4.8%
	103		−47.6s	>150 ns	β−
	104	0+	−44.4s	>300 ns	
	105		−38.6s	>150 ns	β−
39 Y	76		−38.7s	>200 ns	ε?, p?
	77		−46.90s	≈0.06 s	ε, εp
	78	(0+)	−52.5s	50 ms 8	ε
	78m	(5+)	−52.5s	5.7 s 7	ε
	79	(5/2+)	−58.4	14.8 s 6	ε, εp
	80	(4−)	−61.2	30.1 s 5	ε, εp
	80m	(1−)	−61.0	4.8 s 3	IT 81%, ε 19%
	81	(5/2+)	−66.02	70.4 s 10	ε
	82	1+	−68.2	8.30 s 20	ε
	83	9/2+	−72.33	7.08 m 6	ε
	83m	3/2−	−72.26	2.85 m 2	ε 60%, IT 40%
	84	1+	−74.16	4.6 s 2	ε
	84m	(5−)	−74.16	39.5 m 8	ε
	85	(1/2)−	−77.84	2.68 h 5	ε
	85m	9/2+	−77.82	4.86 h 13	ε, IT<2.0×10^{-3}%
	86	4−	−79.28	14.74 h 2	ε
	86m	(8+)	−79.07	48 m 1	IT 99.31%, ε 0.69%
	87	1/2−	−83.019	79.8 h 3	ε
	87m	9/2+	−82.638	13.37 h 3	IT 98.43%, ε 1.57%
	88	4−	−84.299	106.616 d 13	ε
	89	1/2−	−87.702	**100%**	
	89m	9/2+	−86.793	15.28 s 17	IT
	90	2−	−86.488	64.053 h 20	β−
	90m	7+	−85.806	3.19 h 6	IT, β− 1.8×10^{-3}%
39 Y	91	1/2−	−86.345	58.51 d 6	β−
	91m	9/2+	−85.789	49.71 m 4	IT, β−<1.5%
	92	2−	−84.813	3.54 h 1	β−
	93	1/2−	−84.22	10.18 h 8	β−
	93m	7/2+	−83.46	0.82 s 4	IT
	94	2−	−82.349	18.7 m 1	β−
	95	1/2−	−81.207	10.3 m 1	β−
	96	0−	−78.35	5.34 s 5	β−
	96m	(8)+	−78.35	9.6 s 2	β−
	97	(1/2−)	−76.26	3.75 s 3	β−, β−n 0.058%
	97m	(9/2)+	−75.59	1.17 s 3	β−>99.3%, IT<0.7%, β−n<0.08%
	97m	(27/2−)	−72.73	142 ms 8	IT>80%, β−<20%
	98	(0)−	−72.47	0.548 s 2	β−, β−n 0.33%
	98m	(4,5)	−72.06	2.0 s 2	β−>80%, IT<20%, β−n 3.4%
	99	(5/2+)	−70.20	1.470 s 7	β−, β−n 1.9%
	100	1−,2−	−67.29	735 ms 7	β−, β−n 0.92%
	100m	(3,4,5)	−67.29	0.94 s 3	β−
	101	(5/2+)	−64.91	0.45 s 2	β−, β−n 1.5%
	102m		−61.89	0.30 s 1	β−, β−n 4%
	102m		−61.89	0.36 s 4	β−, β−n 4%
	103	(5/2+)	−58.9s	0.23 s 2	β−, β−n 8%
	104		−54.9s	180 ms 60	β−, β−n?
	105		−51.4s	>300 ns	β−?
	106		−46.8s	>150 ns	β−
	107	(5/2+)	−42.7s	≈30 ms	β−
	108		−37.7s	20 ms sy	β−, β−n
40 Zr	78	0+	−41.7s	>200 ns	ε?, εp?
	79		−47.4s	56 ms 30	ε, εp
	80	0+	−56	4.6 s 6	ε, εp
	81	(3/2−)	−58.5	5.5 s 4	ε, εp 0.12%
	82	0+	−64.2s	32 s 5	ε
	83	(1/2−)	−66.46	41.6 s 24	ε, εp
	84	0+	−71.5s	25.9 m 7	ε
	85	7/2+	−73.1	7.86 m 4	ε
	85m	(1/2−)	−72.9	10.9 s 3	IT≤92%, ε>8%
	86	0+	−77.80	16.5 h 1	ε
	87	(9/2)+	−79.348	1.68 h 1	ε
	87m	(1/2)−	−79.012	14.0 s 2	IT
	88	0+	−83.62	83.4 d 3	ε
	89	9/2+	−84.869	78.41 h 12	ε
	89m	1/2−	−84.281	4.161 m 17	IT 93.77%, ε 6.23%
	90	0+	−88.767	**51.45% 40**	
	90m	5−	−86.448	809.2 ms 20	IT
	91	5/2+	−87.890	**11.22% 5**	
	92	0+	−88.454	**17.15% 8**	
	93	5/2+	−87.117	1.53×10^6 y 10	β−
	94	0+	−87.267	**17.38% 28**	
	95	5/2+	−85.658	64.032 d 6	β−
	96	0+	−85.443	>3.9×10^{20} y **2.80% 9**	2β−

TABLE II.1 (cont.)

Nuclide Z El A	Jπ	Δ (MeV)	T½, Γ, or Abundance	Decay Mode
40 Zr 97	1/2+	-82.947	16.744 h 11	β–
98	0+	-81.29	30.7 s 4	β–
99	(1/2+)	-77.77	2.1 s 1	β–
100	0+	-76.60	7.1 s 4	β–
101	(3/2+)	-73.46	2.3 s 1	β–
102	0+	-71.74	2.9 s 2	β–
103	(5/2–)	-68.4	1.3 s 1	β–
104	0+	-66.3s	1.2 s 3	β–
105		-62.4s	0.6 s 1	β–
106	0+	-59.7s	>150 ns	β–?
107		-55.2s	~150 ms	β–
108	0+	-52.2s	80 ms sy	β–, β–n
109		-47.3s	>150 ns	β–, β–n
110	0+	-43.9s	>150 ns	β–
41 Nb 81		-47s	~0.8 s	ε?, εp?, p?
82	0+	-53.0s	50 ms 5	ε
83	(5/2+)	-59.0	4.1 s 3	ε
84	3+	-61.9s	9.5 s 10	ε, εp
85	(9/2+)	-67.1	20.9 s 7	ε
86?		-69.83	56 s 8	ε
86m	(6+)	-69.83	88 s 1	ε
87	(1/2–)	-74.18	3.75 m 9	ε
87m	(9/2+)	-74.18	2.6 m 1	ε
88	(8+)	-76.1	14.55 m 6	ε
88m	(4–)	-76.1	7.78 m 5	ε
89	(9/2+)	-80.65	2.03 h 7	ε
89m	(1/2)–	-80.62	66 m 2	ε
90	8+	-82.656	14.60 h 5	ε
90m	4–	-82.532	18.81 s 6	IT
91	9/2+	-86.632	6.8×10^{2} y 13	ε
91m	1/2–	-86.528	60.86 d 22	IT 96.6%, ε 3.4%
92	(7)+	-86.448	3.47×10^{7} y 24	ε, β–<0.05%
92m	(2)+	-86.313	10.15 d 2	ε
93	9/2+	-87.208	**100%**	
93m	1/2–	-87.177	16.13 y 14	IT
94	(6)+	-86.365	2.03×10^{4} y 16	β–
94m	3+	-86.324	6.263 m 4	IT 99.5%, β– 0.5%
95	9/2+	-86.782	34.991 d 6	β–
95m	1/2–	-86.546	3.61 d 3	IT 94.4%, β– 5.6%
96	6+	-85.604	23.35 h 5	β–
97	9/2+	-85.606	72.1 m 7	β–
97m	1/2–	-84.863	58.7 s 18	IT
98	1+	-83.529	2.86 s 6	β–
98m	(5+)	-83.445	51.3 m 4	β– 99.8%, IT<0.2%
99	9/2+	-82.33	15.0 s 2	β–
99m	1/2–	-81.96	2.6 m 2	β–>96.2%, IT<3.8%
100	1+	-79.94	1.5 s 2	β–
100m	(4+,5+)	-79.44	2.99 s 11	β–
101	(5/2+)	-78.94	7.1 s 3	β–
102m	1+	-76.35	1.3 s 2	β–
102m		-76.35	4.3 s 4	β–
103	(5/2+)	-75.32	1.5 s 2	β–

Nuclide Z El A	Jπ	Δ (MeV)	T½, Γ, or Abundance	Decay Mode
41 Nb 104	(1+)	-72.2	4.9 s 3	β–, β–n 0.06%
104m		-72.0	0.94 s 4	β–, β–n 0.05%
105	(5/2+)	-70.85	2.95 s 6	β–, β–n 1.7%
106		-67.1s	1.02 s 5	β–, β–n 4.5%
107		-64.9s	330 ms 50	β–
108	(2+)	-60.7s	0.193 s 17	β–, β–n 6.2%
109	(5/2)	-58.1s	0.19 s 3	β–, β–n 31%
110		-53.6s	0.17 s 2	β–, β–n 40%
111	(5/2+)	-50.6s	80. ms sy	β–
112	(2+)	-45.8s	>150 ns	β–
113		-42.2s	30 ms sy	β–
42 Mo 83		-47.7s	6 ms +30–3	ε
84	0+	-55.8s	3.7 s +10–8	ε
85	(1/2–)	-59.1s	3.2 s 2	εp 0.14%, ε
86	0+	-64.6	19.6 s 11	ε
87	7/2+	-67.7	14.02 s 26	ε, εp 15%
88	0+	-72.70	8.0 m 2	ε
89	(9/2+)	-75.00	2.11 m 10	ε
89m	(1/2–)	-74.62	190 ms 15	IT
90	0+	-80.167	5.56 h 9	ε
91	9/2+	-82.20	15.49 m 1	ε
91m	1/2–	-81.55	64.6 s 6	ε 50%, IT 50%
92	0+	-86.805	**14.84% 35**	
93	5/2+	-86.803	4.0×10^{3} y 8	ε
93m	21/2+	-84.379	6.85 h 7	IT 99.88%, ε 0.12%
94	0+	-88.410	**9.25% 12**	
95	5/2+	-87.707	**15.92% 13**	
96	0+	-88.790	**16.68% 2**	
97	5/2+	-87.540	**9.55% 8**	
98	0+	-88.112	**24.13% 31**	
99	1/2+	-85.966	2.7489 d 6	β–
100	0+	-86.184	0.78×10^{19} y 8 **9.63% 23**	2β–
101	1/2+	-83.511	14.61 m 3	β–
102	0+	-83.56	11.3 m 2	β–
103	(3/2+)	-80.85	67.5 s 15	β–
104	0+	-80.33	60 s 2	β–
105	(5/2–)	-77.34	35.6 s 16	β–
106	0+	-76.26	8.4 s 5	β–
107	(7/2–)	-72.9	3.5 s 5	β–
108	0+	-71.3s	1.09 s 2	β–
109	(7/2–)	-67.2s	0.53 s 6	β–
110	0+	-65.5s	0.27 s 1	β–
111		-61.1s	200. ms sy	β–
112	0+	-58.8s	>150 ns	β–?
113		-54.1s	100 ms sy	β–
114	0+	-51.3s	80 ms sy	β–
115		-46.3s	60 ms sy	β–, β–n
43 Tc 85		-47.7s	~0.5 s	ε?
86	(0+)	-53.2s	54 ms 7	ε
87	(9/2+)	-59.1s	2.2 s 2	ε
88	(3+)	-62.7s	5.8 s 2	ε

Nuclide Z El A	Jπ	Δ (MeV)	T½, Γ, or Abundance	Decay Mode
43 Tc 88m	(6+)	-62.7s	6.4 s 8	ε
89	(9/2+)	-67.8s	12.8 s 9	ε
89m	(1/2–)	-67.8s	12.9 s 8	ε, IT<0.01%
90m	1+	-71.2	8.7 s 2	ε
90m	(6+)	-70.7	49.2 s 4	ε
91	(9/2)+	-76.0	3.14 m 2	ε
91m	(1/2)–	-75.8	3.3 m 1	ε, IT<1%
92	(8)+	-78.93	4.25 m 15	ε
93	9/2+	-83.603	2.75 h 5	ε
93m	1/2–	-83.211	43.5 m 10	IT 76.6%, ε 23.4%
94	7+	-84.154	293 m 1	ε
94m	(2)+	-84.079	52.0 m 10	ε, IT<0.1%
95	9/2+	-86.017	20.0 h 1	ε
95m	1/2–	-85.978	61 d 2	ε 96.12%, IT 3.88%
96	7+	-85.817	4.28 d 7	ε
96m	4+	-85.783	51.5 m 10	IT 98%, ε 2%
97	9/2+	-87.220	4.21×10^{6} y 16	ε
97m	1/2–	-87.123	91.4 d 8	IT, ε 3.94%
98	(6)+	-86.428	4.2×10^{6} y 3	β–
99	9/2+	-87.323	2.111×10^{5} y 12	β–
99m	1/2–	-87.180	6.0058 h 12	IT, β– 3.7×10^{-3}%
100	1+	-86.016	15.8 s 1	β–, ε 1.8×10^{-3}%
101	9/2+	-86.34	14.22 m 1	β–
102	1+	-84.566	5.28 s 15	β–
102m	(4,5)	-84.566	4.35 m 7	β– 98%, IT 2%
103	5/2+	-84.60	54.2 s 8	β–
104	(3+)	-82.49	18.3 m 3	β–
105	(3/2–)	-82.29	7.6 m 1	β–
106	(1,2)	-79.78	35.6 s 6	β–
107	(3/2–)	-79.1	21.2 s 2	β–
108	(2)+	-76.0	5.17 s 7	β–
109	(5/2+)	-74.54	0.86 s 4	β–, β–n 0.08%
110	(2+)	-70.96	0.92 s 3	β– 99.96%, β–n 0.04%
111	(7/2+,9/2+)	-69.2	290 ms 20	β–, β–n 0.85%
112		-66.0	0.29 s 2	β–, β–n 1.5%
113		-63.7s	170 ms 20	β–, β–n 2.1%
114		-59.7s	150 ms 30	β–, β–n
115		-57.1s	100 ms sy	β–, β–n
116		-52.8s	90 ms sy	β–
117		-49.9s	40 ms sy	β–
118		-45.2s	>150 ns	β–
44 Ru 87		-47.3s	>1.5 μs	ε?
88	0+	-55.6s	1.2 s +3–2	ε, εp
89		-59.5s	1.5 s 2	ε, εp<0.15%
90	0+	-65.3s	11.7 s 9	ε
91	(9/2+)	-68.7s	7.9 s 4	ε
91m	(1/2–)	-68.7s	7.6 s 8	ε>0%, εp>0%, IT
92	0+	-74.4s	3.65 m 5	ε
93	(9/2)+	-77.27	59.7 s 6	ε
93m	(1/2)–	-76.53	10.8 s 3	ε 78%, IT 22%, εp 0.03%

TABLE II.1 (cont.)

Nuclide Z El	A	Jπ	Δ (MeV)	T½, Γ, or Abundance	Decay Mode
44 Ru	94	0+	-82.57	51.8 m 6	ε
	95	5/2+	-83.45	1.643 h 14	ε
	96	0+	-86.072	**5.54% 14**	
	97	5/2+	-86.112	2.791 d 4	ε
	98	0+	-88.225	**1.87% 3**	
	99	5/2+	-87.617	**12.76% 14**	
	100	0+	-89.219	**12.60% 7**	
	101	5/2+	-87.950	**17.06% 2**	
	102	0+	-89.098	**31.55% 14**	
	103	3/2+	-87.259	39.26 d 2	β−
	104	0+	-88.089	**18.62% 27**	
	105	3/2+	-85.928	4.44 h 2	β−
	106	0+	-86.322	373.59 d 15	β−
	107	(5/2)+	-83.9	3.75 m 5	β−
	108	0+	-83.7	4.55 m 5	β−
	109	(5/2+)	-80.85	34.5 s 10	β−
	110	0+	-79.98	11.6 s 6	β−
	111	(5/2+)	-76.67	2.12 s 7	β−
	112	0+	-75.48	1.75 s 7	β−
	113	(5/2+)	-72.20	0.80 s 5	β−
	113m	(11/2−)	-72.07	510 ms 30	β− 92%, IT 8%
	114	0+	-70.5s	0.53 s 6	β−
	115		-66.4	740 ms 80	β−, β−n
	116	0+	-64.4s	400 ms sy	β−?
	117		-60.0s	300 ms sy	β−
	118	0+	-57.9s	>150 ns	β−?
	119		-53.2s	>150 ns	β−
	120	0+	-50.9s	>150 ns	β−
45 Rh	89		-47.7s	>1.5 μs	ε
	90		-53.2s	12 ms +9−4	ε?
	90m		-53.2s	1.0 s +3−2	ε?
	91	(9/2+)	-59.1s	1.47 s 22	ε
	91m	(1/2−)	-59.1s	1.46 s 11	ε
	92	(2+)	-63.4s	0.5 s 4	ε
	92	(≥ 6+)	-63.4s	4.66 s 25	ε
	93	(9/2+)	-69.2s	11.9 s 7	ε
	94m	(8+)	-72.9s	25.8 s 2	ε
	94m	(3+)	-72.9s	70.6 s 6	ε
	95	(9/2)+	-78.3	5.02 m 10	ε
	95m	(1/2)−	-77.8	1.96 m 4	IT 88%, ε 12%
	96	(6+)	-79.68	9.90 m 10	ε
	96m	(3+)	-79.63	1.51 m 2	IT 60%, ε 40%
	97	9/2+	-82.59	30.7 m 6	ε
	97m	1/2−	-82.33	46.2 m 16	ε 94.4%, IT 5.6%
	98	(2)+	-83.17	8.72 m 12	ε
	98m	(5+)	-83.17	3.6 m 2	IT 89%, ε 11%
	99	1/2−	-85.574	16.1 d 2	ε
	99m	9/2+	-85.510	4.7 h 1	ε>99.84%, IT<0.16%
	100	1−	-85.58	20.8 h 1	ε
	100m	(5+)	-85.58	4.6 m 2	IT=98.3%, ε=1.7%
	101	1/2−	-87.41	3.3 y 3	ε
45 Rh	101m	9/2+	-87.25	4.34 d 1	ε 92.8%, IT 7.2%
	102	(1−,2−)	-86.775	207 d 3	ε 78%, β− 22%
	102m	6(+)	-86.634	≈2.9 y	ε 99.77%, IT 0.23%
	103	1/2−	-88.022	**100%**	
	103m	7/2+	-87.982	56.114 m 9	IT
	104	1+	-86.950	42.3 s 4	β− 99.55%, ε 0.45%
	104m	5+	-86.821	4.34 m 3	IT 99.87%, β− 0.13%
	105	7/2+	-87.846	35.36 h 6	β−
	105m	1/2−	-87.716	42.9 s 3	IT
	106	1+	-86.362	29.80 s 8	β−
	106m	(6)+	-86.225	131 m 2	β−
	107	7/2+	-86.86	21.7 m 4	β−
	108	1+	-85.0	16.8 s 5	β−
	108m	(5+)	-85.0	6.0 m 3	β−
	109	7/2+	-85.01	80 s 2	β−
	110m	1+	-82.78	3.2 s 2	β−
	110m	(≥4)	-82.78	28.5 s 15	β−
	111	(7/2+)	-82.36	11 s 1	β−
	112m	1+	-79.74	3.45 s 37	β−
	112m	(4,5,6)	-79.74	6.73 s 15	β−
	113	(7/2+)	-78.68	2.80 s 12	β−
	114	1+	-75.6	1.85 s 5	β−
	114m	(4,5)	-75.6	1.85 s 5	β−
	115	(7/2+)	-74.21	0.99 s 5	β−
	116	1+	-70.7	0.68 s 6	β−
	116m	(6−)	-70.6	0.57 s 5	β−
	117	(7/2+)	-68.9s	0.44 s 4	β−
	118	0+	-65.1s	0.30 s 6	β−
	119		-63.2s	>150 ns	β−
	120		-59.2s	>150 ns	β−?
	121		-57.1s	>150 ns	β−?
	122		-52.9s	≈50 ms	β−?
46 Pd	91		-47.4s	>1 μs	ε?
	92	0+	-55.5s	0.7 s +4−2	ε
	93	(7/2+,9/2+)	-59.7s	1.3 s 2	ε, εp 1.5%
	93m		-59.7s	9.3 s +25−17	ε, IT
	94	0+	-66.3s	9.0 s 5	ε
	95		-70.2s	10 s sy	ε
	95m	(21/2+)	-68.2s	13.3 s 3	ε≥91.3%, IT≤9.7%, εp 0.9%
	96	0+	-76.2	122 s 2	ε
	97	5/2+	-77.8	3.10 m 9	ε
	98	0+	-81.30	17.7 m 3	ε
	99	(5/2)+	-82.19	21.4 m 2	ε
	100	0+	-85.23	3.63 d 9	ε
	101	5/2+	-85.43	8.47 h 6	ε
	102	0+	-87.925	**1.02% 1**	
	103	5/2+	-87.479	16.991 d 19	ε
	104	0+	-89.390	**11.14% 8**	
	105	5/2+	-88.413	**22.33% 8**	
	106	0+	-89.902	**27.33% 3**	
	107	5/2+	-88.368	6.5×10^{6} y 3	β−
46 Pd	107m	11/2−	-88.153	21.3 s 5	IT
	108	0+	-89.524	**26.46% 9**	
	109	5/2+	-87.607	13.7012 h 24	β−
	109m	11/2−	-87.418	4.696 m 3	IT
	110	0+	-88.35	**11.72% 9**	
	111	5/2+	-86.00	23.4 m 2	β−
	111m	11/2−	-85.83	5.5 h 1	IT 73%, β− 27%
	112	0+	-86.34	21.03 h 5	β−
	113	(5/2+)	-83.69	93 s 5	β−
	113m		-83.69	≥100 s	
	113m	(9/2−)	-83.61	0.3 s 1	IT
	114	0+	-83.50	2.42 m 6	β−
	115	(5/2+)	-80.40	25 s 2	β−
	115m	(11/2−)	-80.31	50 s 3	β− 92%, IT 8%
	116	0+	-79.96	11.8 s 4	β−
	117	(5/2+)	-76.53	4.3 s 3	β−
	118	0+	-75.5	1.9 s 1	β−
	119		-71.6s	0.92 s 13	β−
	120	0+	-70.1	0.5 s 1	β−
	121		-66.3s	>150 ns	β−?
	122	0+	-64.7s	>150 ns	β−, β−n
	123		-60.6s	>150 ns	β−
	124	0+	-58.8s	≈0.2 s	β−?
47 Ag	93		-46.8s	>1.5 μs	ε?, p?
	94	(0+)	-53.3s	26 ms +26−9	ε, εp
	94m	(21+)	-53.3s	0.47 s 8	ε, εp
	94m	(7+)	-53.3s	0.59 s 2	ε, εp>0%
	95		-60.1s	2.0 s 1	ε, εp
	96	(8+)	-64.6s	4.40 s 6	ε, εp 8.15%
	96	(2+)	-64.6s	6.9 s 6	ε, εp 18%
	97	9/2+	-70.8	25.9 s 4	ε
	98	(6+)	-73.06	47.5 s 3	ε, εp 1.1×10^{-3}%
	99	(9/2)+	-76.8	124 s 3	ε
	99m	(1/2−)	-76.3	10.5 s 5	IT
	100	(5)+	-78.15	2.01 m 9	ε
	100m	(2)+	-78.13	2.24 m 13	ε, IT
	101	9/2+	-81.2	11.1 m 3	ε
	101m	(1/2)−	-81.0	3.10 s 10	IT
	102	5+	-82.26	12.9 m 3	ε
	102m	2+	-82.26	7.7 m 5	ε 51%, IT 49%
	103	7/2+	-84.79	65.7 m 7	ε
	103m	1/2−	-84.66	5.7 s 3	IT
	104	5+	-85.111	69.2 m 10	ε
	104m	2+	-85.104	33.5 m 20	ε 99.93%, IT<0.07%
	105	1/2−	-87.07	41.29 d 7	ε
	105m	7/2+	-87.04	7.23 m 16	IT 99.66%
	106	1+	-86.937	23.96 m 4	ε 99.5%, β−<1%
	106m	6+	-86.847	8.28 d 2	ε
	107	1/2−	-88.402	**51.839% 8**	
	107m	7/2+	-88.309	44.5 s 8	IT
	108	1+	-87.602	2.37 m 1	β− 97.15%, ε 2.85%
	108m	6+	-87.492	438 y 9	ε 91.3%, IT 8.7%

TABLE II.1 (cont.)

Nuclide Z El	A	Jπ	Δ (MeV)	T½, Γ, or Abundance	Decay Mode
47 Ag	109	1/2−	−88.723	**48.161% 8**	
	109m	7/2+	−88.635	38.0 s 12	IT
	110	1+	−87.461	24.6 s 2	β− 99.7%, ε 0.3%
	110m	6+	−87.343	249.76 d 4	β− 98.64%, IT 1.36%
	111	1/2−	−88.221	7.45 d 1	β−
	111m	7/2+	−88.161	64.8 s 8	IT 99.3%, β− 0.7%
	112	2(−)	−86.62	3.130 h 9	β−
	113	1/2−	−87.03	5.37 h 5	β−
	113m	7/2+	−86.99	68.7 s 16	IT 64%, β− 36%
	114	1+	−84.95	4.6 s 1	β−
	115	1/2−	−84.99	20.0 m 5	β−
	115m	7/2+	−84.95	18.0 s 7	β− 79%, IT 21%
	116	(2)−	−82.57	2.68 m 10	β−
	116m	(5+)	−82.49	8.6 s 3	β− 94%, IT 6%
	117	(1/2−)	−82.27	72.8 s +20−7	β− ≈100%
	117m	(7/2+)	−82.24	5.34 s 5	β− 94%, IT 6%
	118	1(−)	−79.57	3.76 s 15	β−
	118m	4(+)	−79.44	2.0 s 2	β− 59%, IT 41%
	119m	(7/2+)	−78.56	2.1 s 1	β−
	119m	(1/2−)	−78.56	6.0 s 5	β−
	120	3(+)	−75.65	1.23 s 4	β−, β−n<3.0×10^{-3}%
	120m	6(−)	−75.45	0.40 s 3	β− ≈63%, IT≈37%
	121	(7/2+)	−74.7	0.79 s 2	β−, β−n 0.08%
	122	(3+)	−71.2s	0.529 s 13	β−, β−n 0.19%
	122m	(8−)	−71.2s	1.5 s 5	β−, β−n
	123	(7/2+)	−70.0s	0.300 s 5	β−, β−n 0.55%
	124		−66.5s	0.172 s 5	β−, β−n>0.1%
	125	(7/2+)	−64.8s	166 ms 7	β−, β−n
	126		−61.0s	107 ms 12	β−, β−n
	127	(1/2−)	−58.9s	79 ms 3	β−
	128		−54.8s	58 ms 5	β−, β−n
	129	(9/2+)	−52.5s	46 ms +5−9	β−, β−n
	129m	(1/2−)	−52.5s	≈160 ms	β−, β−n
	130		−46.2s	≈50 ms	β−
48 Cd	95		−46.7s	5 ms sy	ε?, εp?
	96	0+	−56.1s	≈1 s	ε?
	97		−60.6s	2.8 s 6	ε, εp
	98	0+	−67.63	9.2 s 3	ε, εp<0.03%
	99	(5/2+)	−69.9s	16 s 3	ε, εp 0.17%, εα<1.0×10^{-4}%
	100	0+	−74.25	49.1 s 5	ε
	101	(5/2+)	−75.7	1.36 m 5	ε
	102	0+	−79.68	5.5 m 5	ε
	103	5/2+	−80.65	7.3 m 1	ε
	104	0+	−83.975	57.7 m 10	ε
	105	5/2+	−84.33	55.5 m 4	ε
	106	0+	−87.132	≥2.6×10^{17} y **1.25% 6**	2ε
	107	5/2+	−86.985	6.50 h 2	ε
	108	0+	−89.252	>1.0×10^{18} y **0.89% 3**	2ε
	109	5/2+	−88.508	461.4 d 12	ε
48 Cd	110	0+	−90.353	**12.49% 18**	
	111	1/2+	−89.257	**12.80% 12**	
	111m	11/2−	−88.861	48.50 m 9	IT
	112	0+	−90.580	**24.13% 21**	
	113	1/2+	−89.049	7.7×10^{15} y 3 **12.22% 12**	β−
	113m	11/2−	−88.786	14.1 y 5	β− 99.86%, IT 0.14%
	114	0+	−90.021	>6.4×10^{18} y **28.73% 42**	2β−
	115	1/2+	−88.090	53.46 h 5	β−
	115m	(11/2)−	−87.910	44.56 d 24	β−
	116	0+	−88.719	3.1×10^{19} y 4 **7.49% 18**	2β−
	117	1/2+	−86.425	2.49 h 4	β−
	117m	(11/2)−	−86.289	3.36 h 5	β−
	118	0+	−86.71	50.3 m 2	β−
	119	3/2+	−83.91	2.69 m 2	β−
	119m	(11/2−)	−83.76	2.20 m 2	β−
	120	0+	−83.97	50.80 s 21	β−
	121	(3/2+)	−81.06	13.5 s 3	β−
	121m	(11/2−)	−80.85	8.3 s 8	β−
	122	0+	−80.73	5.24 s 3	β−
	123	(3/2+)	−77.31	2.10 s 2	β−
	123m	(11/2−)	−76.99	1.82 s 3	β−≤100%, IT
	124	0+	−76.71	1.25 s 2	β−
	125	(3/2+)	−73.36	0.65 s 2	β−
	125m	(11/2−)	−73.31	0.48 s 3	β−
	126	0+	−72.33	0.515 s 17	β−
	127	(3/2+)	−68.52	0.37 s 7	β−
	128	0+	−67.3	0.28 s 4	β−
	129	(3/2+)	−63.2s	0.27 s 4	β−
	130	0+	−61.6	162 ms 7	β−, β−n≈3.5%
	131		−55.3s	68 ms 3	β−, β−n 3.5%
	132	0+	−50.7s	97 ms 10	β−, β−n 60%
49 In	97		−47.0s	5 ms sy	p?, ε?
	98		−53.9s	32 ms +32−11	ε
	98m		−53.9s	1.2 s +12−4	ε
	99	(9/2+)	−61.3s	3.0 s +8−7	ε
	100	(6,7)+	−64.2	5.9 s 2	ε, εp 1.6%
	101		−68.6s	15.1 s 3	ε≈100%, εp
	102	(6+)	−70.7	23.3 s 1	ε, εp 9.3×10^{-3}%
	103	(9/2+)	−74.60	65 s 7	ε
	103m	(1/2−)	−73.97	34 s 2	ε 67%, IT 33%
	104	5,6(+)	−76.11	1.80 m 3	ε
	104m	(3+)	−76.01	15.7 s 5	IT 80%, ε 20%
	105	9/2+	−79.48	5.07 m 7	ε
	105m	(1/2−)	−78.81	48 s 6	IT
	106	7+	−80.61	6.2 m 1	ε
	106m	(3+)	−80.58	5.2 m 1	ε
	107	9/2+	−83.56	32.4 m 3	ε
	107m	1/2−	−82.88	50.4 s 6	IT
	108	7+	−84.116	58.0 m 12	ε
49 In	108m	2+	−84.086	39.6 m 7	ε
	109	9/2+	−86.489	4.2 h 1	ε
	109m	1/2−	−85.839	1.34 m 7	IT
	109m	(19/2+)	−84.387	0.209 s 6	IT
	110	7+	−86.47	4.9 h 1	ε
	110m	2+	−86.41	69.1 m 5	ε
	111	9/2+	−88.396	2.8047 d 5	ε
	111m	1/2−	−87.859	7.7 m 2	IT
	112	1+	−87.996	14.97 m 10	ε 56%, β− 44%
	112m	4+	−87.840	20.56 m 6	IT
	113	9/2+	−89.370	**4.29% 5**	
	113m	1/2−	−88.978	99.476 m 23	IT
	114	1+	−88.572	71.9 s 1	β− 99.5%, ε 0.5%
	114m	5+	−88.382	49.51 d 1	IT 96.75%, ε 3.25%
	115	9/2+	−89.537	4.41×10^{14} y 25 **95.71% 5**	β−
	115m	1/2−	−89.200	4.486 h 4	IT 95%, β− 5%
	116	1+	−88.250	14.10 s 3	β− 99.98%, ε 0.02%
	116m	5+	−88.123	54.29 m 17	β−
	116m	8−	−87.960	2.18 s 4	IT
	117	9/2+	−88.945	43.2 m 3	β−
	117m	1/2−	−88.630	116.2 m 3	β− 52.9%, IT 47.1%
	118	1+	−87.230	5.0 s 5	β−
	118m	5+	−87.170	4.45 m 5	β−
	118m	8−	−87.030	8.5 s 3	IT 98.6%, β− 1.4%
	119	9/2+	−87.704	2.4 m 1	β−
	119m	1/2−	−87.393	18.0 m 3	β− 94.4%, IT 5.6%
	120	1+	−85.74	3.08 s 8	β−
	120m	(8−)	−85.74	47.3 s 5	β−
	120m	(5)+	−85.67	46.2 s 8	β−
	121	9/2+	−85.84	23.1 s 6	β−
	121m	1/2−	−85.53	3.88 m 10	β− 98.8%, IT 1.2%
	122	1+	−83.58	1.5 s 3	β−
	122m	5+	−83.54	10.3 s 6	β−
	122m	8−	−83.29	10.8 s 4	β−
	123	(9/2)+	−83.43	6.17 s 5	β−
	123m	(1/2)−	−83.10	47.4 s 4	β−
	124	3+	−80.88	3.11 s 10	β−
	124m	(8−)	−80.83	3.7 s 2	β−
	125	9/2+	−80.48	2.36 s 4	β−
	125m	1/2(−)	−80.12	12.2 s 2	β−
	126	3(+)	−77.81	1.53 s 1	β−
	126m	(8−)	−77.71	1.64 s 5	β−
	127	(9/2+)	−76.99	1.09 s 1	β−, β−n≤0.03%
	127m	(1/2−)	−76.52	3.67 s 4	β−, β−n 0.69%
	128	(3)+	−74.36	0.84 s 6	β−, β−n<0.05%
	128m	(8−)	−74.02	0.72 s 10	β−, β−n<0.05%
	129	(9/2+)	−72.94	0.61 s 1	β−, β−n 0.25%
	129m	(1/2−)	−72.56	1.23 s 3	β−>99.7%, β−n 2.5%, IT<0.3%
	130	1(−)	−69.89	0.29 s 2	β−, β−n 0.93%
	130m	(10−)	−69.84	0.54 s 1	β−, β−n 1.65%

TABLE II.1 (cont.)

Nuclide Z El A	Jπ	Δ (MeV)	T½, Γ, or Abundance	Decay Mode
49 In 130m	(5+)	−69.49	0.54 s 1	β−, β−n 1.65%
131	(9/2+)	−68.14	0.28 s 3	β−, β−n≤2%
131m	(1/2−)	−67.77	0.35 s 5	β−≥99.98%, β−n≤2%, IT≤0.02%
131m		−63.87	0.32 s 6	β−>99%, IT<1%, β−n 0.03%
132	(7−)	−62.42	0.207 s 6	β−, β−n 6.3%
133	(9/2+)	−57.9s	165 ms 3	β−, β−n 85%
134	(4− to 7−)	−52.0s	140 ms 4	β−, β−n 65%
135		−47.2s	92 ms 10	β−, β−n>0%
50 Sn 99		−47.2s	5 ms sy	ε?, εp?
100	0+	−56.8	0.94 s +54−27	ε, εp<17%
101		−59.6s	3 s 1	ε, εp
102	0+	−64.9	4.5 s 7	ε
103		−67.0s	7.0 s 6	ε, εp
104	0+	−71.6	20.8 s 5	ε
105	(5/2+)	−73.26	34 s 1	ε, εp
106	0+	−77.43	115 s 5	ε
107	(5/2+)	−78.58	2.90 m 5	ε
108	0+	−82.04	10.30 m 8	ε
109	5/2(+)	−82.64	18.0 m 2	ε
110	0+	−85.84	4.11 h 10	ε
111	7/2+	−85.945	35.3 m 6	ε
112	0+	−88.661	**0.97% 1**	
113	1/2+	−88.333	115.09 d 3	ε
113m	7/2+	−88.256	21.4 m 4	IT 91.1%, ε 8.9%
114	0+	−90.561	**0.66% 1**	
115	1/2+	−90.036	**0.34% 1**	
116	0+	−91.528	**14.54% 9**	
117	1/2+	−90.400	**7.68% 7**	
117m	11/2−	−90.085	13.76 d 4	IT
118	0+	−91.656	**24.22% 9**	
119	1/2+	−90.068	**8.59% 4**	
119m	11/2−	−89.979	293.1 d 7	IT
120	0+	−91.105	**32.58% 9**	
121	3/2+	−89.204	27.03 h 4	β−
121m	11/2−	−89.198	43.9 y 5	IT 77.6%, β− 22.4%
122	0+	−89.946	**4.63% 3**	
123	11/2−	−87.821	129.2 d 4	β−
123m	3/2+	−87.796	40.06 m 1	β−
124	0+	−88.237	**5.79% 5**	
125	11/2−	−85.898	9.64 d 3	β−
125m	3/2+	−85.871	9.52 m 5	β−
126	0+	−86.02	2.30×10^{5} y 14	β−
127	(11/2−)	−83.50	2.10 h 4	β−
127m	(3/2+)	−83.49	4.13 m 3	β−
128	0+	−83.33	59.07 m 14	β−
128m	(7−)	−81.24	6.5 s 5	IT
129	(3/2+)	−80.59	2.23 m 4	β−
129m	(11/2−)	−80.56	6.9 m 1	β−, IT<2.0×10^{-3}%
130	0+	−80.14	3.72 m 7	β−
130m	(7−)	−78.19	1.7 m 1	β−
50 Sn 131	(3/2+)	−77.31	56.0 s 5	β−
131m	(11/2−)	−77.07	58.4 s 5	β−, IT≤4.0×10^{-4}%
132	0+	−76.55	39.7 s 8	β−
133	(7/2−)	−70.95	1.45 s 3	β−, β−n 0.08%
134	0+	−66.80	1.050 s 11	β−, β−n 17%
135	(7/2−)	−60.8s	530 ms 20	β−, β−n 21%
136	0+	−56.5s	0.25 s 3	β−, β−n 30%
137		−50.3s	190 ms 60	β−, β−n 58%
51 Sb 103		−56.2s	>1.5 μs	ε?
104		−59.2s	0.44 s +15−11	ε, εp<7%, p<1%
105	(5/2+)	−63.8	1.12 s 16	ε 99%, p 1%
106	(4+)	−66.3s	0.6 s 2	ε
107	(5/2+)	−70.7s	4.0 s 2	ε
108	(4+)	−72.5s	7.4 s 3	ε
109	(5/2+)	−76.26	17.3 s 5	ε
110	(3+,4+)	−77.5s	23.0 s 4	ε
111	(5/2+)	−80.89	75 s 1	ε
112	3+	−81.60	51.4 s 10	ε
113	5/2+	−84.42	6.67 m 7	ε
114	3+	−84.52	3.49 m 3	ε
115	5/2+	−87.00	32.1 m 3	ε
116	3+	−86.821	15.8 m 8	ε
116m	8−	−86.438	60.3 m 6	ε
117	5/2+	−88.645	2.80 h 1	ε, ε 1.7%
118	1+	−87.999	3.6 m 1	ε
118m	8−	−87.749	5.00 h 2	ε
119	5/2+	−89.477	38.19 h 22	ε
119m	(27/2+)	−86.636	0.85 s 9	IT
120	1+	−88.424	15.89 m 4	ε
120m	8−	−88.424	5.76 d 2	ε
121	5/2+	−89.595	**57.21% 5**	
122	2−	−88.330	2.7238 d 2	β− 97.59%, ε 2.41%
122m	(8)−	−88.167	4.191 m 3	IT
123	7/2+	−89.224	**42.79% 5**	
124	3−	−87.620	60.11 d 7	β−
124m	5+	−87.609	93 s 5	IT 75%, β− 25%
124m	(8)−	−87.584	20.2 m 2	IT
125	7/2+	−88.256	2.7586 y 3	β−
126	(8−)	−86.40	12.35 d 6	β−
126m	(5+)	−86.38	19.15 m 8	β− 86%, IT 14%
126m	(3−)	−86.36	≈11 s	IT
127	7/2+	−86.700	3.85 d 5	β−
128	8−	−84.61	9.01 h 4	β−
128m	5+	−84.61	10.4 m 2	β− 96.4%, IT 3.6%
129	7/2+	−84.63	4.40 h 1	β−
129m	(19/2−)	−82.78	17.7 m 1	β− 85%, IT 15%
130	(8−)	−82.29	39.5 m 8	β−
130m	(4,5)+	−82.29	6.3 m 2	β−
131	(7/2+)	−81.99	23.03 m 4	β−
132	(4)+	−79.67	2.79 m 7	β−
132m	(8−)	−79.67	4.10 m 5	β−
133	(7/2+)	−78.94	2.5 m 1	β−
51 Sb 134	(0−)	−74.17	0.78 s 6	β−
134m	(7−)	−74.17	10.07 s 5	β−, β−n 0.09%
135	(7/2+)	−69.7	1.68 s 2	β−, β−n 22%
136	1−	−64.9s	0.923 s 14	β−, β−n 16.3%
137		−60.3s	>150 ns	β−?, β−n?
138		−55.2s	>300 ns	β−?, β−n?
139		−50.3s	>150 ns	β−?
52 Te 105		−52.5s	1 μs sy	α?, ε?
106	0+	−58.2	70 μs +20−10	α
107		−60.5s	3.1 ms 1	α 70%, ε 30%
108	0+	−65.7	2.1 s 1	ε 51%, α 49%, εp 2.4%
109	(5/2+)	−67.61	4.6 s 3	ε 96.1%, εp 9.4%, α 3.9%, εα<5.0×10^{-3}%
110	0+	−72.28	18.6 s 8	ε≈100%, α≈3.0×10^{-3}%
111	(5/2)+	−73.48	19.3 s 4	ε, εp
112	0+	−77.3	2.0 m 2	ε
113	(7/2+)	−78.35	1.7 m 2	ε
114	0+	−81.89	15.2 m 7	ε
115	7/2+	−82.06	5.8 m 2	ε
115m	(1/2)+	−82.04	6.7 m 4	ε≤100%, IT
116	0+	−85.27	2.49 h 4	ε
117	1/2+	−85.10	62 m 2	ε, ε 25%
117m	(11/2−)	−84.80	103 ms 3	IT
118	0+	−87.72	6.00 d 2	ε
119	1/2+	−87.184	16.05 h 5	ε, ε 2.06%
119m	11/2−	−86.923	4.70 d 4	ε, ε 0.41%, IT<8.0×10^{-3}%
120	0+	−89.405	>2.2×10^{16} y **0.09% 1**	2ε
121	1/2+	−88.55	19.16 d 5	ε
121m	11/2−	−88.26	154 d 7	IT 88.6%, ε 11.4%
122	0+	−90.314	**2.55% 12**	
123	1/2+	−89.172	>9.2×10^{16} y **0.89% 3**	ε
123m	11/2−	−88.924	119.2 d 1	IT
124	0+	−90.524	**4.74% 14**	
125	1/2+	−89.022	**7.07% 15**	
125m	11/2−	−88.877	57.40 d 15	IT
126	0+	−90.065	**18.84% 25**	
127	3/2+	−88.281	9.35 h 7	β−
127m	11/2−	−88.193	109 d 2	IT 97.6%, β− 2.4%
128	0+	−88.992	8.8×10^{18} y 4 **31.74% 8**	2β−
129	3/2+	−87.003	69.6 m 3	β−
129m	11/2−	−86.898	33.6 d 1	IT 63%, β− 37%
130	0+	−87.351	>5×10^{23} y **34.08% 62**	2β−
131	3/2+	−85.210	25.0 m 1	β−
131m	11/2−	−85.027	30 h 2	β− 77.8%, IT 22.2%

TABLE II.1 (cont.)

Nuclide Z El A	Jπ	Δ (MeV)	T½, Γ, or Abundance	Decay Mode
52 Te 132	0+	-85.182	3.204 d 13	β–
133	(3/2+)	-82.94	12.5 m 3	β–
133m	(11/2–)	-82.61	55.4 m 4	β– 82.5%, IT 17.5%
134	0+	-82.56	41.8 m 8	β–
135	(7/2–)	-77.83	19.0 s 2	β–
136	0+	-74.43	17.63 s 8	β–, β–n 1.31%
137	(7/2–)	-69.6	2.49 s 5	β–, β–n 2.69%
138	0+	-65.9s	1.4 s 4	β–, β–n 6.3%
139	(7/2–)	-60.8s	>150 ns	β–, β–n
140	0+	-57.0s	>150 ns	β–?, β–n?
141		-51.6s	>150 ns	β–?, β–n?
142	0+	-47.4s	>150 ns	β–?
53 I 108	(1)	-52.7s	36 ms 6	α 91%, ε 9%, p<1%
109	1/2+	-57.6	103 μs 5	p
110		-60.3s	0.65 s 2	ε 83%, α 17%, εp 11%, εα 1.1%
111	(5/2+)	-64.9s	2.5 s 2	ε 99.9%, α≈0.1%
112		-67.1s	3.42 s 11	ε, α≈1.2×10^{-3}%
113	5/2+	-71.13	6.6 s 2	ε, α 3.3×10^{-7}%
114	1+	-72.8s	2.1 s 2	ε, εp
114m	(7)	-72.5s	6.2 s 5	ε 91%, IT 9%
115	(5/2+)	-76.34	1.3 m 2	ε
116	1+	-77.49	2.91 s 15	ε
117	(5/2)+	-80.43	2.22 m 4	ε
118	2–	-80.97	13.7 m 5	ε
118m	(7–)	-80.87	8.5 m 5	ε<100%, IT>0%
119	5/2+	-83.77	19.1 m 4	ε
120	2–	-83.79	81.6 m 2	ε
120m	(7–)	-83.47	53 m 4	ε
121	5/2+	-86.29	2.12 h 1	ε
122	1+	-86.080	3.63 m 6	ε
123	5/2+	-87.943	13.232 h 6	ε
124	2–	-87.365	4.1760 d 3	ε
125	5/2+	-88.836	59.400 d 10	ε
126	2–	-87.910	12.93 d 5	ε 52.7%, β– 47.3%
127	5/2+	-88.983	**100%**	
128	1+	-87.738	24.99 m 2	β– 93.1%, ε 6.9%
129	7/2+	-88.503	1.57×10^{7} y 4	β–
130	5+	-86.932	12.36 h 1	β–
130m	2+	-86.892	8.84 m 6	IT 84%, β– 16%
131	7/2+	-87.444	8.02070 d 11	β–
132	4+	-85.700	2.295 h 13	β–
132m	(8–)	-85.580	1.387 h 15	IT 86%, β– 14%
133	7/2+	-85.887	20.8 h 1	β–
133m	(19/2–)	-84.252	9 s 2	IT
134	(4)+	-84.073	52.5 m 2	β–
134m	(8)–	-83.756	3.52 m 4	IT 97.7%, β– 2.3%
135	7/2+	-83.790	6.57 h 2	β–
136	(1–)	-79.50	83.4 s 10	β–
136m	(6–)	-78.86	46.9 s 10	β–
137	(7/2+)	-76.50	24.5 s 2	β–, β–n 6.97%
138	(2–)	-72.33	6.23 s 3	β–, β–n 5.56%

Nuclide Z El A	Jπ	Δ (MeV)	T½, Γ, or Abundance	Decay Mode
53 I 139	(7/2+)	-68.84	2.280 s 11	β–, β–n 10%
140	(3)	-64.3s	0.86 s 4	β–, β–n 9.3%
141		-60.5s	0.43 s 2	β–, β–n 21.2%
142		-55.7s	≈0.2 s	β–
143		-51.6s	>150 ns	β–?
144		-46.6s	>300 ns	β–?
54 Xe 110	0+	-51.9	105 ms +35–25	α≈64%
110	0+	-51.9	≈0.2 s	ε
111		-54.4s	0.74 s 20	α 8%, ε
112	0+	-60.0	2.7 s 8	ε 99.16%, α 0.84%
113	(5/2+)	-62.09	2.74 s 8	ε≈100%, εp 7%, α≈0.01%, εα≈7.0×10^{-3}%
114	0+	-67.09	10.0 s 4	ε
115	(5/2+)	-68.66	18 s 4	ε, εp 0.34%, α 3.0×10^{-4}%
116	0+	-73.05	59 s 2	ε
117	5/2(+)	-74.19	61 s 2	ε, εp 2.9×10^{-3}%
118	0+	-78.08	3.8 m 9	ε
119	(5/2+)	-78.79	5.8 m 3	ε
120	0+	-82.17	40 m 1	ε
121	(5/2+)	-82.47	40.1 m 20	ε
122	0+	-85.36	20.1 h 1	ε
123	(1/2)+	-85.249	2.08 h 2	ε
124	0+	-87.660	≥1.1×10^{17} y **0.095% 3**	2ε
125	1/2(+)	-87.192	16.9 h 2	ε
125m	9/2(–)	-86.939	56.9 s 9	IT
126	0+	-89.169	**0.089% 1**	
127	1/2+	-88.321	36.4 d 1	ε
127m	9/2–	-88.024	69.2 s 9	IT
128	0+	-89.860	**1.910% 22**	
129	1/2+	-88.697	**26.40% 18**	
129m	11/2–	-88.461	8.88 d 2	IT
130	0+	-89.882	**4.071% 53**	
131	3/2+	-88.415	**21.232% 62**	
131m	11/2–	-88.251	11.934 d 21	IT
132	0+	-89.281	**26.909% 68**	
133	3/2+	-87.644	5.243 d 1	β–
133m	11/2–	-87.410	2.19 d 1	IT
134	0+	-88.124	>5.8×10^{22} y **10.436% 29**	2β– ≥0%
134m	7–	-86.159	290 ms 17	IT
135	3/2+	-86.417	9.14 h 2	β–
135m	11/2–	-85.890	15.29 m 5	IT>99.4%, β–<0.6%
136	0+	-86.425	>2.4×10^{21} y **8.857% 33**	2β–
137	7/2–	-82.379	3.818 m 13	β–
138	0+	-80.15	14.08 m 8	β–
139	3/2–	-75.64	39.68 s 14	β–
140	0+	-72.99	13.60 s 10	β–
141	5/2(–)	-68.33	1.73 s 1	β–, β–n 0.04%

Nuclide Z El A	Jπ	Δ (MeV)	T½, Γ, or Abundance	Decay Mode
54 Xe 142	0+	-65.5	1.250 s 25	β–, β–n 0.21%
143	5/2–	-60.4s	0.511 s 6	β–, β–n 1%
144	0+	-57.3s	0.388 s 7	β–, β–n 3%
145	(3/2–)	-52.1s	188 ms 4	β–
145		-52.1s	188 ms 4	β–n 5%
146	0+	-48.7s	146 ms 6	β–, β–n 6.9%
147		-43.3s	0.10 s +10–5	β–n<8%
55 Cs 112	(0+,3+)	-46.3s	0.5 ms 1	p
113	(3/2+)	-51.7	16.7 μs 7	p, α
114	(1+)	-54.5s	0.57 s 2	ε≈100%, εp 8.7%, εα 0.19%, α 0.02%
115		-59.7s	1.4 s 8	ε, εp≈0.07%
116	(1+)	-62.1s	0.70 s 4	ε, εp>0%, εα>0%
116m	4+,5,6	-62.0s	3.85 s 13	ε, εp>0%, εα>0%
117	(9/2+)	-66.44	8.4 s 6	ε
117m	(3/2+)	-66.29	6.5 s 4	ε
118	2	-68.41	14 s 2	ε, εp<0.04%, εα<2.4×10^{-3}%
118m	6,7,8	-68.41	17 s 3	ε, εp<0.04%, εα<2.4×10^{-3}%
119	9/2+	-72.31	43.0 s 2	ε
119m	3/2(+)	-72.31	30.4 s 1	ε
120	2(+)	-73.89	61.3 s 11	ε, εα 2.0×10^{-5}%, εp 7.0×10^{-6}%
120m	(7–)	-73.89	57 s 6	ε
121	3/2(+)	-77.10	155 s 4	ε
121m	9/2(+)	-77.03	122 s 3	ε 83%, IT 17%
122	1+	-78.14	21.18 s 19	ε
122m	8–	-78.02	3.70 m 11	ε
122m	(5)–	-78.01	0.36 s 2	IT
123	1/2+	-81.04	5.88 m 3	ε
123m	(11/2)–	-80.89	1.64 s 12	IT
124	1+	-81.731	30.8 s 5	ε
124m	(7)+	-81.269	6.3 s 2	IT
125	1/2(+)	-84.088	46.7 m 1	ε
126	1+	-84.34	1.64 m 2	ε
127	1/2+	-86.240	6.25 h 10	ε
128	1+	-85.931	3.66 m 2	ε
129	1/2+	-87.500	32.06 h 6	ε
130	1+	-86.900	29.21 m 4	ε 98.4%, β– 1.6%
130m	5–	-86.737	3.46 m 6	IT 99.84%, ε 0.16%
131	5/2+	-88.060	9.689 d 16	ε
132	2+	-87.156	6.480 d 6	ε 98.13%, β– 1.87%
133	7/2+	-88.071	**100%**	
134	4+	-86.891	2.0652 y 4	β–, ε 3.0×10^{-4}%
134m	8–	-86.753	2.912 h 2	IT
135	7/2+	-87.582	2.3×10^{6} y 3	β–
135m	19/2–	-85.949	53 m 2	IT
136	5+	-86.339	13.04 d 3	β–
136m	8–	-86.339	19 s 2	IT>0%, β–
137	7/2+	-86.546	30.03 y 5	β–
138	3–	-82.887	33.41 m 18	β–

Nuclide Z El A	Jπ	Δ (MeV)	T½, Γ, or Abundance	Decay Mode
55 Cs 138m	6–	–82.808	2.91 m 8	IT 81%, β– 19%
139	7/2+	–80.701	9.27 m 5	β–
140	1–	–77.051	63.7 s 3	β–
141	7/2+	–74.48	24.84 s 16	β–, β–n 0.04%
142	0–	–70.52	1.684 s 14	β–, β–n 0.09%
143	3/2+	–67.67	1.791 s 7	β–, β–n 1.64%
144	1	–63.27	0.994 s 4	β–, β–n 3.2%
144m	(≥4)	–63.27	<1 s	β–
145	3/2+	–60.06	0.594 s 13	β–, β–n 14.3%
146	1–	–55.62	0.321 s 2	β–, β–n 14.2%
147	(3/2+)	–52.02	0.235 s 3	β–, β–n 43%
148		–47.3	146 ms 6	β–, β–n 25.1%
149		–43.8s	>50 ms	β–, β–n
150		–39.0s	>50 ms	β–, β–n
151		–35.2s	>50 ms	β–?, β–n?
56 Ba 114	0+	–45.9	0.43 s +30–15	ε≈100%, εp 20%, α 9.0×10^{-5}%, 12≈3.0×10^{-5}%
115	(5/2+)	–49.0s	0.45 s 5	ε, εp>15%
116	0+	–54.6s	1.3 s 2	ε, εp 3%
117	(3/2)	–57.3s	1.75 s 7	ε, εp>0%, εα>0%
118	0+	–62.4s	5.2 s 2	ε
119	(5/2+)	–64.6	5.4 s 3	ε, εp<25%
120	0+	–68.9	24 s 2	ε
121	5/2(+)	–70.7	29.7 s 15	ε
122	0+	–74.61	1.95 m 15	ε
123	5/2(+)	–75.65	2.7 m 4	ε
124	0+	–79.09	11.0 m 5	ε
125	1/2(+)	–79.67	3.5 m 4	ε
126	0+	–82.67	100 m 2	ε
127	1/2+	–82.82	12.7 m 4	ε
127m	7/2–	–82.74	1.9 s 2	IT
128	0+	–85.40	2.43 d 5	ε
129	1/2+	–85.06	2.23 h 11	ε
129m	7/2+	–85.06	2.16 h 2	ε≤100%, IT
130	0+	–87.262	≥3.5×10^{14} y **0.106% 1**	2ε
131	1/2+	–86.684	11.50 d 6	ε
131m	9/2–	–86.497	14.6 m 2	IT
132	0+	–88.435	>3.0×10^{21} y **0.101% 1**	2ε
133	1/2+	–87.553	3841 d 7	ε
133m	11/2–	–87.265	38.9 h 1	IT 99.99%, ε 9.6×10^{-3}%
134	0+	–88.950	**2.417% 18**	
135	3/2+	–87.851	**6.592% 12**	
135m	11/2–	–87.582	28.7 h 2	IT
136	0+	–88.887	**7.854% 24**	
136m	7–	–86.856	0.3084 s 19	IT
137	3/2+	–87.721	**11.232% 24**	
137m	11/2–	–87.060	2.552 m 1	IT
138	0+	–88.262	**71.698% 42**	
56 Ba 139	7/2–	–84.914	83.06 m 28	β–
140	0+	–83.271	12.752 d 3	β–
141	3/2–	–79.726	18.27 m 7	β–
142	0+	–77.823	10.6 m 2	β–, β–n 0.09%
143	5/2–	–73.94	14.5 s 3	β–
144	0+	–71.77	11.5 s 2	β–, β–n 3.6%
145	5/2–	–67.42	4.31 s 16	β–
146	0+	–65.00	2.22 s 7	β–
147	(3/2+)	–60.6s	0.893 s 1	β–, β–n 0.06%
148	0+	–58.01	0.612 s 17	β–, β–n 0.4%
149		–53.5s	0.344 s 7	β–, β–n 0.43%
150	0+	–50.6s	0.3 s	β–
151		–45.8s	>150 ms	β–?
152	0+	–42.6s	≈0.1 s	β–?
153		–37.6s	≈0.08 s	β–?
57 La 117	(3/2+,3/2–)	–46.5s	23.5 ms 26	p 93.9%, ε 6.1%
117m	(9/2+)	–46.4s	10 ms 5	p 97.4%, ε 2.6%
118		–49.6s	≈1 s	ε?
119		–55.0s	≈2 s	ε?
120m		–57.7s	2.8 s 2	ε, εp>0%
121		–62.4s	5.3 s 2	ε
122		–64.5s	8.6 s 5	ε, εp
123		–68.7s	17 s 3	ε
124m	low	–70.26	<1 s	ε
124m	(7,8–)	–70.26	29 s 1	ε
125		–73.76	64.8 s 12	ε
125m		–73.65	0.4 s 2	IT
126m	(0–,1,2–)	–74.97	<50 s	ε, IT
126m	(5+)	–74.97	54 s 2	ε>0%
127	(11/2–)	–77.90	5.1 m 1	ε
127m	(3/2+)	–77.88	3.7 m 4	ε, IT
128	(5+)	–78.63	5.18 m 14	ε
128m	(1+,2–)	–78.63	<1.4 m	ε
129	3/2+	–81.33	11.6 m 2	ε
129m	11/2–	–81.15	0.56 s 5	IT
130	3(+)	–81.63	8.7 m 1	ε
131	3/2+	–83.77	59 m 2	ε
132	2–	–83.74	4.8 h 2	ε
132m	6–	–83.55	24.3 m 5	IT 76%, ε 24%
133	5/2+	–85.49	3.912 h 8	ε
134	1+	–85.22	6.45 m 16	ε
135	5/2+	–86.65	19.5 h 2	ε
136	1+	–86.04	9.87 m 3	ε
136m	(8+)	–85.81	114 ms 3	IT
137	7/2+	–87.10	6×10^{4} y 2	ε
138	5+	–86.525	1.02×10^{11} y 1 **0.090% 1**	ε 65.6%, β– 34.4%
139	7/2+	–87.231	**99.910% 1**	
140	3–	–84.321	1.6781 d 3	β–
141	(7/2+)	–82.938	3.92 h 3	β–
142	2–	–80.035	91.1 m 5	β–
143	(7/2)+	–78.19	14.2 m 1	β–
57 La 144	(3–)	–74.89	40.8 s 4	β–
145	(5/2+)	–72.99	24.8 s 20	β–
146	2–	–69.12	6.27 s 10	β–
146m	(6–)	–69.12	10.0 s 1	β–
147	(5/2+)	–66.85	4.015 s 8	β–, β–n 0.04%
148	(2–)	–63.13	1.26 s 8	β–, β–n 0.15%
149	(3/2,5/2)	–60.8s	1.05 s 3	β–, β–n 1.43%
150	(3–)	–57.0s	0.51 s 3	β–, β–n 2.7%
151		–54.3s	>150 ns	β–?
152		–50.1s	>150 ns	β–?
153		–46.9s	>100 ns	
154		–42.4s	≈0.1 s	β–?
155		–38.8s	≈0.06 s	β–?
58 Ce 119		–44.0s	≈0.2 s	ε?
120	0+	–49.7s	≈0.25 s	ε?
121		–52.7s	1.1 s 1	ε, εp≈1%
122	0+	–57.8s	≈2 s	ε?, εp?
123	(5/2)	–60.2s	3.8 s 2	ε, εp>0%
124	0+	–64.8s	6 s 2	ε
125	(5/2+)	–66.7s	10.2 s 4	ε, εp
126	0+	–70.82	51.0 s 3	ε
127	(5/2+)	–71.98	31 s 2	ε
128	0+	–75.53	3.93 m 2	ε
129	5/2+	–76.29	3.5 m 5	ε>0%
130	0+	–79.42	22.9 m 5	ε
131	(7/2+)	–79.72	10.2 m 3	ε
131m	(1/2+)	–79.72	5.0 m 10	ε
132	0+	–82.47	3.51 h 11	ε
133	1/2+	–82.42	97 m 4	ε
133m	9/2–	–82.39	4.9 h 4	ε
134	0+	–84.84	3.16 d 4	ε
135	1/2(+)	–84.62	17.7 h 3	ε
135m	(11/2–)	–84.18	20 s 1	IT
136	0+	–86.47	>0.7×10^{14} y **0.185% 2**	2ε
137	3/2+	–85.88	9.0 h 3	ε
137m	11/2–	–85.62	34.4 h 3	IT 99.22%, ε 0.78%
138	0+	–87.57	≥0.9×10^{14} y **0.251% 2**	2ε
139	3/2+	–86.952	137.641 d 20	ε
139m	11/2–	–86.198	54.8 s 10	IT
140	0+	–88.083	**88.450% 18**	
141	7/2–	–85.440	32.508 d 13	β–
142	0+	–84.538	>2.6×10^{17} y **11.114% 17**	2β–
143	3/2–	–81.612	33.039 h 6	β–
144	0+	–80.437	284.91 d 5	β–
145	(3/2–)	–77.10	3.01 m 6	β–
146	0+	–75.68	13.52 m 13	β–
147	(5/2–)	–72.03	56.4 s 10	β–
148	0+	–70.39	56 s 1	β–
149	(3/2–)	–66.70	5.3 s 2	β–

TABLE II.1 (cont.)

Nuclide Z El	A	Jπ	Δ (MeV)	T½, Γ, or Abundance	Decay Mode
58 Ce	150	0+	-64.82	4.0 s 6	β–
	151		-61.5	1.02 s 6	β–
	152	0+	-59.1s	1.4 s 2	β–
	153		-55.3s	>100 ns	β–?
	154	0+	-52.7s	>150 ns	β–?
	155		-48.4s	>300 ns	β–?
	156	0+	-45.4s	≈0.15 s	β–?
	157		-40.7s	≈0.05 s	β–?
59 Pr	121	(3/2–)	-41.6s	1.4 s 8	p
	122		-44.9s	≈0.5 s	ε?
	123		-50.3s	≈0.8 s	ε?
	124		-53.1s	1.2 s 2	ε, εp
	125		-57.9s	3.3 s 7	ε, εp
	126	≥4	-60.3s	3.14 s 22	ε
	126	≥4	-60.3s	3.14 s 22	εp
	127		-64.4s	4.2 s 3	ε
	128	4,5,6	-66.33	2.84 s 9	ε
	129	(11/2–)	-69.77	32 s 3	ε>0%
	130?	(4,5)+	-71.18	40 s 4	ε
	131	(3/2+)	-74.28	94 s 4	ε
	131m	(11/2–)	-74.13	5.7 s 2	IT 96.4%, ε 3.6%
	132	(2)+	-75.21	1.6 m 3	ε
	133	(3/2+)	-77.94	6.5 m 3	ε
	134m	(6–)	-78.51	≈11 m	ε
	134m	2–	-78.51	17 m 2	ε
	135	3/2(+)	-80.94	24 m 2	ε
	136	2+	-81.33	13.1 m 1	ε
	137	5/2+	-83.18	1.28 h 3	ε
	138	1+	-83.13	1.45 m 5	ε
	138m	7–	-82.77	2.12 h 4	ε
	139	5/2+	-84.823	4.41 h 4	ε
	140	1+	-84.695	3.39 m 1	ε
	141	5/2+	-86.021	**100%**	
	142	2–	-83.793	19.12 h 4	β– 99.98%, ε 0.02%
	142m	5–	-83.789	14.6 m 5	IT
	143	7/2+	-83.074	13.57 d 2	β–
	144	0–	-80.756	17.28 m 5	β–
	144m	3–	-80.697	7.2 m 3	IT 99.93%, β– 0.07%
	145	7/2+	-79.632	5.984 h 10	β–
	146	(2)–	-76.71	24.15 m 18	β–
	147	(3/2+)	-75.45	13.4 m 4	β–
	148	1–	-72.53	2.29 m 2	β–
	148m	(4)	-72.44	2.01 m 7	β–
	149	(5/2+)	-71.06	2.26 m 7	β–
	150	(1)–	-68.30	6.19 s 16	β–
	151	(3/2–)	-66.77	18.90 s 7	β–
	152	(4–)	-63.8	3.63 s 12	β–
	153		-61.6	4.28 s 11	β–
	154	(3+,2+)	-58.2	2.3 s 1	β–
	155		-55.8s	>300 ns	β–?
	156		-51.9s	>300 ns	β–?
	157		-49.0s	≈0.3 s	β–?
59 Pr	158		-44.7s	≈0.2 s	β–?
	159		-41.5s	≈0.1 s	β–?
60 Nd	124	0+	-44.5s	0.5 s sy	ε?
	125	(5/2)	-47.6s	0.60 s 15	ε, εp>0%
	126	0+	-52.9s	>200 ns	ε, εp
	127		-55.4s	1.8 s 4	ε, εp
	128	0+	-60.2s	5 s	ε, εp
	129	(5/2+)	-62.2s	7 s 1	ε, εp
	130	0+	-66.60	21 s 3	ε
	131	(5/2)	-67.77	33 s 3	ε, εp
	132	0+	-71.43	94 s 8	ε
	133	(7/2+)	-72.33	70 s 10	ε
	133m	(1/2)+	-72.20	≈70 s	ε, IT
	134	0+	-75.65	8.5 m 15	ε
	135	9/2(–)	-76.21	12.4 m 6	ε
	135m	(1/2+)	-76.15	5.5 m 5	ε>99.97%, IT<0.03%
	136	0+	-79.20	50.65 m 33	ε
	137	1/2+	-79.58	38.5 m 15	ε
	137m	11/2–	-79.06	1.60 s 15	IT
	138	0+	-82.02	5.04 h 9	ε
	139	3/2+	-81.99	29.7 m 5	ε
	139m	11/2–	-81.76	5.50 h 20	ε 88.2%, IT 11.8%
	140	0+	-84.25	3.37 d 2	ε
	141	3/2+	-84.198	2.49 h 3	ε
	141m	11/2–	-83.441	62.0 s 8	IT, ε<0.05%
	142	0+	-85.955	**27.2% 5**	
	143	7/2–	-84.007	**12.2% 2**	
	144	0+	-83.753	2.29×10^{15} y 16 **23.8% 3**	α
	145	7/2–	-81.437	**8.3% 1**	
	146	0+	-80.931	**17.2% 3**	
	147	5/2–	-78.152	10.98 d 1	β–
	148	0+	-77.413	**5.7% 1**	
	149	5/2–	-74.381	1.728 h 1	β–
	150	0+	-73.690	0.79×10^{19} y 7 **5.6% 2**	2β–
	151	3/2+	-70.953	12.44 m 7	β–
	152	0+	-70.16	11.4 m 2	β–
	153	(3/2)–	-67.35	31.6 s 10	β–
	154	0+	-65.7	25.9 s 2	β–
	155		-62.5s	8.9 s 2	β–
	156	0+	-60.5	5.49 s 7	β–
	157		-56.8s	>100 ns	β–?
	158	0+	-54.4s	>50 ns	β–
	159		-50.2s	≈0.7 s	β–?
	160	0+	-47.4s	≈0.3 s	β–?
	161		-43.0s	≈0.2 s	β–?
61 Pm	126		-39.6s	0.5 s sy	ε?
	127		-45.1s	1 s sy	ε?, p?
	128		-48.0s	1.0 s 3	ε, α, εp
	129	5/2–)	-52.9s	2.4 s 9	ε
61 Pm	130	(4,5,6)	-55.5s	2.6 s 2	ε, εp
	131	(11/2)	-59.7s	6.3 s 8	ε
	132	(3+)	-61.7s	6.2 s 6	ε, εp≈5.0×10^{-5}%
	133	(11/2–)	-65.41	15 s 3	ε
	134	(2+)	-66.74	≈5 s	ε
	134m	(5+)	-66.74	22 s 1	ε
	135m	(11/2–)	-69.98	45 s 4	ε
	135m	3/2+,5/2+	-69.98	49 s 3	ε
	136m	(2+)	-71.20	47 s 2	ε
	136m	(5–)	-71.20	107 s 6	ε
	137	11/2–	-74.07	2.4 m 1	ε
	138		-74.94	10 s 2	ε
	138m		-74.92	3.24 m 5	ε
	139	(5/2)+	-77.50	4.15 m 5	ε
	139m	(11/2)–	-77.31	180 ms 20	IT 99.94%, ε 0.06%
	140	1+	-78.21	9.2 s 2	ε
	140m	8–	-78.21	5.95 m 5	ε
	141	5/2+	-80.52	20.90 m 5	ε
	142	1+	-81.16	40.5 s 5	ε
	143	5/2+	-82.968	265 d 7	ε, ε<5.7×10^{-6}%
	144	5–	-81.421	363 d 14	ε
	145	5/2+	-81.274	17.7 y 4	ε, α 3×10^{-7}%
	146	3–	-79.460	5.53 y 5	ε 66%, β– 34%
	147	7/2+	-79.048	2.6234 y 2	β–
	148	1–	-76.872	5.368 d 2	β–
	148m	5–,6–	-76.734	41.29 d 11	β– 95.8%, IT 4.2%
	149	7/2+	-76.071	53.08 h 5	β–
	150	(1–)	-73.60	2.68 h 2	β–
	151	5/2+	-73.395	28.40 h 4	β–
	152	1+	-71.26	4.12 m 8	β–
	152m	4–	-71.11	7.52 m 8	β–
	152m	(8)	-71.11	13.8 m 2	β–≤100%, IT≥0%
	153	5/2–	-70.68	5.25 m 2	β–
	154	(3,4)	-68.50	2.68 m 7	β–
	154m	(0,1)	-68.50	1.73 m 10	β–
	155	5/2–	-66.97	41.5 s 2	β–
	156m	4–	-64.22	26.70 s 10	β–
	157	(5/2–)	-62.4	10.56 s 10	β–
	158		-59.1	4.8 s 5	β–
	159		-56.8s	1.47 s 15	β–
	160		-53.1s	≈2 s	β–?
	161		-50.4s	≈0.7 s	β–?
	162		-46.3s	≈0.5 s	β–?
	163		-43.1s	≈0.2 s	β–?
62 Sm	128	0+	-39.0s	0.5 s sy	ε?, p?
	129	(1/2+)	-42.3s	0.55 10	ε, εp
	130	0+	-47.6s	1 s sy	ε
	131		-50.2s	1.2 s 2	ε, εp>0%
	132	0+	-55.2s	4.0 s 3	ε, εp
	133	(5/2+)	-57.1s	3.7 s 7	ε, εp>0%
	134	0+	-61.5s	9.5 s 8	ε
	135	(3/2+,5/2+)	-62.9	10.3 s 5	ε, εp 0.02%

TABLE II.1 (cont.)

Nuclide Z El A	Jπ	Δ (MeV)	T½, Γ, or Abundance	Decay Mode
62 Sm 136	0+	−66.81	47 s *2*	ε
137	(9/2−)	−68.03	45 s *1*	ε
138	0+	−71.50	3.1 m *2*	ε
139	1/2+	−72.38	2.57 m *10*	ε
139m	11/2−	−71.92	10.7 s *6*	IT 93.7%, ε 6.3%
140	0+	−75.46	14.82 m *12*	ε
141	1/2+	−75.939	10.2 m *2*	ε
141m	11/2−	−75.763	22.6 m *2*	ε 99.69%, IT 0.31%
142	0+	−78.993	72.49 m *5*	ε
143	3/2+	−79.523	8.75 m *8*	ε
143m	11/2−	−78.769	66 s *2*	IT 99.76%, ε 0.24%
144	0+	−81.972	**3.07% *7***	
145	7/2−	−80.658	340 d *3*	ε
146	0+	−81.002	10.3×10^{7} y *5*	α
147	7/2−	−79.272	1.06×10^{11} y *2* **14.99% *18***	α
148	0+	−79.342	7×10^{15} y *3* **11.24% *10***	α
149	7/2−	−77.142	**13.82% *7***	
150	0+	−77.057	**7.38% *1***	
151	5/2−	−74.582	90 y *8*	β−
152	0+	−74.769	**26.75% *16***	
153	3/2+	−72.566	46.284 h *4*	β−
154	0+	−72.462	**22.75% *29***	
155	3/2−	−70.197	22.3 m *2*	β−
156	0+	−69.370	9.4 h *2*	β−
157	(3/2−)	−66.73	8.03 m *7*	β−
158	0+	−65.21	5.30 m *3*	β−
159	5/2−	−62.2	11.37 s *15*	β−
160	0+	−60.4s	9.6 s *3*	β−
161		−57.0s	4.8 s *8*	β−
162	0+	−54.8s	≈2 s	β−?
163		−50.9s	≈1 s	β−?
164	0+	−48.2s	≈0.5 s	β−?
165		−43.8s	≈0.2 s	β−
63 Eu 130	(1+)	−33.9s	0.9 ms *+5−3*	p
131	3/2+	−39.4s	17.8 ms *19*	p 87.9%, ε 12.1%
132		−42.5s	200 ms *sy*	ε
133		−47.3s	≈1 s	ε?
134		−49.8s	0.5 s *2*	ε, εp>0%
135		−54.2s	1.5 s *2*	ε, εp
136m	(7+)	−56.3s	3.3 s *3*	ε, εp 0.09%
136m	(3+)	−56.3s	3.8 s *3*	ε, εp 0.09%
137	(11/2−)	−60.0s	11 s *2*	ε
138	(6−)	−61.75	12.1 s *6*	ε
139	(11/2)−	−65.40	17.9 s *6*	ε
140	1+	−66.99	1.51 s *2*	ε
140m	(5−)	−66.80	125 ms *2*	IT, ε<1%
141	5/2+	−69.93	40.7 s *7*	ε
141m	11/2−	−69.83	2.7 s *3*	IT 87%, ε 13%
142	1+	−71.32	2.34 s *12*	ε
142m	8−	−71.32	1.223 m *8*	ε
63 Eu 143	5/2+	−74.24	2.59 m *2*	ε
144	1+	−75.62	10.2 s *1*	ε
145	5/2+	−77.998	5.93 d *4*	ε
146	4−	−77.122	4.61 d *3*	ε
147	5/2+	−77.550	24.1 d *6*	ε, α 2.2×10^{-3}%
148	5−	−76.30	54.5 d *5*	ε, α 9.4×10^{-7}%
149	5/2+	−76.447	93.1 d *4*	ε
150	5(−)	−74.797	36.9 y *9*	ε
150m	0−	−74.755	12.8 h *1*	β− 89%, ε 11%, IT≤5.0×10^{-8}%
151	5/2+	−74.659	**47.81% *3***	
152	3−	−72.895	13.506 y *6*	ε 72.1%, β− 27.9%
152m	0−	−72.849	9.3116 h *13*	β− 72%, ε 28%
152m	8−	−72.747	96 m *1*	IT
153	5/2+	−73.373	**52.19% *3***	
154	3−	−71.744	8.590 y *3*	β− 99.98%, ε 0.02%
154m	(8−)	−71.599	46.3 m *4*	IT
155	5/2+	−71.825	4.753 y *14*	β−
156	0+	−70.093	15.19 d *8*	β−
157	5/2+	−69.467	15.18 h *3*	β−
158	(1−)	−67.21	45.9 m *2*	β−
159	5/2+	−66.053	18.1 m *1*	β−
160	1	−63.4s	38 s *4*	β−
161		−61.8s	26 s *3*	β−
162		−58.6s	10.6 s *10*	β−
163		−56.6s	6 s *sy*	β−?
164		−53.1s	≈2 s	β−?
165		−50.6s	≈1 s	β−?
166		−46.6s	≈0.4 s	β−
167		−43.6s	≈0.2 s	β−?
64 Gd 134	0+	−41.6s	0.4 s *sy*	ε?
135		−44.2s	1.1 s *2*	ε, εp≈2%
136	0+	−49.1s	≥200 ns	
137	(7/2)	−51.2s	2.2 s *2*	ε, εp
138	0+	−55.8s	4.7 s *9*	ε
139	(9/2−)	−57.5s	5.8 s *9*	ε>0%, εp>0%
139m		−57.5s	4.8 s *9*	ε>0%, εp>0%
140	0+	−61.78	15.8 s *4*	ε
141	1/2+	−63.22	14 s *4*	ε, εp 0.03%
141m	11/2−	−62.85	24.5 s *5*	ε 89%, IT 11%
142	0+	−66.96	70.2 s *6*	ε
143	(1/2)+	−68.2	39 s *2*	ε
143m	(11/2−)	−68.1	110.0 s *14*	ε
144	0+	−71.76	4.47 m *6*	ε
145	1/2+	−72.93	23.0 m *4*	ε
145m	11/2−	−72.18	85 s *3*	IT 94.3%, ε 5.7%
146	0+	−76.093	48.27 d *10*	ε
147	7/2−	−75.363	38.06 h *12*	ε
148	0+	−76.276	70.9 y *10*	α
149	7/2−	−75.133	9.28 d *10*	ε, α 4.3×10^{-4}%
150	0+	−75.769	1.79×10^{6} y *8*	α
151	7/2−	−74.195	124 d *1*	ε, α≈8.0×10^{-7}%
64 Gd 152	0+	−74.714	1.08×10^{14} y *8* **0.20% *1***	α
153	3/2−	−72.890	240.4 d *10*	ε
154	0+	−73.713	**2.18% *3***	
155	3/2−	−72.077	**14.80% *12***	
156	0+	−72.542	**20.47% *9***	
157	3/2−	−70.831	**15.65% *2***	
158	0+	−70.697	**24.84% *7***	
159	3/2−	−68.568	18.479 h *4*	β−
160	0+	−67.949	$>3.1\times10^{19}$ y **21.86% *19***	2β−
161	5/2−	−65.513	3.66 m *5*	β−
162	0+	−64.287	8.4 m *2*	β−
163	(5/2−,7/2+)	−61.5s	68 s *3*	β−
164	0+	−59.7s	45 s *3*	β−
165		−56.5s	10.3 s *16*	β−
166	0+	−54.4s	≈7 s	β−
167		−50.7s	≈3 s	β−?
168	0+	−48.1s	≈0.3 s	β−?
169		−43.9s	≈1 s	β−?
65 Tb 135	(7/2−)		0.94 ms *+33−22*	p
136		−36.0s	0.2 s *sy*	ε?
137		−41.0s	0.6 s *sy*	p?, ε?
138?		−43.6s	≥200 ns	ε, p
139		−48.2s	1.6 s *2*	ε, εp?
140	(7+)	−50.5	2.1 s *4*	ε, εp 0.26%
141	(5/2−)	−54.5	3.5 s *2*	ε
141m		−54.5	7.9 s *6*	ε
142	1+	−57.1s	597 ms *17*	ε, εp 2.2×10^{-3}%
142m	(5−)	−56.8s	303 ms *17*	IT
143	(11/2−)	−60.43	12 s *1*	ε
143m		−60.43	<21 s	ε
144	1+	−62.37	≈1 s	ε
144m	(6−)	−61.97	4.25 s *15*	IT 66%, ε 34%
145	(3/2+)	−65.88	≈20 m	ε?
145m	(11/2−)	−65.88	30.9 s *7*	ε
146	1+	−67.77	8 s *4*	ε
146m	5−	−67.77	23 s *2*	ε
147	(1/2+)	−70.75	1.7 h *1*	ε
147m	(11/2)−	−70.70	1.83 m *6*	ε
148	2−	−70.54	60 m *1*	ε
148m	(9)+	−70.45	2.20 m *5*	ε
149	1/2+	−71.496	4.118 h *25*	ε 83.3%, α 16.7%
149m	11/2−	−71.460	4.16 m *4*	ε 99.98%, α 0.02%
150	(2−)	−71.110	3.48 h *16*	ε, α<0.05%
150m	9+	−70.637	5.8 m *2*	ε
151	1/2(+)	−71.630	17.609 h *1*	ε, α 9.5×10^{-3}%
151m	(11/2−)	−71.530	25 s *3*	IT 93.8%, ε 6.2%
152	2−	−70.72	17.5 h *1*	ε, α<7.0×10^{-7}%
152m	8+	−70.22	4.2 m *1*	IT 78.8%, ε 21.2%
153	5/2+	−71.320	2.34 d *1*	ε
154	0	−70.16	21.5 h *4*	ε, β−<0.1%

TABLE II.1 (cont.)

Nuclide Z El	A	Jπ	Δ (MeV)	T½, Γ, or Abundance	Decay Mode
65 Tb	154m	3−	−70.16	9.4 h *4*	ε 78.2%, IT 21.8%, β−<0.1%
	154m	7−	−70.16	22.7 h *5*	ε 98.2%, IT 1.8%
	155	3/2+	−71.25	5.32 d *6*	ε
	156	3−	−70.098	5.35 d *10*	ε
	156m	(7−)	−70.048	24.4 h *10*	IT
	156m	(0+)	−70.009	5.3 h *2*	IT<100% ε>0%
	157	3/2+	−70.771	71 y *7*	ε
	158	3−	−69.477	180 y *11*	ε 83.4%, β− 16.6%
	158m	0−	−69.367	10.70 s *17*	IT, β−<0.6%, ε<0.01%
	159	3/2+	−69.539	**100%**	
	160	3−	−67.843	72.3 d *2*	β−
	161	3/2+	−67.468	6.906 d *19*	β−
	162	1−	−65.68	7.60 m *15*	β−
	163	3/2+	−64.601	19.5 m *3*	β−
	164	(5+)	−62.1	3.0 m *1*	β−
	165	(3/2+)	−60.7s	2.11 m *10*	β−
	166		−57.8	21 s *6*	β−
	167	(3/2+)	−55.8s	19.4 s *27*	β−
	168	(4−)	−52.5s	8.2 s *13*	β−
	169		−50.1s	≈2 s	β−
	170		−46.3s	≈3 s	β−?
	171		−43.5s	≈0.5 s	β−
66 Dy	138	0+	−34.9s	200 ms sy	ε?
	139	(7/2+)	−37.7s	0.6 s *2*	ε, εp
	141	(9/2−)	−45.3s	0.9 s *2*	ε, εp
	142	0+	−50.0s	2.3 s *3*	ε, εp 0.06%
	143	(1/2+)	−52.3s	5.6 s *10*	ε, εp
	143m	(11/2−)	−52.3s	3.0 s *3*	ε, εp
	144	0+	−56.58	9.1 s *4*	ε, εp
	145	(1/2+)	−58.29	10.5 s *15*	ε
	145m	(11/2−)	−58.29	13.6 s *10*	ε
	146	0+	−62.55	29 s *3*	ε
	146m	(10+)	−59.62	150 ms *20*	IT
	147	1/2+	−64.19	40 s *10*	ε, εp>0%
	147m	11/2−	−63.44	55.7 s *7*	ε 65%, IT 35%
	148	0+	−67.86	3.3 m *2*	ε
	149	(7/2−)	−67.715	4.20 m *14*	ε
	149m	(27/2−)	−65.054	0.490 s *15*	IT 99.3%, ε 0.7%
	150	0+	−69.317	7.17 m *5*	ε 64%, α 36%
	151	7/2(−)	−68.759	17.9 m *3*	ε 94.4%, α 5.6%
	152	0+	−70.124	2.38 h *2*	ε 99.9%, α 0.1%
	153	7/2(−)	−69.150	6.4 h *1*	ε 99.99%, α 9.4×10^{-3}%
	154	0+	−70.398	3.0×10^{6} y *15*	α
	155	3/2−	−69.16	9.9 h *2*	ε
	156	0+	−70.530	**0.06% *1***	
	157	3/2−	−69.428	8.14 h *4*	ε
	158	0+	−70.412	**0.10% *1***	
	159	3/2−	−69.174	144.4 d *2*	ε
	160	0+	−69.678	**2.34% *8***	

Nuclide Z El	A	Jπ	Δ (MeV)	T½, Γ, or Abundance	Decay Mode
66 Dy	161	5/2+	−68.061	**18.91% *24***	
	162	0+	−68.187	**25.51% *26***	
	163	5/2−	−66.386	**24.90% *16***	
	164	0+	−65.973	**28.18% *37***	
	165	7/2+	−63.618	2.334 h *1*	β−
	165m	1/2−	−63.510	1.257 m *6*	IT 97.76%, β− 2.24%
	166	0+	−62.590	81.6 h *1*	β−
	167	(1/2−)	−59.94	6.20 m *8*	β−
	168	0+	−58.6	8.7 m *3*	β−
	169	(5/2−)	−55.6	39 s *8*	β−
	170	0+	−53.7s	≈30 s	β−?
	171		−50.1s	≈6 s	β−
	172	0+	−47.7s	≈3 s	β−
	173		−43.8s	≈2 s	β−?
67 Ho	140	(6−,0−,8+)	−29.3s	6 ms *3*	p
	141	7/2−	−34.4s	4.1 ms *3*	p
	142		−37.5s	≈0.3 s	ε, εp
	143		−42.3s	>200 ns	ε?, εp?
	144		−45.2s	0.7 s *1*	ε, εp
	145		−49.2s	2.4 s *1*	ε
	146	(10+)	−51.6s	3.6 s *3*	ε
	147	(11/2−)	−55.84	5.8 s *4*	ε
	148	(1+)	−58.0	2.2 s *11*	ε
	148m	(6)−	−58.0	9.59 s *15*	ε, εp 0.08%
	149	(11/2−)	−61.69	21.1 s *2*	ε
	149m	(1/2+)	−61.64	56 s *3*	ε
	150	2−	−61.95	72 s *4*	ε
	150m	(9)+	−61.15	23.3 s *3*	ε
	151	(11/2−)	−63.63	35.2 s *1*	ε 78%, α 22%
	151m	(1/2+)	−63.59	47.2 s *10*	α 80%, ε 20%
	152	2−	−63.61	161.8 s *3*	ε 88%, α 12%
	152m	9+	−63.45	50.0 s *4*	ε 89.2%, α 10.8%
	153	11/2−	−65.019	2.01 m *3*	ε 99.95%, α 0.05%
	153m	1/2+	−64.951	9.3 m *5*	ε 99.82%, α 0.18%
	154	2−	−64.644	11.76 m *19*	ε 99.98%, α 0.02%
	154m	8+	−64.644	3.10 m *14*	ε, α<1.0×10^{-3}%, IT?
	155	5/2+	−66.04	48 m *1*	ε
	156	4−	−65.35	56 m *1*	ε
	156m	1−	−65.30	9.5 s *15*	IT
	156m	9+	−65.30	7.8 m *3*	ε 75%, IT 25%
	157	7/2−	−66.83	12.6 m *2*	ε
	158	5+	−66.19	11.3 m *4*	ε
	158m	2−	−66.12	28 m *2*	IT>81%, ε<19%
	158m	(9+)	−66.01	21.3 m *23*	ε≥93%, IT≤7%
	159	7/2−	−67.336	33.05 m *11*	ε
	159m	1/2+	−67.130	8.30 s *8*	IT
	160	5+	−66.39	25.6 m *3*	ε
	160m	2−	−66.33	5.02 h *5*	IT 73%, ε 27%
	160m	(9+)	−66.22	3 s	IT
	161	7/2−	−67.203	2.48 h *5*	ε
	161m	1/2+	−66.992	6.76 s *7*	IT

Nuclide Z El	A	Jπ	Δ (MeV)	T½, Γ, or Abundance	Decay Mode
67 Ho	162	1+	−66.047	15.0 m *10*	ε
	162m	6−	−65.941	67.0 m *7*	IT 62%, ε 38%
	163	7/2−	−66.384	4570 y *25*	ε
	163m	1/2+	−66.086	1.09 s *3*	IT
	164	1+	−64.987	29 m *1*	ε 60%, β− 40%
	164m	6−	−64.847	37.5 m +*15*−*5*	IT
	165	7/2−	−64.905	**100%**	
	166	0−	−63.077	26.83 h *2*	β−
	166m	(7)−	−63.071	1.20×10^{3} y *18*	β−
	167	7/2−	−62.287	3.003 h *18*	β−
	168	3+	−60.07	2.99 m *7*	β−
	168m	(6+)	−60.01	132 s *4*	IT≥99.5%, β−≤0.5%
	169	7/2−	−58.80	4.72 m *10*	β−
	170	(6+)	−56.24	2.76 m *5*	β−
	170m	(1+)	−56.12	43 s *2*	β−
	171	(7/2−)	−54.5	53 s *2*	β−
	172		−51.4s	25 s *3*	β−
	173		−49.1s	≈10 s	β−?
	174		−45.5s	≈8 s	β−?
	175		−42.8s	≈5 s	β−?
68 Er	143		−31.4s	0.2 s sy	ε?
	144	0+	−36.9s	≥200 ns	ε
	145	(11/2−)	−39.7s	0.9 s *3*	ε, εp
	146	0+	−44.7s	1.7 s *6*	ε, εp
	147	(11/2−)	−47.0s	2.5 s *2*	ε, εp>0%
	147m	(1/2+)	−47.0s	≈2.5 s	ε, εp>0%
	148	0+	−51.7s	4.6 s *2*	ε
	149	(1/2+)	−53.74	4 s *2*	ε, εp 7%
	149m	(11/2−)	−53.00	8.9 s *2*	ε 96.5%, IT 3.5%, εp 0.18%
	150	0+	−57.83	18.5 s *7*	ε
	151	(7/2−)	−58.27	23.5 s *13*	ε
	151m	(27/2−)	−55.68	0.58 s *2*	IT 95.3%, ε 4.7%
	152	0+	−60.50	10.3 s *1*	α 90%, ε 10%
	153	(7/2−)	−60.488	37.1 s *2*	α 53%, ε 47%
	154	0+	−62.612	3.73 m *9*	ε 99.53%, α 0.47%
	155	7/2−	−62.215	5.3 m *3*	ε 99.98%, α 0.02%
	156	0+	−64.21	19.5 m *10*	ε≈100%, α 1.7×10^{-5}%
	157	3/2−	−63.42	18.65 m *10*	ε≈100%
	158	0+	−65.30	2.29 h *6*	ε
	159	3/2−	−64.567	36 m *1*	ε
	160	0+	−66.06	28.58 h *9*	ε
	161	3/2−	−65.209	3.21 h *3*	ε
	162	0+	−66.343	**0.139% *5***	
	163	5/2−	−65.174	75.0 m *4*	ε
	164	0+	−65.950	**1.601% *3***	
	165	5/2−	−64.528	10.36 h *4*	ε
	166	0+	−64.932	**33.503% *36***	
	167	7/2+	−63.297	**22.869% *9***	
	167m	1/2−	−63.089	2.269 s *6*	IT
	168	0+	−62.997	**26.978% *18***	

TABLE II.1 (cont.)

Nuclide Z El	A	Jπ	Δ (MeV)	T½, Γ, or Abundance	Decay Mode
68 Er	169	1/2–	–60.929	9.392 d 18	β–
	170	0+	–60.115	**14.910% 36**	
	171	5/2–	–57.725	7.516 h 2	β–
	172	0+	–56.489	49.3 h 3	β–
	173	(7/2–)	–53.7s	1.4 m 1	β–
	174	0+	–51.9s	3.2 m 2	β–
	175	(9/2+)	–48.7s	1.2 m 3	β–
	176	0+	–46.5s	≈20 s	β–?
	177		–42.8s	≈3 s	β–?
69 Tm	145	(11/2–)	–27.9s	3.1 μs 3	p
	146	(5–)	–31.3s	80 ms 10	p, ε
	146m	(8+)	–31.1s	200 ms 10	p, ε
	147	11/2–	–36.4s	0.58 s 3	ε 85%, p 15%
	148m	(10+)	–39.3s	0.7 s 2	ε
	149	(11/2–)	–44.0s	0.9 s 2	ε, εp 0.2%
	150	(6–)	–46.6s	2.2 s 2	ε
	151	(11/2–)	–50.78	4.17 s 10	ε
	151m	(1/2+)	–50.78	6.6 s 14	ε
	152	(2)–	–51.77	8.0 s 10	ε
	152m	(9)+	–51.77	5.2 s 6	ε
	153	(11/2–)	–54.02	1.48 s 1	α 91%, ε 9%
	153m	(1/2+)	–53.97	2.5 s 2	α 92%, ε 8%
	154	(2–)	–54.43	8.1 s 3	α 54%, ε 46%
	154m	(9+)	–54.43	3.30 s 7	α 58%, ε 42%, IT
	155	11/2–	–56.64	21.6 s 2	ε 99.11%, α 0.89%
	155m	1/2+	–56.59	45 s 3	ε>98%, α<2%
	156	2–	–56.84	83.8 s 18	ε 99.94%, α 0.06%
	157	1/2+	–58.71	3.63 m 9	ε
	158	2–	–58.70	3.98 m 6	ε
	158m	(5+)	–58.70	≈20 s	IT?
	159	5/2+	–60.57	9.13 m 16	ε
	160	1–	–60.30	9.4 m 3	ε
	160m	5	–60.23	74.5 s 15	IT 85%, ε 15%
	161	7/2+	–61.90	30.2 m 8	ε
	162	1–	–61.48	21.70 m 19	ε
	162m	5+	–61.48	24.3 s 17	IT 82%, ε 18%
	163	1/2+	–62.735	1.810 h 5	ε
	164	1+	–61.89	2.0 m 1	ε, ε 39%
	164m	6–	–61.89	5.1 m 1	IT≈80%, ε≈20%
	165	1/2+	–62.936	30.06 h 3	ε
	166	2+	–61.89	7.70 h 3	ε
	167	1/2+	–62.548	9.25 d 2	ε
	168	3+	–61.318	93.1 d 2	ε 99.99%, β– 0.01%
	169	1/2+	–61.280	**100%**	
	170	1–	–59.801	128.6 d 3	β– 99.87%, ε 0.13%
	171	1/2+	–59.216	1.92 y 1	β–
	172	2–	–57.380	63.6 h 2	β–
	173	(1/2+)	–56.259	8.24 h 8	β–
	174	(4)–	–53.87	5.4 m 1	β–
	175	(1/2+)	–52.32	15.2 m 5	β–
	176	(4+)	–49.4	1.9 m 1	β–
	177m	(7/2–)	–47.5s	90 s 6	β–≤100%
69 Tm	178		–44.1s	≈30 s	β–?
	179		–41.6s	≈20 s	β–?
70 Yb	148	0+	–30.3s	≈0.25 s	ε?
	149	(1/2+,3/2+)	–33.5s	0.7 s 2	ε, εp=100%
	150	0+	–38.7s	>200 ns	ε?
	151	(1/2+)	–41.5	1.6 s 1	ε, εp
	151m	(11/2–)	–41.5	1.6 s 1	ε=100%, εp, IT?
	152	0+	–46.3	3.04 s 6	ε, εp
	153	7/2–	–47.1s	4.2 s 2	α 60%, ε 40%
	154	0+	–49.93	0.409 s 2	α 92.6%, ε 7.4%
	155	(7/2–)	–50.50	1.793 s 19	α 89%, ε 11%
	156	0+	–53.26	26.1 s 7	ε 90%, α 10%
	157	7/2–	–53.44	38.6 s 10	ε 99.5%, α 0.5%
	158	0+	–56.015	1.49 m 13	ε, α≈2.1×10^{-3}%
	159	5/2(–)	–55.84	1.67 m 9	ε
	160	0+	–58.17	4.8 m 2	ε
	161	3/2–	–57.84	4.2 m 2	ε
	162	0+	–59.83	18.87 m 19	ε
	163	3/2–	–59.30	11.05 m 35	ε
	164	0+	–61.02	75.8 m 17	ε
	165	5/2–	–60.29	9.9 m 3	ε
	166	0+	–61.589	56.7 h 1	ε
	167	5/2–	–60.594	17.5 m 2	ε
	168	0+	–61.575	**0.13% 1**	
	169	7/2+	–60.370	32.018 d 5	ε
	169m	1/2–	–60.346	46 s 2	IT
	170	0+	–60.769	**3.04% 15**	
	171	1/2–	–59.312	**14.28% 57**	
	172	0+	–59.260	**21.83% 67**	
	173	5/2–	–57.556	**16.13% 27**	
	174	0+	–56.950	**31.83% 92**	
	175	(7/2–)	–54.701	4.185 d 1	β–
	176	0+	–53.494	≥1.6×10^{17} y **12.76% 41**	2β–
	176m	(8)–	–52.444	11.4 s 3	IT≥90%, β–≤10%
	177	(9/2+)	–50.989	1.911 h 3	β–
	177m	(1/2–)	–50.658	6.41 s 2	IT
	178	0+	–49.70	74 m 3	β–
	179	(1/2–)	–46.4s	8.0 m 4	β–
	180	0+	–44.4s	2.4 m 5	β–
	181		–40.8s	1 m sy	β–?
71 Lu	150	(2+)	–24.9s	43 ms 5	p 68%, ε 32%
	151	11/2–	–30.2s	80.6 ms 19	p 63.4%, ε 36.6%
	152	(5–,6–)	–33.4s	0.7 s 1	ε, εp 15%
	153	11/2–	–38.4	0.9 s 2	α≈70%, ε≈30%
	154	(2–)	–39.6s	≈2 s	ε?
	154m	(9+)	–39.6s	1.12 s 8	ε≈100%
	155	11/2–	–42.55	68 ms 1	α 90%, ε 10%
	155m	1/2+	–42.53	138 ms 8	α 76%, ε 24%
	156	(2)–	–43.75	494 ms 12	α≈95%, ε≈5%
	156m	9+	–43.75	198 ms 2	α
	157	(1/2+,3/2+)	–46.48	6.8 s 18	α>0%
71 Lu	157m	(11/2–)	–46.46	4.79 s 12	ε 94%, α 6%
	158		–47.21	10.6 s 3	ε 99.09%, α 0.91%
	159		–49.72	12.1 s 10	ε, α 0.1%
	160		–50.27	36.1 s 3	ε, α≤1.0×10^{-4}%
	160m		–50.27	40 s 1	ε≤100%, α
	161	1/2+	–52.56	77 s 2	ε
	161m	(9/2–)	–52.43	7.3 ms 4	IT
	162	(1–)	–52.84	1.37 m 2	ε≤100%
	162m	(4–)	–52.84	1.5 m	ε≤100%
	162m		–52.84	1.9 m	ε≤100%
	163	1/2(+)	–54.79	3.97 m 13	ε
	164	1	–54.64	3.14 m 3	ε
	165	1/2+	–56.44	10.74 m 10	ε
	166	(6–)	–56.02	2.65 m 10	ε
	166m	(3–)	–55.99	1.41 m 10	ε 58%, IT 42%
	166m	(0–)	–55.98	2.12 m 10	ε>80%, IT<20%
	167	7/2+	–57.50	51.5 m 10	ε
	167m	1/2+	–57.50	≥1 m	ε, IT
	168	(6–)	–57.06	5.5 m 1	ε
	168m	3+	–56.84	6.7 m 4	ε>95%, IT<5%
	169	7/2+	–58.077	34.06 h 5	ε
	169m	1/2–	–58.048	160 s 10	IT
	170	0+	–57.31	2.012 d 20	ε
	170m	(4)–	–57.22	0.67 s 10	IT
	171	7/2+	–57.833	8.24 d 3	ε
	171m	1/2–	–57.762	79 s 2	IT
	172	4–	–56.741	6.70 d 3	ε
	172m	1–	–56.699	3.7 m 5	IT
	173	7/2+	–56.886	1.37 y 1	ε
	174	(1)–	–55.575	3.31 y 5	ε
	174m	(6)–	–55.404	142 d 2	IT 99.38%, ε 0.62%
	175	7/2+	–55.171	**97.41% 2**	
	176	7–	–53.387	3.76×10^{10} y 7 **2.59% 2**	β–
	176m	1–	–53.264	3.664 h 19	β– 99.91%, ε 0.1%
	177	7/2+	–52.389	6.6475 d 20	β–
	177m	23/2–	–51.419	160.44 d 6	β– 78.6%, IT 21.4%
	177m	(39/2–)	–49.689	6 m +3−2	β–≤100%, IT
	178	1(+)	–50.343	28.4 m 2	β–
	178m	(9–)	–50.223	23.1 m 3	β–
	179	7/2(+)	–49.064	4.59 h 6	β–
	180	5+	–46.69	5.7 m 1	β–
	181	(7/2+)	–44.7s	3.5 m 3	β–
	182	(0,1,2)	–41.9s	2.0 m 2	β–
	183	(7/2+)	–39.5s	58 s 4	β–
	184	(3+)	–36.4s	20 s 3	β–
72 Hf	153		–27.3s	>60 ns	ε?
	154	0+	–32.7s	2 s 1	ε≈100%, α≈0%
	155		–34.1s	0.89 s 12	ε
	156	0+	–37.9	23 ms 1	α
	157	7/2–	–38.8s	110 ms 6	α 86%, ε 14%
	158	0+	–42.10	2.85 s 7	ε 55.7%, α 44.3%

TABLE II.1 (cont.)

Nuclide Z El A	Jπ	Δ (MeV)	T½, Γ, or Abundance	Decay Mode
72 Hf 159	7/2−	−42.85	5.6 s *4*	ε 65%, α 35%
160	0+	−45.94	13.6 s 2	ε 99.3%, ε 0.7%
161		−46.32	18.2 s 5	ε>99.87%, α<0.13%
162	0+	−49.173	39.4 s *9*	ε 99.99%,
				α 8.0×10⁻³%
163		−49.29	40.0 s *6*	ε, α<1.0×10⁻⁴%
164	0+	−51.82	111 s *8*	ε
165	(5/2−)	−51.64	76 s *4*	ε
166	0+	−53.86	6.77 m *30*	ε
167	(5/2)−	−53.47	2.05 m *5*	ε
168	0+	−55.36	25.95 m *20*	ε
169	(5/2)−	−54.72	3.24 m *4*	ε
170	0+	−56.25	16.01 h *13*	ε
171	7/2(+)	−55.43	12.1 h *4*	ε
171m	1/2(−)	−55.41	29.5 s *9*	IT≤100%, ε
172	0+	−56.40	1.87 y *3*	ε
173	1/2−	−55.41	23.6 h *1*	ε
174	0+	−55.847	2.0×10¹⁵ y *4*	α
			0.16% *1*	
175	5/2(−)	−54.484	70 d 2	ε
176	0+	−54.577	**5.26% *7***	
177	7/2−	−52.890	**18.60% *9***	
177m	23/2+	−51.574	1.09 s *5*	IT
177m	37/2−	−50.150	51.4 m *5*	IT
178	0+	−52.444	**27.28% *7***	
178m	8−	−51.297	4.0 s 2	IT
178m	16+	−49.998	31 y *1*	IT
179	9/2+	−50.472	**13.62% *2***	
179m	1/2−	−50.097	18.67 s *4*	IT
179m	25/2−	−49.366	25.05 d *25*	IT
180	0+	−49.788	**35.08% *16***	
180m	8−	−48.647	5.47 h *4*	IT 99.7%, β− 0.3%
181	1/2−	−47.412	42.39 d *6*	β−
182	0+	−46.059	8.90×10⁶ y *3*	β−
182m	8−	−44.886	61.5 m *15*	β− 58%, IT 42%
183	(3/2−)	−43.29	1.067 h *17*	β−
184	0+	−41.50	4.12 h *5*	β−
184m	8−	−40.23	48 s *10*	β−
185		−38.4s	3.5 m *6*	β−
186	0+	−36.4s	2.6 m *12*	β−
187		−33.0s	30 s *sy*	β−?
188	0+	−30.9s	20 s *sy*	β−
73 Ta 155m	11/2−	−23.7s	12 μs +*4*−*3*	p
156	(2−)	−25.8s	144 ms *24*	p=100%, ε
156m	9+	−25.7s	0.36 s *4*	ε 95.8%, p 4.2%
157	1/2+	−29.6	10.1 ms *4*	α 96.6%, p 3.4%
157m	11/2−	−29.6	4.3 ms *1*	α
157m	(25/2−)	−28.0	1.7 ms *1*	α
158	(2−)	−31.0s	55 ms *15*	α=91%, ε=9%
158m	(9+)	−30.9s	36.7 ms *15*	α 95%, ε 5%
159	(1/2−)	−34.45	0.83 s *18*	ε 66%, α 34%
159m	(11/2−)	−34.38	515 ms *20*	α 55%, ε 45%
73 Ta 160		−35.88	1.55 s *4*	ε 66%, α 34%
160m		−35.88	1.7 s 2	α?
161		−38.73s	2.89 s *12*	ε 95%, α?
162		−39.78	3.57 s *12*	ε 99.93%, α 0.07%
163		−42.54	10.6 s *18*	ε≈99.8%, α≈0.2%
164	(3+)	−43.28	14.2 s *3*	ε
165		−45.86	31.0 s *15*	ε
166	(2)+	−46.10	34.4 s *5*	ε
167	(3/2+)	−48.35	80 s *4*	ε
168	(2−,3+)	−48.39	2.0 m *1*	ε
169	(5/2+)	−50.29	4.9 m *4*	ε
170	(3+)	−50.14	6.76 m *6*	ε
171	(5/2−)	−51.72	23.3 m *3*	ε
172	(3+)	−51.33	36.8 m *3*	ε
173	5/2−	−52.40	3.14 h *13*	ε
174	3+	−51.74	1.14 h *8*	ε
175	7/2+	−52.41	10.5 h 2	ε
176	(1)−	−51.37	8.09 h *5*	ε
177	7/2+	−51.724	56.56 h *6*	ε
178	1+	−50.51	9.31 m *3*	ε
178	(7)−	−50.51	2.36 h *8*	ε
179	7/2+	−50.366	1.82 y *3*	ε
180	1+	−48.936	8.154 h *6*	ε 86%, β− 14%
180m	9−	−48.859	>1.2×10¹⁵ y	2ε?
181	7/2+	−48.442	**99.988% *2***	
182	3−	−46.433	114.43 d *3*	β−
182m	5+	−46.417	283 ms *3*	IT
182m	10−	−45.913	15.84 m *10*	IT
183	7/2+	−45.296	5.1 d *1*	β−
184	(5−)	−42.84	8.7 h *1*	β−
185	(7/2+)	−41.40	49.4 m *15*	β−
186	(2−,3−)	−38.61	10.5 m *3*	β−
186m		−38.61	1.54 m *5*	β−
187		−36.8s	≈2 m	β−?
188		−33.8s	≈20 s	β−
189	(7/2+)	−31.8s	3 s *sy*	β−?
190		−28.7s	0.3 s *sy*	β−?
74 W 158	0+	−23.7s	1.25 ms *21*	α
159		−25.2s	7.3 ms *27*	α≈99.9%, ε≈0.1%
160	0+	−29.4	91 ms *5*	α 87%
161		−30.4s	409 ms *18*	α 73%
162	0+	−34.00	1.36 s *7*	ε 54.8%, α 45.2%
163		−34.91	2.8 s *2*	ε 87%, α 13%
164	0+	−38.23	6.3 s 2	ε 96.2%, α 3.8%
165	(5/2−)	−38.86	5.1 s *5*	ε, α<0.2%
166	0+	−41.89	19.2 s *6*	ε 99.97%, α 0.04%
167	(+)	−42.09	19.9 s *5*	ε 99.96%, α 0.04%
168	0+	−44.89	53 s 2	ε=100%,
				α 3.2×10⁻³%
169	(5/2−)	−44.92	74 s *6*	ε
170	0+	−47.29	2.42 m *4*	ε
171	(5/2−)	−47.09	2.38 m *4*	ε
74 W 172	0+	−49.10	6.6 m *9*	ε
173	5/2−	−48.73	7.6 m *2*	ε
174	0+	−50.23	33.2 m *21*	ε
175	(1/2−)	−49.63	35.2 m *6*	ε
176	0+	−50.64	2.5 h *1*	ε
177	1/2−	−49.70	132 m *2*	ε
178	0+	−50.42	21.6 d *3*	ε
179	(7/2)−	−49.30	37.05 m *16*	ε
179m	(1/2)−	−49.08	6.40 m *7*	IT 99.72%, ε 0.28%
180	0+	−49.645	1.8×10¹⁸ y *2*	α
			0.12% *1*	
181	9/2+	−48.254	121.2 d *2*	ε
182	0+	−48.248	>8.3×10¹⁸ y	α
			26.50% *16*	
183	1/2−	−46.367	>1.3×10¹⁹ y	α
			14.31% *4*	
183m	11/2+	−46.057	5.2 s *3*	IT
184	0+	−45.707	>2.9×10¹⁹ y	α
			30.64% *2*	
185	3/2−	−43.390	75.1 d *3*	β−
185m	11/2+	−43.192	1.67 m *3*	IT
186	0+	−42.509	>2.7×10¹⁹ y	α
			28.43% *19*	
187	3/2−	−39.905	23.72 h *6*	β−
188	0+	−38.667	69.78 d *5*	β−
189	(3/2−)	−35.5	10.7 m *5*	β−
190	0+	−34.3	30.0 m *15*	β−
191		−31.1s	>300 ns	β−?
192	0+	−29.6s	>300 ns	β−?
75 Re 160	(2−)	−16.7s	0.82 ms +*15*−*9*	p 91%, α 9%
161	1/2+	−20.9	0.37 ms *4*	p
161m	11/2−	−20.8	15.6 ms *9*	α 95.2%, p 4.8%
162	(2−)	−22.4s	107 ms *13*	α 94%, ε 6%
162m	(9+)	−22.2s	77 ms 9	α 91%, ε 9%
163	(1/2+)	−26.01	390 ms *72*	ε 68%, α 32%
163m	(11/2−)	−25.89	214 ms *5*	α 66%, ε 34%
164		−27.6s	0.53 s *23*	α≈58%, ε≈42%
165	(1/2+)	−30.66	≈1 s	ε, α
165m	(11/2−)	−30.61	2.1 s *3*	ε 87%, α 13%
166		−31.85s	2.8 s *3*	α≥8%
167	(9/2−)	−34.84s	5.9 s *3*	ε≈99%, α≈1%
167m		−34.84s	3.4 s *4*	α≈100%
168	(5+,6+,7+)	−35.79	4.4 s *1*	ε≈100%,
				α≈5.0×10⁻³%
169	(9/2−)	−38.39	8.1 s *5*	ε, α<0.01%
169m		−38.39	15.1 s *15*	α≈0.2%
170	(5+)	−38.92	9.2 s 2	ε
171	(9/2−)	−41.25	15.2 s *4*	ε
172m	(5)	−41.52	15 s *3*	ε
172m	(2)	−41.52	55 s *5*	ε
173	(5/2−)	−43.55	1.98 m *26*	ε
174		−43.67	2.40 m *4*	ε

TABLE II.1 (cont.)

Nuclide Z El	A	Jπ	Δ (MeV)	T½, Γ, or Abundance	Decay Mode
75 Re	175	(5/2−)	−45.29	5.89 m 5	ε
	176	3+	−45.06	5.3 m 3	ε
	177	5/2−	−46.27	14 m 1	ε
	178	(3+)	−45.65	13.2 m 2	ε
	179	(5/2)+	−46.59	19.5 m 1	ε
	180	(1)−	−45.84	2.44 m 6	ε
	181	5/2+	−46.51	19.9 h 7	ε
	182	7+	−45.4	64.0 h 5	ε
	182m	2+	−45.4	12.7 h 2	ε
	183	5/2+	−45.811	70.0 d 14	ε
	184	3(−)	−44.227	38.0 d 5	ε
	184m	8(+)	−44.039	169 d 8	IT 75.4%, ε 24.6%
	185	5/2+	−43.822	**37.40% 2**	
	186	1−	−41.930	3.7186 d 5	β− 92.53%, ε 7.47%
	186m	(8+)	−41.781	2.0×10^{5} y	IT
	187	5/2+	−41.216	4.12×10^{10} y 11	β−,
				62.60% 2	$\alpha<1.0\times10^{-4}$%
	188	1−	−39.016	17.003 h 3	β−
	188m	(6)−	−38.844	18.59 m 4	IT
	189	5/2+	−37.978	24.3 h 4	β−
	190	(2)−	−35.6	3.1 m 3	β−
	190m	(6−)	−35.4	3.2 h 2	β− 54.4%, IT 45.6%
	191	(3/2+,1/2+)	−34.35	9.8 m 5	β−
	192		−31.7s	16 s 1	β−
	193		−30.3s	30 s sy	β−?
	194		−27.6s	>300 ns	β−
76 Os	162	0+	−14.5s	1.9 ms 2	α
	163		−16.1s	5.5 ms 6	α=100%, ε
	164	0+	−20.5	21 ms 1	α 98%, ε 2%
	165	(7/2−)	−21.6s	71 ms 3	α>60%, ε<40%
	166	0+	−25.44	181 ms 38	α 72%, ε 18%
	167		−26.50	0.81 s 6	α 57%, ε 43%
	168	0+	−29.99	2.1 s 1	α 40%, ε
	169		−30.72	3.40 s 9	ε 88.8%, α 11.2%
	170	0+	−33.93	7.46 s 23	ε 91.4%, α 8.6%
	171	(5/2−)	−34.29	8.3 s 2	ε 98.2%, α 1.8%
	172	0+	−37.24	19.2 s 5	α 1.1%, ε
	173	(5/2−)	−37.44	22.4 s 9	α 0.4%, ε
	174	0+	−40.00	44 s 4	ε 99.98%, α 0.02%
	175	(5/2−)	−40.10	1.4 m 1	ε
	176	0+	−42.10	3.6 m 5	ε
	177	1/2−	−41.95	3.0 m 2	ε
	178	0+	−43.55	5.0 m 4	ε
	179	(1/2−)	−43.02	6.5 m 3	ε
	180	0+	−44.36	21.5 m 4	ε
	181	1/2−	−43.55	105 m 3	ε
	181m	7/2−	−43.50	2.7 m 1	ε=100%, IT≤3%
	182	0+	−44.61	22.10 h 25	ε
	183	9/2+	−43.66	13.0 h 5	ε
	183m	1/2−	−43.49	9.9 h 3	ε 85%, IT 15%
	184	0+	−44.256	$>5.6\times10^{13}$ y	α
				0.02% 1	
76 Os	185	1/2−	−42.809	93.6 d 5	ε
	186	0+	−43.000	2.0×10^{15} y 11	α
				1.59% 3	
	187	1/2−	−41.218	**1.6% 3**	
	188	0+	−41.136	**13.29% 8**	
	189	3/2−	−38.985	**16.21% 5**	
	189m	9/2−	−38.955	5.81 h 6	IT
	190	0+	−38.706	**26.36% 2**	
	190m	(10)−	−37.001	9.9 m 1	IT
	191	9/2−	−36.394	15.4 d 1	β−
	191m	3/2−	−36.320	13.10 h 5	IT
	192	0+	−35.881	**40.93% 19**	
	192m	(10−)	−33.865	5.9 s 1	IT>87%, β−<13%
	193	3/2−	−33.393	30.11 h 1	β−
	194	0+	−32.433	6.0 y 2	β−
	195		−29.7	≈9 m	β−?
	196	0+	−28.28	34.9 m 2	β−
	197			2.8 m 6	β−
77 Ir	164	(9+)	−7.3s	0.11 ms +6−3	p
	165	(1/2+)	−11.6s	<1 μs	p?, α?
	166	(2−)	−13.2s	10.5 ms 22	α 93.1%, p 6.9%
	166m	(9+)	−13.0s	15.1 ms 9	α 98.2%, p 1.8%
	167	1/2+	−17.08	35.2 ms 20	α 48%, p 32%, ε 20%
	167m	11/2−	−16.90	25.7 ms 8	α 80%, ε 20%, p 0.4%
	168		−18.7s	0.161 ms 21	α 82%
	169	(1/2+)	−22.08	0.64 s +46−24	α 50%, ε, p
	169m	(11/2−)	−21.93	0.308 s 22	α 81%
	170		−23.3s	0.87 s +18−12	ε 94.8%, α 5.2%
	170m		−23.3s	0.44 s 6	ε≤64%, IT≤64%,
					α 36%
	171	(1/2+)	−26.43	3.2 s +13−7	ε, α>0%, p
	171m	(11/2−)	−26.43	1.40 s 10	α 58%, ε≤42%,
					p≤42%
	172	(3+)	−27.5s	4.4 s 3	ε 98%, α≈2%
	172m	(7+)	−27.4s	2.0 s 1	ε 77%, α 23%
	173m	(11/2−)	−30.27	2.4 s 9	α 7%, ε
	173m	(3/2+,5/2+)	−30.27	9.0 s 8	ε>93%, α<7%
	174	(3+)	−30.87	7.9 s 6	ε 99.5%, α 0.5%
	174m	(7+)	−30.68	4.9 s 3	ε 97.5%, α 2.5%
	175	(5/2−)	−33.43	9 s 2	ε 99.15%, α 0.85%
	176		−33.86	8.3 s 6	ε 96.9%, α 3.1%
	177	5/2−	−36.05	30 s 2	ε 99.94%, α 0.06%
	178		−36.25	12 s 2	ε
	179	(5/2)−	−38.08	79 s 1	ε
	180	(4,5)	−37.98	1.5 m 1	ε
	181	5/2−	−39.47	4.90 m 15	ε
	182	(5+)	−39.05	15 m 1	ε
	183	5/2−	−40.20	57 m 4	ε
	184	5−	−39.61	3.09 h 3	ε
	185	5/2−	−40.34	14.4 h 1	ε
	186	5+	−39.17	16.64 h 3	ε
	186m	2−	−39.17	1.90 h 5	ε=75%, IT=25%
77 Ir	187	3/2+	−39.716	10.5 h 3	ε
	188	1−	−38.328	41.5 h 5	ε
	189	3/2+	−38.45	13.2 d 1	ε
	190	4−	−36.751	11.78 d 10	ε, $\varepsilon<2.0\times10^{-3}$%
	190m	(1−)	−36.725	1.120 h 3	IT
	190m	(11)−	−36.375	3.087 h 12	ε 91.4%, IT 8.6%
	191	3/2+	−36.706	**37.3% 2**	
	191m	11/2−	−36.535	4.94 s 3	IT
	191m		−34.659	5.5 s 7	IT
	192	4+	−34.833	73.827 d 13	β− 95.13%, ε 4.87%
	192m	1−	−34.777	1.45 m 5	IT 99.98%, β− 0.02%
	192m	(11−)	−34.665	241 y 9	IT
	193	3/2+	−34.534	**62.7% 2**	
	193m	11/2−	−34.454	10.53 d 4	IT
	194	1−	−32.529	19.28 h 13	β−
	194m	(10,11)	−32.339	171 d 11	β−
	195	3/2+	−31.690	2.5 h 2	β−
	195m	11/2−	−31.590	3.8 h 2	β− 95%, IT 5%
	196	(0−)	−29.44	52 s 1	β−
	196m	(10,11−)	−29.03	1.40 h 2	β−≈100%, IT<0.3%
	197	3/2+	−28.27	5.8 m 5	β−
	197m	11/2−	−28.15	8.9 m 3	β− 99.75%, IT 0.25%
	198		−25.8s	8 s 1	β−
	199		−24.40	20 s sy	β−
78 Pt	166	0+	−4.8s	300 μs 100	α
	167		−6.5s	0.9 ms 3	α
	168	0+	−11.0	2.1 ms 2	α≤100%
	169		−12.4s	7.0 ms 2	α≈100%
	170	0+	−16.31	14.0 ms 2	α
	171		−17.47	51 ms 2	α≈98%, ε 2%
	172	0+	−21.10	104 ms 7	α 94%, ε 6%
	173		−21.94	370 ms 13	α 83%, ε
	174	0+	−25.32	0.889 s 17	α 76%, ε 24%
	175	(7/2−)	−25.69	2.53 s 6	α 64%, ε 36%
	176	0+	−28.93	6.33 s 15	ε 62%, α 38%
	177	5/2−	−29.37	10.6 s 4	ε 94.3%, α 5.7%
	178	0+	−32.00	21.1 s 6	ε 92.3%, α 7.7%
	179	1/2−	−32.264	21.2 s 4	ε 99.76%, α 0.24%
	180	0+	−34.44	56 s 2	ε, α≈0.3%
	181	1/2−	−34.37	52.0 s 22	ε, α≈0.08%
	182	0+	−36.17	3.0 m 2	ε 99.96%, α 0.04%
	183	1/2−	−35.77	6.5 m 10	ε, $\alpha\approx1.3\times10^{-3}$%
	183m	(7/2)−	−35.74	43 s 5	ε≈100%,
					$\alpha<4.0\times10^{-4}$%, IT
	184	0+	−37.33	17.3 m 2	ε, α≈0.001%
	185	9/2+	−36.68	70.9 m 24	ε
	185m	1/2−	−36.58	33.0 m 8	ε 99%, IT<2%
	186	0+	−37.86	2.08 h 5	ε, $\alpha\approx1.4\times10^{-4}$%
	187	3/2−	−36.71	2.35 h 3	ε
	188	0+	−37.823	10.2 d 3	ε, $\alpha\ 2.6\times10^{-5}$%
	189	3/2−	−36.48	10.87 h 12	ε

TABLE II.1 (cont.)

Nuclide Z El	A	Jπ	Δ (MeV)	T½, Γ, or Abundance	Decay Mode
78 Pt	190	0+	-37.323	6.5×10^{11} y *3*	α
				0.014% *1*	
	191	3/2-	-35.698	2.862 d *7*	ε
	192	0+	-36.293	**0.782% *7***	
	193	1/2-	-34.477	50 y *6*	ε
	193m	13/2+	-34.327	4.33 d *3*	IT
	194	0+	-34.763	**32.967% *99***	
	195	1/2-	-32.797	**33.832% *10***	
	195m	13/2+	-32.537	4.010 d *5*	IT
	196	0+	-32.647	**25.242% *41***	
	197	1/2-	-30.422	19.8915 h *19*	β-
	197m	13/2+	-30.023	95.41 m *18*	IT 96.7%, β- 3.3%
	198	0+	-29.908	**7.163% *55***	
	199	5/2-	-27.392	30.80 m *21*	β-
	199m	(13/2)+	-26.968	13.6 s *4*	IT
	200	0+	-26.60	12.5 h *3*	β-
	201	(5/2-)	-23.74	2.5 m *1*	β-
	202	0+	-22.6s	44 h *15*	β-
79 Au	169		-1.8s	150 μs sy	α?, p?
	170	(2-)	-3.6s	286 μs *+50-40*	p 89%, α 11%
	171	(1/2+)	-7.56	22 μs *+3-2*	p=100%
	171m	(11/2-)	-7.31	1.09 ms *3*	α 66%, p 36%
	172		-9.3s	6.3 ms *15*	α≤100%, p<2%
	173	(1/2+)	-12.82	25 ms *1*	α 94%, ε, p
	173m	(11/2-)	-12.61	14.0 ms *9*	α 92%, ε, p
	174		-14.2s	139 ms *3*	α>0%
	175	(1/2+)	-17.44	0.1 s sy	α?, ε?
	175m	(11/2-)	-17.44	156 ms *5*	α 94%, ε 6%
	176		-18.5s	0.84 s *+17-14*	α, ε
	177	(1/2+,3/2+)	-21.55	1462 ms *32*	α≤100%, ε
	177m	11/2-	-21.39	1180 ms *12*	α≤100%, ε
	178		-22.33	2.6 s *5*	ε≤60%, α≥40%
	179		-24.95	3.3 s *13*	ε 78%, α 22%
	180		-25.60	8.1 s *3*	ε≤98.2%, α≥1.8%
	181	(3/2-)	-27.87	13.7 s *14*	ε 97.3%, α 2.7%
	182		-28.30	15.6 s *4*	ε 99.87%, α 0.13%
	183	(5/2)-	-30.19	42.8 s *10*	ε 99.45%, α 0.55%
	184	5+	-30.32	20.6 s *9*	α≤0.02%, ε
	184m	2+	-30.25	47.6 s *14*	ε 70%, IT 30%,
					α≤0.02%
	185	5/2-	-31.87	4.25 m *6*	ε 99.74%, α 0.26%
	185m		-31.87	6.8 m *3*	ε<100%, IT
	186	3-	-31.71	10.7 m *5*	ε, α 8.0×10^{-4}%
	187	1/2+	-33.01	8.4 m *3*	ε, α 3.0×10^{-3}%
	187m	9/2-	-32.88	2.3 s *1*	IT
	188	1(-)	-32.30	8.84 m *6*	ε
	189	1/2+	-33.58	28.7 m *3*	ε, α<3.0×10^{-5}%
	189m	11/2-	-33.33	4.59 m *11*	ε
	190	1-	-32.88	42.8 m *10*	ε, α<1.0×10^{-6}%
	190m	(11-)	-32.88	125 ms *20*	IT=100%
	191	3/2+	-33.81	3.18 h *8*	ε
	191m	(11/2-)	-33.54	0.92 s *11*	IT

Nuclide Z El	A	Jπ	Δ (MeV)	T½, Γ, or Abundance	Decay Mode
79 Au	192	1-	-32.78	4.94 h *9*	ε
	192m	(11-)	-32.34	160 ms *20*	IT
	193	3/2+	-33.39	17.65 h *15*	ε
	193m	11/2-	-33.10	3.9 s *3*	IT 99.97%, ε=0.03%
	194	1-	-32.26	38.02 h *10*	ε
	194m	(5+)	-32.15	600 ms *8*	IT
	194m	(11-)	-31.79	420 ms *10*	IT
	195	3/2+	-32.570	186.098 d *47*	ε
	195m	11/2-	-32.251	30.5 s *2*	IT
	196	2-	-31.140	6.1669 d *6*	ε 92.8%, β- 7.2%
	196m	5+	-31.055	8.1 s *2*	IT
	196m	12-	-30.544	9.6 h *1*	IT
	197	3/2+	-31.141	**100%**	
	197m	11/2-	-30.732	7.73 s *6*	IT
	198	2-	-29.582	2.6956 d *3*	β-
	198m	(12-)	-28.770	2.27 d *2*	IT
	199	3/2+	-29.095	3.139 d *7*	β-
	200	1(-)	-27.27	48.4 m *3*	β-
	200m	12-	-26.31	18.7 h *5*	β- 82%, IT 18%
	201	3/2+	-26.401	26 m *1*	β-
	202	(1-)	-24.4	28.8 s *19*	β-
	203	(3/2+)	-23.143	60 s *6*	β-
	204	(2-)	-20.8s	39.8 s *9*	β-
	205	(3/2+)	-18.8s	31 s *2*	β-
80 Hg	171		3.5s	59 μs *+36-16*	α=100%
	172	0+	-1.1	0.25 ms *+35-9*	α
	173		-2.6s	0.6 ms *+5-2*	α=100%
	174	0+	-6.65	2.1 ms *+18-7*	α 99.6%
	175	(7/2-,9/2-)	-8.0	10.8 ms *4*	α
	176	0+	-11.78	20 ms *2*	α=100%
	177	(13/2+)	-12.78	127.3 ms *18*	α 85%, ε 15%
	178	0+	-16.32	0.269 s *3*	α≈70%, ε≈30%
	179		-16.92	1.08 s *9*	α≈53%, ε≈47%,
					εp≈0.15%
	180	0+	-20.24	2.58 s *1*	ε 52%, α 48%
	181	1/2-	-20.66	3.6 s *1*	ε 73%, α 27%,
					εp 0.01%,
					εα 9.0×10^{-6}%
	182	0+	-23.576	10.83 s *6*	ε 84.8%, α 15.2%
	183	1/2-	-23.800	9.4 s *7*	ε 88.3%, α 11.7%,
					εp 2.6×10^{-4}%
	184	0+	-26.35	30.9 s *3*	ε 98.89%, α 1.11%
	185	1/2-	-26.18	49.1 s *10*	ε 94%, α 6%
	185m	13/2+	-26.08	21.6 s *15*	IT 54%, ε 46%,
					α≈0.03%
	186	0+	-28.54	1.38 m *6*	ε 99.98%, α 0.02%
	187	13/2+	-28.12	2.4 m *3*	ε, α>1.2×10^{-4}%
	187m	3/2-	-28.12	1.9 m *3*	ε, α>2.5×10^{-4}%
	188	0+	-30.20	3.25 m *15*	ε, α 3.7×10^{-5}%
	189	3/2-	-29.63	7.6 m *1*	ε, α<3.0×10^{-5}%
	189m	13/2+	-29.63	8.6 m *1*	ε, α<3.0×10^{-5}%
	190	0+	-31.37	20.0 m *5*	ε, α<3.4×10^{-7}%

Nuclide Z El	A	Jπ	Δ (MeV)	T½, Γ, or Abundance	Decay Mode
80 Hg	191	(3/2-)	-30.59	49 m *10*	ε
	191m	13/2+	-30.59	50.8 m *15*	ε
	192	0+	-32.01	4.85 h *20*	ε
	193	3/2-	-31.05	3.80 h *15*	ε
	193m	13/2+	-30.91	11.8 h *2*	ε 92.8%, IT 7.2%
	194	0+	-32.19	444 y *77*	ε
	195	1/2-	-31.00	10.53 h *3*	ε
	195m	13/2+	-30.82	41.6 h *8*	IT 54.2%, ε 45.8%
	196	0+	-31.827	**0.15% *1***	
	197	1/2-	-30.541	64.14 h *5*	ε
	197m	13/2+	-30.242	23.8 h *1*	IT 91.4%, ε 8.6%
	198	0+	-30.954	**9.97% *20***	
	199	1/2-	-29.547	**16.87% *22***	
	199m	13/2+	-29.015	42.67 m *9*	IT
	200	0+	-29.504	**23.10% *19***	
	201	3/2-	-27.663	**13.18% *9***	
	202	0+	-27.346	**29.86% *26***	
	203	5/2-	-25.269	46.595 d *6*	β-
	204	0+	-24.690	**6.87% *15***	
	205	1/2-	-22.288	5.14 m *9*	β-
	206	0+	-20.95	8.15 m *10*	β-
	207	(9/2+)	-16.2	2.9 m *2*	β-
	208	0+	-13.1s	41 m *+5-4*	β-
	209		-8.3s	37 s *8*	β-
	210	0+	-5.1s	>300 ns	β-?
81 Tl	176	(3-,4-,5-)	1.4s	5.2 ms *+30-14*	p=100%
	177	(1/2+)	-3.33	18 ms *5*	α 73%, p 27%
	178		-4.8s	≈60 ms	α?, ε?
	179	(1/2+)	-8.30	0.42 s *6*	α<100%, ε
	179m	(11/2-)	-8.30	1.7 ms *2*	α≤100%, IT, ε
	180		-9.4s	1.5 s *2*	α 7%,
					εSF≈1.0×10^{-4}%, ε
	181	1/2+	-12.801	1.4 ms *5*	ε, α≤10%
	181m	9/2-	-11.951	3.2 s *3*	α
	182	(7+)	-13.35	3.1 s *10*	ε>96%, α<4%
	183	(1/2+)	-16.587	6.9 s *7*	ε>0%, α
	183m	(9/2-)	-15.957	53.3 ms *3*	α 2%, ε, IT
	184	(2+)	-16.89	11 s *1*	ε 97.9%, α 2.1%
	185	(1/2+)	-19.76	19.5 s *5*	ε
	185m	(9/2-)	-19.30	1.93 s *8*	IT, α
	186m	(7+)	-20.2	27.5 s *10*	ε, α=6.0×10^{-3}%
	186m	(10-)	-19.8	2.9 s *2*	IT
	187	(1/2+)	-22.444	≈51 s	ε<100%, α>0%
	187m	(9/2-)	-22.108	15.60 s *12*	ε<99.9%, IT<99.9%,
					α 0.15%
	188m	(2-)	-22.35	71 s *2*	ε
	188m	(7+)	-22.35	71 s *1*	ε
	189	(1/2+)	-24.60	2.3 m *2*	ε
	189m	(9/2-)	-24.34	1.4 m *1*	ε<100%, IT<4%
	190m	2(-)	-24.33	2.6 m *3*	ε
	190m	7(+)	-24.33	3.7 m *3*	ε
	191	(1/2+)	-26.281	?	ε?

TABLE II.1 (cont.)

Nuclide Z El	A	Jπ	Δ (MeV)	T½, Γ, or Abundance	Decay Mode
81 Tl	191m	9/2(−)	−25.982	5.22 m 16	ε
	192	(2−)	−25.87	9.6 m 4	ε
	192m	(7+)	−25.72	10.8 m 2	ε
	193	1/2+	−27.3	21.6 m 8	ε
	193m	9/2−	−27.0	2.11 m 15	IT≤75%, ε≤25%
	194	2−	−26.8	33.0 m 5	ε, α<1.0×10^{-7}%
	194m	(7+)	−26.8	32.8 m 2	ε
	195	1/2+	−28.16	1.16 h 5	ε
	195m	9/2−	−27.67	3.6 s 4	IT
	196	2−	−27.50	1.84 h 3	ε
	196m	(7+)	−27.10	1.41 h 2	ε 95.5%, IT 4.5%
	197	1/2+	−28.34	2.84 h 4	ε
	197m	9/2−	−27.73	0.54 s 1	IT
	198	2−	−27.49	5.3 h 5	ε
	198m	7+	−26.95	1.87 h 3	ε 54%, IT 46%
	199	1/2+	−28.06	7.42 h 8	ε
	200	2−	−27.048	26.1 h 1	ε
	201	1/2+	−27.18	72.912 h 17	ε
	202	2−	−25.98	12.23 d 2	ε
	203	1/2+	−25.761	**29.524% 14**	
	204	2−	−24.346	3.78 y 2	β− 97.1%, ε 2.9%
	205	1/2+	−23.821	**70.476% 14**	
	206	0−	−22.253	4.200 m 17	β−
	206m	(12−)	−19.610	3.74 m 3	IT
	207	1/2+	−21.034	4.77 m 2	β−
	207m	11/2−	−19.686	1.33 s 11	IT
	208	5(+)	−16.750	3.053 m 4	β−
	209	(1/2+)	−13.638	2.161 m 7	β−
	210	(5+)	−9.25	1.30 m 3	β−, β−n 7.0×10^{-3}%
	211		−6.1s	>300 ns	β−?
	212		−1.7s	>300 ns	β−?
82 Pb	178	0+	3.57	0.23 ms 15	α, ε?
	179		2.0s	3 ms sy	α?
	180	0+	−1.94	4.5 ms 11	α≤100%
	181m	(13/2+)	−3.14	45 ms 20	α<100%
	182	0+	−6.83	55 ms +40−35	α≤100%
	183	(3/2−)	−7.57	535 ms 30	α≈90%
	183m	(13/2+)	−7.47	415 ms 20	α≈100%
	184	0+	−11.05	490 ms 25	ε 77%, α 23%
	185m	13/2+	−11.54	4.24 s 17	α≈50%, ε?
	185m	3/2−	−11.54	6.3 s 4	α≈50%, ε?
	186	0+	−14.68	4.82 s 3	ε 60%, α 40%
	187	(3/2−)	−14.980	15.2 s 3	ε 93%, α 7%
	187m	(13/2+)	−14.899	18.3 s 3	ε 88%, α 12%
	188	0+	−17.82	25.1 s 1	ε 90.7%, α 9.3%
	189	(3/2−)	−17.88	51 s 3	ε>99%, α≈0.4%
	190	0+	−20.42	71 s 1	ε 99.6%, α 0.4%
	191	(3/2−)	−20.25	1.33 m 8	ε 99.99%, α 0.01%
	191m	(13/2+)	−20.11	2.18 m 8	ε, α≈0.02%
	192	0+	−22.56	3.5 m 1	ε 99.99%, α 5.9×10^{-3}%
	193	(3/2−)	−22.19	5 m sy	ε
82 Pb	193m	(13/2+)	−22.19	5.8 m 2	ε
	194	0+	−24.21	10.7 m 6	ε, α 7.3×10^{-6}%
	195	3/2−	−23.71	≈15 m	ε
	195m	13/2+	−23.51	15.0 m 12	ε
	196	0+	−25.36	37 m 3	ε≈100%, α≤3.0×10^{-5}%
	197	3/2−	−24.749	8.1 m 17	ε
	197m	13/2+	−24.429	42.9 m 9	ε 81%, IT 19%
	198	0+	−26.05	2.4 h 1	ε
	199	3/2−	−25.23	90 m 10	ε
	199m	(13/2+)	−24.80	12.2 m 3	IT<100%, ε>0%
	200	0+	−26.24	21.5 h 4	ε
	201	5/2−	−25.26	9.33 h 3	ε
	201m	13/2+	−24.63	61 s 2	IT>99%, ε<1%
	202	0+	−25.934	52.5×10^{3} y 28	ε, α<1%
	202m	9−	−23.764	3.53 h 1	IT 90.5%, ε 9.5%
	203	5/2−	−24.787	51.92 h 3	ε
	203m	13/2+	−23.961	6.21 s 8	IT
	203m	29/2−	−21.837	480 ms 7	IT
	204	0+	−25.110	≥1.4×10^{17} y **1.4% 1**	α?
	204m	9−	−22.924	1.14 h 4	IT
	205	5/2−	−23.770	1.73×10^{7} y 7	ε
	206	0+	−23.785	**24.1% 1**	
	207	1/2−	−22.452	**22.1% 1**	
	207m	13/2+	−20.819	0.806 s 6	IT
	208	0+	−21.749	**52.4% 1**	
	209	9/2+	−17.614	3.253 h 14	β−
	210	0+	−14.728	22.20 y 22	β−, α 1.9×10^{-6}%
	211	9/2+	−10.491	36.1 m 2	β−
	212	0+	−7.547	10.64 h 1	β−
	213	(9/2+)	−3.184	10.2 m 3	β−
	214	0+	−0.181	26.8 m 9	β−
	215		4.5s	36 s 1	β−
83 Bi	184m		1.0s	6.6 ms 15	α≈100%
	184m		1.0s	13 ms 2	α≈100%
	185	1/2+	−2.21s	63 μs 3	p 90%, α 10%
	186	(3+)	−3.17	15.0 ms 17	α≈100%
	186m	(10−)	−3.17	9.8 ms 13	α≈100%
	187	(9/2−)	−6.37	32 ms 3	α
	188m	(3+)	−7.20	60 ms 3	α, ε?
	188m	(10−)	−7.20	265 ms 15	α, ε?
	189	(9/2−)	−10.06	674 ms 11	α>50%, ε<50%
	189m	(1/2+)	−9.88	5.0 ms 1	α>50%, ε<50%
	190m	(10−)	−10.9	6.2 s 1	α 70%, ε 30%
	190m	(3+)	−10.9	6.3 s 1	α 90%, ε 10%
	191	(9/2−)	−13.240	12.4 s 4	α 51%, ε 49%
	191m	(1/2+)	−13.000	121 ms 8	α 68%, ε 32%
	192	(3+)	−13.55	34.6 s 9	ε 88%, α 12%
	192m	(10−)	−13.55	39.6 s 4	ε 90%, α 10%
	193	(9/2−)	−15.873	63 s 3	ε 96.2%, α 3.8%
	193m	(1/2+)	−15.566	3.2 s 6	α 84%, ε 16%
83 Bi	194	(3+)	−15.99	95 s 3	ε 99.54%, α 0.46%
	194m	(10−)	−15.99	115 s 4	ε 99.8%, α 0.2%
	194m	(6+,7+)	−15.99	125 s 2	ε
	195	(9/2−)	−18.024	183 s 4	ε 99.97%, α 0.03%
	195m	(1/2+)	−17.623	87 s 1	ε 67%, α 33%
	196	(3+)	−18.01	308 s 12	ε≈100%, α 1.2×10^{-3}%
	196m	(7+)	−17.84	0.6 s 5	IT, ε
	196m	(10−)	−17.74	240 s 3	ε 74.2%, IT 25.8%, α 3.8×10^{-4}%
	197	(9/2−)	−19.688	9.33 m 50	ε, α 1.0×10^{-4}%
	197m	(1/2+)	−19.188	5.04 m 16	α 55%, ε 45%, IT<0.3%
	198	(2+,3+)	−19.37	10.3 m 3	ε
	198m	(7+)	−19.37	11.6 m 3	ε
	198m	10−	−19.12	7.7 s 5	IT
	199	9/2−	−20.80	27 m 1	ε
	199m	(1/2+)	−20.12	24.70 m 15	ε≥98%, IT≤2%, α≈0.01%
	200	7+	−20.37	36.4 m 5	ε
	200m	(2+)	−20.37	31 m 2	ε>90%, IT<10%
	200m	(10−)	−19.94	0.40 s 5	IT
	201	9/2−	−21.42	108 m 3	ε, α<1.0×10^{-4}%
	201m	1/2+	−20.57	59.1 m 6	ε>93%, IT≤6.8%, α≈0.3%
	202	5+	−20.73	1.72 h 5	ε, α<1.0×10^{-5}%
	203	9/2−	−21.54	11.76 h 5	ε
	203m	1/2+	−20.44	305 ms 5	IT
	204	6+	−20.67	11.22 h 10	ε
	205	9/2−	−21.062	15.31 d 4	ε
	206	6(+)	−20.028	6.243 d 3	ε
	207	9/2−	−20.054	32.9 y 14	ε
	208	(5)+	−18.870	3.68×10^{5} y 4	ε
	209	9/2−	−18.258	1.9×10^{19} y 2 **100%**	α
	210	1−	−14.792	5.012 d 5	β−, α 1.3×10^{-4}%
	210m	9−	−14.521	3.04×10^{6} y 6	α
	211	9/2−	−11.858	2.14 m 2	α 99.72%, β− 0.28%
	212	1(−)	−8.117	60.55 m 6	β− 64.06%, α 35.94%
	212m	(8−,9−)	−7.867	25.0 m 2	α 67%, β− 33%, β−α 30%
	212m	≥16	−6.207	7.0 m 3	β−≈100%
	213	9/2−	−5.231	45.59 m 6	β− 97.91%, α 2.09%
	214	1−	−1.20	19.9 m 4	β− 99.98%, α 0.02%
	215	(9/2−)	1.65	7.6 m 2	β−
	215m	(25/2−)	3.00	36.4 s	IT, β−
	216	(1−)	5.87	2.17 m 5	β−≤100%
	217		8.8s	98.5 s 8	β−
	218		13.3s	33 s 1	β−
84 Po	188	0+	−0.54	0.40 ms +20−15	ε<100%, α>0%
	189		−1.42	5 ms 1	α
	190	0+	−4.56	2.46 ms 5	α

TABLE II.1 (cont.)

Nuclide Z El A	Jπ	Δ (MeV)	T½, Γ, or Abundance	Decay Mode
84 Po 191	(3/2–)	–5.05	22 ms 1	α
191m	(13/2+)	–4.92	93 ms 3	α
192	0+	–8.07	33.2 ms 14	α≈99.5%, ε≈0.5%
193m	(13/2+)	–8.36	243 ms +11–10	α≤100%
193m	(3/2–)	–8.36	370 ms +46–40	α≤100%
194	0+	–11.01	0.392 s 4	α≈100%, ε
195	(3/2–)	–11.07	4.64 s 9	α 75%, ε 25%
195m	(13/2+)	–10.84	1.92 s 2	α≈90%, ε≈10%, IT<0.01%
196	0+	–13.47	5.8 s 2	α≈98%, ε≈2%
197	(3/2–)	–13.36	84 s 16	ε 56%, α 44%
197m	(13/2+)	–13.15	32 s 2	α 84%, ε 16%, IT 0.01%
198	0+	–15.47	1.77 m 3	α 57%, ε 43%
199	(3/2–)	–15.22	4.58 m 52	ε 92.5%, α 7.5%
199m	13/2+	–14.90	4.13 m 43	ε 73.5%, α 24%, IT 2.5%
200	0+	–16.95	10.9 m 11	ε 88.9%, α 11.1%
201	3/2–	–16.525	15.3 m 2	ε 98.4%, α 1.6%
201m	13/2+	–16.101	8.9 m 2	IT 56%, ε 41%, α≈2.9%
202	0+	–17.92	44.7 m 5	ε 98.08%, α 1.92%
203	5/2–	–17.31	36.7 m 5	ε 99.89%, α 0.11%
203m	13/2+	–16.67	45 s 2	IT
204	0+	–18.33	3.53 h 2	ε 99.34%, α 0.66%
205	5/2–	–17.51	1.74 h 8	ε 99.96%, α 0.04%
206	0+	–18.182	8.8 d 1	ε 94.55%, α 5.45%
207	5/2–	–17.146	5.80 h 2	ε 99.98%, α 0.02%
207m	19/2–	–15.763	2.79 s 8	IT
208	0+	–17.469	2.898 y 2	α, ε
209	1/2–	–16.366	102 y 5	α 99.52%, ε 0.48%
210	0+	–15.953	138.376 d 2	α
211	9/2+	–12.432	0.516 s 3	α
211m	(25/2+)	–10.970	25.2 s 6	α 99.98%, IT 0.02%
212m	(18+)	–7.447	45.1 s 6	α 99.93%
213	9/2+	–6.653	3.65 μs 4	α
214	0+	–4.470	164.3 μs 20	α
215	9/2+	–0.540	1.781 ms 4	α, β– 2.3×10^{-4}%
216	0+	1.784	0.145 s 2	α
217	(9/2+)	5.901	1.53 s 5	α
218	0+	8.358	3.10 m 2	α 99.98%, β– 0.02%
219		12.8s	≈2 m	α?, β–?
220	0+	15.5s	>300 ns	β–?
85 At 191	(1/2+)		1.7 ms +11–5	α
191m	(7/2–)		2.1 ms +4–3	α
193	(1/2+)	–0.15	28 ms +5–4	α≈100%
193m	(7/2–)	–0.14	21 ms 5	α≈100%
193m	(13/2+)	–0.11	27 ms +4–5	α 24%
194m		–1.2	≈40 ms	α, ε
194m		–1.2	≈250 ms	α, ε, IT
195	(1/2+)	–3.476	328 ms +20	α
195m	(7/2–)	–3.439	147 ms +5	α
85 At 196		–3.92	0.39 s 5	α 94%, ε
197	(9/2–)	–6.34	0.390 s 16	α 96.1%, ε 3.9%
197m	(1/2+)	–6.29	2.0 s 2	α≤100%, IT≤4.0×10^{-3}%, ε
198	(3+)	–6.67	4.2 s 3	α 90%, ε 10%
198m	(10–)	–6.57	1.0 s 2	α 84%, ε 16%
199	(9/2–)	–8.82	6.92 s 13	α 90%, ε 10%
200	(3+)	–8.99	43 s 1	α 57%, ε 43%
200m	(7+)	–8.88	47 s 1	ε≤57%, α 43%
200m	(10–)	–8.65	3.5 s 2	IT≈84%, α≈10.5%, ε≈4.5%
201	(9/2–)	–10.790	89 s 3	α 71%, ε 29%
202	(2,3)+	–10.59	184 s 1	ε 82%, α 18%
202m	(7+)	–10.59	182 s 2	ε 91.3%, α 8.7%
202m	(10–)	–10.20	0.46 s 5	IT 99.7%, ε 0.25%, α 0.1%
203	9/2–	–12.16	7.37 m 13	ε 69%, α 31%
204	7+	–11.88	9.2 m 2	ε 96.2%, α 3.8%
204m	(10–)	–11.29	108 ms 10	IT
205	9/2–	–12.97	26.9 m 8	ε 90%, α 10%
206	(5)+	–12.42	30.6 m 13	ε 99.11%, α 0.89%
207	9/2–	–13.24	1.80 h 4	ε 91.4%, α 8.6%
208	6+	–12.49	1.63 h 3	ε 99.45%, α 0.55%
209	9/2–	–12.880	5.41 h 5	ε 95.9%, α 4.1%
210	(5)+	–11.972	8.1 h 4	ε 99.82%, α 0.18%
211	9/2–	–11.647	7.214 h 7	ε 58.2%, α 41.8%
212	(1–)	–8.621	0.314 s 2	α, ε<0.03%, β–<2.0×10^{-6}%
212m	(9–)	–8.398	0.119 s 3	α>99%, IT<1%
213	9/2–	–6.580	125 ns 6	α
214	1–	–3.380	558 ns 10	α
215	9/2–	–1.255	0.10 ms 2	α
216	1–	2.257	0.30 ms 3	α, β–<6.0×10^{-3}%, ε<3.0×10^{-7}%
217	9/2–	4.396	32.3 ms 4	α 99.99%, β– 7.0×10^{-3}%
218		8.10	1.5 s 3	α 99.9%, β– 0.1%
219		10.397	56 s 3	α≈97%, β–≈3%
220	3	14.35	3.71 m 4	β– 92%, α 8%
221		16.8s	2.3 m 2	β–
222		20.8s	54 s 10	β–
223		23.5s	50 s 7	β–
86 Rn 195		5.07	6 ms +3–2	α
195m		5.12	5 ms +3–2	α
196	0+	1.97	4.4 ms +13–9	α
197	(3/2–)	1.48	65 ms +25–14	α≈100%
197m	(13/2+)	1.48	19 ms +8–4	α≈100%
198	0+	–1.23	65 ms 3	α, ε
199	(3/2–)	–1.52	0.62 s 3	α 94%, ε 6%
199m	(13/2+)	–1.52	0.32 s 2	α 97%, ε 3%
200	0+	–4.01	0.96 s 3	α≈98%, ε≈2%
201	(3/2–)	–4.07	7.1 s 8	α≈80%, ε≈20%
86 Rn 201m	(13/2+)	–3.79	3.8 s 1	α≈90%, ε≈10%, IT≈0%
202	0+	–6.28	10.0 s 3	α 86%, ε 14%
203	(3/2–)	–6.16	44.2 s 16	α 66%, ε 34%
203m	(13/2+)	–5.80	26.9 s 5	α 75%, ε 25%
204	0+	–7.98	1.17 m 18	α 73%, ε 27%
205	5/2–	–7.71	170 s 4	ε 75.4%, α 24.6%
206	0+	–9.12	5.67 m 17	α 62%, ε 38%
207	5/2–	–8.63	9.25 m 17	ε 79%, α 21%
208	0+	–9.65	24.35 m 14	α 62%, ε 38%
209	5/2–	–8.93	28.5 m 10	ε 83%, α 17%
210	0+	–9.598	2.4 h 1	α 96%, ε 4%
211	1/2–	–8.756	14.6 h 2	ε 72.6%, α 27.4%
212	0+	–8.660	23.9 m 12	α
213	(9/2+)	–5.698	19.4 ms 1	α
214	0+	–4.320	0.27 μs 2	α
215	9/2+	–1.169	2.30 μs 10	α
216	0+	0.256	45 μs 5	α
217	9/2+	3.659	0.54 ms 5	α
218	0+	5.218	35 ms 5	α
219	5/2+	8.831	3.96 s 1	α
220	0+	10.613	55.6 s 1	α
221	7/2(+)	14.472	25.7 m 5	β– 78%, α 22%
222	0+	16.374	3.8235 d 4	α
223	7/2	20.3s	24.3 m 4	β–
224	0+	22.4s	107 m 3	β–
225	7/2–	26.5s	4.66 m 4	β–
226	0+	28.8s	7.4 m 1	β–
227		33.0s	20.8 s 7	β–
228	0+	35.4s	65 s 2	β–
87 Fr 199		6.76	12 ms +10–4	α>0%, ε
200	(3+)	6.12	49 ms 4	α
200m	(10–)	6.32	0.57 s +27–14	α
201	(9/2–)	3.60	67 ms 3	α, ε<1%
201m		3.60	19 ms +19–6	α
202	(3+)	3.14	0.23 s +8–4	α≈97%, ε≈3%
202m	(10–)	3.24	0.23 s +14–5	α≈97%, ε≈3%
203	(9/2–)	0.86	0.55 s 2	α≤100%
204	(3+)	0.61	1.7 s 3	α≈80%, ε≈20%
204m	(7+)	0.65	2.6 s 3	α≤100%
204m	(10–)	0.92	≈1 s	α≤100%, IT
205	(9/2–)	–1.310	3.80 s 3	α≤100%
206	(2+,3+)	–1.24	≈16 s	α≈84%, ε≈16%
206m	(7+)	–1.24	15.9 s 1	α 84%, ε 16%
206m	(10–)	–0.71	0.7 s 1	α≈12%, IT
207	9/2–	–2.84	14.8 s 1	α 95%, ε 5%
208	7+	–2.67	59.1 s 3	α 90%, ε 10%
209	9/2–	–3.77	50.0 s 3	α 89%, ε 11%
210	6+	–3.35	3.18 m 6	α 60%, ε 40%
211	9/2–	–4.16	3.10 m 2	α>80%, ε<20%
212	5+	–3.54	20.0 m 6	ε 57%, α 43%
213	9/2–	–3.550	34.6 s 3	α 99.45%, ε 0.55%

TABLE II.1 (cont.)

Nuclide Z El	A	Jπ	Δ (MeV)	T½, Γ, or Abundance	Decay Mode
87 Fr	214	(1−)	−0.958	5.0 ms 2	α
	214m	(8−)	−0.836	3.35 ms 5	α
	215	9/2−	0.318	86 ns 5	α
	216	(1−)	2.98	0.70 μs 2	α, ε<2.0×10^{-7}%
	217	9/2−	4.315	19 μs 3	α
	218	1−	7.059	1.0 ms 6	α
	218m		7.145	22.0 ms 5	α≤100%, IT
	219	9/2−	8.618	20 ms 2	α
	220	1+	11.483	27.4 s 3	α 99.65%, β− 0.35%
	221	5/2−	13.278	4.9 m 2	α, β−<0.1%, ^{14}C 9×10^{-13}%
	222	2−	16.35	14.2 m 3	β−
	223	3/2(−)	18.384	22.00 m 7	β− 99.99%, α 6.0×10^{-3}%
	224	1−	21.66	3.33 m 10	β−
	225	3/2−	23.81	4.0 m 2	β−
	226	1−	27.4	49 s 1	β−
	227	1/2+	29.7	2.47 m 3	β−
	228	2−	33.3s	38 s 1	β−≤100%
	229	(1/2+)	35.82	50.2 s 4	β−
	230		39.6s	19.1 s 5	β−
	231	(1/2+)	42.3s	17.6 s 6	β−
	232		46.4s	5 s 1	β−
88 Ra	202	0+	9.21	0.7 ms +33−3	α
	203	(3/2−)	8.64	1.0 ms +50−5	α≈100%
	203m	(13/2+)	8.64	33 ms +22−10	α≈100%
	204	0+	6.05	59 ms +12−9	α
	205	(3/2−)	5.84	210 ms +60−40	α≤100%, ε
	205m	(13/2+)	5.84	170 ms +60−40	α≤100%, ε
	206	0+	3.57	0.24 s 2	α
	207	(5/2−,3/2−)	3.54	1.3 s 2	α≈90%, ε≈10%
	207m	(13/2+)	4.01	55 ms 10	IT 85%, α 15%, ε≈0.35%
	208	0+	1.71	1.3 s 2	α 95%, ε 5%
	209	5/2−	1.86	4.6 s 2	α≈90%, ε≈10%
	210	0+	0.46	3.7 s 2	α≈96%, ε≈4%
	211	5/2(−)	0.84	13 s 2	α>93%, ε<7%
	212	0+	−0.19	13.0 s 2	α≈85%, ε≈15%
	213	1/2−	0.36	2.74 m 6	α 80%, ε 20%
	213m		2.13	2.1 ms 1	IT≈99%, α≈1%
	214	0+	0.101	2.46 s 3	α 99.94%, ε 0.06%
	215	(9/2+)	2.533	1.55 ms 7	α
	216	0+	3.291	182 ns 10	α, ε<1.0×10^{-8}%
	217	(9/2+)	5.887	1.6 μs 2	α≈100%
	218	0+	6.65	25.2 μs 3	α
	219	(7/2)+	9.394	10 ms 3	α
	220	0+	10.273	18 ms 2	α
	221	5/2+	12.964	28 s 2	α, ^{14}C 1×10^{-12}%
	222	0+	14.321	38.0 s 5	α, ^{14}C 3.0×10^{-8}%
	223	3/2+	17.235	11.43 d 5	α, ^{14}C 8.9×10^{-8}%
	224	0+	18.827	3.6319 d 23	α, ^{14}C 4.0×10^{-9}%
	225	1/2+	21.994	14.9 d 2	β−
88 Ra	226	0+	23.669	1600 y 7	α, ^{14}C 3.2×10^{-9}%
	227	3/2+	27.179	42.2 m 5	β−
	228	0+	28.942	5.75 y 3	β−
	229	5/2(+)	32.56	4.0 m 2	β−
	230	0+	34.52	93 m 2	β−
	231	(5/2+)	38.4s	103 s 3	β−
	232	0+	40.6s	250 s 50	β−
	233		44.8s	30 s 5	β−
	234	0+	47.2s	30 s 10	β−
89 Ac	206m		13.51	11 ms +9−3	α
	206m	(3+)	13.51	22 ms +9−5	α
	206m	(10−)	13.51	33 ms +22−9	α
	207	(9/2−)	11.13	27 ms +11−6	α
	208	(3+)	10.76	95 ms +24−16	ε 1%, α
	208m	(10−)	11.27	25 ms +9−5	IT<10%, ε 1%, α
	209	(9/2−)	8.84	0.10 s 5	α≈99%, ε≈1%
	210		8.79	0.35 s 5	α 91%, ε≈9%
	211		7.20	0.21 s 3	α≈100%
	212		7.28	0.93 s 5	α≈57%, ε≈43%
	213		6.16	0.731 s 17	α≤100%
	214		6.43	8.2 s 2	α≥89%, ε≤11%
	215	9/2−	6.01	0.17 s 1	α 99.91%, ε 0.09%
	216	(1−)	8.12	0.440 ms 16	α
	217	9/2−	8.71	69 ns 4	α≈100%, ε≤2%
	218	(1−)	10.84	1.08 μs 9	α
	219	9/2−	11.57	11.8 μs 15	α
	220	(3−)	13.75	26.4 ms 2	α, ε 5.0×10^{-4}%
	221		14.52	52 ms 2	α
	222	1−	16.621	5.0 s 5	α 99%, ε 1%
	222m		16.621	63 s 3	α≥88%, IT≤10%, ε≥0.7%
	223	(5/2−)	17.826	2.10 m 5	α 99%, ε 1%
	224	0−	20.235	2.78 h 17	ε 90.9%, α 9.1%, β−<1.6%
	225	(3/2−)	21.638	10.0 d 1	α, ^{14}C 5×10^{-10}%
	226	(1)	24.310	29.37 h 12	β− 83%, ε 17%, α 6.0×10^{-3}%
	227	3/2−	25.851	21.772 y 3	β− 98.62%, α 1.38%
	228	3+	28.896	6.15 h 2	β−
	229	(3/2+)	30.75	62.7 m 5	β−
	230	(1+)	33.8	122 s 3	β−
	231	(1/2+)	35.9	7.5 m 1	β−
	232	(1+)	39.1	119 s 5	β−
	233	(1/2+)	41.5s	145 s 10	β−
	234		45.1s	44 s 7	β−
	235		47.7s	≈40 s	β−?
	236		51.5s	≈2 m	β−?
90 Th	209	(5/2−)	16.50	3.8 ms +69−15	α
	210	0+	14.04	9 ms +17−4	α 99%, ε≈1%
	211		13.91	0.04 s +3−1	α
	212	0+	12.09	30 ms +20−10	α, ε≈0.3%
	213		12.12	140 ms 25	α≤100%
90 Th	214	0+	10.71	100 ms 25	α
	215	(1/2−)	10.93	1.2 s 2	α
	216	0+	10.30	0.028 s 2	α, ε≈0.01%
	217	(9/2+)	12.22	0.241 ms 5	α
	218	0+	12.37	109 ns 13	α
	219		14.47	1.05 μs 3	α
	220	0+	14.67	9.7 μs 6	α, ε 2.0×10^{-7}%
	221	(7/2+)	16.938	1.73 ms 3	α
	222	0+	17.20	2.237 ms 13	α
	223	(5/2)+	19.386	0.60 s 2	α
	224	0+	20.00	0.81 s 10	α
	225	(3/2)+	22.310	8.72 m 4	α≈90%, ε≈10%
	226	0+	23.197	30.57 m 10	α
	227	1/2+	25.806	18.68 d 9	α
	228	0+	26.772	1.9116 y 16	α, ^{20}O 1×10^{-11}%
	229	5/2+	29.587	7340 y 160	α
	229m		29.590	13.9 h 30	α
	230	0+	30.864	7.538×10^{4} y 30	α, SF<4.×10^{-11}%
	231	5/2+	33.817	25.52 h 1	β−, α≈4×10^{-11}%
	232	0+	35.448	1.405×10^{10} y 6 100%	α, SF 1.2×10^{-9}%, Ne
	233	1/2+	38.733	21.83 m 4	β−
	234	0+	40.614	24.10 d 3	β−
	235	(1/2+)	44.26	7.2 m 1	β−
	236	0+	46.5s	37.5 m 2	β−
	237	(5/2+)	50.2s	4.7 m 6	β−
	238	0+	52.6s	9.4 m 20	β−
91 Pa	212		21.61	5.1 ms +61−19	α≈100%
	213	(9/2−)	19.66	5.3 ms +40−16	α
	214		19.49	17 ms 3	α
	215		17.87	14 ms 2	α
	216		17.80	105 ms 12	α≈98%, ε≈2%
	217		17.07	3.6 ms 8	α
	217m		18.92	1.2 ms 2	α 73%, IT 27%
	218		18.67	0.113 ms 1	α
	219m	9/2−	18.52	53 ns 10	α
	220?		20.38	0.78 μs 16	α, ε 3.0×10^{-7}%
	221	9/2−	20.38	4.9 μs 8	α
	222		22.12s	3.3 ms 3	α
	223		22.32	5.1 ms 6	α
	224		23.87	0.85 s 2	α
	225		24.34	1.7 s 2	α
	226		26.03	1.8 m 2	α 74%, ε 26%
	227	(5/2−)	26.832	38.3 m 3	α 85%, ε 15%
	228	3+	28.924	22 h 1	ε 98%, α 2%
	229	(5/2+)	29.898	1.50 d 5	ε 99.52%, α 0.48%
	230	(2−)	32.174	17.4 d 5	ε 91.6%, β− 8.4%, α 3.2×10^{-3}%
	231	3/2−	33.426	3.276×10^{4} y 11	α, SF≤3×10^{-10}%
	232	(2−)	35.948	1.31 d 2	β−, ε 3.0×10^{-3}%
	233	3/2−	37.490	26.975 d 13	β−
	234	4+	40.341	6.70 h 5	β−

TABLE II.1 (cont.)

Nuclide Z El	A	Jπ	Δ (MeV)	T½, Γ, or Abundance	Decay Mode
91 Pa	234m	(0−)	40.415	1.17 m *3*	β− 99.84%, IT 0.16%
	235	(3/2−)	42.33	24.44 m *11*	β−
	236	1(−)	45.3	9.1 m *1*	β−
	237	(1/2+)	47.6	8.7 m *2*	β−
	238	(3−)	50.77	2.27 m *9*	β−, SF < 2.6×10^{-6}%
	239	(3/2)	53.3s	1.8 h *5*	β−
	240		56.8s	≈2 m	β−?
92 U	217		22.70	16 ms *+21−6*	α ≤ 100%
	218	0+	21.92	1.5 ms *+73−7*	α
	219		23.21	42 μs *+34−13*	α
	220	0+	23.0s	≈60 ns	α?, ε?
	221		24.6s	≈0.7 μs	α?, ε?
	222	0+	24.3s	1.0 μs *+10−4*	α
	223		25.84	18 μs *+10−5*	α, ε 0.2%
	224	0+	25.71	0.9 ms *3*	α
	225		27.38	84 ms *4*	α
	226	0+	27.33	0.26 s *2*	α
	227	(3/2+)	29.02	1.1 m *1*	α
	228	0+	29.22	9.1 m *2*	α > 95%, ε < 5%
	229	(3/2+)	31.211	58 m *3*	ε ≈ 80%, α ≈ 20%
	230	0+	31.615	20.8 d	α, SF < 1×10^{-10}%
	231	(5/2−)	33.807	4.2 d *1*	ε, α ≈ 4.0×10^{-3}%
	232	0+	34.611	68.9 y *4*	α, Ne 9×10^{-10}%, Mg < 5×10^{-12}%, SF 3×10^{-12}%
	233	5/2+	36.920	1.592×10^{5} y *2*	α, SF < 6.0×10^{-9}%
	234	0+	38.147	2.455×10^{5} y *6* **0.0054%** *5*	α, SF 1.6×10^{-9}%, Mg 1×10^{-11}%, Ne 9×10^{-12}%
	235	7/2−	40.921	7.04×10^{8} y *1* **0.7204%** *6*	α, SF 7.0×10^{-9}%, Ne ≈ $8.\times10^{-10}$%, 28 $8.\times10^{-10}$%
	235m	1/2+	40.921	≈26 m	IT
	236	0+	42.446	2.342×10^{7} y *3*	α, SF 9.4×10^{-8}%, ^{30}Mg
	237	1/2+	45.392	6.75 d *1*	β−
	238	0+	47.309	4.468×10^{9} y *3* **99.2742%** *10*	α, SF 5.5×10^{-5}%
	239	5/2+	50.574	23.45 m *2*	β−
	240	0+	52.715	14.1 h *1*	β−
	241		56.2s	≈5 m	β−?
	242	0+	58.6s	16.8 m *5*	β−
93 Np	225	(9/2−)	31.59	>2 μs	α
	226		32.74s	35 ms *10*	α
	227		32.56	0.51 s *6*	α
	228		33.7s	61.4 s *14*	ε 60%, α 40%
	229		33.78	4.0 m *4*	α 68%, ε 32%
	230		35.24	4.6 m *3*	ε ≤ 97%, α ≥ 3%
	231	(5/2)	35.63	48.8 m *2*	ε 98%, α 2%
	232	(4+)	37.4s	14.7 m *3*	ε

Nuclide Z El	A	Jπ	Δ (MeV)	T½, Γ, or Abundance	Decay Mode
93 Np	233	(5/2+)	37.95	36.2 m *1*	ε, α ≤ 1.0×10^{-3}%
	234	(0+)	39.957	4.4 d *1*	ε
	235	5/2+	41.045	396.1 d *12*	ε, α 2.6×10^{-3}%
	236	(6−)	43.38	154×10^{3} y *6*	ε 87.3%, β− 12.5%, α 0.16%
	236m	1	43.44	22.5 h *4*	ε 52%, β− 48%
	237	5/2+	44.873	2.144×10^{6} y *7*	α, SF ≤ 2×10^{-10}%
	238	2+	47.456	2.117 d *2*	β−
	239	5/2+	49.312	2.356 d *3*	β−
	240	(5+)	52.31	61.9 m *2*	β−
	240m	(1+)	52.31	7.22 m *2*	β− 99.88%, IT 0.12%
	241	(5/2+)	54.26	13.9 m *2*	β−
	242	(1+)	57.4	2.2 m *2*	β−
	242m	(6+)	57.4	5.5 m *1*	β−
	243	(5/2−)	59.88s	1.85 m *15*	β−
	244	(7−)	63.2s	2.29 m *16*	β−
94 Pu	228	0+	36.09	1.1 s *+20−5*	α
	229	(3/2+)	37.40	>2 μs	α
	230	0+	36.93	1.70 m *17*	α 84%, ε 16%
	231	(3/2+)	38.29	8.6 m *5*	ε ≤ 99.8%, α > 0.2%
	232	0+	38.37	33.1 m *8*	ε 80%, α 20%
	233		40.05	20.9 m *4*	ε 99.88%, α 0.12%
	234	0+	40.350	8.8 h *1*	ε ≈ 94%, α ≈ 6%
	235	(5/2+)	42.18	25.3 m *5*	ε, α 2.8×10^{-3}%
	236	0+	42.903	2.858 y *8*	α, SF 1.9×10^{-7}%
	237	7/2−	45.093	45.2 d *1*	ε, α 4.2×10^{-3}%
	237m	1/2+	45.239	0.18 s *2*	IT
	238	0+	46.165	87.7 y *1*	α, SF 1.9×10^{-7}%
	239	1/2+	48.590	24110 y *30*	α, SF $3.\times10^{-10}$%
	240	0+	50.127	6561 y *7*	α, SF 5.7×10^{-6}%
	241	5/2+	52.957	14.290 y *6*	β−, α 2.5×10^{-3}%, SF > $2.\times10^{-14}$%
	242	0+	54.718	3.75×10^{5} y *2*	α, SF 5.5×10^{-4}%
	243	7/2+	57.756	4.956 h *3*	β−
	244	0+	59.806	8.00×10^{7} y *9*	α 99.88%, SF 0.12%
	245	(9/2−)	63.11	10.5 h *1*	β−
	246	0+	65.40	10.84 d *2*	β−
	247		69.0s	2.27 d *23*	β−
95 Am	231		42.4s	≈10 s	ε?, α?
	232		43.4s	79 s *2*	ε ≈ 98%, α ≈ 2%
	233		43.2s	3.2 m *8*	α > 3%, ε
	234		44.5s	2.32 m *8*	ε > 99.96%, α < 0.04%
	235m		44.7s	10.3 m *6*	ε 99.6%, α 0.4%
	236?		46.2s	3.6 m *1*	α 0.004%, ε
	236m		46.2s	0.6 y *2*	ε
	237	5/2(−)	46.57s	73.0 m *10*	ε 99.98%, α 0.03%
	238	1+	48.42	98 m *2*	ε, α 1.0×10^{-4}%
	238m		48.42	3.8 y	α
	239	(5/2)−	49.392	11.9 h *1*	ε 99.99%, α 0.01%
	240	(3−)	51.51	50.8 h *3*	ε, α 1.9×10^{-4}%
	241	5/2−	52.936	432.2 y *7*	α, SF 4×10^{-10}%
	242	1−	55.470	16.02 h *2*	β− 82.7%, ε 17.3%

Nuclide Z El	A	Jπ	Δ (MeV)	T½, Γ, or Abundance	Decay Mode
95 Am	242m	5−	55.518	141 y *2*	IT 99.55%, α 0.45%, SF < 4.7×10^{-9}%
	242m	(2+,3−)	57.670	14.0 ms *10*	SF ≈ 100%, α < 5.0×10^{-3}%, IT
	243	5/2−	57.176	7370 y *40*	α, SF 3.7×10^{-9}%
	244	(6−)	59.881	10.1 h *1*	β−
	244m		59.881	0.90 ms *15*	SF ≤ 100%
	244m	1+	59.967	26 m *1*	β− 99.96%, ε 0.04%
	245	(5/2)+	61.900	2.05 h *1*	β−
	246	(7−)	64.99	39 m *3*	β−
	246m	2(−)	64.99	25.0 m *2*	β−, IT < 0.02%
	247	(5/2)	67.2s	23.0 m *13*	β−
	248		70.6s	≈10 m	β−
	249		73.1s	≈2 m	β−?
96 Cm	232	0+		1 m *?*	SF < 30.3%
	233		47.29		α, ε
	234	0+	46.72	≈2 m	ε?, α?
	235		47.9s	5 m *sy*	ε?, α?
	236	0+	47.9s	≈10 m	ε, α
	237		49.3s	≈20 m	ε?, α?
	238	0+	49.40	2.4 h *1*	ε ≥ 90%, α ≤ 10%
	239	(7/2−)	51.2s	≈2.9 h	ε, α < 0.1%
	240	0+	51.725	27 d *1*	α > 99.5%, ε < 0.5%, SF 3.9×10^{-6}%
	241	1/2+	53.703	32.8 d *2*	ε 99%, α 1%
	242	0+	54.805	162.8 d *2*	α, SF 6.2×10^{-6}%, 34 $1.\times10^{-14}$%
	243	5/2+	57.184	29.1 y *1*	α 99.71%, ε 0.29%, SF 5.3×10^{-9}%
	244	0+	58.454	18.1 y *1*	α, SF 1.4×10^{-4}%
	245	7/2+	61.005	8500 y *100*	α, SF 6.1×10^{-7}%
	246	0+	62.618	4760 y *40*	α 99.97%, SF 0.03%
	247	9/2−	65.534	1.56×10^{7} y *5*	α
	248	0+	67.392	3.48×10^{5} y *6*	α 91.61%, SF 8.39%
	249	1/2(+)	70.750	64.15 m *3*	β−
	250	0+	72.99	≈8.3×10^{3} y	SF ≈ 74%, α ≈ 18%, β− ≈ 8%
	251	(1/2+)	76.65	16.8 m *2*	β−
	252	0+	79.1s	<2 d	β−
97 Bk	235		52.7s	≈20 s	ε?, α?
	236		53.4s	≈42 s	α?, ε?
	236m		46.2s	≥30 d	ε
	237		53.1s	≈1 m	ε?, α?
	238		54.3s	144 s *5*	ε, εSF 0.048%
	240		55.7s	4.8 m *8*	εSF 2.0×10^{-3}%, ε
	241	(7/2+)	56.1s	≈3 m	α?, ε?
	242		57.7s	7.0 m *13*	ε ≤ 100%
	243	(3/2−)	58.691	4.5 h *2*	ε ≈ 99.85%, α ≈ 0.15%
	244	(4−)	60.72	4.35 h *15*	ε 99.99%, α 6.0×10^{-3}%
	245	3/2−	61.815	4.94 d *3*	ε 99.88%, α 0.12%
	246m	2(−)	63.97	1.80 d *2*	ε, α < 0.2%

TABLE II.1 (cont.)

Nuclide Z El A	Jπ	Δ (MeV)	T½, Γ, or Abundance	Decay Mode
97 Bk 247	(3/2−)	65.491	1380 y *250*	α≤100%
248		68.08s	>9 y	α
248m	1(−)	68.08s	23.7 h *2*	β− 70%, ε 30%
249	7/2+	69.850	330 d *4*	β−, α 1.4×10^{-3}%, SF 4.7×10^{-8}%
250	2−	72.951	3.212 h *5*	β−
251	(3/2−)	75.23	55.6 m *11*	β−
252		78.5s	≈2 m	β−?, α?
253		80.9s	≈10 m	β−?
254		84.4s	≈2 m	β−?
98 Cf 237		57.8s	2.1 s *3*	SF≈10%, α?
238	0+	57.2s	21 ms *2*	SF≈100%
239		58.1s	39 s +*37*−*12*	ε, α
240	0+	58.0s	0.96 m *15*	α≈98%, SF≈2%, ε
241		59.4s	3.78 m *70*	ε≈75%, α≈25%
242	0+	59.34	3.7 m *5*	α 80%, ε 20%, SF≤0.01%
243	(1/2+)	60.9s	10.7 m *5*	ε≈86%, α≈14%
244	0+	61.479	19.4 m *6*	α≤100%
245	(1/2+,5/2+)	63.387	45.0 m *15*	ε 64%, α 36%
246	0+	64.092	35.7 h *5*	α, ε<4.0×10^{-3}%, SF 2.5×10^{-4}%
247	(7/2+)	66.137	3.11 h *3*	ε 99.97%, α 0.04%
248	0+	67.240	333.5 d *28*	α, SF 2.9×10^{-3}%
249	9/2−	69.726	351 y *2*	α, SF 5.0×10^{-7}%
250	0+	71.172	13.08 y *9*	α 99.92%, SF 0.08%
251	1/2+	74.135	898 y *44*	α, SF
252	0+	76.034	2.645 y *8*	α 96.91%, SF 3.09%
253	(7/2+)	79.301	17.81 d *8*	β− 99.69%, α 0.31%
254	0+	81.34	60.5 d *2*	SF 99.69%, α 0.31%
255	(7/2+)	84.8s	85 m *18*	β−
256	0+	87.0s	12.3 m *12*	SF, β−<1%, α≈1.0×10^{-6}%
99 Es 240		64.2s	1 s *sy*	α?, ε?
241	(3/2−)	63.8s	8 s +*6*−*5*	α
242		65.0s	13.5 s *25*	α>0%, ε>0%
243		64.8s	21 s *2*	ε≤70%, α≥30%
244		66.0s	37 s *4*	ε 96%, α 4%
245	(3/2−)	66.4s	1.1 m *1*	ε 60%, α 40%
246m		67.9s	7.7 m *5*	ε 90.1%, α 9.9%, ε 3.0×10^{-3}%
247	(7/2+)	68.61s	4.55 m *26*	ε≈93%, α≈7%
247m		68.61s	625 d *84*	α
248	(2−,0+)	70.30s	27 m *5*	ε 99.7%, α≈0.25%
249	7/2+	71.18s	102.2 m *6*	ε 99.43%, α 0.57%
250	(6+)	73.2s	8.6 h *1*	ε>97%, α<3%
250m	1(−)	73.2s	2.22 h *5*	ε≤100%
251	(3/2−)	74.512	33 h *1*	ε 99.5%, α 0.5%
252	(5−)	77.29	471.7 d *19*	α 78%, ε 22%, β−≈0.01%
253	7/2+	79.014	20.47 d *3*	α, SF 8.7×10^{-6}%
254	(7+)	81.992	275.7 d *5*	α≈100%, β− 1.7×10^{-4}%, SF<3.0×10^{-6}%
99 Es 254m	2+	82.072	39.3 h *2*	β− 98%, IT<3%, α 0.32%, ε 0.08%, SF<0.05%
255	(7/2+)	84.09	39.8 d *12*	β− 92%, α 8%, SF 4.1×10^{-3}%
256	(1+,0−)	87.2s	25.4 m *24*	β−
256m	(8+)	87.2s	7.6 h	β−
257		89.4s	7.7 d *2*	β−, SF
258		92.7s	3 m *sy*	ε?, α?
100Fm 242	0+	68.4s	0.8 ms *2*	SF≤100%
243	(7/2+)	69.3s	0.18 s +*8*−*4*	α≤100%
244	0+	69.0s	3.3 ms *5*	SF≤100%
245		70.2s	4.2 s *13*	α≤100%, SF≤0.1%
246	0+	70.14	1.1 s *2*	α 92%, SF 8%, ε≤1%
247	(7/2+)	71.6s	29 s *1*	α≥50%, ε≤50%
247m	(1/2+)	71.6s	4.3 s *4*	α≤100%
248	0+	71.91	36 s *2*	α 93%, ε 7%, SF 0.1%
249	(7/2+)	73.6s	2.6 m *7*	ε 67%, α 33%
250	0+	74.07	30 m *3*	α>90%, ε<10%, SF 6.9×10^{-3}%
250m	0+	74.07	1.8 s *1*	IT≥80%, α<20%, SF≤8.2×10^{-5}%, ε
251	(9/2−)	75.987	5.30 h *8*	ε 98.2%, α 1.8%
252	0+	76.817	25.39 h *4*	α, SF 2.3×10^{-3}%
253	(1/2)+	79.350	3.00 d *12*	ε 88%, α 12%
254	0+	80.904	3.240 h *2*	α 99.94%, SF 0.06%
255	7/2+	83.799	20.07 h *7*	α, SF 2.4×10^{-5}%
256	0+	85.486	157.6 m *13*	SF 91.9%, α 8.1%
257	(9/2+)	88.590	100.5 d *2*	α 99.79%, SF 0.21%
258	0+	90.4s	370 μs *43*	SF≤100%
259		93.7s	1.5 s *3*	SF
260	0+	95.6s	≈4 ms	SF
101Md 245	(1/2−)	75.3s	0.90 ms *25*	α, SF
245m		75.6s	0.35 s +*23*−*16*	α, ε
246m		76.3s	1.0 s *4*	SF, α>0%, ε>0%
247		76.0s	1.12 s *22*	α≤100%
248		77.1s	7 s *3*	ε 80%, α 20%, SF≤0.05%
249		77.3s	24 s *4*	α>60%, ε≤40%
250		78.6s	52 s *6*	ε 93%, α 7%
251		79.0s	4.0 m *5*	ε≥90%, α≤10%
252		80.6s	2.3 m *8*	ε≤100%
253	(1/2−)	81.3s	6 m +*12*−*3*	ε≤100%, α
254m		83.5s	10 m *3*	ε≤100%
254m		83.5s	28 m *8*	ε≤100%
255	(7/2−)	84.843	27 m *2*	ε 92%, α 8%, SF<0.15%
256	(1−)	87.62	77 m *2*	ε 90.8%, α 9.2%, SF<3%
257	(7/2−)	88.996	5.52 h *5*	ε 85%, α 15%, SF<1%
258		91.688	51.5 d *3*	α, SF
258m		91.688	57.0 m *9*	ε≥70%, SF
101Md 259		93.6s	96 m *3*	SF≈100%, α<1.3%
260		96.6s	31.8 d *5*	SF≥42%, α≤25%, ε≤23%, β−≤10%
261		98.5s	40 m *sy*	α?
262		101.4s	3 m *sy*	SF?, α?
102No 248	0+	80.7s	<2 μs	SF?
249		81.8s	54 μs +*15*−*10*	SF
250	0+	81.5s	6 μs *1*	SF≤100%, α 0.1%, ε 1.0×10^{-3}%
251	(7/2+)	82.9s	0.78 s *2*	α≤100%, SF≤8%, ε
251m	(1/2+)	83.0s	0.93 s *6*	α≤100%
252	0+	82.88	2.27 s *14*	α 58%, ε 23%, SF 19%
252m		82.88	26 d *7*	α
253	(9/2−)	84.5s	1.62 m *15*	α≤100%, ε
254	0+	84.72	51 s *10*	α 90%, ε 10%, SF 0.17%
254m	0+	84.72	0.28 s *4*	IT>80%
255	(1/2+)	86.85	3.1 m *2*	α 61%, ε 39%
256	0+	87.824	2.91 s *5*	α 99.47%, SF 0.53%
257	(7/2+)	90.24	25 s *3*	α≤100%, SF≤1.5%
258	0+	91.5s	1.2 ms *2*	SF≤100%
259		94.1s	58 m *5*	α 75%, ε 25%, SF<10%
260	0+	95.6s	106 ms *8*	SF
261		98.5s		β−, α
262	0+	100.0s	≈5 ms	SF
263		103.0s	20 m *sy*	α?, SF?
264	0+	104.6s	1 m *sy*	α?
103 Lr 251		87.9s		ε?, α?
252		88.8s	0.36 s +*11*−*7*	α≈90%, ε≈10%, SF<1%
253		88.7s	0.57 s +*7*−*6*	α, SF<2%
253m		88.7s	1.5 s +*3*−*2*	α 90%, SF<2%
254		89.8s	13 s *3*	α 76%, ε 24%
255		90.1s	22 s *4*	α 85%, ε<30%, SF≤0.1%
256		91.9s	27 s *3*	α 85%, ε 15%, SF<0.03%
257		92.7s	0.646 s *25*	α≤100%, SF≤0.03%
258		94.8s	4.1 s *3*	α>95%, SF<5%
259		95.85s	6.2 s *3*	α 78%, SF 22%
260		98.3s	180 s *30*	α 80%, ε<40%, SF<10%
261		99.6s	39 m *12*	SF
262		102.1s	≈4 h	SF<10%, ε, α
263		103.7s	5 h *sy*	α?
264		106.2s	10 h *sy*	α?, SF?
265		107.9s	10 h *sy*	α?, SF?
266		111.1s	1 h *sy*	α?, SF?
104 Rf 253m		93.8s	48 μs +*17*−*10*	SF≤100%, α
253m		93.8s	≈1.8 s	SF≈50%, α≈50%

TABLE II.1 (cont.)

Nuclide Z El	A	Jπ	Δ (MeV)	T½, Γ, or Abundance	Decay Mode
104 Rf	254	0+	93.3s	23 μs *3*	SF≤100%
	255	(9/2−)	94.4s	1.64 s *11*	SF 52%, α 48%
	255m		94.4s	0.8 s *+5−2*	α≤100%
	256	0+	94.24	6.4 ms *2*	SF 99.68%, α 0.32%
	257	(1/2+)	95.9s	4.7 s *3*	α<100%, SF≤1.4%, ε>0%
	257m		95.9s	3.9 s *4*	α<100%, SF≤1.4%, ε>0%
	258	0+	96.4s	12 ms *2*	SF 87%, α 13%
	259		98.40s	3.2 s *6*	α 92%, SF 8%
	260	0+	99.1s	21 ms *1*	SF≤100%, α?
	261		101.32	65 s *10*	α>80%, ε<15%, SF<10%
	262	0+	102.4s	2.3 s *4*	SF≤100%, α<3%
	263		104.8s	10 m *2*	SF≈100%, α
	264	0+	106.2s	1 h *sy*	α?
	265		108.7s	≈13 h	α?
	266	0+	109.9s	10 h *sy*	α? SF?
	267?		113.2s	2.3 h *+980−17*	SF
	268	0+	115.2s	6 h *sy*	α?, SF?
105 Db	255		100.0s	1.6 s *+6−4*	α≈80%, SF≈20%
	256		100.7s	1.6 s *+5−3*	α≈64%, ε≈36%, SF≈0.02%
	257		100.3s	1.50 s *+19−15*	α>94%, SF<6%
	257m		100.3s	0.76 s *+15−11*	α≥87%, SF≤13%
	258		101.7s	4.0 s *10*	α 67%, ε 33%, SF<1%
	258m		101.7s	20 s *10*	ε≈100%
	259		102.1s	0.51 s *16*	α
	260		103.7s	1.52 s *13*	α≥90.4%, SF≤9.6%, ε<2.5%
	261		104.4s	1.8 s *4*	α≥82%, SF≤18%
	262		106.3s	35 s *5*	α≈67%, SF
	263		107.1s	27 s *+10−7*	SF 56%, α 41%, ε 3%
	264		109.4s	3 m *sy*	α?
	265		110.5s	15 m *sy*	α?
	266		112.7s	20 m *sy*	α?, SF?
	267?		114.0s	73 m *+350−33*	SF>0%
	268?		116.9s	32 h *+11−7*	SF
	269		118.7s	3 h *sy*	α?, SF?
106 Sg	258	0+	105.4s	2.9 ms *+13−7*	SF≤100%, α?
	259	(1/2+)	106.7s	0.48 s *+28−13*	α 90%, SF<20%
	260	0+	106.58	3.6 ms *9*	α 50%, SF 50%
	261		108.2s	0.23 s *6*	α≈100%, SF<1%
	262	0+	108.4s	6.9 ms *+38−18*	SF≥78%, α≤22%
	263		110.2s	1.0 s *2*	α>70%, SF<30%
	263m		110.2s	0.12 s	α, IT
	264	0+	110.8s	0.4 s *sy*	α?
	265	(9/2+)	112.82	8 s *3*	SF≤57%, α≥43%
	266	0+	113.7s	21 s *+20−12*	SF≥50%, α≤50%
	268	0+	117.0s	30 s *sy*	α?, SF?
	270	0+	121.4s	10 m *sy*	α?, SF?
	271?		124.3s	2.4 m *+43−10*	α 50%, SF 50%
106 Sg	272	0+	125.9s	1 h *sy*	α?, SF?
	273		128.8s	1 m *sy*	SF?
107 Bh	260		113.6s	0.3 ms *sy*	α≤100%
	261		113.3s	12 ms *+5−3*	α 95%, SF<10%
	262m		114.5s	8.0 ms *21*	α≤100%
	262m		114.5s	102 ms *26*	α≤100%
	263		114.6s	0.2 ms *sy*	α?
	264		116.1s	0.44 s *+60−16*	α≤100%
	265		116.6s	0.9 s *+7−3*	α
	266?		118.2s	1.7 ms *+82−8*	α
	267?		118.9s	17 s *+14−6*	α
	271?		125.9s	40 s *sy*	α?
	272?		128.6s	10 s *+12−4*	α
	273		130.1s	90 m *sy*	α?, SF?
	274		132.7s	90 m *sy*	α?, SF?
	275		134.4s	40 m *sy*	SF?
108 Hs	263		119.8s		α≤100%
	264	0+	119.60	≈0.8 ms	α≈50%, SF≈50%
	265		121.2s	2.0 ms *+3−2*	α≈100%, SF≤1%
	266	0+	121.2s	2.3 ms *+13−6*	α≈100%, SF<1.4%
	267		122.8s	52 ms *+13−8*	α≥80%, SF<20%
	267m		122.8s	0.80 s *+380−37*	α>0%
	269?		124.9s	9.7 s *+97−33*	α
	270?	0+	125.4s	3.6 s *+8−14*	α
	271		128.2s	40 s *sy*	α?, SF?
	272	0+	129.5s	40 s *sy*	α?, SF?
	273	(3/2+)	132.3s	50 s *sy*	α?
	274	0+	133.3s	1 m *sy*	α?, SF?
	275?		136.0s	0.15 s *+27−6*	α
	276	0+	137.1s	1 h *sy*	α?, SF?
109 Mt	265		126.8s	2 m *sy*	α?
	266?		127.9s	1.7 ms *+18−16*	α≤100%
	267		127.9s	10 ms *sy*	α?
	268?		129.2s	21 ms *+8−5*	α
	270?		131.0s	4.96 ms	α
	271		131.5s	5 s *sy*	α?
	272		133.9s	10 s *sy*	α?, SF?
	273		135.0s	20 s *sy*	α?, SF?
	274		137.4s	20 s *sy*	α?, SF?
	275?		138.5s	9.7 ms *+460−44*	α
	276?		140.8s	0.72 s *+87−25*	α
	279		145.5s	6 m *sy*	α?, SF?
110 Ds	267?		134.5s	3 μs *+6−2*	α
	268	0+	133.9s	100 μs *sy*	
	269?		135.2s	179 μs *+245−66*	α
	270	0+	134.8s	0.10 ms *+14−4*	α, SF<0.2%
	270m		135.9s	6.0 ms *+82−22*	α>70%, IT≤30%
	271		136.1s	1.63 ms *+44−29*	α
	271m		136.1s	69 ms *+56−21*	α>0%, IT?
	272	0+	136.3s	1 s *sy*	SF
	273		138.7s	0.17 ms *+17−6*	α
110 Ds	274	0+	139.3s	2 s *sy*	α?, SF?
	275		141.8s	2 s *sy*	α?
	276	0+	142.6s	5 s *sy*	α?, SF?
	277		145.0s	5 s *sy*	α?
	278	0+	145.8s	10 s *sy*	α?, SF?
	279?		148.0s	0.18 s *+5−3*	SF 90%, α 10%
	281?		151.0s	11.1 s *+50−27*	SF
111 Rg	272?		143.1s	3.8 ms *+14−8*	α
	273		143.2s	5 ms *sy*	α?
	274?		145.1s	6.4 ms *+307−29*	α
	275		145.4s	10 ms *sy*	α?
	276		147.6s	100 ms *sy*	α?, SF?
	277		148.6s	1 s *sy*	α?, SF?
	278		150.5s	1 s *sy*	α?, SF?
	279?		151.3s	0.17 s *+81−8*	α
	280?		153.2s	3.6 s *+43−13*	α
	281		154.0s	1 m *sy*	α?, SF?
	282		156.0s	4 m *sy*	α?, SF?
	283		156.9s	10 m *sy*	α?, SF?
112	277		152.7s	0.69 ms *+69−24*	α
	278	0+	153.1s	10 ms *sy*	α?, SF?
	279		155.1s	0.1 s *sy*	α?, SF?
	280	0+	155.6s	1 s *sy*	α?, SF?
	282?		158.1s	0.50 ms *+33−14*	SF
	283?		160.0s	4.0 s *+13−7*	α≥99%, SF≤1%
	284?		160.6s	97 ms *+31−19*	SF
	285?		162.2s	29 s *+13−7*	α
113	278?			0.24 ms *+114−11*	α
	283?		164.4s	100 ms *+490−45*	α
	284?		165.9s	0.48 s *+58−17*	α
	285		166.5s	2 m *sy*	α?, SF?
	286		168.1s	5 m *sy*	α?, SF?
	287		168.6s	20 m *sy*	α?, SF?
114	286?		171.3s	0.16 ms *+7−3*	SF 60%, α 40%
	287?		172.9s	0.51 s *+18−10*	α
	288?		173.0s	0.8 s *+27−16*	α
	289?		174.4s	2.6 s *+12−7*	α
115	287?		178.1s	32 ms *+155−14*	α
	288?		179.3s	87 ms *+105−30*	α
	289		180s	10 s *sy*	α?, SF?
	290		180.8s	10 s *sy*	α?, SF?
	291		181.1s	1 m *sy*	α?, SF?
116	290	0+	185.0s	15 ms *+26−6*	α
	291		186.3s	6.3 ms *+116−25*	α
	292?	0+	186.1s	18.0 ms *+16−6*	α
	293?			61 ms *+57−20*	α
117	291		192.4s	10 ms *sy*	α?, SF?
	292		193.3s	50 ms *sy*	α?, SF?
118	294?	0+		1.8 ms *+84−8*	α>0%, SF?

APPENDIX III

THE 2006 VALUES OF THE PHYSICAL CONSTANTS RECOMMENDED BY CODATA[1]

Quality	Symbol	Value	Unit
Avogadro constant	N_A	$6.022\ 141\ 79(30) \times 10^{23}$	mol^{-1}
Molar Volume (ideal gas) (at 273.15K, 101325Pa)	V_m	$22.413\ 996(39) \times 10^{3}$	$m^3 \cdot mol^{-1}$
Speed of light in vacuum	c	$2.997\ 924\ 58 \times 10^{8}$	$m \cdot s^{-1}$
Farady constant	F	$9.648\ 533\ 99(24) \times 10^{4}$	$C \cdot mol^{-1}$
Elementary charge	e	$1.602\ 176\ 487(40) \times 10^{-19}$	C
Planck constant	h	$6.626\ 068\ 96(33) \times 10^{-34}$	$J \cdot s$
		$4.135\ 667\ 33(10) \times 10^{-21}$	$MeV \cdot s$
	$\hbar$	$6.582\ 118\ 99(16) \times 10^{-22}$	$MeV \cdot s$
Composite constants	hc	1240	$MeV \cdot fm$
	$\hbar c$	197.326 9631 (49)	$MeV \cdot fm$
	e^2	1.439 964 416 (51)	$MeV \cdot fm$
Fine structure constant $e^2/\hbar c$	α	1/137.035 999 679(94)	
Boltzmann constant	k	$1.380\ 6504(24) \times 10^{-23}$	$J \cdot K^{-1}$
		$8.617\ 343(15) \times 10^{-11}$	$MeV \cdot K^{-1}$
Stefan-Boltzmann constant	σ	$5.670\ 400(40) \times 10^{-8}$	$J \cdot s^{1} \cdot m^{-2} \cdot K^{-4}$
Electron mass	m_e	$9.109\ 382\ 15(45) \times 10^{-31}$	kg
	$m_e c^2$	0.510 998 910(13)	MeV

[1]P. J. Mohr, B. N. Taylor and D.B. Newell, *Physics Today* **60** (July 2007) 52.

Quality	Symbol	Value	Unit
Proton mass	m_p	$1.672\ 621\ 637(83) \times 10^{-27}$	kg
		1.007 276 466 77(10)	u
	m_pc^2	938.272 013(23)	MeV
	m_p/m_e	1836.152 672 47(80)	
Atomic mass unit $m(^{12}C)/12$	u	$1.660\ 538\ 782(83) \times 10^{-27}$	kg
		931.494 028 (23)	MeV
Neutron mass	m_nc^2	939.565 346(23)	MeV
Deuteron mass	m_dc^2	1875.612 793(47)	MeV
Electron classical radius, e^2/m_ec^2	r_e	2.817 940 2894(58)	fm(10^{-15}m)
Compton wavelength of electron, h/m_ec	λ_{ec}	$2.426\ 310\ 2175(33) \times 10^{-12}$	m
Reduced Compton wavelength of Electron, $\lambda_{ec}/2\pi$, $\hbar/m_ec$	$\bar{\lambda}_{ec}$	$386.159\ 264\ 59(53) \times 10^{-15}$	m
Bohr radius	a_1	$0.529\ 177\ 208\ 59(36) \times 10^{-10}$	m
Bohr magneton, $e\hbar/2m_e$	μ_B	$9.274\ 009\ 15(23) \times 10^{-24}$	$J{\cdot}T^{-1}$
		$5.788\ 381\ 7555(79) \times 10^{-5}$	$eV{\cdot}T^{-1}$
Nuclear magneton, $e\hbar/2m_p$	μ_N	$5.050\ 783\ 24(13) \times 10^{-27}$	$J{\cdot}T^{-1}$
		$3.152\ 451\ 2326(45) \times 10^{-8}$	$eV{\cdot}T^{-1}$
Electron magnetic moment	μ_e	$-928.\ 476\ 377\ (23) \times 10^{-26}$	$J{\cdot}T^{-1}$
	μ_e/μ_B	-1.001 159 652 181 11(74)	
Proton magnetic moment/μ_N	μ_p/μ_N	2.792 847 356(23)	
Neutron magnetic moment/μ_N	μ_n/μ_N	-1.913 042 73(45)	
Rydberg constant	R_∞	10 973 731.568 527(73)	m^{-1}
	$R_\infty hc$	13.605 691 93(34)	eV
Standard acceleration of gravity	g	9.806 65	$m{\cdot}s^{-2}$
Newtonian constant of gravitation	G	$6.674\ 28(67) \times 10^{-11}$	$m^3{\cdot}kg^{-1}{\cdot}s^{-2}$
Temperature analogous to 1eV	eV	$1.160\ 4505(20) \times 10^4$	K
1 million electron-volts	MeV	$1.602\ 176\ 487(40) \times 10^{-13}$	J

APPENDIX IV

TABLES OF ATOMIC-ELECTRON BINDING ENERGIES AND X-RAY ENERGIES AND INTENSITIES

The following tables are reprinted from *Table of Radioactive Isotopes*, E. Browne and Richard Firestone, V. S. Shirley, ed., Wiley, New York (1986) with the permission of the authors and editor.

IV.1 ATOMIC-ELECTRON BINDING ENERGIES

The binding energies given in Table IV.1 are those reported by Larkins,[1] mainly from the compilations of Sevier[2] for $Z \leq 83$, and of Porter and Freedman,[3] for $Z \geq 84$. All binding energies listed are for solid systems referenced to the Fermi level, excepting those for Ne, Cl, Ar, Br, Kr, Xe, and Rn. These latter binding energies are for vapor-phase systems referenced to the vacuum level.

The binding energies are accurate to within at least 1 to 2 eV for most of the subshells in the lighter elements, and for the outer orbitals in the heavier elements.

(*continued on page 759*)

[1] F. B. Larkins, *At. Data Nucl. Data Tables* **20** (1977) 313.

[2] K. D. Sevier, *Low Energy Electron Spectrometry*, Wiley, New York (1972).

[3] F. T. Porter and M. S. Freedman, *J. Phys. Chem. Ref. Data* **7** (1978) 1267.

[4] D. A. Shirley, R. L. Martin, S. P. Kowalczyk, F. R. McFeely, and L. Ley, *Phys. Rev.* **B15** (1977) 544.

[5] J. A. Bearden and A. F. Burr, *Rev. Mod. Phys.* **39** (1967) 125.

TABLE IV.1
Atomic-electron binding energies

El	K	L_1	L_2	L_3	M_1	M_2	M_3	M_4	M_5	N_1	N_2	N_3	N_4	N_5
1 H	0.0136													
2 He	0.0246													
3 Li	0.0548	0.0053												
4 Be	0.1121	0.0080												
5 B	0.1880	0.0126	0.0047	0.0047										
6 C	0.2838	0.0180	0.0064	0.0064										
7 N	0.4016	0.0244	0.0092	0.0092										
8 O	0.5320	0.0285	0.0071	0.0071										
9 F	0.6854	0.0340	0.0086	0.0086										
10 Ne	0.8701	0.0485	0.0217	0.0216										
11 Na	1.0721	0.0633	0.0311	0.0311	0.0007									
12 Mg	1.3050	0.0894	0.0514	0.0514	0.0021									
13 Al	1.5596	0.1177	0.0732	0.0727	0.0007	0.0055	0.0055							
14 Si	1.8389	0.1487	0.0995	0.0989	0.0076	0.0030	0.0030							
15 P	2.1455	0.1893	0.1362	0.1353	0.0162	0.0099	0.0099							
16 S	2.4720	0.2292	0.1654	0.1642	0.0158	0.0080	0.0080							
17 Cl	2.8224	0.2702	0.2016	0.2000	0.0175	0.0068	0.0068							
18 Ar	3.2060	0.3263	0.2507	0.2486	0.0292	0.0159	0.0158							
19 K	3.6074	0.3771	0.2963	0.2936	0.0339	0.0178	0.0178							
20 Ca	4.0381	0.4378	0.3500	0.3464	0.0437	0.0254	0.0254							

El	K	L_1	L_2	L_3	M_1	M_2	M_3	M_4	M_5	N_1	N_2	N_3	N_4	N_5
21 Sc	4.4928	0.5004	0.4067	0.4022	0.0538	0.0323	0.0323	0.0066	0.0066					
22 Ti	4.9664	0.5637	0.4615	0.4555	0.0603	0.0346	0.0346	0.0037	0.0037					
23 V	5.4651	0.6282	0.5205	0.5129	0.0665	0.0378	0.0378	0.0022	0.0022					
24 Cr	5.9892	0.6946	0.5837	0.5745	0.0741	0.0425	0.0425	0.0023	0.0023					
25 Mn	6.5390	0.7690	0.6514	0.6403	0.0839	0.0486	0.0486	0.0033	0.0033					
26 Fe	7.1120	0.8461	0.7211	0.7081	0.0929	0.0540	0.0540	0.0036	0.0036					
27 Co	7.7089	0.9256	0.7936	0.7786	0.1007	0.0595	0.0595	0.0029	0.0029					
28 Ni	8.3328	1.0081	0.8719	0.8547	0.1118	0.0681	0.0681	0.0036	0.0036					
29 Cu	8.9789	1.0961	0.9510	0.9311	0.1198	0.0736	0.0736	0.0016	0.0016					
30 Zn	9.6586	1.1936	1.0428	1.0197	0.1359	0.0866	0.0866	0.0081	0.0081					
31 Ga	10.3671	1.2977	1.1423	1.1154	0.1581	0.1068	0.1029	0.0174	0.0174	0.0015	0.0008	0.0008		
32 Ge	11.1031	1.4143	1.2478	1.2167	0.1800	0.1279	0.1208	0.0287	0.0287	0.0050	0.0023	0.0023		
33 As	11.8667	1.5265	1.3586	1.3231	0.2035	0.1464	0.1405	0.0412	0.0412	0.0085	0.0025	0.0025		
34 Se	12.6578	1.6539	1.4762	1.4358	0.2315	0.1682	0.1619	0.0567	0.0567	0.0120	0.0056	0.0056		
35 Br	13.4737	1.7820	1.5960	1.5499	0.2565	0.1893	0.1815	0.0701	0.0690	0.0273	0.0052	0.0046		
36 Kr	14.3256	1.9210	1.7272	1.6749	0.2921	0.2218	0.2145	0.0950	0.0938	0.0275	0.0147	0.0140		
37 Rb	15.1997	2.0651	1.8639	1.8044	0.3221	0.2474	0.2385	0.1118	0.1103	0.0293	0.0148	0.0140		
38 Sr	16.1046	2.2163	2.0068	1.9396	0.3575	0.2798	0.2691	0.1350	0.1331	0.0377	0.0199	0.0199		
39 Y	17.0384	2.3725	2.1555	2.0800	0.3936	0.3124	0.3003	0.1596	0.1574	0.0454	0.0256	0.0256	0.0024	0.0024
40 Zr	17.9976	2.5316	2.3067	2.2223	0.4303	0.3442	0.3305	0.1824	0.1800	0.0513	0.0287	0.0287	0.0030	0.0030
41 Nb	18.9856	2.6977	2.4647	2.3705	0.4684	0.3784	0.3630	0.2074	0.2046	0.0581	0.0339	0.0339	0.0032	0.0032
42 Mo	19.9995	2.8655	2.6251	2.5202	0.5046	0.4097	0.3923	0.2303	0.2270	0.0618	0.0348	0.0348	0.0018	0.0018
43 Tc	21.0440	3.0425	2.7932	2.6769	0.5440	0.4449	0.4250	0.2564	0.2529	0.0680	0.0389	0.0389	0.0020	0.0020
44 Ru	22.1172	3.2240	2.9669	2.8379	0.5850	0.4828	0.4606	0.2836	0.2794	0.0749	0.0431	0.0431	0.0020	0.0020
45 Rh	23.2199	3.4119	3.1461	3.0038	0.6271	0.5210	0.4962	0.3117	0.3070	0.0810	0.0479	0.0479	0.0025	0.0025

TABLE IV.1 (cont.)

El	K	L_1	L_2	L_3	M_1	M_2	M_3	M_4	M_5	N_1	N_2	N_3	N_4	N_5
46 Pd	24.3503	3.6043	3.3303	3.1733	0.6699	0.5591	0.5315	0.3400	0.3347	0.0864	0.0511	0.0511	0.0015	0.0015
47 Ag	25.5140	3.8058	3.5237	3.3511	0.7175	0.6024	0.5714	0.3728	0.3667	0.0952	0.0626	0.0559	0.0033	0.0033
48 Cd	26.7112	4.0180	3.7270	3.5375	0.7702	0.6507	0.6165	0.4105	0.4037	0.1076	0.0669	0.0669	0.0093	0.0093
49 In	27.9399	4.2375	3.9380	3.7301	0.8256	0.7022	0.6643	0.4508	0.4431	0.1219	0.0774	0.0774	0.0162	0.0162
50 Sn	29.2001	4.4647	4.1561	3.9288	0.8838	0.7564	0.7144	0.4933	0.4848	0.1365	0.0886	0.0886	0.0239	0.0239
51 Sb	30.4912	4.6983	4.3804	4.1322	0.9437	0.8119	0.7656	0.5369	0.5275	0.1520	0.0984	0.0984	0.0314	0.0314
52 Te	31.8138	4.9392	4.6120	4.3414	1.0060	0.8697	0.8187	0.5825	0.5721	0.1683	0.1102	0.1102	0.0398	0.0398
53 I	33.1694	5.1881	4.8521	4.5571	1.0721	0.9305	0.8746	0.6313	0.6194	0.1864	0.1227	0.1227	0.0496	0.0496
54 Xe	34.5644	5.4528	5.1037	4.7822	1.1487	1.0021	0.9406	0.6894	0.6767	0.2133	0.1455	0.1455	0.0695	0.0675
55 Cs	35.9846	5.7143	5.3594	5.0119	1.2171	1.0650	0.9976	0.7395	0.7255	0.2308	0.1723	0.1616	0.0788	0.0765
56 Ba	37.4406	5.9888	5.6236	5.2470	1.2928	1.1367	1.0622	0.7961	0.7807	0.2530	0.1918	0.1797	0.0925	0.0899
57 La	38.9246	6.2663	5.8906	5.4827	1.3613	1.2044	1.1234	0.8485	0.8317	0.2704	0.2058	0.1914	0.0989	0.0989
58 Ce	40.4430	6.5488	6.1642	5.7234	1.4346	1.2728	1.1854	0.9013	0.8833	0.2896	0.2233	0.2072	0.1100	0.1100
59 Pr	41.9906	6.8348	6.4404	5.9643	1.5110	1.3374	1.2422	0.9511	0.9310	0.3045	0.2363	0.2176	0.1132	0.1132
60 Nd	43.5689	7.1260	6.7215	6.2079	1.5753	1.4028	1.2974	0.9999	0.9777	0.3152	0.2433	0.2246	0.1175	0.1175
61 Pm	45.1840	7.4279	7.0128	6.4593	1.6500	1.4714	1.3569	1.0515	1.0269	0.3310	0.2420	0.2420	0.1204	0.1204
62 Sm	46.8342	7.7368	7.3118	6.7162	1.7228	1.5407	1.4198	1.1060	1.0802	0.3457	0.2656	0.2474	0.1290	0.1290
63 Eu	48.5190	8.0520	7.6171	6.9769	1.8000	1.6139	1.4806	1.1606	1.1309	0.3602	0.2839	0.2566	0.1332	0.1332
64 Gd	50.2391	8.3756	7.9303	7.2428	1.8808	1.6883	1.5440	1.2172	1.1852	0.3758	0.2885	0.2709	0.1405	0.1405
65 Tb	51.9957	8.7080	8.2516	7.5140	1.9675	1.7677	1.6113	1.2750	1.2412	0.3979	0.3102	0.2850	0.1470	0.1470
66 Dy	53.7885	9.0458	8.5806	7.7901	2.0468	1.8418	1.6756	1.3325	1.2949	0.4163	0.3318	0.2929	0.1542	0.1542
67 Ho	55.6177	9.3942	8.9178	8.0711	2.1283	1.9228	1.7412	1.3915	1.3514	0.4357	0.3435	0.3066	0.1610	0.1610
68 Er	57.4855	9.7513	9.2643	8.3579	2.2065	2.0058	1.8118	1.4533	1.4093	0.4491	0.3662	0.3200	0.1767	0.1676
69 Tm	59.3896	10.1157	9.6169	8.6480	2.3068	2.0898	1.8845	1.5146	1.4677	0.4717	0.3859	0.3366	0.1796	0.1796
70 Yb	61.3323	10.4864	9.9782	8.9436	2.3981	2.1730	1.9498	1.5763	1.5278	0.4872	0.3967	0.3435	0.1981	0.1849
71 Lu	63.3138	10.8704	10.3486	9.2441	2.4912	2.2635	2.0236	1.6394	1.5885	0.5062	0.4101	0.3593	0.2048	0.1950
72 Hf	65.3508	11.2707	10.7394	9.5607	2.6009	2.3654	2.1076	1.7164	1.6617	0.5381	0.4370	0.3804	0.2238	0.2137
73 Ta	67.4164	11.6815	11.1361	9.8811	2.7080	2.4687	2.1940	1.7932	1.7351	0.5655	0.4648	0.4045	0.2413	0.2293
74 W	69.5250	12.0998	11.5440	10.2068	2.8196	2.5749	2.2810	1.8716	1.8092	0.5950	0.4916	0.4253	0.2588	0.2454
75 Re	71.6764	12.5267	11.9587	10.5353	2.9317	2.6816	2.3673	1.9489	1.8829	0.6250	0.5179	0.4444	0.2737	0.2602

El	K	L_1	L_2	L_3	M_1	M_2	M_3	M_4	M_5	N_1	N_2	N_3	N_4	N_5
76 Os	73.8708	12.9680	12.3850	10.8709	3.0485	2.7922	2.4572	2.0308	1.9601	0.6543	0.5465	0.4682	0.2894	0.2728
77 Ir	76.1110	13.4185	12.8241	11.2152	3.1737	2.9087	2.5507	2.1161	2.0404	0.6901	0.5771	0.4943	0.3114	0.2949
78 Pt	78.3948	13.8805	13.2726	11.5638	3.2976	3.0270	2.6453	2.2015	2.1211	0.7240	0.6076	0.5191	0.3307	0.3138
79 Au	80.7249	14.3528	13.7336	11.9187	3.4249	3.1478	2.7430	2.2911	2.2057	0.7588	0.6437	0.5454	0.3520	0.3339
80 Hg	83.1023	14.8393	14.2087	12.2839	3.5616	3.2785	2.8471	2.3849	2.2949	0.8030	0.6810	0.5769	0.3785	0.3593
81 Tl	85.5304	15.3467	14.6979	12.6575	3.7041	3.4157	2.9566	2.4851	2.3893	0.8455	0.7213	0.6090	0.4066	0.3862
82 Pb	88.0045	15.8608	15.2000	13.0352	3.8507	3.5542	3.0664	2.5856	2.4840	0.8936	0.7639	0.6445	0.4352	0.4129
83 Bi	90.5259	16.3875	15.7111	13.4186	3.9991	3.6963	3.1769	2.6876	2.5796	0.9382	0.8053	0.6789	0.4636	0.4400
84 Po	93.1000	16.9280	16.2370	13.8100	4.1520	3.8440	3.2930	2.7940	2.6800	0.9870	0.8510	0.7150	0.4950	0.4690
85 At	95.7240	17.4820	16.7760	14.2070	4.3100	3.9940	3.4090	2.9010	2.7810	1.0380	0.8970	0.7510	0.5270	0.4990
86 Rn	98.3970	18.0480	17.3280	14.5100	4.4730	4.1500	3.5290	3.0120	2.8840	1.0900	0.9440	0.7900	0.5580	0.5300
87 Fr	101.1300	18.6340	17.8990	15.0250	4.6440	4.3150	3.6560	3.1290	2.9940	1.1480	0.9990	0.8340	0.5970	0.5670
88 Ra	103.9150	19.2320	18.4840	15.4440	4.8220	4.4830	3.7850	3.2480	3.1050	1.2080	1.0550	0.8790	0.6360	0.6030
89 Ac	106.7560	19.8460	19.0810	15.8700	4.9990	4.6550	3.9150	3.3700	3.2190	1.2690	1.1120	0.9240	0.6760	0.6400
90 Th	109.6500	20.4720	19.6930	16.3000	5.1820	4.8310	4.0460	3.4910	3.3320	1.3300	1.1680	0.9670	0.7130	0.6770
91 Pa	112.5960	21.1050	20.3140	16.7330	5.3610	5.0010	4.1740	3.6060	3.4420	1.3830	1.2170	1.0040	0.7430	0.7080
92 U	115.6020	21.7580	20.9480	17.1680	5.5480	5.1810	4.3040	3.7260	3.5500	1.4410	1.2710	1.0430	0.7790	0.7370
93 Np	118.6690	22.4270	21.6000	17.6100	5.7390	5.3660	4.4350	3.8490	3.6640	1.5010	1.3280	1.0850	0.8160	0.7710
94 Pu	121.7910	23.1040	22.2660	18.0570	5.9330	5.5470	4.5630	3.9700	3.7750	1.5590	1.3800	1.1230	0.8460	0.7980
95 Am	124.9820	23.8080	22.9520	18.5100	6.1330	5.7390	4.6980	4.0960	3.8900	1.6200	1.4380	1.1650	0.8800	0.8290
96 Cm	128.2410	24.5260	23.6510	18.9700	6.3370	5.9370	4.8380	4.2240	4.0090	1.6840	1.4980	1.2070	0.9160	0.8620
97 Bk	131.5560	25.2560	24.3710	19.4350	6.5450	6.1380	4.9760	4.3530	4.1270	1.7480	1.5580	1.2490	0.9550	0.8980
98 Cf	134.9390	26.0100	25.1080	19.9070	6.7610	6.3450	5.1160	4.4840	4.2470	1.8130	1.6200	1.2920	0.9910	0.9300
99 Es	138.3960	26.7820	25.8650	20.3840	6.9810	6.5580	5.2590	4.6170	4.3680	1.8830	1.6830	1.3360	1.0290	0.9650
100 Fm	141.9260	27.5740	26.6410	20.8680	7.2080	6.7760	5.4050	4.7520	4.4910	1.9520	1.7490	1.3790	1.0670	1.0000
101 Md	146.5260	28.3870	27.4380	21.3560	7.4400	7.0010	5.5520	4.8890	4.6150	2.0240	1.8160	1.4240	1.1050	1.0340
102 No	149.2080	29.2210	28.2550	21.8510	7.6780	7.2310	5.7020	5.0280	4.7410	2.0970	1.8850	1.4690	1.1450	1.0700
103 Lr	152.9700	30.0830	29.1030	22.3590	7.9300	7.4740	5.8600	5.1760	4.8760	2.1800	1.9630	1.5230	1.1920	1.1120
104	156.2880	30.8810	29.9860	22.9070	8.1610	7.7380	6.0090	5.3360	5.0140	2.2370	2.0350	1.5540	1.2330	1.1490

TABLE IV.1 (cont.)

El	N_6	N_7	O_1	O_2	O_3	O_4	O_5	O_6	O_7	P_1	P_2	P_3	P_4	P_5
48 Cd			0.0022	0.0022	0.0022									
49 In			0.0001	0.0008	0.0008									
50 Sn			0.0009	0.0011	0.0011									
51 Sb			0.0067	0.0021	0.0021									
52 Te			0.0116	0.0023	0.0023									
53 I			0.0136	0.0033	0.0033									
54 Xe			0.0234	0.0134	0.0121									
55 Cs			0.0227	0.0131	0.0114									
56 Ba			0.0291	0.0166	0.0146									
57 La			0.0323	0.0144	0.0144									
58 Ce	0.0001	0.0001	0.0378	0.0198	0.0198									
59 Pr	0.0020	0.0020	0.0374	0.0223	0.0223									
60 Nd	0.0015	0.0015	0.0375	0.0211	0.0211									
61 Pm	0.0040	0.0040	0.0380	0.0220	0.0220									
62 Sm	0.0055	0.0055	0.0374	0.0213	0.0213									
63 Eu			0.0318	0.0220	0.0220									
64 Gd	0.0001	0.0001	0.0361	0.0203	0.0203									
65 Tb	0.0026	0.0026	0.0390	0.0254	0.0254									
66 Dy	0.0042	0.0042	0.0629	0.0263	0.0263									
67 Ho	0.0037	0.0037	0.0512	0.0203	0.0203									
68 Er	0.0043	0.0043	0.0598	0.0294	0.0294									
69 Tm	0.0053	0.0053	0.0532	0.0323	0.0323									
70 Yb	0.0063	0.0063	0.0541	0.0234	0.0234									
71 Lu	0.0069	0.0069	0.0568	0.0280	0.0280	0.0046	0.0046							
72 Hf	0.0171	0.0171	0.0649	0.0381	0.0306	0.0066	0.0066							
73 Ta	0.0275	0.0256	0.0711	0.0449	0.0364	0.0057	0.0057							
74 W	0.0379	0.0358	0.0771	0.0468	0.0356	0.0061	0.0061							
75 Re	0.0481	0.0457	0.0828	0.0456	0.0346	0.0035	0.0035							
76 Os	0.0538	0.0510	0.0837	0.0580	0.0454									
77 Ir	0.0640	0.0610	0.0952	0.0630	0.0505	0.0038	0.0038							

El	N_6	N_7	O_1	O_2	O_3	O_4	O_5	O_6	O_7	P_1	P_2	P_3	P_4	P_5
78 Pt	0.0745	0.0711	0.1017	0.0653	0.0510	0.0021	0.0021							
79 Au	0.0878	0.0841	0.1078	0.0717	0.0537	0.0025	0.0025							
80 Hg	0.1040	0.0999	0.1203	0.0840	0.0650	0.0098	0.0078							
81 Tl	0.1231	0.1188	0.1363	0.0996	0.0730	0.0153	0.0131							
82 Pb	0.1412	0.1363	0.1473	0.1048	0.0830	0.0218	0.0192			0.0031	0.0007	0.0007		
83 Bi	0.1624	0.1571	0.1593	0.1168	0.0930	0.0265	0.0244			0.0080	0.0027	0.0027		
84 Po	0.1840	0.1780	0.1760	0.1320	0.1020	0.0340	0.0300			0.0090	0.0040	0.0010		
85 At	0.2060	0.1990	0.1920	0.1440	0.1130	0.0410	0.0370			0.0130	0.0060	0.0010		
86 Rn	0.2290	0.2220	0.2080	0.1580	0.1230	0.0480	0.0430			0.0160	0.0080	0.0020		
87 Fr	0.2580	0.2490	0.2290	0.1780	0.1380	0.0600	0.0550			0.0240	0.0140	0.0070		0.0038
88 Ra	0.2870	0.2790	0.2510	0.1970	0.1530	0.0720	0.0660			0.0310	0.0200	0.0120		0.0047
89 Ac	0.3160	0.3070	0.2720	0.2170	0.1680	0.0840	0.0760			0.0370	0.0240	0.0150	0.0044	0.0054
90 Th	0.3440	0.3350	0.2900	0.2360	0.1800	0.0940	0.0870			0.0410	0.0240	0.0170	0.0055	0.0059
91 Pa	0.3660	0.3550	0.3050	0.2450	0.1880	0.0970	0.0900	0.0073		0.0430	0.0270	0.0170	0.0046	0.0056
92 U	0.3890	0.3790	0.3240	0.2570	0.1940	0.1040	0.0950	0.0085		0.0440	0.0270	0.0170	0.0046	0.0057
93 Np	0.4140	0.4030	0.3380	0.2740	0.2060	0.1090	0.1010	0.0097		0.0470	0.0290	0.0180	0.0046	0.0058
94 Pu	0.4360	0.4240	0.3500	0.2830	0.2130	0.1130	0.1020	0.0070		0.0460	0.0290	0.0160		0.0054
95 Am	0.4610	0.4460	0.3650	0.2980	0.2190	0.1160	0.1060	0.0079	0.0066	0.0480	0.0290	0.0160		0.0055
96 Cm	0.4840	0.4700	0.3830	0.3130	0.2290	0.1240	0.1100	0.0129	0.0113	0.0500	0.0300	0.0160	0.0045	0.0061
97 Bk	0.5110	0.4950	0.3990	0.3260	0.2370	0.1300	0.1170	0.0140	0.0122	0.0520	0.0320	0.0160	0.0044	0.0062
98 Cf	0.5380	0.5200	0.4160	0.3410	0.2450	0.1370	0.1220	0.0105	0.0087	0.0540	0.0330	0.0170		0.0057
99 Es	0.5640	0.5460	0.4340	0.3570	0.2550	0.1420	0.1270	0.0113	0.0094	0.0570	0.0350	0.0170		0.0058
100 Fm	0.5910	0.5720	0.4520	0.3730	0.2620	0.1490	0.1330	0.0170	0.0147	0.0590	0.0360	0.0170	0.0042	0.0065
101 Md	0.6180	0.5970	0.4710	0.3890	0.2720	0.1540	0.1370	0.0129	0.0105	0.0610	0.0370	0.0170		0.0059
102 No	0.6450	0.6240	0.4900	0.4060	0.2300	0.1610	0.1420	0.0136	0.0111	0.0630	0.0380	0.0180		0.0060
103 Lr	0.6800	0.6580	0.5160	0.4290	0.2960	0.1740	0.1540	0.0199	0.0170	0.0710	0.0440	0.0210	0.0039	0.0069
104	0.7250	0.7010	0.5350	0.4480	0.3190	0.1900	0.1710	0.0260	0.0228	0.0820	0.0550	0.0330	0.0050	0.0075

TABLE IV.2

Notations for x-ray transitions

Classical designation (Siegbahn notation)	Associated initial–final shell vacancies
K_{α_1}	K–L_3
K_{α_2}	K–L_2
K_{α_3}	K–L_1
K_{β_1}	K–M_3
K_{β_2}	K–N_2N_3
K_{β_3}	K–M_2
K_{β_4}	K–N_4N_5
K_{β_5}	K–M_4M_5
$KO_{2,3}$	K–O_2O_3
$KP_{2,3}$	K–P_2P_3
L_{α_1}	L_3–M_5
L_{α_2}	L_3–M_4
L_{β_1}	L_2–M_4
$L_{\beta_{2,15}}$	L_3–N_4N_5
L_{β_3}	L_1–M_3
L_{β_4}	L_1–M_2
L_{β_5}	L_3–O_4O_5
L_{β_6}	L_3–N_1
L_{γ_1}	L_2–N_4
L_{γ_2}	L_1–N_2
L_{γ_3}	L_1–N_3
L_{γ_6}	L_2–O_4
L_{η}	L_2–M_1
L_{l}	L_3–M_1

Group designation	Associated transitions
$K'_{\beta 1}$	$K_{\beta 1} + K_{\beta 3} + K_{\beta 5}$
$K'_{\beta 2}$	$K_{\beta 2} + K_{\beta 4} + \ldots$
L_{α}	$L_{\alpha 1} + L_{\alpha 2}$
L_{β}	$L_{\beta 1} + L_{\beta 2,15} + L_{\beta 3} + L_{\beta 4} + L_{\beta 5} + L_{\beta 6}$
L_{γ}	$L_{\gamma 1} + L_{\gamma 2} + L_{\gamma 3} + L_{\gamma 6}$

TABLE IV.3

Atomic shell	Associated x-rays
K	$K_{\alpha 1}$, $K_{\alpha 2}$, $K_{\alpha 3}$, $K_{\beta 1}$, $K_{\beta 2}$, $K_{\beta 3}$, $K_{\beta 4}$, $K_{\beta 5}$, $KO_{2,3}$, $KP_{2,3}$
L_1	$L_{\beta 3}$, $L_{\beta 4}$, $L_{\gamma 2}$, $L_{\gamma 3}$
L_2	$L_{\beta 1}$, L_{η}, $L_{\gamma 1}$, $L_{\gamma 6}$
L_3	$L_{\alpha 1}$, $L_{\alpha 2}$, $L_{\beta 2,15}$, $L_{\beta 5}$, $L_{\beta 6}$, L_{l}

Uncertainties may be as large as 10 or 20 eV for the inner orbitals in the high-Z elements, and changes in chemical state can lead to substantial shifts in the binding energies of nonvalence shells.[4] Bearden and Burr[5] reevaluated existing data on x-ray emission wavelengths, and discussed binding energies determined from atomic energy level differences.

IV.2 X-RAY ENERGIES AND INTENSITIES

Table IV.4 lists energies and intensities for x-rays with intensities greater than 0.001 per 100 primary vacancies in the K, L_1, and L_3 atomic shells. The first column shows the Siegbahn notations for the x-ray transitions (the associations with initial and final atomic-shell vacancies are given in Table IV.2). The following columns give, for each element, the x-ray energies in keV (boldface) rounded to the nearest eV, and their corresponding intensities directly below. Intensities for the L x-rays are totals from both primary and secondary atomic-shell vacancies.

X-ray energies have been determined from the differences between the corresponding atomic-shell binding energies reported by Larkins.[6] Energies of complex x-ray transitions, for example, $L_{\beta 2,15}$, are unweighted averages of those for the single-line components.

X-ray intensities have been determined from the experimental relative emission probabilities of Salem et al.,[7] and the atomic yields of Krause.[8] The theoretical emission probabilities of Scofield[9] have occasionally been used whenever experimental values were not available.

The relative intensities of x-rays from the same initial atomic shells are independent of the processes creating the shell vacancies. Table IV.4 may therefore be used to separate experimentally unresolved or complex x-ray intensities from the photon tables of the *Table of Radioactive Isotopes*. Table IV.3 shows the initial atomic shells and their associated x-rays.

[6] F. B. Larkins, *At. Data Nucl. Data Tables* **20** (1977) 313.

[7] S. I. Salem, S. L. Panossian, and R. A. Krause, *At. Data Nucl. Data Tables* **14** (1974) 91.

[8] M. O. Krause, *J. Phys. Chem. Ref. Data* **8** (1979) 307.

[9] J. H. Scofield, *At. Data Nucl. Data Tables* **14** (1974) 121.

TABLE IV.4

X-ray energies and intensities (per 100 K-shell vacancies)

	$_5$B	$_6$C	$_7$N	$_8$O	$_9$F	$_{10}$Ne	$_{11}$Na	$_{12}$Mg	$_{13}$Al	$_{14}$Si	$_{15}$P	$_{16}$S	$_{17}$Cl	$_{18}$Ar	$_{19}$K
$K_{\alpha1}$	**0.183** 0.11 *5*	**0.277** 0.19 *8*	**0.392** 0.35 *14*	**0.525** 0.55 *22*	**0.677** 0.9 *4*	**0.849** 1.20 *12*	**1.041** 1.53 *16*	**1.254** 2.0 *2*	**1.487** 2.6 *3*	**1.740** 3.3 *3*	**2.010** 4.1 *4*	**2.308** 5.0 *5*	**2.622** 6.1 *6*	**2.957** 7.3 *7*	**3.314** 8.5 *9*
$K_{\alpha2}$	**0.183** 0.056 *23*	**0.277** 0.09 *4*	**0.392** 0.17 *7*	**0.525** 0.28 *11*	**0.677** 0.43 *17*	**0.848** 0.60 *6*	**1.041** 0.77 *8*	**1.254** 1.00 *10*	**1.486** 1.29 *13*	**1.739** 1.64 *17*	**2.009** 2.04 *21*	**2.307** 2.49 *25*	**2.621** 3.0 *3*	**2.955** 3.6 *4*	**3.311** 4.3 *4*
$K_{\beta1}$									**1.554** 0.0155 *16*	**1.836** 0.056 *6*	**2.136** 0.122 *12*	**2.464** 0.229 *23*	**2.816** 0.38 *4*	**3.190** 0.58 *6*	**3.590** 0.79 *8*
$K_{\beta3}$									**1.554** 0.0079 *8*	**1.836** 0.028 *3*	**2.136** 0.062 *6*	**2.464** 0.116 *12*	**2.816** 0.192 *20*	**3.190** 0.30 *3*	**3.590** 0.40 *4*
$L_{\beta1}$														**0.251** 0.011 *3*	**0.296** 0.013 *4*
$L_{\beta3}$														**0.310** 0.0038 *13*	**0.359** 0.0050 *17*
$L_{\beta4}$														**0.310** 0.0024 *9*	**0.359** 0.0010 *5*

	$_{20}$Ca	$_{21}$Sc	$_{22}$Ti	$_{23}$V	$_{24}$Cr	$_{25}$Mn	$_{26}$Fe	$_{27}$Co	$_{28}$Ni	$_{29}$Cu	$_{30}$Zn	$_{31}$Ga	$_{32}$Ge	$_{33}$As	$_{34}$Se
$K_{\alpha1}$	**3.692** 9.8 *4*	**4.091** 11.3 *5*	**4.511** 12.8 *6*	**4.952** 14.5 *7*	**5.415** 16.4 *7*	**5.899** 18.3 *8*	**6.404** 20.2 *9*	**6.930** 22.1 *10*	**7.478** 24.0 *11*	**8.048** 26.0 *12*	**8.639** 28.0 *10*	**9.252** 29.8 *11*	**9.886** 31.3 *11*	**10.544** 32.7 *12*	**11.222** 34.1 *12*
$K_{\alpha2}$	**3.688** 4.93 *22*	**4.086** 5.68 *25*	**4.505** 6.4 *3*	**4.945** 7.3 *3*	**5.405** 8.3 *4*	**5.888** 9.3 *4*	**6.391** 10.2 *5*	**6.915** 11.2 *5*	**7.461** 12.2 *6*	**8.028** 13.3 *6*	**8.616** 14.3 *5*	**9.225** 15.2 *6*	**9.855** 16.1 *6*	**10.508** 16.8 *6*	**11.182** 17.6 *6*
$K_{\beta1}$	**4.013** 1.02 *5*	**4.461** 1.22 *6*	**4.932** 1.42 *6*	**5.427** 1.64 *7*	**5.947** 1.84 *8*	**6.490** 2.14 *10*	**7.058** 2.40 *11*	**7.649** 2.65 *12*	**8.265** 2.88 *13*	**8.905** 3.10 *14*	**9.572** 3.39 *12*	**10.264** 3.70 *13*	**10.982** 3.98 *14*	**11.726** 4.25 *15*	**12.496** 4.54 *16*
$K_{\beta2}$												**10.366** 0.0314 *11*	**11.101** 0.097 *4*	**11.864** 0.194 *7*	**12.652** 0.323 *12*
$K_{\beta3}$	**4.013** 0.519 *23*	**4.461** 0.62 *3*	**4.932** 0.72 *3*	**5.427** 0.84 *4*	**5.947** 0.94 *4*	**6.490** 1.09 *5*	**7.058** 1.23 *6*	**7.649** 1.36 *6*	**8.265** 1.48 *7*	**8.905** 1.59 *7*	**9.572** 1.74 *6*	**10.260** 1.90 *7*	**10.975** 2.05 *7*	**11.720** 2.19 *8*	**12.490** 2.34 *8*
$K_{\beta5}$							**7.108** 0.00127 *7*	**7.706** 0.00188 *11*	**8.329** 0.00264 *15*	**8.977** 0.00365 *21*	**9.651** 0.00504 *25*	**10.350** 0.0063 *3*	**11.074** 0.0078 *4*	**11.826** 0.0095 *5*	**12.601** 0.0116 *6*
$L_{\alpha1}$		**0.396** 0.026 *7*	**0.452** 0.063 *16*	**0.511** 0.12 *3*	**0.572** 0.19 *5*	**0.637** 0.26 *7*	**0.704** 0.33 *8*	**0.776** 0.41 *10*	**0.851** 0.50 *13*	**0.929** 0.60 *15*	**1.012** 0.65 *13*	**1.098** 0.70 *14*	**1.188** 0.81 *16*	**1.282** 0.87 *17*	**1.379** 0.98 *20*
$L_{\alpha2}$		**0.396** 0.0028 *7*	**0.452** 0.0070 *18*	**0.511** 0.013 *3*	**0.572** 0.021 *5*	**0.637** 0.029 *7*	**0.704** 0.037 *9*	**0.776** 0.045 *11*	**0.851** 0.056 *14*	**0.929** 0.066 *17*	**1.012** 0.072 *15*	**1.098** 0.077 *16*	**1.188** 0.090 *18*	**1.282** 0.096 *19*	**1.379** 0.108 *22*

$L_{\beta 1}$	**0.350** 0.016 *4*	**0.400** 0.020 *5*	**0.458** 0.050 *12*	**0.518** 0.096 *24*	**0.581** 0.15 *4*	**0.648** 0.20 *5*	**0.717** 0.25 *6*	**0.791** 0.31 *8*	**0.868** 0.34 *9*	**0.949** 0.39 *10*	**1.035** 0.42 *11*	**1.125** 0.46 *12*	**1.219** 0.49 *12*	**1.317** 0.52 *13*	**1.420** 0.58 *15*
$L_{\beta 3}$	**0.412** 0.0062 *19*	**0.468** 0.0075 *23*	**0.529** 0.009 *3*	**0.590** 0.010 *3*	**0.652** 0.012 *4*	**0.720** 0.014 *4*	**0.792** 0.016 *5*	**0.866** 0.018 *5*	**0.940** 0.020 *6*	**1.022** 0.021 *6*	**1.107** 0.023 *7*	**1.195** 0.024 *7*	**1.294** 0.025 *7*	**1.386** 0.027 *8*	**1.492** 0.029 *9*
$L_{\beta 4}$	**0.412** 0.0039 *12*	**0.468** 0.0048 *15*	**0.529** 0.0056 *17*	**0.590** 0.0067 *20*	**0.652** 0.0079 *24*	**0.720** 0.009 *3*	**0.792** 0.010 *3*	**0.866** 0.012 *4*	**0.940** 0.013 *4*	**1.022** 0.014 *4*	**1.107** 0.015 *5*	**1.191** 0.016 *5*	**1.286** 0.016 *5*	**1.380** 0.018 *5*	**1.486** 0.019 *6*
$L_{\beta 6}$		**0.402** 0.0017 *4*	**0.456** 0.0018 *5*	**0.513** 0.0022 *6*		**0.640** 0.0023 *6*	**0.708** 0.0022 *6*	**0.779** 0.0022 *6*	**0.855** 0.0022 *6*		**1.020** 0.0021 *4*	**1.114** 0.0027 *5*	**1.212** 0.0033 *7*	**1.315** 0.0038 *8*	**1.424** 0.0045 *9*
$L_{\gamma 3}$												**1.297** 0.0012 *4*	**1.412** 0.0042 *13*	**1.524** 0.0047 *15*	**1.648** 0.0051 *16*
L_{η}		**0.353** 0.020 *5*	**0.401** 0.022 *6*	**0.454** 0.026 *7*	**0.510** 0.025 *6*	**0.568** 0.026 *7*	**0.628** 0.028 *7*	**0.693** 0.028 *7*	**0.760** 0.026 *7*	**0.831** 0.028 *7*	**0.907** 0.029 *7*	**0.984** 0.030 *8*	**1.068** 0.031 *8*	**1.155** 0.031 *8*	**1.245** 0.034 *9*
L_{ℓ}		**0.348** 0.026 *7*	**0.395** 0.029 *8*	**0.446** 0.034 *9*	**0.500** 0.033 *9*	**0.556** 0.038 *10*	**0.615** 0.040 *11*	**0.678** 0.043 *11*	**0.743** 0.045 *12*	**0.811** 0.048 *13*	**0.884** 0.047 *10*	**0.957** 0.048 *10*	**1.037** 0.052 *11*	**1.120** 0.053 *11*	**1.204** 0.056 *12*

	$_{35}$Br	$_{36}$Kr	$_{37}$Rb	$_{38}$Sr	$_{39}$Y	$_{40}$Zr	$_{41}$Nb	$_{42}$Mo	$_{43}$Tc	$_{44}$Ru	$_{45}$Rh	$_{46}$Pd	$_{47}$Ag	$_{48}$Cd	$_{49}$In
$K_{\alpha 1}$	**11.924** 35.6 *13*	**12.651** 36.8 *13*	**13.395** 38.0 *14*	**14.165** 39.1 *14*	**14.958** 40.1 *14*	**15.775** 41.0 *12*	**16.615** 41.8 *12*	**17.479** 42.6 *12*	**18.367** 43.3 *12*	**19.279** 44.0 *12*	**20.216** 44.6 *13*	**21.177** 45.1 *13*	**22.163** 45.6 *13*	**23.174** 46.1 *13*	**24.210** 45.3 *13*
$K_{\alpha 2}$	**11.878** 18.4 *7*	**12.598** 19.0 *7*	**13.336** 19.7 *7*	**14.098** 20.3 *7*	**14.883** 20.9 *8*	**15.691** 21.4 *6*	**16.521** 21.9 *6*	**17.374** 22.4 *6*	**18.251** 22.8 *6*	**19.150** 23.2 *7*	**20.074** 23.5 *7*	**21.020** 23.9 *7*	**21.990** 24.2 *7*	**22.984** 24.5 *7*	**24.002** 24.5 *7*
$K_{\alpha 3}$													**21.708** 0.00100 *4*	**22.693** 0.00115 *4*	**23.702** 0.00135 *5*
$K_{\beta 1}$	**13.292** 4.84 *17*	**14.111** 5.12 *19*	**14.961** 5.39 *19*	**15.836** 5.63 *20*	**16.738** 5.89 *21*	**17.667** 6.15 *17*	**18.623** 6.35 *18*	**19.607** 6.61 *19*	**20.619** 6.80 *19*	**21.657** 6.99 *20*	**22.724** 7.18 *20*	**23.819** 7.35 *21*	**24.943** 7.52 *21*	**26.095** 7.69 *22*	**27.276** 7.85 *22*
$K_{\beta 2}$	**13.469** 0.484 *18*	**14.311** 0.676 *24*	**15.185** 0.85 *3*	**16.085** 1.00 *4*	**17.013** 1.13 *4*	**17.969** 1.25 *4*	**18.952** 1.33 *4*	**19.965** 1.45 *4*	**21.005** 1.54 *4*	**22.074** 1.64 *5*	**23.172** 1.72 *5*	**24.299** 1.79 *5*	**25.455** 1.88 *5*	**26.644** 1.98 *6*	**27.863** 2.09 *6*
$K_{\beta 3}$	**13.284** 2.50 *9*	**14.104** 2.64 *10*	**14.952** 2.78 *10*	**15.825** 2.91 *10*	**16.726** 3.04 *11*	**17.653** 3.17 *9*	**18.607** 3.28 *9*	**19.590** 3.41 *10*	**20.599** 3.51 *10*	**21.634** 3.61 *10*	**22.699** 3.71 *10*	**23.791** 3.81 *11*	**24.912** 3.90 *11*	**26.060** 3.99 *11*	**27.238** 4.07 *12*
$K_{\beta 4}$							**18.982** 0.0010 *5*	**19.998** 0.0015 *7*	**21.042** 0.0023 *11*	**22.115** 0.0032 *16*	**23.217** 0.0043 *21*	**24.349** 0.006 *3*	**25.511** 0.007 *3*	**26.702** 0.008 *4*	**27.924** 0.010 *5*
$K_{\beta 5}$	**13.404** 0.0139 *7*	**14.231** 0.0162 *8*	**15.089** 0.0186 *9*	**15.971** 0.0215 *11*	**16.880** 0.0244 *12*	**17.816** 0.0275 *12*	**18.780** 0.0305 *14*	**19.771** 0.0341 *15*	**20.789** 0.0377 *17*	**21.836** 0.0418 *19*	**22.911** 0.0446 *20*	**24.013** 0.0496 *22*	**25.144** 0.0547 *25*	**26.304** 0.060 *3*	**27.493** 0.065 *3*
$KO_{2,3}$															**27.939** 0.0170 *18*
$L_{\alpha 1}$	**1.481** 1.09 *22*	**1.581** 1.20 *24*	**1.694** 1.3 *3*	**1.806** 1.4 *3*	**1.923** 1.5 *3*	**2.042** 1.66 *25*	**2.166** 1.8 *3*	**2.293** 1.9 *3*	**2.424** 2.0 *3*	**2.558** 2.1 *3*	**2.697** 2.3 *3*	**2.839** 2.4 *4*	**2.984** 2.5 *4*	**3.134** 2.6 *4*	**3.287** 2.8 *4*

TABLE IV.4 (cont.)

	$_{35}$Br	$_{36}$Kr	$_{37}$Rb	$_{38}$Sr	$_{39}$Y	$_{40}$Zr	$_{41}$Nb	$_{42}$Mo	$_{43}$Tc	$_{44}$Ru	$_{45}$Rh	$_{46}$Pd	$_{47}$Ag	$_{48}$Cd	$_{49}$In
continued															
$L_{\alpha 2}$	1.480 0.121 24	1.580 0.13 3	1.693 0.15 3	1.805 0.16 3	1.920 0.17 3	2.040 0.19 3	2.163 0.20 3	2.290 0.21 3	2.420 0.23 3	2.554 0.24 4	2.692 0.25 4	2.833 0.26 4	2.978 0.28 4	3.127 0.29 5	3.279 0.31 5
$L_{\beta 1}$	1.526 0.64 16	1.632 0.69 17	1.752 0.75 19	1.872 0.81 20	1.996 0.86 21	2.124 0.90 14	2.257 0.93 14	2.395 1.00 15	2.537 1.07 16	2.683 1.14 17	2.834 1.20 18	2.990 1.28 19	3.151 1.39 21	3.317 1.51 23	3.487 1.63 25
$L_{\beta 2,15}$					2.078 0.0044 9	2.219 0.0116 18	2.367 0.056 9	2.518 0.100 15	2.675 0.150 23	2.836 0.20 3	3.001 0.24 4	3.172 0.28 4	3.348 0.32 5	3.528 0.38 6	3.714 0.43 7
$L_{\beta 3}$	1.601 0.029 9	1.707 0.030 9	1.827 0.031 9	1.947 0.032 10	2.072 0.035 11	2.201 0.038 10	2.335 0.049 12	2.473 0.048 12	2.617 0.050 13	2.763 0.052 13	2.916 0.053 13	3.073 0.054 14	3.234 0.059 15	3.402 0.062 16	3.573 0.065 16
$L_{\beta 4}$	1.593 0.020 6	1.699 0.020 6	1.818 0.022 7	1.936 0.022 7	2.060 0.025 7	2.187 0.026 7	2.319 0.034 9	2.456 0.034 9	2.598 0.035 9	2.741 0.035 9	2.891 0.035 9	3.045 0.036 9	3.203 0.038 10	3.367 0.039 10	3.535 0.041 10
$L_{\beta 6}$	1.523 0.0052 11	1.647 0.0060 12	1.775 0.0069 14	1.902 0.0079 16	2.035 0.0088 18	2.171 0.0100 15	2.312 0.0110 17	2.458 0.0119 18	2.609 0.0128 20	2.763 0.0139 21	2.923 0.0149 23	3.087 0.0157 24	3.256 0.0166 25	3.430 0.018 3	3.608 0.019 3
$L_{\gamma 1}$					2.153 0.012 3	2.304 0.030 5	2.462 0.041 6	2.623 0.055 8	2.791 0.068 10	2.965 0.083 13	3.144 0.109 17	3.329 0.137 21	3.520 0.147 23	3.718 0.161 25	3.922 0.18 3
$L_{\gamma 2}$	1.777 0.0013 4	1.906 0.0018 6	2.050 0.0022 7	2.196 0.0027 8	2.347 0.0031 9	2.503 0.0036 9	2.664 0.0050 13	2.831 0.0051 13	3.004 0.0055 14	3.181 0.0059 15	3.364 0.0062 16	3.553 0.0065 17	3.743 0.0073 19	3.951 0.0080 20	4.160 0.0087 22
$L_{\gamma 3}$	1.777 0.0052 16	1.907 0.0054 17	2.051 0.0058 18	2.196 0.0061 19	2.347 0.0066 20	2.503 0.0072 19	2.664 0.0095 24	2.831 0.0095 24	3.004 0.010 3	3.181 0.011 3	3.364 0.011 3	3.553 0.011 3	3.750 0.012 3	3.951 0.013 3	4.160 0.014 4
L_{η}	1.339 0.036 9	1.435 0.037 9	1.542 0.039 10	1.649 0.04 1	1.762 0.041 10	1.876 0.041 6	1.996 0.041 6	2.120 0.043 7	2.249 0.044 7	2.382 0.046 7	2.519 0.046 7	2.660 0.048 7	2.806 0.051 8	2.957 0.054 8	3.112 0.056 9
L_{ℓ}	1.293 0.060 13	1.383 0.063 14	1.482 0.067 14	1.582 0.069 15	1.686 0.073 16	1.792 0.078 13	1.902 0.082 14	2.016 0.085 15	2.133 0.089 15	2.253 0.092 16	2.377 0.095 16	2.503 0.097 17	2.634 0.101 17	2.767 0.107 18	2.905 0.112 19

	$_{50}$Sn	$_{51}$Sb	$_{52}$Te	$_{53}$I	$_{54}$Xe	$_{55}$Cs	$_{56}$Ba	$_{57}$La	$_{58}$Ce	$_{59}$Pr	$_{60}$Nd	$_{61}$Pm	$_{62}$Sm	$_{63}$Eu	$_{64}$Gd
$K_{\alpha 1}$	25.271 45.7 10	26.359 46.0 10	27.472 46.2 11	28.612 46.4 11	29.782 46.6 11	30.973 46.7 11	32.194 46.7 11	33.442 46.8 11	34.720 47.0 11	36.026 47.1 11	37.361 47.2 10	38.725 47.3 10	40.118 47.5 10	41.542 47.6 10	42.996 47.5 10
$K_{\alpha 2}$	25.044 24.7 6	26.111 24.9 6	27.202 25.0 6	28.317 25.2 6	29.461 25.3 6	30.625 25.5 6	31.817 25.6 6	33.034 25.7 6	34.279 25.9 6	35.550 26.1 6	36.847 26.2 6	38.171 26.3 6	39.522 26.4 6	40.902 26.6 6	42.309 26.6 6
$K_{\alpha 3}$	24.735 0.00154 5	25.793 0.00179 6	26.875 0.00203 6	27.981 0.00227 7	29.112 0.00262 8	30.270 0.00296 9	31.452 0.00334 11	32.658 0.00373 12	33.894 0.00422 13	35.156 0.00472 15	36.443 0.00531 16	37.756 0.00580 18	39.097 0.00678 21	40.467 0.00727 22	41.864 0.00824 25
$K_{\beta 1}$	28.486 7.99 18	29.726 8.09 18	30.995 8.21 18	32.295 8.34 19	33.624 8.42 19	34.987 8.53 19	36.378 8.63 19	37.801 8.70 19	39.258 8.83 20	40.748 8.9 2	42.272 8.97 19	43.827 9.08 19	45.414 9.15 19	47.038 9.21 19	48.695 9.30 19

$K_{\alpha2}$	**25.044** 24.7 *6*	**26.111** 24.9 *6*	**27.202** 25.0 *6*	**28.317** 25.2 *6*	**29.461** 25.3 *6*	**30.625** 25.5 *6*	**31.817** 25.6 *6*	**33.034** 25.7 *6*	**34.279** 25.9 *6*	**35.550** 26.1 *6*	**36.847** 26.2 *6*	**38.171** 26.3 *6*	**39.522** 26.4 *6*	**40.902** 26.6 *6*	**42.309** 26.6 *6*
$K_{\alpha3}$	**24.735** 0.00154 *5*	**25.793** 0.00179 *6*	**26.875** 0.00203 *6*	**27.981** 0.00227 *7*	**29.112** 0.00262 *8*	**30.270** 0.00296 *9*	**31.452** 0.00334 *11*	**32.658** 0.00373 *12*	**33.894** 0.00422 *13*	**35.156** 0.00472 *15*	**36.443** 0.00531 *16*	**37.756** 0.00580 *18*	**39.097** 0.00678 *21*	**40.467** 0.00727 *22*	**41.864** 0.00824 *25*
$K_{\beta1}$	**28.486** 7.99 *18*	**29.726** 8.09 *18*	**30.995** 8.21 *18*	**32.295** 8.34 *19*	**33.624** 8.42 *19*	**34.987** 8.53 *19*	**36.378** 8.63 *19*	**37.801** 8.70 *19*	**39.258** 8.83 *20*	**40.748** 8.9 *2*	**42.272** 8.97 *19*	**43.827** 9.08 *19*	**45.414** 9.15 *19*	**47.038** 9.21 *19*	**48.695** 9.30 *19*
$K_{\beta2}$	**29.111** 2.19 *5*	**30.393** 2.28 *5*	**31.704** 2.37 *5*	**33.047** 2.47 *6*	**34.419** 2.55 *6*	**35.818** 2.64 *6*	**37.255** 2.73 *6*	**38.726** 2.81 *6*	**40.228** 2.84 *6*	**41.764** 2.88 *7*	**43.335** 2.93 *6*	**44.942** 2.98 *6*	**46.578** 3.02 *6*	**48.249** 3.05 *6*	**49.959** 3.11 *6*
$K_{\beta3}$	**28.444** 4.15 *9*	**29.679** 4.20 *9*	**30.944** 4.26 *10*	**32.239** 4.32 *10*	**33.562** 4.36 *10*	**34.920** 4.42 *10*	**36.304** 4.47 *10*	**37.720** 4.51 *10*	**39.170** 4.57 *10*	**40.653** 4.61 *10*	**42.166** 4.65 *10*	**43.713** 4.69 *10*	**45.293** 4.73 *10*	**46.905** 4.76 *10*	**48.551** 4.81 *10*
$K_{\beta4}$	**29.176** 0.012 *6*	**30.460** 0.013 *6*	**31.774** 0.015 *7*	**33.120** 0.017 *8*	**34.496** 0.019 *9*	**35.907** 0.021 *10*	**37.349** 0.023 *11*	**38.826** 0.025 *12*	**40.333** 0.027 *13*	**41.877** 0.028 *14*	**43.451** 0.030 *15*	**45.064** 0.032 *16*	**46.705** 0.034 *17*	**48.386** 0.036 *18*	**50.099** 0.038 *19*
$K_{\beta5}$	**28.711** 0.070 *3*	**29.959** 0.071 *3*	**31.236** 0.075 *3*	**32.544** 0.081 *3*	**33.881** 0.086 *4*	**35.252** 0.091 *4*	**36.652** 0.100 *4*	**38.085** 0.105 *4*	**39.551** 0.110 *5*	**41.050** 0.116 *5*	**42.580** 0.121 *5*	**44.145** 0.130 *5*	**45.741** 0.136 *6*	**47.373** 0.141 *6*	**49.038** 0.146 *6*
$KO_{2,3}$	**29.199** 0.049 *5*	**30.489** 0.092 *10*	**31.812** 0.147 *15*	**33.166** 0.212 *22*	**34.552** 0.29 *3*	**35.972** 0.35 *4*	**37.425** 0.40 *4*	**38.910** 0.45 *5*	**40.423** 0.42 *4*	**41.968** 0.42 *4*	**43.548** 0.42 *4*	**45.162** 0.42 *4*	**46.813** 0.42 *4*	**48.497** 0.42 *4*	**50.219** 0.45 *5*
$L_{\alpha1}$	**3.444** 2.9 *3*	**3.605** 3.0 *3*	**3.769** 3.2 *3*	**3.938** 3.4 *4*	**4.106** 3.6 *4*	**4.286** 3.8 *4*	**4.466** 4.1 *4*	**4.651** 4.3 *5*	**4.840** 4.6 *5*	**5.033** 4.9 *5*	**5.230** 5.2 *3*	**5.432** 5.4 *3*	**5.636** 5.7 *3*	**5.846** 6.0 *3*	**6.058** 6.3 *4*
$L_{\alpha2}$	**3.435** 0.32 *3*	**3.595** 0.34 *4*	**3.759** 0.36 *4*	**3.926** 0.38 *4*	**4.093** 0.40 *4*	**4.272** 0.43 *4*	**4.451** 0.45 *5*	**4.634** 0.48 *5*	**4.822** 0.51 *5*	**5.013** 0.54 *6*	**5.208** 0.57 *3*	**5.408** 0.60 *3*	**5.610** 0.63 *4*	**5.816** 0.67 *4*	**6.026** 0.70 *4*
$L_{\beta1}$	**3.663** 1.75 *18*	**3.843** 1.84 *19*	**4.029** 1.96 *20*	**4.221** 2.07 *21*	**4.414** 2.16 *22*	**4.620** 2.32 *24*	**4.828** 2.47 *25*	**5.042** 2.6 *3*	**5.263** 2.8 *3*	**5.489** 3.0 *3*	**5.722** 3.12 *23*	**5.961** 3.31 *24*	**6.206** 3.5 *3*	**6.457** 3.7 *3*	**6.713** 3.9 *3*
$L_{\beta2,15}$	**3.905** 0.46 *5*	**4.101** 0.52 *6*	**4.302** 0.58 *6*	**4.508** 0.64 *7*	**4.714** 0.71 *7*	**4.934** 0.77 *8*	**5.156** 0.84 *9*	**5.384** 0.91 *10*	**5.613** 0.97 *10*	**5.851** 1.03 *11*	**6.090** 1.10 *7*	**6.339** 1.15 *7*	**6.587** 1.21 *7*	**6.844** 1.27 *8*	**7.102** 1.32 *8*
$L_{\beta3}$	**3.750** 0.113 *23*	**3.933** 0.114 *23*	**4.121** 0.114 *23*	**4.314** 0.116 *23*	**4.512** 0.115 *23*	**4.717** 0.116 *23*	**4.927** 0.118 *24*	**5.143** 0.120 *24*	**5.363** 0.120 *24*	**5.593** 0.120 *24*	**5.829** 0.121 *18*	**6.071** 0.119 *18*	**6.317** 0.122 *18*	**6.571** 0.125 *19*	**6.832** 0.126 *19*
$L_{\beta4}$	**3.708** 0.071 *14*	**3.886** 0.070 *14*	**4.070** 0.069 *14*	**4.258** 0.070 *14*	**4.451** 0.069 *14*	**4.649** 0.069 *14*	**4.852** 0.070 *14*	**5.062** 0.071 *14*	**5.276** 0.071 *14*	**5.497** 0.071 *14*	**5.723** 0.072 *11*	**5.956** 0.071 *11*	**6.196** 0.073 *11*	**6.438** 0.075 *11*	**6.687** 0.077 *12*
$L_{\beta5}$								**5.483** 0.0091 *9*							**7.243** 0.0108 *6*
$L_{\beta6}$	**3.792** 0.0205 *21*	**3.980** 0.0225 *23*	**4.173** 0.0245 *25*	**4.371** 0.026 *3*	**4.569** 0.029 *3*	**4.781** 0.031 *3*	**4.994** 0.034 *4*	**5.212** 0.036 *4*	**5.434** 0.039 *4*	**5.660** 0.042 *4*	**5.893** 0.0454 *25*	**6.128** 0.049 *3*	**6.370** 0.053 *3*	**6.617** 0.058 *3*	**6.867** 0.063 *4*
$L_{\gamma1}$	**4.132** 0.206 *22*	**4.349** 0.226 *24*	**4.572** 0.25 *3*	**4.802** 0.28 *3*	**5.034** 0.30 *3*	**5.281** 0.33 *4*	**5.531** 0.36 *4*	**5.792** 0.39 *4*	**6.054** 0.43 *5*	**6.327** 0.46 *5*	**6.604** 0.50 *4*	**6.892** 0.54 *4*	**7.183** 0.58 *4*	**7.484** 0.62 *5*	**7.790** 0.67 *5*
$L_{\gamma2}$	**4.376** 0.016 *3*	**4.600** 0.016 *3*	**4.829** 0.017 *4*	**5.065** 0.018 *4*	**5.307** 0.018 *4*	**5.542** 0.019 *4*	**5.797** 0.020 *4*	**6.060** 0.020 *4*	**6.326** 0.021 *4*	**6.599** 0.021 *4*	**6.883** 0.022 *4*	**7.186** 0.022 *4*	**7.471** 0.023 *4*	**7.768** 0.023 *4*	**8.087** 0.024 *4*
$L_{\gamma3}$	**4.376** 0.025 *5*	**4.600** 0.025 *5*	**4.829** 0.026 *5*	**5.065** 0.027 *6*	**5.307** 0.027 *6*	**5.553** 0.028 *6*	**5.809** 0.028 *6*	**6.075** 0.029 *6*	**6.342** 0.030 *6*	**6.617** 0.030 *6*	**6.901** 0.031 *5*	**7.186** 0.031 *5*	**7.489** 0.032 *5*	**7.795** 0.033 *5*	**8.105** 0.034 *6*

TABLE IV.4 (cont.)

	$_{65}$Tb	$_{66}$Dy	$_{67}$Ho	$_{68}$Er	$_{69}$Tm	$_{70}$Yb	$_{71}$Lu	$_{72}$Hf	$_{73}$Ta	$_{74}$W	$_{75}$Re	$_{76}$Os	$_{77}$Ir	$_{78}$Pt	$_{79}$Au
$K_{\alpha 1}$	**44.482** 47.5 *10*	**45.998** 47.5 *10*	**47.547** 47.5 *10*	**49.128** 47.5 *10*	**50.742** 47.4 *10*	**52.389** 47.4 *10*	**54.070** 47.3 *10*	**55.790** 47.1 *10*	**57.535** 47.2 *10*	**59.318** 47.0 *10*	**61.141** 46.9 *10*	**63.000** 46.7 *10*	**64.896** 46.7 *10*	**66.831** 46.5 *10*	**68.806** 46.4 *10*
$K_{\alpha 2}$	**43.744** 26.7 *6*	**45.208** 26.8 *6*	**46.700** 26.9 *6*	**48.221** 27.0 *6*	**49.773** 27.2 *6*	**51.354** 27.2 *6*	**52.965** 27.3 *6*	**54.611** 27.3 *6*	**56.280** 27.3 *6*	**57.981** 27.4 *6*	**59.718** 27.4 *6*	**61.486** 27.4 *6*	**63.287** 27.4 *6*	**65.122** 27.4 *6*	**66.991** 27.5 *6*
$K_{\alpha 3}$	**43.288** 0.0092 *3*	**44.743** 0.0102 *3*	**46.224** 0.0111 *3*	**47.734** 0.0126 *4*	**49.274** 0.0135 *4*	**50.846** 0.0145 *4*	**52.443** 0.0159 *5*	**54.080** 0.0173 *5*	**55.735** 0.0192 *6*	**57.425** 0.0206 *6*	**59.150** 0.0224 *7*	**60.903** 0.0242 *7*	**62.693** 0.0261 *8*	**64.514** 0.0298 *9*	**66.372** 0.0326 *10*
$K_{\beta 1}$	**50.384** 9.44 *19*	**52.113** 9.58 *20*	**53.877** 9.68 *20*	**55.674** 9.77 *20*	**57.505** 9.86 *20*	**59.383** 9.99 *20*	**61.290** 10.1 *2*	**63.243** 10.20 *21*	**65.222** 10.30 *21*	**67.244** 10.30 *21*	**69.309** 10.40 *21*	**71.414** 10.60 *21*	**73.560** 10.60 *22*	**75.749** 10.70 *22*	**77.982** 10.70 *22*
$K_{\beta 2}$	**51.698** 3.15 *7*	**53.476** 3.20 *7*	**55.293** 3.24 *7*	**57.142** 3.28 *7*	**59.028** 3.32 *7*	**60.962** 3.38 *7*	**62.929** 3.42 *7*	**64.942** 3.48 *7*	**66.982** 3.53 *7*	**69.067** 3.58 *7*	**71.195** 3.63 *7*	**73.363** 3.71 *8*	**75.575** 3.75 *8*	**77.831** 3.81 *8*	**80.130** 3.84 *8*
$K_{\beta 3}$	**50.228** 4.88 *10*	**51.947** 4.95 *10*	**53.695** 5.0 *1*	**55.480** 5.06 *10*	**57.300** 5.11 *10*	**59.159** 5.18 *10*	**61.050** 5.21 *10*	**62.985** 5.28 *11*	**64.948** 5.32 *11*	**66.950** 5.35 *11*	**68.995** 5.42 *11*	**71.079** 5.48 *11*	**73.202** 5.52 *11*	**75.368** 5.56 *11*	**77.577** 5.57 *11*
$K_{\beta 4}$	**51.849** 0.040 *20*	**53.634** 0.042 *21*	**55.457** 0.045 *22*	**57.313** 0.047 *23*	**59.210** 0.049 *24*	**61.141** 0.051 *25*	**63.114** 0.05 *3*	**65.132** 0.06 *3*	**67.181** 0.06 *3*	**69.273** 0.06 *3*	**71.409** 0.07 *3*	**73.590** -0.07 *3*	**75.808** 0.07 *3*	**78.073** 0.08 *4*	**80.382** 0.08 *4*
$K_{\beta 5}$	**50.738** 0.156 *6*	**52.475** 0.166 *7*	**54.246** 0.176 *7*	**56.054** 0.186 *8*	**57.898** 0.195 *8*	**59.780** 0.204 *8*	**61.700** 0.213 *9*	**63.662** 0.222 *9*	**65.652** 0.232 *9*	**67.685** 0.241 *10*	**69.760** 0.25 *1*	**71.875** 0.259 *10*	**74.033** 0.268 *11*	**76.233** 0.276 *11*	**78.476** 0.285 *11*
$KO_{2,3}$	**51.970** 0.42 *4*	**53.762** 0.42 *4*	**55.597** 0.42 *4*	**57.456** 0.42 *4*	**59.357** 0.42 *4*	**61.309** 0.41 *4*	**63.286** 0.44 *5*	**65.316** 0.46 *5*	**67.376** 0.49 *5*	**69.484** 0.51 *5*	**71.636** 0.54 *6*	**73.819** 0.56 *6*	**76.054** 0.57 *6*	**78.337** 0.60 *6*	**80.660** 0.62 *6*
$L_{\alpha 1}$	**6.273** 6.7 *4*	**6.495** 7.1 *4*	**6.720** 7.4 *4*	**6.949** 7.7 *4*	**7.180** 8.1 *5*	**7.416** 8.3 *4*	**7.656** 8.6 *4*	**7.899** 8.9 *4*	**8.146** 9.3 *4*	**8.398** 9.7 *4*	**8.652** 10.1 *5*	**8.911** 10.4 *5*	**9.175** 10.9 *5*	**9.443** 11.2 *5*	**9.713** 11.6 *5*
$L_{\alpha 2}$	**6.239** 0.74 *4*	**6.458** 0.78 *4*	**6.680** 0.82 *5*	**6.905** 0.86 *5*	**7.133** 0.90 *5*	**7.367** 0.93 *4*	**7.605** 0.97 *5*	**7.844** 1.00 *5*	**8.088** 1.04 *5*	**8.335** 1.08 *5*	**8.586** 1.13 *5*	**8.840** 1.17 *5*	**9.099** 1.22 *6*	**9.362** 1.26 *6*	**9.628** 1.30 *6*
$L_{\beta 1}$	**6.977** 4.1 *3*	**7.248** 4.4 *3*	**7.526** 4.7 *3*	**7.811** 4.9 *4*	**8.102** 5.2 *4*	**8.402** 5.4 *3*	**8.709** 5.7 *3*	**9.023** 6.0 *3*	**9.343** 6.2 *3*	**9.672** 6.5 *4*	**10.010** 6.7 *4*	**10.354** 7.0 *4*	**10.708** 7.2 *4*	**11.071** 7.5 *4*	**11.443** 7.8 *4*
$L_{\beta 2,15}$	**7.367** 1.38 *8*	**7.636** 1.45 *9*	**7.910** 1.49 *9*	**8.186** 1.55 *9*	**8.468** 1.59 *10*	**8.752** 1.62 *8*	**9.044** 1.75 *9*	**9.342** 1.90 *10*	**9.646** 2.05 *10*	**9.955** 2.19 *11*	**10.268** 2.31 *12*	**10.590** 2.44 *13*	**10.912** 2.57 *13*	**11.242** 2.69 *14*	**11.576** 2.82 *14*
$L_{\beta 3}$	**7.097** 0.127 *19*	**7.370** 0.131 *20*	**7.653** 0.132 *20*	**7.940** 0.133 *20*	**8.231** 0.137 *21*	**8.537** 0.139 *21*	**8.847** 0.144 *22*	**9.163** 0.147 *22*	**9.488** 0.151 *23*	**9.819** 0.159 *24*	**10.159** 0.152 *23*	**10.511** 0.131 *20*	**10.868** 0.118 *18*	**11.235** 0.109 *16*	**11.610** 0.099 *15*
$L_{\beta 4}$	**6.940** 0.078 *12*	**7.204** 0.081 *12*	**7.471** 0.083 *12*	**7.746** 0.085 *13*	**8.026** 0.088 *13*	**8.313** 0.091 *14*	**8.607** 0.096 *15*	**8.905** 0.100 *15*	**9.213** 0.104 *16*	**9.525** 0.112 *17*	**9.845** 0.109 *16*	**10.176** 0.096 *15*	**10.510** 0.088 *13*	**10.854** 0.083 *13*	**11.205** 0.078 *12*
$L_{\beta 5}$							**9.240** 0.0103 *5*	**9.554** 0.0268 *12*	**9.875** 0.0372 *17*	**10.201** 0.0483 *22*	**10.532** 0.091 *4*	**10.871** 0.138 *6*	**11.211** 0.179 *8*	**11.562** 0.222 *10*	**11.916** 0.268 *12*
$L_{\beta 6}$	**7.116** 0.068 *4*	**7.374** 0.074 *4*	**7.635** 0.079 *4*	**7.909** 0.087 *5*	**8.176** 0.092 *5*	**8.456** 0.098 *5*	**8.738** 0.103 *5*	**9.023** 0.108 *5*	**9.316** 0.114 *5*	**9.612** 0.121 *6*	**9.910** 0.132 *6*	**10.217** 0.143 *7*	**10.525** 0.152 *7*	**10.840** 0.160 *7*	**11.160** 0.170 *8*
$L_{\gamma 1}$	**8.105** 0.71 *5*	**8.426** 0.76 *6*	**8.757** 0.82 *6*	**9.088** 0.88 *7*	**9.437** 0.93 *7*	**9.780** 0.99 *6*	**10.144** 1.04 *6*	**10.516** 1.10 *6*	**10.895** 1.16 *7*	**11.285** 1.22 *7*	**11.685** 1.29 *8*	**12.096** 1.35 *8*	**12.513** 1.41 *8*	**12.942** 1.47 *9*	**13.382** 1.56 *9*

$L_{\gamma 2}$	**8.398** 0.025 4	**8.714** 0.025 4	**9.051** 0.026 4	**9.385** 0.026 4	**9.730** 0.028 5	**10.090** 0.029 5	**10.460** 0.030 5	**10.834** 0.031 5	**11.217** 0.033 5	**11.608** 0.035 6	**12.009** 0.034 6	**12.421** 0.030 5	**12.841** 0.028 5	**13.273** 0.027 4	**13.709** 0.025 4
$L_{\gamma 3}$	**8.423** 0.035 6	**8.753** 0.037 6	**9.088** 0.038 6	**9.431** 0.039 6	**9.779** 0.040 7	**10.143** 0.041 7	**10.511** 0.044 7	**10.890** 0.045 7	**11.277** 0.047 8	**11.675** 0.050 8	**12.082** 0.049 8	**12.500** 0.043 7	**12.924** 0.039 6	**13.361** 0.037 6	**13.807** 0.034 6
$L_{\gamma 6}$							**10.344** 0.0063 6	**10.733** 0.0167 16	**11.130** 0.030 3	**11.538** 0.047 4	**11.955** 0.081 8	**12.385** 0.112 11	**12.820** 0.145 14	**13.270** 0.180 17	**13.731** 0.209 20
L_{η}	**6.284** 0.095 7	**6.534** 0.099 7	**6.789** 0.103 8	**7.058** 0.106 8	**7.310** 0.110 8	**7.580** 0.114 6	**7.857** 0.119 6	**8.139** 0.124 7	**8.428** 0.130 7	**8.724** 0.136 7	**9.027** 0.142 8	**9.337** 0.148 8	**9.650** 0.155 8	**9.975** 0.163 9	**10.309** 0.172 9
L_{ℓ}	**5.546** 0.28 3	**5.743** 0.30 3	**5.943** 0.32 3	**6.151** 0.34 3	**6.341** 0.36 3	**6.545** 0.37 3	**6.753** 0.39 4	**6.960** 0.41 4	**7.173** 0.43 4	**7.387** 0.46 4	**7.604** 0.49 4	**7.822** 0.52 5	**8.042** 0.55 5	**8.266** 0.58 5	**8.494** 0.61 6

	$_{80}$Hg	$_{81}$Tl	$_{82}$Pb	$_{83}$Bi	$_{84}$Po	$_{85}$At	$_{86}$Rn	$_{87}$Fr	$_{88}$Ra	$_{89}$Ac	$_{90}$Th	$_{91}$Pa	$_{92}$U	$_{93}$Np	$_{94}$Pu
$K_{\alpha 1}$	**70.818** 46.3 9	**72.873** 46.3 9	**74.969** 46.2 9	**77.107** 46.2 9	**79.290** 46.1 9	**81.517** 46.1 9	**83.787** 46.0 9	**86.105** 45.8 9	**88.471** 45.7 9	**90.886** 45.5 9	**93.350** 45.4 9	**95.863** 45.3 9	**98.434** 45.1 9	**101.059** 45.1 9	**103.734** 45.1 9
$K_{\alpha 2}$	**68.894** 27.5 6	**70.832** 27.6 6	**72.805** 27.7 6	**74.815** 27.7 6	**76.863** 27.7 6	**78.948** 27.9 6	**81.069** 27.9 6	**83.231** 27.9 6	**85.431** 28.0 6	**87.675** 28.1 6	**89.957** 28.1 6	**92.282** 28.1 6	**94.654** 28.2 6	**97.069** 28.3 6	**99.525** 28.4 6
$K_{\alpha 3}$	**68.263** 0.0358 11	**70.184** 0.0395 12	**72.144** 0.0428 13	**74.138** 0.0474 14	**76.172** 0.196 6	**78.242** 0.0571 17	**80.349** 0.0616 19	**82.496** 0.0675 20	**84.683** 0.0732 22	**86.910** 0.0791 24	**89.178** 0.085 3	**91.491** 0.091 3	**93.844** 0.099 3	**96.242** 0.105 3	**98.687** 0.114 3
$K_{\beta 1}$	**80.255** 10.70 22	**82.574** 10.70 22	**84.938** 10.70 22	**87.349** 10.70 21	**89.807** 10.70 21	**92.315** 10.70 21	**94.868** 10.60 21	**97.474** 10.70 21	**100.130** 10.70 21	**102.841** 10.70 21	**105.604** 10.70 21	**108.422** 10.70 22	**111.298** 10.70 22	**114.234** 10.70 22	**117.228** 10.70 22
$K_{\beta 2}$	**82.473** 3.87 8	**84.865** 3.90 8	**87.300** 3.91 8	**89.784** 3.93 8	**92.317** 3.95 8	**94.900** 3.97 8	**97.530** 3.98 8	**100.214** 4.01 8	**102.948** 4.04 8	**105.738** 4.07 8	**108.582** 4.10 8	**111.486** 4.13 8	**114.445** 4.15 8	**117.463** 4.17 8	**120.540** 4.18 8
$K_{\beta 3}$	**79.824** 5.59 11	**82.115** 5.59 11	**84.450** 5.58 11	**86.830** 5.59 11	**89.256** 5.57 11	**91.730** 5.58 11	**94.247** 5.56 11	**96.815** 5.58 11	**99.432** 5.59 11	**102.101** 5.61 11	**104.819** 5.61 11	**107.595** 5.64 11	**110.421** 5.65 11	**113.303** 5.65 11	**116.244** 5.44 11
$K_{\beta 4}$	**82.733** 0.08 4	**85.134** 0.09 4	**87.580** 0.09 4	**90.074** 0.09 4	**92.618** 0.09 4	**95.211** 0.10 5	**97.853** 0.10 5	**100.548** 0.10 5	**103.295** 0.11 5	**106.098** 0.11 5	**108.955** 0.11 5	**111.870** 0.12 6	**114.844** 0.12 6	**117.876** 0.12 6	**120.969** 0.13 6
$K_{\beta 5}$	**80.762** 0.294 12	**83.093** 0.303 12	**85.470** 0.312 12	**87.892** 0.321 13	**90.363** 0.330 13	**92.883** 0.339 14	**95.449** 0.349 14	**98.069** 0.358 14	**100.738** 0.362 15	**103.462** 0.371 15	**106.239** 0.380 15	**109.072** 0.389 16	**111.964** 0.397 16	**114.912** 0.405 16	**117.918** 0.413 16
$KO_{2,3}$	**83.028** 0.64 7	**85.444** 0.67 7	**87.911** 0.70 7	**90.421** 0.73 8	**92.983** 0.76 8	**95.595** 0.78 8	**98.257** 0.81 8	**100.972** 0.84 9	**103.740** 0.86 9	**106.563** 0.89 9	**109.442** 0.90 9	**112.380** 0.93 10	**115.377** 0.95 10	**118.429** 0.97 10	**121.543** 0.99 10
$KP_{2,3}$		**85.530** 0.0059 6	**88.003** 0.0165 17	**90.522** 0.031 3	**93.095** 0.049 5	**95.717** 0.070 7	**98.389** 0.094 10	**101.118** 0.114 12	**103.899** 0.132 13	**106.738** 0.146 15	**109.630** 0.160 16	**112.575** 0.156 16	**115.580** 0.159 16	**118.646** 0.162 17	**121.768** 0.157 16
$L_{\alpha 1}$	**9.989** 12.0 5	**10.268** 12.4 5	**10.551** 12.8 5	**10.839** 13.2 5	**11.130** 13.6 5	**11.426** 13.9 5	**11.726** 14.2 5	**12.031** 14.6 5	**12.339** 14.9 6	**12.651** 15.3 6	**12.968** 15.6 7	**13.291** 16.2 8	**13.618** 16.8 8	**13.946** 17.3 8	**14.282** 17.9 9
$L_{\alpha 2}$	**9.899** 1.35 5	**10.172** 1.39 5	**10.450** 1.44 5	**10.731** 1.48 6	**11.016** 1.52 6	**11.306** 1.56 6	**11.598** 1.59 6	**11.896** 1.63 6	**12.196** 1.67 6	**12.500** 1.71 6	**12.809** 1.75 8	**13.127** 1.82 8	**13.442** 1.89 9	**13.761** 1.94 9	**14.087** 2.00 10
$L_{\beta 1}$	**11.824** 8.0 4	**12.213** 8.3 4	**12.614** 8.5 5	**13.024** 8.8 5	**13.443** 9.1 5	**13.875** 9.4 5	**14.316** 9.7 5	**14.770** 9.9 5	**15.236** 10.2 6	**15.711** 10.4 6	**16.202** 10.7 11	**16.708** 10.5 11	**17.222** 10.3 11	**17.751** 10.3 10	**18.296** 10.3 10

TABLE IV.4 (cont.)

	$_{80}$Hg	$_{81}$Tl	$_{82}$Pb	$_{83}$Bi	$_{84}$Po	$_{85}$At	$_{86}$Rn	$_{87}$Fr	$_{88}$Ra	$_{89}$Ac	$_{90}$Th	$_{91}$Pa	$_{92}$U	$_{93}$Np	$_{94}$Pu
continued															
$L_{\beta 2,15}$	**11.915** 2.94 *13*	**12.261** 3.06 *13*	**12.611** 3.18 *14*	**12.967** 3.28 *14*	**13.328** 3.40 *15*	**13.694** 3.52 *15*	**14.066** 3.65 *16*	**14.443** 3.74 *16*	**14.825** 3.86 *17*	**15.212** 3.97 *17*	**15.605** 4.09 *21*	**16.008** 4.27 *22*	**16.410** 4.45 *23*	**16.817** 4.59 *24*	**17.235** 4.77 *25*
$L_{\beta 3}$	**11.992** 0.097 *15*	**12.390** 0.094 *14*	**12.794** 0.095 *14*	**13.211** 0.096 *15*	**13.635** 0.107 *16*	**14.073** 0.101 *15*	**14.519** 0.106 *16*	**14.978** 0.106 *16*	**15.447** 0.110 *17*	**15.931** 0.111 *17*	**16.426** 0.116 *20*	**16.931** 0.113 *19*	**17.454** 0.122 *21*	**17.992** 0.124 *21*	**18.541** 0.135 *23*
$L_{\beta 4}$	**11.561** 0.077 *12*	**11.931** 0.077 *12*	**12.307** 0.080 *12*	**12.691** 0.083 *13*	**13.084** 0.095 *14*	**13.488** 0.092 *14*	**13.898** 0.099 *15*	**14.319** 0.102 *15*	**14.749** 0.109 *16*	**15.191** 0.113 *17*	**15.641** 0.122 *21*	**16.104** 0.121 *21*	**16.577** 0.134 *23*	**17.061** 0.140 *24*	**17.557** 0.16 *3*
$L_{\beta 5}$	**12.275** 0.315 *12*	**12.643** 0.362 *13*	**13.015** 0.411 *15*	**13.393** 0.458 *17*	**13.778** 0.506 *19*	**14.168** 0.556 *21*	**14.565** 0.605 *23*	**14.967** 0.683 *25*	**15.375** 0.71 *3*	**15.790** 0.76 *3*	**16.209** 0.81 *4*	**16.639** 0.87 *4*	**17.069** 0.94 *4*	**17.505** 1.17 *6*	**17.950** 1.06 *5*
$L_{\beta 6}$	**11.481** 0.180 *7*	**11.812** 0.190 *7*	**12.142** 0.200 *8*	**12.480** 0.210 *8*	**12.823** 0.220 *8*	**13.169** 0.230 *9*	**13.520** 0.239 *9*	**13.877** 0.251 *9*	**14.236** 0.263 *10*	**14.601** 0.273 *10*	**14.970** 0.284 *13*	**15.350** 0.300 *14*	**15.727** 0.318 *15*	**16.109** 0.333 *16*	**16.498** 0.349 *17*
$L_{\gamma 1}$	**13.830** 1.63 *10*	**14.291** 1.71 *10*	**14.765** 1.78 *10*	**15.248** 1.87 *11*	**15.742** 1.95 *11*	**16.249** 2.05 *12*	**16.770** 2.15 *13*	**17.302** 2.25 *13*	**17.848** 2.34 *14*	**18.405** 2.42 *14*	**18.980** 2.5 *3*	**19.571** 2.5 *3*	**20.169** 2.5 *3*	**20.784** 2.5 *3*	**21.420** 2.5 *3*
$L_{\gamma 2}$	**14.158** 0.025 *4*	**14.625** 0.026 *4*	**15.097** 0.027 *4*	**15.582** 0.029 *5*	**16.077** 0.033 *5*	**16.585** 0.033 *5*	**17.104** 0.036 *6*	**17.635** 0.038 *6*	**18.177** 0.041 *7*	**18.734** 0.044 *7*	**19.304** 0.048 *9*	**19.888** 0.048 *9*	**20.487** 0.055 *10*	**21.099** 0.059 *11*	**21.724** 0.067 *12*
$L_{\gamma 3}$	**14.262** 0.034 *6*	**14.738** 0.033 *5*	**15.216** 0.034 *6*	**15.709** 0.035 *6*	**16.213** 0.040 *6*	**16.731** 0.038 *6*	**17.258** 0.040 *7*	**17.800** 0.041 *7*	**18.353** 0.044 *7*	**18.922** 0.045 *7*	**19.505** 0.048 *9*	**20.101** 0.047 *9*	**20.715** 0.052 *9*	**21.342** 0.054 *10*	**21.981** 0.059 *11*
$L_{\gamma 6}$	**14.199** 0.248 *23*	**14.683** 0.28 *3*	**15.178** 0.31 *3*	**15.685** 0.34 *3*	**16.203** 0.38 *4*	**16.735** 0.41 *4*	**17.280** 0.44 *4*	**17.839** 0.47 *4*	**18.412** 0.50 *5*	**18.997** 0.52 *5*	**19.599** 0.54 *7*	**20.217** 0.53 *7*	**20.844** 0.53 *7*	**21.491** 0.53 *7*	**22.153** 0.53 *7*
L_{η}	**10.647** 0.180 *10*	**10.994** 0.188 *10*	**11.349** 0.196 *11*	**11.712** 0.207 *11*	**12.085** 0.218 *12*	**12.466** 0.228 *12*	**12.855** 0.238 *13*	**13.255** 0.247 *13*	**13.662** 0.255 *14*	**14.082** 0.266 *14*	**14.511** 0.28 *3*	**14.953** 0.28 *3*	**15.400** 0.27 *3*	**15.861** 0.28 *3*	**16.333** 0.28 *3*
L_{ℓ}	**8.722** 0.65 *6*	**8.953** 0.68 *6*	**9.184** 0.72 *6*	**9.420** 0.75 *7*	**9.658** 0.79 *7*	**9.897** 0.82 *7*	**10.137** 0.86 *7*	**10.381** 0.89 *8*	**10.622** 0.93 *8*	**10.871** 0.98 *8*	**11.118** 1.02 *9*	**11.372** 1.08 *10*	**11.620** 1.14 *10*	**11.871** 1.20 *11*	**12.124** 1.25 *11*

TABLE IV.4 (cont.)

	$_{95}$Am	$_{96}$Cm	$_{97}$Bk	$_{98}$Cf	$_{99}$Es	$_{100}$Fm	$_{101}$Md	$_{102}$No	$_{103}$Lr	104
$K_{\alpha1}$	**106.472** 44.9 *9*	**109.271** 44.8 *9*	**112.121** 44.6 *9*	**115.032** 44.4 *9*	**118.012** 44.3 *9*	**121.058** 44.2 *9*	**125.170** 44.1 *9*	**127.357** 44.0 *9*	**130.611** 43.8 *9*	**133.381** 43.6 *9*
$K_{\alpha2}$	**102.030** 28.5 *6*	**104.590** 28.5 *6*	**107.185** 28.7 *6*	**109.831** 28.7 *6*	**112.531** 28.8 *6*	**115.285** 28.9 *6*	**119.088** 29.0 *6*	**120.953** 29.0 *6*	**123.867** 29.1 *6*	**126.302** 29.2 *6*
$K_{\alpha3}$	**101.174** 0.123 *4*	**103.715** 0.132 *4*	**106.300** 0.145 *4*	**108.929** 0.158 *5*	**111.614** 0.171 *5*	**114.352** 0.184 *6*	**118.139** 0.201 *6*	**119.987** 0.218 *7*	**122.887** 0.235 *7*	**125.407** 0.252 *8*
$K_{\beta1}$	**120.284** 10.70 *21*	**123.403** 10.60 *21*	**126.580** 10.70 *21*	**129.823** 10.70 *22*	**133.137** 10.70 *22*	**136.521** 10.80 *22*	**140.974** 10.70 *22*	**143.506** 10.70 *22*	**147.110** 10.80 *22*	**150.279** 10.80 *22*
$K_{\beta2}$	**123.680** 4.19 *8*	**126.889** 4.19 *8*	**130.152** 4.22 *9*	**133.483** 4.26 *9*	**136.887** 4.27 *9*	**140.362** 4.31 *9*	**144.906** 4.30 *9*	**147.531** 4.32 *9*	**151.227** 4.35 *9*	**154.494** 4.36 *9*
$K_{\beta3}$	**119.243** 5.64 *11*	**122.304** 5.63 *11*	**125.418** 5.67 *11*	**128.594** 5.70 *11*	**131.838** 5.71 *11*	**135.150** 5.74 *12*	**139.525** 5.73 *12*	**141.977** 5.75 *12*	**145.496** 5.78 *12*	**148.550** 5.79 *12*
$K_{\beta4}$	**124.127** 0.13 *6*	**127.352** 0.13 *6*	**130.630** 0.14 *7*	**133.979** 0.14 *7*	**137.399** 0.14 *7*	**140.892** 0.15 *7*	**145.456** 0.15 *7*	**148.100** 0.15 *7*	**151.818** 0.16 *8*	**155.097** 0.16 *8*
$K_{\beta5}$	**120.989** 0.421 *17*	**124.124** 0.429 *17*	**127.316** 0.437 *18*	**130.573** 0.449 *18*	**133.904** 0.454 *18*	**137.304** 0.457 *18*	**141.774** 0.465 *19*	**144.323** 0.472 *19*	**147.944** 0.479 *19*	**151.113** 0.486 *19*
$KO_{2,3}$	**124.723** 1.0 *1*	**127.970** 1.02 *10*	**131.274** 1.04 *11*	**134.646** 1.05 *11*	**138.090** 1.06 *11*	**141.608** 1.08 *11*	**146.195** 1.09 *11*	**148.865** 1.10 *11*	**152.607** 1.12 *11*	**155.904** 1.13 *12*
$KP_{2,3}$	**124.955** 0.158 *16*	**128.210** 0.169 *17*	**131.524** 0.170 *17*	**134.908** 0.162 *17*	**138.363** 0.163 *17*	**141.889** 0.163 *17*	**146.490** 0.164 *17*	**149.171** 0.165 *17*	**152.926** 0.174 *18*	**156.236** 0.183 *19*
$L_{\alpha1}$	**14.620** 18.2 *9*	**14.961** 18.5 *9*	**15.308** 18.8 *9*	**15.660** 19.0 *9*	**16.016** 19.2 *9*	**16.377** 19.4 *11*	**16.741** 19.6 *11*	**17.110** 19.7 *11*	**17.483** 19.8 *11*	**17.893** 19.8 *11*
$L_{\alpha2}$	**14.414** 2.04 *10*	**14.746** 2.08 *10*	**15.082** 2.1 *1*	**15.423** 2.12 *10*	**15.767** 2.15 *10*	**16.116** 2.17 *13*	**16.467** 2.19 *13*	**16.823** 2.20 *13*	**17.183** 2.22 *13*	**17.571** 2.22 *13*
$L_{\beta1}$	**18.856** 10.4 *11*	**19.427** 10.6 *11*	**20.018** 10.7 *11*	**20.624** 10.8 *11*	**21.248** 11.0 *11*	**21.889** 11.2 *11*	**22.549** 11.4 *12*	**23.227** 11.5 *12*	**23.927** 11.7 *12*	**24.650** 12.0 *12*
$L_{\beta2,15}$	**17.655** 4.9 *3*	**18.081** 5.0 *3*	**18.509** 5.1 *3*	**18.946** 5.2 *3*	**19.387** 5.3 *3*	**19.834** 5.3 *3*	**20.286** 5.4 *3*	**20.744** 5.5 *3*	**21.207** 5.5 *4*	**21.716** 5.6 *4*
$L_{\beta3}$	**19.110** 0.137 *24*	**19.688** 0.142 *24*	**20.280** 0.142 *24*	**20.894** 0.145 *25*	**21.523** 0.15 *3*	**22.169** 0.15 *3*	**22.835** 0.15 *3*	**23.519** 0.15 *3*	**24.223** 0.15 *3*	**24.872** 0.15 *3*
$L_{\beta4}$	**18.069** 0.16 *3*	**18.589** 0.17 *3*	**19.118** 0.18 *3*	**19.665** 0.19 *3*	**20.224** 0.21 *4*	**20.798** 0.21 *4*	**21.386** 0.23 *5*	**21.990** 0.24 *5*	**22.609** 0.24 *5*	**23.143** 0.26 *5*
$L_{\beta5}$	**18.399** 1.11 *5*	**18.853** 1.16 *6*	**19.312** 1.19 *6*	**19.777** 1.23 *6*	**20.249** 1.26 *6*	**20.727** 1.29 *8*	**21.210** 1.32 *8*	**21.700** 1.35 *8*	**22.195** 1.38 *8*	**22.727** 1.41 *8*
$L_{\beta6}$	**16.890** 0.361 *17*	**17.286** 0.373 *18*	**17.687** 0.385 *18*	**18.094** 0.396 *19*	**18.501** 0.407 *19*	**18.916** 0.419 *24*	**19.332** 0.430 *25*	**19.754** 0.44 *3*	**20.179** 0.45 *3*	**20.670** 0.46 *3*
$L_{\gamma1}$	**22.072** 2.6 *3*	**22.735** 2.7 *3*	**23.416** 2.7 *3*	**24.117** 2.8 *3*	**24.836** 2.8 *3*	**25.574** 2.9 *3*	**26.333** 3.0 *3*	**27.110** 3.1 *3*	**27.911** 3.1 *3*	**28.753** 3.2 *3*
$L_{\gamma2}$	**22.370** 0.073 *13*	**23.028** 0.079 *14*	**23.698** 0.082 *15*	**24.390** 0.087 *16*	**25.099** 0.093 *17*	**25.825** 0.096 *20*	**26.571** 0.102 *21*	**27.336** 0.109 *23*	**28.120** 0.109 *23*	**28.846** 0.116 *24*
$L_{\gamma3}$	**22.643** 0.062 *11*	**23.319** 0.065 *12*	**24.007** 0.065 *12*	**24.718** 0.068 *12*	**25.446** 0.071 *13*	**26.195** 0.071 *15*	**26.963** 0.073 *15*	**27.752** 0.075 *16*	**28.560** 0.074 *15*	**29.327** 0.076 *16*
$L_{\gamma6}$	**22.836** 0.54 *7*	**23.527** 0.55 *7*	**24.241** 0.57 *7*	**24.971** 0.59 *8*	**25.723** 0.60 *8*	**26.492** 0.63 *8*	**27.284** 0.65 *8*	**28.094** 0.68 *9*	**28.929** 0.70 *9*	**29.796** 0.73 *9*
L_{η}	**16.819** 0.28 *3*	**17.314** 0.29 *3*	**17.826** 0.30 *3*	**18.347** 0.30 *3*	**18.884** 0.31 *3*	**19.433** 0.31 *3*	**19.998** 0.32 *3*	**20.577** 0.33 *3*	**21.173** 0.34 *3*	**21.825** 0.35 *4*
L_{ℓ}	**12.377** 1.31 *12*	**12.633** 1.36 *12*	**12.890** 1.40 *13*	**13.146** 1.45 *13*	**13.403** 1.49 *14*	**13.660** 1.53 *15*	**13.916** 1.57 *15*	**14.173** 1.61 *16*	**14.429** 1.64 *16*	**14.746** 1.68 *16*

INDEX